Methods of Organic Synthesis

Alexander Düfert

Methods of Organic Synthesis

Principles, Mechanisms and Applications

 Springer

Alexander Düfert
Ludwigshafen, Germany

ISBN 978-3-662-70962-7   ISBN 978-3-662-70963-4   (eBook)
https://doi.org/10.1007/978-3-662-70963-4

This book is a translation of the original German edition "Organische Synthesemethoden" by Alexander Düfert, published by Springer-Verlag GmbH, DE in 2023. The translation was done with the help of an artificial intelligence machine translation tool. A subsequent human revision was done primarily in terms of content, so that the book will read stylistically differently from a conventional translation. Springer Nature works continuously to further the development of tools for the production of books and on the related technologies to support the authors.

Translation from the first language edition: "Organische Synthesemethoden" by Alexander Düfert, © 2023 Published by Springer-Verlag GmbH. All Rights Reserved.

This Springer imprint is published by the registered company Springer-Verlag GmbH, DE, part of Springer Nature.
The registered company address is: Heidelberger Platz 3, 14197 Berlin, Germany

If disposing of this product, please recycle the paper.

The ability to prepare molecules efficiently and selectively is important to many areas of organic chemistry and related science. This is true whether it is for drug discovery or manufacture, the preparation of agrochemicals, or the invention of new electronic materials. Also included are various aspects of chemical biology and the development of new sensory materials. To become well-versed in the field of organic synthesis, a student must gain exposure to a variety of subjects. Alexander Düfert's book, Methods of Organic Synthesis, covers a broad range of topics that every student of the field should be familiar with. Chemistry is an ever-changing field. New synthetic reactions are constantly being invented, many of which are significant improvements over those previously employed by chemists. Yet, decades-old, time-tested traditional techniques are still of tremendous importance, particularly on scale.

Düfert's book starts with a review of many important fundamental topics in Chap. 1. This includes a review of A-values and simple molecular orbital considerations. This chapter also surveys the anomeric effect, allylic strain, Baldwin's Rules, the Thorpe-Ingold effect, and acid-base chemistry. The book then discusses polar reactions, such as carbonyl chemistry in Chap. 2, and the means to selectively prepare geometrically pure alkenes in Chap. 3. The ability to manipulate the oxidation state and interconvert alcohols and carbonyl compounds is detailed in Chap. 4. It is also essential to have an introduction to Woodward-Hoffman rules and learn about pericyclic reactions, which are discussed in Chap. 5. Transition metal-catalyzed processes have revolutionized how molecules are assembled. Thus, familiarity with key transformations such as the Suzuki-Miyaura, Negishi, and Heck coupling processes, the three Nobel Prize-winning reactions, and the Sonogashira coupling reaction are covered. Chapter 6 includes these and metal-catalyzed C-H activation reactions, which are increasingly important as the techniques are improved. Also included is a discussion of related carbon-heteroatom bond-forming methods, including older ones such as Ullmann and Goldberg processes that employ copper catalysts. The remarkable applicability of olefin metathesis to preparing large and small ring compounds has made this a must-have topic. This is covered in Chap. 7, along with enyne and alkyne metathesis. Also included is a discussion of applications in both academia and industry. Nobel Prize-winning chemistry is again covered in Chaps. 8 and 9. In the former, organocatalysis is described, including the different reaction modes, types of catalysts, and many applications. In the latter, click reactions are featured. This includes well-known processes such as combining alkynes and azides to form triazoles and thiol-ene reactions and Staudinger ligations. Their utility is then illustrated through the applications of these methods. There has been a dramatic rise in the application of radical chemistry to synthetic chemistry in recent years. Once considered untameable, this field of chemistry has grown exponentially through selective generation and mechanistically guided incorporation into catalytic processes. Two topics of great current interest (no pun intended) that are covered are electrochemistry and photoredox catalysis. These exciting areas of chemistry are covered in Chap. 10. In Chap. 11, the book's final chapter, a detailed description of the principles of planning synthetic routes is discussed. This includes retrosynthetic analysis and the differences in considerations for academic synthesis versus that carried out in industry. Finally, a discussion of how to evaluate and optimize synthetic routes is presented.

Overall, this book guides the student through fundamental topics and a variety of important (old and new) areas of synthetic chemistry. It thoroughly describes each area and illustrates its utility in modern-day synthetic chemistry with applications. This book is a welcome addition to the field and should be invaluable for both classes and self-study. Dr. Düfert is to be congratulated for this significant contribution to the field of organic synthesis.

October 2024

Stephen L. Buchwald
Camille Dreyfus Professor of Chemistry
Massachusetts Institute of Technology
Cambridge, MA, USA

Introduction

The information about any type of reaction can nowadays be easily found through a simple internet search. Do you then really need another textbook on organic chemistry? The key advantage of a textbook usually lies in its *contextualization* of said information. The comparison of the different methods and their advantages and disadvantages allow a practitioner to make a preselection for a given synthetic challenge (*Which olefination method is suitable for a given substrate? Is it perhaps worth taking a look at a metathesis or a CC cross-coupling? What levers do I have to influence the stereoselectivity of an aldol reaction? Which substrates are suitable for an asymmetric dihydroxylation and which functionalities could interfere?*).

While a few standard monographs cover a good range of the topics contained in the pages at hand, the entire breadth of subjects is so far not found in the available textbooks. Additionally certain information, especially the mechanisms and examples of applications in complex settings, are highly relevant and should be up-to-date to assess a method's utility.

Basic knowledge of certain reaction mechanisms or of principles of reactivity is presumed in many places, as there are already excellent textbooks and online sources available covering that field. This book is accordingly targeting Ph.D. students and industry practitioners whose challenges are more of an advanced nature.

Due to the desired level of detail, the goal of a comprehensive textbook can never be fully met, which is why a selection was necessary both in terms of thematic scope and highlighted methods. The reactions were chosen based on importance and utility for the synthesis of complex target molecules, mostly natural products. This cuts short some areas, despite them being in the focus of current research efforts, as only few applications in natural product syntheses have been reported (yet). The chosen reaction conditions also mirror this aim. This results in a certain mix of discussed examples and methods, where both decades-old established and relatively new synthetic methods are covered. This has the advantage that the presented tools of the trade should not become outdated too quickly and should not have lost significant relevance even after several years. On average, it takes ten or more years for a new method to become broadly established.

The chapters are either organized by the key functionalities, such as CC multiple bonds, or by the type of synthetic transformation, for example a metathesis. An attempt was made to avoid redundancies and, if necessary, to refer to the chapters where the supplementary content is located. A Shi epoxidation, for example, is an oxidation (Chap. 4), but also falls under derivatization of a double bond (Chap. 3) and an organocatalytic method (Chap. 8). Best efforts were made to allow for thematic homogeneity and easy flow of reading.

As this is a textbook, no claim to completeness in terms of the current state of research can be made. The phrase of "typical" or "common" conditions or substrates can therefore only attempt to cover 90–95% of the published possibilities, if even that. There are numerous exemptions in the recent literature, which is why current review articles should be consulted for individual topics.

As the book is aiming to support the chemist in problem solving, the chapters are mostly structured in a similar manner: The relevance of the methods in the overall context is explained. This is typically followed by an in-depth discussion of the mechanism based on up-to-date evidence. This understanding plays a key role: Obstacles can either be avoided in advance or unexpected results can be rationalized afterwards. The sections are concluded by a selection of currently used or historically important methods, along with several examples of their application.

All the examples presented are chosen in such a way that they can clearly convey the synthetic principle (supported by the use of a highlighting color) and at the same time have as complex a molecular architecture as possible to demonstrate the limits of the methods.

Almost all syntheses come from an academic context. This is not necessarily intentional, but simply determined by the fewer number of publications covering industrial syntheses. Given that, an attempt was made to also cover industrial challenges, which mostly stem from pharmaceutical chemistry.

Based on personal experience and as a conscious choice, an emphasis was placed on certain details: 1. Literature references for all statements herein. 2. For well-established topics, a selection of review articles that allow the reader an easy entry for further in-depth study. 3. Detailed reaction conditions for the reader to develop a "gut feeling" about each method.

While only a limited number of people were involved at the beginning of this project, this has greatly expanded towards the end, and their contributions should be acknowledged here: Particular mention should be made of the participation of Daniel B. Werz, who was the first reviewer for every single page of this textbook in the nearly 1 1/2 decades of the writing process. He not only positively influenced the thematic orientation of this book in our many discussions but was also able to catch the most glaring errors. Further thanks go to Richard Dehn, Christian Ducho, Volker Hickmann, Uli Kazmaier, Michael Rack, Christian Raith, Thomas Schaub, Mathias Schelwies, Christiane Schotten and Grigory

Shevchenko for careful proofreading and all professional suggestions. The Springer team deserves praise for supporting my vision up to the publication of the initial German edition, as well as this revised English version.

The book is dedicated to my academic mentors, which helped shape my perspective and approach to chemistry in all its facets. Lutz Tietze, Steve Ley, Daniel Werz and Steve Buchwald instilled a deep fascination for all things chemical with a special place for complex syntheses. My tenure in the R&D department of BASF subsequently added a desire to see how things could work on a larger scale and how to convert a laboratory synthesis into an industrial process.

If you notice any inconsistencies, errors or the like—please do not hesitate to get in touch (alexander.duefert@uni-saarland.de). I hope you enjoy the read, receive answers to all your questions and can use it to help you tackling your synthetic challenges!

Ludwigshafen, Germany   Alexander Düfert
Summer 2025

Abbreviations and Acronyms

Abbreviation	Name	Common Use
9-BBN	9-borabicyclo[3.3.1]nonane	–
Ac	acetyl	–
acac	acetylacetonate	ligand
Ad	adamantyl	–
AIBN	azobis(isobutyronitrile)	radical initiator
Ar	aryl	–
BAIB	bis(acetato-O)phenyliodine, bis(acetoxy)iodobenzene	oxidizing agent
BHT	3,5-di-*tert*-butyl-4-hydroxytoluene	radical stabilizer
BINAL, BINAL-H	2,2'-dihydroxy-1,1'-binaphthyl aluminum hydride	chiral reducing agent
BINAP	2,2'-bis(diphenylphosphino)-1,1'-binaphthyl	ligand
BINOL	1,1'-bi-2,2'-naphthol	ligand
bipy	2,2'-bipyridine	ligand
Bn	benzyl	protecting group
Boc	*tert*-butoxycarbonyl	protecting group
BOM	benzyloxymethyl	protecting group
BOP	1H-benzotriazolyloxytris (dimethylamino) phosphonium hexafluorophosphate	carboxylic acid activation
box	2,2-bis(oxazolinyl)propane	ligand
brsm	based on recovered starting material	–
Bz	benzoyl	protecting group
CAN	cerium ammonium nitrate, $Ce(NH_4)_2(NO_3)_6$	oxidizing agent
catBH	catecholborane	–
Cbz	benzyloxycarbonyl	protecting group
CDI	carbonyl diimidazole	OH activation, carboxylic acid activation
CM	(alkene) cross metathesis	–
COD	1,5-cyclooctadiene	ligand
Cp	cyclopentadiene	ligand
Cp*	pentamethyl cyclopentadiene	ligand

Abbreviation	Name	Common Use
CSA	camphorsulfonic acid	acid
Cy, cHex	cyclohexyl	–
DABCO	1,4-diazabicyclo[2.2.2]octane	base
dba	dibenzylideneacetone	ligand
DBU	1,8-diazabicyclo[5.4.0]undec-7-ene	base
DCC	*N,N'*-dicyclohexyl carbodiimide	carboxylic acid activation
DCE	1,2-dichloroethane	solvent
DDQ	2,3-dichloro-5,6-dicyano-1,4-benzoquinone	oxidizing agent
DEAD	diethyl azodicarboxylate	OH activation
DEG	diethylene glycol	solvent, ligand
DET	diethyl tartrate	ligand
DHQ	dihydroquinine (Cinchona alkaloid)	ligand, base, catalyst
DHQD	dihydroquinidine (Cinchona alkaloid)	ligand, base, catalyst
DIAD	diisopropyl azodicarboxylate	OH activation
DIBAL	diisobutylaluminum hydride	reducing agent
DIC	*N,N'*-diisopropyl carbodiimide	carboxylic acid activation
DIPEA	diisopropylethylamine	base
DIPT	diisopropyl tartrate	ligand
DMAc	*N,N*-dimethylacetamide	solvent
DMAP	*N,N*-dimethyl-4-aminopyridine	carboxylic acid activation, additive
DMDO	dimethyl dioxirane	epoxidation reagent
DME	1,2-dimethoxyethane	solvent
DMP	Dess-Martin periodinane	oxidizing agent
DMPU	dimethyl propylene urea	solvent, ligand
dppb	bis(diphenylphosphino)butane	ligand
dppe	bis(diphenylphosphino)ethane	ligand
dppf	1,1'-bis(diphenylphosphino)ferrocene	ligand
dppp	bis(diphenylphosphino)propane	ligand
EDC, EDCl	1-ethyl-3-(3-dimethylaminopropyl)carbodiimide	carboxylic acid activation
EDG	electron-donating group	–
EWG	electron-withdrawing group	–
Fmoc	9-fluorenylmethoxycarbonyl	protecting group
HATU	*O*-(7-azabenzotriazol-1-yl)-*N,N,N',N'*-tetramethyluronium hexafluorophosphate	carboxylic acid activation
HBTU	2-(1H-benzotriazol-1-yl)-1,1,3,3-tetramethyluronium hexafluorophosphate	carboxylic acid activation
HetAr, Het	heteroaryl	–
hfacac	hexafluoroacetylacetonate	ligand
HFIP	hexafluoroisopropanol	solvent
HMPA	hexamethylphosphoramide	solvent, ligand
HOAt	1-hydroxy-7-azabenzotriazole	carboxylic acid activation, additive
HOBt	1-hydroxybenzotriazole	carboxylic acid activation, additive

Abbreviation	Name	Common Use
HWE	Horner-Wadsworth-Emmons (olefination)	–
IBX	2-iodoxybenzoic acid	oxidizing agent
Imid / Im	imidazoyl	base
Ipc	isopinocampheyl	–
KHMDS	potassium hexamethyldisilazide	base
LA	Lewis acid	–
LDA	lithium diisopropylamide	base
LiHMDS	lithium hexamethyldisilazide	base
LiTMP	lithium tetramethylpiperidine	base
*m*CPBA	*meta*-chloroperbenzoic acid	oxidizing agent
MEM	(2-methoxyethoxy)methyl	protecting group
Mes	mesityl	–
MOM	methoxymethyl	protecting group
MoOPH	oxodiperoxymolybdenum(pyridine)hexamethylphosphoric triamide	oxidizing agent
Ms	mesyl, methanesulfonyl	–
MS	molecular sieve	–
MSA	methanesulfonic acid	–
MTBE	methyl-*tert*-butyl ether	solvent
NaHMDS	sodium hexamethyldisilazide	base
NBS	*N*-bromosuccinimide	halogenation reagent
NCS	*N*-chlorosuccinimide	halogenation reagent
NFSI	*N*-fluorobenzenesulfonimide	halogenation reagent
NIS	*N*-iodosuccinimide	halogenation reagent
NMM	*N*-methylmorpholine	–
NMO	*N*-methylmorpholine *N*-Oxide	oxidizing agent
PCC	pyridinium chlorochromate	oxidizing agent
PDC	pyridinium dichromate	oxidizing agent
PEG	polyethylene glycol	ligand
PHAL	phthalazine	–
phen	9,10-phenanthroline	ligand
PHOX	phosphine oxazoline	ligand
Phth	phthaloyl	–
pin	pinacol	–
Piv	pivaloyl	–
PMB (MPM)	*p*-methoxybenzyl	protecting group
PMHS	polymethylhydrosiloxane	reducing agent
PMP	*p*-methoxyphenyl	–
PPTS	pyridinium *p*-toluenesulfonate	acid
PTC	phase transfer catalysis	–
PTSA	*p*-toluenesulfonic acid, TsOH	acid

Abbreviation	Name	Common Use
Py (Pyr)	pyridine	base, ligand
pybox	2,6-bis(oxazolinyl)pyridine	ligand
RAMP	*R*-1-amino-2-methoxymethylpyrrolidine	auxiliary
RCAM	alkyne ring-closing metathesis	–
RCM	(alkene) ring-closing metathesis	–
rds	rate determining step	–
S, (S)	solvent (if not sulfur)	–
salen	bis(salicylidene)ethylenediamine	ligand
SAMP	*S*-1-amino-2-methoxymethylpyrrolidine	auxiliary
SEM	2-(trimethylsilyl)ethoxymethyl	protecting group
SET	single electron transfer	–
T3P	propanephosphonic acid anhydride	carboxylic acid activation, amidation
TBAF	tetra-*n*-butylammonium fluoride	halogenation reagent, F^- source
TBAI	tetra-*n*-butylammonium iodide	halogenation reagent, I^- source
TBDPS	*tert*-butyldiphenylsilyl	protecting group
TBHP	*tert*-butyl hydroperoxide	oxidizing agent
TBS (TBDMS)	*tert*-butyldimethylsilyl	protecting group
TC	thiophene-2-carboxylate	ligand
TEMPO	2,2,6,6-tetramethylpiperidine-1-oxyl	radical agent
Teoc	2-(trimethylsilyl)ethoxycarbonyl	protecting group
TES	triethylsilyl	protecting group
Tf, OTf	triflate, trifluoromethanesulfonate	–
TFA	trifluoroacetic acid	acid
TFAA	trifluoroacetic anhydride	–
THP	2-tetrahydropyranyl	protecting group
TIPS	triisopropylsilyl	protecting group
TMEDA	*N,N,N',N'*-tetramethylethylenediamine	–
TMS	trimethylsilyl	protecting group
TMSE	2-(trimethylsilyl)ethyl	protecting group
Tol	*p*-tolyl	–
TPAP	tetra-*n*-propylammonium perruthenate, $(n\text{Pr})_4\text{N RuO}_4$	–
Tr	trityl, triphenylmethyl	–
Troc	2,2,2-trichloroethoxycarbonyl	protecting group
Ts, Tos	tosyl, *p*-toluenesulfonyl	–

Contents

1   Stereoelectronic Effects   1
    1.1   Applications to Structure and Reactivity   6
        1.1.1   The Anomeric Effect   7
        1.1.2   1,3- and 1,2-Allylic Strain   13
        1.1.3   Fürst-Plattner Rule   16
        1.1.4   Cyclization Reactions and Baldwin's Rules   18
        1.1.5   Thorpe-Ingold Effect   24
        1.1.6   Acidity of Organic Compounds   27
    References   33

2   Carbonyl Chemistry   37
    2.1   Additions to Carbonyls   40
        2.1.1   1,2-Induction   41
        2.1.2   1,3-Induction   49
        2.1.3   4- Versus 6-Membered Transition States   53
    2.2   Formation and Reactivity of Enolates   55
        2.2.1   Lithium Enolates and Silyl Enol Ethers   56
        2.2.2   Boron, Titanium, Tin, and Magnesium Enolates   63
        2.2.3   Alternative Enolization Approaches   71
    2.3   Principles of Aldol Chemistry   74
        2.3.1   Diastereoselective Aldol Reactions   76
        2.3.2   Enantioselective Aldol Reactions   85
        2.3.3   Enantioselective Aldol Reactions—Catalytic Methods   97
        2.3.4   Aldol-Type Addition Reactions   105
    2.4 α-Functionalization Reactions   108
        2.4.1 α-Alkylation—Selectivity and Kinetics   109
        2.4.2   Diastereoselective Alkylations   112
        2.4.3   Enantioselective Alkylations   116
        2.4.4   Hydroxylation, Amination, and Halogenation   123

2.5	1,2- versus 1,4-Addition		127
	2.5.1	1,2-Addition	129
	2.5.2	1,4-Addition	145
2.6	Interconversion of Carboxylic Acid Derivatives		152
	References		161

3 Synthesis and Derivatization of Alkenes and Alkynes ... 179
3.1	Synthesis of Alkenes		180
	3.1.1	Olefination Using Phosphorus Reagents	181
	3.1.2	Olefination by Sulfones	197
	3.1.3	Transition Metal-Mediated Olefinations	205
	3.1.4	Eliminations	218
	3.1.5	Other Methods	225
3.2	Synthesis of Alkynes		226
	3.2.1	Construction of Alkynes Starting from Carbonyl Compounds	227
	3.2.2	Other Approaches to Alkyne Formation	234
3.3	Derivatization of CC-Multiple Bonds		236
	3.3.1	Oxidation of Olefins	237
	3.3.2	Oxidation of Alkynes	286
	3.3.3	Reductions of Alkenes and Alkynes	299
	References		302

4 Oxidation and Reduction ... 319
4.1	Oxidation		320
	4.1.1	Oxidations of Alcohols and Carbonyl Compounds	321
	4.1.2	Selective Oxidations of Primary or Secondary Alcohols in Polyols	346
	4.1.3	Other Oxidations	350
4.2	Reduction		362
	4.2.1	Metal Hydrides	363
	4.2.2	Solvated Metals	383
	4.2.3	Reductive Defunctionalization	390
	4.2.4	Hydrogen as a Reducing Agent	401
	4.2.5	Asymmetric Reductions	427
	References		440

5 Pericyclic Reactions ... 457
5.1	Theoretical and Mechanistic Principles		459
	5.1.1	Conservation of Symmetry and Correlation Diagrams	462
	5.1.2	Perturbation Theory and Frontier Orbital Interactions	468

5.2 Cycloadditions 476
    5.2.1 Reactivity and Selectivity 477
    5.2.2 Diels-Alder Reactions 489
    5.2.3 1,3-Dipolar Cycloadditions 509
    5.2.4 [2+2]-Cycloadditions 518
    5.2.5 Cheletropic Reactions 528
    5.2.6 Further Cycloadditions 536
5.3 Sigmatropic Rearrangements 537
    5.3.1 Mechanism and Selectivity 538
    5.3.2 Modern Variants and Synthetic Applications 552
5.4 Electrocyclic Reactions 556
    5.4.1 Mechanism and Selectivity 556
    5.4.2 Torquoselectivity 558
    5.4.3 Variations of Electrocyclic Reactions 563
5.5 Group Transfer Reactions 571
    5.5.1 Ene Reactions 571
    5.5.2 Diimide Reductions 576
References 578

6 Transition Metal-Catalyzed Coupling Reactions 589
6.1 Mechanistic Aspects 594
    6.1.1 Formation of the Catalytically Active Species 602
    6.1.2 Oxidative Addition 605
    6.1.3 Transmetalation 610
    6.1.4 Reductive Elimination 621
    6.1.5 Off-Cycle Species and Inhibition of the Catalytic Cycle .. 624
6.2 CC Cross-Couplings 626
    6.2.1 Synthesis of the Organometallic Reagents 628
    6.2.2 Characteristics of the Various Cross-Coupling Methods 633
    6.2.3 Modern Developments and Selected Applications 647
6.3 The Mizoroki–Heck Reaction and Related Couplings 661
    6.3.1 Mechanism of the Heck Reaction 662
    6.3.2 Modern Variants of the Heck Reaction and Application
           in Synthesis 669
    6.3.3 Carbopalladation of Alkynes 672
6.4 C-X Coupling Reactions 674
    6.4.1 Buchwald–Hartwig-Type Couplings 675
    6.4.2 Asymmetric Allylic Alkylation 686
    6.4.3 Wacker Oxidation 692

6.5   CH Activation 694

      6.5.1   Mechanisms of CH Activation 696

      6.5.2   Reactivity and Selectivity 700

References 705

7   Metathesis 723

7.1   Alkene Metathesis 724

      7.1.1   Mechanism 726

      7.1.2   Modes of Alkene Metathesis 729

      7.1.3   Modern Applications 733

7.2   Enyne Metathesis 740

7.3   Alkyne Metathesis 744

7.4   Application in the Academic and Industrial Sector 749

References 753

8   Organocatalysis 757

8.1   Reaction Modes, Catalyst Classes, and Applications 760

      8.1.1   Enamine Catalysis 764

      8.1.2   Iminium Ion Catalysis 768

      8.1.3   Umpolung Reactions 776

      8.1.4   Hydrogen Bond-Based Activation and Bifunctional
              Catalysis 781

      8.1.5   Ionic Interactions—Phase Transfer Catalysis and Chiral
              Counterions 787

      8.1.6   Other Mechanisms and Methods 796

References 805

9   Click Reactions 813

9.1   Metal-Catalyzed Huisgen Cycloaddition 814

      9.1.1   Mechanism of the Catalyzed Cycloaddition 815

      9.1.2   Cu-Free Azide-Alkyne Cycloadditions 817

9.2   Thiol-Ene and Thiol-Yne Reactions 818

9.3   Staudinger Ligation 819

9.4   Selected Applications 820

References 822

10 Modern Radical and Redox Chemistry 825

10.1 Electrosynthesis 826

      10.1.1 Basic Principles of Reactivity and Selectivity 828

      10.1.2 Practical Aspects and Applications 832

10.2   VIS-Photoredox Catalysis 835
      10.2.1 Mechanistic Principles 836
      10.2.2 Selected Applications of Photoredox Catalysis 841
References 844

11 Principles of Synthetic Planning 847
  11.1   Retrosynthesis 849
      11.1.1 Nomenclature 852
      11.1.2 Classic Approaches 854
      11.1.3 Control of Stereochemistry 865
      11.1.4 Advanced Concepts 874
  11.2   Requirements of Academic and Industrial Syntheses 883
  11.3   Evaluation and Optimization of Synthesis Routes 894
References 907

Index 915

About the Author

Alexander Düfert studied chemistry at the Universities of Göttingen and Cambridge, UK. After completing his doctorate in organic synthesis under Lutz F. Tietze, he completed a postdoctoral research stay with Steve Buchwald at the Massachusetts Institute of Technology in the field of homogeneous Pd catalysis. After joining BASF's research department, focusing on heterogeneous hydrogenation, he held various positions within the company. Since 2019, he has also been an external lecturer at Saarland University, teaching industrial organic synthesis.

Stereoelectronic Effects**1**

Not just since the thalidomide scandal of the 1950s/60s has the importance of the three-dimensional structure of molecules been in focus for the identification, production, and application of chemical substances. The properties of constitutional isomers and stereoisomers can differ drastically — even enantiomers display different properties in a chiral environment like the human body (see Fig. 1.1) [1, 2].

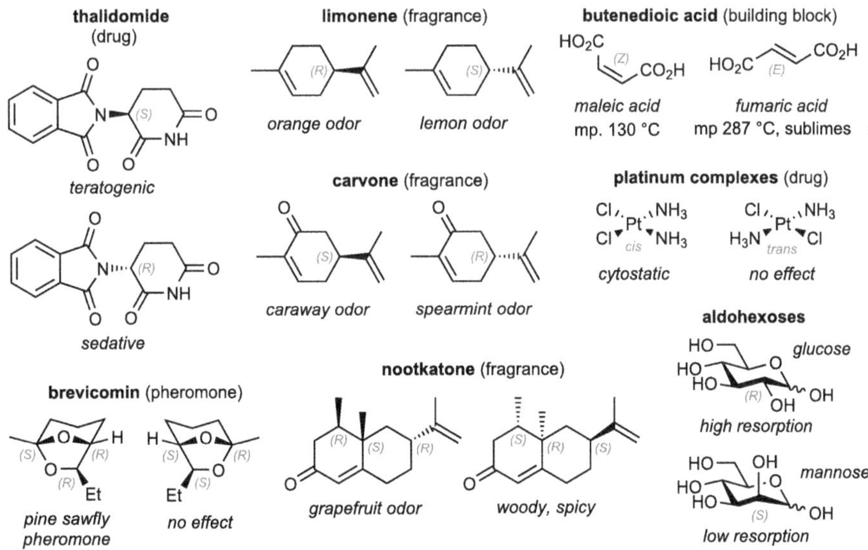

Fig. 1.1 Properties of selected enantio- and diastereomers [1]

A. Düfert, *Methods of Organic Synthesis*,
https://doi.org/10.1007/978-3-662-70963-4_1

Access to target structures with the desired constitution, configuration, and conformation requires methods whose selectivity is often based on the stereoinformation of the corresponding starting materials through substrate control. Additionally, this innate selectivity can be enhanced or even overwritten by the skillful choice of catalysts and reagents.

The chemo-, regio-, and stereoselectivity of organic reactions can be influenced either by steric effects, electronic effects, or a combination of both. Figure 1.2 shows examples of reactions whose selectivity or reactivity is primarily based on either steric or electronic factors: The diastereoselective Michael addition of the cuprate to enone **1** delivers ketone **2** with high facial selectivity and an α-positioned methyl group. In the tritylation of the C5 hydroxy group in the glucose derivative **3**, the protection as carbohydrate **4** occurs solely at the primary OH functionality. In contrast to these sterically controlled reactions, the rate of saponification of differently substituted benzoic acid esters is only influenced by the electronic nature of the *para*-positioned substituent. The stronger the acceptor, the faster the reaction (see Fig. 1.2) [3–5].

R	k_{rel}
NH_2	0.04
OMe	0.24
H	1.0
Br	3.8
NO_2	59

Fig. 1.2 Examples of sterically and electronically controlled reactions [3–5]

Especially in the conformational equilibria of carbo- and heterocycles, the preference of one isomer is governed by steric effects. For example, the relative energies of different conformers and transition states of cyclohexane and cyclohexene result from a combination of angle strain, torsional strain, and van der Waals repulsion (see Fig. 1.3) [6–8].

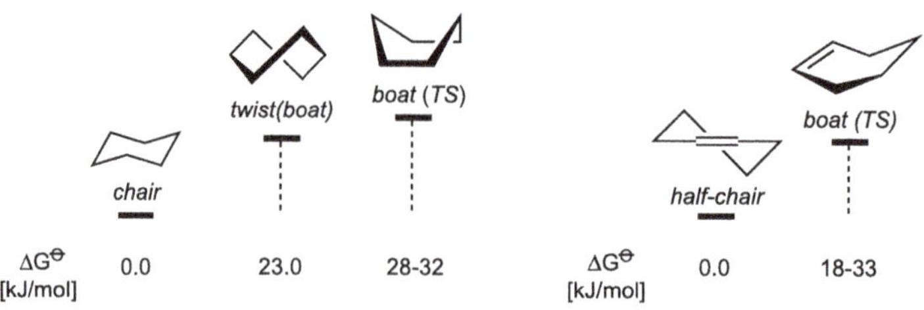

Fig. 1.3 Selected conformers and transition states of cyclohexane and cyclohexene [8–10]

In contrast to the conformers shown in Fig. 1.3, the steric repulsion of axial groups in substituted cyclohexanes can be easily detected directly via spectroscopic methods[I]: The equilibrium of the conformers with either axially- or equatorially-oriented substituent (K_{eq}) can be determined and the free energy ΔG^o can be calculated based on this ratio. In the literature, this is usually referred to as the *A-value* and is often used as an approximation for the steric demand of a functional group (see Table 1.1).

Table 1.1 A-values of common functional groups [11–13]. For OH and NH_2, the values are given in aprotic/protic medium

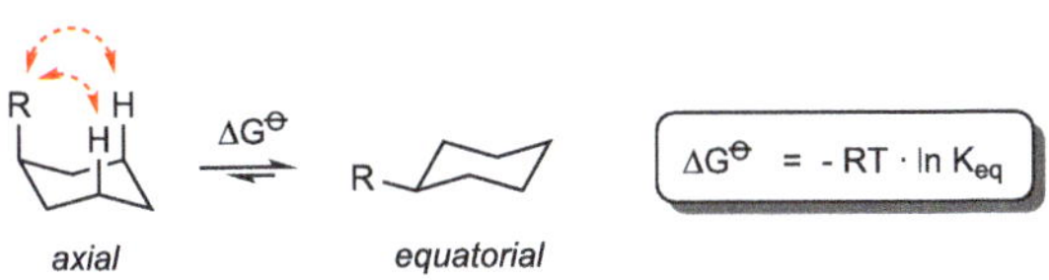

Substituent	ΔG^o [kcal/mol]	Ratio eq./ax.
H	0.0	50 : 50
F	0.26	61 : 39
Cl	0.43	67 : 33
Br	0.38	66 : 34
I	0.43	67 : 33
Me	1.8	95 : 5
*i*Pr	2.2	98 : 2
*t*Bu	>4.5	>99.95 : <0.05
Ac	1.2	88 : 12
CO_2Me	1.27	90 : 10
OMe, OAc	0.6	73 : 27
OH	0.52 (aprot.) / 0.87 (prot.)	71 : 29 / 81 : 19
NH_2	1.2 (aprot.) / 1.6 (prot.)	88 : 12 / 95 : 6
Ph	2.9	99 : 1
$HC=CH_2$	1.7	95 : 5
$C\equiv CH$	0.5	70 : 30
CN	0.17	57 : 43
CF_3	2.2	98 : 2
TMS	2.5	99 : 1

Stereoelectronic effects are more than just the cumulative sum of steric and electronic interactions as exemplified by the reactions in Fig. 1.2. A stereoelectronic effect refers to the influence of a molecule's electronic composition on its three-dimensional structure and reactivity. It encompasses an electronic interaction that depends on the correct geometry and which results in a stabilizing effect due to an increased delocalization of electron density.

[I] However, in this case already a minor influence by stereoelectronic effects can also be calculated by quantum-mechanic methods. See D. S. Ribeiro, R. Rittner, *J. Org. Chem.* **2003**, *68*, 6780–6787.

Localized orbitals (natural bond orbitals, *NBO*) can be used to describe stereoelectronic interactions and roughly correspond to the Lewis structure-like bonding pattern of electron pairs. Primary orbital interactions (from atomic orbitals) form the basis of valence bonds through linear combination. Secondary orbital interactions of bonding orbitals generally lead to a stabilizing effect and lower the total energy of the system [14]. This is referred to as *conjugation*. In addition to the π, π interaction, σ, π and σ, σ interactions also exist. Free electron pairs of an atom X are labelled as n_X. The key to their stabilizing effect likewise rests on the distribution of electron density [15, 16]. *Hyperconjugation* refers to the delocalization of electrons from σ bonds into adjacent molecular orbitals. The distinction between conjugation and hyperconjugation is an academic definition, and both are based on donor-acceptor interactions of orbitals.

Stabilizing donor-acceptor interactions (E_{STAB}) is most pronounced when the geometric overlap is large (e.g., antiperiplanar alignment of the orbitals in $\sigma \to \sigma^*$), the orbitals have the correct symmetry, and the energy difference ΔE_i is minimal. Quantum-mechanically, the orbital interaction is expressed by the Fock matrix F_{ij}, which is proportional to the overlap S^2 (see Fig. 1.4) [17, 18].

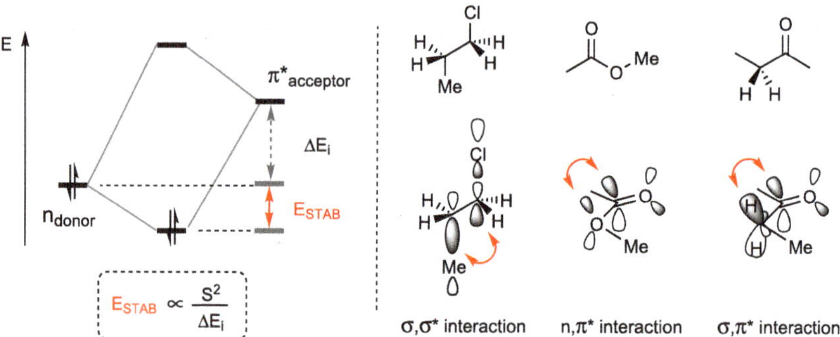

Fig. 1.4 Orbital scheme of a donor-acceptor interaction and types of conjugation

A donor-acceptor interaction leads to a change in bonding. For example, in an $n_X \to \pi^*_{CO}$ interaction, the bond order of a carbonyl group is reduced by increased population of the π^*_{CO} orbital, while an $n_O \to \sigma^*_{CX}$ interaction has the opposite effect. This can be directly observed and confirmed: The CO bond distance is modified accordingly and the change in bonding can be monitored, for example, by IR spectroscopy through a shift of the CO absorption band to lower or higher wavenumbers (see Fig. 1.5) [19–21].

Due to their relative energetic state, the best acceptors are empty p-orbitals, followed by π^*- and σ^*-orbitals. The acceptor capacity is determined by the polarization of the orbital, which correlates with the electronegativity of the heteroatom in CX bonds, and with the energetic position of the orbital.[II] The acceptor capacity of σ^*_{CX}-orbitals generally increases

---

[II] Trends are sometimes difficult to predict, for example with σ(C-Halogen)-orbitals [18].

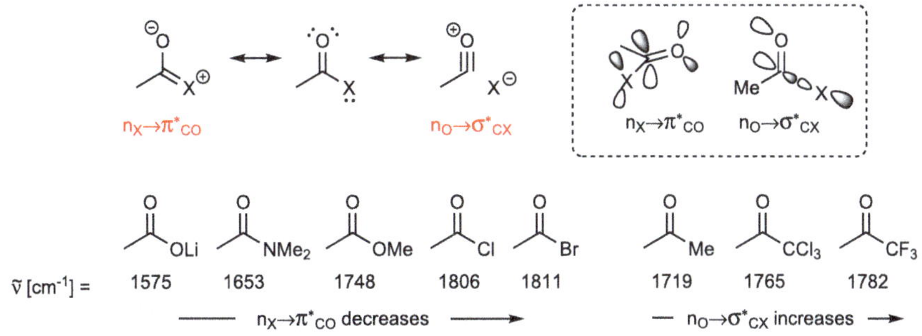

Fig. 1.5 Resonance structures of carbonyl compounds and IR stretching vibration of various carbonyl derivatives [22–25]

within a period (due to polarization) and when moving to higher homologues (due to the energetic position of the orbital). Particularly ΔE counterintuitively leads to the acceptor capacity being more pronounced for C-Br and C-Cl compared to C-F bonds (see Fig. 1.6) [18, 26]. This trend can be observed, for example, in the strength of the anomeric effect, even though steric and electrostatic factors also play a minor role in this case (see below). The stronger the polarization of the orbital, the higher the anisotropy of the bond and thus the orbital coefficients: A strong interaction at the C-terminus of a CX bond results in a weak interaction at the X-terminus. An $n \rightarrow \sigma^*_{CO}$-interaction is therefore not the same as an $n \rightarrow \sigma^*_{OC}$-interaction. This anisotropy applies to both donor and acceptor capacities [18].

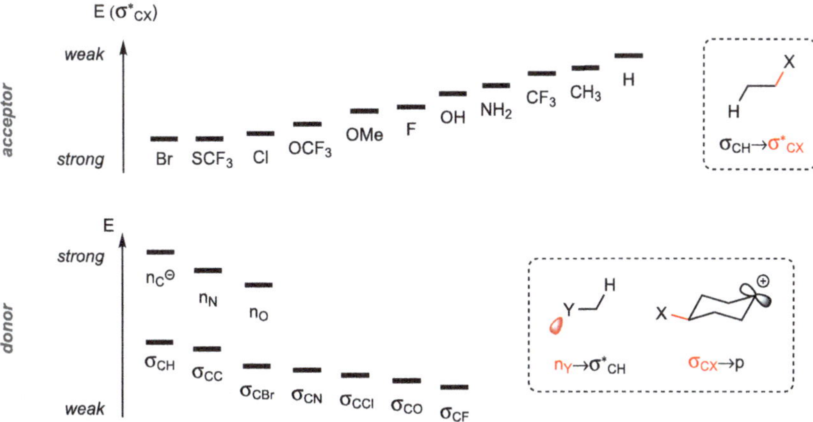

Fig. 1.6 Acceptor and donor capacities of σ^*_{CX}-, σ_{CX}- and n-orbitals [16, 18, 26]

The trend in donor capacities is approximately opposite to that of acceptor capacities. The more electronegative the heteroatom in the CX bond, the lower the energetic level of the σ-orbital and the weaker its donor properties. High polarizability leads to an increase in donor capacity. Since the hybridization and the electronic properties of other substituents can strongly influence the donor properties, the order shown in Fig. 1.6 can — unlike the more robust acceptor capacity — change readily. Particularly the sequence of the relative energetic position of the CC versus the CH bond is the subject of intense discussions [27]. Lone pairs of nitrogen or oxygen atoms are significantly higher in energy and thus enable a stronger donor-acceptor interaction.

The interplay of the donor/acceptor capacity of a single bond and the correct spatial arrangement, which results in a good overlap and thus a strong interaction, can for example be eminently demonstrated by the relative reactivity of phosphites [28]. The rate of the Michael addition of structurally rigid phosphites **7-9** to **5** can be directly correlated with the presence of strong $n_P \rightarrow \sigma^*_{CO}$-interactions: The more electron density transferred to the acceptor orbitals, the more slowly the phosphite reacts (see Fig. 1.7).

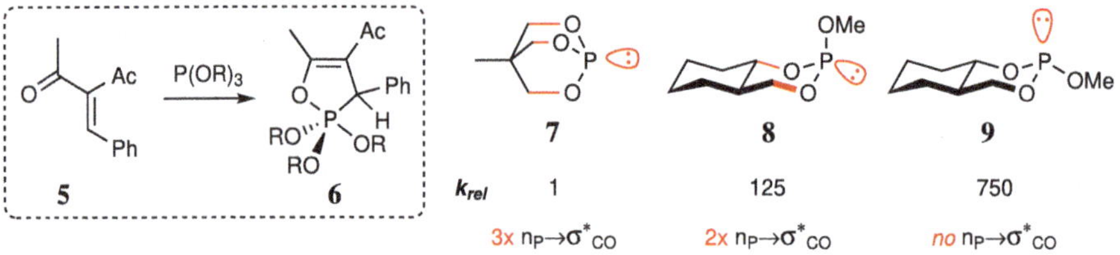

Fig. 1.7 Reactivity of different phosphites with 3-benzylidene-2,4-pentanedione [28]

The concept of (hyper)conjugation and donor/acceptor strengths can rationalize the observed preferences in the conformation of molecules (see this chapter) as well as the latter's impact on the corresponding reactivity: For example in the torquoselectivity of pericyclic reactions, the addition of nucleophiles to carbonyl compounds (Felkin-Anh selectivity), or the selective epoxidation of polyenes.

1.1 Applications to Structure and Reactivity

The most common occurrences of stereoelectronic effects are encountered in the conformational preferences of acyclic molecules. In addition to the classical steric repulsion, the donor and acceptor interactions of σ- and π-bonds determine which structure the molecules adopt in their ground state (see Fig. 1.8) [29].

In the case of butane, $\sigma_{CH} \rightarrow \sigma^*_{CC}$- and $\sigma_{CC} \rightarrow \sigma^*_{CH}$-interactions can lead to a stabilization of the staggered structure. Despite the strong steric repulsion of the methyl groups, the synclinal alignment is therefore energetically almost as favorable as the antiperiplanar

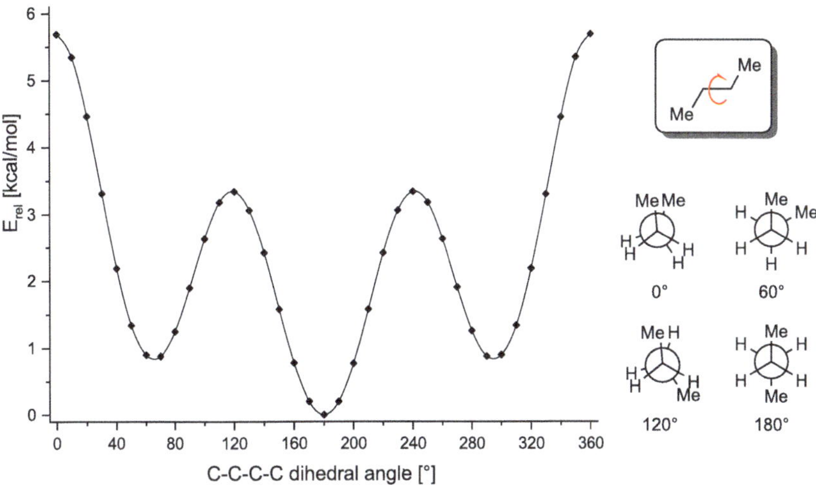

Fig. 1.8 Calculated energy profile of the conformations of butane (B3LYP/6-31G*)

conformation. In the eclipsed conformation, only synperiplanar interactions can stabilize the molecule (E_{STAB}), but due to the less favorable geometric arrangement, they have a significantly weaker influence ($S^2_{anti} \gg S^2_{syn}$, cf. Fig. 1.4).

In the case of vicinal disubstituted alkanes and alkenes, the ground state energies of the conformers or diastereomers can differ to an even larger degree. It has been shown experimentally that in alkanes, those substituents that display the highest acceptor capacity of the σ^*_{CX}-orbitals prefer to adopt a gauche conformation (so-called *gauche effect*) [30–32]. The dominating influence of these interactions can for example be observed in difluoroethane or difluorodiazene (see Fig. 1.9).

Considering the acceptor strengths outlined in Fig. 1.6, dichloromethane and dibromomethane should show an even stronger gauche effect. However, the hyperconjgation is countermanded by steric and electrostatic effects. A synclinal arrangement of the halogen substituents is preferred only in polar solvents, which stabilize the more dipolar gauche conformation [38]. The orbital interactions are thus often a dominant, but not the only factor rationalizing the observed products and transition structures and reactivities of chemical reactions.

1.1.1 The Anomeric Effect

The *anomeric effect* originally referred to the thermodynamic preference of polar groups to adopt an axial position at the anomeric center of glucopyranoses. The underlying stereoelectronic effect can be applied to all geminally disubstituted systems X–C–Y–C (X,Y =

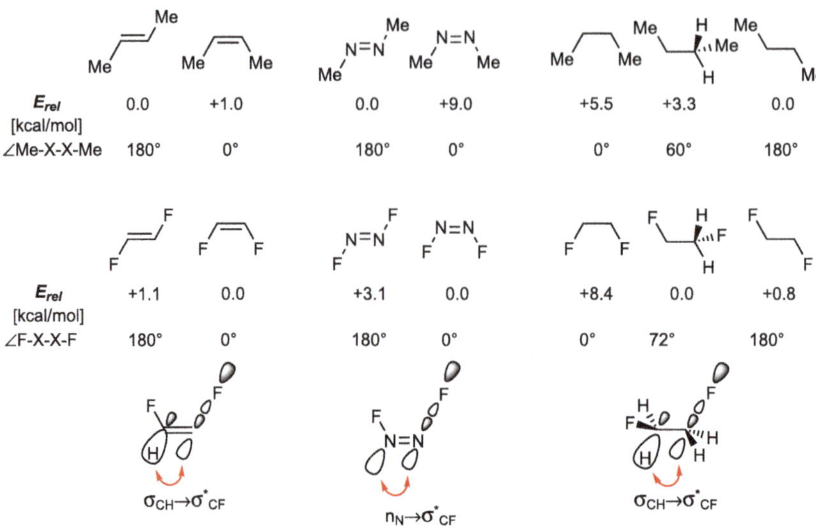

Fig. 1.9 Ground state energies of Me- and F-substituted molecules in the gas phase and dominant donor/acceptor interactions [33–37]

heteroatoms with free valence electrons; mostly O, N, S, F) and describes the preference for a synclinal conformation around the C-Y bond (so-called *generalized* anomeric effect, see Fig. 1.10) [39–42].

Fig. 1.10 Preferred ground state conformations of substituted tetrahydropyrans and geminally disubstituted methylenes bearing heteroatoms

The anomeric effect influences the ratio of configurational isomers in equilibrium as well as the conformational equilibrium. The lower the energy of the σ^*_{CX} acceptor orbital, the more of the axial isomer is observed. In the case of the glucose derivatives **10** and **11**, the more stable α anomer results. β-Haloxylopyranoses even prefer the *all*-axial conformers **13** and **15** by virtue of the anomeric effect (see Fig. 1.11).

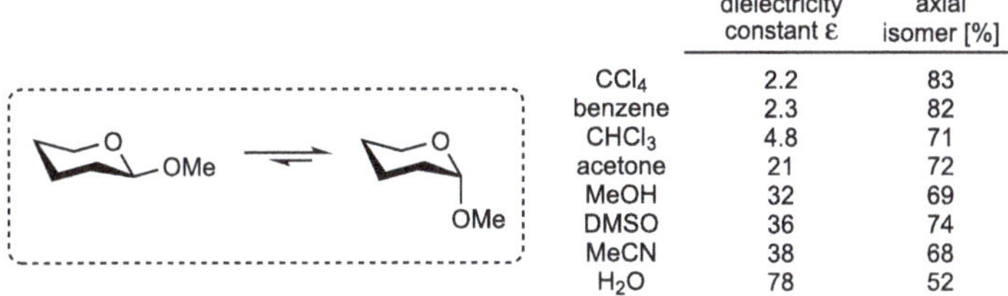

Fig. 1.11 Equilibrium ratios of the anomers (**10**, **11**) or conformers (**12-15**) of gluco- and xylopyranoses [42–44]

The strength of the anomeric effect is depending significantly on the solvent system: Polar solvents reduce the energy difference between the isomers, while a nonpolar medium enhances the anomeric effect (see Fig. 1.12) [45, 46].

	dielectricity constant ε	axial isomer [%]
CCl$_4$	2.2	83
benzene	2.3	82
CHCl$_3$	4.8	71
acetone	21	72
MeOH	32	69
DMSO	36	74
MeCN	38	68
H$_2$O	78	52

Fig. 1.12 Solvent dependence of the anomeric equilibrium [45]

The magnitude of the anomeric effect $\Delta\Delta G^{\ominus}$ can be approximated by the energy difference of the isomers in tetrahydropyrans *vs.* cyclohexanes in chemical equilibrium, neglecting certain structural differences [47]. In cyclohexanes, the ratio of isomers is solely determined by steric interactions. The difference to the observed ratio in tetrahydropyrans is thus of a stereoelectronic nature and correctly reflects the trends qualitatively (see Fig. 1.13). The magnitude of the anomeric stabilization, depending on the substituent and solvent, is significant, with 1–2.5 kcal/mol [39, 40, 48].

Fig. 1.13 Quantitative estimation of the anomeric effect according to Franck [47]

Two explanations are cited as the cause of the conformational preference of the anomeric effect: an $n \rightarrow \sigma^*$ interaction or a dipole-dipole interaction [39, 40, 42, 49]. The explanation via a conjugative interaction was derived from X-ray crystallographic analyses, in which a longer axial CX bond is observed compared to the equatorial isomer. The partial shift of the electron density of the lone electron pair at the oxygen into the vacant antibonding orbital σ^*_{CX} explains the preference of the α-isomer, which is energetically most stabilized in the axial position by the optimal overlap.

As a mesomeric Lewis structure of **20**, a zwitterionic species **21** can be formulated by shifting the electron density of the free electron pairs, which should lead to shortened CO and lengthened CX bonds. In chloro-substituted dioxanes a shorter C-O and a longer C-Cl bond are indeed observed as expected for axial compared to equatorial substituents. This is predicted assuming a hyperconjugation, leading to a strengthening of the CO and weakening of the CX bond [50]. The same trends are found in a variety of tetrahydropyrans (see Fig. 1.14) [42].

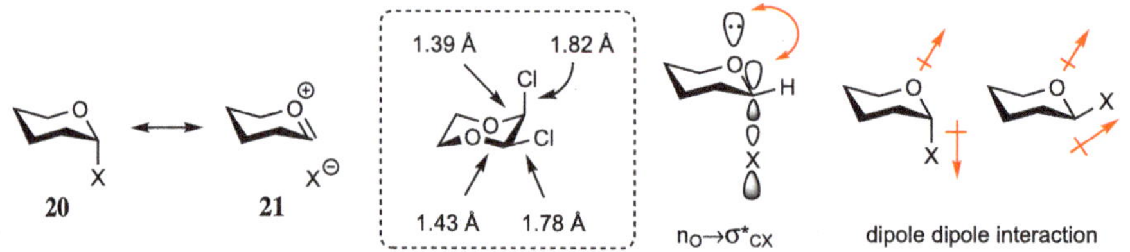

Fig. 1.14 Mesomeric structures, bond lengths in 1,4-dioxane and postulated interactions [40]

The reduction of electrostatic interaction of the dipoles in the polarized covalent bonds constitutes an alternative explanation. The orientation of the dipoles in the axial isomer leads to a minimization of the total energy and also favors the α conformation. Both hyperconjugation and dipole-dipole interaction can be substantiated by experimental observations. While the changes in bond distances can be explained with an $n_O \rightarrow \sigma^*_{CX}$ interaction, the considerable solvent dependence can be well rationalized by the electrostatic model. Nevertheless, hyperconjugation has become mostly accepted as the presumed cause of the anomeric effect, with contributions from electrostatic interactions also potentially influencing the energetic equilibrium.

In addition to the impact of the anomeric effect on the structure, also a significant deviation in reactivity can result. In the direct comparison of the reactivity of α- and β-anomers, surprisingly the β-isomer is typically converted faster, although the α-isomer should have the more labile CX bond due to the $n_O \rightarrow \sigma^*_{CX}$ interaction. This so-called *kinetic* anomeric effect, however, is based on the lower activation energy $\Delta G^{\ddagger}$ of the β-stereoisomer due to the stabilization of the α-isomer by the anomeric effect [42, 51, 52]. Thus, in the structurally more rigid oxadecalin **22**, Kirby *et al.* observed a nearly threefold increase in rate of the α- compared to the β-anomer in the reaction to **23** (see Fig. 1.15) [53].

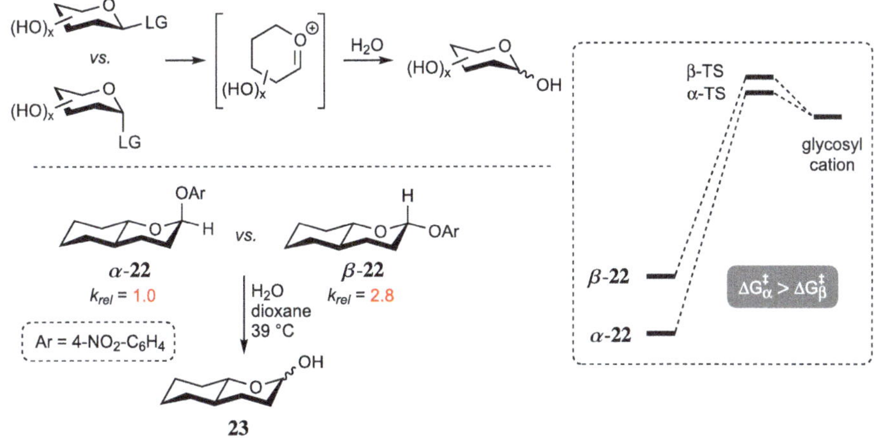

Fig. 1.15 Hydrolysis of tetrahydropyrans and kinetic anomeric effect [53]

Especially in the syntheses of spiroketals, the anomeric effect is commonly used for a diastereoselective access to the desired products. Spiroketals are often found as structural units in secondary metabolites from marine, bacterial or plant sources [54, 55]. Several possible isomers can be constructed, but only one experiences a double anomeric stabilization (see Fig. 1.16).

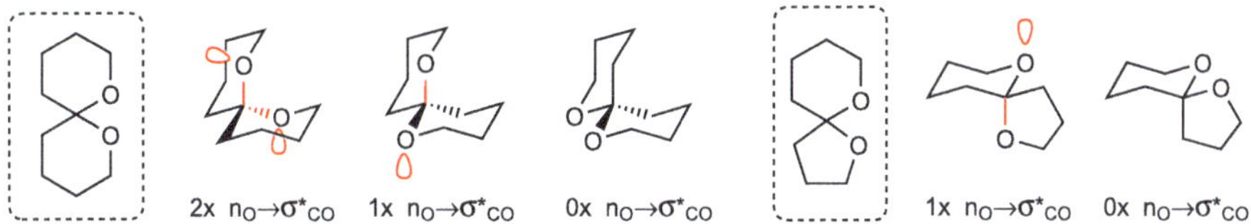

Fig. 1.16 Isomers of [6,6]- and [5,6]-spiroketals and possible stabilizing $n_O \rightarrow \sigma^*_{CO}$ interactions (highlighted in red)

The multiple anomeric stabilization of spiroketals was pivotal in the synthetic efforts of the Smith group towards the marine natural product spirastrellolide B, which contains a [6,6,5]-bispiroketal motif [56]. Based on a cyclization developed by De Brabander, alkyne **24** was converted to the hemiketal [57]. Following the PMB-deprotection with DDQ, **25** was reacted to the desired bisketal **26** in the presence of PPTS. In the three-step one-pot reaction, the triple anomeric-stabilized product could be obtained in 38% in a diastereomeric ratio of 7:1 (see Fig. 1.17).

In Ley's synthesis of spongistatin 1, a 1:1 mixture of the *R/S*-diastereomers of **27** was used, which seems counterintuitive but did yield great merit in the approach to the marine macrolide. While the *S*-configured starting material exclusively afforded the double anomerically-stabilized spiroketal **28**, the corresponding *R*-configured diastereomer of **27** gave a 3:1 ratio of the double (**29**) to the single anomerically-stabilized cyclization product

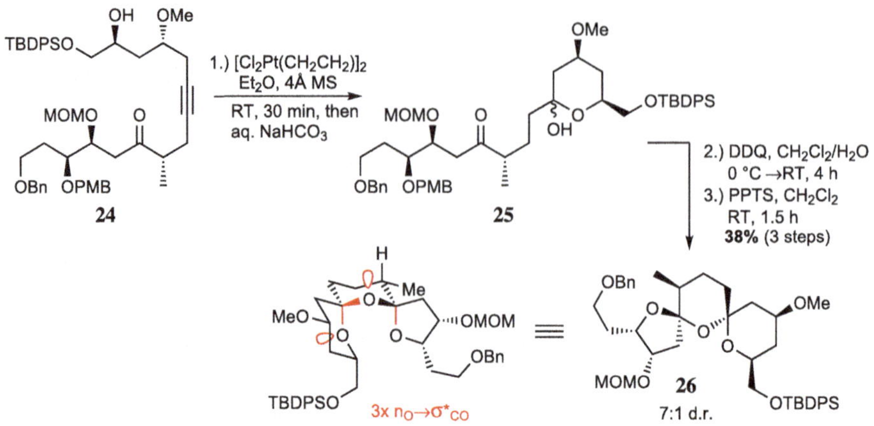

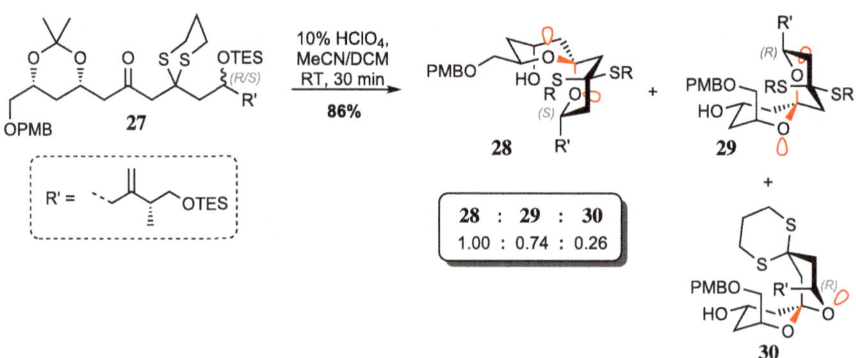

Fig. 1.17 A.B. Smith's synthesis of the C(26)-C(40) fragment of spirastrellolide B (**26**) [56]

(**30**). In contrast to the clear preference for one diastereomer **29** *vs.* **30** due to stereoelectronic interactions, the substitution pattern can counteract the preferred formation of the double anomerically-stabilized product **29** through steric interactions. This was intended, as **30** represents the desired product. The product mixture was subsequently used to access the northern and southern hemisphere of the target structure: **28** afforded the AB fragment, while **30** was converted to the CD framework. **29** could then also be transformed to **30** by epimerization, so that the structural elements of the two diastereomers **28** and **30** are ultimately found in the complex natural product (see Fig. 1.18) [58].

Fig. 1.18 Synthesis of the AB- (**28**) and CD-fragments (**30**) of spongistatin 1 according to Ley [58]

Beyond the impacts of anomeric effect discussed so far, the preference of cationic substituents in tetrahydropyrans to reside in the equatorial position is referred to as the *reverse anomeric effect* (see Fig. 1.19). Its existence remains a topic of controversial discussion still today, especially since the observed β-preference can often be attributed to steric or electrostatic interactions. A supposed $n_O \rightarrow \sigma^*_{CN^+}$ molecular orbital interaction would even predict an enhanced "regular" anomeric effect. Therefore, its existence has not been confirmed with certainty thus far [41, 49, 52].

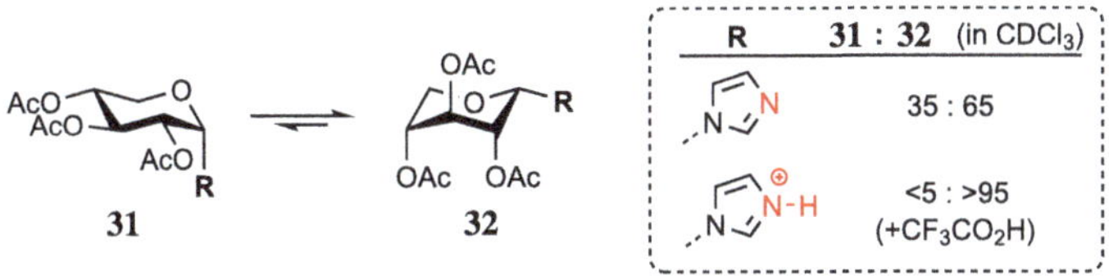

Fig. 1.19 Enhanced preference for the equatorial conformer due to protonation in glycosylimidazoles [59]

The *exo-anomeric effect* is also frequently encountered. However, it represents a special case of the generalized anomeric effect, whereby the arrangement of the R-substituent in OR groups on tetrahydropyrans is preferably synclinal, analogous to acyclic systems (see Fig. 1.10) [40, 42]. The anomeric effect not only encompasses oxygen as a donor but also other heteroatoms. It is typically observed in cyclic compounds such as dithianes or nitrogen-containing heterocycles as well as in acyclic molecules, for example in (hemi-)ketals.

1.1.2 1,3- and 1,2-Allylic Strain

The titular allylic strain refers to the destabilization of a molecule or a conformation based on a van der Waals repulsion between a substituent at a double bond and one in an allylic position. Depending on the position of the substituents at the double bond, it is referred to as 1,2- or 1,3-allylic strain, which is also abbreviated to $A^{1,2}$- or $A^{1,3}$-strain (see Fig. 1.20) [60, 61].

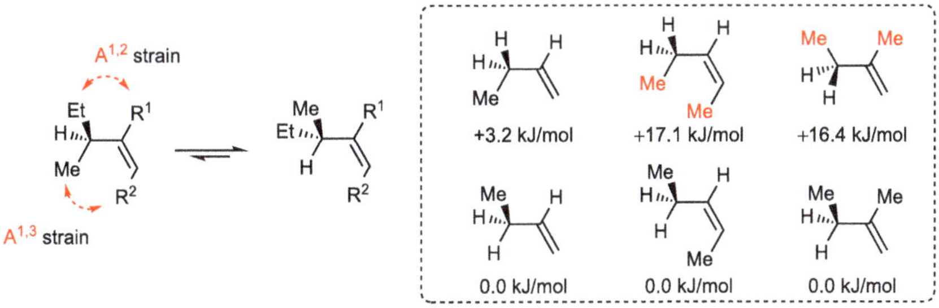

Fig. 1.20 Definition of the allylic strain and calculated energies of conformers in the gas phase (B3LYP/6-31G*)

As is evident from the calculated energy of the conformers shown in Fig. 1.20, the magnitude of the 1,3-allylic strain is significantly greater than the anomeric effect. In fact, the $A^{1,3}$-strain is commonly referred to as the strongest stereoelectronic effect. As governed by the steric repulsion, rotation around the CC-single bond gives the preferred conformer, in which this interaction is minimized. This allows to selectively promote reaction pathways that lead to the exclusive formation of a desired stereoisomer. With an energetic difference of about 17 kJ/mol, impressive selectivities beyond 99.9 : 0.1 can be achieved.

Danishefsky *et al.* used an 1,3-allylic strain to control a hetero-Diels-Alder reaction in their synthesis of the natural product grandisin A (**38**) [62]. The BF$_3$-mediated construction of the *cis*-linked pyranone **37** could be achieved starting from the racemic dehydropiperidine **33**. In the transition state **34**, the desired diastereomer was formed by avoiding the 1,3-allylic strain between the vinyl substituent and the TBS-protected hydroxy group, with only the *endo*-product **36** being isolated. In the diastereomorphic transition state **35**, said repulsion destabilizes the approach of the activated aldehyde to the β-face. **38** was obtained after deprotection in nine further steps (see Fig. 1.21).

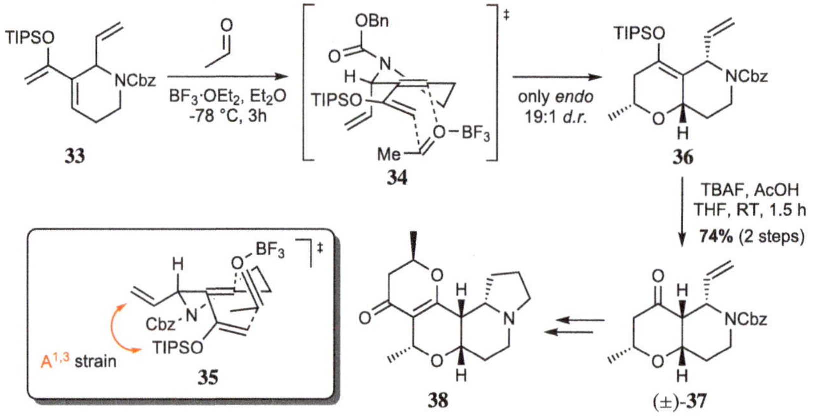

TIPSO
NCbz
33
BF₃·OEt₂, Et₂O
-78 °C, 3h
OBn
O
N
H
TIPSO
Me
O–BF₃
34
only endo
19:1 d.r.
TIPSO
NCbz
O
H
36
TBAF, AcOH
THF, RT, 1.5 h
74% (2 steps)
H
Cbz-N
TIPSO
A^1,3 strain
35
O
H
H
N
O
H
38
O
H
NCbz
O
H
(±)-37

Fig. 1.21 Danishefsky's synthesis of grandisin A (**38**) [62]

The members of the spinosyn class of natural products possess potent anti-insecticidal properties, the main component of which, spinosyn A, is marketed along with other metabolites as the commercial pesticide Spinosad. The synthetic access to the 5/6/5-ring skeleton of spinosyn A by the Roush group was realized using a transannular Diels-Alder reaction [63]. In the domino reaction, the macrolactone was closed in the first step by a Horner-Wadsworth-Emmons reaction of **39**, followed by the cycloaddition of **40** to **42**. The two transition states **41** and **43** differ in their conformation. This leads to the avoidance of a 1,3-allylic strain occurring only in **43** to furnish the desired diastereomer **42** via the $A^{1,3}$-free and thus preferred transition state **41** (see Fig. 1.22).

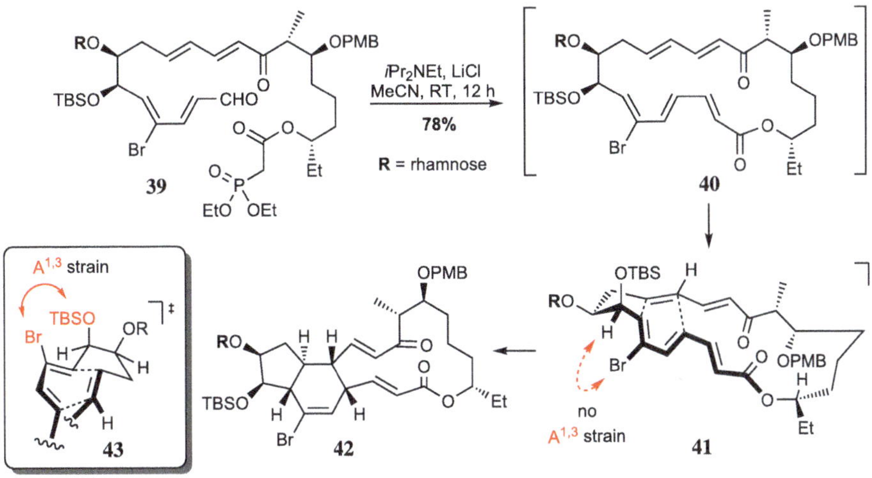

Fig. 1.22 Synthesis of spinosyn A by the Roush group [63]

The 1,2-allylic strain is less commonly used for reaction control, even though the strength of the repulsive interaction also allows for a good facial differentiation of the double bond. Evans and co-workers used this to selectively introduce a stereogenic center in the polyether antibiotic lonomycin A [64]. The hydroboration of the substituted tetrahydropyran **44** to **45** preferably proceeds via transition state **46**. The approach to the α-face, in which an $A^{1,2}$ interaction is avoided, then affords the observed diastereomer as the main product. In **47**, the allylic strain destabilizes the transition state (see Fig. 1.23).

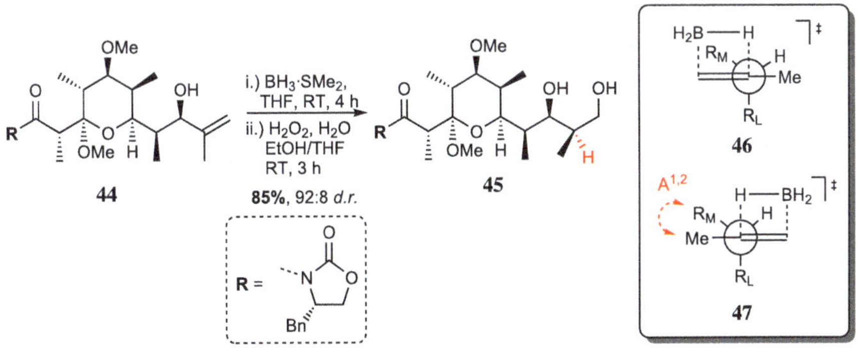

Fig. 1.23 Evans' synthesis of lonomycin A [64]

In hydroborations, the energy differences between the transition states caused by the 1,2-allylic strain are in the range of 1.5–2.5 kcal/mol, often achieving diastereomeric ratios of >95 : 5 [65].

1.1.3　Fürst-Plattner Rule

The Fürst-Plattner rule or *trans*-diaxial effect describes the facial selectivity or regiose-lectivity in addition reactions to cyclohexene derivatives that preferably adopt a half-chair conformation. Beyond substituted cyclohexenes, this can additionally include epoxides or aziridines derived from cyclohexenes, cyclic oxo- or azacarbenium ions, as well as bromo-nium or iodonium species (see Fig. 1.24) [66].

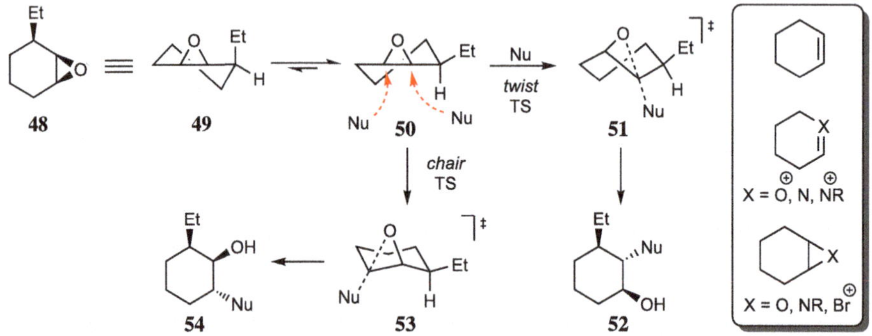

Fig. 1.24 Rationale of the Fürst-Plattner rule

Mechanistically, the observed selectivity is based on two competing transition states (**51** *vs.* **53**), the relative energetic position of which determines the product distribution. Starting from the epoxide **48**, two conformers can be adopted. In the depicted case the pre-equilibrium lies on the side of the conformer with (pseudo-)equatorial orientation of the ethyl group (**50**). The nucleophilic attack on **50** can occur on both sides of the epoxide. Due to the $n_{Nu} \rightarrow \sigma^*_{CO}$-interaction, the nucleofugal oxygen atom is aligned at an 180° angle with regard to the nucleophile, as one would expect in a nucleophilic substitution. This results in a twist-like (**52**) transition state for the reaction to **53** and a chair-like transition state (**54**) for the reaction to **54**. In analogy to the relative stability of the conformers in cyclohexane (see Fig. 1.3), the chair-like transition state is strongly preferred, resulting in **54** being obtained as major product.

The ratio of the obtained isomers can also display a limited solvent-dependence. The more polar a solvent, the later the transition state should occur, which is why the energy difference between the twist- and the chair-like transition state should be enhanced. With increasing polarity of the solvent a growing preference of the double (**58**) over the single anomerically-stabilized spiroketal **57** can indeed be observed (see Fig. 1.25) [67].

In the synthesis of (–)-quinocarcin by Zhu *et al.*, the stereocenter in the C-ring and thus the control of the stereochemistry of the C/D-ring was achieved by an addition reaction governed by the effects outlined in the Fürst-Plattner rule [68]. The hydroxy group in **59** was cleaved in the presence of the strongly oxophilic acid Hf(OTf)$_4$ with concomitant deprotection of the

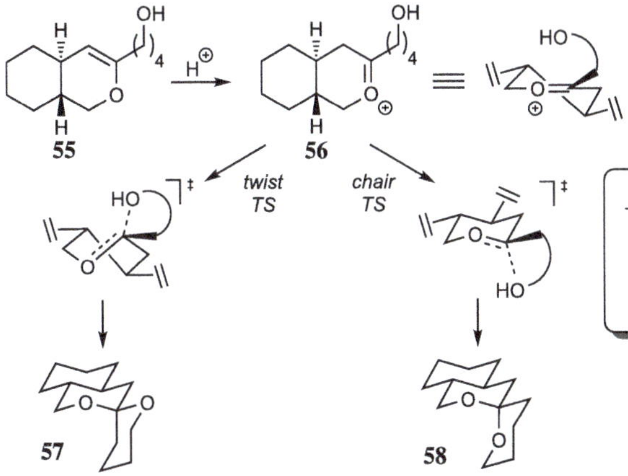

solvent	ε	57 / 58
benzene	2.3	1 : 1.5
CHCl$_3$	4.8	1 : 1.6
CH$_2$Cl$_2$	9.1	1 : 1.7
MeCN	37.5	1 : 2.6

Fig. 1.25 Solvent-dependence of diastereoselectivity according to the Fürst-Plattner rule [67]

silyl ether to the azacarbenium species **60**. The attack of the thioethanolate on the rigid half-chair can occur either from the α- or the β-face. **62** was obtained as the only diastereomer in 88%, presumably via the chair-like transition state **61**. The corresponding diastereomer **64**, which would result from the twist-like transition state **63**, was not observed (see Fig. 1.26).

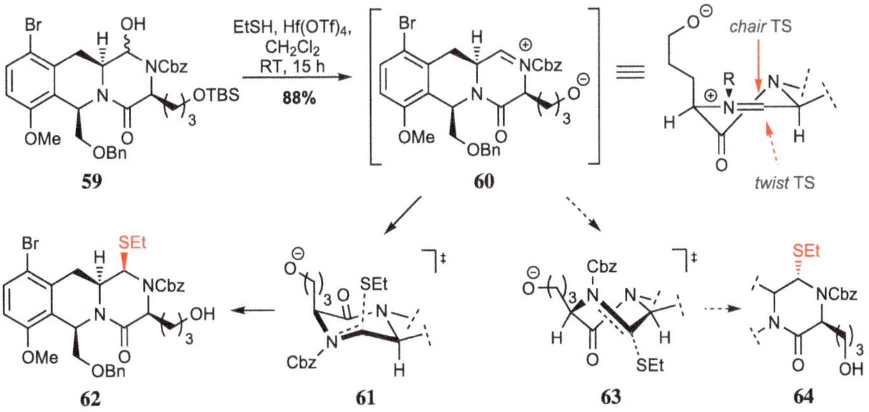

Fig. 1.26 Application of the Fürst-Plattner rule in Zhu's synthesis of (–)-quinocarcin [68]

Singaram and co-workers at Dow utilized the selectivity of the epoxide-opening to **67** according to the Fürst-Plattner rule to separate a diastereomeric mixture of limonene epoxide by kinetic resolution [69]. The obtained products are various β-amino alcohols, which can thus be easily accessed from the respective terpenes and might serve as chiral ligands. A

1:1 mixture of **65** and **66** could be cleaved by secondary amines, with only **65** reacting. In the half-chair form, a chair-like transition state leads to an opening of **65** at the lower substituted C-atom of the epoxide. **66** can only be opened at the higher substituted terminus in the chair-like transition state and therefore does not react (see Fig. 1.27).

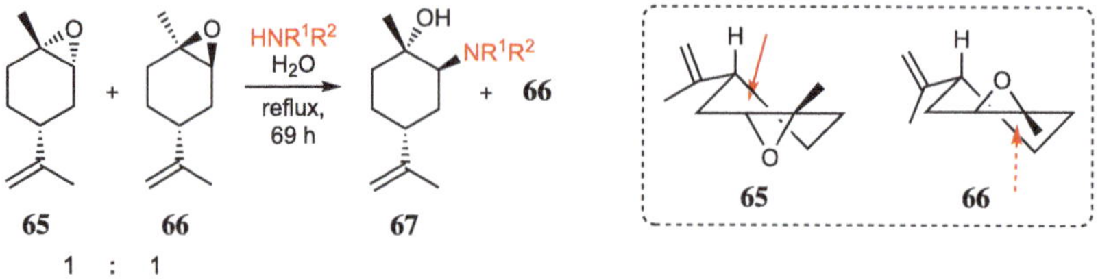

Fig. 1.27 Kinetic resolution via epoxide opening [69]

1.1.4 Cyclization Reactions and Baldwin's Rules

Natural products, their derivatives, or pharmaceuticals contain a plethora of carbo- and heterocyclic structures. In addition to a holistic nomenclature of all cyclization reactions, the British chemist Jack E. Baldwin developed a set of empirical guidelines which ring closing reactions should be preferred [70]. *Baldwin's rules* have become a fundamental tool in the hands of synthetic chemists to predict the relative facility of a given transformation. Subsequent investigations could show that the corresponding predicted selectivities as expressed by the activation energies of competing cyclization pathways can be rationalized by stereoelectronic parameters (see Fig. 1.28) [71–73].

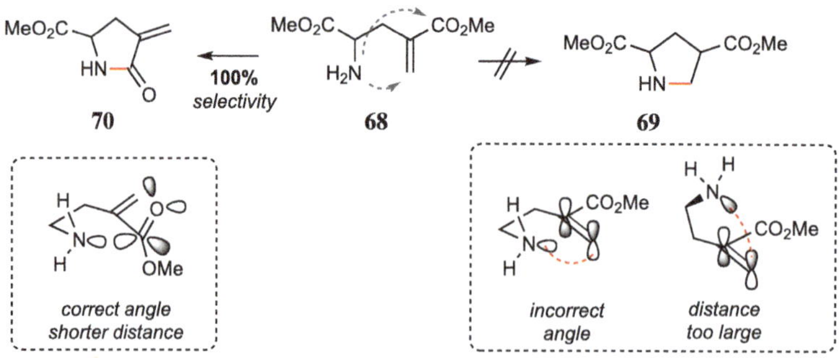

Fig. 1.28 Selectivity of the cyclization of β'-aminoacrylates [74]

For parallel reaction trajectories, the interaction of the attacking nucleophile with the orbitals of the broken bonds is maximized (σ_{CX}^*, π_{CX}^*). The nomenclature for cyclization reactions consists of

- the resulting ring size (3–7)
- the location of the involved broken bond (*endo*, *exo*) and
- the hybridization of the reacting atom in the ring (*tet*, *trig*, *dig*)

(see Fig. 1.29).

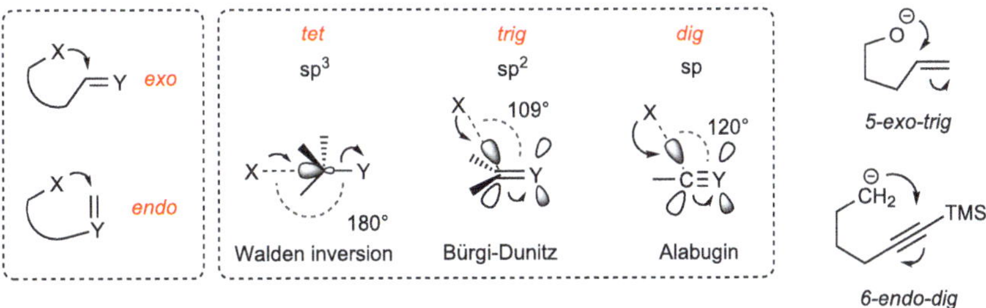

Fig. 1.29 Nomenclature of the Baldwin rules and favored reaction angles

Baldwin postulated for common cyclizations whether they are **favored** or **disfavored**. The Baldwin rules refer to the *kinetics* of a reaction. This categorization covers a general, empirical assessment of the feasibility of a reaction based on the involved orbitals, bond lengths, and molecular structure. A disfavored reaction can still occur, but it proceeds significantly slower than a competing favored reaction. Despite *endo-tet* reactions by definition not being cyclizations, their transition states should proceed in a cyclic manner and can thus also be contained within the frame of Baldwin's rules. Based on the seminal work by Baldwin, subsequent experimental and theoretical studies have been published over the years for their confirmation and extension (see Fig. 1.30) [75–77].

Baldwin's rules are not strictly adhered to and show quite a few exceptions. Higher homologues (X, Y ≥2. period) and pericyclic reactions are not covered. The larger atomic radii and longer bonds allow greater flexibility for atoms of the second and higher periods and therefore they often do not follow Baldwin's rules. Pericyclic processes, on the other hand, are governed by orbital symmetries (Woodward-Hoffmann rules). In the case of thermodynamically controlled, reversible cyclizations, the thermodynamically favored product is also formed, regardless of the prediction of Baldwin's rules. Cationic reactions can also only be anticipated to a certain degree under these guidelines [78].

Although not shown in Fig. 1.30, Baldwin also commented on cyclizations of seven-membered systems, however few subsequent studies have been reported. In the original publication, all 7-*exo/endo* reactions were classified as favored, with 7-*endo-tet* as a disfavored exception.

For the opening of epoxides, the reactions are often categorized as following an *exo-tet*- and *endo-tet* mode. This is not correct according to Baldwin's original definition: both paths proceed *exo-tet*. To be more precise, alcohols with an *exo*- or *endo*-CC bond compared to

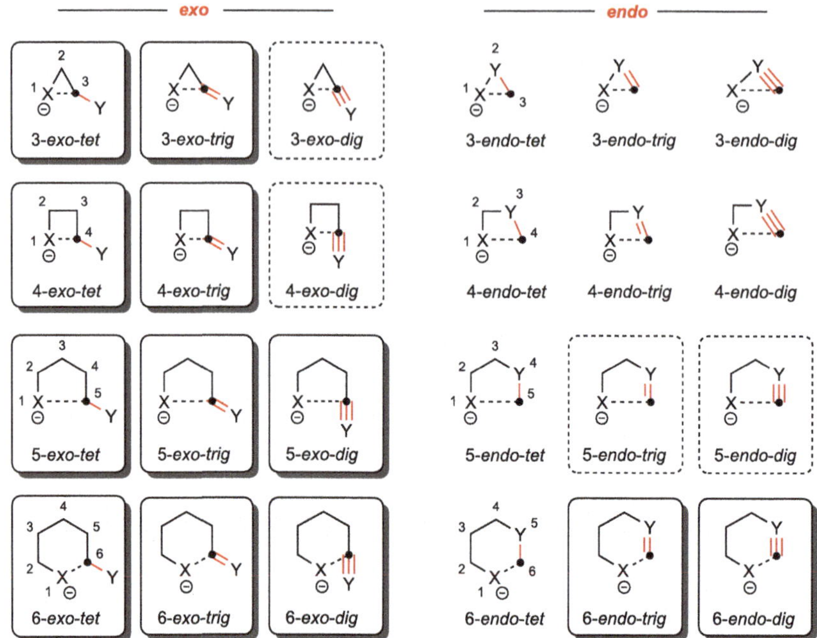

Fig. 1.30 Extended Baldwin's rules for nucleophilic and radical ring closure reactions [72]. Solid boxes indicate favored cyclizations, dashed boxes indicate borderline cases that require additional support. Reactions without a box are disfavored

the CC bond of the epoxide are formed [78]. Depending on the reaction conditions, the selectivity can be modified for a given substrate (see Fig. 1.31).

Fig. 1.31 Examples of epoxide openings [79–83]. Dashed arrows indicate the reaction to the minor isomer

The regioselectivity of the oxirane opening can be rationalized by the relative rate of formation of the corresponding cyclic ethers, as all *exo-tet* reactions are favored according to Baldwin. The influence of ring size on different cyclizations was systematically investigated by the groups of Illuminati and Calli: The cyclization rate can be attributed to the activation energy and the likelyhood of the ring termini coming into proximity. Since these ring closures entail a cyclic, product-like transition state, the activation energy approximately reflects the ring strain of the product. This varies greatly depending on the ring size and, apart from 3- and 4-membered rings, is particularly high for medium-sized rings. On the other hand, the probability of both termini meeting decreases with increasing chain length. As a result of these cumulative effects, five-membered rings form the fastest, followed by 6- or 4-membered (hetero)cycles (see Fig. 1.32) [84]. Deviations from the expected regioselectivity are often due to the stabilization of partial charges, which energetically favor the transition state of the "wrong" regioisomer.

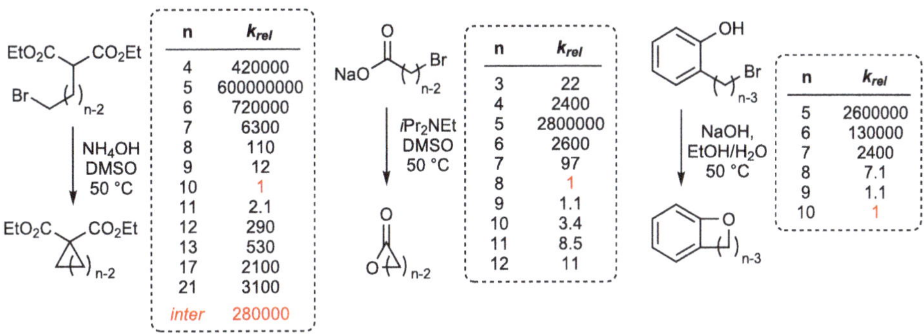

Fig. 1.32 Relative rates of different ring closure reactions with ring size n [85–87]

Endo-tet reactions are disfavored according to Baldwin's rules. This classification is consistent with the observation that for reactions to proceed intra- and no longer intermolecularly, a cyclic transition state of a minimum ring size is required [88]. For example, Eschenmoser and co-workers were able to elegantly demonstrate the intermolecular reaction path in 6-*endo-tet* reactions. In crossover-experiments of deuterated and undeuterated substrates, the statistically expected isomer ratio of 1:2:1 resulted, which would only be obtained by following an intermolecular pathway (see Fig. 1.33) [89].

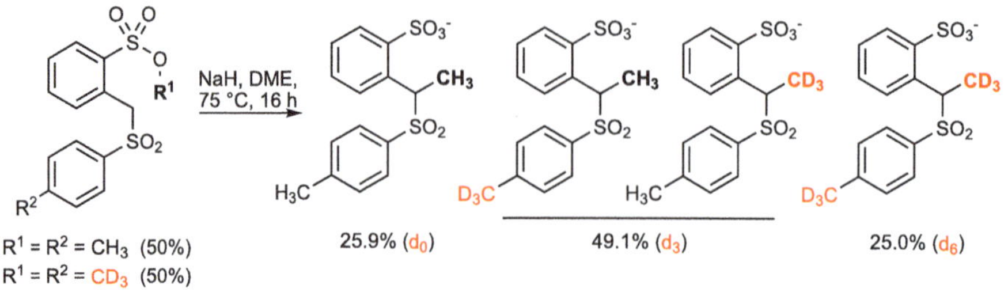

Fig. 1.33 Crossover-experiment according to Eschenmoser *et al.* [89]

Macrocyclic natural products typically comprise 12 or more ring atoms, which is why they are not covered by Baldwin's rules. Their formation is competing with that of the corresponding dimers or oligomers due to the low thermodynamic driving force of the ring closure. As can be seen on the left side of Fig. 1.32, the rate of formation in large rings hardly shows any correlation to the chain length of the substrates. However, if one compares the kinetics of the *inter*molecular conversion of malonic esters with the intramolecular reaction ($n > 6$), it is evident that the rate of the intermolecular process is significantly higher under the chosen conditions [85]. There are several approaches to counteract this innate selectivity and direct a reaction to favor the respective cyclization: Working under (pseudo-)diluted conditions as well as the preformation of the ring structure by complexing the reactive termini (see Fig. 1.34) [90–94].

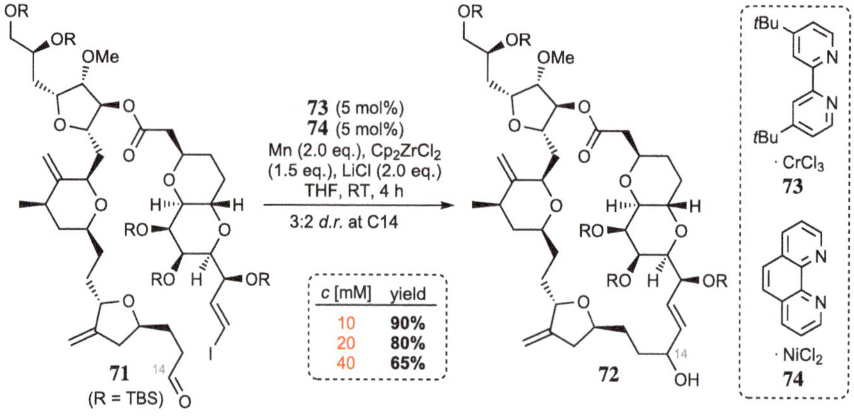

Fig. 1.34 Influence of the concentration on the yield of the Nozaki-Hiyama-Kishi reaction in the synthesis of halichondrin B [95]

If the ring closure is irreversible (kinetic control), it is often necessary to add the reactant to the reaction mixture slowly in order to avoid using large amounts of solvents (*semibatch* process): This keeps the concentration of the added reactant low, as it is constantly consumed and the rate of reaction increases significantly relative to the reaction volume. This procedure is also referred to as pseudo-dilution, infinite dilution or simulated high dilution. Biphasic conditions similarly result in low concentrations of the substrates or reagents in the boundary layer [84, 90].

Structural features of the substrate, such as the ability to form hydrogen bonds, can additionally favor an intra- over an intermolecular pathway. Apart from the ring size, the substitution pattern has a decisive influence on the kinetics of cyclization. It can either favor or hinder the adoption of the required reactive conformation (see Fig. 1.35).

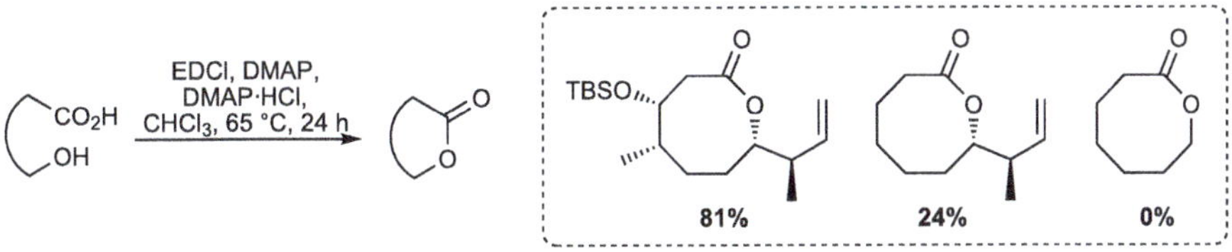

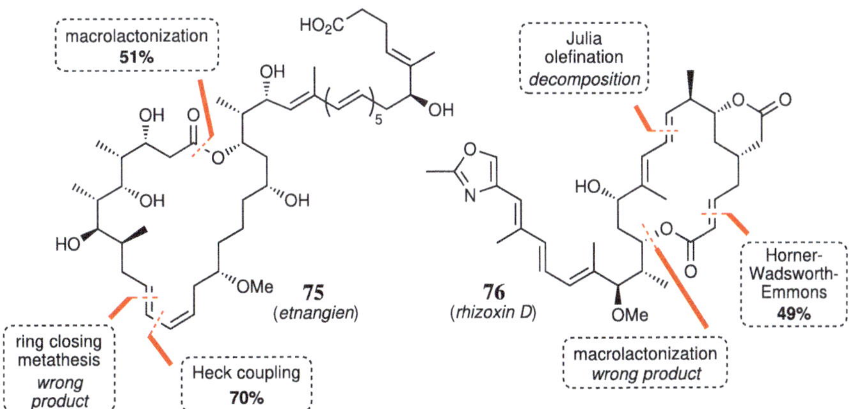

Fig. 1.35 Cyclization of differently substituted 7-hydroxyheptanoic acids [96]

Macrocycles can theoretically be constructed using a variety of approaches, however in practice, some methods are used predominantly to accomplish a ring closure. This can often be related to the presence of certain functionalities such as ester or amide groups, which is why macrolactonizations and -lactamizations are used very widely by synthetic chemists (see Sect. 2.6). In addition, alkene and alkyne ring-closing metatheses (see Chap. 7) as well as Pd-mediated coupling reactions (see Chap. 6) are frequently employed for CC-bond formations (see Fig. 1.36) [92, 94, 97].

Fig. 1.36 Macrocyclization approaches in the total syntheses of etnangien (**75**) and rhizoxin D (**76**) [98, 99]

1.1.5   Thorpe-Ingold Effect

The significance of the substitution pattern in cyclization reactions was already illustrated by the examples shown in Fig. 1.35. The *Thorpe-Ingold effect* or geminal dialkyl effect comes to bear for a certain substitution arrangement, which is used advantageously for improving slow ring closure reactions: It describes an acceleration of a cyclization when a quaternary carbon atom is contained in the backbone of the acyclic precursor (see Fig. 1.37) [100].

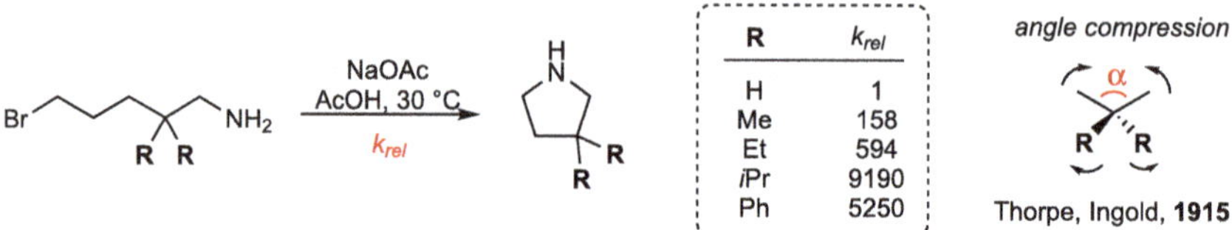

Fig. 1.37 Acceleration of cyclizations by geminal disubstitution [101]

The original rationalization for this rate enhancement was initially attributed to a change of the bonding angles at the quaternary carbon atom. Due to mutual steric repulsion of the alkyl substituents their bonding angle widens. As a result, the reactive termini show a smaller angle α as they are pushed closer together, and react more rapidly by virtue of their proximity. While this so-called angle compression has been observed by X-ray structural analyses, it is small and therefore should contribute only to a limited degree to the rate acceleration.

The potential causes of the effect have been investigated by many groups over the years and can be attributed to the interplay of several factors. The postulated aspects include (i.) a change in the bonding angles of the substrate and product (angle compression in the substrate and ring strain in the product), (ii.) a faster ring closure reaction due to thermodynamic or kinetic factors (e.g., destabilization of the substrate and reduction of the activation enthalpy) and (iii.) other causes (change in the rate-determining step, solvent effects, etc.). The weight of these contributions seems to vary depending on the substrate, foremost being determined by the ring size of the resulting product [100, 102, 103].

Especially for the second item, there are a multitude of interpretations and resulting consequences, which either rely on thermodynamic or kinetic arguments (see Fig. 1.38) [100].

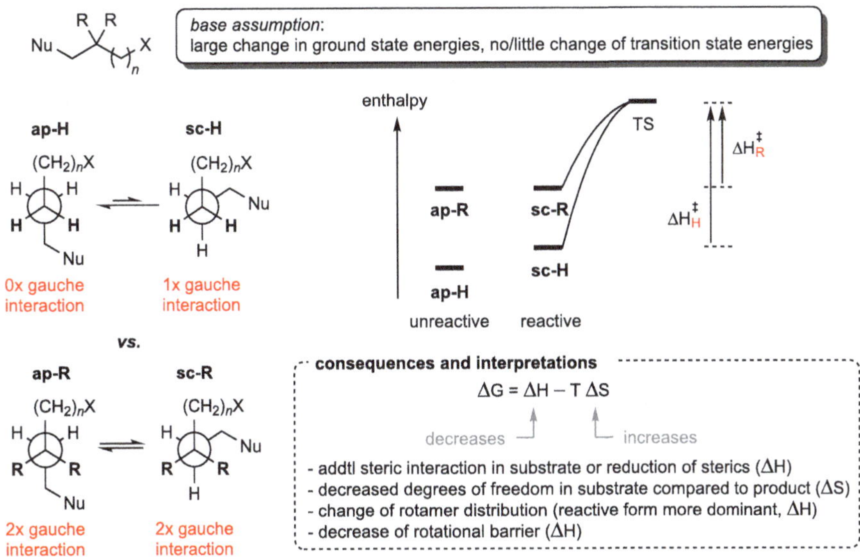

Fig. 1.38 A more detailed overview of the rationalizations by thermodynamic and kinetic factors [100]

By introducing substituents, the increased *gauche* interactions in the substrate relative to the product can change both the enthalpic and entropic terms of a reaction's driving force: a stronger reduction of interactions ($\Delta H^{\ominus}$) upon substitution as well as a smaller entropy increase ($\Delta S^{\ominus}$) ensues, as the rotation is hindered by the added substituents. If the ring closure is the rate-determining step, these arguments can also be applied to the transition state, leading to a modification of the terms $\Delta H^{\ddagger}$ and $\Delta S^{\ddagger}$. Alternatively, if it is assumed that only one conformer would lead to ring closure, a change in the rotamer distribution in equilibrium can populate the reactive form more strongly (reactive synclinal *vs.* unproductive antiperiplanar conformer), which would change $\Delta H^{\ddagger}$. A third hypothesis assumes that the substitution *reduces* the rotational barriers in the substrate and transition state, thus allowing a faster conversion of the rotamers (so-called "facilitated transition hypothesis", $\Delta H^{\ddagger}$) [100].

The gem-dialkyl effect is particularly pronounced for normal rings, while it has been observed to be negligibly small for medium and large heterocycles (see Fig. 1.39).

ring size	k_{Me}/k_H
6	38.5
9	6.6
10	1.1
11	0.6
16	1.2

Fig. 1.39 Dependence of the cyclization rate on the ring size and substitution pattern [104]

The Thorpe-Ingold effect also plays a role in reversible ring openings. In contrast to cyclization reactions, the introduction of substituents results in a *slowing down* of the ring cleavage. This is attributed to the acceleration of the reverse reaction (cyclization, k_{-1}) by the substituents [100]. It is particularly relevant in the hydrolysis of cyclic ketal protecting groups, as this can influence the stability of the protected ketone under acidic conditions in a desired selective deprotection (see Fig. 1.40).

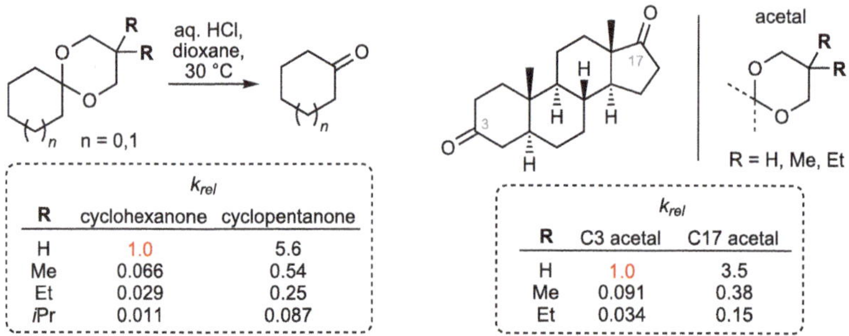

Fig. 1.40 Influence of the ketal substituents and the ring size on the hydrolysis rate of *O,O*-ketals [105, 106]

Gem-dialkyl substitution patterns are frequently encountered in the cyclic backbone of natural products. As the substitutents are an integral part of the target structure, their presence can support the synthesis of the secondary metabolites by exploiting the Thorpe-Ingold effect. The situation is different when this element only has a transitive character and is removed during the synthesis after the desired cyclization has been achieved. *S,S*-acetals constitute prime examples in this regard, as they can be removed under reductive and oxidative conditions and represent masked ketones [107]. Two examples of this approach can be found in the syntheses of panaginsene and quinolizidine (–)-217. In the complex radical reaction of **77**, a diradical intermediate was obtained after the isolation of the hydrazone tosylate upon N_2 elimination, which cyclized to **78**. In addition to the increased yield, an improved diastereoselectivity was observed upon introduction of the propanedithiane. In Lee's synthesis of the alkaloid quinolizidine (–)-217, the dithiane was initially used in an Umpolung to form the acyclic precursor **79**. In the presence of the Jørgensen organocatalyst **81**, the desired piperidine was afforded in nearly quantitative yield (see Fig. 1.41) [108, 109].

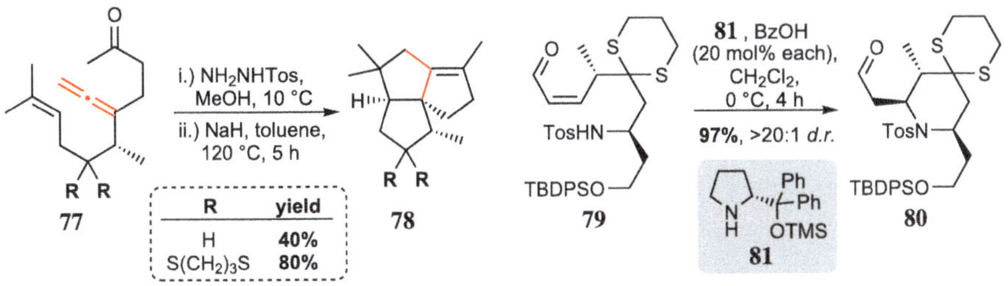

Fig. 1.41 Use of temporary dithianes to exploit the Thorpe-Ingold effect in the total syntheses of panaginsene and quinolizidine (−)-217 [108, 109]

1.1.6 Acidity of Organic Compounds

The acidity of organic compounds is likewise governed by stereoelectronic principles. The acidity of a molecule, whether it is CH-, OH- or NH-acidity, is essential in understanding basic reactivities of fundamental biological processes (see Fig. 1.42).

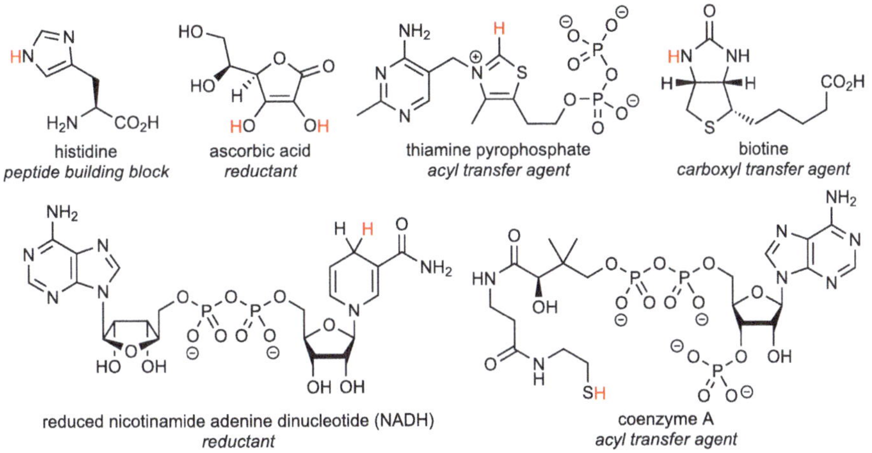

Fig. 1.42 Selection of biologically relevant Brønsted acids

In the conversion of carbonyl compounds, understanding the acidity of different C-H bonds is paramount to predicting the desired chemo-, regio- and enantioselectivity for synthetic applications. Other fields such as organocatalysis or organometallic chemistry also require an intimate knowledge of acid/base equilibria. For example, Buchwald and co-workers observed a strong slowdown in the Suzuki cross-coupling of unprotected NH-acidic nitrogen heterocycles **82** to give **83** when moving from indoles to indazoles and benzimidazoles. Inhibition studies showed that under biphasic conditions, the pK_a values of the starting materials correlate with the lower rate. The cause for the decrease was attributed to the formation of dimeric Pd-heterocycle complexes (**85**) when using substrates of higher acidity than the water of the reaction medium. These off-cycle species tie up the majority of the catalyst in an unreactive Pd reservoir with the deprotonated azole (see Fig. 1.43) [110].

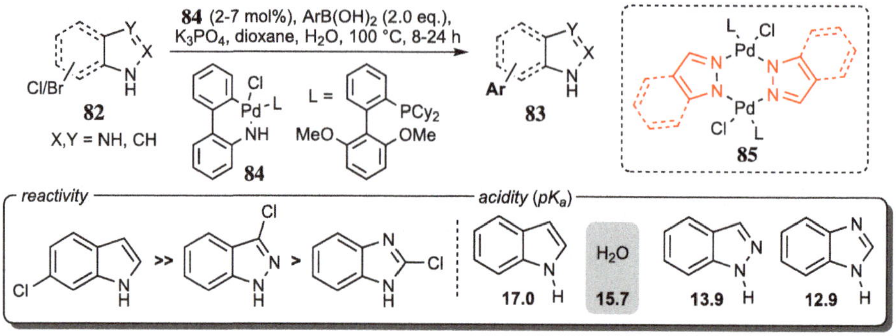

Fig. 1.43 Suzuki cross-coupling of unprotected azoles after Buchwald *et al.* [110]

The definition of *Lewis* acidity/basicity is particularly important with regard to transition metal complexes [111]. When considering deprotonations, the *Brønsted-Lowry* definition [112, 113] is sufficient: acids are capable of donating protons (H-donor), while a base accepts protons (H-acceptor). The equilibrium of formation in carbanions or carbocations, as well as the stabilization of the negative charge in the conjugate base determines the acidity of a substance, among other factors.

This ability to stabilize the anion in the reprotonation/deprotonation equilibrium is referred to as *thermodynamic* acidity. The *kinetic* acidity, on the other hand, describes the rate of deprotonation of a given hydrogen atom.

Since most C-H bonds display very low acidity, strong bases are required for an efficient deprotonation. The solvent also influences the pK_a values. For a deprotonation to occur, the medium must therefore have a higher pK_a value than the molecule to be deprotonated. For this reason, most pK_a values are measured in polar aprotic solvents such as DMSO or cyclohexylamine [114]. For simplicity's sake, the term "pK_a-value" will be used for all solvents, even though the term only refers to the deprotonation equilibrium in water.

Since the 60s and 70s of the last century, the various influences on the acidity of hydrocarbons, heterocycles, and carbocycles have been intensively researched, primarily by the groups of Bordwell and Streitwieser [115–117]. It was shown that in addition to the solvent, inductive and mesomeric effects of the substituents, hybridization effects as well as the orbital or proton orientation in the substrate can have a significant impact on the pK_a values.

The choice of solvent greatly influences the deprotonation equilibrium. Apart from the pK_a of the solvents, the stabilization of the reactive intermediate by the reaction medium plays a crucial role. In addition, the influence of the counterion also should be considered (see Fig. 1.44) [118].

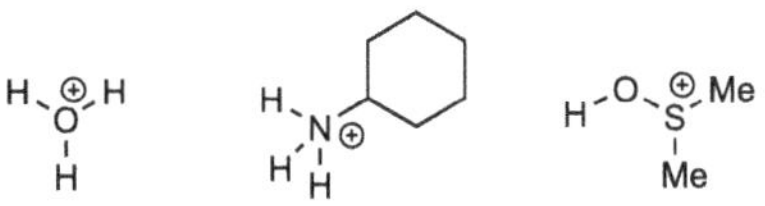

Fig. 1.44 Protonated water, cyclohexylamine, and DMSO as counterions

Both the oxonium (**86**) and the cyclohexylamine cation (**87**) can stabilize the conjugate base of an organic acid through hydrogen bonds while DMSO cannot. For this reason, substances in water show a higher tendency to dissociate and the pK_a value is lower by several orders of magnitude. The lower the stabilization of the carbanion by polarization of the solvent, the *higher* the pK_a value. As a consequence, molecules in vacuum display the lowest tendency to dissociate, where no external stabilization is possible (Table 1.2).

Table 1.2 Influence of the solvent on the acidity of H_2O

Medium	pK_a (H_2O)
H_2O	15.7
DMSO	32
Vacuum^a	279

^{a} Calculated pK_a value [119]

Although DMSO is rarely used as solvent, it is routinely chosen for pK_a measurements as relative trends can be compared. In addition, a set of standardized reference systems was established, which allows the measured pK_a values to be converted to other solvents [120].

Both negative inductive and mesomeric effects support stabilization of the carbanion, which corresponds to a decrease in the pK_a-value as delocalization of the negative charge occurs. Consequently, substituents with high electronegativity and/or the possibility of resonance stabilization (–I or –M effect) likewise lead to a lower pK_a value. Substrates with extensive π systems (nitrile, nitro, acetyl, etc.) as acceptors therefore portray a pronounced acidity in the α-position of the acceptor group. The stronger the effects, the lower the pK_a value (see Fig. 1.45).

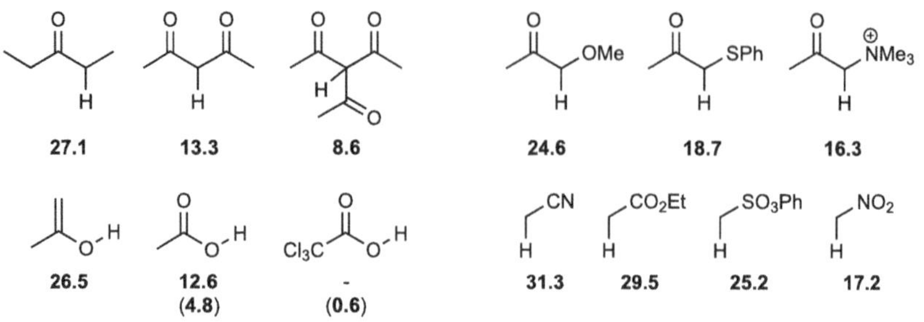

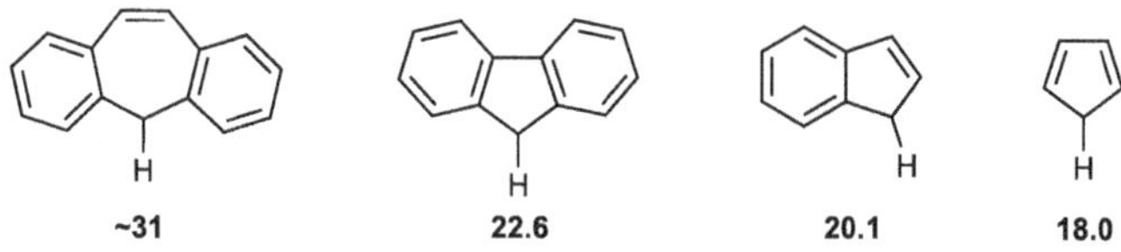

Fig. 1.45 Effects of mesomeric and inductive effects on the acidities of substituted carbonyls (in DMSO). Values in brackets correspond to measurements in water

The trends in acidity of various fluorene and cyclopentadienyl derivatives can also be understood through the aromaticity of the corresponding carbanions. The dibenzocyclo-heptatriene anion is antiaromatic (8π electrons) in nature, the carbanion is strongly desta-bilized and the corresponding hydrocarbon is therefore only slightly acidic. In contrast, the cyclopentadiene anion has a pK_a value comparable to phenols by virtue of its highly aromatic character (see Fig. 1.46).

Fig. 1.46 Acidities of fluorene and cycloheptadiene derivatives (in DMSO)

The *hybridization* is interpreted as the organization of the electrons of a molecule or atom, in which the minimization of energy results in a specific spatial geometry. The average probability of electrons residing in s-orbitals is significantly higher near the nucleus than for p-electrons of the same shell. Due to the stabilizing electron-nucleus interaction a negative charge on atoms with a *high* s-character is stabilized significantly better than in those with a low s-character. For this reason, the acidity of differently hybridized C-atoms is observed in the order:

$$pK_a \qquad C_{sp} < C_{sp^2} < C_{sp^3}$$

This explains both the low pK_a values of terminal alkynes and the high acidity of cyclo-propane derivatives compared to "normal" sp^3-hybridized C-atoms: these carbon atoms dis-play an sp^2-like character and deviate from the ideal tetrahedral structure with H-C-C and H-C-H angles of 118° and 114°, respectively [121]. However, the hybridization and derived qualitative trends should only be considered in the comparison of differently hybridized atoms of the same period: When progressing to higher homologues (alcohols vs. thiols), they show a more pronounced acidity. This trend is due to the increased polarizability when moving to atoms of a higher period, which leads to a higher stabilization of negative charges. When staying within a main group, acidity and electronegativity exhibit a good correlation (see Fig. 1.47).

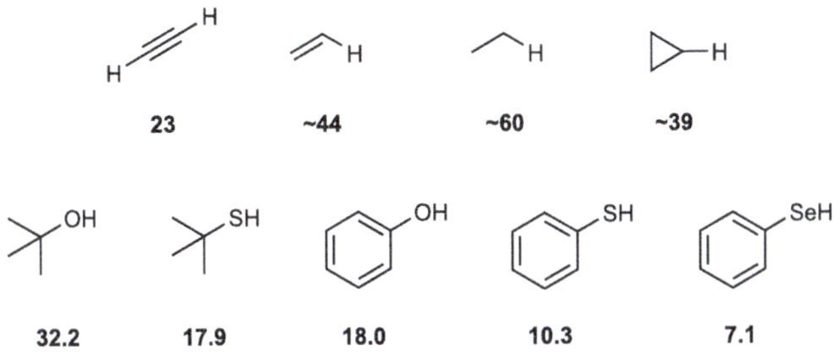

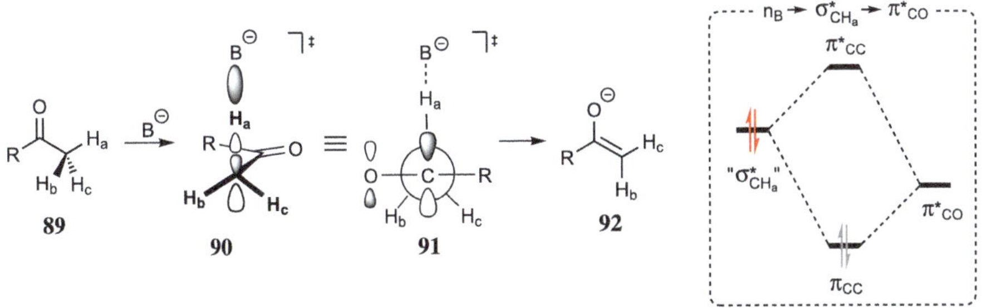

Fig. 1.47 Effects of hybridization on the acidity of C-H bonds. Acidities of various homologues of the chalcogenides (in DMSO)

Since orbital interactions require a certain spatial orientation for good overlap during deprotonation as well as in the conjugate base, *conformational* effects also have a significant influence on acidity. Although all groups with low-lying acceptor orbitals (p, π^* or σ^*) can *per se* stabilize anions, an anti- (or syn-)periplanar arrangement is required to enable a *strong* interaction. For example, the deprotonation of hydrogen in α-position of ketones is facilitated by a synclinal position between the σ^*_{CH} bond and the π^*_{CO} orbital at the C-terminus of the carbonyl functionality (**90, 91**). An overlap of the orbitals in the transition state leads to a stabilizing interaction and thus a lowering of the activation barrier (see Fig. 1.48).

Fig. 1.48 Transition state of the deprotonation of the H_a proton showing a synperiplanar arrangement and energetic scheme of the interactions in the transition state

In the transition state the breaking $\sigma^*_{CH_a}$ bond is populated with electron density ($n_B \rightarrow \sigma^*_{CH}$) and can distribute this through interaction with the π^*_{CO} orbital ($\sigma^*_{CH_a} \rightarrow \pi^*_{CO}$) during formation of the new CC double bond. A geometric fixation of protons in the α-position of acceptor groups can thus either significantly increase the acidity (**93**) or, in contrast, lead to a molecule showing little to no acidity at this position (**94, 96**) — solely depending on the spatial orientation of the CH bonds (see Fig. 1.49).

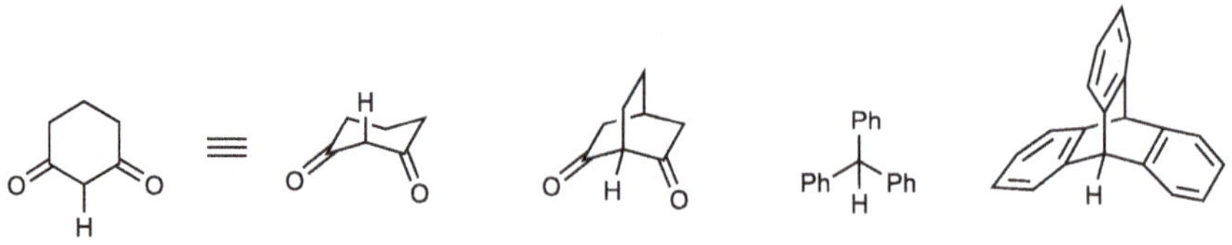

93[122]   **94**[123]   **95**   **96**

10.3   not detectable   30.6   47.7

Fig. 1.49 Effects of the orientation of different CH-bonds on the pK_a values (in DMSO)

Besides a thermodynamic effect, orbital interactions can also affect the kinetic acidity. In the cyclic thioethers **97** and **98**, a significant preference for deprotonation of those H-atoms antiperiplanar to the σ^*_{CS} acceptor orbitals should occur, as these bonds are weakened by $\sigma_{CH} \rightarrow \sigma^*_{CS}$ interactions. This was indeed observed: In **97**, deprotonation of the equatorial hydrogen atoms occurs up to 35 times faster (see Fig. 1.50) [124, 125].

$H_e : H_a = 35$ $H_e : H_a = 8.6$

97   **98**

Fig. 1.50 Ratio of the kinetic acidities of equatorial and axial protons in cyclic thioethers [124, 125]

When comparing the acidity of nitromethane with phenol, their pK_a-values are almost identical. However, when measuring the rate of deprotonation, the carbanion derived from phenol is formed more rapidly by several orders of magnitude (see Fig. 1.51).

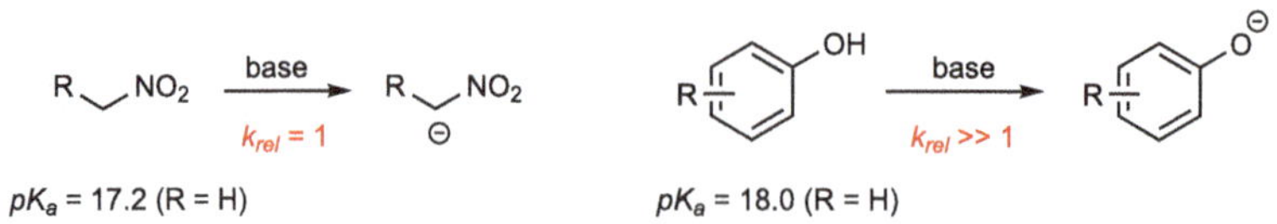

$pK_a = 17.2$ (R = H) $pK_a = 18.0$ (R = H)

Fig. 1.51 Comparison of the kinetic acidities of nitroalkanes and phenols

This phenomenon, which is referred to as "nitroalkane anomaly", can also be understood in terms of their kinetic acidity. In deprotonation equilibria that lie far on the product side, the reaction can become so rapid that the rate of reaction becomes limited by the diffusion of reactants and products. Especially CH-acidic molecules whose carbanions are resonance-stabilized are deprotonated several orders of magnitude more slowly than compounds in which only inductive effects come into play — even if both show comparable pK_a-values. This can be explained by the difference of inertia of atomic nuclei in comparison to electrons: A structural reorganization of the bonding situation in the carbanion (rehybridization $sp^3 \rightarrow sp^2$, shortening of the CC bond by forming partial double bond character) during the deprotonation requires an energy expenditure before the mesomeric stabilization of the anion can come into effect. This energy barrier of rehybridization slows down the deprotonation (see Fig. 1.52).

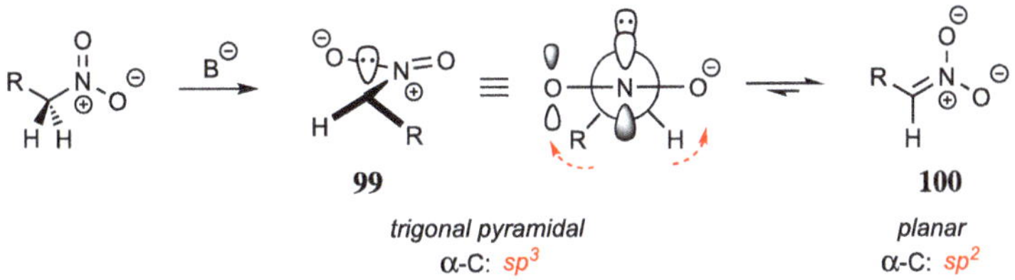

99

trigonal pyramidal
α-C: sp^3

100

planar
α-C: sp^2

Fig. 1.52 Molecular reorganization in the stabilization of nitroalkane anions

In an ideal transition state, these processes occur in a concerted manner. However, the reluctance of masses to move from their position (*principle of least motion*) [126] and to enable resonance stabilization results in a slightly asynchronous sequence of bond breaking and charge delocalization (*principle of non-perfect synchronization*) [127]. This delay of resonance stabilization by the π-system has also been hypothesized to take place in the formation of enolates from ketones [128]. In general, the kinetic acidity of a given molecule is smaller the greater the structural reorganization during the deprotonation process. Especially carbanions seem to be susceptible to this effect as opposed to N- or O-based anions.

References

1. R. Bentley, *Chem. Rev.* **2006**, *106*, 4099–4112.
2. J. M. Finefield, D. H. Sherman, M. Kreitman, R. M. Williams, *Angew. Chem. Int. Ed.* **2012**, *51*, 4802–4836.
3. J.-B. Farcet, M. Himmelbauer, J. Mulzer, *Org. Lett.* **2012**, *14*, 2195–2197.
4. M. K. Singh, R. Xu, S. Moebs, A. Kumar, Y. Queneau, S. J. Cowling, J. W. Goodby, *Chem. Eur. J.* **2013**, *19*, 5041–5049.
5. C. K. Hancock, C. P. Falls, *J. Am. Chem. Soc.* **1961**, *83*, 4214–4216.
6. N. Leventis, S. B. Hanna, C. Sotiriou-Leventis, *J. Chem. Educ.* **1997**, *74*, 813–814.
7. R. R. Saucrs, *J. Chem. Educ.* **2000**, *77*, 332.
8. F. A. L. Anet, D. I. Freedberg, J. W. Storer, K. N. Houk, *J. Am. Chem. Soc.* **1992**, *114*, 10969–10971.

9. M. Squillacote, R. S. Sheridan, O. L. Chapman, F. A. L. Anet, *J. Am. Chem. Soc.* **1975**, *97*, 3244–3246.

10. K. Kakhiani, U. Lourderaj, W. Hu, D. Birney, W. L. Hase, *J. Phys. Chem. A* **2009**, *113*, 4570–4580.

11. J. A. Hirsch, *Top. Stereochem.* **1967**, *1*, 199–222.

12. E. L. Eliel, M. Manoharan, *J. Org. Chem.* **1981**, *46*, 1959–1962.

13. Y. Carcenac, P. Diter, C. Wakselman, M. Tordeux, *New J. Chem.* **2006**, *30*, 442–446.

14. T. K. Brunck, F. Weinhold, *J. Am. Chem. Soc.* **1979**, *101*, 1700–1709.

15. I. V. Alabugin, *Stereoelectronic Effects: A Bridge Between Structure and Reactivity*, Wiley, 1st ed., **2016**.

16. I. V. Alabugin, K. M. Gilmore, P. W. Peterson, *WIREs Comput. Mol. Sci.* **2011**, *1*, 109–141.

17. S. Z. Vatsadze, Y. D. Loginova, G. dos Passos Gomes, I. V. Alabugin, *Chem. Eur. J.* **2017**, *23*, 3225–3245.

18. I. V. Alabugin, T. A. Zeidan, *J. Am. Chem. Soc.* **2002**, *124*, 3175–3185.

19. F. Bohlmann, *Angew. Chem.* **1957**, *69*, 641–642.

20. J.-H. Lii, K.-H. Chen, N. L. Allinger, *J. Phys. Chem. A* **2004**, *108*, 3006–3015.

21. E. Juaristi, G. Cuevas, *Acc. Chem. Res.* **2007**, *40*, 961–970.

22. L. J. Bellamy in *The Infrared Spectra of Complex Molecules, Vol. II*, Springer, 2nd ed., **1980**, Chapter 5, pp. 128–194.

23. L. S. Barreto, K. A. Mort, R. A. Jackson, O. L. Alves, *J. Non-Cryst. Solids* **2002**, *303*, 281–290.

24. C. Gallina, C. Giordano, *Synthesis* **1989**, 466–468.

25. B. Bogdanov, T. B. McMahon, *Int. J. Mass Spectrom.* **2002**, *219*, 593–613.

26. I. V. Alabugin, M. Manoharan, *J. Org. Chem.* **2004**, *69*, 9011–9024.

27. M. Spiniello, J. M. White, *Org. Biomol. Chem.* **2003**, *1*, 3094–3101.

28. K. Taira, W. L. Mock, D. G. Gorenstein, *J. Am. Chem. Soc.* **1984**, *106*, 7831–7835.

29. R. Chen, Y. Shen, S. Yang, Y. Zhang, *Angew. Chem. Int. Ed.* **2020**, *59*, 14198–14210.

30. S. Wolfe, *Acc. Chem. Res.* **1970**, *5*, 102–111.

31. Y.-P. Chang, T.-M. Su, T.-W. Li, I. Chao, *J. Phys. Chem. A* **1997**, *101*, 6107–6117.

32. K. A. Brameld, B. Kuhn, D. C. Reuter, M. Stahl, *J. Chem. Inf. Model* **2008**, *48*, 1–24.

33. N. C. Craig, L. G. Piper, V. L. Wheeler, *J. Phys. Chem.* **1971**, *75*, 1453–1460.

34. C.-H. Hu, H. F. Schaefer, *J. Phys. Chem.* **1995**, *99*, 7507–7513.

35. G. D. Smith, R. L. Jaffe, *J. Phys. Chem.* **1996**, *100*, 18718–18724.

36. T. Yamamoto, D. Kaneno, S. Tomoda, *J. Org. Chem.* **2008**, *73*, 5429–5435.

37. L. Goodman, H. Gu, V. Pophristic, *J. Phys. Chem. A* **2005**, *109*, 1223–1229.

38. R. K. Sreeruttun, P. Ramasami, *Phys. Chem. Liq.* **2006**, *44*, 315–328.

39. I. V. Alabugin, L. Kuhn, N. V. Krivoshchapov, P. Mehaffy, M. G. Medvedev, *Chem. Soc. Rev.* **2021**, *50*, 10212–10252.

40. E. Juaristi, G. Cuevas, *Tetrahedron* **1992**, *48*, 5019–5087.

41. C. L. Perrin, *Tetrahedron* **1995**, *51*, 11901–11935.

42. *The Anomeric Effect and Associated Stereoelectronic Effects, ACS Symposium Series, Vol. 539*, (Ed.: G. R. J. Thatcher), **1993**.

43. P. L. Durette, D. Horton, *Adv. Carbohydr. Chem. Biochem.* **1971**, *26*, 49–125.

44. G. Kothe, P. Luger, H. Paulsen, *Acta Cryst. B* **1979**, *B35*, 2079–2087.

45. R. U. Lemieux, A. A. Pavia, J. C. Martin, K. A. Watanabe, *Can. J. Chem.* **1969**, *47*, 4427–4439.

46. C. Wang, F. Ying, W. Wu, Y. Mo, *J. Org. Chem.* **2014**, *79*, 1571–1581.

47. R. W. Franck, *Tetrahedron* **1983**, *39*, 3251–3252.

48. P. Deslongchamps, D. D. Rowan, N. Pothier, G. Sauvé, J. K. Saunders, *Can. J. Chem.* **1981**, *59*, 1105–1121.

49. C. M. Filloux, *Angew. Chem. Int. Ed.* **2015**, *54*, 8880–8894.

50. C. Romers, C. Altona, H. R. Buys, E. Havinga, *Topics Stereochem.* **1969**, *4*, 39–97.

51. I. Cumpstey, *Org. Biomol. Chem.* **2012**, *10*, 2503–2508.

52. S. Chandrasekhar, *Arkivoc* **2005**, 37–66.

53. S. Chandrasekhar, A. J. Kirby, R. J. Martin, *J. Chem. Soc. Perkin Trans. 2* **1983**, 1619–1626.

54. K. T. Mead, B. N. Brewer, *Curr. Org. Chem.* **2003**, *7*, 227–256.

55. B. R. Raju, A. K. Saikia, *Molecules* **2008**, *13*, 1942–2038.

56. X. Wang, T. J. Paxton, N. Li, A. B. Smith, *Org. Lett.* **2012**, *14*, 3998–4001.

57. B. Liu, J. K. De Brabander, *Org. Lett.* **2006**, *8*, 4907–4910.

58. M. Ball, M. J. Gaunt, D. F. Hook, A. S. Jessiman, S. Kawahara, P. Orsini, A. Scolaro, A. C. Talbot, H. R. Tanner, S. Yamanoi, S. V. Ley, *Angew. Chem. Int. Ed.* **2005**, *44*, 5433–5438.

59. H. Paulsen, Z. Györgydeák, M. Friedmann, *Chem. Ber.* **1974**, *107*, 1590–1613.

60. R. W. Hoffmann, *Chem. Rev.* **1989**, *89*, 1841–1860.

61. F. Johnson, *Chem. Rev.* **1968**, *68*, 375–413.

62. D. J. Maloney, S. J. Danishefsky, *Angew. Chem. Int. Ed.* **2007**, *46*, 7789–7792.

63. S. M. Winbush, D. J. Mergott, W. R. Roush, *J. Org. Chem.* **2008**, *73*, 1818–1829.

64. D. A. Evans, A. M. Ratz, B. E. Huff, G. S. Sheppard, *J. Am. Chem. Soc.* **1995**, *117*, 3448–3467.

65. K. Houk, N. G. Rondan, Y.-D. Wu, J. T. Metz, M. N. Paddon-Row, *Tetrahedron* **1984**, *40*, 2257–2274.

66. A. Fürst, P. A. Plattner, *Helv. Chim. Acta* **1949**, *32*, 275–283.

67. P. Deslongchamps in *The Anomeric Effect and Associated Stereoelectronic Effects* (Ed.: G. R. J. Thatcher), **1993**, Chapter III. Intramolecular Strategies and Stereoelectronic Effects, p. 38 ff.

68. Y.-C. Wu, M. Liron, J. Zhu, *J. Am. Chem. Soc.* **2008**, *130*, 7148–7152.

69. W. Chrisman, J. N. Camara, K. Marcellini, B. Singaram, C. T. Goralski, D. L. Hasha, P. R. Rudolf, L. W. Nicholson, K. K. Borodychuk, *Tetrahedron Lett.* **2001**, *42*, 5805–5807.

70. J. E. Baldwin, *J. Chem. Soc. Chem. Comm.* **1976**, 734–736.

71. C. D. Johnson, *Acc. Chem. Res.* **1993**, *26*, 476–482.

72. K. Gilmore, R. K. Mohamed, I. V. Alabugin, *WIREs Comput. Mol. Sci.* **2016**, *6*, 487–514.

73. K. Gilmore, I. V. Alabugin, *Chem. Rev.* **2011**, *111*, 6513–6556.

74. J. E. Baldwin, J. Cutting, W. Dupont, L. Kruse, L. Silberman, R. C. Thomas, *J. Chem. Soc. Chem. Comm.* **1976**, 736–738.

75. A. L. J. Beckwith, C. J. Easton, A. K. Serelis, *J. Chem. Soc. Chem. Commun.* **1980**, 482–483.

76. I. V. Alabugin, K. Gilmore, M. Manoharan, *J. Am. Chem. Soc.* **2011**, *133*, 12608–12623.

77. I. V. Alabugin, V. I. Timokhin, J. N. Abrams, M. Manoharan, R. Abrams, I. Ghiviriga, *J. Am. Chem. Soc.* **2008**, *130*, 10984–10995.

78. I. V. Alabugin, K. Gilmore, *Chem. Commun.* **2013**, *49*, 11246–11250.

79. C. An, J. A. Jurica, S. P. Walsh, A. T. Hoye, A. B. Smith, *J. Org. Chem.* **2013**, *78*, 4278–4296.

80. B. Thirupathi, P. P. Reddy, D. K. Mohapatra, *J. Org. Chem.* **2011**, *76*, 9835–9840.

81. K. C. Nicolaou, M. E. Duggan, C.-K. Hwang, P. K. Somers, *J. Chem. Soc. Chem. Commun.* **1985**, 1359–1362.

82. K. C. Nicolaou, C. V. C. Prasad, P. K. Somers, C.-K. Hwang, *J. Am. Chem. Soc.* **1989**, *111*, 5335–5340.

83. Y. Morimoto, Y. Nishikawa, C. Ueba, T. Tanaka, *Angew. Chem. Int. Ed.* **2006**, *45*, 810–812.

84. G. Illuminati, L. Mandolini, *Acc. Chem. Res.* **1981**, *14*, 95–102.

85. M. A. Casadei, C. Calli, L. Mandolini, *J. Am. Chem. Soc.* **1984**, *106*, 1051–1056.

86. C. Calli, G. Illuminati, L. Mandolini, P. Tamborra, *J. Am. Chem. Soc.* **1977**, *99*, 2591–2597.

87. G. Illuminati, L. Mandolini, B. Masci, *J. Am. Chem. Soc.* **1975**, *97*, 4960–4966.

88. P. Beak, *Acc. Chem. Res.* **1992**, *25*, 215–222.

89. L. Tenud, S. Farooq, J. Seibl, A. Eschenmoser, *Helv. Chim. Acta* **1970**, *53*, 2059–2069.

90. V. Martí-Centelles, M. D. Pandey, M. I. Burguete, S. V. Luis, *Chem. Rev.* **2015**, *115*, 8736–8834.

91. C. M. Madsen, M. H. Clausen, *Eur. J. Org. Chem.* **2011**, 3107–3115.
92. X. Yu, D. Sun, *Molecules* **2013**, *18*, 6230–6268.
93. I. Saridakis, D. Kaiser, N. Maulide, *ACS Cent. Sci.* **2020**, *6*, 1869–1889.
94. A. Fürstner, *Acc. Chem. Res.* **2021**, *54*, 861–874.
95. K. Namba, Y. Kishi, *J. Am. Chem. Soc.* **2005**, *127*, 15382–15383.
96. M. B. Andrus, A. B. Argade, *Tetrahedron Lett.* **1996**, *37*, 5049–5052.
97. M. Cordes, M. Kalesse, *Top. Heterocycl. Chem.* **2014**, *36*, 369–427.
98. P. Li, J. Li, F. Arikan, W. Ahlbrecht, M. Dieckmann, D. Menche, *J. Org. Chem.* **2010**, *75*, 2429–2444.
99. J. A. Lafontaine, D. P. Provencal, C. Gardelli, J. W. Leahy, *J. Org. Chem.* **2003**, *68*, 4215–4234.
100. M. E. Jung, G. Piizzi, *Chem. Rev.* **2005**, *105*, 1735–1766.
101. R. Brown, N. Van Gulick, *J. Org. Chem.* **1956**, *21*, 1046–1049.
102. S. M. Bachrach, *J. Org. Chem.* **2008**, *73*, 2466–2468.
103. J. Kostal, W. L. Jorgensen, *J. Am. Chem. Soc.* **2010**, *132*, 8766–8773.
104. C. Galli, G. Giovannelli, G. Illuminati, L. Mandolini, *J. Org. Chem.* **1979**, *44*, 1258–1261.
105. M. S. Newman, R. J. Harper, *J. Am. Chem. Soc.* **1958**, *80*, 6350–6355.
106. S. W. Smith, M. S. Newman, *J. Am. Chem. Soc.* **1968**, *90*, 1249–1253.
107. M. Yus, C. Nájera, F. Foubelo, *Tetrahedron* **2003**, *59*, 6147–6212.
108. S. Geum, H.-Y. Lee, *Org. Lett.* **2014**, *16*, 2466–2469.
109. H. Choi, J. Hong, K. Lee, *Eur. J. Org. Chem.* **2020**, *6*, 689–692.
110. M. A. Düfert, K. L. Billingsley, S. L. Buchwald, *J. Am. Chem. Soc.* **2013**, *135*, 12877–12885.
111. W. B. Jensen, *Chem. Rev.* **1978**, *78*, 1–22.
112. J. Brønsted, *Recl. Trav. Chim. Pays-Bas* **1923**, *42*, 718–728.
113. T. Lowry, *Chem. Ind.* **1923**, *42*, 43–47.
114. A. Streitwieser, J. H. Hammons, E. Ciuffarin, J. I. Brauman, *J. Am. Chem. Soc.* **1967**, *89*, 59–62.
115. F. G. Bordwell, *Acc. Chem. Res.* **1988**, *21*, 456–463.
116. https://organicchemistrydata.org/hansreich/resources/pka/.
117. *Unless stated otherwise, all pK_a values were taken from the Bordwell tables.*
118. C. Lambert, P. von Rague Schleyer, *Angew. Chem. Int. Ed. Engl.* **1994**, *33*, 1129–1140.
119. J. E. Bartmess, J. A. Scott, R. T. McIver, *J. Am. Chem. Soc.* **1979**, *101*, 6046–6056.
120. D. A. Bors, M. J. Kaufman, A. Streitwieser, *J. Am. Chem. Soc.* **1985**, *107*, 6975–6982.
121. D. Cremer, J. Gauss, *J. Am. Chem. Soc.* **1986**, *108*, 7467–7477.
122. E. M. Anrett, J. A. Andersen, *J. Am. Chem. Soc.* **1987**, *109*, 809–812.
123. N. H. Werstiuk, *Can. J. Chem.* **1988**, *66*, 2958–2960.
124. E. L. Eliel, A. A. Hartmann, A. G. Abatjoglou, *J. Am. Chem. Soc.* **1974**, *96*, 1807–1816.
125. G. Barbarella, P. Dembech, A. Garbesi, F. Bernardi, A. Bottoni, A. Fava, *J. Am. Chem. Soc.* **1978**, *100*, 200–202.
126. J. Hine, *Adv. Phys. Org. Chem.* **1977**, *15*, 1–61.
127. C. F. Bernasconi, *Acc. Chem. Res.* **1992**, *25*, 9–16.
128. Z. Zhong, T. S. Snowden, M. D. Best, E. V. Anslyn, *J. Am. Chem. Soc.* **2004**, *126*, 3488–3495.

Carbonyl Chemistry

2

The carbonyl group represents one of the most important and versatile functionalities in organic chemistry. It encompasses not only ketones/aldehydes but also the entire spectrum of interconvertible carboxylic acid derivatives. In addition, CC-multiple bonds and C-heteroatom single bonds are likewise accessible using reductive methods. Furthermore, Umpolung reactions and α- or β-functionalizations are part of the broad canon of carbonyl chemistry.

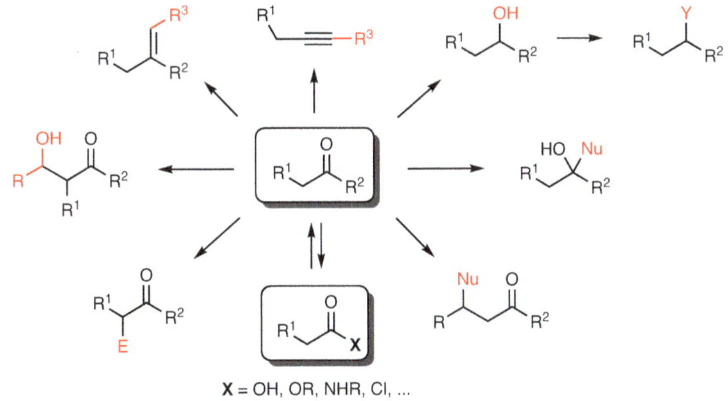

Due to the opposite polarization of the oxygen and carbon atom, both nucleophilic and electrophilic reactions can be entertained. The electrophilicity of the carbon atom is based on a positive polarization, which can also be expressed by a zwitterionic valence structure.

A. Düfert, *Methods of Organic Synthesis*,
https://doi.org/10.1007/978-3-662-70963-4_2

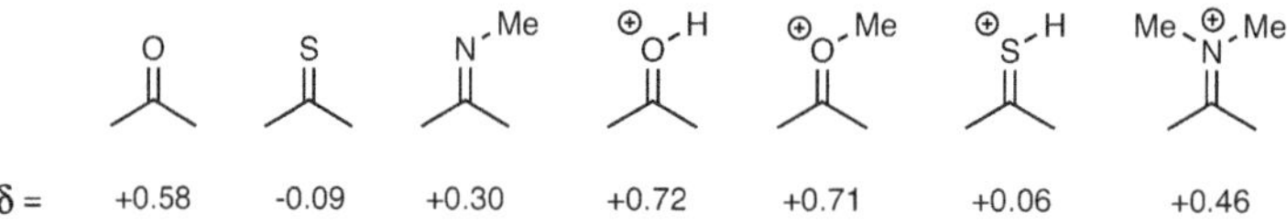

Depending on the substituents R^1 and R^2, the valence structure can either be stabilized or destabilized. A decrease in energy of the π^* antibonding orbital promotes a nucleophilic addition to the carbonyl group. The higher the partial charge localized at the carbon atom, for example by attaching electron-withdrawing substituents, the higher the reactivity. If in contrast the contribution of the zwitterionic valence structure is reduced through dissipation of the charge by introducing electron-donating substituents, e.g., with amides and esters, a nucleophilic attack would occur more slowly. On the other hand, carboxylic acid derivatives with good acceptor substituents display an acceleration in their conversion with nucleophiles (see Fig. 2.1).

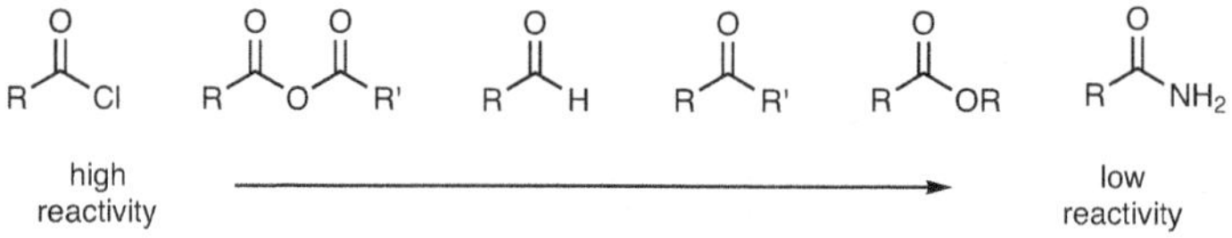

Fig. 2.1 Reactivity of various carboxylic acid derivatives towards nucleophilic attack

It primarily depends on the properties of the available leaving groups whether the carbonyl system is reformed from the tetrahedral intermediate under elimination of a substituent. Since the dominant interaction (see Sect. 2.1) includes the π^* orbital, the substituents and the heteroatom of the C=X bond influence the electron density or polarization to a significant degree (see Fig. 2.2).

Fig. 2.2 Partial charge of the carbon atom in different carbonyl derivatives (natural population analysis, B3LYP/6-31G*)

The change in partial charge is likewise correlated with a change in the MO coefficients at the C and O atoms. In the π orbital, an increased electron density is located at the oxygen atom, whereas in the π^* orbital the orbital coefficient is higher at the carbon atom (see Fig. 2.3). The reactivity of a carbonyl group is governed by both the partial charge (influence on energy level, "hardness" of the atom according to HSAB) and the MO coefficients (influence on orbital overlap in the transition state, see Fig. 2.4).

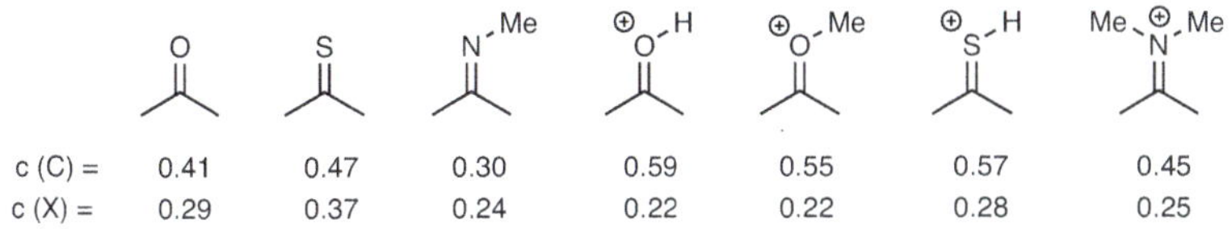

Fig. 2.3 Orbital coefficients of the C=X bond of the π^* orbital in different carbonyl derivatives (B3LYP/6-31G*). c indicates the contribution of the atomic orbitals to the π^*_{CX} (LUMO) in percent. By definition, $\sum_i c_i^2 = 1$

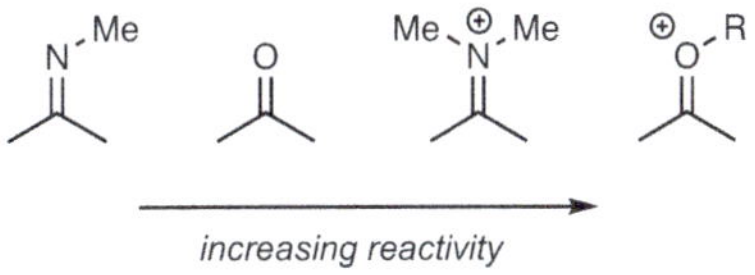

Fig. 2.4 Reactivity of different carbonyl derivatives towards nucleophilic attack

Based on studies of various crystal structures by *H.B. Bürgi* and *J.D. Dunitz* in the 1970s, a reaction trajectory following a minimum energy path was postulated for the addition of nucleophiles to the carbonyl group (see Fig. 2.5) [1].

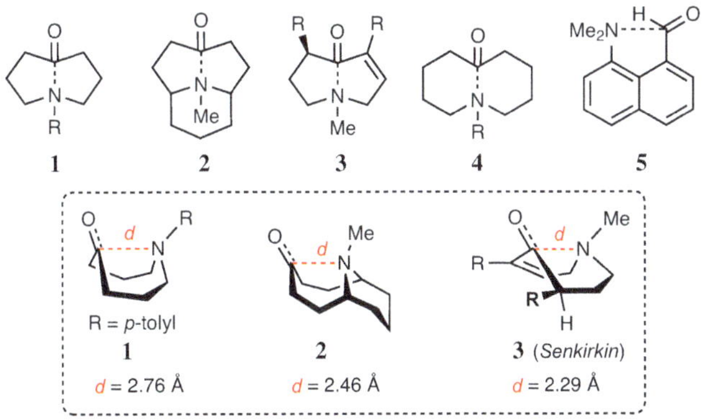

Fig. 2.5 Compounds investigated via crystal structure analysis. Preferred conformations of **1–3** with indication of the C-N distance [2–4]

Various carbonyls were examined that bear amino functionalities at strictly defined distances to the CO group [4–8]. With decreasing distance, an increasing tendency towards pyramidalization of the CO functionality could be observed, i.e., a partial rehybridization of the sp^2 carbon atom to a tetrahedral structure. The attack angle of the amino groups with respect to the CO bond was observed to be $(105 \pm 5)°$ and represents the path of least energy when approaching the carbonyl group.

This attack angle of approx. 107° has become known as the *Bürgi-Dunitz trajectory*. It is derived from the orbital interaction of the nucleophile with the carbonyl group. $n_{Nu} \rightarrow \pi^*_{CO}$ is the dominant interaction in the formation of the new σ_{Nu-C} bond. The approach of the nucleophile following the Bürgi-Dunitz trajectory is energetically particularly favorable for achieving maximum overlap and interaction of the orbitals between nucleophile and carbonyl: the orbital lobes of the C atom in the π^*_{CO} orbital are oriented at approx. 107° relative to the the C-O axis (see Fig. 2.6) [8–10].

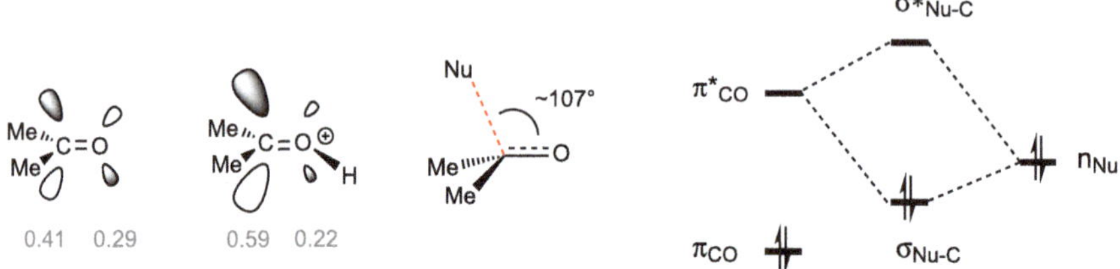

Fig. 2.6 Comparison of the CO orbitals of the LUMO of acetone and protonated acetone including orbital coefficients. Bürgi-Dunitz trajectory and dominant Nu-π^*_{CO} interaction in the nucleophilic attack on carbonyl groups

2.1   Additions to Carbonyls

The addition of nucleophiles to the CO bond constitutes a fundamental transformation, especially in the context of aldol-type chemistry, and is based on the electrophilicity of the carbonyl carbon atom. The basics of selectivity can also be transferred to other nucleophiles beyond enolates such as organometallic compounds or metal hydrides. The structural complexity of a given substrate places a particular emphasis on controlling the stereoselectivity especially in the later stages of a (natural product) synthesis: the later an unselective step occurs, the higher is the wasted effort of assembling the advanced intermediates because of the lack of control (see Fig. 2.7).

The facial selectivity in an asymmetric reaction can be based on several stereoinductors: chiral auxiliaries, chiral catalysts, as well as the 1,*n*-induction of persistent stereogenic centers. Chiral auxiliaries and catalysts primarily rely on steric effects for stereoinduction, allowing the prediction of the stereoselectivity if the transition structures are known. Under substrate control, more complex models are needed to understand the diastereoselectivity governed by a 1,2-, 1,3- and 1,4-induction relying on stereogenic centers present in the starting materials.

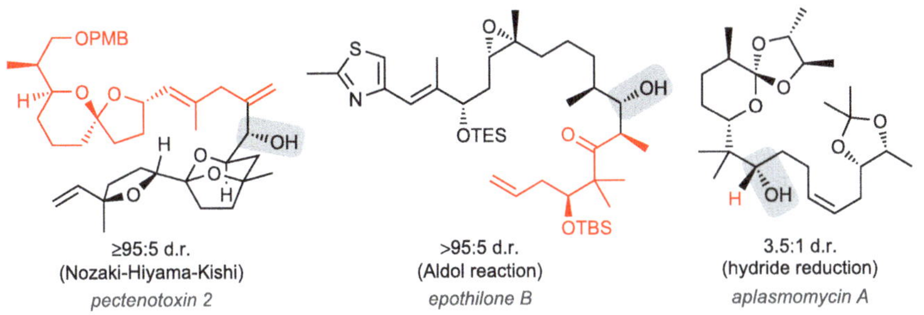

≥95:5 d.r.
(Nozaki-Hiyama-Kishi)
pectenotoxin 2

>95:5 d.r.
(Aldol reaction)
epothilone B

3.5:1 d.r.
(hydride reduction)
aplasmomycin A

Fig. 2.7 Facial selectivity in carbonyl additions of complex intermediates. The nucleophile-derived fragments are highlighted in red [11–13]

2.1.1 1,2-Induction

The addition of nucleophiles to aldehydes or ketones bearing α-chiral centers displays an intrinsic preference for one product diastereomer. Steric interactions, typically expressed by a functionality's A-value (cf. Table 1.1, Chap. 1), can explain a trend when comparing a set of substrates. However, electronic effects might induce an impact that runs counter to steric effects, thereby changing the outcome of the reaction (see Fig. 2.8).

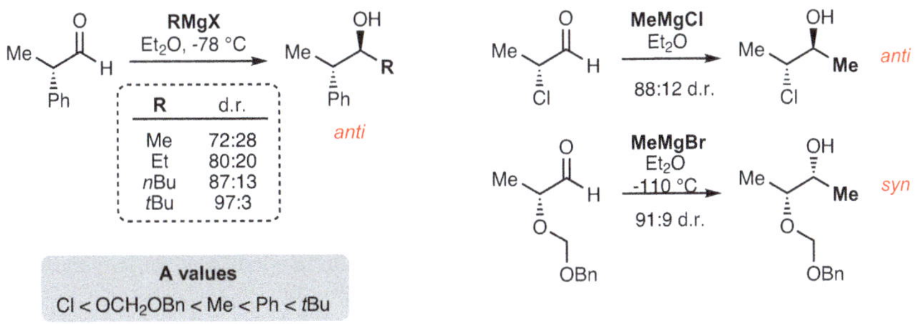

R	d.r.
Me	72:28
Et	80:20
nBu	87:13
tBu	97:3

A values

$Cl < OCH_2OBn < Me < Ph < tBu$

Fig. 2.8 Examples of nucleophilic additions to α-chiral aldehydesb [14–17]

The best selectivities can be achieved through 1,2-induction of persistent stereogenic centers, which can be easily understood through its proximity to the newly formed stereocenter. Depending on the steric and electronic environment of the substrate, these trends are intensified or weakened, making it obvious that an underlying fundamental principle must exist which can rationalize the observed isomer ratios [17]. While initially purely steric (Cram, Karabatsos) [18–20] or electrostatic models (Cornforth) [21] were considered, it was subsequently realized that stereoelectronic factors play an equally important role (Felkin-Anh) [22, 23].

The first investigations date back to research undertaken by Cram. [18] Substituents were divided according to their steric demand into large (R_L), medium (R_M), and small (R_S, cf. **6**). Electronic factors were not considered, the model was based purely on minimizing steric interactions in the transition state. As a consequence it incorrectly predicts the selectivity for polar substituents and is only mentioned for the sake of completeness. Cram and co-workers also postulated a model considering rigid, intramolecularly chelated intermediates of type **8**. It is still in use today in a slightly modified form to predict the facial preference, giving **9** as major product (so-called *Cram chelate model* **7**, see Fig. 2.9) [19, 24].

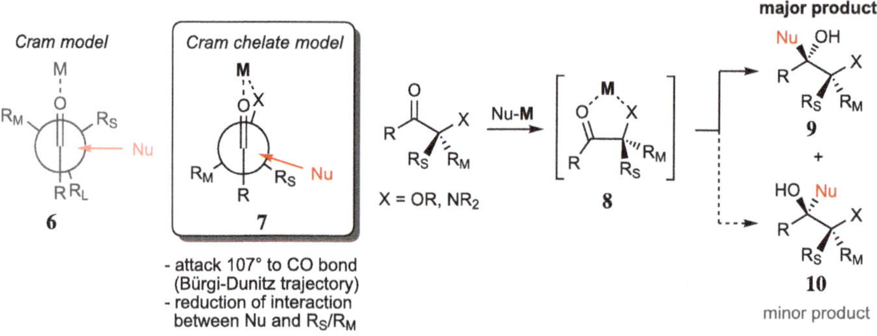

Fig. 2.9 Mnemonic for additions to α-chiral carbonyls according to the Cram chelate model

Fürstner and his team employed a carbonyl addition in their synthesis of the presumed structure of the macrolide chagosensin, in which the stoichiometric use of $MgBr_2$ induces the formation of an intramolecular chelate, thus leading to a reaction trajectory according to the Cram chelate model. The formation of the homoallylic alcohol **12** starting from **11** was achieved by reaction with allyl trimethylsilane. The postulated intermediate **13** explains the high selectivity via addition of the nucleophile to the *Si*-face of the aldehyde. The use of two equivalents of $MgBr_2$ was necessary to prevent the neighboring groups (MOM) from competing for coordination of the Lewis acid. As a consequence, the α-positioned benzyl ether group could determine the facial selectivity as the dominant complexing partner (see Fig. 2.10) [25].

In the absence of chelating substituents or complexing Lewis acids, the reaction follows an acyclic transition state. According to a model by H. Felkin, the reacting substrate should assume a staggered and not an eclipsed conformation to minimize steric interactions of the α-positioned substituents (R_L, R_M, R_S) with R. Acceptor substituents (R_X), e.g. chlorine, are aligned perpendicular with regard to the carbonyl double bond. While the initially postulated reaction trajectory proceeded at a $90°$ angle to the CO bond (as originally proposed by Cram), theoretical studies by Anh and Eisenstein were able to reconcile the experimentally observed Bürgi-Dunitz trajectory [8] with the Felkin model [23, 26]. The result is the now widely accepted *Felkin-Anh model* (see Fig. 2.11).

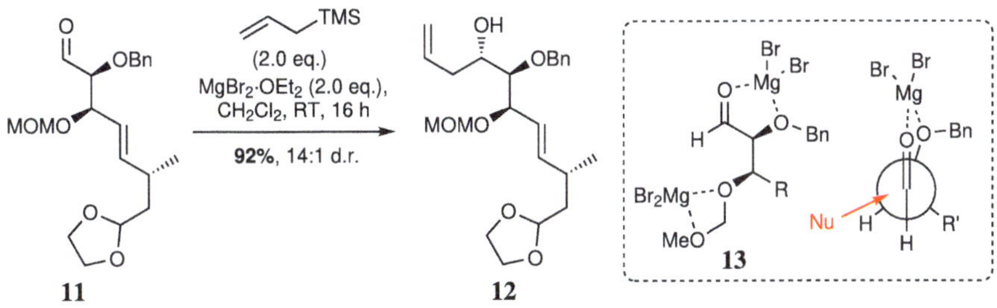

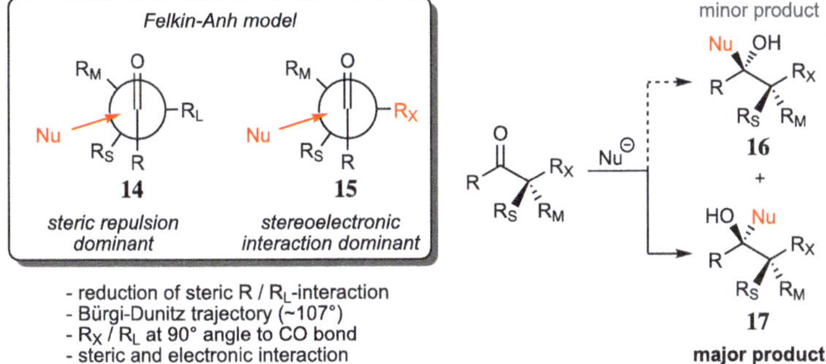

Fig. 2.10 Excerpt of the synthesis of chagosensin according to Fürstner *et al.* [25]

Fig. 2.11 Selectivity of carbonyl additions according to the Anh-Eisenstein-modified Felkin model

In the Felkin-Anh model, the observed selectivity is based both on the minimization of steric factors and the stabilization of the transition states by orbital interaction between the nucleophile (n_{Nu}) and the π^*_{CO} as well as the $\sigma^*_{C\text{-}R_L}$ or $\sigma^*_{C\text{-}R_X}$ orbitals. By including interactions of both antibonding orbitals - of the carbonyl and the electron-withdrawing group R_X - in the transition state, the LUMO of the electrophile is generated anew, subsequently decreasing the activation barrier of the addition (see Fig. 2.12) [26].

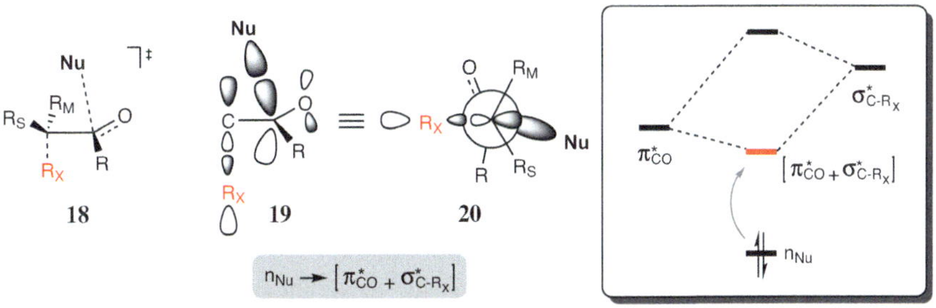

Fig. 2.12 Dominant orbital interactions of the Felkin-Anh model [26]. The substituents R_M and R_S in **19** are omitted for clarity

The lower the energy of the $\sigma^*_{C\text{-}R_X}$ orbital, the closer the interacting orbitals are in their energetic state and the greater the stabilization of the transition state. Strong acceptor substituents thus lead to a large stabilization and allow for high selectivities. If the α-polar substituent is also the most sterically demanding ($R_X = R_L$), both the steric and the polar Felkin-Anh model predict the same product (**22**) as the major isomer. Strongly electron-withdrawing substituents such as chlorine are assumed to adopt the position as largest substituent R_L. However, when the size of R_M increases, steric effects counteract the stereoelectronic impact of the acceptor, which can lead to a reduction in selectivity (see Table 2.1) [15, 27–29].

Table 2.1 Influence of the substituent size of R_M on the selectivity of the addition to aldehydes/ketones [27, 28]

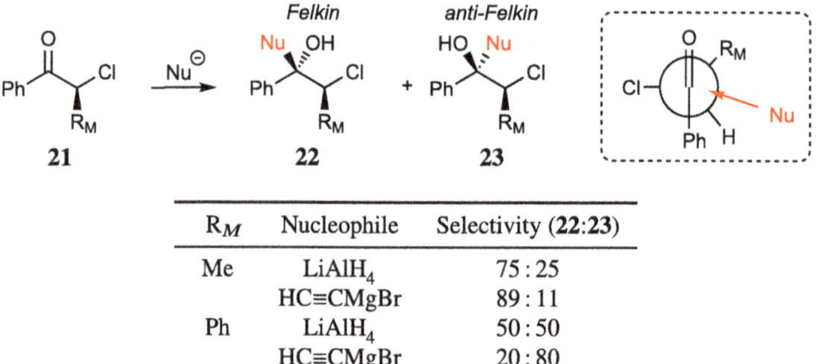

R_M	Nucleophile	Selectivity (**22**:**23**)
Me	LiAlH$_4$	75 : 25
	HC≡CMgBr	89 : 11
Ph	LiAlH$_4$	50 : 50
	HC≡CMgBr	20 : 80

This was impressively demonstrated in A. B. Smith's approach to rapamycin: The vinyl residue of the substrate **24** is sterically so demanding that it can overpower the stereoelectronic preference of the protected α-hydroxy group. Further increasing the steric bulk of the α-hydroxy protecting group cannot reverse this. Instead, it counterintuitively even increases the selectivity of the addition. The authors attributed this observation to the competing transition states **27–29**. Due to the size of the vinyl group, only the transition states **27, 28** with R' in the antiperiplanar position are energetically viable, thereby preventing the pathway to the polar Felkin product (via **29**). Upon increasing the size of the protecting group the steric repulsion of the nucleophile with OR in **28** is enhanced, which further favors **27**. As a consequence, a more pronounced facial selectivity is observed (see Fig. 2.13) [30].

The addition of nucleophiles to aldehydes is frequently employed in synthetic endeavors to access natural products. A reliable prediction of the expected selectivity in an envisioned reaction is therefore indispensable in order to prevent a change of the synthetic route at an advanced stage. Two illustrative examples of addition reactions affording the Felkin-Anh product are encountered in the syntheses of the polyether laurenidificin and macrolide Sch725674 (see Fig. 2.14) [31, 32].

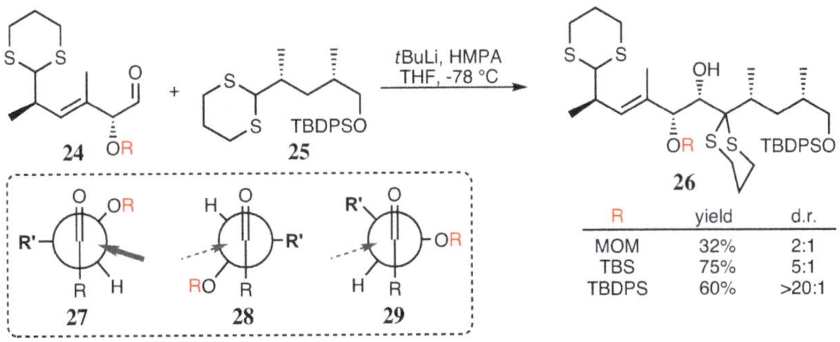

Fig. 2.13 Carbonyl addition in the synthesis of rapamycin [30]

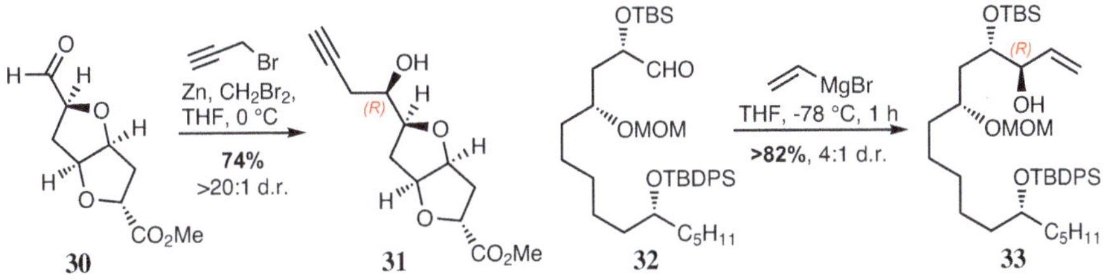

R	yield	d.r.
MOM	32%	2:1
TBS	75%	5:1
TBDPS	60%	>20:1

Fig. 2.14 Felkin-Anh selectivity in Kobayashi's and Kumar's syntheses of laurenidificin and Sch725674 [31, 32]

While the OTBS group in **32** prevents chelation and thus only allows a classic Felkin-Anh addition, chelate formation in **30** with reversed selectivity according to the Cram chelate model was observed under the chosen reaction conditions. Formation of the isolated diastereomer **31** can only be explained by proceeding through an open transition state according to the Felkin-Anh model, which implies negligible chelate formation with the α-positioned alkoxy group.

The competition of open and chelate intermediates in the addition can typically be well controlled by judicious choice of reaction conditions and by adjusting the steric demand of the substituents. Alkyl ethers (OR = OBn, OMOM, OMEM, etc.) can complex metal ions by serving as a basic ligand, while sterically demanding silyl ethers (OR = OTBS, OTIPS, OTBDPS) are hardly capable thereof or not at all. With ketones, chelates react significantly faster than the corresponding mono-complexed substrates. Their occurrence largely depends on the employed solvent [33, 34]. This is consistent with the Curtin-Hammett principle: The acceleration of the kinetically controlled reaction provides the product ratio **34:35** according to the energy difference of the transition states $\Delta\Delta G^{\ddagger}$ (see Fig. 2.15) [35, 36]. In the absence of a pre-equilibrium, the facial selectivity reflects the difference in activation energies $\Delta G^{\ddagger}_{\text{Chelate}} - \Delta G^{\ddagger}_{\text{acyclic}}$.

The choice of solvent and cation significantly influences the complexing behavior and subsequently the stereoselectivity. Occasionally, the polarity of the solvent correlates with the

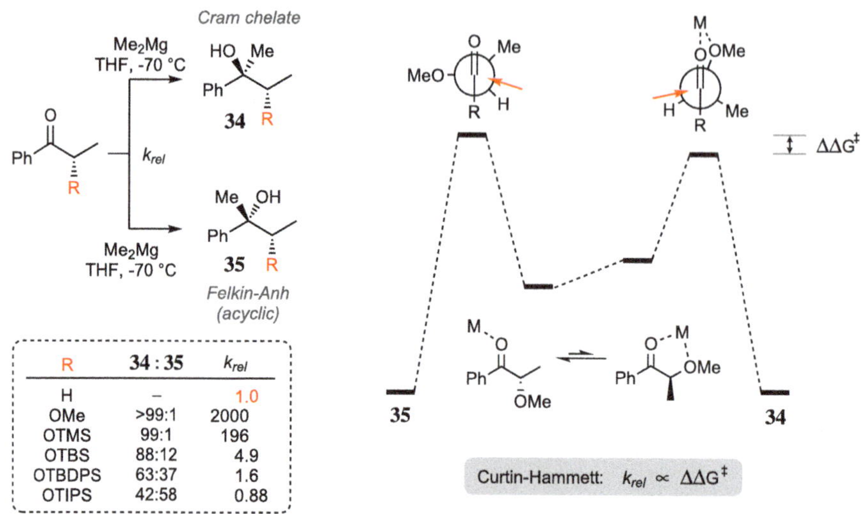

Fig. 2.15 Kinetics and selectivity of carbonyl additions depending on chelation [33] Energy scheme of a Curtin-controlled reaction with a pre-equilibrium displaying low chelate pathway contribution [35]

ability to form chelates, as solvents that are too polar can compete with internal complexation. Unfortunately this is not universally applicable and many exceptions exist [17]. Regarding the utilized counterions, magnesium shows a high tendency for complexation whereas it is seldom observed with lithium (see Fig. 2.16) [37].

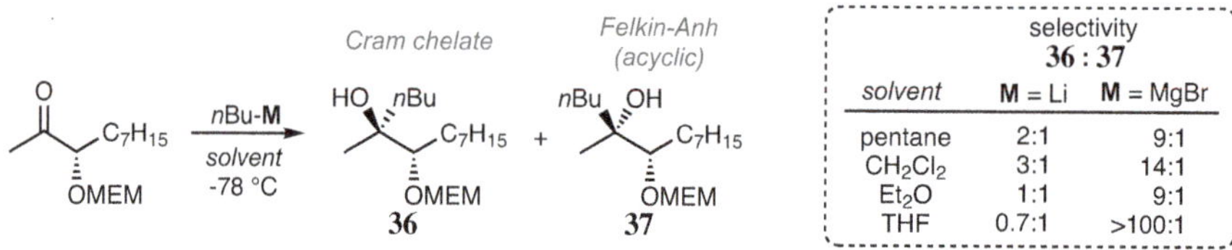

Fig. 2.16 Dependence of the selectivity on the counterion and solvent [37]

Further examples of complexing transition states can typically be found when using Cu and Zn reagents as well as with Ti intermediates. However, when the utilizing these metals, the employed combination of organometallic reagents and/or salts can exhibit a pronounced difference in product ratios. In general, ketones react more selectively than aldehydes, both on an acyclic or chelate pathway [17, 24]. It has been postulated that the higher reactivity of aldehydes cannot be much further enhanced through complexation, thus providing a similar reactivity and therefore selectivity both with and without chelation (see Fig. 2.17, cf. Sect. 4.2.1 for a detailed look at hydride nucleophiles) [34].

Isobe and co-workers used alkynyl magnesiates as nucleophiles in their synthesis of the highly functionalized dideoxytetrodotoxin. Upon addition of the Li acetylide to **38**, no facial selectivity was observed. However, the propargyl alcohol **39** could be obtained with high

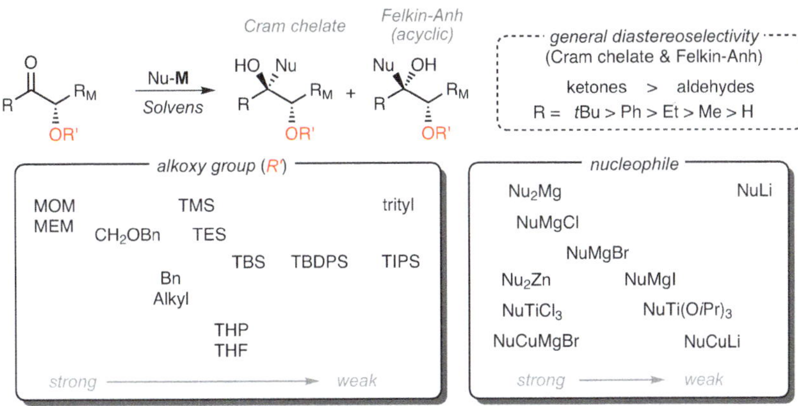

Fig. 2.17 Qualitative assessment of the influence of reaction conditions on the degree of chelation [17, 24, 35]

selectivity when switching to the Mg homologue. It was postulated that the addition proceeds via an attack on the *Re*-face of the aldehyde in intermediate **40** (see Fig. 2.18) [38].

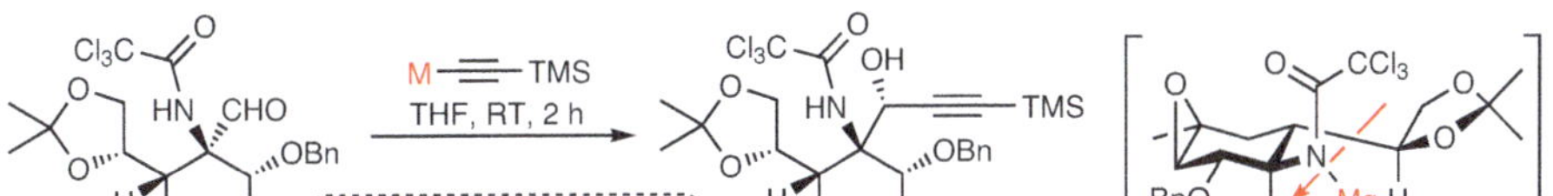

Fig. 2.18 Isobe's synthesis of dideoxytetrodotoxin [38]

In addition to the discussed exceptions to the Cram-chelate model, there are also several examples where the product distribution shows a selectivity contrary to the prediction of the Felkin-Anh model. Mehta, Le Noble, and Wipf were able to achieve reversed selectivity in additions to rigid cyclic ketones [39–41]. The observed product distributions were rationalized either by a dominant orbital interaction deviating from the Felkin-Anh model (*Cieplak* model, **43**, **45**) or electrostatic effects (**47**) (see Fig. 2.19) [42].

The model proposed by Cieplak relies on stereoelectronic factors to explain a selectivity similar to the Felkin-Anh model. However, according to the Cieplak model, hyperconjugation of polar substituents at the α-stereogenic center leads to a stabilization of the transition states through interaction with the newly forming C-Nu bond through $\sigma_{C\text{-}R_X} \rightarrow \sigma^*_{C\text{-}Nu}$ interactions, when the substituent with the best donor properties aligns in an antiperiplanar fashion to the nucleophile (cf. Fig. 2.20) [43, 44].

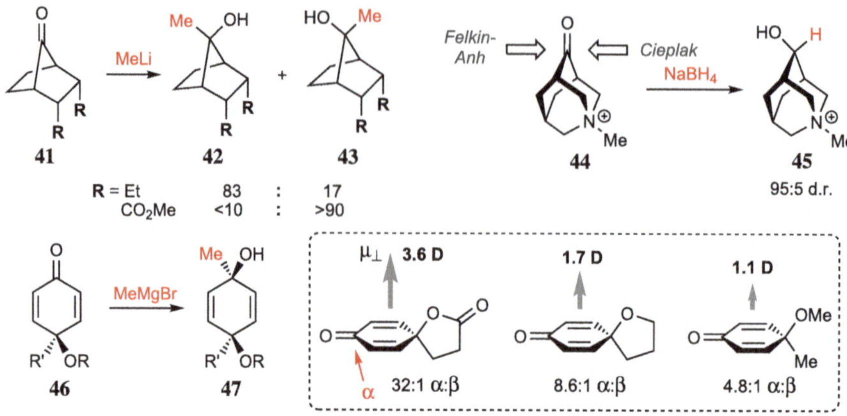

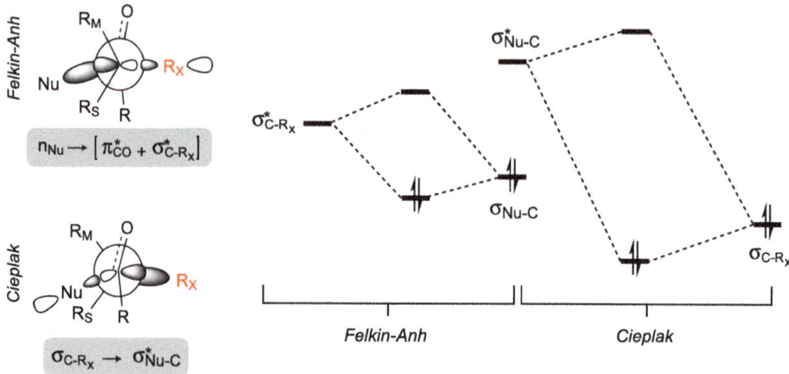

Fig. 2.19 Reactions with *anti*-Felkin-Anh products as major products [39–41]

Fig. 2.20 Comparison of the dominant orbital interactions in the Cieplak and Felkin-Anh model [26, 45]

The Cieplak model explains the selectivities for **43** and **45** shown in Fig. 2.19. Furthermore, cyclohexanones usually react to the *anti*-Felkin-Anh products, which should be favored according to Cieplak. Nevertheless, the Felkin-Anh model has been widely accepted to predict the 1,2-selectivity in additions to carbonyl compounds although the Cieplak model is considered for some observed exceptions. The proposed orbital interactions that constitute the mechanistic basis of the Cieplak model are still being discussed controversially until the present day [46, 47]. Evans also proposed a modified variant of the electrostatic Cornforth model [48]. The occurrence of reactions that do not correspond to the predictions of these models makes it clear that either not all relevant factors were considered or they were weighted incorrectly. The models presented so far are thus purely qualitative in nature, with the Felkin-Anh model being able to most often predict the experimentally observed selectivities.

Quantitative methods have also been proposed, with the contributions of Tomoda (EFOE model, *exterior frontier orbital extension*) and Dannenberg (PPFMO model, *polarized π-frontier molecular orbital*) being particularly noteworthy [49, 50]. The disadvantage of these methods lies in the inherent necessity of theoretical calculations to predict the major product, requiring a significant effort as opposed to the "simple" Felkin-Anh model.

In addition to the conversion of aldehydes or ketones, the Felkin-Anh and Cram-chelate models can also be used for related transformations, for example in the hydroboration of α-chiral olefins (see Sect. 3.3).

2.1.2 1,3-Induction

If the inducing element moves away by one methylene unit from the newly forming stereogenic center, its influence and thus the stereochemical selectivity should decrease. In order to enable a high diastereoselectivity, rigid transition states with low degrees of freedom are required for these substrates. Accordingly, polar functional groups capable of forming chelates are preferred as β-substituents, for example when using organometallic reagents as nucleophiles.

The stereoinduction under chelate control was investigated by Cram as early as the 1950s (cf. Fig. 2.9) [51, 52]. A deeper understanding of 1,3-stereoinduction was achieved three decades later through fundamental work by Reetz. It was subsequently refined and extended to include acyclic reactions without the involvement of chelates by the Evans group (see Fig. 2.21) [53–56].

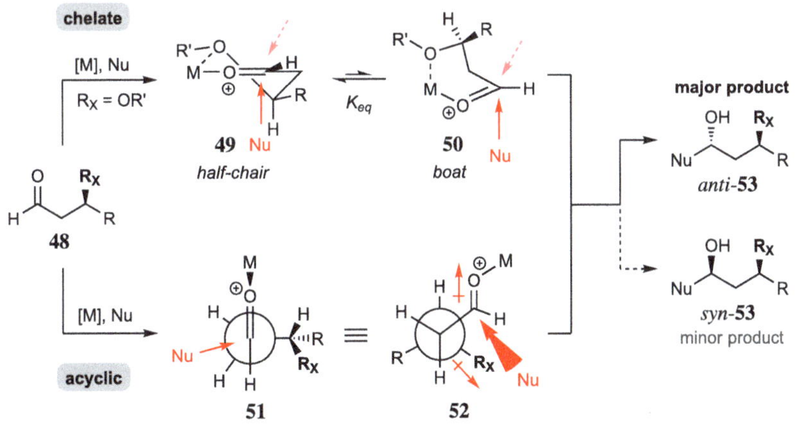

Fig. 2.21 Evans model for 1,3-induction in aldehydes [55, 56]

According to the Reetz model, 1,3-chelates adopt a half-chair conformation. The attack on the carbonyl function preferably occurs under formation of a chair-like transition state

($\rightarrow$ Fürst-Plattner rule, cf. Sect. 1.1). The energetically most favorable half-chair conformation **59** affords the *anti*-diol **53** (the alternative, energetically higher conformation with R in axial position is omitted). Based on studies by Evans and co-workers the boat-like transition state **50** was proposed to be additionally feasible and potentially even energetically more favorable, which likewise predicts *anti*–**53** to be the major product [55].

In the absence of chelate intermediates, the 1,3-model showing a selectivity based on electrostatic interactions is used (via **51** or **52**). The β-stereogenic center is oriented perpendicularly with regard to the carbonyl group, in the same manner as the orientation of the substituents R_L or R_X in the Felkin-Anh model (**51**). The polar group R_X, typically a protected hydroxy group, can be arranged synclinal or antiperiplanar to the C_α-carbonyl bond to minimize the dipoles in the staggered conformation. The conformation in which R shows the least steric repulsion with the carbonyl group is preferred, which is antiperiplanar to C_α-CO bond (**52**). *Anti*–**53** is thus predicted to be the major product, both under chelate control as well as following an acyclic reaction pathway [56].

Álvarez and co-workers employed a Lewis acid-mediated Mukaiyama-Aldol reaction utilizing a protected β-hydroxyaldehyde in their synthesis of the polyhydroxylated side chain of oscillariolide and phormidolides A-C. The use of non-chelating $BF_3 \cdot OEt_2$ as an activator suggests an acyclic reaction pathway. The expected 1,3-*anti*-configured hydroxyketone **55** was afforded as major product in moderate yield with excellent diastereoselectivity (see Fig. 2.22) [57].

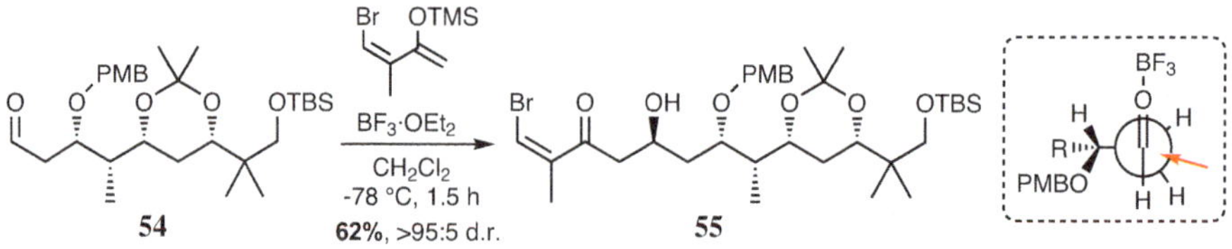

Fig. 2.22 Synthesis of the polyketide side chain of oscillariolide and phormidolides A-C according to Álvarez *et al.* [57]

In the case of α, β-chiral carbonyl substrates, the facial induction of the two stereocenters can either reinforce each other (*matched*) or direct towards opposite products (*mismatched*). Evans developed a qualitative model to predict the expected diastereoselectivities for substrates which bear polar β-substituents (1,3-induction) and an α-stereogenic center (1,2-induction). Depending on the relative configuration, this results in either chelate intermediates (*syn*) or an acyclic reaction path (*anti*), yielding **58** or **61** via **56/57** or **59/60**, respectively. Under chelate control, the *syn*-substrate provides high selectivities, while under non-chelating conditions, the *anti*-configured reactant is similarly selective (see Fig. 2.23) [55, 56].

The Roush group reported the first total synthesis of the macrocyclic cytotoxin tedanolide. In the aldol reaction of ketone **62** with aldehyde **63**, the preferred diastereomer could be reliably predicted by the Evans model. The reaction proceeded through an unchelated aldehyde intermediate (see Sect. 2.3 for aldol transition states) and the β-hydroxyketone

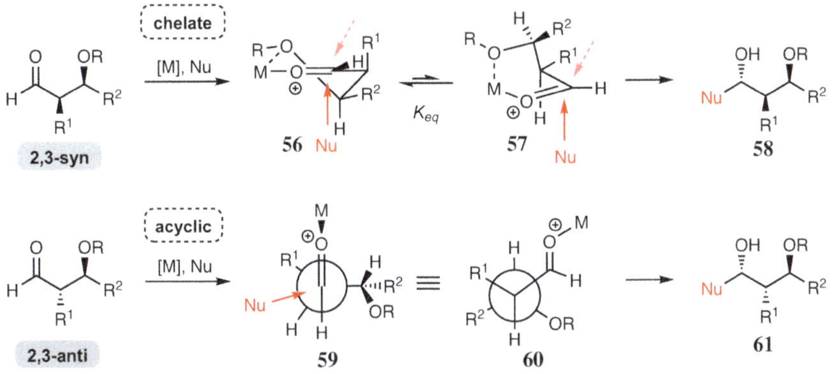

Fig. 2.23 *Matched* 1,2-/1,3-induction in additions to aldehydes. [55, 56]

64 was obtained in an acceptable yield without traces of the undesired diastereomer (see Fig. 2.24) [58].

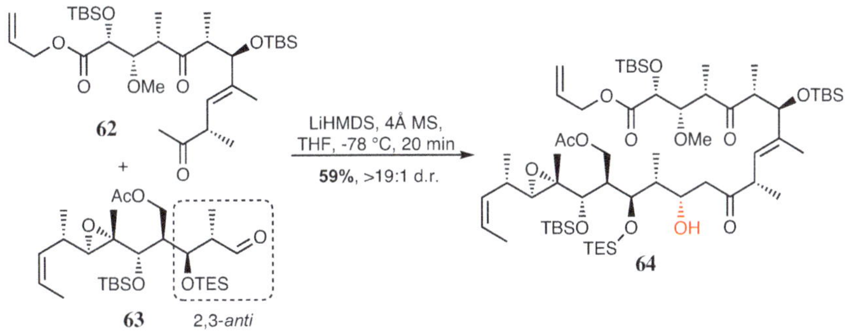

Fig. 2.24 *Matched* aldol reaction in the synthesis of tedanolide by Roush *et al.* [58]

When combining chelating Lewis acids with 2,3-*anti*-configured aldehydes or monodentate Lewis acids with 2,3-*syn*-carbonyls, the steric and electronic interactions at the two stereocenters each direct towards opposite products. The smaller the steric demand of the nucleophile, the more dominant the 1,3-induction becomes (*anti*-Felkin product via **67**). As the size increases, the interactions with the α-stereogenic center become the major deciding factor for determining the diastereoselectivity of the addition (Felkin product via **65**, see Table 2.2) [56].

This interplay could also be demonstrated in the syntheses of the natural product psymberin and of a fragment of spiroketal metabolite spirangien A: When employing the sterically demanding silyl enol ether as a nucleophile, the facial selectivity is determined by the α-stereogenic center of the aldehyde (**69**). In Kalesse's approach to **70**, the Li enolate is ster-

Table 2.2 Influence of nucleophile size on the facial selectivity [56]. Groups showing strong steric repulsion in the transition state (iPr$\leftrightarrow$CHO, Me$\leftrightarrow$Nu) are highlighted in red

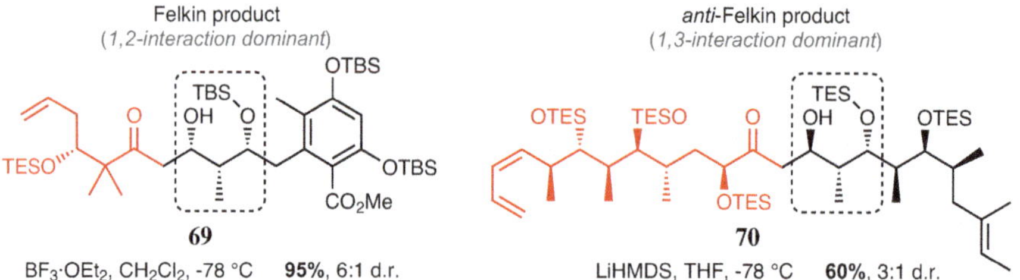

R	Selectivity (**66:68**)	
	CH$_2$Cl$_2$	toluene
tBu	96:4	88:12
iPr	56:44	32:68
Me	17:83	6:94

ically much less demanding and thus 1,3-induction dominates, affording the energetically less favorable addition product according to the Felkin-Anh model (see Fig. 2.25) [59, 60].

Fig. 2.25 Syntheses of psymberin and spirangien A by the Floreancig and Kalesse groups. The nucleophile fragment is highlighted in red [59, 60]

The diastereomeric ratio can occasionally show a low to even opposite preference for an addition to one of the diastereotopic carbonyl faces according to the herein presented transition state models — despite an optimally matched situation as depicted in Fig. 2.23 [61]. In addition, if chiral nucleophiles are used in conjunction with electrophiles bearing persistent stereocenters, further effects or directing elements for double or higher stereodifferentiation need to be considered [62, 63]. The higher the complexity of the substrates, the more an extensive literature search with comparable substrates and model studies lend themselves to the synthetic planning.

2.1.3   4-Versus 6-Membered Transition States

In addition to the influence of the electrophile on the diastereoselectivity, the structure of the nucleophile also plays a pivotal role for the stereochemical course of a reaction. Apart from the presence of stereogenic centers, the question of the stoichiometry of the nucleophile likewise can result in significant implications in the formulation of the transition structures.

For example, early theoretical studies of the solvolysis of carbonyl compounds implied that a transition state involving **two** equivalents of H_2O is approximately 41 kcal/mol more stable than a transition state containing only one molecule of water [64].

The postulate of transition states requiring two reagent equivalents is also applied in the addition of organometallic compounds to carbonyls [65]. In Grignard reactions, the exact mechanism was long disputed due to the Schlenk equilibria occurring in solution (see Fig. 2.26) [66, 67].

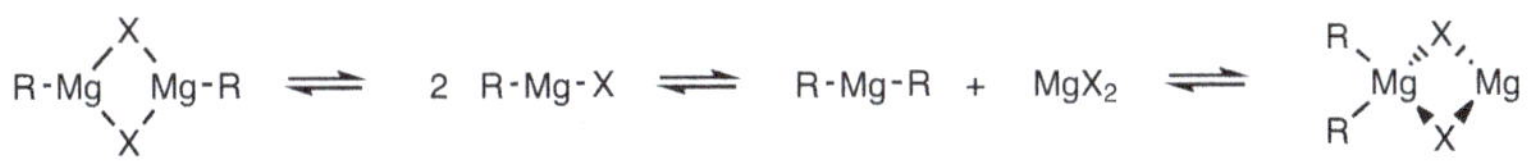

Fig. 2.26 Schlenk equilibrium in polar-aprotic media. Coordinated solvents are not depicted [67]

Transition states incorporating one and two equivalents of organomagnesium reagents have both been discussed [68]. Current theoretical studies favor a bimetallic mechanism under formation of four-centered transition states (see Fig. 2.27) [69, 70] However, since in Grignard reactions not only one individual species exists in solution, there is no generally accepted and proven mechanism. Consequently, both an alternative mononuclear or radical mechanism cannot be ruled out.

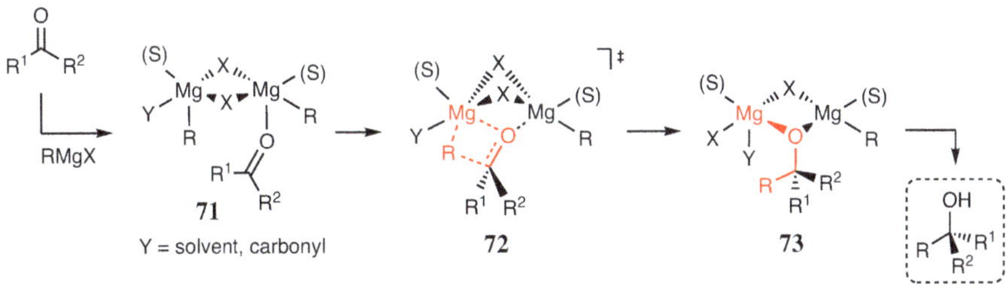

Fig. 2.27 Postulated mechanism of the Grignard addition to carbonyls [69, 70]

The situation is different with organoaluminium species, for which reactions via bimetallic, 6-membered transition states were assumed early on [65]. Organozinc reagents can also follow a monometallic, 4-membered pathway in additions to aldehydes. However, this reaction was observed to proceed very sluggishly. In contrast, catalytic amounts of metal salts or organometallic compounds significantly enhance the rate of reaction. Their presence enables bimetallic, 6-membered transition states, which are predicted to have a significantly lower activation energy than the 4-membered analogs. Furthermore, the introduction of chiral substituents on the catalyst can enable an asymmetric reaction, which has led to a multitude of reports on asymmetric additions of organozinc reagents (see Fig. 2.28) [71, 72].

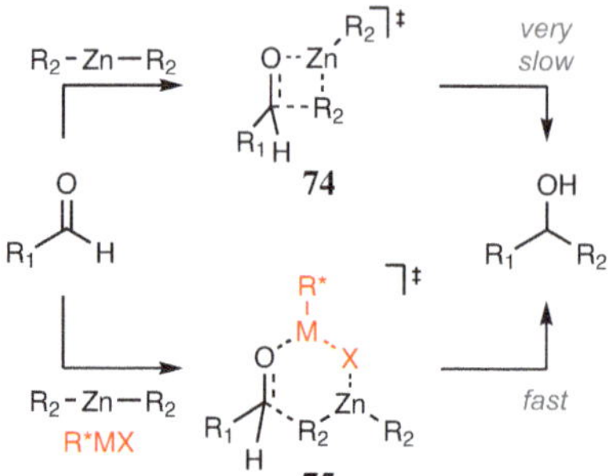

Fig. 2.28 Competing pathways in the addition of organozinc reagents to aldehydes [73]

This principle of bimetallic catalysis can also be advantageously used in many further transformations, including asymmetric reactions exploiting nonlinear effects through aggregation of the metal species [74].

The Ziegler-Natta polymerization constitutes an example of a monometallic, 4-membered transition state. According to the accepted mechanism, which has been ascertained via both theoretical and crystallographic means, only one equivalent of the metal species is required to synthesize the desired polymers (see Fig. 2.29) [75–77].

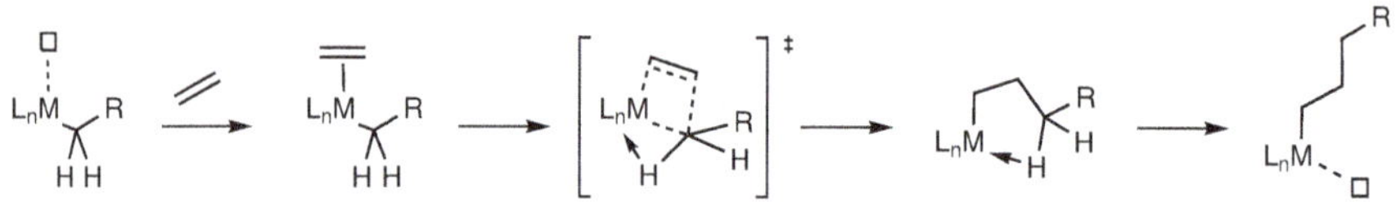

Fig. 2.29 Mechanism of Ziegler-Natta polymerization. □ indicates a free coordination site at the metal center [77]

As elaborated by the illustrated examples, the stoichiometry of a reagent can directly affect the involved transition states and thus the reaction mechanism. Without a doubt, the depiction of a reaction without exact and detailed description of the substrates/reagents including stoichiometry allows for a quicker overview of the big picture. However, even when "just" optimizing the reaction conditions and not pursuing fundamental mechanistic studies, it is useful to keep the exact stoichiometry in mind.

2.2   Formation and Reactivity of Enolates

Apart from a direct addition to the carbonyl CO bond, the α-functionalization constitutes the second common class of derivatizations of ketones, amides, and esters. These transformations employ metal enolates or the corresponding homologues, whereby the "classic" application is the ubiquitous aldol reaction. Furthermore, the carbonyl-derived nucleophiles can react with a considerable range of electrophilic coupling partners, which significantly extends the spectrum of accessible products beyond a mere CC bond formation (see Fig. 2.30).

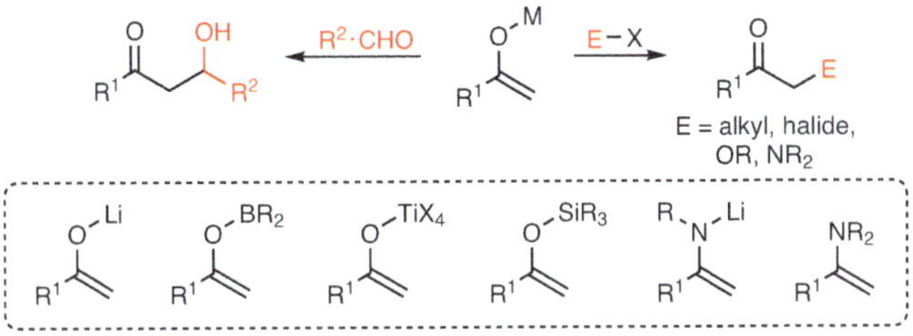

Fig. 2.30 Functionalization of enolates and common enolate equivalents

The individual enolate species significantly influences the chemo-, regio-, and diastereoselectivity of the transformation, which is why their selection and method of preparation can have a dramatic impact. The most commonly used enolates comprise lithium, silyl, and boron enolates. However, there exists a wide range of metals as counterions beyond this first line of reagents (based on Ti, Sn, Zr, Na). In addition to the base-mediated preparation of enolates from the corresponding aldehydes or ketones, several variants of the Reformatsky reaction and the Birch reduction can analogously be used to generate the required nucleophile. This is further supplemented by organocatalytic approaches, which proceed through the intermediacy of enamines as reactive species (see Sect. 8.1).

The reactivity of an enolate equivalent towards an electrophile can be roughly estimated as follows (see Fig. 2.31).

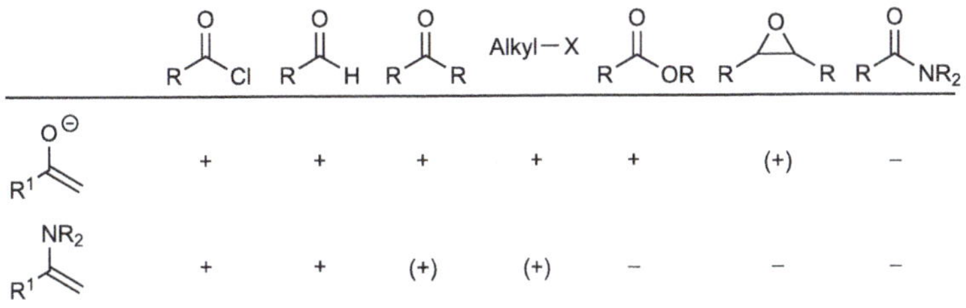

	$\underset{R\;\;\;Cl}{O}$	$\underset{R\;\;\;H}{O}$	$\underset{R\;\;\;R}{O}$	Alkyl—X	$\underset{R\;\;\;OR}{O}$	$\underset{R\quad\;R}{O}$	$\underset{R\;\;\;NR_2}{O}$
$\underset{R^1}{O^\ominus}$	+	+	+	+	+	(+)	–
$\underset{R^1}{NR_2}$	+	+	(+)	(+)	–	–	–

Fig. 2.31 Reactivity of enolates or enamines towards various electrophiles

2.2.1 Lithium Enolates and Silyl Enol Ethers

The methods referred to as "classic" enolate formation rely on ketones and strong bases as reactants for the deprotonation of the α-acidic CH bonds. A careful combination of a base with the respective starting material is a prerequisite for a selective transformation – both in terms of the compatibility of the substrate with the base and the deprotonation of the desired CH bond in the presence of several acidic protons. Depending on the reaction conditions, drastic differences in the product distribution can thus be encountered.

Only those reagents should be used as base that i.) have very high (kinetic) basicities and ii.) display a low nucleophilicity. The given basicity is particularly decisive, as in the case of slow deprotonations both the starting material and the deprotonated species coexist and can react (see Fig. 2.32).

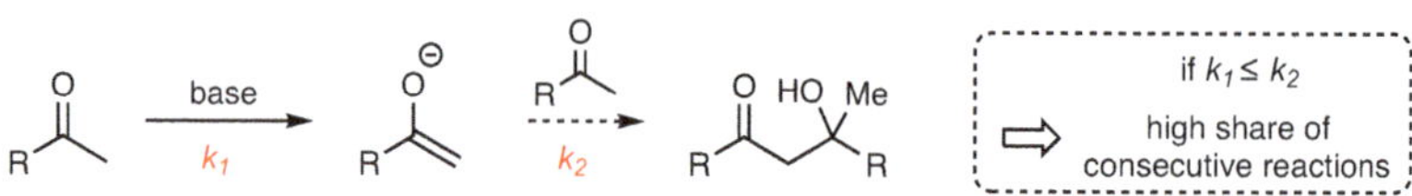

Fig. 2.32 Possible homoaldol reaction under slow substrate deprotonation

The rate of deprotonation (k_1) should be significantly greater than the rate of an aldol reaction, condensation, or alkylation (k_2) to suppress unwanted side reactions. A pK_a difference between the base and enolate precursor of 5 units or more serves as a guideline — assuming that the deprotonation rate correlates with the thermodynamic basicity ($\rightarrow$ Sect. 1.1). Efficient deprotonation is possible with many reagents in the case of ketones, but with aldehydes, this can often only be achieved by resorting to strong bases like LDA, as the homocondensation proceeds very quickly (k_2 large). Since nucleophiles can add to the electrophilic C-atom of the CO bond, bases with low nucleophilicity are required to avoid this addition of the base. Many organolithium reagents do not meet this criterion and are unsuitable for enolate formation. Although there are also sterically hindered, non-nucleophilic organolithium bases (**85–87**), these are rarely encountered due to their laborious preparation [78–82]. Lithium amide bases (**81–83**) have advanced to become the reagents of choice, as they meet both of the above criteria and their preparation is easy to carry out both in the laboratory and on an industrial scale (see Fig. 2.33) [83].

The deprotonation with LDA affords HNiPr$_2$ as byproduct upon reaction with acidic CH bonds. Its presence can significantly reduce the reactivity or selectivity of a reaction by reprotonating the enolate, which is why amine-free Li-enolate solutions often show an improved reaction profile [87]. If sterically more hindered (LiTMP, LOBA) and less aggregated bases (LiHMDS, KHMDS) are used instead, this typically can prevent enolate reprotonation and affords improved results.

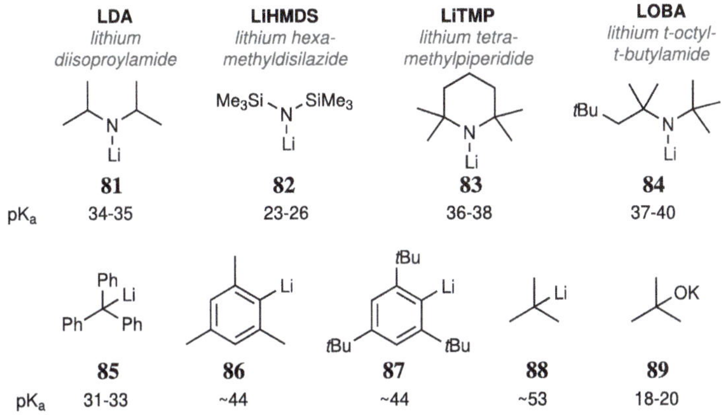

LDA
*lithium
diisoproylamide*

LiHMDS
*lithium hexa-
methyldisilazide*

LiTMP
*lithium tetra-
methylpiperidide*

LOBA
*lithium t-octyl-
t-butylamide*

81   **82**   **83**   **84**

pK$_a$ 34-35   23-26   36-38   37-40

85   **86**   **87**   **88**   **89**

pK$_a$ 31-33   ~44   ~44   ~53   18-20

Fig. 2.33 Selected bases and their pK$_a$ values [83–86]

The **regioselectivity** of the deprotonation step can also be controlled by the choice of reaction conditions. In unsymmetrical ketones, the proton on the sterically less hindered side is deprotonated significantly faster than on the sterically more encumbered side if both possess comparable thermodynamic acidity. Two products are possible, the *kinetic* product (**91**, less substituted double bond) and the *thermodynamic* product (**92**, more substituted double bond). By using a stoichiometric excess of base (> 1.0 eq.) and low temperatures, the sterically more accessible and thus kinetically more acidic proton is removed first (lower activation energy $\Delta G^{\ddagger}$). When using substoichiometric amounts of base, the reaction becomes reversible: The rapidly formed kinetic product **91** reacts with the carbonyl substrate **90** present in traces and thus equilibrates to the thermodynamic enolate **92** (lower formation enthalpy ΔG, see Figs. 2.34 and 2.35).

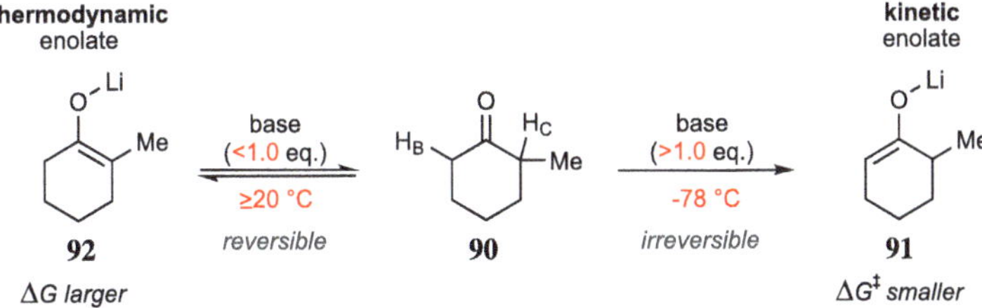

Fig. 2.34 Formation of the kinetic and thermodynamic products

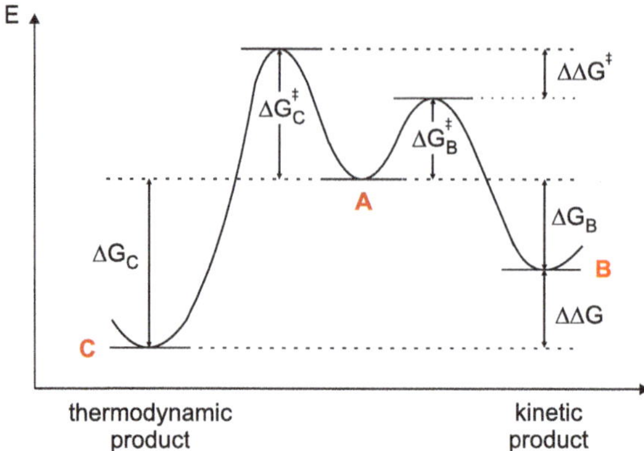

Fig. 2.35 Energy profile of the reaction to thermodynamic and kinetic enolate

By utilizing sterically hindered bases (**81-84**), the preference for the removal of the kinetically more acidic proton H_B is further exacerbated, as the steric interaction disproportionately increases when deprotonating H_C. A higher reaction temperature, on the other hand, accelerates the establishment of the thermodynamic equilibrium (see Table 2.3) [81, 88–91].

Table 2.3 Regioselectivity of enolate formation under kinetic and thermodynamic control [81, 88–91]

Base	Temperature	Selectivity (91 : 92)
LDA	−78°C	99 : 1
LiHMDS	−78°C	95 : 5
MesLi	−78°C	90 : 10
Ph$_3$CLi	−78°C	86 : 14
NaH	≥20°C	26 : 74
LDA	≥20°C	20 : 80
Ph$_3$CLi	≥20°C	10 : 90

Under kinetic conditions, the same principles apply, as can be seen from the product ratio across a range of different ketones. Only a few substrates direct to the more substituted enolate. The atypical regioselectivity of the urethane-substituted ketone is presumably caused by the strongly electron-withdrawing group in the α-position (see Fig. 2.36).

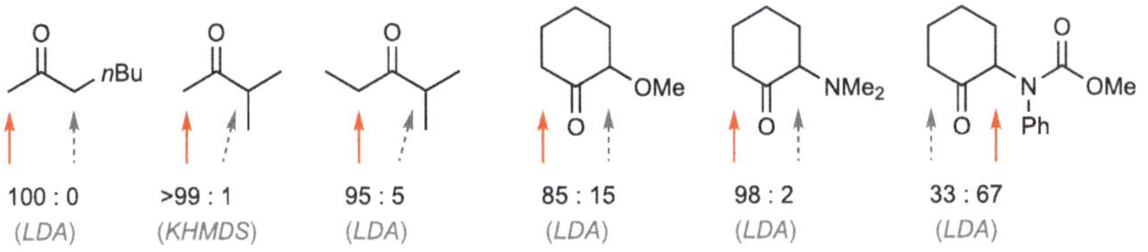

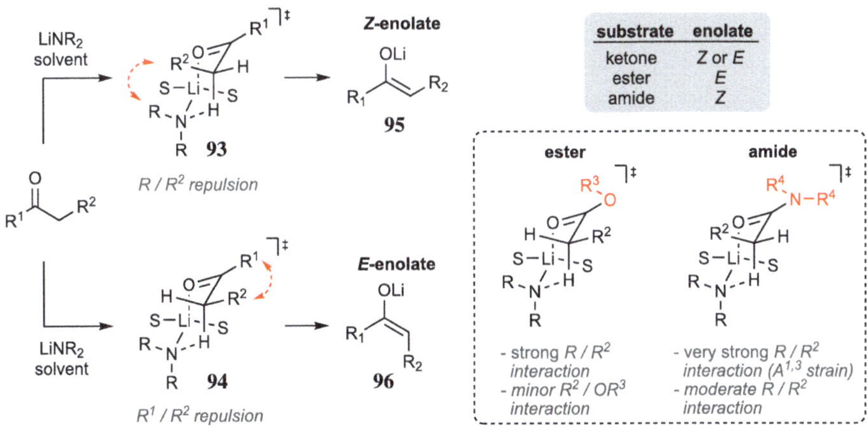

Fig. 2.36 Regioselectivity in the enolate formation of different substrates under kinetic control [90, 92–95]

During deprotonation, acyclic systems with two acidic α-protons can either afford an E- or a Z-configured enolate. Mechanistically, the deprotonation steps with lithium amide bases proceed through six-membered, chair-like transition states (termed *Ireland model*) due to the 90° angle between the α-CH and CO bond (overlap of the σ_{CH}^* and π^* orbitals, see Sect. 1.1) [96, 97]. In **93**, a 1,3-diaxial interaction occurs between R and R^2. For **94**, a 1,2-*gauche* interaction between R^1 and R^2 results. A closer look at the steric interactions in **93** and **94** reveals the possibility to obtain both products when using ketones as substrates, highly depending on the size of R, R^1 and R^2. However, often the E-enolate is formed as major product. In esters, the alkoxy group can point away from R^2, so that the repulsion is minimized in **94**. In amides, due to the planar structure of the amide bond in **94**, a 1,3-allylic strain of the amine substituents with R^2 occurs, which strongly destabilizes this transition state. Esters thus preferentially afford E-enolates, while amides yield Z-enolates (see Fig. 2.37).

Fig. 2.37 Ireland model for ketones, esters, and amides [96, 97] S = Solvent

Since both isomers can in principle be formed from ketones, control of the diastereoselectivity through modification of the reaction conditions can often be achieved. Sterically demanding bases and large R^2 groups enhance the tendency to form the E-enolate (see Fig. 2.38).

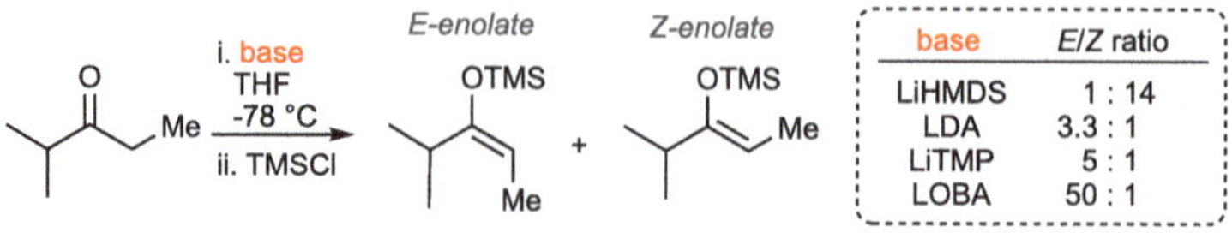

Fig. 2.38 Dependence of diastereoselectivity on the steric demand of the base [98, 99]

If a given ketone primarily forms Z-enolates due to steric interactions, a method introduced by the Collum group can be used for the highly selective access to the E-diastereomer, provided sterically more hindered bases are no viable option. When NEt$_3$ is added to a solution of LiHMDS, eight-membered transition states (**97**) (as opposed to six-membered) have been postulated to explain the observed selectivity. They can avoid a strong R^1/R^2 interaction due to significantly greater conformational flexibility (see Table 2.4) [100].

Table 2.4 E-selective enolate formation from ketones according to Collum *et al.* [100]

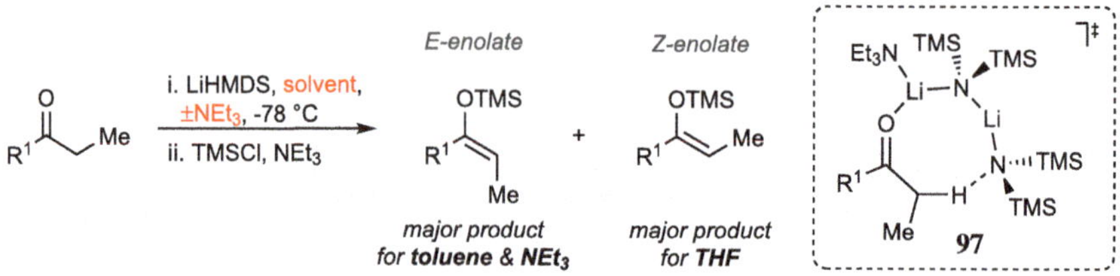

R^1	Selectivity $E:Z$	
	NEt$_3$, toluene	THF
Et	140 : 1	1 : 4
iPr	100 : 1	1 : 14
cHex	80 : 1	1 : 30
Ph	3.5 : 1	1 : >100
OMe	22 : 1	1 : 12

The dependence of the transition structures and thus the diastereoselectivity on the stoichiometry of the reagents and the solvent, as documented in Table 2.4, illustrates the variability of "simple" enolate formations with respect to the choice of reaction conditions. Especially LDA and LiHMDS show significant selectivity differences when comparing them for the same substrate, while LiTMP or LOBA have kinetics, stoichiometries and thus transition states comparable to those of LDA [101, 102].

The properties of the organolithium compounds are strongly influenced by their tendency to aggregate. In view of this, a change of the solvent system can also be considered to alter the product selectivity in addition to changing the steric demand of the substrate or the base. The solvent thus acts not only as reaction medium but also as a Lewis base for stabilizing the lithium counterions. By choice of the solvent, both the structures of the intermediate metal species and their reactivities can be modified (see Fig. 2.39). Spectroscopic studies have provided evidence that enolate aggregates can directly undergo α-functionalization reactions. The reactivity of these complexes increases with decreasing degree of aggregation, which can be attributed to a combination of two effects: a lower steric shielding of less aggregated species and a higher negative charge density at the bonding partner of the metal [87, 102, 103].

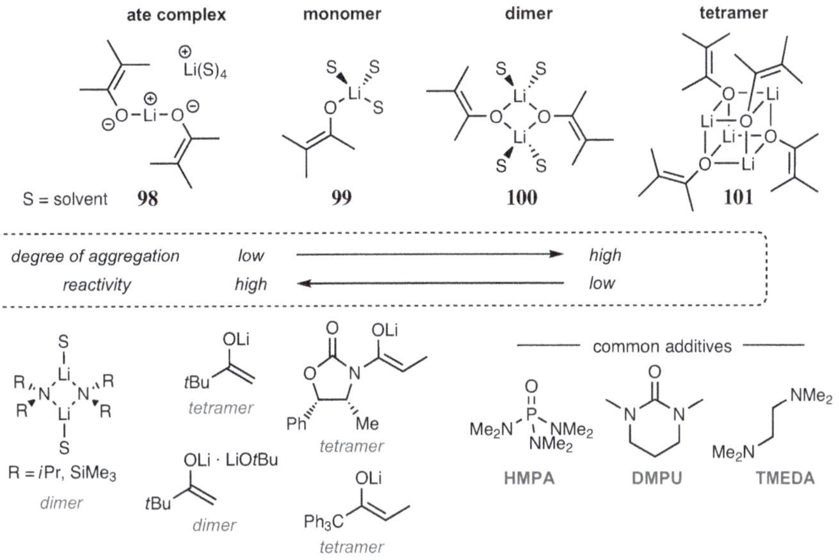

Fig. 2.39 Structures of organolithium aggregates and dominant structure of selected compounds in THF solution [87, 102, 103]

In addition to the effects of the solvent system on the deprotonation of ketones, the diastereoselectivity in the enolization of esters can also be influenced by varying the solvent. In the presence of strongly chelating ligands such as HMPA or DMPU, the preferred enolate changes from E- to Z-configured products in the deprotonation of esters and ketones (see Fig. 2.40).

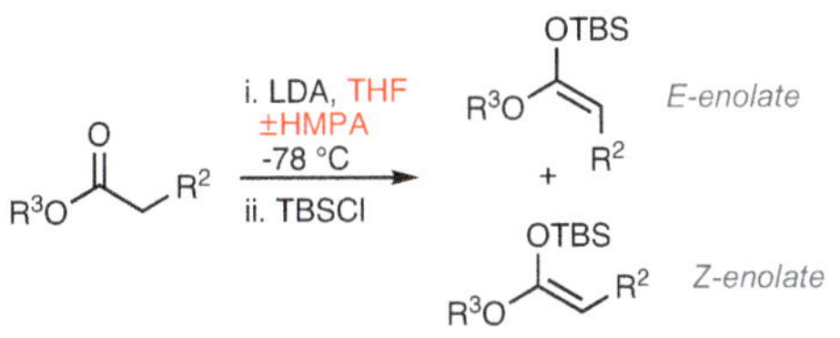

R^3	R^2	ratio E-/Z-enolate	
		THF	THF, 23% HMPA
Me	Et	91 : 9	16 : 48
Me	Ph	29 : 71	5 : 95
Me	tBu	97 : 3	9 : 91
Et	Me	94 : 6	15 : 85
tBu	Et	95 : 5	23 : 77

Fig. 2.40 Influence of HMPA on diastereoselectivity [96, 97]

The mechanistic background of selectivity changes can be extremely complex, often misunderstood, and tends to be non-transferrable [104]. Research indicates that each substrate/solvent combination can potentially possess a different dominant mechanism [105–107]. The strength of the cation-solvent interaction increases significantly when using HMPA or DMPU, which is why Collum's investigations were able to evidence a changed stoichiometry in the transition states of Li enolates (see Fig. 2.41) [102, 106]. In addition to the occurrence of mono-coordinated or ate-transition states (**103, 104**), a later transition state

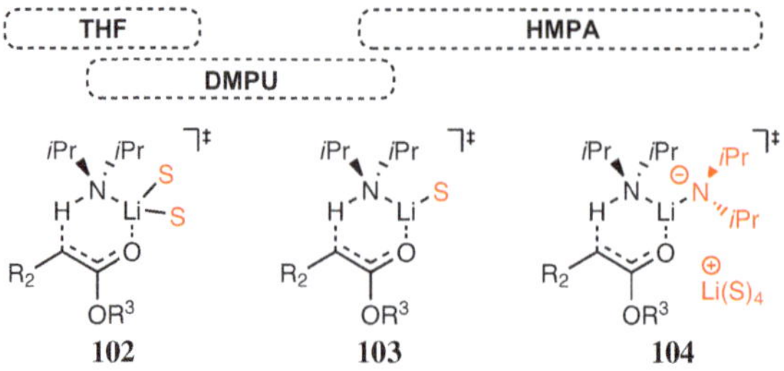

Fig. 2.41 Solvent dependence of the transition states in the enolate formation of esters [106]

with lower steric interactions of R^2 and *i*Pr (in LDA), as well as an acyclic pathway have been postulated to rationalize the observed solvent effects [96, 97].

The formation of **silyl enol ethers** is achieved by trapping the *in situ*-formed lithium enolate with silyl monochlorides, thereby maintaining enolate regio- and stereoselectivity. The use of enol silanes as nucleophiles in the aldol reaction enables access to complementary product selectivities (open transition states, see Sect. 2.3). The decreased reactivity also allows a catalyzed, and thus possibly more selective, transformation as uncatalyzed background reactions hardly occur. Furthermore, this allows the use of amine-free conditions for derivatizations. TMSCl, TBSCl or TESCl are commonly employed as silylation reagent. If amine-free lithium enolates are desired, the route via TMS enolates can be taken as well: when they react with methyl lithium, tetramethylsilane and the desired lithium species are afforded (see Fig. 2.42) [88, 108].

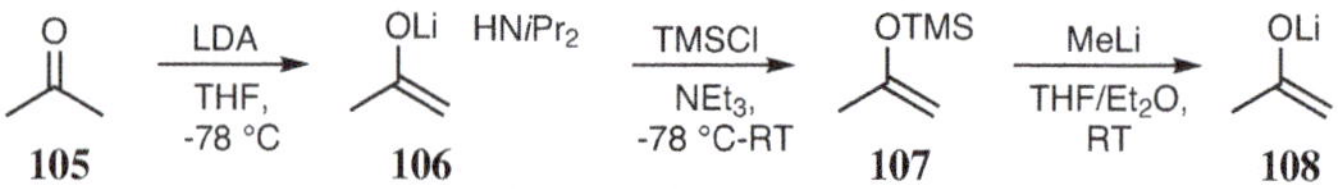

Fig. 2.42 Formation of amine-free lithium enolates according to Stork *et al.* [88, 108]

Thus, *E*- or *Z*-enolates can be accessed from the corresponding ketones and esters with the approaches described above. Amides exclusively direct to *Z*-enolates (see Fig. 2.43).

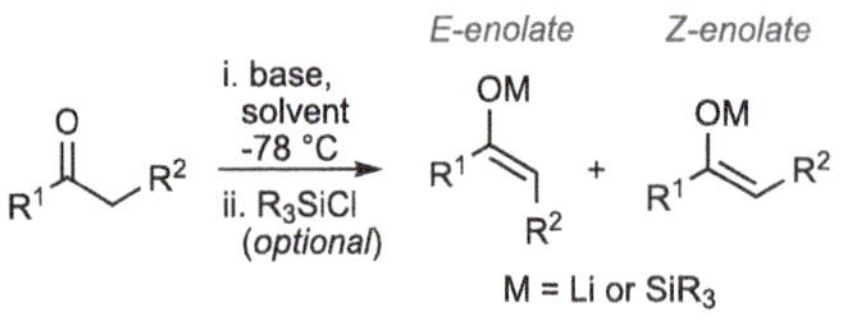

	E-enolate	*Z*-enolate
ketone	LiHMDS NEt$_3$, toluene	LiHMDS, possibly LDA THF, ±HMPA/DMPU
ester	LiHMDS NEt$_3$, toluene	LDA or LiTMP THF, HMPA/DMPU
amide	–	LDA or LiTMP THF

Fig. 2.43 Commonly used conditions for the diastereoselective enolate formation

2.2.2 Boron, Titanium, Tin, and Magnesium Enolates

Strong bases are required for a quantitative and rapid deprotonation. However, a drawback is the partial lack of selectivity in the presence of multiple acidic CH bonds and the risk of epimerizing base-labile stereogenic centers. Weak bases, on the other hand, can be used much more selectively, but the small pK_a difference between the α-acidic carbonyl position and the corresponding amine only allows a partial and slow deprotonation. This opens up the possibility of competing (undesired) autocondensation reactions (cf. Fig. 2.32) and renders them insufficent to produce useful quantities of the desired enolate. Their use is still feasible if the carbonyl function can be activated, for example by virtue of an added Lewis acid, thereby increasing the rate of deprotonation. For instance, the coordination of BF_3 leads to a significant increase in the acidity of the α-CH bond. For acetaldehyde, a decrease in the pK_a value by 24 units was estimated (see Fig. 2.44) [109].

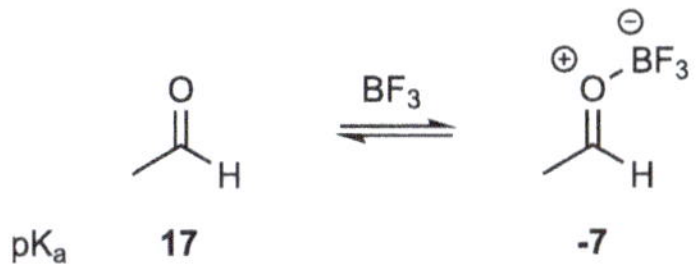

Fig. 2.44 Calculated change in acidity (in H_2O) by coordination of BF_3 [109]

The so-called *soft enolization* employs a combination of weak bases, mostly tertiary amines, and a suitable Lewis acid to efficiently and selectively form the desired enolate. As opposed to alkali metals with spherical symmetry at the metal center, the range of potential Lewis acids allows for the use of metals with a defined coordination geometry at the metal and can pose an additional element for stereocontrol in the subsequent conversion of the enolate (see Fig. 2.45) [110].

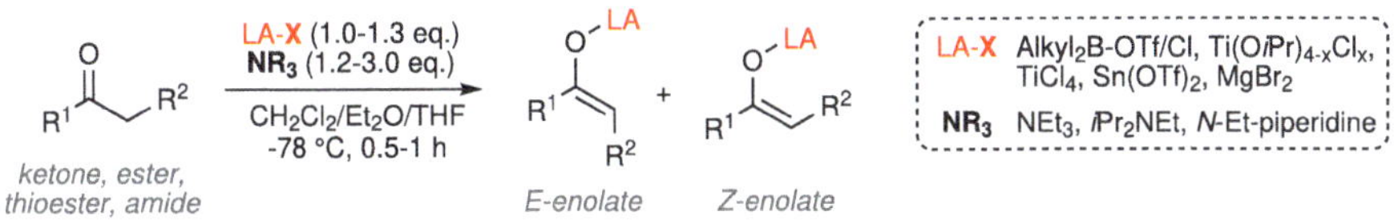

Fig. 2.45 Typical reaction conditions in soft enolizations [111–113]

The Lewis acid is used stoichiometrically or in slight excess and forms the desired metal enolate by displacement of a leaving group (X = OTf, Cl, Br, I). The proper choice of the Lewis acid, base, as well as the reaction temperature permits a high degree of control with regard to the diastereoselectivity of the reaction [112, 114]. An obvious challenge with the combination of a Lewis acid and base lies in the irreversible formation of salts of type **109**.

If the formation of the adduct is irreversible, the order of addition must be strictly adhered to in order to prevent its occurrence. Otherwise no coordination of the Lewis acid to the carbonyl group can take place and the reaction to **111** is inhibited (see Fig. 2.46).

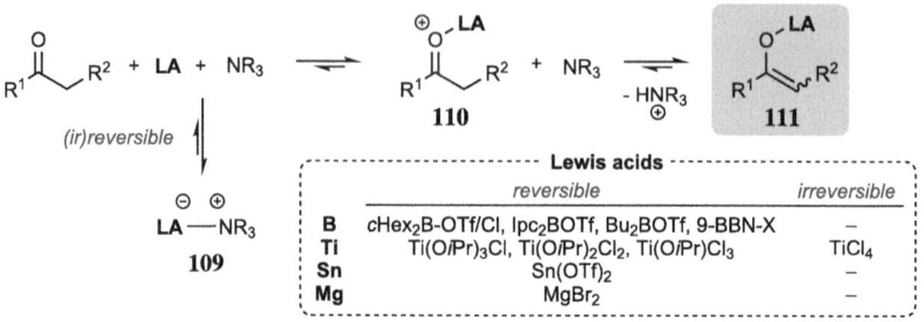

Fig. 2.46 Principle of soft enolization and reversibility of Lewis acid adduct formation [110, 111, 115]

Lewis acids based on boron, tin, and titanium allow complete deprotonation of the ketone and (in principle) full formation of the corresponding enolate. Lithium and magnesium salts enolize ketones only partially, but can achieve full conversion if subsequent steps are irreversible [110, 111]. Esters, thioesters, and amides can also be employed in addition to ketones, as these are acidic enough to afford the enolate using soft enolization. **Boron enolates** are the most commonly encountered nucleophiles that are accessible via soft enolization. The stereoselectivity is highly dependent on the boron species (leaving group, alkyl substituents) as well as the amine base. In principle, both E- and Z-diastereomers can be obtained but the degree of selectivity can vary. Based on initial investigations by Mukaiyama and co-workers, a general model for predicting the regio- and diastereoselectivity in the enolate formation could be developed through subsequent studies. With ketones, the steric environment of the substrate controls the regioselectivity, while the amine and the leaving group on the borane determine the E-/Z-ratio (see Fig. 2.47) [114, 116–118].

The non-complexed electron pair of the carbonyl oxygen and the BX bond are aligned in an antiperiplanar fashion in the initial complex **112** due to a $n_O \rightarrow \sigma^*_{BX}$ interaction (generalized anomeric effect), leading to steric repulsion between the leaving group X and R^1 in **113** or CH_2R^2 in **114**. Boron chlorides partially polarize the α-C-atom on the side of the carbonyl group where L_2BCl is coordinated (**112**). Due to this polarization, the lone pair of the carbonyl oxygen *cis* to a group that can better stabilize this negative charge preferably bonds the boron, which typically is the less substituted side (Me > Et > iPr > tBu). The equilibrium thus shifts towards **114** as long as R^1 is more highly substituted than CH_2R^2. The substituent R^2 avoids a steric clash with Cl in **114** and aligns synclinally with regards to R^1. Small amines then allow deprotonation to the E-enolate, without negating the preference for **114** due to steric repulsion between amine and Lewis acid.

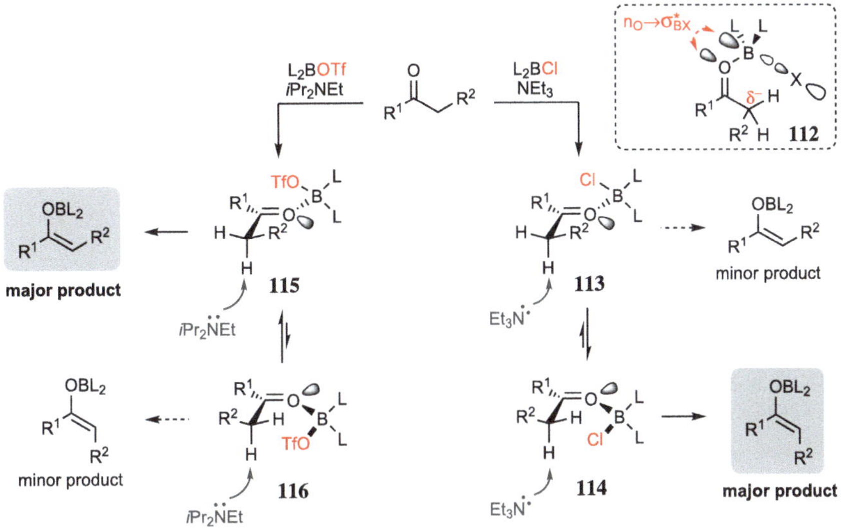

Fig. 2.47 Origin of regio- and diastereoselectivity in boron enolate formation [114]

When using a boron triflate the equilibrium lies on the side of the *trans*-complex **115**, which preferentially affords the Z-boron enolate. The polarization effect of the α-C-atom decreases when increasing the CO-B-X angle. The larger steric demand of the triflate widens the dihedral angle and negates the stereoelectronic preference for deprotonation of one side. By virtue of the weaker stereoelectronic influence, deprotonation occurs *trans* to the Lewis acid due to the steric repulsion of the base with the OTf group. The equilibrium **115 / 116** is governed by purely steric aspects, so deprotonation is possible on both sides of the ketone. Sterically demanding bases ensure enolate formation on the sterically less crowded side via **115** by deprotonating the CH bond at a 90° angle to the CO bond (see Fig. 1.48, Sect. 1.1). Removal of the other proton of the methylene group following rotation of the CC bond would result in a 1,2-*gauche*-interaction between R^1 and R^2 in the transition state (not shown). As a result, the Z-enolate is formed under base control. However, if R^1 becomes too large (e.g., R = tBu), the selectivity decreases as the equilibrium progressively shifts towards **116**. As an alternative rationalization, complete displacement of OTf was proposed, which would alleviate a steric interaction between the ketone substituents and the alkyl groups on boron. The sterically very demanding base would then force the boron onto the more hindered carbonyl side as the base more easily deprotonates the less crowded side and thus favor formation of the Z-enolate [110].

The size of the substituents L on boron can further support these trends in diastereoselectivity: large groups reduce the CO-B-X angle, which results in a stronger polarization and increased E-selectivity in L₂BCl/NEt₃. As a consequence, primarily cyclohexyl- and sometimes isopinocampheyl-substituted boron chlorides constitute the reagents of choice. With boron triflates, the steric interactions and thus the diastereoselectivity are enhanced by large substituents. The size of the substituents on the boron triflate was observed to have a less pronounced influence than with the corresponding chlorides in several studies,[117, 119] so that a wider range of Lewis acids such as cHex$_2$BOTf, Ipc$_2$BOTf or nBu$_2$BOTf might be required for obtaining the Z-boron enolates with high geometric purity (see Table 2.5).

Table 2.5 Influence of borane and amine on the diastereoselectivity of enolization. Calculated from the product ratio in reaction with benzaldehyde [117, 118, 120, 121]

R^1	Borane	Amine	E/Z-Ratio
Et	9-BBN-B-OTf/Cl/I	iPr$_2$NEt/NEt$_3$	1 : 99
	cHex$_2$BOTf	iPr$_2$NEt	7 : 93
	cHex$_2$BCl	NEt$_3$	79 : 21
Ph	cHex$_2$BOTf	iPr$_2$NEt	1 : 99
	cHex$_2$BCl	NEt$_3$	99 : 1
iPr	Ipc$_2$BOTf	iPr$_2$NEt	5 : 95
	cHex$_2$BCl	NEt$_3$	97 : 3
iBu	Ipc$_2$BOTf	iPr$_2$NEt	3 : 97
	cHex$_2$BCl	NEt$_3$	88 : 12
tBu	cHex$_2$BI	NEt$_3$	3 : 97
	cHex$_2$BCl	NEt$_3$	97 : 3
cHex	cHex$_2$BCl	NEt$_3$	99 : 1

The regioselectivity of enolization is the same for both leaving groups and is usually very selective: while the triflate leads to deprotonation of the sterically more accessible side due to steric interaction with the substrate, the chloride directs the base to the less substituted group due to better stabilization of the partial charge. As a result, the kinetic enolate with the less substituted double bond is formed [114, 122, 123].

Carbonyl derivatives with lower acidity such as esters or amides require slightly more tailored conditions. Brown and co-workers investigated the use of boron iodides, which increase reactivity through the enhanced nucleofugality of the leaving group I$^-$ [124, 125]. Masamune later developed a methodology that reliably allows the formation of both E and Z enolates [119]. When using benzyl esters, an acceptable E:Z ratio of either 90:10 or 10:90 can be achieved just through switching of the base (see Fig. 2.48).

In contrast to thioesters, ketones, and imides, boron enolates derived from esters can undergo equilibration at temperatures around 0 °C, allowing their E/Z ratio to be subsequently altered [112].

The soft enolization with boron-based Lewis acids proceeds under extremely mild conditions, allowing their use with highly functionalized substrates. An example can be found

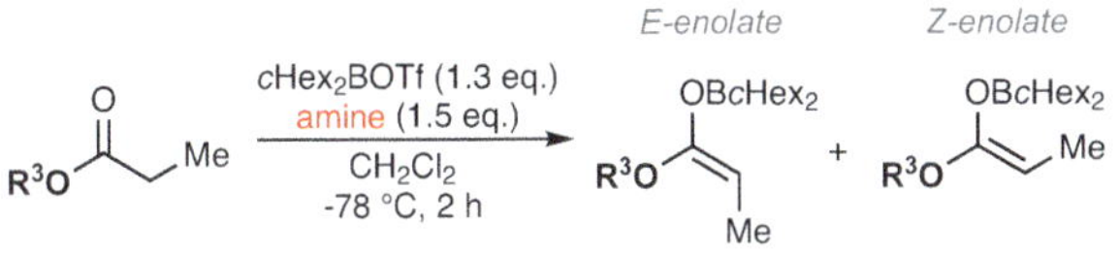

OR^3	E/Z ratio *i*Pr$_2$NEt	E/Z ratio NEt$_3$
OMe	5 : 95	<5 : 95
OEt	<5 : 95	40 : 60
OCH$_2$Cy	<5 : 95	<5 : 95
O*i*Pr	15 : 85	30 : 70
OBn	10 : 90	90 : 10

Fig. 2.48 Diastereoselectivity of boron enolate formation from carboxylic acid esters. [a] Calculated from the product ratio in reaction with *i*-butyraldehyde [119]

in Evans' synthesis of the macrocyclic spiroketal spongistatin 2, in which the AB and CD fragments were coupled by an aldol reaction. After the formation of the *E*-enolate from **118** by *c*Hex$_2$BCl and NEt$_3$, aldehyde **117** was added to provide the β-hydroxyketone **119** with high selectivity and in good yield (see Fig. 2.49) [126].

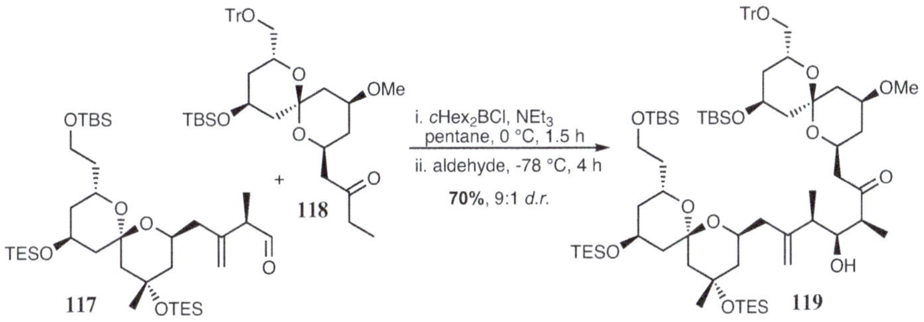

Fig. 2.49 Synthesis of spongistatin 2 by Evans *et al.* [126]

In addition to boron enolates, **titanium enolates** are also employed, although complex syntheses have become less reliant on them over the years. The range of possible derivatizations is broader than with any other metal under soft enolization conditions. Apart from aldol reactions, Michael additions and α-alkylations can also be realized with excellent diastereoselectivities. Titanium enolates can be prepared from ketones, esters, thioesters, and amides bearing oxazolidinone- or thiazolidinone groups, preferentially forming the *Z*-enolate with these substrates [115, 127]. Ketones and esters remain challenging, as they might either lead to more pronounced homocondensation reactions ($k_1 \leq k_2$, cf. Fig 2.32) or they can display insufficient acidity for deprotonation. This renders some starting materials poorly suited for enolization under these conditions (see Fig. 2.50) [111].

The diastereoselectivity of the titanium-mediated enolization is not influenced by the nature of the tertiary amine bases. While *i*Pr$_2$NEt, NEt$_3$ and *N*-ethylpiperidine provide comparable results, no enolization is observed with DBU and tetramethylguanidine. Polar, aprotic solvents are employed as reaction medium, typically favoring dichloromethane. As Lewis acids, TiCl$_4$ or TiCl$_3$O*i*Pr are used, as only these Lewis acids allow a complete

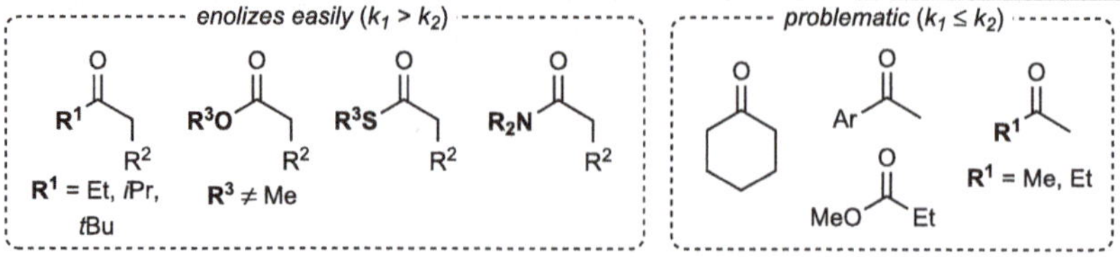

Fig. 2.50 Substrate scope of titanium-mediated soft enolization [111]

conversion to the enolate. Upon increasing substitution of the titanium with alkoxy groups, the Lewis-acidic character of the metal decreases (see Fig. 2.51) [111].

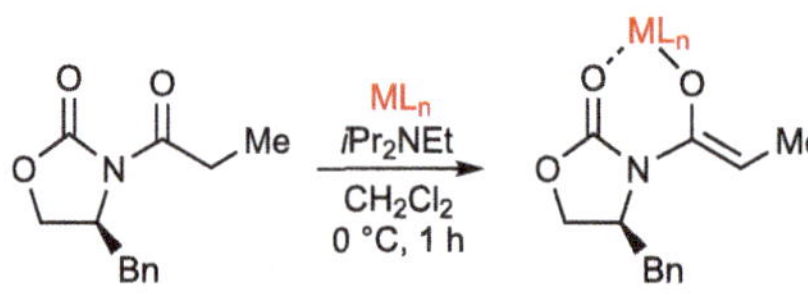

ML_n	enolate formation
TiCl$_4$, TiCl$_3$(OiPr)	100%
TiCl$_2$(OiPr)$_2$	70%
AlCl$_3$	70%
MgBr$_2$·OEt$_2$	25%
TiCl(OiPr)$_3$	10%
Et$_2$AlCl	0%
SnCl$_4$	0%

Fig. 2.51 Enolization with different Lewis acids [111, 128]

The biggest advantage of soft enolization can be attributed to its pronounced chemoselectivity. In the presence of several, differently acidic CH groups, typically the most acidic is reliably deprotonated. Since titanium tends to form chelates, this can be used advantageously to selectively form the desired enolate if chelate formation is viable, e.g., in β-ketocarbonyl substrates (see Fig. 2.52).

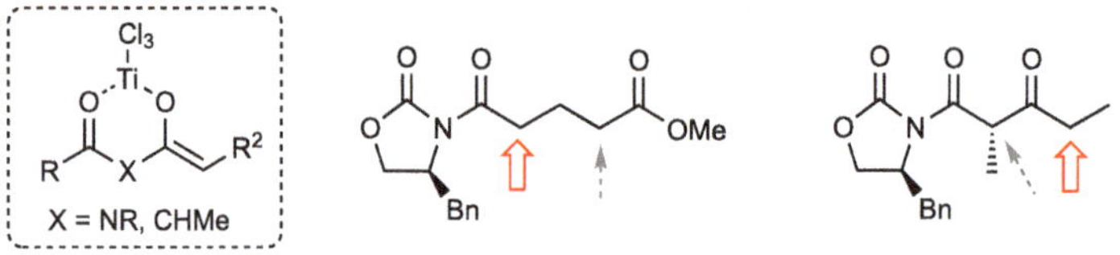

Fig. 2.52 Chemoselective enolization mediated by TiCl$_4$/NEt$_3$ with substrates bearing multiple acidic sites [129, 130]

Apart from being used as a nucleophile, a number of radical reactions can also be carried out with Ti-enolates which exploits its partially diradical character. The majority of the reactions utilizes the "classic" reactivity of carbonyl derivatives and results in their α-functionalization [131, 132].

A noteworthy example of a large-scale application of titanium enolates can be found in the process development of the AKT inhibitor ipatasertib (**123**). The access to the required quantities for clinical trials was enabled by the diastereoselective alkylation of a titanium enolate, which was developed by researchers at Genentech and Array BioPharma. Following deprotonation by TiCl$_4$/iPr$_2$NEt, oxazolidinone amide **120** was reacted with an iminium electrophile derived from **122** to afford **121** in 96% with excellent diastereoselectivity. The crude alkylation product was then used directly in the next step (see Fig. 2.53) [133].

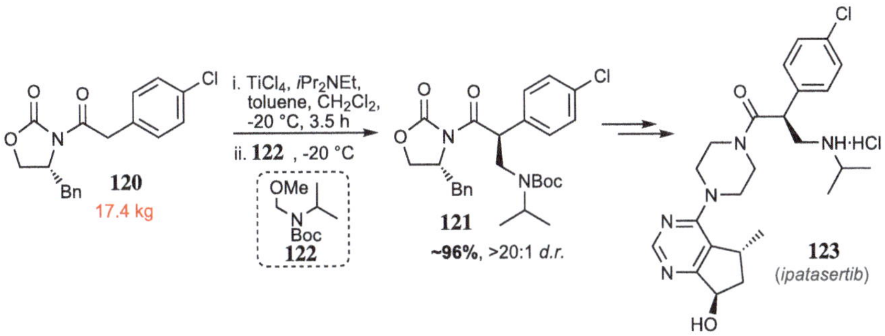

Fig. 2.53 α-Alkylation in the synthesis of ipatasertib (**123**) [133]

Tin enolates are the softest method across the range of enolization approaches, even when compared to Lewis acids based on boron and titanium [134]. However, only relatively acidic substrates such as ketones are viable due to the low Lewis acidity of the metal center. Esters are unreactive under these conditions, but the introduction of thiazolidinethiones (**124**) as ester equivalents could indirectly overcome this issue [135]. Since unsubstituted oxazolidinones (R^2 = alkyl) are not acidic enough for enolization, α-substituted, activated derivatives were developed (**125**), which can be subsequently converted into a number of useful products such as unnatural amino acids (see Fig. 2.54) [113, 136, 137].

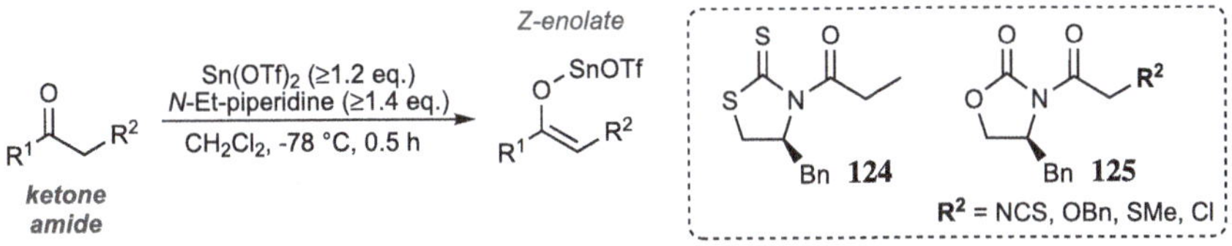

Fig. 2.54 Substrate scope of tin-mediated soft enolizations [113, 134–137]

In analogy to titanium enolates, tin enolates also yield the thermodynamically more stable Z-enolate. N-Ethylpiperidine is the most commonly employed base (in some cases iPr$_2$NEt is used instead), as other tertiary amines can result in considerable homocondensation of the ketone (see Table 2.6) [134].

A. B. Smith and co-workers used the addition of a tin enolate to an aldehyde in their second-generation synthesis of phorboxazole A, a complex marine macrolactone. After deprotonation of the acetate donor **129**, aldehyde **130** was slowly added at low temperatures,

Table 2.6 Influence of the base on the formation of the aldol (**127**) and the homocondensation product (**128**) [134]

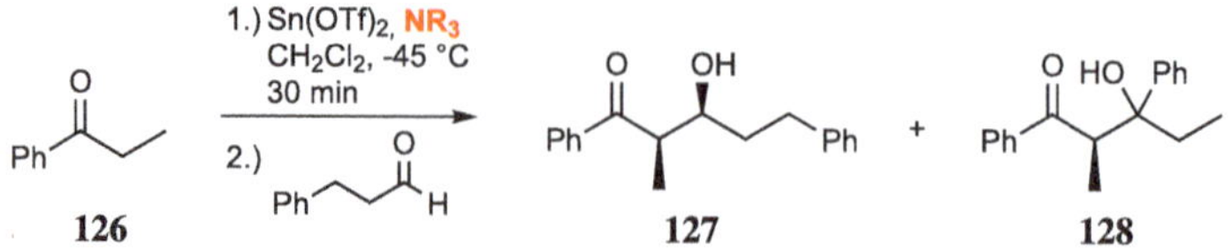

NR₃	Yield	
	127 [%]	128 [%]
pyridine	0	0
DBU	0	0
NEt₃	50	15
N-methyl morpholine	22	65
N-ethyl piperidine	80	traces

allowing for the isolation of the β-hydroxyamide **131** in excellent yield with very good facial differentiation (see Fig. 2.55) [138].

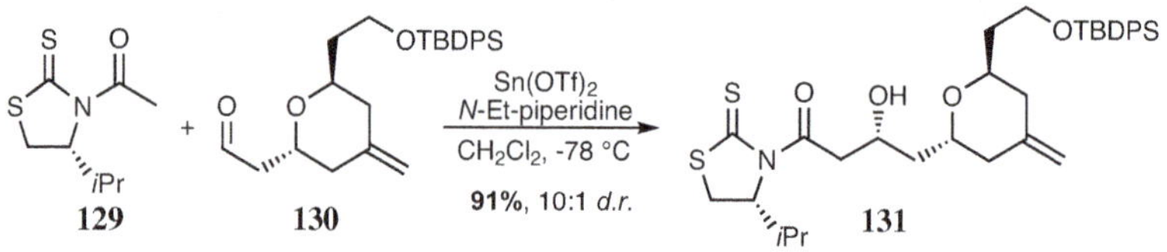

Fig. 2.55 Excerpt from A. B. Smith's synthesis of phorboxazole A [138]

The mild conditions in tin-mediated enolizations lead to a low reactivity compared to other enolate species. Although this indicates an excellent tolerance towards various functional groups that are labile under other methods, it ultimately limits the usefulness of the methodology.

Based on the propensity of various Lewis acids for enolizations as depicted in Fig. 2.51, it is evident that even magnesium salts have the ability to increase the acidity of carbonyl derivatives upon coordination. The formation of **magnesium enolates** under soft conditions is, however, limited to select substrate classes such as pyrazoles, [139, 140] imines [141] and certain oxazolidinone-derived amides. Instead of relying on soft methods, strong Mg-amide bases can alternatively provide the corresponding enolates under classical conditions.

Partial deprotonation under soft conditions can still result in full conversion if the formed Mg-enolate can irreversibly react in a rapid, subsequent derivatization. Under these conditions, only substoichiometric amounts of Mg salts are required for the reaction. Based on work by Fürstner[142], Evans developed an aldol variant in the presence of catalytic amounts of magnesium salt by adding TMSCl, whereby the enolate is converted to a silyl ether. In contrast to previous methods, *anti*-aldol reactions using oxazolidinone auxiliaries are thus possible [143]. An application of this method can be found in Micalizio's approach to the GH ring framework of pectenotoxin 2 and a simplified macrolactone thereof (**132**, see Fig. 2.56) [144].

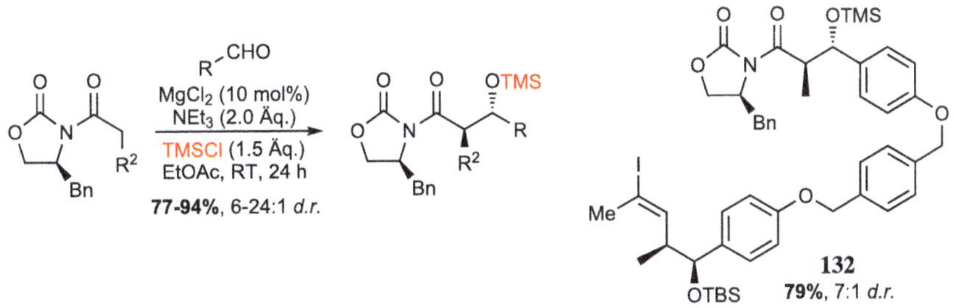

Fig. 2.56 Mg-catalyzed *anti*-aldol reactions and their application [143, 144]

2.2.3   Alternative Enolization Approaches

In addition to enolizations under classical or soft conditions, various transformations can also generate enolates or their equivalent. These methods particularly play a significant role in the context of domino sequences. These approaches typically comprise the Michael addition, Birch reduction, or Reformatsky reaction. Furthermore, many reactions can also be conducted in the presence of organocatalysts (see Fig. 2.57).

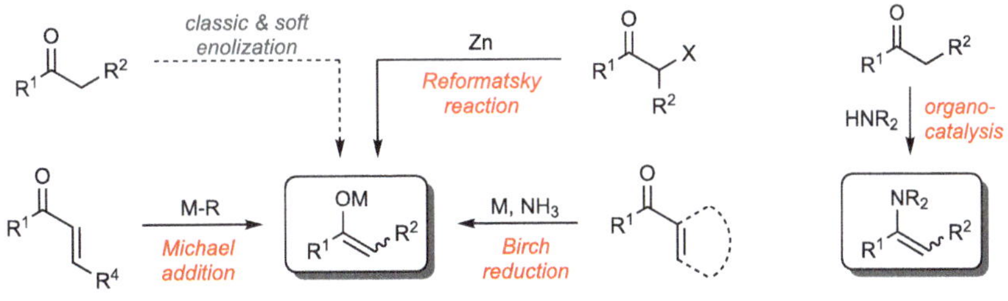

Fig. 2.57 Various routes to enolates and enamines

The *Michael addition* of cuprates (see Sect. 2.5) to α, β-unsaturated carbonyls is a common method for the formation of enolates. However, the diastereoselectivity of a subsequent aldol or alkylation reaction highly depends on the metal being used (see Fig. 2.58). After the addition, either the corresponding silyl enol ethers can be formed or the copper enolates can be converted to a different metal enolate by adding organometallic reagents or salts to achieve a transmetalation. Chiu *et al.* used [CuH(PPh$_3$)]$_6$ (*Stryker's reagent*) as a hydride

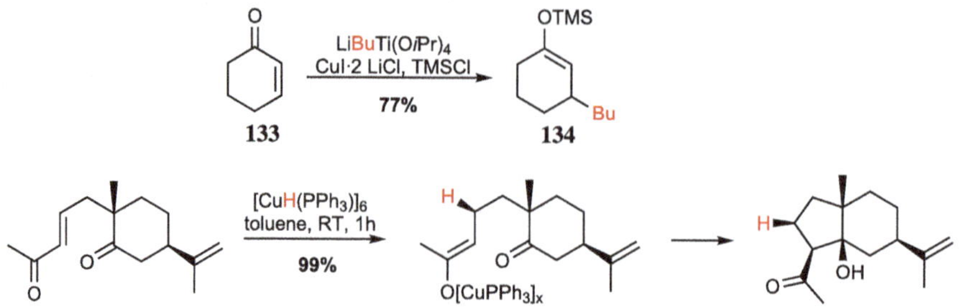

Fig. 2.58 Copper-catalyzed Michael additions to enones [145, 146]

source in their synthesis of lucinone to subsequently assemble the basic framework of the natural product (**137**) in an intermolecular aldol reaction from **136** [145].

Cuprates form copper enolates with good diastereoinduction. Zinc enolates, on the other hand, which result from the Cu-catalyzed reaction in the presence of stoichiometric amounts of organozinc species, do not display sufficient facial differentiation. Copper enolates can be converted into the corresponding lithium enolates or silyl enol ethers by transmetalation (see Fig. 2.59).

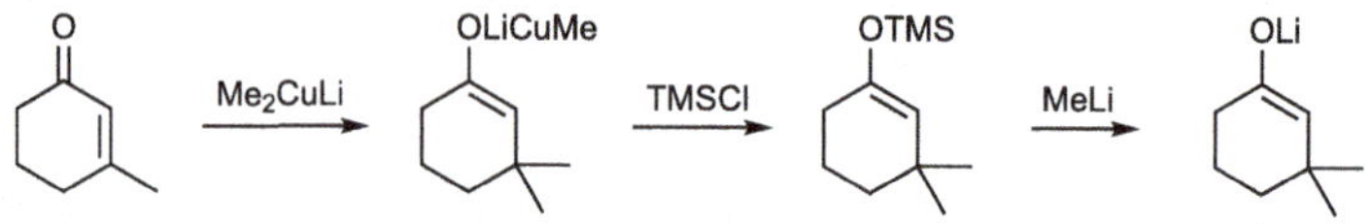

Fig. 2.59 Transmetalation of copper enolates

The 1,4-addition to cyclopentenones, followed by the reaction of the intermediate enolate with an electrophile has been used extensively in the synthesis of various prostaglandins. The Noyori group's approach relied on the Cu-mediated addition of **143** to **142**, followed by transmetalation of the copper to the tin enolate, and a concluding attachment of the allyl iodide **145** in the α-position (see Fig. 2.60) [147].

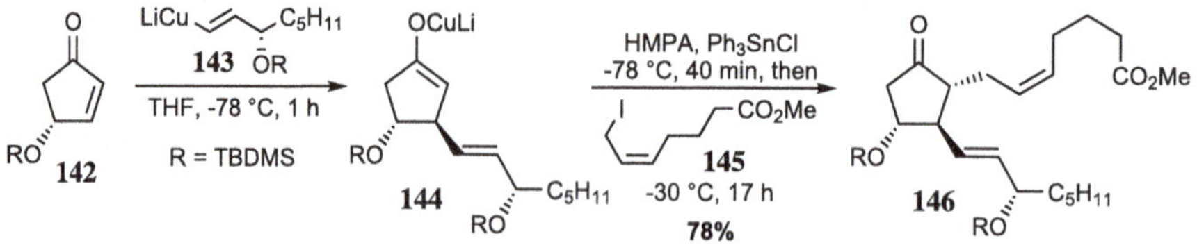

Fig. 2.60 Noyori's approach to prostaglandins [147]

The *Birch reduction* is primarily suited for aryl ketones as substrates; enones can also be reduced to the respective enolates under Birch conditions (see Fig. 2.61) [148]. Based on the mechanism of the reduction, the reaction of acceptor-substituted arenes results in 3-substituted cyclohexadienes (**151**), while donor-substituted systems give cyclohexadienes with the respective donor groups at position 2 (**152**) (see Sect. 4.2 for details on the mechanism) [149].

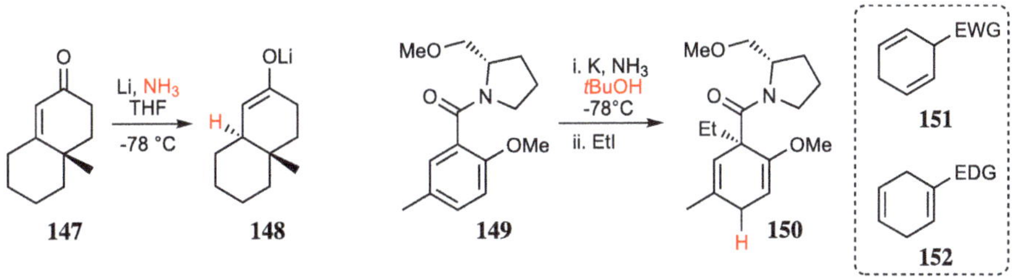

Fig. 2.61 Birch reduction of enones and aryl amides [148]

The conversion of electron-rich α-halo carbonyls (esters, amides, rarely ketones) in the presence of activated, low-valent metals (Zn, Sm, Ti, Co, In, Fe) and their subsequent reaction with electrophiles is termed the *Reformatsky reaction* [150–153]. An advantage of the reaction compared to a classic α-deprotonation can be attributed to its neutral conditions. The less pronounced nucleophilicity of zinc and samarium enolates does increase the chemoselectivity of a subsequent conversion, but the diastereoselectivities and yields in aldol reactions are significantly lower than with lithium enolates. An equilibrium exists between a dimeric form (**155**, with esters) or a C_α-metalated carbonyl (**154**, with ketones) as reactive species. When electrophiles such as aldehydes or ketones are added, they are coordinated to the metal (**156**) and can then be converted accordingly (see Fig. 2.62)

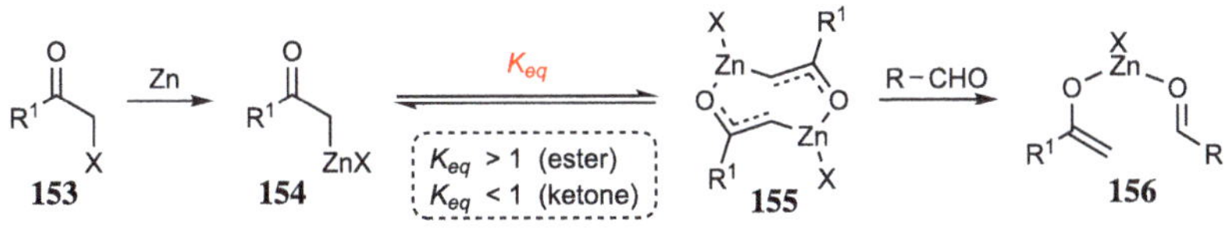

Fig. 2.62 Activation and monomer/dimer equilibrium in the Reformatsky reaction [151]

Compared to the generation and conversion of "classic" enolates, the Reformatsky variant stands out for its unsurpassed chemo- and regioselectivity during the enolate formation, even with α-branched and sterically very demanding substrates. In addition, the α-halo starting materials do not undergo unwanted side reactions and a conversion usually takes place in a one-pot process, in which the metal (salt), the α-halo carbonyl and the aldehyde/ketone can be simply mixed. In comparison to the original protocol, two modifications in particular led to a significantly improved functional group tolerance and an increased diastereoinduction: The pre-activation of zinc (I_2, CuX_2, Zn-Cu, Rieke-Zn, K, Na/Li-Naphthalide) and the use of metal salts with more favorable redox potentials such as SmI_2 [150]. The advantage of using Sm is that the metal enolates display a sufficient stereodifferentiation in aldol reactions whereby the corresponding zinc homologues do not. Secondly, the conditions are particularly suitable for intramolecular processes, as a classic synthesis of **159** by Heathcock and co-workers could demonstrate [154]. An intermolecular example involving a less reactive α, α-disubstituted amide (**160**) was successfully employed in Mulzer's approach to pasteurestin A and B (see Fig. 2.63) [155].

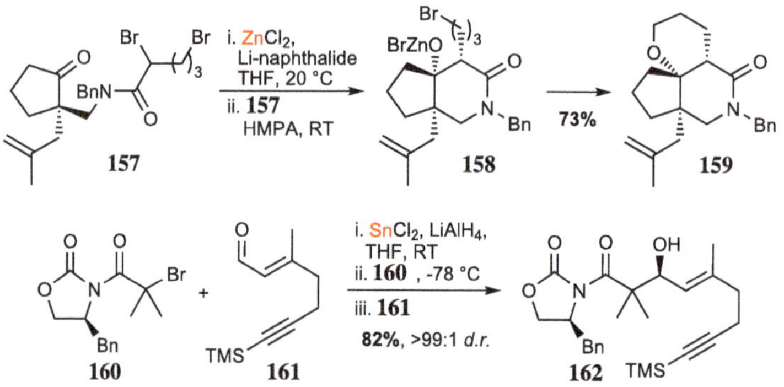

Fig. 2.63 Applications of the Reformatsky reaction [154, 155]

Apart from metal enolates, enamines can also be utilized for the functionalization of carbonyl derivatives. The amines used allow metal-free, *organocatalytic* reaction conditions. The slightly lower reactivity towards electrophiles compared to enolates (see Fig. 2.31) does limit the range of possible reactions somewhat, but in return, asymmetric reactions using chiral catalysts are operationally simple and generally show high enantioselectivities. The detailed methods and their applications will be covered in-depth in Chap. 8.

2.3 Principles of Aldol Chemistry

The aldol reaction is amongst the most common methods for CC bond formation and plays a pivotal role in the construction of stereogenic centers, especially for acyclic substrates. The reaction provides β-hydroxycarbonyls as a structural motif, in which concurrently an additional stereocenter can be introduced in the α-position. The introduction of C2 or C3 units bears close resemblance to the enzymatic biosynthesis of the natural product class of polyketides, which is why they represent obvious targets for the multiple use of the aldol reaction (see Fig. 2.64) [156, 157].

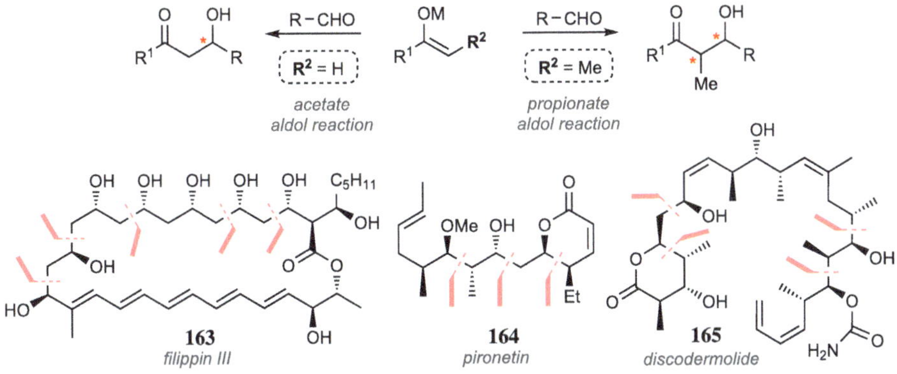

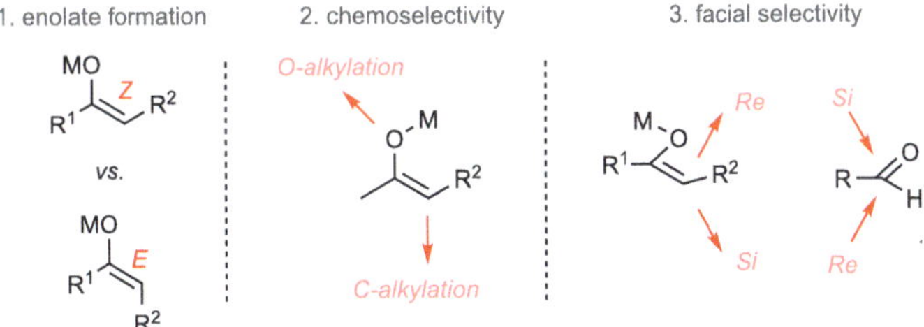

Fig. 2.64 Propionate and acetate aldol reactions and their applications in the synthesis of polyketides [158–160]

The diastereo- and enantioselective construction of stereocenters in the aldol reaction is based on the cumulative selectivity of several factors (see Fig. 2.65) [161]:

1. regio- and diastereoselectivity during the enolization
2. chemoselectivity in reaction of enolates with electrophiles (*C-versus O*-selectivity)
3. facial or *syn-/anti*-selectivity upon addition to carbonyl compounds
4. loss of selectivity upon work-up or during removal of auxiliaries

Fig. 2.65 Diastereo- and enantioselectivity-determining factors of the aldol reaction

The higher the selectivity of the individual steps, the greater the selectivity of the overall reaction. While the previous part dealt with the different aspects of enolate formation, this section will cover the last two topics of the above list. The chemoselectivity of aldol reaction (item 2) directs exclusively to the formation of β-hydroxycarbonyls except for very few examples, thereby affording the CC-bond formation products. This preference can only be reversed under very specific conditions to perform an *O*-alkylation (see Sect. 2.4).

2.3.1 Diastereoselective Aldol Reactions

The stereoselectivity of the Aldol reaction is determined by the facial preference of both the enolate and the electrophile (see Fig. 2.65). For diastereoselective reactions in propionate aldol reactions, two combinations (*Re-Re*/*Si-Si* or *Re-Si*/*Si-Re*) can each deliver the desired *syn*- or *anti*-configured product. In enantioselective reactions, only a single nucleophile/electrophile combination out of these leads to the desired enantiomers.

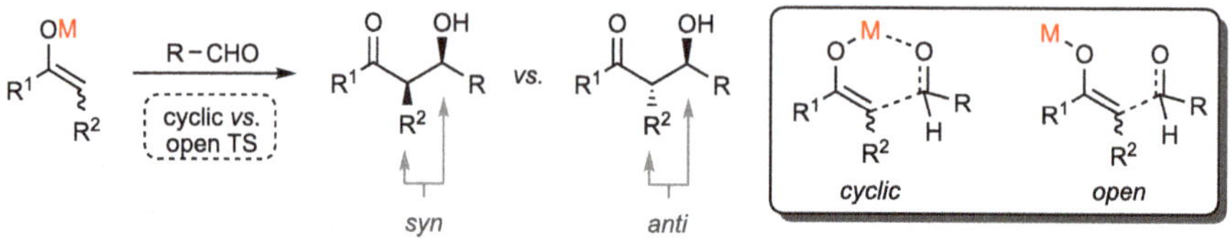

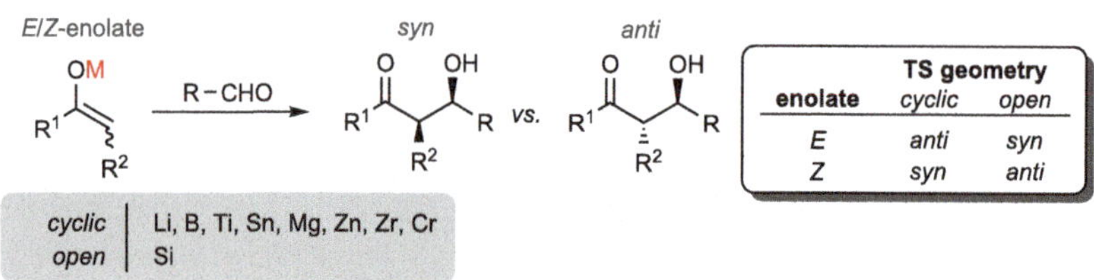

Fig. 2.66 Diastereoselectivity of the Aldol reaction and possible transition states

The aldol reaction proceeds either through a cyclic or an acyclic transition state (see Fig. 2.66). The pathway can be controlled by choice of the appropriate metal enolate used and be further modified by certain additives. The vast majority of enolates directs towards a cyclic transition structure under standard conditions, while silyl enol ethers prefer an acyclic path. The geometry of the enolate determines under kinetic control whether the aldol product preferably yields a *syn*- or *anti*-configured substitution pattern (see Fig. 2.67) [161, 162].

Fig. 2.67 Influence of transition state geometry on diastereoselectivity

Based on detailed mechanistic studies, H. E. Zimmerman and M. D. Traxler postulated that cyclic transition states typically form a chair-like structure [163]. This allows for very rigid and highly organized transition states with good stereoinduction by the enolate. The *Zimmerman-Traxler* model thus correlates the configuration of the product with the double bond geometry of the metal enolate in case of cyclic transition states as the dominant pathway (see Fig. 2.68).

The stereocontrol in the chair-like transition state results from the preferred equatorial arrangement of the substituents to minimize steric interactions. For *E*-enolates, the two transition states **166** and **167** are conceivable: the aldehyde is approached from the *Si*- or *Re*-

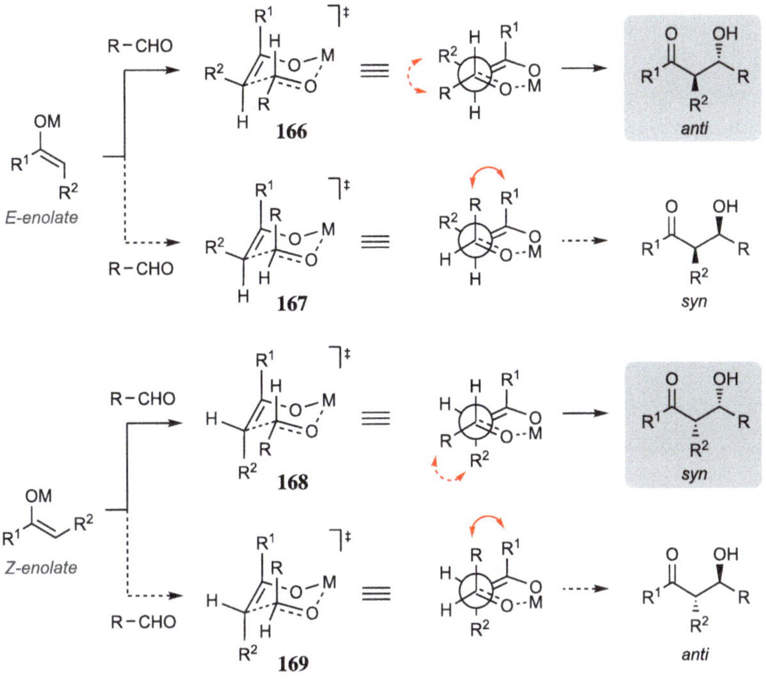

Fig. 2.68 Zimmerman-Traxler transition states for *E*- and *Z*-enolates [161, 162]

face, resulting in an orientation of the substituent R of the aldehyde in either equatorial and axial position. In **166**, a weak 1,2-*gauche*-interaction of R and R^2 ensues; in contrast, **167** shows a strong 1,3-diaxial repulsion (R/R^1) which significantly destabilizes the transition state. **166** is therefore energetically preferred and primarily the *anti*-product is isolated. For the *Z*-enolate, a preference for **168** over **169** can be derived accordingly to avoid an unfavorable 1,3-diaxial interaction. The *syn*-configured β-hydroxycarbonyl is consequently obtained as the major product.

The steric influence of the enolate substituents R^1 and R^2 plays a key role in the stereo-selectivity of the aldol reaction, while the aldehyde does not seem to have a dominant influence. The correlation between high diastereoselectivity and the steric demand of R^1 applies for enolates derived from ketones, amides, and esters [162]. Apart from aldehydes, ketones are also suitable as electrophile. They are however less reactive than the former, leading to longer reaction times and lower yields. In addition, the exchange of a hydrogen atom for an alkyl or aryl group gives rise to poor discrimination of the two substituents in the transition state and by extension a decreased differentiation of the enantio-/diastereotopic π-faces of the electrophile. Ketones therefore typically express a lower stereoselectivity in the ratio of the respective products [164]. The selectivities for *E*- and *Z*-enolates shown in Figs. 2.67 and 2.68 do not strictly apply in all cases. Enolates derived from aldehydes

(R^1 = H) often show only minor stereoinduction, as the dominant 1,3-diaxial repulsion of R and R^1 in the transition state significantly decreases. In acetate aldol reactions (R^2 = H), lower stereoselectivities are likewise encountered. This is due to the absence of R^2, thereby rendering boat-like transition states energetically viable, which dilute the clear $Z \rightarrow syn$-, $E \rightarrow anti$-relation of the Zimmerman-Traxler model [165]. It was also observed that Z-enolates display higher stereoselectivities than E-enolates if the steric demand of R^1 is low (see Table 2.7) [166].

Table 2.7 Diastereoselectivity of the aldol reaction of various enolates [98]

R^1	enolate geometry Z/E	ratio syn/anti
H	100:0	50:50
	0:100	65:35
Et	30:70	64:36
	66:34	77:23
iPr	>98:2	90:10
	32:68	58:42
	0:100	45:55
tBu	>98:2	>98:2
1-adamantyl	>98:2	>98:2
Ph	>98:2	88:12
mesityl	8:92	8:92
	87:13	88:12

The cause of this disparity between the E- and Z-enolate selectivity is not provided by the Zimmerman-Traxler model. Dubois postulated a distortion of the chair-like transition states (approx. 90° angle between C=O and C=C bonds) to rationalize these differences (see Fig. 2.69) [167].

In a direct comparison of the transition states of the E- and Z-enolates according to Zimmerman-Traxler and Dubois, strong steric interactions are observed in both cases for the E-enolate, with the R/R^1 interaction being dominant (cf. **170** and **171**). For the Z-enolate, the preferred transition state **172** is largely devoid of destabilizing interactions, while **173** shows a 1,3-diaxial or 1,2-*gauche* interaction. This significant discrepancy would strongly favor **172** over **173** for Z-enolates and explain their increased stereoselectivity.

As possible alternative explanation, twist or boat conformers can also be considered as reactive transition structures. The smaller the substituents at the metal center, the more energetically feasible a boat transition state (**174**) becomes, which can lead to deviating stereoselectivities (see Fig. 2.70) [168–170].

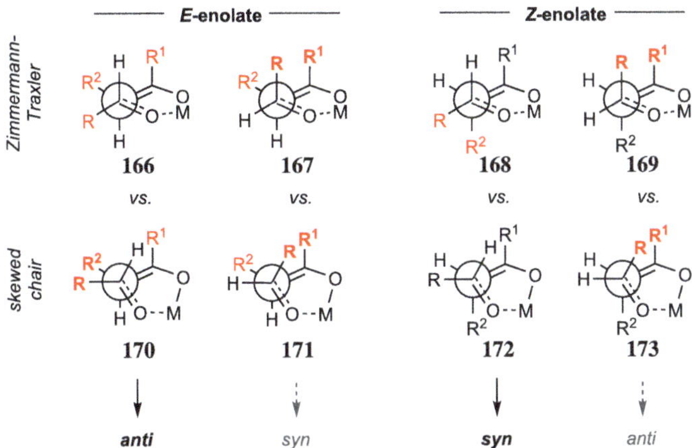

Fig. 2.69 Comparison of the classic Zimmerman-Traxler transition states and the distorted transition states according to Dubois [167]. Groups highlighted in red indicate steric interactions. Emboldened groups display strong interactions

Fig. 2.70 *anti*-selective aldol addition proceeding via a boat transition state **174**

Further theories have also been postulated to account for the transition states of aldol reactions. "Modern" variants of transition state rationalizations often cover deviations very well; however, the classic Zimmerman-Traxler model can adequately predict the outcome of the majority of reactions and thus has led to its broad acceptance. In addition, a deviation from the predicted selectivity can also indicate a thermodynamic equilibration of the enolates or aldol products.

The aldol reaction of lithium enolates shows a lower stereoselectivity than the corresponding boron homologues at a defined enolate geometry (cf. Fig. 2.71).

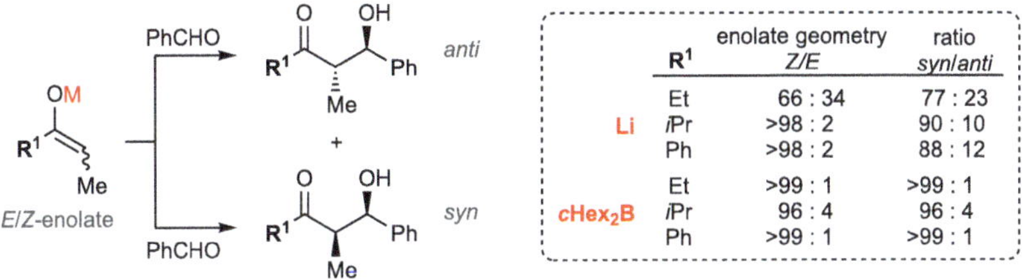

	R^1	enolate geometry Z/E	ratio syn/anti
Li	Et	66 : 34	77 : 23
	*i*Pr	>98 : 2	90 : 10
	Ph	>98 : 2	88 : 12
cHex₂B	Et	>99 : 1	>99 : 1
	*i*Pr	96 : 4	96 : 4
	Ph	>99 : 1	>99 : 1

Fig. 2.71 Comparison of the diastereoselectivity of lithium and boron enolates [98, 171]

When comparing the typical bond lengths of the metal-oxygen species, a correlation between the diastereoselectivities of the aldol reaction and the M-O distance of the metal enolate is observed (cf. Fig. 2.72).

stereoinduction enolate → aldol product		metal	M-O-distance [Å]
		Li	1.9-2.0
		B	1.4-1.5
B > Ti > Mg, Sn^{II}, Zn ≥ Li		Ti	1.6-1.7
		Mg	2.0-2.1
		Sn	1.9-2.0
		Zn	1.9-2.2

Fig. 2.72 Qualitative stereoinduction of metal enolates and bond lengths for M-O bonds [98, 115, 136, 143, 171–173]

The high diastereoselectivity is attributed to the very short B-O-distance and thus a more compact transition state, resulting in a superior stereoinduction by the enolate geometry. Therefore, even with only moderate *E/Z*-selectivities in the enolization, almost identical *syn-/anti*-ratios are found in the derived products [98, 171]. Due to the lower nucleophilicity of the B-O-bond compared to the Li-O-bond, boron-mediated aldol reactions are less prone to the formation of by-products.

Since both *E*- and *Z*-boron enolates are accessible for ketones and esters, (cf. Table 2.5, Fig. 2.47, → Sect. 2.2), this class of metal enolates has enjoyed great popularity for highly functionalized substrates due to its outstanding selectivity. In Kang's synthesis of the C1-C10 fragment of the polyether ionophores monensin B and laidlomycin, the *E*-configured boron enolate derived from **178** was used to obtain the desired *anti*-aldol product **180**. The stereotransfer via a Zimmerman-Traxler transition state and the addition to aldehyde **179** according to the Felkin-Anh model predicted the correct facial selectivity for both aldehyde and ketone (see Fig. 2.73) [174].

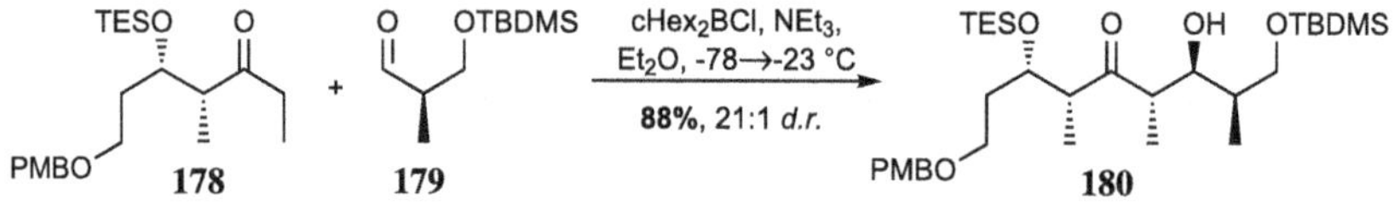

Fig. 2.73 Synthesis of the C1-C10 fragment of monensin B and laidlomycin by Kang *et al.* [174]

Titanium enolates provide very good yields with a preference for the formation of the *syn*-product under mild conditions. The selectivities typically only fall slightly short compared with boron enolates [115]. Zinc enolates are rarely utilized in simple diastereoselective reactions, but are one of the most frequently encountered metal enolates in enantioselective variants alongside silyl enol ethers, boron and titanium enolates. Other than ketones, the mild conditions only allow for amides bearing thiazolidinethione groups, typically intended as an auxiliary, to be used efficiently as substrates (see Fig. 2.74).

Other enolates that are resorted to in individual cases rely on zirconium, copper or chromium as the metal center. Before the widespread use of boron enolates and silyl enol

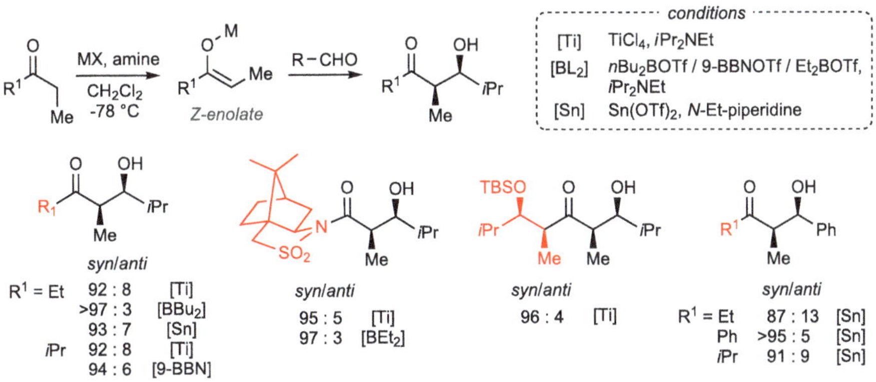

Fig. 2.74 *syn*-selective aldol additions of various metal enolates [115, 134, 135]

ethers, zirconium enolates were also common nucleophiles with good diastereoselectivities in aldol reactions [175–177]. However, they have fallen out of fashion, partly due to the growing popularity of alternative methods such as organocatalysis that display a distinct preparative simplicity.

The use of *silyl enol ethers* as a nucleophilic component in aldol reactions was discovered early on by Mukaiyama,[178, 179] but the full potential of the *Mukaiyama* aldol reaction when employing them under enantioselective, catalytic conditions was not fully realized until the 1990s [180, 181]. With few exceptions, silyl enol ethers are unreactive towards aldehydes. The silicon atom of the silyl enol ethers is not sufficiently Lewis-acidic for complexation and activation of the aldehyde. The reaction only takes place in the presence of an exogenous activator as catalyst, usually a Lewis acid. The primary products are silylated β-hydroxycarbonyls, which are typically desilylated in the work-up (see Fig. 2.75) [181].

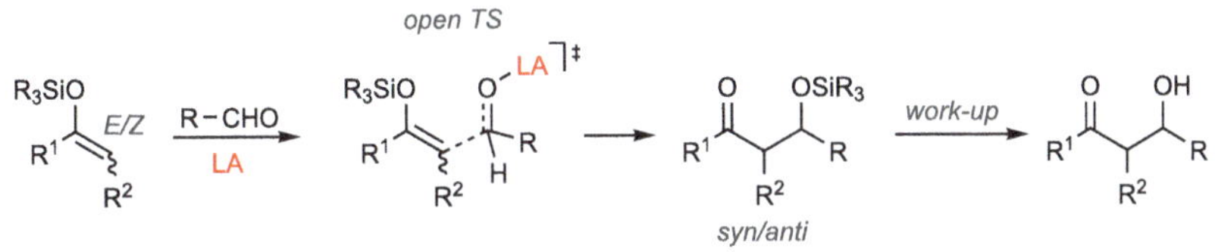

Fig. 2.75 Lewis acid-catalyzed Mukaiyama aldol reaction

The catalyzed reaction is much faster than the uncatalyzed background reaction due to the subdued reactivity in the absence of an activator ($k_{cat} \gg k_{uncat}$). This forms the basis of an asymmetric reaction with high selectivity, provided the (chiral) activator allows for a strong enantioinduction. Activation by Lewis acids constitutes the most common variant of the Mukaiyama aldol reaction. However, the reaction can also be realized by increasing the reactivity of the silyl enol ether instead of the aldehyde component, for example by adding Lewis bases to Lewis acidic silyl substituents or by removing the silyl group via addition of fluoride sources (see Fig. 2.76).

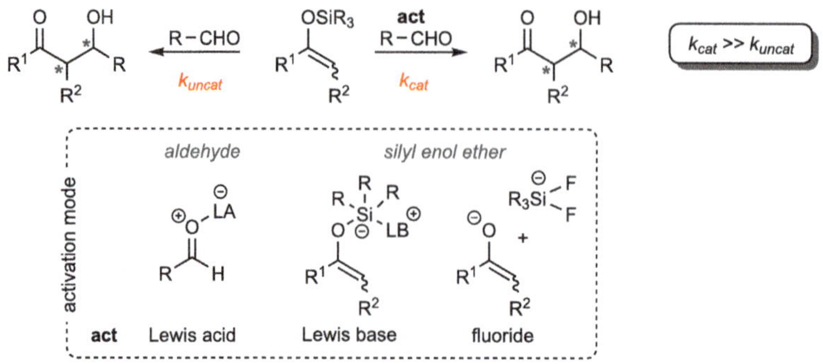

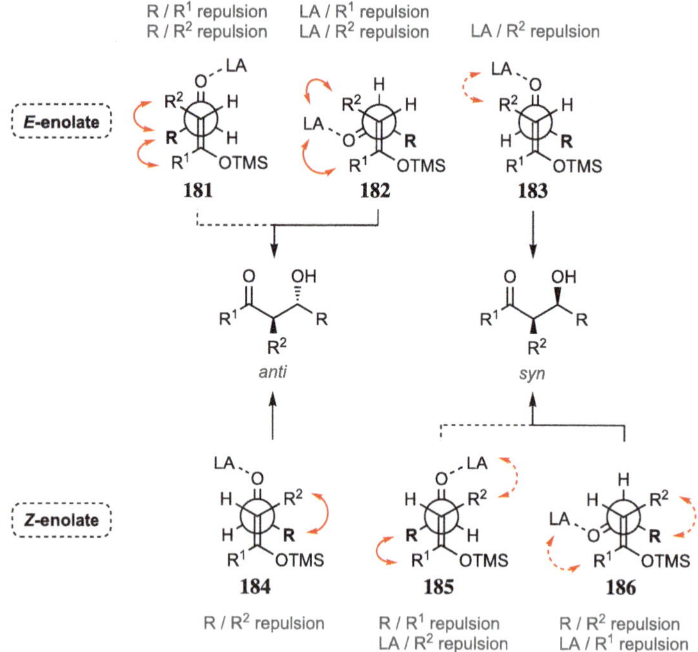

Fig. 2.76 Competition of the catalyzed and uncatalyzed aldol reaction and possible activation modes

Originally, stoichiometric amounts of activators were needed, but current variants only require catalytic amounts. Unlike the other metal enolates, the Mukaiyama aldol reaction typically proceeds via acyclic transition structures and not via cyclic Zimmerman-Traxler transition states (see Fig. 2.77) [182–185].

Fig. 2.77 Postulated transition states of the Mukaiyama aldol reaction [184]

In the transition states, the oxygen atoms orient themselves in an anticlinal fashion to each other to avoid an electrostatic build-up. The steric demand of the substituents R^1, R^2 and R as well as the Lewis acid govern the strength of the repulsive interactions and the relative energetic position of the transition states. Based on extensive experimental and theoretical studies, most Lewis acids tend to direct towards an antiperiplanar alignment of the aldehyde CO- and enol ether CC-bond (**181**, **183–185**), showing ratios of 1.5:1 to 4:1 [184]. This is also reflected in the simple stereoselectivity of the reaction: In some cases, high diastereoselectivities are attainable in the absence of further directing factors – but this is not a general trend. Often, the Mukaiyama aldol reaction is even stereoconvergent, i.e., there is no correlation between the enolate geometry and the diastereoselectivity of the process (see Fig. 2.78) [186].

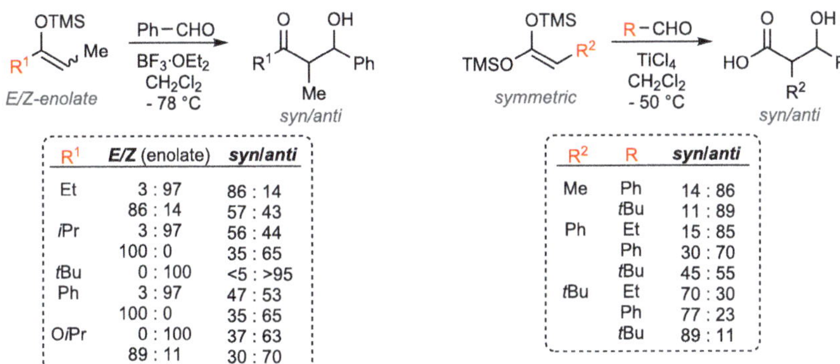

R¹	E/Z (enolate)	syn/anti
Et	3 : 97	86 : 14
	86 : 14	57 : 43
iPr	3 : 97	56 : 44
	100 : 0	35 : 65
tBu	0 : 100	<5 : >95
Ph	3 : 97	47 : 53
	100 : 0	35 : 65
OiPr	0 : 100	37 : 63
	89 : 11	30 : 70

R²	R	syn/anti
Me	Ph	14 : 86
	tBu	11 : 89
Ph	Et	15 : 85
	Ph	30 : 70
	tBu	45 : 55
tBu	Et	70 : 30
	Ph	77 : 23
	tBu	89 : 11

Fig. 2.78 Comparison of the diastereoselectivity of various silyl enol ethers [182, 187]

In the presence of functional groups that can form a chelate with the Lewis acid, however, the Mukaiyama aldol reaction is characterized by a very good facial differentiation. Especially β-alkoxy and silyloxy groups are used for efficient diastereoinduction. Regardless of the configuration of the silyl enol ether nucleophile, the reaction is *syn*-selective for these substrates, which is accounted for by the transition states **187** and **188** (see Fig. 2.79) [186].

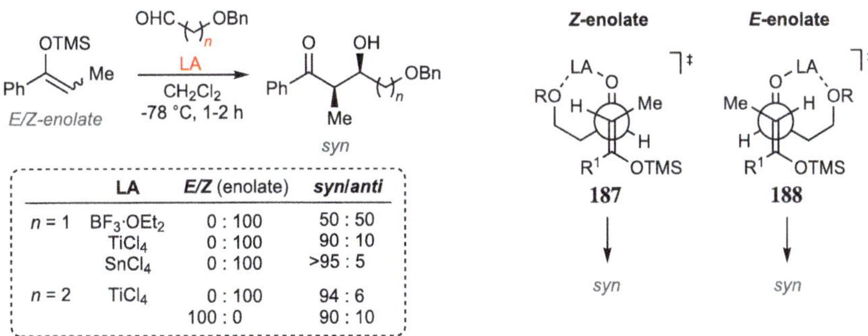

	LA	E/Z (enolate)	syn/anti
n = 1	BF₃·OEt₂	0 : 100	50 : 50
	TiCl₄	0 : 100	90 : 10
	SnCl₄	0 : 100	>95 : 5
n = 2	TiCl₄	0 : 100	94 : 6
		100 : 0	90 : 10

Fig. 2.79 Diastereoselectivity and transition states of the Mukaiyama aldol reaction with chelating aldehydes [186, 188]

The size of the substituents R^1 and R^2 determines which diastereomer is formed: If R^2 is small and R^1 sterically demanding, good *anti*-selectivity is observed, regardless of the double bond geometry of the enol ether (dominant R-R^2 interaction). When R^2 increases in size, the *syn*-product is preferably obtained. With chelating aldehydes, the *syn*-product is furnished [186].

If substrates bearing stereogenic centers (α- or β-chiral aldehydes, chiral silyl enol ethers) are utilized, typically good diastereoselectivities are achieved. Kiyooka and co-workers synthesized the C1-C13 fragment of the polyketide (+)-discodermolide incorporating several aldol reactions. The concluding CC-bond formation to give **191** was realized with excellent diastereoselectivity by coupling the silyl enol ether **190** with aldehyde **189** in the presence of $TiCl_4$. The facial differentiation of the aldehyde was achieved under chelating conditions by a 1,3-stereoinduction of the β-stereocenter ($\rightarrow$ Sect. 2.1, Fig. 2.21), presumably via transition state **192** (see Fig. 2.80) [189].

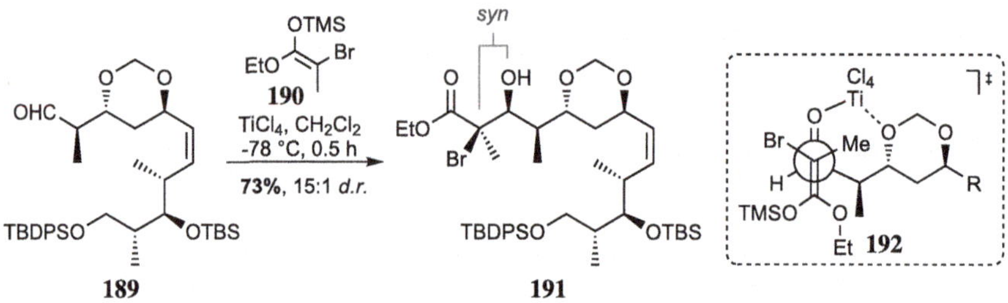

Fig. 2.80 Synthesis of the C1-C13 fragment of (+)-discodermolide [189]

Apart from the "classic enolates" – d^2 systems relying on an Umpolung - a functionalization at the γ-position in α, β- or β, γ-unsaturated carbonyl compounds can occur instead of a reaction at the α-position. This type of reactivity is referred to as *vinylogous* aldol reactions (see Fig. 2.81).

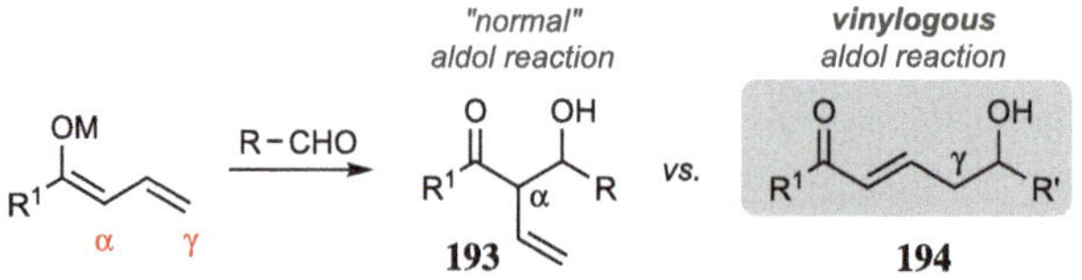

Fig. 2.81 Normal and vinylogous aldol reaction

Most vinylogous metal enolates only display a limited γ-selectivity. In contrast, a hallmark of β, γ-unsaturated silyl enol ethers like **195** is their good vinylogous reactivity, which was observed in pioneering studies by T. Mukaiyama (see Fig. 2.82) [190, 191].

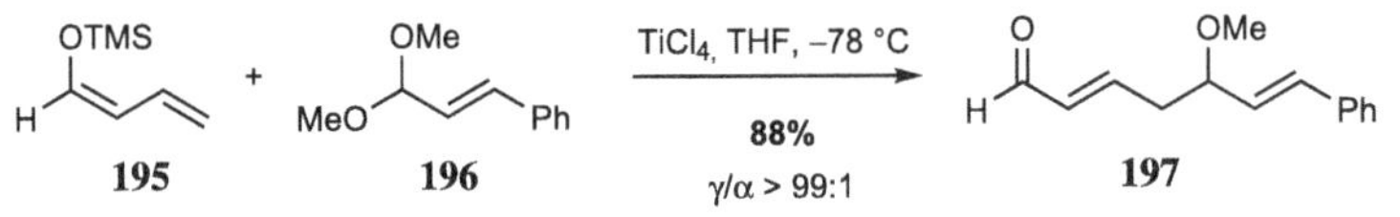

Fig. 2.82 Vinylogous alkylation by Mukaiyama [190]

Apart from a few exceptions, vinylogous aldol reactions do not possess simple diastereo-selectivity [192, 193]. However, there exist now several asymmetric methods with both good enantio- and double diastereoselectivities (see below). Based on the selectivities described above, the following guidelines for the diastereoselectivity of kinetically – controlled aldol reactions of metal enolates and silyl enol ethers can be formulated:

- If the enolate does not contain a large substituent R^1, the diastereoselectivity is low. Z-enolates tend to be slightly more stereoselective than E-enolates in these cases.
- Boron enolates show the best stereoinduction, lithium enolates often the least.
- Enolates derived from aldehydes are not very diastereoselective. The reaction with ketones as electrophiles also shows minor stereoinduction and provides only moderate yields.

Apart from a simple diastereoselectivity (*syn/anti*) furnishing racemic mixtures, a double diastereoselectivity can also be exploited in enantioselective transformations.

2.3.2 Enantioselective Aldol Reactions

A core advantage of the aldol reaction is reflected in its ability to not only reliably provide the relative (*syn/anti*) configuration, but also to allow access to the newly formed stere-ocenters in an enantioselective manner. Hardly any class of reaction has been probed so intensively in terms of its reaction mechanism, substrate scope, as well as its chemo-, regio-, and diastereoselectivity, and used as widely in total synthesis compared to other synthetic methods [157, 162, 172, 180, 181, 186, 194–197].

To introduce one or more stereogenic elements, the stereoinduction must either be pro-vided by the substrates themselves (enolate, aldehyde) or by an exogenous inductor. The starting materials typically used for the assembly of advanced intermediates in total synthe-ses often already contain persistent stereogenic centers in α- or β-position, which can be capitalized on for a facial differentiation of enantio- or diastereotopic surfaces. Carboxylic acid derivatives can play a special role, as an auxiliary can be attached as the corresponding amides or esters, which can be cleaved off and recycled after the aldol reaction. As a third category, chiral reagents can be used in the enolization step, for example chiral boranes in the case of boron enolates or or metal enolates bearing suitable ligands at the metal center (Ti,

Sn). Finally, chiral Lewis acids lend themselves as catalysts specifically in the Mukaiyama aldol reaction. Organocatalysts can furthermore be used to enable an asymmetric transformation (see Fig. 2.83).

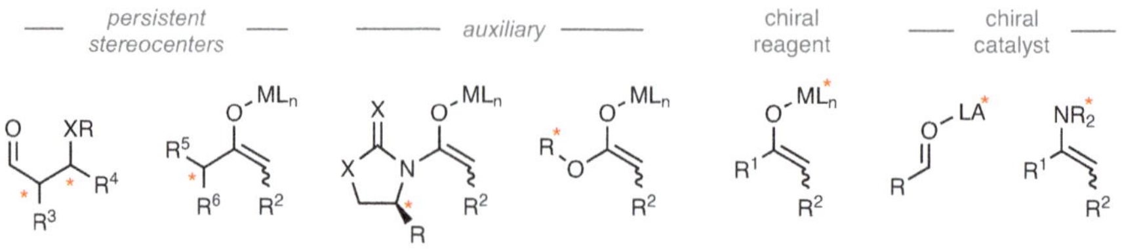

Fig. 2.83 Approaches to stereoinduction in enantioselective aldol reactions

The selective addition to α- or β-chiral aldehydes has already been extensively discussed in Sect. 2.1. The inclination for a *Re*- or *Si*-approach of the electrophile can be accounted for by the stereoelectronic preference for one face by the starting material. In this "double" stereoselectivity, one major diastereomer is obtained, with the absolute configuration then being determined by the stereocenter present in the starting materials. This provides an enantiomer instead of a racemic mixture as product. The inclusion of models for 1,2- (Felkin-Anh-, Cram-Chelate model) or 1,3-induction (Evans/Reetz models) into the Zimmerman-Traxler model leads to transition states of the type **200** or **201**, depending on the geometry of the enolate (see Fig. 2.84) [198].

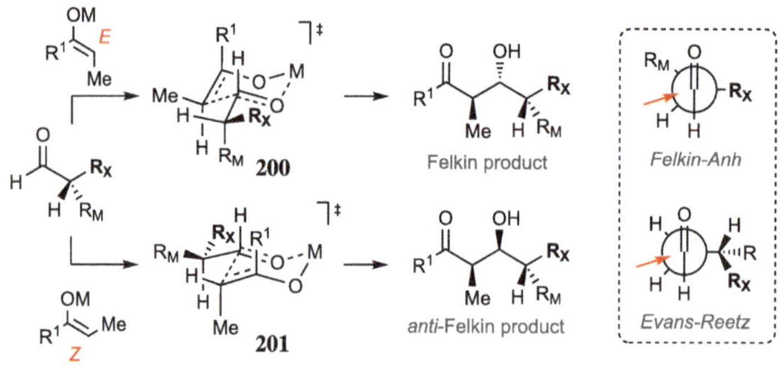

Fig. 2.84 Aldol reaction of chiral aldehydes and integrated Zimmerman-Traxler transition states [198]

Access to natural product analogs is particularly relevant in the context of structure-activity relationship (SAR) studies to determine the influence of individual molecular fragments on the bioactivity of a drug. The Taylor group synthesized an epothilone D analog and used the aldol addition to an α-chiral aldehyde as a key step to undertake an SAR-assessment. The soft enolization starting from ketone **202** selectively gave the Z-titanium enolate, which added to **203** to yield the *anti*-Felkin product **204** with excellent stereoinduction (see Fig. 2.85) [199].

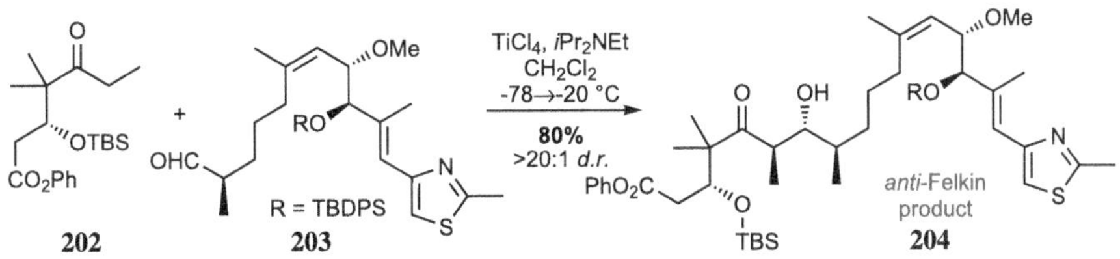

Fig. 2.85 Synthesis of an epothilone analog by Taylor *et al.* [199]

The principle works analogously with enolates: The stereogenic center of the nucleophile orients itself in the transition state by minimizing repulsive interactions so that ideally one face of the enolate is completely shielded. This shielding is referred to as *passive stereoinduction* or the shielding substituent termed *passive volume* (see Fig. 2.86).

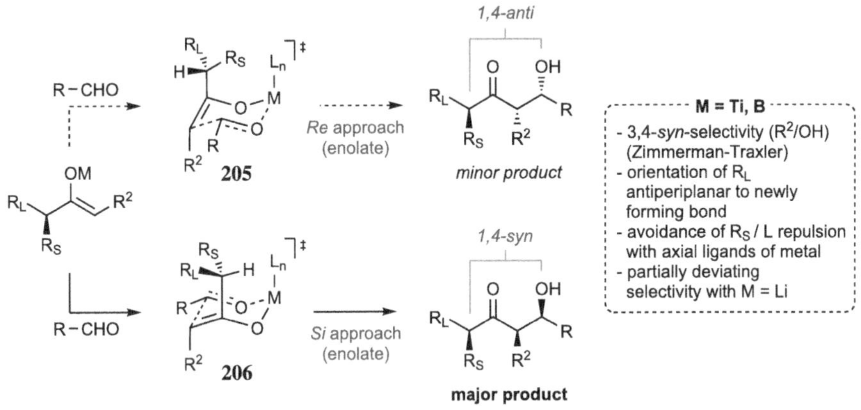

Fig. 2.86 Double diastereoselectivity in the presence of α-stereogenic centers on the enolate [115]

This rather robust principle could also be used in the synthesis of the ionophore zincophorin by Cossy and co-workers to control the facial selectivity of a late-stage aldol addition. The enolate formation from **207** proceeded as predicted Z-selectively in the presence of $TiCl_4/iPr_2NEt$. The α-stereogenic center of the enolate enabled the formation of the β-hydroxyketone **209** with excellent 1,4-*syn*-selectivity in the subsequent addition. The final steps to the natural product included the reduction of the ketone at C13 to the corresponding alcohol, a global desilylation with HF-pyridine, and the saponification of the ester (see Fig. 2.87) [200].

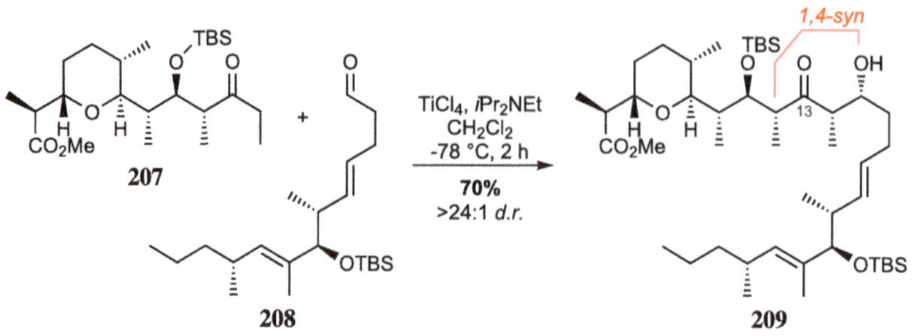

Fig. 2.87 Cossy's approach to zincophorin [200]

The conformation of a stereogenic center in α- or β-position can be locked in the transition state not only by a minimization of steric interactions but also through complexation, given the substituents are Lewis-basic. Whether coordination takes place can then be controlled by judicious choice of the metal. The HSAB principle[201] and the coordination sphere of the metal accordingly influence the tendency to form chelates: titanium, for example, is rather oxophilic, whereas tin prefers to bind to soft ligands such as sulfur. Heathcock investigated this phenomenon using enolates of type **210**. The formation of a chelate complex and the shielding of the *Re*- or *Si*-face of the enolate occurs in accordance with the principles discussed in the Cram chelate model (**211**; cf. Fig. 2.9, Sect. 2.1). If the metal is not coordinating, the stereogenic center is aligned in such a way so that the dipoles are oriented in an antiperiplanar manner (**212**) in order to negate the build-up of dipoles and minimize the energy. By variation of the reaction conditions to allow or prevent complexation, a complementary selectivity can thus be achieved with the identical substrate (cf. Fig. 2.88) [202].

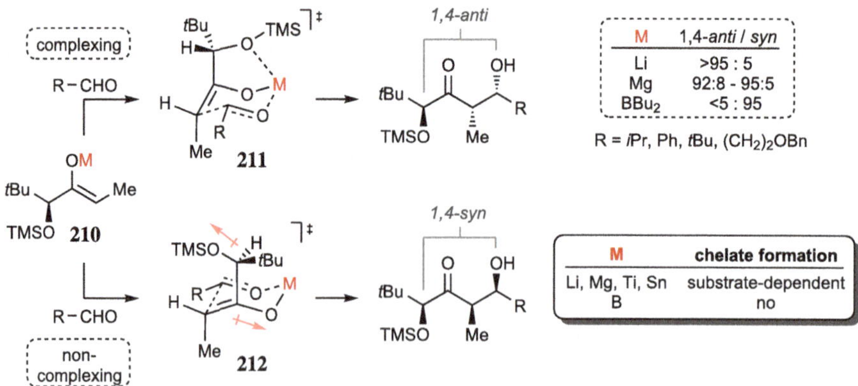

Fig. 2.88 Influence of the metal in metal enolates on the product distribution, postulated transition states and coordination tendency of common metal ions with basic α/β-substituents [202]

Menche and co-workers examined the possibility of a diastereodivergent aldol reaction with protected β-hydroxyketones, solely by varying the reaction conditions. In addition to an investigation of various metal enolates, the effect of different protecting groups was also elucidated in order to be able to subsequently access different derivatives from the aldol products **214** and **215**. The competition of chelating (via a β-analog of **211**) and steric interactions (via **206**) was postulated as the stereodetermining effect in the transition states. Interestingly, the Li enolate yielded the diastereomer which is routinely obtained under non-chelating conditions. Using the Sn and Ti enolates, however, the complementary products (under chelate control) were isolated [203]. The method was subsequently applied successfully in the total synthesis of the macrolide etnangien (via **214**, R = TBS) (see Fig. 2.89) [204].

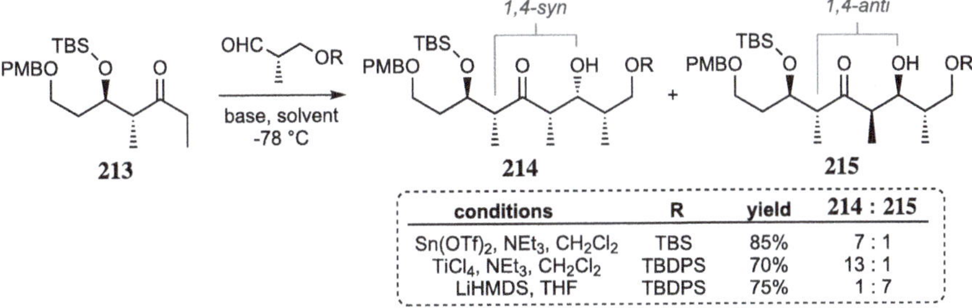

conditions	R	yield	214 : 215
Sn(OTf)$_2$, NEt$_3$, CH$_2$Cl$_2$	TBS	85%	7 : 1
TiCl$_4$, NEt$_3$, CH$_2$Cl$_2$	TBDPS	70%	13 : 1
LiHMDS, THF	TBDPS	75%	1 : 7

Fig. 2.89 Control of 1,4-selectivity through variation of the reaction conditions [203, 204]

There is a myriad of further methods and their application in the synthesis of complex natural products that rely on an intramolecular chelate formation for control of the product distribution beyond the examples already mentioned [205]. Certain enolate-metal combina-

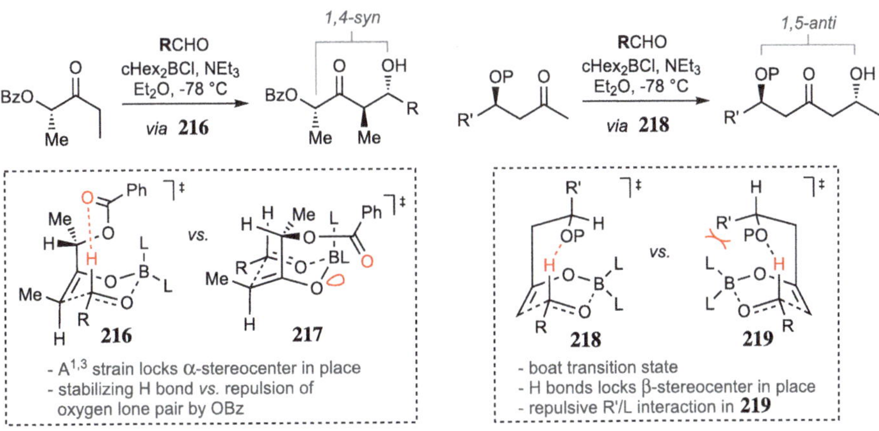

Fig. 2.90 1,4- and 1,5-selective aldol reactions of chiral enolates and their transition states [206, 207, 209, 210]

tions warrant a closer look at the specific postulated transition structures, as they do not fall under the general schemes already discussed. Particularly Paterson and co-workers developed a variety of methods using boron enolates, [206, 207] which are specifically tailored for use in the synthesis of complex polyketides and have found wide application in this field (see Fig. 2.90) [205, 208].

Although great progress has been made in developing catalytic aldol reactions (see below), one of the most reliable methods for stereoinduction remains the use of *chiral auxiliaries* (see Fig. 2.91) [211–213].

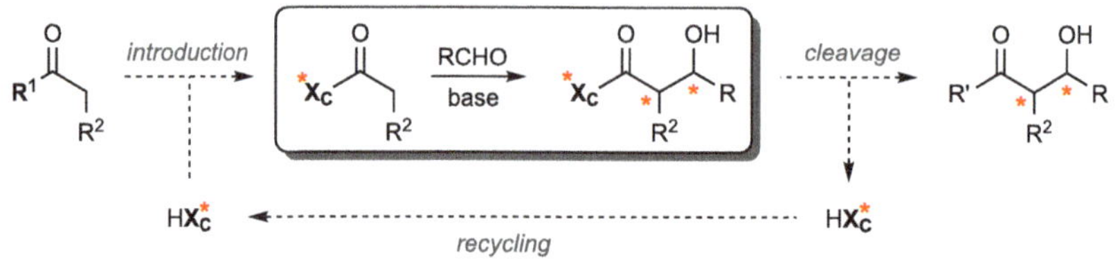

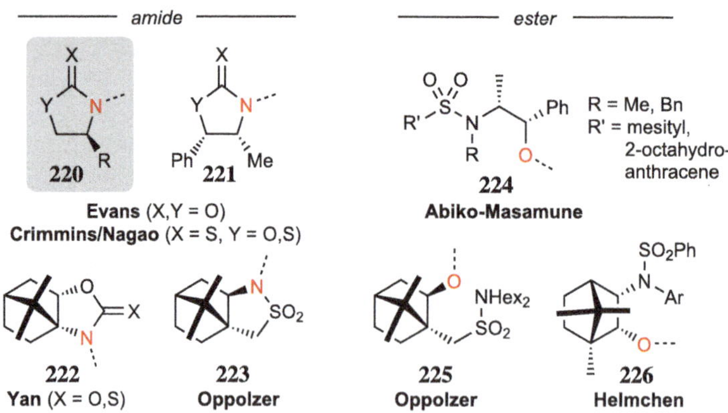

Fig. 2.91 Principle of stereoinduction by auxiliaries

The chiral auxiliary (X_C) should be readily available, its removal should occur without racemization of the desired product, and if possible, it should be recyclable. The majority of auxiliaries are based on abundant natural sources: derivatives of camphor (Helmchen auxiliary, Oppolzer auxiliaries) [214–217] as well as amino acid-, pseudoephedrine- (Evans-, Crimmins-, Yan auxiliaries) [218–220] or epinephrine-derived structures (Abiko-Masamune auxiliary) [221]. These amines or alcohols can be introduced as corresponding amides or esters in the carbonyl substrates (see Fig. 2.92).

Fig. 2.92 Common auxiliaries in aldol reactions

Particularly oxazolidinones (**220**), which were introduced by Evans and co-workers, have established themselves as gold standard in the field of auxiliary-controlled aldol reactions [222]. These auxiliary have also been used in the α-functionalization of carbonyl compounds as well as asymmetric cycloadditions. In addition, the systems of Crimmins and Abiko-Masamune (**224**) have proven to be valuable tools in the synthesis of complex intermediates, showing complementary reactivity in their introduction, removal, and enantioinduction compared to the Evans auxiliary. Their synthesis relies on molecules of the chiral pool as source of stereogenic information (see Fig. 2.93) [219, 223].

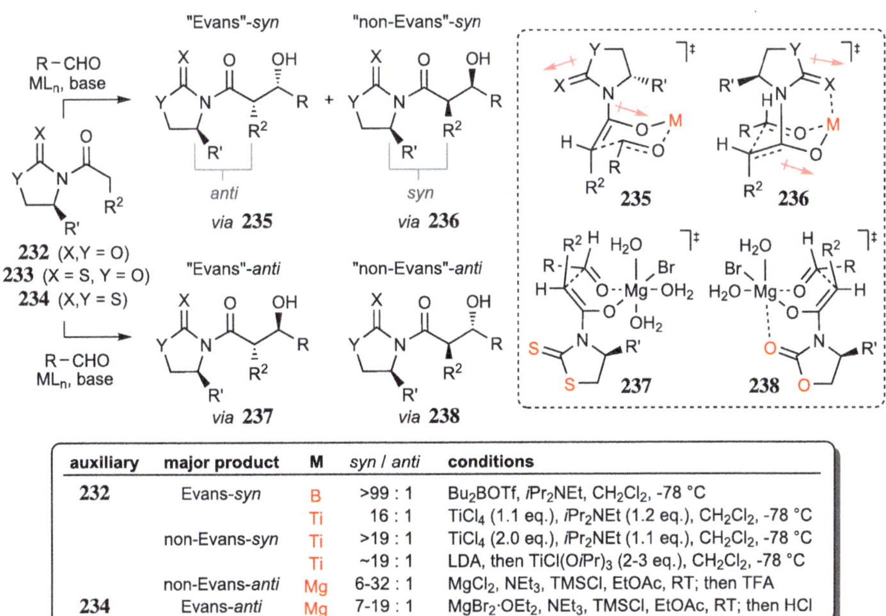

Fig. 2.93 Synthesis of Evans- (**229**) and Crimmins-type auxiliaries (**230, 231**) [219, 223]

When using oxazolidinones and thiazolidinones as auxiliaries, the steric shielding of one of the diastereotopic faces accounts for their stereoinduction, in analogy to the facial discrimination by α-stereogenic enolates. Since amides selectively lead to the formation

auxiliary	major product	M	syn / anti	conditions
232	Evans-*syn*	B	>99 : 1	Bu$_2$BOTf, *i*Pr$_2$NEt, CH$_2$Cl$_2$, -78 °C
		Ti	16 : 1	TiCl$_4$ (1.1 eq.), *i*Pr$_2$NEt (1.2 eq.), CH$_2$Cl$_2$, -78 °C
	non-Evans-*syn*	Ti	>19 : 1	TiCl$_4$ (2.0 eq.), *i*Pr$_2$NEt (1.1 eq.), CH$_2$Cl$_2$, -78 °C
		Ti	~19 : 1	LDA, then TiCl(O*i*Pr)$_3$ (2-3 eq.), CH$_2$Cl$_2$, -78 °C
	non-Evans-*anti*	Mg	6-32 : 1	MgCl$_2$, NEt$_3$, TMSCl, EtOAc, RT; then TFA
234	Evans-*anti*	Mg	7-19 : 1	MgBr$_2$·OEt$_2$, NEt$_3$, TMSCl, EtOAc, RT; then HCl

Fig. 2.94 Auxiliary-controlled propionate aldol reaction [115, 143, 218, 224–226]

of Z-enolates, only the corresponding *syn*-configured aldol products can be obtained via Zimmerman-Traxler transition states. Based on the seminal work of the Evans group, the major products are referred to as "*Evans*"-*syn*- and "*non-Evans*"-*syn*-product [120, 176, 218]. Depending on the ability of the metal and the auxiliary to coordinate in the transition state, both *syn*-diastereomers can be obtained by careful choice of the reaction conditions (via **235** or **236**). The complementary *anti*-configured β-hydroxyimides can be accessed by using the catalytic MgX$_2$/NEt$_3$ methodology (see Fig. 2.56): The auxiliary controls which of the *anti*-diastereomers results as the main product, which is likewise determined by the ability to coordinate in the postulated boat-like transition states **237** and **238** (see Fig. 2.94).

The selective access to the Evans and non-Evans *syn*-products by simply varying the equivalents of TiCl$_4$ can be understood in the context of a change in the coordination sphere of the titanium and thus the coordination geometry of the auxiliary (**235** *versus* **236**). In the reaction of oxazolidinones (X,Y = O; **232**) and oxazolidinethiones (X = S, Y = O; **233**), a second equivalent of TiCl$_4$ can abstract a chloride to enable a chelate transition state **242** to the non-Evans *syn*-product. In the case of thiazolidinethiones (X,Y = S; **234**), the coordination of the thioimide is possible in the transition state, even in the presence of only one equivalent of amine (see Fig. 2.95) [219, 224, 227]. In the presence of an excess of TiCl$_4$, a Mukaiyama-type open transition state is conceivable instead of a cyclic transition state. Under these conditions the non-Evans *syn*-product **240** is obtained.

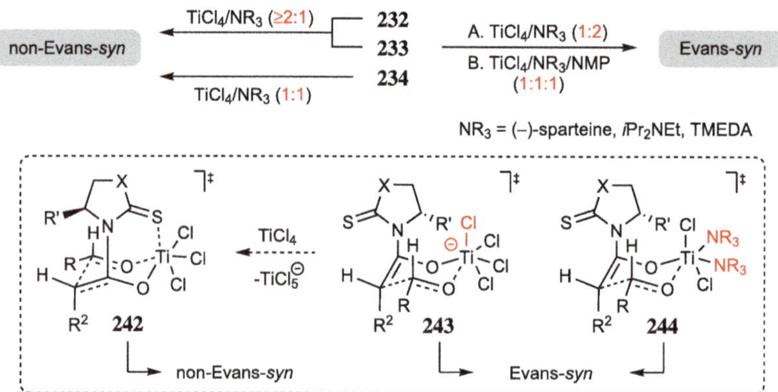

Fig. 2.95 Access to the Evans/non-Evans *syn*-aldol product by variation of the reagent stoichiometry [219, 224, 227]

The variation of the amine/TiCl$_4$ ratio has been successfully used in several natural product syntheses for the selective construction of the desired Evans or non-Evans *syn*-product (see Fig. 2.96) [227–230].

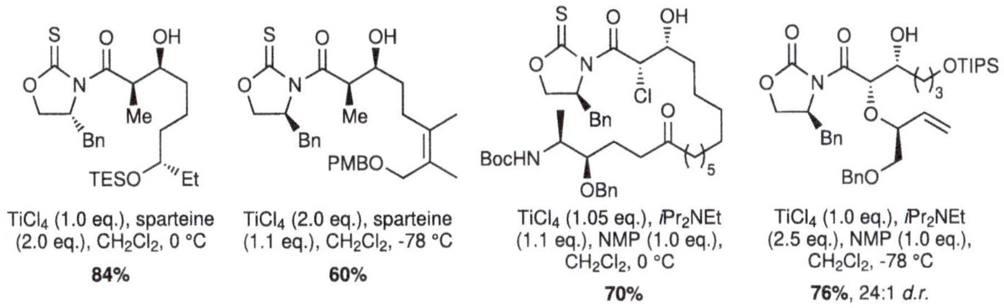

Fig. 2.96 Application of the Crimmins *syn*-aldol methodology [227–230]

An extension of this methodology by Crimmins allowed the access to Evans *anti*-products of glycol derivatives. This required the aldehyde to be complexed by an additional equivalent of TiCl$_4$ before initiation of the reaction upon addition of the Ti enolate. The *anti*-selectivity was rationalized by postulating acyclic transition states [231]. This is consistent with findings by Heathcock and co-workers that the utilization of an additional equivalent of a Lewis acid in boron-mediated aldol reactions preferentially yields the *anti*-product [232]. Acyclic transition states were also used to explain the observed selectivity [233, 234].

When using tin enolates under soft enolization conditions (Sn(OTf)$_2$, *N*-ethylpiperidine), coordination of the auxiliary should occur accordingly in the transition states. Oxazolidinethiones and thiazolidinethiones are used almost exclusively in these cases by virtue of the high affinity of tin for sulfur, as opposed to oxazolidinones [235, 236].

If substrates containing a 1,3-oxygenated framework such as β-keto-imides are used to synthesize polyketides, the methodologies relying on titanium/tin/boron enolates already

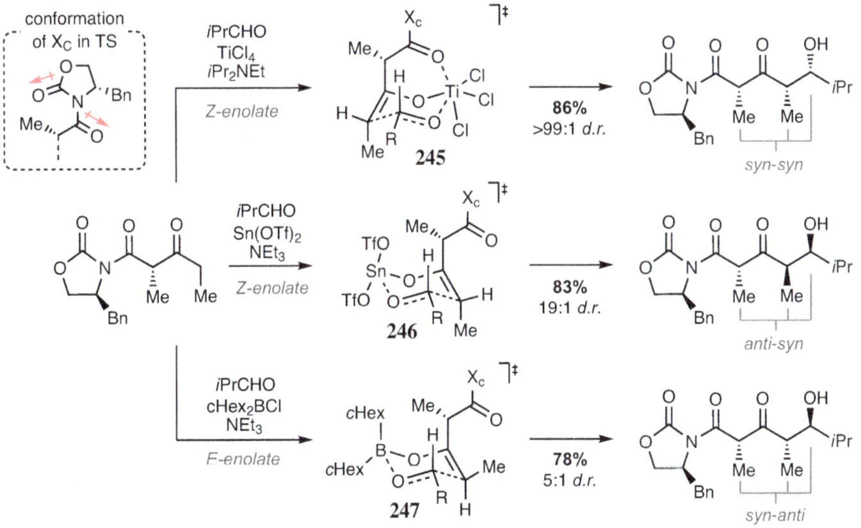

Fig. 2.97 Aldol reaction of β-ketoimides [129, 237]

covered above can be applied. No epimerization of the strongly acidic α-stereogenic center takes place, which is attributed to an unfavorable 1,3-allylic strain between the benzyl group of the auxiliary and the methyl group during the deprotonation. The kinetically – controlled enolization selectively removes the proton in the γ-position (see Fig. 2.52). Depending on the enolate geometry and the ability to complex the auxiliary in the transition state (**245–247**), almost all possible diastereomeric products can be furnished by the same carbonyl substrate (see Fig. 2.97) [129, 237].

Following the aldol reaction, Evans and Crimmins auxiliaries can be cleaved under mild conditions and converted into a variety of functional groups [222]. The amount of undesired hydrolysis of the **endo**cyclic amide bond with concomitant cleavage of the oxazolidinone (**248**) depends on the specific nucleophile. Lithium hydroperoxide very selectively leads to scission of the desired exocyclic bond (see Fig. 2.98) [238]. Thiazolidinethiones and thiazolinones can be converted under even milder conditions, which is why they are sometimes preferred to the more frequently employed oxazolidinones [219].

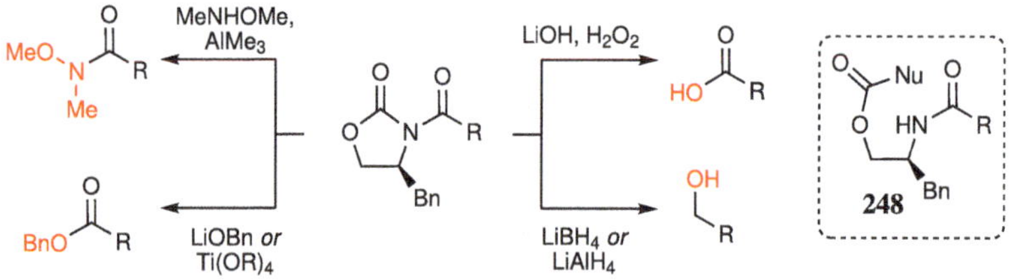

Fig. 2.98 Cleavage of the Evans auxiliary. Formation of the undesired cleavage product (**248**) through fissure of the endocyclic amide bond [222]

In addition to imides, esters can also be used for selective propionate aldol reactions when alcohol-based auxiliaries are employed. Abiko and Masamune introduced auxiliaries derived from epinephrine, which can afford both the *syn*- and *anti*-configured aldol products. Depending on which auxiliary is used, the desired β-hydroxyesters can be isolated via the corresponding *E*- or *Z*-enolate presumably proceeding through Zimmerman-Traxler transition states (see Fig. 2.99) [239]. Fuwa and co-workers used the Abiko-Masamune protocol in their synthesis of the marine macrolide lyngbyaloside B to afford the *anti*-aldol product **253**. Despite the larger steric bulk of the alkyl chain in **251** and a β-stereocenter in the aldehyde **252**, the addition product was obtained with the expected configuration in good diastereoselectivity. The ester was subsequently reduced to the corresponding alcohol using DIBAL and further derivatized [240].

The ester bond is surprisingly stable: the alcohol can be cleaved under mild conditions only by activation of the carbonyl group with Lewis acids or by resorting to the sulfur analog (C=S instead of C=O) [212].

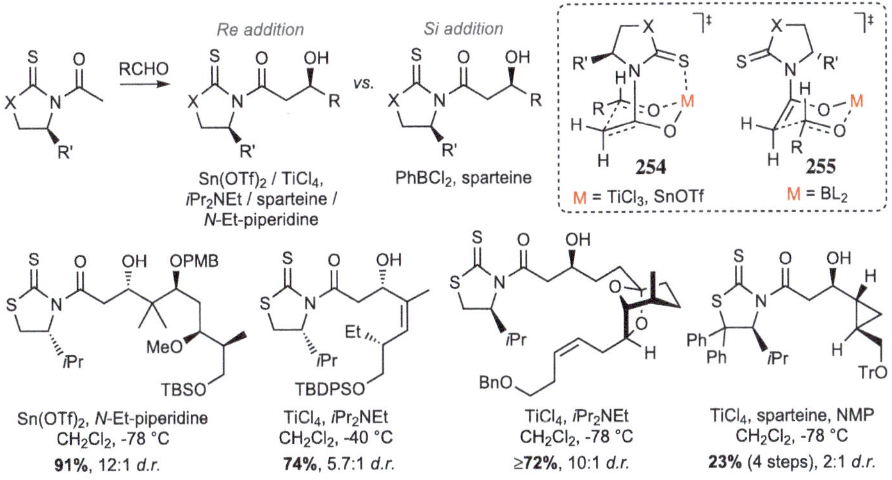

Fig. 2.99 *syn-* and *anti*-selective aldol reaction according to Abiko and Masamune and application in the total synthesis of lyngbyaloside B [239, 240]

While the propionate aldol addition allows for high stereodiscrimination due to the presence of an alkyl group R^2, the respective acetate aldol reactions lack good stereoinduction, as requisite dominant interactions in the transition state are negligible in the absence of this substituent. Theoretical studies have indicated the preference of a boat-shaped instead of a chair-like transition state, resulting in altered selectivities in several cases [165].

The methods presented so far for propionate derivatives are therefore only partially suitable for selective reactions of methyl ketones. While oxazolidinones offer only moderate selectivities, sulfur-based auxiliaries can enable high diastereoselectivities. The Nagao group reported early on a method that uses a combination of thiazolidinones and Sn(OTf)$_2$/N-Et-piperidine that has found wide application [235, 241]. In addition, modified oxazolidine- or

Fig. 2.100 Acetate aldol reaction, postulated transition states and select applications [241, 246–249]

thiazolidinethiones can be successfully used as boron or titanium enolates [242–245]. The selectivity is governed by the complexation behavior of the metal atom in the transition state (**254** *vs.* **255**, see Fig. 2.100) [241].

Although these approaches provide good results, auxiliary-based methods are overshadowed by chiral Lewis acid-catalyzed Mukaiyama aldol reactions in total synthesis (see below). Furthermore, α- and β-chiral aldehydes can likewise afford good stereoinduction, especially in the presence of chelating Lewis acids.

Instead of stereoinduction relying on an auxiliary, stoichiometric amounts of chiral reagents can be used, for example metal enolates attached to chiral ligands at the metal center. Given their compact transition states resulting in high selectivities, particularly boron enolates based on C_2-symmetric B-reagents have become broadly established (see Fig. 2.101).

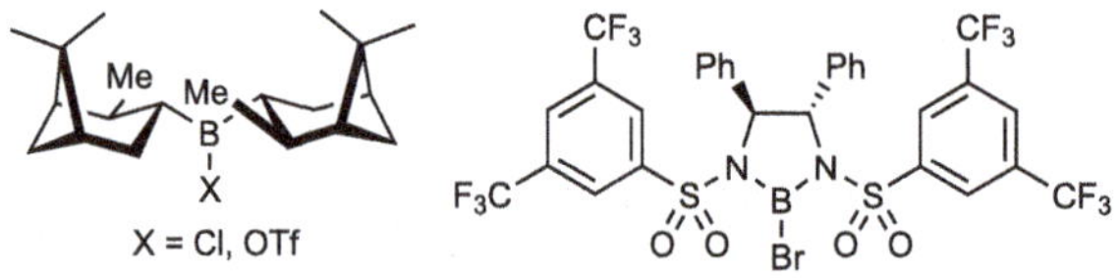

Fig. 2.101 Boron reagents developed by Paterson (**256**), Corey (**257**) and Masamune (**258**)

The most frequently used reagents are based on isopinocampheylborane (Ipc$_2$B-X) [121]. The enantioselectivities of the corresponding *Z*-boron enolates are good and the *syn/anti*

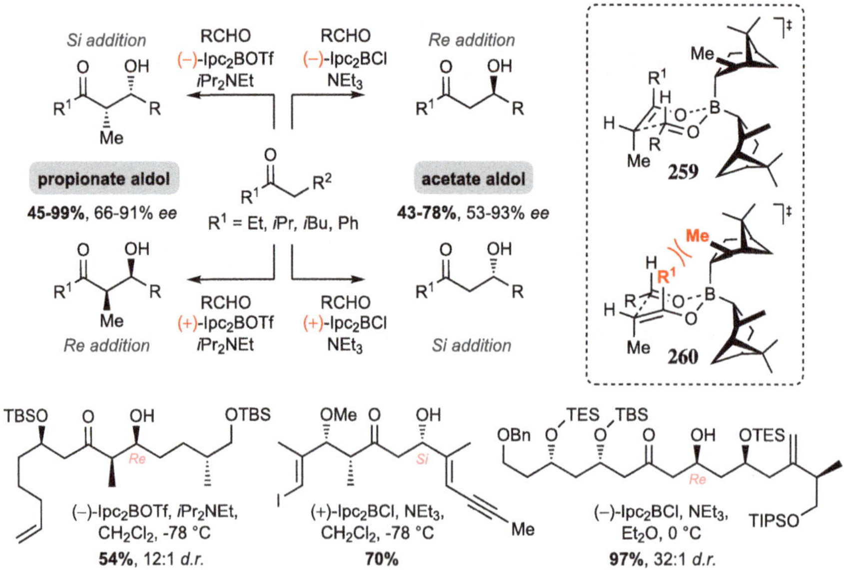

Fig. 2.102 Asymmetric aldol reaction with Ipc$_2$B-enolates, transition states for propionates and application in natural product syntheses [121, 250, 252–254]

ratios are excellent [250, 251]. The facial discrimination arises from a destabilizing R^1/Me interaction, which is avoided in the favored Zimmerman-Traxler transition state **259**. The *E*-enolate, accessible with Ipc_2BCl/NEt_3, only provides moderate *anti*-diastereoselectivities and a low enantiomeric excess. Acetate aldol reactions of complex intermediates, on the other hand, can be carried out with good stereoinduction, but surprisingly the facial selectivity of the aldehyde reverses, possibly due to switching to a boat-like transition state (see Fig. 2.102) [252].

The methodologies developed by Corey (**257**) and Masamune (**258**) similarly show good enantioselectivities, but are less frequently used due to their narrower substrate scope [255–258]. The discrimination of the π-faces in the transition states is rationalized by a steric interaction with the stereogenic centers in the α- or β-position.

In summary, specific methods are available for ketones, amides and esters to selectively obtain the various desired diastereomers. Only few broadly applicable methods exist to afford the *anti*-configured addition products for amides and esters. With ketones, a facial selectivity is typically achieved by capitalizing on existing stereogenic centers in the enolates or aldehydes, necessitating a very substrate-specific approach (see Fig. 2.103) [212].

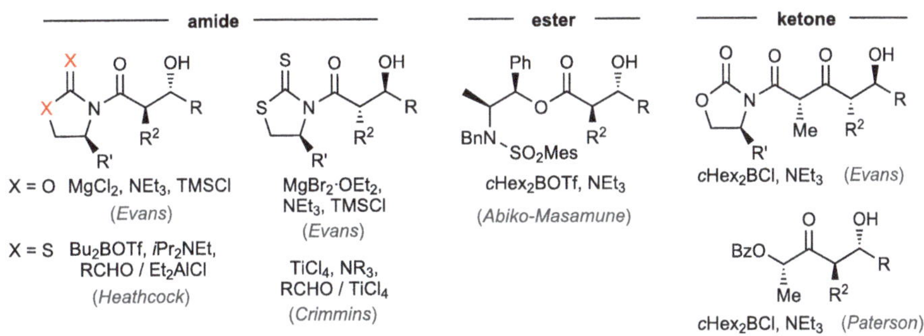

Fig. 2.103 Non-catalytic, *anti*-selective asymmetric aldol methods [143, 206, 233, 234, 237, 239]

2.3.3   Enantioselective Aldol Reactions — Catalytic Methods

The foundation of modern, catalytic aldol chemistry was laid by Hayashi and Ito, who first reported a catalytic, enantioselective domino-aldol/cyclization reaction using chiral phosphinoferrocenes in 1986 (see Fig. 2.104) [259, 260].

Aldol reactions utilizing catalytic amounts of a stereoinductor constitute a promising approach compared to the use of chiral auxiliaries or reagents. Accordingly, this field has evolved rapidly and led to the development of methods with impressive selectively that allow for the application in densely functionalized substrates.

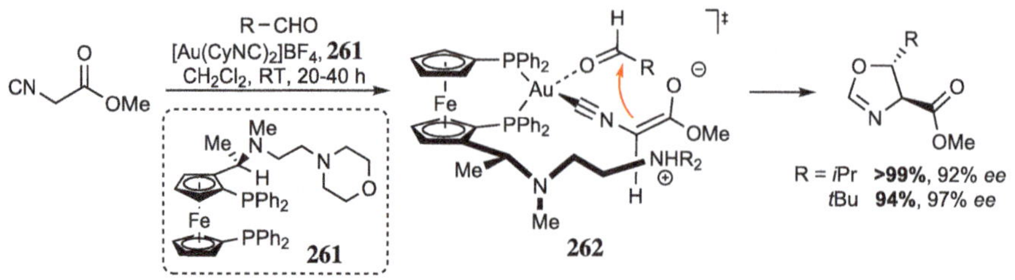

Fig. 2.104 Gold(I)-catalyzed enantioselective aldol reaction for the construction of oxazoles according to Hayashi and Ito [259, 260]

When moving from stoichiometric use of chiral inductors to catalytic variants, the envisioned reaction should only occur in the presence of the chiral catalyst as activator to avoid unselective background reactions ($k_{cat} \gg k_{uncat}$). Since most enolates are already too reactive even without the addition of an activator, only certain substrates are suitable for this approach as competing reaction pathways can hardly be suppressed (see Fig. 2.105).

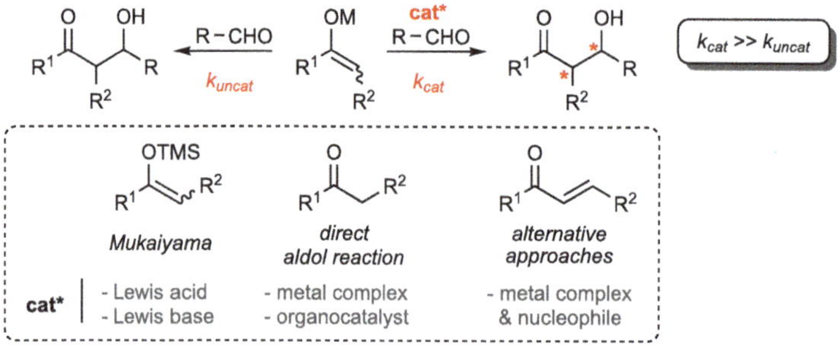

Fig. 2.105 Principle of catalyzed reactions and common enolate equivalents

The aldol reaction of silyl enol ethers requires the presence of an activator, such as a Lewis acid or base, to proceed. This makes them ideal substrates for catalytic approaches and most catalytic, asymmetric aldol reactions are based on silyl enol ethers as the enolate component [186, 261]. Following initial studies by Mukaiyama and Kobayashi using stoichiometric amounts of a chiral inductor, subsequent research was conducted soon after in the groups of Yamamoto, Corey, Mikami, Keck, Carreira and others (see Fig. 2.106) [262–272].

While the Mukaiyama/Kobayashi protocol directs an approach of the aldehyde to the *Si*-side via TS **266** using the *in situ*-formed Sn-pyrrolidine catalyst, the tryptophan-derived oxazaborolidine **264** (via TS **267**) and Ti-catalyst **265** induce a *Re*-face attack.

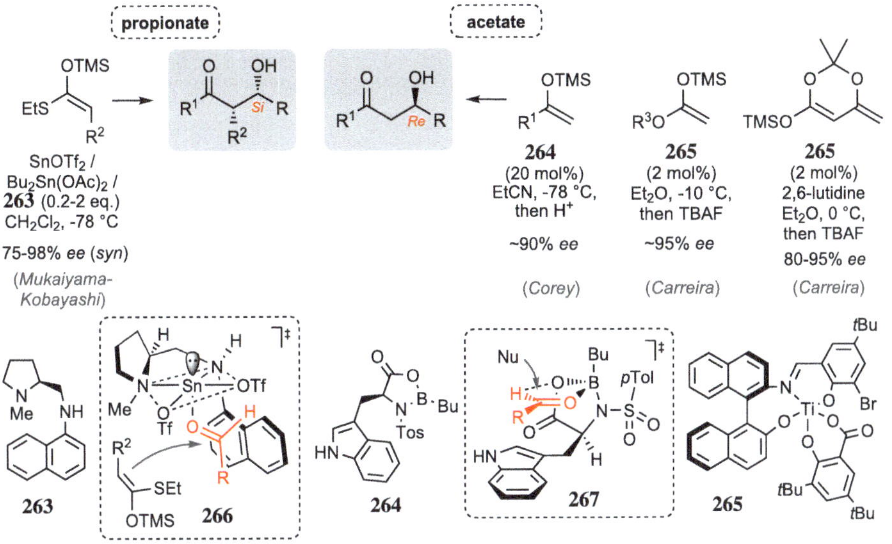

Fig. 2.106 Asymmetric-catalytic Mukaiyama aldol reactions [262–267]

Especially the Mukaiyama/Kobayashi variant has established itself as one of the standard methods of asymmetric Mukaiyama aldol reactions and been used extensively in natural product syntheses. The methods developed by Corey and Carreira for normal and vinylogous acetate aldol reactions have also proven themselves to be suitable for complex substrates (see Fig. 2.107) [197]. Apart from Carreira's Ti-salicylate imine complex **265**, vinylogous silyl enol ethers can also be converted selectively to the aldol products in the presence of copper-BINOL complexes [273]. Mechanistic studies on the latter catalysts indicated the formation of copper enolates as intermediates as opposed to the direct conversion of the silyl enol ethers via the Mukaiyama aldol mechanism [274].

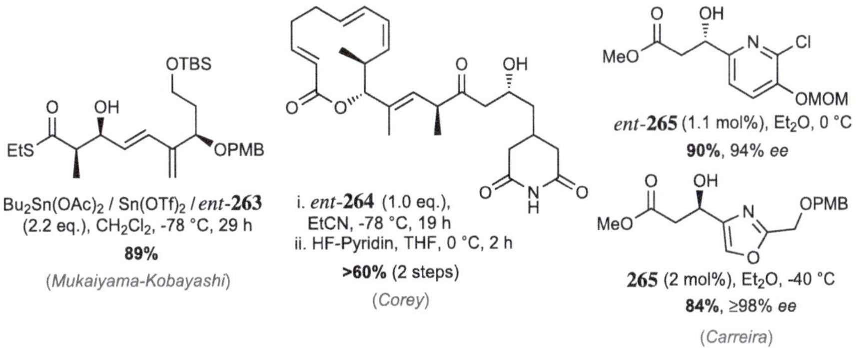

Fig. 2.107 Applications of asymmetric Mukaiyama aldol reactions [275–278]

The Evans group intensively studied copper-, tin-, scandium- and zinc-catalyzed aldol reactions with stereodiscrimation by bisoxazoline-based ligands. Since bidentate complexation of the metal center by the substrate is a prerequisite for good stereoinduction, all substrates carry an oxygen atom in the α-position (OBn, ketone, ester). Both normal and vinylogous acetate as well as propionate aldol reactions can be carried out with TMS-ethers derived from thioesters. However, the corresponding acetate aldol reactions display a broader substrate scope for these nucleophiles (see Fig. 2.108) [279–285].

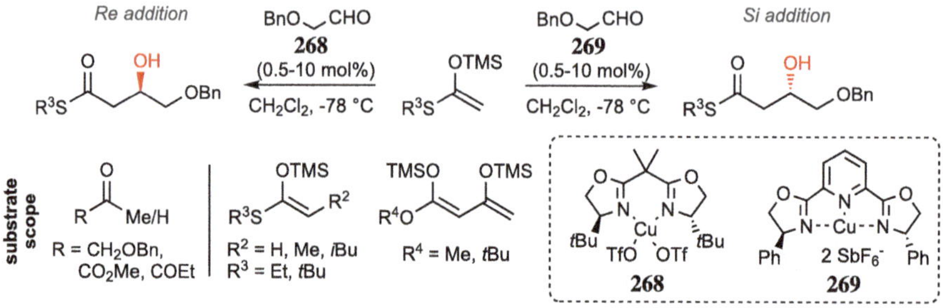

Fig. 2.108 Cu-box/pybox-catalyzed asymmetric Mukaiyama aldol reaction and substrate spectrum of the reaction

The facial discrimination of the electrophile can be influenced by the judicious choice of ligand, metal, and counterion. Cu(II) as a d^8-metal prefers a square-planar environment with weak binding of further axial ligands. When OTf is used as a counterion with bisoxazoline ligands (*box*) (**267**), it occupies an equatorial position, like the coordinated aldehyde group of the substrate. This forms a chelate complex **270** by concomitant association of the OBn group, in which the *t*Bu groups on the ligand only allows for access to the *Re*-face. In the case of **269**, the reverse selectivity is achieved: With non-coordinating SbF$_6$ as a counterion, no competition for binding sites at the metal center can occur. The tridentate pyridine bisoxazoline ligand (*pybox*) forces the substrate into the structure shown in **271**, again with weak binding by the α-OBn group in an axial position. The use of Cu-box complexes with SbF$_6^-$ counterion (not shown in Fig. 2.108) provides the opposite stereoselectivity to **268**. Here, in the absence of further competing ligands, the substrate occupies an equatorial position with both functional groups, so the attack can only occur on the *Si*-face (see Fig. 2.109) [281].

Despite their high conceptual relevance, the Cu-box/pybox-catalyzed Mukaiyama aldol reaction has so far only been used sporadically in syntheses due to the rather limited substrate scope, as very specific substituents are required in the α-position of the electrophile [286–288].

Instead of resorting to Lewis acids, the use of Lewis bases can also facilitate the reaction of silyl enol ethers with aldehydes [181]. The combination of electron-poor, Lewis-acidic chlorosilanes with chiral Lewis bases was developed by the Denmark group. Initial research

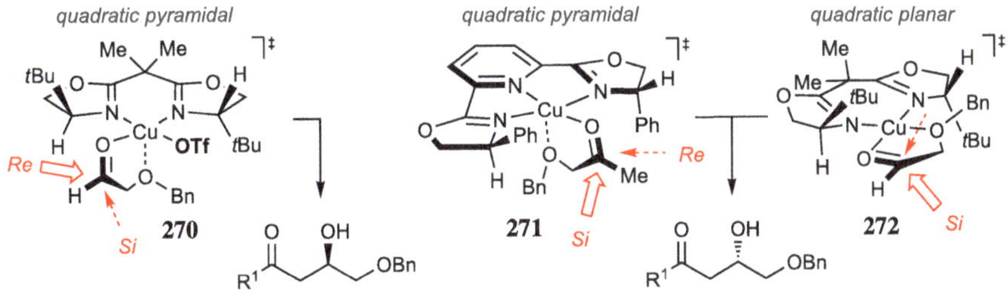

Fig. 2.109 Transition states of the Cu/Sn-box/pybox-catalyzed aldol reaction and major products [281]

aimed at activating the silyl enol ether, while further studies used the combination of $SiCl_4$ and chiral bases to form a chiral Lewis acid *in situ* to activate the aldehyde (see Fig. 2.110) [289–293].

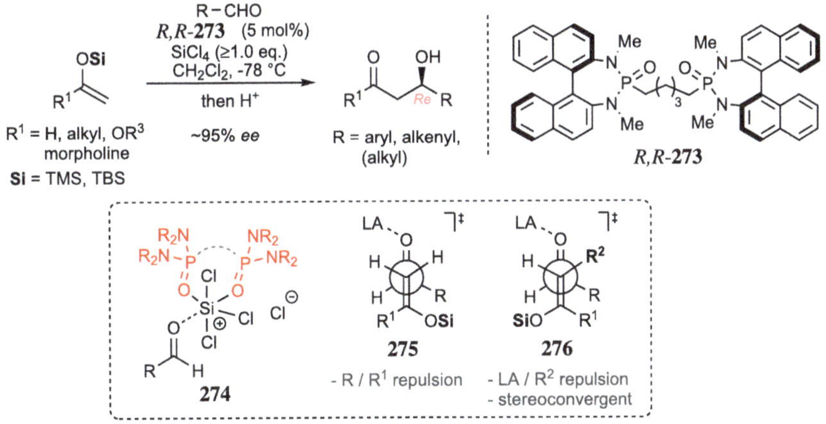

Fig. 2.110 Tetrachlorosilane-mediated, Lewis base-catalyzed Mukaiyama aldol reaction developed by Denmark [294] Activation of the aldehyde and postulated TSs for acetate and propionate aldol reactions

Bifunctional Lewis bases were employed since two equivalents of base are required when utilizing chlorosilanes. The chelation then ensures good transfer of the stereoinformation from the binaphthyl backbone to allow for efficient facial discrimination of the substrate (**274**). The transition states are acyclic in nature, as expected in Mukaiyama aldol reactions. In the case of the acetate aldol reaction, the sterically bulky Lewis acid defines the structural arrangement in the transition state (via **275**) to avoid the very unfavorable R/R^1 repulsion, thus leading to addition of the nucleophile to the *Re*-face of the aldehyde. In propionate aldol reactions, the reaction is stereoconvergent: the *E/Z* ratio of the silyl enol ether influences the facial selectivity and thus the enantiomeric excess only to a limited extent. In the preferred transition state **276**, the LA/R^2 repulsion is minimized.

This approach can enable both normal and vinylogous aldol reactions, the aldehyde scope usually focuses on aromatic or α,β-unsaturated carbonyl derivatives. The methodology has been successfully applied to some total syntheses, but as with many catalytic variants, the substrates only display a limited complexity (see Fig. 2.111).

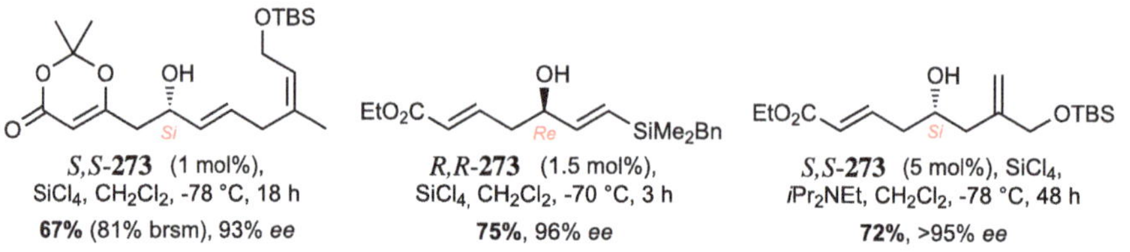

Fig. 2.111 Applications of the base-catalyzed, SiCl$_4$-mediated Mukaiyama aldol reaction [287, 295, 296]

In addition to the use of activated enolate equivalents such as silyl enol ethers, some methods allow the direct use of ketones (or aldehydes), from which the enolate or enamine is formed *in situ*. These so-called **direct** catalytic aldol reactions have the significant advantage that no previous steps or additional reagents are needed to activate the electrophilic carbonyl for the reaction. However, this is offset by the more limited substrate scope: Since two carbonyl components react with each other in an aldol addition, the catalyst must be able to distinguish between them in order to enable a selective product formation. This requires the use of two substrates with different reactivity to avoid a statistical product distribution including unwanted homocoupling. Aldehydes without α-acidic protons or with α-branching are typically employed as electrophiles, displaying very slow to negligible enolate formation. The regio- and diastereoselectivity of the enolization is controlled by the use of methyl aryl ketones or symmetric ketones, usually cyclohexanone or cyclopentanone. The individual reaction steps are normally reversible (see Fig. 2.112).

While transition metal catalysts dominated during the early phase of research in the field, they have been almost completely replaced by secondary amines (amongst others) as catalysts due to the rapid recent developments in organocatalysis. This following section will only cover significant developments relying on metal catalysis; for an overview on the topic of organocatalysis, please refer to Chap. 8 and the relevant literature [297–300]. The metal catalysts used for a direct reaction are modeled after enzymatic aldol reactions. They activate both the enolate and the aldehyde (**277, 278**). Depending on the catalyst species, the deprotonation is achieved by an external base or by the catalyst itself. The systems developed by Shibasaki and Trost represent important milestones of this principle and are based on bi- and tridentate alcohols as chiral inductors.

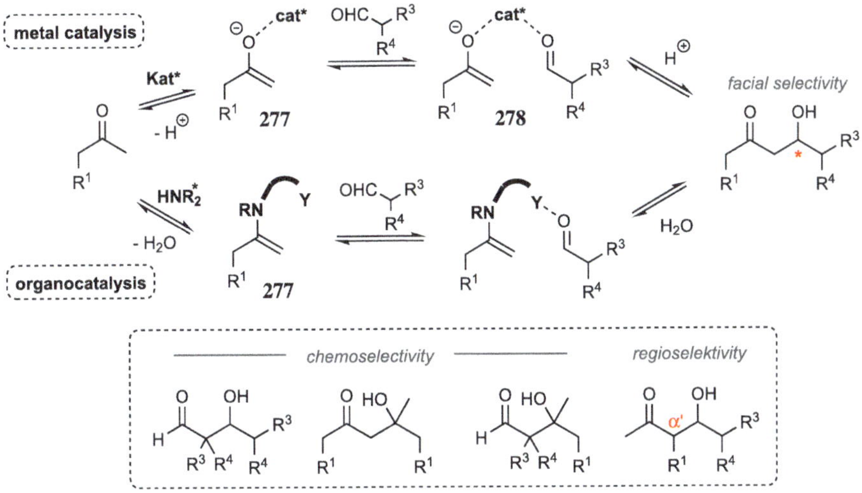

Fig. 2.112 Principle of the direct aldol reaction and possible products of unselective processes

A bimetallic lithium-lanthanum cluster of type **279** was introduced by the Shibasaki group, which contains both Lewis-acidic and Lewis-basic sites (see Fig. 2.113) [301, 302].

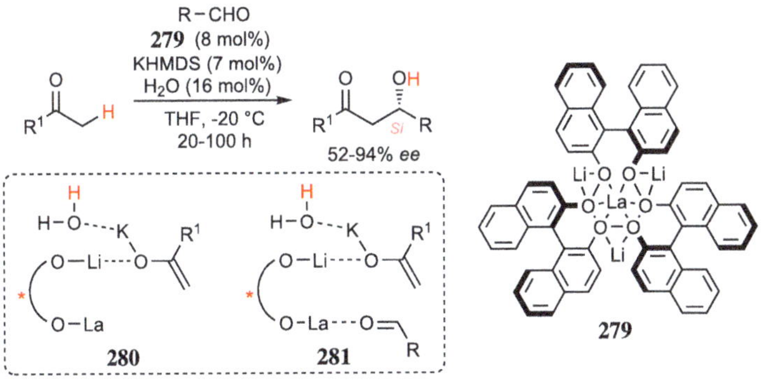

Fig. 2.113 Direct, asymmetric aldol reaction according to Shibasaki *et al.* [301, 302]

Because of these, the bifunctional cluster is capable of coordinating and activating both the enolate (**280**) and the aldehyde (**281**). The addition of catalytic amounts of water and KHMDS as base significantly reduce the reaction time, as the endogenous alkoxy group can only slowly deprotonate the ketone. The substrate scope of the direct aldol reaction with this catalyst remains much narrower compared to catalytic Mukaiyama reactions, even though subsequent improvements reported the use of α-hydroxyketones as enolate precursors [301, 303]. The Shibasaki group disclosed the method's successful application in the total synthesis of several natural products [304, 305].

Trost and co-workers also developed a direct aldol reaction of aryl ketones in the presence of homoleptic, bimetallic zinc complexes [306]. The mechanism involves both Lewis-acidic and -basic coordination sites, as postulated for the heteroleptic lanthanum complex **279** and homoleptic Zn complexes also studied by Shibasaki [307]. Upon addition of weakly coordinating ligands, the reaction rate increased significantly, with triphenylphosphine sulfide affording the best results in terms of yield and *ee*. Et$_2$Zn and **282** form the active catalyst **283**. The complex likewise fulfills a bifunctional role by activating the enolate as well as the aldehyde (**284**, see Fig. 2.114) [306, 308].

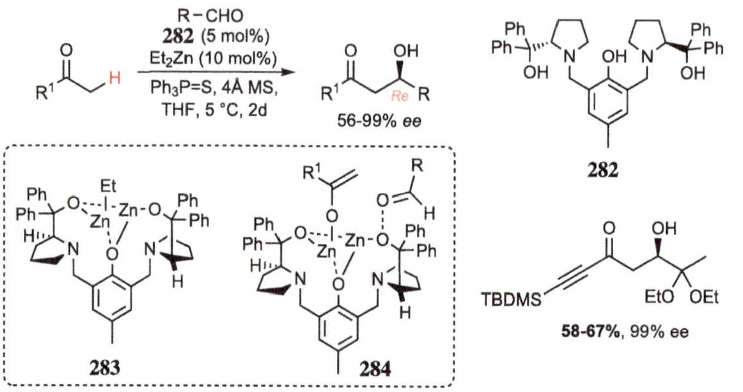

Fig. 2.114 Direct asymmetric aldol reaction by Trost and application in the synthesis of fostriecin [306, 308, 309]

Like the aldol methodology developed by Shibasaki,[307] α-hydroxyketones could be subsequently also introduced as substrates and the method was successfully utilized in the total synthesis of (+)-boronolide [310, 311].

A third approach employs α,β-unsaturated carbonyl compounds as substrates, which can be converted *in situ* to enolates by 1,4-addition of hydride equivalents. The **reductive** catalytic aldol reaction was mainly developed by the Krische group based on pioneering studies by Revis and Morken [312–315] (Fig. 2.115).

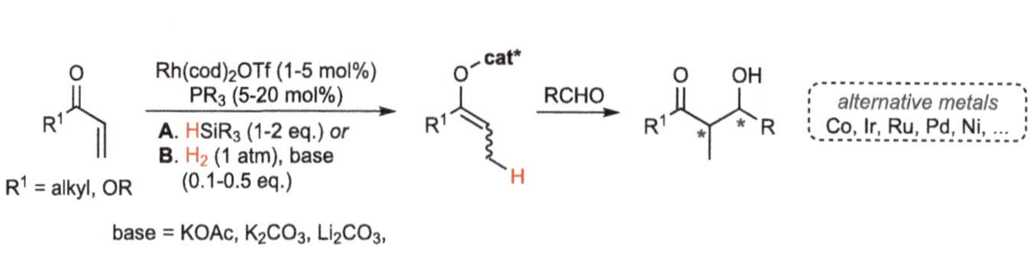

Fig. 2.115 Reductive, catalytic aldol reaction [315]

A hallmark of this approach lies in the fact that enolate precursors with low acidity like esters can also be used, and that enolates derived from unsymmetrical ketones react regioselectively. The stereoinduction (relative and absolute) and the occurence of unwanted reductions of the carbonyl group or the olefin however remain its Achilles' heel. Silanes or hydrogen in combination with an inorganic base serve as reducing agents. The catalyst is usually a rhodium complex (see Fig. 2.116) [315, 316].

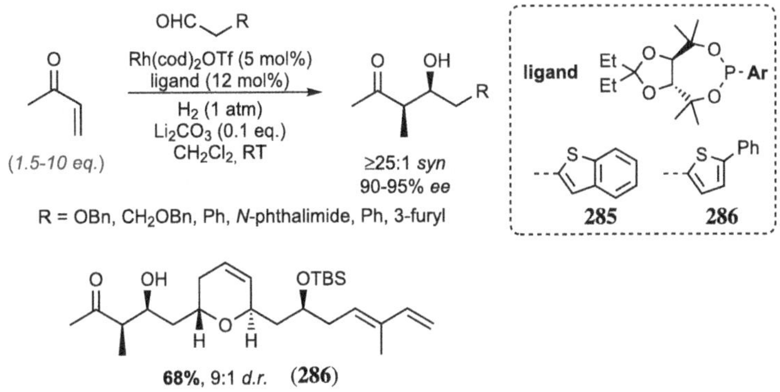

Fig. 2.116 Reductive, asymmetric aldol reaction according to Krische *et al.* and application in the synthesis of swinholide A [316, 317]

2.3.4 Aldol-Type Addition Reactions

By resorting to related nucleophiles and electrophiles (instead of enolates and aldehydes) that display a similar reactivity, the class of aldol-type reactions was explored. The *Mannich reaction* and the *Nitroaldol/Henry reaction* have been studied intensively and catalytic, asymmetric variants have similarly been developed. Both transformations, as well as other related reactions, are now dominated by the use of organocatalytic methods, akin to the catalytic aldol reaction. Further details can be found in Chap. 8 and the review articles cited below.

A synthetically common variant of the aldol reaction can be found in the *Mannich reaction*, in which an imine or iminium salt (instead of an aldehyde/ketone) reacts with the enolate to afford β-aminoketones. The product of the Mannich reaction, often referred to as *Mannich base*, allows easy access to 1,3-difunctionalized structural motifs, especially β-amino alcohols (see Fig. 2.117) [318–324].

Since the transformation constitutes an equilibrium reaction whose reactive tautomers **287–289** are only present in small amounts, long reaction times ensue, which can result in a significant amount of unwanted by-products. While aldehydes are reacted *in situ* with primary or secondary amines in the presence of enolizable carbonyl compounds under "classical" conditions, asymmetric Mannich variants often resort to preformed electrophiles such as imines or iminium ions as coupling partners to improve the traditionally low regio-, diastereo- and chemoselectivity.

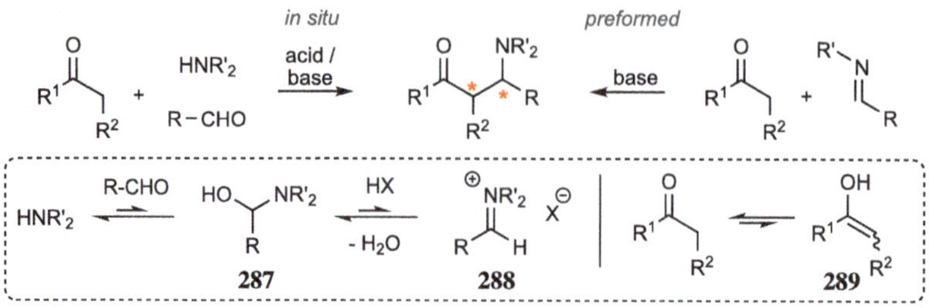

Fig. 2.117 General Mannich reaction

Under organocatalysis, either non-covalent systems are used that activate the reactants via hydrogen bond interactions, or secondary amines which operate via enamine intermediates. For selected organocatalysts, the origin of the observed diastereo- and enantioselectivity has also been investigated in detail, allowing a good understanding of the operative mechanism and the transition states (e.g., **290** for proline) [325, 326].

Metal-based catalysts, as in the direct catalytic aldol reaction, are typically bifunctional and silyl enol ethers serve as enolate equivalents. The most efficient systems commonly rely on similar catalysts employed for aldol reactions such as polyol-metal complexes of the Trost-type, Cu- or Pd-bisphosphines, Cu/Mg or La-box/pybox complexes, and a variety of other catalyst classes. The nitrogen of the aldimine is often Boc- or Tos-substituted, as these can be subsequently removed under relatively mild conditions to further functionalize the resultant amine (see Fig. 2.118) [318–320].

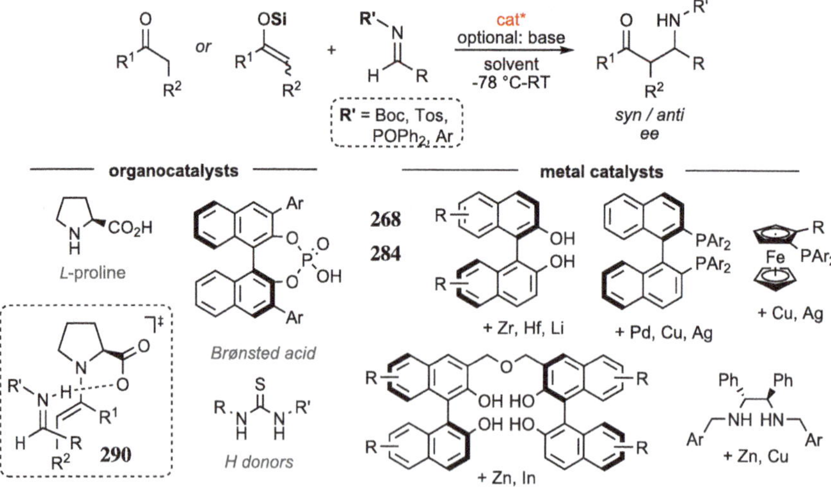

Fig. 2.118 Overview of catalyst classes used in the asymmetric Mannich reaction [318–320]

Since preformed iminium salts or imines allow the accommodation of high complexity in the substrates, the Mannich reaction — especially the intramolecular variant — is often used for the construction of alkaloids and other nitrogen-containing natural products [327]. The selected examples cover the range of possible target frameworks and reaction modes: intermolecular to intramolecular, diastereo-, enantioselective or auxiliary-controlled conversions to vinylogous Mannich reactions are all conceivable (see Fig. 2.119).

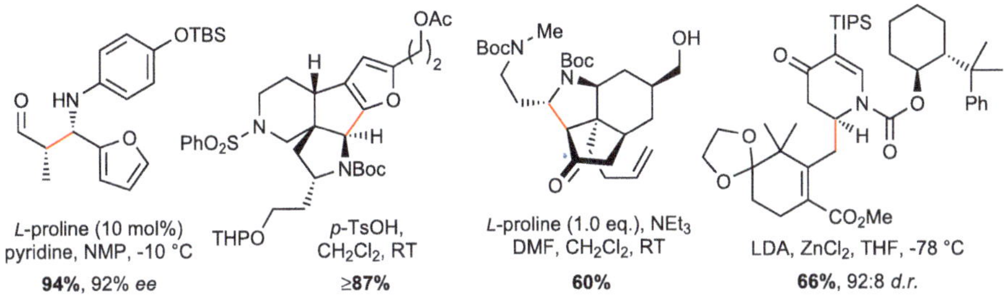

Fig. 2.119 Examples of the Mannich reaction in the synthesis of natural products [328–331] The newly formed CC bond has been highlighted

In addition to the classic Mannich reaction, there also exist enantioselective variants of the vinylogous Mannich reaction,[332] as well as the nitro-Mannich reaction [321, 333–335].

In the *Henry reaction*, nitroalkanes are deprotonated in the presence of stoichiometric or catalytic amounts of a base and are subsequently reacted with carbonyl compounds to form β-nitro alcohols. For diastereoselective Henry reactions, the requisite base typically comprises metal alkoxides and non-nucleophilic amines. Furthermore, the use of a variety of inorganic salts and even heterogeneous bases such as Al_2O_3 or SiO_2 have also been reported (see Fig. 2.120) [336–338].

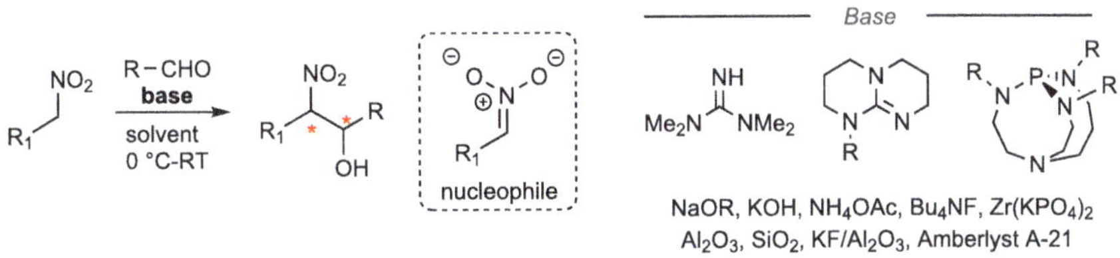

Fig. 2.120 Reaction conditions of the common Henry reaction [336–338]

The reaction is often carried out in the nitroalkane reagent as solvent. The control of the *syn/anti*-diastereoselectivity is a fundamental challenge of the Henry reaction due to the reversibility of the steps. The β-nitro alcohol can subsequently be converted further by oxidative or reductive means. The required, often harsh reaction conditions remain a drawback of this transformation. The reaction mechanism proceeds similarly to that of the aldol reaction, with initial deprotonation of the nitroalkane and concluding addition of the nucleophile to the aldehyde. For uncharged or doubly charged nitronates, both acyclic and cyclic transition states have been postulated to rationalize the origin of diastereoselectivity [339].

Asymmetric Henry variants often rely on the same catalysts that facilitate the enantioselective aldol reaction [340, 341]. Initial methods used the hetero- and homoleptic metal clusters introduced by Shibasaki (**279**) and Trost (**284**, transition state **293**) as well as Cu-pybox ligands (**292**, via transition state **294**) [342–345]. Apart from the outlined methods, various copper-amine ligand systems have been investigated as catalysts in the asymmetric nitroaldol reaction. The examined chiral ligands range from (–)-sparteine, tridentate pybox ligands to diamine chelate ligands [341]. Initial studies following an organocatalytic approach employed tetramethylguanidine, since it was already known as a catalytically active base in the analogous Henry reaction. Subsequent work could elaborate on the scope of the base, proving the viability of chiral guanidines, thioureas as well as phase transfer systems (**291**, Maruoka and Cinchona ammonium salts; see Fig. 2.121, cf. Chap. 8) [346].

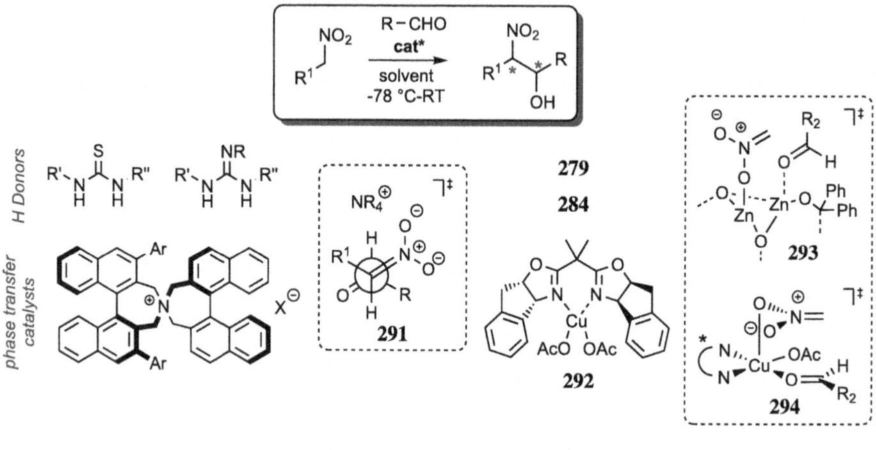

Fig. 2.121 Asymmetric nitroaldol reaction. Selected catalysts and TS [341, 344–346]

Applications of the Henry reaction to the synthesis of complex natural products are rather rare and usually limited to structurally simple substrates [347].

2.4 α-Functionalization Reactions

The derivatization of ketones, esters, or amides in the α-position of the carbonyl group introduces another handle for the use of this versatile functionality. Due to the innate reactivity of the carbonyl group, this type of reaction typically proceeds by formation of an enolate or enamine and subsequent coupling with electrophilic reagents. Alternatively, carbonyl derivatives with a leaving group in the α-position can be converted via substitution of the nucleofuge by nucleophiles. The introduction of an alkyl group constitutes the most often encountered functionalization, but the formation of a C-O or C-N bond is also very common (see Fig. 2.122).

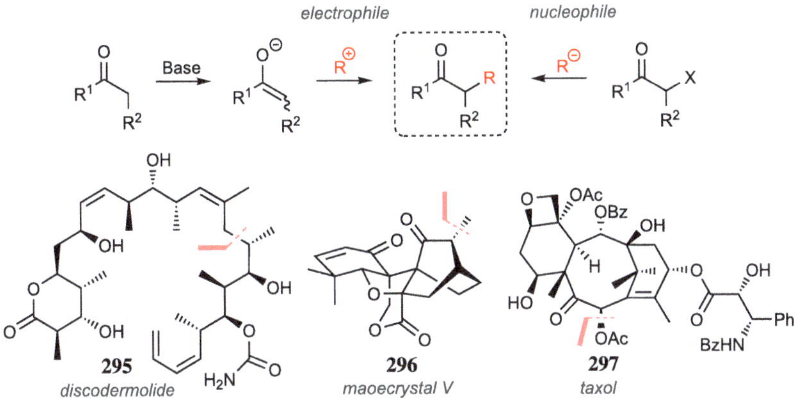

Fig. 2.122 Approaches to the α-functionalization of carbonyls and examples of alkylations (**295**, **296**) or hydroxylations (**297**) in natural product syntheses [348–350]

The reaction of acyclic starting materials is less common than the derivatization of cyclic ketones, lactones, or lactams due to the greater structural degrees of freedom and potentially poorer stereoselectivity. However, facial stereodifferentiation of prochiral enolates can still be high in acyclic substrates if additional stereoelectronic control factors (allylic strain, steric shielding, hindered rotation around single bonds, chelate formation) are present. Aldehydes are often too reactive under basic conditions to specifically obtain the α-functionalized product in acceptable yield due to an unwanted homocondensation. Alkylations can be carried out both inter- and intramolecularly with good chemo- and stereoselectivities [351, 352].

2.4.1 α-Alkylation — Selectivity and Kinetics

Enolates act as ambident nucleophiles, in which both a C- as well as an O-functionalization is viable [353, 354]. The chemoselectivity can be influenced by judicious choice of the reaction conditions (counterion, solvent system) as well as the structure of the electrophile (leaving group, degree of substitution) (see Fig. 2.123).

The trends observed for these influencing factors can be rationalized both by steric interactions (aggregation behavior) and by HSAB theory [356–359]. Rapid and efficient formation of the enolates is a prerequisite for a selective reaction, with the resulting enolate geometry playing an important role for diastereo- and enantioselective alkylations.

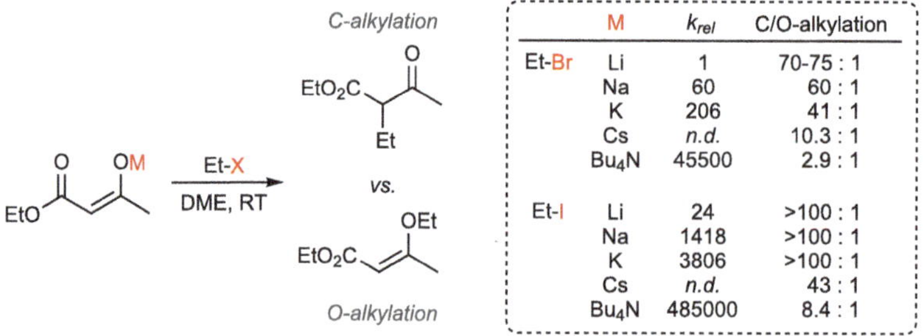

Fig. 2.123 Influence of reaction conditions on chemoselectivity [355]

Large, poorly coordinating cations usually allow solvent-separated ion pairs or even "naked" anions [360]. This effect can also be enhanced by the addition of crown ethers [353]. The observed build-up of O-alkylated products is mirrored by an increasing trend for enolate-counterion dissociation. Lithium enolates exhibit a high degree of covalent bonding with a small ionic radius, which causes hard electrophiles to react at the oxygen according to HSAB theory. The tighter coordination reduces reactivity and also produces strongly aggregated species in solution (cf. Section 2.2). High aggregation favors the functionalization of the C-terminus by steric shielding of the enolate oxygen. Conversely, weakly coordinating ammonium ions form associative complexes only to a very limited degree and therefore O-alkylations increasingly occur (see Fig. 2.124) [361].

Fig. 2.124 Reaction rate and ratio of C-/O-alkylation in the reaction of ethyl acetoacetate [361]

	M	k_{rel}	C/O-alkylation
Et-Br	Li	1	70-75 : 1
	Na	60	60 : 1
	K	206	41 : 1
	Cs	n.d.	10.3 : 1
	Bu$_4$N	45500	2.9 : 1
Et-I	Li	24	>100 : 1
	Na	1418	>100 : 1
	K	3806	>100 : 1
	Cs	n.d.	43 : 1
	Bu$_4$N	485000	8.4 : 1

Polar, aprotic solvents support the formation of solvent-separated ion pairs or even naked anions and increase the reactivity of the enolate oxygen towards electrophiles [362]. In general, an increase in the rate of reaction is accompanied by an increase in O-alkylation (see Fig. 2.125) [363–365].

Polar protic solvents such as water, alcohols or ammonia also lead to ion separation, but with concomitant reduction of the reaction rate, as they can coordinate not only the metal cation but also the enolate anion to provide some steric shielding. Most commonly, either polar aprotic solvents (HMPA, DMSO, DMF) or chelating agents (crown ethers, cryptands) as additives are used to promote O-alkylation [353].

solvent	k_{rel}	C/O-alkylation
EtOH	0.04	2.1 : 1
MeCN	1.0	1 : 3.4
DMSO	2.0	1 : 4.2
DMF	2.4	1 : 4.5
DMA	3.9	1 : 5.9
Me$_2$NCONMe$_2$	8.9	1 : 6.7
NMP	11.6	1 : 7.1
HMPA	14.0	1 : 7.7

Fig. 2.125 Reaction rates and chemoselectivity of the alkylation of malonic esters [364, 365]

Variation of the electrophile likewise exerts an influence on the reactivity and chemoselectivity of the transformation. Both the nucleofugality as well as the hardness of the leaving group and the degree of substitution of the electrophile play a decisive role. Hard leaving groups with high charge density preferentially react charge-controlled via the oxygen terminus. Soft, easily polarizable leaving groups with diffuse charge typically follow an orbital-controlled pathway under formation of the *C*-alkylation product. Since the alkylations proceed in a S$_N$2 fashion, the alkylating agents must possess sufficient reactivity towards an attack by the nucleophile. Benzylic and allylic alkyl halides are the most reactive, followed by primary alkyl reagents. The reactions of substrates with a secondary substitution pattern occur significantly slower and only with moderate yields. Tertiary halides almost exclusively afford the elimination product (see Fig. 2.126) [353, 366–368].

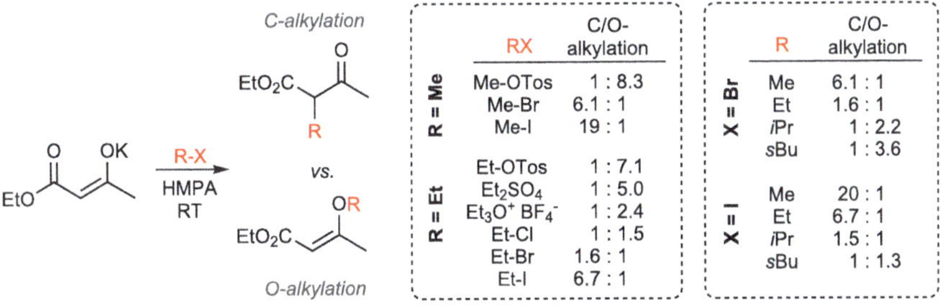

	RX	C/O-alkylation
R = Me	Me-OTos	1 : 8.3
	Me-Br	6.1 : 1
	Me-I	19 : 1
R = Et	Et-OTos	1 : 7.1
	Et$_2$SO$_4$	1 : 5.0
	Et$_3$O$^+$ BF$_4^-$	1 : 2.4
	Et-Cl	1 : 1.5
	Et-Br	1.6 : 1
	Et-I	6.7 : 1

	R	C/O-alkylation
X = Br	Me	6.1 : 1
	Et	1.6 : 1
	iPr	1 : 2.2
	sBu	1 : 3.6
X = I	Me	20 : 1
	Et	6.7 : 1
	iPr	1.5 : 1
	sBu	1 : 1.3

Fig. 2.126 Influence of the leaving group and the steric environment of the alkyl residue on the chemoselectivity of alkylation [353, 369]

The trends in chemoselectivity and reaction rate can be summarized as follows: for an *O*-alkylation, preferably tosylates should be used in combination with ammonium enolates in DMSO as solvent. Lithium enolates with alkyl, allyl or benzyl iodides preferentially provide the *C*-alkylation product in etheric solvents (see Fig. 2.127).

Boron enolates, like silyl enol ethers or titanium enolates, are rarely used for α-alkylations. While in the case of boron enolates this is attributed to the lack of reactivity towards alkylating

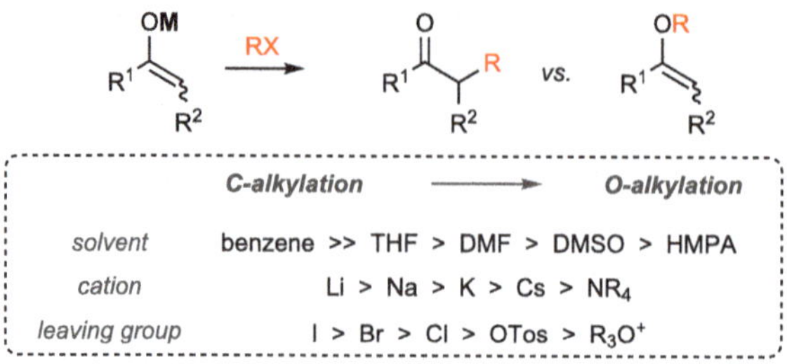

Fig. 2.127 Trends in reaction conditions in the chemoselectivity of alkylations

agents, the use of silyl enol ethers eliminates the possibility of activating the electrophile with Lewis acids — the most common activation mode in the aldol reaction for these enolate equivalents. Alternatively, fluorine sources such as TBAF can be utilized to generate the naked enolate, which is highly amenable towards α-functionalizations [351].

Methylene groups can also be doubly alkylated. The extent of multiple alkylation is primarily determined by the degree of aggregation of the metal enolates: The monoalkylated product is less aggregated in the case of Li enolates and can therefore react more quickly with another equivalent of the electrophile, provided the product can be deprotonated again to the enolate under the reaction conditions [370]. Under basic conditions and thermodynamic control, dihalogenated alkanes can afford cyclic products, presuming an intramolecular reaction is preferred (see Fig. 2.128; cf. Baldwin's rules, Sect. 1.1.4) [371].

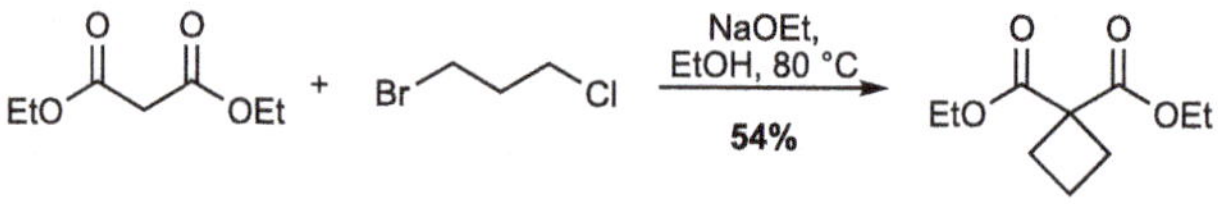

Fig. 2.128 Inter-/intramolecular bis-alkylation of diethyl malonate [371]

2.4.2 Diastereoselective Alkylations

Prochiral enolates inevitably result in racemic mixtures in an α-functionalization, as the enantiomorphic transition states are isoenergetic and thus do not permit a differentiation of the *Re*- or *Si*-face. The situation is different when stereogenic elements are already present in the starting material, enabling a *diastereoselective* reaction. Stereoinduction can be achieved in these cases by using chiral enolates with persistent stereogenic centers.

Bimolecular substitution reactions do not undergo a Zimmerman-Traxler transition state, as the nucleophile and nucleofuge must adopt a 180° angle due to the requisite orbital interactions. Since the stereodifferentiating enolate-electrophile transition state is thus acyclic in nature, rigid transition states with few conformational degrees of freedom usually provide the best diastereoselectivity. The desired facial discrimination of the electrophile can therefore be most easily achieved either by steric shielding, the formation of a chelate in the presence of Lewis basic substituents, or by using cyclic substrates (conformational control) (see Fig. 2.129).

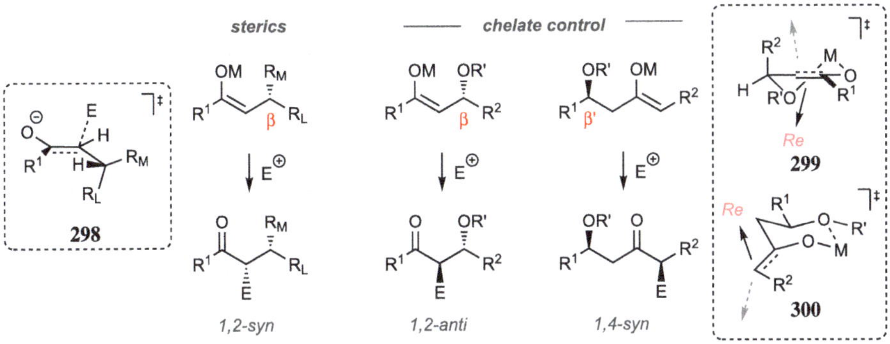

Fig. 2.129 Diastereoselectivity of alkylation reactions in the presence of stereogenic centers and relevant transition states [372, 373]

In acyclic enolates, only moderate diastereoselectivity is usually observed upon minimizing conformational interactions. For example, to prevent the occurrence of an 1,3-allylic strain, sterically demanding substituents R_L are aligned in an antiperiplanar orientation with regard to the approaching electrophile (**298**) [372, 374]. Significantly improved selectivities can be achieved with more rigid, chelated enolates. **299** assumes a half-chair conformation (cf. Fürst-Plattner rule, Sect. 1.1), wherein the attack on the electrophile proceeds via the *Re*-face [373]. With basic $β'$-substituents, a chair (**300**) or boat transition state can be adopted. In this case, the exocyclic enolate functionality becomes more accessible from one face as determined by the substituents and the conformation of the ring (see Fig. 2.130).

In the presence of acidic $β$-hydroxy/amino-carbonyl compounds, chelate formation also furnishes 1,2-*anti*-configured products, but in this case two equivalents of base are required for the double deprotonation [377, 378]. The selectivity of the bottom-right $β'$-alkoxyketone shown in Fig. 2.130 was rationalized by a chair-like transition state such as **300**. Due to the presence of a shielding $α$-methyl group, the steric repulsion in the allylation transition state leads to the *anti*-configured instead of the *syn*-product [348].

Cyclic substrates, whether they are ketones, esters or amides, typically display an excellent facial selectivity. In medium rings, the energetically most favored conformation is determined by the substituents. The attack on the electrophile then takes place via the sterically more accessible face to give the alkylated products diastereoselectively. Especially six-membered hetero- or carbocycles have been extensively studied due to their rigid molecular framework and the good predictability of the diastereoselectivity [351, 352]. With regard to the double bond of the enolate, it can be positioned either within (*endocyclic*) or outside the ring (*exocyclic*).

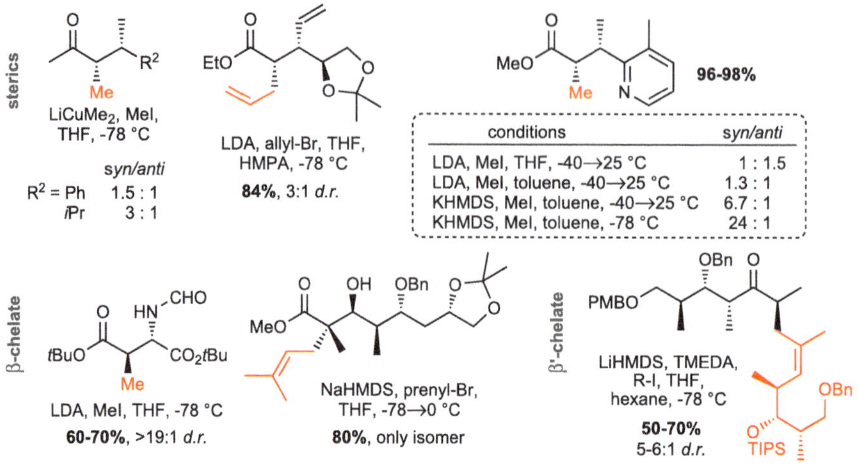

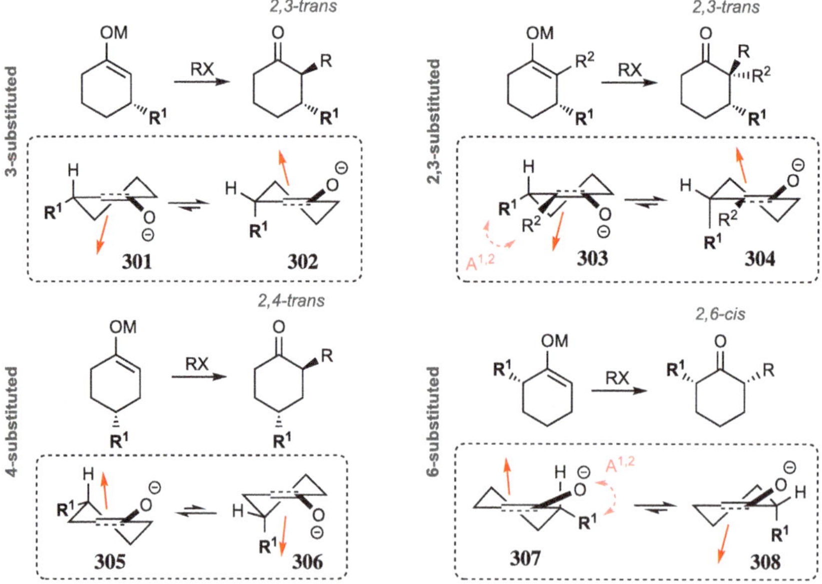

Fig. 2.130 Examples of alkylations of enolates with persistent stereogenic centers under steric or chelate control in total syntheses [348, 372, 375–378]

In the *endocyclic* case, six-membered carbonyls adopt a half-chair conformation. The substitution pattern determines the reactive face and the addition to the electrophile proceeds according to the Fürst-Plattner model via a chair-like transition state [373]. Depending on the substitution pattern, different facial preferences are predicted (see Fig. 2.131). In some cases,

Fig. 2.131 Diastereoselectivity of the alkylation of variously substituted, endocyclic enolates derived from six-membered, cyclic substrates [351, 352]

the dominant steric interaction determining the reactive conformation might be governed by the avoidance of either a 1,3-diaxial interaction (**301/302**; **305/306**) or a 1,2-allylic strain (**303/304**; **307/308**). Figure 2.131 provides some examples of the selectivity in differently substituted six-membered ketones, lactones and lactams to illustrate this point.

Especially with substituents in pseudoaxial or -equatorial positions, there exists only a small energy difference between the two half-chair conformers, so that additional substituents or steric interactions can easily tip the balance for or against a conformer. An overview of substrates with the substitution patterns mentioned in Fig. 2.131 (and beyond) are depicted in Fig. 2.132.

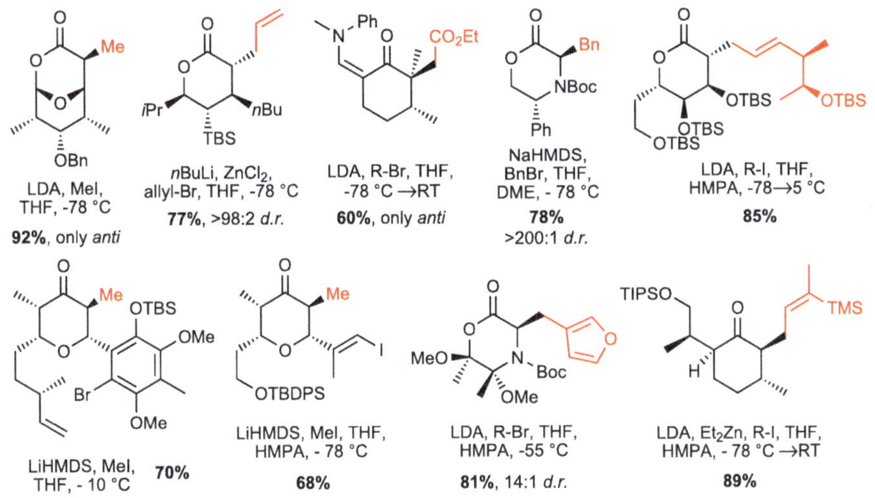

Fig. 2.132 Examples of alkylation reactions in complex six-membered carbonyls [138, 379–386]

Moving from six-membered enolates to larger cyclic systems, the consideration of the viable conformers presents an effective basis for predicting the nucleophile's facial preference. The diastereoselectivity of intramolecular, ring closing alkylation reactions can often be rationalized by analyzing the possible conformers of the newly formed ring system in the products [387–389]. Of course, the more complex the substitution pattern in specific cases, the more challenging a reliable prediction can become (see Fig. 2.133).

In *exocyclic*, cyclohexane-derived substrates devoid of substituents, there exists a slight preference for alkylation at the equatorial position (via **312**). In the trajectory leading to the axial product, the electrophile must approach through the β-positioned hydrogen atoms, while in the equatorial reaction path, the approach takes place more parallel to the α-positioned H atoms. If additional substituents are introduced that create a passive volume by shielding one of the two faces, this can lead to a significant enhancement of facial discrim-

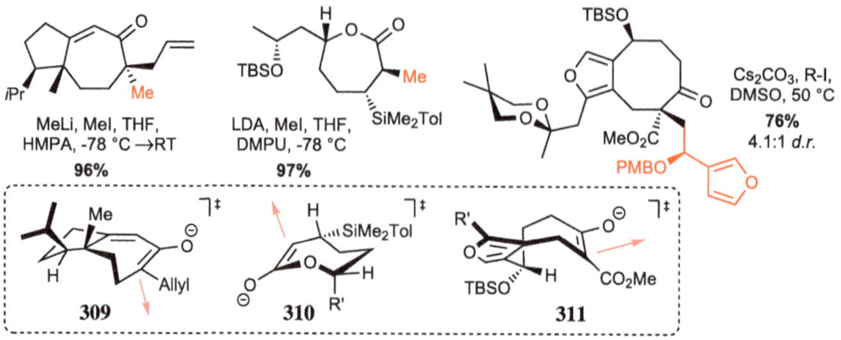

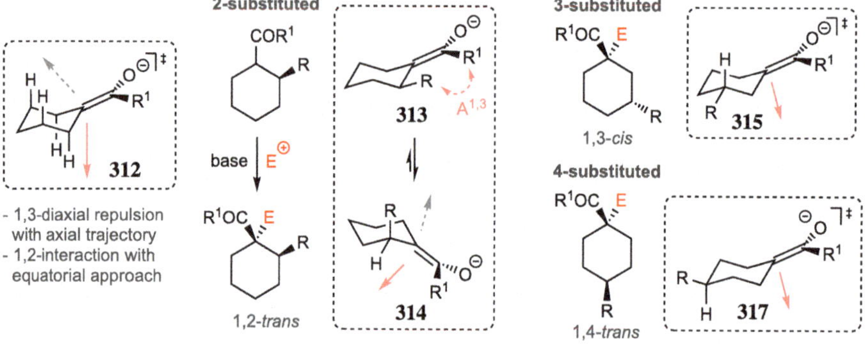

Fig. 2.133 Alkylation of 7- and 8-membered carbonyls and postulated TSs [390–392]

ination. These substituents might additionally serve as a conformational anchor to favor a certain conformation (see Fig. 2.134) [393–397].

Fig. 2.134 Diastereoselectivity of alkylation reactions in exocyclic systems

2.4.3 Enantioselective Alkylations

Beyond diastereoselective functionalization reactions, the electrophile can also be attached *enantioselectively*. Different strategies are followed depending on the carbonyl derivative in question: While ketones are enantioselectively alkylated either organocatalytically or using hydrazone-based auxiliaries, carboxylic acid derivatives typically require chiral amine auxiliaries or phase transfer catalysis. In addition, there exist methods based on chiral lithium bases, in which enantioinduction is realized by the counterion, as well as transition metal-mediated allylations using allyl carboxylates as substrates [351, 398–400].

Auxiliary-induced asymmetric alkylations can be roughly divided into two substrate classes: Nitrogen analogs of ketones or aldehydes (hydrazones, imines, sulfonimides, etc.) and carboxylic acid derivatives (amides, esters). The most widely applied methods on the basis of auxiliary control rely on hydrazones of type **318** (SAMP/RAMP, **319**) and ephedrine-based amides (**321**). Additionally, a plethora of other auxiliary-based alkylation reactions exist (see Fig. 2.135) [401–406].

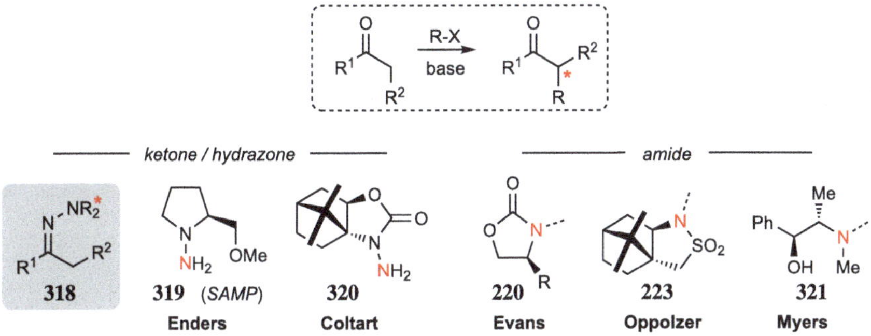

Fig. 2.135 Common auxiliaries in asymmetric alkylations [351, 398, 399]

In contrast to enolates derived from carboxylic acids, ketones and aldehydes present particularly challenging substrates, as they show a comparatively low nucleophilicity and higher tendency to epimerize. The *S*- or *R*-amino methoxymethyl pyrrolidines (*SAMP/RAMP*) introduced by the Enders group constitute an elegant way of functionalizing substrates where alternative methods typically fail to induce a stereoselectivity [407, 408].

In the alkylation step, the intermediate metal hydrazone **323** is shielded on one side through coordination of the cation by the methoxy group in the transition state **325**, resulting in high stereoselectivity (see Fig. 2.136) [407–409].

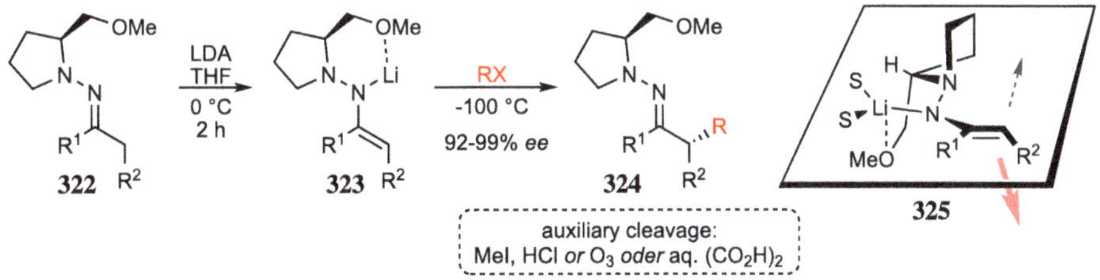

Fig. 2.136 SAMP-mediated alkylations and transition state of the reaction

Disadvantages of the SAMP/RAMP methodology include the weak acidity of the corresponding hydrazones, which necessitates a deprotonation by a strong base and extended reaction times, as well as the very low alkylation temperatures. Despite its weaknesses, **319** has proven itself in accessing advanced intermediates for total synthesis, as impressively demonstrated in Hayashi's approach to an epimer of amphidinolide N. The hydrazone **326** was deprotonated and successively alkylated with **327** and **328**. Neither the ester, thioester, nor the epoxide functionality were affected under these conditions. The mild work-up using aqueous oxalic acid then furnished the bis-alkylated ketone **329**. This densely functionalized intermediate was subsequently converted to the desired natural product in another 11 steps comprising a macrocyclization, Grieco olefination, and several deprotections (see Fig. 2.137) [410].

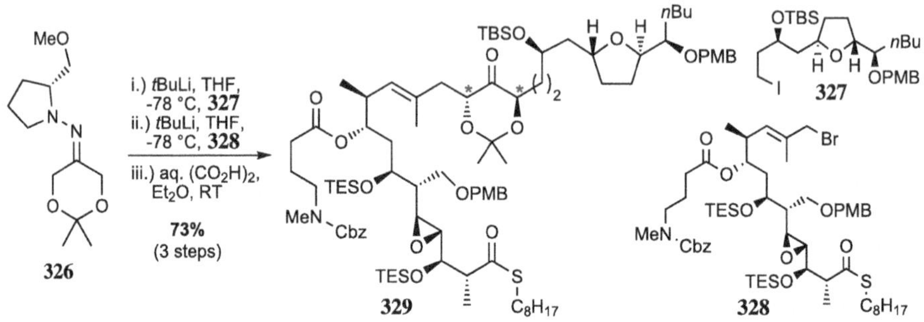

Fig. 2.137 Asymmetric RAMP-mediated bisalkylation according to Hayashi *et al.* [410]

SAMP is produced on an industrial scale in four or six steps from proline (**330**), RAMP is accessible from *R*-glutamic acid (**333**) (see Fig. 2.138) [408, 411].

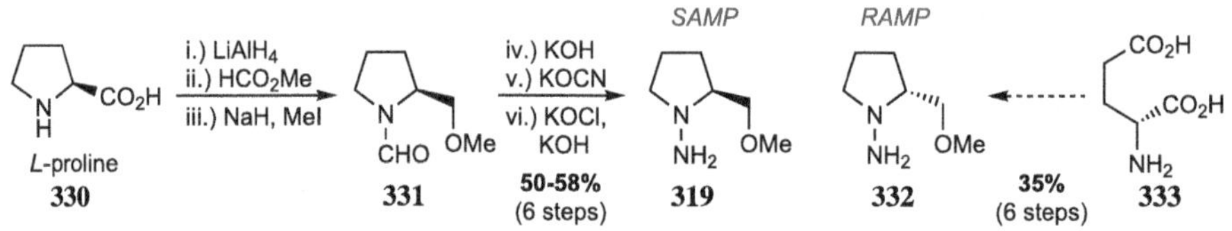

Fig. 2.138 Synthesis of SAMP and RAMP [408, 411]

The cyclic aminocarbamates developed by Coltart allow somewhat milder conditions in the deprotonation compared to the SAMP/RAMP protocol, which is due to a polarization of the α-positioned protons of the derived hydrazones. In addition, the alkylation does not require temperatures quite that low, even if it is typically carried out at -40 to -78 °C [412]. The facial differentiation is achieved according to the same principle as with the Enders system: The rotation of the auxiliary is prevented by intramolecular chelate formation in the postulated transition state **334** and the substituents of the carbamate sterically shield one face (see Fig. 2.139) [413].

The oxazolidinones developed by Evans are primarily applied for enantioselective aldol reactions. However, they can also be used in asymmetric alkylations. Alkyl bromides provide excellent selectivities for most substrates in the range of 95-99:1 *d.r.*. In contrast, methyl

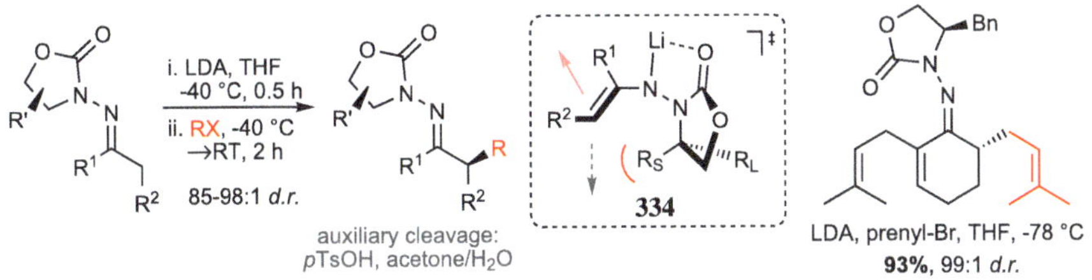

Fig. 2.139 Asymmetric hydrazone-mediated alkylation by Coltart *et al.*, postulated transition state and application in the synthesis of clusianone [412–414]

iodide does not allow such good discrimination of the enantiotopic faces of the enolate due to its lower steric demand [415]. Orthoesters, which normally tend to prefer S_N1 reactions, can also be used as electrophiles when employing $TiCl_4$/iPr$_2$NEt and provide outstanding diastereoselectivities. These conditions also work very well for the "classic" alkylating agents such as alkyl halides (see Fig. 2.140) [111, 416].

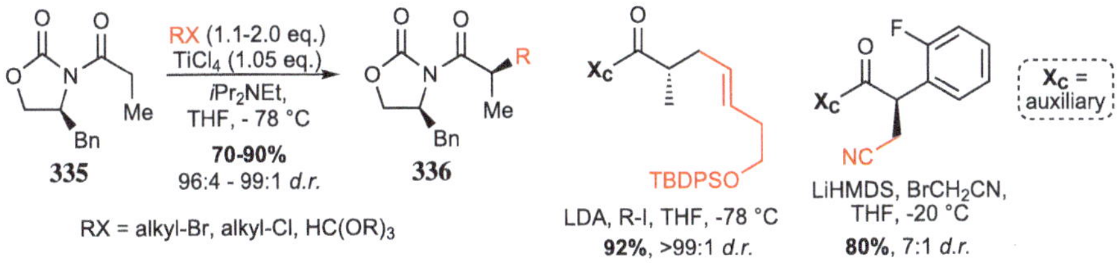

Fig. 2.140 Asymmetric alkylation developed by Evans *et al.* and examples of oxazolidinone-controlled alkylations [111, 417, 418]

The Myers group developed a broadly applicable method with regard to the substrate scope and the possibility for further derivatization of the alkylation product. In addition to alkyl halides, epoxides can also be used as electrophiles. Based on a pseudoephedrine derivative as stereoinductor, the addition of two equivalents of base deprotonates both the α-acidic proton of the amide and the hydroxy group of the auxiliary. The diastereoselectivity in the reaction of the dianion was rationalized using a model in which the free hydroxy group of the pseudoephedrine can block one of the two enolate faces due to the coordination of the solvent-complexed lithium (see Fig. 2.141) [419, 420].

Alkyl halides and epoxides show opposite diastereoselectivity with respect to the diastereotopic π-faces of the enolate. When using epoxides, a coordination and activation of the epoxide by the lithium ion was postulated (**340**) [421, 422]. The range of viable derivatizations was extensively investigated by Myers and alternative conditions for acid- or base-labile substrates were developed [420]. Some examples of the application of this method are depicted in Fig. 2.142.

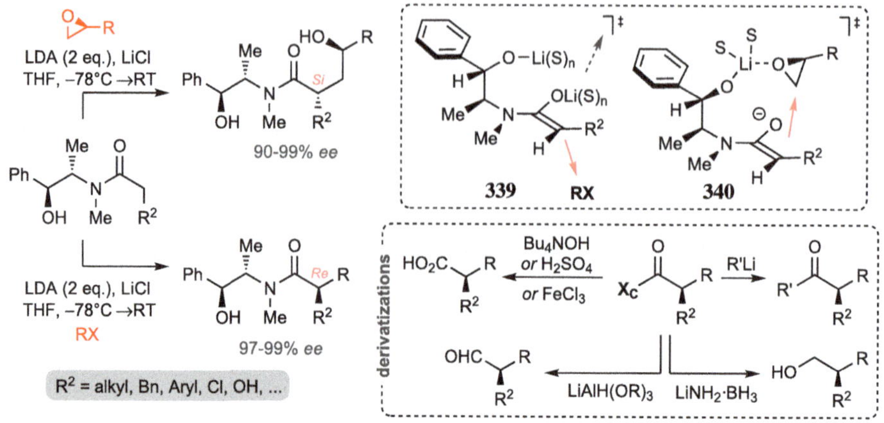

Fig. 2.141 Asymmetric alkylation according to Myers and co-workers [420, 421]

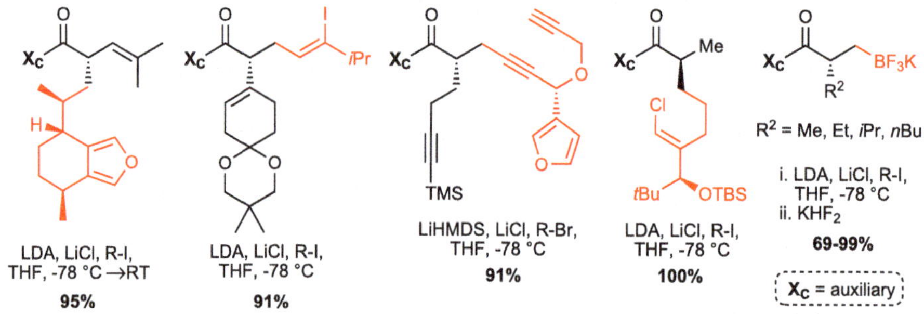

Fig. 2.142 Examples of the Myers alkylation in natural product syntheses [423–427]

There are only a few examples of catalytic, enantioselective alkylation reactions. The available methods offer only a limited substrate scope, both with regard to the carbonyl component or the electrophiles [400]. Initial work by Jacobsen and Evans with tin enolates and N-acyl thiazolidinethiones could deliver good enantiomeric excess, but the restricted substrate scope has so far prevented the widespread utilization in advanced intermediates [428–430].

A comprehensively studied and widely applicable α-functionalization is the transition metal-catalyzed asymmetric allylation, also known as *Tsuji-Trost allylation* ($\rightarrow$ Sect. 6.4). Acyclic or carboxylated allylic alcohol serves as the electrophile component. It can also be directly attached to the ketone or ester that serves as enolate precursor. This approach has the advantage that under the employed reaction conditions the enolate is generated in the absence of a base, which also allows its use for base-labile substrates. The catalyst cleaves the acyl or carboxyl group to form the reactive metal allyl system **341** *in situ*. Attachment of chiral ligands to the the metal center in the electrophilic reagent **341** enables an enantioinduction during the conversion with the enolate. While initial methods heavily relied on Pd complexes as catalysts, Ir-, Rh- and Ru-derived systems have now become firmly established (see Fig. 2.143) [399, 431–434].

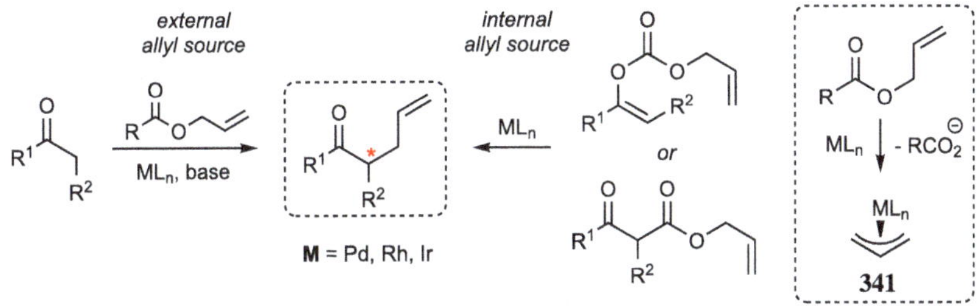

Fig. 2.143 Generalized scheme of the transition metal-catalyzed asymmetric allylation

Chelate ligands in the allylation are typically based on phosphorus or nitrogen atoms as basic donors. Despite the attractiveness of the concept also for CC-bond formation, most applications in complex syntheses focus on the formation of C-heteroatom bonds by using oxygen or nitrogen nucleophiles. Nevertheless, there have been some target molecules in which the α-allylation provides the most efficient entry to the selective construction of one or more stereocenters (see Fig. 2.144).

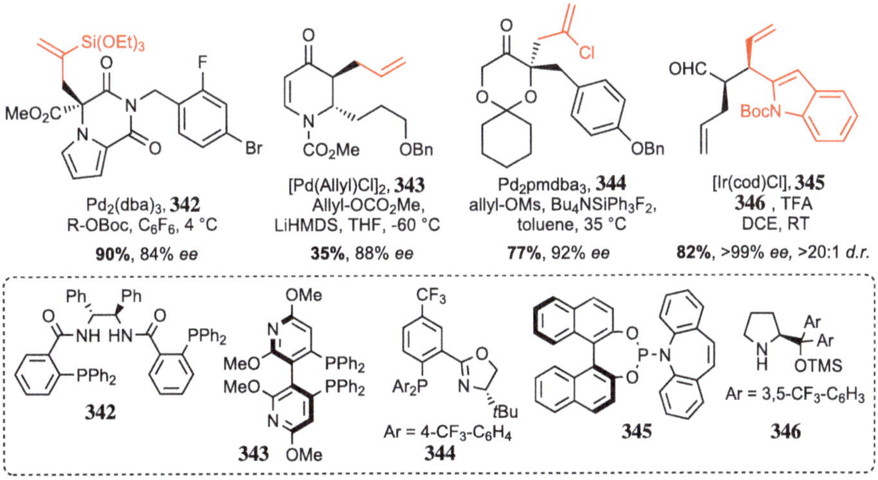

Fig. 2.144 Asymmetric allylation in the synthesis of complex natural products [435–438]

The desymmetrization of meso-compounds via selective deprotonation of enantiotopic protons using chiral bases was already investigated early on by the research groups of

Koga and Simpkins [439, 440]. Based on seminal work by Shioiri, Zakarian disclosed in 2011 an alkylation of various aryl acetic acids using a chiral Li-amide base [441, 442]. The enantioinducing base is connected to the deprotonated substrate via the jointly shared counterion. In addition to aryl acetic acids, vinyl acetic acid could also be successfully used (see Fig. 2.145).

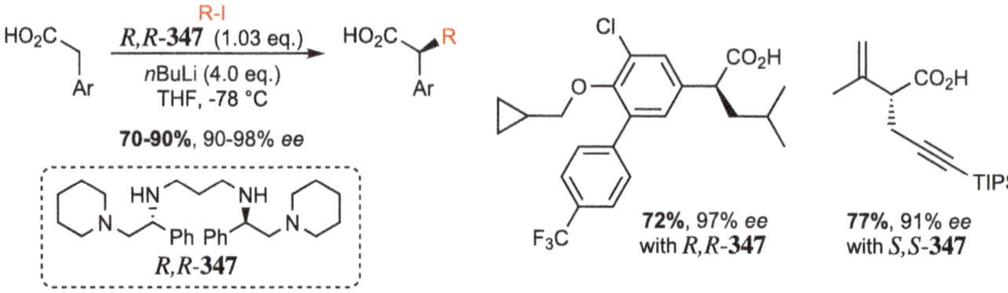

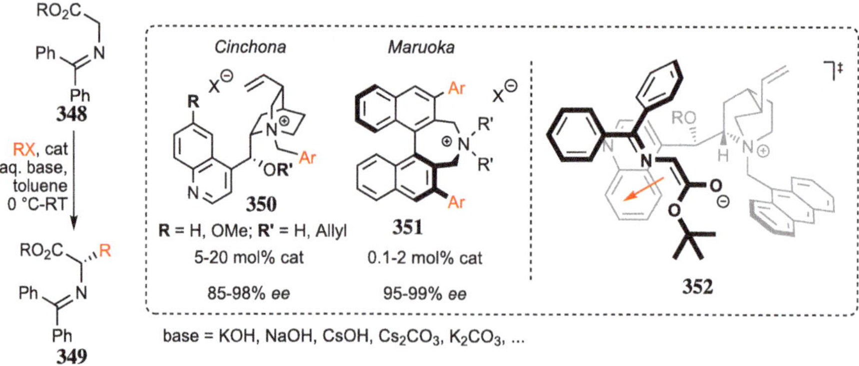

Fig. 2.145 Alkylation introduced by Zakarian *et al.* and selected applications [442, 443]

In addition to homogeneous metal catalysis, the asymmetric alkylation can also be carried out in a biphasic solvent system in the presence of a phase transfer catalyst [444–449]. In addition to work by O'Donnell, Corey, and Lygo using Cinchona alkaloid-derived catalysts (**350**), the symmetric binaphthyl ammonium salts (**351**) developed by Maruoka have proven particularly effective. Schiff base-protected amino acid esters (**348**) serve as substrates in the majority of methods and applications (see Fig. 2.146, cf. Sect. 8.1.5).

Fig. 2.146 Asymmetric phase transfer catalyst-mediated catalytic alkylation. Comparison of reaction conditions and postulated transition state [445, 448, 449]

A transition state has been postulated for catalysts based on Cinchona alkaloids. The substrate undergoes a π/π interaction with the quinoline part of the Cinchona alkaloid, locking the enolate, while the bicyclic backbone and the anthracenyl side chain shield its back face (**352**) [445, 450].

2.4.4 Hydroxylation, Amination, and Halogenation

Apart from alkyl substituents, functional groups based on oxygen, nitrogen, or halogens can also be introduced via α-functionalizations by resorting to suitable heteroatom electrophiles. The most common types of electrophilic reagents are based on molybdenum peroxides and oxaziridines (hydroxylation), azodicarboxylates (amination), and *N*-organo halides (halogenations, see Fig. 2.147).

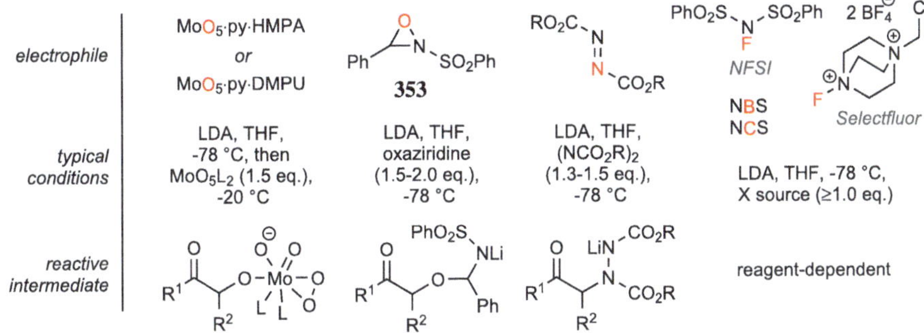

Fig. 2.147 Reagents, conditions, and postulated intermediates in α-functionalizations [451–455]

Hydroxylations constitute one of the largest classes of oxidation reactions at the α-carbon atom. The first broadly applicable method was developed by Vedejs using molybdenum oxodiperoxy-pyridine-HMPA (MoO₅·py·HMPA, *MoOPH*) as an oxygen source. Due to the toxicity of the HMPA, MoOPD (MoO₅·py·DMPU) was developed as an alternative reagent, in which the HMPA was replaced by the less toxic DMPU [451, 456]. Even though *N*-sulfonyl oxaziridines pioneered by Davis (**353**, *Davis reagent*) have now become the *de facto* method of choice in total synthesis, MoOPH is still used occasionally (see Fig. 2.148). Beyond these approaches, silyl enol ethers can also be converted into the corresponding α-hydroxyketones by *m*CPBA via intermediate epoxidation and subsequent acidic work-up (*Rubottom oxidation*) [457, 458].

Following the introduction of oxaziridine **353** as oxygen source, enantioselective hydroxylations could also be carried out, either with the aid of the Evans auxiliary or Enders' SAMP/RAMP methodology, to obtain the oxidation product in very good to excellent enantiomeric excesses of up to 98% *ee* [452, 463, 464]. Apart from a substrate-controlled reaction, the reagent might also serve as the source of stereogenic information: F. A. Davis and co-workers developed a series of chiral oxaziridines (**354, 355**) and investigated their applicability for asymmetric hydroxylation reactions [465, 466]. A strong dependence of the enan-

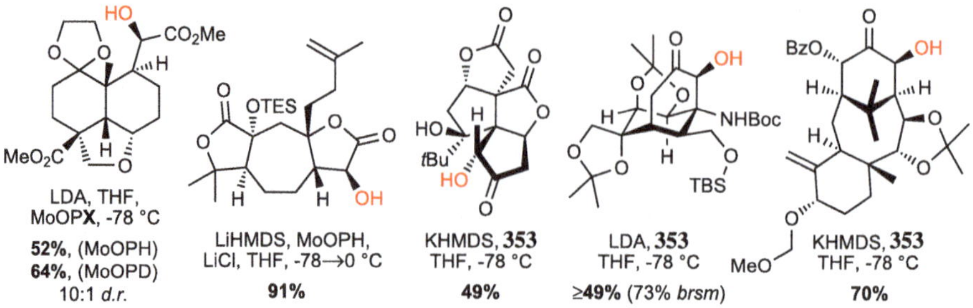

Fig. 2.148 Substrate-directed α-hydroxylations of complex intermediates [456, 459–462]

tiomeric excess on the counterion of the enolate species and its aggregation was observed, both in the auxiliary- and the reagent-controlled enantioselective reactions. Lithium enolates give only low *ee* values, while Na-enolates afford superior results. The selectivity could be further improved by varying the base or by the addition of HMPA. Especially chiral oxaziridines have established themselves as method of choice in asymmetric oxidation reaction to access complex secondary metabolites (see Fig. 2.149).

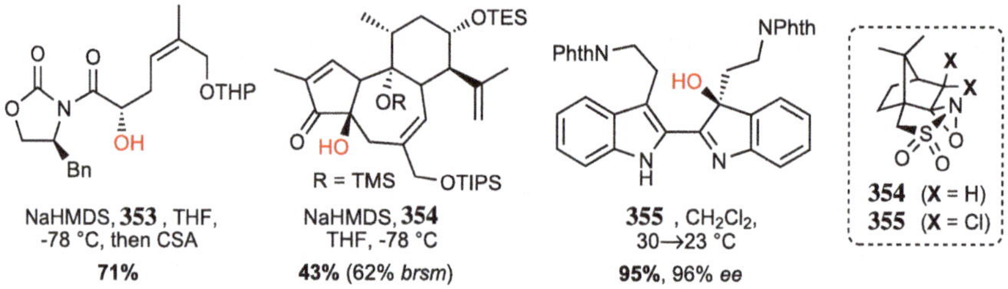

Fig. 2.149 Examples of auxiliary- or reagent-controlled enantioselective α-hydroxylations [467–469]

So far, there have only been isolated studies on the *catalytic* α-hydroxylation of carbonyl compounds. In addition to organocatalytic approaches with nitrosamines as an oxygen source (see Chap. 8), transition metal catalysts have been successfully used under CH activation with O_2, although the substrate spectrum is largely limited to aryl alkyl ketones or β-oxoketones [470, 471].

In contrast to α-hydroxylation reactions of carbonyl compounds, a highly selective entry to chiral 1,2-oxygenated products can be gained from the corresponding alkenes, either by direct dihydroxylation or via an intermediate epoxide. This allows easy access to vicinally bis-oxygenated motifs in an enantioselective manner (see Sect. 3.3). Starting from these 1,2-diols, one of the hydroxy groups can then be subsequently oxidized to the desired α-oxygenated ketone or aldehyde, which is why this route is particularly used when high enantiomeric excesses and the avoidance of a stoichiometric stereoinductor (auxiliary, reagent) are key criteria [472].

The lack of suitable electrophilic nitrogen sources limits the substrate scope of the α-*amination*. Diazo compounds of type **356**, azides (**358**) or hypervalent iminoiodanes (**357**) typically serve as electrophilic reagent. A much more common approach is the nucleophilic substitution of α-halo carbonyl compounds, as the innate nucleophilicity of amines can thus be easily exploited (see Figs. 2.150 and 2.151) [453, 454, 473].

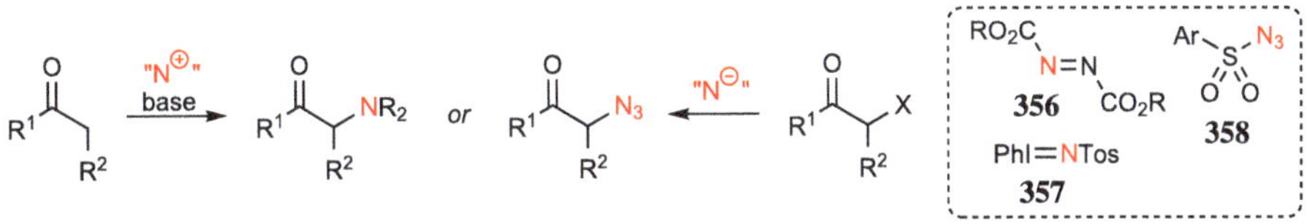

Fig. 2.150 Approaches to the construction of α-aminocarbonyl compounds

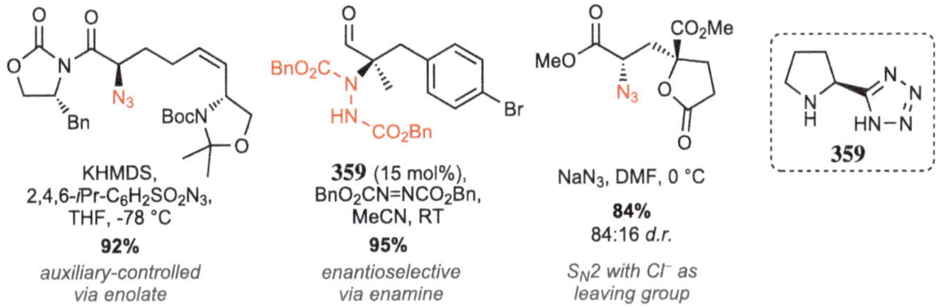

Fig. 2.151 Synthetic applications of α-aminations [474–476]

Halides play a crucial role in pharmaceuticals and agrochemicals for modulating lipophilicity, bioavailability, and pharmacokinetics [477]. The controlled α-*halogenation* of ketones, esters, and amides, whether in a diastereo- or enantioselective fashion, is therefore crucial for the efficient access to bioactive target compounds that contain a C_{sp^3}-X bond. While fluorine has become almost ubiquitous in pharmaceuticals, there are hardly any naturally occurring secondary metabolites that contain fluorine or iodine (see Fig. 2.152) [478–480].

α-halo carbonyls can easily be constructed by reacting enolates with a suitable electrophilic halogen source, often affording high yields [455]. Due to their innate reactivity, hardly any α-iodo or α-bromo carbonyl derivatives are found in nature. Their synthesis is usually of interest for use as an intermediate for a subsequent nucleophilic substitution (see Fig. 2.153).

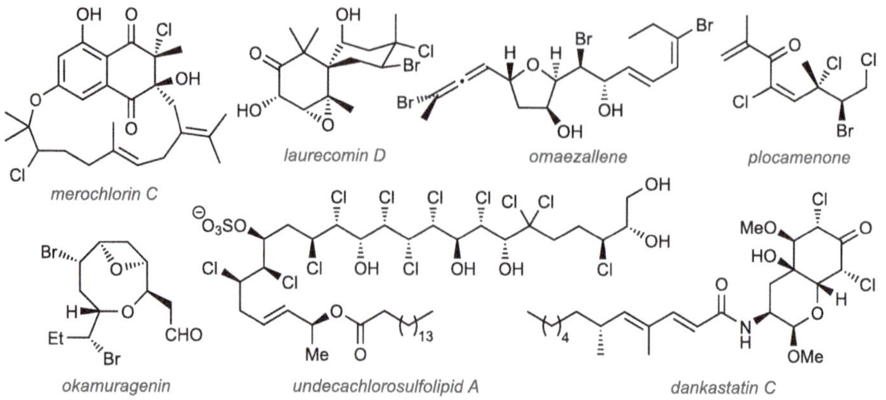

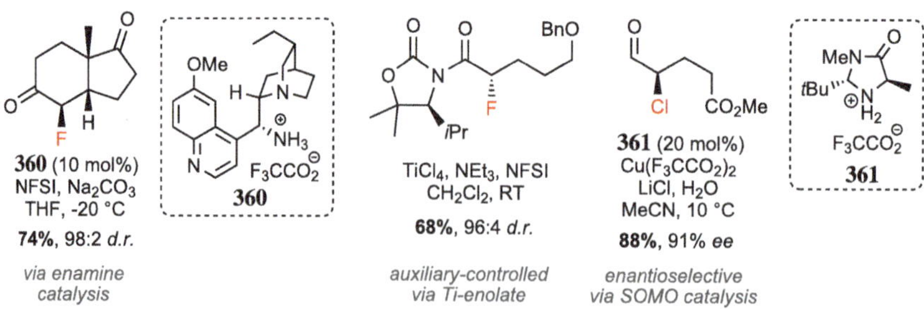

Fig. 2.152 Examples of halogenated natural products [478]

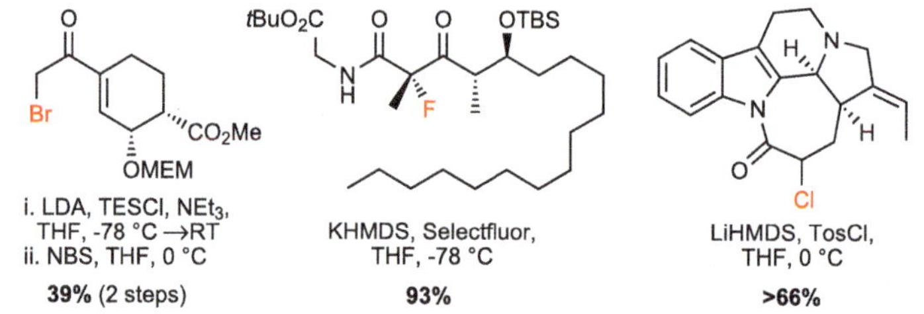

Fig. 2.153 α-Halogenations in natural product syntheses [481–483]

The situation is rather different when moving to chlorine and fluorine. The development of asymmetric variants has experienced a rapid expansion for these halides in recent years, driven by the ongoing discovery of new chlorine-containing natural products and the increasing use of fluorine in pharmaceuticals and agrochemicals (see Fig. 2.154) [484–486].

Fig. 2.154 Enantioselective methods for the introduction of F and Cl [487–489]

2.5   1,2- versus 1,4-Addition

The addition of nucleophiles to α,β-unsaturated carbonyl compounds can result in the bond formation occurring either at the C-atom of the carbonyl group or in the β-position. The preference can be greatly influenced by the choice of reagents (see Fig. 2.155).

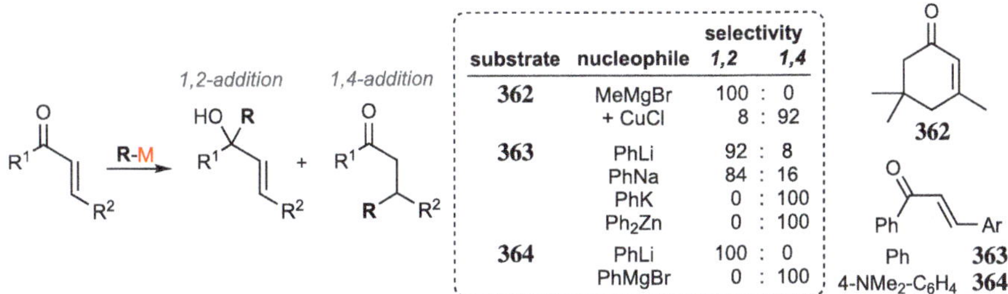

Fig. 2.155 Influence of the nucleophile on chemoselectivity [490, 491]

The selectivity of the addition is governed by the stereoelectronic properties of the substrates (acceptor and nucleophile). The reaction is mediated by a mixture of Coulomb and orbital interactions (see Klopman-Salem equation, Sect. 5.1.2). If the Coulombic contribution dominates, a nucleophile is classified as "hard" according to Pearson, while orbital-controlled nucleophiles are referred to as "soft". Hard electrophiles possess low steric bulk, are charged and have a high-lying LUMO. Soft electrophiles, on the other hand, are sterically demanding, seldom charged, and have a low-lying acceptor orbital. α,β-Unsaturated carbonyls carry a larger orbital coefficient at the β-terminus in the LUMO. The partial positive charge, however, is higher on the carbonyl carbon than in the β-position. Coordination of Lewis acids at the oxygen atom further lowers the LUMO, intensifies the charges, and promotes a 1,2-addition (see Fig. 2.156) [492].

Fig. 2.156 Frontier orbital coefficients, effective charges derived from the electrostatic potentials, and π charges of free and Li-coordinated acrolein [492–494]

Conjugated carbonyls are *ambident* electrophiles: with hard nucleophiles the 1,2-addition products are typically received due to a dominant Coulomb interaction, while with soft nucleophiles a Michael addition is observed [26]. The more electron-deficient the carbonyl group, the stronger the tendency for 1,2-addition. This leads to acid chlorides being exclusively attacked at the carbonyl C-atom, whereby the tendency shifts towards the Michael product when going from ketones via esters to amides. As with the 1,2-addition, the Michael addition is an established tool to effectuate CC-, CN- or CS-bond formation in total synthesis (see Fig. 2.157).

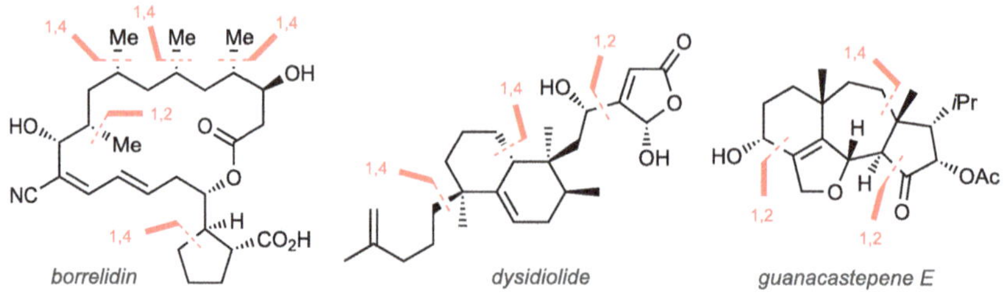

borrelidin   *dysidiolide*   *guanacastepene E*

Fig. 2.157 Examples of natural products in whose syntheses both 1,4- and 1,2-additions were used [495–497]

The product ratio for these substrates can be steered to follow the desired pathway by prudent choice of the reaction conditions, especially the organometallic reagent. A further degree of freedom can be gained when considering the ability of many organometallic nucleophiles to transfer their substituents to other catalytically active metals. This transmetalation enables further modulation of their chemoselectivity. A classic example exploiting this is the addition of catalytic amounts of Cu(I) salts to Grignard reagents. The transmetalation of a hard nucleophile to a soft copper center shifts the reaction to follow a 1,4-addition. However, in order to avoid challenges with mixed chemoselectivity and competing reaction paths, most 1,2-additions do not utilize α,β-conjugated carbonyl derivatives as starting materials.

A wide range of 1,2- or 1,4-selective methods is available (see Fig. 2.158): Li- and Mg-based organometallic reagents typically display high 1,2-selectivity, while the corresponding copper species exclusively add to the β-position. Their reliability makes them the reagent of choice for 1,4-additions under stoichiometric, catalytic, and enantioselective conditions. Organozinc nucleophiles show almost no reactivity towards aldehydes. In the presence of a suitable catalyst, the reaction is however efficient and usually 1,2-selective. Under Cr-catalysis (Nozaki-Hiyama-Kishi conditions), a preference for addition to the CO bond of enals can be observed in the few reported examples. In allylation reactions, almost all variants are 1,2-selective, only intramolecular reactions lead to the Michael product under certain conditions. Metal enolates add selectively to the CO bond, while silyl enol ethers result in a preference for 1,4-addition (both F-mediated additions and under organocatalysis). Hydride reductions typically tend to favor 1,2-addition. The *Luche* reduction (1,2-selective) or the use of copper hydrides (1,4-selective) yield complementary results for α,β-unsaturated carbonyls and are routinely applied. Further, more specialized catalysts or nucleophiles with a differing selectivities in CC- and CH-couplings have been developed, but their use has not yet been widely established.

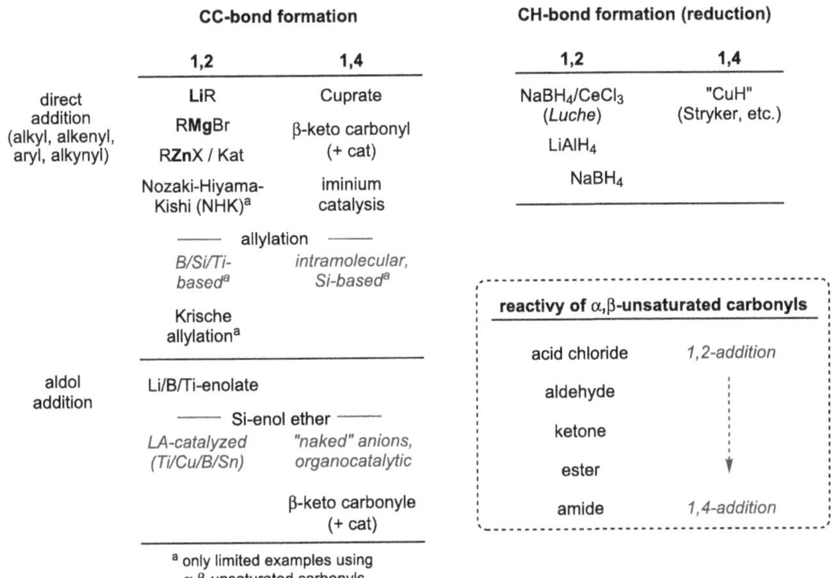

Fig. 2.158 Overview of the selectivity of reagents and methods for CC- and CH-bond formation with α,β-unsaturated carbonyl compounds [197, 498–503]

2.5.1 1,2-Addition

Organometallic reagents based on lithium and magnesium react charge-controlled with the carbonyl group of Michael systems in the absence of additives or catalysts. The usually encountered lower functional group tolerance can be tempered by modifying their reactivity and thus selectivity. The use of more polar solvents or internal coordination of basic substituents under chelate formation allows for fine-tuning of the yield. Wipf's synthesis of the alkaloid cycloclavine employed a new TEMPO-lithium carbamate (**366**), as the use of the corresponding Boc-carbamate resulted in the attack on both the amide and enone moiety in **365**. The concluding steps comprised an intramolecular Diels-Alder reaction/aromatization of the furan with the cyclohexene fragment upon thermolysis of the carbamate, followed by reduction of the lactam. The desired natural product was thus obtained in only eight steps (see Fig. 2.159) [504].

The addition of HMPA to organolithium reagents leads to the conversion of the contact ion pair into a solvent-separated ion pair ("naked" anion) by the strong complexation of the cation, with a change in chemoselectivity towards the Michael addition [505]. Organomagnesium compounds can also be used for selective 1,2-additions to enals or enones. However, their application in densely functionalized substrates is limited to a few published examples, as the Nozaki-Hiyama-Kishi reaction is usually preferred in these cases (see below). Britton and Paterson reported the reaction of an *in situ*-generated vinyl magnesium halide

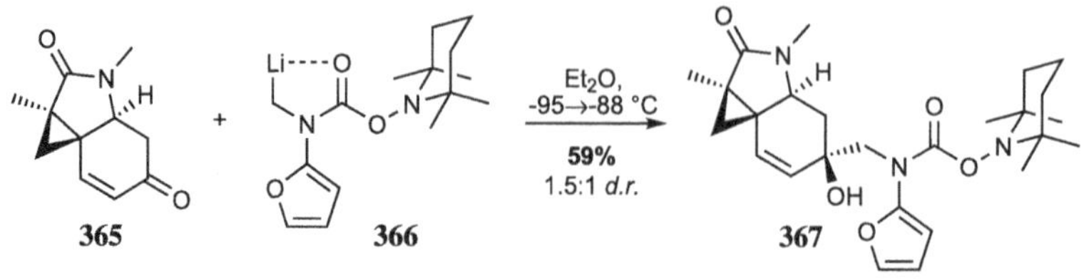

Fig. 2.159 Synthesis of (–)-cycloclavine by Wipf *et al.* [504]

(derived from **368**) with an aldehyde in the presence of an α,β-unsaturated ester (**369**). The diastereoselectivity can be rationalized by a Cram chelate transition state, which favors the addition to the α-chiral aldehyde via its *Si*-face (see Fig. 2.160) [506].

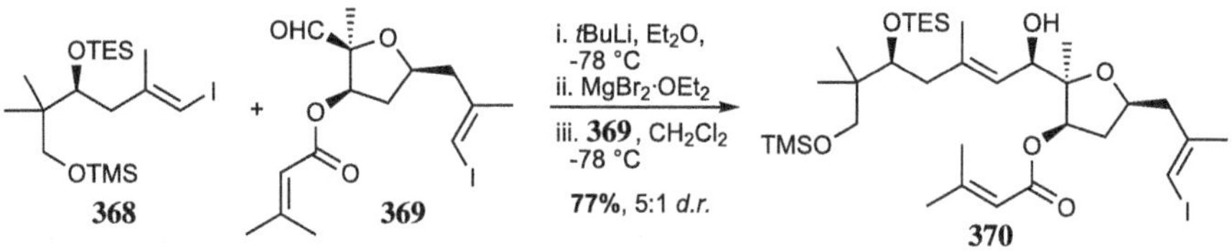

Fig. 2.160 Synthesis of phormidolide A by Britton and Paterson [506]

Organozinc compounds are mostly inert towards addition to aldehydes or ketones, which is why they are routinely used as nucleophiles in cross-coupling reactions in the presence of a variety of functional groups (ketones, esters, nitriles, etc.). This makes them ideal substrates for catalytic transformations, as no background reaction for an unselective product formation occurs. To enable a reaction for dialkylzinc, usually both the aldehyde and the zinc species need to be activated by coordination of a Lewis acid and Lewis base, respectively. Diarylzinc, on the other hand, reacts with aldehydes even in the absence of additives. The first asymmetric, catalytic addition of organozinc compounds to aldehydes was reported by the Noyori group already in 1986 [507]. This seminal work laid the foundation for subsequent research in the field of asymmetric addition reactions and is one of the first reactions with a pronounced nonlinear effect, as high enantiomeric excess in the product is achieved even with low *ee* of the ligand (see Fig. 2.161) [508, 509].

The origin of the nonlinear effect can be understood by the formation of dimeric Zn-ligand complexes. Homodimers (**373**) of two ligands with identical configuration can easily dissociate to release the active species **373**. Heterodimers (**372**) of (+)- and (–)-DAIB, on the other hand, are more stable and show very limited dissociation. Thus they can temporarily remove the minor enantiomer of the ligand even at low enantiomeric excess until the reactive species **374** is enantiomerically strongly enriched. Several viable transition states have been postulated, with theoretical studies pointing to **375** as the dominant transition structure [510–512]. The change of the coordination sphere from the linear ZnR$_2$ to the tetrahedral species

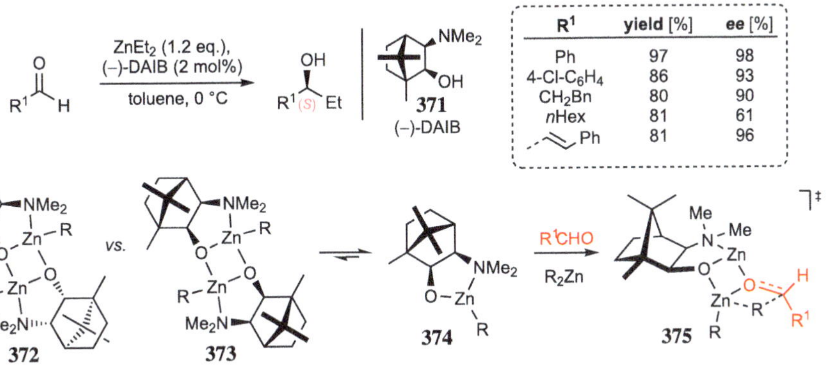

R^1	yield [%]	*ee* [%]
Ph	97	98
4-Cl-C_6H_4	86	93
CH_2Bn	80	90
*n*Hex	81	61
⁄\\Ph	81	96

Fig. 2.161 Asymmetric addition of organozinc compounds to aldehydes by Noyori *et al.* [507, 509]

in **372–374** reduces the Zn-C bond order, and increases the nucleophilicity of the transferred alkyl group R [72].

The formation of organozinc compounds can occur via several routes, depending on the nature of the group to be transferred. Most reagents show the composition ZnR_2 (homoleptic), whereas the substituents differ in heteroleptic organozinc species. Polyfunctional organozinc reagents are accessible by direct insertion into alkyl iodides, by transmetalation of Li/Mg/B or from alkynes by deprotonation of the terminal C_{sp}H-bond (see Sect. 6.4) [513, 514]. Alkynyl residues can alternatively be introduced *in situ* by $Zn(OTf)_2/NEt_3$ in the presence of the parent alkyne, since the terminal CH bond is acidic enough to form the alkynyl-Zn(OTf) under these conditions [515].

There are various well-established methods for asymmetric additions to carbonyl substrates, which differ depending on the type of substituent to be transferred [516, 517]. First studies were undertaken in the groups of Noyori and Oguni, and most methods rely on the use of commercially available dialkyl or diarylzinc reagents ($ZnMe_2$, $ZnEt_2$, Zn(iPr)$_2$, $ZnPh_2$). Functionalized organozinc nucleophiles are indispensable for coupling more complex fragments, such as can be found foremost in the area of total synthesis. Knochel and co-workers extensively investigated methods for accessing polyfunctional organozinc compounds and enabled the incorporation of functional groups such as ketones, esters, nitriles or even NH-acidic amines in the nucleophile (see Fig. 2.162) [513].

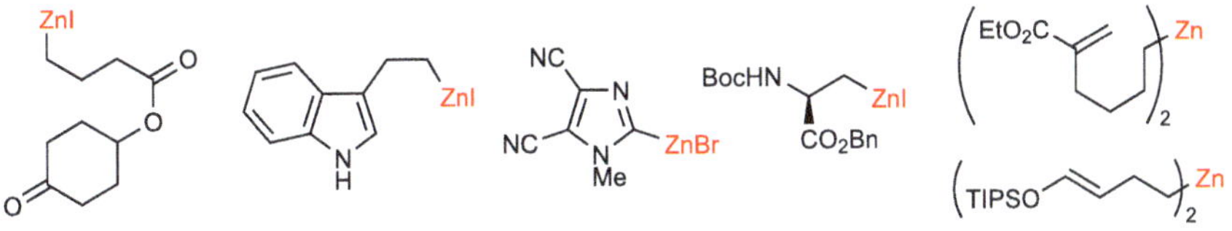

Fig. 2.162 Examples of *in situ* generated complex organozinc reagents [513]

To enable an enantioinduction in the addition reaction, *N,O*- or *O,O*-complexing chiral ligands are routinely added. Figure 2.163 provides a selection of methods used in the synthesis of complex intermediates: In addition to ephedrines, Schiff bases such as salen can also be employed. Many Zn-based methods further rely on the addition of stoichiometric amounts of Ti(O*i*Pr)$_4$, often in combination with sulfonamide ligands. For the alkynylation of aldehydes two established protocols are often referred to, which use the ProPhenol ligand developed by Trost (**282**) and the conditions introduced by Carreira (**379** as ligand). Especially the Carreira system has proven to be extremely robust and suitable for the reaction of very complex substrates, despite the use of mostly stoichiometric amounts of ligands.

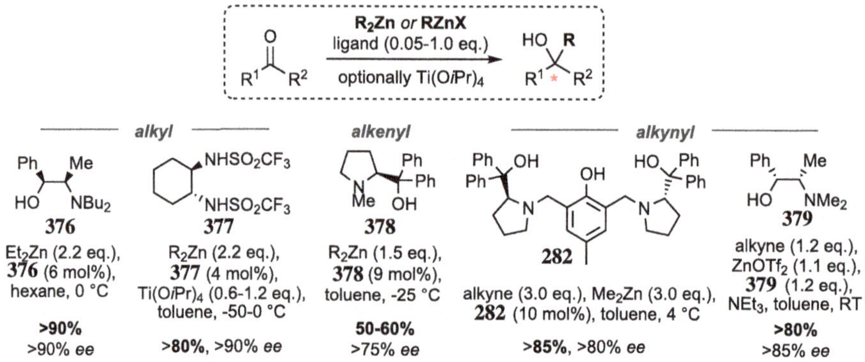

Fig. 2.163 Selection of Zn-mediated asymmetric carbonyl addition reactions [518–524]

The mechanism of the Ti-mediated 1,2-addition has been extensively studied and deviates significantly from the mechanism postulated for most Zn/ligand combinations, which is supposed to proceed via analogous species of **374** and **375**. Instead, the nucleophilic residue is transferred to the bimetallic complex **380**, presumably via a Ti(O*i*Pr)$_3$R species. After coordination of the substrate to **381**, a four-membered transition state **382** is assumed. The exact coordination environment of the Ti atoms has not yet been fully elucidated and seems to vary depending on the ligand (see Fig. 2.164) [525, 526].

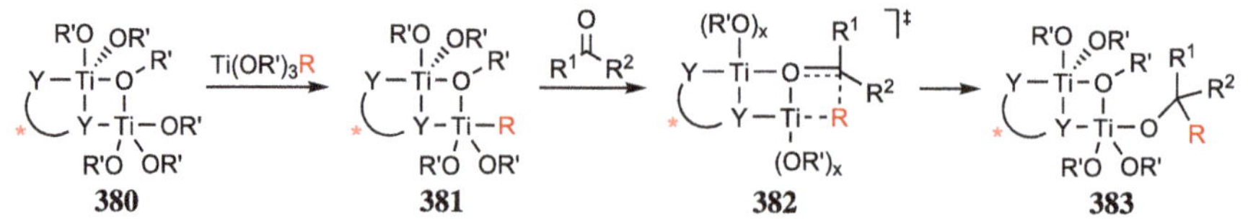

Fig. 2.164 Postulated mechanism in the presence of Ti(O*i*Pr)$_4$ [525, 526]

While aldehydes undergo additions under very mild conditions, ketones or esters usually require stronger activation, which is why the use of Ti(OiPr)$_4$ is often observed for these classes of substrates. In the synthesis of complex natural products, Zn-mediated carbonyl additions are usually limited to aldehydes as reactants (see Fig. 2.165).

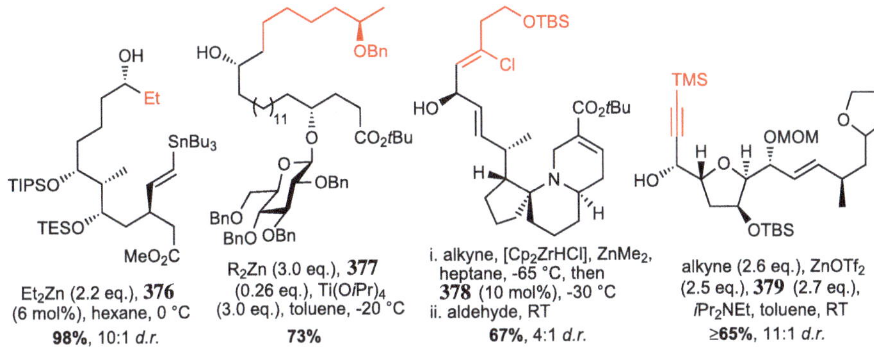

Fig. 2.165 Applications of Zn-mediated additions to aldehydes [25, 527–529]

Organochromium reagents were used by Nozaki and Hiyama in the late 1970s to promote the addition of allylic halides to aldehydes. Later work by Kishi and Nozaki showed that the addition of nickel(II) salts further allowed the conversion of less reactive vinyl and aryl halides. The *Nozaki-Hiyama-Kishi reaction* (NHK reaction) is extremely selective for aldehydes and can be carried out in the presence of ketones, esters, amides, nitriles, acetals, alcohols, and olefins [530]. The chromium(II) salts must be used stoichiometrically to enable the oxidative addition into the carbon-halogen bond as a reducing agent (conditions A). Due to its high toxicity, methods have been developed requiring only catalytic amounts of chromium by using an additional reducing agent (manganese metal, conditions B) [531–533]. Under Cr/Ni catalysis, asymmetric variants have also been reported that rely on chiral ligands such as sulfonamide **384** (see Fig. 2.166) [501].

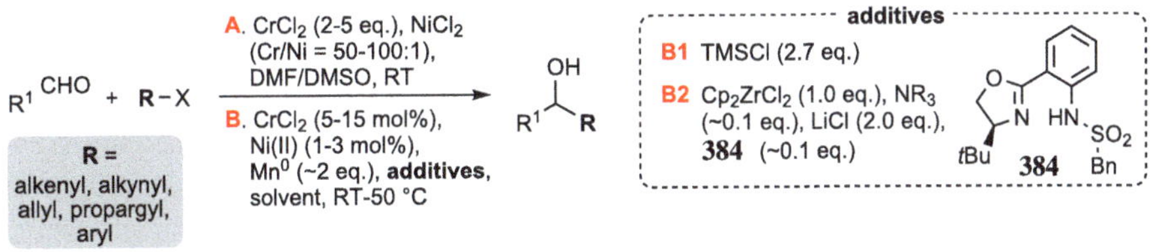

Fig. 2.166 Typical conditions of the NHK reaction using stoichiometric (A) or catalytic amounts of CrCl$_2$ (B)

The driving force of the reaction lies in the formation of stable metal alkoxylates or silyl enol ethers. Under stoichiometric conditions, at least two equivalents of $CrCl_2$ are required in the absence of a secondary reductant, as Cr(II) is a one-electron donor and the oxidative insertion into the carbon-halogen bond ($Ni^0 \rightarrow Ni^{II}$) is a two-electron process. There are several interdependent catalytic cycles, which is why a finely tuned reaction system with suitable redox potentials is required. Initially, Cr(II) is oxidized to Cr(III) and reduces Ni(II) to Ni(0). The low-valent Ni species can insert into the CX bond of the electrophile (**385**) and the vinyl-Ni species transmetalates to form the Cr(III)-vinyl complex **386**. After addition to the aldehyde, Cr(III)-alkoxylate **387** is obtained. The addition of a trapping agent (TMSCl or Cp_2ZrCl_2) releases the Cr(III) intermediate to form **388**, which can again enter the primary or secondary catalytic Cr-cycle. The work-up subsequently affords the desired alcohol from **388** either via hydrolysis or by using a fluoride source. Under stoichiometric conditions, the regeneration of Cr(II) in the primary Cr-cycle does not take place and the secondary cycle stops at **387**, from which the product is likewise obtained by hydrolysis (see Fig. 2.167) [534].

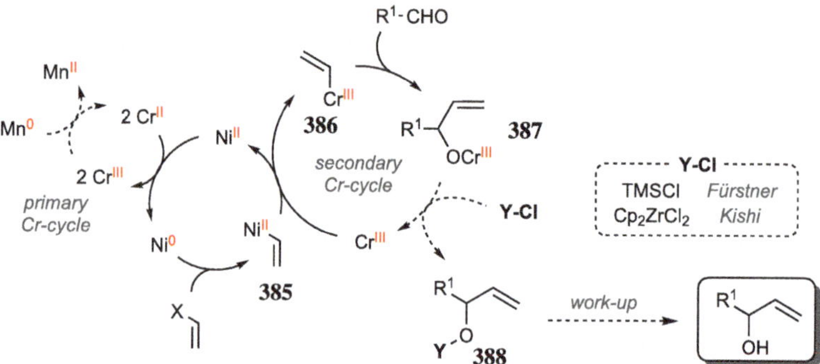

Fig. 2.167 Postulated mechanism of the NHK coupling [534] Dashed steps do not take place under stoichiometric conditions

The reaction generally tolerates traces of water due to the low polarity of the C-Cr bond. However, its slow protolysis and the competing reduction of various functional groups by $Cr(II)/H_2O$ still require aprotic reaction conditions. $CrCl_2$ is only slightly soluble in organic solvents such as THF or DME, but the addition of LiCl increases its solubility. More commonly, the use of more polar solvents such as DMF or DMSO obviates the use of LiCl as additive.

The addition to α-chiral aldehydes typically proceeds without noticeable involvement of chelated intermediates, favoring the Felkin-Anh addition products. Intramolecular reactions and α,β-chiral substrates may however deviate from this preference. To explain the simple diastereoselectivity, a reaction via a Zimmerman-Traxler-like transition state was postulated (see below). When using crotylchromium reagents, the *anti*-configured addition products are isolated, regardless of the configuration of the employed crotyl halides. This outcome was rationalized by the existence of a pre-equilibrium to the more reactive E-isomer prior to the addition [530].

The Nozaki-Hiyama-Kishi coupling has established itself as one of the standard methods for the preparation of allylic alcohols from densely functionalized substrates [535, 536]. In addition to alkenyl, alkynyl, and aryl halides, alkyl iodides like iodoform can also be converted. With alkenyl halides and α,β-unsaturated aldehydes, the configuration of the double bond remains intact over the course of the reaction, and epimerizations of CH-acidic α-carbonyl positions are extremely rare [530]. Despite the development of catalytic conditions by Fürstner and Kishi, a superstoichiometric use of $CrCl_2$ is still common practice in total syntheses (see Fig. 2.168). An electrochemical variant has also been reported by Reisman, Baran and Blackmond [537].

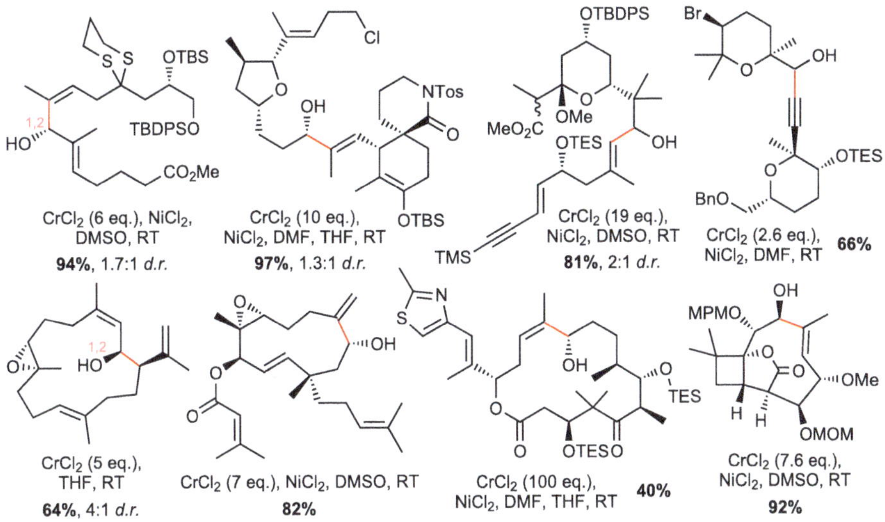

Fig. 2.168 Application of the Nozaki-Hiyama-Kishi reaction in total syntheses [538–545]

Kishi more recently disclosed a Ni-catalyzed NHK-type reaction that employs $Cp_2Zr(II)$ for the activation of alkyl iodides, presumably proceeding via **392** and **393**. Thioesters constitute the electrophilic component, allowing for the one-pot synthesis of ketones without the need for Weinreb amides (*vide infra*) [546]. Eisai employed this novel method in their gram-scale approach to the anticancer drug candidate E7130 (see Fig. 2.169) [547].

As opposed to the previously discussed 1,2-selective methods, the nucleophilic addition of allyl groups seems to allow only a very limited range of accessible product structures and thus, a narrow application scope. The allyl group is in fact a very useful structural element, in which the double bond can be further derivatized by oxidative processes or an alkene

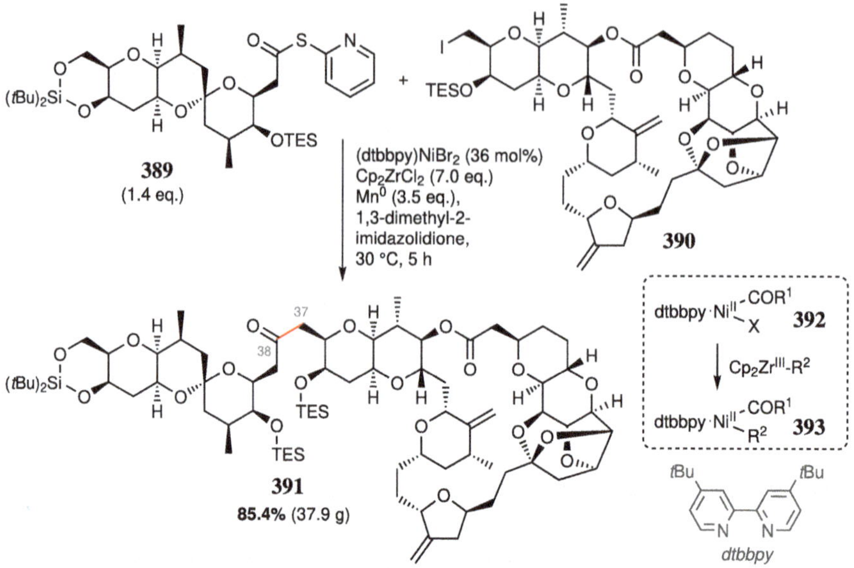

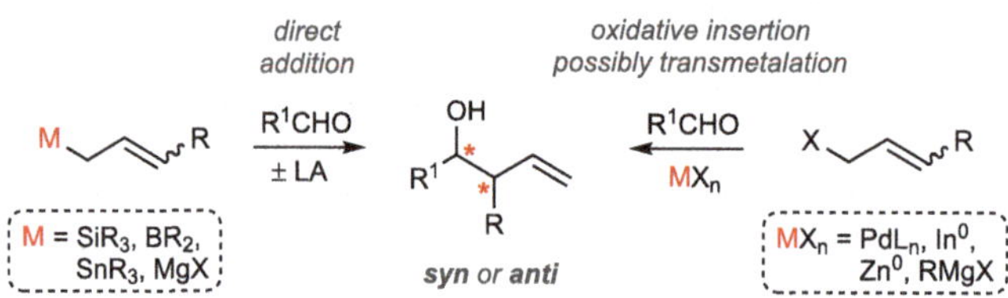

Fig. 2.169 Eisai's approach to E7130 [547]

metathesis. From this perspective, an *allylation* is typically used to introduce a stereocenter diastereo- or enantioselectively and subsequently convert the resulting homoallylic alcohol. In addition to the direct use of allylic organometallic reagents (allyl silanes, stannanes, boranes) under activation of the carbonyl group with Lewis acids, the nucleophiles can also be formed *in situ* from allyl halides and metal salts (see Fig. 2.170) [499, 548, 549].

Fig. 2.170 Approaches to the allylation of carbonyl derivatives

The principles of stereoinduction by internal or external control elements that have been previously discussed for aldol reactions also apply to allylations: In addition to simple diastereoselectivity (*syn/anti*) when using crotyl nucleophiles, stereoinduction can occur through chiral carbonyl electrophiles, chiral nucleophiles, or through chiral Lewis acids and chiral ligands. Both diastereoselective (especially allyl boranes) and stereoconvergent reactions (allyl silanes, allyl stannanes) have been reported. Cyclic (**394, 395**) or acyclic transition states (**396, 397**, ...) have been postulated to rationalize the observed stereoselectivity (see Fig. 2.171) [499].

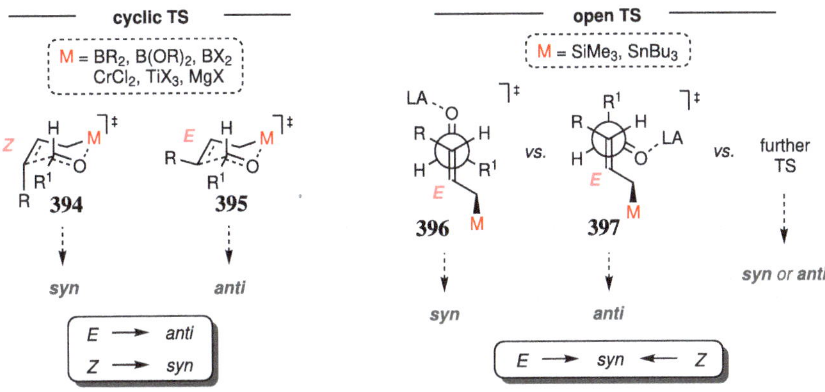

Fig. 2.171 Postulated transition states of the allylation of aldehydes [499]

Boron-based allyl reagents usually exhibit excellent stereoinduction due to the rigid, cyclic transition states. Both crotyl boranes and boronates are synthetically easily accessible and commercially available. The NHK reaction was first carried out with allyl halides as nucleophilic precursors, so that allyl chromium reagents can also be directly used in the NHK coupling. A chair-like transition state is assumed, but since the Cr reagent equilibrates to the *E*-configured diastereomer under reaction conditions, the reaction becomes *anti*-selective (via **395**). Sn- and especially Si-based nucleophiles have also been developed and established in synthesis. The diastereoselectivity of the reaction is usually *syn*-selective, regardless of the configuration of the starting material. The rationale can be derived from the acyclic transistion states: The orientation of the starting materials can either be antiperiplanar with respect to the CC- and CO-double bonds (cf. **396**) or aligned in a (+)- or (−)-synclinal manner (cf. **397**), with an antiperiplanar orientation delivering the *syn*-configured products. Depending on the substrate in question, other transition states with different arrangements may also become energetically viable. For a *quantitative* prediction of selectivity, several competing pathways need to be considered and included in the calculation [550, 551]. The facial discrimination of α-chiral carbonyl substrates in the reaction with various organometallic compounds (M = B, Si, Sn, In, Zn, Ti) is usually based on the minimization of steric and electronic interactions according to the Felkin-Anh model.

Unlike other Grignard reagents, allyl magnesium nucleophiles are often not very selective or counterintuitively display the reverse stereoselectivity compared to other nucleophiles in the conversion of carbonyl compounds. Especially the addition to α-chiral aldehydes is extremely difficult to predict (Felkin-Anh, Cram selectivity) and in many cases only affords moderate selectivities. The reasons for this have been extensively studied and seem to be due to a combination of different causes (chelate formation, diffusion control, reversibility of the reaction, etc.) (see Fig. 2.172) [35].

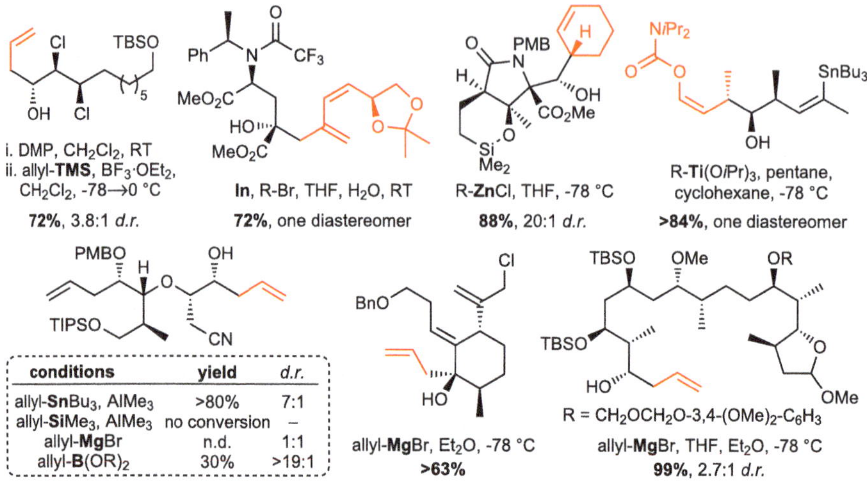

Fig. 2.172 Examples of diastereoselective allylations of complex substrates [552–558]

The reactivity of the various allyl reagents has been well studied using theoretical and experimental means. Organozinc compounds have the highest nucleophilicity, followed by stannanes, silanes, and boronates (see Fig. 2.173) [559, 560]. While stannanes add to carbonyls at only slightly elevated temperatures, allylsilanes do not react at all in the absence of strong Lewis acids. The reactivity of neutral B-allyl reagents does not match the order of their nucleophilicity shown in Fig. 2.173 — they are significantly more reactive. The boron atom activates the carbonyl group by coordination which in turn activates the boron center, so that the reactivity is in fact closer to that of anionic boranate complexes.

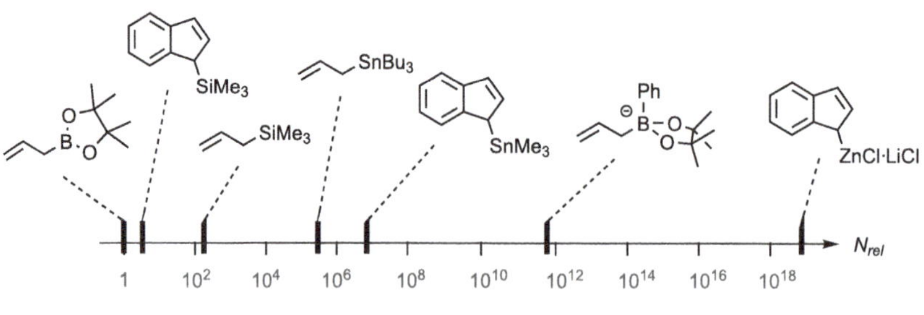

Fig. 2.173 Relative nucleophilicity of various allyl reagents [559, 560]

The stoichiometric use of chiral allyl reagents quickly enabled an asymmetric allylation reaction. Especially the allyl boranes introduced by the groups of Roush and Brown have shown themselves to be very reliable even in advanced stages of synthetic endeavors. Subsequent developments by Leighton, Corey and others were also able to make valuable contributions and have proven their worth (see Fig. 2.174).

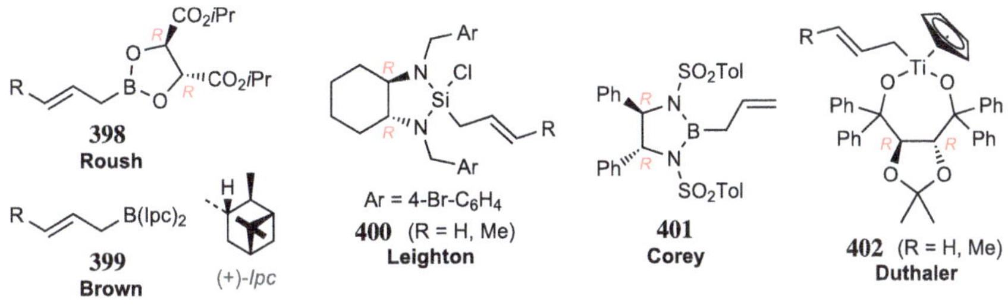

Fig. 2.174 Chiral allylation and crotylation reagents

The high diastereoselectivity of chiral allylation and crotylation reagents is usually attributed to the adoption of chair-like transition states. The Roush allyl boronates (**398**) show a destabilizing e^-/e^- repulsion in the energetically disfavored transition state **404**, which is avoided in the favored transition state **403** [561]. In the Brown allylation, the facial differentiation in the addition of **398** to the aldehyde is based on the minimization of steric interactions ($A^{1,3}$ strain, cf. Fig. 2.102) [562]. The chlorosilanes **400** developed by Leighton can also enable cyclic transition states by coordination of the Lewis-acidic silane to the carbonyl bond (see Fig. 2.175) [563].

While there are numerous methods with broad substrate scope available for the enantioselective allylation of aldehydes with chiral boron reagents, there are significantly fewer examples for the allylation of ketones. These are usually limited to the conversion of aromatic or α,β-unsaturated ketones. Some challenges of asymmetric, catalytic variants are the occasionally unpredicted stereoselectivity and the competing stereoinduction of the chiral substrates when bearing stereogenic centers, which is more the rule than the exception in complex natural product syntheses.

In addition to the protocols relying on a stoichiometric use of chiral allylation reagents, the development of asymmetric, catalytic conditions have also been reported to successfully furnish high stereoinduction. Allyl silanes (*Hosomi-Sakurai allylation*),[568] -stannanes or -boronates are typically utilized as nucleophiles, which add to carbonyl group in the presence of chiral Lewis acids as activators [549, 569, 570]. Due to the myriad of published methods, only the synthetically more frequently used approaches can be highlighted herein. The first widely applied catalyst system based on acyloxyboranes was reported by Yamamoto and co-workers (CAB catalyst, **407**) [571, 572]. The use of Ti-BINOL complexes was independently developed by the groups of Umani-Ronchi and Keck [573, 574]. The reaction now known as

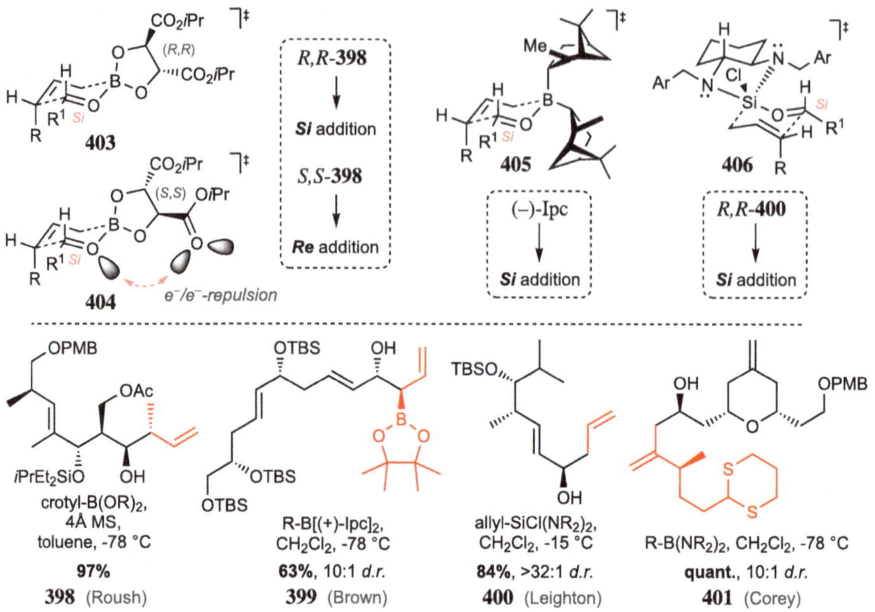

Fig. 2.175 Postulated transition states of the chiral allylation reagents **398–400** and selected applications [561–567]

Keck allylation has proven itself in highly complex substrates and is regarded as a cornerstone of asymmetric allylation methods [575]. More recent studies from the laboratories of the Schaus and Hall groups have introduced chiral diols as stereoinductors for the addition of allyl nucleophiles to carbonyls [576–579]. For this purpose, either BINOL derivatives (**408**) or aliphatic biaryl diols (e.g., **409**) are used as ligands in combination with $SnCl_4$ (see Fig. 2.176).

The mechanisms of the listed methodologies are more or less well understood. While only basic considerations were published for Yamamoto's CAB system which rationalize the stereoselectivity on the basis of an acyclic transition state,[572, 580] in-depth studies were undertaken for the elucidation of the underlying mechanism for the allylation methods of Keck, Schaus and Hall. The mechanistic details of the Keck allylation are only partially understood today and sometimes contradict each other. While an intermediate Ti-allyl-aldehyde complex was postulated to explain the observed selectivity (**410**) [580], mechanistic investigations indicated an **inter**molecular allylation involving a dimeric, bimetallic complex, which has not been further characterized to date [549, 581–583]. The addition of undried molecular sieve as a water reservoir was identified as being critical for an efficient and selective reaction [584]. In the Schaus system, an alkoxy group of the allylboronate is substituted by the diol **408** to give **411**. In the presence of a Brønsted acid for activation of the boronate, the homoallylic alcohol can be obtained, presumably via a cyclic transition state

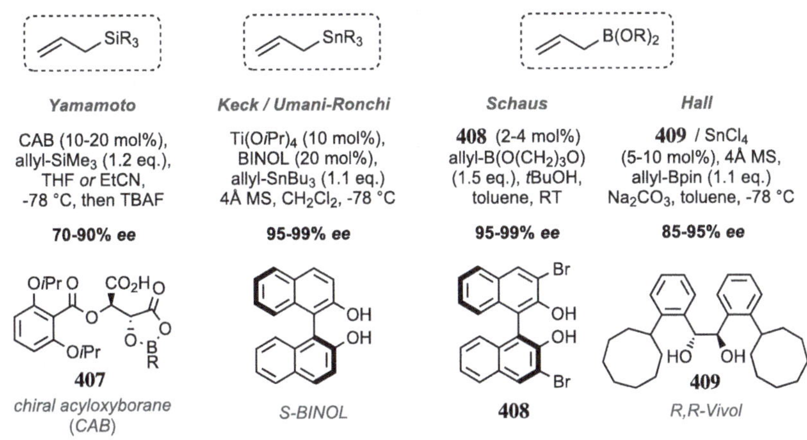

Fig. 2.176 Selection of asymmetric, catalytic allylation reactions

412. The release of the chiral diol from the addition product constitutes the rate-determining step of the catalytic cycle, which is why the addition of *t*BuOH leads to an overall acceleration of the reaction [577]. The combination of weakly acidic diols such as Vivol (**409**) with Lewis acidic metal complexes increases the acidity of the hydroxy protons (**413**) upon coordination to the metal center. In the postulated transition state **414** a Brønsted acid-mediated activation occurs similar to the Schaus allylation, whereby the staggered complex **413** presumably activates the boronate reagent by coordination to one of the alkoxy substituents of the boron atom (see Fig. 2.177) [579].

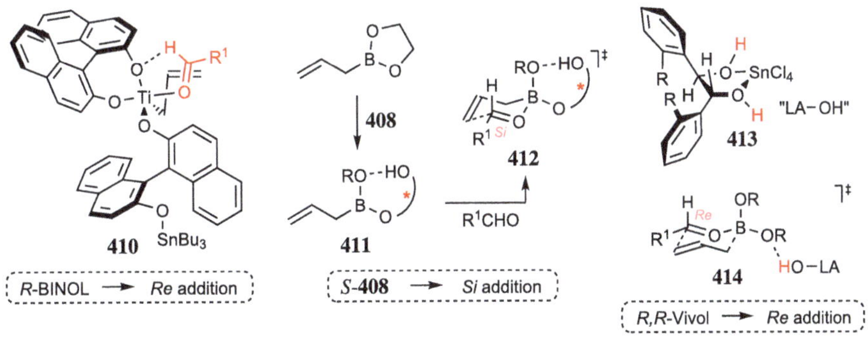

Fig. 2.177 Postulated key intermediates and transition states in the reaction mechanisms of the allylations by Keck, Schaus, and Hall [577, 579, 580]

In addition to simple allyl nucleophiles, more complex allyl moieties can also be introduced via the corresponding organosilanes, -stannanes, and -boronates. In combination with the highlighted enantioselective allylation methods, a high complexity and density of functional groups can thus be efficiently incorporated in the products (see Fig. 2.178).

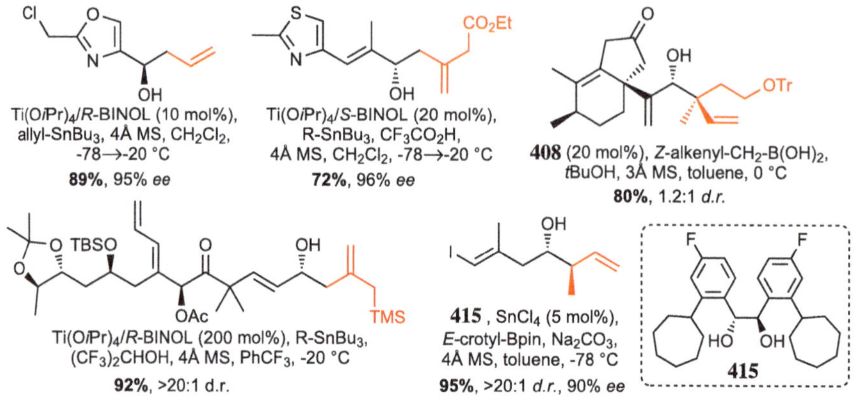

Fig. 2.178 Use of asymmetric allylations in natural product syntheses [585–589]

Beyond these, further approaches for the enantioselective addition of allyl nucleophiles exist. Allyltrichlorosilanes can also undergo asymmetric allylation reactions in the presence of chiral Lewis bases. However, the method pioneered by Denmark and Hayashi is rarely used to access complex intermediates due to the high reactivity of the chlorosilanes and their difficile handling. Organocatalytic conditions with chiral phosphoric acids have been successfully used as well in enantioselective reactions [590]. In addition, copper complexes can similarly be utilized for the allylation of carbonyls by boronate-based nucleophiles (Shibasaki, Kanai, Hoveyda) [569].

π-allyl complexes can serve as electrophiles, the most prominent example possibly being the Tsuji-Trost reaction. In contrast, their use as nucleophiles is just as desirable and can be achieved in several ways [591]. Covalently bonded allyl metal reagents such as allyl stannanes are possibly the most common approach to utilize the innate reactivity of the allyl anion. Alternatively, Pd-π-allyl complexes, which are accessible from allyl electrophiles in the presence of a Pd catalyst, can be transmetalated *in situ* with stoichiometric Et$_2$Zn to form a nucleophilic allyl zinc species. The third approach is a transition metal-catalyzed transfer hydrogenation. This methodology was extensively investigated in the Krische group and successfully used as a key step in several natural product syntheses. The *Krische allylation* is based on the CH activation of an alcohol in the presence of iridium complexes and the transfer of the proton to the leaving group of the nucleophile donor. The charm of the method lies in the fact that only allyl acetates are needed as masked nucleophiles. The preparation and isolation of organometallic reagents is thus omitted, which increases the overall atom economy (see Fig. 2.179) [592].

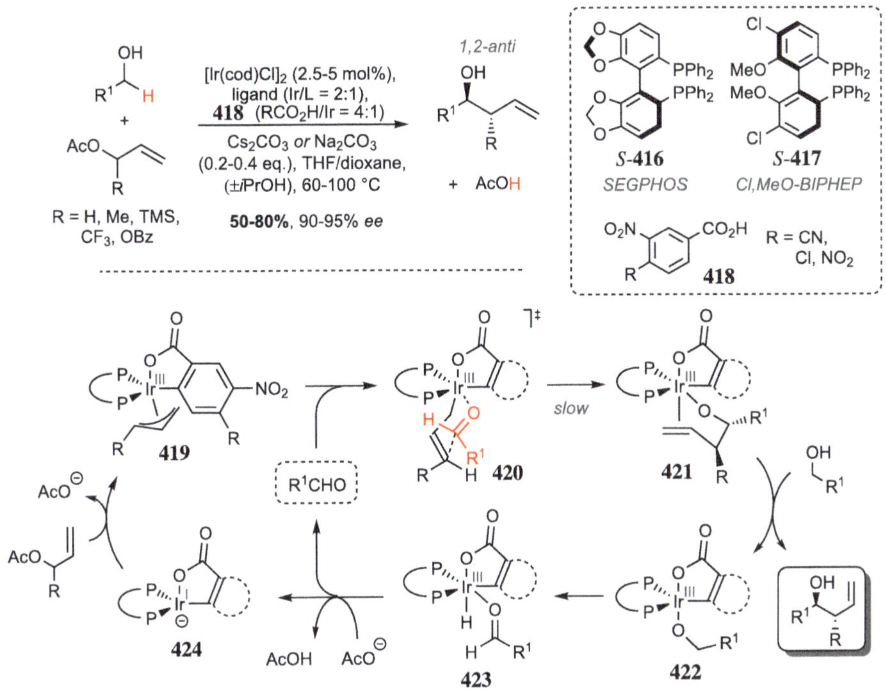

Fig. 2.179 Reaction conditions and postulated mechanism of the Krische allylation of alcohols [592, 593]

The mechanism was studied in depth by Krische and co-workers [593]. After the formation of the catalytically active species **419** from [Ir(cod)Cl]$_2$, an equivalent of alcohol is first oxidized for initiating catalytic turnover. This presumably occurs by substitution of the allyl substituent to form **422**, from which the requisite aldehyde is generated (via **423**→**423**→**424**). Starting the catalytic cycle from **419**, the just-generated aldehyde is coordinated and the allyl residue adds to the carbonyl group via a chair-like transition state **420**. In the case of unsymmetric allylic acetates, the η^3-allyl ligand is in equilibrium with the σ-bound allyl complex (σ-π-σ-equilibration), in which the R group is favored to lie at the opposite end of the allyl system where it is bound to the metal center [594]. The homoallyl alcoholate resulting from the addition chelates the metal center through bidentate attachment (**421**), and the desired product can be liberated by coordination of one equivalent of the alcoholic substrate. The alcohol is subsequently oxidized to the aldehyde and the Ir-hydride species **423** can reductively eliminate the carbonyl, while a base takes up the proton. The 16-electron Ir(I) complex **424** can regenerate the initial species **419** by reaction with an allyl acetate to enable reentry into the catalytic cycle. The addition to the carbonyl group via **420** was identified as the rate-determining step.

If aldehydes are used instead of alcohols, an external reducing agent is required to access the intermediates **421–423** in the catalytic cycle. For this purpose, isopropanol is added in superstoichiometric amounts, as the derived acetone is too unreactive for allylation under the reaction conditions. For more challenging substrates, the catalytically active species **419** is used directly instead of the precatalyst [Ir(cod)Cl]$_2$ [592]. Ketones are mostly inert towards allylation with the Krische protocol, so aldehydes can be selectively converted in the presence of other carbonyl functionalities. A strong preference for primary over secondary alcohols is observed based on the different redox potentials and the sluggishness of ketone allylations. Nitrogen-containing heterocycles and a range of sensitive functionalities such as halides, nitriles, ketones, (sulfon)amides and thioethers are also tolerated without noticeable degradation (see Fig. 2.180) [595, 596].

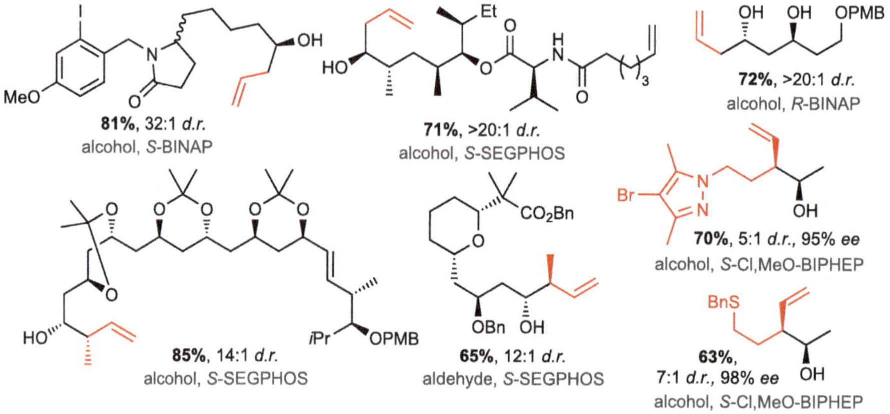

Fig. 2.180 Examples of the Krische allylation indicating the oxidation state of the substrate and the ligand used [597–602]

Beyond the combination of carbonyl and organometallic compounds typically used in the Krische allylation, the substrate scope can also be extended to include dienes, allenes, propargyls and enynes, and include other transition metals such as Rh [603, 604].

In addition to the formation of CC bonds by addition to carbonyls, hydrides can be similarly used as nucleophiles under reductive conditions. Most common metal hydrides direct towards a 1,2-reduction, especially mild reagents such as borohydrides. In case a desired reduction of a complex intermediate is unselective and/or overhydrogenation cannot be suppressed, this calls for tailor-made solutions. The only method specifically developed for the 1,2-reduction of enones and enals is the *Luche reduction* ($NaBH_4/CeCl_3$), in which the carbonyl group is activated by $CeCl_3$ (see Sect. 4.2.1). The 1,4-selective counterpart to the Luche reduction is the use of copper hydrides, which will be discussed in more detail in the following section.

2.5.2   1,4-Addition

The seminal work by Kharasch revealed early on that the regioselectivity in the conversion of enones can be shifted in favor of the Michael addition product by the addition of catalytic amounts of copper salts (see Fig. 2.155) [490]. The combination of copper, classified as "soft" according to HSAB theory, with different secondary cations (Li, Mg, Zn, etc.) and additional ligands to give cuprates made possible a large array of impressive Michael additions with highly functionalized substrates (see Fig. 2.181).

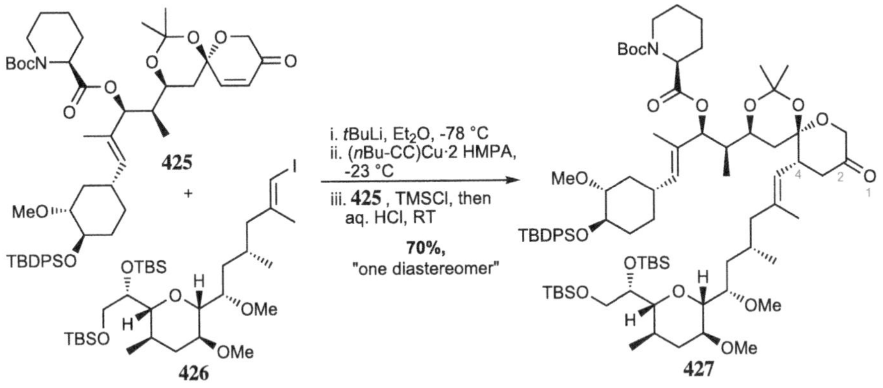

Fig. 2.181 Michael addition in Ireland's synthesis of FK-506 [605]

The stoichiometry, structures, and aggregation state of the standard Cu(I) reagents used for 1,4-additions are exceedingly complex and can vary greatly under the reaction conditions, highly depending on the solvent and additives. The nucleophilic organocopper(I) reagents can be described as neutral organocopper species RCu, as well as organocuprates of the composition R_2CuM (homocuprate) or $RCu(X)M$ (heterocuprate). Cuprates of the type R_2CuLi are commonly referred to as *Gilman* cuprates, whereas the corresponding Mg derivatives are termed *Normant* cuprates. Organocuprates which result from the reaction of CuCN with one or two equivalents of organolithium reagents were originally categorized as cuprates of lower and higher order. However, recent studies have shown that the majority of higher-order cyanocuprates does not contain a triple-coordinated, dianionic species, but consist only of an R_2Cu^- species with CN^- outside of the coordination sphere of the copper atom. They are therefore now referred to as cyano-Gilman cuprates. The structures of the cuprates are typically linear, but during the addition they take on a bent structure to alter the orbital symmetry and allow the involved orbitals (d_{xz}-π^*) to interact. Alkynyl and cyanide ligands can be additionally coordinated at the π-system in an η^2 mode by the secondary cation (**428**). In solution, cuprates mainly exist as dimeric species (**429**, see Fig. 2.182) [606, 607]. The behavior of magnesium cuprates differs considerably from the extensively studied lithium cuprates, as they are less stable and are more hesitant to form heteronuclear species with cuprate anions [608].

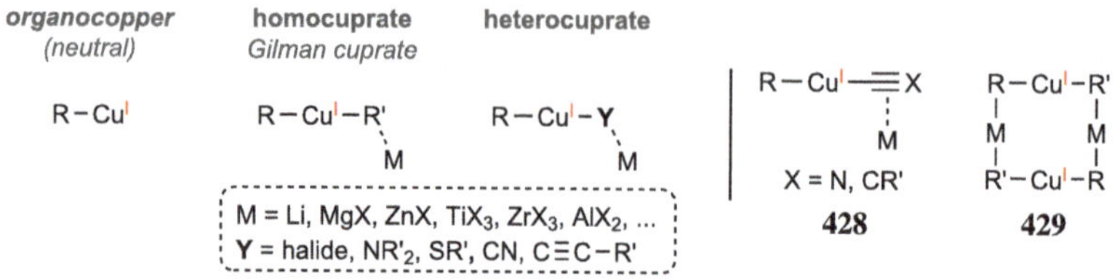

Fig. 2.182 Overview of reagent classes and structures of organocopper compounds (coordinated solvents not shown) [606]

The C-C bond formation involves the nucleophilic attack of the Cu(I) center on the electrophile (oxidative addition) and the reductive elimination of the resulting Cu(III) species [609]. The addition is controlled by the orbital interaction of the cuprate with the electrophile and the coordination of the other Lewis-acidic cations to the substrate. The reductive elimination is often rate-determining as well as the regio- and stereochemistry-defining step of the sequence (see Fig. 2.183) [606].

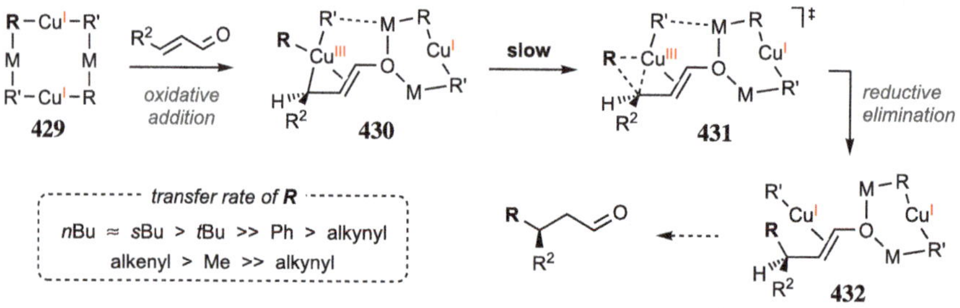

Fig. 2.183 Mechanism of the 1,4-addition with cuprates and relative transfer rate of different ligands from the Cu atom [606, 610–613]

The extensive experimental and theoretical studies indicate a reversible formation of the Cu(III)-enone complex **430**[609] from the dimeric precursor **429** as the dominant mechanism, while the secondary cation M (e.g., Li) complexes the oxygen of the carbonyl substrate, thereby activating it. The unimolecular transfer of the R ligand via transition state **431** under concomitant regeneration of the Cu(I) species affords **432**, which furnishes the desired addition product. The details of the aggregate structures of complexes **430–432** remain to date unclear [606, 610]. In homocuprates, only one ligand R is transferred, while the other is lost as [R'Cu]. Heterocuprates can selectively transfer only the desired R group when containing a non-transferable ligand Y. Cyanide or alkyne substituents are typically used for this purpose, as the latter display a negligible transfer tendency in the presence of other substituents such as alkenyl, alkyl or aryl groups [611–613].

The reactivity of cuprates is not only determined by the ligands but also by the identity of the secondary cation M. Zinc cuprates show a high tolerance towards functional groups, but the reaction may require additional activation to proceed. Al-cuprates are more reactive and allow the conversion of even challenging substrates, while Mg-cuprates, due to their high reactivity, are only suitable for use in uncatalyzed reactions [614].

With regard to the substrate, the stereoelectronic properties of the carbonyl can likewise enhance or diminish the reactivity and selectivity. The presence of β-substituents as well as α,β-unsaturated esters and amides slow down the addition. While an optimal combination of reactivity and 1,4-selectivity is observed with ketones, a partial 1,2-addition can also occur with more reactive enals. Among the transferrable cuprate substituents, alkyl groups are particularly 1,4-selective, followed by the slightly less reactive alkenyl and aryl groups. Alkynes are mostly inert and are therefore often used as "dummy" ligands.

The development of methods to increase reactivity, selectivity and atom economy has brought about some important modifications: The addition of TMS-X (X = Br, Cl) increases the reactivity and 1,4-selectivity in the Michael addition. The cause of this effect is still controversially discussed [606]. Recent experimental studies have indicated a silylation of the Cu(III)-enone complex **430** to **435** prior to the transfer of the ligand R and reductive elimination as the cause (see Fig. 2.184) [615–617].

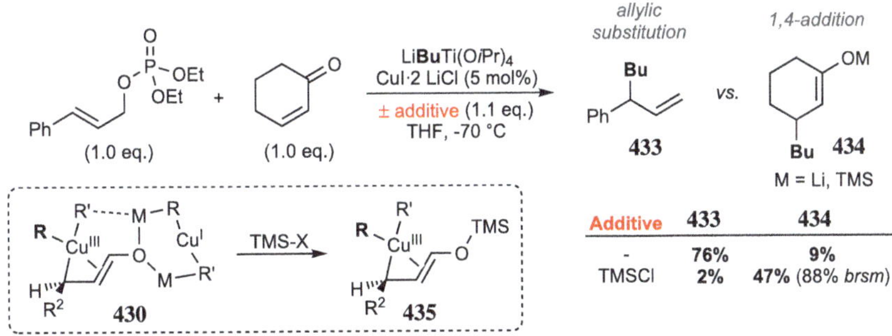

Fig. 2.184 Effect of added TMSCl on chemoselectivity [146, 615–617]

In the presence of BF_3, a significant rate acceleration of the 1,4-addition is also observed. Smith and co-workers used this to enable the sluggish conversion of the Gilman cuprate with the β-disubstituted cyclic enone in their synthesis of the sesquiterpene modhephene [618]. The effect was postulated to rely on the activation of the Cu(III)-enone complex **430** by BF_3, by bringing the charge distribution of **436** closer to the transition state of the reductive elimination (see Fig. 2.185) [619]. Nevertheless, a complexation of the carbonyl group cannot be ruled out as an alternative explanation [606].

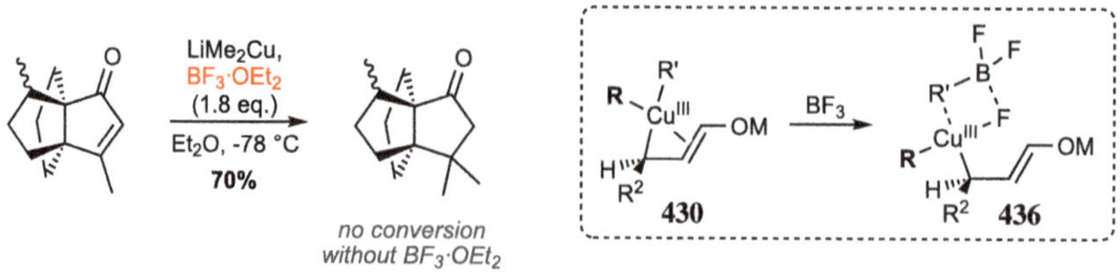

Fig. 2.185 Activation through BF$_3$ addition in Cu-mediated Michael additions [618, 619]

The achievable complexity of the organometallic nucleophiles is primarily determined by the functional group tolerance during their formation. Accordingly, organolithium and organomagnesium compounds are structurally limited. Zinc- or zirconium-based protocols could however be used in some complex natural product syntheses, allowing the preparation of organometallic reagents that permit an efficient access to the desired Michael products in (see Fig. 2.186) [620].

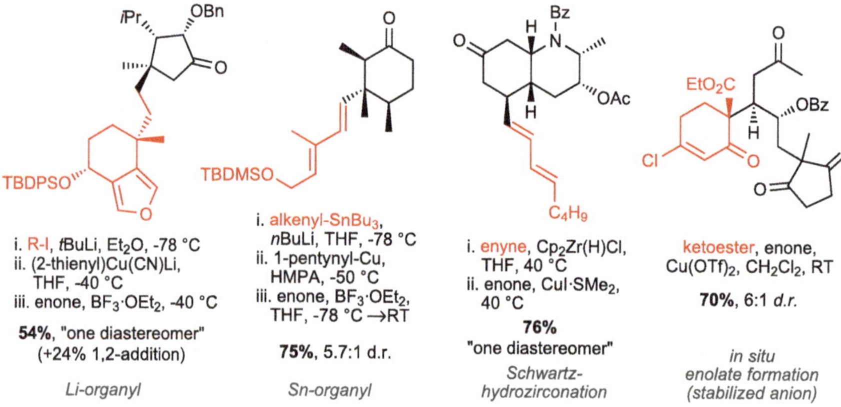

Fig. 2.186 Use of Cu-mediated Michael addition in total syntheses and indication of the precursors of the Cu organyl [497, 621–623]

Enantioselective variants of Cu-catalyzed 1,4-additions were also reported early on in addition to the developed diastereoselective methods. The possibility of a conversion relying on catalytic amounts of copper initiated the race to identify the ideal ligands for this transformation, which would enable both high yields as well as good stereoinduction. Particularly monodentate phosphoramidites as well as bidentate phosphines (P,P), (mostly) dimeric phosphites (P,P), iminophosphines (P,N) and N-heterocyclic carbenes (NHCs) are frequently encountered ligand classes in this context. The nature of the organometallic nucleophile plays an equally pivotal role, as its reactivity must be adapted to the substrate to avoid significant (unselective) background reactions. Thus, there is no universally valid method, but rather different substrate classes (α,β-unsaturated aldehydes, ketones, esters, amides, etc.) require tailor-made solutions (see Fig. 2.187) [614, 624, 625].

	R-ZnX, R_2Zn	AlR_3, AlR_2X, $AlRX_2$	R-MgX
α,β-unsaturated substrates	enal enone (cyclic & acyclic) nitroalkene	β-substituted enone (cyclic & acyclic)	α,β-unsat. ester α,β-unsat. thioester β-substituted enone
typical ligands	phosphoramidite iminophosphine phosphite NHC	phosphite phosphoramidite	bis-phosphine

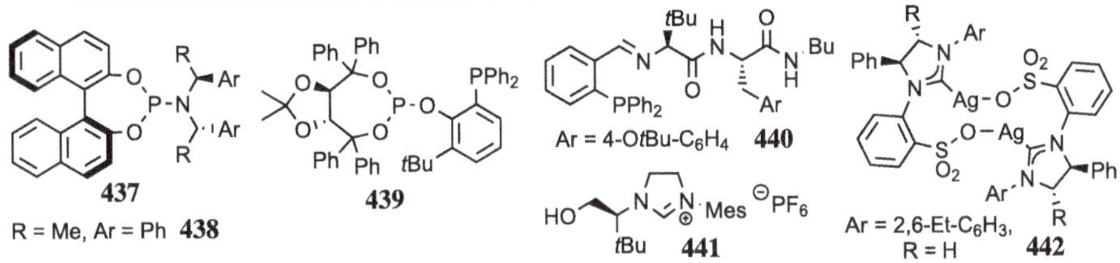

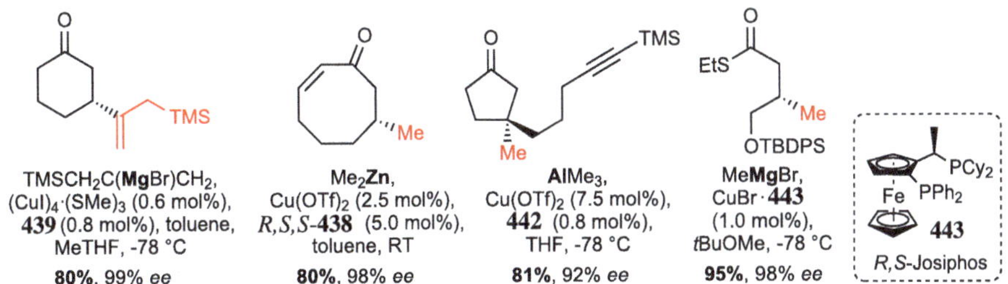

Fig. 2.187 Guide for matching the reaction conditions in asymmetric 1,4-additions to the substrates [614, 624, 625]

The typically employed copper precatalyst is $Cu(OTf)_2$, but Cu(I)-thiophene carboxylate (CuTC) as well as Cu(I) halides and Cu acetylacetonate are also utilized. Since most methods use cyclohexenone or cyclopentenone as reference substrates, cyclic enones lend themselves as target motifs for use of asymmetric 1,4-additions in the context of total synthesis. Outside of this structural pattern, there are hardly any examples of enantioselective methods to access complex intermediates (see Fig. 2.188) [614, 625, 626].

Fig. 2.188 Examples of 1,4-additions in total syntheses [627–630]

In addition to copper, other metals (Ni, Ru, Rh, Pd, La, Al, etc.) as well as organocatalysts (see Sect. 8.1) are capable of enabling an enantioselective reaction in Michael additions. With C-nucleophiles, acidic dicarbonyl compounds such as β-keto-carbonyls play a prominent role. In addition, thiols or other heteroatom nucleophiles (N, O, B, etc.) can also be used in a 1,4-addition with α,β-unsaturated carbonyls (see Fig. 2.189) [503, 631].

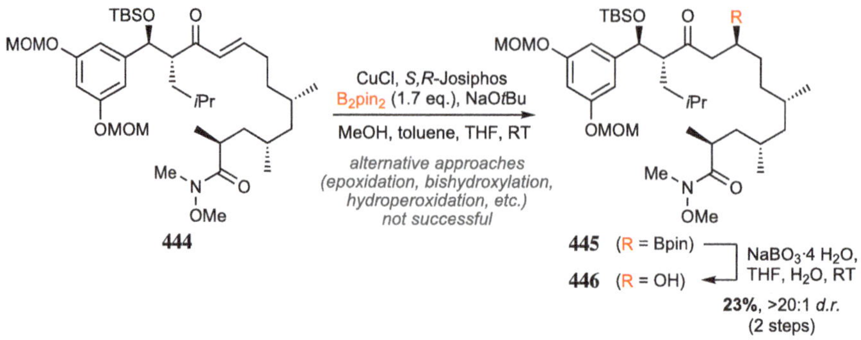

Fig. 2.189 1,4-Borylation and subsequent oxidation in the synthesis of baulamycin A according to Williams *et al.* [632]

In addition to the CC- and C-heteroatom-coupling with various nucleophiles, α,β-unsaturated carbonyls can also be converted into the corresponding saturated products by reducing them with metal hydrides. While a multitude of methods exist for the reduction of non-conjugated carbonyl compounds to the respective alcohols or aldehydes (see Sect. 4.2), methods for selective 1,4-reduction are encountered significantly less often. Due to their outstanding reaction profile in Michael additions, 1,4-selective reductions typically rely on copper hydrides as reductants [500, 633–635]. The first studies employed stoichiometric amounts of copper hydrides, of which the hexameric $[\text{CuH(PPh}_3)]_6$ (*Stryker's reagent*) has seen the most synthetic use (see Fig. 2.190).

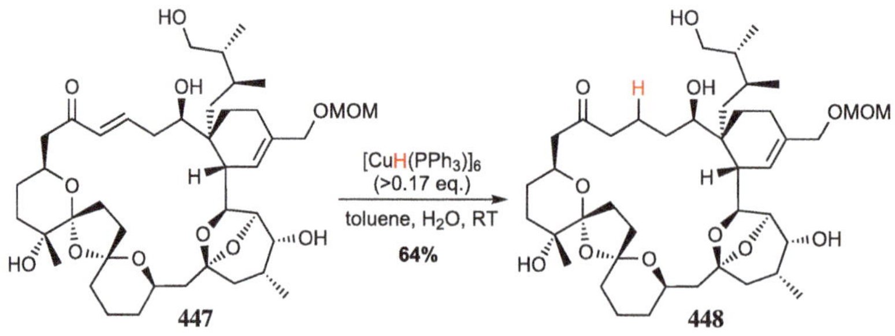

Fig. 2.190 Excerpt from Hirama's synthesis of (+)-pinnatoxin A [636]

The tolerance towards potentially affected functional groups (acetals, epoxides, silyl ethers, alkenes, alkynes, esters, free alcohols) is high. Even ketones are usually inert under the reaction conditions. When using catalytic amounts of Cu and the hydride being formed *in situ*, silanes are employed as stoichiometric reductants, with phosphines and occasionally *N*-heterocyclic carbenes serving as ligands. Depending on the silane reductant, the presence of alcohols additives and the work-up, either silyl enol ethers (**449**) or the free carbonyl (**450**) are isolated (see Fig. 2.191) [633, 635].

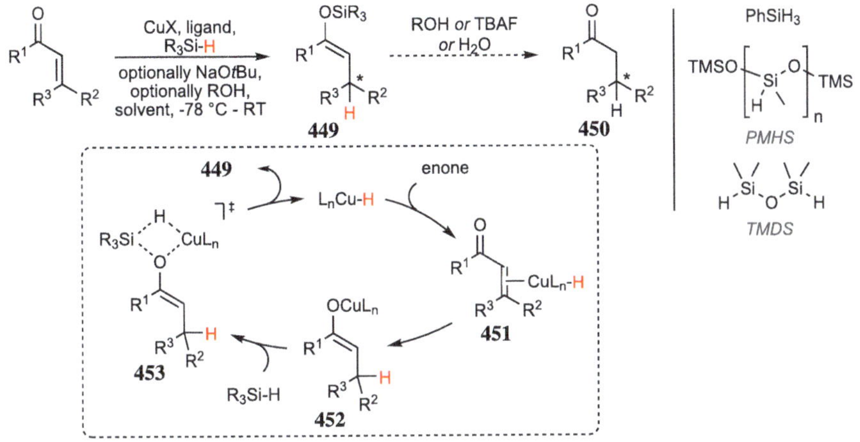

Fig. 2.191 Typical reaction conditions and postulated mechanism of the 1,4-reduction [633, 635]

Starting from the catalytically active CuH species, the postulated mechanism is initiated by formation of the enone-π-complex **451**. The transfer of the hydride to the β-C-atom yields an enolate species **452**. Following the rate-determining σ-bond metathesis via the four-membered transition state **453**, the copper hydride is regenerated and the silyl enol ether **449** is received. The acceleration of the reaction by addition of sterically demanding alcohols (e.g., *t*BuOH) can be attributed to the formation of the Cu-*t*butanolate, which proceeds by substitution of the enolate directly from **452** instead of via **453** [637, 638]. Theoretical studies suggest the formation of a *C*-bound instead of an *O*-bound copper enolate, with the β-hydride transfer as the rate-determining step [639]. Hydrogen or stannanes can also be used as an alternative hydride source.

In addition to PPh$_3$, the use of chiral ligands can enable an asymmetric reaction. Significant contributions were reported by the groups of Buchwald (BINAP derivatives) and Lipshutz (DTBM-SEGPHOS) [640, 641]. Subsequent studies showed that the use of sterically bulky ligands can alter the selectivity in α-substituted enones towards favoring a 1,2-reduction and aryl alkyl ketones can be converted to the corresponding alcohols. Enantioselective, reductive aldol reactions (cf. Sect. 2.3.3) have also been accomplished with variants of these reaction conditions (see Fig. 2.192) [634, 635].

For complex substrates, the 1,4-reduction is typically conducted with achiral catalysts to avoid competing stereoinduction from stereocenters in the substrate and the external ligand.

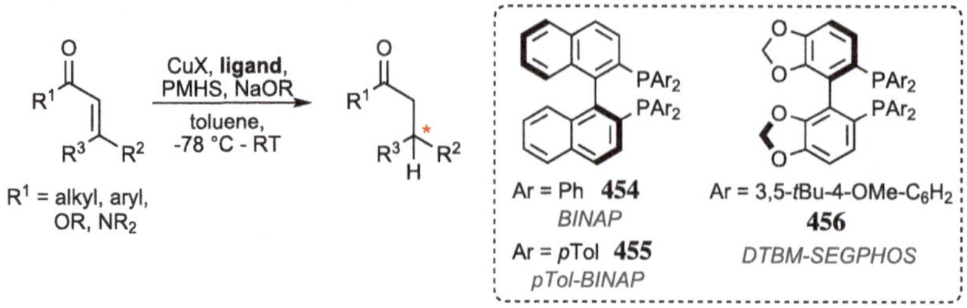

Fig. 2.192 Asymmetric 1,4-reductions [635]

However, occasional examples under asymmetric reaction control have been reported (see Fig. 2.193).

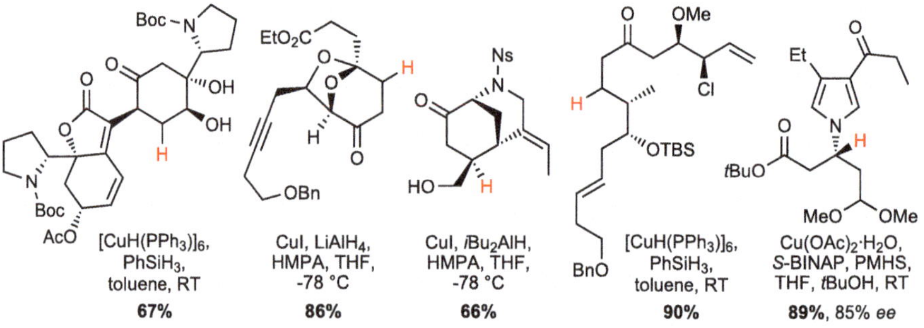

Fig. 2.193 Syntheses employing CuH-mediated 1,4-reductions [248, 642–645]

Some alternative, but so far rarely encountered methods for selective 1,4-reduction include the use of $Co(acac)_2$/DIBAL, Raney nickel, $Na_2S_2O_4$, and selectrides as well as heterogeneous Pd-catalyzed hydrogenations [646–649].

Another important reaction displaying strong similarities with the mechanisms and catalytic systems discussed so far is the allylic substitution, which is catalyzed by Cu, Ir, Pd, or Mo [614, 650–652]. It will be discussed in more detail in Sect. 6.4.

2.6　Interconversion of Carboxylic Acid Derivatives

The derivatization of carbonyl compounds by aldol chemistry, α-functionalization, and 1,2-/1,4-addition was covered in the previous sections. However, a seemingly simple but incredibly important class of reactions should not go unmentioned: the interconversion of various carbonyls, notably carboxylic acid derivatives. While acid chlorides and to a limited extent also anhydrides are rarely the target of a synthesis due to their high innate reactivity, esters and amides in particular represent strategically relevant functional groups (see Fig. 2.194).

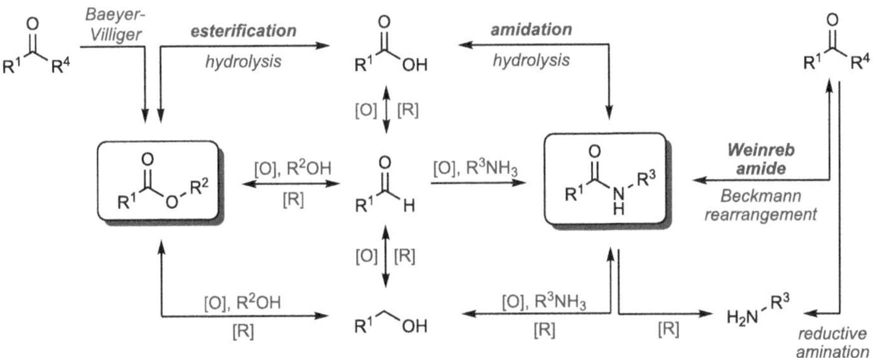

Fig. 2.194 Established routes between various carbonyl derivatives

Many classes of natural products contain ester bonds. Especially in the retrosynthetic analysis of macrocyclic metabolites, the formation of the ester bond is an obvious step. Amide groups are present in approximately a quarter of all commercially available active pharmaceutical ingredients and in medicinal chemistry, amide formation is one of the most common reactions in the synthesis of new drug candidates [653, 654]. The formation of amides in peptides, especially in solid-phase synthesis, represents a separate field of expertise, which over time has moved its focus from the initial question of the feasibility of a coupling to improving its atom economy [655, 656].

The formation of carboxylic acids, esters, or amides from ketones by oxidative means will be discussed in more detail in Chap. 4, as will the reduction of carbonyl derivatives to the corresponding alcohols or amines.

The direct conversion of carboxylic acids with alcohols can be carried out under thermal conditions in the presence of catalytic amounts of a Brønsted acid (*Fischer* esterification), if the resulting water byproduct is continuously removed from the reaction equilibrium (molecular sieve, Dean-Stark trap). Unfortunately, the approach is rarely compatible with more sensitive functional groups. The straightforward reaction of amines with carboxylic acids only results in the formation of the salt of the two reactants, an ammonium carboxylate. To enable sufficient yield under mild conditions, either the carboxylic acid or the respective coupling partner need to be activated. With few exceptions, the former approach dominates (carboxylic acid activation) and the reagents for coupling can be categorized by the respective activated intermediates (see Fig. 2.195).

The formation of acid chlorides (**457**) and mixed anhydrides (**458**) as well as pyridinium esters (**461**) is preferably chosen to access esters from carboxylic acids, while amides are usually obtained by reaction with carbodiimides via **459**, phosphorus(V) reagents (via **460**) or by activated esters (**462**). Other frequently employed methods are the reaction with sulfonyl

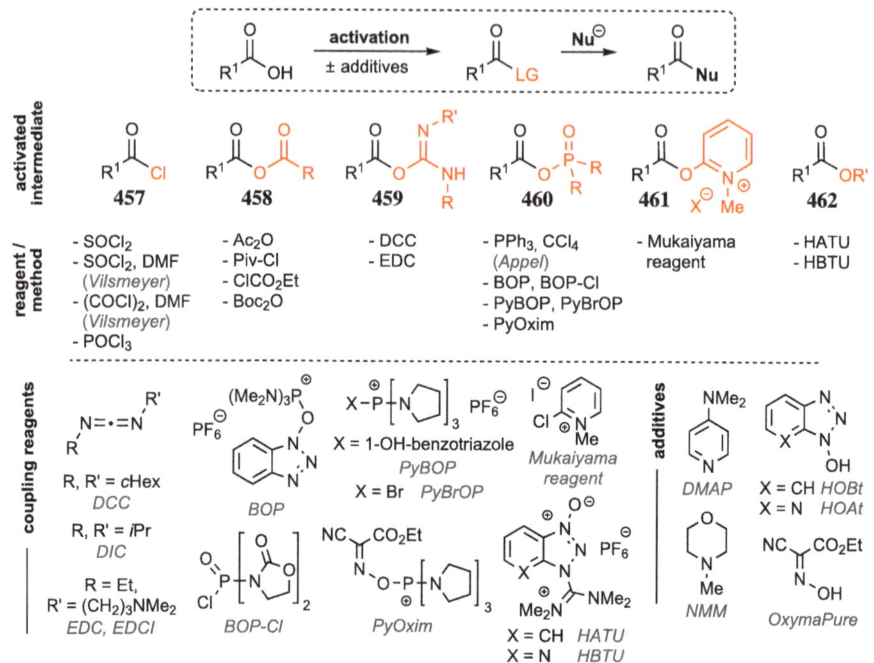

Fig. 2.195 Overview of common approaches to the activation of carboxylic acids [655–657]

chlorides (e.g., tosyl chloride) to form the mixed sulfonic/carboxylic acid anhydride and activation by carbonyldiimidazole (CDI), which generates the reactive imidazole amide. The imidazole can be subsequently displaced quite easily as an excellent nucleofuge (see Fig. 2.196).

The activated carboxylic acid can proceed through the intermediacy of the corresponding labile amides and esters by specific additives, in which said additive acts as a leaving group. Especially DMAP enjoys great popularity and is utilized particularly often. In peptide couplings the α-acidic stereocenter can undergo unwanted epimerizations. Carbodiimide- and phosphorus(V)-based approaches often make use of alcohol additives such as hydroxybenzotriazole or oximes, since loss of stereointegrity can usually be avoided under these conditions (see Fig. 2.197).

A much less frequently used approach for esterifications is the reaction of the free carboxylic acid with an activated coupling partner via substitution of a leaving group. Classic examples can be found in the Mitsunobu reaction (conditions A) or the displacement of a mesylate/tosylate (conditions B, see Fig. 2.198) [662, 663].

Despite its importance, the mechanism of the Mitsunobu reaction is still the subject of active discussion. A postulated path proceeds via the formation of the phosphonium ion **464**, which subsequently activates the alcohol via **465** for nucleophilic substitution by the carboxylate [662]. The nature of the ester groups on the azodicarboxylate influences the

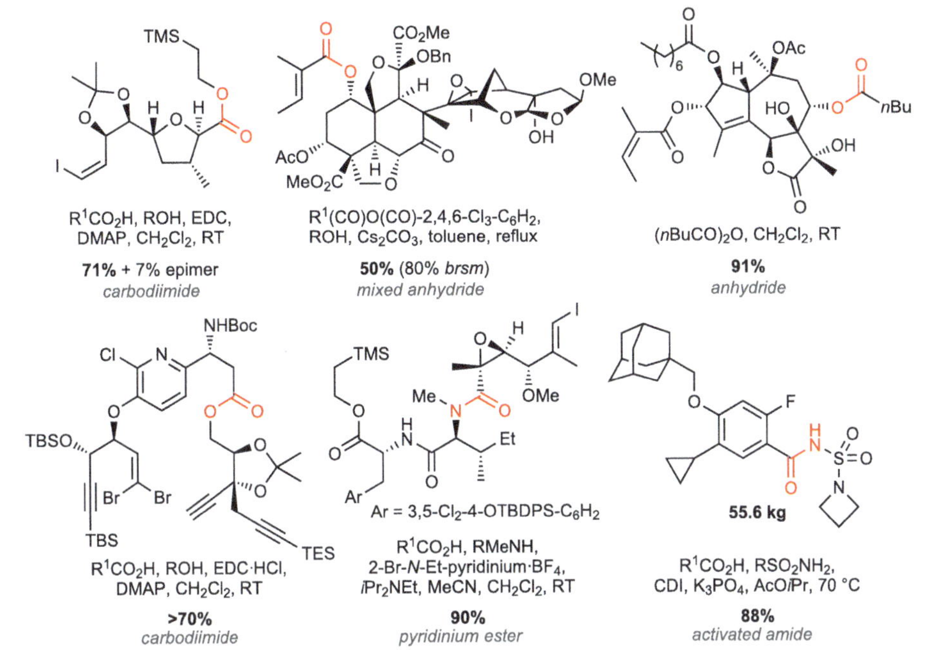
R^1CO_2H, ROH, EDC, DMAP, CH_2Cl_2, RT
71% + 7% epimer
carbodiimide
$R^1(CO)O(CO)$-2,4,6-Cl_3-C_6H_2, ROH, Cs_2CO_3, toluene, reflux
50% (80% brsm)
mixed anhydride
$(nBuCO)_2O$, CH_2Cl_2, RT
91%
anhydride
R^1CO_2H, ROH, EDC·HCl, DMAP, CH_2Cl_2, RT
>70%
carbodiimide
Ar = 3,5-Cl_2-4-OTBDPS-C_6H_2
R^1CO_2H, RMeNH, 2-Br-N-Et-pyridinium·BF_4, iPr_2NEt, MeCN, CH_2Cl_2, RT
90%
pyridinium ester
55.6 kg
R^1CO_2H, RSO_2NH_2, CDI, K_3PO_4, AcOiPr, 70 °C
88%
activated amide

Fig. 2.196 Examples of esterifications or amidations in advanced intermediates [25, 277, 658–661]

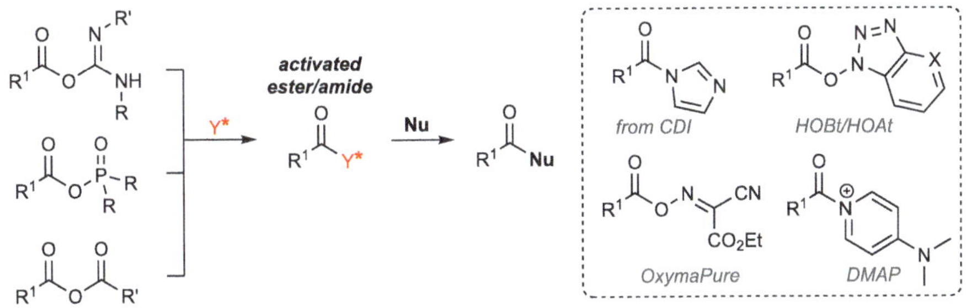
activated ester/amide
Nu
from CDI
HOBt/HOAt
OxymaPure
DMAP

Fig. 2.197 Modus operandi of additives in the formation of activated esters or amides

thermal stability and the reactivity of the intermediate **464** towards nucleophiles possessing a broad acidity range. The phosphine oxide, which is obtained stoichiometrically as a by-product, is often difficult to remove from the crude product by chromatographic means. For this challenge, the use of polymer-bound PPh_3 or the variation of the phosphine species can provide a solution [662, 663]. There are now also methods available in which the phosphine is required only in catalytic amounts (see Sect. 8.1). Methyl esters in particular can be obtained by converting the carboxylic acid in the presence of diazomethane. The highly toxic profile of the reagent and its explosiveness does however counterbalance the simple

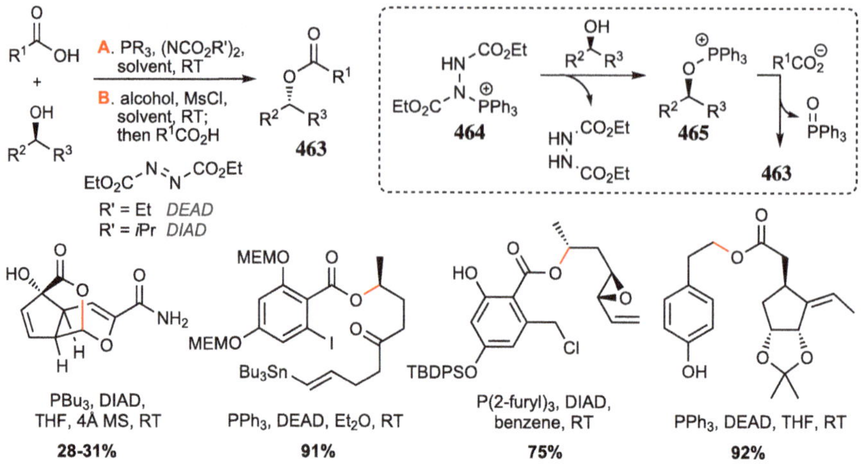

Fig. 2.198 Esterification by activation of an alcohol. Postulated mechanism of the Mitsunobu reaction and applications [662, 664–667]

handling and high yields. Therefore, it is only rarely encountered. As a safer alternative, TMS-diazomethane can be used as a solution in methanol, from which diazomethane is generated *in situ* by methanolytic desilylation (see Fig. 2.199) [668, 669].

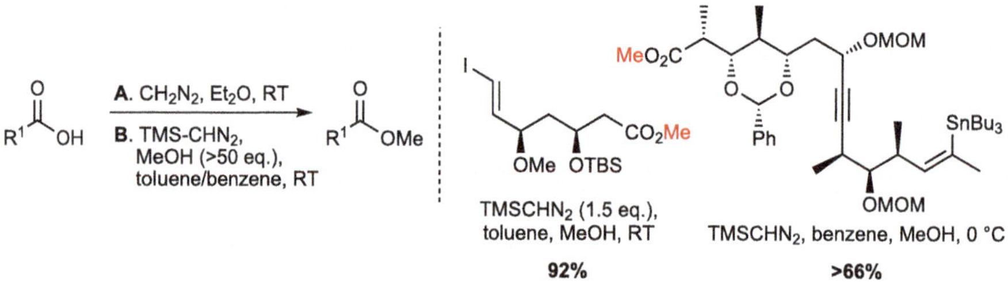

Fig. 2.199 Synthesis of methyl esters using diazomethane [555, 668, 670]

For amidations on an industrial scale (>100 mmol), an analysis of the most frequently utilized coupling methods has been reported. Since cost, process safety, as well as atom economy play a more central role on large scale, methods that are not always considered "first choice" for complex substrates dominate in an industrial context. In addition to EDC, thionyl chloride, CDI and oxalyl chloride are thus used preferably (see Fig. 2.200) [657, 671].

In addition to these classic approaches, more innovative routes to esters and amides have also been exploited. Particular attention is paid to catalytic methods, thereby avoiding

method	share
carbodiimide	27%
- EDC	21%
- DCC	5%
acid chloride	27%
- SOCl$_2$	16%
- (COCl)$_2$	8%
CDI	14%
carboxylic anhydride	13%
P(V) anhydride	6%
others	13%

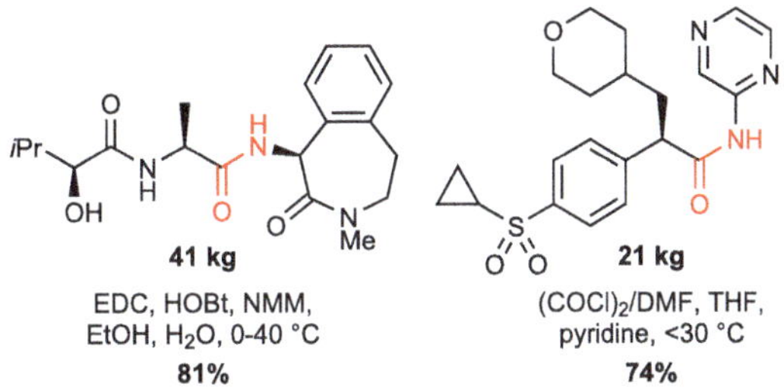

41 kg

EDC, HOBt, NMM,
EtOH, H$_2$O, 0-40 °C

81%

21 kg

(COCl)$_2$/DMF, THF,
pyridine, <30 °C

74%

Fig. 2.200 Use of amidation methods on an industrial scale [657, 672, 673]

the stoichiometric use of activation reagents [674, 675]. In this context, boric, borinic, and boronic acids have established themselves as a viable alternative. The existence of the originally postulated, activated species **466** has come under question in a comprehensive, combined experimental and theoretical study. Instead, dimeric, bridged ate-complexes **467** were proposed as operative intermediates, which result from the reaction with an amine **468** (see Fig. 2.201) [675–677].

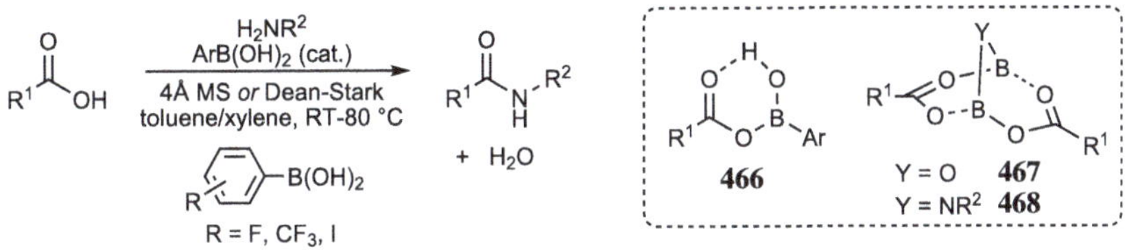

Fig. 2.201 Direct amidation through boronic acid catalysis [676, 677]

Another variant is the use of primary alcohols, which are oxidized *in situ* under the loss of H$_2$ and then afford the corresponding esters or amides with an equivalent of an alcohol or an amine, respectively. The reaction proceeds in three steps: Dehydrogenation of the primary alcohol to the aldehyde (**470**) and release of H$_2$, nucleophilic attack of an alcohol or amine on the hemiacetal (**471**), and dehydrogenation of the hemiacetal to the ester or amide. This acceptorless alcohol dehydrogenation requires tailor-made catalysts, which are able to liberate hydrogen before a coupling partner (R^{2}OH, R^{2}NH$_2$) coordinates. The first efficient catalytic systems were developed in the Milstein group and were based on Ru complexes bearing PNP-pincer ligands (**469**) [678, 679]. The spectrum of viable catalysts has since been expanded to additionally include Ir-, Rh- as well as Fe- and Co-complexes [680–682] (Fig. 2.202).

The esterification in macrolactones entails an added complexity. Their synthesis is hampered further by the general thermodynamic difficulty of forming medium and large ring structures.

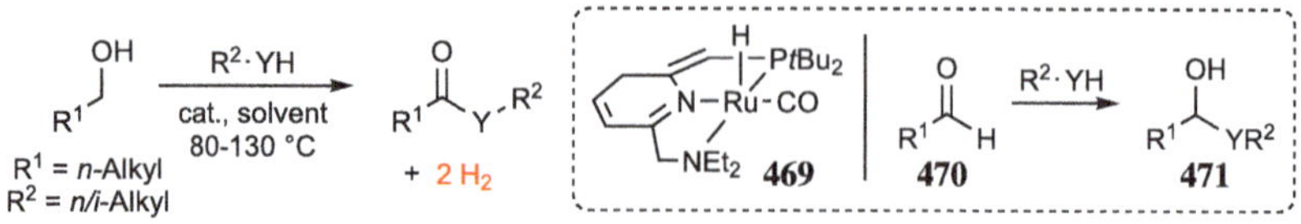

Fig. 2.202 Acceptorless alcohol dehydrogenation to esters and amides and Milstein catalyst **469**

Macrolactonizations can be achieved either by activating the acid or the hydroxy group in α, ω-hydroxycarboxylic acids. The activation of the acid is the more common approach and there is a wide range of methods available (see Fig. 2.203). In total syntheses, the conditions developed by Yamaguchi and Shiina have advanced to be the primary methods of choice, both of which proceed via the formation of mixed anhydrides. Alcohols, on the other hand, can either be activated under Mitsunobu conditions or by conversion to the corresponding mesylates/tosylates [683, 684].

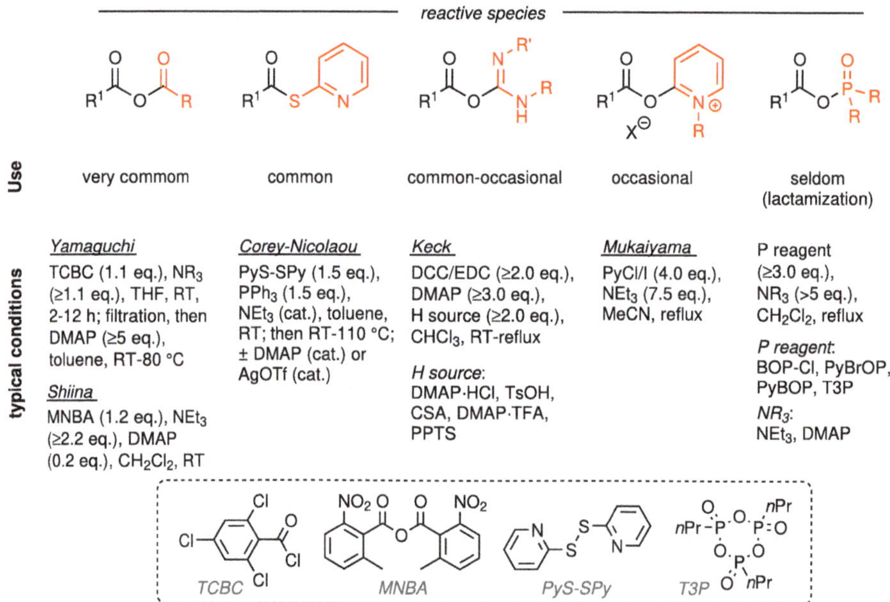

Fig. 2.203 Reactive intermediates of activated carboxylic acids and frequently used methods in macrolactonizations [683, 684]

Unfortunately, it is often not possible to predict which lactonization method affords the highest yield. Mixed anhydrides (Yamaguchi, Shiina) have been relied on most often, but epimerizations have been reported for some substrates bearing α-acidic protons. The wide range of available options regularly leads to extensive investigation of which conditions most efficiently deliver the complex target molecule (see Fig. 2.204).

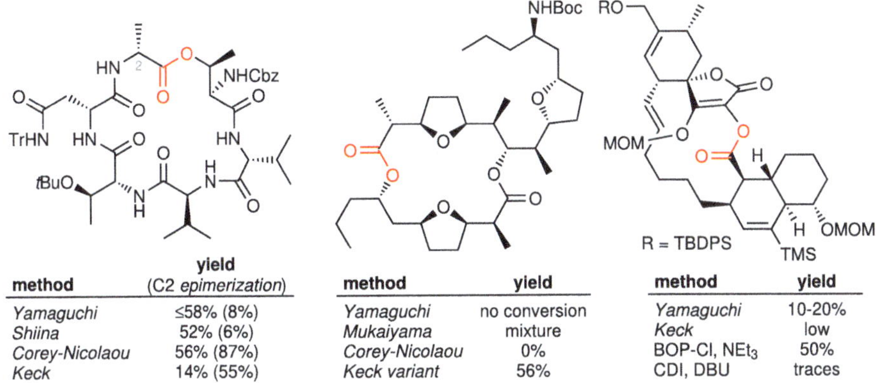

method	yield (C2 *epimerization*)
Yamaguchi	≤58% (8%)
Shiina	52% (6%)
Corey-Nicolaou	56% (87%)
Keck	14% (55%)

method	yield
Yamaguchi	no conversion
Mukaiyama	mixture
Corey-Nicolaou	0%
Keck variant	56%

method	yield
Yamaguchi	10-20%
Keck	low
BOP-Cl, NEt₃	50%
CDI, DBU	traces

Fig. 2.204 Comparison of different lactonization methods in the syntheses of LI-F04a, pamamycin-607 and (–)-chlorothricolide [685–687]

While macrolactones are a common structural motif in natural products, macrolactams are encountered much less frequently. The variety of methods available for amide formation have therefore been investigated only to a limited extent in a densely functionalized setting. Nevertheless, some impressive examples of macrolactamizations in the synthesis of bioactive compounds have been reported (see Fig. 2.205).

Fig. 2.205 Lactamizations in complex macrocycles [688, 689]

To synthesize ketones or aldehydes, nucleophiles can be added to acyl donors. However, when esters or carboxylic acid chlorides react with organolithium or organomagnesium reagents, the conversion cannot be halted selectively upon addition of one equivalent of the nucleophile. Therefore not the carbonyl derivative, but rather the corresponding alcohol is obtained. Methoxy-methylamides (**472**, *Weinreb amides*) allow acylation of organometallic reagents by stabilizing the tetrahedral intermediate **473**,[690] which is why the desired carbonyl is isolated after an acidic work-up (see Fig. 2.206) [691–693].

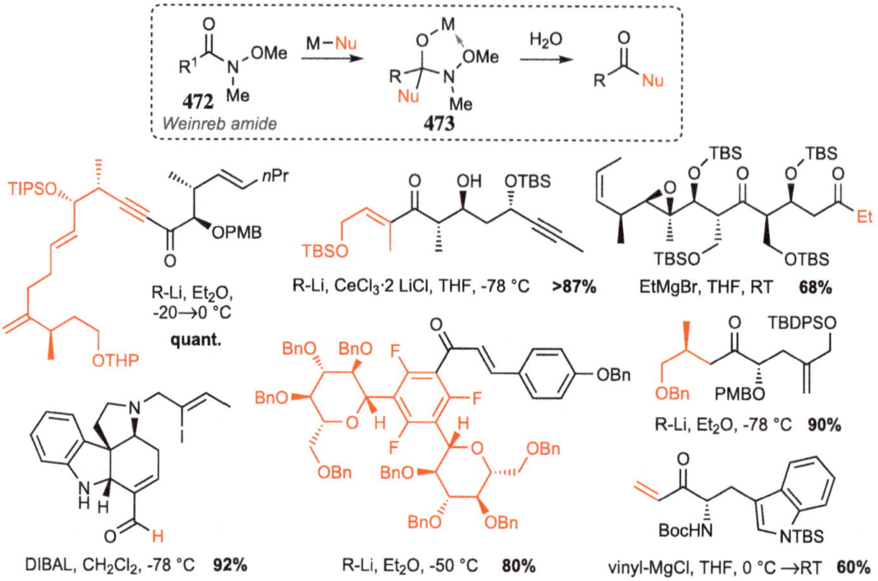

Fig. 2.206 Reactivity of Weinreb amides [694–700]

Weinreb amides can be prepared from all common carboxylic acid derivatives: esters, carboxylic acids, and acid chlorides. The synthesis can be reliably used even on a kilogram scale and usually affords high yields above 90% (see Fig. 2.207).

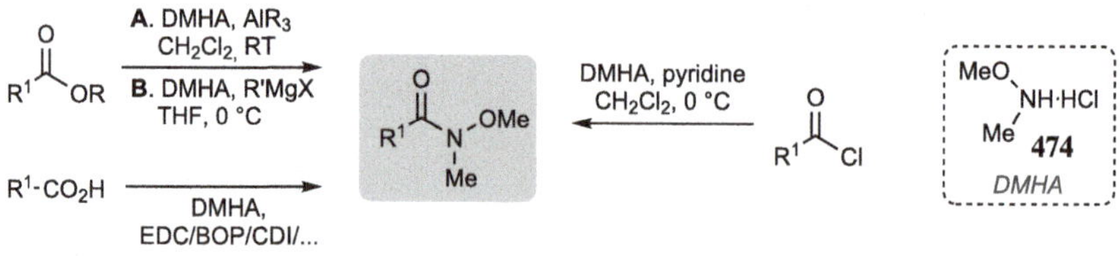

Fig. 2.207 Formation of Weinreb amides [692]

Sulfonamides are a class of substances related to amides in terms of stability, physico-chemical properties, and three-dimensional structure. While the sulfonamide functionality does not occur in natural products, pharmaceuticals in particular often rely on it as amide analog in a wide range of active ingredients, for example in celecoxib, sildenafil, and pirox-icam (see Fig. 2.208) [701] Since sulfonamides are not derived from carboxylic acids, their synthesis and relevance will not be discussed further here. Nevertheless, their formation from sulfonic acids utilizes chemistry closely related to that of carboxylic acids.

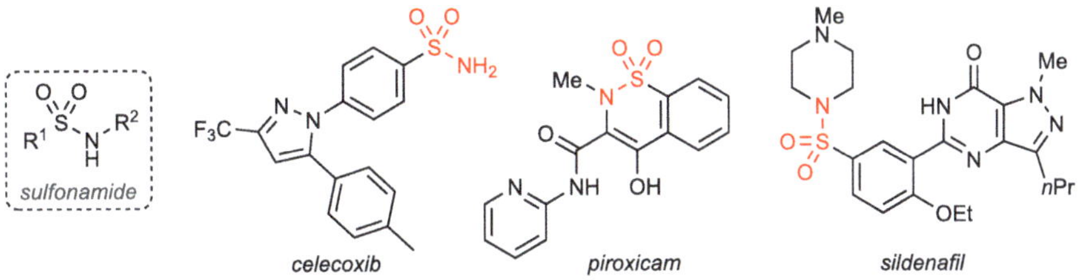

Fig. 2.208 Examples of sulfonamide drugs

References

1. H. B. Bürgi, J. D. Dunitz, E. Shefter, *J. Am. Chem. Soc.* **1973**, *95*, 5065–5067.
2. M. Kaftory, J. D. Dunitz, *Acta Cryst. B* **1975**, *B31*, 2912–2914.
3. M. Kaftory, J. D. Dunitz, *Acta Cryst. B* **1975**, *B31*, 2914–2916.
4. G. I. Birnbaum, *J. Am. Chem. Soc.* **1974**, *96*, 6165–6168.
5. J. A. Wunderlich, *Acta Cryst. B* **1967**, *23*, 846–855.
6. S. R. Hall, F. R. Ahmed, *Acta Cryst. B* **1968**, *24*, 337–346.
7. S. R. Hall, F. R. Ahmed, *Acta Cryst. B* **1968**, *24*, 346–355.
8. H. B. Bürgi, J. D. Dunitz, *Acc. Chem. Res.* **1983**, *16*, 153–161.
9. H. B. Bürgi, J. D. Dunitz, J. M. Lehn, G. Wipf, *Tetrahedron* **1974**, *30*, 1563–1572.
10. H. B. Bürgi, *Angew. Chem. Int. Ed. Engl.* **1975**, *14*, 460–473.
11. O. Kubo, D. P. Canterbury, G. C. Micalizio, *Org. Lett.* **2012**, *14*, 5748–5751.
12. H. J. Martin, M. Drescher, J. Mulzer, *Angew. Chem. Int. Ed.* **2000**, *39*, 581–583.
13. M. A. Avery, S. C. Choudhry, O. P. Dhingra, B. D. Gray, M. Kang, S. Kuo, T. R. Vedananda, J. D. White, A. J. Whittle, *Org. Biomol. Chem.* **2014**, *12*, 9116–9132.
14. K. Maruoka, T. Itoh, H. Yamamoto, *J. Am. Chem. Soc.* **1985**, *107*, 4573–4576.
15. G. Frenking, K. F. Köhler, M. T. Reetz, *Tetrahedron* **1991**, *47*, 9005–9018.
16. W. C. Still, J. A. Schneider, *Tetrahedron Lett.* **1980**, *21*, 1035–1038.
17. A. Mengel, O. Reiser, *Chem. Rev.* **1999**, *99*, 1191–1223.
18. D. J. Cram, F. A. A. Elhafez, *J. Am. Chem. Soc.* **1952**, *74*, 5828–5835.
19. D. J. Cram, K. R. Kopecky, *J. Am. Chem. Soc.* **1959**, *81*, 2748–2755.
20. G. J. Karabatsos, *J. Am. Chem. Soc.* **1967**, *89*, 1367–1371.
21. J. W. Cornforth, R. H. Cornforth, K. K. Mathew, *J. Chem. Soc.* **1959**, 112–127.
22. M. Chérest, H. Felkin, N. Prudent, *Tetrahedron Lett.* **1968**, *9*, 2199–2204.
23. N. T. Anh, O. Eisenstein, *Nouveau J. Chim.* **1977**, *1*, 61–70.
24. M. T. Reetz, *Angew. Chem. Int. Ed. Engl.* **1984**, *23*, 556–569.
25. M. Heinrich, J. J. Murphy, M. K. Ilg, A. Letort, J. Flasz, P. Philipps, A. Fürstner, *Angew. Chem. Int. Ed.* **2018**, *57*, 13575–13581.
26. N. T. Anh, *Top. Curr. Chem.* **1980**, *88*, 145–162.
27. A. U. Rahman, Z. Shah in *Stereoselective Synthesis in Organic Chemistry*, Springer Verlag, **1993**, pp. 168–169.

28. F. Toda, K. Tanaka, K. Mori, *Chem. Lett.* **1983**, *12*, 827–830.

29. G. E. Keck, E. P. Boden, *Tetrahedron Lett.* **1984**, *25*, 265–268.

30. A. B. Smith, III, S. M. Condon, J. A. McCauley, J. L. Leazer, J. W. Leahy, R. E. Maleczka, *J. Am. Chem. Soc.* **1997**, *119*, 947–961.

31. S. Kobayashi, T. Yokoi, T. Inoue, Y. Hori, T. Saka, T. Shimomura, A. Masuyama, *J. Org. Chem.* **2016**, *81*, 1484–1498.

32. B. M. Sharma, A. Gontala, P. Kumar, *Eur. J. Org. Chem.* **2016**, 1215–1226.

33. X. Chen, E. R. Hortelano, E. L. Eliel, S. V. Frye, *J. Am. Chem. Soc.* **1992**, *114*, 1778–1784.

34. J. A. Read, Y. Yang, K. A. Woerpel, *Org. Lett.* **2017**, *19*, 3346–3349.

35. N. D. Bartolo, J. A. Read, E. M. Valentín, K. A. Woerpel, *Chem. Rev.* **2020**, *120*, 1513–1619.

36. J. I. Seeman, *Chem. Rev.* **1983**, *83*, 83–134.

37. W. C. Still, J. H. McDonald, *Tetrahedron Lett.* **1980**, *21*, 1031–1034.

38. M. Asai, T. Nishikawa, N. Ohyabu, N. Yamamoto, M. Isobe, *Tetrahedron* **2001**, *57*, 4543–4558.

39. G. Mehta, F. A. Khan, *J. Am. Chem. Soc.* **1990**, *112*, 6140–6142.

40. J. M. Hahn, W. J. Le Noble, *J. Am. Chem. Soc.* **1992**, *114*, 1916–1917.

41. P. Wipf, Y. Kim, *J. Am. Chem. Soc.* **1994**, *116*, 11678–11688.

42. P. Wipf, J.-K. Jung, *Chem. Rev.* **1999**, *99*, 1469–1480.

43. A. S. Cieplak, *J. Am. Chem. Soc.* **1981**, *103*, 4540–4552.

44. A. S. Cieplak, *Chem. Rev.* **1999**, *99*, 1265–1336.

45. A. S. Cieplak, B. D. Tait, C. R. Johnson, *J. Am. Chem. Soc.* **1989**, *111*, 8447–8462.

46. G. Frenking, K. F. Köhler, M. T. Reetz, *Angew. Chem. Int. Ed. Engl.* **1991**, *30*, 1146–1149.

47. B. W. Gung, *Tetrahedron* **1996**, *52*, 5263–5301.

48. D. A. Evans, S. J. Siska, V. J. Cee, *Angew. Chem. Int. Ed.* **2003**, *42*, 1761–1765.

49. S. Tomoda, *Chem. Rev.* **1999**, *99*, 1243–1263.

50. J. J. Dannenberg, *Chem. Rev.* **1999**, *99*, 1225–1241.

51. T. J. Leitereg, D. J. Cram, *J. Am. Chem. Soc.* **1968**, *90*, 4019–4026.

52. M. T. Reetz, *Acc. Chem. Res.* **1993**, *26*, 462–468.

53. M. T. Reetz, A. Jung, *J. Am. Chem. Soc.* **1983**, *105*, 4833–4835.

54. D. A. Evans, A. H. Hoveyda, *J. Org. Chem.* **1990**, *55*, 5190–5192.

55. D. A. Evans, B. D. Allison, M. G. Yang, C. E. Masse, *J. Am. Chem. Soc.* **2001**, *123*, 10840–10852.

56. D. A. Evans, M. J. Dart, J. L. Duffy, M. G. Yang, *J. Am. Chem. Soc.* **1996**, *118*, 4322–4343.

57. A. Gil, J. Lamariano-Merketegi, A. Lorente, F. Albericio, M. Álvarez, *Org. Lett.* **2016**, *18*, 4485–4487.

58. J. R. Dunetz, L. D. Julian, J. S. Newcom, W. R. Roush, *J. Am. Chem. Soc.* **2008**, *130*, 16407–16416.

59. S. Wan, F. Wu, J. C. Rech, M. E. Green, R. Balachandran, W. S. Horne, B. W. Day, P. E. Floreancig, *J. Am. Chem. Soc.* **2011**, *133*, 16668–16679.

60. M. Lorenz, M. Kalesse, *Org. Lett.* **2008**, *10*, 4371–4374.

61. Z. A. Kasun, X. Gao, R. M. Lipinski, M. J. Krische, *J. Am. Chem. Soc.* **2015**, *137*, 8900–8903.

62. D. A. Evans, M. J. Dart, J. L. Duffy, D. L. Rieger, *J. Am. Chem. Soc.* **1995**, *117*, 9073–9074.

63. D. A. Evans, M. G. Yang, M. J. Dart, J. L. Duffy, *Tetrahedron Lett.* **1996**, *37*, 1957–1960.

64. I. H. Williams, D. Spangler, D. A. Femec, G. M. Maggiora, R. L. Schowen, *J. Am. Chem. Soc.* **1983**, *105*, 31–40.

65. E. C. Ashby, J. T. Laemmle, *Chem. Rev.* **1975**, *75*, 521–546.

66. W. Schlenk, W. Schlenk jr., *Chem. Ber.* **1929**, *62*, 920–924.

67. F. W. Walker, E. C. Ashby, *J. Am. Chem. Soc.* **1969**, *91*, 3845–3850.

68. C. G. Swain, H. B. Boyles, *J. Am. Chem. Soc.* **1951**, *73*, 870–872.

69. R. M. Peltzer, J. Gauss, O. Eisenstein, M. Cascella, *J. Am. Chem. Soc.* **2020**, *142*, 2984–2994.

70. S. Yamazaki, S. Yamabe, *J. Org. Chem.* **2002**, *67*, 9346–9353.

71. K. Soai, S. Niwa, *Chem. Rev.* **1992**, *92*, 833–856.

72. L. Pu, H.-B. Yu, *Chem. Rev.* **2001**, *101*, 757–824.

73. E. J. Corey, F. J. .Hannon, *Tetrahedron Lett.* **1987**, *28*, 5233–5236.

74. T. Satyanarayana, S. Abraham, H. B. Kagan, *Angew. Chem. Int. Ed.* **2009**, *48*, 456–494.

75. P. Cossee, *J. Catal.* **1964**, *3*, 80–88.

76. E. J. Arlman, P. Cossee, *J. Catal.* **1964**, *3*, 99–104.

77. P. C. A. Guerra, L. Cavallo, *Acc. Chem. Res.* **2004**, *37*, 231–241.

78. H. O. House, B. M. Trost, *J. Org. Chem.* **1965**, *30*, 1341–1348.

79. P. J. Reider, R. S. E. Conn, P. Davis, V. J. Grenda, A. J. Zambito, E. J. J. Grabowski, *J. Org. Chem.* **1987**, *52*, 3326–3334.

80. A. K. Beck, M. S. Hoekstra, D. Seebach, *Tetrahedron Lett.* **1977**, *18*, 1187–1190.

81. D. Seebach, T. Weller, G. Protschuk, A. K. Beck, M. S. Hoekstra, *Helv. Chim. Acta* **1981**, *64*, 716–735.

82. M. Yoshifuji, T. Nakamura, N. Inamoto, *Tetrahedron Lett.* **1987**, *28*, 6325–6328.

83. R. K. Henderson, A. P. Hill, A. M. Redman, H. F. Sneddon, *Green Chem.* **2015**, *17*, 945–949.

84. R. R. Fraser, T. S. Mansour, S. Savard, *J. Org. Chem.* **1985**, *50*, 3232–3234.

85. R. R. Fraser, T. S. Mansour, *J. Org. Chem.* **1984**, *49*, 3442–3443.

86. A. Streitwieser, A. Facchetti, L. Xie, X. Zhang, E. C. Wu, *J. Org. Chem.* **2012**, *77*, 985–990.

87. D. Seebach, *Angew. Chem. Int. Ed. Engl.* **1988**, *27*, 1624–1654.

88. G. Stork, P. F. Hudrlik, *J. Am. Chem. Soc.* **1968**, *90*, 4462–4464.

89. D. Caine, B. J. L. Huff, *Tetrahedron Lett.* **1966**, *7*, 4695–4700.

90. H. O. House, L. J. Czuba, M. Gall, H. D. Olmstead, *J. Org. Chem.* **1969**, *34*, 2324–2336.

91. M. T. Reetz, H. Haning, *Tetrahedron Lett.* **1993**, *34*, 7395–7398.

92. R. D. Clark, C. H. Heathcock, *Tetrahedron Lett.* **1974**, *15*, 2027–2030.

93. C. A. Brown, *J. Org. Chem.* **1974**, *39*, 3913–3918.

94. C. Kowalski, X. Creary, A. J. Rollin, M. C. Burke, *J. Org. Chem.* **1978**, *43*, 2601–2608.

95. M. E. Garst, J. N. Bonfiglio, D. A. Grudoski, J. Marks, *J. Org. Chem.* **1980**, *45*, 2307–2315.

96. R. E. Ireland, R. H. Mueller, A. K. Willard, *J. Am. Chem. Soc.* **1976**, *98*, 2868–2877.

97. R. E. Ireland, P. Wipf, J. D. Armstrong III., *J. Org. Chem.* **1991**, *56*, 650–657.

98. C. H. Heathcock, C. T. Buse, W. A. Kleschick, M. C. Pirrung, J. E. Sohn, J. Lampe, *J. Org. Chem.* **1980**, *45*, 1066–1081.

99. P. L. Hall, J. H. Gilchrist, D. B. Collum, *J. Am. Chem. Soc.* **1991**, *113*, 9571–9574.

100. P. F. Godenschwager, D. B. Collum, *J. Am. Chem. Soc.* **2008**, *130*, 8726–8732.

101. P. F. Godenschwager, D. B. Collum, *J. Am. Chem. Soc.* **2007**, *129*, 12023–12031.

102. D. B. Collum, A. J. McNeil, A. Ramirez, *Angew. Chem. Int. Ed.* **2007**, *46*, 3002–3017.

103. B. L. Lucht, D. B. Collum, *Acc. Chem. Res.* **1999**, *41*, 1035–1042.

104. D. B. Collum, personal communication, **2021**.

105. R. A. Woltornist, D. B. Collum, *J. Am. Chem. Soc.* **2021**, *143*, 17452–17464.

106. X. Sun, D. B. Collum, *J. Am. Chem. Soc.* **2000**, *122*, 2452–2458.

107. L. Xie, K. M. Isenberger, G. Held, L. M. Dahl, *J. Org. Chem.* **1997**, *62*, 7516–7519.

108. G. Stork, P. F. Hudrlik, *J. Am. Chem. Soc.* **1968**, *90*, 4464–4465.

109. J. Ren, C. J. Cramer, R. R. Squires, *J. Am. Chem. Soc.* **1999**, *121*, 2633–2634.

110. D. A. Evans, J. T. Shaw, *unpublished Angew. Chem. Int. Ed. Review* **2005**.

111. D. A. Evans, F. Urpi, T. C. Somers, J. S. Clark, M. T. Bilodeau, *J. Am. Chem. Soc.* **1990**, *112*, 8215–8216.

112. A. Abiko, *Acc. Chem. Res.* **2004**, *37*, 387–395.

113. D. A. Evans, A. E. Weber, *J. Am. Chem. Soc.* **1986**, *108*, 6757–6761.

114. J. M. Goodman, I. Paterson, *Tetrahedron Lett.* **1992**, *33*, 7223–7226.

115. D. A. Evans, D. L. Rieger, M. T. Bilodeau, F. Urpi, *J. Am. Chem. Soc.* **1991**, *113*, 1047–1049.

116. T. Mukaiyama, T. Inoue, *Chem. Lett.* **1976**, *5*, 559–562.

117. H. C. Brown, R. K. Dhar, R. K. Bakshi, P. K. Pandiarajan, B. Singaram, *J. Am. Chem. Soc.* **1989**, *111*, 3441–3442.

118. H. C. Brown, K. Ganesan, R. K. Dhar, *J. Org. Chem.* **1993**, *58*, 147–153.

119. A. Abiko, J.-F. Liu, S. Masamune, *J. Org. Chem.* **1996**, *61*, 2590–2591.

120. D. A. Evans, J. V. Nelson, E. Vogel, T. R. Taber, *J. Am. Chem. Soc.* **1981**, *103*, 3099–3111.

121. I. Paterson, J. M. Goodman, M. A. Lister, R. C. Schumann, C. K. McClure, R. D. Norcross, *Tetrahedron* **1990**, *46*, 4663–4684.

122. H. C. Brown, R. K. Dhar, K. Ganesan, B. Singaram, *J. Org. Chem.* **1992**, *57*, 499–504.

123. T. Inoue, T. Mukaiyama, *Bull. Chem. Soc. Jpn.* **1980**, *53*, 174–178.

124. K. Ganesan, H. C. Brown, *J. Org. Chem.* **1994**, *59*, 2336–2340.

125. H. C. Brown, K. Ganesan, *Tetrahedron Lett.* **1992**, *33*, 3421–3424.

126. D. A. Evans, B. W. Trotter, B. Coté, P. J. Coleman, L. C. Dias, A. N. Tyler, *Angew. Chem. Int. Ed. Engl.* **1997**, *36*, 2744–2747.

127. Y. Tanabe, N. Matsumoto, S. Funakoshi, N. Manta, *Synlett* **2001**, 1959–1961.

128. D. A. Evans, *unpublished results*.

129. D. A. Evans, J. S. Clark, R. Metternich, V. J. Novack, G. S. Sheppard, *J. Am. Chem. Soc.* **1990**, *112*, 866–868.

130. D. A. Evans, M. Bilodeau, T. C. Somers, J. Clardy, D. Cherry, Y. Kato, *J. Org. Chem.* **1991**, *56*, 5750–5752.

131. C. Heras, A. Gómez-Palomino, P. Romea, F. Urpí, J. M. Bofill, I. de P. R. Moreira, *J. Org. Chem.* **2017**, *82*, 8909–8916.

132. I. de P. R. Moreira, J. M. Bofill, J. M. Anglada, J. G. Solsona, J. Nebot, P. Romea, F. Urpí, *J. Am. Chem. Soc.* **2008**, *130*, 3242–3243.

133. T. Remarchuk, F. St-Jean, D. Carrera, S. Savage, H. Yajima, B. Wong, S. Babu, A. Deese, J. Stults, M. W. Dong, D. Askin, J. W. Lane, K. L. Spencer, *Org. Process Res. Dev.* **2014**, *18*, 1652–1666.

134. T. Mukaiyama, R. W. Stevens, N. Iwasawa, *Chem. Lett.* **1982**, *13*, 353–356.

135. T. Mukaiyama, N. Iwasawa, *Chem. Lett.* **1982**, *13*, 1903–1906.

136. A. Abdel-Magid, L. N. Pridgen, D. S. Eggleston, I. Lantos, *J. Am. Chem. Soc.* **1986**, *108*, 4595–4602.

137. D. L. Boger, T. Honda, *Tetrahedron Lett.* **1993**, *34*, 1567–1570.

138. A. B. Smith, III, T. M. Razler, J. P. Ciavarri, T. Hirose, T. Ishikawa, *Org. Lett.* **2005**, *7*, 4399–4402.

139. C. Kashima, K. Fukusaka, K. Takahashi, *J. Het. Chem.* **1997**, *34*, 1559–1565.

140. C. Kashima, I. Fukuchi, K. Takahashi, K. Fukusaka, A. Hosomi, *Heterocycles* **1998**, *47*, 357–365.

141. K. Hayashi, H. Kogiso, S. Sano, Y. Nagao, *Synlett* **1996**, 1203–1205.

142. A. Fürstner, *Chem. Eur. J.* **1998**, *4*, 567–570.

143. D. A. Evans, J. S. Tedrow, J. T. Shaw, C. W. Downey, *J. Am. Chem. Soc.* **2002**, *124*, 392–393.

144. N. F. O'Rourke, M. A, H. N. Higgs, A. Eastman, G. C. Micalizio, *Org. Lett.* **2017**, *19*, 5154–5157.

145. P. Chiu, C. P. Szeto, Z. Geng, K. F. Cheng, *Tetrahedron Lett.* **2001**, *42*, 4091–4093.

146. M. Arai, B. H. Lipshutz, E. Nakamura, *Tetrahedron* **1992**, *48*, 5709–5718.

147. M. Suzuki, A. Yanagisawa, R. Noyori, *J. Am. Chem. Soc.* **1985**, *107*, 3348–3349.

148. A. G. Schultz, W. P. Malachowski, Y. Pan, *J. Org. Chem.* **1997**, *62*, 1223–1229.

149. H. E. Zimmerman, P. A. Wang, *J. Am. Chem. Soc.* **1990**, *112*, 1280–1281.

150. R. Ocampoa, W. R. Dolbier, *Tetrahedron* **2004**, *60*, 9325–9374.

151. A. Fürstner, *Synthesis* **1989**, 571–590.

152. S. Choppin, L. Ferreiro-Medeiros, M. Barbarotto, F. Colobert, *Chem. Soc. Rev.* **2013**, *42*, 937–949.

153. H. Pellissier, *Beilstein J. Org. Chem.* **2018**, *14*, 325–344.

154. R. B. Ruggeri, C. H. Heathcock, *J. Org. Chem.* **1987**, *52*, 5746–5749.

155. M. Kögl, L. Brecker, R. Warrass, J. Mulzer, *Angew. Chem. Int. Ed.* **2007**, *46*, 9320–9322.

156. I. Paterson, J. P. Scott, *J. Chem. Soc. Perkin Trans. 1* **1999**, 1003–1014.

157. B. Schetter, R. Mahrwald, *Angew. Chem. Int. Ed.* **2006**, *45*, 7506–7525.

158. S. ichi Kiyooka, M. A. Hena, T. Yabukami, K. Murai, F. Goto, *Tetrahedron Lett.* **2000**, *41*, 7511–7516.

159. M. T. Crimmins, A.-M. R. Dechert, *Org. Lett.* **2009**, *11*, 1635–1638.

160. I. Paterson, G. J. Florence, K. Gerlach, J. P. Scott, *Angew. Chem. Int. Ed.* **2000**, *39*, 377–340.

161. D. A. Evans, J. M. Takacs, L. R. McGee, M. D. Ennis, D. J. Mathre, J. Bartroli, *Pure Appl. Chem.* **1981**, *53*, 1109–1127.

162. D. A. Evans, J. V. Nelson, T. R. Taber, *Top. Stereochem.* **1982**, *13*, 1–115.

163. H. E. Zimmerman, M. D. Traxler, *J. Am. Chem. Soc.* **1957**, *79*, 1920–1923.

164. S. Adachi, T. Harada, *Eur. J. Org. Chem.* **2009**, 3661–3671.

165. P. Romea, F. Urpí in *Modern Methods in Stereoselective Aldol Reactions* (Ed.: R. Mahrwald), Wiley-VCH, **2013**, Chapter 1, pp. 1–81.

166. J.-E. Dubois, P. Fellmann, *Tetrahedron Lett.* **1975**, *14*, 1225–1228.

167. P. Fellmann, J.-E. Dubois, *Tetrahedron* **1978**, *34*, 1343–1347.

168. R. W. Hoffmann, K. Ditrich, S. Froech, D. Cremer, *Tetrahedron* **1985**, *41*, 5517–5524.

169. C. Gennari, R. Todeschini, M. G. Beretta, G. Favini, C. Scolastico, *J. Am. Chem. Soc.* **1986**, *51*, 612–616.

170. Y. Li, K. N. Houk, *J. Am. Chem. Soc.* **1989**, *111*, 1236–1240.

171. P. V. Ramachandran, W. Xu, H. C. Brown, *Tetrahedron Lett.* **1997**, *38*, 769–772.

172. C. H. Heathcock in *Asymmetric synthesis*, *Vol. 3* (Ed.: J. D. Morrison), Academic press, **1984**, Chapter 2, pp. 111–212.

173. M. Yasuda, K. Chiba, A. Baba, *J. Am. Chem. Soc.* **2000**, *122*, 7549–7555.

174. S. Kang, W. Lee, B. Jung, H.-S. Lee, S. H. Kang, *Asian J. Org. Chem.* **2015**, *4*, 567–572.

175. Y. Yamamoto, K. Maruyama, *Tetrahedron Lett.* **1980**, *21*, 4607–4610.

176. D. A. Evans, L. R. McGee, *J. Am. Chem. Soc.* **1981**, *103*, 2876–2878.

177. S. Yamago, D. Machii, E. Nakamura, *J. Org. Chem.* **1991**, *56*, 2098–2106.

178. T. Mukaiyama, K. Narasaka, K. Banno, *Chem. Lett.* **1972**, *2*, 1011–1014.

179. T. Mukaiyama, K. Banno, K. Narasaka, *J. Am. Chem. Soc.* **1974**, *96*, 7503–7509.

180. J. Matsuo, M. Murakami, *Angew. Chem. Int. Ed.* **2013**, *52*, 9109–9118.

181. G. L. Beutner, S. E. Denmark, *Angew. Chem. Int. Ed.* **2013**, *52*, 9086–9096.

182. C. H. Heathcock, K. T. Hug, L. A. Flippin, *Tetrahedron Lett.* **1984**, *25*, 5973–5976.

183. C. H. Heathcock, S. K. Davidsen, K. T. Hug, L. A. Flippin, *J. Org. Chem.* **1986**, *51*, 3027–3037.

184. S. E. Denmark, W. Lee, *Chem. Asian J.* **2008**, *3*, 327–341.

185. J. M. Lee, P. Helquist, O. Wiest, *J. Am. Chem. Soc.* **2012**, *134*, 14973–14981.

186. R. Mahrwald, *Chem. Rev.* **1999**, *99*, 1095–1120.

187. J.-E. Dubois, G. Axiotis, E. Bertounesque, *Tetrahedron Lett.* **1984**, *25*, 4655–4658.

188. M. T. Reetz, K. Kesseler, A. Jung, *Tetrahedron* **1984**, *40*, 4327–4336.

189. K. A. Shahid, J. Mursheda, M. Okazaki, Y. Shuto, F. Goto, S. Kiyooka, *Tetrahedron Lett.* **2002**, *43*, 6377–6381.

190. T. Mukaiyama, A. Ishida, *Chem. Lett.* **1975**, *4*, 319–322.

191. G. Casiraghi, F. Zanardi, G. Appendino, G. Rassu, *Chem. Rev.* **2000**, *100*, 1929–1972.

192. F. von der Ohe, R. Brückner, *New. J. Chem.* **2000**, *24*, 659–669.

193. C. S. López, R. Álvarez, B. Vaz, O. N. Faza, Á. R. de Lera, *J. Org. Chem.* **2005**, *70*, 3654–3659.

194. *Modern Aldol Reactions*, (Ed.: R. Mahrwald), Wiley-VCH, **2004**.

195. *Modern Methods in Stereoselective Aldol Reactions*, (Ed.: R. Mahrwald), Wiley-VCH, **2013**.

196. *Modern Enolate Chemistry: From Preparation to Applications in Asymmetric Synthesis*, (Ed.: M. Braun), Wiley-VCH, **2016**.

197. S. B. J. Kan, K. K.-H. Ng, I. Paterson, *Angew. Chem. Int. Ed.* **2013**, *52*, 9097–9108.

198. W. R. Roush, *J. Org. Chem.* **1991**, *51*, 4151–4157.

199. J. D. Frein, R. E. Taylor, D. L. Sackett, *Org. Lett.* **2009**, *11*, 3186–3189.

200. M. Defosseux, N. Blanchard, C. Meyer, J. Cossy, *J. Org. Chem.* **2004**, *69*, 4626–4647.

201. R. G. Pearson, *J. Am. Chem. Soc.* **1963**, *85*, 3533–3539.

202. N. A. V. Draanen, S. Arseniyadis, M. T. Crimmins, C. H. Heathcock, *J. Org. Chem.* **1991**, *56*, 2499–2506.

203. F. Arikan, J. Li, D. Menche, *Org. Lett.* **2008**, *10*, 3521–3524.

204. P. Li, J. Li, F. Arikan, W. Ahlbrecht, M. Dieckmann, D. Menche, *J. Am. Chem. Soc.* **2009**, *131*, 11678–11679.

205. L. C. Dias, E. C. Polo, E. C. de Lucca, M. A. B. Ferreira in *Modern Methods in Stereoselective Aldol Reactions* (Ed.: R. Mahrwald), Wiley-VCH, **2013**, Chapter 5, pp. 293–375.

206. I. Paterson, D. J. Wallace, C. J. Cowden, *Synthesis* **1998**, 639–652.

207. I. Paterson, K. R. Gibson, R. M. Oballa, *Tetrahedron Lett.* **1996**, *37*, 8585–8588.

208. L. C. Dias, A. M. Aguilar, *Chem. Soc. Rev.* **2008**, *37*, 451–469.

209. D. A. Evans, P. J. Coleman, B. Cote, *J. Org. Chem.* **1997**, *62*, 788–790.

210. R. S. Paton, J. M. Goodman, *J. Org. Chem.* **2008**, *73*, 1253–1263.

211. D. A. Evans, G. Helmchen, M. Rüping in *Asymmetric Synthesis - The Essentials* (Eds.: M. Christmann, S. Bräse), Wiley-VCH, **2007**, Chapter 1, pp. 3–9.

212. N. D. V. Thanh, *Tetrahedron* **2020**, *76*, 130618.

213. A. Nazari, M. M. Heravi, V. Zadsirjan, *J. Organomet. Chem.* **2021**, *932*, 121629.

214. G. Helmchen, U. Leikauf, I. Taufer-Knöpfel, *Angew. Chem. Int. Ed. Engl.* **1985**, *24*, 874–875.

215. W. Oppolzer, J. Marco-Contelles, *Helv. Chim. Acta* **1986**, *69*, 1699–1703.

216. W. Oppolzer, J. Blagg, I. Rodriguez, E. Walther, *J. Am. Chem. Soc.* **1990**, *112*, 2767–2772.

217. W. Oppolzer, C. Starkemann, I. Rodriguez, G. Bernardinelli, *Tetrahedron Lett.* **1991**, *32*, 61–64.

218. D. A. Evans, J. Bartroli, T. L. Shih, *J. Am. Chem. Soc.* **1981**, *103*, 2127–2129.

219. M. T. Crimmins, B. W. King, E. A. Tabet, K. Chaudhary, *J. Org. Chem.* **2001**, *66*, 894–902.

220. T. H. Yan, C. W. Tan, H. C. Lee, H. C. Lo, T. Y. Huang, *J. Am. Chem. Soc.* **1993**, *115*, 2613–2621.

221. A. Abiko, J.-F. Liu, S. Masamune, *J. Am. Chem. Soc.* **1997**, *119*, 2586–2587.

222. D. J. Ager, I. Prakash, D. R. Schaad, *Chem. Rev.* **1996**, *96*, 835–875.

223. J. R. Gage, D. A. Evans, *Org. Synth.* **1990**, *68*, 77–82.

224. M. T. Crimmins, B. W. King, E. A. Tabet, *J. Am. Chem. Soc.* **1997**, *119*, 7883–7884.

225. M. Nerz-Stormes, E. R. Thornton, *J. Org. Chem.* **1991**, *56*, 2489–2498.

226. D. A. Evans, C. W. Downey, J. T. Shaw, J. S. Tedrow, *Org. Lett.* **2002**, *4*, 1127–1130.

227. M. T. Crimmins, J. She, *Synlett* **2004**, 1371–1374.

228. H. J. Kim, R. Pongdee, Q. Wu, L. Hong, H. wen Liu, *J. Am. Chem. Soc.* **2007**, *129*, 14582–14584.

229. P. A. Clarke, J. Winn, *Tetrahedron Lett.* **2011**, *52*, 1469–1472.

230. D.-S. Yu, W.-X. Xu, L.-X. Liu, P.-Q. Huang, *Synlett* **2008**, 1189–1192.

231. M. T. Crimmins, P. J. McDougall, *Org. Lett.* **2003**, *5*, 591–594.

232. H. Danda, M. M. Hansen, C. H. Heathcock, *J. Org. Chem.* **1990**, *55*, 173–181.

233. M. A. Walker, C. H. Heathcock, *J. Org. Chem.* **1991**, *56*, 5747–5750.

234. B. C. Raimundo, C. H. Heathcock, *Synlett* **1995**, 1213–1214.

235. Y. Nagao, Y. Hagiwara, T. Kumagai, M. Ochiai, T. Inoue, K. Hashimoto, E. Fujita, *J. Org. Chem.* **1986**, *51*, 2391–2393.

236. R. M. Rzasa, H. A. Shea, D. Romo, *J. Am. Chem. Soc.* **1998**, *120*, 591–592.

237. D. A. Evans, H. P. Ng, J. S. Clark, D. L. Rieger, *Tetrahedron* **1992**, *48*, 2127–2142.

238. D. A. Evans, T. C. Britton, J. A. Ellman, *Tetrahedron Lett.* **1987**, *28*, 6141–6144.

239. T. Inoue, J.-F. Liu, D. C. Buske, A. Abiko, *J. Org. Chem.* **2002**, *67*, 5250–5256.

240. H. Fuwa, Y. Okuaki, N. Yamagata, M. Sasaki, *Angew. Chem. Int. Ed.* **2015**, *54*, 868–873.

241. Y. Nagao, S. Yamada, T. Kumagai, M. Ochiai, E. Fujita, *J. Chem. Soc. Chem. Commun.* **1985**, 1418–1419.

242. T.-H. Yan, A.-W. Hung, H.-C. Lee, C.-S. Chang, W.-H. Liu, *J. Org. Chem.* **1995**, *60*, 3301–3306.

243. N. R. Guz, A. J. Phillips, *Org. Lett.* **2002**, *4*, 2253–2256.

244. Y. Zhang, A. J. Phillips, T. Sammakia, *Org. Lett.* **2004**, *6*, 23–25.

245. Y. Zhang, T. Sammakia, *J. Org. Chem.* **2006**, *71*, 6262–6265.

246. K. K. Pulukuri, T. K. Chakraborty, *Org. Lett.* **2014**, *16*, 2284–2287.

247. S. C. D. Kennington, J. M. Romo, P. Romea, F. Urpí, *Org. Lett.* **2016**, *18*, 3018–3021.

248. J. Ren, J. Wang, R. Tong, *Org. Lett.* **2015**, *17*, 744–747.

249. J. D. White, C. M. Lincoln, J. Yang, W. H. C. Martin, D. B. Chan, *J. Org. Chem.* **2008**, *73*, 4139–4150.

250. I. Paterson, R. M. Oballa, *Tetrahedron Lett.* **1997**, *38*, 8241–8244.

251. I. Paterson, D. J. Wallace, K. R. Gibson, *Tetrahedron Lett.* **1997**, *38*, 8911–8914.

252. A. Bernardi, C. Gennari, J. M. Goodman, I. Paterson, *Tetrahedron: Asymmetry* **1995**, *6*, 2613–2636.

253. G. Symkenberg, M. Kalesse, *Angew. Chem. Int. Ed.* **2014**, *53*, 1795–1798.

254. C. M. Neuhaus, M. Liniger, M. Stieger, K.-H. Altmann, *Angew. Chem. Int. Ed.* **2013**, *52*, 5866–5870.

255. S. Masamune, T. Sato, B. M. Kim, T. A. Wollmann, *J. Am. Chem. Soc.* **1986**, *108*, 8279–8281.

256. E. J. Corey, S. S. Kim, *J. Am. Chem. Soc.* **1990**, *112*, 4976–4977.

257. E. J. Corey, D.-H. Lee, *Tetrahedron Lett.* **1993**, *34*, 1737–1740.

258. M. T. Reetz, E. Rivadeneira, C. Niemeyer, *Tetrahedron Lett.* **1990**, *31*, 3863–3866.

259. Y. Ito, M. Sawamura, T. Hayashi, *J. Am. Chem. Soc.* **1986**, *108*, 6405–6406.

260. T. Hayashi, M. Sawamura, Y. Ito, *Tetrahedron* **1992**, *48*, 1999–2012.

261. Y. Yamashita, T. Yasukawa, W.-J. Yoo, T. Kitanosono, S. Kobayashi, *Chem. Soc. Rev.* **2018**, *47*, 4388–4480.

262. S. Kobayashi, H. Uchiro, Y. Fujishita, I. Shiina, T. Mukaiyama, *J. Am. Chem. Soc.* **1991**, *113*, 4247–4252.

263. S. Kobayashi, H. Uchiro, I. Shiina, T. Mukaiyama, *Tetrahedron* **1993**, *49*, 1761–1772.

264. E. J. Corey, C. L. Cywin, T. D. Roper, *Tetrahedron Lett.* **1992**, *33*, 6907–6910.

265. E. J. Corey, T.-P. Loh, T. D. Roper, M. D. Azimioara, M. C. Noe, *J. Am. Chem. Soc.* **1992**, *114*, 8290–8292.

266. E. M. Carreira, R. A. Singer, W. Lee, *J. Am. Chem. Soc.* **1994**, *116*, 8837–8838.

267. R. A. Singer, E. M. Carreira, *J. Am. Chem. Soc.* **1995**, *117*, 12360–12361.

268. K. Furuta, T. Maruyama, H. Yamamoto, *J. Am. Chem. Soc.* **1991**, *113*, 1041–1042.

269. E. R. Parmee, O. Tempkin, S. Masamune, A. Abiko, *J. Am. Chem. Soc.* **1991**, *113*, 9365–9366.

270. K. Mikami, S. Matsukawa, *J. Am. Chem. Soc.* **1993**, *115*, 7039–7040.

271. K. Mikami, S. Matsukawa, *J. Am. Chem. Soc.* **1994**, *116*, 4077–4078.

272. G. E. Keck, D. Krishnamurthy, *J. Am. Chem. Soc.* **1995**, *117*, 2363–2364.

273. J. Krüger, E. M. Carreira, *J. Am. Chem. Soc.* **1998**, *120*, 837–838.

274. B. L. Pagenkopf, J. Krüger, A. Stojanovic, E. M. Carreira, *Angew. Chem. Int. Ed.* **1998**, *37*, 3124–3126.

275. B. M. Trost, W.-J. Bai, C. E. Stivala, C. Hohn, C. Poock, M. Heinrich, S. Xu, J. Rey, *J. Am. Chem. Soc.* **2018**, *140*, 17316–17326.

276. K. Micoine, A. Fürstner, *J. Am. Chem. Soc.* **2010**, *132*, 14064–14066.

277. A. G. Myers, P. C. Hogan, A. R. Hurd, S. D. Goldberg, *Angew. Chem. Int. Ed.* **2002**, *41*, 1062–1067.

278. A. B. Smith, III, K. P. Minbiole, P. R. Verhoest, M. Schelhaas, *J. Am. Chem. Soc.* **2001**, *123*, 10942–10953.

279. D. A. Evans, J. A. Murry, M. C. Kozlowski, *J. Am. Chem. Soc.* **1996**, *118*, 5814–5815.

280. D. A. Evans, M. C. Kozlowski, C. S. Burgey, D. W. C. MacMillan, *J. Am. Chem. Soc.* **1997**, *119*, 7893–7894.

281. D. A. Evans, M. C. Kozlowski, J. A. Murry, C. S. Burgey, K. R. Campos, B. T. Connell, R. J. Staples, *J. Am. Chem. Soc.* **1999**, *121*, 669–685.

282. D. A. Evans, C. S. Burgey, M. C. Kozlowski, S. W. Tregay, *J. Am. Chem. Soc.* **1999**, *121*, 686–699.

283. D. A. Evans, D. W. C. MacMillan, K. R. Campos, *J. Am. Chem. Soc.* **1997**, *119*, 10859–10860.

284. D. A. Evans, C. E. Masse, J. Wu, *Org. Lett.* **2002**, *4*, 3375–3378.

285. D. A. Evans, M. C. Kozlowski, J. S. Tedrow, *Tetrahedron Lett.* **1996**, *37*, 7481–7484.

286. P. L. DeRoy, A. B. Charette, *Org. Lett.* **2003**, *5*, 4163–4165.

287. D. L. Aubele, S. Wan, P. E. Floreancig, *Angew. Chem. Int. Ed.* **2005**, *44*, 3485–3488.

288. D. M. Black, R. Davis, B. D. Doan, T. C. Lovelace, A. Millar, J. F. Toczko, S. Xie, *Tetrahedron: Asymmetry* **2008**, *19*, 2015–2019.

289. S. E. Denmark, S. B. D. Winter, X. Su, K.-T. Wong, *J. Am. Chem. Soc.* **1996**, *118*, 7404–7405.

290. S. E. Denmark, R. A. Stavenger, *Acc. Chem. Res.* **2000**, *33*, 432–440.

291. S. E. Denmark, Y. Fan, *J. Am. Chem. Soc.* **2002**, *124*, 4233–4235.

292. S. E. Denmark, T. Wynn, G. L. Beutner, *J. Am. Chem. Soc.* **2002**, *124*, 13405–13407.

293. S. E. Denmark, G. L. Beutner, T. Wynn, M. D. Eastgate, *J. Am. Chem. Soc.* **2005**, *127*, 3774–3789.

294. S. E. Denmark, B. M. Eklov, P. J. Yao, M. D. Eastgate, *J. Am. Chem. Soc.* **2009**, *131*, 11770–11787.

295. S. E. Denmark, S. Fujimori, *J. Am. Chem. Soc.* **2005**, *127*, 8971–8973.

296. L. Fang, H. Xue, J. Yang, *Org. Lett.* **2008**, *10*, 4645–4648.

297. B. M. Trost, C. S. Brindle, *Chem. Soc. Rev.* **2010**, *39*, 1600–1632.

298. G. Guillena in *Modern Methods in Stereoselective Aldol Reactions* (Ed.: R. Mahrwald), Wiley-VCH, **2013**, Chapter 3, pp. 155–268.

299. M. Shibasaki, S. Matsunaga, N. Kumagai in *Modern Aldol Reactions* (Ed.: R. Mahrwald), Wiley-VCH, **2004**, Chapter 6, pp. 197–227.

300. S. Saito, H. Yamamoto, *Acc. Chem. Res.* **2004**, *37*, 570–579.

301. N. Yoshikawa, Y. M. A. Yamada, J. Das, H. Sasai, M. Shibasaki, *J. Am. Chem. Soc.* **1999**, *121*, 4168–4178.

302. Y. M. A. Yamada, N. Yoshikawa, H. Sasai, M. Shibasaki, *Angew. Chem. Int. Ed. Engl.* **1997**, *36*, 1871–1873.

303. N. Yoshikawa, T. Suzuki, M. Shibasaki, *J. Org. Chem.* **2002**, *67*, 2556–2565.

304. D. Sawada, M. Kanai, M. Shibasaki, *J. Am. Chem. Soc.* **2000**, *122*, 10521–10532.

305. D. Sawada, M. Shibasaki, *Angew. Chem. Int. Ed.* **2000**, *39*, 209–213.

306. B. M. Trost, H. Ito, *J. Am. Chem. Soc.* **2000**, *122*, 12003–12004.

307. N. Kumagai, S. Matsunaga, N. Yoshikawa, T. Ohshima, M. Shibasaki, *Org. Lett.* **2001**, *3*, 1539–1542.

308. B. M. Trost, E. R. Silcoff, H. Ito, *Org. Lett.* **2001**, *3*, 2497–2500.

309. B. M. Trost, M. U. Frederiksen, J. P. N. Papillon, P. E. Harrington, S. Shin, B. T. Shireman, *J. Am. Chem. Soc.* **2005**, *127*, 3666–3667.

310. B. M. Trost, H. Ito, E. R. Silcoff, *J. Am. Chem. Soc.* **2001**, *123*, 3367–3368.

311. B. M. Trost, V. S. C. Yeh, *Org. Lett.* **2002**, *4*, 3513–3516.

312. C. C. Meyer, E. Ortiz, M. J. Krische, *Chem. Rev.* **2020**, *120*, 3721–3748.

313. A. Revis, T. K. Hilty, *Tetrahedron Lett.* **1987**, *28*, 4809–4812.

314. S. J. Taylor, M. O. Duffey, J. P. Morken, *J. Am. Chem. Soc.* **2000**, *122*, 4528–4529.

315. A. E. Russell, N. O. Fuller, S. J. Taylor, P. Aurriset, J. P. Morken, *Org. Lett.* **2004**, *6*, 2309–2312.

316. C. Bee, S. B. Han, A. Hassan, H. Iida, M. J. Krische, *J. Am. Chem. Soc.* **2008**, *130*, 2746–2747.

317. I. Shin, S. Hong, M. J. Krische, *J. Am. Chem. Soc.* **2016**, *138*, 14246–14249.

318. S. Salim, N. A. Harry, K. K. Krishnan, G. Anilkumar, *Asian J. Org. Chem.* **2018**, *7*, 613–633.

319. B. Karimi, D. Enders, E. Jafari, *Synthesis* **2013**, *45*, 2769–2812.

320. A. Ting, S. E. Schaus, *Eur. J. Org. Chem.* **2007**, 5797–5815.

321. G. K. Friestad, A. K. Mathies, *Tetrahedron* **2007**, *63*, 2541–2569.

322. M. Tramontini, *Synthesis* **1973**, 703–775.

323. M. Tramontini, L. Angiolini, *Tetrahedron* **1990**, *46*, 1791–1837.

324. M. Arend, B. Westermann, N. Risch, *Angew. Chem. Int. Ed.* **1998**, *37*, 1044–1070.

325. S. Bahmanyar, K. N. Houk, *Org. Lett.* **2003**, *5*, 1249–1251.

326. P. H.-Y. Cheong, C. Y. Legault, J. M. Um, N. Celebi-Ölcüm, K. N. Houk, *Chem. Rev.* **2011**, *111*, 5042–5137.

327. Y. Shi, Q. Wang, S. Gao, *Org. Chem. Frontiers* **2018**, *5*, 1049–1066.

328. Y. Hayashi, T. Urushima, M. Shin, M. Shoji, *Tetrahedron* **2005**, *61*, 11393–11404.

329. T. Nagata, M. Nakagawa, A. Nishida, *J. Am. Chem. Soc.* **2003**, *125*, 7484–7485.

330. B. M. Williams, D. Trauner, *Angew. Chem. Int. Ed.* **2016**, *55*, 2191–2194.

331. Z. Bian, C. C. Marvin, S. F. Martin, *J. Am. Chem. Soc.* **2013**, *135*, 10886–10889.

332. M. Sickert, C. Schneider, *Angew. Chem. Int. Ed.* **2008**, *47*, 3631–3634.

333. K. Yamada, S. J. Harwood, H. Gröger, M. Shibasaki, *Angew. Chem. Int. Ed.* **1999**, *38*, 3504–3506.

334. N. Nishiwaki, K. R. Knudsen, K. V. Gothelf, K. A. Jørgensen, *Angew. Chem. Int. Ed.* **2001**, *40*, 2992–2995.

335. J. C. Anderson, G. P. Howell, R. M. Lawrence, C. S. Wilson, *J. Org. Chem.* **2005**, *70*, 5665–5670.

336. F. A. Luzzio, *Tetrahedron* **2001**, *57*, 915–945.

337. D. Simoni, F. P. Invidiata, S. Manfredini, R. Ferroni, I. Lampronti, M. Roberti, G. P. Pollini, *Tetrahedron Lett.* **1997**, *38*, 2749–2752.

338. P. B. Kisanga, J. G. Verkade, *J. Org. Chem.* **1999**, *64*, 4298–4303.

339. B. Lecea, A. Arrieta, I. Morao, F. P. Cossío, *Chem. Eur. J.* **1997**, *3*, 20–28.

340. C. Palomo, M. Oiarbide, A. Mielgo, *Angew. Chem. Int. Ed.* **2004**, *43*, 5442–5444.

341. C. Palomo, M. Oiarbide, A. Laso, *Eur. J. Org. Chem.* **2007**, 2561–2574.

342. H. Sasai, T. Suzuki, S. Arai, T. Arai, M. Shibasaki, *J. Am. Chem. Soc.* **1992**, *114*, 4418–4420.

343. T. Arai, Y. M. A. Yamada, N. Yamamoto, H. Sasai, M. Shibasaki, *Chem. Eur. J.* **1996**, *2*, 1368–1372.

344. B. M. Trost, V. S. C. Yeh, *Angew. Chem. Int. Ed.* **2002**, *41*, 861–863.

345. D. A. Evans, D. Seidel, M. Rueping, H. W. Lam, J. T. Shaw, C. W. Downey, *J. Am. Chem. Soc.* **2003**, *125*, 12692–12693.

346. Y. Alvarez-Casao, E. Marques-Lopez, R. P. Herrera, *Symmetry* **2011**, *3*, 220–245.

347. B. M. Trost, V. S. C. Yeh, H. Ito, N. Bremeyer, *Org. Lett.* **2002**, *4*, 2621–2623.

348. S. S. Harried, C. P. Lee, G. Yang, T. I. H. Lee, D. C. Myles, *J. Org. Chem.* **2003**, *68*, 6646–6660.

349. P. Lu, Z. Gu, A. Zakarian, *J. Am. Chem. Soc.* **2013**, *135*, 14552–14555.

350. H. Kusama, R. Hara, S. Kawahara, T. Nishimori, H. Kashima, N. Nakamura, K. Morihira, I. Kuwajima, *J. Am. Chem. Soc.* **2000**, *122*, 3811–3820.

351. B. M. Stoltz, N. B. Bennett, D. C. Duquette, A. F. G. Goldberg, Y. Liu, M. B. Loewinger, C. M. Reeves in *Comp. Org. Synth. II, Vol. 3* (Ed.: I. Marek), Elsevier, **2014**, Chapter 3.01, pp. 1–55.

352. J. A. Marco, M. Carda, J. Murga, E. Falomir in *Comp. Chirality, Vol. 2* (Ed.: J. Mulzer), Elsevier, **2012**, Chapter 2.14, pp. 398–440.

353. L. M. Jackman, B. C. Lange, *Tetrahedron Lett.* **1977**, *33*, 2737–2769.

354. R. Gompper, H.-H. Vogt, *Chem. Ber.* **1981**, *114*, 2866–2883.

355. M. W. Rathke, D. F. Sullivan, *Synth. Commun.* **1973**, *3*, 67–72.

356. F. Méndez, J. L. Gázquez, *J. Am. Chem. Soc.* **1994**, *116*, 9298–9301.

357. S. Damoun, G. Van de Woude, K. Choho, P. Geerlings, *J. Phys. Chem. A* **1999**, *103*, 7861–7866.

358. M. Elango, R. Parthasarathi, V. Subramanian, P. K. Chattaraj, *Int. J. Quantum. Chem.* **2006**, *106*, 852–862.

359. H. Mayr, M. Breugst, A. R. Ofial, *Angew. Chem. Int. Ed.* **2011**, *50*, 6470–6505.

360. J. H. Exner, E. C. Steiner, *J. Am. Chem. Soc.* **1974**, *96*, 1782–1787.

361. F. Guibé, P. Sarthou, G. Bram, *Tetrahedron* **1974**, *30*, 3139–3151.

362. W. J. Le Noble, H. F. Morris, *J. Org. Chem.* **1969**, *34*, 1969–1973.

363. H. E. Zaugg, *J. Am. Chem. Soc.* **1961**, *83*, 837–840.

364. A. L. Kurts, P. I. Demyanov, I. P. Beletskaya, O. A. Reutov, *J. Org. Chem. USSR (Engl.)* **1973**, *9*, 1341.

365. A. L. Kurts, S. M. Sakembaeva, I. P. Beletskaya, O. A. Reutov, *Dokl. Akad. Nauk. SSSR (Engl.)* **1973**, *211*, 590.

366. A. Streitwieser, *Chem. Rev.* **1956**, *56*, 571–752.

367. J. M. Conia, *Bull. Soc. Chim. Fr.* **1950**, 533–537.

368. J. M. Conia, *Bull. Soc. Chim. Fr.* **1950**, 537–541.

369. A. L. Kurts, N. K.Genkina, A. Macias, L. P. Beletskaya, O. A. Reutov, *Tetrahedron* **1971**, *27*, 4777–4785.

370. A. Streitwieser, Y.-J. Kim, D. Z.-R. Wang, *Org. Lett.* **2001**, *3*, 2599–2601.

371. R. P. Mariella, R. Raube, *Org. Synth.* **1953**, *33*, 23–24.

372. I. Fleming, J. J. Lewis, *J. Chem. Soc. Perkin Trans. 1* **1992**, 3257–3266.

373. H. O. House, M. J. Umen, *J. Org. Chem.* **1973**, *38*, 1000–1003.

374. R. W. Hoffmann, *Chem. Rev.* **1989**, *89*, 1841–1860.

375. S. Ghosh, S. Sinha, M. G. B. Drew, *Org. Lett.* **2006**, *8*, 3781–3784.

376. A. C. Spivey, L. Shukla, J. F. Hayler, *Org. Lett.* **2007**, *9*, 891–894.

377. D. Seebach, D. Wasmuth, *Angew. Chem. Int. Ed. Engl.* **1981**, *20*, 971.

378. D. R. Williams, C. M. Rojas, S. L. Bogen, *J. Org. Chem.* **1999**, *64*, 736–746.

379. J. S. Yadav, P. N. Lakshmi, *Synlett* **2010**, 1033–1036.

380. F. Marion, S. Calvet, C. Courillon, M. Malacria, *Tetrahedron Lett.* **2002**, *43*, 3369–3371.

381. M. Pena-López, M. M. Martínez, L. A. Sarandeses, J. P. Sestelo, *Chem. Eur. J.* **2009**, *15*, 910–916.

382. J. F. Dellaria, B. D. Santarsiero, *Tetrahedron Lett.* **1988**, *29*, 6079–6082.

383. C. Mckay, T. J. Simpson, C. L. Willis, A. K. Forrest, P. J. O'Hanlon, *Chem. Commun.* **2000**, 1109–1110.

384. A. B. Smith, III, E. F. Mesaros, E. A. Meyer, *J. Am. Chem. Soc.* **2006**, *128*, 5292–5299.

385. D. J. Dixon, C. I. Harding, S. V. Ley, D. M. G. Tilbrook, *Chem. Commun.* **2003**, 468–469.

386. G. B. Dudley, D. A. Engel, I. Ghiviriga, H. Lam, K. W. C. Poon, J. A. Singletary, *Org. Lett.* **2007**, *9*, 2839–2842.

387. M. K. Leong, V. S. Mastryukov, J. E. Boggs, *J. Mol. Struct.* **1998**, *445*, 149–160.

388. S. Saebo, J. E. Boggs, *J. Mol. Struct.* **1982**, *87*, 365–373.

389. E. L. Eliel, S. H. Wilen, *Stereochemistry of Organic Compounds*, John Wiley & Sons, **1994**.

390. G. B. Dudley, D. S. Tan, G. Kim, J. M. Tanskib, S. J. Danishefsky, *Tetrahedron Lett.* **2001**, *42*, 6789–6791.

391. I. Fleming, S. K. Ghosh, *J. Chem. Soc. Perkin Trans. 1* **1998**, 2733–2748.

392. L. A. Paquette, I. Efremov, *J. Am. Chem. Soc.* **2001**, *123*, 4492–4501.

393. A. P. Krapcho, E. A. Dundulis, *J. Org. Chem.* **1980**, *45*, 3236–3245.

394. J. A. Hogg, *J. Am. Chem. Soc.* **1948**, *70*, 161–164.

395. F. Plavac, C. H. Heathcock, *Tetrahedron Lett.* **1979**, *20*, 2115–2118.

396. R. D. Clark, *Org. Synth.* **1979**, *9*, 325–331.

397. H. O. House, T. M. Bare, *J. Org. Chem.* **1968**, *33*, 943–949.

398. M. C. Kohler, S. E. Wengryniuk, D. M. Coltart in *Stereoselective Synthesis of Drugs and Natural Products* (Eds.: V. Andrushko, N. Andrushko), John Wiley & Sons, **2013**, Chapter 7, pp. 183–213.

399. R. Cano, A. Zakarian, G. P. McGlacken, *Angew. Chem. Int. Ed.* **2017**, *56*, 9278–9290.

400. T. B. Wright, P. A. Evans, *Chem. Rev.* **2022**, *122*, 9196–9242.

401. R. Schmierer, G. Grotemeier, G. Helmchen, A. Selim, *Angew. Chem. Int. Ed. Engl.* **1981**, *20*, 207–208.

402. G. Helmchen, R. Wierzchowski, *Angew. Chem. Int. Ed. Engl.* **1984**, *23*, 60–61.

403. W. Oppolzer, *Tetrahedron* **1987**, *43*, 1969–2004.

404. W. Oppolzer, R. Moretti, S. Thomi, *Tetrahedron Lett.* **1989**, *40*, 5603–5606.

405. D. Romo, A. I. Meyers, *Tetrahedron* **1991**, *47*, 9503–9569.

406. D. Enders, M. Klatt, *Synthesis* **1996**, 1403–1418.

407. D. Enders, H. Eichenauer, *Angew. Chem. Int. Ed. Engl.* **1976**, *15*, 549–551.

408. A. Job, C. F. Janeck, W. Bettray, R. Peters, D. Enders, *Tetrahedron* **2002**, *58*, 2253–2329.

409. D. Enders, T. Hundertmark, R. Lazny, *Synlett* **1998**, 721–722.

410. K. Ochiai, S. Kuppusamy, Y. Yasui, K. Harada, N. R. Gupta, Y. Takahashi, T. Kubota, J. Kobayashi, Y. Hayashi, *Chem. Eur. J.* **2016**, *22*, 3287–3291.

411. D. Enders, P. Fey, H. Kipphardt, *Org. Synth.* **1987**, *65*, 173–182.

412. D. Lim, D. M. Coltart, *Angew. Chem. Int. Ed.* **2008**, *47*, 5207–5210.

413. E. H. Krenske, K. N. Houk, D. Lim, S. E. Wengryniuk, D. M. Coltart, *J. Org. Chem.* **2010**, *75*, 8578–8584.

414. M. R. Garnsey, D. Lim, J. M. Yost, D. M. Coltart, *Org. Lett.* **2010**, *12*, 5234–5237.

415. D. A. Evans, M. D. Ennis, D. J. Mathre, *J. Am. Chem. Soc.* **1982**, *104*, 1737–1739.

416. M. T. Reetz, *Angew. Chem. Int. Ed. Engl.* **1982**, *21*, 96–108.

417. K. Ohtsuki, K. Matsuo, T. Yoshikawa, C. Moriya, K. Tomita-Yokotani, K. Shishido, M. Shindo, *Org. Lett.* **2008**, *10*, 1247–1250.

418. M. Prashad, D. Har, L. Chen, H.-Y. Kim, O. Repic, T. J. Blacklock, *J. Org. Chem.* **2002**, *67*, 6612–6617.

419. A. G. Myers, B. H. Yang, H. Chen, J. L. Gleason, *J. Am. Chem. Soc.* **1994**, *116*, 9361–9362.

420. A. G. Myers, B. H. Yang, H. Chen, L. McKinstry, D. J. Kopecky, J. L. Gleason, *J. Am. Chem. Soc.* **1997**, *119*, 6496–6511.

421. A. G. Myers, L. McKinstry, *J. Org. Chem.* **1996**, *61*, 2428–2440.

422. D. Askin, R. P. Volante, K. M. Ryan, R. A. Reamer, I. Shinkai, *Tetrahedron Lett.* **1988**, *29*, 4245–4248.

423. I. T. Chen, I. Baitinger, L. Schreyer, D. Trauner, *Org. Lett.* **2014**, *16*, 166–169.

424. T. Ling, E. Griffith, K. Mitachi, F. Rivas, *Org. Lett.* **2013**, *15*, 5790–5793.

425. S. Levin, R. R. Nani, S. E. Reisman, *J. Am. Chem. Soc.* **2011**, *133*, 774–776.

426. C. Cui, W.-M. Dai, *Org. Lett.* **2018**, *20*, 3358–3361.

427. G. A. Molander, I. Shin, L. Jean-Gérard, *Org. Lett.* **2010**, *12*, 4384–4387.

428. A. G. Doyle, E. N. Jacobsen, *J. Am. Chem. Soc.* **2005**, *127*, 62–63.

429. A. G. Doyle, E. N. Jacobsen, *Angew. Chem. Int. Ed.* **2007**, *46*, 3701–3705.

430. D. A. Evans, R. J. Thomson, *J. Am. Chem. Soc.* **2005**, *127*, 10506–10507.

431. U. Kazmaier, *Org. Chem. Front.* **2016**, *3*, 1541–1560.

432. B. M. Trost, *Org. Process Res. Dev.* **2012**, *16*, 185–194.

433. R. A. Fernandes, J. L. Nallasivam, *Org. Biomol. Chem.* **2019**, *17*, 8647–8672.

434. Q. Cheng, H.-F. Tu, C. Zheng, J.-P. Qu, G. Helmchen, S.-L. You, *Chem. Rev.* **2019**, *119*, 1855–1969.

435. B. M. Trost, M. Osipov, G. Dong, *Org. Lett.* **2010**, *12*, 1276–1279.

436. B.-L. Lei, Q.-S. Zhang, W.-H. Yu, Q.-P. Ding, C.-H. Ding, X.-L. Hou, *Org. Lett.* **2014**, *16*, 1944–1947.

437. B. I. Estipona, B. P. Pritchett, R. A. Craig, B. M. Stoltz, *Tetrahedron* **2016**, *72*, 3707–3712.

438. X. Liang, T.-Y. Zhang, C.-Y. Meng, X.-D. Li, K. Wei, Y.-R. Yang, *Org. Lett.* **2018**, *20*, 4575–7578.

439. R. Shirai, M. Tanaka, K. Koga, *J. Am. Chem. Soc.* **1986**, *108*, 543–545.

440. C. M. Cain, R. P. C. Cousins, G. Coumbarides, N. S. Simpkins, *Tetrahedron* **1990**, *46*, 523–544.

441. A. Ando, T. Shioiri, *J. Chem. Soc. Chem. Commun.* **1987**, 656–658.

442. C. E. Stivala, A. Zakarian, *J. Am. Chem. Soc.* **2011**, *133*, 11936–11939.

443. W. Zhang, Z. Zhang, J.-C. Tang, J.-T. Che, H.-Y. Zhang, J.-H. Chen, Z. Yang, *J. Am. Chem. Soc.* **2020**, *142*, 19487–19492.

444. M. J. O'Donnell, *Acc. Chem. Res.* **2004**, *37*, 506–517.

445. B. Lygo, B. I. Andrews, *Acc. Chem. Res.* **2004**, *37*, 518–525.

446. T. Ooi, K. Maruoka, *Acc. Chem. Res.* **2004**, *37*, 526–533.

447. T. Ooi, K. Maruoka, *Angew. Chem. Int. Ed.* **2007**, *46*, 4222–4266.

448. J. Tan, N. Yasuda, *Org. Process Res. Dev.* **2015**, *19*, 1731–1746.

449. K. Maruoka, *Org. Process Res. Dev.* **2008**, *12*, 679–697.

450. E. J. Corey, F. Xu, M. C. Noe, *J. Am. Chem. Soc.* **1997**, *119*, 12414–12415.

451. E. Vedejs, D. A. Engler, J. E. Telschow, *J. Org. Chem.* **1978**, *43*, 188–196.

452. F. A. Davis, B. C. Chen, *Chem. Rev.* **1992**, *92*, 919–934.

453. E. Ciganek, *Org. React.* **2008**, *72*, 1–366.

454. E. Erdik, *Tetrahedron* **2004**, *60*, 8747–8782.

455. R. C. Larock, L. Zhang in *Comprehensive Organic Transformations: A Guide to Functional Group Preparations* (Ed.: R. C. Larock), John Wiley & Sons, **2018**.

456. J. C. Anderson, S. C. Smith, *Synlett* **1990**, 107–108.

457. G. M. Rubottom, M. A. Vazquez, D. R. Pelegrina, *Tetrahedron Lett.* **1974**, *15*, 4319–4322.

458. G. M. Rubottom, J. M. Gruber, R. K. Boeckman, M.Ramaiah, J. B. Medwid, *Tetrahedron Lett.* **1978**, *19*, 4603–4606.

459. Y.-X. Han, Y.-L. Jiang, Y. Li, H.-X. Yu, B.-Q. Tong, Z. Niu, S.-J. Zhou, S. Liu, Y. Lan, J.-H. Chen, Z. Yang, *Nature Comm.* **2017**, *8*, 14233.

460. R. M. Demoret, M. A. Baker, M. Ohtawa, S. Chen, C.-C. Lam, S. Khom, M. Roberto, S. Forli, K. N. Houk, R. A. Shenvi, *J. Am. Chem. Soc.* **2020**, *142*, 18599–18618.

461. S. Xu, M. A. Ciufolini, *Org. Lett.* **2015**, *17*, 2424–2427.

462. L. A. Paquette, H.-L. Wang, Z. Su, M. Zhao, *J. Am. Chem. Soc.* **1998**, *120*, 5213–5225.

463. D. Enders, V. Bhushan, *Tetrahedron Lett.* **1988**, *29*, 2437–2440.

464. D. A. Evans, M. M. Morrissey, R. L. Dorow, *J. Am. Chem. Soc.* **1985**, *107*, 4346–4348.

465. F. A. Davis, A. C. Sheppard, B. C. Chen, M. S. Haque, *J. Am. Chem. Soc.* **1990**, *112*, 6679–6690.

466. F. A. Davis, M. C. Weismiller, *J. Org. Chem.* **1990**, *55*, 3715–3717.

467. J. D. White, R. G. Carter, K. F. Sundermann, *J. Org. Chem.* **1999**, *64*, 684–685.

468. D. Urabe, T. Asaba, M. Inoue, *Bull. Chem. Soc. Jpn.* **2016**, *89*, 1137–1144.

469. S. Han, K. C. Morrison, P. J. Hergenrother, M. Movassaghi, *J. Org. Chem.* **2014**, *79*, 473–486.

470. G. J. Chuang, W. Wang, E. Lee, T. Ritter, *J. Am. Chem. Soc.* **2011**, *133*, 1760–1762.

471. A. S.-K. Tsang, A. Kapat, F. Schoenebeck, *J. Am. Chem. Soc.* **2016**, *138*, 518–526.

472. J. Streuff, *Synlett* **2013**, *24*, 276–280.

473. L. A. T. Allen, R.-C. Raclea, P. Natho, P. J. Parsons, *Org. Biomol. Chem.* **2020**, *19*, 498–513.

474. S. Derrer, J. E. Davies, A. B. Holmes, *J. Chem. Soc. Perkin Trans. 1* **2000**, 2943–2956.

475. N. S. Chowdari, C. F. Barbas, III, *Org. Lett.* **2005**, *7*, 867–870.

476. S. Kortet, A. Claraz, P. M. Pihko, *Org. Lett.* **2020**, *22*, 3010–3013.

477. J. Wang, M. Sánchez-Roselló, J. L. Aceña, C. del Pozo, A. E. Sorochinsky, S. Fustero, V. A. Soloshonok, H. Liu, *Chem. Rev.* **2014**, *114*, 2432–2506.

478. G. W. Gribble, *Mar. Drugs* **2015**, *13*, 4044–4136.

479. G. W. Gribble, *Environ. Sci. & Pollut. Res.* **2000**, *7*, 37–49.

480. L. Wang, X. Zhou, M. Fredimoses, S. Liao, Y. Liu, *RSC Adv.* **2014**, *4*, 57350–57376.

481. A. B. Smith III, J. R. Empfield, R. A. Rivero, H. A. Vaccaro, J. J. W. Duan, M. M. Sulikowski, *J. Am. Chem. Soc.* **1992**, *114*, 9419–9434.

482. Y. Sun, R. Zhou, H. Xu, D. Wang, X. Su, C. Wang, Y. Ding, L. Wang, Y. Chen, *Tetrahedron* **2019**, *75*, 1808–1818.

483. M. Jarret, A. Tap, V. Turpin, N. Denizot, C. Kouklovsky, E. Poupon, L. Evanno, G. Vincent, *Eur. J. Org. Chem.* **2020**, 6340–6351.

484. R. T. Thornbury, F. D. Toste, *Org. React.* **2020**, *100*, 749–800.

485. K. D. Dykstra, N. Ichiishi, S. W. Krska, P. F. Richardson in *Fluorine in Life Sciences: Pharmaceuticals, Medicinal Diagnostics, and Agrochemicals, Vol. IV* (Eds.: G. Haufe, F. R. Leroux), Academic Press, **2019**, Chapter 1, pp. 1–90.

486. K. Shibatomi, A. Narayama, *Asian J. Org. Chem.* **2013**, *2*, 812–823.

487. P. Kwiatkowski, T. D. Beeson, J. C. Conrad, D. W. C. MacMillan, *J. Am. Chem. Soc.* **2011**, *133*, 1738–1741.

488. J. Alvarado, A. T. Herrmann, A. Zakarian, *J. Org. Chem.* **2014**, *79*, 6206–6220.

489. B. Kang, S. Chang, S. Decker, R. Britton, *Org. Lett.* **2010**, *12*, 1716–1719.

490. M. S. Kharasch, P. O. Tawney, *J. Am. Chem. Soc.* **1941**, *63*, 2308–2316.

491. H. Gilman, R. H. Kirby, *J. Am. Chem. Soc.* **1941**, *63*, 2046–2048.

492. J.-M. Lefour, *Tetrahedron* **1978**, *34*, 2597–2605.

493. K. N. Houk, R. W. Strozier, *J. Am. Chem. Soc.* **1973**, *95*, 4094–4096.

494. K. B. Wiberg, R. E. Rosenberg, P. R. Rablen, *J. Am. Chem. Soc.* **1991**, *113*, 2890–2898.

495. S. Hanessian, Y. Yang, S. Giroux, V. Mascitti, J. Ma, F. Raeppel, *J. Am. Chem. Soc.* **2003**, *125*, 13784–13792.

496. D. Demeke, C. J. Forsyth, *Org. Lett.* **2000**, *2*, 3177–3179.

497. A. K. Miller, C. C. Hughes, J. J. Kennedy-Smith, S. N. Gradl, D. Trauner, *J. Am. Chem. Soc.* **2006**, *128*, 17057–17062.

498. B. H. Lipshutz, S. Sengupta, *Org. React.* **1992**, *41*, 135–631.

499. M. Yus, J. C. González-Gómez, F. Foubelo, *Chem. Rev.* **2013**, *113*, 5595–5598.

500. B. H. Lipshutz, *Synlett* **2009**, 509–524.

501. Q. Tian, G. Zhang, *Synthesis* **2016**, *48*, 4038–4049.

502. T. Ohshima in *Comp. Chirality, Vol. 4* (Ed.: M. Shibasaki), Elsevier, **2012**, Chapter 4.19, pp. 355–377.

503. E. Reyes, U. Uria, J. L. Vicario, L. Carrillo, *Org. React.* **2016**, *90*, 1–898.

504. S. R. McCabe, P. Wipf, *Angew. Chem. Int. Ed.* **2017**, *56*, 324–327.

505. H. J. Reich, *J. Org. Chem.* **2012**, *77*, 5471–5491.

506. N. Y. S. Lam, G. Muir, V. R. Challa, R. Britton, I. Paterson, *Chem. Commun.* **2019**, *55*, 9717–9720.

507. M. Kitamura, S. Suga, K. Kawai, R. Noyori, *J. Am. Chem. Soc.* **1986**, *108*, 6071–6072.

508. N. Oguni, Y. Matsuda, T. Kaneko, *J. Am. Chem. Soc.* **1988**, *110*, 7877–7878.

509. M. Kitamura, S. Okada, S. Suga, R. Noyori, *J. Am. Chem. Soc.* **1989**, *111*, 4028–4036.

510. S. Itsuno, J. M. J. Frechet, *J. Org. Chem.* **1987**, *52*, 4140–4142.

511. M. Yamakawa, R. Noyori, *J. Am. Chem. Soc.* **1995**, *117*, 6327–6335.

512. B. Goldfuss, K. N. Houk, *J. Org. Chem.* **1998**, *63*, 8998–9006.

513. P. Knochel, J. J. A. Perea, P. Jones, *Tetrahedron* **1998**, *54*, 8275–8319.

514. P. Knochel, N. Millot, A. L. Rodriguez, *Org. React.* **2001**, *58*, 417–745.

515. D. E. Frantz, R. Fässler, C. S. Tomooka, E. M. Carreira, *Acc. Chem. Res.* **2000**, *33*, 373–381.

516. C. M. Binder, B. Singaram, *Org. Prep. Proc. Int.* **2011**, *43*, 139–208.

517. T. Bauer, *Coord. Chem. Rev.* **2015**, *299*, 83–150.

518. K. Soai, S. Yokoyama, K. Ebihara, T. Hayasaka, *J. Chem. Soc. Chem. Commun.* **1987**, 1690–1691.

519. H. Takakashi, T. Kawakita, M. Ohno, M. Yoshioka, S. Kobayashi, *Tetrahedron* **1992**, *48*, 5691–5700.

520. K. Soai, K. Takahashi, *J. Chem. Soc. Perkin Trans. 1* **1994**, 1257–1258.

521. B. M. Trost, A. H. Weiss, A. J. von Wangelin, *J. Am. Chem. Soc.* **2006**, *128*, 8–9.

522. B. M. Trost, M. J. Bartlett, A. H. Weiss, A. J. von Wangelin, V. S. Chan, *Chem. Eur. J.* **2012**, *18*, 16498–16509.

523. D. Boyall, D. E. Frantz, E. M. Carreira, *Org. Lett.* **2002**, *4*, 2605–2606.

524. N. K. Anand, E. M. Carreira, *J. Am. Chem. Soc.* **2001**, *123*, 9687–9688.

525. P. J. Walsh, *Acc. Chem. Res.* **2003**, *36*, 739–749.

526. K.-H. Wu, H.-M. Gau, *Organometallics* **2004**, *23*, 580–588.

527. D. A. Evans, W. C. Black, *J. Am. Chem. Soc.* **1992**, *114*, 2260–2262.

528. J. Mlynarski, J. Ruiz-Caro, A. Fürstner, *Chem. Eur. J.* **2004**, *10*, 2214–2222.

529. D. Trauner, J. B. Schwarz, S. J. Danishefsky, *Angew. Chem. Int. Ed.* **1999**, *38*, 3542–3545.

530. K. Takai, *Org. React.* **2004**, *64*, 253–612.

531. A. Fürstner, N. Shi, *J. Am. Chem. Soc.* **1996**, *118*, 12349–12357.

532. K. Namba, Y. Kishi, *Org. Lett.* **2004**, *6*, 5031–5033.

533. K. Namba, J. Wang, S. Cui, Y. Kishi, *Org. Lett.* **2005**, *7*, 5421–5424.

534. W. Harnying, A. Kaiser, A. Klein, A. Berkessel, *Chem. Eur. J.* **2011**, *17*, 4765–4773.

535. A. Gil, F. Albericio, M. Álvarez, *Chem. Rev.* **2017**, *117*, 8420–8446.

536. K. ichi Takao, A. Ogura, K. Yoshida, S. Simizu, *Synlett* **2020**, 421–433.

537. Y. Gao, D. E. Hill, W. Hao, B. J. McNicholas, J. C. Vantourout, R. G. Hadt, S. E. Reisman, D. G. Blackmond, P. S. Baran, *J. Am. Chem. Soc.* **2021**, *143*, 9478–9488.

538. M. Nakata, J. Ohashi, K. Ohsawa, T. Nishimura, M. Kinoshita, K. Tatsuta, *Bull. Chem. Soc. Jpn.* **1993**, *66*, 3464–3474.

539. K. Kong, Z. Moussa, C. Lee, D. Romo, *J. Am. Chem. Soc.* **2011**, *133*, 19844–19856.

540. K. Suenaga, H. Hoshino, T. Yoshii, K. Mori, H. Sone, Y. Bessho, A. Sakakura, I. Hayakawa, K. Yamada, H. Kigoshi, *Tetrahedron* **2006**, *62*, 7687–7698.

541. I. C. González, C. J. Forsyth, *J. Am. Chem. Soc.* **2000**, *122*, 9099–9108.

542. W. C. Still, D. Mobilio, *J. Org. Chem.* **1983**, *48*, 4785–4786.

543. K. Takao, K. Tsunoda, T. Kurisu, A. Sakama, Y. Nishimura, K. Yoshida, K. Tadano, *Org. Lett.* **2015**, *17*, 756–759.

544. J. T. Njardarson, K. Biswas, S. J. Danishefsky, *Chem. Commun.* **2002**, 2759–2761.

545. K. Takao, N. Hayakawa, R. Yamada, T. Yamaguchi, U. Morita, S. Kawasaki, K. Tadano, *Angew. Chem. Int. Ed.* **2008**, *47*, 3426–3429.

546. Y. Ai, N. Ye, Q. Wang, K. Yahata, Y. Kishi, *Angew. Chem. Int. Ed.* **2017**, *56*, 10791–10795.

547. Y. Kaburagi, K. Kira, K. Yahata, K. Iso, Y. Sato, F. Matsuura, I. Ohashi, Y. Matsumoto, M. Isomura, T. Sasaki, T. Fukuyama, Y. Miyashita, H. Azuma, D. Iida, T. Ishida, W. Itano, M. Matsuda, M. Matsukura, N. Murai, S. Nagao, M. Seki, A. Yamamoto, Y. Yamamoto, N. Yoneda, Y. Watanabe, A. Kamada, A. Kayano, K. Tagami, O. Asano, T. Owa, Y. Kishi, *Org. Lett.* **2024**, *26*, 2837–2842.

548. S. E. Denmark, N. G. Almstead in *Modern Carbonyl Chemistry* (Ed.: J. Otera), Wiley-VCH, **2000**, Chapter 10, pp. 299–401.

549. S. E. Denmark, J. Fu, *Chem. Rev.* **2003**, *103*, 2763–2793.

550. L. F. Tietze, T. Kinzel, S. Schmatz, *J. Am. Chem. Soc.* **2006**, *128*, 11483–11495.

551. L. M. Wolf, S. E. Denmark, *J. Am. Chem. Soc.* **2013**, *135*, 4743–4756.

552. T. Yoshimitsu, N. Fukumoto, R. Nakatani, N. Kojima, T. Tanaka, *J. Org. Chem.* **2010**, *75*, 5425–5437.

553. J.-M. Huang, K.-C. Xu, T.-P. Loh, *Synthesis* **2003**, 755–764.

554. L. R. Reddy, P. Saravanan, E. J. Corey, *J. Am. Chem. Soc.* **2004**, *126*, 6230–6231.

555. E. de Lemos, F.-H. Porée, A. Commercon, J.-F. Betzer, A. Pancrazi, J. Ardisson, *Angew. Chem. Int. Ed.* **2007**, *46*, 1917–1921.

556. M. T. Crimmins, J. M. Ellis, K. A. Emmitte, P. A. Haile, P. J. McDougall, J. D. Parrish, J. L. Zuccarello, *Chem. Eur. J.* **2009**, *15*, 9223–9234.

557. M. Kito, T. Sakai, H. Shirahama, M. Miyashita, F. Matsuda, *Synlett* **1997**, 219–220.

558. M. Kita, H. Oka, A. Usui, T. Ishitsuka, Y. Mogi, H. Watanabe, M. Tsunoda, H. Kigoshi, *Angew. Chem. Int. Ed.* **2015**, *54*, 14174–14178.

559. F. Corral-Bautista, L. Klier, P. Knochel, H. Mayr, *Angew. Chem. Int. Ed.* **2015**, *54*, 12497–12500.

560. C. García-Ruiz, J. L.-Y. Chen, C. Sandford, K. Feeney, P. Lorenzo, G. Berionni, H. Mayr, V. K. Aggarwal, *J. Am. Chem. Soc.* **2017**, *139*, 15324–15327.

561. B. W. Gung, X. Xue, W. R. Roush, *J. Am. Chem. Soc.* **2002**, *124*, 10692–10697.

562. H. Lachance, D. G. Hall, *Org. React.* **2008**, *73*, 1–573.

563. X. Zhang, K. N. Houk, J. L. Leighton, *Angew. Chem. Int. Ed.* **2005**, *44*, 938–941.

564. A. B. Smith, III, C. M. Adams, S. A. L. Barbosa, A. P. Degnan, *J. Am. Chem. Soc.* **2003**, *125*, 350–351.

565. E. M. Flamme, W. R. Roush, *Org. Lett.* **2005**, *7*, 1411–1414.

566. H. Guo, M. S. Mortensen, G. A. O'Doherty, *Org. Lett.* **2008**, *10*, 3149–3152.

567. D. R. Williams, S. Patnaik, S. V. Plummer, *Org. Lett.* **2003**, *5*, 5035–5038.

568. J. H. Lee, *Tetrahedron* **2020**, *33*, 131351.

569. M. Yus, J. C. González-Gómez, F. Foubelo, *Chem. Rev.* **2011**, *111*, 7774–7854.

570. H.-X. Huo, J. R. Duvall, M.-Y. Huang, R. Hong, *Org. Chem. Front.* **2014**, *1*, 303–320.

571. K. Furuta, M. Mouri, H. Yamamoto, *Synlett* **1991**, 561–562.

572. K. Ishihara, M. Mouri, Q. Gao, T. Maruyama, K. Furuta, H. Yamamoto, *J. Am. Chem. Soc.* **1993**, *115*, 11490–11495.

573. A. L. Costa, M. G. Piazza, E. Tagliavini, C. Trombini, A. Umani-Ronchi, *J. Am. Chem. Soc.* **1993**, *115*, 7001–7002.

574. G. E. Keck, K. H. Tarbet, L. S. Geraci, *J. Am. Chem. Soc.* **1993**, *115*, 8467–8468.

575. T. A. Fattaha, A. Saeed, *New. J. Chem.* **2017**, *41*, 12804–14821.

576. S. Lou, P. N. Moquist, S. E. Schaus, *J. Am. Chem. Soc.* **2006**, *128*, 12660–12661.

577. D. S. Barnett, P. N. Moquist, S. E. Schaus, *Angew. Chem. Int. Ed.* **2009**, *48*, 8679–8682.

578. V. Rauniyar, D. G. Hall, *Angew. Chem. Int. Ed.* **2006**, *45*, 2426–2428.

579. V. Rauniyar, H. Zhai, D. G. Hall, *J. Am. Chem. Soc.* **2008**, *130*, 8481–8490.

580. E. J. Corey, T. W. Lee, *Chem. Commun.* **2001**, 1321–1329.

581. G. E. Keck, D. Krishnamurthy, M. C. Grier, *J. Org. Chem.* **1993**, *58*, 6543–6544.

582. S. E. Denmark, S. Hosoi, *J. Org. Chem.* **1994**, *59*, 5133–5135.

583. J. W. Faller, D. W. I. Sams, X. Liu, *J. Am. Chem. Soc.* **1996**, *118*, 1217–1218.

584. M. Kurosu, M. Lorca, *Synlett* **2005**, 1109–1112.

585. J. P. Vitale, S. A. Wolckenhauer, N. M. Do, S. D. Rychnovsky, *Org. Lett.* **2005**, *7*, 3255–3258.

586. G. E. Keck, R. L. Giles, V. J. Cee, C. A. Wager, T. Yu, M. B. Kraft, *J. Org. Chem.* **2008**, *73*, 9675–9691.

587. E. P. Farney, S. S. Feng, F. Schäfers, S. E. Reisman, *J. Am. Chem. Soc.* **2018**, *140*, 1267–1270.

588. Y. Lu, S. K. Woo, M. J. Krische, *J. Am. Chem. Soc.* **2008**, *133*, 13876–13879.

589. M. Penner, V. Rauniyar, L. T. Kaspar, D. G. Hall, *J. Am. Chem. Soc.* **2009**, *131*, 14216–14217.

590. D. M. Sedgwick, M. N. Grayson, S. Fustero, P. Barrio, *Synthesis* **2018**, *50*, 1935–1957.

591. K. Spielmann, G. Niel, R. M. de Figueiredo, J.-M. Campagne, *Chem. Soc. Rev.* **2018**, *47*, 1159–1173.

592. S. W. Kim, W. Zhang, M. J. Krische, *Acc. Chem. Res.* **2017**, *50*, 2371–2380.

593. I. S. Kim, M.-Y. Ngai, M. J. Krische, *J. Am. Chem. Soc.* **2008**, *130*, 14891–14899.

594. S. B. Han, X. Gao, M. J. Krische, *J. Am. Chem. Soc.* **2010**, *132*, 9153–9156.

595. J. Feng, Z. A. Kasun, M. J. Krische, *J. Am. Chem. Soc.* **2016**, *138*, 5467–5478.

596. R. S. Doerksen, C. C. Meyer, M. J. Krische, *Angew. Chem. Int. Ed.* **2019**, *58*, 14055–14064.

597. M. Shimomura, M. Sato, H. Azuma, J. Sakata, H. Tokuyama, *Org. Lett.* **2020**, *22*, 3313–3317.

598. M. Yang, W. Peng, Y. Guo, T. Ye, *Org. Lett.* **2020**, *22*, 1776–1779.

599. F. Della-Felice, A. M. Sarotti, R. A. Pilli, *J. Org. Chem.* **2017**, *82*, 9191–9197.

600. S. B. Han, A. Hassan, I. S. Kim, M. J. Krische, *J. Am. Chem. Soc.* **2010**, *132*, 15559–15561.

601. G. Wang, M. J. Krische, *J. Am. Chem. Soc.* **2016**, *138*, 8088–8091.

602. C. C. Meyer, N. P. Stafford, M. J. Cheng, M. J. Krische, *Angew. Chem. Int. Ed.* **2021**, *60*, 10542–10546.

603. M. Holmes, L. A. Schwartz, M. J. Krische, *Chem. Rev.* **2018**, *118*, 6026–6052.

604. F. Panahi, F. Bauer, B. Breit, *Acc. Chem. Res.* **2023**, *56*, 3676–3693.

605. R. E. Ireland, J. L. Gleason, L. D. Gegnas, T. K. Highsmith, *J. Org. Chem.* **1996**, *61*, 6856–6872.

606. N. Yoshikai, E. Nakamura, *Chem. Rev.* **2012**, *112*, 2339–2372.

607. A. Alexakis in *Transition Metals for Organic Synthesis*, *Vol. 1* (Eds.: M. Beller, C. Bolm), Wiley-VCH, 2nd ed., **2004**, Chapter 3.8, pp. 553–562.

608. S. Weske, T. Auth, F. Kraft, A. Winkler, S. Schneider, K. Koszinowski, *Organometallics* **2024**, *43*, 3216–3225.

609. I. M. DiMucci, J. T. Lukens, S. Chatterjee, K. M. Carsch, C. J. Titus, S. J. Lee, D. Nordlund, T. A. Betley, S. N. MacMillan, K. M. Lancaster, *J. Am. Chem. Soc.* **2019**, *141*, 18508–18520.

610. S. Woodward, *Chem. Soc. Rev.* **2000**, *29*, 393–401.

611. W. H. Mandeville, G. M. Whitesides, *J. Org. Chem.* **1974**, *39*, 400–405.

612. G. H. Posner, C. E. Whitten, J. J. Sterling, D. J. Brunelle, *Tetrahedron Lett.* **1974**, *15*, 2591–2594.

613. H. O. House, C.-Y. Chu, J. M. Wilkins, M. J. Umen, *J. Org. Chem.* **1975**, *40*, 1460–1469.

614. A. Alexakis, J. E. Bäckvall, N. Krause, O. Pàmies, M. Diéguez, *Chem. Rev.* **2008**, *108*, 2796–2823.

615. S. H. Bertz, S. Cope, M. Murphy, C. A. Ogle, B. J. Taylor, *J. Am. Chem. Soc.* **2007**, *129*, 7208–7209.

616. M. Eriksson, A. Johansson, M. Nilsson, T. Olsson, *J. Am. Chem. Soc.* **1996**, *118*, 10904–10905.

617. D. E. Frantz, D. A. Singleton, *J. Am. Chem. Soc.* **2000**, *122*, 3288–3295.

618. A. B. Smith, III, P. J. Jerris, *J. Am. Chem. Soc.* **1981**, *103*, 194–195.

619. E. Nakamura, M. Yamanaka, S. Mori, *J. Am. Chem. Soc.* **2000**, *122*, 1826–1827.

620. P. Siengalewicz, J. Mulzer, U. Rinner in *Comp. Chirality*, *Vol. 2* (Ed.: J. Mulzer), Elsevier, **2012**, Chapter 2.15, pp. 441–471.

621. G. B. Dudley, K. S. Takaki, D. D. Cha, R. L. Danheiser, *Org. Lett.* **2000**, *2*, 3407–3410.

622. G. Barbe, A. B. Charette, *J. Am. Chem. Soc.* **2008**, *130*, 13873–13875.

623. W. Kaplan, H. R. Khatri, P. Nagorny, *J. Am. Chem. Soc.* **2016**, *138*, 7194–7198.

624. T. Thaler, P. Knochel, *Angew. Chem. Int. Ed.* **2009**, *48*, 645–648.

625. M. Hayashi, R. Matsubara, *Tetrahedron Lett.* **2017**, *58*, 1793–1805.

626. Z. Wang, *Org. Chem. Front.* **2020**, *7*, 3815–3841.

627. A. Cernijenko, R. Risgaard, P. S. Baran, *J. Am. Chem. Soc.* **2016**, *138*, 9425–9428.

628. C. Ferrer, P. Fodran, S. Barroso, R. Gibson, E. C. Hopmans, J. S. Damsté, S. Schouten, A. J. Minnaard, *Org. Biomol. Chem.* **2013**, *11*, 2482–2492.

629. C. Peng, P. Arya, Z. Zhou, S. A. Snyder, *Angew. Chem. Int. Ed.* **2020**, *59*, 13521–13525.

630. B. ter Horst, B. L. Feringa, A. J. Minnaard, *Org. Lett.* **2007**, *9*, 3013–3015.

631. C. Hui, F. Pu, J. Xu, *Chem. Eur. J.* **2017**, *23*, 4023–4036.

632. J. R. Thielman, D. H. Sherman, R. M. Williams, *J. Org. Chem.* **2020**, *85*, 3812–3823.

633. C. Deutsch, N. Krause, B. H. Lipshutz, *Chem. Rev.* **2008**, *108*, 2916–2927.

634. A. V. Malkov, K. Lawson, *Top. Organomet. Chem.* **2016**, *58*, 207–220.

635. A. J. Jordan, G. Lalic, J. P. Sadighi, *Chem. Rev.* **2016**, *116*, 8318–8372.

636. S. Sakamoto, H. Sakazaki, K. Hagiwara, K. Kamada, K. Ishii, T. Noda, M. Inoue, M. Hirama, *Angew. Chem. Int. Ed.* **2004**, *43*, 6505–6510.

637. G. Hughes, M. Kimura, S. L. Buchwald, *J. Am. Chem. Soc.* **2003**, *125*, 11253–11258.

638. B. H. Lipshutz, J. M. Servesko, B. R. Taft, *J. Am. Chem. Soc.* **2004**, *126*, 8352–8353.

639. H. Liu, W. Zhang, L. He, M. Luo, S. Qin, *RSC Adv.* **2014**, *4*, 5726–5733.

640. D. H. Appella, Y. Moritani, R. Shintani, E. M. Ferreira, S. L. Buchwald, *J. Am. Chem. Soc.* **1999**, *121*, 9473–9474.

641. B. H. Lipshutz, J. M. Servesko, T. B. Petersen, P. P. Papa, A. A. Lover, *Org. Lett.* **2004**, *6*, 1273–1275.

642. S. Jeon, J. Lee, S. Park, S. Han, *Chem. Sci.* **2020**, *11*, 10928–10932.

643. J. M. Smith, J. Moreno, B. W. Boal, N. K. Garg, *J. Org. Chem.* **2015**, *80*, 8954–8967.

644. I. Paterson, E. A. Anderson, S. M. Dalby, J. H. Lim, J. Genovino, P. Maltas, C. Moessner, *Angew. Chem. Int. Ed.* **2008**, *47*, 3016–3020.

645. M. Movassaghi, A. E. Ondrus, *Org. Lett.* **2005**, *7*, 4423–4426.

646. T. Ikeno, T. Kimura, Y. Ohtsuka, T. Yamada, *Synlett* **1999**, 96–98.

647. A. F. Barrero, E. J. Alvarez-Manzaneda, R. Chahboun, R. Meneses, *Synlett* **1999**, 1663–1666.

648. R. S. Dhillon, R. P. Singh, D. Kaur, *Tetrahedron Lett.* **1995**, *36*, 1107–1108.

649. A. Haskel, E. Keinan in *Handbook of Organopalladium Chemistry for Organic Synthesis, Vol. 2* (Ed.: E. Negishi), John Wiley & Sons, **2002**, Chapter VII.2.3, pp. 2767–2782.

650. Q. Cheng, H.-F. Tu, C. Zheng, J.-P. Qu, G. Helmchen, S.-L. You, *Chem. Rev.* **2019**, *119*, 1855–1969.

651. L. Milhau, P. J. Guiry, *Top. Organomet. Chem.* **2011**, *38*, 95–153.

652. C. Moberg, *Org. React.* **2014**, *84*, 1–73.

653. S. D. Roughley, A. M. Jordan, *J. Med. Chem.* **2011**, *54*, 3451–3479.

654. J. S. Carey, D. Laffan, C. Thomson, M. T. Williams, *Org. Biomol. Chem.* **2006**, *4*, 2337–2374.

655. A. El-Faham, F. Albericio, *Chem. Rev.* **2011**, *111*, 6557–6602.

656. K. Hollanders, B. U. W. Maes, S. Ballet, *Synthesis* **2019**, *51*, 2261–2277.

657. J. R. Dunetz, J. Magano, G. A. Weisenburger, *Org. Process Res. Dev.* **2016**, *20*, 140–177.

658. G. E. Veitch, E. Beckmann, B. J. Burke, A. Boyer, C. Ayats, S. V. Ley, *Angew. Chem. Int. Ed.* **2007**, *46*, 7633–7635.

659. M. Ball, S. P. Andrews, F. Wierschem, E. Cleator, M. D. Smith, S. V. Ley, *Org. Lett.* **2007**, *9*, 663–666.

660. L. Liao, J. Zhou, Z. Xu, T. Ye, *Angew. Chem. Int. Ed.* **2016**, *55*, 13263–13266.

661. A. Stumpf, Z. K. Cheng, D. Beaudry, R. Angelaud, F. Gosselin, *Org. Process Res. Dev.* **2019**, *23*, 1829–1840.

662. K. C. K. Swamy, N. N. B. Kumar, E. Balaraman, K. V. P. P. Kumar, *Chem. Rev.* **2009**, *109*, 2551–2651.

663. D. L. Hughes, *Org. React.* **1992**, *42*, 335–656.

664. A. B. Smith, III, G. A. Sulikowski, K. Fujimoto, *J. Am. Chem. Soc.* **1989**, *111*, 8039–8041.

665. A. Kalivretenos, J. K. Stille, L. S. Hegedus, *J. Org. Chem.* **1991**, *56*, 2883–2894.

666. R. M. Garbaccio, S. J. Stachel, D. K. Baeschlin, S. J. Danishefsky, *J. Am. Chem. Soc.* **2001**, *123*, 10903–10908.
667. A. B. Smith, III, Q. Han, P. A. S. Breslin, G. K. Beauchamp, *Org. Lett.* **2005**, *7*, 5075–5078.
668. N. Hashimoto, T. Aoyama, T. Shioiri, *Chem. Pharm. Bull.* **1981**, *29*, 1475–1478.
669. E. Kühnel, D. D. P. Laffan, G. C. Lloyd-Jones, T. M. del Campo, I. R. Shepperson, J. L. Slaughter, *Angew. Chem. Int. Ed.* **2007**, *46*, 7075–7078.
670. M. Kretschmer, M. Dieckmann, P. Li, S. Rudolph, D. Herkommer, J. Troendlin, D. Menche, *Chem. Eur. J.* **2013**, *19*, 15993–16018.
671. J. Magano, *Org. Process Res. Dev.* **2022**, *26*, 1562–1689.
672. Y. J. Pu, R. K. Vaid, S. K. Boini, R. W. Towsley, C. W. Doecke, D. Mitchell, *Org. Process Res. Dev.* **2009**, *13*, 310–314.
673. N. A. Magnus, T. M. Braden, J. Y. Buser, A. C. DeBaillie, P. C. Heath, C. P. Ley, J. R. Remacle, D. L. Varie, T. M. Wilson, *Org. Process Res. Dev.* **2012**, *16*, 830–835.
674. V. R. Pattabiraman, J. W. Bode, *Nature* **2011**, *480*, 471–479.
675. M. Todorovic, D. M. Perrin, *Pept. Sci.* **2020**, *112*, e24210.
676. K. Ishihara, S. Ohara, H. Yamamoto, *J. Org. Chem.* **1996**, *61*, 4196–4197.
677. S. Arkhipenko, M. T. Sabatini, A. S. Batsanov, V. Karaluka, T. D. Sheppard, H. S. Rzepa, A. Whiting, *Chem. Sci.* **2018**, *9*, 1058–1072.
678. J. Zhang, G. Leitus, Y. Ben-David, D. Milstein, *J. Am. Chem. Soc.* **2005**, *127*, 10840–10841.
679. C. Gunanathan, Y. Ben-David, D. Milstein, *Science* **2007**, *317*, 790–792.
680. G. E. Dobereiner, R. H. Crabtree, *Chem. Rev.* **2010**, *110*, 681–703.
681. R. H. Crabtree, *Chem. Rev.* **2017**, *117*, 9228–9246.
682. M. Trincado, J. Bösken, H. Grützmacher, *Coord. Chem. Rev.* **2021**, *443*, 213967.
683. A. Parenty, X. Moreau, G. Niel, J.-M. Campagne, *Chem. Rev.* **2013**, *113*, PR1–PR40.
684. M. Cordes, M. Kalesse, *Top. Heterocycl. Chem.* **2014**, *36*, 369–427.
685. J. R. Cochrane, D. H. Yoon, C. S. P. McErlean, K. A. Jolliffe, *Beilst. J. Org. Chem.* **2012**, *8*, 1344–1351.
686. E. J. Jeong, E. J. Kang, L. T. Sung, S. K. Hong, E. Lee, *J. Am. Chem. Soc.* **2002**, *124*, 14655–14662.
687. W. R. Roush, R. J. Sciotti, *J. Am. Chem. Soc.* **1998**, *120*, 7411–7419.
688. B. Ma, B. Banerjee, D. N. Litvinov, L. He, S. L. Castle, *J. Am. Chem. Soc.* **2010**, *132*, 1159–1171.
689. Z. J. Song, D. M. Tellers, M. Journet, J. T. Kuethe, D. Lieberman, G. Humphrey, F. Zhang, Z. Peng, M. S. Waters, D. Zewge, A. Nolting, D. Zhao, R. A. Reamer, P. G. Dormer, K. M. Belyk, I. W. Davies, P. N. Devine, D. M. Tschaen, *J. Org. Chem.* **2011**, *76*, 7804–7815.
690. M. Adler, S. Adler, G. Boche, *J. Phys. Org. Chem.* **2005**, *18*, 193–209.
691. S. Nahm, S. M. Weinreb, *Tetrahedron Lett.* **1981**, *22*, 3815–3818.
692. S. Balasubramaniam, I. S. Aidhen, *Synthesis* **2008**, *23*, 3707–3738.
693. R. Senatore, L. Ielo, S. Monticelli, L. Castoldi, V. Pace, *Synthesis* **2019**, *51*, 2792–2808.
694. M. Barbazanges, C. Meyer, J. Cossy, *Org. Lett.* **2008**, *10*, 4489–4492.
695. A. Fürstner, P. Karier, F. Ungeheuer, A. Ahlers, F. Anderl, C. Wille, *Angew. Chem. Int. Ed.* **2019**, *58*, 248–253.
696. M. Pérez, C. del Pozo, F. Reyes, A. Rodríguez, A. Francesch, A. M. Echavarren, C. Cuevas, *Angew. Chem. Int. Ed.* **2004**, *43*, 1724–1727.
697. G. Sirasani, R. B. Andrade, *Org. Lett.* **2011**, *13*, 4736–4737.
698. T. C. Ho, H. Kamimura, K. Ohmori, K. Suzuki, *Org. Lett.* **2016**, *18*, 4488–4490.
699. D. A. Evans, L. Kværnø, T. B. Dunn, A. Beauchemin, B. Raymer, J. A. Mulder, E. J. Olhava, M. Juhl, K. Kagechika, D. A. Favor, *J. Am. Chem. Soc.* **2008**, *130*, 16295–16309.
700. Y. Tsunematsu, S. Nishimura, A. Hattori, S. Oishi, N. Fujii, H. Kakeya, *Org. Lett.* **2015**, *17*, 258–261.
701. K. A. Scott, J. T. Njardarson, *Top. Curr. Chem.* **2018**, *376*, 5.

Synthesis and Derivatization of Alkenes and Alkynes

3

CC-multiple bonds may be considered one of the most versatile functional groups in organic chemistry, along with carbonyl derivatives. Olefins constitute a platform functionality that enables the selective introduction of 1,2-difunctionalized segments in a carbon backbone through a variety of oxidative methods. Furthermore, their stereoconfiguration is a handle that allows for subsequent diastereo- and enantioselective conversions to proceed. For this reason, it is not surprising that in addition to a wide range of olefin derivatizations, an equivalent effort has gone into accessing alkenes not only efficiently, but diastereoselectively as well.

Alkynes offer two particularly efficient means for their derivatization. Either the triple bond is leveraged for the oxidative or reductive construction of a specific alkene, or, in the case of terminal alkynes, the high $C_{sp}H$ acidity is exploited to use them as nucleophiles in addition or substitution reactions.

Carbonyl derivatives can be employed for the construction of both alkenes and alkynes. Since alkenes can also be oxidatively cleaved to afford carbonyl compounds, olefins can be considered to represent masked carbonyls.

A. Düfert, *Methods of Organic Synthesis*,
https://doi.org/10.1007/978-3-662-70963-4_3

3.1   Synthesis of Alkenes

The diastereoselective access to olefins often relies on the derivatization of the corresponding carbonyl compounds by phosphorus or sulfur ylides, especially in academic syntheses. A second common approach encompasses the use of transition metal complexes. These methods notwithstanding, eliminations or pericyclic reactions are encountered as well, albeit to a lesser degree (see Fig. 3.1).

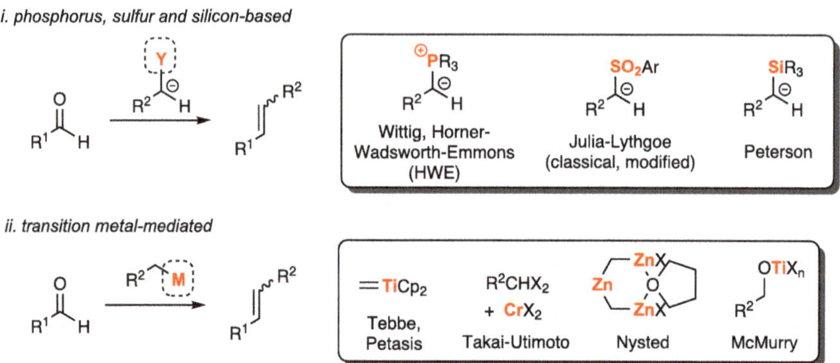

Fig. 3.1 Common reagents for the construction of CC double bonds from carbonyl derivatives

In an industrial context, the use of ylides is often prohibitive due to the considerable amount of by-products received [1]. Depending on the reaction scale (gram, kilogram, or ton scale), olefin metathesis can be a suitable complementary method, as it often shows the highest atom economy. Details on metathesis reactions will not be further elaborated here but can be found in Chap. 7.

The stereoselectivity in the formation of 1,2-disubstituted alkenes can usually be well controlled. Trisubstituted alkenes on the other hand are typically obtained much less selectively, as steric effects usually do not favor one of the products or one of the transition states as clearly. In the case of electronic and steric interactions directing to different products, "atypical" diastereomeric mixtures can also be received.

An overview of the most frequently employed methods and their diastereoselectivity can be found in Table 3.1.

Apart from the reaction conditions of the alkene formation step, one of the defining differences between the individual methods also is expressed in the preparation of the corresponding olefination reagents. The lack of functional group tolerance in the synthesis of sulfur- and phosphorus-derived ylides, for example, might preclude its application with certain substrates or necessitate alternative routes in lieu of the standard approach. A prominent example of this can be found in the Horner–Wadsworth–Emmons (HWE) reaction, wherein the olefination reagent may be accessed by means of an Arbuzov reaction. Furthermore, the synthesis of S-ylides is often much simpler and milder than the access to the respective P-ylides.

Table 3.1 Overview of common olefination methods [2–5]

reaction	*E/Z*-selective	remarks
Wittig	*E* or *Z*	often *Z*-selective; *E/Z* ratio dependent on ylide (stabilized, non-stabilized) and can be influenced by salt additives and solvent
Wittig (*Schlosser modification*)	*E*	*E*-selective with non-stabilized ylides
HWE (*standard, Masamune-Roush*)	*E*	sterically demanding substrates are more reactive than in classic Wittig; phosphonate group influences stereoselectivity
HWE (*Still-Gennari, Ando*)	*Z*	*Z*-selective even with trisubstituted alkenes
classic Julia-Lythgoe	rather *E*	multistep process
modified Julia-Lythgoe	*E* or *Z*	selectivity can be controlled by conditions
Tebbe, Petasis	-	methenylation; also reacts with esters, amides
Takai-Utimoto	*E*	high excess of Cr reagent necessary; neutral conditions; $k_{\text{aldehyde}} \gg k_{\text{ketone}}$
Nysted	-	methenylation; almost neutral conditions
McMurry	not selective	mostly used for homocoupling; particularly suitable for sterically demanding substrates; easily reducible groups in the substrate are affected / degraded
Metathesis	*E* or *Z*	mostly *E*-selective, also *Z*-selective catalysts known
Peterson	*E* or *Z*	stereoselectivity can be reversed by acidic/basic conditions
Corey-Winter	*E* or *Z*	completely stereospecific; particularly suitable for sterically demanding substrates
Bamford-Stevens & Shapiro	not selective	often forms sterically least hindered olefin
Barton-Kellogg	not selective	particularly suitable for tetrasubstituted olefins
Ramberg-Bäcklund	*Z*	with KO*t*Bu as base, *E*-alkenes are major product

3.1.1 Olefination Using Phosphorus Reagents

The conversion of carbonyl derivatives with phosphorus ylides was discovered in the 1950s by Wittig and Schöllkopf [6]. The so-called *Wittig* reaction has been continuously improved since its inception and has become one of the most important tools for the selective construction of double bonds, especially since no shift of the formed double bond occurs over the course of the transformation [7, 8]. Based on Wittig's seminal contributions, further variants such as the Schlosser modification and the Horner–Wadsworth–Emmons reaction (HWE) were subsequently introduced. They differ from the classic Wittig reaction in the type of phosphorus reagents used (phosphonium ylide *vs.* phosphonate anion) and the reaction conditions (see Fig. 3.2).

The classic Wittig reaction displays the broadest substrate scope: ylides with alkyl, aryl, vinyl substituents, or other electron-withdrawing groups (EWG = ester, ketone, nitrile, sulfonate, etc.) in vinylic position (R') can be employed. The Schlosser modification, on

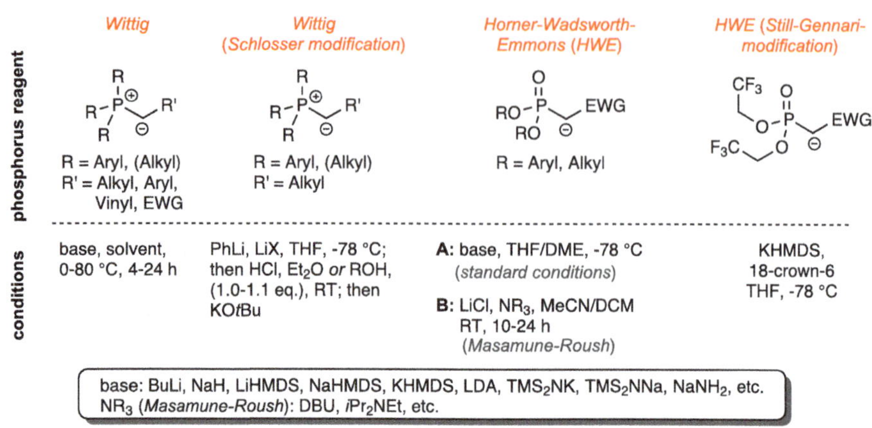

Fig. 3.2 Comparison of olefination methods mediated by phosphorus reagents

the other hand, is only used with alkyl-substituted ylides, as it offers the complementary diastereoselectivity compared to the classic Wittig conditions for these substrates.

Phosphonium ylides are categorized into three classes depending on the vinylic substituents on the α-carbon, which also correlate with their reactivity: [3] Alkyl-substituted ylides are considered *non-stabilized* and are already sensitive to atmospheric moisture, while aryl or vinyl-substituted ylides are referred to as *semi-stabilized* and are less susceptible to hydrolysis. If the ylide bears an electron-withdrawing group in the α-position, it is termed as *stabilized*; the latter ylides are usually stable against hydrolytic decomposition.

The driving force of the olefination relies on the cleavage of a weak P-C bond to form a significantly more stable phosphine oxide. The exact mechanism is complex and depends on several factors: the nature of the ylide (non-stabilized *vs.* stabilized), the presence of soluble salts in the reaction medium, as well as the nature of the substituents on the P-atom [3, 8]. The Schlosser modification and the HWE reaction proceed via related reaction paths (see below).

The mechanism of the Wittig reaction under classic conditions has been intensively studied for decades, both experimentally as well as by theoretical means. The steps described below apply to olefinations in which the resulting salt does **not** dissolve in the reaction medium. This almost exclusively applies to lithium salts, which result during the ylide generation from bases like BuLi or LiHMDS and the counterion of the phosphonium precursor.

The now generally accepted [3] mechanism of the Wittig reaction involves only three discrete steps and is shown in Fig. 3.3: The initial $[\pi 2s+\pi 2a]$-cycloaddition of the substrates via transition state **1** to an oxaphosphetane **2** is followed by a rearrangement of the oxaphosphetane (**2**→**3**). In the last step, a [2+2]-cycloreversion of the oxaphosphetane via **4** occurs, affording the alkene and an equivalent of a phosphine oxide as by-product. The sequence is irreversible and the product is obtained under kinetic control [9]. The relevance of betaines (**5**) has been in focus for a long time — their involvement as intermediates in the salt-free Wittig reaction could however ultimately be excluded [10, 11].

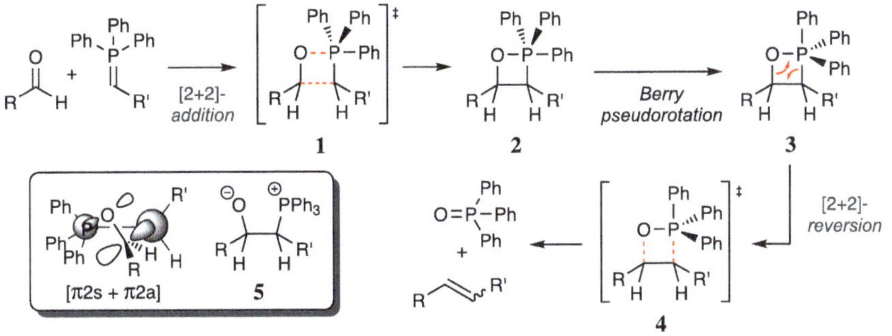

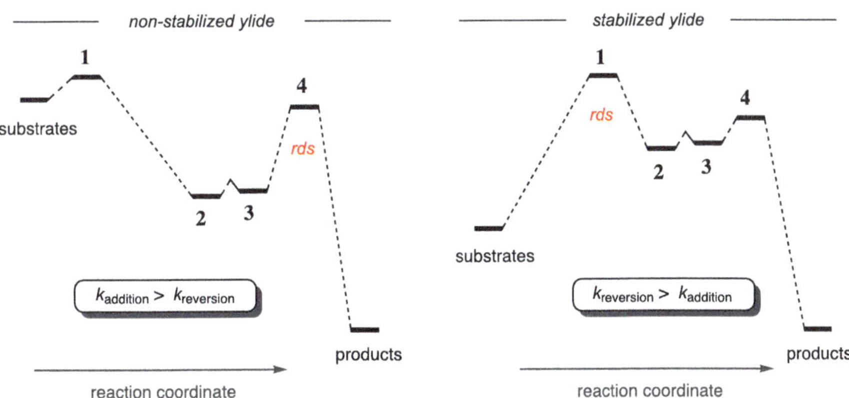

Fig. 3.3 Mechanism of the Li-salt-free Wittig olefination [3, 8]

The attack on the phosphorus and liberation of the leaving group always occurs in the apical position in trigonal-bipyramidal systems (*Westheimer* rule) [12, 13]. This implies that an intermittent pseudorotation of the oxaphosphetane **2** to **3** must proceed to subsequently furnish the products in the concluding cycloreversion [14].

For non-stabilized ylides, the initial cycloaddition is an early transition state with a low energy barrier. The cycloreversion displays a high activation energy and is rate-determining. For stabilized ylides, it is the opposite (see Fig. 3.4) [15].

Fig. 3.4 Energy profile of the Li-salt-free Wittig olefination of non-stabilized and stabilized ylides [15]

The decomposition of the oxaphosphetane is stereospecific: 1,2-disubstituted *cis*-oxaphosphetanes exclusively yield the *Z*-alkene in salt-free reactions, while the *trans*-intermediate leads to the *E*-product [10, 16, 17]. Non-stabilized ylides preferentially form the *Z*-alkene, stabilized ylides on the other hand give the *E*-configured product (see Fig. 3.5).

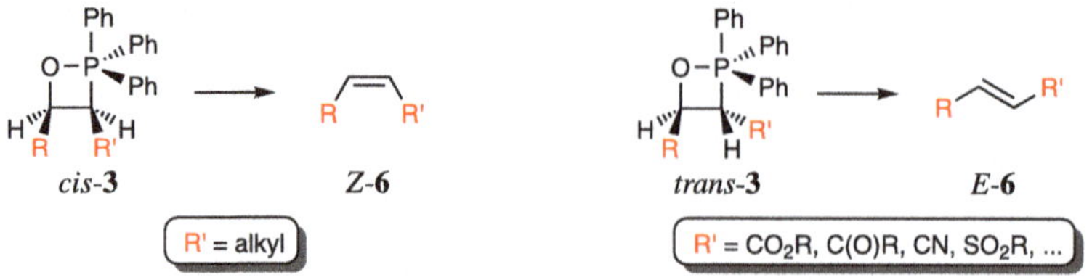

Fig. 3.5 Stereospecificity of the oxaphosphetane decomposition [10, 16, 17]

Since the formation is irreversible with only few exceptions (e.g., with aromatic alde-hydes), [16, 18] the stereoinducing step must be the initial cycloaddition due to the stereo-specificity of the sequence. The stereoselectivity of the reaction is thus entirely depen-dent on the relative energies of the *cis-* and *trans*-selective transition states en route to the oxaphosphetane **2** [19]. In an early transition state, the steric interactions are maximized when the reactants approach each other, while in a late transition state, the steric interaction in the oxaphosphetane **2** plays a prominent role, as the transition state is very similar to this intermediate.

In non-stabilized ylides, a skewed transition state is adopted to avoid a strong 1,2-repulsion (see Fig. 3.6). While steric interactions are minimized in **7**, an almost planar transition state **8** must be adopted due to a 1,3-diaxial interaction of the substituent R in pseu-doaxial position, which renders it energetically less favorable than **7**. In an early transition state, the *cis*-oxaphosphetane **2** is thus the major product via **7**. In stabilized ylides, the 1,2-interaction of eclipsed substituents is significantly reduced, as the bond angles between the substituents widen due to the more advanced rehybridization ($C_{sp^2} \rightarrow C_{sp^3}$ and $P_{sp^3} \rightarrow P_{dsp^3}$).

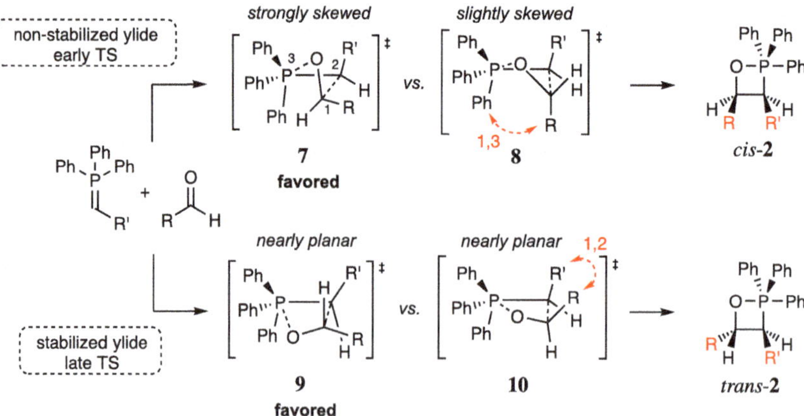

Fig. 3.6 Origin of selectivity in the formation of *E*- and *Z*-alkenes in the salt-free Wittig reaction [20, 21]

As a consequence, (almost) planar structures are energetically preferred. Since in **9** the 1,2-repulsion is diminished compared to **10**, *E*-alkenes are preferentially formed via *trans*-**2**. As an alternative rationalization, the formation of stabilizing dipole-dipole interactions in the transition state might also result in the pronounced *E*-selectivity when using ylides with strong electron-withdrawing groups [20].

Since the transition state of semi-stabilized ylides is located between the two discussed cases (early and late transition state) on the reaction coordinate, the energetic differences between the path to *cis*- and *trans*-**2** are small. This often leads to the formation of a product mixture with a similar *E/Z* ratio [3, 20, 21].

The dependence of the diastereoselectivity on the nature of substituents on the phosphorus further illustrates the importance of steric interactions in the transition state. The smaller the substituent on phosphorus, the smaller the 1,3-diaxial interaction and the higher the proportion of *E*-olefins. A geminal disubstitution on the aldehyde, on the other hand, increases the repulsion with the P-substituents (see Table 3.2) [21].

Table 3.2 Influence of phosphorus substituents on the diastereoselectivity of Wittig olefinations [21]

| | PhCH$_2$CH$_2$CHO | PhCH$_2$C(**Me**)$_2$CHO |
R	*Z/E*-**12**	*Z/E*-**12**
Ph	16 : 1	>99 : 1
*t*Bu	17 : 1	99 : 1
*i*Pr	1 : 4.6	1 : 1
Et	1 : 2.3	5.7 : 1

The presence of Li salts can give rise to deviating *E/Z* ratios [22, 23]. This is particularly the case with semi-stabilized ylides, which often show no clear preference for a diastereomer, and aromatic aldehydes, whose reaction can be partially reversible (see Table 3.3). The observed trends can depend on the concentration of the lithium salt. The higher the concentration of LiX, the more the *E/Z* ratio can differ. The presence of salts enables the stabilization of betaine intermediates, but whether this is the cause of the changed diastereoselectivity is not entirely understood: The effect of salts in the solvent on the mechanism has not been fully elucidated [3]. In addition to the ylide used, the choice of base also plays a major role (see Fig. 3.7).

Table 3.3 Change in the diastereoselectivity of the Wittig reaction upon addition of LiX [23]

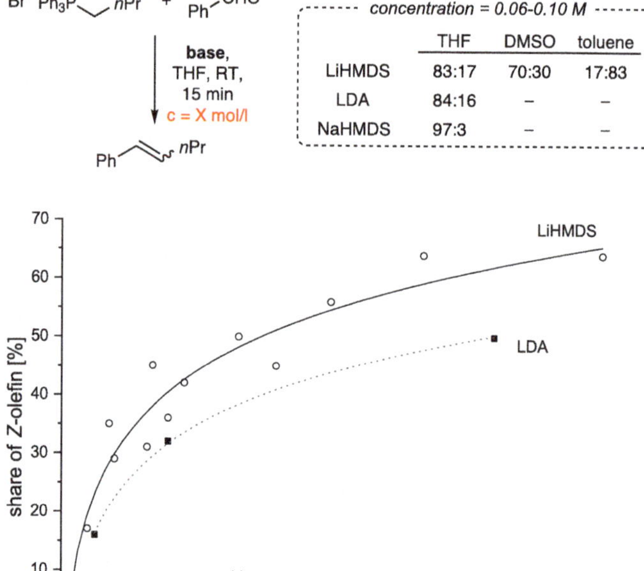

	Z/E-Ratio	
LiX	**R = Me**	**R = Et**
–	87 : 13	96 : 4
LiCl	81 : 19	90 : 10
LiBr	61 : 39	86 : 14
LiI	58 : 42	83 : 17
Li(BPh$_4$)	50 : 50	52 : 48

Fig. 3.7 Influence of bases, solvents, and reaction concentration on the diastereomer ratio of the Wittig reaction in the presence of LiBr [22]

In the rare cases where the *E/Z* ratio of the alkenes does not reflect the ratio of the oxaphosphetanes (which would result assuming an irreversible formation), this deviation is referred to as *stereochemical drift*. The phenomenon may be based on the fact that the formation of oxaphosphetanes can be partially **reversible**. Under partial thermodynamic equilibration, a higher proportion of *trans*-oxaphosphetane is obtained, resulting in *E*-enriched product mixtures. Assuming an irreversible cycloaddition step, the possible influence of Li$^+$ on the relative positions of the stereodirecting transition states **7** and **9** poses a further, alternative explanation for the stereochemical drift [15, 19].

Typically, non-nucleophilic, strong metal bases are employed (e.g., BuLi, LDA, KHMDS, TMS$_2$NK, NaH) for the deprotonation of phosphonium salts to remove the weakly acidic protons in the α-position. Both the counterion and the solvent may influence the diastereomeric ratio of the products (see Fig. 3.8).

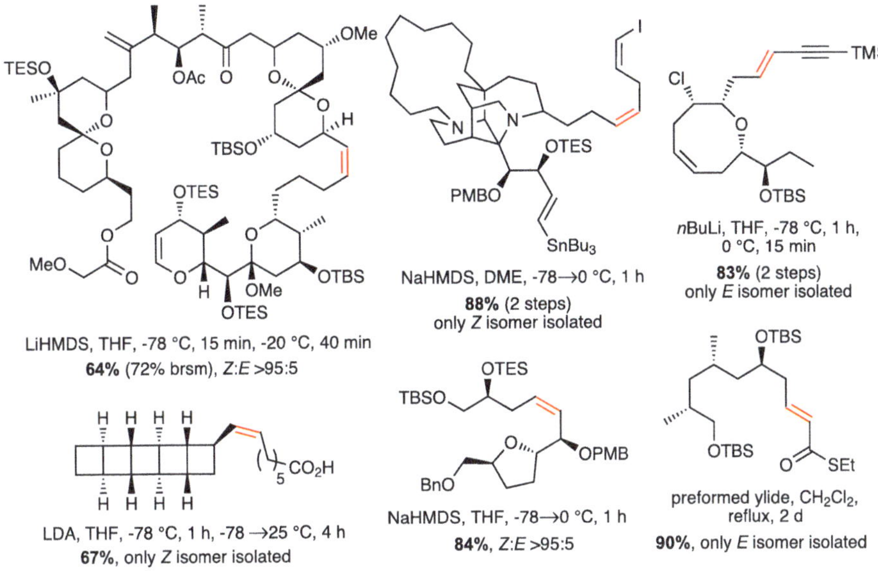

Fig. 3.8 Applications of the Wittig reaction in natural product synthesis [24–29]

The Wittig reaction also sometimes plays an important role on an industrial scale despite its poor atom economy and considerable waste generation. Apart from its use in the production of pharmaceuticals on a kg scale, vitamin A acetate is produced commercially via a Wittig reaction. The C$_{15}$ salt **15** derived from β-ionone can be converted into retinol acetate **16** on a kiloton scale. The strong exothermicity of the reaction is successfully controlled by utilizing a dose-controlled reagent addition with sufficient heat exchange capacity. The triphenylphosphine oxide is recycled following its separation in a wash column, thus enabling the economic sustainability of the process (see Fig. 3.9) [30–32].

Apart from the classic Wittig conditions, which show the selectivities described above for unstabilized, semi-stabilized, and stabilized ylides, the Schlosser group disclosed a variant in the 1960s by that selectively provides *E*-alkenes [33]. The selectivity of the *Schlosser modification* is based on the use of an additional equivalent of a lithium base to remove the proton in α-position to the phosphorus atom in the initial *cis*-oxaphosphetane and, after reprotonation, to afford the more stable *trans*-oxaphosphetane.

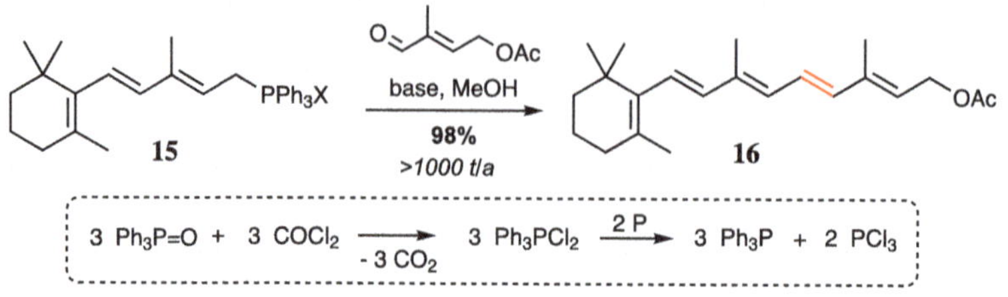

Fig. 3.9 BASF's production process of vitamin A acetate including phosphine oxide recycling [30–32]

According to Schlosser's studies, the zwitterionic betaine **17** is formed in the presence of Li salts. To avoid the dissociation of the α-lithiated betaine (**18**) to the free anion, a high concentration of soluble lithium halide is required. LiBr is typically used as an additive. The α-lithio species can epimerize to the corresponding *trans*-diastereomer **19**, which is protonated to the *trans*-betaine **21** via transition state **20** in the presence of strictly stoichiometric amounts of acid or alcohol. The addition of KO*t*Bu sequesters the LiBr and the oxaphosphetane (*trans*-**3**) provides the *E*-alkene with high selectivity (see Fig. 3.10) [34, 35].

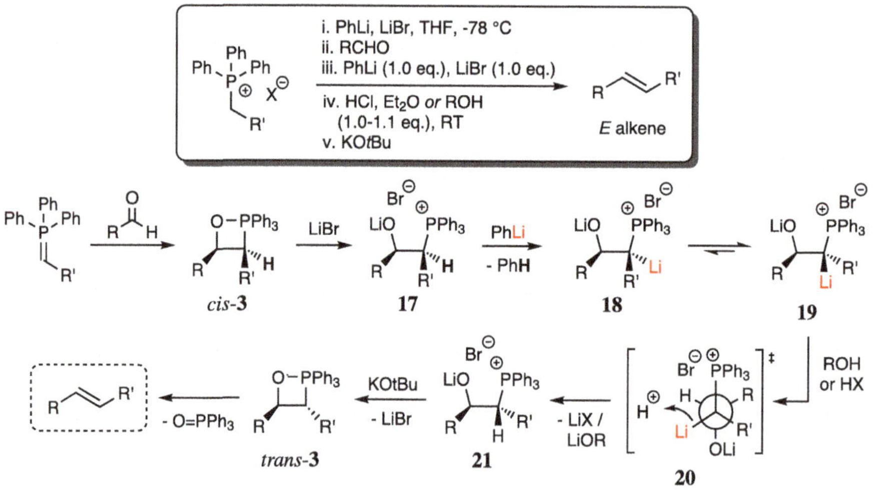

Fig. 3.10 Schlosser modification of the Wittig olefination — reaction conditions and postulated mechanism [34]

Whether the mechanism proposed by Schlosser corresponds to the actual pathway has not been conclusively clarified, as many of the original postulates on the classic Wittig reaction also had to be revised over time in light of new mechanistic results [3].

The nature of the second equivalent of the alkyl- or aryllithium base should have no influence on the selectivity from the point of view of basicity, as its purpose is to enable a complete and rapid deprotonation of the betaine **17**. Conversely, only PhLi consistently delivers high *E*-selectivities. The aggregation behavior of the different organolithium reagents and intermediates **17-21** seems to indicate that only slightly aggregating lithium species guarantee an efficient formation of **18** (see Table 3.4) [34].

Table 3.4 Influence of the Li-base on the diastereoselectivity of the Schlosser-modified Wittig reaction [34]

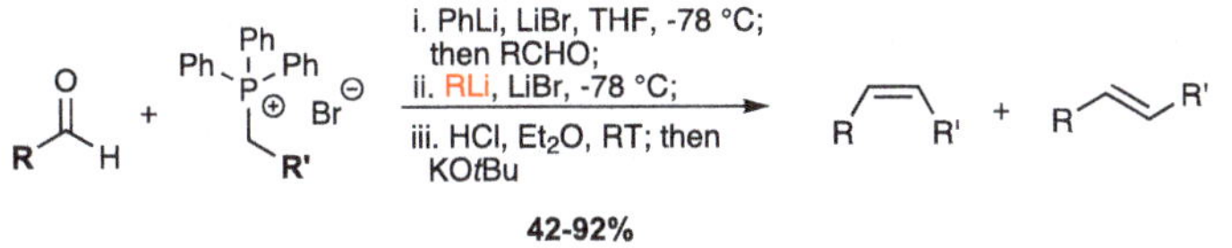

i. PhLi, LiBr, THF, -78 °C; then RCHO;
ii. RLi, LiBr, -78 °C;
iii. HCl, Et₂O, RT; then KO*t*Bu

42-92%

R	R'		Z/E-Product ratio			
		PhLi	MeLi	*n*BuLi	*s*BuLi	*t*BuLi
*n*Bu	*n*Hex	<0.5:99.5	3:97	50:50	33:67	17:83
*n*Pr	Ph	<0.5:99.5	1.5:98.5	12:88	6:94	23:77
CH₂Bn	H₂C=C-Me	<0.5:99.5	2:98	16:84	12:88	36:64
*n*Pent	*i*Pr	<0.5:99.5	2:98	16:84	–	13:87
*n*Pent	*t*Bu	94:6	98:2	97:3	–	76:24

Shiina and co-workers used a Schlosser-modified Wittig reaction in their enantioselective synthesis of the macrolide (+)-eushearilide. The central double bond in **23** could be constructed under typical conditions (PhLi, LiBr) from the phosphonium salt **22** and the protected ω-hydroxy-aldehyde in high yield, with only the *E*-isomer being isolated. Following concluding steps to the natural product (oxidation, Mukaiyama aldol, macrolactonization, and esterification), studies on its antimicrobial activity could be successfully conducted (see Fig. 3.11) [36].

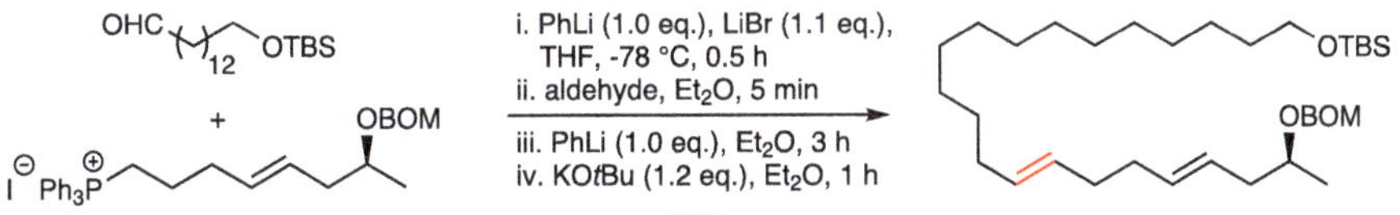

i. PhLi (1.0 eq.), LiBr (1.1 eq.), THF, -78 °C, 0.5 h
ii. aldehyde, Et₂O, 5 min
iii. PhLi (1.0 eq.), Et₂O, 3 h
iv. KO*t*Bu (1.2 eq.), Et₂O, 1 h

76%

22   **23**

Fig. 3.11 Synthesis of eushearilide according to Shiina *et al.* [36]

The Wittig reaction was further improved with the primary goal to obviate the need for stoichiometric amounts of base and phosphine [37, 38]. O'Brien and co-workers reported the first catalytic Wittig reaction, in which the phosphine oxide can be reduced back to the respective phosphine by the use of silanes (**24**→**25**), thus forming the ylide **27** *in situ*. The choice of phosphine and base are crucial for the successful deprotonation of **26**, since the pKa values of the reagents must be matched. The method is applicable to non-stabilized, semi-stabilized, and stabilized ylides (see Fig. 3.12) [38, 39].

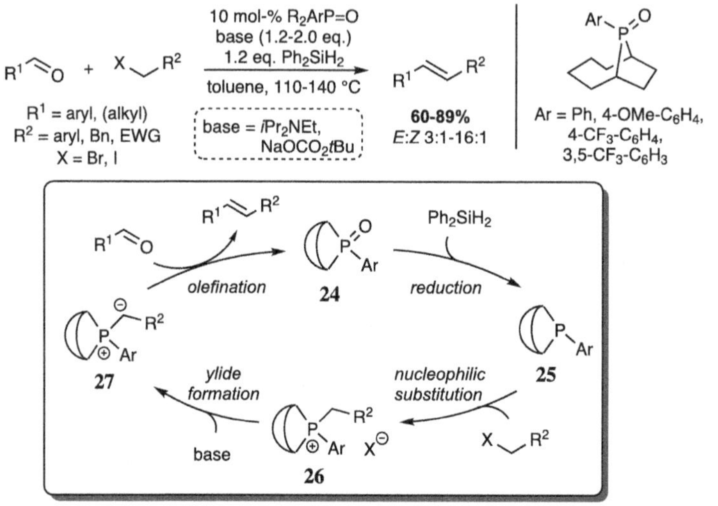

Fig. 3.12 Catalytic Wittig reaction according to O'Brien and co-workers [38]

By judicious choice of the phosphonium ylides and the reaction conditions, the *E/Z* ratio of the resulting olefins can thus be controlled within certain limitations (see Table 3.5).

Table 3.5 Factors governing the diastereoselectivity in Wittig olefinations [8, 40, 41]

preference for *Z*-alkenes	preference for *E*-alkenes
Li-salt-free conditions	Schlosser modification
sterically demanding aldehydes	small substituents on phosphorus or aldehyde
ethers as solvents	protic solvents (not applicable for stabilized ylides)

While the Wittig olefination is often used in synthetic projects targeting a natural product, one of the biggest disadvantages of the reaction lies in the stoichiometric formation of phosphine oxide, which often proves difficult to separate from the product. In addition, the reluctance of hindered ketones to afford the desired product under Wittig conditions and the sometimes challenging access to the phosphonium salt often results in the preference for alternative methods.

The *Horner–Wadsworth–Emmons* reaction (*HWE*) uses stabilized phosphonate carbanions instead of phosphonium ylides [8]. It generally proceeds at lower temperatures or under less harsh conditions than the classic Wittig reaction, which enhances its tolerance towards most major functional groups. The more reactive substrates also allow successful conversion of sterically very hindered substrates. The Horner–Wadsworth–Emmons reaction, like the Wittig reaction of stabilized phosphonium ylides, is highly *E*-selective, but unlike the Wittig reaction, *Z*-selective HWE variants exist. The Wittig and HWE olefination can thus be considered complimentary with regard to their scope. The HWE reaction is limited to the use of phosphonates whose substituent can stabilize a carbanion (EWG = COO^-, CO_2Me, CN, aryl, vinyl, SO_2R, $P(O)(OR)_2$, SR, OR, and NR_2). In the absence of such substituents, the desired alkene is isolated only in minor quantities. Unlike in the Wittig reaction, the separation of the phosphorus by-product is unproblematic in the HWE reaction, as the formed phosphoric acid diester is water-soluble (see Fig. 3.13).

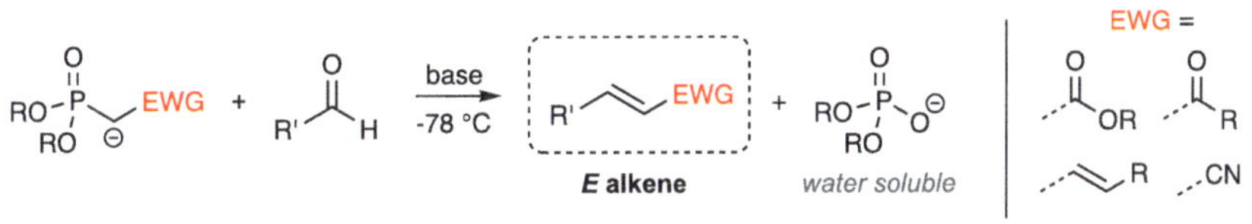

Fig. 3.13 General reaction scheme of the Horner–Wadsworth–Emmons reaction

In contrast to the Wittig reaction, the HWE olefination is almost completely reversible and closely resembles a classic aldol reaction in the initial addition step (see Fig. 3.14). After deprotonation of the acidic α-position, a β-ketophosphoric acid ester ylide is generated *in situ* (**28**). A carbonyl compound can be approached either via the *Re-* (**29**) or *Si*-face (**30**). In contrast to **29**, **30** experiences a steric repulsion of the carbonyl substituent R^2 with the alkoxy groups on the phosphonate ester, which renders the formation of the *cis*-product **33** slower than the corresponding *trans*-derivative **31**. Both reaction paths are characterized by a low activation barrier and are reversible, resulting in a pre-equilibrium before the ensuing cyclization. The rate-determining formation to the oxaphosphetane proceeds faster for the *trans*-oxaphosphetane **34** than for the *cis*-intermediate **32** ($k_{anti} > k_{syn} \gg k_{trans} > k_{cis}$) by virtue of the weaker steric repulsion in the *trans*-transition state. Following the generation of the oxaphosphetane, a pseudorotation at the P-atom as outlined by the Westheimer rule occurs (not shown in Fig. 3.14). The concluding retro-cycloaddition is irreversible, which results in the preferential formation of *E*-alkenes in the HWE olefination [8, 42, 43].

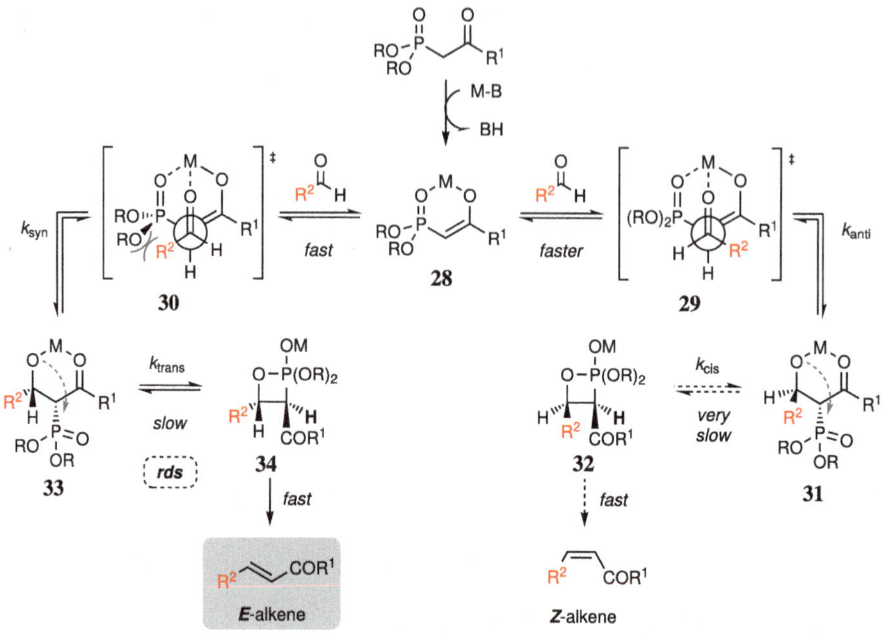

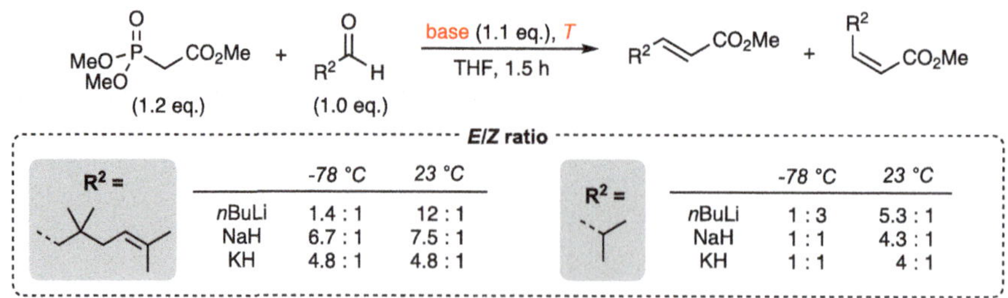

Fig. 3.14 Postulated mechanism of the HWE olefination [42]

The stereoselectivity can be modified by three factors: The choice of the cation in the base, the temperature, as well as the substituents on the phosphorus atom. The solvent can further impart a limited influence on the isomer ratio of the product [44].

More electronegative cations and an elevated temperature may increase the reversibility of the equilibria, thus increasing the difference of $k_{syn,anti}$ versus $k_{cis,trans}$, enabling a more efficient formation of the kinetically preferred *trans*-oxaphosphetane **34**. High temperatures and the switch from K to Na to Li thus increase the proportion of the *E*-stereoisomer. The more sterically demanding the substituent on the aldehyde, the more pronounced is the preference to afford the *E*-alkene (see Fig. 3.15) [44].

R² =		-78 °C	23 °C	R² =		-78 °C	23 °C
	*n*BuLi	1.4 : 1	12 : 1		*n*BuLi	1 : 3	5.3 : 1
	NaH	6.7 : 1	7.5 : 1		NaH	1 : 1	4.3 : 1
	KH	4.8 : 1	4.8 : 1		KH	1 : 1	4 : 1

Fig. 3.15 Influence of temperature and base on the diastereomer ratio of the HWE olefination [44]

The *E/Z* ratio can be further influenced by changing the electronic properties of the phosphorus through variation of the substituents. The stronger the electron-withdrawing properties, the smaller the activation barrier for the oxaphosphetane formation should become ($k_{cis,trans}$ increases). As the cyclization rate increases, the influence of the upstream equilibrium decreases in equal measure, so that the selectivity of the initial addition step (**28**→**31**, **33**) is dominated by the more stable *anti*-configured intermediate **31**. As a result, Z-alkenes are preferred (see Table 3.6).

Table 3.6 Influence of the phosphorus substituents on the diastereoselectivity [45]

R	yield	*E/Z* ratio
PhS	25	only E
2,6-Me$_2$-C$_6$H$_3$	62	85 : 15
Ph	80	69 : 31
2,4-F$_2$-C$_6$H$_3$	68	59 : 41
2,6-F$_2$-C$_6$H$_3$	80	35 : 65
CH$_2$CBr$_3$	62	71 : 29
CH$_2$CCl$_3$	37	65 : 35
CH$_2$CF$_3$	56	51 : 49
CH(CF$_3$)$_2$	55	12 : 88

Still and Gennari exploited this handle in their HWE modification to specifically obtain the complementary Z-alkene [46, 47]. Another Z-selective variant was introduced by the Ando group, which is also widely used (see Fig. 3.16) [48–50].

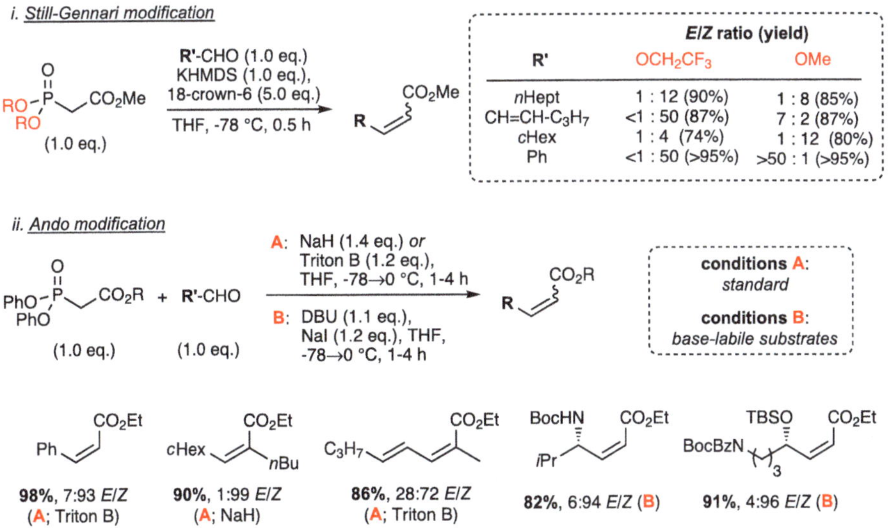

Fig. 3.16 Horner–Wadsworth Emmons modifications according to Still/Gennari and Ando [46, 48, 50]

The observed Z-selectivity can be rationalized by the modified electronic properties of the phosphonate group. In the Still–Gennari modification, the electrophilicity of the electron-poor phosphorus is greatly enhanced, while sequestering the cation from the coordination sphere of the oxygen anion by the added crown ether. As a result, the rate of formation of the oxaphosphetane increases to such a degree that the addition reaction of the ylide in turn becomes the rate-determining step ($k_{trans} > k_{cis} > k_{anti} > k_{syn}$). In addition, the formation of oxaphosphetane shows only a very limited to no reversibility, thus removing the addition product from a possible equilibrium [46]. A similar rationale is used for the diarylphosphonate esters in the Ando modification to explain their Z preference [49].

In order to expand the substrate scope to include very base-sensitive substrates, further improvements of the HWE reaction have been reported in addition to Z-selective variants. Especially the method introduced by Masamune and Roush accommodates the coupling of aldehydes or phosphonates with easily epimerizable stereocenters or substrates that tend to decompose (see Fig. 3.17). The addition of stoichiometric LiCl increases the acidity of the phosphonate, enabling even weak amine bases to be used for the formation of the anion and the successful coupling reaction. Depending on the base, the reaction proceeds faster or slower. In general, high yields and excellent E/Z selectivities are achieved [51].

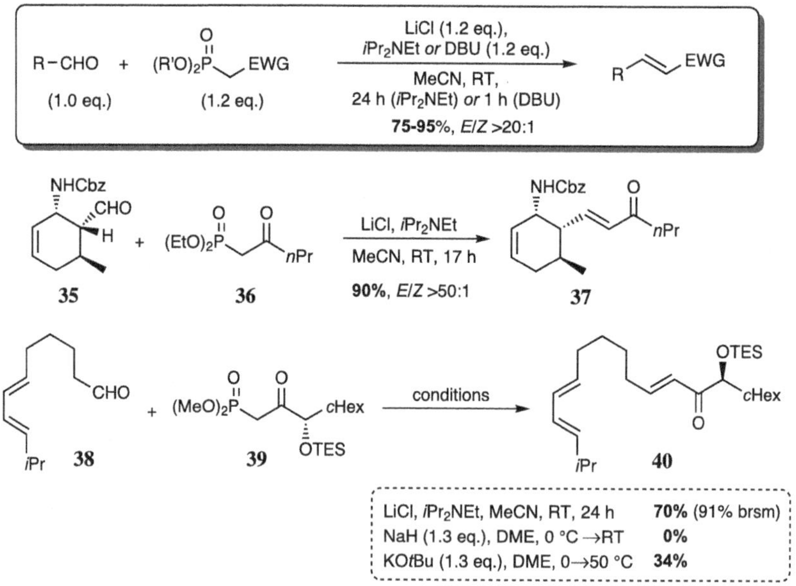

Fig. 3.17 Typical reaction conditions of the Masamune–Roush modification and selected synthetic examples [51]

While **35** easily epimerizes in the presence of bases such as NaH, this is not observed in the presence of LiCl/iPr$_2$NEt. Phosphonate **39** also possesses an epimerizable stereocenter and the aldehyde tends to undergo an aldol homocondensation. If classic coupling conditions are used instead of the Masamune–Roush conditions, the yield of the desired product **40** thus drastically decreases.

In addition, the Paterson group developed another variant relying on Ba(OH)$_2$ as a base, which also converts base-sensitive substrates to the corresponding olefins in high yield [52].

The Horner–Wadsworth–Emmons reaction finds wide application in synthetic studies, with the different modifications often allowing a tailor-made solution to address a challenging coupling [53–55]. Figure 3.18 provides a selection of intermediates that were used as key fragments in total syntheses, including the modification used, if applicable.

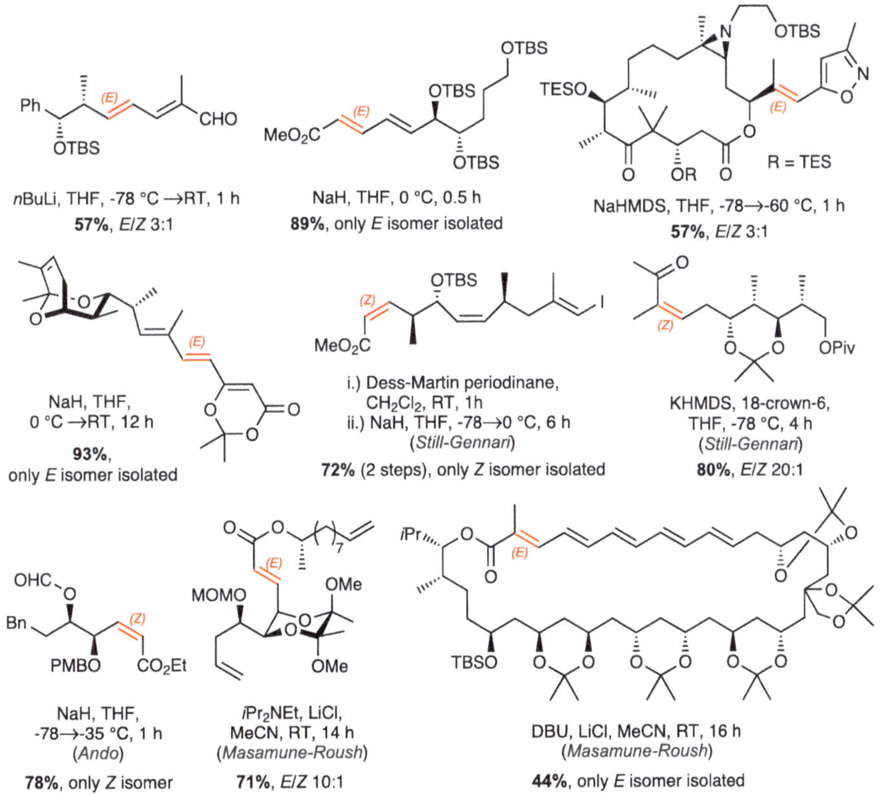

nBuLi, THF, -78 °C →RT, 1 h
57%, *E/Z* 3:1

NaH, THF, 0 °C, 0.5 h
89%, only *E* isomer isolated

R = TES
NaHMDS, THF, -78→-60 °C, 1 h
57%, *E/Z* 3:1

NaH, THF, 0 °C →RT, 12 h
93%, only *E* isomer isolated

i.) Dess-Martin periodinane, CH$_2$Cl$_2$, RT, 1h
ii.) NaH, THF, -78→0 °C, 6 h (*Still-Gennari*)
72% (2 steps), only *Z* isomer isolated

KHMDS, 18-crown-6, THF, -78 °C, 4 h (*Still-Gennari*)
80%, *E/Z* 20:1

NaH, THF, -78→-35 °C, 1 h (*Ando*)
78%, only *Z* isomer

iPr$_2$NEt, LiCl, MeCN, RT, 14 h (*Masamune-Roush*)
71%, *E/Z* 10:1

DBU, LiCl, MeCN, RT, 16 h (*Masamune-Roush*)
44%, only *E* isomer isolated

Fig. 3.18 Application of HWE olefination in total syntheses of natural products [56–63]

Synthesis of the Phosphorus Reagents

Phosphonium salts can be prepared by reacting trialkyl or triaryl phosphines with alkyl halides. Primary alkyl iodides and benzyl bromides can be converted to the corresponding reagents upon moderate heating. Primary alkyl bromides, chlorides, or branched alkyl halides require elevated temperatures. Non-stabilized alkyl phosphonium halides necessitate strictly anhydrous conditions, as they tend to hydrolyze in the presence of H_2O. β-functionalized phosphonium salts can be synthesized by 1,2-addition of a nucleophilic ylide and subsequent trapping of the anion with an electrophile (see Fig. 3.19).

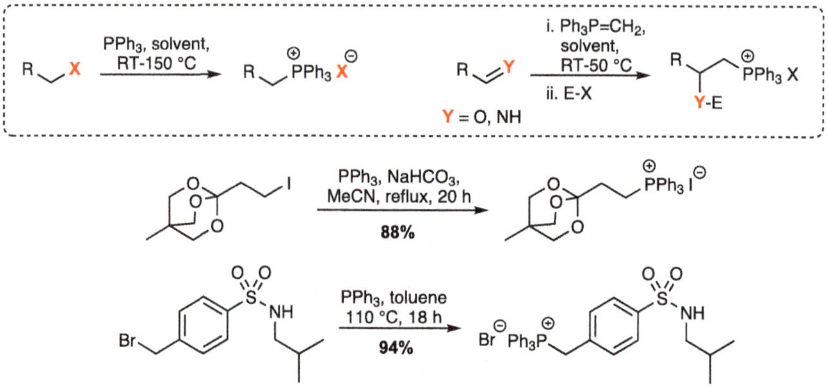

Fig. 3.19 General access to phosphonium reagents and representative examples [64]

The precursors for the Horner–Wadsworth–Emmons olefination are obtained by the conversion of trialkyl phosphites with alkyl halides (*Arbuzov* reaction, see Fig. 3.20) [65]. Secondary halides tend to eliminate, but the resulting phosphonate can be further derivatized by alkylation [66, 67]. In the presence of Lewis acids as catalyst, the reaction can often be carried out at room temperature. Elevated temperatures are required for uncatalyzed reactions.

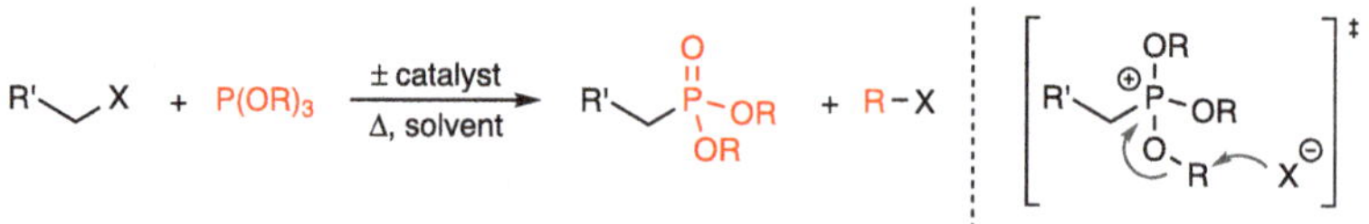

Fig. 3.20 Arbuzov reaction including postulated transition state [65]

Mechanistically, the reaction proceeds in two steps: A substitution of the halide by the phosphite $P(OR)_3$ results in the liberation of X^-, which acts as base and subsequently attacks one of the substituents via **41** [65].

The Dougherty group developed an alternative to the Arbuzov reaction by resorting to carboxylic acid chlorides as starting materials (see Fig. 3.21) [68]. According to the postulated mechanism, α-ketophosphonate **42** is initially formed, followed by its conversion to the hydrazone **43**. In the presence of a base (KO*t*Bu), the diazo compound **44** affords the corresponding phosphonate. The reaction provides moderate to good yields, only α-branched carboxylic acids tend to decompose under the reaction conditions.

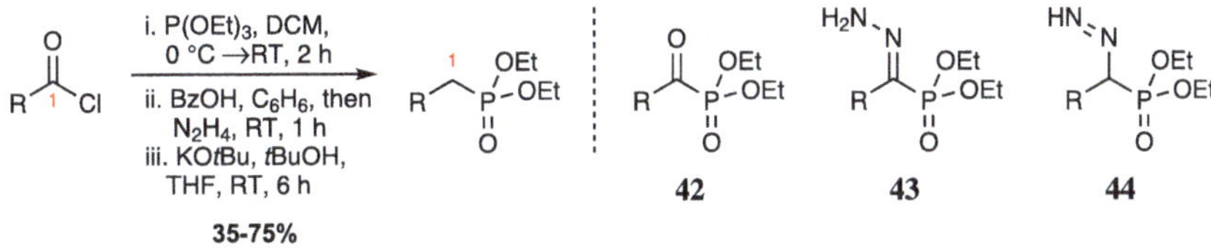

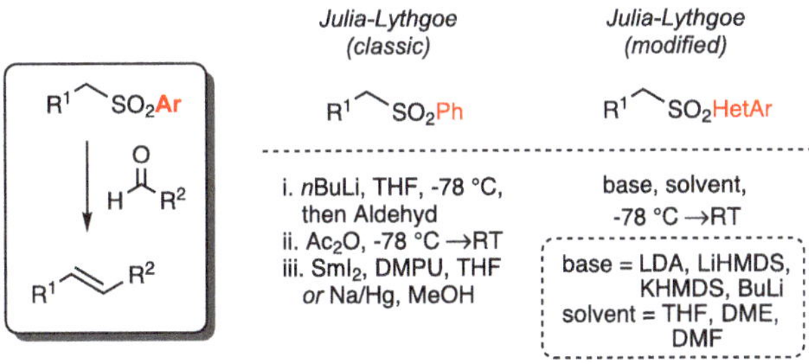

Fig. 3.21 Reductive deoxygenation of acylphosphonates according to Dougherty *et al.* [68]

3.1.2   Olefination by Sulfones

The second carbonyl olefination method considered "standard" relies on sulfur-derived carbanions. The *Julia–Lythgoe* reaction allows access to primarily *E*-configured alkenes. Under classic conditions, it is a multi-step synthesis sequence. However, modern variants can deliver the corresponding olefins in a one-pot setup and allow the construction of *E*- or *Z*-olefins, depending on the reaction conditions and the individual sulfur species (see Fig. 3.22) [4, 69].

Fig. 3.22 Julia–Lythgoe olefination: Modifications and typical reaction conditions [70]

The classic Julia olefination begins with the addition of the sulfone anion to the aldehyde. The resulting β-hydroxysulfone is intercepted by an acylation reagent. Said acylated intermediate is normally isolated. In the concluding step, the reductive cleavage to the respective olefin is mediated by one-electron reductants. Typical acylation reagents include acetic anhydride as well as benzoyl, mesyl, and tosyl chloride. Sodium amalgam in methanol was initially employed as a reductant, but SmI_2 in the presence of DMPU has now established itself as a non-basic standard reductant. Further variants utilize RMgX, Bu_3SnH, Li/Na in NH_3, $Na_2S_2O_4$, etc [69].

The olefination sequence comprises three discrete steps: An aldol-like addition of the sulfonate carbanion **45** to the carbonyl component yields β-hydroxysulfonate **46**, followed by acylation to **47** (see Fig. 3.23). The addition of sulfones to carbonyl compounds is reversible and can result in low yields due to an unfavorable equilibrium position. Intercepting **46** by acylation or silylation shifts this equilibrium towards **47** [71]. In the concluding elimination, a dependence of the mechanism on the choice of the reducing agent has been observed (see Fig. 3.24) [69, 72, 73].

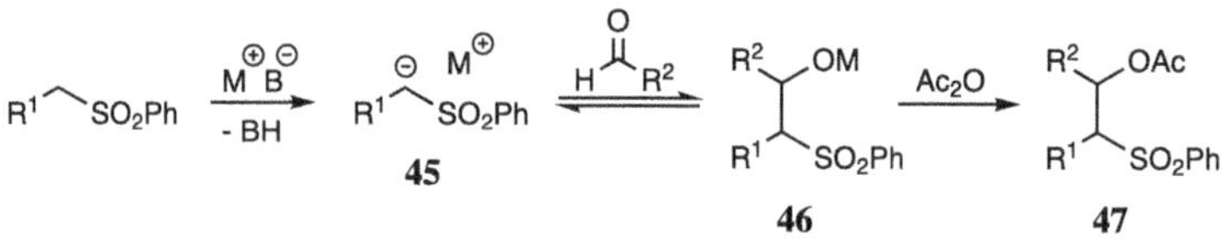

Fig. 3.23 Primary steps of the classical Julia–Lythgoe olefination [69]

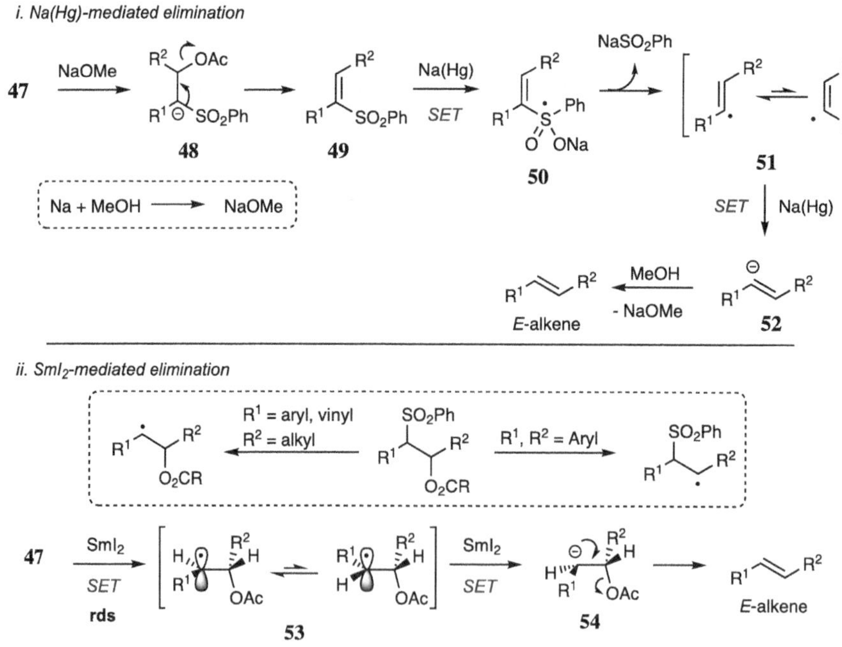

Fig. 3.24 Postulated mechanism of elimination in the Julia–Lythgoe olefination [69, 72, 73]

When using Na-amalgam for the elimination step, the α-position to the sulfone group is deprotonated first by *in situ*-generated NaOMe. Following the elimination of AcO^- to the vinyl sulfonate (**48→49**), an intermediate vinyl radical **51** is obtained by single electron transfer (SET), whose equilibrium favors the thermodynamically more stable *trans*-diastereomer. After another SET, the respective anion is reprotonated to give the *E*-alkene. With SmI_2 as a reducing agent, the formation of the primary radical can result in the cleavage of either the sulfone group or the acylated hydroxy functionality, depending on the individual substrate. The chemoselectivity of the elimination (desulfonylation *vs.* decarboxylation) is determined by the relative stability of the carbon-centered radicals [74]. Both paths are viable, even if the C-S cleavage usually dominates [73]. Following the first electron transfer in the desulfonylation path, the radical **53** equilibrates to the *trans*-isomer. After transfer of the second electron, the anion **54** forms the corresponding alkene with concomitant cleavage of the carboxylate group [72]. The frequent use of additives (HMPA, DMPU) with SmI_2 increases the reduction potential of Sm(II) and is often necessary, as typically no elimination occurs in their absence [71].

A dependence of the stereoselectivity can be observed with regard to the degree of branching in the α-position of the double bond. The cause was postulated to be based on the increasing steric repulsion of R^1 and R^2 in **51** during the equilibration of the radicals. The effect is limited, the reaction remains *E*-selective even at low branching (see Fig. 3.25) [75].

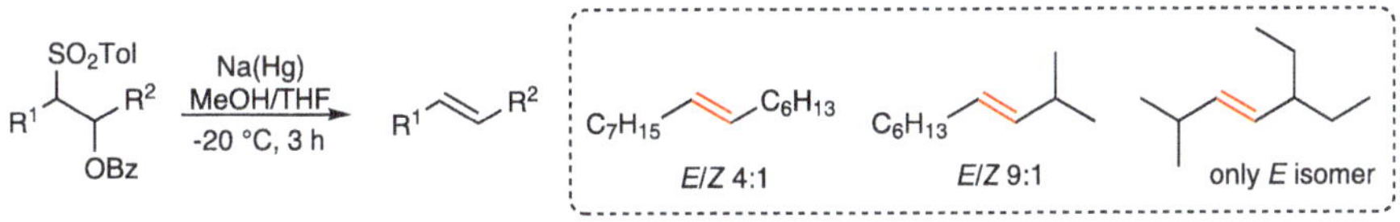

Fig. 3.25 Dependence of the diastereoselectivity on the substitution pattern [75]

The classic Julia–Lythgoe olefination suffers from its generally low tolerance towards various functional groups. This can be partially compensated by switching to SmI_2 as reductant and selecting the appropriate reaction conditions, e.g., the acylation reagent. A rather impressive example can be found in the synthesis of the kedarcidin aglycone **56** from **55** by

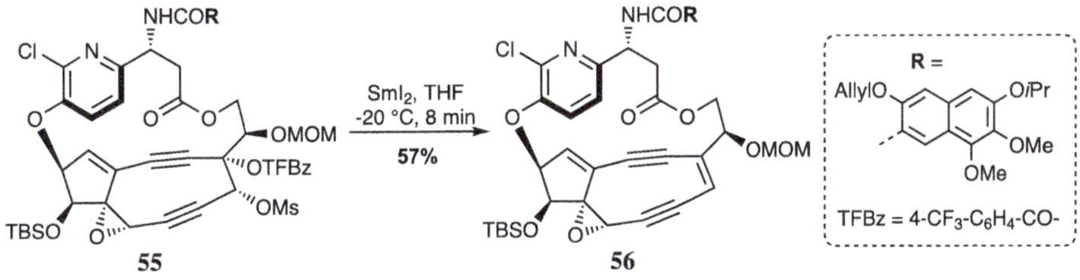

Fig. 3.26 Construction of the enediyne moiety in the total synthesis of the protected kedarcidin aglycone **56** [76]

the Hirama group, in which the epoxide, the propargylic pyridyl ether, the enyne, as well as the ester functionalities remain intact (see Fig. 3.26) [76].

The Julia–Lythgoe olefination has gained popularity in recent decades as the variety of methods for introducing sulfones in complex intermediates has increased. The non-toxicity of the sulfur by-products and a high degree of regio- and stereoselectivity are hallmarks of this method. The purification challenges of the classic Wittig reaction are also not relevant, since the by-products can be easily removed chromatographically or by crystallization. The yields of SmI_2-mediated reactions are usually superior to those employing Na-amalgam. Even though reductions with SmI_2 display slightly deviating diastereoselectivities, the stereo-specificity is not dependent on the reaction temperature. Particularly mono-, geminally and vicinally disubstituted alkenes can be reliably synthesized with the Julia–Lythgoe olefination (see Fig. 3.27) [71].

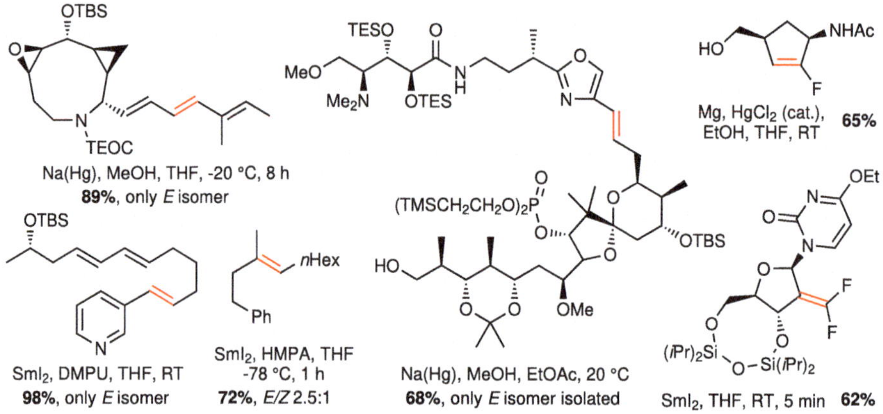

Fig. 3.27 Synthetic applications of the classic Julia–Lythgoe olefination [77–82]. The conditions and yields refer to the elimination step

While the Julia olefination provides reliable access to E-configured alkenes, it necessitates three separate steps. Although these can be carried out as a one-pot variant, the yields usually improve significantly when the intermediates are isolated. The group around S. Julia was able to develop a greatly improved one-step variant by resorting to heteroaryl-substituted sulfones. This *modified* Julia olefination allows the preparation of olefins from sulfones and aldehydes in one step. The modified variant is also referred to as *Julia–Kocienski* olefination when N-Ph-tetrazolyl sulfones are used (see Fig. 3.28) [4, 70, 83, 84].

The mechanism of the modified Julia olefination has been extensively studied by both experimental and theoretical means [4, 84]. The steps are usually stereospecific, which results in the initial addition of the deprotonated sulfone to the aldehyde being the stereodetermining step. For PT-sulfones, the irreversibility of the addition has been demonstrated, while the remaining steps are at least partially reversible. The addition product (**57, 60**) then undergoes a *Smiles rearrangement* [70, 85, 86]. The spirocycle **58** was postulated to be a transition state instead of a stable intermediate for PT-sulfones [87]. Its nature (intermediate *vs.* TS) can thus potentially vary from substrate to substrate (BT, PT, TBT, Pyr). Following the Smiles

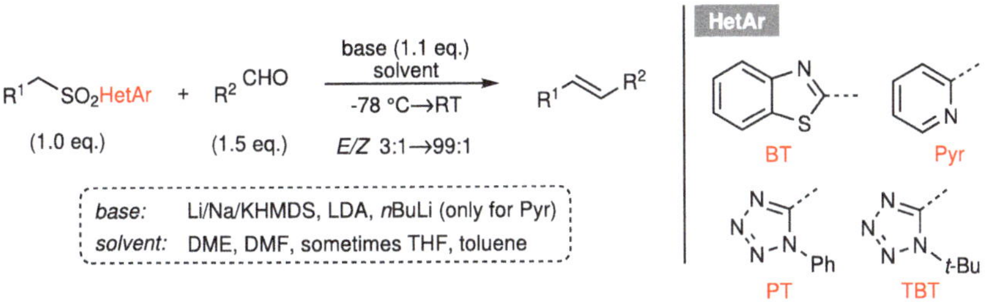

Fig. 3.28 Typical conditions and common heteroarylsulfones of the modified Julia olefination [83]

rearrangement, **59** decomposes stereospecifically to the E-alkene by elimination of SO_2 and the hydroxyarene. The sulfone group and OAr are aligned in an antiperiplanar fashion in **59**, **62**. Due to the higher energy of **57** and **58** caused by the synclinal arrangement of R^1 and R^2, the path to the E-alkene is slower and mostly irreversible. Nevertheless, it is the requisite path for the formation of the observed major diastereomer. The selectivity of the reaction to the Z-alkene (via **60-62**) is determined by the rate of the initial addition, which for **60** usually proceeds slower than for **57** ($k_{anti} > k_{syn}$). The ratio of **57** to **60** is thus reflected in the E/Z ratio of the products and the E-alkene results as the main product (see Fig. 3.29) [4, 70].

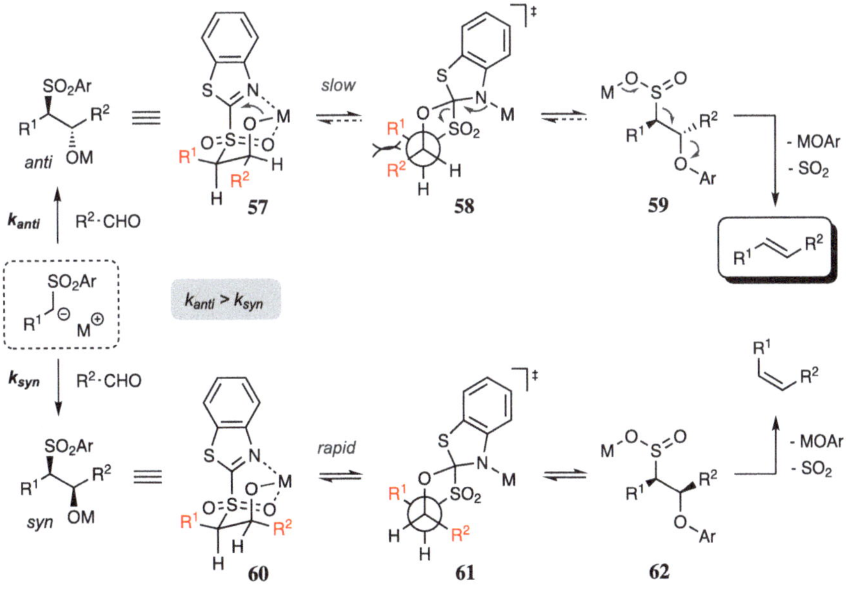

Fig. 3.29 Mechanism of the modified Julia olefination [4, 87]

From a practical point of view, the base either is allowed to deprotonate the sulfone prior to the addition of the aldehyde (*premetalation*) or it is added to a mixture of sulfone and aldehyde (*Barbier conditions*). The tendency of sulfones to undergo homocondensation reactions is one of the challenges of the modified Julia olefination, and depends on the individual heteroaryl substituent. This competing reaction can result in drastically diminished yields. Barbier conditions can partially alleviate this issue and are thus preferred in these instances, since the deprotonated sulfone can directly react to the alkene in the presence of the aldehyde.

Benzothiazolyl-(*BT*)-sulfones consistently provide high yields and *E*-diastereoselectivities with α-branched sulfones. In contrast, *n*-alkyl-BT-sulfones afford the corresponding products in low yield due to the enhanced rate of homocondensation. They can particularly play to their strengths in terms of yield and selectivity with α, β-unsaturated aldehydes. Aromatic and allylic BT-sulfones yield low diastereoselectivities and might even afford *Z*-configured olefins as major products. This can be rationalized by the addition to **57** and **60** becoming reversible. PT- and especially TBT-sulfones are suitable for base-labile substrates (see Fig. 3.30) [4, 70, 83, 84, 88].

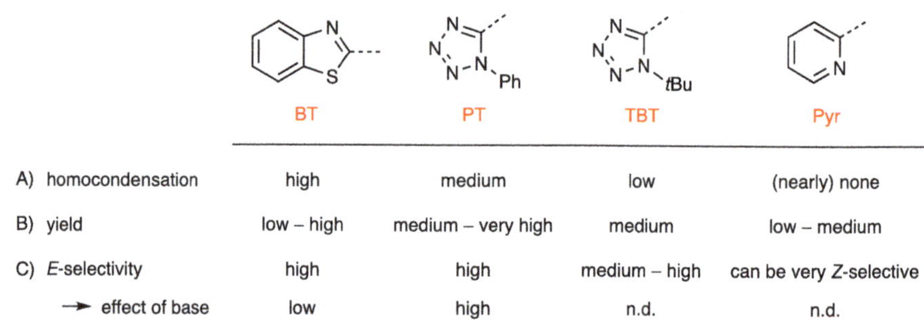

	BT	PT	TBT	Pyr
A) homocondensation	high	medium	low	(nearly) none
B) yield	low – high	medium – very high	medium	low – medium
C) *E*-selectivity	high	high	medium – high	can be very *Z*-selective
→ effect of base	low	high	n.d.	n.d.

Fig. 3.30 Comparison of the various heteroaryl substituents of the sulfone [4, 70, 83, 88]

A difference between BT- and PT-sulfones is the dependence of the observed diastereoselectivity on the base. BT-sulfones can only be influenced to a limited extent, PT-sulfones allow more leeway in that regard through judicious choice of the cation. All sulfones are prone to solvent effects on the *E/Z* ratio: A polar solvent (DME, DMF) and large counterion (K^+) generally favor the *E*-alkene (see Table 3.7) [89].

Table 3.7 Change of the *E/Z* ratio depending on the solvent and base [89]

	toluene	Et$_2$O	THF	DME
LiHMDS	51 : 49	61 : 39	69 : 31	72 : 28
NaHMDS	65 : 35	65 : 35	73 : 27	89 : 11
KHMDS	77 : 23	89 : 11	97 : 3	99 : 1

The cause of the base effect was postulated to be the competition of an open versus a closed transition state in the addition to the aldehyde (see Fig. 3.31) [83].

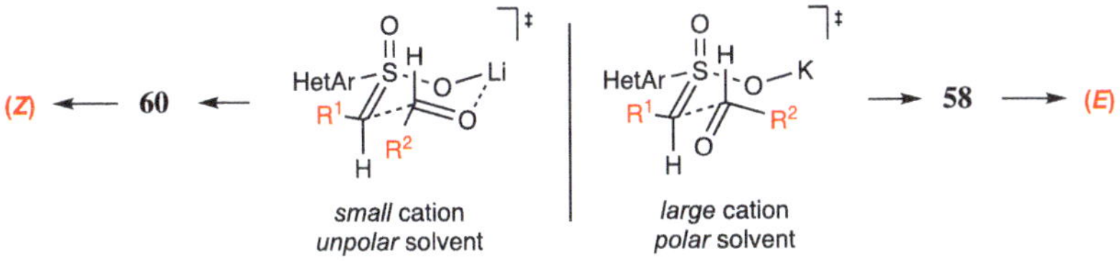

Fig. 3.31 Postulated transition states in the addition of sulfones to aldehydes [83]

The significant influence of the solvent also had to be taken into account in the total synthesis of the bacterial metabolite U-106305. For optimizing the construction of one of the central double bonds, the simplified model substrate **65** was devised (see Fig. 3.32). After extensive screening of a wide range of solvents, the critical coupling of **63** and **64** could be carried out with an acceptable diastereomeric ratio of *E/Z* 4:1 using a DMF/THF mixture. Alternative olefination methods yielded the desired product only in negligible yield [90].

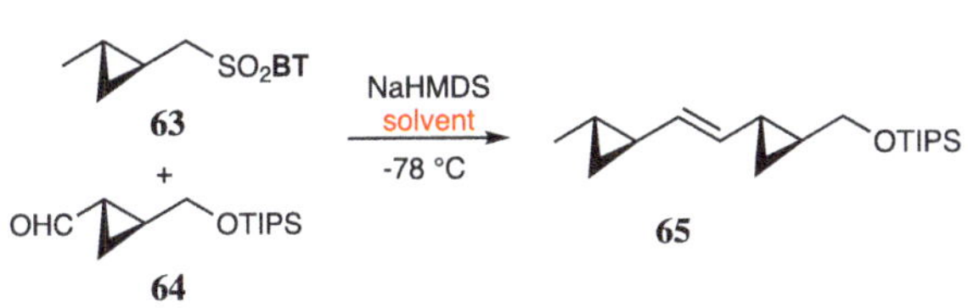

solvent	E/Z
toluene	1 : 10
CH$_2$Cl$_2$	1 : 10
Et$_2$O	1 : 7.7
THF	1.1 : 1
DME	2.4 : 1
DMF	3.5 : 1

Fig. 3.32 Influence of the solvent on the modified Julia olefination in the synthesis of U-106305 [90]

The choice of KHMDS as base can lead to high selectivity at the expense of yield. If the latter is also critical, NaHMDS might offer the optimum balance between yield and diastereoselectivity [89].

In contrast to the classic Julia olefination, the dependence of the diastereoselectivity on the steric bulk of the substrates occurs only to a very limited extent, especially with PT-sulfones (see Fig. 3.33) [89].

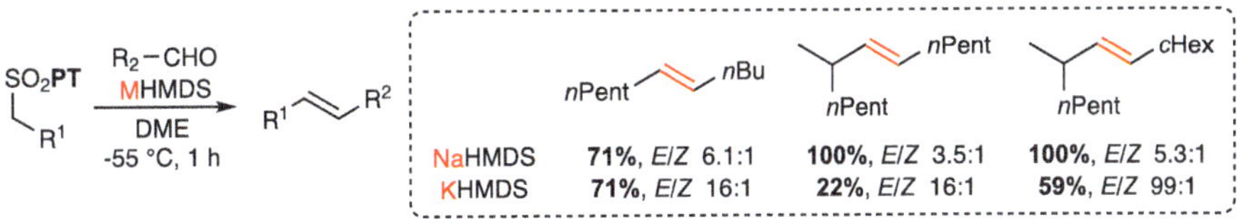

	*n*Pent / *n*Bu	*n*Pent / *n*Pent	*n*Pent / *c*Hex
NaHMDS	71%, *E/Z* 6.1:1	100%, *E/Z* 3.5:1	100%, *E/Z* 5.3:1
KHMDS	71%, *E/Z* 16:1	22%, *E/Z* 16:1	59%, *E/Z* 99:1

Fig. 3.33 Substrate dependence of stereoselectivity [89]

In the preparation of active ingredients or drugs, the modified Julia olefination also lends itself to effect CC linkages on a larger scale. Bristol-Myers Squibb investigated an approach to new statins. In the process development of BMS-644950, a PT-sulfone (**67**) was used in the critical coupling reaction to convert the aldehyde **66** on a kg scale to the advanced intermediate **68**. The desired HMGR inhibitor, obtained after a concluding global deprotection, possesses the potential to allow access to a new generation of statins (see Fig. 3.34) [91].

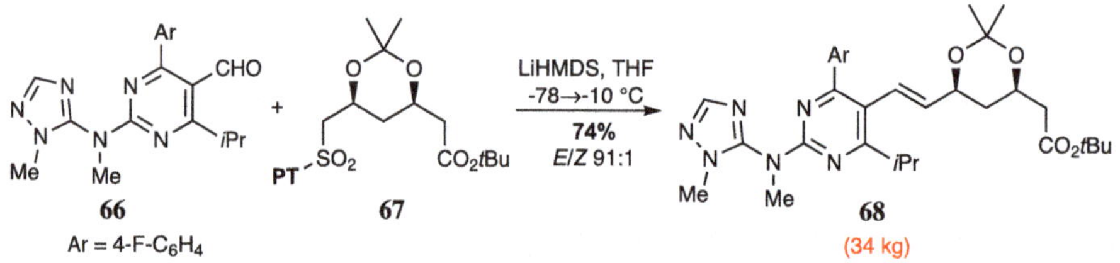

Fig. 3.34 Construction of the central double bond in the industrial synthesis of BMS-644950 [91]

Due to the preparatively simple procedure and the high tolerance towards functional groups, the reaction has found its way into the canon of standard CC coupling methods of complex natural product syntheses (see Fig. 3.35) [4, 83].

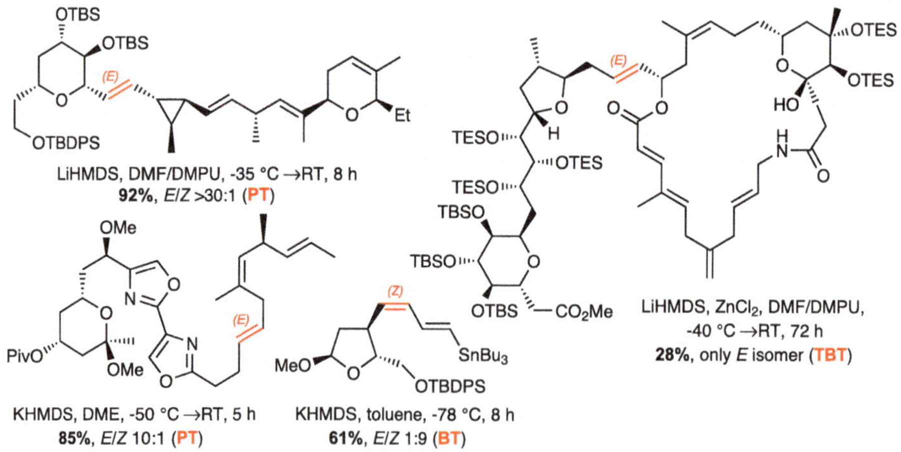

Fig. 3.35 Examples of the modified Julia olefination in total synthesis [92–95]

Synthesis of the Sulfones

The (hetero)aryl-substituted sulfones can be easily prepared in two steps by *S*-alkylation/*S*-oxidation from the corresponding thiols via the respective thioether (see Fig. 3.36) [4, 83].

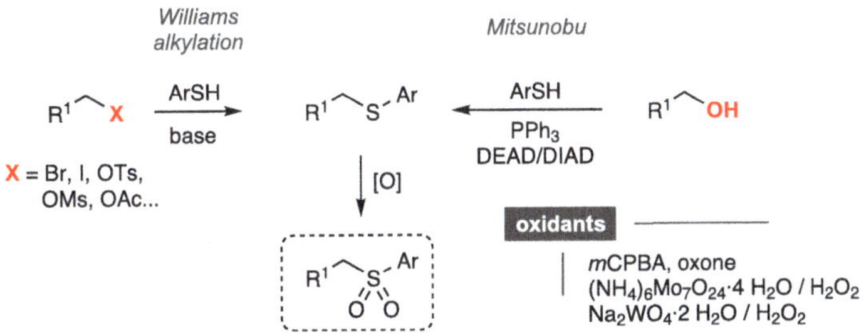

Fig. 3.36 Preparation procedures of sulfones from thiols

The alkylation is achieved under basic conditions by nucleophilic substitution of a suitable leaving group. As a second option, the thioether can be synthesized from an alcohol under Mitsunobu conditions. Especially the second approach is now used as a standard route to access the desired sulfones (see Fig. 3.37).

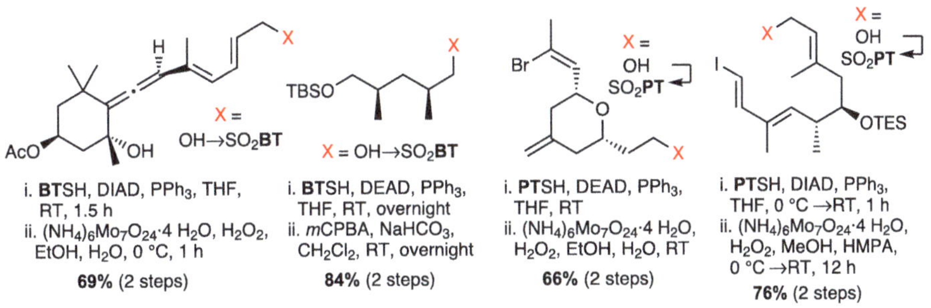

Fig. 3.37 Introduction of the requisite heteroaryl sulfones for the modified Julia olefination in various total syntheses [96–99]

3.1.3 Transition Metal-Mediated Olefinations

In addition to the Wittig and Julia olefinations and their variants, the construction of alkenes mediated by transition metal-based reagents forms the third major group of olefination methods frequently used in complex syntheses (Fig. 3.38).

Reactions based on transition metals possess certain advantages over P- and S-based methods: Unlike Wittig and Julia olefinations, titanium-mediated olefinations are not basic,

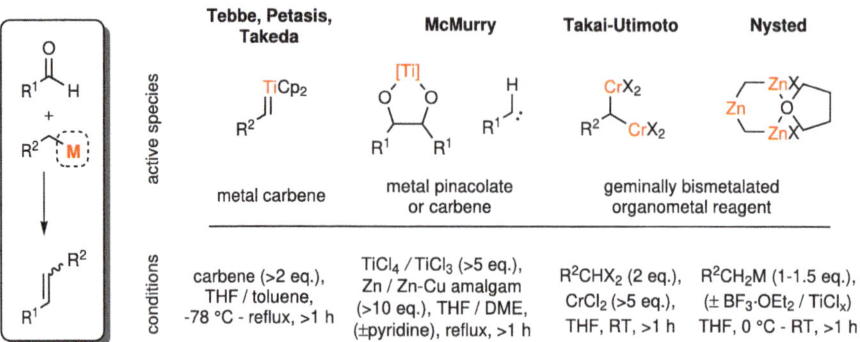

Fig. 3.38 Overview of different metal-mediated olefinations [100–106]

thus epimerization in the α-position of the carbonyl groups can be avoided. Carboxylic acid derivatives as well as sterically highly hindered carbonyls can also be converted due to the low steric demand of the metal reagents, whereas the previously discussed methods typically fail to convert these substrates (see Table 3.8) [100, 107].

Table 3.8 Reactivity of transition metal-based olefination reagents. [100, 101, 104, 107] Relative reaction rates **within a method** are indicated with +++ to (+). *Notes:* [a] formation of the Ti-enolate. [b] reactivity unknown/not published. [c] in the presence of Ti-salts. [d] only intramolecular

	Tebbe	Petasis	Takeda	McMurry	Takai-Utimoto	Nysted
aldehyde	+++	+++	+++	++	+++	+++
ketone	++	++	++	++	+	(+) / ++ [c]
ester	+	+	+	+	–	– / + [c]
lactone	+	+	+	–	–	– / + [c]
carbonate	(+)	(+)	– [b]	–	–	– [b]
amide	(+)	(+)	(+)	(+) [d]	–	– [b]
anhydride	– [a]	(+)	– [b]	–	–	– [b]
methenylation	+	+	–	–	±	+
alkylidenation	–	±	+	+	+	–

In contrast, the preparation of the organometallic reagents constitutes a distinct disadvantage of transition metal-mediated olefinations, which is significantly less universally applicable than the corresponding Wittig or Julia variants due to the more limited functional group tolerance. However, "simple" reactions such as methenylations can usually be carried out with excellent yields, often superior to other methods.

Tebbe, Petasis, and Takeda Olefinations

Olefinations based on Ti-alkylidenes are typically utilized for the methenylation of unreactive substrates such as esters and even amides [101, 108]. The most widely applied methods were developed in the groups of Tebbe, Petasis and Takeda and differ primarily in the nature of the Ti-precursors (Fig. 3.39).

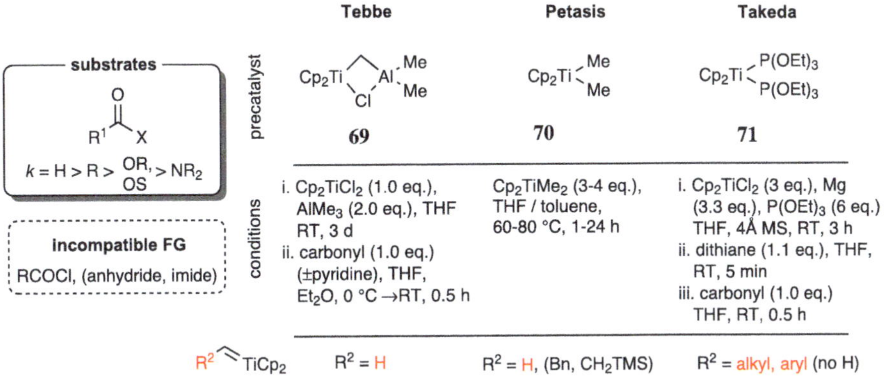

Fig. 3.39 Typical reaction conditions and substrate scope in Ti-mediated olefinations [101, 109, 110]

Comparing the efficiency of a methenylation of ketones based on either a Wittig or Tebbe olefination, Ti-mediated methods often provide superior yields. This difference is particularly noticeable in sterically crowded environments, as the sterically less demanding Ti-alkylidenes possess lower activation barriers (Fig. 3.40) [111, 112].

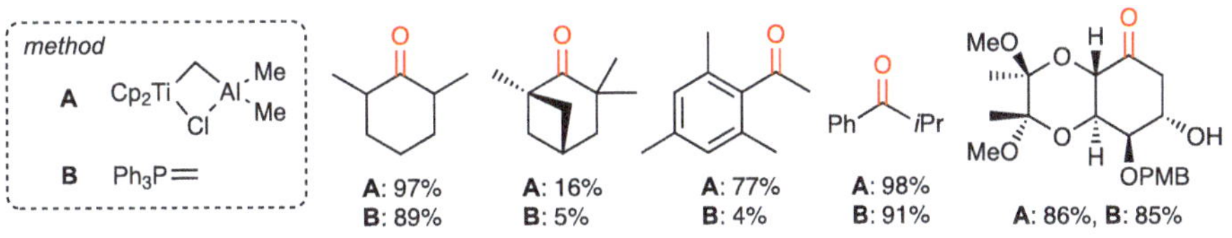

Fig. 3.40 Comparison of the olefination of ketones by the Tebbe or Wittig reagent [111, 112]

Titanium-mediated reagents are based on metal-alkylidenes as active species. Ti-alkylidenes are electron-rich (Schrock carbenes) and exhibit a nucleophilic character. They rapidly react with carbonyl groups at the electrophilic C-atom. Even substrates that possess several reactive functional groups usually are highly chemoselective. Mechanistically, the transformation resembles the Wittig reaction: An initial, concerted [2+2]-cycloaddition (via **72**) forms an oxatitanacyclobutane **73**, which subsequently decomposes to afford the desired alkene and a titanium oxide species. The driving force of the reaction is derived from the formation of the highly stable Ti-O double bond, ensuring the reactions' irreversibility. The formation of the Ti-alkylidene has been shown to be rate-determining for the Petasis olefination. By analogy, the Takeda method likely mirrors this trend. However, since the substrate also influences the individual rates of all steps, the cycloaddition cannot be ruled out to constitute the rate-determining step for some starting materials. The Tebbe olefination deviates slightly from the general mechanism presented here: In the absence of a Lewis base (e.g., pyridine, THF), the olefination was postulated to proceed through a bimetallic intermediate **75** instead of the oxatitanacyclobutane **73** (see Fig. 3.41) [107, 113, 114].

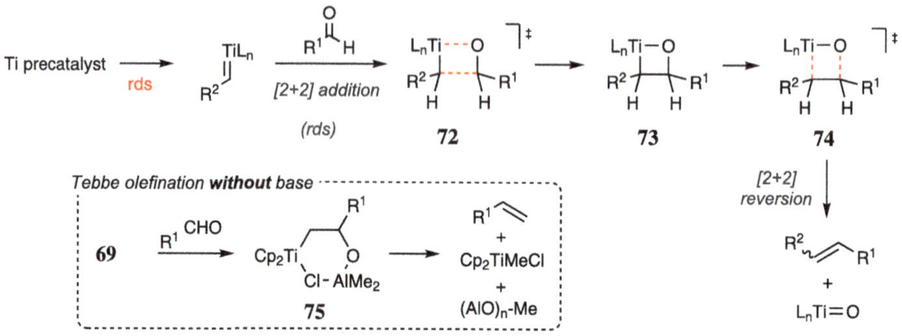

Fig. 3.41 Postulated mechanism of the Tebbe, Petasis and Takeda olefination [113, 114]

Ti-alkylidenes are also reactive via a classic metathesis pathway — this reaction is however usually significantly slower than the respective olefination [107] The metathesis is therefore not relevant in most cases as an undesired competitive reaction.

The individual methods introduced by Tebbe, Petasis, and Takeda differ primarily in the preparative approach to the active species from the corresponding precursors. Only methenylations are possible with the Tebbe olefination ($R^2 = H$). Even though the Petasis variant allows alkylidenations to a limited extent, the method is mainly used for methenylations as well. In contrast, the Takeda olefination only yields acceptable results with alkylidenations, thus presenting a complementary product scope to the other methods. Since the requisite dithianes are easily accessible from the corresponding carbonyls, a wide range of Ti-alkylidenes can be prepared for their use in the Takeda olefination (see Fig. 3.42).

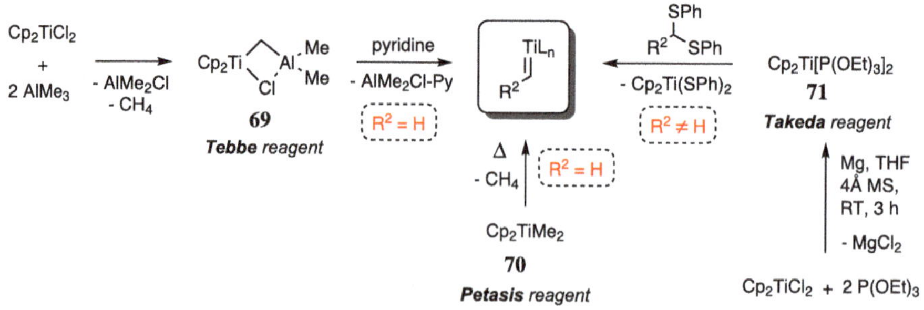

Fig. 3.42 Preparation of active metal reagents for the different olefination methods [113]

While the Tebbe reaction is very sensitive to the presence of air and moisture, the reagent is very reactive already at low temperatures. In contrast, Petasis and Takeda olefination require elevated temperatures. In general, all olefinations operate with an excess of the titanium carbene reagents.

As a by-product of the Tebbe olefination, a Lewis-acidic aluminum compound is generated. When using Cp_2TiMe_2 (**70**), methane is generated as the only by-product, which proves particularly advantageous with acid-sensitive substrates such as β-lactones (**76**) (see Fig. 3.43) [115].

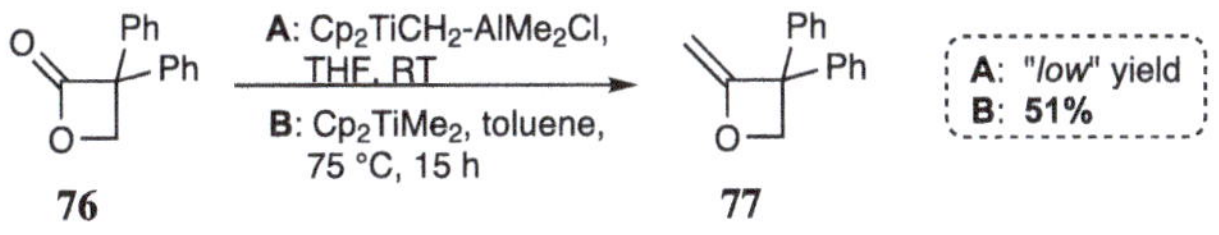

Fig. 3.43 Dependence of the yield on the type of methenylation reagent [115]

The Takeda olefination is rarely used, possibly due to the higher preparative effort to the reagent and the inability to facilitate methenylations. On the other hand, the Tebbe and Petasis olefinations are regularly encountered in complex natural product syntheses (see Fig. 3.44).

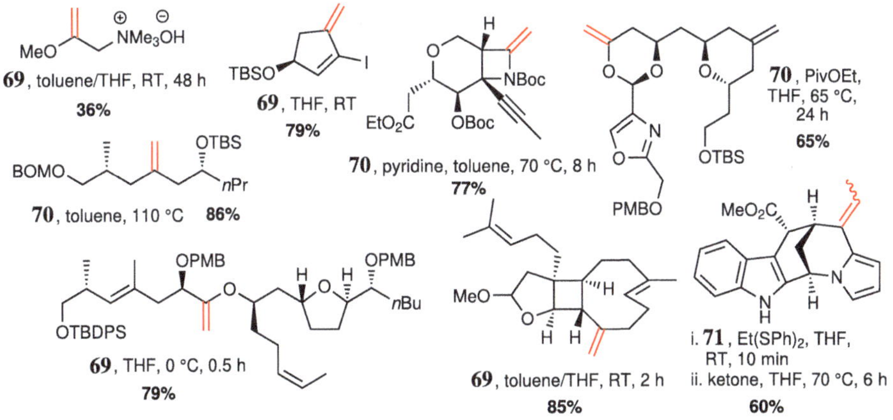

Fig. 3.44 Applications of the Tebbe (Cp_2TiCH_2-$AlMe_2Cl$, **69**), [116–119] Petasis (Cp_2TiMe_2, **70**) [120–122] and Takeda reagents ($Cp_2Ti[P(OEt)_3]_2$, **71**) [123]

To prevent the formation of by-products as well as the decomposition of the desired products by metathesis, other carboxylic acid derivatives can be added as a sacrificial reagent, which allow excess Ti-alkylidenes to be scavenged after the desired reaction has reached completion. Suitable additives are sterically more hindered structural analogs of the substrate, so as not to interfere with olefination. In a study for the industrial scale synthesis of the antiemetic aprepitant by Merck, $PhMe_2COAc$ was chosen as optimal sacrificial additive (see Fig. 3.45). Due to the scale, it was further required to recycle the Ti reagent. To facilitate this and to improve the yield and purity of the Petasis olefination, it was observed that the presence of trace $Cp_2TiMeCl$ allowed for a much cleaner reaction. This finding was rationalized by the formation of the oxo-bridged dimer $Cp_2MeTi\text{-}O\text{-}TiCp_2Me$ from the primary product $Cp_2Ti{=}O$. The select addition of small amounts of Cp_2TiCl_2 facilitates the requisite reaction to the Ti dimer and allows the recycle loop to be closed: $Cp_2MeTi\text{-}O\text{-}TiCp_2Me$ can be converted to Cp_2TiCl_2, from which the Petasis reagent can be generated again *in situ*. The optimized conditions allowed the conversion of **78** to **79** in excellent yield without chromatographic purification on a 227 kg scale [124].

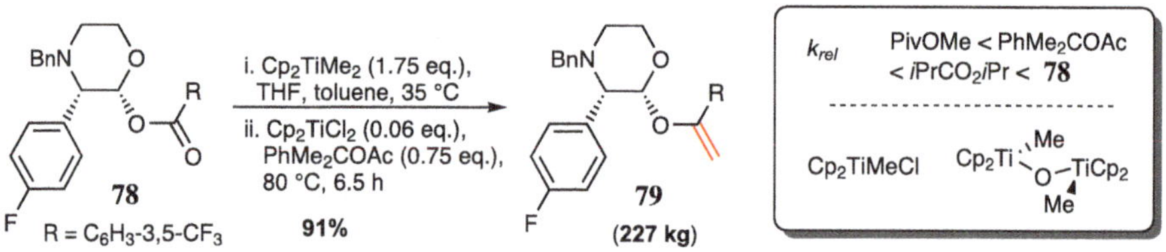

Fig. 3.45 Petasis olefination on an industrial scale [124]

1,1-Bimetallic Reagents

In addition to the classic methods by Tebbe, Petasis, and Takeda that rely on preformed Ti reagents, organotitanium compounds for carbonyl olefination can also be produced *in situ* by adding Ti-salts to the appropriate Mg or Zn salts. Since an alkene metathesis is usually not observed under these conditions, the active reagents are assumed to be 1,1-bimetallic compounds, which cannot facilitate this type of reaction. However, the details of the mechanisms have not yet been elucidated (Fig. 3.46) [100, 107].

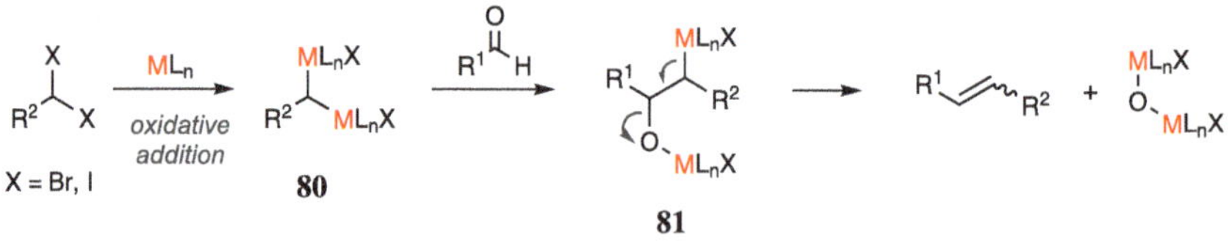

Fig. 3.46 General reaction mechanism of olefination using 1,1-bimetallic reagents [105]

Alkyl halides usually serve as the carbon component for the metal reagent. The standard conditions utilize Zn salts in combination with $TiCl_4$ and $PbCl_2$ and are referred to as Takai olefination (not to be confused with the Takai–Utimoto olefination, see below) [100]. The related Lombardo olefination uses similar conditions (see Fig. 3.47) [125].

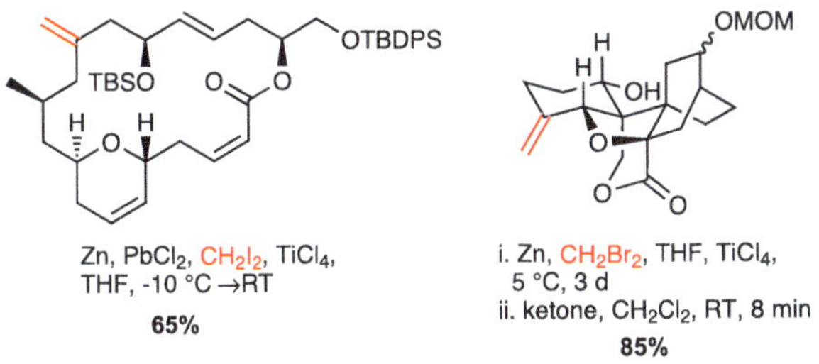

Fig. 3.47 Applications of the Takai and Lombardo olefination with ketones [126, 127]

When using metallic Zn, Mg, or other elemental metals in combination with Ti complexes, an oxidative addition affords organozinc or Grignard reagents, which are subsequently transmetalated to yield either a Tebbe reagent analog or the corresponding Ti-alkylidene (see Fig. 3.48) [100, 128].

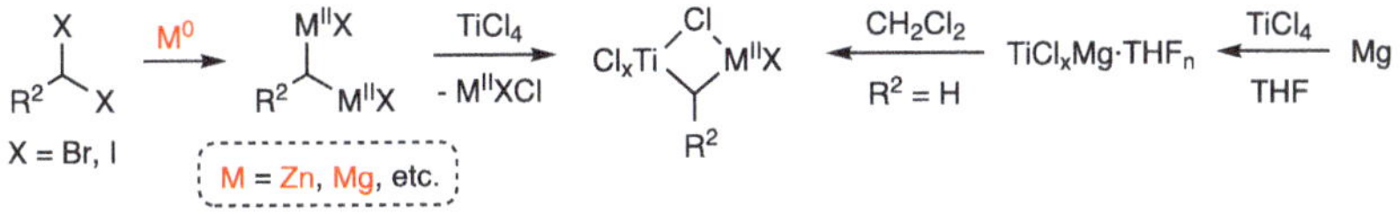

Fig. 3.48 Possible intermediates of olefinations in the presence of Ti-salts and Zn-/Mg organyls [100, 107, 128]

A method developed in Yan's group exploits the formation of the Tebbe-like reactive Ti-Mg complex from $CH_2Cl_2/TiCl_4/Mg/THF$ to synthesize the corresponding olefins from aldehydes, ketones, and esters [128, 129].

Apart from titanium-based reagents as mediators, chromium salts, organozirconium species as heavier Ti homologs, or even organozinc compounds such as the Nysted reagent **83** may be used. The Cr-based method developed by and named after Takai and Utimoto allows easy access to 1,2-disubstituted alkenes, foremost vinyl halides, boronates, and stannanes [130–132]. *E*-Configured alkenes are usually received from the aldehyde substrates. Many common functional groups (esters, amides, nitriles, silyl ethers, acetals, thioethers, alkynes, stannanes) are tolerated under the reaction conditions. Even though the Nysted reagent (**83**) can only introduce a CH_2 unit, it possesses a similar tolerance for a range of sensitive functionalities comparable to the Takai–Utimoto olefination while not necessitating stoichiometric amounts of chromium. The use of $HC_2(ZnI)_2$ is possible under identical conditions as well (see Fig. 3.49) [133].

The Takai–Utimoto olefination has become the method of choice for obtaining *E*-configured vinyl halides, as these compounds provide easy access to complex, conjugated polyenes via subsequent cross-coupling reactions. However, its substrate scope is by no means limited to halides only (see Table 3.9) [132].

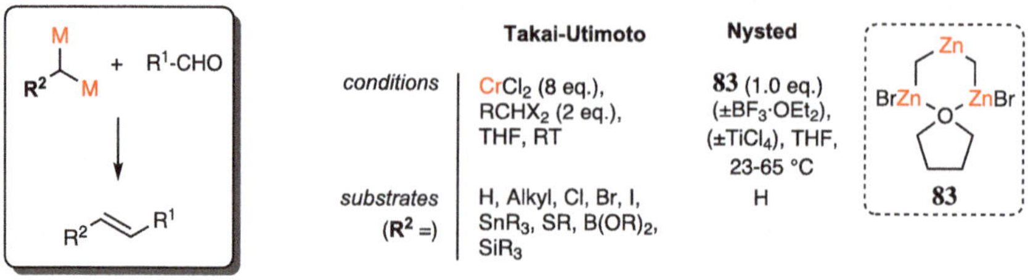

Fig. 3.49 Reaction conditions and substrate scope of Takai–Utimoto and Nysted olefinations [106, 131, 132]

Table 3.9 Substrate scope of the Takai–Utimoto olefination [132]. [a] LiI as additive. [b] boron pinacolate. [c] yield refers only to E-diastereomer

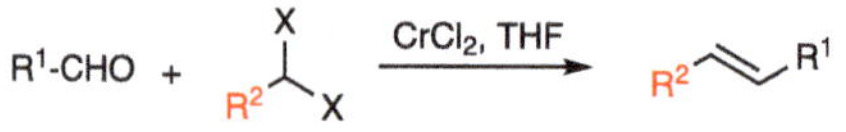

R^2	X	yield [%]	E-diastereoselectivity [%]
H	I	70–92	–
Alkyl	I	73–97	88–98
Cl	Cl	75	>90
Br	Br	55–73	81–95
I	I	66–80	80–92
SiR_3	Br	70–80	only E
SnR_3	Br [a]	60[c]	n.d.
SR	Cl	68–83	70–80
$B(OR)_2$ [b]	Cl [a]	78–91	87–97

Following its initial formation from R^2CHX_2, the mechanism of the conversion of **84** with the carbonyl substrate proceeds through a pseudo-chair transition state **85**. It has been postulated that an additional equivalent of the Cr salt participates in the transition state. After

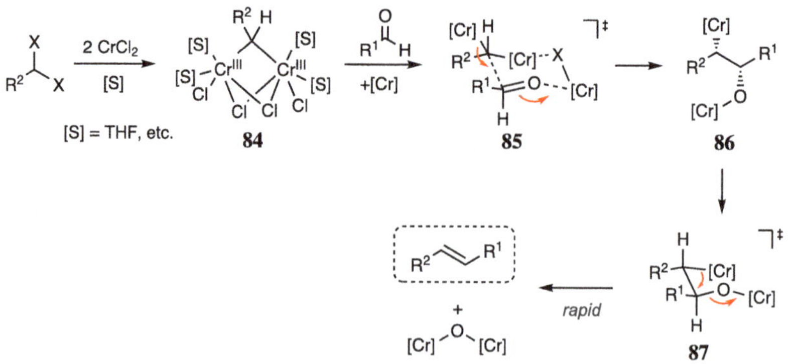

Fig. 3.50 Postulated mechanism of the Takai–Utimoto olefination [134–136]

obtaining the β-chromohydroxy-organochromium intermediate **86**, the desired alkene can be formed via a rapid *syn*-elimination to afford the *E*-diastereomer as the preferred product (see Fig. 3.50) [134–136].

1,1-Dizinc compounds such as the Nysted reagent **83** or $HC_2(ZnI)_2$ are able to efficiently olefinate aldehydes, while the corresponding ketones are usually converted too slowly to make the method synthetically useful. This normally prevents ketone groups in the substrate to be affected under the reaction conditions, but some ketones do show a sufficient reactivity. If their conversion is desired, Lewis acids such as BF_3 etherate are routinely added to activate the carbonyl function [106]. The zinc reagents can also be prompted to convert more reactive functional groups such as ketones or even esters under forced conditions by adding Ti-salts. The increased reactivity was rationalized by the formation of Ti-alkylidenes as the operative species under the reaction conditions [137]. Thus, *trans*-decalin **88** could be converted to the respective alkene **89** in high yield in Li's approach to the indole sesquiterpenoid sespenine (see Fig. 3.51) [138].

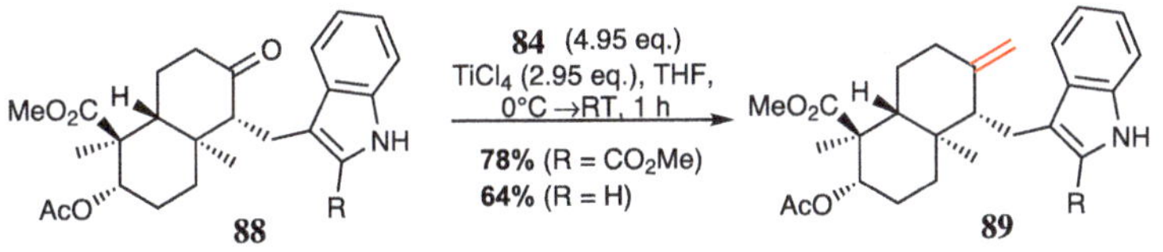

Fig. 3.51 Excerpt from the synthesis of sespenin by Li *et al.* [138]

As with other methods, the chemoselective conversion of a carbonyl function in the presence of other reactive groups remains the biggest challenge for metal-mediated olefinations. As decomposition reactions of the starting material are a common occurrence, no generalized substrate scope can be given. However, the reactivities provided in Table 3.8 might serve as good initial guess.

Ketoester **90** constitutes a key intermediate in Vatèle's synthetic route to the class of plakortolides and needed to be converted to the corresponding olefin. While a Wittig and Takai olefination only led to a decomposition of the starting material, the Tebbe and Petasis reagents could deliver the desired product **91**, albeit in low yield or in the presence of the vinyl ether **93**. The Nysted reagent however successfully afforded **91** in high yield when further modulated by the addition of Ti-salts. As expected, the preferred formation of the vinyl ether from the ester group (**90**→**92**) was negligible (see Table 3.10) [139].

The Takai–Utimoto method is primarily employed for the synthesis of vinyl halides. Even though alkylidenations are rather rare, some impressive examples can be found in terms of chemo- and regioselectivity. Figure 3.52 shows a selection of (natural product) syntheses in which the Takai–Utimoto or Nysted reagent was successfully used.

Table 3.10 Comparison of various methenylation methods with a ketoester [139]

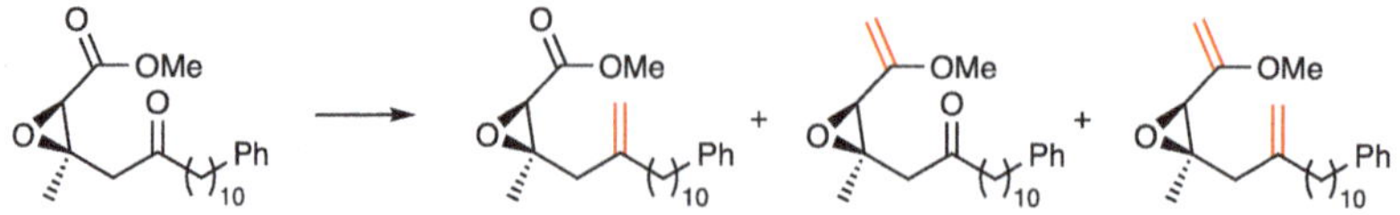

Method	Conditions	Conversion [%]	Yield (in %) 91 / 92 / 93
Wittig	Ph$_3$P=CH$_2$, -78 °C→RT, 3 h	100	– / – / –
Tebbe	Cp$_2$TiCH$_2$AlMe$_2$Cl, -78 °C→RT, 14 h	100	23 / – / –
Petasis	Cp$_2$TiMe$_2$, 80 °C, 21 h	78	34 / 3 / 41
Petasis	Cp$_2$TiMe$_2$, Cp$_2$TiCl$_2$ (kat.), 80 °C, 6 h	79	31 / 3 / 27
Takai	Zn, CH$_2$Br$_2$, TiCl$_4$, 0 °C→RT, 4 h	100	– / – / –
Nysted	**83**, TiCl$_4$, 0→15 °C, 1 h	100	35 / – / –
Nysted	**83**, Ti(OiPr)$_2$Cl$_2$, 0→15 °C, 0.25 h	85	70 / – / –

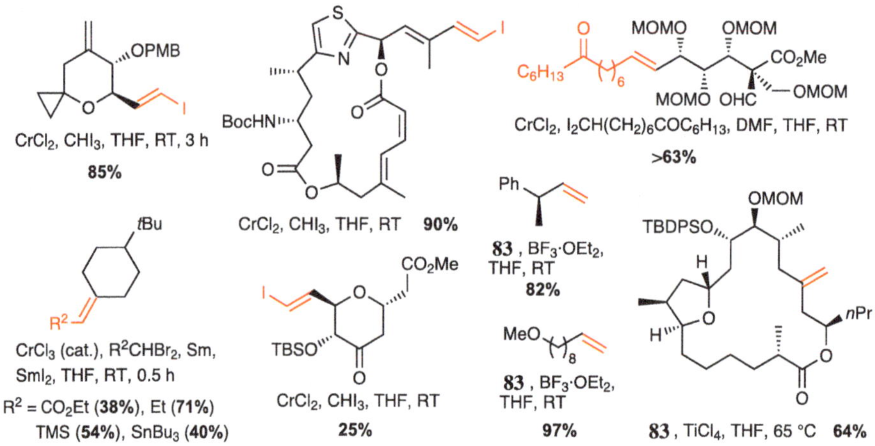

Fig. 3.52 Synthesis examples of the Takai–Utimoto olefination and the Nysted reagent [106, 140–145]

McMurry Coupling

The McMurry coupling was independently discovered by the groups of Mukaiyama, Tyrlik, and McMurry and further studied in-depth by McMurry. The reaction relies on low-valent titanium species to afford alkenes from the corresponding carbonyl compounds. The McMurry coupling can be conducted either as a homocoupling or using two different carbonyl substrates. The reactive reagent is obtained by exposing Ti chlorides to a suitable reducing agent, usually elemental metals (Zn, Zn-Cu, Li, etc., see Fig. 3.53) [102, 146–148].

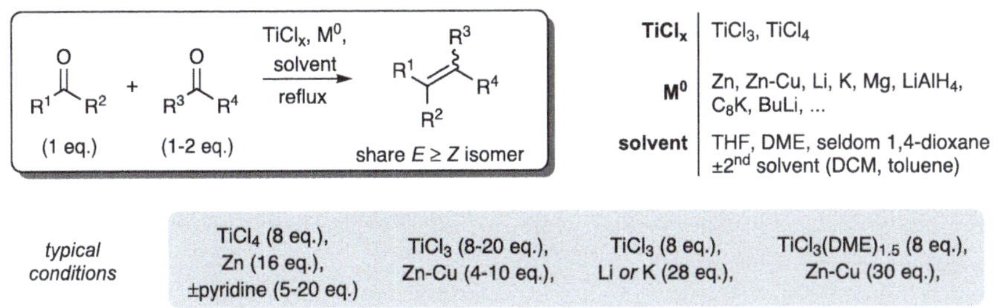

Fig. 3.53 General reaction scheme of the McMurry coupling [146, 149]

The reactions are carried out in etheric solvents (THF, DME), usually under reflux. For certain titanium-activator combinations, additives such as pyridine ($TiCl_4$/Zn) or NEt_3 ($TiCl_3$/$LiAlH_4$) and alkaline earth halides ($TiCl_3$/Li) are used to increase selectivity and suppress the formation of pinacol by-products.

In general, there is a strong dependence of the yield on the combination of the Ti source, the reducing agent, and the solvent. Frequently encountered conditions including their typical stoichiometry are listed in Fig. 3.53. $TiCl_4$/Zn/pyridine constitutes the most widely utilized system, followed by $TiCl_3$/Li. Newer developments, which are commonly cited as more reproducible and broadly applicable, are $TiCl_3(DME)_{1.5}$/Zn-Cu and $TiCl_3$/C_8K (see Fig. 3.54) [146].

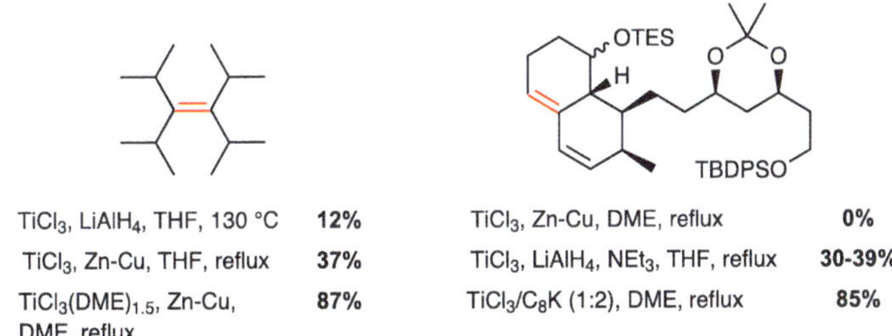

Fig. 3.54 Effect of using $TiCl_3(DME)_{1.5}$/Zn-Cu or $TiCl_3$/C_8K on the yield [150, 151]

The McMurry coupling has some disadvantages compared to other olefination methods. Not all functional groups are stable under the reaction conditions — particularly functionalities prone to reduction can be affected: 1,2-diols, nitro groups, quinones, oximes, α-halogenated ketones, sometimes also epoxides and allylic/benzylic alcohols. Secondly, the *E/Z*-geometry cannot be controlled for practical purposes by the reaction conditions. Although there is usually a (weak) *E*-preference, the diastereoselectivity tends to be strongly substrate-dependent (Fig. 3.55) [152].

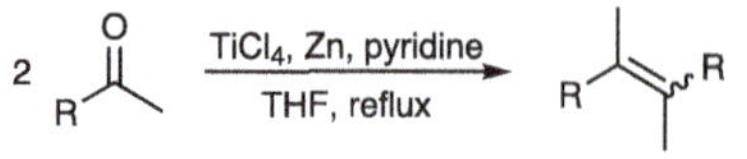

Fig. 3.55 Influence of substrate structure on the diastereoselectivity of olefinization [152]

R	yield [%]	E/Z
Et	65	1.1 : 1
nPr	61	1.4 : 1
iPr	46	1.5 : 1
neopentyl	48	only E
tBu	62	only E

Thirdly, the *inter*molecular McMurry coupling of different carbonyls typically affords a statistical mixture of products. To obtain a significant improvement of the desired product beyond the statistical outcome, a significant excess of one of the starting materials is required (see Fig. 3.56) [153].

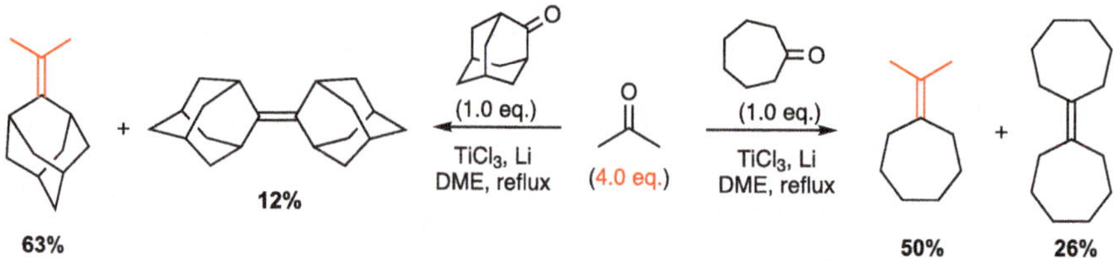

Fig. 3.56 Product distribution in the McMurry coupling [153]

Formaldehyde remains an unsuitable substrate for intermolecular couplings so far, thus precluding the McMurry coupling as a viable option for methenylations.

Despite all these disadvantages, the McMurry coupling remains the method of choice for the construction of tri- or tetrasubstituted alkenes. Furthermore, the simplicity of the procedure without the prerequisite to prepare the active reagents (phosphonates, sulfonates, Ti-alkylidenes, etc.) constitutes an advantage that should not be underestimated. The mechanism of the coupling appears to be substrate-dependent and its details are not fully understood, [146] as a suspension of the Ti species and dissolved substrate are usually present under the reaction conditions. This renders the elucidation of the heterogeneous mechanism immensely difficult.

The identity of the involved Ti species is described as "*low-valent titanium*". Neither the exact oxidation state of the metal nor the question of whether it is always a heterogeneous or sometimes also a homogeneous system has been conclusively clarified. Most studies postulate a reaction at the surface of heterogeneous Ti particles in the oxidation states 0 to +2 or +3, sometimes also an ensemble of Ti centers of different oxidation states. However, the involved species apparently can vary with the used Ti/reducing agent combination [154–156].

The fate of the organic component follows two possible paths: Either a Ti pinacolate or a carbene have been postulated as the reactive intermediate, which are formed through one- or two-electron reductions (see Fig. 3.57) [146, 157].

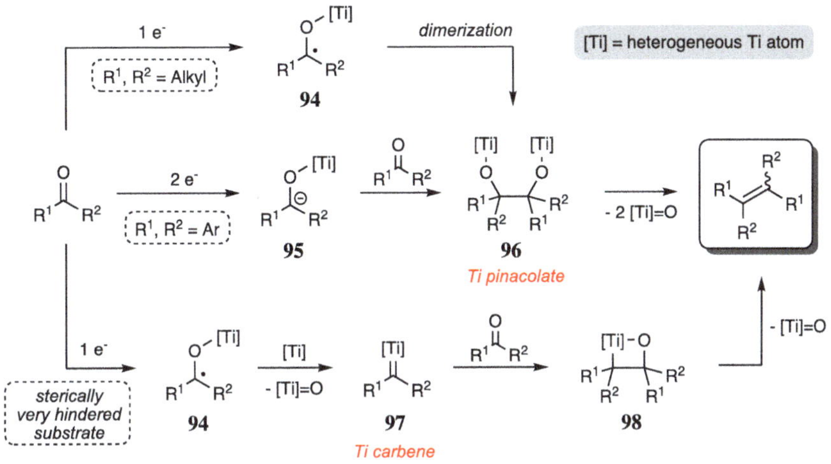

Fig. 3.57 Postulated mechanism of the McMurry coupling [146, 157]

For aliphatic ketones, a one-electron reduction (SET) to **94** followed by a dimerization to the Ti pinacolate **96** is assumed. More reactive aromatic ketones can be reduced twice. The anion **95** would then afford **96** in a nucleophilic addition to the ketone, which provides the alkene under the formation of the oxygenated Ti species. The pinacolate path is supported by the isolation of pinacols at low temperatures, whereas at higher temperature the alkene is obtained as the only product under identical conditions. In the reaction of sterically hindered substrates, it is assumed that carbene **97** constitutes the reactive species instead of Ti pinacolates following an initial one-electron reduction to **94**. The subsequent mechanism probably proceeds in analogy to a classic coupling with homogeneous Ti-alkylidenes [158–160]. The driving force of the reaction lies in the formation of extremely stable titanium-oxygen bonds.

The McMurry coupling is typically employed in the construction of complex di- or tetramers for use as functional materials, for example, as molecular switches, cavitands, or complexing agents. In the context of total synthesis, repeating units that can be coupled in this way are rarely present. Moreover, the intermolecular coupling of different carbonyl derivatives in a ∼1:1 ratio is possible, but often burdened with moderate yields, unless alkyl ketones are reacted with aryl-substituted ketones or aldehydes. For this reason, the reaction is traditionally used in an intramolecular fashion for the access to highly substituted alkenes (see Fig. 3.58).

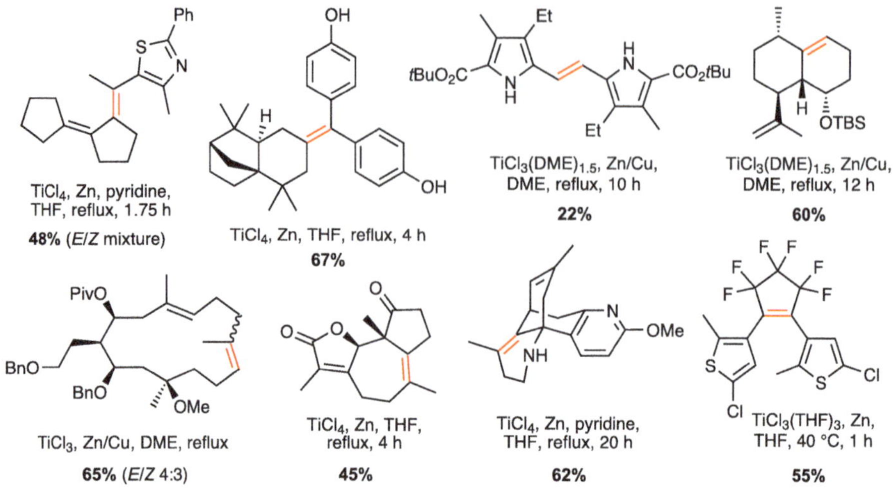

Fig. 3.58 Use of the McMurry coupling in the synthesis of complex molecules [161–167]

3.1.4 Eliminations

Historically, olefins were primarily prepared via a selective elimination. A disadvantage of a synthesis via cleavage of leaving groups such as halides or selenium oxide (see below) is the inherent possibility of forming various diastereo- and regioisomers (see Fig. 3.59).

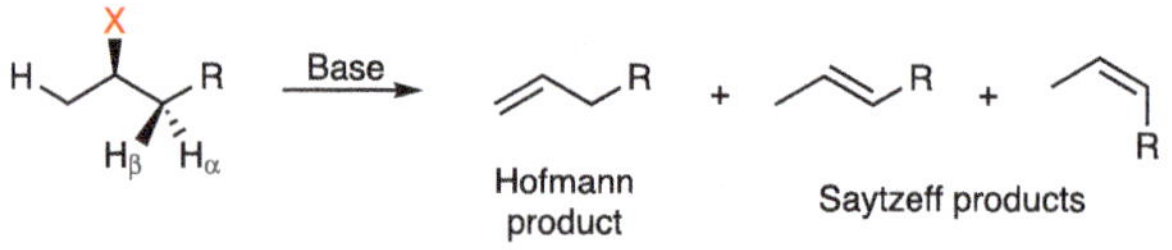

Fig. 3.59 Possible elimination products

The ratio of the Hofmann to Saytzeff product can usually be shifted towards the more substituted Saytzeff products under thermodynamic control. Despite the mentioned challenges, eliminations are still used very successfully on a case-by-case basis to introduce the desired double bond in a synthesis with minimal effort. The trick lies in the use of two vicinal functional groups that can jointly be removed in a chemo- and diastereoselective manner. Prominent examples can be found in the Peterson and Corey–Winter olefination (see Fig. 3.60).

The *Peterson olefination* involves the addition of silyl carbanions to carbonyl compounds followed by elimination of the resulting β-hydroxysilanes to the corresponding olefin [168–170]. The β-hydroxysilane intermediates are usually isolated when carbanions without stabilizing, electron-withdrawing functionalities are used. The second step, the elimination, is

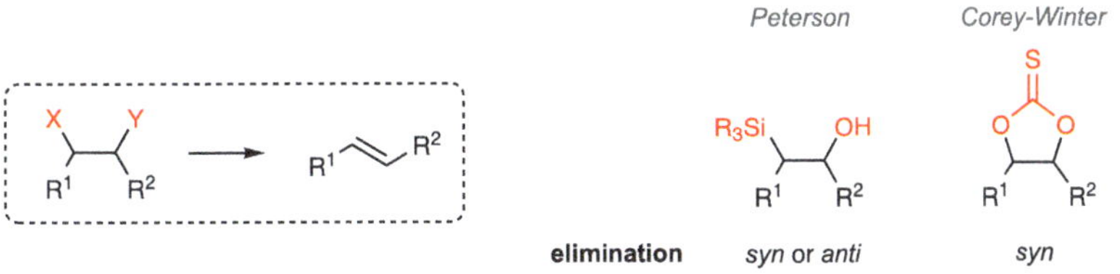

Fig. 3.60 Eliminations of substrates with vicinal functional group

stereospecific and dependent on the reaction conditions used. It is also referred to as *Peterson elimination* (see Fig. 3.61).

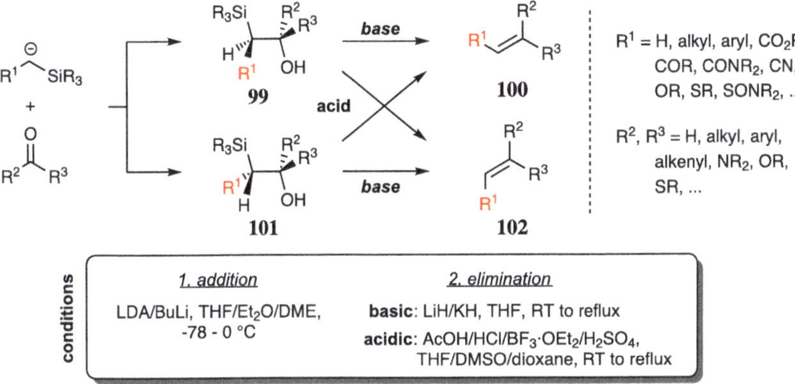

Fig. 3.61 Reaction sequence of the Peterson olefination including common reaction conditions [168–170]

The addition of the carbanion to the carbonyl can afford the two diastereomers **99** and **101**. Depending on the elimination conditions (basic or acidic), one of two stereospecific pathways is followed (synperiplanar under basic or antiperiplanar elimination under acidic conditions). The stereochemistry of the products **100** and **102** is thus determined by the aldehyde/ketone addition in the first step. If the initial carbonyl addition does not display high selectivities, no strong preference of a diastereomer is obtained in the desired alkene due to the stereospecificity of the subsequent steps.

The exact mechanism of the Peterson olefination has not been conclusively elucidated. There are two postulated paths under basic conditions, which proceed either stepwise or in a concerted manner (see Fig. 3.62). In the stepwise mechanism, which seems to be preferred in the literature as the more likely case, the carbanion addition can result in the *syn*-betaine **103** or the opposite diastereomer (not shown). The rapid formation of the strong O-Si bond then leads to the Z-alkene either upon elimination of the silanol functionality (via **104**) or through a cleavage of the oxasiletane **105**. The path is assumed to proceed so rapidly

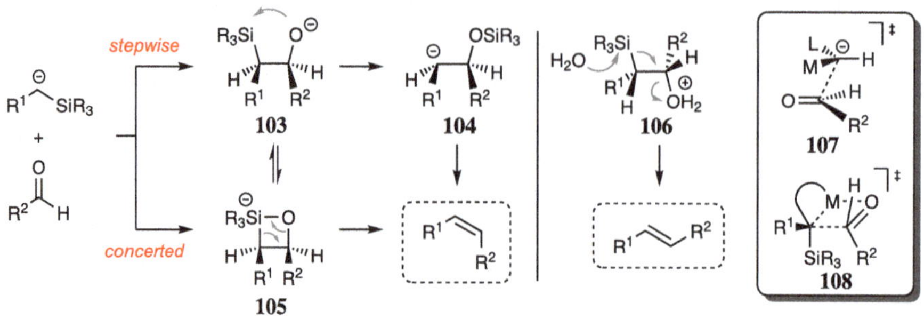

Fig. 3.62 Postulated mechanism under basic and acidic conditions [168, 169]. For simplification, only one diastereomer after the addition is depicted

that no intermediate rotation around the central C-C bond occurs, thereby preserving the stereospecificity of the reaction. In the concerted case the oxasiletane would be obtained directly via a [2+2]-cycloaddition, similar to the mechanism of the Wittig reaction. Under acidic conditions, the β-hydroxysilane is isolated and converted to the E-alkene by an acid-catalyzed *anti*-elimination [168, 169]. The initial addition is governed by steric factors according to a model proposed by Bassindale (via **107**) [171]. A small difference in the steric demand of the substituents thus results in diastereomeric mixtures with little to no clear preference (usually 1:1 to a maximum of 3:1). The use of a chelating functionality in one of the substituents of the silyl anion constitutes a more promising variant, whereby the reaction is assumed to proceed via a more rigid butterfly transition state **108** [172, 173].

The higher basicity and nucleophilicity of the silyl carbanions used in the Peterson olefination signifies one advantage the method possesses compared to the classic Wittig variants. On contrast to phosphorus-based anions, reactions of sterically hindered substrates can thus proceed more easily with the respective silyl homologues. Secondly, the by-product hexamethyldisiloxane can be separated relatively easily by distillation. This increased reactivity of the starting materials is offset by a diminished tolerance towards sensitive electrophilic functionalities — the chemoselectivity is thus significantly lower than with a corresponding Wittig variant. Another disadvantage of the Peterson olefination lies in the sometimes lacking stereocontrol of the addition with the consequence of a diastereomeric mixture of the alkenes. The commercial availability of the respective phosphorus reagents is also significantly better in comparison.

A multitude of studies and improved methodologies have been dedicated to enhance the diastereoselectivity of the addition step. A preparatively complex but atom-efficient approach is the isolation and chromatographic separation of the primary addition products. The *syn*- and *anti*-β-hydroxysilane can then subsequently be converted stereospecifically to the same olefin under complementary basic and acidic conditions (cf. Fig. 3.61). Beyond the laborious isolation, the most common strategy for an increased facial differentiation relies on the introduction of basic functionalities for chelate formation, thereby reducing the degrees of freedom. Furthermore, stereoelectronic effects such as the 1,3-allylic strain can also be exploited for improved control (Fig. 3.63).

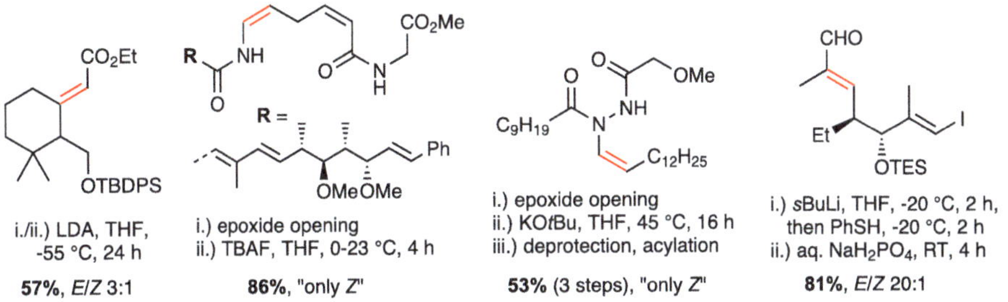

Fig. 3.63 Examples of Peterson olefinations in natural product syntheses [174–177]. Step *i* is the formation of the β-hydroxysilane, *ii* the elimination to the alkene

A.B. Smith and co-workers made use of a Peterson olefination in their total synthesis of the marine spiroketal metabolites calyculin A and B, which differ in the *E/Z*-stereoconfiguration of the tetraene moiety (see Fig. 3.64). To gain access to both diastereomers, the group resorted to a Peterson elimination of the corresponding *syn*- and *anti*-β-hydroxysilanes in the concluding steps of the synthesis. To form both secondary metabolites, the highly complex ketone was reacted with TMS-acetonitrile under basic conditions to afford the desired products in high yield. After chromatographic separation of the diastereomers, the respective calyculins A and B were obtained in the presence of HF by global deprotection of the 7 (!) silyl ether protecting groups [178].

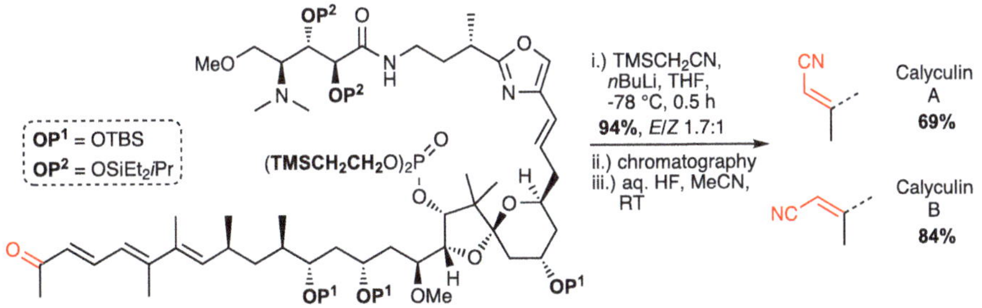

Fig. 3.64 Total synthesis of calyculin A and B by Smith *et al.* [178]

The *Corey–Winter olefination* is defined as the two-step conversion of vicinal diols to alkenes via the formation of cyclic thiocarbonates [179]. This method, developed in the 1960s, has the charm of being stereospecific and very chemoselective. Particularly sterically strained cycloalkenes can be synthesized stereoselectively with this approach. However, the extended reaction times and elevated temperatures, which are required for the elimination step to proceed efficiently, come at the expense of a possible impairment of sensitive functional groups. Due to the mechanism, the diols must necessarily be able to assume a synperiplanar orientation. This allows the selective introduction of a double bond even in structurally fixed ring systems in the presence of several other reactive functionalities (see Fig. 3.65).

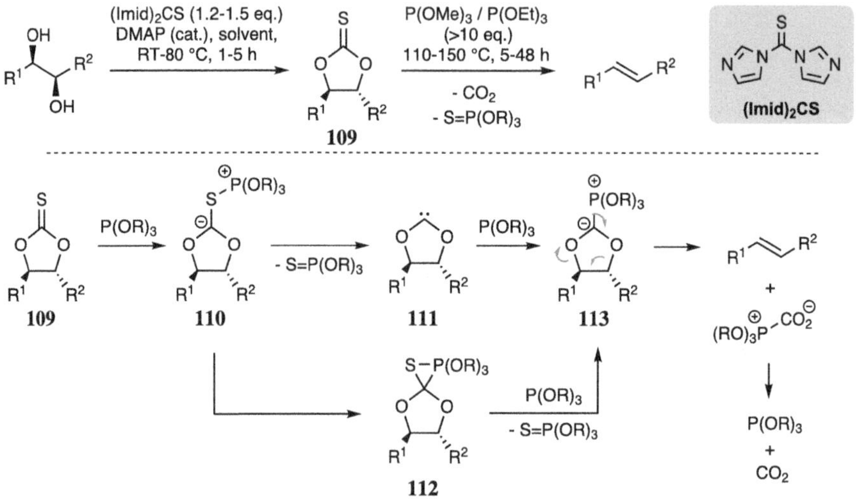

Fig. 3.65 General reaction conditions and postulated mechanism of the Corey–Winter olefination [179]

The driving force of the reaction is the formation of a stable P=S bond and the loss of CO_2. Following the initial formation of the thionocarbonate **109** with thiocarbonyldiimidazole or thiophosgene, it is converted to the corresponding ylide **110** in the presence of trialkyl phosphites. The transformation can proceed either via a carbene (**111**) or a phosphathiirane **112**. After the addition of another equivalent of $P(OR)_3$, a cycloreversion to the corresponding alkene takes place with elimination of $(RO)_3PCO_2$, from which CO_2 is ultimately released [179]. Despite its advantages, the Corey–Winter olefination is not as widely used as alternative methods, except in those selected applications that can fully exploit the method's strengths. These examples elegantly incorporate the Corey–Winter olefination into the synthetic scheme (see Fig. 3.66).

Selenium oxide-mediated eliminations are not often used due to the toxicity of selenium and its smell, [184] but there exists ample precedent in the synthesis of complex natural products where it can be very elegantly applied if alternative methods do not perform well [185, 186]. One of the most frequently used SeO-mediated eliminations was introduced by Grieco and co-workers. The method converts primary and occasionally secondary alcohols

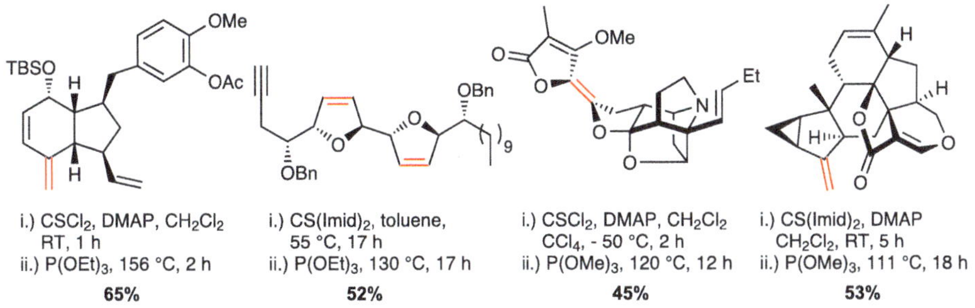

Fig. 3.66 Examples of the Corey–Winter olefination [180–183]

to the corresponding selenides **117** in the presence of an aryl-substituted selenocyanate. The desired olefin is subsequently obtained under oxidative conditions (H_2O_2, *m*CPBA, etc.). The reaction mechanistically resembles a Mitsunobu inversion. After the activation of the arylselenide (**115**), the phosphine is transferred to the alcohol to afford the activated intermediate **116**. The selenide **117** is then obtained via nucleophilic substitution with inversion of configuration. Following an oxidation to the selenoxide, the elimination of the selenol via **118** yields the desired alkene (see Fig. 3.67) [187, 188].

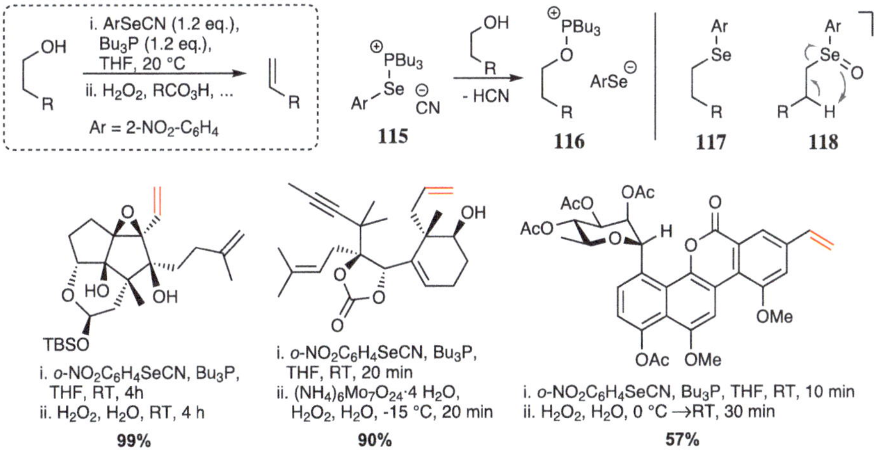

Fig. 3.67 Conditions of the Grieco elimination, postulated intermediates and synthetic examples [188–191]

An impressive example can be found in Cramer's synthesis of the marine secondary metabolite fijiolide A. The oxidative elimination in various solvents only yielded the allylic alcohol through a [2,3]-Wittig rearrangement (see Fig. 5.143, Chap. 5). A biphasic system (toluene/30% H_2O_2) allowed for the selective elimination with an acceptable yield of 67%. The global desilylation with HF·pyridine afforded the desired natural product after a total of 36 steps (18 steps of the longest linear sequence see Fig. 3.68) [192].

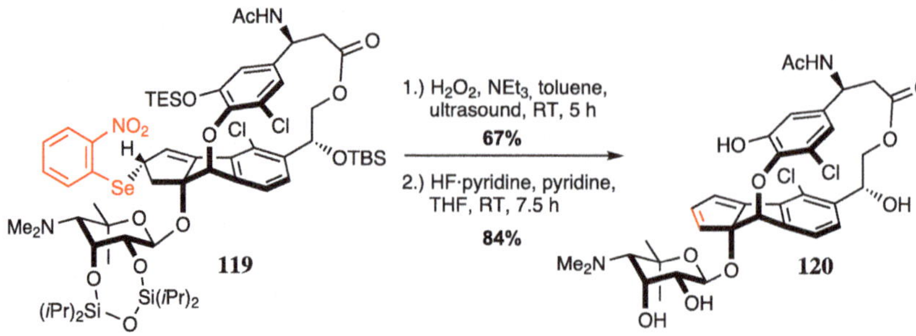

Fig. 3.68 Synthesis of fijiolide A (**120**) according to Cramer *et al.* [192]

Further olefination methods based on the elimination of alcohols were developed in the groups of Martin and Burgess [193, 194]. The reagents both rely on the activation of a hydroxy group followed by its elimination. Due to the neutral reaction conditions, base-labile functionalities remain unaffected, as the requisite equivalent of base is already contained in the reagent (see Fig. 3.69).

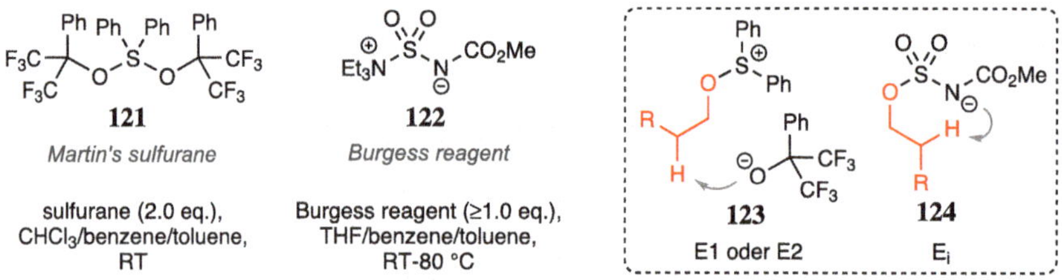

Fig. 3.69 Olefinations by Martin and Burgess and postulated intermediates [193–195]

In both cases, the sulfur of the reagents activates the alcohol. While a basic alkoxy group is displaced by the substrate in Martin's sulfurane, the sulfamate ester salt derived from the Burgess reagent releases NEt_3 as endogeneous base. Upon reductive elimination of the second alkoxy substituent, the alkene formation from **123** proceeds either via an E1 or E2 mechanism with antiperiplanar alignment [193, 196]. The Burgess reagent in turn enables the olefination via **124**, which is postulated to occur in an intramolecular fashion and synperiplanar orientation (E_i mechanism) [193, 195].

The reactions are particularly suitable for secondary and tertiary alcohols as substrates, as these are more easily prone to elimination reactions. Primary hydroxy groups without acidic β-H atoms can undergo a substitution instead of elimination in the presence of Martin's sulfurane. Densely functionalized alcohols can be converted with these reagents, but substrates possessing multiple hydroxy groups often suffer from a lack of differentiation and should generally be avoided (see Fig. 3.70) [195].

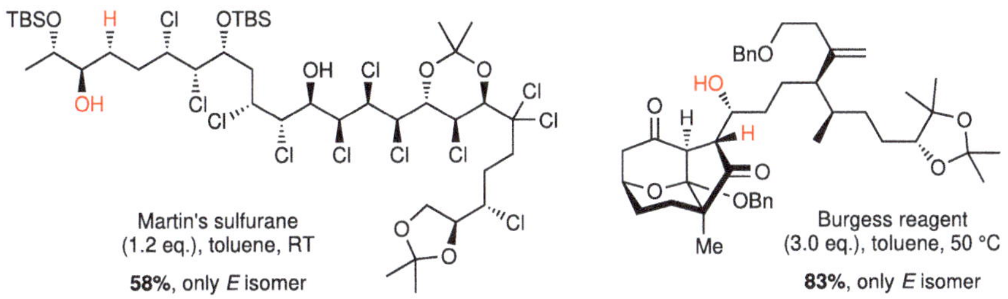

Fig. 3.70 Applications of the olefination methods by Martin and Burgess [197, 198]

In addition to these approaches, the straightforward formation of the respective mesylate, tosylate, or halide and subsequent elimination can be used to obtain an alkene under basic conditions. Despite the mild conditions of the Martin and Burgess methods, **all** possible competing reaction pathways to the various regioisomers and diastereomers need to be considered in the presence of multiple viable H-atoms. The use of a "simple" elimination without resorting to 1,2-difunctionalized substrates (Peterson, Corey–Winter) is especially recommended for substrates where no isomer mixtures can or should be obtained, for example, with primary alcohols or in the formation of α,β-unsaturated alkenes.

3.1.5   Other Methods

In addition to the methods described so far, there are other niche approaches to prepare olefins. An efficient, yet rarely used concept relies on a sequence incorporating a pericyclic reaction as a key step. This category comprises the *Ramberg-Bäcklund rearrangement*, which is discussed in more detail in Sect. 5.5. The *Barton–Kellogg* reaction or olefination is particularly suited for the synthesis of unsymmetrical, sterically highly strained tetrasubstituted alkenes [199–201]. In the reaction of thioketones with diazo compounds, two carbonyl derivatives or equivalents are converted to alkenes via an intermittent thiirane (see Fig. 3.71).

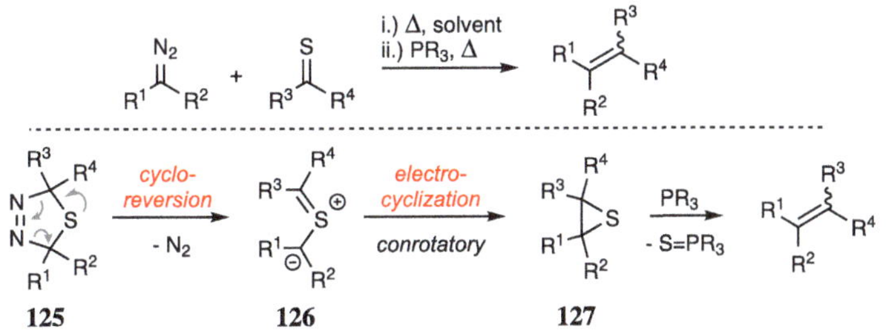

Fig. 3.71 Postulated mechanism of the Barton–Kellogg reaction [199]

Mechanistically, a sequence of pericyclic reactions takes place: Following an initial [3+2]-cycloaddition, N_2 is eliminated in the cycloreversion **125→126**. The conrotatory electrocyclization of **127** leads to the formation of a thiirane **127**. The sulfur is excised at elevated temperatures in the presence of phosphines with formation of a thiophosphine upon transformation of **127** to the respective alkene. The reaction is completely stereospecific from the stage of cycloaddition (cycloreversion, electrocyclization, S-extrusion) [199, 202]. The addition thus determines the stereochemistry of the alkene. So far, there is no comprehensive methodological study following the initial publications that explores the substrate scope and selectivity.

The reaction is used almost exclusively for the synthesis of highly substituted, unsymmetrical alkenes. The advantage over a McMurry coupling lies in the possibility of using two different building blocks, which means that it is not limited to a homocoupling and avoids statistical mixtures.

In Li's total synthesis of clostrubin, a variant of the Barton–Kellogg olefination is employed for the construction of the central olefinic bond. The stabilized diazo compound **128** was converted to the reactive metal carbene in the presence of the rhodium salt and reacted with thioketone **129**. After the *in situ* formation of the episulfide **130**, Cu powder was used for the reductive formation of the central double bond in **131**, and the desired alkene was obtained in 85% yield (Fig. 3.72) [203].

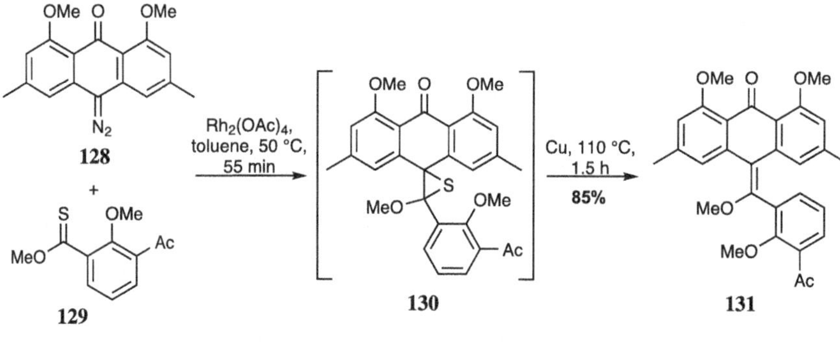

Fig. 3.72 Use of a modified Barton–Kellogg reaction in the synthesis of clostrubin [203]

3.2   Synthesis of Alkynes

By virtue of their easy derivatization, alkynes display a similar versatility as platform functionality as carbonyl compounds or olefins do and are therefore important intermediates in synthetic endeavors. The increasing importance of Cu-catalyzed azide-alkyne cycloadditions (Click Chemistry, → Chap. 9) has also brought the CC-triple bond back into the focus of research. The accessible repertoire of transformations to build internal or terminal alkynes can largely be traced back to the conversion of two functional groups: aldehydes/ketones or CC-triple bonds (see Fig. 3.73).

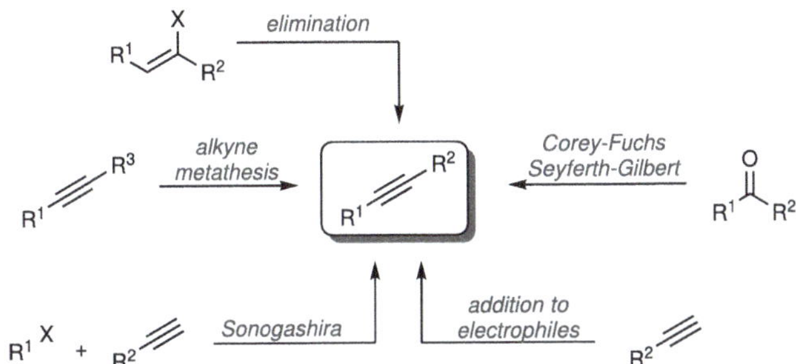

Fig. 3.73 Routes for the synthesis of alkynes

The reaction starting from carbonyl compounds specifically synthesizes the desired triple bond *de novo*. As alternatives, the deprotonation of an alkyne and reaction with an electrophile, an alkyne metathesis or a Sonogashira cross-coupling (see Sects. 7.3 and 6.2 for these reactions) rely on the derivatization of an already existing CC-triple bond.

3.2.1 Construction of Alkynes Starting from Carbonyl Compounds

The dominant approach in total syntheses lies in the use of aldehydes/ketones as starting material, also due to the wide arsenal of available methods for the construction and deriva-

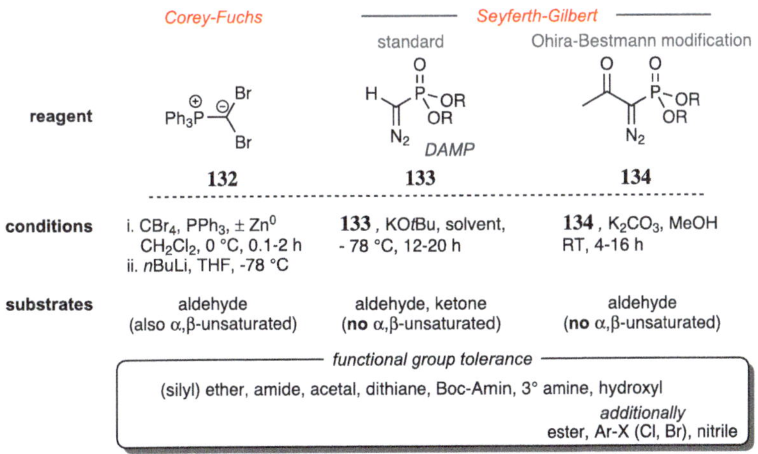

Fig. 3.74 Comparison of the Corey–Fuchs and Seyferth–Gilbert alkynylation [204–206]

tization of CO double bonds (see Chaps. 4, 2). The complementary methods starting from carbonyls comprise the *Corey–Fuchs reaction* and the *Seyferth–Gilbert homologation* (see Fig. 3.74) [204].

Both alkynylation methods tolerate a range of functionalities, but these often require protection (silyl/benzyl ether, acetal, Boc-amine). The Ohira–Bestmann variant also allows the use of esters, aryl halides, and nitriles. In some cases hydroxyl groups can be employed directly without prior protection without showing a negative impact on the yield.

Mechanistically, the Corey–Fuchs reaction is based on two discrete reactions, which also can be carried out separately with intermittent steps in between. In the first step, the phosphonium ylide **132** is formed from PPh_3 and CBr_4 via **135**, which leads to the geminally dibrominated alkene **136** analogous to the mechanism of the Wittig reaction. The subsequent step affords alkyne **137** in the presence of butyllithium by elimination of HBr and Br-Li exchange, which either yields a terminal alkyne under acidic work-up or an internal alkyne in the presence of electrophiles (see Fig. 3.75) [205]. If an equivalent of NaHMDS is used as the base, isolable 1-bromoalkynes (**138**) are obtained and can be converted to **137** upon exposure to another equivalent of *n*BuLi.

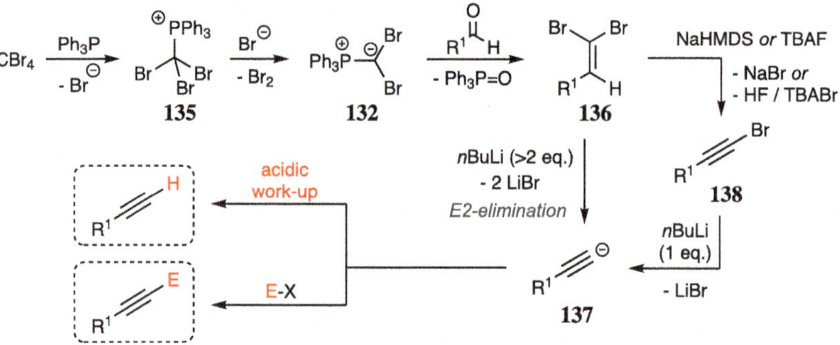

Fig. 3.75 Mechanism of the Corey–Fuchs reaction and possible products [205, 207, 208]

If the dibromoalkene **136** is isolated, the conversion to the corresponding alkyne can be carried out later in the synthesis. Other transformations such as cross-coupling reactions can also be realized instead. Such interrupted Corey–Fuchs reactions occur quite often in total syntheses, as the tolerance towards functional groups is in large part limited by the strong base in the elimination step (see Fig. 3.76).

A striking example can be found in Overman's access to (+)-pumiliotoxin B. Following the formation of the dibromoalkene, four transformations were carried out before the desired alkyne **139** was synthesized in the concluding step [209]. Instead of introducing CC-triple bond, the dibromoalkene can also serve as handle for other transformations: In the presence of catalytic amounts of $Pd(PPh_3)_4$ and Bu_3SnH as a reducing agent, one of the two C-Br bonds is replaced by a C-H bond: After the reduction, a Pd-catalyzed Heck reaction ensues with a

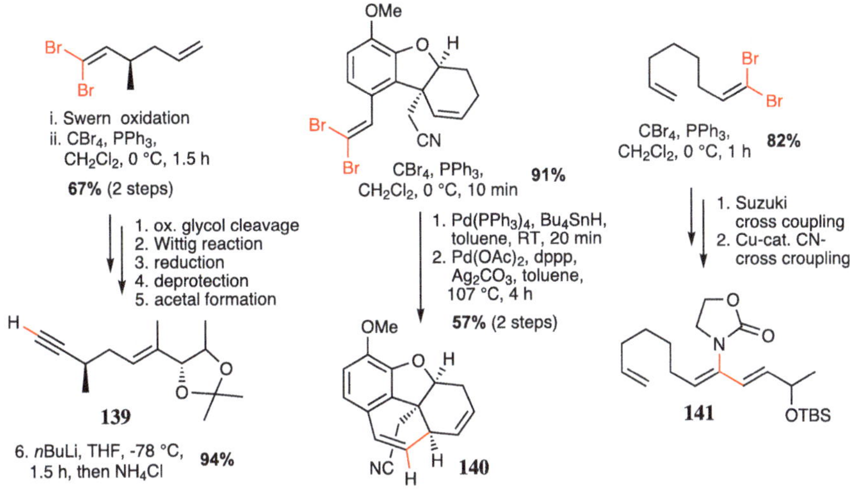

Fig. 3.76 Examples of interrupted Corey–Fuchs reactions in the total syntheses of pumiliotoxin B, codeine/morphine, and *Galbulimima* alkaloid 13 [209–211]

concomitant shift of the endocyclic double bond and formation of the C-ring to **140**, which ultimately allowed the synthesis of codeine and morphine [210]. In the last example, the C-Br functionality was converted to **141** in successive CC and CN cross-coupling reactions catalyzed by Pd and Cu complexes, respectively. In a further ten steps, Galbulimima alkaloid 13 could be obtained [211].

One advantage of the Corey–Fuchs reaction over the Seyferth–Gilbert homologation (both under standard conditions as well as the Ohira–Bestmann variant) is the ability to convert α,β-unsaturated aldehydes and ketones to the corresponding enynes without isomerization of the double bond (see Fig. 3.77). For sensitive substrates, NEt_3 can also be added in the first step to suppress potential side reactions [207].

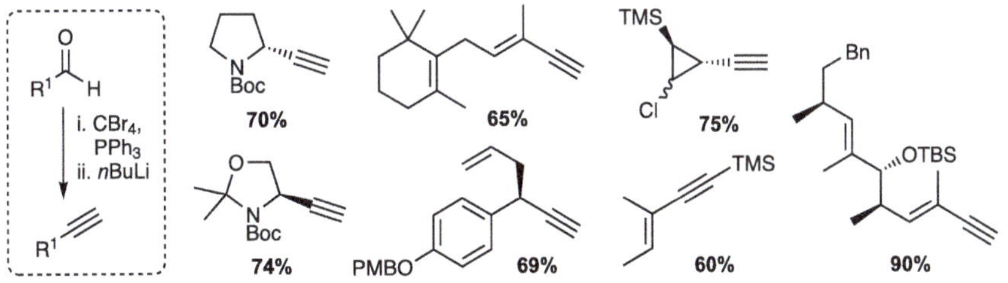

Fig. 3.77 Substrate scope of the Corey–Fuchs alkynylation [206, 212, 213]

As in the Wittig reaction, the phosphine oxide by-product is difficult to separate from the product by chromatographic means. One approach for improvement is the replacement of PPh$_3$ with P(OiPr)$_3$, as the corresponding O=P(OiPr)$_3$ by-product can be easily separated as an oil in the work-up [214]. The *Seyferth–Gilbert homologation* relies on a diazo compound (**133**) as carbon source. Under the classic reaction conditions, a strong base is required for the formation of the initial anion **142**. After reversible deprotonation, **142** adds to an aldehyde or a ketone and the primary product **143** rapidly cyclizes to the oxaphosphetane **144**, analogous to the intermediate of the Horner–Wadsworth–Emmons olefination. Following its decomposition to diazoalkene **145**, a rearrangement was postulated, which comprises a 1,2-shift of the substituents on the original carbonyl C-atom and a subsequent elimination of N$_2$ (see Fig. 3.78) [215]. In the Ohira–Bestmann modification, **134** is used as an alkynylation reagent instead. A weak base in alcoholic solvents suffices to irreversibly cleave the acetyl group via attack by methoxide, affording the anion **142**. Under classic conditions, KOtBu serves as the base. The larger potassium counterion (compared to nBuLi in the Corey–Fuchs reaction) leads to a faster decay of anion **143** and largely avoids the formation of undesired by-products.

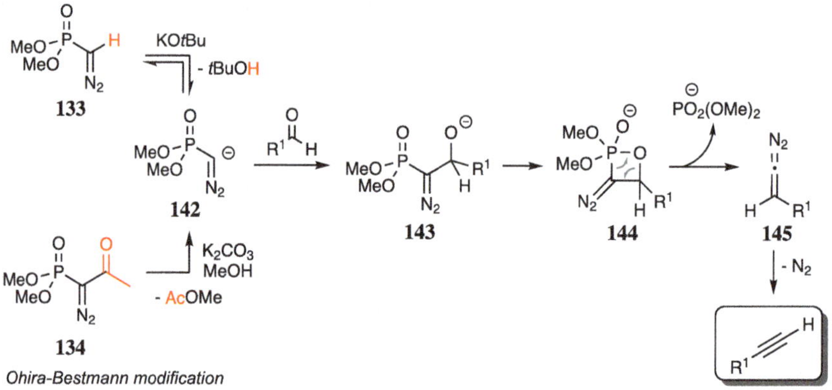

Fig. 3.78 Postulated mechanism of the Seyferth–Gilbert homologation [215]

Dialkylketones are unreactive under the reaction conditions and α,β-unsaturated aldehydes afford the corresponding products in greatly diminished yields, which is why the Seyferth–Gilbert homologation is considered unsuitable for these substrates. The reaction requires aprotic solvents, as the diazo compound **145** can otherwise react to the corresponding enol ether with loss of dinitrogen (see Fig. 3.79) [205, 206].

The Ohira–Bestmann modification uses the acetyl analog of DAMP (**134**), which forms **142** *in situ* in a slightly basic medium. The low concentration of methoxide anions prevents premature decomposition of **142** and allows the conversion of aldehydes to alkynes already by addition of the weak base K$_2$CO$_3$.

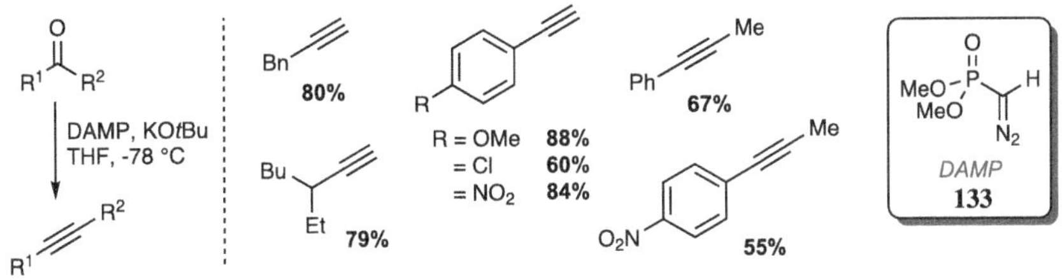

Fig. 3.79 Substrate scope of the classic Seyferth–Gilbert homologation [215]

Ketones are not reactive via the Ohira–Bestmann modification. α,β-Unsaturated aldehydes suffer a 1,4-addition by MeO$^-$ and thus afford homopropargylic ethers **146** instead of the respective enyne. A hallmark of the Ohira–Bestmann modification is the possibility of converting esters, nitriles and α-chiral aldehydes to the respective alkynes, while only necessitating a weak base for the conversion (see Fig. 3.80) [216, 217].

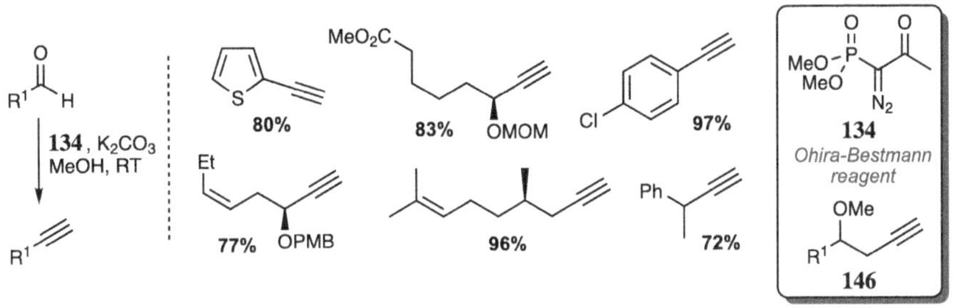

Fig. 3.80 Substrate scope of the Ohira–Bestmann modification [216, 217]

One of the disadvantages of the Seyferth–Gilbert homologation is the lack of commercial availability of the diazo reagents **133** and **134**. They must be prepared individually prior to the alkynylation (see Fig. 3.81).

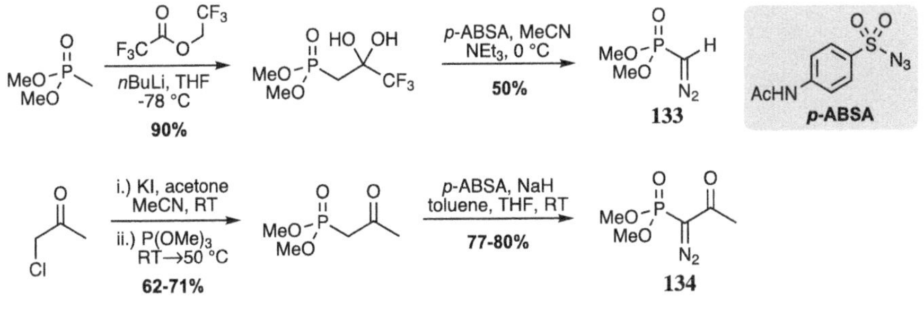

Fig. 3.81 Synthesis of the diazo reagents [218, 219]

Although the Ohira–Bestmann modification is the most commonly used alkynylation method, a direct comparison underlines that it may not always be the method of choice. Depending on the starting materials, the classic Seyferth–Gilbert homologation or the Corey–Fuchs reaction can yield better results (see Fig. 3.82).

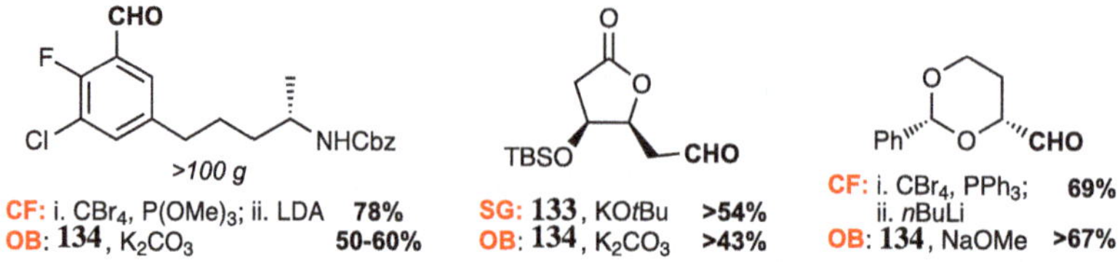

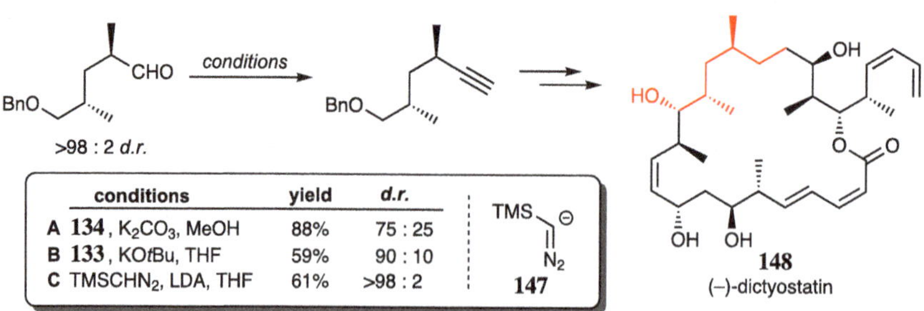

Fig. 3.82 Comparison of various alkynylation methods of selected aldehydes [220–222]. CF = Corey–Fuchs; SG = Seyferth–Gilbert; OB = Ohira–Bestmann

For α-chiral carbonyl compounds, the stereointegrity during the alkynylation plays a particularly important role. While most publications describe the Ohira–Bestmann modification as the reaction with the lowest tendency towards epimerization, this is not true for all substrates. Particularly the use of the Colvin reagent TMSC(Li)N_2 (**147**) can often fully avoid the loss of stereoinformation under the mild conditions, while other methods may tend to epimerize the substrate [204, 223, 224]. The commercial availability of the reagents TMSCHN_2 and LDA is another advantage of the *Colvin rearrangement*.

Gennari's synthesis of a (–)-dictyostatin fragment assessed the various alkynylation methods and impressively demonstrated the superiority of the Colvin rearrangement (see Fig. 3.83) [225]. While the Ohira–Bestmann modification counterintuitively performed worst in terms of maintaining the diastereomeric ratio, a *d.r.* of 90:10 could be achieved under classic Seyferth–Gilbert conditions. The Colvin rearrangement afforded the desired alkyne diastereomerically pure in moderate yield.

conditions	yield	d.r.
A **134**, K_2CO_3, MeOH	88%	75 : 25
B **133**, KO*t*Bu, THF	59%	90 : 10
C TMSCHN_2, LDA, THF	61%	>98 : 2

Fig. 3.83 Synthesis of a (–)-dictyostatin fragment by Gennari *et al.* [225]

Modern Methods and Synthetic Applications

One of the main disadvantages of the Corey–Fuchs reaction and the various Seyferth–Gilbert variants is the need to isolate the intermediates or prepare the diazo reagents prior to the alkynylation. To remedy both shortcomings, various one-pot procedures have been developed.

Rassat and co-workers reported a variant using a different CBr_2 source. By resorting to the Wittig salt **149**, whose tribromine congener **135** is formed as an intermediate under classic Corey–Fuchs reaction conditions, the alkyne can be isolated directly in the presence of KO*t*Bu (see Fig. 3.84) [226]. Schmidt *et al.* used the one-pot method for the synthesis of the polyacetylene natural product atractylodemayne A, in which they obtained the key intermediate **150** in one step with acceptable yield [227].

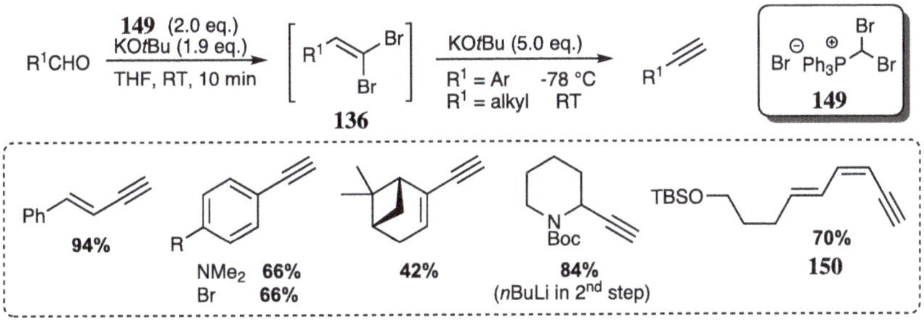

Fig. 3.84 Modification of the Corey–Fuchs reaction as a one-pot variant [226]

The Ohira–Bestmann modification was also expanded to proceed as a one-pot variant. In the groups of Roth and Meffre, two methods were investigated that primarily differ in the nature of the azide donor. First, diazaphosphonate **134** is generated *in situ*, followed by the conversion to the alkyne after the addition of the aldehyde (see Fig. 3.85) [228, 229].

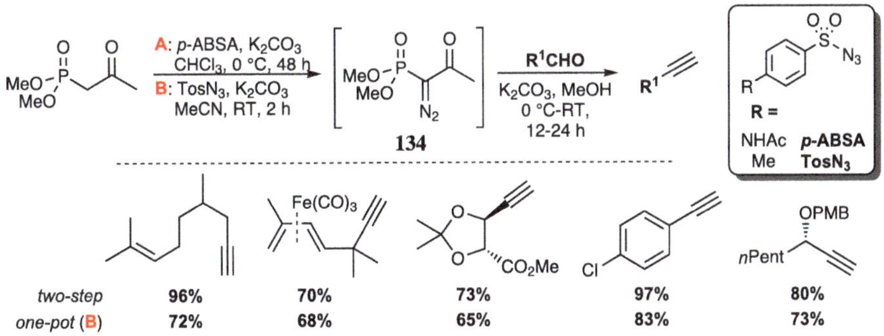

Fig. 3.85 *In situ* variants of the Ohira–Bestmann modification according to Meffre (**A**) and Roth (**B**) [228, 229]

In a direct comparison between the sequential, two-step variant and the one-pot procedure according to Roth, the one-step protocol was observed to perform only slightly inferior in terms of conversion with most substrates. Nevertheless, if yield is the pivotal factor of a given alkynylation, the two-step procedure is accordingly recommended [229].

As illustrated by these aspects (tolerance towards complex functionalities, stereointegrity of substrate and product, purification of the product, availability of the reagents), each of the methods possesses its strengths and weaknesses. Depending on the substrate, it may make sense to consult the published examples to select the appropriate method (see Fig. 3.86).

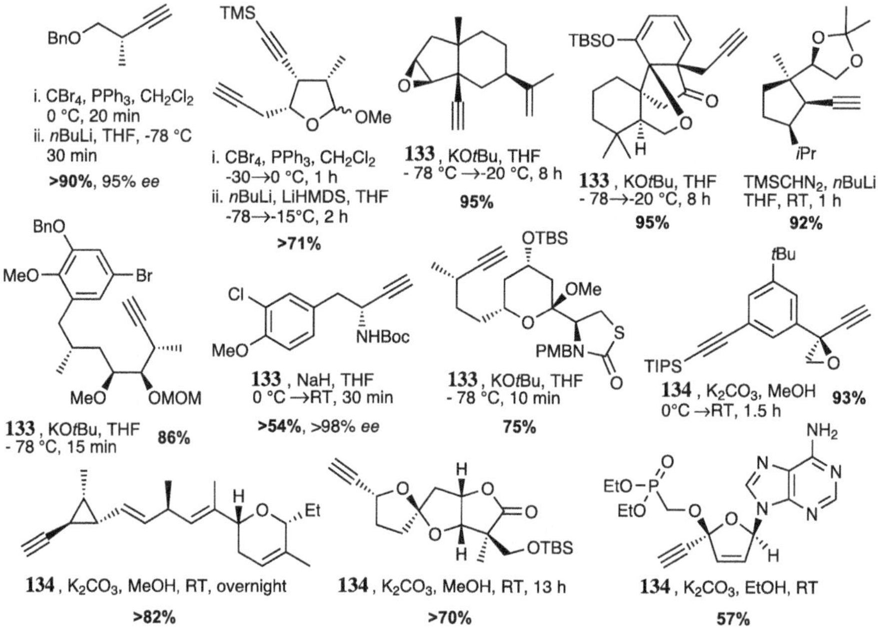

Fig. 3.86 Applications of various alkynylation methods in total synthetic projects [224, 230–240]

3.2.2 Other Approaches to Alkyne Formation

In addition to the standard alkynylation methods described thus far (Corey–Fuchs, Seyferth–Gilbert, Ohira–Bestmann) and alkyne metathesis (→ Sect. 7.3), elimination reactions of olefins in the presence of a base constitute a third approach to install a CC-triple bond. The single or double functionalized alkene substrates can usually be easily accessed from the corresponding ketones by enolization and subsequent O-functionalization. A Wittig reaction might likewise prove suitable for their synthesis, depending on the desired functionality (see Fig. 3.87).

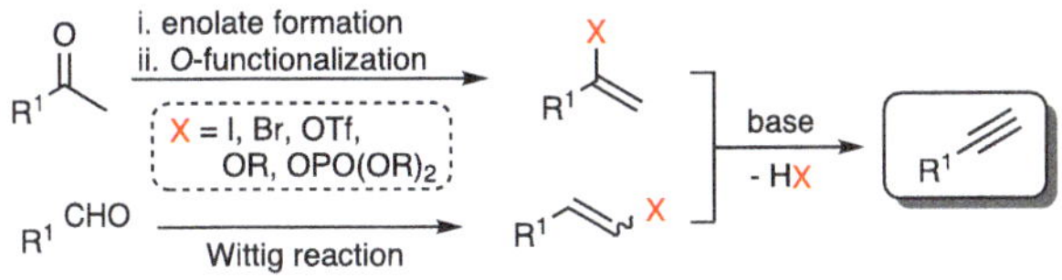

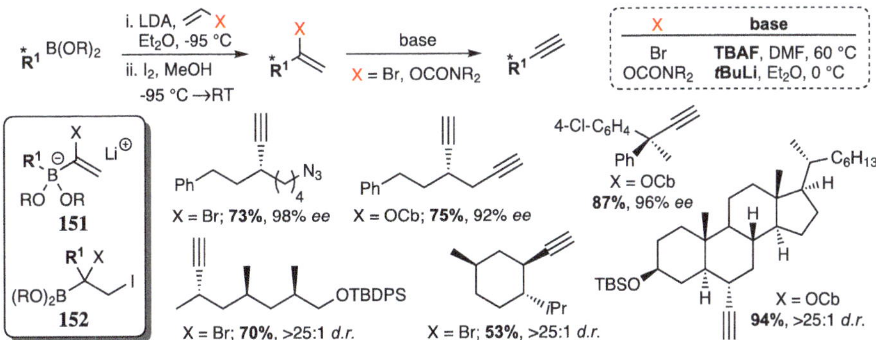

Fig. 3.87 Routes to alkynes via elimination reactions

Thadani and co-workers developed a mild elimination of Z-iodoalkenes upon exposure to TBAF [241]. A variant introduced by Li only necessitates LiCl as a solution in DMF at room temperature as a "base" to convert alkenyl triflates to the respective alkynes [242]. Pyridine or other nitrogen bases can also be used (see Fig. 3.88).

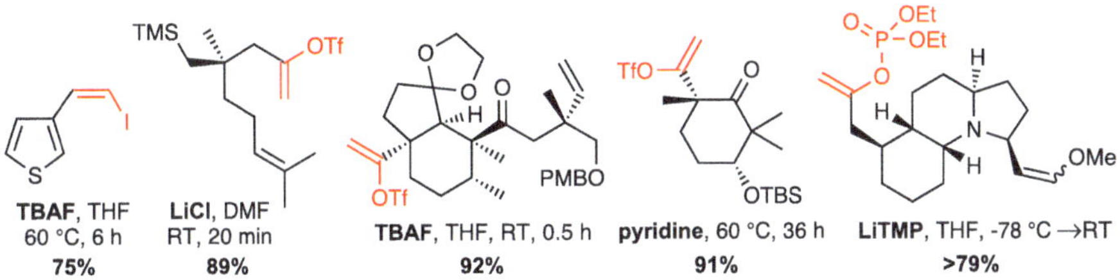

Fig. 3.88 Elimination of differently substituted alkenes to give the respective alkyne [241–245]

Aggarwal *et al.* published an alternative sequence to provide alkenyl bromides and carbamates, which can also be eliminated to give alkynes. After the formation of the boronates **151**, an I_2-mediated rearrangement to **152** takes place, which subsequently forms the geminal disubstituted olefin upon loss of I^- and $MeOB(OR)_2$. Depending on the leaving group, the chiral alkyne is then obtained with retention of the stereoconfiguration in the substituent R^1 either in the presence of TBAF or *t*BuLi (see Fig. 3.89) [246].

Fig. 3.89 Synthesis of chiral alkynes reported by Aggarwal *et al.* [246]

The Harusawa group developed a method to convert aldehydes and ketones even under neutral conditions to the corresponding alkynes. The approach involves the formation of cyanophosphates (**153**), which are transformed into tetrazoles (**154**) by azides. Following the cleavage of $PO_2(OEt)_2^-$, tetrazole (**155**) is obtained. In analogy to the Seyferth–Gilbert homologation, this intermediate decomposes to an alkylidene carbene (**156**), which rearranges to furnish the CC-triple bond [247]. Both aldehydes and ketones can be converted to afford terminal or internal alkynes under these conditions (see Fig. 3.90) [248].

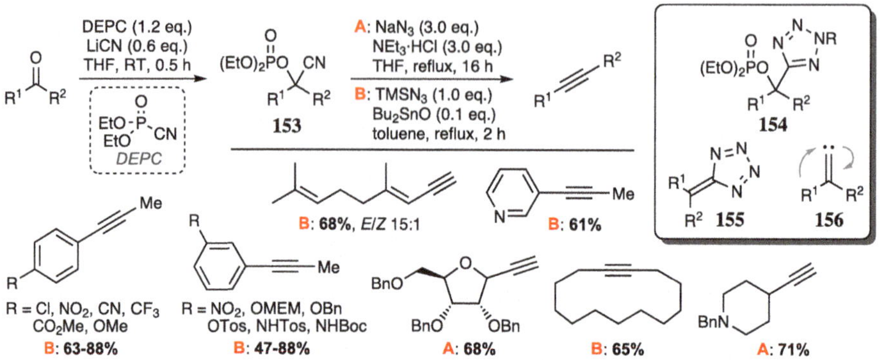

Fig. 3.90 Preparation of alkynes according to Harusawa *et al.* [248]

3.3   Derivatization of CC-Multiple Bonds

The alkene or alkyne functionalities, if not itself contained in the skeletal framework of a synthetic target, constitute a convenient handle to access a plethora of functional groups through a variety of transformations (see Fig. 3.91) [249].

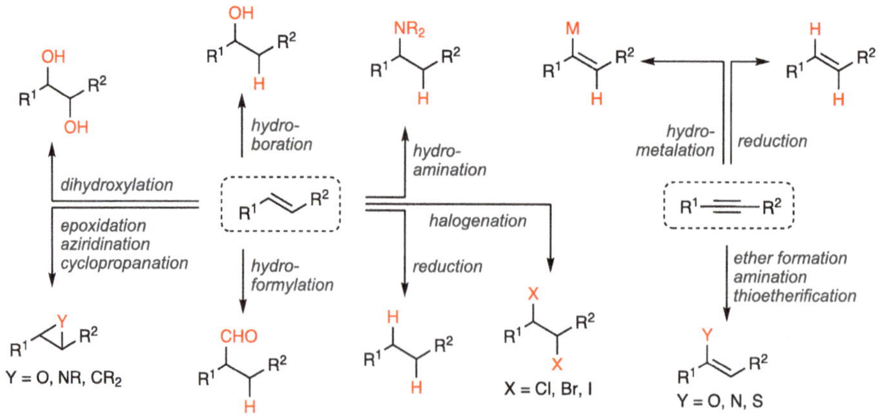

Fig. 3.91 Selected functionalizations of alkenes and alkynes

These are usually oxidative processes, which involve the introduction of heteroatoms, even if some reactions are redox-neutral with respect to the olefin carbon atoms [250]. Reductive methods, mainly referring to hydrogenations, are extensively treated in Sect. 4.2 and will not be covered here. A number of different reactions, which also serve for the derivatization of olefins and alkynes, are also discussed separately: metatheses (Chap. 7), cyclopropanations (Sect. 5.2), dipolar cycloadditions (e.g., ozonolysis, Sect. 5.2) as well as Heck reactions (Sect. 6.3).

3.3.1   Oxidation of Olefins

The addition of oxygen to olefinic double bonds accounts for the largest proportion of derivatization reactions encountered in total syntheses, as the addition products in turn serve as an ubiquitous access for further functionalities. The diastereo- and enantioselective variants of epoxidation, dihydroxylation, and hydroboration make up the lion's share — also because they were among the first asymmetric methods developed and reliably allow excellent discrimination of enantio- and diastereotopic functionalities [251].

In alkene derivatizations, the double bond usually comprises the nucleophilic component, while the oxidizing agent assumes the role of an electrophile [252, 253]. As a result, electron-rich double bonds can be chemoselectively converted given the absence of further directing effects (see Fig. 3.92).

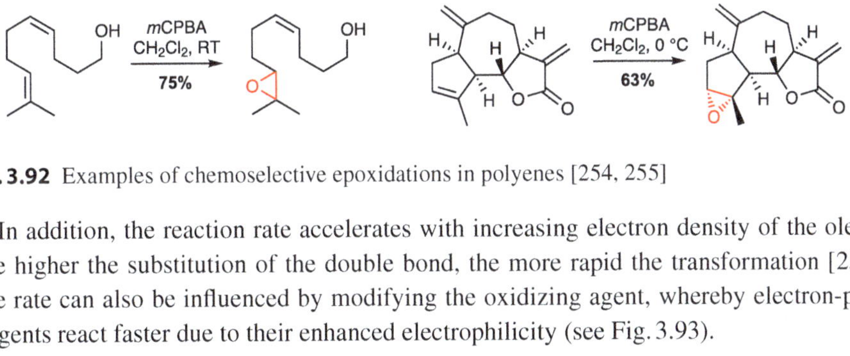

Fig. 3.92 Examples of chemoselective epoxidations in polyenes [254, 255]

In addition, the reaction rate accelerates with increasing electron density of the olefin. The higher the substitution of the double bond, the more rapid the transformation [256]. The rate can also be influenced by modifying the oxidizing agent, whereby electron-poor reagents react faster due to their enhanced electrophilicity (see Fig. 3.93).

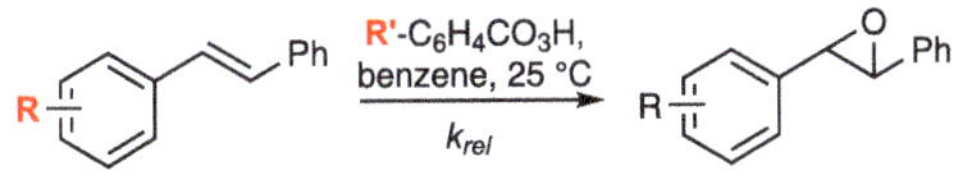

$PhCO_3H$ (R' = H) with *subst. stilbenes*		$R'\text{-}C_6H_4CO_3H$ with stilbene (R = H)	
R	k_{rel}	**R'**	k_{rel}
4-OMe	4.7	4-NO$_2$	10.4
4-Me	2.3	4-Cl	2.4
H	1.0	H	1.0
4-Cl	0.6	4-Me	0.6
4-NO$_2$	0.2	4-OMe	0.3

Fig. 3.93 Dependence of the reaction rate on the electronic properties of the olefin and oxidizing agent [257]

Depending on the mechanism of an asymmetric method, exceedingly high enantioselectivities can only be achieved with certain substrates. Some substrates do not react at all or only very slowly in diastereoselective transformations. However, a clever combination of the available transformations allows almost all substrates to be oxidized to the desired products in high yields and enantioselectivities — provided the functional groups are compatible (see Fig. 3.94).

	R^1 (terminal)	R^1=R^2 (trans)	R^1,R^2 (cis)	enone R^1/R^2	allylic alcohol R^1/R^2	R^1,R^2,R^3	R^1,R^2,R^3,R^4
rac. epoxidation (*m*CPBA, dioxirane)	+	+	+	(–)[a]	+	+	(+)[b]
V-mediated epoxidation	–	–	–	–	+	–	–
base, peroxide/H_2O_2				+			
Sharpless epoxidation	–	–	–	–	+	–	–
Jacobsen epoxidation	(–)[c]	(–)[c]	+	–	+	+	(+)[b]
Jacobsen hydrolytic kinetic resolution (HKR)	+[d]	+[d]	+[d]	+[d]	+[d]	(+)[b,d]	(+)[b,d]
Shi epoxidation	+	+	+	(–)[a]	+	+	(+)[b]
aziridination	+	+	+	+	+	(+)[b,c]	(+)[b,c]
rac. dihydroxylation	+	+	+	(+)[b]	+	(+)[b]	(+)[b]
Sharpless dihydroxylation	+	+	(+)[c]	+	+	+	(+)[b,c]

[a] requires strong oxidant; [b] very slow; [c] low enantioselectivity; [d] starting from epoxide

Fig. 3.94 Feasibility of oxidative functionalizations for various substrates

Epoxidation and Aziridination

Epoxidations are usually carried out in the presence of a catalyst, with peroxides, peracids and dioxiranes comprising the most common suitable oxygen sources. Typical catalysts are

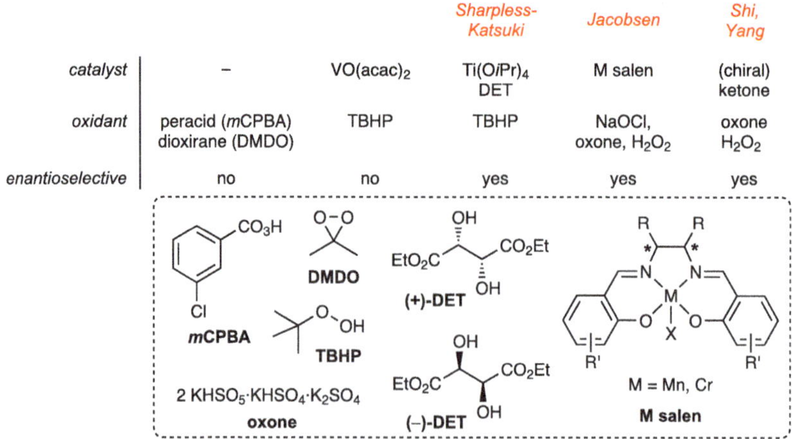

Fig. 3.95 Overview of common catalysts and oxygen sources in epoxidations

usually based on transition metals such as vanadium, titanium, and manganese, although organocatalytic methods in the presence of ketones have also been developed (see Fig. 3.95).

The uncatalyzed reaction can be carried out in the presence of peracids or dioxiranes. The reaction with peracids is occasionally also referred to as the *Prilezhaev* reaction. Despite *m*-chloroperbenzoic acid (*m*CPBA) often being the reagent of choice, other peracids may be similarly utilized. Oxone, whose active species is derived from peroxomonosulfuric acid ($KHSO_5$), is just one of the possible oxygenation reagents.

The epoxidation with dioxiranes or peracids is a bimolecular, stereospecific reaction. A *spiro* arrangement has been postulated for the concerted transition states (**157**, **158**) [256, 258–260]. While the interaction $\pi_{CC} \rightarrow \sigma_{OO}^*$ proceeds in a symmetric fashion with respect to the O-O axis, the transfer of electron density from oxygen to the alkene ($n_O \rightarrow \pi_{CC}^*$) is maximized in a *spiro* geometry [261]. After the transfer of oxygen, a ketone or a carboxylic acid is obtained as by-products alongside the desired epoxides (see Fig. 3.96).

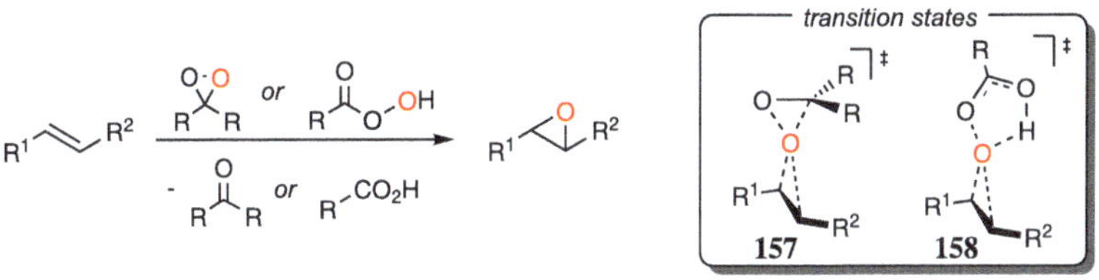

Fig. 3.96 Transition states of the epoxidation with dioxiranes and peracids [256, 258–260]

The reaction with peracids displays a limited dependence on the solvent. Polar, aprotic solvents yield high reaction rates, whereby protic solvents form strong hydrogen bonds with the peracid reagent, prevent intramolecular H-bond formation, and significantly slow down the reaction (see Fig. 3.97) [256].

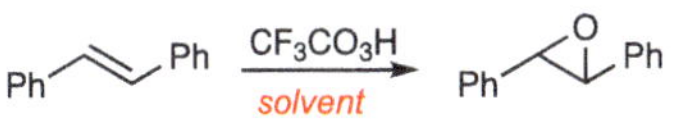

solvent	k_{rel}
CCl_4	0.8
$CHCl_3$	1.9
CH_2Cl_2	1.9
DCE	1.7
benzene	1.0
*t*BuOH	0.02

solvent	k_{rel}
CCl_4	0.7
DCE	1.0
benzene	1.0
$AcOH, H_2O$	0.04
$AcOH, Ac_2O$	0.3

Fig. 3.97 Solvent dependence of the Prilezhaev reaction [262]

The reaction of dioxiranes with *Z*-configured alkenes is almost an order of magnitude faster than the conversion of the respective *E*-alkenes. This is attributed to a possible steric repulsion of the substituents R in the dioxirane with the substituents on the alkene in the transition state **157**. In a *Z*-configured olefin, the R-R$^{1/2}$ interaction can be completely circumvented, whereas in *E*-alkenes, there always exists a repulsion with a substituent R^1

or R^2. In contrast, reactions with peracids show hardly any negative impact on the yield due to steric factors [259].

Electron-poor double bonds necessitate stronger peracids as oxidants. In Corey's synthesis of ($\pm$)-bilobalide, treatment of the diene **159** with *m*CPBA afforded only the monoepoxide of the trisubstituted double bond. To successfully convert the carboxylate-substituted dihydrofuran, the bis-epoxide **160** required the use of 3,5-dinitroperbenzoic acid, which led to the isolation of the desired product in high yield [263]. Wender employed the same reagent in his synthesis of the decalin warburganal (**161**→**162**) in the presence of an additional radical inhibitor [264] (see Fig. 3.98) [265].

Epoxidation by peracids intrinsically results in *acidic* conditions due to the corresponding acid by-products. To protect the newly formed epoxides as well as to stabilize other acid-labile groups, bases or buffer salts can be added, as shown in the reactions in Fig. 3.98.

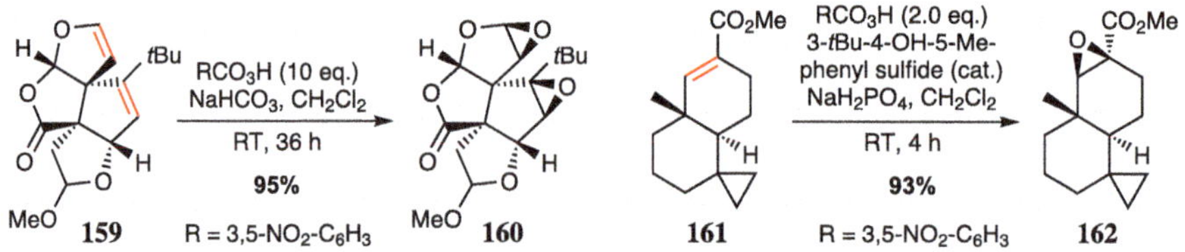

Fig. 3.98 Epoxidation of electron-poor double bonds in total syntheses [263, 265]

Dioxiranes constitute another group of oxygen transfer reagents that can be reacted under *neutral* conditions, with ketones as the only by-product. The higher reactivity allows the reactions to proceed at low temperatures, which further contributes to a high yield. Acetone-derived dimethyldioxirane (DMDO) is the most commonly employed reagent of that class, but in principle all ketones, including chiral ones, can be used (*vide infra*). The reactivity of the respective dioxiranes increases with decreasing electron density of the carbonyl group in the corresponding ketones (see Fig. 3.99).

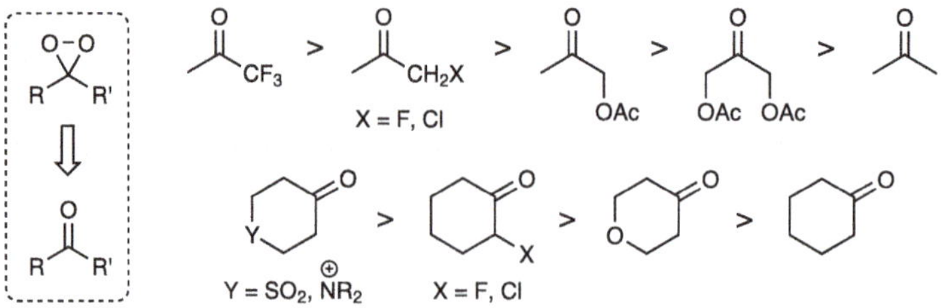

Fig. 3.99 Reactivity of dioxiranes derived from different ketones [266]

The dioxirane reagent is either prepared *in situ* or it can be used in a ready-made form. Due to their reactive nature, they are exclusively isolated as a solution in those ketones from which they are derived. Typical oxidants utilized in the preparation are peracids, especially oxone (see Fig. 3.100).

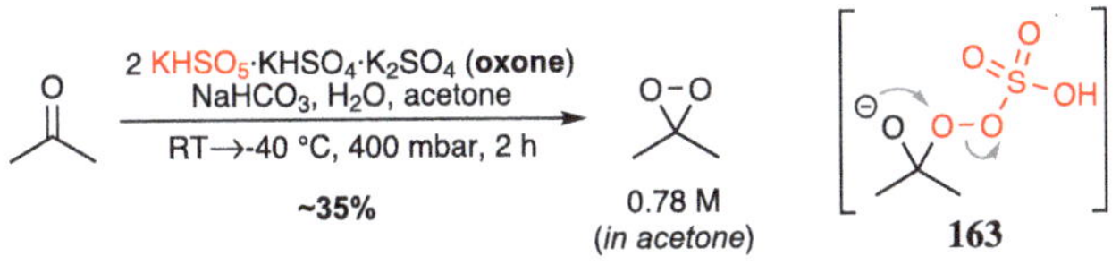

Fig. 3.100 Preparation of DMDO on kg scale and postulated intermediate [259, 267]

Mixtures with various cosolvents, such as acetone/CH_2Cl_2 or in an aqueous solution with H_2O_2 as the oxidant are suitable as well. This increases the reactivity of the dioxirane but limits its stability, so it necessitates a preparation *in situ* [259]. Yang's synthesis of triptolide **166** offers a representative example in this regard (see Fig. 3.101) [268].

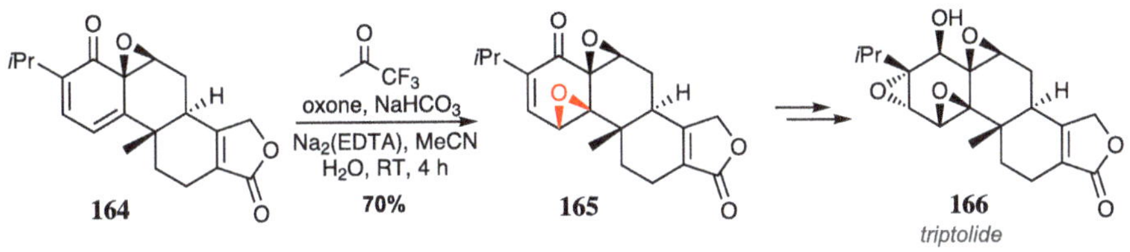

Fig. 3.101 Epoxidation using *in situ*-generated dioxirane [268]

Metal salts are ideally suited to catalyze epoxidations, with vanadium-based systems being particularly dominant for achiral reactions. Allylic alcohols lend themselves as substrates by virtue of the hydroxy group's ability to coordinate the metal center, thereby allowing an intramolecular reaction to proceed with optimal distance to the peroxo group (see Fig. 3.102) [269].

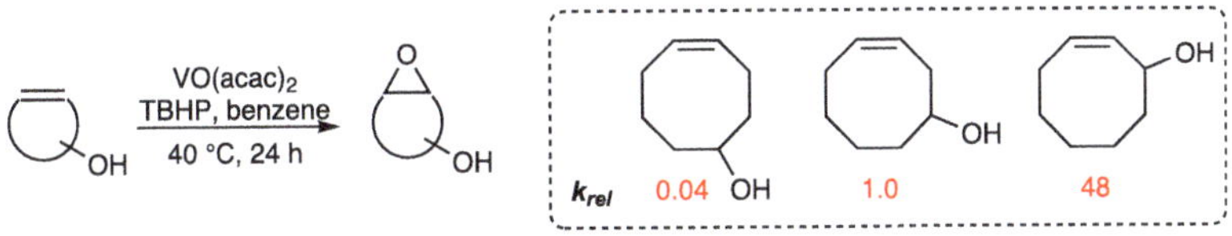

Fig. 3.102 Influence of the OH position of enols on the reaction with VO(acac)$_2$ [269]

The mechanism of V-catalyzed epoxidation is complex. Especially the ligand environment of the vanadium remains to be fully elucidated in the individual intermediates during the activation step and in the catalytic cycle. When using VO(acac)$_2$ (**167**), a VO(acac)$_2$-peroxo and subsequently an alkoxy complex **168** are initially formed in the activating oxidation of the vanadium (IV→V). **168** is afterwards converted to the catalytically active peroxo complex **169** in the presence of allylic alcohols and *t*butyl hydroperoxide (TBHP). If VO(O*i*Pr)$_3$ is employed instead, **169** is obtained directly. The ligands of the V complex are either O*t*Bu (from TBHP), OAc (as a degradation product of acac) or O*i*Pr. Following the transfer of the oxygen of the η^2-peroxo ligand via **170**, **171** is received, which reenters the catalytic cycle upon elimination of the product/*t*BuOH by substitution with another equivalent of the substrate and oxidant (see Fig. 3.103) [270–272].

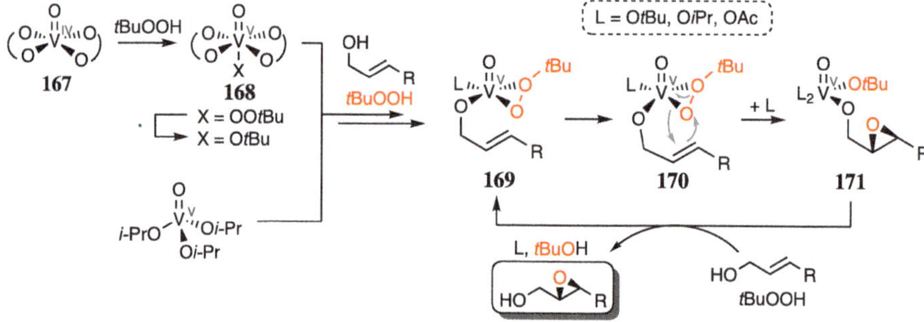

Fig. 3.103 Postulated mechanism of vanadium-mediated epoxidation of allylic alcohols [270–272]

The transfer of oxygen to alkenes under the control of directing groups is a frequently encountered and highly reliable concept in synthesis [273]. In epoxidations, this approach is also referred to as the *Henbest* effect. Given the absence of an external stereoinduction by chiral reagents, allylic alcohols in particular can preferentially afford one specific

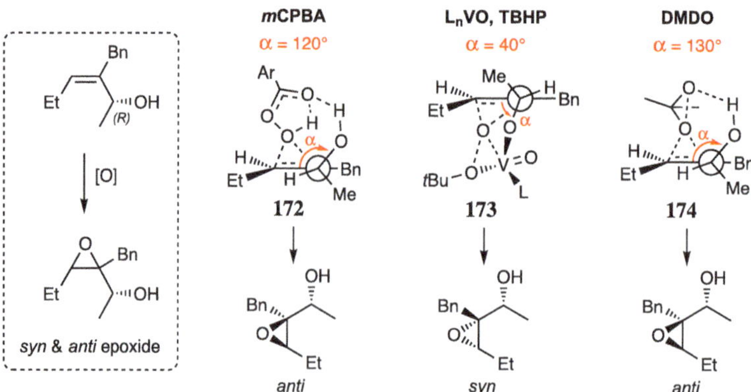

Fig. 3.104 Transition states in epoxidations of allylic alcohols [273, 274]

diastereomer solely due to the stereoelectronic effects occurring in the transition state such as allylic strain (cf. Sect. 1.1.2). Depending on the reagent in question, a different dihedral angle between the directing OH group and the CC double bond in the transition state is required. This can lead to a reversal of the stereoselectivity in the major product even when starting from the same substrate (see Fig. 3.104).

*m*CPBA and DMDO preferentially approach the depicted substrate from the β-face to avoid the 1,3-allylic strain between the allylic methyl group and the ethyl residue at the requisite dihedral angle of 120–130°. With VO(acac)$_2$, the preferred angle of 40° leads to the addition taking place from the α-side to circumvent an A^{1,2} interaction between Me- and Bn-substituent. This interplay of A^{1,3} and A^{1,2} interactions can be exploited to obtain a desired diastereomer with high selectivity from a substrate by simply exchanging the epoxidation reagents (see Fig. 3.105).

			syn:anti selectivity		
*m*CPBA	45:55	95:5	90:10	48:52	95:5
DMDO	57:43	67:33	87:13	51:49	76:24
VO(acac)$_2$, TBHP	5:95	71:29	33:67	10:90	86:14

Fig. 3.105 Diastereoselectivity of different classes of allylic alcohols [273]

Other metal catalysts, for example, Ti(O*i*Pr)$_4$ or M-Salen, follow the same mechanistic path as V-systems and show preferred dihedral angles in the range of 40° < α < 90° [273].

As Trost and co-workers observed in their synthesis of the macrolide *des*-epoxy-amphidinolide N, the stereoselectivity can also be extremely substrate-dependent in rare cases. The simplified model substrate **175** was utilized to optimize the chemo- and diastereo-selectivity of the epoxidation with the most common epoxidation methods. While DMDO only oxidized the geminal disubstituted olefin at the 6,6'-position, the other reagents preferably or exclusively yielded the desired epoxide at the allylic double bond. The addition of metal catalysts was required to obtain a moderate to good diastereoselectivity. However, when the optimized conditions were transferred to the key intermediate in the total synthesis, no turnover was obtained with VO(acac)$_2$. Ti(O*i*Pr)$_4$ did provide the desired chemoselec-tivity, but only a 1:1 mixture of the diastereomers of the highly unstable product **176** was afforded (see Fig. 3.106) [275].

Enones are difficult to epoxidize due to their electron-deficient nature. However, it is possible to convert the CC double bond to the desired oxirane by the action of hydrogen

175

conditions	4,5- *vs.* 6,6'-epoxide	*d.r.*
DMDO	0:100	-
*m*CPBA	91:9	1.6 : 1
VO(acac)$_2$, TBHP, 5Å MS	100:0	3 : 1
Ti(O*i*Pr)$_4$, TBHP, 4Å MS	100:0	<1 : >20

176

Ti(O*i*Pr)$_4$, TBHP, CH_2Cl_2, 4Å MS, RT, 1 h

76%, 1:1 *d.r.*

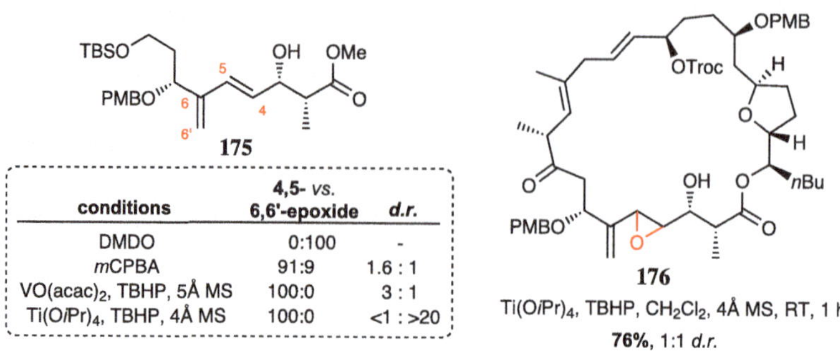

Fig. 3.106 Allylic epoxidation in the attempted synthesis of amphidinolide N [275]

peroxide, TBHP or peracids if a base is added. The tolerance of functional groups is naturally limited by the basic medium, although esters and many protective groups typically remain intact. The purported mechanism was postulated to proceed in a 1,4-addition of the oxidizing agent via **177**. The enolate **178** can subsequently furnish the desired epoxide by elimination of RO$^-$ (see Fig. 3.107) [276].

Weitz-Scheffer	H_2O_2, NaOH
Meth-Cohn	*n*BuLi/TBHP *or* KHMDS/Ph_3COOH
Yadav	DBU, TBHP

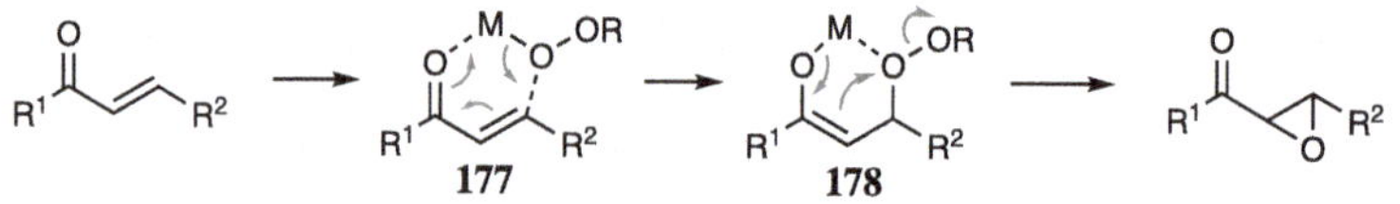

Fig. 3.107 Methods for the epoxidation of α, β-unsaturated ketones and proposed mechanism [276–278]

The tolerance toward various functional groups when using peracids, *m*CPBA and VO complexes can be rationalized by the oxidation potential of the corresponding functionalities. Amines, thiols, thioethers, and very electron-rich olefins are labile, while alkynes, aldehydes, and ketenes usually remain untouched. Ketones, epoxides, esters, nitriles, amides, and halogens (Br, Cl, F) are inert to the reaction conditions (see Fig. 3.108) [258].

In addition to the possibility of using the inherent stereochemistry of the starting material for diastereoselective epoxidation (substrate or active control), there are numerous ways to construct the heterocycle in an asymmetric manner by relying on chiral reagents or catalysts (exogeneous stereoinduction) [250].

The *Sharpless epoxidation* constitutes the by far most frequently employed asymmetric method [290, 291]. It relies on Ti(O*i*Pr)$_4$ and tartrate esters (diethyl tartrate, *DET*; diisopropyl tartrate, *DIPT*) as a catalyst and *t*butyl hydroperoxide as the source of oxygen. Allylic alcohols are required as substrates to enable a covalent interaction of the catalyst with the substrate, in analogy to the reaction with vanadium complexes (see Fig. 3.109) [252].

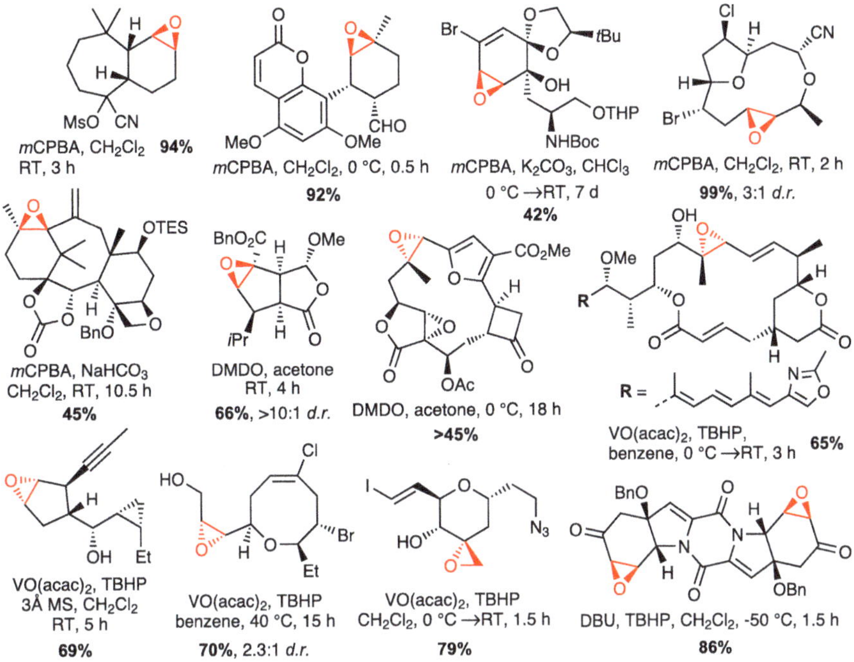

Fig. 3.108 Applications of diastereoselective epoxidations in the synthesis of complex natural products [144, 279–289]

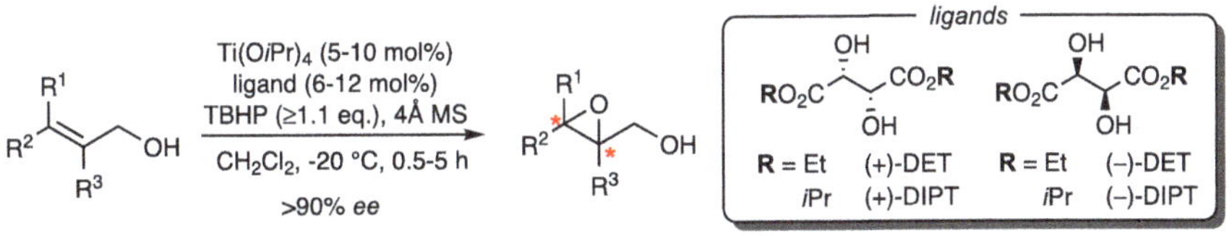

Fig. 3.109 Reaction conditions of the Sharpless epoxidation [252, 291]

The typical reaction conditions entail 5–10% Ti(OiPr)$_4$ and a slight excess of 20% of the tartrate ligand with regard to titanium. Too much ligand significantly reduces the turnover and the enantiomeric excess. The requisite addition of 3–5 Å molecular sieves is due to the high moisture-sensitivity of the Ti complex, as small amounts of H$_2$O can be produced by some undesired side reactions. The potential loss of efficiency/rate in the epoxidation is thus avoided. While DET remains the ligand of choice, DIPT often displays superior chemo- and enantioselectivity at the cost of a lower reaction rate. Especially in the resolution of racemic α-substituted allylic alcohols, DIPT is preferred (*vide infra*) [291].

The postulated mechanism accounts for the peculiar effect of the specific metal/ligand stoichiometry. It has been studied in-depth by both experimental as well as theoretical means and strongly resembles the catalytic cycle already described for vanadium complexes (cf.

Fig. 3.103). In the presence of an equivalent of tartrate, Ti(O*i*Pr)$_4$ is converted to the active dimeric species **179**. If an excess of tartrate is present, a monomeric, catalytically inactive Ti(tartrate)$_2$ complex can be reversibly formed. Even though the exact nature of the dimeric complex **179** has not yet been elucidated [292], the pseudo-C_2-symmetric structure **179** represents the most widely accepted proposal [293]. The left half of the complex does not actively participate but rather only provides steric shielding of the lower left quadrant and the "correct" positioning of the allylic alcohol during the reaction. The coordination of the *tert*-butyl peroxide and allylic alcohol affords the initial peroxo species **180**. In analogy to the V complex **170**, the transfer of oxygen to the CC double bond takes place via the spiro-shaped transition state **181** [294]. The desired product and *t*BuOH are finally obtained upon substitution in **182** by another equivalent of the substrate and TBHP (see Fig. 3.110). The experimentally determined rate law shows a dependence on the concentration of the complex **179** and the substrate/reagent. The presence of alcohols (*i*PrOH, *t*BuOH) has an inhibitory effect on the transformation [252, 293, 295].

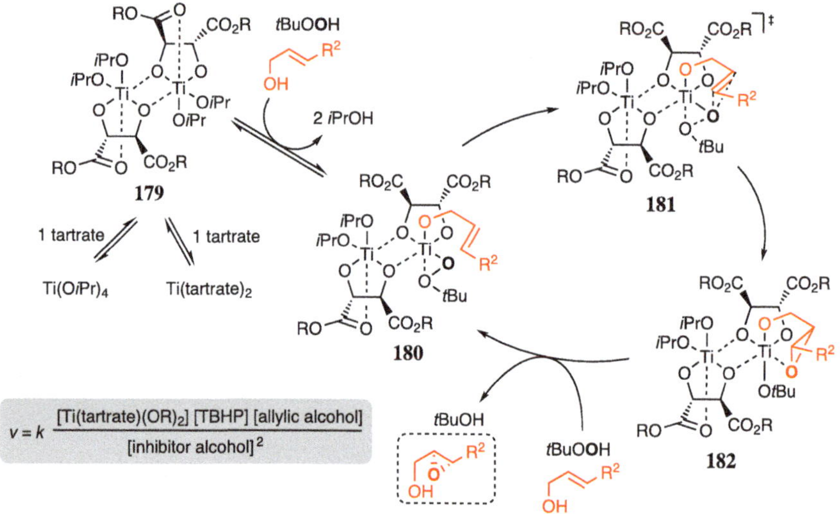

Fig. 3.110 Mechanism of the Sharpless epoxidation with (+)-tartrate [252, 293, 295]

Depending on the configuration of the ligand ((+)- or (−)-tartrate), an epoxidation occurs either from the *Re*- or the *Si*-face, respectively. Based on empirical data, a mnemonic for the selectivity was composed, which can also be derived from the mechanism shown in Fig. 3.110: If the allylic alcohol is placed in a plane with the OH group in the southeast corner, the addition of oxygen occurs from β-face with (−)-tartrate, while (+)-tartrate favors the approach from below (see Fig. 3.111) [290].

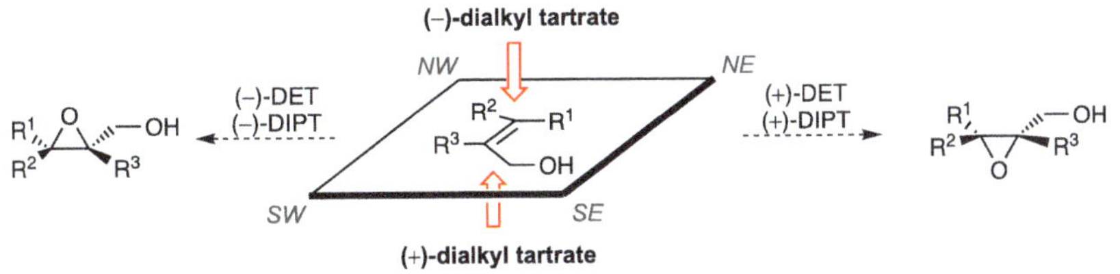

Fig. 3.111 Mnemonic of facial selectivity in the Sharpless epoxidation of allylic alcohols [290]

Disubstituted Z-configured olefins only lead to low enantioselectivity, as the substituent R^1 experiences a steric repulsion with the uncoordinated ester group of the second metal center comprising the left catalyst hemisphere (see Fig. 3.112).

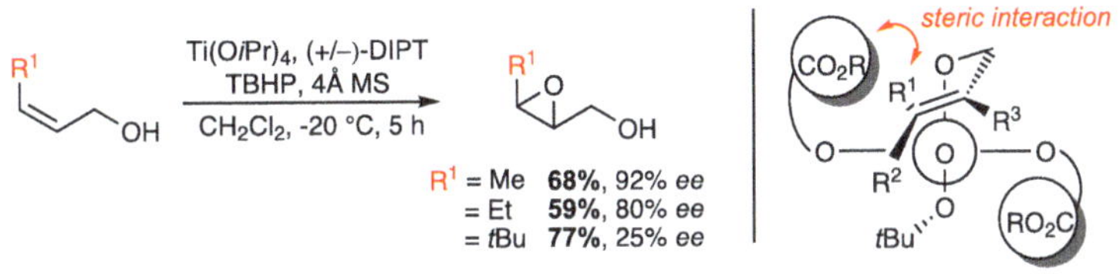

Fig. 3.112 Dependence of enantioselectivity on the size of R^1. Schematic representation of **180** viewed along the distal O_{Peroxo}-Ti bond [252]

The larger the steric demand of R^1, the lower the observed *ee*-values. The highest enantioselectivities are achieved with disubstituted E-configured alkenes (see Fig. 3.113) [296]

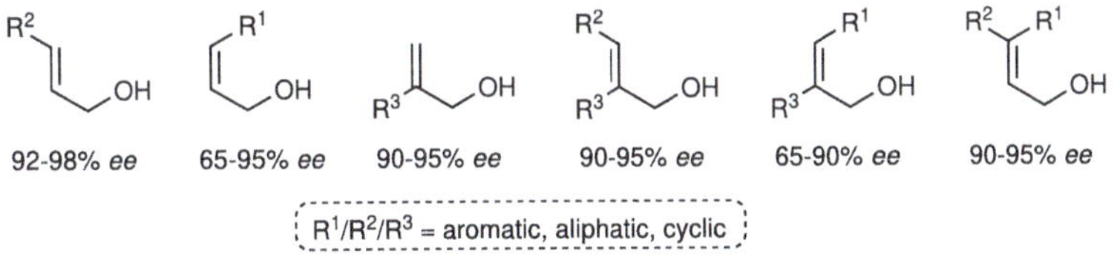

Fig. 3.113 Enantioselectivity of various substrate classes [296]

Homo- and bishomoallylic alcohols show the *reverse* facial selectivity compared to allylic alcohols. (−)-Tartrates thus approach from the α-face with the identical placement of the CH_2OH group in the SO quadrant. The enantioselectivity in the conversion of these substrates also tends to be typically lower than with allylic alcohols. As with the corresponding vanadium catalysts (see Fig. 3.102), the conversion is significantly slower as well [296].

Chiral allylic alcohols possess an inherent facial selectivity, provided the stereoinducing element is not positioned too remote from the double bond. Especially α-chiral substrates favor one of the possible diastereomers. Since the Ti-tartrate system displays a strong stereoinducing effect, this can either reinforce the substrate guidance (*matched*) or counteract it (*mismatched*).

Given an absence of chiral ligands, the epoxidation of the homochiral allylic alcohol **184** shows a slight substrate-induced preference for the formation of the *anti*-configured diastereomer**185** under V- and Ti-catalysis. The use of (–)-DIPT enhances the preference of the substrate further to afford a very high *matched* diastereoselectivity. In contrast, the substrate control is superseded by the dominating ligand induction in the presence of (+)-DIPT and yields *syn*-**185** with a significantly lower stereoselectivity (see Fig. 3.114) [297].

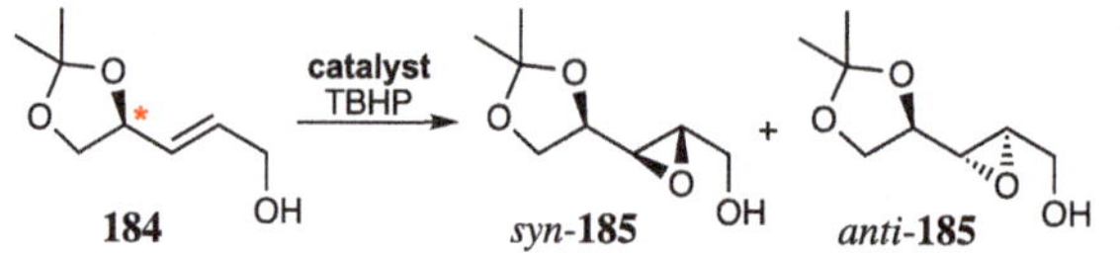

catalyst	syn / anti
VO(acac)$_2$	1 : 1.8
Ti(OiPr)$_4$	1 : 2.3
Ti(OiPr)$_4$, (–)-DIPT	1 : 90
Ti(OiPr)$_4$, (+)-DIPT	22 : 1

Fig. 3.114 Substrate control, *matched*- and *mismatched*-reagent control [297]

Matched substrates react significantly faster than *mismatched* reactants. In this way, racemic mixtures can be separated in a kinetic resolution (see Fig. 3.115). When, for example, using (+)-DIPT, R^5 either shields the substrate's α-face and slows down the reaction (*mismatched*, R^4 = H, R^5 = cHex), or it enhances its preference (*matched*, R^4 = cHex, R^5 = H), depending on the configuration of the stereocenter [291].

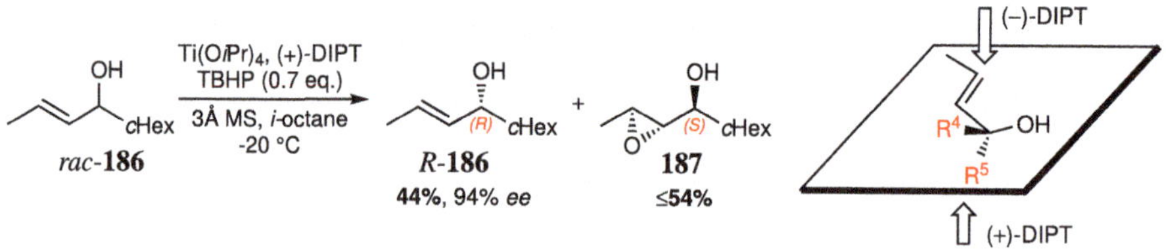

Fig. 3.115 Kinetic resolution via Sharpless epoxidation [291]

Olefins lacking a hydroxy group in the α-position are inert to Ti-mediated epoxidation reactions. In general, the reaction displays excellent chemoselectivity, so that in the presence of commonly sensitive functionalities (alkyne, acetal, amide, ester, silyl ether, ketone, halogens) only the allylic alcohol is converted to epoxy alcohol (see Fig. 3.116) [298].

While the Sharpless epoxidation necessitates a hydroxy group to enable a covalent attachment to the catalytically active center, there are several methods that can enantioselectively epoxidize unfunctionalized alkenes. The *Jacobsen epoxidation* is based on the use of chiral Mn salen complexes in the presence of NaOCl, iodosobenzene or H_2O_2 as oxidants. In

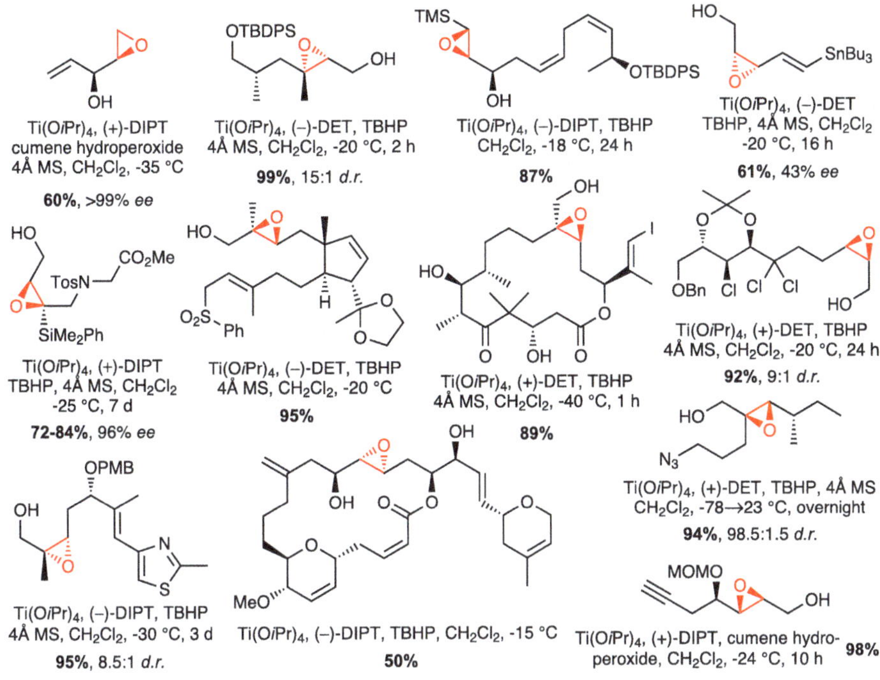

Fig. 3.116 Examples of Sharpless epoxidation in total syntheses [299–310]

addition, imidazoles and *N*-oxides are often used as basic additives that can coordinate the metal center and modify its reactivity (see Fig. 3.117) [311–313].

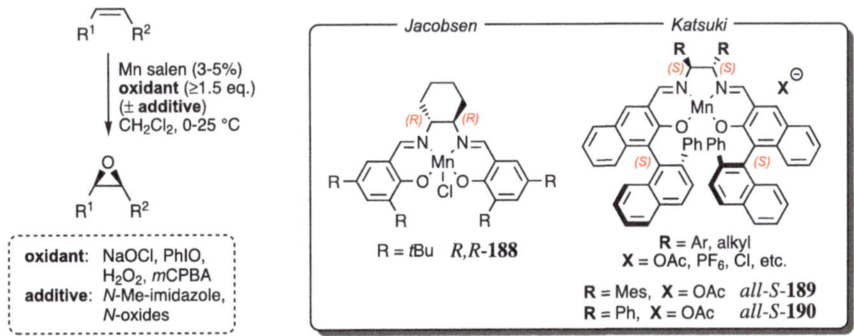

Fig. 3.117 Typical reaction conditions of the Jacobsen epoxidation and common catalysts [311]

Both enantiomers of the catalyst introduced by Jacobsen (**188**) are commercially available by virtue of their short and modular synthesis. The more complex catalysts **189-190**, which were developed in the Katsuki group, often show a superior selectivity. They are however not commercially available and must be individually synthesized in several steps, which may be why they are utilized less frequently. The mechanistic details of the transformation are not yet fully understood [311, 314]. The formation of a Mn=O species **191** has been experimentally proven, [315] which is assumed to oxidize the alkene without prior coordination to the metal center. It is controversely debated whether the nature of the oxygen transfer constitutes a concerted process (via **193**), a diradical course (via **192**) or involves a metalaoxetane **194**. Theoretical and experimental studies suggest that the mechanism of many substrates is diradical, as acyclic Z-alkenes can also form *trans*-epoxides, depending on the oxidizing agent and counterion in question [316, 317]. The preference for Z-olefins can be attributed to the postulated trajectory of the alkene to the oxygen of the Mn=O species: In the transition state, the alkene approaches at a right angle to the R_L-C=C-R_S plane (*side-on* attack).

Katsuki proposed an approach via the Mn-N_{imine} bond (R_S points towards Ar-*t*Bu substituent), while Jacobsen rationalizes the reaction to occur via a trajectory over the chiral backbone of the salen ligand (R_S points towards axial H atom) [313, 318]. Both postulates account for the reduced stereoinduction with *E*-alkenes, which would experience a steric repulsion between one of the substituents (R_L, R_S) and the salen ligand (see Fig. 3.118) [319].

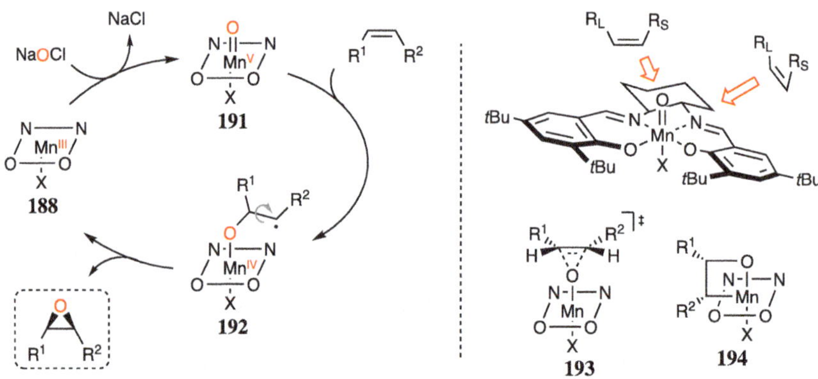

Fig. 3.118 Postulated mechanism of the Jacobsen epoxidation [311]

The reaction is not stereo*specific*, which may possibly be due to a diradical mechanism, as the radical intermediate **192** can rotate around the CC single bond. Depending on the

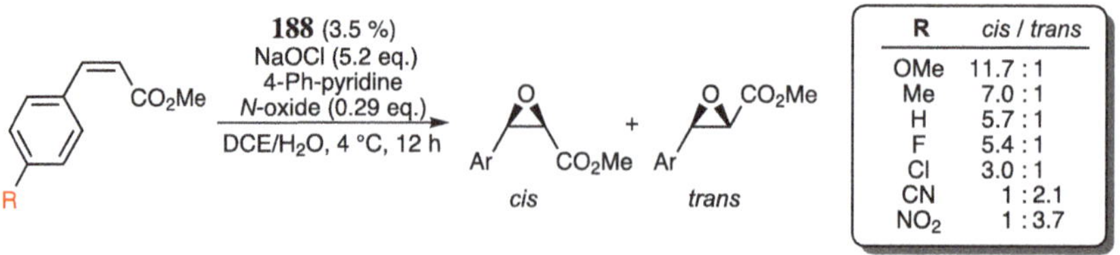

R	cis / trans
OMe	11.7 : 1
Me	7.0 : 1
H	5.7 : 1
F	5.4 : 1
Cl	3.0 : 1
CN	1 : 2.1
NO₂	1 : 3.7

Fig. 3.119 Stereoselectivity in the epoxidation of cinnamic acid esters [318]

substrate and the reaction conditions, mixtures of *cis-* and *trans*-epoxides are obtained. Particularly conjugated polyenes and stilbene analogs stabilize the radical intermediate **192** and allow isomerization to the *trans*-epoxide (see Fig. 3.119) [318, 320].

In addition to the previously discussed impact of the electron density at the olefinic double bond, *Z*-configured olefins react significantly faster than *E*-olefins. Thus, in conjugated substrates, a double bond may be selectively epoxidized (see Fig. 3.120) [320].

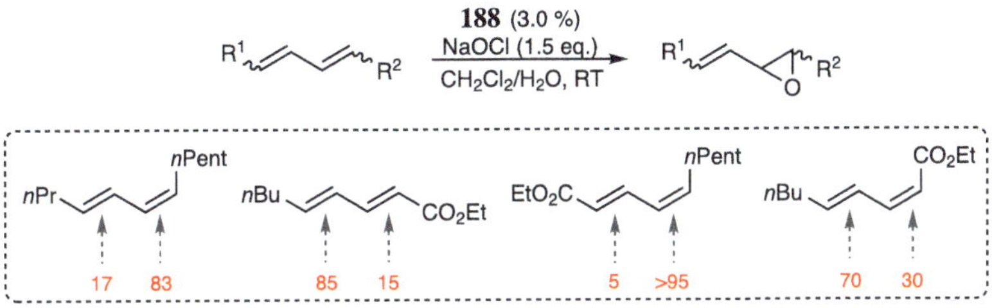

Fig. 3.120 Regioselectivity of the Jacobsen epoxidation of dienes [320]

The substrate scope of the Jacobsen olefination can be considered to be complementary to the Sharpless method, as particularly disubstituted, *Z*-configured alkenes lead to a high enantioselectivity. *E*-configured and terminal alkenes, on the other hand, can be converted to the corresponding epoxides with significantly lower stereoselectivity. The catalysts of type **189** introduced by Katsuki are more suited for these substrates and generally afford a higher enantiomeric excess (see Fig. 3.121) [250].

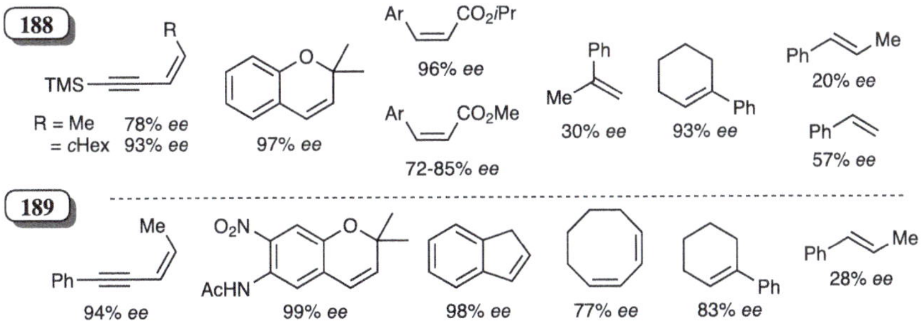

Fig. 3.121 Substrate scope of the Jacobsen epoxidation with catalysts **188** and **189** [318, 321, 322]

Since the substrate scope is limited due to the low enantioselectivities encountered with terminal and *E*-alkenes, the potential of the Jacobsen epoxidation could only be demonstrated sporadically in the synthesis of complex natural products (see Fig. 3.122).

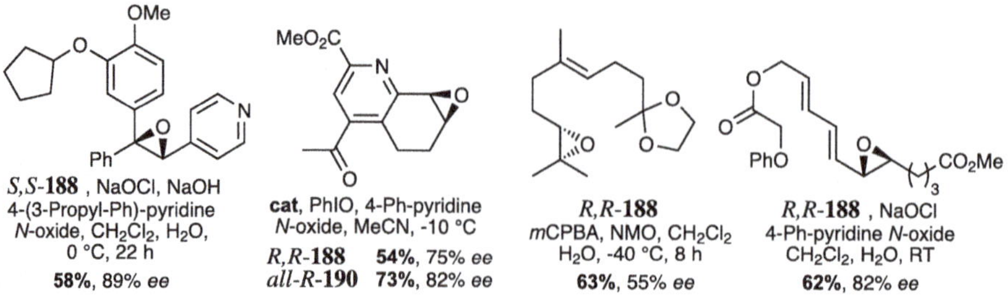

Fig. 3.122 Applications of the Jacobsen epoxidation in natural product syntheses [320, 323–325]

An alternative approach to synthesize the desired oxirane with terminal alkenes is via the formation of the racemic epoxide, followed by a kinetic resolution (*Jacobsen (hydrolytic) kinetic resolution, HKR*). If the manganese in the active center is replaced by cobalt or chromium, these salen complexes become excellent catalysts for the asymmetric cleavage of epoxides (see Fig. 3.123) [326].

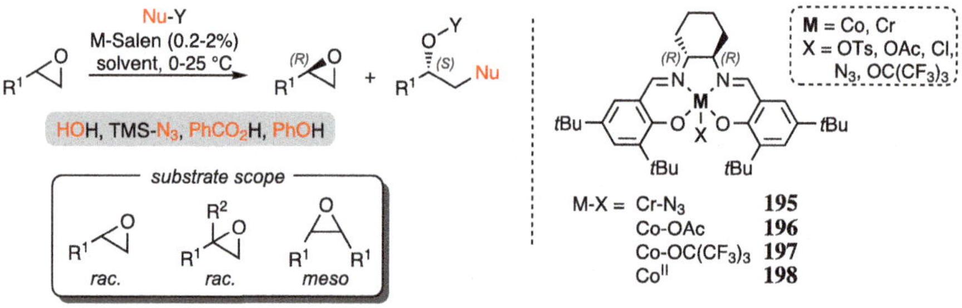

Fig. 3.123 General scheme of the kinetic resolution of epoxides by Jacobsen [326]

The reaction has been successfully applied to terminal epoxides, which show low selectivity as substrates in the classic Jacobsen epoxidation, as well as 1,1-disubstituted and *meso*-epoxides. Terminal alkenes and the corresponding epoxides are readily available on an industrial scale. Since only a yield of a maximum of 50% can be achieved in a resolution, the reactants should be as inexpensive as possible. Terminal alkenes meet this requirement, which is why the Jacobsen HKR is often used in the early stages of a synthesis with structurally simple substrates. *Meso*-epoxides, on the other hand, can theoretically yield 100% by virtue of their prochiral nature. For this reason, they are ideally suited as substrates, even if their requisite symmetric structure rarely provides an opportunity for application in natural product syntheses.

A variety of nucleophiles are eligible for opening the epoxide, provided they possess sufficient acidity. N-based nucleophiles are typically converted in the presence of Cr salenes, while the reaction of oxygen nucleophiles (H_2O, carboxylic acids, phenols) with epoxides is most efficiently catalzed by cobalt complexes. The most synthetically versatile and widely used reactants are H_2O and $TMSN_3$ (see Fig. 3.124) [326].

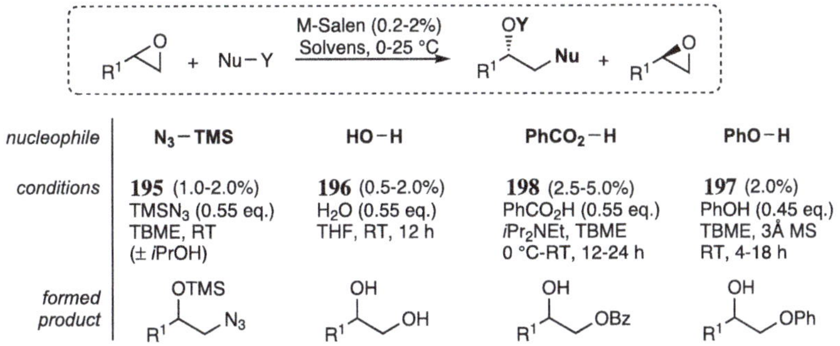

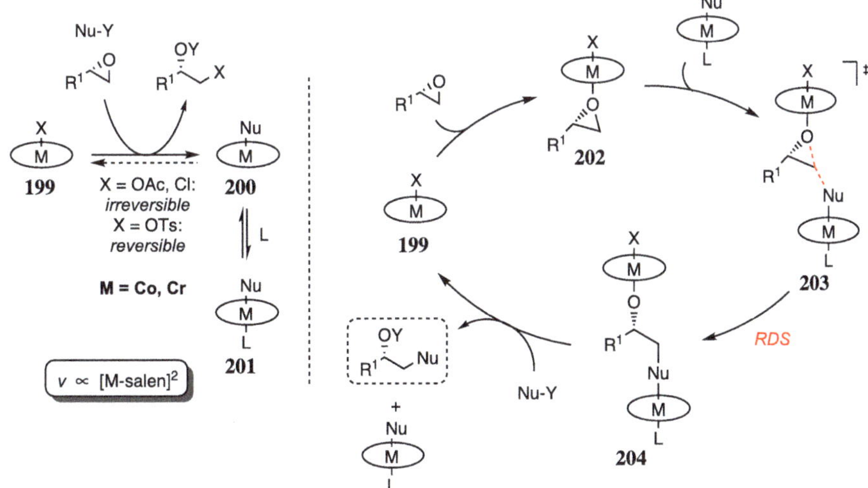

Fig. 3.124 Conditions of the Jacobsen (H)KR with different nucleophiles [327–331]

As opposed to the mechanism of the Jacobsen epoxidation, an epoxide cleavage requires **two** equivalents of the metal catalyst for the elementary step of the nucleophilic attack ($v \propto [\text{M-salen}]^2$). The nucleophile does not attack the epoxide directly but is activated and transferred by a second salen complex in a cooperative manner. One of the requisite salen complexes thus acts as a Lewis acid. As an initial step, the metal salen **199** reacts with one equivalent of the epoxide and nucleophile to **200**, which is reversible in the case of the tosylate. The reactive nucleophile **201** is obtained by the coordination of a second ligand. In the catalytic cycle, the epoxide coordinates the metal salen, thus polarizing and activating the C-O bond. The opening of the coordinated oxirane in the epoxy complex **202** is postulated to be the rate-determining step, which occurs via the transition state **203**. The bridged dimetallic complex **204** finally eliminates the desired product and the resulting salens **199** and **201** can reenter the catalytic cycle (see Fig. 3.125) [332–334].

Fig. 3.125 Postulated mechanism of the Jacobsen HKR [332–334]

The high difference in reaction rates of the two epoxide enantiomers is rationalized by the cooperative interactions of the two metal centers in the transition state. A dominant stereoinduction by only one metal complex, which either activates the epoxide (**202**) or the nucleophile (**201**), could be ruled out by careful control experiments. Thus, only the "double" matched case in the rate-determining step would result in the highest reaction rate, leading to the observed high selectivities of up to 500:1. For example, the opening of the *R*-configured epoxide is preferred for the hydrolytic resolution in the presence of *S,S*-**196**. Of the diastereomeric transition states, *S,S-R-S,S*-**205** is thus the energetically most favorable path, in which Lewis acid, epoxide, and nucleophile display a reinforcing, cooperative interaction (see Fig. 3.126) [332].

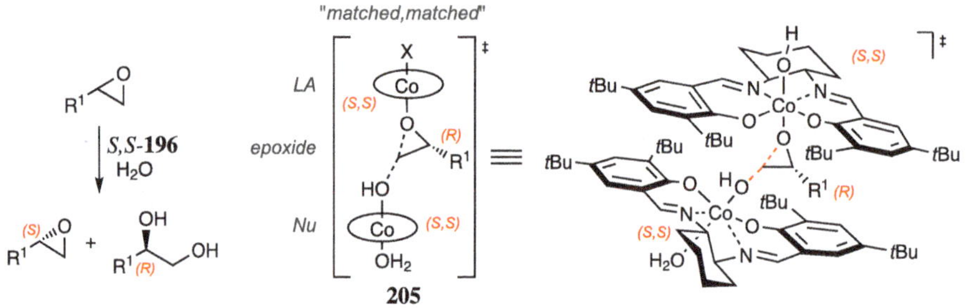

Fig. 3.126 Origin of the stereoselectivity in the Jacobsen HKR [332]

Due to the typically low cost of the substrates and a maximum theoretical yield of 50%, a chiral resolution is typically used with structurally simple building blocks for the introduction of the first stereoelement. While the functional group tolerance of the Jacobsen kinetic resolution is expected to largely correspond to that displayed by the Jacobsen epoxidation, this has not yet been shown for more complex intermediates (see Fig. 3.127).

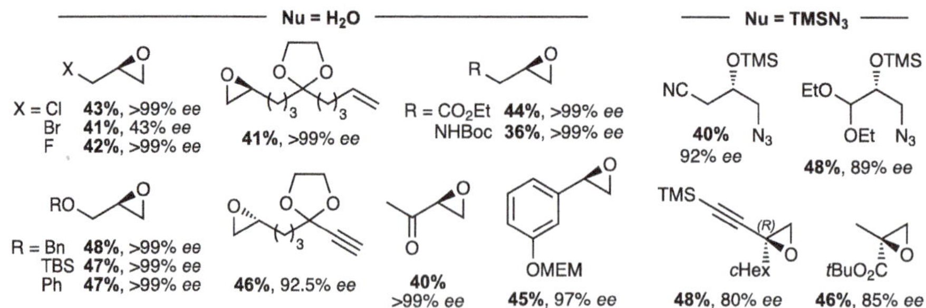

Fig. 3.127 Substrate scope of the Jacobsen (H)KR [327–329, 335, 336]

In addition to the metal salens developed by Jacobsen, ring opening reactions based on different metal catalysts and further nucleophiles have been reported by other researchers as well [337, 338].

The asymmetric epoxidation mediated by chiral dioxiranes has also found widespread use [339–342]. The mechanistic principles are identical to the details depicted in Figs. 3.96 and 3.99, only differing in fact that the ketone and the derived dioxirane are chiral in nature. The probably most frequently encountered ketone from this class of reagents was developed by Shi and co-workers based on fructose (**206, 207**). The *Shi epoxidation* typically uses oxone as an oxidizing agent in a buffered medium (see Fig. 3.128).

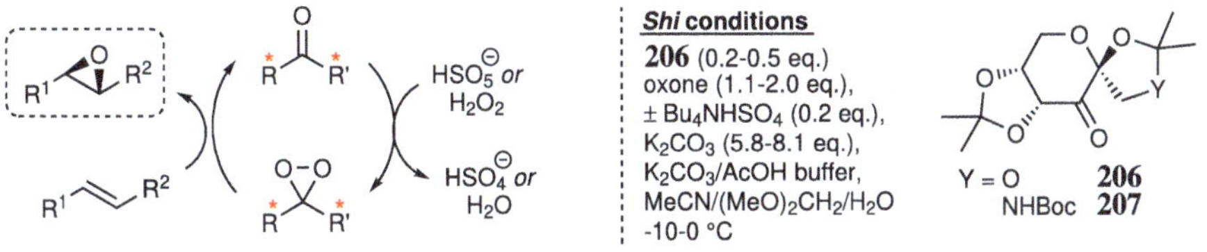

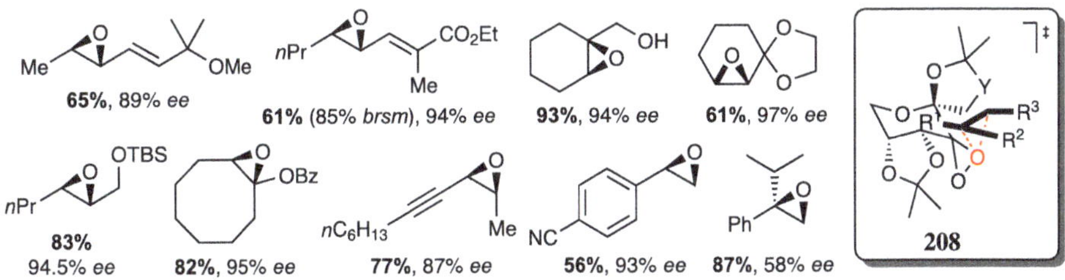

Fig. 3.128 Catalytic cycle and typical conditions of the Shi epoxidation [340]

The basicity of the reaction medium is pivotal: if the pH value falls below 10, the turnover of the reaction rapidly decreases due to an oxidative degradation of the catalyst. In addition to minimizing the requisite amount of catalyst for an efficient turnover, the use of a buffer at high pH can furthermore facilitate a reduction of the amount of oxidant while slightly increasing the enantioselectivity [343]. Suitable substrates include *E*- and *Z*-configured alkenes as well as trisubstituted and terminal olefins (see Fig. 3.129).

Fig. 3.129 Substrate scope of the asymmetric epoxidation with **206, 207** [344–346]

The epoxidation preferably proceeds via a *spiro* transition state **208**. In contrast to the Jacobsen epoxidation, *E*-configured and trisubstituted alkenes also afford high enantioselectivities. This is rationalized by the steric arrangement of the substituents of the dioxirane reagent, which allows these substrates to come into proximity to the reactive center without incurring a destabilizing steric repulsion.

The reaction shows a comparable tolerance towards functional groups as the epoxidation with DMDO, but the higher pH value can negatively affect base-labile groups. Despite the high catalyst loading, the reaction is regularly used with complex intermediates (see Fig. 3.130).

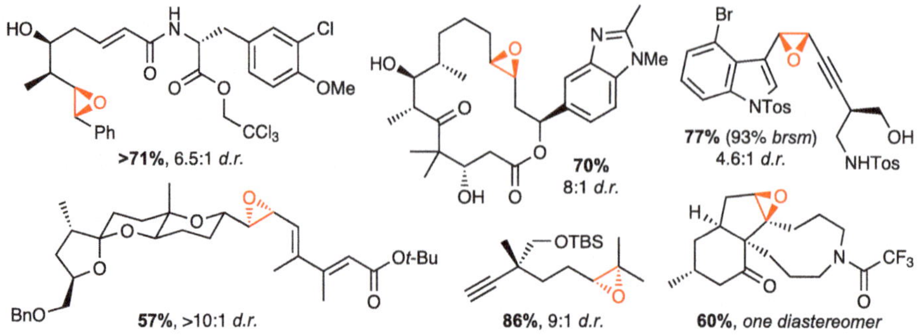

Fig. 3.130 Applications of the Shi epoxidation in natural product syntheses [347–352]

In the presence of nitriles as solvents, H_2O_2 can also be used as oxidant [353]. In addition to carbohydrate-derived catalysts, systems based on BINOL and biphenyl as well as quaternary cinchona and peptide derivatives have also been successfully developed [340, 342].

A comparison of the epoxidation methods in terms of their tolerance towards the major functional groups shows why total syntheses routinely resort to this class of reactions: the lion's share of potentially sensitive functionalities frequently remains intact while many functional groups are generally unaffected (see Table 3.11).

Table 3.11 Comparison of the functional group tolerance of different epoxidation methods. Dioxirane also includes the Shi epoxidation [252, 258, 311, 312, 326, 329, 339–341, 354]. *Notes:* [a] sp^2 and sp^3. [b] Br, Cl, F, rarely I. [c] 3° amines are tolerated

functionality	*m*CPBA	dioxirane	VO(acac)$_2$	Sharpless	Jacobsen	Jacobsen HKR
acetal, ketal	+	+	+	+	+	+
alcohol	+	+	+	+		
aldehyde	(−)	(+)		+		
alkene, e-rich	(−)	(−)	+	+	−	+
alkene, e-poor	(+)	(+)	+	+	(−)	+
alkyne	(+)	(+)	+	+	+	+
amide	+	+	+	+	+	+
amine		−[c]		−		
azide	+	+	+	+		
ester	+	+	+	+	+	+
ether	+	+	+	+	+	+
halogen [a,b]	+	+	+	+	+	+
ketone	(+)	+	+	+	(+)	+
nitrile	+	+	+	+	+	+
silyl ether	+	+	+	+	+	+
sulfon(amide)	+	+	+	+	+	
sulfoxide	−	(−)		+		
thiol	−	−	−	−		

The addition of oxygen to alkenes is well documented and has long been established, while the introduction of nitrogen was developed with significant delay: The challenge of the **aziridination** lies in the increased reactivity of the nitrogen precursors used for the reaction as well as the stability of the starting materials and the product under the reaction conditions (see Fig. 3.131) [355, 356].

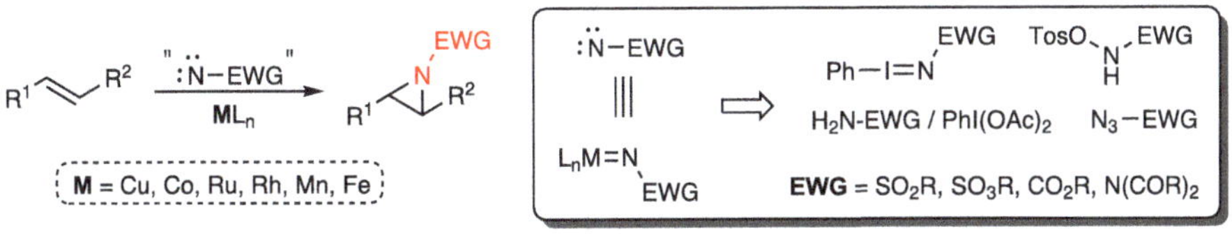

Fig. 3.131 General scheme of the aziridination of alkenes [355]

Since the olefin constitutes the nucleophile, the electrophilic nitrogen component requires an additional activation, e.g., through careful choice of the attached substituents. If used in conjunction with a metal catalyst, the reactive N-donors achieve high chemoselectivity and improved stability when incorporated into a metal nitrene species. Even though the nitrene precursors often display a suboptimal tolerance towards functional groups, the variability of

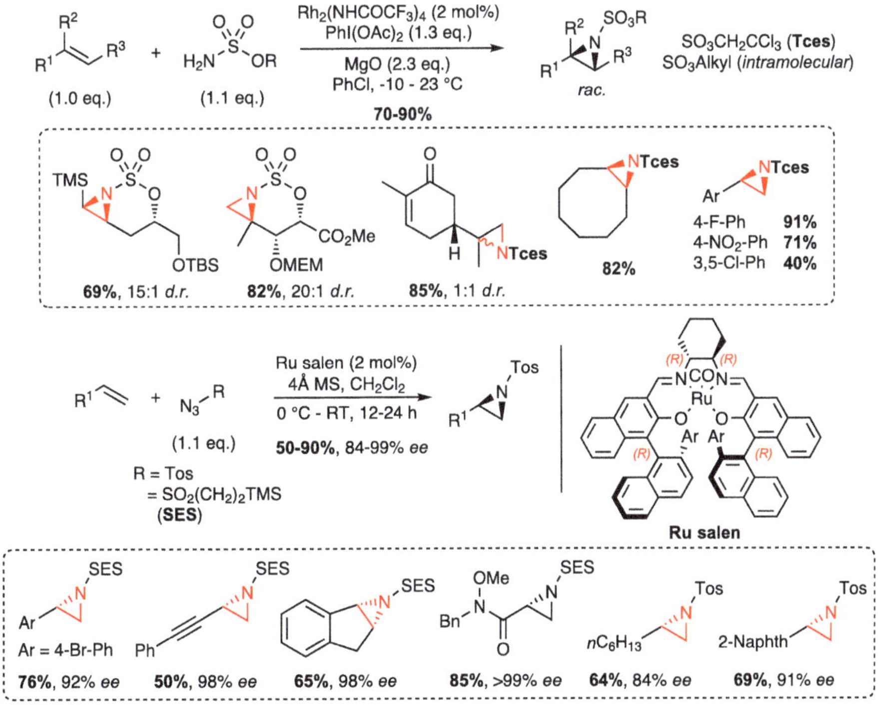

Fig. 3.132 Diastereo- and enantioselective aziridination methods [357, 358]

the nitrogen substituents (Boc, Troc, Tos, etc.) permits an adjustment of the reactivity over a wide range (see Fig. 3.132).

Despite the fact that syntheses rarely incorporate an aziridination, there are some individual cases in which the method was used to great effect in the access to natural products (see Fig. 3.133).

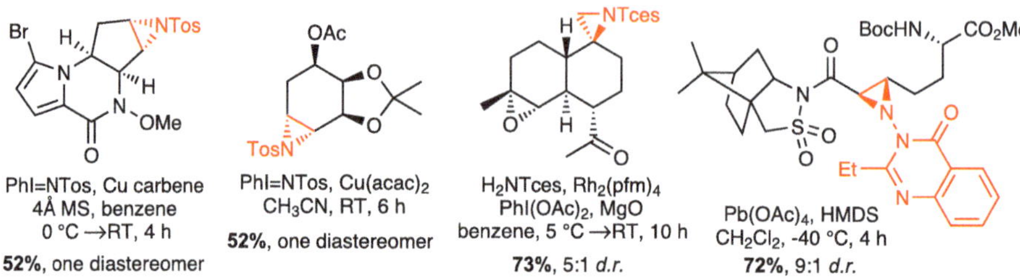

Fig. 3.133 Applications of diastereoselective aziridinations in complex synthesis intermediates [359–362]

Aside from the direct transformation starting from an olefin (Sharpless, Jacobsen, Shi, etc.), an epoxide or aziridine can also be constructed from a carbonyl compound and a sulfur ylide (*Corey–Chaykovsky* reaction, see Fig. 3.134) [363, 364]. Furthermore, a cyclopropane is obtained if the sulfur ylide is combined with an alkene [365]. In addition to the stoichiometric use of sulfur ylides, catalytic variants have been reported as well (see Sect. 8.1.6) [366, 367].

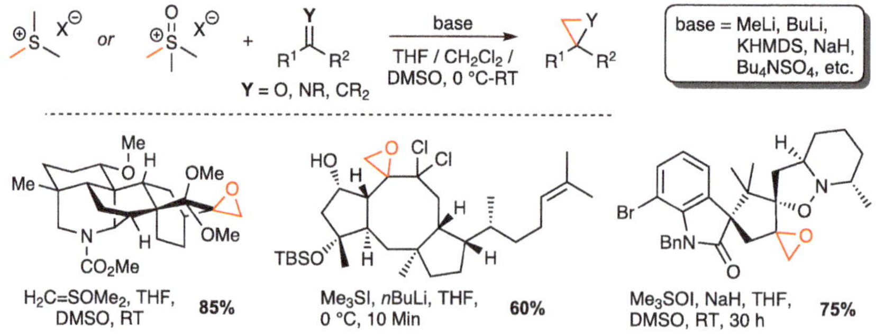

Fig. 3.134 Conditions of the Corey–Chaykovsky reaction and examples of complex intermediates. [368–370]

Dihydroxylation and Aminohydroxylation

The second major class of oxidative derivatizations encompasses the addition of two hydroxy groups to the double bond. The mild and exceptionally selective reaction typically relies on a highly oxidized metal complex to facilitate the oxygen transfer. Common metal salts are osmium-, manganese- and ruthenium-based. It was discovered early on that the addition of certain tertiary amines as ligands significantly accelerates the reaction. A chiral variant of the amine component can also be used to effect an enantioselective reaction (see Fig. 3.135) [250, 253, 371–373].

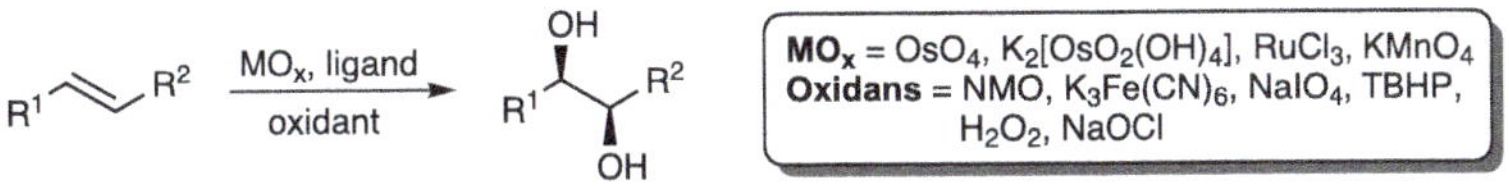

Fig. 3.135 Generic scheme for the dihydroxylation of alkenes

While early methods required stoichiometric amounts of the metal salt, the addition of co-oxidants such as $NaIO_4$ or NMO rendered the methods catalytic. The reaction is stereospecific, furnishing *syn*-diols as major products. The most commonly used precatalyst is an osmium oxide, which is similarly employed in the asymmetric variant introduced by Sharpless. Depending on the catalytic species, an oxidative cleavage of the resulting vicinal diols by the secondary oxidants is the most commonly occurring side reaction.

The addition step of OsO_4 to the alkene is assumed to proceed via a concerted [3+2]-cycloaddition according to the latest publications [373–377]. After extensive mechanistic studies, an alternative [2+2]-cycloaddition to give the corresponding metalaoxetane was ruled out as unlikely due to the high reaction barriers. Whether the equilibrium between OsO_4 and **209** lies outside or inside the catalytic cycle has not been clarified yet. However, the coordination of the amine to the metal center is essential to minimize the activation barrier of the subsequent rate-determining cycloaddition [377, 378]. In the pericyclic transition state **210** following the generation of the trigonal-bipyramidal catalyst **209**, oxygens in the axial and an equatorial positions are transferred, respectively. The stereospecificity of the conversion is rooted in the cycloaddition step, which yields the osmium glycolate **211**. The

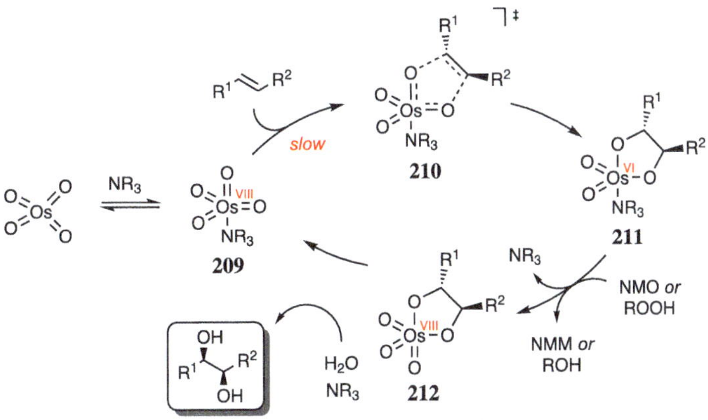

Fig. 3.136 Postulated mechanism of Os-catalyzed dihydroxylation in the presence of *tert*-amines [253, 373, 377]

exact mechanism from **211** is prone to variations in its details, depending on the reaction conditions and the oxidant (N-oxide, $K_3Fe(CN)_6$) in question. In a homogeneous reaction, the co-oxidant (N-oxide, peroxide) reoxidizes **211** to the trioxo-Os(VIII)-glycolate **212**, whose subsequent hydrolysis regenerates **209** upon coordination of the amine and concomitant release of the desired diol (see Fig. 3.136) [253, 373]. Also the existence of a second cycle involving a reaction of **211** with another equivalent of alkene prior to glycolate hydrolysis has been proposed under homogeneous conditions [379, 380].

Under biphasic conditions ($K_3Fe(CN)_6$, K_2CO_3, tBuOH/H$_2$O), a hydrolysis of the glycolate ester **211** was postulated to proceed prior to the reoxidation of the catalyst (Os^{VI}→ Os^{VIII}) in the aqueous phase [373]. The mechanism presented in Fig. 3.136 does not contain and explain all influences or experimental trends [378, 381]. Despite the gaps in the understanding of the mechanistic details that can change with the reaction conditions, the postulated catalytic cycle is considered to be an acceptable approximation.

The reaction displays a broad tolerance towards a variety of functional groups. Typically, either NMO or $K_3Fe(CN)_6$ are used as co-oxidants (see Fig. 3.137) [382, 383].

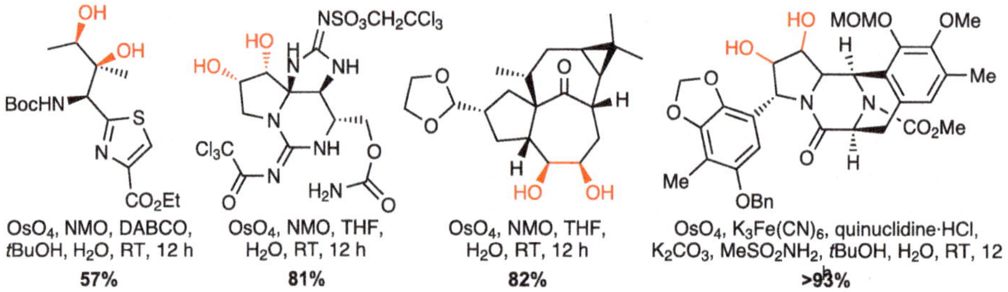

Fig. 3.137 Examples of Os-catalyzed dihydroxylations in natural product syntheses [324, 384–386]

The use of osmium tetroxide comes with some significant disadvantages: The toxicity of the catalyst and its volatility imply a significant safety hazard; in addition, osmium is relatively expensive. Despite the high selectivity and yield of Os-mediated dihydroxylations, methods relying on alternative catalysts have been developed to overcome the associated disadvantages of using Os [372, 387]. These methods are commonly based either on catalytic amounts of $RuCl_3$ with $NaIO_4$ as co-oxidant or stoichiometric $KMnO_4$. The $RuCl_3$-mediated reaction is particularly successful with substrates that possess insufficient reactivity in dihydroxylations in the presence of OsO_4 (see Fig. 3.138) [388, 389].

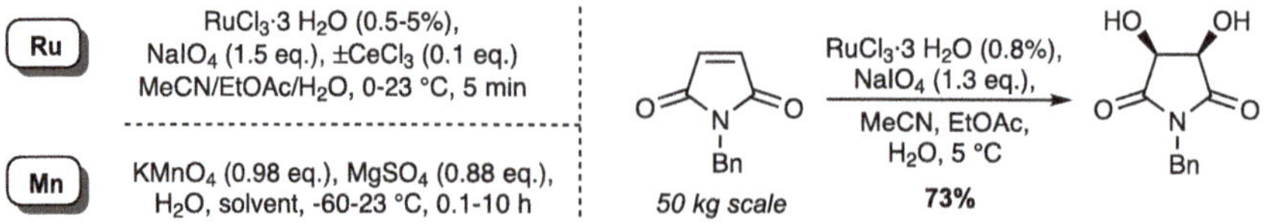

Fig. 3.138 Conditions of osmium-free dihydroxylations and their application on an industrial scale [372, 389, 390]

The enantioselective variant was first introduced by the group of K.B. Sharpless and has been extensively optimized by intensive research on aspects such as the ligands, the co-oxidant, and the solvent system [250, 253, 373]. Dimeric structures based on Cinchona alkaloids are the standard ligand class being used. The linker-bridged dihydroquinidine (DHQD) and dihydroquinine (DHQ) behave like quasi-enantiomers and facilitate opposite enantioinduction in the substrates (see Fig. 3.139).

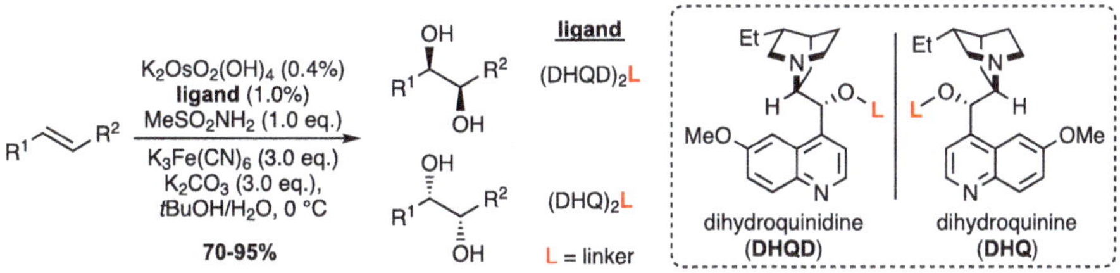

Fig. 3.139 Optimized reaction conditions of the Sharpless dihydroxylation [253]

Non-volatile $K_2OsO_2(OH)_4$ serves as the Os source and $K_3Fe(CN)_6$ as an inorganic co-oxidant. This mixture including the ligand and K_2CO_3 is commercially available, called AD Mix α ((DHQ)$_2$PHAL) and AD Mix β ((DHQD)$_2$PHAL), which is recommended as a starting point on small scale and standard substrates. The addition of one equivalent $MeSO_2NH_2$ significantly accelerates the reaction by facilitating the hydrolysis of the osmium glycolate **211**. In the case of terminal olefins it does not effect any change and sometimes might even slow down the hydrolysis step, rendering its addition counterproductive for this class of

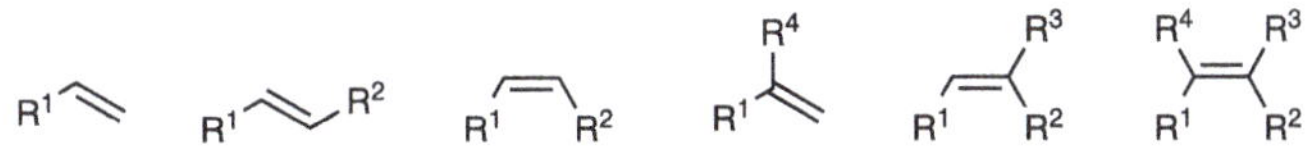

R^1, R^2, R^3, R^4	$R^1\diagup$	$R^1\diagup R^2$	$R^1 \diagup R^2$	$R^4 \atop R^1\diagup$	$R^4 \quad R^3 \atop R^1\diagup R^2$	$R^4 \quad R^3 \atop R^1\diagup R^2$
aromatic	DPP, PHAL	DPP, PHAL		DPP, PHAL	PHAL, DPP, AQN	PYR, PHAL
aliphatic	AQN	AQN		AQN		
others	PYR (branched)		IND (acyclic) PYR, DPP, AQN (cyclic)	PYR (branched)		
enantiomeric excess	**70-97%**	**90-99%**	**20-80%**	**70-97%**	**90-99%**	**20-97%**

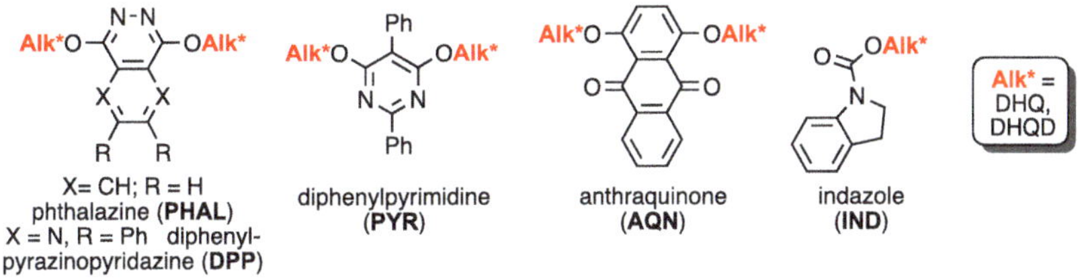

Fig. 3.140 Recommended linker L and expected enantioselectivity [250, 253]

alkenes. Depending on the type of substrate, dimeric ligands with different linkers L are employed to maximize the enantiomeric excess (see Fig. 3.140) [250, 253].

Based on the empirical results with a wide variety of substrates a mnemonic was developed, which predicts the steric requirements for high enantioselectivity and the expected absolute stereochemistry (see Fig. 3.141).

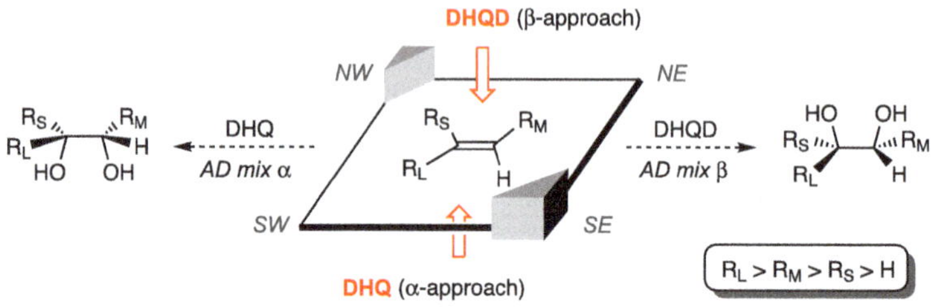

Fig. 3.141 Mnemonic of the Sharpless dihydroxylation [253]

Under the optimized reaction conditions, the conversion takes place in a biphasic solvent system with $K_3Fe(CN)_6$ as a co-oxidant. The mechanism closely resembles that outlined for the dihydroxylation in the absence of stereoinducing ligands, with the steps being identical up to the formation of the glycolate complex **211** via **209**. The hydrolysis of the Os(VI) ester takes place prior to the outer-sphere oxidation of Os(VI) to Os(VIII) and occurs in the aqueous phase. This has the advantage that a secondary catalytic cycle occurring under homogeneous conditions with lower enantioselectivity is effectively prevented [391]. The hydrolysis of **211** is accelerated in the presence of methanesulfonamide, which possibly acts as a nucleophile under slightly basic conditions. The [3+2]-cycloaddition was postulated as the enantioinducing step, in which the ligand arms form a U-shaped pocket (**213**) and shield one face of the substrate (see Fig. 3.142) [253, 373].

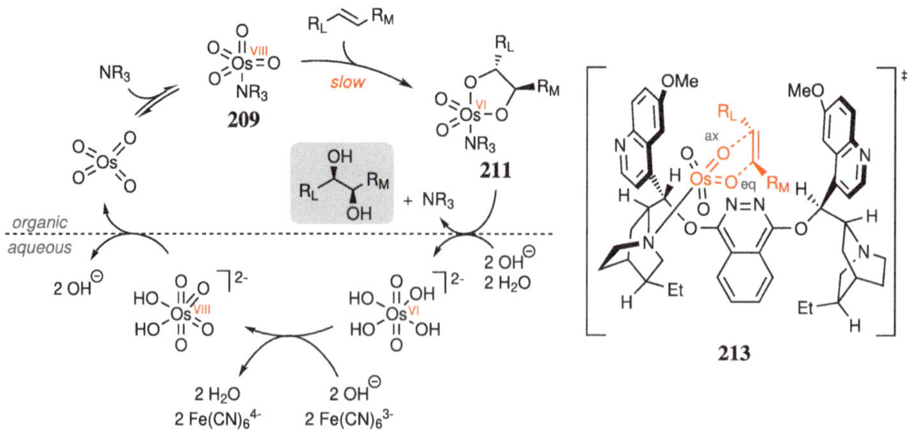

Fig. 3.142 Postulated mechanism of the Sharpless dihydroxylation and transition state of the cycloaddition with $(DHQ)_2PHAL$ [253, 392]

A wide range of olefins is suitable as substrates. High enantioselectivity is expected in those cases where the substrate can closely adhere to the structure of the alkene depicted in the menomic (cf. Fig. 3.141). Thus, disubstituted Z-configured olefins only achieve moderate enantioselectivity. Terminal alkenes usually show high to very high enantioinduction, but substrates with small substituentes such as allylic derivatives or unbranched alkenes result in *ee* values in the range of 70–80% [253]. As a consequence, the intermediates used in natural product syntheses for the dihydroxylation are usually terminal or disubstituted E-alkenes (see Fig. 3.143).

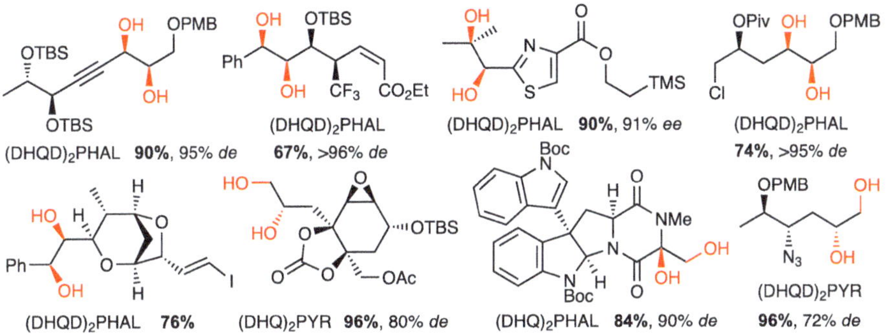

Fig. 3.143 Applications of the Sharpless dihydroxylation in total syntheses [393–400]

Functional groups such as acetals, alcohols, alkynes, amides, esters, halogens, nitriles, ketones, and sulfonamides are tolerated. Even substrates with thioethers can be converted to the corresponding diols without significantly oxidizing the sulfur moiety to the respective sulfoxide [373].

The asymmetric Os-catalyzed *aminohydroxylation* was also developed in the Sharpless group and is closely related to the dihydroxylation reaction. Salts of halogenated sulfonamides, carbamates, and amides are used as the nitrogen source, the oxygen stems from the H_2O used as a cosolvent (see Fig. 3.144) [401–403].

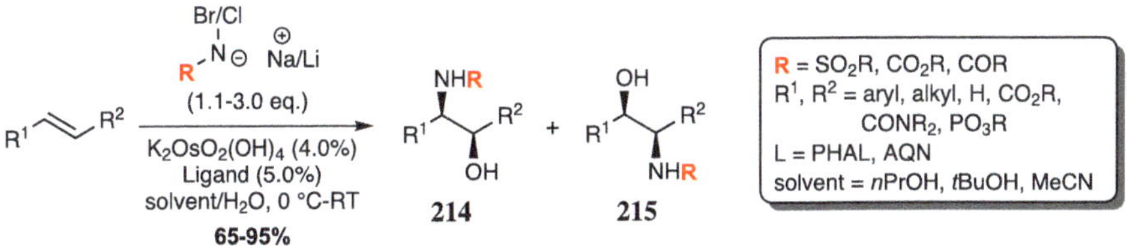

Fig. 3.144 Typical conditions of the Sharpless aminohydroxylation

The reaction with carbamates and amides generally yields higher enantioselectivities than with sulfonamides. Furthermore, the latter is not suited for converting terminal alkenes. The regioselectivity can also be influenced by judicious choice of the reaction conditions with carbamates and amides (see Table 3.12).

Table 3.12 Comparison of different N sources in the Sharpless aminohydroxylation [401, 404, 405]

N source	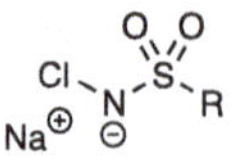	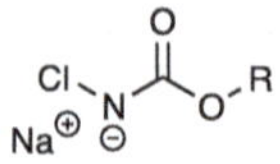	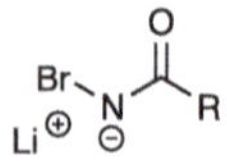
R	Tol, Me, Ph	Bn, Et, *t*Bu	Me, Pr, Ph
ee values	75–95%	85–99%	85–99%
regioselectivity	≥5:1	≥3:1	≥5:1
substrates	cinnamic acid deriv.	term. alkenes	term. alkenes
	acrylates	acrylates	acrylates
		cinnamic acid deriv.	cinnamic acid deriv.
N equivalents	3.0	3.0	1.1
anion formation	isolated	*in situ*	isolated
N deprotection	*harsh*	*mild*	*medium-harsh*
	e.g. Li/NH$_3$;	e.g. LiOH/MeOH;	e.g. HCl, Δ;
	Red-Al	TBAF/THF;	NaBH$_4$/THF
		H$^+$ (Boc); H$_2$/Pd-C (Cbz)	

The challenge of aminohydroxylation reactions lies the selective formation of the desired regioisomer **214** or **215**. In unsymmetrical alkenes, the oxygen atom is usually added to the electron-poorer terminus of the double bond (see Fig. 3.145). Mechanistically, the inherent preference during the cycloaddition for one of the addition pathways accounts for the observed regioselectivity. Due to its kinship with the dihydroxylation reaction, this step may proceed similarly via a [2+2] or [3+2] cycloaddition (via **217/218** or **219/220**). For the aminohydroxylation, a competition of both paths hasn't yet been investigated extensively, neither by theoretical or experimental means, leaving this question unresolved thus far.

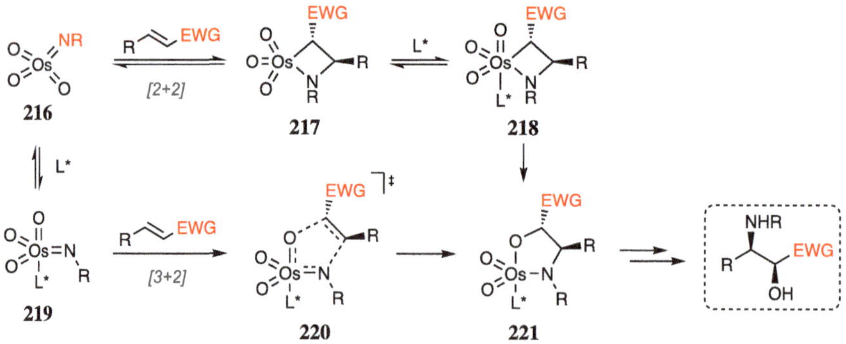

Fig. 3.145 Origin of regioselectivity and transition states of a possible [2+2]- or [3+2]-cycloaddition [403, 404]

A reversal of the regioselectivity could be obtained by resorting to (Alk*)$_2$AQN as ligand and by changing the solvent system (see Fig. 3.146).

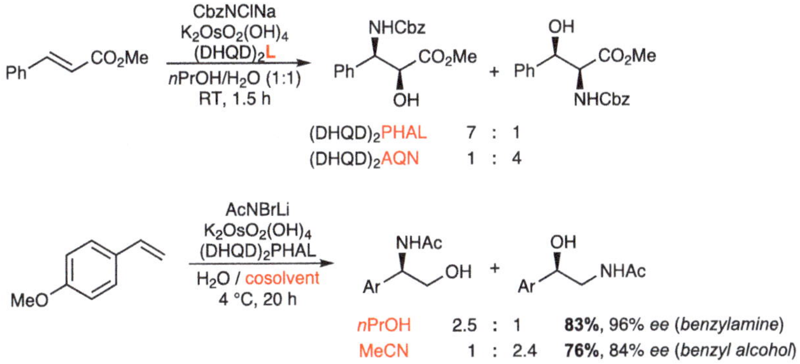

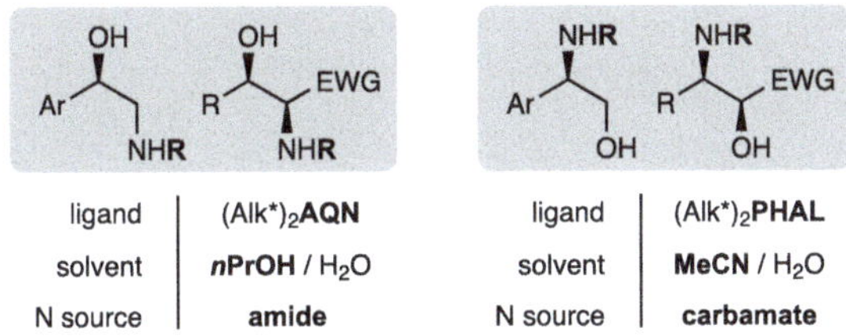

Fig. 3.146 Influence of the reaction conditions on the regioselectivity [406, 407]

The regioselectivity is only partially influenced by the steric environment of the substrate, but can be controlled by the choice of ligand, the solvent system, and the nitrogen source [404]. In the case of carbamate reagents, a change in regioselectivity often correlates with a decrease in enantioselectivity when using AQN-bridged ligands. In contrast, if *N*-halogen amides constitute the nitrogen source, high *ee*-values for the formation of the desired benzyl alcohols can be achieved (see Fig. 3.147) [408].

Fig. 3.147 Handles enabling control of the regioselectivity [404]

The mechanism of the reaction progresses analogously to the Sharpless dihydroxylation under homogeneous conditions. Two catalytic cycles exist, which are interconnected by the shared intermediates **221** and **223**. The primary catalytic cycle proceeds through an initial cycloaddition of **216** to **221**, followed by oxidation to **223** and the concluding rate-determining hydrolysis of the product. This cycle affords high *ee*-values, while the secondary catalytic cycle displays a significantly lower enantioselectivity. A branching into the less selective cycle can be prevented by accelerating the hydrolysis of **223**, for example, by increasing the concentration of H_2O. In addition, the size of the substituent on the nitrogen reagent influences the enantioselectivity, as the hydrolysis of **222** is similarly delayed by sterically demanding groups: the smaller the steric bulk of the nitrogen source, the higher the *ee*-values (see Fig. 3.148) [409].

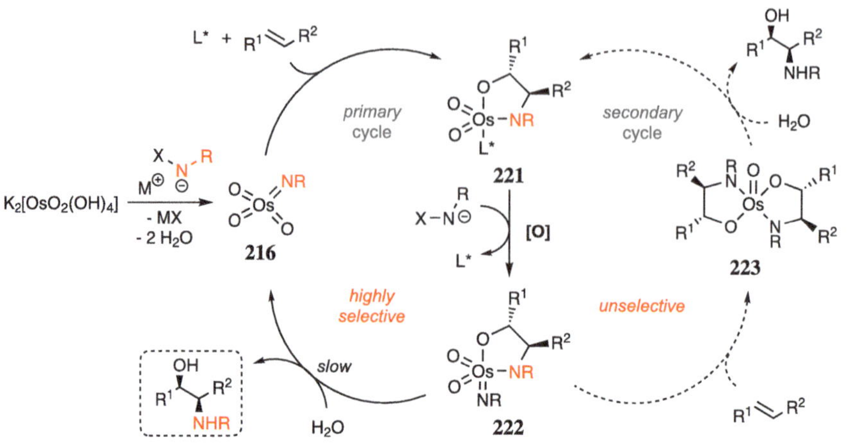

Fig. 3.148 Postulated mechanism of the Sharpless aminohydroxylation [401, 404]

Especially carbamates are the nitrogen donor of choice in more complex syntheses. Even though amides show a comparable substrate scope as carbamates and high enantioselectivities, the advantage of the latter lies in the relatively simple access to the deprotected amino alcohols (see Fig. 3.149) [410].

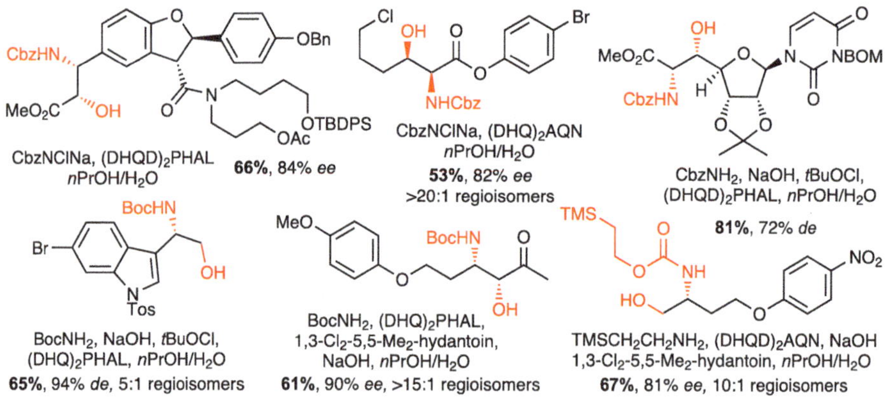

Fig. 3.149 Examples of Sharpless aminohydroxylations using $K_2OsO_2(OH)_4$ as precatalyst [411–416]

Halofunctionalization and Dihalogenation

The reactions of alkenes with elemental bromine or iodine are classic examples of ionic dihalogenations and have been used in a multitude of synthetic endeavors to achieve a 1,2-functionalized substitution pattern. The intermediate halonium compounds can be easily cleaved with a variety of nucleophiles based on halogens, amines, alcohols, and carboxylic acids (see Fig. 3.150).

Fig. 3.150 Scheme of the halofunctionalization of alkenes

The conversion of olefins with halogens to **224** or **225** is irreversible for X = F, Cl. In contrast, brominations and iodinations are reversible equilibrium reactions [417, 418]. In the derivatization of alkenes with *dihalogens*, an olefin-X_2 charge-transfer complex is obtained as the initial step. Cyclic halonium compounds (**224**) lead to a *trans*-substitution pattern in the product due to the stereospecific S_N2 attack of the nucleophile. Not all halogens form haliranium ions. The tendency is particularly pronounced for large, highly polarizable atoms (I>Br>Cl≫F). In contrast, the ability to stabilize carbocations in benzylic or allylic position degrades the observed *trans*-selectivity. Acyclic halonium ions (**225**) can rotate around their central CC single bond, given a sufficient lifetime and low rotational barrier. As they can intrinsically favor one diastereomer, a high selectivity does not necessarily indicate the involvement of cyclic intermediates (see Fig. 3.151) [419].

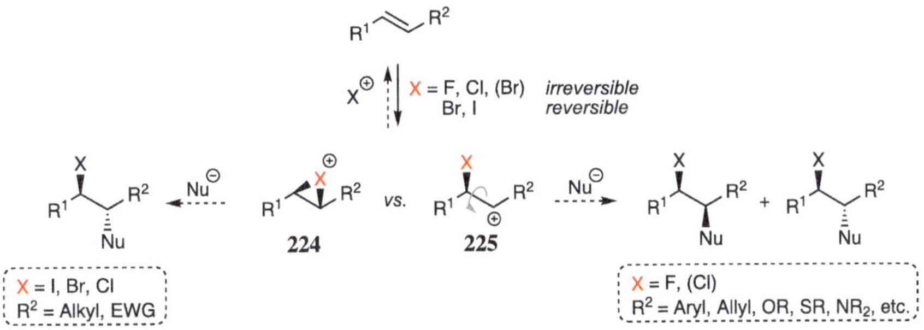

Fig. 3.151 Reaction paths and resulting stereochemistry of halofunctionalizations

The product distribution in the dihalogenation of different olefins can usually be rationalized by assuming a strictly stereospecific mechanism. However, functionalizations involving fluorine and chlorine as well as substrates bearing electron-rich or aryl substituents in the α-position routinely yield mixtures of diastereomers (see Fig. 3.152).

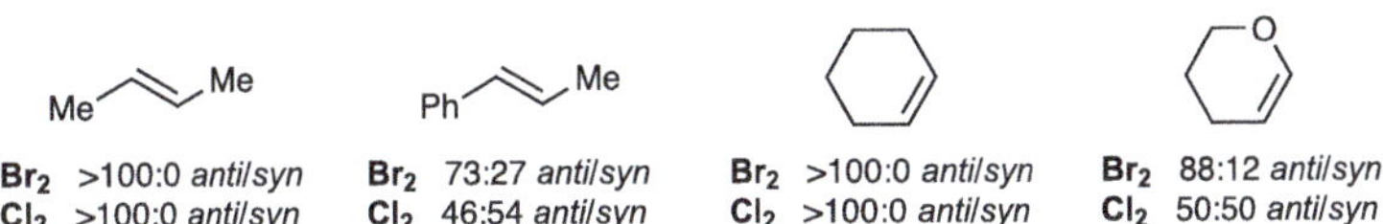

Fig. 3.152 Stereoselectivity in the dihalogenation of various olefins [420–424]

Since the olefinic double bond constitutes the nucleophilic component, electrophilic halo-gen sources are needed to form the positively charged halonium ions. Given a dihalogenation by elemental halogen, the corresponding counterion would consist of the halides Cl^-, Br^-, or I^-. These can subsequently act as a nucleophile and intercept the cationic haliranium ion. Different electrophilic reagents are thus required for different substrates and modes of halofunctionalization (see Fig. 3.153).

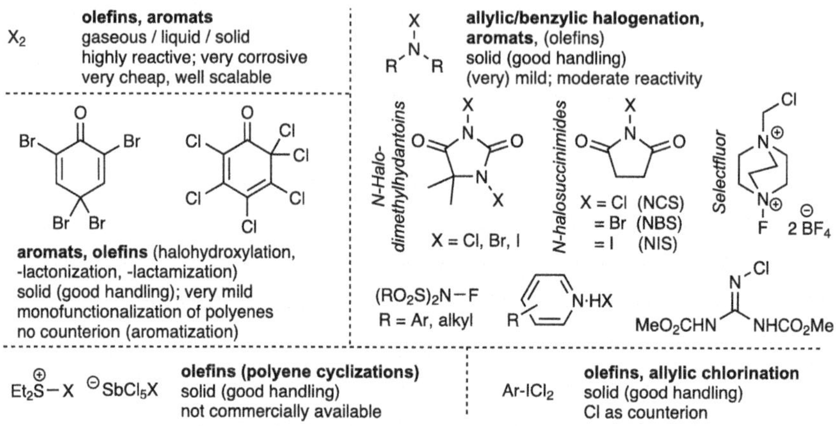

Fig. 3.153 Comparison of different halogenation reagents

In an industrial context, inexpensive but highly reactive halogen sources such as elemental dihalogens or thionyl chloride are typically preferred. In the synthesis of fine chemicals, active pharmaceutical ingredients, or densely functionalized natural products, the complexity of the substrates, the length of the synthetic sequence, and the increasing demands on chemoselectivity mandates a focus on high selectivity rather than cost of the halogen source. The requisite reagents are significantly milder, but often decidedly more expensive (see Fig. 3.154).

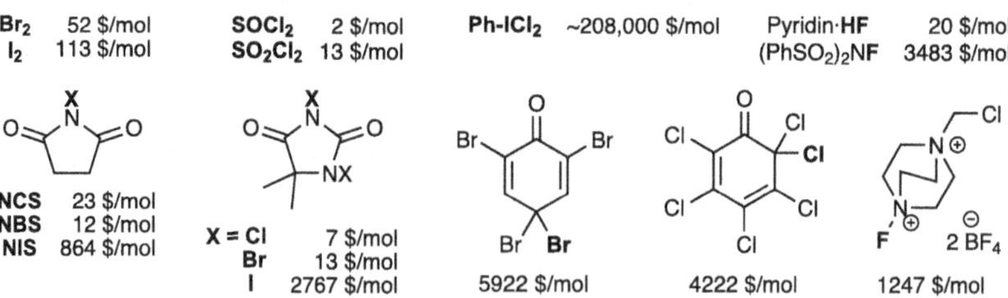

Fig. 3.154 Costs of common halogenation reagents per electrophilic halogen equivalent. Prices taken from Sigma-Aldrich or Thermo-Fisher in 2/2024, the price of the largest pack size was used.) [425]

Dihalogenations are encountered less frequently, as the tolerance towards functional groups is usually limited. However, the development of particularly mild chlorinations and brominations, including enantioselective variants (*vide infra*), has led to an increasing number of successful total syntheses of halogenated natural products in recent years (see Fig. 3.155).

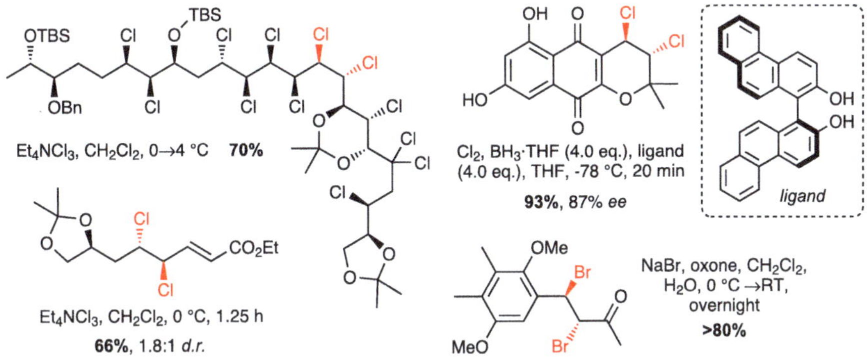

Fig. 3.155 Examples of dihalogenations in total syntheses [306, 426, 427]

Difluorinations and fluorofunctionalizations enjoy a special status due to the safety-related challenges in handling fluorine gas. This necessitates the development of reagents and methods dedicated exclusively for the introduction of fluorine [428, 429]. Hypervalent iodine can efficiently transfer fluorine to alkenes, either stoichiometrically or in a catalytic fashion (see Fig. 3.156) [430–433].

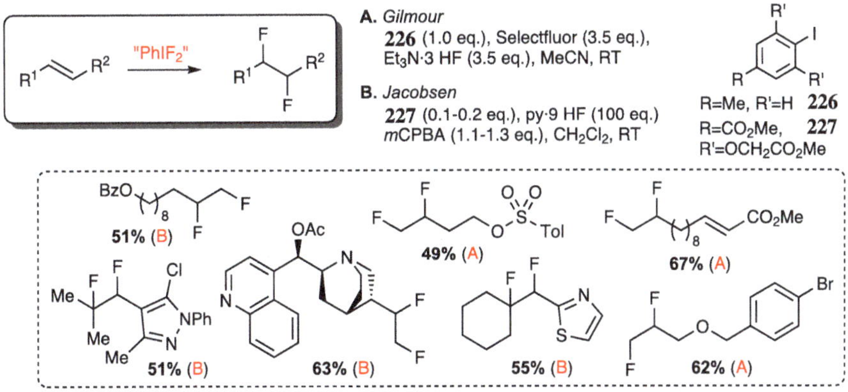

Fig. 3.156 Catalytic difluorinations of terminal and internal alkenes [431, 432]

In addition to dihalogenations, the interception of halonium intermediates with O- and N-nucleophiles can provide rapid access to a range of valuable products [434–437]. In these cases, a low concentration of halogens is required, such as reagents that slowly release X^+ *in situ*. Halo-nitrogen compounds and cyclohexadienone derivatives are particularly successful as electrophilic halogenation reagents in these transformations.

In addition to the question of chemoselectivity, two constitutional isomers **228** or **229** can be formed with unsymmetric olefins, depending on the regioselectivity of the nucleophilic attack. In the absence of dominant steric factors, the Markovnikov product is usually obtained preferentially (see Fig. 3.157).

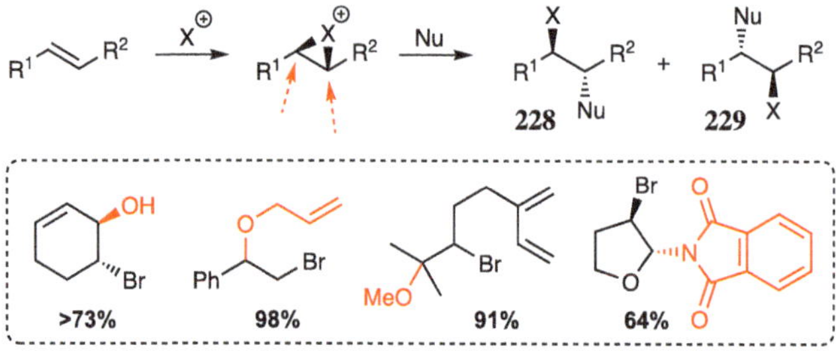

Fig. 3.157 Regioselectivity of intermolecular halofunctionalizations [434, 438]

In *intermolecular* halofunctionalizations, the nucleophile is typically used in excess, for example, as a solvent. *Intramolecular* reactions usually afford one isomer as a major product, as different ring sizes are formed significantly faster than others due to stereoelectronic factors. Lactonizations and the formation of halohydrins have been extensively used in synthetic endeavors (see Fig. 3.158).

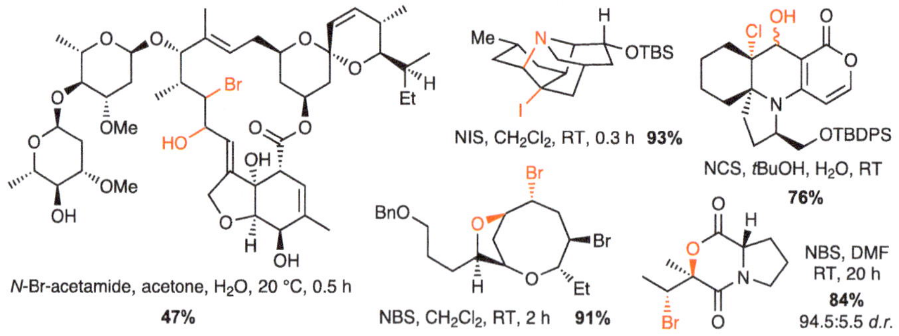

Fig. 3.158 Diastereoselective halofunctionalizations of complex synthesis intermediates [439–443]

With the exception of fluorinations, stoichiometric dihalogenations proceed stereospecifically due to the cyclic nature of the haliranium intermediates, resulting in *trans*-configured products. In the case of catalytic variants, some *syn*-selective methods have been reported.

Denmark and co-workers developed a Se-catalyzed dichlorination in the presence of pyridinium-N-fluoride as an oxidant and lutidine-N-oxide as a Lewis basic additive to further accelerate the turnover. The catalytically active species is $PhSeCl_3$, which adds to alkenes and forms the cyclic selenium ion **231**. Following the S_N2 addition of a chloride and cleavage of the seleniranium intermediate, the $PhSeCl_2$ substituent in **232** is displaced by another chloride and the catalyst $PhSeCl_3$ is subsequently regenerated. The isolated product is *syn*-configured due to the twofold inversion. The diastereoselectivities are usually excellent and the yields acceptable to good (see Fig. 3.159) [444].

Fig. 3.159 *syn*-Selective dichlorination by Denmark *et al.* [444]

The requirements of an efficient enantioselective, catalytic halogenation represent a high hurdle in their development. Most enantioselective halofunctionalizations are based on halolactonizations (X = I, Br) and are limited to the formation of 5- or 6-membered heterocycles as the kinetics of those cyclizations are very favorable compared to other reaction modes. Intermolecular methods have been developed especially for cinnamic acid derivatives, but the substrate scope in turn limits their applicability [445]. Apart from affording different regioisomers, the (partial) reversibility in the formation of the haliranium intermediates can also result in racemization. The haliranium ions can transfer the halogen atom to another alkene if their formation is reversible, thus rendering them configurationally unstable. In the case of dihalogenations (Nu = X), symmetric alkene substrates ($R^1 = R^2$) exclusively yield *meso*-compounds as products since an inversion center is present in the molecule (see Fig. 3.160) [445–448].

Initial studies on enantioselective dihalogenations of allylic alcohols relied on dimeric Cinchona alkaloids as ligands and hypervalent iodoarenes as halide source. The Burns group developed a Ti-catalyzed variant in which a wide variety of allylic alcohols could be successfully converted, providing high enantiomeric excess as rationalized by the postulated

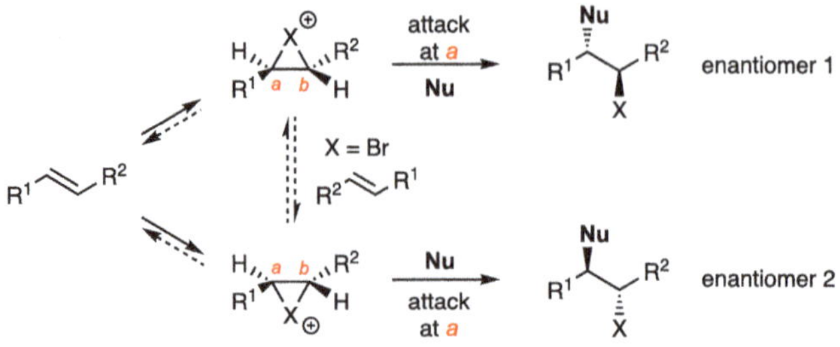

Fig. 3.160 Challenges in asymmetric halofunctionalizations [446]

transition state **233** (see Fig. 3.161) [449]. In addition to dihalogenations, other halofunctionalizations are also possible via this approach [450].

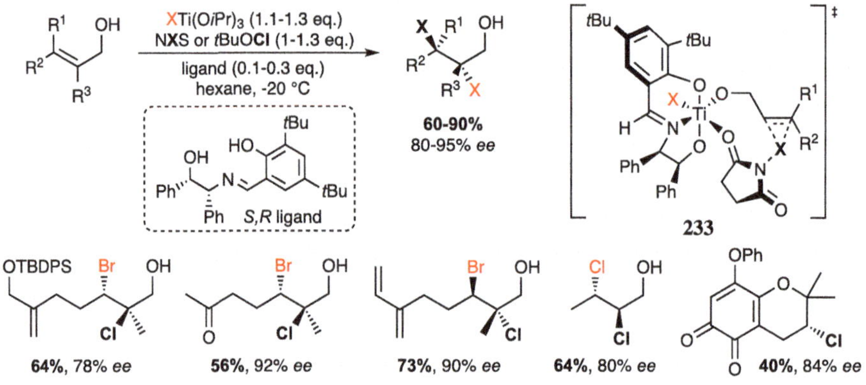

Fig. 3.161 Enantioselective dihalogenation developed by Burns *et al.* and postulated transition state
233 [449, 451–453]

Hydroboration

The efficient conversion of alkenes (and alkynes) by boranes was developed in the 1950s
by the group of H.C. Brown and is referred to as hydroboration [454]. The reaction can
also be carried out in the presence of transition metal catalysts, thus enabling asymmetric,
catalytic variants. The resulting alk(en)ylated boranes are useful intermediates and can be
further transformed subsequently, for example, by transmetalation or direct reaction with
various electrophiles. The most frequently employed derivatization is the basic oxidation to
the corresponding saturated alcohols (see Fig. 3.162).

A wide range of boranes can be used: While diborane, B_2H_6, or BH_3 complexes react
with three equivalents of the olefin to form the trimeric trialkylboranes, with dialkylboranes
or boronic acid diesters only one equivalent of the substrate is consumed to afford the
alkyl borane products. Particularly $BH_3 \cdot SMe_2$, 9-borabicyclo[3.3.1]nonane (9-BBN) and

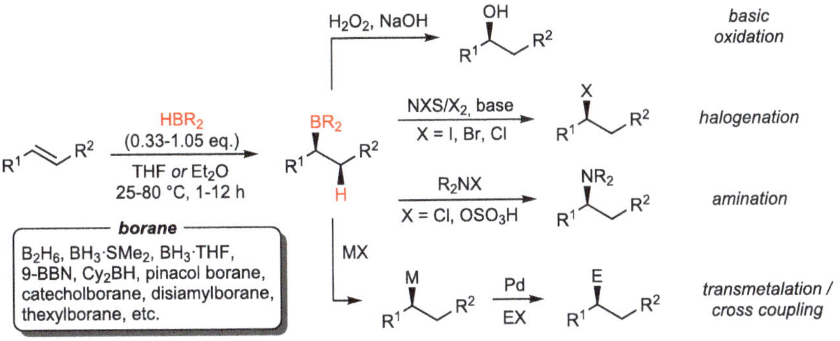

Fig. 3.162 Hydroboration of alkenes and common subsequent transformations of alkylboranes [455, 456]

catecholborane have established themselves as the most frequently used reagents and are (partially) considered complementary in their respective applications (see Table 3.13) [457].

Table 3.13 Characteristics of different boranes [457]

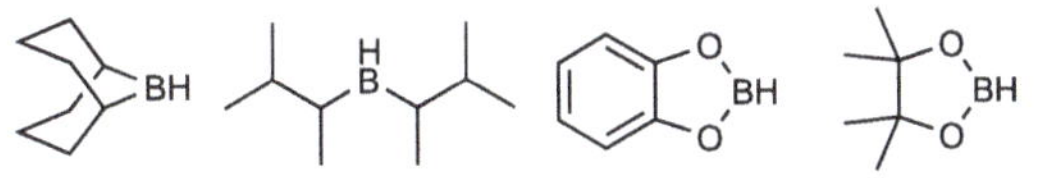

	B_2H_6 $BH_3 \cdot THF$ $BH_3 \cdot SMe_2$	9-BBN 9-borabicyclo-[3.3.1]nonane	Sia₂BH disiamylborane	catBH catecholborane	pinBH pinacol borane
reactivity	high	high	high	low	low
regioselectivity	low-medium	very high	very high	medium-high	medium-high
monofunctionalization of polyenes	bad	good	good	mediochre	mediochre
preference alkene *vs* alkyne	none	alkene	n.d.	alkyne	n.d.
used catalytically	no	no	no	yes	yes

Catecholborane and pinacol borane necessitate elevated temperatures of 80–100 °C due to their diminished reactivity. Alcohols, aldehydes as well as 1° and 2° amines must generally be protected during hydroborations. By virtue of their tolerance toward a variety of functional groups, particularly 9-BBN and disiamylborane are suitable for chemoselective reactions of complex intermediates. Catechol- and pinacol borane are resorted to in the catalyzed variant due to their low reactivity and acceptable tolerance towards various functionalities (see Table 3.14) [458].

The uncatalyzed reaction generally favors a C-B bond formation at the less substituted terminus of the double bond based on steric and electronic effects (*anti*-Markovnikov product). Particularly 9-BBN and disiamylborane are highly selective in this regard due to their higher steric demand (see Fig. 3.163).

Table 3.14 Reaction rate of various boranes towards various functional groups in the absence of catalysts [458–461]

	BH_3	9-BBN	catecholborane
alkene	++++	++++	+
alkyne	++++	+++	++
ketone	++++	+++	+
ester	+	−	+
amide	+	+	+
carboxylic acid	++++	−	++
acid chloride	−	++++	++
halogen	−	−	−
epoxide	+	+	+
nitrile	+++	−	+

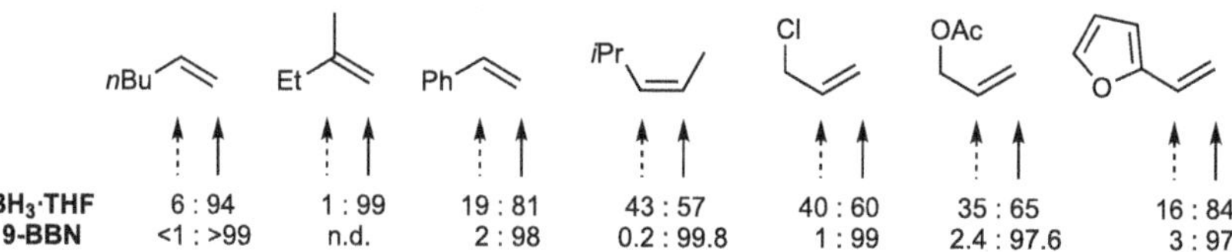

	nBu	Et	Ph	iPr	Cl	OAc	furyl
BH_3·THF	6 : 94	1 : 99	19 : 81	43 : 57	40 : 60	35 : 65	16 : 84
9-BBN	<1 : >99	n.d.	2 : 98	0.2 : 99.8	1 : 99	2.4 : 97.6	3 : 97

Fig. 3.163 Regioselectivity of the hydroboration of various alkenes. The figures refer to the percentage of C-B bond formation at the indicated carbon [457]

The necessity to conduct the reaction in etheric solvents for efficiency's sake is rooted in the dimerization behavior of boranes. The dimeric boranes (**234**) or Lewis pairs such as BH_3·SMe_2 initially have to dissociate to the free borane HBR_2. This is greatly facilitated in solvents such as THF or Et_2O, supposedly by enabling the formation of weakly bound intermediates such as HBR_2·THF complex [462–464]. Depending on the reactivity of the alkene substrate, either the formation of the borane monomer or the subsequent attack on the alkene can be rate-determining ($k_1 > k_2$ or $k_1 < k_2$). The transition state has been postulated based on kinetic and theoretical investigations to adopt a 4-membered, distorted structure **235** [465]. Despite its resemblance to the thermally forbidden $[\pi 2s + \sigma 2s]$-addition, the transition state is symmetry-allowed, as the p-orbital of the boron is involved in the orbital interaction [458]. The tendency to yield *anti*-Markovnikov products can be attributed to steric and electronic factors, with the influence of steric factors usually dominating: BR_2 adds to the sterically less crowded terminus of the CC bond to minimize a repulsive interaction. Furthermore, in terminal alkenes, the unsubstituted terminus is negatively polarized and can thus interact more strongly with the positively polarized boron. While trialkylboranes are stable at room temperature, the hydroboration product **236** can revert back to the starting materials approximately above 100 °C. The reaction is therefore partially reversible at elevated temperatures, depending on the individual alkene and boron reagent (see Fig. 3.164) [466].

Depending on the reaction conditions that **236** is obtained under, either a retention or inversion of the newly formed stereocenter may result. For the basic oxidation, an S_N2

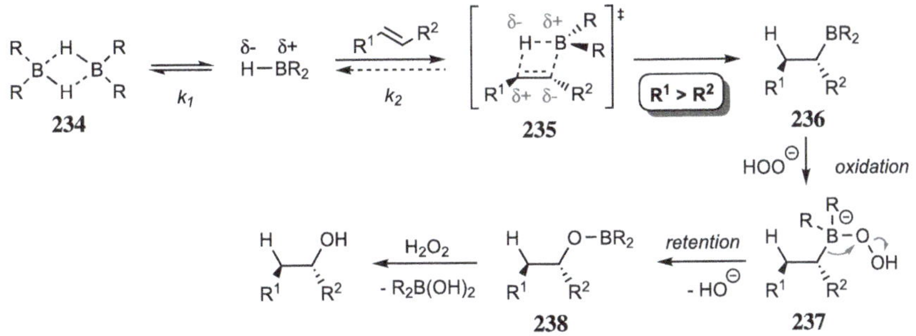

Fig. 3.164 Postulated mechanism of hydroboration and subsequent oxidation in basic medium [456, 462, 465]

attack by HOO^- via **237** has been postulated with stereoretentive transfer of the alkyl substituent at boron to give **238**. The concluding oxidation step affords the desired alcohol upon exposure to another equivalent of H_2O_2 [456].

The catalyzed process relies on the use of less reactive boronic acid esters. The resulting slower background reactions allow for the desired reaction to be significantly accelerated by the catalyst and thus achieve a high overall selectivity. Catecholborane or occasionally pinacol borane are typically used. Common catalysts are often based on rhodium or iridium complexes, but there are also methods employing Ru, La, Ti, Zr, Co, or Fe. In the presence of a catalyst, the hydroboration can display different chemoselectivity as well in addition to changed regioselectivity (see Fig. 3.165) [467–472].

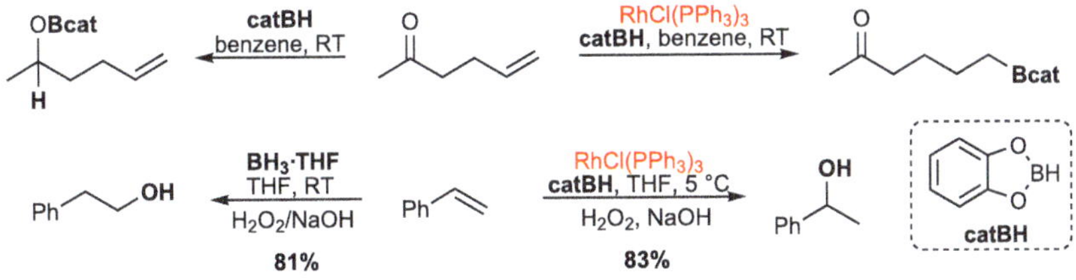

Fig. 3.165 Impact of catalyzed hydroboration on the chemo- and regioselectivity [473, 474]

The deviating chemo- and regioselectivity can be rationalized by the differences in the mechanism of the catalyzed hydroboration. In the presence of the Wilkinson catalyst [RhCl(PPh$_3$)$_3$] (**239**), the borane oxidatively adds in the initial step. Following the coordination of the alkene *trans* to the equatorial Cl ligand in **241**, the olefin is inserted into the metal-hydride bond and **243** is obtained. The steric and electronic properties of the alkene, the identity of the borane (cf. Fig. 3.163) as well as the catalyst determine the regioselectivity of the insertion and thus of the final product. While the uncatalyzed hydroboration exclusively affords the *anti*-Markovnikov product, the catalyzed version can be induced to furnish either the Markovnikov or *anti*-Markovnikov alkyl boranes. Unfunctionalized terminal olefins typically provide the unbranched product (catalyzed and uncatalyzed). In the case of vinyl arenes, carefully controlled (oxygen-free) conditions induce the branched isomer to be the major product in the presence of most catalysts while the linear isomer predominates in the noncatalyzed hydroboration. The reductive elimination, which was postulated to be the rate-determining step based on theoretical studies, affords the addition product and regenerates the catalytically active species **240**. Depending on the substrate and the reaction conditions, the initial steps of the catalytic cycle (**240**→**241**→**242**) may be reversible (see Fig. 3.166) [470, 471].

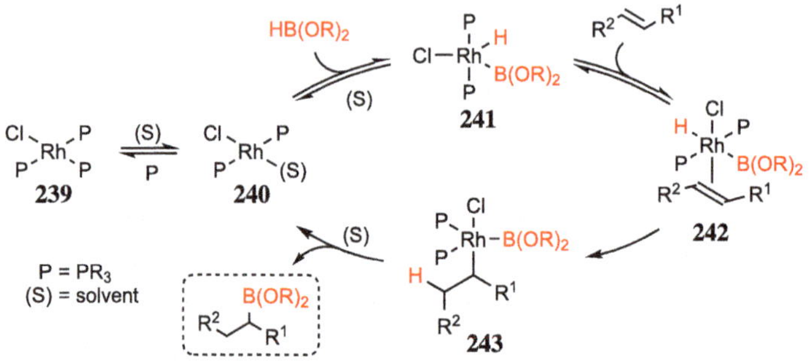

Fig. 3.166 Postulated mechanism of catalytic hydroboration [470, 471]

Despite the great advantages of a catalytic reaction, such as the possibility of an enantioselective, catalytic variant, the transformation utilizing the most common catalyst/organoborane combination (Rh-cat/catecholborane) faces a number of challenges: The reaction is extremely sensitive to the presence of water and oxygen. Unwanted oxidation of the catalyst or ligand can significantly influence the regioselectivity (and enantioselectivity). Iridium catalysts are somewhat more robust against oxidative degradation in this respect [468]. Difficulties arise with unreactive substrates, and thus the major limitation of the catalytic hydroboration is the substrate scope. The most commonly employed substrates are vinyl arenes, 1,1-disubstituted olefins and cyclohexene derivatives [471].

The uncatalyzed reaction with 9-BBN is mainly based on steric interactions to control the regioselectivity. Under catalytic conditions, Lewis basic functionalities can additionally be exploited for a directed borane transfer and even overcome the inherent steric induction of the substrate [468].

In the synthesis of complex natural products, the uncatalyzed variant is typically used for diastereoselective transformations. While the hydroboration is regularly employed in combination with a subsequent basic oxidation to the corresponding alcohol, variations with a successive cross-coupling of the respective alkyl borane addition products have also been routinely reported (see Fig. 3.167).

Beyond diastereoselective reactions, chiral organoboranes have furthermore enabled an entry into enantioselective hydroborations. Reagents such as α-pinene-derived isopinocampheylboranes (IpcBH$_2$, Ipc$_2$BH) or the C_2-symmetric Masamune borane (**245**) provide the greatest enantioinduction [483, 484]. Geminal disubstituted alkenes remain challenging.

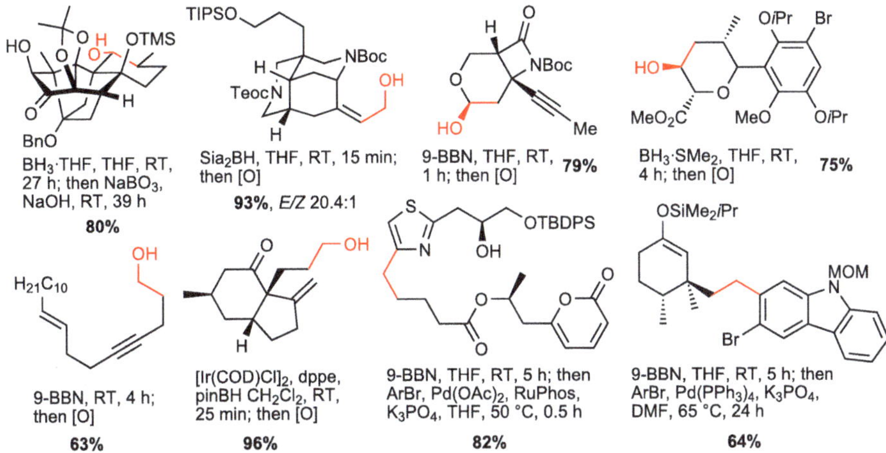

Fig. 3.167 Applications of hydroboration in total syntheses [475–482]. Unless otherwise stated, oxidation was carried out using H_2O_2/NaOH

However, Soderquist and co-workers disclosed boranes **246**, **247**, which were primarily developed for this class of substrates (see Table 3.15) [485].

Table 3.15 Enantioselectivity (in *% ee*) of the hydroboration of different olefin classes [485, 486]

alkene	IpcBH$_2$	Ipc$_2$BH	Masamune	**246**	**247**
	73	14	99.5	96	95
	24	99.1	97.6	32	84
	53	15	97.6	74	–
	–	32	1.5	38	52
	5	–	–	78	66

When focusing on the catalytic variants of enantioselective hydroborations, a wide range of ligand classes was examined, and moderate to excellent enantiomeric excess was achieved, typically in combination with Rh-precatalysts. While styrene derivatives usually constitute model substrates for comparing various methods, their application on more complex, and thus representative, substrates is reported much less frequently. The most common ligand classes are based on atropisomeric P,N- or P,P-scaffolds such as BINAP or QUINAP (see Fig. 3.168) [469–472]. In addition to Rh catalysts, reactions mediated by other metals have also been published sporadically, such as systems relying on Fe or Co [468].

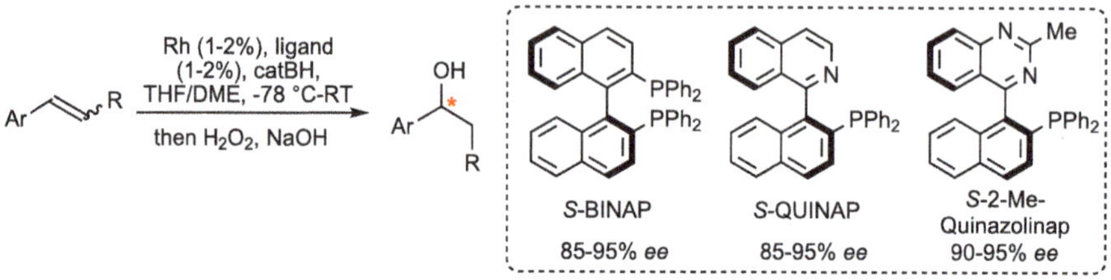

Fig. 3.168 Rh-catalyzed, enantioselective hydroboration of vinyl arenes [470]

Enantioselective hydroborations are rarely encountered in a complex synthetic environment. The facial induction of existing stereocenters in the substrate can often be exploited to allow for a diastereoselective hydroboration (see Fig. 3.169).

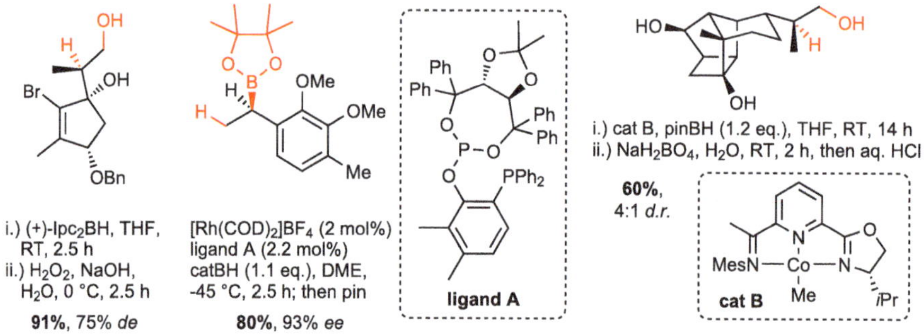

Fig. 3.169 Examples of enantio- and diastereoselective hydroborations in natural product syntheses [487–489]

Other Reactions

Further transformations of olefins beyond the reactions already discussed include hydroaminations, hydroformylations, or ozonolysis, among others.

An ozonolysis requires specialized equipment to generate the ozone gas, which inherently limits the broad applicability of the reaction. Given that the formation of the various possible products can be controlled by judicious choice of the reaction conditions, it has nonetheless established itself as a standard method in total syntheses (see Fig. 3.170) [490].

Reductive conditions during work-up yield aldehydes, ketones, or alcohols depending on the exact reaction conditions. Oxidative conditions lead to ketones or carboxylic acids

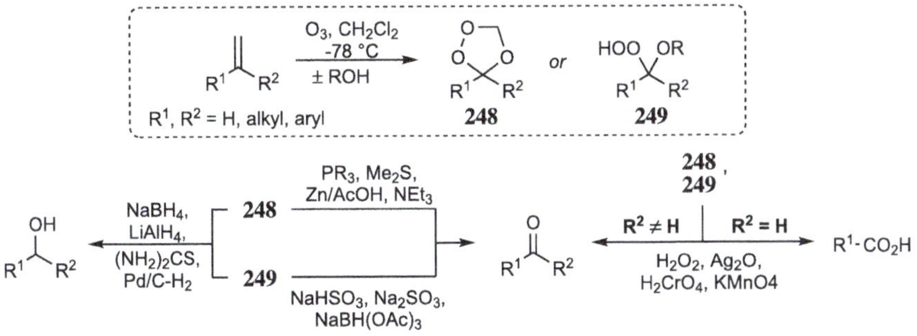

Fig. 3.170 Conditions of ozonolysis and dependence of the products on the reaction conditions of the work-up [490, 491]

[491]. If the work-up is carried out in the absence of reagents to modify the oxidation state, the resulting peroxy intermediates **248**, **249** collapse to provide aldehydes/ketones and carboxylic acids. The mode of fragmentation is determined to a large degree by the properties of the solvent, whereby the heterolytic cleavage is often accelerated by basic conditions [492].

Mechanistically, the ozonolysis constitutes a pericyclic reaction. More precisely, it can be classified as a sequence of a 1,3-dipolar cycloaddition, followed by a cycloreversion. Its mechanism was correctly recognized by Criegee: alkenes act as dipolarophiles and initially form molozonide **250**, which subsequently fragments to **251**. Finally, the carbonyl oxide and aldehyde can recombine to the 1,2,4-trioxolane **248** (sometimes referred to as secondary ozonide). In the presence of nucleophiles such as alcohols, the hydroperoxyacetal **249** is obtained instead of the 1,2,4-trioxolane. Starting from these labile intermediates, the formation of the target product can be similarly steered by choosing the requisite work-up conditions (see Fig. 3.171) [490, 493].

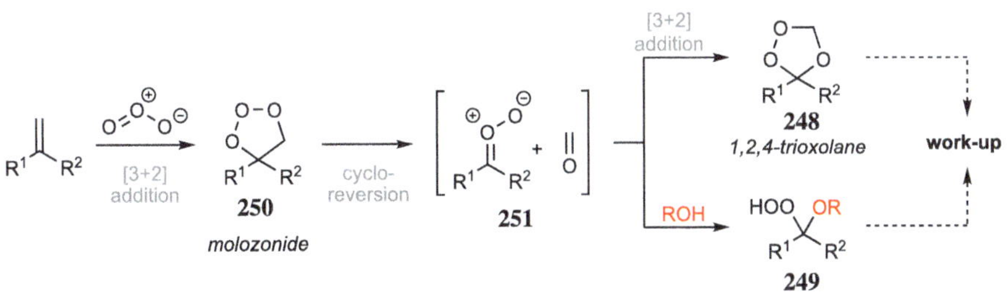

Fig. 3.171 Criegee mechanism of ozonolysis [490, 493]

The more electron-rich the double bond, the faster the reaction of the alkene substrates proceeds. The intrinsically high reactivity of ozone can limit the usefulness of an ozonolysis when working with densely functionalized substrates or polyenes, as many functional groups can be oxidized with O_3 and an "overoxidation" of the substrate may thus occur. The use of an indicator dye is an elegant way to monitor the reaction. Upon reaching the end of the desired transformation, the added dye is decomposed by the ozone and indicates that the reaction needs to be terminated, lest an overoxidation ensues (see Fig. 3.172) [494].

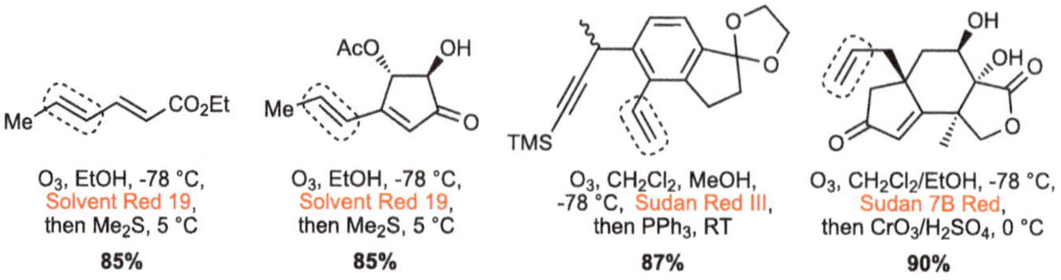

Fig. 3.172 Selective ozonolysis in the presence of an indicator dye [494–496]

In the synthesis of complex natural products, an ozonolysis is typically employed rather early in a synthetic sequence to avoid affecting arene moieties or other oxidation-prone functional groups. Nevertheless, some impressive examples have been reported wherein the transformation could be used to great effect in highly functionalized intermediates (see Fig. 3.173).

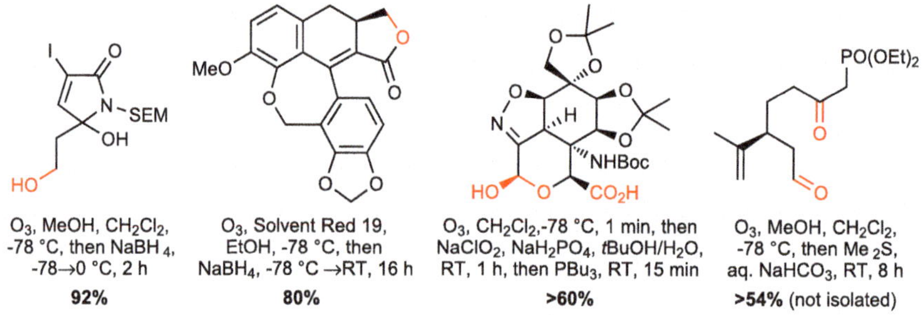

Fig. 3.173 Ozonolysis of complex intermediates in total syntheses [497–500]

In addition to the challenges of chemoselectivity, process safety plays a critical role especially on larger scale. Extensive calorimetric studies to identify the energy content of all intermediates, especially the 1,2,4-trioxolane, as well as a dose-controlled work-up enable a safe reaction in the kg range [490, 501].

Ragan and co-workers required larger amounts of an indanone-derived aldehyde for a reductive amination, the substrate of which was retrosynthetically disconnected to allylic alcohol **252**. A direct ozonolysis with reductive work-up (Me$_2$S) yielded a mixture of dimeric hemiacetals. Following the formation of the hydroperoxyacetal **253**, the bisulfite adduct **254** was specifically isolated to circumvent this product mixture and drive the reaction to completion. **254** could be used directly in the subsequent reductive amination as the protected aldehyde. The synthetic sequence was successfully scaled up to 2.3 kg (see Fig. 3.174) [502].

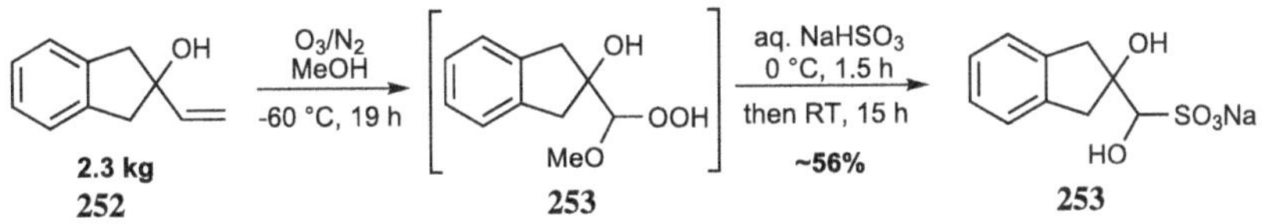

Fig. 3.174 Large scale application of an ozonolysis [502]

The **hydroformylation** or oxo synthesis assumes a particularly prominent role in an industrial environment and is even used on a million-ton scale, for example, in the synthesis of butanal from propene. It is one of the few homogeneously catalyzed processes for the production of bulk chemicals. The regioselectivity of the process (*n*- vs. *iso*-aldehyde) is a pivotal criterion and can be well controlled by choosing the proper conditions. Typically, industrial processes are catalyzed by Rh- or Co-phosphines. The pressure and temperature may vary in the range of 15–350 bar and 60–200 °C, respectively. If a high selectivity has first priority, process conditions with lower temperatures and pressures are resorted to, yet often at the expense of the turnover rate. If a stereocenter is introduced over the course of the transformation, the possibility of an enantioselective hydroformylation opens up, for example, as in the case of the conversion of terminal alkenes to *iso*-aldehydes (see Fig. 3.175) [503–505].

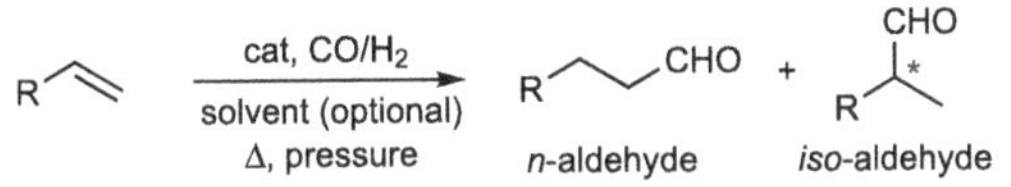

	$[HRhCO(PPh_3)_3]$	$[HRh(CO)_4]$	$[HCo(CO)_3PR_3]$	$[HCo(CO)_4]$
pressure [bar]	15-20	200-300	50-100	200-350
temperature [°C]	80-120	100-140	160-200	110-180
n/i ratio	92:8	50:50	88:12	80:20
hydrogenation	low	low	high	medium

Fig. 3.175 General reaction scheme of hydroformylation and comparison of reaction conditions typically employed at large scale [503]

The co-catalyzed process is still used on an industrial scale despite the harsher, more energy-intensive conditions. One of its main benefits lies in the heightened resistance of the catalyst towards catalyst poisons that may, for example, be contained in the raw materials. Especially mixtures of internal and branched long-chain alkenes are successfully hydroformylated under these conditions [504]. If the corresponding alcohols are the required products, a Co-mediated process is more suitable since the catalyst tends to partially hydrogenate the aldehyde hydroformylation products during the process [503]. In addition to PPh_3 or phosphites as ligands favored for large-scale applications, academic methods typically rely on bidentate ligands in conjunction with Rh-precatalysts (see Fig. 3.176).

BiPhePhos was originally developed by Union Carbide for the hydroformylation. Derived from this ligand framework, other companies commercialized processes based on variations thereof [504]. In the academic field, the hydroformylation is often encountered as part of

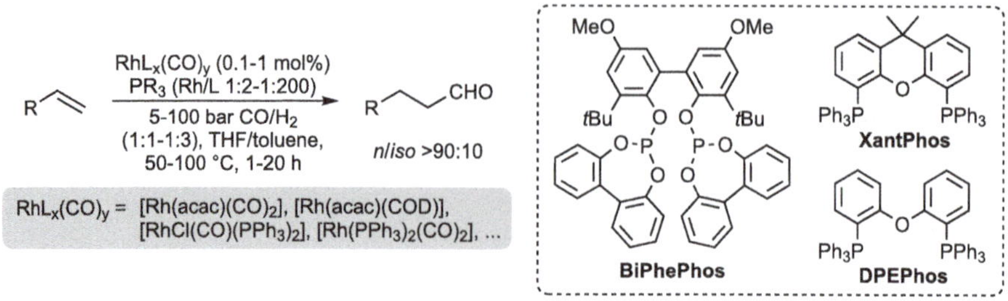

Fig. 3.176 Typical reaction conditions of hydroformylation [504]

a domino reaction, foremost in domino-hydroformylation/amination sequences [506]. In order to access the enantiomerically enriched products, the structure of the ligands lends itself to the switch to chiral phosphines. This has been shown to significantly alter or even reverse the *n/iso*-selectivity in some instances (see Fig. 3.177).

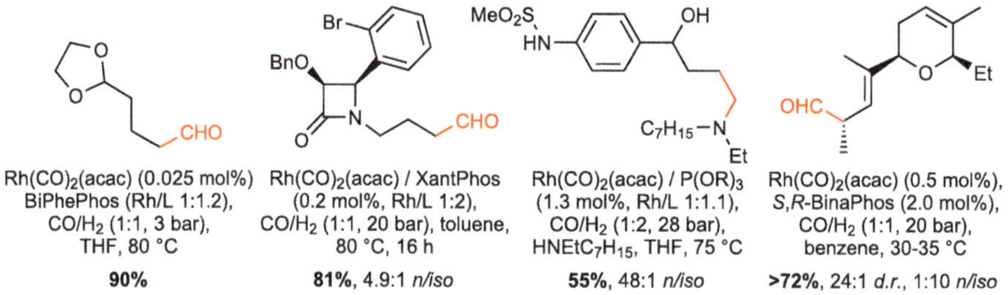

Fig. 3.177 Applications of hydroformylation [92, 507–509]

The **hydroamination** constitutes another class of derivatization reactions, which entails the addition of amines to alkenes or alkynes. It is thermally forbidden as a direct [2+2]-cycloaddition and thus incurs a high reaction barrier, but metal-mediated variants allow an efficient transformation. The reaction is challenging insofar as amines generally also embody good ligands for a variety of catalysts. The majority of methods can be classified according to the two dominant approaches: Either via activation of the alkene (**255**), which is usually mediated by late transition metals. Alternatively, the amine component can be activated (**256**), typically achieved in the presence of early transition metals, actinides, and lanthanides (see Fig. 3.178) [510–514].

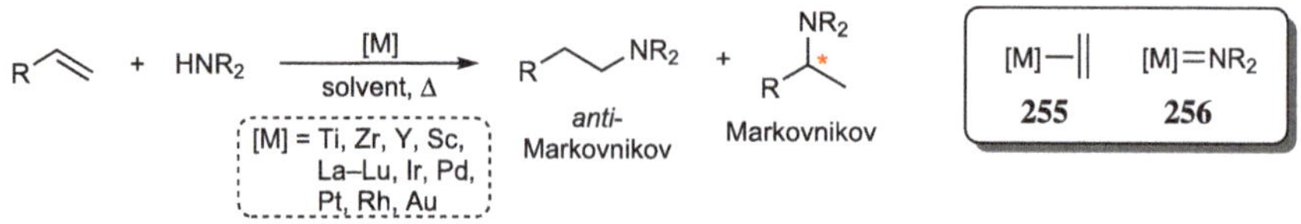

Fig. 3.178 Reaction scheme of hydroamination

Many catalysts display a very specific substrate scope whose conversion they can efficiently promote; only few are suitable for a wider range of reactants. Especially intramolecular hydroaminations have been employed in the synthesis of various alkaloid frameworks and pharmacologically active molecules. Most applications rely on achiral catalysts, even though a substrate-controlled reaction can afford enantiomerically pure products (see Fig. 3.179) [511].

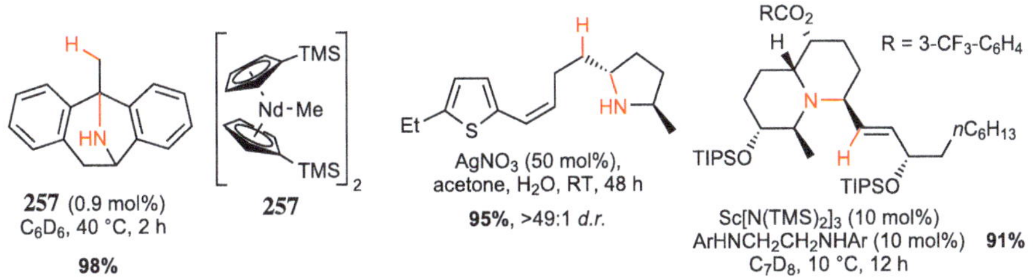

Fig. 3.179 Examples of hydroaminations in alkaloid syntheses [515–517]

The asymmetric hydroamination of olefins developed by Buchwald uses chiral Cu complexes, mirroring the conditions usually employed in copper hydride-mediated 1,4-additions. Suitable substrates include styrene derivatives as well as terminal alkenes, vinyl silanes, and boranes, as well as dienes or allenes. The amine component is derived from hydroxylamine esters, in which the carboxylate constitutes the leaving group. NH_2 as well as primary and secondary amines can be transferred. Silanes serve as the hydride source (see Fig. 3.180) [518].

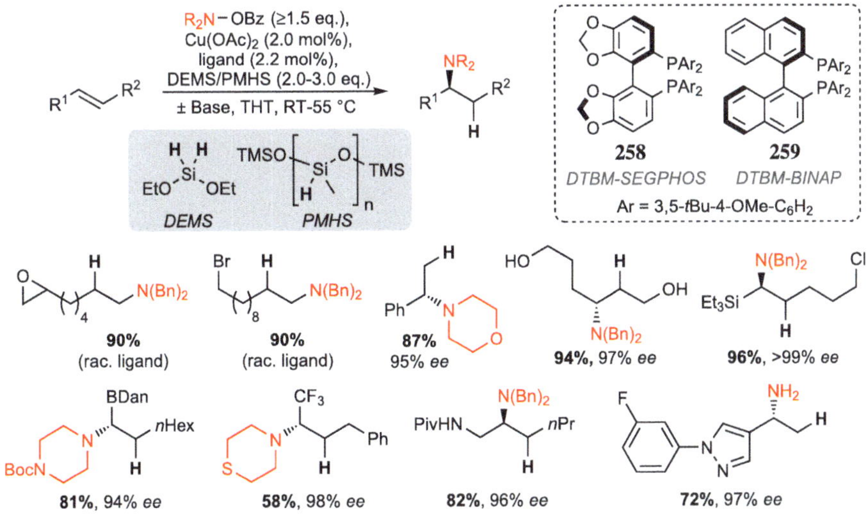

Fig. 3.180 Asymmetric hydroamination developed by Buchwald *et al.* [518]

After the hydrocupration of the alkene by a Cu-H species, the chiral alkyl copper intermediate reacts with the electrophilic amination reagent and subsequently dissociates the hydroamination product. The silane regenerates the active CuH species. In the case of terminal alkenes, the *anti*-Markovnikov product is obtained exclusively, whereas the hydroamination of styrenes selectively affords the benzylamine Markovnikov-regioisomer. The CuH-mediated reaction is thermodynamically driven by the energy gain of the Si-X bond formation from the hydrosilane [518].

In analogy to the hydrofunctionalization of alkynes (*vide infra*), radical **hydrofunctionalization** reactions of alkenes, which are routinely referred to as hydrogen atom transfer (HAT) reactions, have also been disclosed. While these are now mostly mediated by photocatalytic processes (cf. Chap. 10), also aerobic oxidations employing Mn-, Fe- and Co-catalysts have been reported (see Fig. 3.181) [519]. Despite some impressive examples, they have however yet to reach their full potential in synthetic applications.

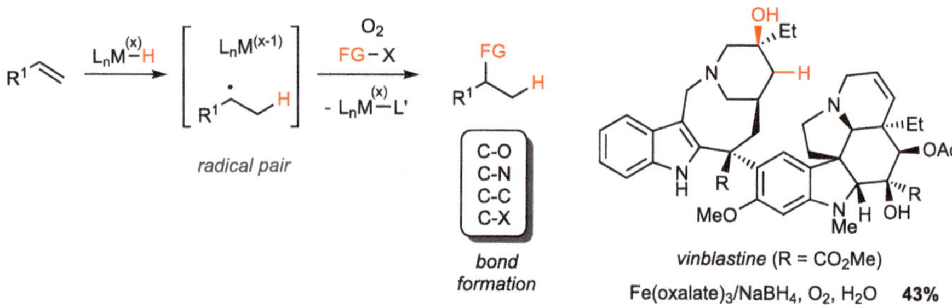

Fig. 3.181 General scheme of aerobic hydrofunctionalizations and application in Boger's synthesis of vinblastine [519, 520]

Further Transformations of Derivatized Olefins

The products obtained by the above methods allow the rapid access to a variety of downstream compounds with only a few reactions. Epoxides can be cleaved by nucleophiles (azide, carboxylate, alkoxylate, thiolate, chloride, etc.), 1,2-diols can be converted to epoxides or alkenes, and aziridines can be derived from amino alcohols (see Fig. 3.182).

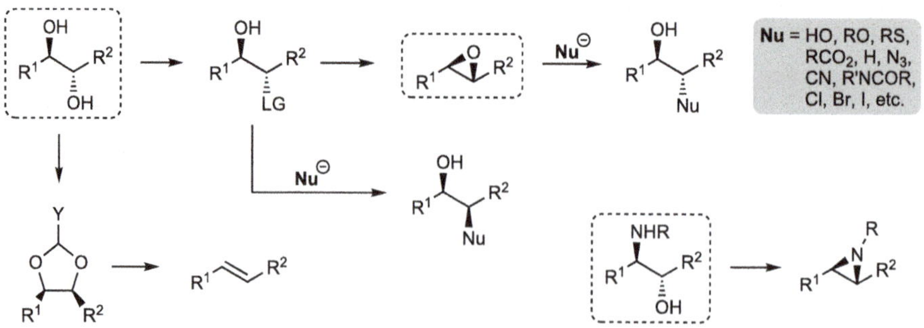

Fig. 3.182 Further reactions of primary products of olefin oxidations [250]

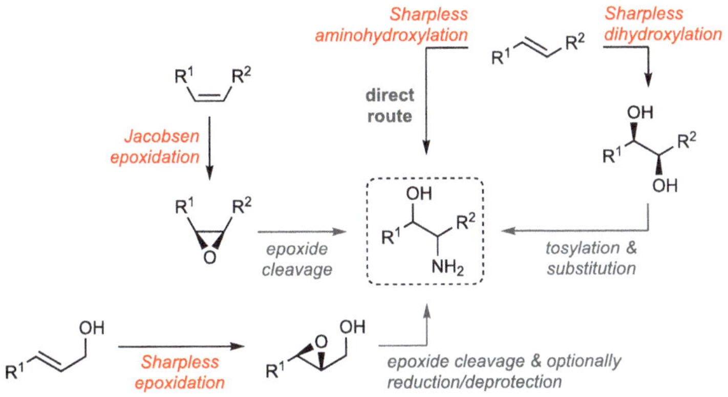

Fig. 3.183 Various asymmetric routes to 1,2-amino alcohols [250]

Apart from 1,2-diols and epoxides, particularly hydroxy esters, amino alcohols and amino esters can be obtained via different routes using the outlined approaches (see Fig. 3.183). Despite the fact that methods for the direct routes exist — this additional effort of a multi-step, alternative synthesis route can be justified if (a.) the starting material of an initially obvious route is not available or can only be produced with great effort, (b.) high *ee* values are not afforded by an asymmetric method, but can be with an alternative approach, and (c.) if incompatibilities with functionalities occur in a given route. In an industrial context, additional criteria such as process safety, efficient use of available equipment, or the large-scale availability of the raw materials need also be considered.

The entry to *syn*- or *anti*-1,2-chloroalcohols can be realized, for example, by selective opening of diastereomerically pure epoxides. Carreira and co-workers synthesized the densely-funtionalized secondary metabolite undecachlorosulfolipid A, which, in addition to the enantioselective introduction of several 1,2-dichloro moieties, also comprised the selective access to two *syn*-1,2-chlorohydroxy units from the corresponding epoxides. The requisite oxirane in **260** was synthesized using a Sharpless dihydroxylation. The epoxide could be opened diastereoselectively and the advanced intermediate **261** was isolated with acceptable yield (see Fig. 3.184) [306].

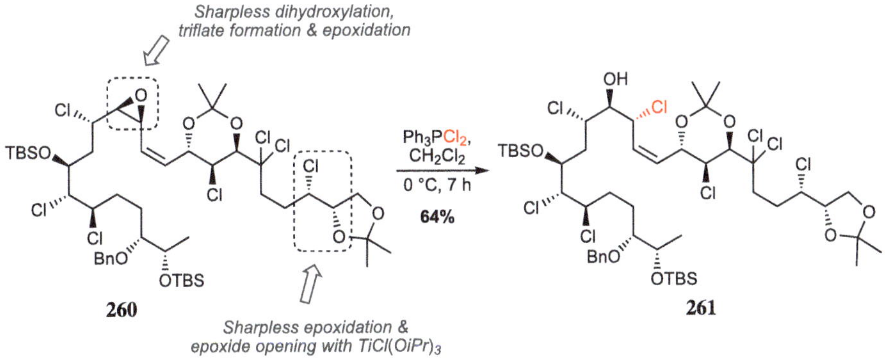

Fig. 3.184 Epoxide opening in the synthesis of undecachlorosulfolipid A [306]

3.3.2   Oxidation of Alkynes

The derivatizations of alkynes possess a narrower range than the broad spectrum of transformations available for alkenes. Although the reactivity of the corresponding CC double and triple bonds is similar, only a small proportion of methods for alkenes have been transferred to alkynes and established themselves as a standard in synthesis. Furthermore, new methods such as alkyne metathesis have been developed to supplement the repertoire of feasible transformations (see Fig. 3.185).

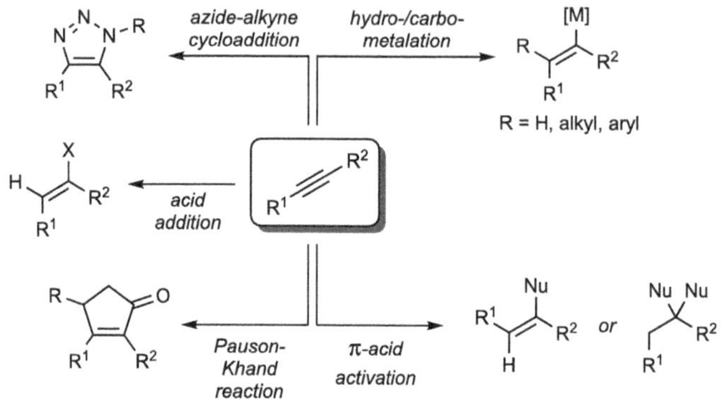

Fig. 3.185 Overview of various derivatizations of alkynes

While hydrometalations are redox-neutral with respect to the CC-triple bond, they are included among the oxidative derivatizations since they are usually used to selectively oxidize one of the carbon atoms of the alkyne. A carbometalation is preferably carried out with palladium complexes, but organoaluminum reagents can also transfer an alkyl or aryl group to generate a vinyl aluminum species. Starting from these intermediates, the carbon-metal bonds can be further derivatized by transmetalations, cross-couplings, or direct reactions with nucleophiles or electrophiles. The azide/alkyne cycloaddition and metathesis variants involving the alkyne are covered more extensively in Sects. 5.2 and 7.3, respectively.

When the alkyne serves as nucleophile, its reactivity generally correlates with the electron density of the triple bond. Nevertheless, steric factors may prevent a reaction despite auspicious boundary conditions, for example, due to energetically very unfavorable transition states or highly strained products.

Hydrometalation, Carbometalation, and Dimetalation

Alkynes can be efficiently converted to the corresponding vinyl derivatives in the presence of (transition) metal complexes [521–523]. The coordination of the metal center to the Lewis basic triple bond activates it for the transfer of a variety of metal-bound substituents. Depending on the reagents used, the addition of a catalyst may enable the conversion of otherwise inert substrates to the respective metal vinyl species. Some of these are stable and can be isolated (M = Si, Sn, B), whereas more reactive organometallic compounds (M = Zr, Pd, Al) can be further derivatized *in situ* upon exposure to a variety of (electrophilic) reaction partners.

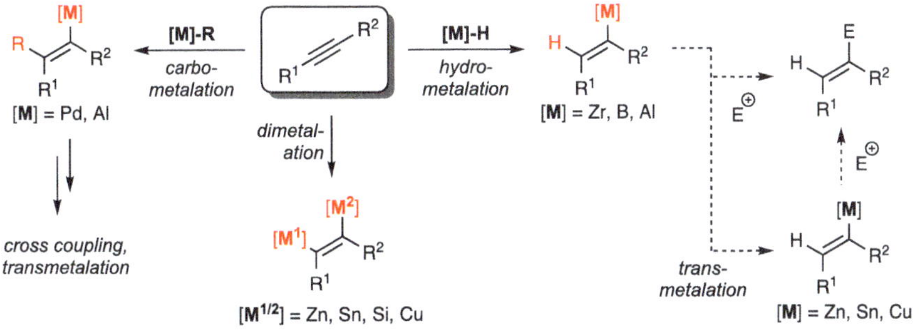

The addition of (almost) all organometallic reagents to alkynes proceeds in a concerted manner, resulting in stereospecific *syn*-addition reactions.

The addition of transition metal hydrides to alkynes proceeds without the need for a catalyst and affords metal vinyl intermediates. Of all the hydrometalation reactions, the **hydrozirconation** with zirconocenes is probably the most extensively used transformation for the diastereoselective derivatization of alkynes [524–526]. The standard Schwartz reagent (Cp$_2$ZrClH) is usually freshly prepared given its limited stability [527]. Various one-pot procedures (conditions B) have been introduced over the years that rely on an *in situ* Zr activation (see Fig. 3.186) [528].

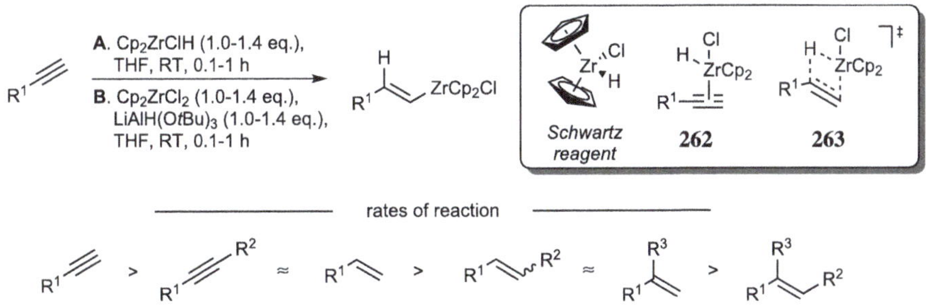

Fig. 3.186 Common conditions in hydrozirconations, their mechanism, and relative reactivity of alkynes/alkenes [525]

Mechanistically, the reaction is typically initiated by the coordination of the alkyne to the metal center to afford the π-complex **262**, followed by a *syn*-addition via transition state **263**. The reactivity of different CC-multiple bonds is largely determined by steric factors, so that the reaction rate of substituted alkynes and alkenes decreases with an increasing degree of substitution. Alkynes generally tend to be more reactive than olefins due to their higher electron density and reduced steric demand [525].

The oxophilic, hard Lewis acidity of the organozirconocene reagents limits the range of functionalities that remain stable under the reaction conditions. Aldehydes, ketones, amides, nitriles, epoxides, as well as various esters are similarly reactive as the alkyne and thus should be avoided in the substrate. Silyl-, benzyl- and *t*butyl ethers and esters are tolerated, as are unprotected hydroxy groups, sulfonamides, carbamates, acetals, halides, and even alkenes, if strictly stoichiometric conditions are adhered to. For NH- or OH-acidic functionalities, an additional equivalent of zirconocene hydrochloride is consumed for their deprotonation [526].

The regioselectivity is primarily determined by steric interactions, resulting in the C-Zr bond formation to occur at the sterically less hindered terminus with unsymmetric and terminal alkynes. If only a moderate regioselectivity is obtained under kinetic control, the addition of small amounts of zirconocene can facilitate an equilibration to the thermodynamically preferred product, further increasing regioselectivity (see Table 3.16) [529]. In the hydrozirconation of propargyl alcohols, the presence of $ZnCl_2$ can furnish the higher substituted regioisomer with excellent selectivity due to a directed reaction enabled by the hydroxy group [530].

Table 3.16 Regioselectivity of hydrozirconations under kinetic and thermodynamic control by subsequent addition of Cp_2ZrClH [529]

		ratio **264 : 265**	
R_L	R_S	conditions i.	conditions i.+ii.
*n*Bu	H	>98 : 2	–
Et	Me	55 : 45	89 : 11
*n*Pr	Me	69 : 31	91 : 9
*i*Bu	Me	55 : 45	>95 : 5
*i*Pr	Me	84 : 16	>98 : 2
*t*Bu	Me	>98 : 2	–

A challenge of vinyl zirconocenes is their limitation to the reaction with sterically less demanding electrophiles if one wants to achieve synthetically useful reaction rates [531]. When successful, however, the corresponding addition products are isolated without erosion of the *E/Z* ratio. Particularly, the reaction with halogenation reagents (Br_2, I_2, NCS, NBS, $PhICl_2$) is used for the diastereoselective access to vinyl halides (see Fig. 3.187).

Furthermore, carbon monoxide and isonitriles are also successfully converted by vinyl zirconocenes to give the acyl derivatives or nitriles/aldehydes, respectively. With carbon monoxide, an insertion of CO into the C-Zr bond occurs to generate an acyl zirconium species. These can be subsequently converted to the corresponding aldehydes, carboxylic acids, esters, or acid chlorides [525].

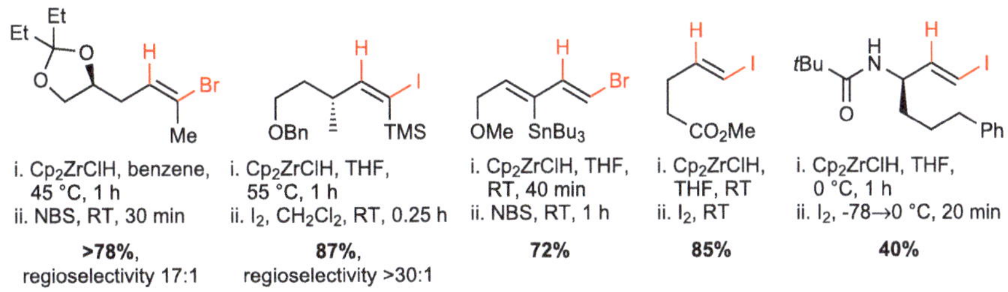

Fig. 3.187 Examples of the hydrozirconation/halogenation of alkynes [532–536]

Since the highly nucleophilic organozirconocenes are very sensitive to the steric environment of electrophilic component, their broad applicability is based on the ease of transmetalation to other alkenyl metal complexes that do not suffer from these steric constraints in the same manner. Especially Zn, Sn, and Cu offer a wide range of applications in cross-couplings, as well as 1,2- and 1,4-addition reactions (see Fig. 3.188) [531].

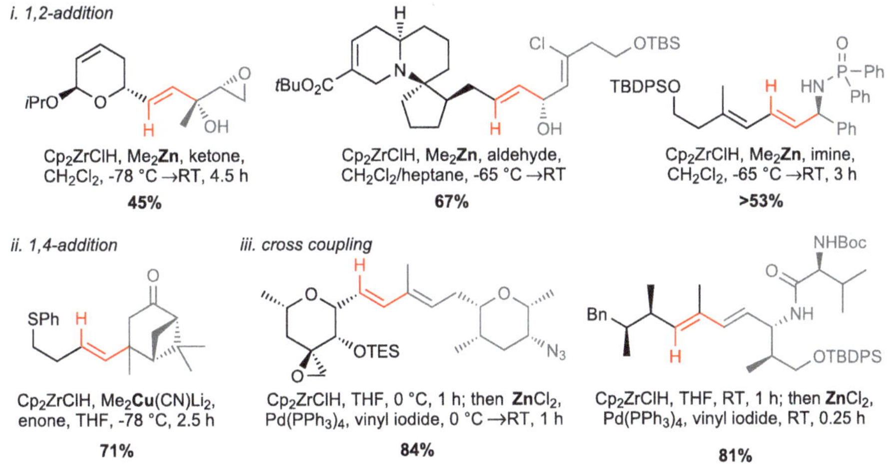

Fig. 3.188 Examples of hydrozirconation/transmetalation sequences. The alkenyl functionality derived from the alkyne is marked in red, and the reaction partner of the alkenyl zirconocene is gray [537–542]

The **hydroboration** of alkynes shows a higher regioselectivity with unsymmetric substrates compared to the reaction of olefins. The boron adds *syn*-selectively via the 4-membered transition state **266** to terminal and internal alkynes at the less substituted terminus. In the absence of olefinic double bonds, the range of tolerated functional groups corresponds to the hydroboration already discussed for alkenes (see Fig. 3.189) [467, 543].

The reaction is encountered much less frequently with alkynes than with alkenes, as substrates that bear both alkene and alkyne moieties are derivatized more rapidly at CC double bond with most borane reagents (see Table 3.14). Disiamylborane is a notable exception,

Fig. 3.189 Hydroboration of alkynes and subsequent transformations

even enabling enynes to be selectively converted (see Fig. 3.190) [544]. Dicyclohexylborane and thexylborane also preferentially add to the triple bond in enynes [545].

	$nHex$⟍	nBu⟍	Et⟍Et	Et⟍Et	Et≡Et
9-BBN	1.00	0.15	0.006	0.01	0.007
Sia$_2$BH	1.00	3.45	0.02	0.002	2.08

Fig. 3.190 Relative reactivity of various hydroboration reagents [544]

Mechanistically, the reactions of both alkenes and alkynes are closely related: The rate-determining step in the uncatalyzed hydroboration of alkynes is similarly assumed to be the dissociation of the dimeric borane [462]. The stereospecific *syn*-addition exclusively affords *E*-alkenylboranes. These are subsequently either oxidatively converted to ketones or participate in Suzuki cross-couplings to yield the substituted olefins (see Fig. 3.191).

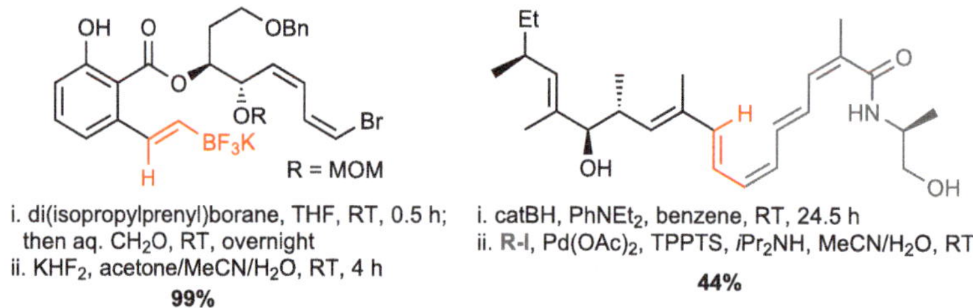

i. di(isopropylprenyl)borane, THF, RT, 0.5 h;
   then aq. CH$_2$O, RT, overnight
ii. KHF$_2$, acetone/MeCN/H$_2$O, RT, 4 h
99%

i. catBH, PhNEt$_2$, benzene, RT, 24.5 h
ii. R-I, Pd(OAc)$_2$, TPPTS, *i*Pr$_2$NH, MeCN/H$_2$O, RT
44%

Fig. 3.191 Selected hydroborations of alkynes [546, 547]

A particularly elegant variant was employed by the Fürstner group in their synthesis of the alkaloid xestocyclamine A (**272**). The enyne **268** was hydroborated by 6 equivalents of 9-BBN at the terminal alkene and internal alkyne to give **269**. The selective protonolysis of the more labile vinyl borane moiety yielded the *Z*-alkene **270**, which in the presence of a Pd complex closed the 11-membered macrocycle by means of an intramolecular alkyl-Suzuki cross-coupling to afford **271** in the one-pot reaction in 48%. The concluding reduction of the amide functionality with concomitant deprotection of the silyl ether provided the natural product **272** (see Fig. 3.192) [548].

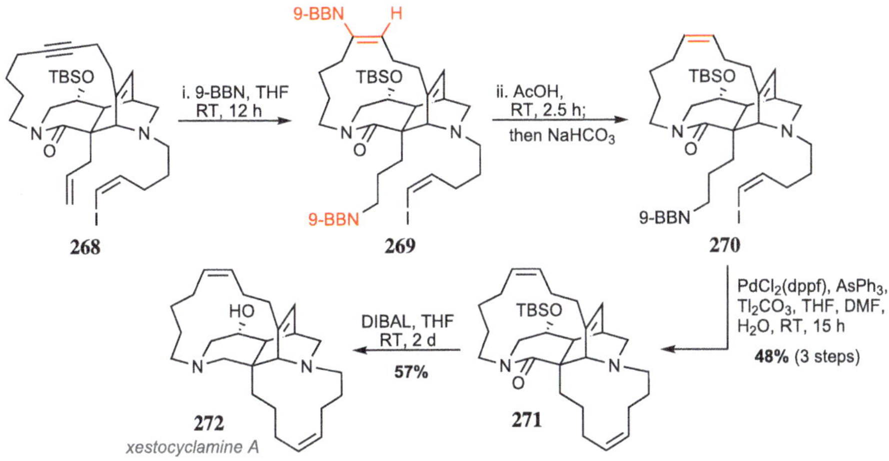

Fig. 3.192 Synthesis of xestocyclamine A (**272**) reported by Fürstner *et al.* [548]

The hydroboration of alkynes can also be mediated by transition metal catalysts [549]. In addition to Rh and Ir catalysts, Cu complexes bearing carbene or phosphine ligands have been disclosed as efficient catalysts in recent years [550]. With terminal alkynes as substrates, Cu/NHC catalysts facilitate an attack of boron at the more substituted terminus of the triple bond, contrary to the expected regioselectivity obtained with the majority of catalysts or in the absence thereof. Depending on the reaction conditions, either the β- or the α-alkenyl borane is isolated (see Table 3.17) [551].

Table 3.17 Influence of NHC ligands on the regioselectivity of the catalyzed hydroboration [551, 552]

R	β-selective	α-selective
CH$_2$OR, CH$_2$NR$_2$	**276** (89:11)	**273** (83:17–98:2)
aryl	**274** (>93:7)	**275** (78:22–96:4)
alkyl	**274** (>81:19)	–

The mechanism of the Cu-catalyzed hydroboration is postulated to comprise an initial boracupration of the alkyne by **277** with subsequent stereoretentive protonolysis of the intermediate vinyl metalates. Following the formation of the C-B and C-Cu bonds, the regioisomers **278** or **279** can furnish the corresponding β-/α-selective products. The Cu-alkoxide complex **280** resulting from methanolysis regenerates the catalytically active species **277** by a concluding reaction with B_2pin_2 (see Fig. 3.193) [553, 554].

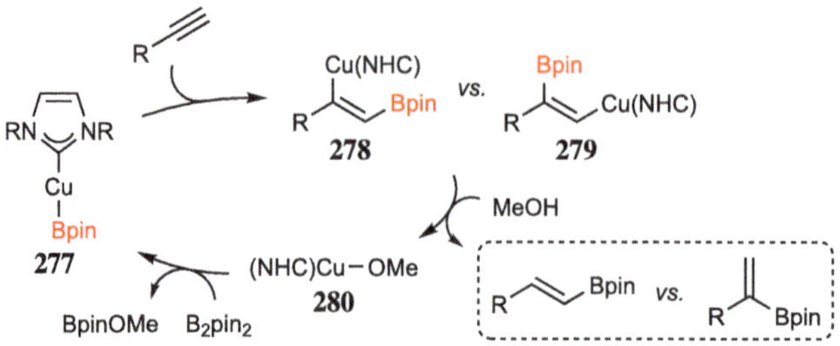

Fig. 3.193 Postulated mechanism of the Cu/NHC-catalyzed hydroboration [553, 554]

In Yadav's synthesis of the cytotoxic natural product EBC-23, a highly regioselective hydroboration was chosen as a key step to construct the central ketone functionality. The Cu-catalyzed derivatization of **281** afforded the desired vinyl borane **282** with excellent yield and selectivity. In the ensuing basic oxidation, the ketone **283** was also obtained in high yield. The carbonyl group subsequently served as a convenient handle to introduce the spirocyclic ketal function contained in the total synthesis of the desired secondary metabolite (see Fig. 3.194) [555].

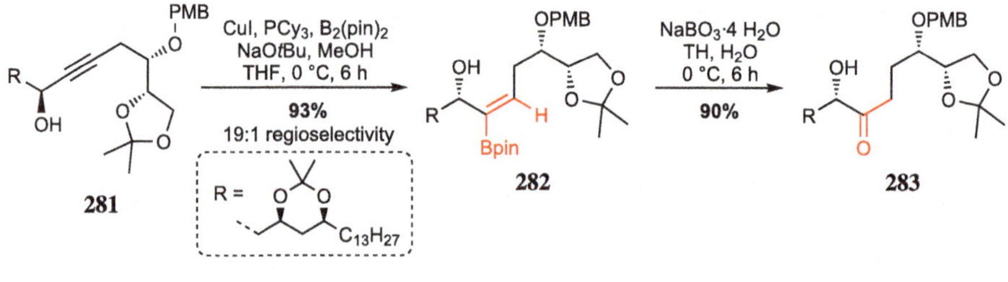

Fig. 3.194 Synthesis of EBC-23 according to Yadav *et al.* [555]

In hydrometalations, the use of alanes plays a limited role [521]. The drawbacks of the narrower substrate scope due to the high reactivity of the organoaluminum reagents and their limited commercial availability likely account for the scarce occurrence of *hydroaluminations* in complex syntheses.

As opposed to hydrometalations, the presence of stoichiometric amounts of trialkylaluminium or Pd catalysis in conjunction with aryl-/alkyl-halides and -sulfonates allow for a **carbometalation** to take place, in which CC- and C-metal bonds are jointly formed. The challenge of the regioselectivity can be overcome by directing groups or the use of symmetric alkynes. With unsymmetric substrates, directing groups are routinely used to avoid obtaining a mixture of regioisomers. Its inducing effect to ensure a selective transformation can be both steric and electronic in nature (passive reaction control) as well as encompass chelating functionalities (active control).

The construction of heterocyclic frameworks can be realized most efficiently by means of a *carbopalladation*. Since no β-H-elimination can occur in the primary products of metalation (vinyl palladium complexes), they can be subsequently converted with a plethora of electrophiles, nucleophiles or even other CC-multiple bonds in a domino reaction (see Fig. 3.195) [556–558].

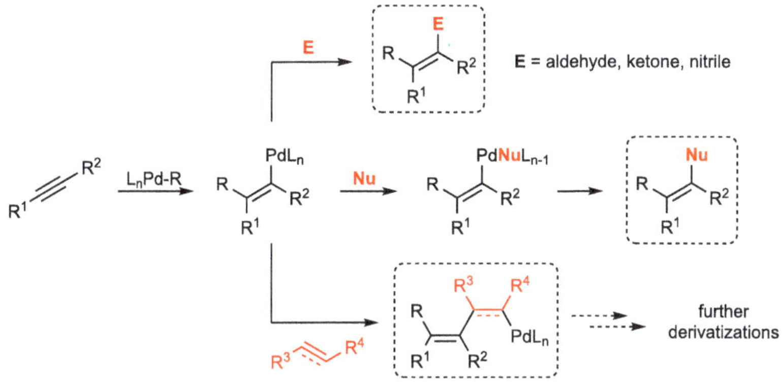

Fig. 3.195 Carbopalladation of alkynes and subsequent transformations

While the carbopalladation constitutes an elemental step in the Heck reaction of olefins, there exist several examples where carbopalladations of alkynes represent a key step in total syntheses. Anderson and co-workers exploited a triple carbopalladation to access the 5/5/7/6/5-fused molecular skeleton in their approach to the secondary metabolite rubriflordilactone A (**286**, see Fig. 3.196). After the initial oxidative addition of Pd into the vinylic C-Br bond, the reaction proceeds smoothly along the two alkyne functionalities and terminates at the initial alkene unit to afford **285** in 91% yield. The final four-step sequence of the total synthesis comprised the deoxygenation of the TBS-protected alcohol, construction of the F-ring, and attachment of the unsaturated γ-butyrolactone moiety [559].

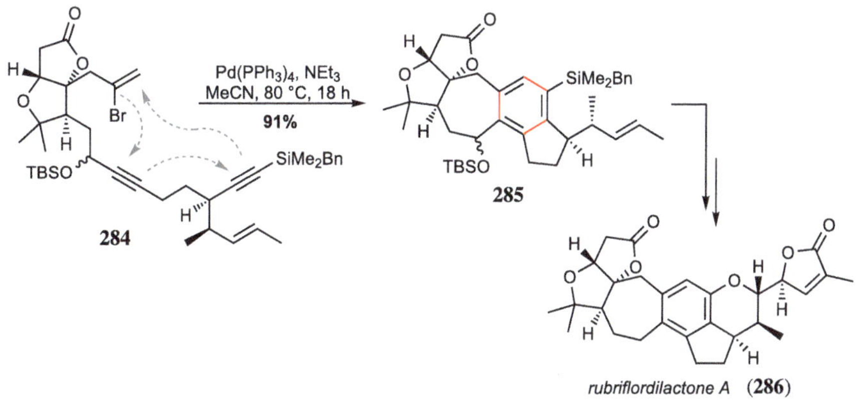

Fig. 3.196 Key step in Anderson's synthesis of rubriflordilactone A (**286**) [559]

Even though a *carboalumination* can be directly carried out by alanes, the reaction is typically conducted in the presence of zirconocene chloride as catalyst [560, 561]. Similar to the Tebbe reagent, the catalytically active species **287** was postulated to be a heteroleptic, bimetallic complex (see Fig. 3.197).

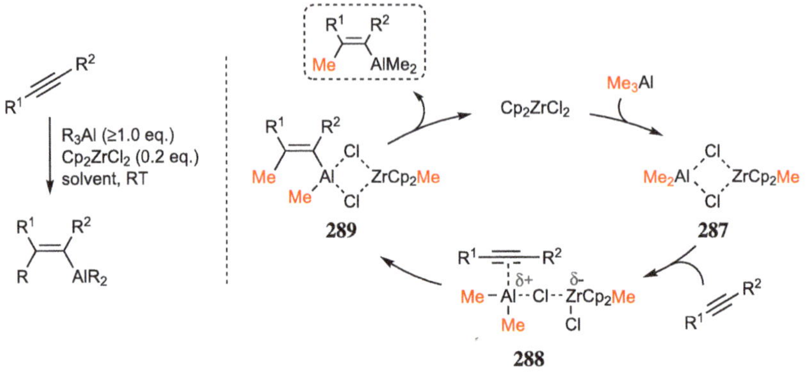

Fig. 3.197 Typical reaction conditions and postulated mechanism of the Zr-catalyzed carboalumination [560]

In analogy to the organometallic intermediates of a hydrometalation, the generated vinyl alanes can subsequently act as nucleophilic component in the direct reaction with electrophiles or in cross-couplings (see Fig. 3.198).

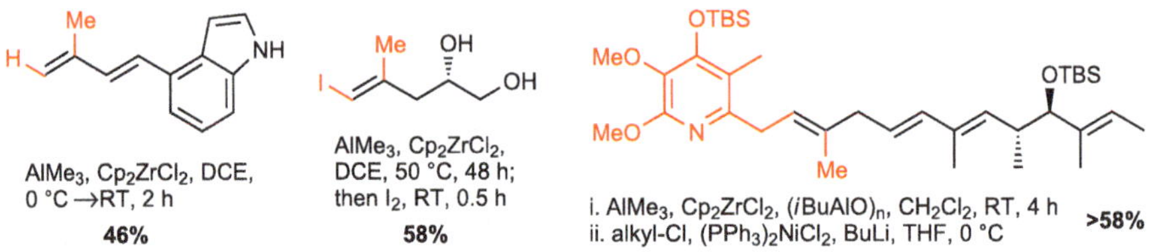

Fig. 3.198 Applications of the carboalumination [562–564]

The **dimetalation** of alkynes can be carried out by a variety of viable metals. Especially distannanes or silastannanes are routinely obtained as stable products in the presence of a Pd-catalyst. Alternatively, Cu- or In-catalyzed vinyl metalates can be formed *in situ* and then reacted with assorted electrophiles such as halogens, epoxides, or carbonyl derivatives (see Fig. 3.199).

π-Acid Catalysis

In contrast to olefins, which preferentially react as nucleophiles, CC-triple bonds can furthermore be activated for a nucleophilic attack through an interaction with late transition metals (**290**), most commonly Au and Pt [567–572]. The primary addition products **291**

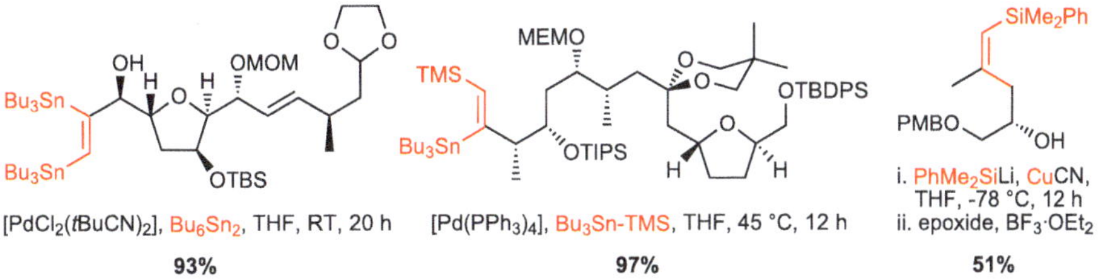

Fig. 3.199 Products of the dimetalation of alkynes obtained in total syntheses [93, 565, 566]

can subsequently be further derivatized via different paths, depending on the stereoelectronic structure of the substrate and the catalyst: by substituting the metal with hydrogen (protodemetalation) or in a straightforward reaction with various electrophiles. Especially vinyl gold compounds can display properties possessed by a cationic or carbenoid species with the corresponding reactivity, which has led to intensive mechanistic investigations and a controversial discussion regarding the "true" nature of these species (see Fig. 3.200) [568, 573, 574].

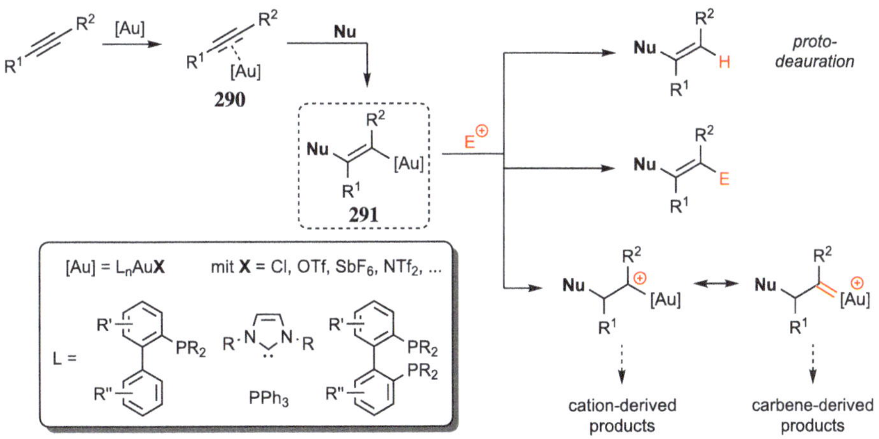

Fig. 3.200 Activation of alkynes and subsequent reactions in the presence of Au complexes

The addition of the nucleophile to **290** proceeds in an *anti* fashion with regard to the alkyne face coordinating the Au catalyst, affording *trans*-substituted metal vinyl species. Starting from the primary intermediate **291**, a protodeauration, protodehalogenation, or direct reaction with electrophiles can ensue. This type of oxidative functionalization mediated by π-acidic transition metals can also be carried out in the presence of alkenes or other sensitive functionalities, with high selectivity for the CC-triple bond. In the absence of alkynes, allenes or occasionally olefins can also be derivatized (see Fig. 3.201).

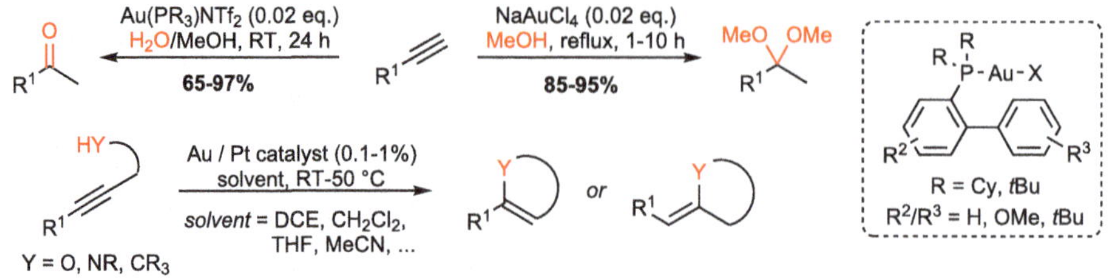

Fig. 3.201 Derivatization of alkynes to ketones, acetals, as well as enol ethers or enamines [569, 570, 575, 576]

Modern Au precatalysts often rely on dialkylbiphenylphosphines as ligands (see Fig. 3.201) and weakly coordinating counterions or silver salts (AgBF$_4$, AgOTf, AgSBF$_6$) as additives for the abstraction of halide ligands. NTf$_2$, SbF$_6$, OTf, and Cl constitute the most common counterions and can impart a decisive influence on the reactivity and selectivity of a reaction with Au catalysts [577]. By virtue of their negligible oxophilicity, particularly Au- and Pt-based systems can activate unsaturated CC bonds in the presence of H$_2$O, alcohols, or other oxygen-containing functionalities. Occasionally, the stronger Lewis acidity of gold complexes can be detrimental for a selective reaction of very densely functionalized substrates, in which cases milder Pt(II) salts are better suited [578]. Apart from oxygen nucleophiles, nitrogen- and carbon-based nucleophiles can also add to the CC-triple bond (see Fig. 3.202) [571, 579].

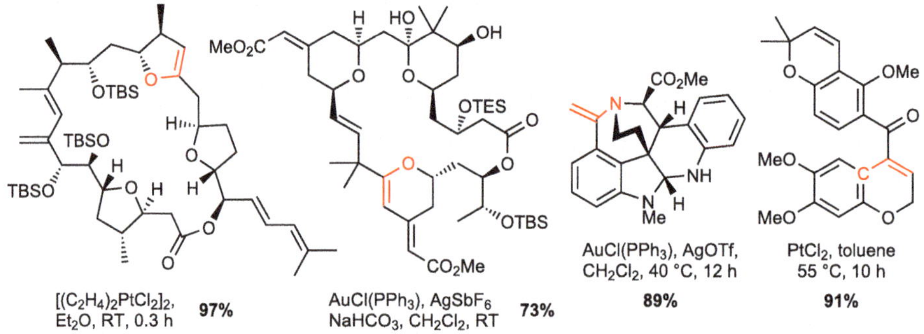

Fig. 3.202 Products of π-acid-catalyzed reactions in total syntheses [580–583]. The former triple bond and the nucleophile are marked in red

In addition to these approaches (hydroalkoxylation, hydroamination, hydrocarboxylation, ketone, and acetal formation), the reaction of a CC-triple bond with alkenes represents a common reaction mode, which furnishes complex rearrangement products, derived from either a cationic or carbenoid behavior (see Fig. 3.203).

The most common reaction motifs exploiting these competing modes are the Conia-ene reaction and enyne cycloisomerizations [578, 584]. The majority of examples correspond to an intramolecular reaction, with the product distribution correlating with the Baldwin rules for ring closure reactions (see Sect. 1.1). Mechanistically, in the Conia-ene reaction, an attack of the enol or enolate on the activated alkyne (**292**) is assumed to occur, forming

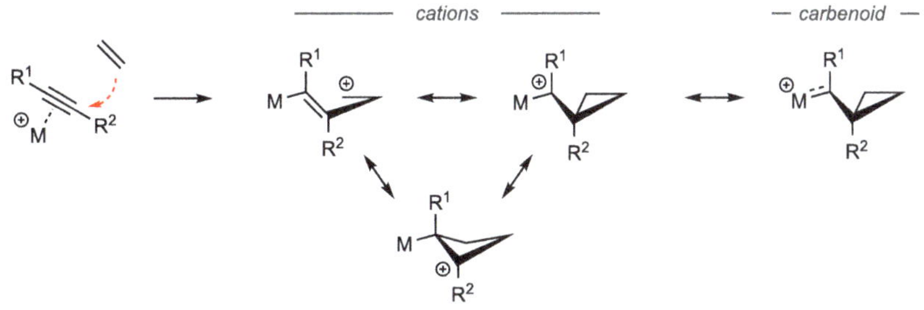

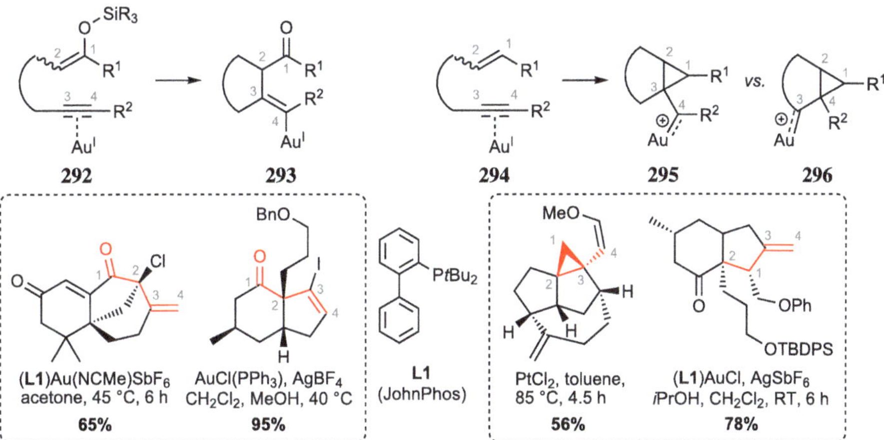

Fig. 3.203 Competition of cationic and carbene intermediates in the reaction of enynes [579]

the β,γ-unsaturated ketone intermediate **293** [585]. In contrast, the reaction of an alkene as a nucleophile with the CC-triple bond affords two regioisomers **295** or **296**, depending on the orientation of the alkyne in the addition. These can subsequently behave as either Au cation or carbene, as represented by the valence structures depicted in Fig. 3.203 (see Fig. 3.204) [578].

Fig. 3.204 Postulated intermediates of the catalyzed Conia-ene reaction and enyne-cycloisomerization as well as applications in natural product syntheses [578, 585–589]

Subsequent isomerizations or bond formations with nucleophiles starting from **293**, **295** or **296** generate an almost infinite variety of possible derivatives. Numerous applications in the synthesis of complex natural products impressively demonstrate the range of highly functionalized structures that can be accessed by this approach [571, 579, 590, 591].

Subsequently, seminal studies have also been reported in the field of *asymmetric* Au-catalysis. Cationic Au(I) catalysts, which have proven particularly effective by virtue of their reactivity and selectivity, form linear, doubly coordinated complexes. The reacting substrate is thus at a maximum distance from the inducing ligand. The direct attack on the alkyne without coordination with the metal center (outer-sphere mechanism) moves an enantioinducing ligand even further away from the newly forming stereogenic element.

For these reasons, the realization of high enantioselectivities was unattainable for π-acid-catalyzed reactions for a rather extended period. Approaches to overcome these challenges have been sought in the use of sterically very demanding ligands, dimeric catalysts, and enantioinducing counterions (see Fig. 3.205) [592–594].

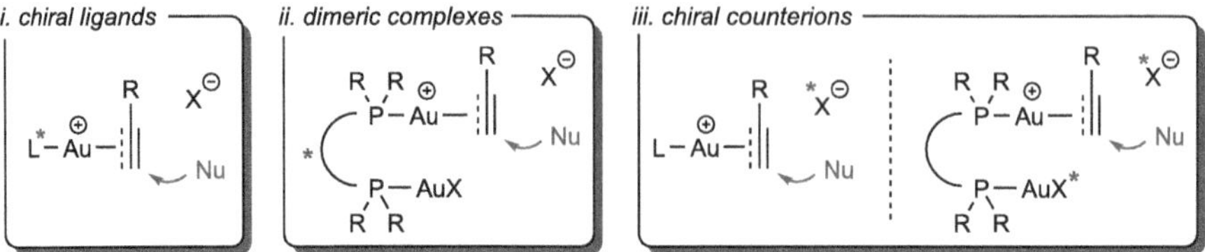

Fig. 3.205 Approaches to enantioselective Au-catalyzed transformations

The majority of methodological work in the field of enantioselective π-acid catalysis utilizes allenes as substrates. As the alternative hydrofunctionalization of CC-triple bonds introduces new stereoelements only in the α- or β-position of the resulting double bond, asymmetric methods using alkynes, therefore, focus primarily on enyne cycloisomerizations. Apart from methodological studies, there have been only occasional applications in the synthesis of complex hetero- or carbocycles.

Fürstner and co-workers disclosed a highly selective access to the active drug GSK-1360707 (**299**) originally developed by Glaxo–Smith–Kline. The [6,3]-fused molecular framework of the antidepressant was synthesized via an Au-catalyzed enyne cycloisomerization in the presence of the chiral phosphoramidite L1 as a ligand. The monodentate ligand can rotate freely around the Au-P bond when incorporated into linear Au(I) complexes and induces a high enantioselectivity of 95% in the conversion of **297** to **298** through its TADDOL backbone. The final stages to the active drug entailed a hydrogenation of the enamine and the acid-catalyzed deprotection of the Cbz-amine (see Fig. 3.206) [595].

The Ohno group pursued the total synthesis of the pain-relieving alkaloid (+)-conolidine. The key step of their approach relied on a domino hydroamination/Conia-ene reaction to convert the enediyne **300** to indole **301**. The activation of both alkyne functionalities in the presence of the bimetallic, bridged Au complex afforded a high enantioselectivity, albeit only a moderate yield. As competing reactions in the second step, a 7-endo-dig attack is viable instead of the desired 6-exo-dig cyclization, which diminishes the yield of piperidine **301**, similar to the unwanted desilylation of the enol ether. The desired natural product was obtained in two further steps starting from **301** [596].

Other Reactions

In addition to azide-alkyne cycloadditions (click chemistry) and the alkyne or enyne metathesis discussed in subsequent chapters, additional reactions for the derivatization of alkynes

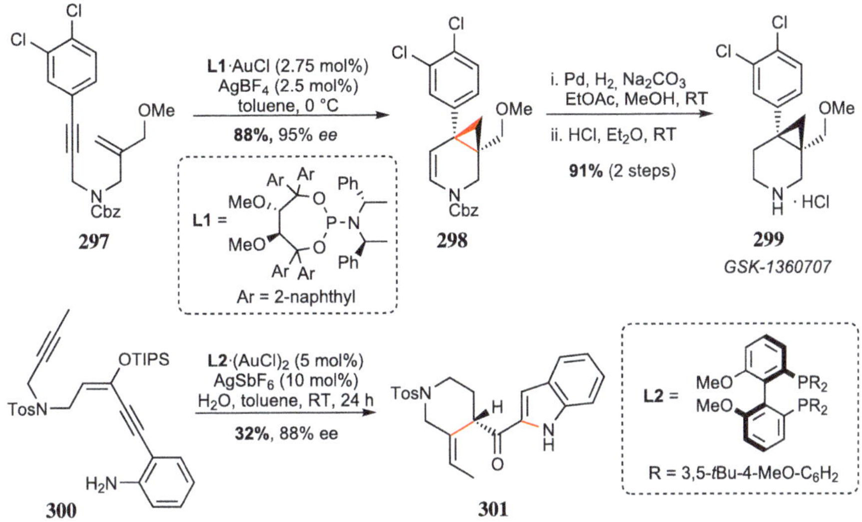

Fig. 3.206 Enantioselective Au-catalyzed alkyne activations according to Fürstner and Ohno. Newly formed bonds are highlighted [595, 596]

comprise classic transformations such as the Glaser coupling [597, 598] or the Pauson–Khand reaction (see Fig. 3.185), [599–601] which will not be explored here in further detail.

3.3.3   Reductions of Alkenes and Alkynes

A variety of reductive methods will be discussed in the following chapter. The selective reduction of CC bonds is routinely exploited synthetically to introduce a stereocenter. In some cases, the reactive site resulting from a CC coupling by Wittig olefination or metathesis is subsequently masked/removed by a reduction to the respective alkane. With *alkynes*, partial reductions are commonly used to selectively obtain either an *E*- or *Z*-configured alkene (see Table 3.18) [602].

Table 3.18 Methods for the diastereoselective reduction of alkynes

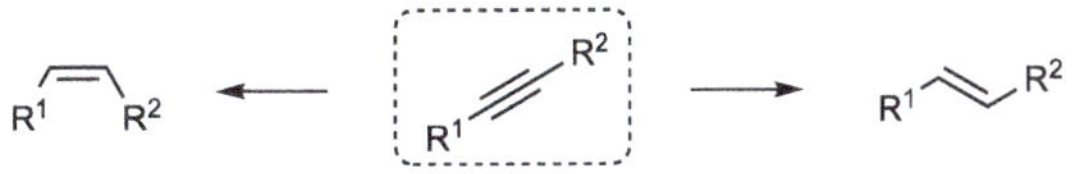

Z-alkene	*E*-alkene
Lindlar reduction & related methods	Birch reduction
diimide reduction	Ru *trans*-hydrogenation
hydrometalation and C-M-protonolysis	

The methods listed in Table 3.18 demonstrate the breadth of possibilities for partially reducing alkynes diastereoselectively. If Z-alkenes are desired, a hydrogenation in the presence of the **Lindlar** catalyst can be considered as the first choice when reducing alkynes. Beyond this method, there are also complementary, heterogeneous conditions (P-2 Ni catalyst, H_2 or $Zn(Cu/Ag)$, MeOH) that have been successfully used in total syntheses on complex intermediates (see Fig. 3.207).

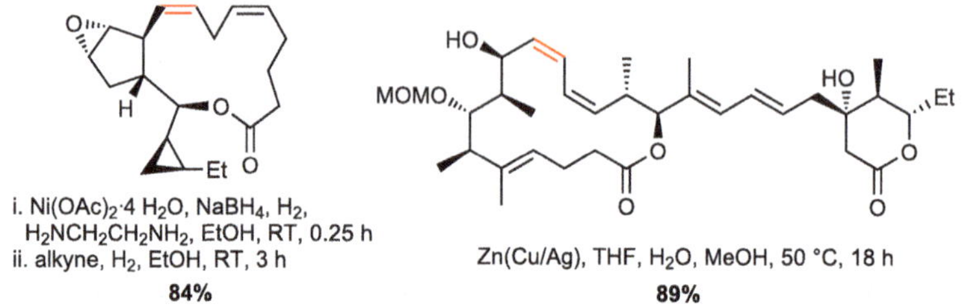

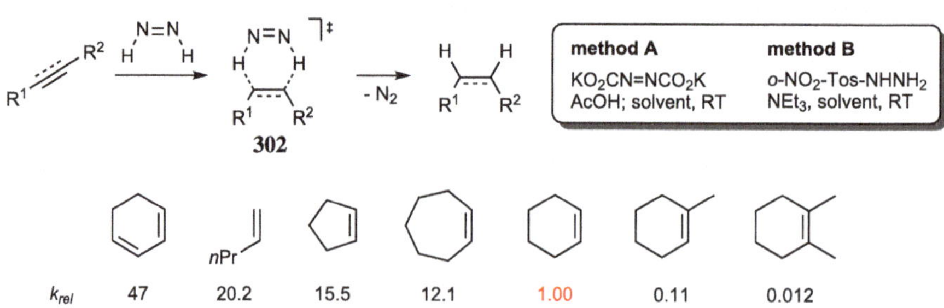

Fig. 3.207 Alternative methods to the Lindlar catalyst in the partial reduction of alkynes [287, 603]

Diimide formally constitutes an H_2 transfer agent, whose driving force in the reduction is derived from the N_2 release, thereby ensuring high turnovers. A hallmark of diimide reductions is the exceptionally high selectivity for nonpolar multiple bonds: alkynes and olefins are selectively reduced in the presence of carbonyls (ketones, esters), imines, nitriles, nitro groups as well as halogens, amines, peroxides or disulfides without affecting the more polar functionalities [604].

Diimide is not stable and must be generated *in situ*. This can either be accomplished via oxidative or thermal means starting from hydrazine, or — more commonly — by using azodicarboxylate in an acidic medium or *ortho*-nitro-tosylhydrazine in the presence of NEt$_3$. The reaction mechanism resembles a concerted *syn*-addition of H_2 to the acceptor triple or double bond, thereby stereospecifically affording Z-alkenes from alkyne substrates. The attack usually occurs from the sterically less hindered side (see Fig. 3.208) [604, 605].

Fig. 3.208 Mechanism of diimide reduction, common reaction conditions and relative reactivity of various alkenes [604]

The steric environment of the olefinic or acetylenic multiple bonds impacts the rate of reduction. Terminal functionalities react more rapidly than internal systems and the higher the degree of substitution, the slower the reaction proceeds. Conjugated systems are more reactive than isolated multiple bonds. Z-configured alkenes are converted more quickly than E-alkenes. Alkynes possess an enhanced reactivity compared to alkenes. Despite this preference for acetylenic bonds, a halt at the olefinic stage is often difficult to achieve. Nevertheless, diimide reductions have also been successfully reported in total syntheses on advanced intermediates that bear several CC-multiple bonds (see Fig. 3.209).

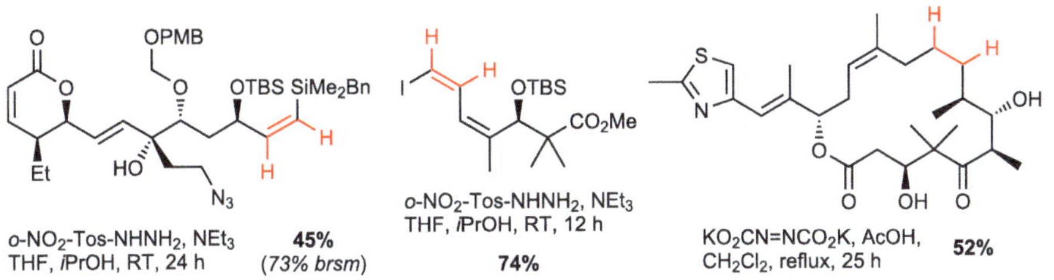

Fig. 3.209 Examples of selective diimide reductions of enynes and polyenes [606–608]

The third approach for the selective access to Z-configured olefins from alkynes is a stereospecific **hydrometalation**, followed by protonolysis of the carbon-metal bond. In addition to vinyl boranes (see Fig. 3.192), vinyl silanes and stannanes can also be introduced via this method under very mild conditions and subsequently reduced to the corresponding alkenes (see Fig. 3.210).

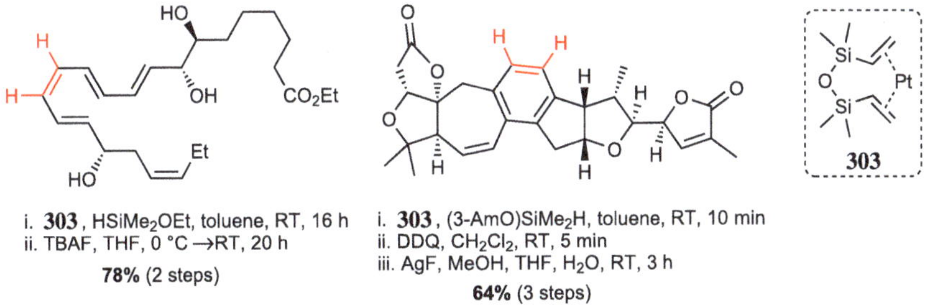

Fig. 3.210 Hydrosilylation/protodesilylation sequences in total syntheses [500, 609]

Radical reductions are particularly suitable for the diastereoselective synthesis of E-alkenes from alkynes. In addition to the **Birch reduction** (see Sect. 4.2), radical hydrometalations can also preferentially yield *trans*-configured alkenyl derivatives, which can be converted to E-olefins [523, 610]. Terminal alkynes cannot be reduced under classic Birch conditions (Na, NH$_3$) because the intermittent Na acetylides resulting from deprotonation of the C$_{sp}$-H bond are inert towards a reduction of the triple bond. In such cases, Li/amines

(Benkeser reaction) are recommended as an alternative method [611, 612]. Given the limited compatibility with more sensitive functional groups in the reduction by solvated metals, a Ru-catalyzed method for ***trans*-hydrogenation** and *trans*-hydrometalation was developed in the Fürstner group derived from previous studies by Trost and co-workers (see Fig. 3.211) [613].

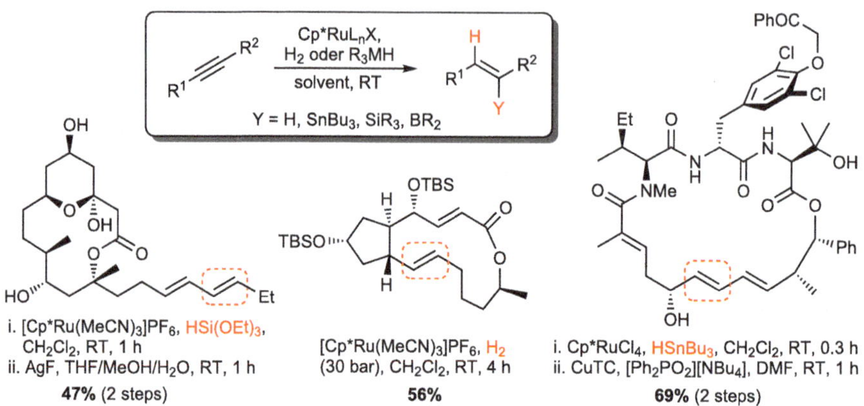

Fig. 3.211 Ru-catalyzed *trans*-functionalization of alkynes [613–616]

References

1. S. D. Roughley, A. M. Jordan, *J. Med. Chem.* **2011**, *54*, 3451–3479
2. *Modern Carbonyl Olefination*, (Ed.: T. Takeda), Wiley-VCH, **2004**
3. P. A. Byrne, D. G. Gilheany, *Chem. Soc. Rev.* **2013**, *42*, 6670–6696
4. P. R. Blakemore, *J. Chem. Soc. Perkin Trans. 1* **2002**, 2563–2585
5. R. J. K. Taylor, G. Casy, *Org. React.* **2003**, *62*, 359–475
6. G. Wittig, U. Schöllkopf, *Chem. Ber.* **1954**, *87*, 1318–1330
7. K. C. Nicolaou, M. W. Härter, J. L. Gunzner, A. Nadin, *Liebigs Ann.* **1997**, 1283–1301
8. B. E. Maryanoff, A. B. Reitz, *Chem. Rev.* **1989**, *89*, 863–927
9. E. Vedejs, G. P. Meier, K. A. J. Snoble, *J. Am. Chem. Soc.* **1981**, *103*, 2823–2831
10. E. Vedejs, C. F. Marth, R. Ruggeri, *J. Am. Chem. Soc.* **1988**, *110*, 3940–3948
11. Z. Chen, Y. Nieves-Quinones, J. R. Waas, D. A. Singleton, *J. Am. Chem. Soc.* **2014**, *136*, 13122–13125
12. F. H. Westheimer, *Acc. Chem. Res.* **1968**, *1*, 70–78
13. C. R. Hall, T. D. Inch, *Tetrahedron* **1980**, *36*, 2059–2095
14. J. G. López, A. M. Ramallal, J. González, L. Roces, S. García-Granda, M. J. Iglesias, P. Oña-Burgos, F. L. Ortiz, *J. Am. Chem. Soc.* **2012**, *134*, 19504–19507
15. P. A. Byrne, D. G. Gilheany, *J. Am. Chem. Soc.* **2012**, *134*, 9225–9239
16. B. E. Maryanoff, A. B. Reitz, M. S. Mutter, R. R. Whittle, R. A. Olofson, *J. Am. Chem. Soc.* **1986**, *108*, 7664–7678

17. F. Bangerter, M. Karpf, L. A. Meier, P. Rys, P. Skrabal, *J. Am. Chem. Soc.* **1998**, *120*, 10653–10659

18. M. Schlosser, K. F. Christmann, *Angew. Chem. Int. Ed. Engl.* **1965**, *4*, 689–690

19. P. A. Byrne, J. Muldoon, Y. Ortin, H. Müller-Bunz, D. G. Gilheany, *Eur. J. Org. Chem.* **2014**, 86–98

20. R. Robiette, J. Richardson, V. K. Aggarwal, J. N. Harvey, *J. Am. Chem. Soc.* **2006**, *128*, 2394–2409

21. E. Vedejs, C. F. Marth, *J. Am. Chem. Soc.* **1988**, *110*, 3948–3958

22. A. B. Reitz, S. O. Nortey, A. D. Jordan, M. S. Mutter, B. E. Maryanoff, *J. Org. Chem.* **1986**, *51*, 3302–3308

23. M. Schlosser, K. F. Christmann, *Justus Liebigs Ann. Chem.* **1967**, *708*, 1–35

24. D. A. Evans, B. W. Trotter, P. J. Coleman, B. Côté, L. C. Dias, H. A. Rajapakse, A. N. Tyler, *Tetrahedron* **1999**, *55*, 8671–8726

25. V. Mascitti, E. J. Corey, *J. Am. Chem. Soc.* **2004**, *126*, 15664–15665

26. M. H. Becker, P. Chua, R. Downham, C. J. Douglas, N. K. Garg, S. Hiebert, S. Jaroch, R. T. Matsuoka, J. A. Middleton, F. W. Ng, L. E. Overman, *J. Am. Chem. Soc.* **2007**, *129*, 11987–12002

27. H. S. Kim, T. Kim, J. Ahn, H. Yun, C. Lim, J. Jang, J. Sim, H. An, Y.-J. Surh, J. Lee, Y.-G. Suh, *J. Org. Chem.* **2018**, *83*, 1997–2005

28. H. Fuwa, S. Matsukida, T. Miyoshi, Y. Kawashima, T. Saito, M. Sasaki, *J. Org. Chem.* **2016**, *81*, 2213–2227

29. V. K. Mishra, P. C. Ravikumar, M. E. Maier, *J. Org. Chem.* **2016**, *81*, 9728–9737

30. H. Pommer, *Angew. Chem. Int. Ed. Engl.* **1977**, *16*, 423–429

31. G. L. Parker, L. K. Smith, I. R. Baxendale, *Tetrahedron* **2016**, *72*, 1645–1652

32. W. Bonrath, B. Gao, P. Houston, T. McClymont, M.-A. Müller, C. Schäfer, C. Schweiggert, J. Schütz, J. A. Medlock, *Org. Process Res. Dev.* **2023**, *27*, 1557–1584

33. M. Schlosser, K. F. Christmann, *Angew. Chem. Int. Ed. Engl.* **1966**, *5*, 126

34. Q. Wang, D. Deredas, C. Huynh, M. Schlosser, *Chem. Eur. J.* **2003**, *9*, 570–574

35. M. Schlosser, K.-F. Christmann, A. Piskala, *Chem. Ber.* **1970**, *103*, 2814–2820

36. T. Tonoi, R. Kawahara, T. Inohana, I. Shiina, *J. Antibiotics* **2016**, *69*, 697–701

37. M.-L. Schirmer, S. Adomeit, T. Werner, *Org. Lett.* **2015**, *17*, 3078–3081

38. E. E. Coyle, B. J. Doonan, A. J. Holohan, K. A. Walsh, F. Lavigne, E. H. Krenske, C. J. O'Brien, *Angew. Chem. Int. Ed.* **2014**, *53*, 12907–12911

39. C. J. O'Brien, Z. S. Nixon, A. J. Holohan, S. R. Kunkel, J. L. Tellez, B. J. Doonan, E. E. Coyle, F. Lavigne, L. J. Kang, K. C. Przeworski, *Chem. Eur. J.* **2013**, *19*, 15281–15289

40. M. Schlosser, B. Schaub, J. De Oliveira-Neto, S. Jeganathan, *Chimia* **1986**, *40*, 244–245

41. M. Edmonds, A. Abell in *Modern Carbonyl Olefination* (Ed.: T. Takeda), Wiley-VCH, **2004**, Chapter 1, pp. 1–17

42. K. Ando, *J. Org. Chem.* **1999**, *64*, 6815–6821

43. P. Brandt, P.-O. Norrby, I. Martin, T. Rein, *J. Org. Chem.* **1998**, *63*, 1280–1289

44. S. K. Thompson, C. H. Heathcock, *J. Org. Chem.* **1990**, *55*, 3386–3388

45. J. Motoyoshiya, T. Kusaura, K. Kokin, S. Yokoya, Y. Takaguchi, S. Narita, H. Aoyama, *Tetrahedron* **2001**, *57*, 1715–1721

46. W. C. Stil, C. Gennari, *Tetrahedron Lett.* **1983**, *24*, 4405–4408

47. I. Janickia, , P. Kiełbasińskia, *Adv. Synth. Catal.* **2020**, *362*, 2552–2596

48. K. Ando, *J. Org. Chem.* **1997**, *62*, 1934–1939

49. K. Ando, *J. Org. Chem.* **1998**, *63*, 8411–8416

50. K. Ando, T. Oishi, M. Hirama, H. Ohno, T. Ibuka, *J. Org. Chem.* **2000**, *65*, 4745–4749

51. M. A. Blanchette, W. Choy, J. T. Davis, A. P. Essenfeld, S. Masamune, W. R. Roush, T. Sakai, *Tetrahedron Lett.* **1984**, *25*, 2183–2186

52. I. Paterson, K.-S. Yeung, J. B. Smaill, *Synlett* **1993**, 774–777

53. J. Á Bisceglia, L. R. Orelli, *Curr. Org. Chem.* **2012**, *16*, 2206–2230

54. K. Kobayashi, K. Tanaka III, H. Kogen, *Tetrahedron Lett.* **2018**, *59*, 568–582

55. D. Roman, M. Sauer, C. Beemelmanns, *Synthesis* **2021**, *53*, 2713–2739

56. C. Poock, M. Kalesse, *Org. Lett.* **2017**, *19*, 4536–4539

57. M. Morita, Y. Kobayashi, *J. Org. Chem.* **2018**, *83*, 3906–3914

58. K. C. Nicolaou, D. Rhoades, Y. Wang, R. Bai, E. Hamel, M. Aujay, J. Sandoval, J. Gavrilyuk, *J. Am. Chem. Soc.* **2017**, *139*, 7318–7334

59. B. Chandrasekhar, S. Athe, P. P. Reddya, S. Ghosh, *Org. Biomol. Chem.* **2015**, *13*, 115–124

60. J. S. Yadav, S. Dhara, S. S. Hossain, D. K. Mohapatra, *J. Org. Chem.* **2012**, *77*, 9628–9633

61. R. P. Singh, V. K. Singh, *J. Org. Chem.* **2004**, *69*, 3425–3430

62. D. J. Dixon, A. C. Foster, S. V. Ley, *Org. Lett.* **2000**, *2*, 123–125

63. S. D. Rychnovsky, U. R. Khire, G. Yang, *J. Am. Chem. Soc.* **1997**, *119*, 2058–2059

64. M. B. van Niel, B. Fauber, S. Gaines, J. Killen, S. Ward, *Int. Patent WO2014090712 A1* **2015**

65. A. K. Bhattacharya, G. Thyagarajan, *Chem. Rev.* **1981**, *81*, 415–430

66. R. D. Clark, L. G. Kozar, C. H. Heathcock, *Synthesis* **1975**, 635–636

67. P. Savignac, F. Mathey, *Tetrahedron Lett.* **1976**, *17*, 2829–2832

68. S. M. A. Kedrowski, D. A. Dougherty, *Org. Lett.* **2010**, *12*, 3990–3993

69. R. Durmeunier, I. E. Markó in *Modern Carbonyl Olefination* (Ed.: T. Takeda), Wiley-VCH, **2004**, Chapter 3, pp. 104–150

70. K. Plesniak, A. Zarecki, J. Wicha, *Top. Curr. Chem.* **2007**, *275*, 163–250

71. D. A. Alonso, C. Nájera, *Org. React.* **2008**, *72*, 367–656

72. G. E. Keck, K. A. Savin, M. E. Weglarz, *J. Org. Chem.* **1995**, *60*, 3194–3204

73. E. O. Volz, G. W. O'Neil, *J. Org. Chem.* **2011**, *76*, 8428–8432

74. M. Szostak, M. Spain, D. J. Procter, *Chem. Soc. Rev.* **2013**, *42*, 9155–9183

75. P. J. Kocienski, B. Lythgoe, I. Waterhouse, *J. Chem. Soc. Perkin Trans. 1* **1980**, 1045–1050

76. K. Ogawa, Y. Koyama, I. Ohashi, I. Sato, M. Hirama, *Angew. Chem. Int. Ed.* **2009**, *48*, 1110–1113

77. G. Kim, M. Y. Chu-Moyer, S. J. Danishefsky, G. K. Schulte, *J. Am. Chem. Soc.* **1993**, *115*, 30–39

78. J. M. Storvick, E. Ankoudinova, B. R. King, H. V. Epps, G. W. O'Neil, *Tetrahedron Lett.* **2011**, *52*, 5858–5861

79. I. E. Markó, F. Murphy, L. Kumps, A. Ates, R. Touilleaux, D. Craig, S. Caballeres, S. Dolan, *Tetrahedron* **2001**, *57*, 2609–2619

80. N. Tanimoto, S. W. Gerritz, A. Sawabe, T. Noda, S. A. Filla, S. Masamune, *Angew. Chem. Int. Ed. Engl.* **1994**, *33*, 673–675

81. A. Toyota, A. Nishimura, C. Kaneko, *Tetrahedron Lett.* **1998**, *39*, 4687–4690

82. J. S. Sabol, J. R. McCarthy, *Tetrahedron Lett.* **1992**, *33*, 3101–3104

83. B. Chatterjee, S. Bera, D. Mondal, *Tetrahedron: Asymm.* **2014**, *25*, 1–55

84. G. Sakaine, Z. Leitis, R. Ločmele, G. Smits, *Eur. J. Org. Chem.* **2023**, *26*, e202201217

85. W. E. Truce, E. M. Kreider, W. W. Brand, *Org. React.* **1970**, *18*, 99–215

86. C. M. Holden, M. F. Greaney, *Chem. Eur. J.* **2017**, *23*, 8992–9008

87. L. Legnani, A. P. N. P. Caramella, L. Toma, G. Zanoni, G. Vidari, *J. Org. Chem.* **2015**, *80*, 3092–3100

88. P. J. Kocienski, A. Bell, P. R. Blakemore, *Synlett* **2000**, 365–366

89. P. R. Blakemore, W. J. Cole, P. J. Kocienski, A. Morley, *Synlett* **1998**, 26–28

90. A. B. Charette, H. Lebel, *J. Am. Chem. Soc.* **1996**, *118*, 10327–10328

91. L. A. Hobson, O. Akiti, S. S. Deshmukh, S. Harper, K. Katipally, C. J. Lai, R. C. Livingston, E. Lo, M. M. Miller, S. Ramakrishnan, L. Shen, J. Spink, S. Tummala, C. Wei, K. Yamamoto, J. Young, R. L. Parsons, *Org. Process Res. Dev.* **2010**, *14*, 441–458

92. P. Liu, E. N. Jacobsen, *J. Am. Chem. Soc.* **2001**, *123*, 10772–10773

93. F. Anderl, S. Größl, C. Wirtz, A. Fürstner, *Angew. Chem. Int. Ed.* **2018**, *57*, 10712–10717

94. D. R. Williams, D. A. Brooks, M. A. Berliner, *J. Am. Chem. Soc.* **1999**, *121*, 4924–4925

95. V. S. Enev, W. Felzmann, A. Gromov, S. Marchart, J. Mulzer, *Chem. Eur. J.* **2012**, *18*, 9651–9668

96. N. Furuichi, H. Hara, T. Osaki, H. Mori, S. Katsumura, *Angew. Chem. Int. Ed.* **2002**, *41*, 1023–1026

97. R. Bellingham, K. Jarowicki, P. Kocienski, V. Martin, *Synthesis* **1996**, 285–296

98. A. B. Smith, I. G. Safonov, R. M. Corbett, *J. Am. Chem. Soc.* **2001**, *123*, 12426–12427

99. G. J. Florence, J. Wlochal, *Chem. Eur. J.* **2012**, *18*, 14250–14254

100. R. C. Hartley, G. J. McKiernan, *J. Chem. Soc. Perkin Trans. 1* **2002**, 2763–2793

101. T. Takeda, A. Tsubouchi in *Modern Carbonyl Olefination* (Ed.: T. Takeda), Wiley-VCH, **2004**, Chapter 4, pp. 151–199

102. J. E. McMurry, *Chem. Rev.* **1989**, *89*, 1513–1524

103. A. Fürstner, B. Bogdanovic, *Angew. Chem. Int. Ed. Engl.* **1996**, *35*, 2442–2469

104. M. Ephritikhine, C. Villiers in *Modern Carbonyl Olefination* (Ed.: T. Takeda), Wiley-VCH, **2004**, Chapter 6, pp. 223–285

105. S. Matsubara, K. Oshima in *Modern Carbonyl Olefination* (Ed.: T. Takeda), Wiley-VCH, **2004**, Chapter 5, pp. 200–222

106. S. Matsubara, M. Sugihara, K. Utimoto, *Synlett* **1998**, 313–315

107. R. C. Hartley, J. Li, C. A. Main, G. J. McKiernan, *Tetrahedron* **2007**, *63*, 4825–4864

108. C. Müller, M. Cokoja, F. E. Kühn, *Science of Synthesis* **2014**, *2*, 1–29

109. S. H. Pine, G. Kim, V. Lee, *Org. Synth.* **1990**, *69*, 72–79

110. Y. Horikawa, M. Watanabe, T. Fujiwara, T. Takeda, *J. Am. Chem. Soc.* **1997**, *119*, 1127–1128

111. S. H. Pine, G. S. Shen, H. Hoang, *Synthesis* **1991**, 165–167

112. Y. Sun, M. Nitz, *J. Org. Chem.* **2012**, *77*, 7401–7410

113. D. L. Hughes, J. F. Payack, D. Cai, T. R. Verhoeven, P. J. Reider, *Organometallics* **1996**, *15*, 663–667

114. E. C. Meurer, L. Silva Santos, R. A. Pilli, M. N. Eberlin, *Org. Lett.* **2003**, *5*, 1391–1394

115. L. M. Dollinger, A. R. Howell, *J. Org. Chem.* **1996**, *61*, 7248–7249

116. R. J. Capon, C. Skene, E. H. Liu, E. Lacey, J. H. Gill, K. Heiland, T. Friedel, *Nat. Prod. Res.* **2004**, *18*, 305–309

117. S. Kobayashi, R. S. Reddy, Y. Sugiura, D. Sasaki, N. Miyagawa, M. Hirama, *J. Am. Chem. Soc.* **2001**, *123*, 2887–2888

118. C. S. Mushti, J.-H. Kim, E. J. Corey, *J. Am. Chem. Soc.* **2006**, *128*, 14050–14052

119. Y. Kawashima, A. Toyoshima, H. Fuwa, M. Sasaki, *Org. Lett.* **2016**, *18*, 2232–2235

120. S. Diethelm, E. M. Carreira, *J. Am. Chem. Soc.* **2013**, *135*, 8500–8503

121. A. B. Smith, T. M. Razler, J. P. Ciavarri, T. Hirose, T. Ishikawa, R. M. Meis, *J. Org. Chem.* **2008**, *73*, 1192–1200

122. L.-S. Deng, X.-P. Huang, G. Zhao, *J. Org. Chem.* **2006**, *71*, 4625–4635

123. T. Gerfaud, C. Xie, L. Neuville, J. Zhu, *Angew. Chem. Int. Ed.* **2011**, *50*, 3954–3957

124. J. F. Payack, M. A. Huffman, D. Cai, D. L. Hughes, P. C. Collins, B. K. Johnson, I. F. Cottrell, L. D. Tuma, *Org. Process Res. Dev.* **2004**, *8*, 256–259

125. L. Lombardi, *Org. Synth.* **1987**, *65*, 81–89

126. I. Paterson, C. de Savi, M. Tudge, *Org. Lett.* **2001**, *3*, 213–216

127. F. Peng, S. J. Danishefsky, *J. Am. Chem. Soc.* **2012**, *134*, 18860–18867

128. T.-H. Yan, C.-C. Tsai, C.-T. Chien, C.-C. Cho, P.-C. Huang, *Org. Lett.* **2004**, *6*, 4961–4963

129. T.-H. Yan, C.-T. Chien, C.-C. Tsai, K.-W. Lin, Y.-H. Wu, *Org. Lett.* **2004**, *6*, 4965–4967

130. K. Takai, K. Nitta, K. Utimoto, *J. Am. Chem. Soc.* **1986**, *108*, 7408–7410

131. A. Fürstner, *Chem. Rev.* **1999**, *99*, 991–1045

132. L. A. Wessjohann, G. Scheid, *Synthesis* **1999**, 1–36
133. M. Sada, S. Komagawa, M. Uchiyama, M. Kobata, T. Mizuno, K. Utimoto, K. Oshima, S. Matsubara, *J. Am. Chem. Soc.* **2010**, *132*, 17452–17458
134. D. M. Hodgson, L. T. Boulton, G. N. Maw, *Tetrahedron Lett.* **1994**, *35*, 2231–2234
135. K. Takai, N. Shinomiya, H. Kaihara, N. Yoshida, T. Moriwake, K. Utimoto, *Synlett* **1995**, 963–964
136. D. Werner, R. Anwander, *J. Am. Chem. Soc.* **2018**, *140*, 14334–14341
137. A. Haahr, Z. Rankovic, R. C. Hartley, *Tetrahedron Lett.* **2011**, *52*, 3020–3022
138. Y. Sun, P. Chen, D. Zhang, M. Baunach, C. Hertweck, A. Li, *Angew. Chem. Int. Ed.* **2014**, *53*, 9012–9016
139. B. Barnych, J.-M. Vatèle, *Synlett* **2011**, 1912–1916
140. A. K. Ghosh, G. C. Reddy, S. Kovela, N. Relitti, V. K. Urabe, B. E. Prichard, M. S. Jurica, *Org. Lett.* **2018**, *20*, 7293–7297
141. C.-X. Zhuo, A. Fürstner, *J. Am. Chem. Soc.* **2018**, *140*, 10514–10523
142. K. Ikeuchi, M. Hayashi, T. Yamamoto, M. Inai, T. Asakawa, Y. Hamashima, T. Kan, *Eur. J. Org. Chem.* **2013**, 6789–6792
143. S. Matsubara, M. Horiuchi, K. Takai, K. Utimoto, *Chem. Lett.* **1995**, *24*, 259–260
144. K. C. Nicolaou, D. Rhoades, S. M. Kumar, *J. Am. Chem. Soc.* **2018**, *140*, 8303–8320
145. C. Aïssa, R. Riveiros, J. Ragot, A. Fürstner, *J. Am. Chem. Soc.* **2003**, *125*, 15512–15520
146. T. Takeda, A. Tsubochi, *Org. React.* **2013**, *82*, 1–470
147. F. T. Ladipo, *Curr. Org. Synth.* **2006**, *10*, 965–980
148. A. Bongso, R. Roswanda, Y. M. Syah, *RSC Adv.* **2022**, *12*, 15885–15909
149. T. Takeda, A. Tsubochi in *Science of Synthesis, 47a: Category 6, Compounds with All-Carbon* (Ed.: A. de Meijere), Thieme, **2010**, Chapter 47.1.1.5, pp. 247–345
150. J. E. McMurry, T. Lectka, J. G. Rico, *J. Org. Chem.* **1989**, *54*, 3748–3749
151. D. L. J. Clive, K. S. K. Murthy, A. G. H. Wee, J. S. Prasad, G. V. J. Da Silva, M. Majewski, P. C. Anderson, C. F. Evans, R. D. Haugen, L. D. Heerze, J. R. Barrie, *J. Am. Chem. Soc.* **1990**, *112*, 3018–3028
152. D. Lenoir, H. Burghard, *J. Chem. Res. (S)* **1980**, 396–397
153. J. E. McMurry, M. P. Fleming, K. L. Kees, L. R. Krepski, *J. Org. Chem.* **1978**, *43*, 3255–3266
154. R. Dams, M. Malinowski, I. Westdorp, H. Geise, *J. Org. Chem.* **1982**, *47*, 248–259
155. K. G. Pierce, M. A. Barteau, *J. Org. Chem.* **1995**, *60*, 2405–2410
156. L. E. Aleandri, S. Becke, B. Bogdanovíc, D. J. Jones, J. Rozière, *J. Organomet. Chem.* **1994**, *472*, 97–112
157. M. Ephritikhine, *Chem. Commun.* **1998**, 2549–2554
158. C. Villiers, M. Ephritikhine, *Angew. Chem. Int. Ed. Engl.* **1997**, *36*, 2380–2382
159. C. Villiers, M. Ephritikhine, *Chem. Eur. J.* **2001**, *7*, 3043–3051
160. C. Villiers, A. Vandais, M. Ephritikhine, *J. Organomet. Chem.* **2001**, *617–618*, 744–747
161. R. S. Muthyala, S. Sheng, K. E. Carlson, B. S. Katzenellenbogen, J. A. Katzenellenbogen, *J. Med. Chem.* **2003**, *46*, 1589–1602
162. H. Xie, D. A. Lee, D. M. Wallace, M. O. Senge, K. M. Smith, *J. Org. Chem.* **1996**, *61*, 8508–8517
163. E. J. Corey, A. Palani, *Tetrahedron Lett.* **1997**, *38*, 2397–2400
164. W. G. Dauben, T. Wang, R. W. Stephens, *Tetrahedron Lett.* **1990**, *31*, 2393–2396
165. B. Venkataiah, C. Ramesh, N. Ravindranath, B. Das, *Phytochemistry* **2003**, *63*, 383–386
166. V. Rajendran, S.-B. Rong, A. Saxena, B. P. Doctor, A. P. Kozikowski, *Tetrahedron Lett.* **2001**, *42*, 5359–5361
167. L. N. Lucas, J. van Esch, R. M. Kellogg, B. L. Feringa, *Tetrahedron Lett.* **1999**, *40*, 1775–1778
168. L. F. van Staden, D. Gravestock, D. J. Ager, *Chem. Soc. Rev.* **2002**, *31*, 195–200
169. N. Kano, T. Kawashima in *Modern Carbonyl Olefination* (Ed.: T. Takeda), Wiley-VCH, **2004**, Chapter 2, pp. 18–103

170. D. J. Ager, *Org. React.* **1990**, *39*, 1–110
171. A. R. Bassindale, R. J. Ellis, J. C.-Y. Lau, P. G. Taylor, *J. Chem. Soc. Perkin Trans. 2* **1986**, 593–597
172. R. Waschbüsch, J. Carran, P. Savignac, *Tetrahedron* **1996**, *52*, 14199–14216
173. L. F. van Staden, B. Bartels-Rahm, J. S. Field, N. D. Emslie, *Tetrahedron* **1998**, *54*, 3255–3278
174. B. J. Moritz, D. J. Mack, L. Tong, R. J. Thomson, *Angew. Chem. Int. Ed.* **2014**, *53*, 2988–2991
175. T. Chakraborty, P. Laxman, *Tetrahedron Lett.* **2003**, *44*, 4989–4992
176. R. E. Beveridge, R. A. Batey, *Org. Lett.* **2013**, *15*, 3086–3089
177. F. Glaus, K.-H. Altmann, *Angew. Chem. Int. Ed.* **2015**, *54*, 1937–1940
178. A. B. Smith, G. K. Friestad, J. J.-W. Duan, J. Barbosa, K. G. Hull, M. Iwashima, Y. Qiu, P. G. Spoors, E. Bertounesque, B. A. Salvatore, *J. Org. Chem.* **1998**, *63*, 7596–7597
179. E. Block, *Org. React.* **1984**, *30*, 457–566
180. H. Araki, M. Inoue, T. Suzuki, T. Yamori, M. Kohno, K. Watanabe, H. Abe, T. Katoh, *Chem. Eur. J.* **2007**, *13*, 9866–9881
181. T. Kawamata, A. Yamaguchi, M. Nagatomo, M. Inoue, *Chem. Eur. J.* **2018**, *24*, 18907–18912
182. M. Brüggemann, A. I. McDonald, L. E. Overman, M. D. Rosen, L. Schwink, J. P. Scott, *J. Am. Chem. Soc.* **2003**, *125*, 15284–15285
183. B. Du, C. Yuan, T. Yu, L. Yang, Y. Yang, B. Liu, S. Qin, *Chem. Eur. J.* **2014**, *20*, 2613–2622
184. L. A. Wessjohann, U. Sinks, *J. prakt. Chem.* **1998**, *340*, 189–203
185. F. Bihelovic, Z. Ferjancic, *Angew. Chem. Int. Ed.* **2016**, *55*, 2569–2572
186. D. L. Re, Y. Zhou, J. Mucha, L. F. Jones, L. Leahy, C. Santocanale, M. Krol, P. V. Murphy, *Chem. Eur. J.* **2015**, *21*, 18109–18121
187. P. A. Grieco, S. Gilman, M. Nishizawa, *J. Org. Chem.* **1976**, *41*, 1485–1486
188. A. Krief, A.-M. Laval, *Bull. Soc. Chim. Fr.* **1997**, *134*, 869–874
189. K. Du, M. J. Kier, A. L. Rheingold, G. C. Micalizio, *Org. Lett.* **2018**, *20*, 6457–6461
190. A. Letort, D.-L. Long, J. Prunet, *J. Org. Chem.* **2016**, *81*, 12318–12331
191. X. Cai, K. Ng, H. Panesar, S.-J. Moon, M. Paredes, K. Ishida, C. Hertweck, T. G. Minehan, *Org. Lett.* **2014**, *16*, 2962–2965
192. C. Heinz, N. Cramer, *J. Am. Chem. Soc.* **2015**, *137*, 11278–11281
193. R. J. Arhart, J. C. Martin, *J. Am. Chem. Soc.* **1972**, *94*, 5003–5010
194. E. M. Burgess, H. R. Penton, E. A. Taylor, *J. Org. Chem.* **1973**, *38*, 26–31
195. P. Hjerrild, T. Tørring, T. B. Poulsen, *Nat. Prod. Rep.* **2020**, *37*, 1043–1064
196. Z. Ma, J. Jiang, S. Luo, Y. Cai, J. M. Cardon, B. M. Kay, D. H. Ess, S. L. Castle, *Org. Lett.* **2014**, *16*, 4044–4047
197. P. Sondermann, E. M. Carreira, *J. Am. Chem. Soc.* **2019**, *141*, 10510–10519
198. Z. Lu, X. Zhang, Z. Guo, Y. Chen, T. Mu, A. Li, *J. Am. Chem. Soc.* **2018**, *140*, 9211–9218
199. R. M. Kellogg, S. Wassenaar, J. Buter, *Tetrahedron Lett.* **1970**, *54*, 4689–4692
200. D. H. R. Barton, B. J. Willis, *J. Chem. Soc. D* **1970**, 1225–1226
201. D. H. R. Barton, F. S. Guziec, I. Shahak, *J. Chem. Soc. Perkin Trans. 1* **1974**, 1794–1799
202. N. P. Neureiter, F. G. Bordwell, *J. Am. Chem. Soc.* **1959**, *81*, 578–580
203. M. Yang, J. Li, A. Li, *Nat. Commun.* **2015**, *6*, 6445–6450
204. D. Habrant, V. Rauhala, A. M. P. Koskinen, *Chem. Soc. Rev.* **2010**, *39*, 2007–2017
205. M. M. Heravi, S. Asadi, N. Nazari, B. M. Lashkariani, *Curr. Org. Chem.* **2015**, *19*, 2196–2219
206. F. Eymery, B. Iorga, P. Savignac, *Synthesis* **2000**, 185–213
207. D. Grandjean, P. Pale, J. Chuche, *Tetrahedron Lett.* **1994**, *35*, 3529–3530
208. M. Okutani, Y. Mori, *J. Org. Chem.* **2009**, *74*, 442–444
209. N.-H. Lin, L. E. Overman, M. H. Rabinowitz, L. A. Robinson, M. J. Sharp, J. Zablocki, *J. Am. Chem. Soc.* **1996**, *118*, 9062–9072
210. B. M. Trost, W. Tang, *J. Am. Chem. Soc.* **2002**, *124*, 14542–14543

211. M. Movassaghi, D. K. Hunt, M. Tjandra, *J. Am. Chem. Soc.* **2006**, *128*, 8126–8127
212. J. Reyes, N. Winter, L. Spessert, D. Trauner, *Angew. Chem. Int. Ed.* **2018**, *57*, 15587–15591
213. X. Zeng, F. Zeng, E. Negishi, *Org. Lett.* **2004**, *6*, 3245–3248
214. Y.-Q. Fang, O. Lifchits, M. Lautens, *Synlett* **2008**, 413–417
215. J. C. Gilbert, U. Weerasooriya, *J. Org. Chem.* **1982**, *47*, 1837–1845
216. S. Müller, B. Liepold, G. J. Roth, H. J. Bestmann, *Synlett* **1996**, 521–522
217. M. Dhameja, J. Pandey, *Asian J. Org. Chem.* **2018**, *7*, 1502–1523
218. D. G. Brown, E. J. Velthuisen, J. R. Commerford, R. G. Brisbois, T. R. Hoye, *J. Org. Chem.* **1996**, *61*, 2540–2541
219. J. Pietruszka, A. Witt, *Synthesis* **2006**, 4266–4268
220. J. Martynow, R. Hanselmann, E. Duffy, A. Bhattacharjee, *Org. Process Res. Dev.* **2019**, *23*, 1026–1033
221. G. V. Ramakrishna, R. A. Fernandes, *J. Org. Chem.* **2019**, *84*, 14127–14132
222. S. Mahapatra, R. G. Carter, *J. Am. Chem. Soc.* **2013**, *135*, 10792–10803
223. K. Miwa, T. Aoyama, T. Shioiri, *Synlett* **1994**, 107–108
224. D. Tymann, U. Bednarzick, L. Iovkova-Berends, M. Hiersemann, *Org. Lett.* **2018**, *20*, 4072–4076
225. C. Monti, O. Sharon, C. Gennari, *Chem. Commun.* **2007**, 4271–4273
226. P. Michel, D. Gennet, A. Rassat, *Tetrahedron Lett.* **1999**, *40*, 8575–8578
227. B. Schmidt, S. Audörsch, *Org. Lett.* **2016**, *18*, 1162–1165
228. P. Meffre, S. Hermann, P. Durand, G. Reginato, A. Riu, *Tetrahedron* **2002**, *58*, 5159–5162
229. G. J. Roth, B. Liepold, S. G. Müller, H. J. Bestmann, *Synthesis* **2004**, 59–62
230. S. Gahalawat, S. K. Pandey, *Org. Biomol. Chem.* **2016**, *14*, 9287–9293
231. M. Mohammad, V. Chintalapudi, J. M. Carney, S. J. Mansfield, P. Sanderson, K. E. Christensen, E. A. Anderson, *Angew. Chem. Int. Ed.* **2019**, *58*, 18177–18181
232. P. Klahn, A. Duschek, C. Liébert, S. F. Kirsch, *Org. Lett.* **2012**, *14*, 1250–1253
233. Z. Lv, B. Chen, C. Zhang, G. Liang, *Chem. Eur. J.* **2018**, *24*, 9773–9777
234. I. E. Wrona, A. Gozman, T. Taldone, G. Chiosis, J. S. Panek, *J. Org. Chem.* **2010**, *75*, 2820–2835
235. M. Nahrwold, T. Bogner, S. Eissler, S. Verma, N. Sewald, *Org. Lett.* **2010**, *12*, 1064–1067
236. B. D. Williams, A. B. Smith, III., *J. Org. Chem.* **2014**, *79*, 9284–9296
237. M. D. Clay, A. G. Fallis, *Angew. Chem. Int. Ed.* **2005**, *44*, 4039–4042
238. S. Hanessian, T. Focken, X. Mi, R. Oza, B. Chen, D. Ritson, R. Beaudegnies, *J. Org. Chem.* **2010**, *75*, 5601–5618
239. Y. Hara, T. Honda, K. Arakawa, K. Ota, K. Kamaike, H. Miyaoka, *J. Org. Chem.* **2018**, *83*, 1976–1987
240. G. H. Shen, J. H. Hong, *Carbohyd. Res.* **2018**, *463*, 47–106
241. M. Beshai, B. Dhudshia, R. Mills, A. N. Thadani, *Tetrahedron Lett.* **2008**, *49*, 6794–6796
242. X. Yang, D. Wu, Z. Lu, H. Sun, A. Li, *Org. Biomol. Chem.* **2016**, *14*, 5591–5594
243. M. Zeng, S. K. Murphy, S. B. Herzon, *J. Am. Chem. Soc.* **2017**, *139*, 16377–16388
244. C. L. Chapelain, *Org. Biomol. Chem.* **2017**, *15*, 6242–6256
245. S. Chu, S. Wallace, M. D. Smith, *Angew. Chem. Int. Ed.* **2014**, *53*, 13826–13829
246. Y. Wang, A. Noble, E. L. Myers, V. K. Aggarwal, *Angew. Chem. Int. Ed.* **2016**, *55*, 4270–4274
247. R. Knorr, *Chem. Rev.* **2004**, *104*, 3795–3849
248. H. Yoneyama, M. Numata, K. Uemura, Y. Usami, S. Harusawa, *J. Org. Chem.* **2017**, *82*, 5538–5556
249. J. R. Coombs, J. P. Morken, *Angew. Chem. Int. Ed.* **2016**, *55*, 2636–2649
250. C. Bonini, G. Righi, *Tetrahedron* **2002**, *58*, 4981–5021
251. K. B. Sharpless, *Angew. Chem. Int. Ed.* **2002**, *41*, 2024–2032
252. T. Katsuki, V. S. Martin, *Org. React.* **1995**, *48*, 1–299

253. H. C. Kolb, M. S. Van Nieuwenhze, K. B. Sharpless, *Chem. Rev.* **1994**, *94*, 2483–2547

254. A. Armstrong, A. G. Draffan, *J. Chem. Soc. Perkin Trans. 1* **2001**, 2861–2873

255. C. Li, X. Yu, X. Lei, *Org. Lett.* **2010**, *12*, 4284–4287

256. D. I. Metelitsa, *Russ. Chem. Rev.* **1972**, *41*, 807–821

257. B. M. Lynch, K. H. Pausacker, *J. Chem. Soc.* **1955**, 1525–1531

258. W. Adam, C. R. Saha-Müller, C.-G. Zhao, *Org. React.* **2002**, *61*, 219–516

259. R. W. Murray, *Chem. Rev.* **1989**, *89*, 1187–1201

260. D. V. Deubel, *J. Org. Chem.* **2001**, *66*, 3790–3796

261. A. Düfert, D. B. Werz, *J. Org. Chem.* **2008**, *73*, 5514–5519

262. N. N. Schwartz, J. H. Blumbergs, *J. Org. Chem.* **1964**, *29*, 1976–1979

263. E. J. Corey, W. Su, *J. Am. Chem. Soc.* **1987**, *109*, 7534–7536

264. Y. Kishi, M. Aratani, H. Tanino, T. Fukuyama, T. Goto, S. Inoue, S. Sugiura, H. Kakoi, *Chem. Commun.* **1972**, 64–65

265. P. A. Wender, S. L. Eck, *Tetrahedron Lett.* **1982**, *23*, 1871–1874

266. D. Yang, *Acc. Chem. Res.* **2004**, *37*, 497–505

267. H. Mikula, D. Svatunek, D. Lumpi, F. Glöcklhofer, C. Hametner, J. Fröhlich, *Org. Process Res. Dev.* **2013**, *17*, 313–316

268. D. Yang, X.-Y. Ye, M. Xu, *J. Org. Chem.* **2000**, *65*, 2208–2217

269. T. Itoh, K. Jitsukawa, K. Kaneda, S. Teranishi, *J. Am. Chem. Soc.* **1979**, *101*, 159–169

270. C. K. Sams, K. A. Jørgensen, *Acta Chem. Scand.* **1995**, *49*, 839–847

271. E. P. Talsi, V. D. Chinakov, V. P. Babenko, K. I. Zamaraev, *J. Mol. Cat.* **1993**, *81*, 235–254

272. M. Vandichel, K. Leus, P. Van Der Voort, M. Waroquier, V. Van Speybroeck, *J. Catal.* **2012**, *294*, 1–18

273. W. Adam, T. Wirth, *Acc. Chem. Res.* **1999**, *32*, 703–710

274. W. Adam, R. Paredes, A. K. Smerza, L. A. Veloza, *Eur. J. Org. Chem.* **1998**, 349–354

275. B. M. Trost, W.-J. Bai, C. E. Stivala, C. Hohn, C. Poock, M. Heinrich, S. Xu, J. Rey, *J. Am. Chem. Soc.* **2018**, *140*, 17316–17326

276. C. Clark, P. Hermans, O. Meth-Cohn, C. Moore, H. C. Taljaard, G. van Vuuren, *J. Chem. Soc. Chem. Commun.* **1986**, 1378–1380

277. E. Weitz, A. Scheffer, *Chem. Ber.* **1921**, *54*, 2344–2353

278. V. K. Yadav, K. K. Kapoor, *Tetrahedron* **1995**, *51*, 8573–8584

279. K. Matsunaga, N. Saito, H. Kogen, K. Takatori, *Org. Lett.* **2019**, *21*, 6054–6057

280. B. Cheng, G. Volpin, J. Morstein, D. Trauner, *Org. Lett.* **2018**, *20*, 4358–4361

281. M. E. Assal, P. A. Peixoto, R. Coffinier, T. Garnier, D. Deffieux, K. Miqueu, J.-M. Sotiropoulos, L. Pouységu, S. Quideau, *J. Org. Chem.* **2017**, *82*, 11816–11828

282. J. Clarke, K. J. Bonney, M. Yaqoob, S. Solanki, H. S. Rzepa, A. J. P. White, D. S. Millan, D. C. Braddock, *J. Org. Chem.* **2016**, *81*, 9539–9552

283. S. J. Danishefsky, J. J. Masters, W. B. Young, J. T. Link, L. B. Snyder, T. V. Magee, D. K. Jung, R. C. A. Isaacs, W. G. Bornmann, C. A. Alaimo, C. A. Coburn, M. J. Di Grandi, *J. Am. Chem. Soc.* **1996**, *118*, 2843–2859

284. D. J. Tao, Y. Slutskyy, M. Muuronen, A. Le, P. Kohler, L. E. Overman, *J. Am. Chem. Soc.* **2018**, *140*, 3091–3102

285. N. Toelle, H. Weinstabl, T. Gaich, J. Mulzer, *Angew. Chem. Int. Ed.* **2014**, *53*, 3859–3862

286. C. M. Neuhaus, M. Liniger, M. Stieger, K.-H. Altmann, *Angew. Chem. Int. Ed.* **2013**, *52*, 5866–5870

287. V. Hickmann, A. Kondoh, B. Gabor, M. Alcarazo, A. Fürstner, *J. Am. Chem. Soc.* **2011**, *133*, 13471–13480

288. G. Kim, T. ik Sohn, D. Kim, R. S. Paton, *Angew. Chem. Int. Ed.* **2014**, *53*, 272–276

289. H. F. Zipfel, E. M. Carreira, *Chem. Eur. J.* **2015**, *21*, 12475–12480

290. T. Katsuki, K. B. Sharpless, *J. Am. Chem. Soc.* **1980**, *102*, 5974–5976
291. Y. Gao, J. M. Klunder, R. M. Hanson, H. Masamune, S. Y. Ko, K. B. Sharpless, *J. Am. Chem. Soc.* **1987**, *109*, 5765–5780
292. A. S. Fernandes, P. Maître, T. C. Correra, *J. Phys. Chem. A* **2019**, *123*, 1022–1029
293. M. G. Finn, K. B. Sharpless, *J. Am. Chem. Soc.* **1991**, *113*, 113–126
294. Y.-D. Wu, D. K. W. Lai, *J. Am. Chem. Soc.* **1995**, *117*, 11327–11336
295. S. S. Woodard, M. G. Finn, K. B. Sharpless, *J. Am. Chem. Soc.* **1991**, *113*, 106–113
296. R. A. Johnson, K. B. Sharpless, *Comp. Org. Syn.* **1991**, *7*, 389–436
297. S. Y. Ko, A. W. M. Lee, S. Masamune, L. A. Reed, III, K. B. Sharpless, F. J. Walker, *Tetrahedron* **1990**, *46*, 245–264
298. M. M. Heravi, T. B. Lashaki, N. Poorahmed, *Tetrahedron: Asymmetry* **2015**, *26*, 405–495
299. S. M. Kaplan, P. E. Floreancig, *Angew. Chem. Int. Ed.* **2018**, *57*, 15866–15870
300. D. Lee, H. Kondo, Y. Kuwayama, K. Takahashi, S. Arima, S. Omura, M. Ohtawa, T. Nagamitsu, *Tetrahedron* **2019**, *75*, 3178–3185
301. K. Nishimura, T. Sakaguchi, Y. Nanba, Y. Suganuma, M. Morita, S. Hong, Y. Lu, B. Jun, N. G. Bazan, M. Arita, Y. Kobayashi, *J. Org. Chem.* **2018**, *83*, 154–166
302. K. C. Nicolaou, G. Bellavance, M. Buchman, K. K. Pulukuri, *J. Am. Chem. Soc.* **2017**, *139*, 15636–15639
303. M.-C. Lamas, M. Malacria, S. Thorimbert, *Eur. J. Org. Chem.* **2011**, 2777–2780
304. H. Miyaoka, Y. Isaji, Y. Kajiwara, I. Kunimune, Y. Yamada, *Tetrahedron Lett.* **1998**, *39*, 6503–6506
305. K. C. Nicolaou, N. P. King, M. R. V. Finlay, Y. He, F. Roschangar, D. Vourloumis, H. Vallberg, F. Sarabia, S. Ninkovic, D. Hepworth, *Bioorg. Med. Chem.* **1999**, *7*, 665–697
306. C. Nilewski, N. R. Deprez, T. C. Fessard, D. B. Li, R. W. Geisser, E. M. Carreira, *Angew. Chem. Int. Ed.* **2011**, *50*, 7940–7943
307. J. Mulzer, A. Mantoulidis, E. Öhler, *Tetrahedron Lett.* **1997**, *38*, 7725–7728
308. B. M. Gallagher, H. Zhao, M. Pesant, F. G. Fang, *Tetrahedron Lett.* **2005**, *46*, 923–926
309. T. Müller, M. Göhl, I. Lusebrink, K. Dettner, K. Seifert, *Eur. J. Org. Chem.* **2012**, 2323–2330
310. G. Sabitha, P. Gopal, C. N. Reddy, J. S. Yadav, *Tetrahedron Lett.* **2009**, *50*, 6298–6302
311. E. M. McGarrigle, D. G. Gilheany, *Chem. Rev.* **2005**, *105*, 1563–1602
312. T. Flessner, S. Doye, *J. Prakt. Chem.* **1999**, *341*, 436–444
313. T. Katsuki, *Synlett* **2003**, 281–297
314. T. Linker, *Angew. Chem. Int. Ed. Engl.* **1997**, *36*, 2060–2062
315. D. Feichtinger, D. A. Plattner, *Angew. Chem. Int. Ed. Engl.* **1997**, *36*, 1718–1719
316. W. Adam, K. J. Roschmann, C. R. Saha-Möller, D. Seebach, *J. Am. Chem. Soc.* **2002**, *124*, 5068–5073
317. T. Bogaerts, S. Wouters, P. Van Der Voort, V. Van Speybroeck, *ChemCatChem* **2015**, *7*, 2711–2719
318. E. N. Jacobsen, L. Deng, Y. Furukawa, L. E. Martínez, *Tetrahedron* **1994**, *50*, 4323–4334
319. B. D. Brandes, E. N. Jacobsen, *J. Org. Chem.* **1994**, *59*, 4378–7380
320. S. Chang, N. H. Lee, E. N. Jacobsen, *J. Org. Chem.* **1993**, *58*, 6939–6941
321. W. Zhang, J. L. Loebach, S. R. Wilson, E. N. Jacobsen, *J. Am. Chem. Soc.* **1990**, *112*, 2801–2803
322. H. Sasaki, R. Irie, T. Hamada, K. Suzuki, T. Katsuki, *Tetrahedron* **1994**, *50*, 11827–11838
323. J. E. Lynch, W.-B. Choi, H. R. O. Churchill, R. P. Volante, R. A. Reamer, R. G. Ball, *J. Org. Chem.* **1997**, *62*, 9223–9228
324. T. Mori, S. Higashibayashi, T. Goto, M. Kohno, Y. Satouchi, K. Shinko, K. Suzuki, S. Suzuki, H. Tohmiya, K. Hashimoto, M. Nakata, *Chem. Asian J.* **2008**, *3*, 984–1012
325. J. Justicia, A. G. Campaña, B. Bazdi, R. Robles, J. M. Cuerva, J. E. Oltra, *Adv. Synth. Catal.* **2008**, *350*, 571–576

326. E. N. Jacobsen, *Acc. Chem. Res.* **2000**, *33*, 421–431
327. J. F. Larrow, S. E. Schaus, E. N. Jacobsen, *J. Am. Chem. Soc.* **1996**, *118*, 7420–7421
328. H. Lebel, E. N. Jacobsen, *Tetrahedron Lett.* **1999**, *40*, 7303–7306
329. S. E. Schaus, B. D. Brandes, J. F. Larrow, M. Tokunaga, K. B. Hansen, A. E. Gould, M. E. Furrow, E. N. Jacobsen, *J. Am. Chem. Soc.* **2001**, *124*, 1307–1315
330. E. N. Jacobsen, F. Kakiuchi, R. G. Konsler, J. F. Larrow, M. Tokunaga, *Tetrahedron Lett.* **1997**, *38*, 773–776
331. J. M. Ready, E. N. Jacobsen, *J. Am. Chem. Soc.* **1999**, *121*, 6086–6087
332. D. D. Ford, L. P. C. Nielsen, S. J. Zuend, E. N. Jacobsen, *J. Am. Chem. Soc.* **2013**, *135*, 15595–15608
333. L. P. C. Nielsen, C. P. Stevenson, D. G. Blackmond, E. N. Jacobsen, *J. Am. Chem. Soc.* **2004**, *126*, 1360–1362
334. M. P. Mower, D. G. Blackmond, *ACS Catal.* **2018**, *8*, 5977–5982
335. M. K. Gurjar, L. M. Krishna, B. V. N. B. S. Sarma, M. S. Chorghade, *Org. Process Res. Dev.* **1998**, *2*, 422–424
336. B. B. Snider, J. Zhou, *Org. Lett.* **2006**, *8*, 1283–1286
337. I. M. Pastor, M. Yus, *Curr. Org. Chem.* **2005**, *9*, 1–29
338. J. A. Kalow, A. G. Doyle, *J. Am. Chem. Soc.* **2010**, *132*, 3268–3269
339. Y. Shi, *Acc. Chem. Res.* **2004**, *37*, 488–496
340. O. A. Wong, Y. Shi, *Chem. Rev.* **2008**, *108*, 3958–3987
341. M. Frohn, Y. Shi, *Synthesis* **2000**, 1979–2000
342. Y. Zhu, Q. Wang, R. G. Cornwall, Y. Shi, *Chem. Rev.* **2014**, *114*, 8199–8256
343. Z.-X. Wang, Y. Tu, M. Frohn, Y. Shi, *J. Org. Chem.* **1997**, *62*, 2328–2329
344. H. Tian, X. She, H. Yu, L. Shu, Y. Shi, *J. Org. Chem.* **2002**, *67*, 2435–2446
345. M. Frohn, M. Dalkiewicz, Y. Tu, Z.-X. Wang, Y. Shi, *J. Org. Chem.* **1998**, *63*, 2948–2953
346. Y. Zhu, Y. Tu, H. Yu, Y. Shi, *Tetrahedron Lett.* **1998**, *39*, 7819–7822
347. D. W. Hoard, E. D. Moher, M. J. Martinelli, B. H. Norman, *Org. Lett.* **2002**, *4*, 1813–1815
348. F. Cachoux, T. Isarno, M. Wartmann, K.-H. Altmann, *ChemBioChem* **2006**, *7*, 54–57
349. A. Iwata, S. Inuki, S. Oishi, N. Fujii, H. Ohno, *J. Org. Chem.* **2011**, *76*, 5506–5512
350. A. B. Smith, S. P. Walsh, M. Frohn, M. O. Duffey, *Org. Lett.* **2005**, *7*, 139–142
351. T. Barton, D. Siegel, *Synthesis* **2012**, *44*, 2770–2778
352. Y.-R. Yang, Z.-W. Lai, L. Shen, J.-Z. Huang, X.-D. Wu, J.-L. Yin, K. Wei, *Org. Lett.* **2012**, *12*, 3430–3433
353. L. Shu, Y. Shi, *Tetrahedron Lett.* **1999**, *40*, 8721–8724
354. K. B. Sharpless, T. R. Verhoeven, *Aldrichimica Acta* **1979**, *12*, 63–74
355. L. Degennaro, P. Trinchera, R. Luisi, *Chem. Rev.* **2014**, *114*, 7881–7929
356. H. Pellissier, *Tetrahedron* **2010**, *66*, 1509–1555
357. K. Guthikonda, P. M. Wehn, B. J. Caliando, J. Du Bois, *Tetrahedron* **2006**, *62*, 11331–11342
358. H. Kawabata, K. Omura, T. Uchida, T. Katsuki, *Chem. Asian J.* **2007**, *2*, 248–256
359. B. M. Trost, G. Dong, *J. Am. Chem. Soc.* **2006**, *128*, 6054–6055
360. J. S. Yadav, G. Satheesh, C. V. S. R. Murthy, *Org. Lett.* **2010**, *12*, 2544–2547
361. G. F. Keaney, J. L. Wood, *Tetrahedron Lett.* **2005**, *46*, 4031–4034
362. C. M. Diaper, A. Sutherland, B. Pillai, M. N. G. James, P. Semchuk, J. S. Blancharde, J. C. Vederas, *Org. Biomol. Chem.* **2005**, *3*, 4402–4411
363. E. J. Corey, M. Chaykovsky, *J. Am. Chem. Soc.* **1965**, *87*, 1353–1364
364. M. M. Heravi, S. Asadi, N. Nazari, B. M. Lashkariani, *Curr. Org. Synth.* **2016**, *13*, 308–333
365. G. L. Beutner, D. T. George, *Org. Process Res. Dev.* **2023**, *27*, 10–41
366. V. K. Aggarwal, C. L. Winn, *Acc. Chem. Res.* **2004**, *37*, 611–620
367. E. M. McGarrigle, E. L. Myers, O. Illa, M. A. Shaw, S. L. Riches, V. K. Aggarwal, *Chem. Rev.* **2007**, *107*, 5841–5883

368. K. G. M. Kou, B. X. Li, J. C. Lee, G. M. Gallego, T. P. Lebold, A. G. DiPasquale, R. Sarpong, *J. Am. Chem. Soc.* **2016**, *138*, 10830–10833

369. Z. G. Brill, H. K. Grover, T. J. Maimone, *Science* **2016**, *352*, 1078–1082

370. K. Kong, J. A. Enquist, M. E. McCallum, G. M. Smith, T. Matsumaru, E. Menhaji-Klotz, J. L. Wood, *J. Am. Chem. Soc.* **2013**, *135*, 10890–10893

371. M. Schröder, *Chem. Rev.* **1980**, *80*, 187–213

372. C. J. R. Bataille, T. J. Donohoe, *Chem. Soc. Rev.* **2011**, *40*, 114–128

373. M. C. Noe, M. A. Letavic, S. L. Snow, S. W. McCombie, *Org. React.* **2005**, *66*, 109–625

374. S. Dapprich, G. Ujaque, F. Maseras, A. Lledós, D. G. Musaev, K. Morokuma, *J. Am. Chem. Soc.* **1996**, *118*, 11660–11661

375. M. Torrent, L. Deng, M. Duran, M. Sola, T. Ziegler, *Organometallics* **1997**, *16*, 13–19

376. U. Pidun, C. Boehme, G. Frenking, *Angew. Chem. Int. Ed. Engl.* **1996**, *35*, 2817–2820

377. G. Ujaque, F. Maseras, A. Lledós, *Eur. J. Org. Chem.* **2003**, 833–839

378. A. J. DelMonte, J. Haller, K. N. Houk, K. B. Sharpless, D. A. Singleton, T. Strassner, A. A. Thomas, *J. Am. Chem. Soc.* **1997**, *119*, 9907–9908

379. J. S. M. Wai, I. Marko, J. S. Svendsen, M. G. Finn, E. N. Jacobsen, K. B. Sharpless, *J. Am. Chem. Soc.* **1123**, *111*, 1123–1125

380. A. A. Hussein, Y. Ma, G. A. I. Moustafa, *Catal. Sci. Technol.* **2022**, *12*, 880–893

381. D. W. Nelson, A. Gypser, P. T. Ho, H. C. Kolb, T. Kondo, H.-L. Kwong, D. V. McGrath, A. E. Rubin, P.-O. Norrby, K. P. Gable, K. B. Sharpless, *J. Am. Chem. Soc.* **1997**, *119*, 1840–1858

382. V. VanRheenen, R. C. Kelly, D. Y. Cha, *Tetrahedron Lett.* **1976**, *17*, 1973–1976

383. M. Minato, K. Yamamoto, J. Tsuji, *J. Org. Chem.* **1990**, *55*, 766–768

384. J. V. Mulcahy, J. R. Walker, J. E. Merit, A. Whitehead, J. Du Bois, *J. Am. Chem. Soc.* **2016**, *138*, 5994–6001

385. H. Tang, N. Yusuff, J. L. Wood, *Org. Lett.* **2001**, *3*, 1563–1566

386. F. Kawagishi, T. Toma, T. Inui, S. Yokoshima, T. Fukuyama, *J. Am. Chem. Soc.* **2013**, *135*, 13684–13687

387. T. Achard, S. Bellemin-Laponnaz, *Eur. J. Org. Chem.* **2021**, 877–896

388. T. K. M. Shing, E. K. W. Tam, V. W.-F. Tai, I. H. F. Chung, Q. Jiang, *Chem. Eur. J.* **1996**, *2*, 50–57

389. B. Plietker, M. Niggemann, *J. Org. Chem.* **2005**, *70*, 2402–2405

390. M. Couturier, B. M. Andresen, J. B. Jorgensen, J. L. Tucker, F. R. Busch, S. J. Brenek, P. Dubé, D. J. am Ende, J. T. Negri, *Org. Process Res. Dev.* **2002**, *6*, 42–48

391. H.-L. Kwong, C. Sorato, Y. Ogino, H. Chen, K. B. Sharpless, *Tetrahedron Lett.* **1990**, *31*, 2999–3002

392. E. J. Corey, A. Guzman-Perez, M. C. Noe, *Tetrahedron Lett.* **1995**, *36*, 3481–3484

393. U. Ramulu, S. Rajaram, D. Ramesh, K. S. Babu, *Tet. Asymm.* **2015**, *26*, 928–934

394. J.-L. Chen, Z.-W. You, F.-L. Qing, *J. Fluorine Chem.* **2013**, *155*, 143–150

395. F. Yokokawa, H. Sameshima, D. Katagiri, T. Aoyama, T. Shioiri, *Tetrahedron* **2002**, *58*, 9445–9458

396. S. Raghavan, V. V. Kumar, *Tetrahedron* **2013**, *69*, 4835–4844

397. A. B. Smith III, S. Dong, R. J. Fox, J. B. Brenneman, J. A. Vanecko, T. Maegawa, *Tetrahedron* **2011**, *67*, 9809–9828

398. K. Kuwata, M. Suzuki, Y. Inami, K. Hanaya, T. Sugai, M. Shoji, *Tetrahedron Lett.* **2014**, *55*, 2856–2858

399. S. Sato, A. Hirayama, H. Ueda, H. Tokuyama, *Asian J. Org. Chem.* **2017**, *6*, 54–58

400. D. J. Mergott, S. A. Frank, W. R. Roush, *Proc. Nat. Acad. Sci.* **2004**, *101*, 11955–11959

401. J. A. Bodkin, M. D. McLeod, *J. Chem. Soc. Perkin Trans. 1* **2002**, 2733–2746

402. D. Nilov, O. Reiser, *Adv. Synth. Catal.* **2002**, *344*, 1169–1173

403. K. Muñiz, *Chem. Soc. Rev.* **2004**, *33*, 166–174

404. H. C. Kolb, K. B. Sharpless in *Transition Metals for Organic Synthesis* (Eds.: M. Beller, C. Bolm), Wiley-VCH, **2004**, Chapter 2.6, pp. 309–336

405. P. G. M. Wuts, T. W. Greene, *Greene's Protective Groups in Organic Synthesis*, John Wiley & Sons, 4th ed., **2007**

406. B. Tao, G. Schlingloff, K. B. Sharpless, *Tetrahedron Lett.* **1998**, *39*, 2507–2510

407. M. Bruncko, G. Schlingloff, K. B. Sharpless, *Angew. Chem. Int. Ed. Engl.* **1997**, *36*, 1483–1486

408. K. L. Reddy, K. B. Sharpless, *J. Am. Chem. Soc.* **1998**, *120*, 1207–1213

409. J. Rudolph, P. C. Sennhenn, C. P. Vlaar, K. B. Sharpless, *Angew. Chem. Int. Ed. Engl.* **1996**, *35*, 2810–2813

410. M. M. Heravi, T. B. Lashaki, B. Fattahi, V. Zadsirjan, *RSC Adv.* **2018**, *8*, 6634–6659

411. W. Kurosawa, T. Kan, T. Fukuyama, *J. Am. Chem. Soc.* **2003**, *125*, 8112–8113

412. C. E. Masse, A. J. Morgan, J. S. Panek, *Org. Lett.* **2000**, *2*, 2571–2573

413. S. Hirano, S. Ichikawa, A. Matsuda, *Angew. Chem. Int. Ed.* **2005**, *44*, 1854–1856

414. C.-G. Yang, J. Wang, X.-X. Tang, B. Jiang, *Tet. Asymm.* **2002**, *13*, 383–394

415. J. A. Bodkin, G. B. Bacskay, M. D. McLeod, *Org. Biomol. Chem.* **2008**, *6*, 2544–2553

416. M. Harding, J. A. Bodkin, C. A. Hutton, M. D. McLeod, *Synlett* **2005**, 2829–2831

417. R. S. Brown, *Acc. Chem. Res.* **1997**, *30*, 131–137

418. G. Bellucci, C. Chiappe, R. Bianchini, *Ind. Chem. Library* **1995**, *7*, 128–151

419. M.-F. Ruasse, *Adv. Phys. Org. Chem.* **1993**, *28*, 207–291

420. J. H. Rolston, K. Yates, *J. Am. Chem. Soc.* **1969**, *91*, 1469–1476

421. M. L. Poutsma, *J. Am. Chem. Soc.* **1965**, *87*, 2172–2183

422. R. C. Fahey, C. Schubert, *J. Am. Chem. Soc.* **1965**, *87*, 5172–5179

423. S. Winstein, *J. Am. Chem. Soc.* **1942**, *64*, 2792–2795

424. L. Crombie, R. D. Wyvill, *J. Chem. Soc. Perkin Trans. 1* **1985**, 1971–1978

425. As of April 26th 2020 from Acros, Sigma-Aldrich, TCI and Matrix Scientific websites. The largest batch size was used for price calculation.

426. S. A. Snyder, Z.-Y. Tang, R. Gupta, *J. Am. Chem. Soc.* **2009**, *131*, 5744–5745

427. Z. Wu, S. R. Harutyunyan, A. J. Minnaard, *Chem. Eur. J.* **2014**, *20*, 14250–14255

428. I. G. Molnár, C. Thiehoff, M. C. Holland, R. Gilmour, *ACS Catal.* **2016**, *6*, 7167–7173

429. J. R. Wolstenhulme, V. Gouverneur, *Acc. Chem. Res.* **2014**, *47*, 3560–3570

430. S. Hara, J. Nakahigashi, K. Ishi-i, M. Sawaguchi, H. Sakai, T. Fukuhara, N. Yoneda, *Synlett* **1998**, 495–496

431. S. M. Banik, J. W. Medley, E. N. Jacobsen, *J. Am. Chem. Soc.* **2016**, *138*, 5000–5003

432. I. G. Molnár, R. Gilmour, *J. Am. Chem. Soc.* **2016**, *138*, 5004–5007

433. J. C. Sarie, C. Thiehoff, R. J. Mudd, C. G. Daniliuc, G. Kehr, R. Gilmour, *J. Org. Chem.* **2017**, *82*, 11792–11798

434. J. Rodriguez, J.-P. Dulcère, *Synthesis* **1993**, 1177–1205

435. S. A. Snyder, D. S. Treitler, A. P. Brucks, *Aldrichimica Acta* **2011**, *44*, 27–40

436. S. Ranganathan, K. M. Muraleedharan, N. K. Vaish, N. Jayaraman, *Tetrahedron* **2004**, *60*, 5273–5308

437. S. R. Chemler, M. T. Bovino, *ACS Catal.* **2013**, *3*, 1076–1091

438. S. Knapp, P. J. Kukkola, S. Sharma, S. Pietranico, *Tetrahedron Lett.* **1987**, *28*, 5399–5402

439. T. L. Shih, H. Mrozik, J. Ruiz-Sanchez, M. H. Fisher, *J. Org. Chem.* **1989**, *54*, 1459–1463

440. D. C. Beshore, A. B. Smith, *J. Am. Chem. Soc.* **2007**, *129*, 4148–4149

441. J. J. Swidorski, J. Wang, R. P. Hsung, *Org. Lett.* **2006**, *8*, 777–780

442. B. S. Dyson, J. W. Burton, T. Sohn, B. Kim, H. Bae, D. Kim, *J. Am. Chem. Soc.* **2012**, *134*, 11781–11790

443. S. Terashima, S. Jew, *Tetrahedron Lett.* **1977**, *18*, 1005–1008

444. A. J. Cresswell, S. T.-C. Eey, S. E. Denmark, *Nature Chem.* **2015**, *7*, 146–152

445. S. E. Denmark, W. E. Kuester, M. T. Burk, *Angew. Chem. Int. Ed.* **2012**, *51*, 10938–10953

446. A. J. Cresswell, S. T.-C. Eey, S. E. Denmark, *Angew. Chem. Int. Ed.* **2015**, *54*, 15642–15682

447. J. Bock, S. Guria, V. Wedek, U. Hennecke, *Chem. Eur. J.* **2020**, *27*, 4517–4530

448. K. D. Ashtekar, A. Jaganathan, B. Borhan, D. C. Whitehead, *Org. React.* **2021**, *105*, 1–266

449. M. L. Landry, N. Z. Burns, *Acc. Chem. Res.* **2018**, *51*, 1260–1271

450. F. J. Seidl, C. Min, J. A. Lopez, N. Z. Burns, *J. Am. Chem. Soc.* **2018**, *140*, 15646–15650

451. M. L. Landry, D. X. Hu, G. M. McKenna, N. Z. Burns, *J. Am. Chem. Soc.* **2016**, *138*, 5150–5158

452. D. X. Hu, F. J. Seidl, C. Bucher, N. Z. Burns, *J. Am. Chem. Soc.* **2015**, *137*, 3795–3798

453. M. L. Landry, G. M. McKenna, N. Z. Burns, *J. Am. Chem. Soc.* **2019**, *141*, 2867–2871

454. H. C. Brown, *Organic Synthesis via Boranes*, John Wiley & Sons, **1975**

455. H. C. Brown in *Comprehensive Organometallic Chemistry*, *Vol. 7* (Eds.: G. Wilkinson, F. G. A. Stone, E. W. Abel), Elsevier, **1982**, Chapter 45.1, pp. 111–142

456. H. C. Brown, B. Singararn, *Pure Appl. Chem.* **1987**, *59*, 879–894

457. M. Zaidlewicz in *Kirk-Othmer Encyclopedia of Chemical Technology*, *Vol. 13*, Wiley-VCH, **2005**, Chapter Hydroboration, pp. 631–684

458. M. Zaidlewicz in *Comprehensive Organometallic Chemistry*, *Vol. 7* (Eds.: G. Wilkinson, F. G. A. Stone, E. W. Abel), Elsevier, **1982**, Chapter 45.2, pp. 143–160

459. H. C. Brown, P. Heim, N. M. Yoon, *J. Am. Chem. Soc.* **1970**, *92*, 1637–1346

460. H. C. Brown, S. Krishnamurthy, N. M. Yoon, *J. Org. Chem.* **1976**, *41*, 1778–1791

461. C. F. Lane, G. W. Kabalka, *Tetrahedron* **1976**, *32*, 981–990

462. H. C. Brown, J. Chandrasekharan, K. K. Wang, *Pure Appl. Chem.* **1983**, *55*, 1387–1414

463. K. K. Wang, H. C. Brown, *J. Am. Chem. Soc.* **1982**, *104*, 7148–7155

464. H. C. Brown, J. Chandrasekharan, *J. Am. Chem. Soc.* **1984**, *106*, 1863–1865

465. K. Houk, N. G. Rondan, Y.-D. Wu, J. T. Metz, M. N. Paddon-Row, *Tetrahedron* **1984**, *40*, 2257–2274

466. S. E. Wood, B. Rickborn, *J. Org. Chem.* **1983**, *48*, 555–562

467. I. Beletskaya, A. Pelter, *Tetrahedron* **1997**, *53*, 4957–5026

468. J. M. Brown, B. N. Nguyen in *Science of Synthesis: Stereoselective Synthesis*, *Vol. 1* (Ed.: J. G. de Vries), Thieme, **2011**, Chapter 1.7, pp. 295–324

469. S. J. Geier, C. M. Vogels, S. A. Westcott, *ACS Symposium Series* **2016**, *1236*, 209–225

470. A.-M. Carroll, T. P. O. Sullivan, P. J. Guiry, *Adv. Synth. Catal.* **2005**, *349*, 609–631

471. C. M. Crudden, D. Edwards, *Eur. J. Org. Chem.* **2003**, 4695–4712

472. K. Burgess, M. J. Ohlmeyer, *Chem. Rev.* **1991**, *91*, 1179–1191

473. D. Männig, H. Nöth, *Angew. Chem. Int. Ed. Engl.* **1985**, *24*, 878–879

474. J. Zhang, B. Lou, G. Guo, L. Dai, *J. Org. Chem.* **1991**, *56*, 1670–1672

475. K. Masuda, M. Koshimizu, M. Nagatomo, M. Inoue, *Chem. Eur. J.* **2016**, *22*, 230–236

476. T. Suto, Y. Yanagita, Y. Nagashima, S. Takikawa, Y. Kurosu, N. Matsuo, T. Sato, N. Chida, *J. Am. Chem. Soc.* **2017**, *139*, 2952–2955

477. S. Diethelm, E. M. Carreira, *J. Am. Chem. Soc.* **2015**, *137*, 6084–6096

478. H.-L. Qin, J. S. Panek, *Org. Lett.* **2008**, *10*, 2477–2479

479. J. Adrian, C. B. W. Stark, *J. Org. Chem.* **2016**, *81*, 8175–8186

480. F. W. W. Hartrampf, D. Trauner, *J. Org. Chem.* **2017**, *82*, 8206–8212

481. C.-X. Zhuo, A. Fürstner, *Angew. Chem. Int. Ed.* **2016**, *55*, 6051–6056

482. A. E. Goetz, A. L. Silberstein, M. A. Corsello, N. K. Garg, *J. Am. Chem. Soc.* **2014**, *136*, 3036–3039

483. H. C. Brown, B. Singaram, *Acc. Chem. Res.* **1988**, *21*, 287–293

484. S. P. Thomas, V. K. Aggarwal, *Angew. Chem. Int. Ed.* **2009**, *48*, 1896–1898

485. A. Z. Gonzalez, J. G. Román, E. Gonzalez, J. Martinez, J. R. Medina, K. Matos, J. A. Soderquist, *J. Am. Chem. Soc.* **2008**, *130*, 9218–9219

486. S. Masamune, B. M. Kim, J. S. Petersen, T. Sato, S. J. Veenstra, T. Imai, *J. Am. Chem. Soc.* **1985**, *107*, 4549–4551

487. A. Richter, C. Hedberg, H. Waldmann, *J. Org. Chem.* **2011**, *76*, 6694–6702

488. S. Werle, T. Fey, J. M. Neudörfl, H.-G. Schmalz, *Org. Lett.* **2007**, *9*, 3555–3558

489. H. Wolleb, E. M. Carreira, *Angew. Chem. Int. Ed.* **2017**, *56*, 10890–10893

490. T. J. Fisher, P. H. Dussault, *Tetrahedron* **2017**, *73*, 4233–4258

491. S. G. Van Ornum, R. M. Champeau, R. Pariza, *Chem. Rev.* **2006**, *106*, 2990–3001

492. G. Y. Ishmuratov, Y. V. Legostaeva, L. P. Botsman, G. A. Tolstikov, *Russ. J. Org. Chem.* **2010**, *46*, 1593–1621

493. R. Criegee, *Angew. Chem. Int. Ed. Engl.* **1975**, *14*, 745–752

494. T. Veysoglu, L. A. Mitscher, J. K. Swayze, *Synthesis* **1980**, 807–810

495. E. O. Onyango, P. A. Jacobi, *J. Org. Chem.* **2012**, *77*, 7411–7427

496. Y. S. Cho, D. A. Carcache, Y. Tian, Y.-M. Li, S. J. Danishefsky, *J. Am. Chem. Soc.* **2004**, *126*, 14358–14359

497. R. S. Coleman, M. C. Walczak, E. L. Campbell, *J. Am. Chem. Soc.* **2005**, *127*, 16038–16039

498. L. F. Tietze, S.-C. Duefert, J. Clerc, M. Bischoff, C. Maaß, D. Stalke, *Angew. Chem. Int. Ed.* **2013**, *52*, 3191–3194

499. T. Maehara, K. Motoyama, T. Toma, S. Yokoshima, T. Fukuyama, *Angew. Chem. Int. Ed.* **2017**, *56*, 1549–1552

500. P. Yang, M. Yao, J. Li, Y. Li, A. Li, *Angew. Chem. Int. Ed.* **2016**, *55*, 6964–6968

501. T. Hida, J. Kikuchi, M. Kakinuma, H. Nogusa, *Org. Process Res. Dev.* **2010**, *14*, 1485–1489

502. J. A. Ragan, D. J. am Ende, S. J. Brenek, S. A. Eisenbeis, R. A. Singer, D. L. Tickner, J. J. Teixeira, Jr., B. C. Vanderplas, N. Weston, *Org. Process Res. Dev.* **2003**, *7*, 155–160

503. H. Bahrmann, H. Bach, G. D. Frey in *Ullmann's Encyclopedia of Industrial Chemistry*, Wiley-VCH, **2013**, Chapter *Oxo Synthesis*

504. R. Franke, D. Selent, A. Börner, *Chem. Rev.* **2012**, *112*, 5675–5732

505. S. Chakrabortty, A. A. Almasalma, J. G. de Vries, *Catal. Sci. Technol.* **2021**, *11*, 5388–5411

506. P. H. Gehrtz, V. Hirschbeck, B. Ciszek, I. Fleischer, *Synthesis* **2016**, *48*, 1573–1596

507. C. J. Cobley, C. H. Hanson, M. C. Lloyd, S. Simmonds, W. J. Peng, *Org. Process Res. Dev.* **2011**, *15*, 284–290

508. S. Dekeukeleire, M. D'hooghe, C. Müller, D. Vogt, N. De Kimpe, *New J. Chem.* **2010**, *34*, 1079–1083

509. J. R. Briggs, J. Klosin, G. T. Whiteker, *Org. Lett.* **2005**, *7*, 4795–4798

510. T. E. Müller, K. C. Hultzsch, M. Yus, F. Foubelo, M. Tada, *Chem. Rev.* **2008**, *108*, 3795–3892

511. A. L. Reznichenko, K. C. Hultzsch, *Org. React.* **2015**, *88*, 1–554

512. T. E. Müller, M. Beller, *Chem. Rev.* **1998**, *98*, 675–703

513. E. Bernoud, C. Lepori, M. Mellah, E. Schulz, J. Hannedouche, *Catal. Sci. Technol.* **2015**, *5*, 2017–2037

514. K. C. Hultzsch, *Org. Biomol. Chem.* **2005**, *3*, 1819–1824

515. G. A. Molander, E. D. Dowdy, *J. Org. Chem.* **1999**, *64*, 6515–6517

516. T. Jiang, T. Livinghouse, *Org. Lett.* **2010**, *12*, 4271–4273

517. J. D. Ha, J. K. Cha, *J. Am. Chem. Soc.* **1999**, *121*, 10012–10020

518. R. Y. Liu, S. L. Buchwald, *Acc. Chem. Res.* **2020**, *53*, 1229–1243

519. S. W. M. Crossley, C. Obradors, R. M. Martinez, R. A. Shenvi, *Chem. Rev.* **2016**, *116*, 8912–9000

520. H. Ishikawa, D. A. Colby, S. Seto, P. Va, A. Tam, H. Kakei, T. J. Rayl, I. Hwang, D. L. Boger, *J. Am. Chem. Soc.* **2009**, *131*, 4904–4916

521. M. Zaidlewicz, A. Wolan, M. Budny, *Comp. Org. Syn. II* **2014**, *8*, 877–963

522. A. P. Dobbs, F. K. I. Chio, *Comp. Org. Syn. II* **2014**, *8*, 964–998

523. M. Alami, A. Hamze, O. Provot, *ACS Catal.* **2019**, *9*, 3437–3466

524. Z. Song, T. Takahashi, *Comp. Org. Syn. II* **2014**, *8*, 838–876
525. P. Wipf, H. Jahn, *Tetrahedron* **1996**, *52*, 12853–12910
526. P. Wipf, C. Kendall, *Topics Organomet. Chem.* **2004**, *8*, 1–25
527. S. L. Buchwald, S. J. LaMaire, R. B. Nielsen, B. T. Watson, S. M. King, *Org. Synth.* **1993**, *71*, 77–82
528. Y. Zhao, V. Snieckus, *Org. Lett.* **2013**, *16*, 390–393
529. D. W. Hart, T. F. Blackburn, J. Schwartz, *J. Am. Chem. Soc.* **1975**, *97*, 679–680
530. D. Zhang, J. M. Ready, *J. Am. Chem. Soc.* **2007**, *129*, 12088–12089
531. D. L. J. Pinheiro, P. P. de Castro, G. W. Amarante, *Eur. J. Org. Chem.* **2018**, 4828–4844
532. A. B. Smith, S. S.-Y. Chen, F. C. Nelson, J. M. Reichert, B. A. Salvatore, *J. Am. Chem. Soc.* **1997**, *119*, 10935–10946
533. A. Arefolov, N. F. Langille, J. S. Panek, *Org. Lett.* **2001**, *3*, 3281–3284
534. M.-Z. Cai, Y. Wang, P.-P. Wang, *J. Organomet. Chem.* **2008**, *693*, 2954–2958
535. T. Kim, S. I. Lee, S. Kim, S. Y. Shim, D. H. Ryu, *Tetrahedron* **2019**, *75*, 130593
536. X.-Y. Bai, W.-W. Zhang, Q. Li, B.-J. Li, *J. Am. Chem. Soc.* **2018**, *140*, 506–514
537. D. E. Chavez, E. N. Jacobsen, *Angew. Chem. Int. Ed.* **2001**, *40*, 3667–3670
538. D. Trauner, J. B. Schwarz, S. J. Danishefsky, *Angew. Chem. Int. Ed.* **1999**, *38*, 3542–3545
539. K. A. Babiak, J. R. Behling, J. H. Dygos, K. T. McLaughlin, J. S. Ng, V. J. Kalish, S. W. Kramer, R. L. Shone, *J. Am. Chem. Soc.* **1990**, *112*, 7441–7442
540. B. H. Lipshutz, M. R. Wood, *J. Am. Chem. Soc.* **1994**, *116*, 11689–11702
541. C. F. Thompson, T. F. Jamison, E. N. Jacobsen, *J. Am. Chem. Soc.* **2001**, *123*, 9974–9983
542. T. Hu, J. S. Panek, *J. Am. Chem. Soc.* **2002**, *124*, 11368–11378
543. D. S. Matteson, *Stereodirected Synthesis with Organoboranes*, Springer, **1995**
544. M. Zaidlewicz in *Comprehensive Organometallic Chemistry*, Vol. 7 (Eds.: G. Wilkinson, F. G. A. Stone, E. W. Abel), Elsevier, **1982**, Chapter 45.4, pp. 199–227
545. G. Zweifel, G. M. Clark, N. L. Polston, *J. Am. Chem. Soc.* **1971**, *93*, 3395–3399
546. G. A. Molander, F. Dehmel, *J. Am. Chem. Soc.* **2004**, *126*, 10313–10318
547. A. K. Mapp, C. H. Heathcock, *J. Org. Chem.* **1999**, *64*, 23–27
548. Z. Meng, A. Fürstner, *J. Am. Chem. Soc.* **2020**, *142*, 11703–11708
549. I. J. Munslow in *Modern Reduction Methods* (Eds.: P. G. Andersson, I. J. Munslow), Wiley-VCH, **2008**, Chapter 15, pp. 363–385
550. J. Chen, J. Guo, Z. Lu, *Chin. J. Chem.* **2018**, *36*, 1075–1109
551. H. Jang, A. R. Zhugralin, Y. Lee, A. H. Hoveyda, *J. Am. Chem. Soc.* **2011**, *133*, 7859–7871
552. Y. Lee, H. Jang, A. H. Hoveyda, *J. Am. Chem. Soc.* **2009**, *131*, 18234–18235
553. B. M. Trost, Z. T. Ball, *Synthesis* **2005**, 853–887
554. R. Corberán, N. W. Mszar, A. H. Hoveyda, *Angew. Chem. Int. Ed.* **2011**, *50*, 7079–7082
555. D. V. Reddy, G. Sabitha, T. P. Rao, J. S. Yadav, *Org. Lett.* **2016**, *18*, 4202–4205
556. S. Cacchi, G. Fabrizi in *Handbook of Organopalladium Chemistry for Organic Synthesis* (Ed.: E. Negishi), John Wiley & Sons, **2002**, Chapter IV.2.5, pp. 1335–1359
557. V. Gevorgyan, Y. Yamamoto in *Handbook of Organopalladium Chemistry for Organic Synthesis* (Ed.: E. Negishi), John Wiley & Sons, **2002**, Chapter IV.2.6, pp. 1361–1367
558. A. Düfert, D. Werz, *Chem. Eur. J.* **2016**, *22*, 16718–16732
559. S. S. Goh, G. Chaubet, B. Gockel, M.-C. A. Cordonnier, H. Baars, A. W. Phillips, E. A. Anderson, *Angew. Chem. Int. Ed.* **2015**, *54*, 12618–12621
560. E. Negishi, D. Y. Kondakov, *Chem. Soc. Rev.* **1996**, *25*, 417–426
561. L. V. Parfenova, L. M. Khalilov, U. M. Dzhemilev, *Russ. Chem. Rev.* **2012**, *81*, 524–548
562. K. K. Anantoju, B. S. Mohd, T. C. Maringanti, *Tetrahedron Lett.* **2017**, *58*, 1499–1500
563. T. R. Hoye, M. E. Danielson, A. E. May, H. Zhao, *J. Org. Chem.* **2010**, *75*, 7052–7060
564. B. H. Lipshutz, B. Amorelli, *J. Am. Chem. Soc.* **2009**, *131*, 1396–1397

565. M. Heinrich, J. J. Murphy, M. K. Ilg, A. Letort, J. T. Flasz, P. Philipps, A. Fürstner, *J. Am. Chem. Soc.* **2020**, *142*, 6409–6422
566. D. R. Williams, R. De, M. W. Fultz, D. A. Fischer, A. Morales-Ramos, D. Rodríguez-Reyes, *Org. Lett.* **2020**, *22*, 4118–4122
567. A. Fürstner, P. W. Davies, *Angew. Chem. Int. Ed.* **2007**, *46*, 3410–3449
568. A. Fürstner, *Chem. Soc. Rev.* **2009**, *38*, 3208–3221
569. A. S. K. Hashmi, *Chem. Rev.* **2007**, *107*, 3180–3211
570. A. Corma, A. Leyva-Pérez, M. J. Sabater, *Chem. Rev.* **2011**, *111*, 1657–1712
571. R. Dorel, A. M. Echavarren, *Chem. Rev.* **2015**, *115*, 9028–9072
572. C. C. Chintawar, A. K. Yadav, A. Kumar, S. P. Sancheti, N. T. Patil, *Chem. Rev.* **2021**, *121*, 8478–8558
573. A. S. K. Hashmi, *Angew. Chem. Int. Ed.* **2008**, *47*, 6754–6756
574. G. Seidel, A. Fürstner, *Angew. Chem. Int. Ed.* **2014**, *53*, 4807–4811
575. Y. Fukuda, K. Utimoto, *J. Org. Chem.* **1991**, *56*, 3729–3731
576. A. Leyva, A. Corma, *J. Org. Chem.* **2009**, *74*, 2067–2074
577. M. Jia, M. Bandini, *ACS Catal.* **2015**, *5*, 1638–1652
578. E. Jiménez-Núñez, A. M. Echavarren, *Chem. Rev.* **2008**, *108*, 3326–3350
579. A. Fürstner, *Acc. Chem. Res.* **2014**, *47*, 925–938
580. G. Valot, C. S. Regens, D. P. O'Malley, E. Godineau, H. Takikawa, A. Fürstner, *Angew. Chem. Int. Ed.* **2013**, *52*, 9534–9538
581. B. M. Trost, G. Dong, *Nature* **2008**, *456*, 485–488
582. S. L. Crawley, R. L. Funk, *Org. Lett.* **2006**, *8*, 3995–3998
583. S. J. Pastine, D. Sames, *Org. Lett.* **2003**, *5*, 4053–4055
584. D. Hack, M. Blümel, P. Chauhan, A. R. Philipps, D. Enders, *Chem. Soc. Rev.* **2015**, *44*, 6059–6093
585. J. J. Kennedy-Smith, S. T. Staben, F. D. Toste, *J. Am. Chem. Soc.* **2004**, *126*, 4526–4527
586. N. Huwyler, E. M. Carreira, *Angew. Chem. Int. Ed.* **2012**, *51*, 13066–13069
587. S. T. Staben, J. J. Kennedy-Smith, D. Huang, B. K. Corkey, R. L. LaLonde, F. Toste, *Angew. Chem. Int. Ed.* **2006**, *45*, 5991–5994
588. T. D. Michels, M. S. Dowling, C. D. Vanderwal, *Angew. Chem. Int. Ed.* **2012**, *51*, 7572–7576
589. S. M. Canham, D. J. France, L. E. Overman, *J. Am. Chem. Soc.* **2010**, *132*, 7876–7877
590. A. S. K. Hashmi, M. Rudolph, *Chem. Soc. Rev.* **2008**, *37*, 1766–1775
591. D. Pflästerer, A. S. K. Hashmi, *Chem. Soc. Rev.* **2016**, *45*, 1331–1367
592. W. Zi, F. D. Toste, *Chem. Soc. Rev.* **2016**, *45*, 4567–4589
593. Y. Li, W. Li, J. Zhang, *Chem. Eur. J.* **2016**, *23*, 467–512
594. G. Cera, M. Bandini, *Isr. J. Chem.* **2013**, *53*, 848–855
595. H. Teller, A. Fürstner, *Chem. Eur. J.* **2011**, *17*, 7764–7767
596. S. Naoe, Y. Yoshida, S. Oishi, N. Fujii, H. Ohno, *J. Org. Chem.* **2016**, *81*, 5690–5698
597. P. Siemsen, R. C. Livingston, F. Diederich, *Angew. Chem. Int. Ed.* **2000**, *39*, 2632–2357
598. K. S. Sindhu, G. Anilkumar, *RSC Adv.* **2014**, *4*, 27867–27887
599. J. Blanco-Urgoiti, L. Añorbe, L. Pérez-Serrano, G. Domínguez, J. Pérez-Castells, *Chem. Soc. Rev.* **2004**, *33*, 32–42
600. J. D. Ricker, L. M. Geary, *Top. Catal.* **2017**, *60*, 609–619
601. S. E. Gibson, N. Mainolfi, *Angew. Chem. Int. Ed.* **2005**, *44*, 3022–3037
602. C. Oger, L. Balas, T. Durand, J.-M. Galano, *Chem. Rev.* **2013**, *113*, 1313–1350
603. D. Mailhol, J. Willwacher, N. Kausch-Busies, E. E. Rubitski, Z. Sobol, M. Schuler, M.-H. Lam, S. Musto, F. Loganzo, A. Maderna, A. Fürstner, *J. Am. Chem. Soc.* **2014**, *134*, 15718–15729
604. D. J. Pasto, R. T. Taylor, *Org. React.* **1991**, *40*, 91–155
605. C. E. Miller, *J. Chem. Ed.* **1965**, *42*, 254–259

606. B. M. Trost, B. Biannic, C. S. Brindle, B. M. O'Keefe, T. J. Hunter, M.-Y. Ngai, *J. Am. Chem. Soc.* **2015**, *137*, 11594–11597

607. E. O. Onyango, J. Tsurumoto, N. Imai, K. Takahashi, J. Ishihara, S. Hatakeyama, *Angew. Chem. Int. Ed.* **2007**, *46*, 6703–6705

608. J. D. White, R. G. Carter, K. F. Sundermann, M. Wartmann, *J. Am. Chem. Soc.* **2001**, *123*, 5407–5413

609. J. E. Tungen, L. Gerstmann, A. Vik, R. De Matteis, R. A. Colas, J. Dalli, N. Chiang, C. N. Serhan, M. Kalesse, T. V. Hansen, *Chem. Eur. J.* **2019**, *25*, 1476–1480

610. R. Damrauer, *J. Org. Chem.* **2006**, *71*, 9165–9171

611. E. M. Kaiser, *Synthesis* **1972**, 391–415

612. L. Brandsma, W. F. Nieuwenhuizen, J. W. Zwikker, U. Mäeorg, *Eur. J. Org. Chem.* **1999**, 775–779

613. A. Fürstner, *J. Am. Chem. Soc.* **2019**, *141*, 11–24

614. A. ElMarrouni, R. Lebeuf, J. Gebauer, M. Heras, S. Arseniyadis, J. Cossy, *Org. Lett.* **2012**, *14*, 314–317

615. M. Fuchs, A. Fürstner, *Angew. Chem. Int. Ed.* **2015**, *54*, 3978–3982

616. Z. Meng, L. Souillart, B. Monks, N. Huwyler, J. Herrmann, R. Müller, A. Fürstner, *J. Org. Chem.* **2018**, *83*, 6977–6994

The oxidative or reductive conversion of functional groups presumably represents the most frequently applied category of synthetic transformations. Even if a chemist may not automatically think of it this way, the formation of a radical by abstraction of an H-atom should also formally be considered a reduction. The addition of HX to a CC-double bond, on the other hand, is a redox reaction, as one of the two C-atoms is oxidized while the other is reduced by the formation of a C-H bond (see Fig. 4.1).

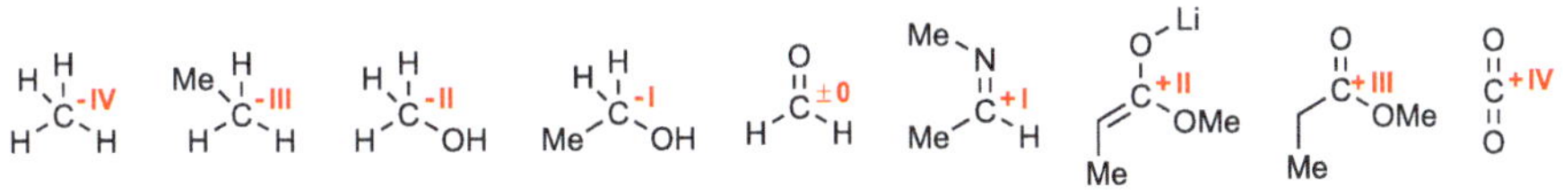

Fig. 4.1 Oxidation states of carbon in organic compounds

The challenge in the synthesis of highly functionalized molecules, in addition to the construction of the functionalities themselves, to a large degree centers around a possible impairment of each synthesis step by the substrates' functional groups. This may range from a simple lack of a conversion to the decomposition of the starting material. For this reason, the choice of compatible conditions or the protection of critical functionalities is essential for the success of a synthesis (see Fig. 4.2).

The oxidations and reductions discussed in this chapter mainly cover the transformation of C-O bonds, i.e., the formation and conversion of alcohols and carbonyl derivatives, which are the most frequently encountered modifications in a molecule's oxidation state.

© The Author(s), under exclusive license to Springer-Verlag GmbH, DE, part of Springer Nature 2026
A. Düfert, *Methods of Organic Synthesis*,
https://doi.org/10.1007/978-3-662-70963-4_4

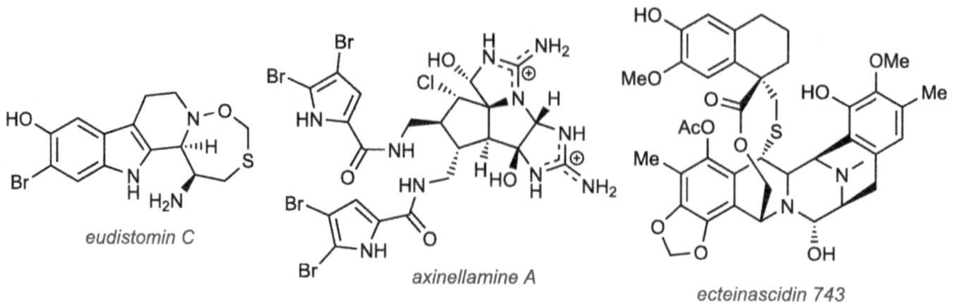

eudistomin C

axinellamine A

ecteinascidin 743

Fig. 4.2 Selected complex natural substances with functionalities susceptible to reductions or oxidations

4.1   Oxidation

An oxidation is defined as the donation of electrons. This may entail, among others, an increase in the bond order or the transferal of a heteroatom such as oxygen or nitrogen to a carbon atom.

The oxidation reactions shown in Fig. 4.3 comprise a selection of transformations considered standard in organic chemistry. In addition to the already discussed transformations of alkenes (epoxidation, dihydroxylation, aminohydroxylation, etc.), a significant part of oxidations in synthesis focus on the manipulation of alcohols to form carbonyl compounds. α-Functionalizations of carbonyls also constitute a formal oxidation. The oxidation of sulfur (sulfide $\rightarrow$ sulfoxide $\rightarrow$ sulfone) and nitrogen compounds (amine $\rightarrow$ N-oxide) is also beyond the scope of this overview. For these, the interested reader is kindly referred to the relevant monographs and review articles [1–4].

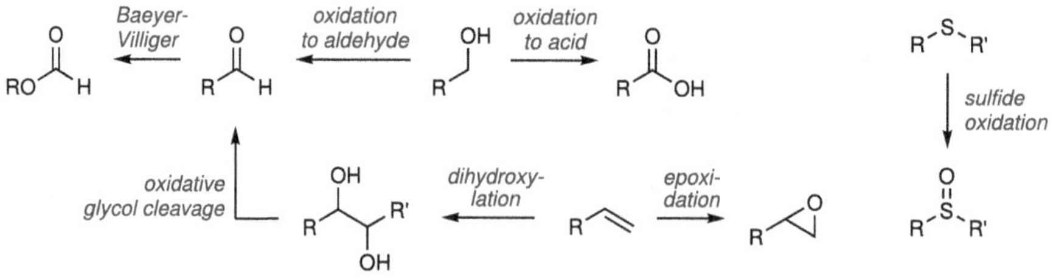

Fig. 4.3 Examples of oxidation reactions

There are hardly any syntheses of complex natural products which do not necessitate a modification of the intermediates' oxidation states. Accordingly, examples of oxidation reactions can be found in a myriad of synthetic projects (see Fig. 4.4). The reasons for choosing a respective method are rarely apparent from the literature and in only a few cases are the most efficient reaction conditions compared with lower yielding methods.

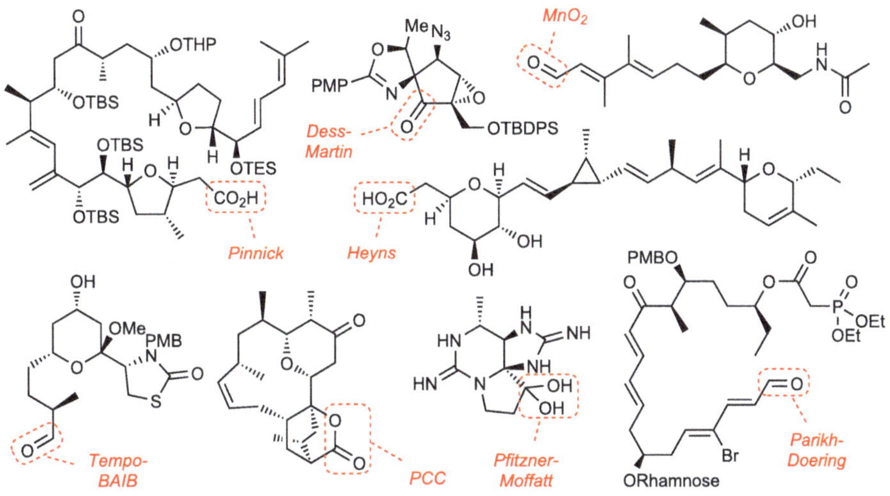

Fig. 4.4 Selection of various oxidations in natural product syntheses [5–12]

The different oxidation methods can be roughly categorized by their primary oxidant: Chromium(VI) reagents (*Jones* reagent, *Collins* reagent, PCC, PDC), activated DMSO (*Swern*, *Parikh–Doering*, *Pfitzner–Moffatt*, *Corey–Kim*), hypervalent iodine compounds (IBX, *Dess–Martin* periodinane), TEMPO or nitrosyl radicals, Ru reagents (RuO$_4$, TPAP/ NMO), Mn oxidants (MnO$_2$, KMnO$_4$), and others. The choice of a suitable method is typically based on the tolerance of a substrate to the reagents used, primarily focusing on whether the functional groups could be acid- or base-labile (see Fig. 4.5). As a second step, it is prudent to review if a desired regioselectivity might also be achieved in addition to the chemoselectivity, for example in the oxidation of a primary in the presence of a secondary alcohol.

4.1.1 Oxidations of Alcohols and Carbonyl Compounds

The conversion of alcohols to aldehydes or ketones comprises by far the largest group of oxidations. While Cr(VI)-based oxidants were the first selective reagents in the 1940s to provide access to carbonyl derivatives from alcohols, this was quickly followed by the development of a large number of other, equally selective or superior methods (see Tables 4.1, 4.2) [13–18].

During the oxidation of alcohols and carbonyl compounds, an unwanted epimerization of the product can easily occur due to the acidity of the α-CH protons of carbonyls, especially

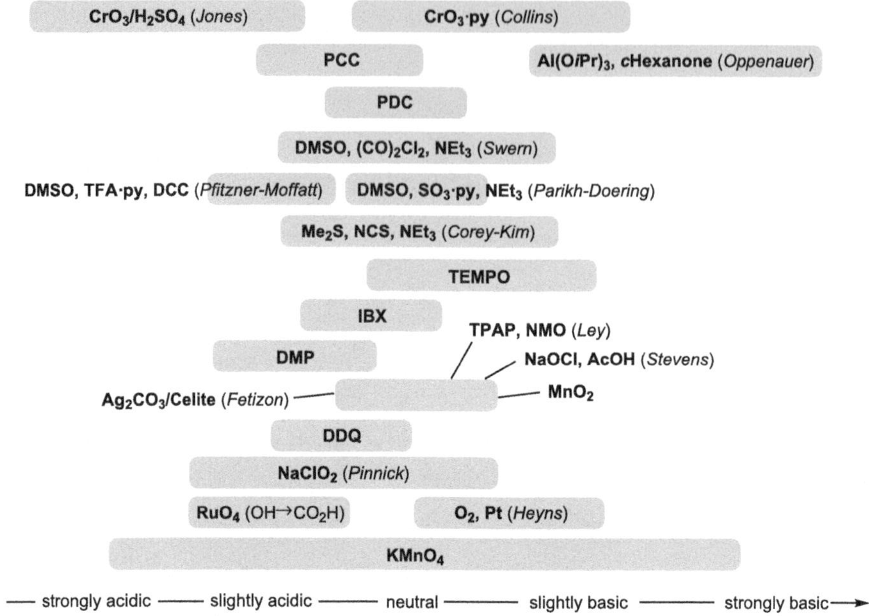

Fig. 4.5 Qualitative classification of the most common oxidation methods as acidic/neutral/basic [13–18]

under basic conditions. When substrates or products displaying pronounced α-acidity are encountered, several methods should be compared to determine which one best retains the stereointegrity.

Normally, oxidations do not tolerate amines, as primary and secondary amines can form an imine and tertiary amines can be converted into the corresponding N-oxide by the oxidant. However, for some methods exceptions exist for tertiary and sterically hindered secondary amines (*vide infra*).

Chromium-based Methods

The first method using Cr(VI) salts was developed in the Jones group and relied on a diluted sulfuric acid solution of CrO_3, furnishing ketones from secondary alcohols. Subsequent studies saw the introduction of oxidants such as PCC (pyridinium chlorochromate), PDC (pyridinium dichromate), and the *Collins/Sarett* reagent [15].

The innate toxicity of the oxidants constitutes the major drawback of all methods based on hexavalent chromium. As a consequence, these methods are still occasionally utilized on a small scale with highly functionalized substrates today, but they have never entered large-scale use, for example in medicinal chemistry.

Table 4.1 Common methods for the oxidation of alcohols to aldehydes/ketones

$CrO_3 \cdot 2$ Py (*Collins*)	requires large excess of oxidant
PCC ($CrO_3Cl \cdot HPy$)	addition of NaOAc for buffering; molecular sieve accelerates the reaction
PDC ($Cr_2O_7 \cdot HPy$) in CH_2Cl_2	can be accelerated with molecular sieve and organic acids as additives
DMSO, oxalyl chloride, NEt_3 (*Swern*)	sterically demanding bases can suppress side reactions; formed Cl^- can act as a nucleophile
DMSO, $SO_3 \cdot py$, NEt_3 (*Parikh-Doering*)	use of iPr_2NEt instead of NEt_3 can suppress epimerizations
Me_2S, NCS, NEt_3 (*Corey-Kim*)	not used as often due to the odor
DMSO, DCC, py·TFA (*Pfitzner-Moffatt*)	tolerates thiols as starting material; some acids beyond py·TFA also possible
TEMPO, NaOCl or $PhI(OAc)_2$, KBr	can be accelerated by a phase transfer catalyst, but may afford overoxidation; use of $NaHCO_3$ to correct pH; olefins incompatible with NaOCl, use of $PhI(OAc)_2$ instead
IBX	explosive, tolerates H_2O; poorly soluble, DMSO is usually used as solvent
DMP (*Dess-Martin*)	presence of H_2O or tBuOH accelerates the reaction
TPAP, NMO (*Ley*)	requires 4Å MS, almost all functional groups are tolerated (except $1°$, $2°$ amines & sulfides)
NaOCl, AcOH (*Stevens*)	favors secondary over primary alcohols
$NaBrO_3$, $KBrO_3$	favors secondary over primary alcohols
MnO_2	selective for allylic and benzylic alcohols; reproducibility strongly depends on MnO_2 source; $2°$ amines are usually tolerated; $BaMnO_4$ can be used as an alternative oxidant
$Al(OiPr)_3$, cyclohexanone (*Oppenauer*)	only oxidizes secondary alcohols
Ag_2CO_3/Celite (*Fetizon*)	very nonpolar reagents necessary; polar functionalities (also in substrate) inhibit the oxidation
DDQ	selective for allylic and benzylic alcohols; the more electron-rich the benzylic or allylic position, the faster the reaction

The oxidation with chromium trioxide or chromium acid derivatives is postulated to proceed via the general mechanism shown in Fig. 4.6. After the formation of a chromate ester, the proton at the alcohol moiety in the intermediate is removed by an external base as the rate-determining step. The liberated chromium(IV) species[19] can convert another equivalent of the alcohol to form the carbonyl compound. The chromium(II) acid is assumed to subsequently react with a Cr(VI) species to ultimately be transformed to its final oxidation state III, whose salts precipitate from the reaction medium [20, 21]. In the presence of water the desired aldehyde can generate the respective hydrate in equilibrium, which is oxidized to the carboxylic acid by a second equivalent of the oxidant. Apart from the *Jones* oxidation, which is carried out in aq. H_2SO_4/acetone, oxidations of primary alcohols to aldehydes by

Table 4.2 Common methods for the oxidation of alcohols/aldehydes to carboxylic acids

CrO_3, conc. H_2SO_4 (*Jones*)	acid-sensitive groups can be affected, decomposition partly limited by biphasic conditions; for reaction to acid, cat. CrO_3 with H_5IO_6 also applicable (Zhao modification)
PDC ($Cr_2O_7 \cdot HPy$) in DMF	only oxidizes saturated alcohols, benzylic/allylic alcohols only oxidized to the aldehyde
TEMPO, NaOCl or $PhI(OAc)_2$ or $NaClO_2$	phase transfer catalyst affords oxidation to the acid; olefins incompatible with NaOCl, use $PhI(OAc)_2$ or $NaClO_2$ instead
RuO_4	can also be used catalytically (RuO_2 or $RuCl_3$) with $NaIO_4$ as oxidant; does not tolerate amines and can oxidize many other functional groups (double/triple bonds, ethers, aromats, etc.); RuO_4^- is a possible alternative for more sensitive substrates
$KMnO_4$	can be performed under basic/acidic conditions; aldehydes are oxidized faster than alcohols; seldom used for difficult substrates; alkenes are not tolerated
O_2, Pt (Heyns)	can selectively oxidize 1° alcohols to the acid in the presence of 2° alcohols; neutral to slightly basic conditions necessary (usually by base or buffer); amines, S are strong cat poisons; aldehydes can be oxidized in the presence of 1° alcohols

chromium reagents must therefore be carried out under *anhydrous* conditions. Secondary alcohols can easily be converted to ketones, and a stringent exclusion of water is not necessary in these cases.

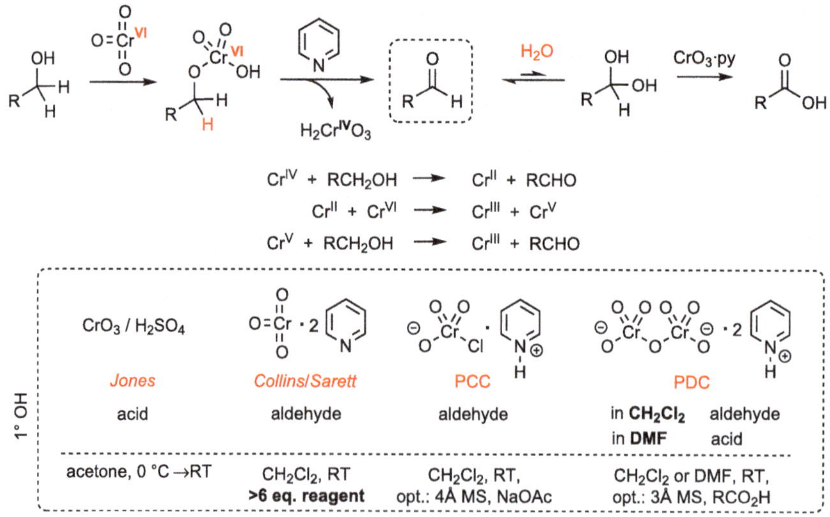

Fig. 4.6 Schematic mechanism of Cr(VI)-mediated oxidations (depicted using the Collins reagent), product scope and comparison of typical reaction conditions [14, 15, 21]

Anhydrous chromium reagents tolerate a variety of functional groups. Often, even sensitive functionalities such as sulfides stay intact. Although the *Jones* oxidation is carried out in the presence of a strong acid, many acid-labile groups often remain unaffected due to the biphasic conditions (aq. H_2SO_4/acetone). PCC is at most slightly acidic, while the *Collins* oxidation is slightly basic. The PCC oxidation can be conducted under almost neutral conditions by the addition of a weak base (NaOAc) to enhance the resilience of acid-labile protecting groups. Only PDC is quasi-neutral (see Fig. 4.5). The addition of molecular sieves to PCC or PDC expedites the reaction, and an organic acid or Ac_2O can also be added to PDC to promote faster product formation [22, 23].

The Collins reagent ($CrO_3 \cdot 2$ Py) is hygroscopic, which makes an anhydrous handling challenging. It can also be generated *in situ*, which is referred to as the *Sarett* reagent. In contrast, PCC and PDC possess little or no hygroscopicity, respectively, and can thus be easily handled as solids.

Deslongchamps *et al.* used a Collins oxidation in their synthesis of the diterpene (+)-ryanodol to form a lactol from two proximal hydroxy groups [24]. The transformation was carried out in the presence of a secondary alcohol functionality, which remained intact under the reaction conditions (see Fig. 4.7).

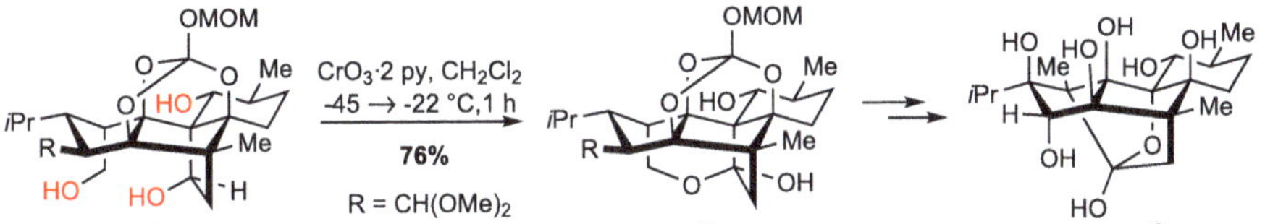

Fig. 4.7 Deslongchamps' synthesis of (+)-ryanodol (**3**) [24]

Apart from its conversion to aldehydes, primary alcohols can also be directly oxidized to the respective carboxylic acids using hexavalent chromium. Either the Jones reagent or PDC in DMF as solvent is suitable for this purpose. With the Jones oxidation, aldehydes are converted to the corresponding acids faster than alcohols can react, enabling the selective oxidation of an aldehyde in the presence of an alcohol (see Fig. 4.8).

Barring the Jones oxidation, which directly converts 1° alcohols to acids, the latter can also be obtained from aldehydes or 1° alcohols with PDC in **DMF** as solvent. Furthermore, 2° alcohols are oxidized to ketones and lactols to lactones. However, due to the mild conditions, it is usually not possible to oxidize 1° alcohols selectively in the presence of 2° alcohols, aldehydes, and lactols. The oxidation of allylic and benzylic 1° alcohols usually ceases after formation of the aldehyde [27]. An extension of the method was introduced by Zhao

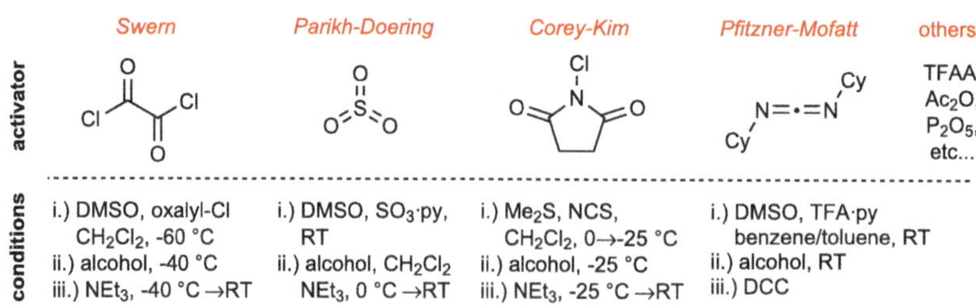

Fig. 4.8 Chemoselective oxidation of an aldehyde [25] and concomitant deprotection/oxidation [26]

in 1998, in which CrO_3 is used catalytically with H_5IO_5 as stoichiometric or secondary[l] oxidant [28].

The *Jones* oxidation is preparatively simple as no inert conditions (O_2, H_2O) need to be used, but the acidity of the reagents typically prevent its application on complex substrates. The *Collins* oxidation is preparatively somewhat more demanding due to the need of rigorously anhydrous conditions, the reagents are however comparatively cheap. PCC and PDC usually afford the highest yields, but the preparation of the reagents or their price can be considered a disadvantage compared to the other Cr(VI) methods.

Activated DMSO as Oxidant

DMSO can form a highly reactive dimethylsulfonium salt in the presence of a suitable activator, which reacts with alcohols to yield aldehydes or ketones with concomitant liberation of dimethyl sulfide (see Fig. 4.9) [17, 29]. With DMSO-based methods, no overoxidation to the acid is observed, nor are diols converted to lactols or lactones. The reactive species

	Swern	Parikh-Doering	Corey-Kim	Pfitzner-Mofatt	others
activator					TFAA, Ac$_2$O, P$_2$O$_5$, etc...
conditions	i.) DMSO, oxalyl-Cl CH$_2$Cl$_2$, -60 °C ii.) alcohol, -40 °C iii.) NEt$_3$, -40 °C →RT	i.) DMSO, SO$_3$·py, RT ii.) alcohol, CH$_2$Cl$_2$ NEt$_3$, 0 °C →RT	i.) Me$_2$S, NCS, CH$_2$Cl$_2$, 0→-25 °C ii.) alcohol, -25 °C iii.) NEt$_3$, -25 °C →RT	i.) DMSO, TFA·py benzene/toluene, RT ii.) alcohol, RT iii.) DCC	

Fig. 4.9 Overview of different oxidations with activated DMSO [14]

---

[l] The primary oxidant conducts the actual oxidation of the substrate and can be used stoichiometrically or catalytically. A secondary oxidant serves as reoxidant of the (stoichiometric or catalytic) primary oxidant and does not directly interact with the substrate.

first needs to be generated *in situ*. The order of addition of all reagents is therefore pivotal for an efficient conversion and high yields. In addition, the preformation precludes the unwanted (decomposition) reaction of the electrophilic activators with the alcohol, as the full conversion of the reagents to the dimethylsulfonium salt can be ensured before coming into contact with the substrate.

Strong activators such as oxalyl chloride or trifluoroacetic anhydride can oxidize alcohols even at very low temperatures, which can be particularly advantageous for highly functionalized substrates to suppress side reactions. Milder activators like $SO_3 \cdot py$ or DCC do however necessitate room temperature or slightly below to provide acceptable yields. All methods commonly share the instability of its activated DMSO reagent, which is why a reaction with these activators results in the decomposition of the oxidant at higher temperatures. This suggests that an increase of the reaction temperature to convert even unreactive alcohols seldom leads to an increase in yield. In these cases, alternative activators should rather be employed or a decrease of the temperature considered (see Fig. 4.10).

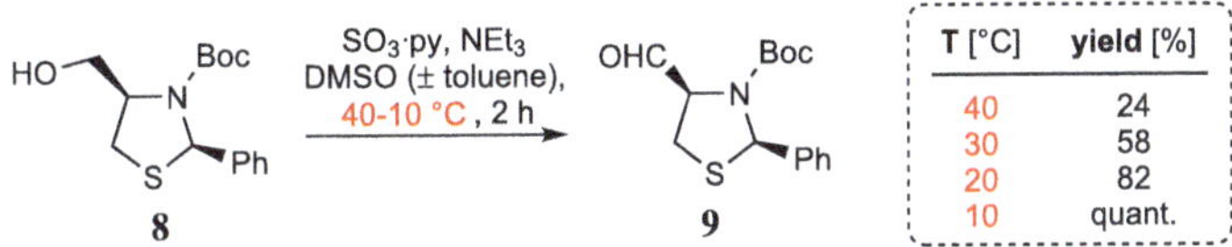

T [°C]	yield [%]
40	24
30	58
20	82
10	quant.

Fig. 4.10 Influence of the reaction temperature on the yield of the Parikh–Doering oxidation [30].

The mechanism of the oxidation has been studied in-depth and is shown in Fig. 4.11. The nucleophilic DMSO, or Me_2S in the case of the Corey–Kim oxidation, initially attacks the electrophilic activator and forms the dimethylsulfonium salt **10**. Following the substitution of the leaving group by the substrate yields species **11**, which is deprotonated by the external base. The sulfur ylide **12** finally removes the proton of the alcohol in an intramolecular fashion, affording the aldehyde or ketone and releasing dimethyl sulfide. While the formation of **10** with $SO_3 \cdot py$, DCC, NCS, etc. occurs in one step, the reaction with oxalyl chloride under Swern conditions is a two-step procedure. First, DMSO displaces a chloride in the oxalyl chloride and reacts to **13**. This activated DMSO species then quickly decomposes under release of gaseous CO and CO_2 to afford the active reagent, chlorodimethylsulfonium chloride [14, 29, 31–33].

The activated DMSO variants do not tolerate primary and most secondary amines, as these are better nucleophiles than the respective alcohol functionality and thus will preferentially add to the DMSO intermediate (in addition to converting the product to an imine). Sterically hindered secondary amines may be tolerated, but this is strongly substrate-dependent.

The corresponding methylthiomethyl ethers of the alcohol starting material (**15**) constitute a frequently occurring by-product. There are several possible reaction paths which furnish **15**. Alkylation of methylene sulfonium salts **14**, which can be formed by dissociation

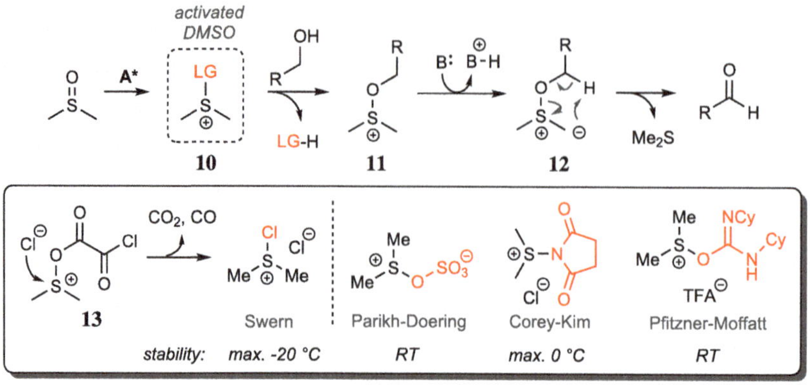

Fig. 4.11 Mechanism of oxidation with activated DMSO and formation of the reactive species **10** with oxalyl chloride [31]

of the ylides from **10** (so-called *Pummerer rearrangement*) [34], is postulated to be a major contributor. Methylthiomethyl ether formation can be largely suppressed by judicious choice of the activator and the reaction conditions. Under Swern conditions, the presence of electrophilic or nucleophilic chlorine enables another side reaction: nucleophilic functionalities such as indoles or the product aldehydes/ketones can be chlorinated by attacking the sulfonium salt (**16**). Chlorinations can be suppressed by strictly stoichiometric use of oxalyl chloride or by resorting to alternative activators (e.g., Ac_2O, $SO_3 \cdot py$). Benzylic alcohols can also be chlorinated to **17** starting from the sulfonium salt **11** under elimination of DMSO (see Fig. 4.12) [14, 31].

Fig. 4.12 Possible side reactions. LG = leaving group, B = base [14, 31]

The use of triethylamine can facilitate the α-epimerization of carbonyl compounds, the isomerization of alkenes, and the formation of α, β-unsaturated carbonyl compounds in the presence of good leaving groups in β-position. These side reactions can be largely suppressed by the use of sterically demanding bases (e.g., $i\mathrm{Pr}_2\mathrm{NEt}$) (see Fig. 4.13). Under Parikh–Doering conditions, base-induced side reactions also do not occur to a significant extent [29].

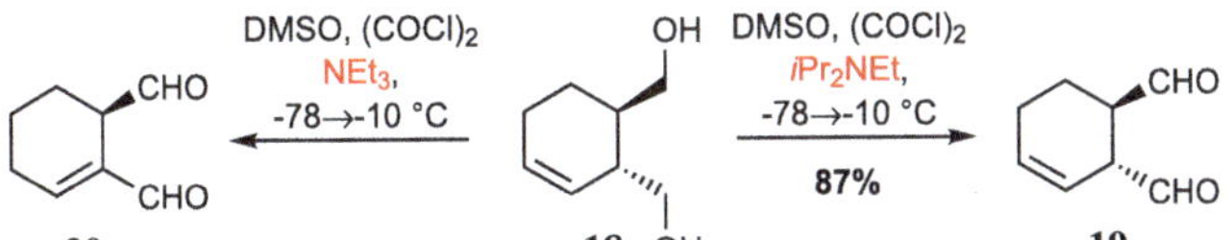

Fig. 4.13 Influence of the base in the Swern oxidation [35]

In contrast to chromium reagents, the reaction conditions can only be varied to a limited extent. The pH of the reaction mixture can for example be modulated, as the necessary reagents or intermediates intrinsically possess a certain innate acidity or basicity. The pivotal determinant is the choice of the activator, after which the extent of side reactions is mainly controlled by the temperatures and reaction times of the individual steps. Since activated DMSO (**10**) and the activated alcohols (**11**) behave slightly acidic, an extended activation time prior to addition of the base can impair very acid-labile functionalities (*Swern, Parikh–Doering, Corey–Kim*). Other, rather rarely used activators are trifluoroacetic anhydride, which affords good yields especially with sterically demanding substrates, as well as acetic anhydride (*Albright–Goldman conditions*; also good for sterically hindered alcohols), phosphorus pentoxide and thionyl chloride [14, 32].

During the synthesis of the sesterterpenoid nitidasin by the Trauner group, various oxidation methods were employed. Besides a classic Swern oxidation, a PCC-mediated lactone formation from a diol, a Dess–Martin, and a Ley oxidation were utilized to construct the tetracyclic 5/8/6/5-linked framework of the natural product (see Fig. 4.14) [36].

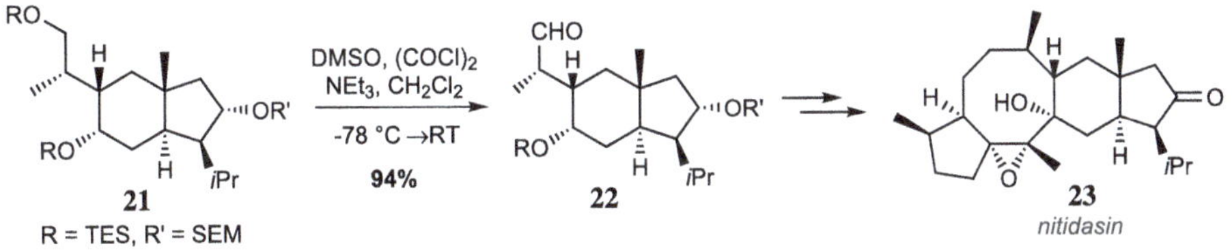

Fig. 4.14 Excerpt of the synthesis of nitidasin (**23**) by Trauner [36]

Under Pfitzner–Moffatt conditions, even thiol groups are often tolerated in the starting material. Basic functionalities in the substrate can react with the TFA·py used, which is why additional equivalents of the acid activator are needed in their presence. The challenging separation of the formed alkyl urea by-product resulting from the EDC reagent can be considered a drawback of Pfitzner–Moffatt oxidations.

Hypervalent Iodine Compounds

There are a number of common methods that utilize hypervalent iodine as oxidant. Although iodine exists in oxidation states +III and +V in primary oxidants,[II] the most important methods rely on I^V. Iodoxybenzoic acid (*IBX*, **24**) and the periodinane introduced by Dess and Martin **25** (*DMP*) are the reagents of choice. Bisacetoxyiodobenzene (**26**, *BAIB* or *PIDA*) is usually employed as secondary oxidant and is used in the oxidation with TEMPO, TPAP, or similar catalytic primary oxidants (see Fig. 4.15) [37–41].

24 IBX

25 Dess-Martin periodinane (DMP)

26 BAIB / PIDA

Fig. 4.15 Common oxidants based on hypervalent iodine

The Swern and Dess–Martin oxidation typically are the first choice for more complex substrates, followed by TEMPO- and TPAP-mediated oxidations (see below). Apart from the mild conditions of the reactions, this can additionally be attributed to the fact that an extensive body of syntheses has already been carried out with these reagents and the substrate scope as well as the tolerance towards critical functional groups are well known. The success of DMP over the less frequently encountered IBX largely relies on the significant difference in solubility: While DMP forms a homogeneous solution in many standard solvents, IBX can only be fully dissolved in DMSO. Due to this challenge IBX was used only sporadically and its advantages and substrate scope investigated accordingly. The observation that a suspension of IBX in an "unsuitable" solvent (CH_2Cl_2, EtOAc, MeCN) also affords the desired products and exhibits a reactivity complementary to DMP for certain substrates has since rekindled interest in this oxidant (see Fig. 4.16) [42, 43].

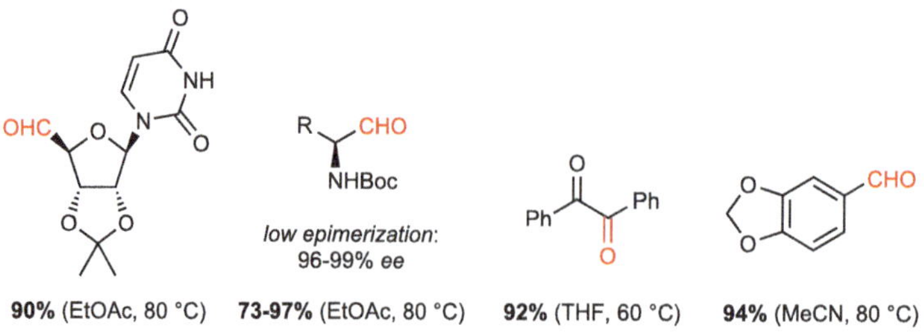

low epimerization: 96–99% ee

90% (EtOAc, 80 °C)   **73-97%** (EtOAc, 80 °C)   **92%** (THF, 60 °C)   **94%** (MeCN, 80 °C)

Fig. 4.16 Oxidation of alcohols by IBX in DMSO-free suspensions [43–45]

[II] The nomenclature and identity of the oxidation state of iodine is controversely discussed in the literature. Most publications cite the oxidation states indicated herein. Alternatively, it has also been suggested to refer to tri- (λ^3) and pentavalent iodine (λ^5).

IBX and DMP are synthesized from iodobenzoic acid (**27**) (see Fig. 4.17) [46, 47]. IBX has been reported to display explosive behavior and shock-sensitivity, irrespective of the preparation method [48, 49]. An explosion can be prevented by adding benzoic acid and isophthalic acid, affording "stabilized IBX" (SIBX) [50]. The synthesis of IBX is nowadays almost exclusively carried out with oxone (2 $KHSO_5 \cdot KHSO_4 \cdot K_2SO_4$) as the milder conditions are safer, notwithstanding the slightly diminished yields compared to the bromate route (see Fig. 4.17).

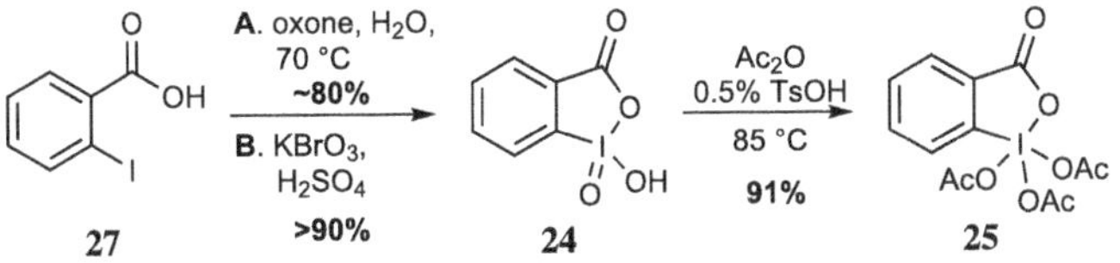

Fig. 4.17 Synthesis of IBX and DMP [46, 47]

The postulated mechanism of the alcohol oxidation is based on the highly electrophilic nature of hypervalent iodine, making it susceptible to nucleophilic attacks. In the oxidation with DMP (**25**), the alcohol first substitutes an acetate at the iodine (**28**), then the α-proton can be removed and the final iodine(III) species (**29**, OR = OAc) is obtained upon elimination of the desired aldehyde and another equivalent of AcOH. Although it is assumed that an external base is required to remove the proton, this has not yet been conclusively proven. In the presence of pyridine or acetic acid, no acceleration of the reaction occurs. Furthermore, the reaction displays first-order kinetics in the presence of an excess of alcohol [49, 51]. Both observations indicate the rate-determining deprotonation to be intramolecular. The reaction using IBX (**24**) proceeds similarly. The only difference is found in the nature of the substituent on the iodoso compound **29** (OR = OH) after the oxidation. The initial substitution reaction occurs rapidly in both cases, and the subsequent decay to the iodinane **29** represents the rate-determining step in DMP. With IBX, the alcohol-coordinated iodinane can form the diastereomers **31** and **32**, which are in equilibrium. The conformer **32** with an axially bound alcohol is the reactive species. Whether the hypervalent twisting (**31**→**32**) or the intramolecular elimination to **29** constitute the rate-determining step has not yet been conclusively determined (see Fig. 4.18) [45, 52, 53].

In the presence of an excess of the alcohol substrate, a double substitution of an acetate to **30** can occur for the Dess–Martin periodinane. This bis-alcoholate species reacts significantly faster than **28**, yet an additional equivalent of the substrate is consumed by remaining attached to the λ^3-iodinane **29** (OR = OCH$_2$R'). To exploit this reactivity, an alcohol without α-protons (e.g., tBuOH) can be added stoichiometrically. This leads to formation of alkoxyiodinane **34**, thereby significantly accelerating the overall oxidation as this intermediate is more reactive than the parent periodinane **25** [49]. A slight excess of substrate or another alcohol is thus beneficial, whereas an excess of reagent counterintuivtely does not

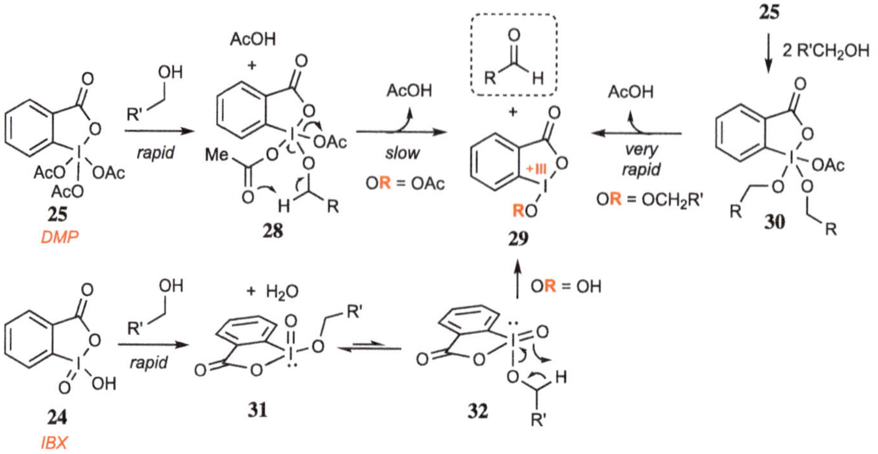

Fig. 4.18 Mechanism of oxidation with DMP and IBX [49, 52–54]

offer any advantages due to the conversion of **28** not depending on the concentration of the periodinane reagent. Another method to increase the reaction rate is the *strictly stoichiometric* addition of water. This forms acetoxyiodinane **33**, which similarly shows a faster conversion of alcohols to aldehydes or ketones than DMP itself (see Fig. 4.19) [54].

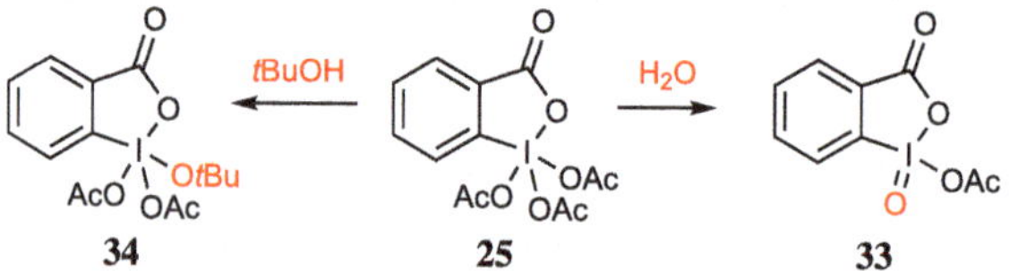

Fig. 4.19 Acceleration of the Dess–Martin oxidation by addition of *t*BuOH or H_2O by formation of the more reactive species **33**, **34** [49, 54, 55]

The effectiveness of the oxidation with DMP strongly depends on the quality of the (self-) prepared synthesized reagent. The incomplete formation of **25** from **24**, a partial hydrolysis to **33** or a complete hydrolysis to IBX are probably the causes of fluctuating yields in the oxidation [54, 56].

Ghosh *et al.* used a Dess–Martin oxidation in their synthesis of the aglycones of lycoperdinoside A and B to convert the primary alcohol to the aldehyde **35** by oxidation, followed by a Z-selective Still–Gennari modification of the HWE olefination. In the later stages of the synthesis, the Dess–Martin oxidation was used again for the three-step access to a methyl ester (DMP, Pinnick, diazomethane). In addition to the complexity of the polyene **36**, the tolerance of the vinyl iodide in **35** under the reaction conditions is especially notable, as the periodinane could easily lead to its oxidation (*comproportionation* of the oxidation states).

The iodide remains untouched and the oxidation affords the aldehyde without epimerization of the α-methyl group (see Fig. 4.20) [57].

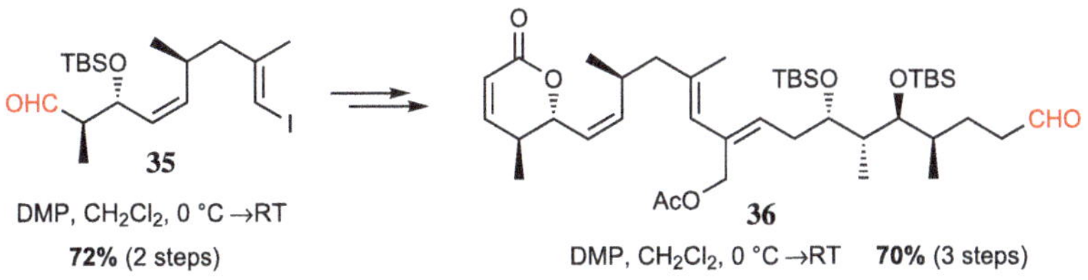

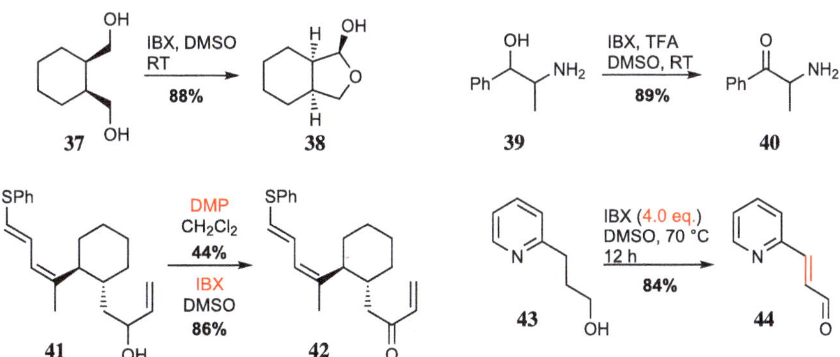

Fig. 4.20 Intermediates in the synthesis of the lycoperdinoside aglycone by Ghosh [57]

Primary amines are usually not tolerated in the Dess–Martin oxidation, whereas secondary and tertiary amines routinely are. 1, 2-diols are cleaved to the respective aldehydes. To protect acid-labile functionalities, pyridine can be added to neutralize the acetic acid by-product [14].

Since the discovery by Frigerio *et al.* that IBX dissolves in DMSO, it has developed into a very mild method complementary to DMP oxidations. Alcohols can be converted to the corresponding carbonyl compounds under quasi-neutral conditions [45]. Unlike DMP, IBX can oxidize 1,2-diols without CC-bond cleavage and usually does not oxidize aminals to amides or lactols to lactones (**37**→**38**). Primary amines or secondary amines can be preserved by adding acid (e.g., TFA) prior to mixing with the oxidant (**39**→**40**). Pyridines also remain unaffected (**43**→**44**) and sulfides are even less susceptible to oxidation with IBX than with DMP (**41**→**42**). Another important reaction is the dehydrogenation of carbonyl compounds to α, β-unsaturated aldehydes or ketones (**43**→**44**). The method developed by Nicolaou uses an excess of IBX for oxidation and subsequent dehydrogenation of the intermediate carbonyl derivative (see Fig. 4.21) [14, 42].

Fig. 4.21 Examples of selective oxidations with IBX [45, 58–60]

The access to the marine macrolide amphidinolide E was accomplished by Lee and co-workers by means of a late macrocyclization. The requisite carboxylic acid was synthesized in two steps using IBX and a subsequent Pinnick oxidation. While a Dess–Martin oxidation led to the modification of the polyene side chain, the triene terminus in **45** was completely preserved under the quasi-neutral conditions with IBX. The final steps starting from **46** consisted of a TIPS deprotection, macrolactonization, and the removal of the acetal and MOM groups (see Fig. 4.22) [61].

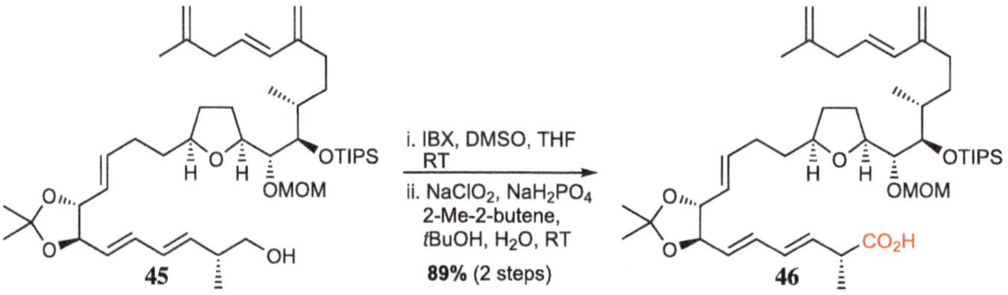

Fig. 4.22 Late oxidation in Lee's synthesis of amphidinolide E [61]

In addition to oxidation with stoichiometric reagent, a variety of methods using catalytic amounts of hypervalent iodanes have also been developed [38]. Apart from *o*-iodobenzoic acid as a catalyst, which in combination with oxone displays only moderate chemoselectivity, the sodium salt of *o*-iodobenzenesulfonic acid (**47**) was reported to be an effective reagent by Ishihara *et al.* (see Fig. 4.23) [62].

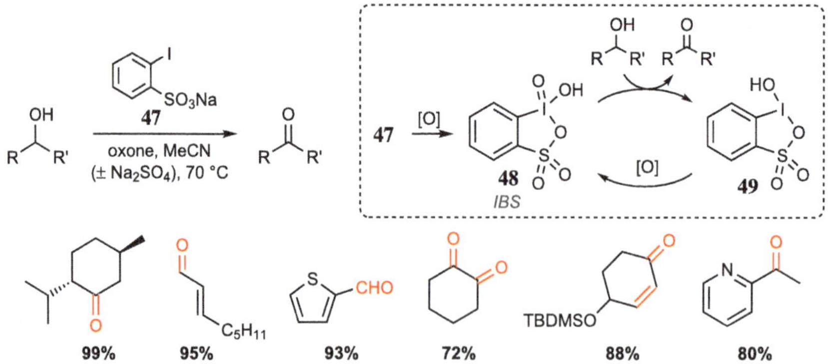

Fig. 4.23 Oxidation of alcohols to aldehydes, ketones with IBS according to Ishihara [62]

The active catalyst, iodoxybenzenesulfonic acid (*IBS*, **48**), is formed *in situ* from **47** with oxone. After the oxidation of the alcohol, **49** can then be reoxidized with oxone and reenter the catalytic cycle. Already 1 mol-% of IBS is sufficient for efficient turnover. Na_2SO_4 can additionally be added for acid-sensitive substrates.

Despite the charm of high atom economy and the use of cost-effective secondary oxidants under catalytic conditions, these have not been widely established in natural product synthesis yet. The Dess–Martin oxidation typically is the method of choice for complex syntheses.

TEMPO-mediated Oxidations

Although the conversion of alcohols to the corresponding aldehydes by stoichiometric oxoammonium salts was already reported in the 1960s, the true potential of the reaction remained unrecognized for long thereafter. It was not until the late 1980s that the Anelli group introduced the selective oxidation of alcohols by *catalytic* amounts of nitroxyl radicals using an additional, stoichiometric oxidizing agent, the secondary oxidant NaOCl [63]. Based on this seminal work, it was quickly realized that a large number of other stoichiometric oxidants are also suitable for the reaction (see Fig. 4.24) [64–68].

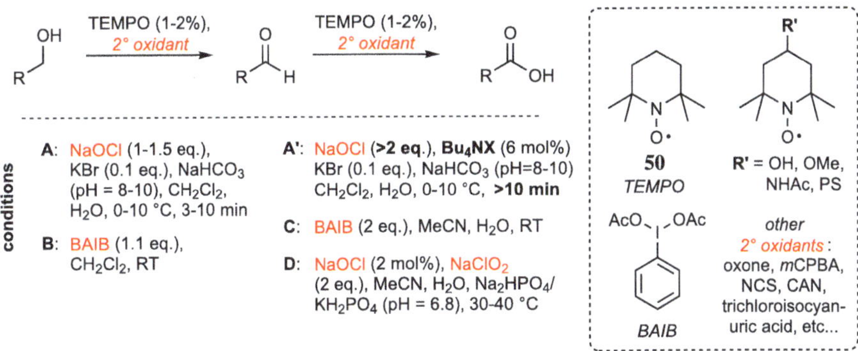

Fig. 4.24 Overview of the TEMPO-catalyzed oxidations of alcohols to aldehydes and carboxylic acids [63, 69–71]

The oxidation to aldehydes or ketones in the presence of catalytic amounts of TEMPO typically utilizes either NaOCl (Anelli conditions A, A') or BAIB as a secondary oxidant (B, C). If superstoichiometric quantities of NaOCl are used, the consecutive reaction to the carboxylic acid proceeds very slowly (conditions A'). However, an efficient conversion to the acid is observed upon addition of a quaternary ammonium salt as a phase transfer catalyst. Added $NaHCO_3$ is commonly used with Anelli conditions to adjust the reaction mixture to a slightly basic pH (8–10), as the oxidation proceeds rather slowly in an acidic environment. The use of BAIB has the advantage that slightly acidic to neutral conditions are retained throughout the conversion. This prevents degradation of base-labile groups and affords only iodobenzene as by-product, which is easy to separate. For the selective transformation to the respective acids instead of the aldehyde, an increased amount of secondary oxidant, the presence of water, and (usually) an increased reaction time are required. Zhao *et al.* also introduced chlorite ($NaClO_2$) as a secondary oxidant for the oxidation of alcohols to the acid (conditions D) [71]. $NaClO_2$ serves a dual role: It generates NaOCl, which is the secondary oxidant for TEMPO in the reaction to the aldehyde, and also constitutes the *primary* oxidant of the subsequent conversion to the carboxylic acid.

Nitroxyl radicals (**50**) as well as oxoammonium salts (**51**) qualify as reactive species in this transformation. Since oxoammonium salts are however significantly stronger oxidizing agents, they are assumed to be the primary oxidant of the reaction. TEMPO is reduced to hydroxylamine **52** in the process and needs to be reoxidized to **51** by the secondary oxidant. Due to the acid-mediated equilibrium reaction between the oxoammonium salt **51**, hydroxylamine **52**, and the nitroxyl radical **50**, either **50** or **52** is reoxidized depending on the secondary oxidant.

After passing through the catalytic cycle, NaOCl reoxidizes **50** to **51** via HOCl as active species, which is generated from NaOCl. At elevated pH, HOCl is deprotonated and the reaction significantly slows down. The presence of catalytic amounts of KBr under Anelli conditions (**A, A'**) likely leads to the formation of HOBr instead of HOCl, which is the actual secondary oxidant [72]. In contrast to NaOCl, BAIB does not regenerate the corresponding oxoammonium salt **51** from the nitroxyl radical [69]. Instead, TEMPO disproportionates in the presence of a proton source (AcOH) to the oxoammonium salt **51** and hydroxylamine **52**, which is reoxidized to the nitroxyl radical under release of AcOH. The acetic acid required for the initial reoxidation can be generated by the reaction of the alcohol with BAIB (see Fig. 4.25) [13, 14, 64, 68].

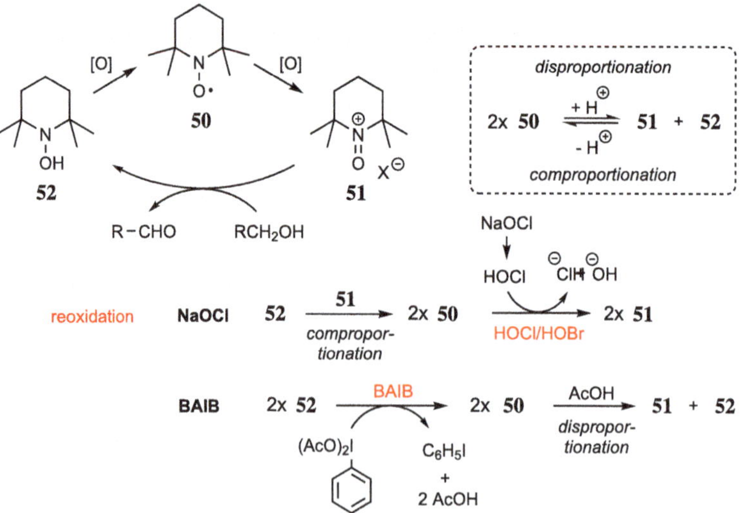

Fig. 4.25 Mechanism of TEMPO-mediated oxidation — catalytic cycle and reoxidation of hydroxylamine [64]

Depending on whether acidic or basic conditions are chosen, a different reaction path may dominate: In basic media, the zwitterionic species **53** is formed preferentially, in which the α-proton of the coordinated alcohol is removed intramolecularly to furnish the desired aldehyde. In contrast, a bimolecular hydride transfer is assumed in acidic media. The reaction proceeds significantly faster under basic conditions, as the α-deprotonation is comparatively slow at acidic pH (rate-determining step). When the utilized reagent combination promotes the consecutive reaction of the aldehyde to the respective carboxylic acid, a small amount of the hydrate is formed from the aldehyde in equilibrium, irrespective of whether acidic or basic conditions are used. This hydrate intermediate is then oxidized to the carboxylic acid by reacting with another equivalent of the oxoammonium salt (**conditions A', C**, Fig. 4.24) or $NaClO_2$ (**conditions D**) (see Fig. 4.26) [64, 73].

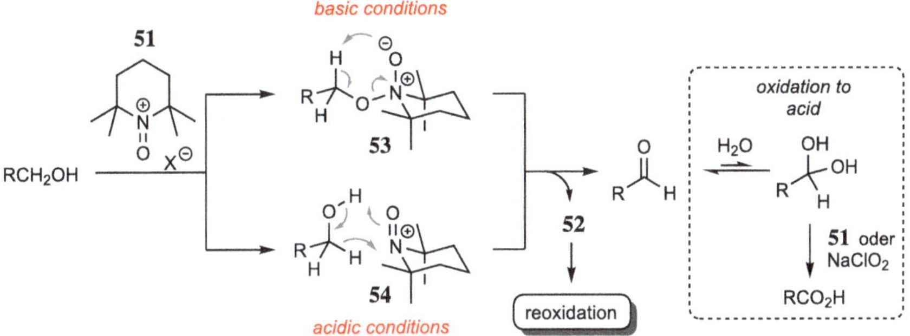

Fig. 4.26 Postulated mechanism of TEMPO-mediated oxidation — transition states and oxidation to acid [73, 74]

The pH thus assumes a pivotal role for the chemoselectivity of the transformation: In a basic medium, primary alcohols are oxidized faster than secondary ones, while under acidic conditions the relative reaction rates reverse. Substrates with multiple hydroxy groups can thus be converted selectively, depending on the chosen conditions [73]. The paths postulated by Semmelhack and Wiberg via **53**, **54** are widely cited as the operative mechanism [67]. Recent theoretical and experimental studies additionally indicate an intermolecular hydride transfer as a viable alternative under basic conditions, proceeding analogously through **54** [75, 76].

The stereointegrity of acidic protons can always be affected in sensitive substrates by any reaction under basic or to some extent acidic conditions. Using 2-methylbutanol as model substrate, the least amount of racemization was observed when resorting to the only slightly basic conditions of the TEMPO oxidation compared to other methods (see Fig. 4.27) [77].

The high selectivity of primary over secondary alcohols under basic conditions can also be further exploited in consecutive oxidations: Primary hydroxy groups can even be oxidized to the respective acids while keeping secondary alcohols intact (see Sect. 4.1.2) [78, 79].

oxidant	yield [%]	ee [%]
PCC	45	55
SO$_3$·py (*Parikh-Doering*)	56	87
Swern	63	90
Dess-Martin	75	90
TEMPO, NaOCl, KBr	82	93

Fig. 4.27 Comparison of various oxidation protocols in the synthesis of 2-methylbutanal [77]

In Kishi's synthesis of the framework of mycolactone metabolites, a selective oxidation of a terminal alcohol to the aldehyde was used in the presence of a secondary hydroxy group to access the macrocyclic lactone of this class of natural products. *N*-Chlorosuccinimide served as the secondary oxidizing agent in this case (see Fig. 4.28) [80].

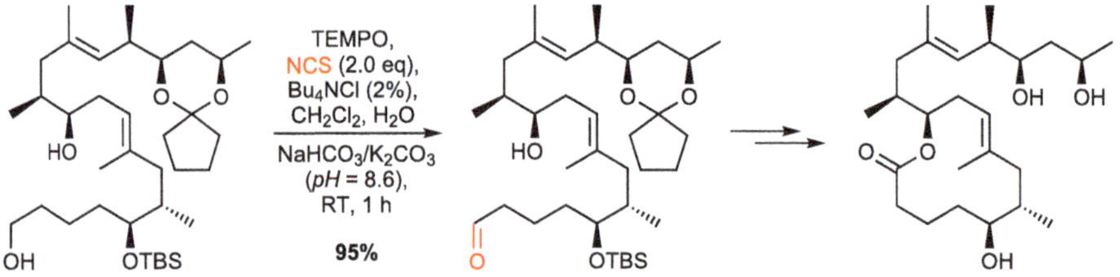

Fig. 4.28 Synthesis of the mycolactone structure (**59**) by Kishi *et al.* [80]

The chlorination of electron-rich arenes and alkenes by HOCl is one of the most commonly encountered side reactions. This can be prevented by using BAIB or, when oxidizing to the respective acid, NaClO$_2$. The latter is a weaker chlorinating agent than the HOCl generated *in situ* from NaOCl. As a last resort, the oxoammonium salt can also be used stoichiometrically. To avoid an undesired overoxidation to the acid, anhydrous conditions are required to prevent the formation of the hydrate, for example by using BAIB in CH$_2$Cl$_2$ (conditions **B**, Fig. 4.24). In addition, less than two equivalents of secondary oxidant should be used (see Fig. 4.29).

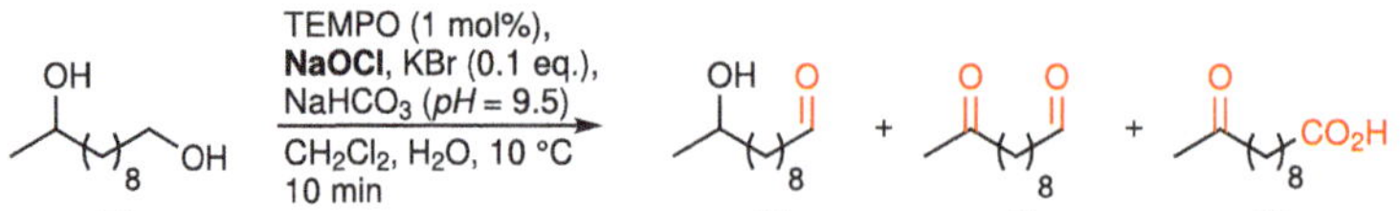

	61	62	63
1.1 eq. NaOCl	68%	10%	–
2.2 eq. NaOCl	–	69%	14%
3.6 eq. NaOCl, 0.05 eq. (nC$_8$H$_{17}$)$_3$NMeCl	–	–	57%

Fig. 4.29 Single or multiple oxidation of 1,10-undecandiol [78]

An advanced variant of the classic TEMPO-mediated oxidation was developed by the Stahl group. Under aerobic conditions, molecular oxygen serves as secondary oxidant. The mechanism proposed by Stahl assumes *in situ*-generated Cu-imidazolate complex **64** to bind the substrate after reduction of oxygen (**65**) and furnish the desired product via the Cu-TEMPO-alcohol complex **67**. An alternative aerobic mechanism was also suggested based on DFT calculations [81]. This extremely mild method tolerates free amino and hydroxy groups, thioethers as well as nitro and iodine functionalities (see Fig. 4.30) [82, 83]. Under anaerobic conditions, a proton-coupled electron transfer between TEMPOH and Cu(II) was suggested as operative mechanism to regenerate the nitroxyl radical [84].

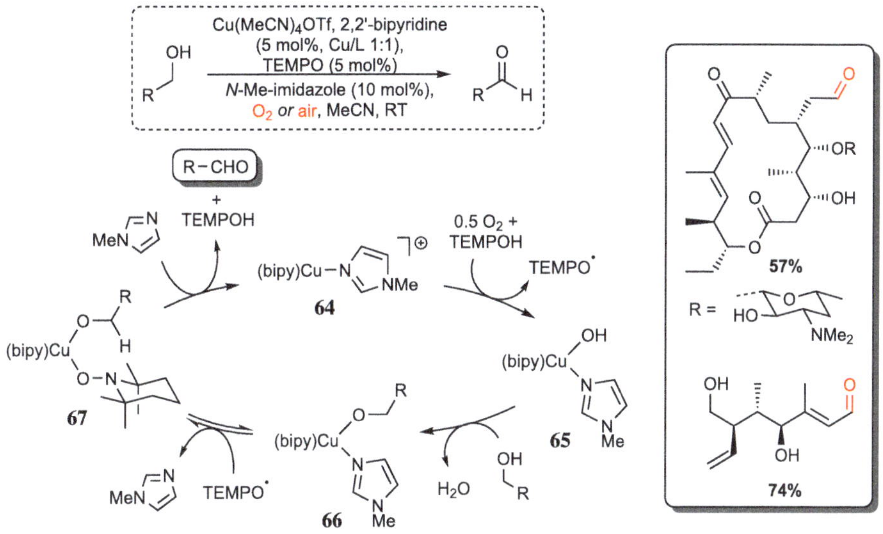

Fig. 4.30 Cu/TEMPO-mediated aerobic oxidation according to Stahl *et al.* and synthetic applications [82, 83, 85, 86]

Oxidation with Ruthenium Oxides

Ruthenium oxides are efficient oxidants, which can be used for the formation of aldehydes/ketones as well as carboxylic acids [13, 14, 87]. RuO_4 is an extremely strong oxidant that also enables difficult transformations beyond the oxidation of alcohols to acids, such as the conversion of ethers into esters and the oxidative degradation of aromatic systems. Selective oxidations of 2° alcohols to ketones are seldom successful with ruthenium tetroxide due to the high reactivity, since it requires precise control of the reaction. It is therefore primarily used for the formation of carboxylic acids. Ruthenium derivatives in lower oxidation states than 8+ are milder than RuO_4 as oxidants. Perruthenates (RuO_4^-) are very mild and selective in the presence of large organic counterions such as tetraalkylammonium and highly soluble in common organic solvents. The tetrapropylammonium perruthenate (*TPAP*) reagent developed by Ley has thus been widely established even for densely functionalized substrates (*Ley oxidation*). In the presence of water the oxidation to the respective carboxylic acid can also be achieved with ammonium perruthenates [88]. Otherwise the acid can similarly be afforded by resorting to alternative oxidants such as RuO_4 or $RuCl_3/NaIO_4$ (see Fig. 4.31) [13]

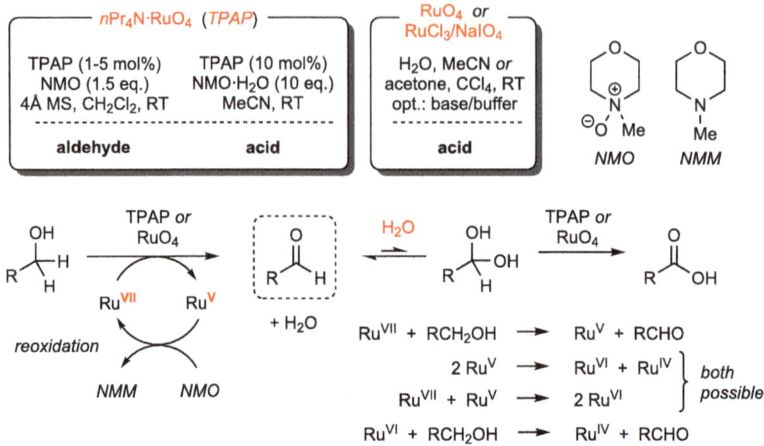

Fig. 4.31 Ruthenium oxide-mediated oxidations and postulated mechanism of the transformation [89, 90]

The mechanism of oxidation is complex, as ruthenium complexes of oxidation states $+8, +7, +6, +5$, and $+4$ are all capable of stoichiometrically oxidizing alcohols to aldehydes or ketones [89]. Some details remain unknown, but a recently published in-depth mechanistic study was able to shed more light on the role of the various species. The reactive oxidant is a high-valent Ru complex, which is subsequently reduced to insoluble RuO_2. For oxidation with TPAP, it was observed that only two-electron transfers take place, which is why reoxidation of Ru through oxidation states V and VII is assumed — Ru(V) is a weak oxidant compared to Ru(VI) or Ru(VII) [89]. Traces of colloidal $Ru(IV)O_2$ are essential for an efficient conversion, as the oxidation steps were indicated to take place on the surface of the heterogeneous particles. NMO has a dual role: In addition to the reoxidation of Ru(V) to Ru(VII), the disproportionation of Ru(V) to Ru(VI) and RuO_2 is also inhibited, thereby stabilizing the active catalyst by preventing a rapid decomposition to RuO_2 [90].

In analogy to chromium-mediated oxidations, a Ru ester was postulated as active intermediate (**68**, cf. Fig. 4.6). In the oxidation with *stoichiometric* amounts of TPAP, a diruthenate of type **69** was derived from the rate law to be an oxidatively competitive species [91, 92]. Under *catalytic* conditions, the rate law shows a first order in perruthenate and alcohol, but no dependence on NMO, which is why diruthenates apparently play no role [90]. The postulated spent Ru(V) species is stabilized by the solvent (**70**) before reoxidation by NMO to RuO$_4^-$ can occur [93]. In the presence of water, NMO stabilizes the formation of hydrates (**71**) and can thus facilitate oxidation to the acid (see Fig. 4.32) [88].

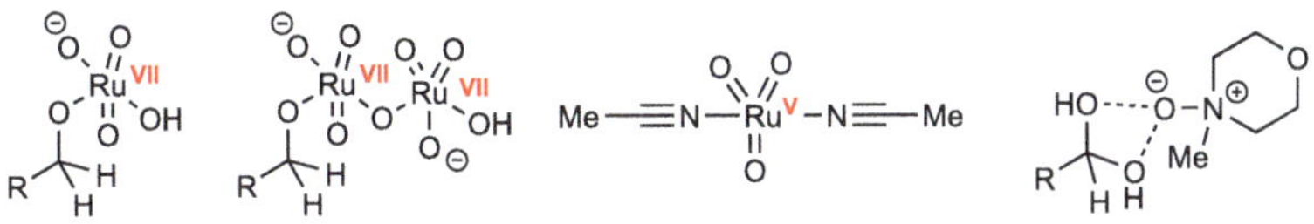

Fig. 4.32 Postulated intermediates of TPAP-catalyzed oxidations [88, 91, 93]

Overoxidation to the acid can also occur with TPAP in the presence of water. Additionally, it was shown that the addition of water leads to a decrease in the reaction rate [91]. Therefore, crushed molecular sieves are routinely added to the reaction as water scavengers to prevent both effects.

The oxidation with TPAP is widely used in natural product synthesis. The obtained aldehydes and ketones can serve two purposes: Either they are used for further derivatization and subsequently converted to alkenes or alcohols or the CO functionality is present in the target molecule themselves (see Fig. 4.33).

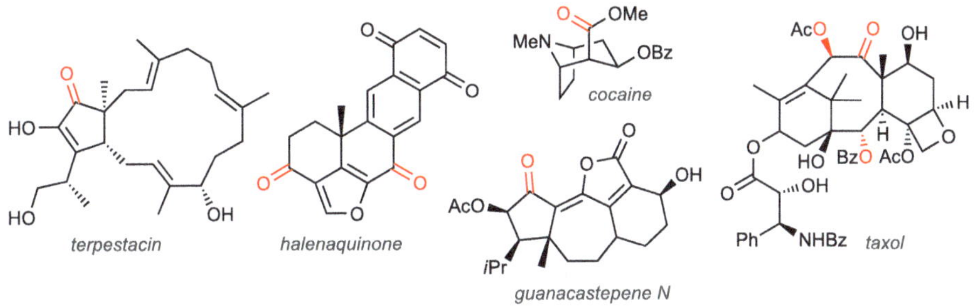

Fig. 4.33 Examples of natural product syntheses using TPAP-catalyzed oxidations [94–98]

In the synthesis of carboxylic acids, either RuO$_4$ or a low-valent Ru salt (RuCl$_3$, RuO$_2$) is used in conjunction with a secondary oxidant. RuO$_4$ possesses high toxicity, similar to OsO$_4$, and can lead to the undesired oxidation of a variety of functional groups (alkenes, 1,2-diols, ethers, electron-rich arenes) as side reaction. Since low-valent Ru, which is routinely generated as a by-product after oxidation, can easily be complexed by the carboxylic acid products, MeCN or acetone are preferably used as solvents. They can competitively complex Ru in low oxidation states and stabilize it for reoxidation (*Sharpless* modification, see Fig. 4.34) [99, 100]. The addition of a base or a buffer can be used to avoid decomposition reactions of acid-labile functionalities.

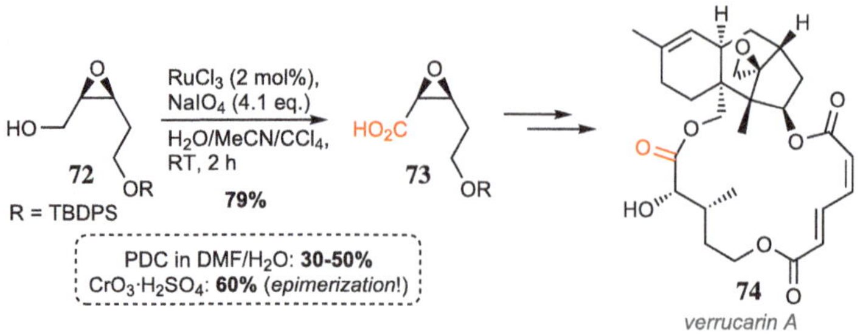

Fig. 4.34 RuCl$_3$-catalyzed oxidation in Still's synthesis of verrucarin A (**74**) [101]

A RuCl$_3$-catalyzed oxidation of a primary alcohol to the acid was used in Still's synthesis of the macrocyclic trislactone verrucarin A (**74**) [101]. While many oxidants only yielded traces of the desired carboxylic acid, the Sharpless modification allowed **73** to be obtained in 79% yield. PDC in DMF furnished the acid in moderate 30–50%. While a Jones oxidation **73** led to an improved yield of 60%, it was however accompanied by an epimerization of the product to the *trans*-configured epoxide.

Other Methods for the Oxidation of alcohols to carbonyl derivatives

In addition to the previously outlined oxidations, a multitude of further, less frequently employed methods exist. Even though they may not be encountered as often in the context of total syntheses, these reactions are suitable for a range of diverse applications due to their specific reaction conditions, the substrate scope or their typical side reactions.

The heterogeneous aerobic oxidation of alcohols to acids in the presence of transition metal catalysts is one of the mildest methods to introduce this functionality [13, 102–104]. Under classical conditions, finely dispersed or supported platinum is used in an aqueous solution of the substrate (*Heyns* oxidation). Even though the aerobic oxidation was historically restricted to the reaction of carbohydrates, all primary alcohols can be converted to the respective acids. Jacobsen and co-workers used the Pt-mediated oxidation of a primary alcohol as the final step in their synthesis of the secondary metabolite ambruticin (**76**, see Fig. 4.35) [105].

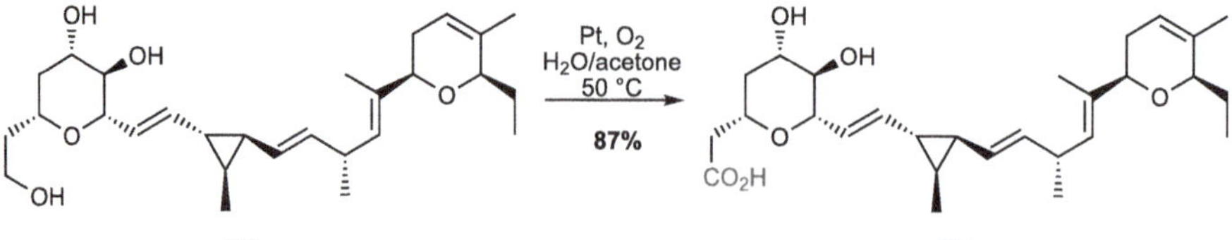

Fig. 4.35 Synthesis of ambruticin (**76**) by Jacobsen employing a Pt/O$_2$ oxidation as the final step [105]

The oxidation is usually carried out under slightly basic conditions ($pH = 8$–10). If the pH is too acidic, the reaction slows down drastically, which is why a base usually has to be continuously dosed to ensure turnover or a buffer has to be used to neutralize the formed carboxylic acid. The mechanism can be rationalized by a dehydration of the alcohol, although the exact elemental steps and their order have not been elucidated in detail. Following the adsorption of the substrate on the metal surface, hydrogen is transferred from the alcohol to Pt and subsequently forms water with molecular oxygen. The hydrate analog of the aldehyde resulting from the equilibrium with H_2O can then be dehydrated a second time to afford the carboxylic acid. It is assumed that the oxidation follows Langmuir–Hinshelwood–Hougen–Watson kinetics, which require all reactants and intermediates to be adsorbed on the surface during the course of the reaction (see Fig. 4.36) [102–104]. Nevertheless, an intermittent desorption cannot be ruled out. In a study with isotopically labeled $H_2^{18}O$ and $^{18}O_2$, it was also shown that only a small part of the oxygen in the product comes from O_2, but instead can be traced back to H_2O [106].

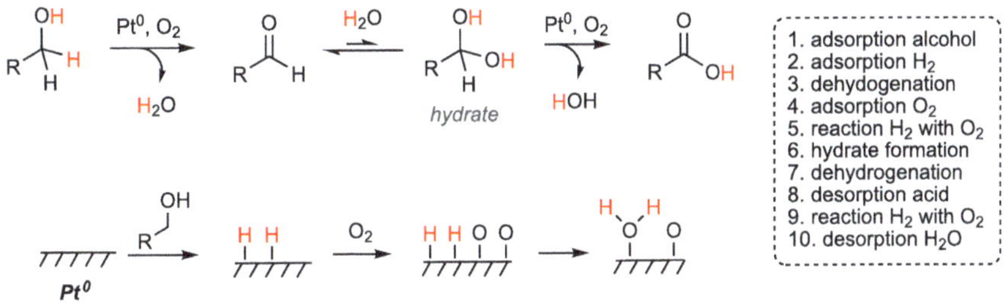

Fig. 4.36 Postulated mechanism of the Pt-catalyzed oxidation to acids [102, 104]

When using heterogeneous catalysts, a certain degree of deactivation is always observed. The metal catalyst can undergo a "natural" deactivation through sintering processes or metal leaching. An excess of oxygen can additionally oxidize the elemental metal, thereby losing its catalytic properties. Apart from these processes, the catalyst can be poisoned by certain compounds that form a strong bond with the active centers and thus irreversibly block them. Especially amines, sulfur-containing functionalities, and phosphorus compounds are known to deactivate noble metal catalysts [107]. While the acid formed as a product itself could also potentially pose a problem, the maintenance of a slightly basic pH during reaction can effectively prevent this.

The most outstanding criterion of oxidations mediated by O_2 under noble metal catalysis is the selectivity towards primary hydroxy groups. The method allows even individual primary alcohols to be selectively oxidized in the presence of other primary alcohols that possess only a slightly different steric environment (see Fig. 4.37). Furthermore, aldehydes are oxidized faster than alcohols, making possible the selective oxidation of an aldehyde in the presence of primary alcohols. Apart from platinum metal, other supported systems based on Pd, Cu, Au, or Co have been introduced as catalysts in recent years [108].

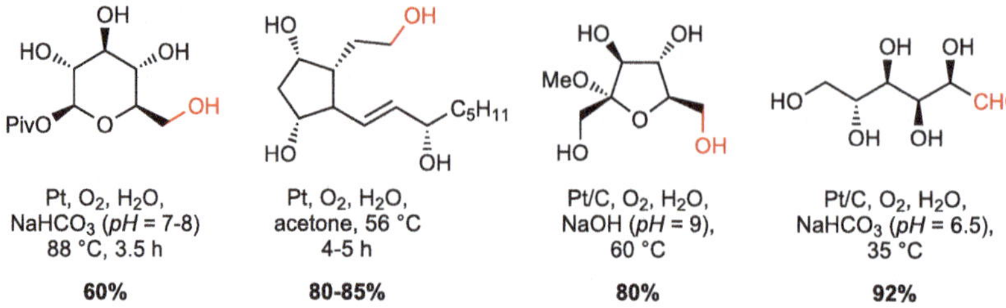

Pt, O$_2$, H$_2$O,
NaHCO$_3$ (pH = 7-8)
88 °C, 3.5 h

60%

Pt, O$_2$, H$_2$O,
acetone, 56 °C
4-5 h

80-85%

Pt/C, O$_2$, H$_2$O,
NaOH (pH = 9),
60 °C

80%

Pt/C, O$_2$, H$_2$O,
NaHCO$_3$ (pH = 6.5),
35 °C

92%

Fig. 4.37 Selective oxidations in the presence of other reactive functional groups. The highlighted functionality was oxidized to the corresponding acid [109–111]

Aqueous KMnO$_4$ constitutes another reagent for the synthesis of carboxylic acids from alcohols [13, 112]. It is an extremely strong oxidant and due to its high reactivity it is widely regarded to be incompatible with a variety of functional groups, foremost alkenes, 1,2-diols, sulfides, and thiols as well as benzyl groups, which are oxidized to ketones or acids. A hallmark of the oxidation with KMnO$_4$ lies in its adaptability to a range of different media: it can be carried out from strongly basic to strongly acidic conditions, so that it can be specifically adjusted to tolerate certain functional groups. Nevertheless, the oxidation is routinely carried out in a basic medium, as the rate of reaction is usually significantly higher compared to neutral or acidic conditions [113]. Since KMnO$_4$ is unstable in aqueous solution and decomposes to MnO$_2$, an excess of the reagent is usually used for the oxidation. To improve solubility, crown ether additives or biphasic systems in the presence of phase transfer catalysts can be employed (see Fig. 4.38).

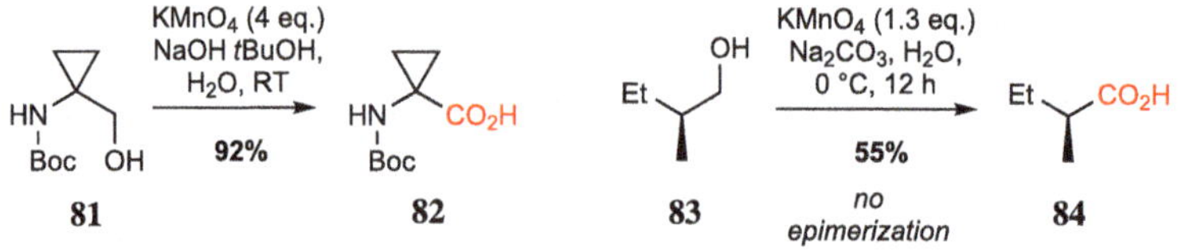

KMnO$_4$ (4 eq.)
NaOH tBuOH,
H$_2$O, RT

92%

81   **82**

KMnO$_4$ (1.3 eq.)
Na$_2$CO$_3$, H$_2$O,
0 °C, 12 h

55%

no epimerization

83   **84**

Fig. 4.38 Oxidations of alcohols to carboxylic acid with KMnO$_4$ [114, 115]

If a one-step oxidation of alcohols to the corresponding carboxylic acid affords only low yields or is incompatible with certain functional groups, a two-step protocol can be chosen instead. NaClO$_2$ is widely employed for the oxidation of aldehydes to carboxylic acids (*Pinnick* oxidation) [116]. The HOCl by-product generated from ClO$_2^-$ can autocatalytically decompose ClO$_2^-$ to chlorine dioxide (ClO$_2$) and Cl$^-$ [117]. The redox pair HOCl/Cl$^-$ is a stronger oxidant than ClO$_2^-$/HOCl, HOCl can easily undergo side reactions with double bonds and it oxidizes ClO_2^-. Consequently, the Pinnick oxidation benefits from the addition of a HOCl scavenger for its efficient removal to prevent undesired side reactions [118]. 2-Methyl-2-butene is the most common additive, but H_2O_2, resorcinol, or sulfamic acid also have been reported to be effective. NaH_2PO_4 is often added as a buffer, but transformations can also be carried out without it and still afford the product in high yield (see Fig. 4.39) [116].

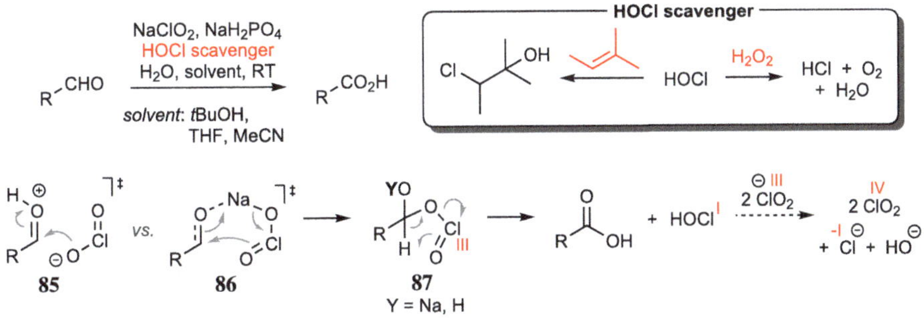

Fig. 4.39 Mechanism of the Pinnick oxidation [118–121]

The oxidation is initiated either by a nucleophilic attack of the chlorite on the protonated aldehyde (**85**) or by a direct reaction of the unprotonated carbonyl with $NaClO_2$ (**86**) [119, 120]. The resulting intermediate **87** rapidly decomposes, presumably intramolecularly, to the carboxylic acid and hypochlorous acid. HOCl must then be rapidly degraded by reaction with a suitable HOCl scavenger. The reaction is also amenable to kg-scale (see Fig. 4.40) [122].

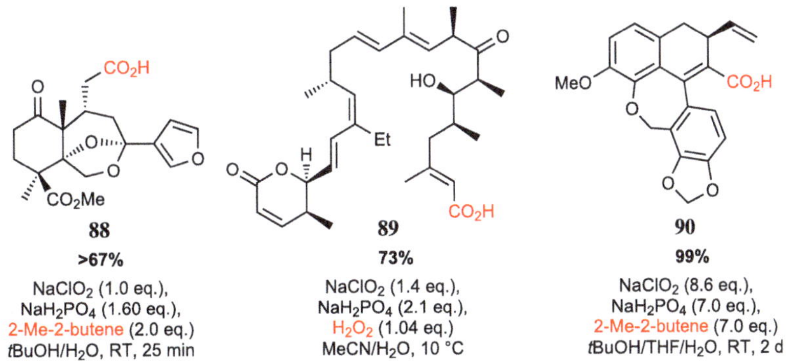

Fig. 4.40 Products of a Pinnick oxidation in the syntheses of complex secondary metabolites [123–125]

4.1.2   Selective Oxidations of Primary or Secondary Alcohols in Polyols

Primary alcohols are sterically less crowded compared to secondary alcohols. In the absence of dominant electronic factors, many reagents therefore display a higher reaction rate with primary alcohols. These include TPAP, PCC, DMP, IBX, or the Swern oxidation, which can achieve high chemoselectivities. Especially TEMPO-mediated reactions can very efficiently convert a primary alcohol in the presence of a secondary one to the aldehyde (see Fig. 4.41).

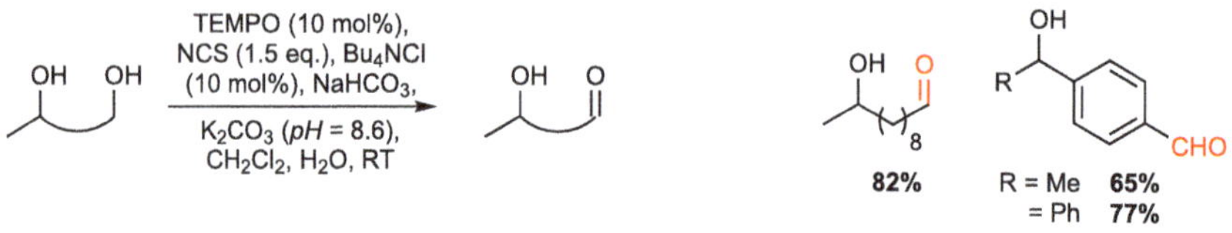

Fig. 4.41 Selective oxidations of primary alcohols [126]

For the oxidation of a scaffold bearing both axial and equatorial hydroxyl groups, nitroxide-radical-based reagents should be the first choice if oxidation of the equatorial hydroxyl group is needed. Stevens or Dess–Martin reagents should be employed for the preferential oxidation of the axial hydroxyl group [127].

Silver salts (Ag_2O, Ag_2CO_3) have long been known to stoichiometrically oxidize alcohols to aldehydes/ketones or the corresponding carboxylic acids. The reagents, similar to activated MnO_2, are strongly dependent on the surface structure of the metal nanoparticles. Fétizon and co-workers hence introduced Ag_2CO_3 adsorbed on Celite as an efficient oxidant, as a high active surface can be ensured by the support [14]. The efficiency of the *Fétizon* reagent is highly correlated with the steric environment of the substrate. For example, **94** is oxidized much more rapidly than **95**, as the α-proton is more accessible. This allows primary alcohols to be selectively oxidized in the presence of secondary alcohols with Ag_2CO_3/Celite (see Fig. 4.42).

In the first step, an adsorption of the substrate on the surface occurs (**91**). Following the electron transfer from the substrate to the active Ag centers (**92**), the protonated aldehyde and a proton are likely formed, which are adsorbed on the reduced metal. The carbonate finally decomposes to CO_2 and H_2O. The silver carbonate is converted to Ag^0 by the oxidation of the alcohol to the aldehyde, which necessitates stoichiometric amounts of the reagent. This makes the method expensive, but the conditions are so mild that many functional groups are tolerated and more than make up for the costs [128].

The reaction rate of the oxidation is highly dependent on the choice of solvent. The presence of polar functionalities, both in the solvent and in the substrate, can significantly slow down the oxidation, as the chemisorption of the hydroxy group (**91**) is competitively inhibited. Therefore, toluene or hydrocarbons are used as standard solvents [128].

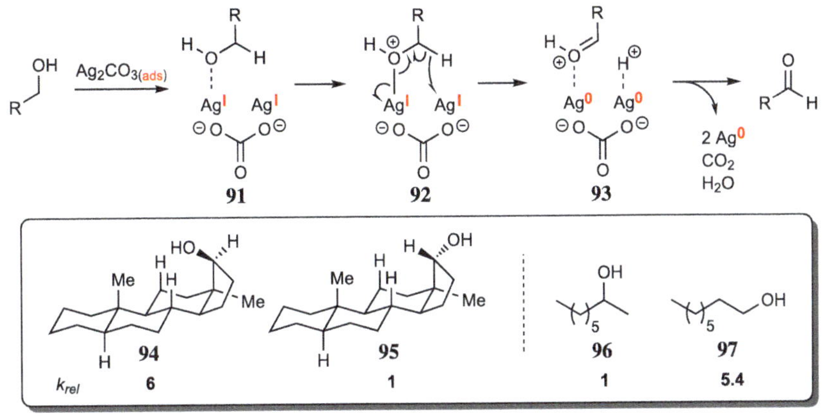

Fig. 4.42 Postulated mechanism of oxidation with Ag_2CO_3/Celite and relative reaction rates of select substrates [128]

The oxidation potential of aldehydes is higher than that of ketones, which is why the oxidation of **secondary** alcohols is thermodynamically preferred [129]. Very mild reagents can therefore react faster with secondary alcohols barring steric factors. The Corey–Kim oxidation or PDC, for example, show a certain preference for secondary alcohols.

The method of choice for a selective oxidation of secondary alcohols was introduced by Stevens *et al.* relying on NaOCl in acetic acid. The Stevens oxidation uses technical grade sodium hypochlorite, which makes the method very cost-effective. Its mechanism has not yet been elucidated [130]. Apart from NaOCl, other electrophilic halogen sources such as Br_2, NBS or $KBrO_3$ can also selectively oxidize secondary alcohols (see Fig. 4.43) [131].

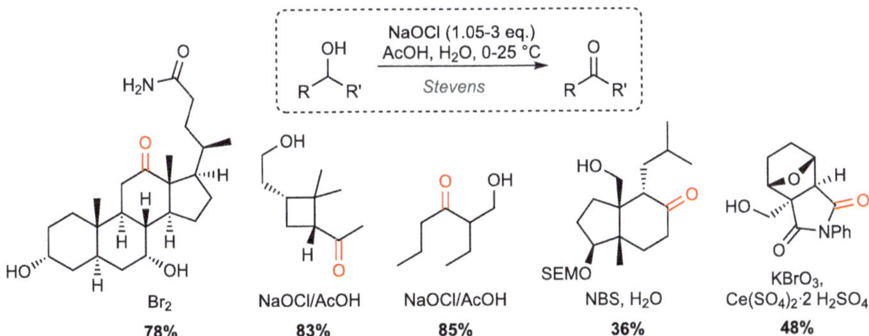

Fig. 4.43 Selective oxidation of secondary alcohols [130, 131]

The conversion of a secondary alcohol to a ketone via organoaluminum-mediated transfer hydrogenation (**98**) is referred to as *Oppenauer* oxidation [132, 133]. It is conducted in the presence of a "sacrificial carbonyl", which serves as the secondary stoichiometric oxidant. The reaction is a thermodynamically controlled and the product formation is determined by the redox equilibrium between the oxidant and the substrate. It has lost much of its preparative significance due to the newer and usually milder methods outlined above. However, given the basic reaction conditions, it is suitable for the oxidation of acid-sensitive substrates and is still occasionally used, even on an industrial scale [18]. Based on the redox potentials, secondary alcohols are oxidized more easily than primary alcohols. The oxidant must therefore possess the lowest reduction potential for a selective oxidation of polyols. Carbonyl compounds with low reduction potential, such as aromatic aldehydes, are therefore particularly suitable oxidizing agents (see Fig. 4.44) [132]. Apart from aluminum salts, which are commonly added stoichiometrically, other promoters can also be used substoichiometrically or catalytically [134–136].

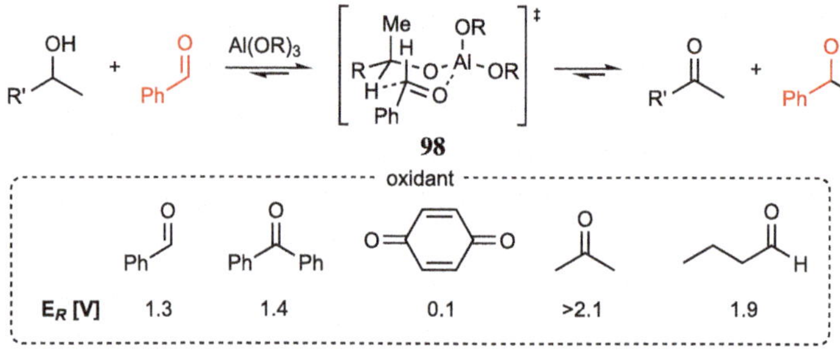

Fig. 4.44 Mechanism of the Oppenauer oxidation and reduction potentials of selected carbonyl compounds [132]

If a hydroxy group of a polyol resides in the **allylic** or **benzylic** position, one can resort to the available methods that selectively enable an oxidation of these activated functionalities. Manganese(IV) oxide and DDQ are considered prime reagents for this transformation. $BaMnO_4$ can be utilized analogously to MnO_2 as it has a comparable reactivity profile (see Fig. 4.45) [14, 112, 137, 138].

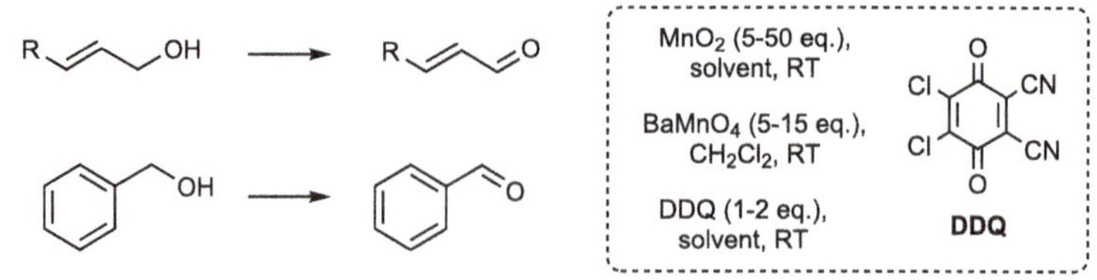

Fig. 4.45 Oxidations of allylic and benzylic alcohols [14, 112, 137, 138]

Even though manganese dioxide especially lends itself for allylic oxidations, it possesses some peculiarities that can limit its efficiency [137, 138]. There are many methods for preparing activated MnO_2, but depending on the conditions of the preparation they provide an oxidant with varying efficiency. This leads to the oxidations often not being reproducible if MnO_2 from a source other than the published one is used and also results in the addition of a large excess of the oxidant. Polar solvents should generally be avoided due to a competitive adsorption versus the substrate. Hence hydrocarbons, halogenated hydrocarbons, and ethers are routinely employed (see Fig. 4.46).

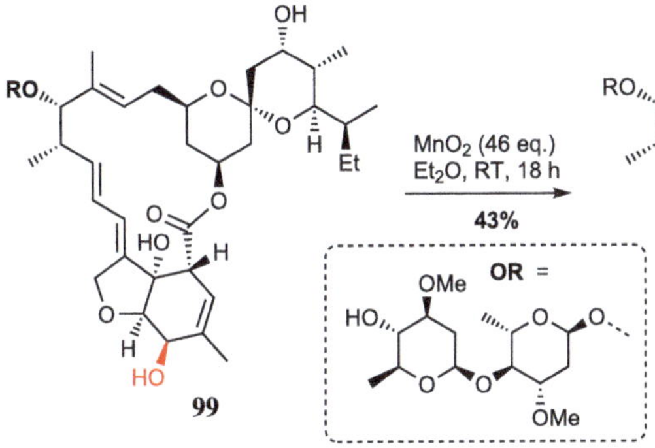

Fig. 4.46 Derivatization of avermectin B_2a (**99**) [139]

The reaction should not be carried out at higher temperatures, as the oxidation potential of manganese dioxide increases significantly with heat. Thus normally inert functional groups can be affected at elevated temperatures. Allylic alcohols are on average oxidized slightly faster than benzylic hydroxy groups [14].

Electron-poor quinones and specifically DDQ can oxidize allylic and benzylic hydroxy groups to the corresponding aldehyde or ketone. Quinones mechanistically act as hydride acceptors and convert the alcohol (**101**) to a stabilized adduct (**103**). This intermediate furnishes the desired carbonyl (**104**) and a hydroquinone (**105**) upon elimination of H^+. The rate-determining step is postulated to be the intramolecular H-transfer **103**→**104** (see Fig. 4.47) [140, 141].

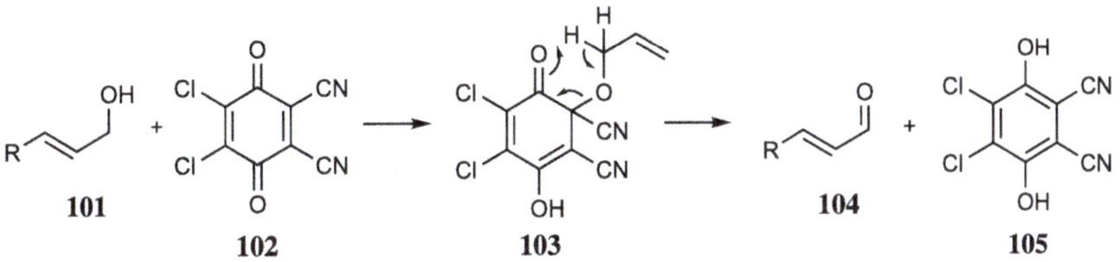

Fig. 4.47 Postulated mechanism of DDQ oxidation [140]

The most common use for DDQ is the oxidative cleavage of PMB protecting groups, which can be removed orthogonally to other protecting groups with this reagent.

4.1.3 Other Oxidations

Apart from the oxidation of alcohols to ketones and aldehydes, there are a number of other reactions that have not yet been covered: the oxidation of ketones and aldehydes to esters and amides, allylic or benzylic oxidations, and the one- or multi-step formal dehydrogenation of carbonyls [18]. The α-functionalization of carbonyl compounds and the derivatization of CC-multiple bonds can be included as well. They have already been described in detail in the two previous chapters (see Fig. 4.48).

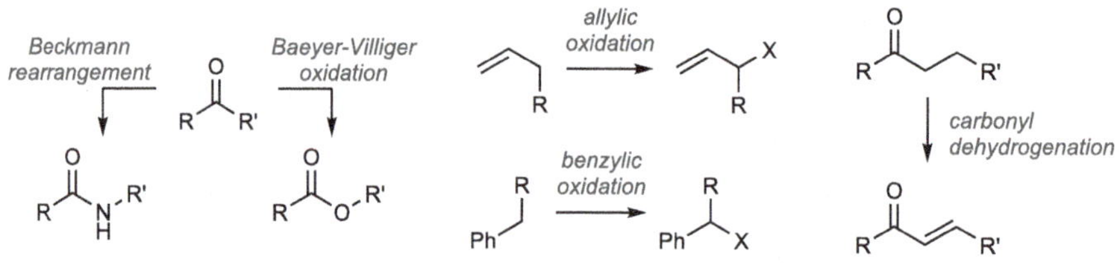

Fig. 4.48 Oxidation of carbonyl derivatives as well as allylic and benzylic oxidation

In this context, the selective formation of carboxylic acid esters from aldehydes or ketones mediated by peroxycarboxylic acids is one of the most important methodologies (see Fig. 4.49) [142–144].

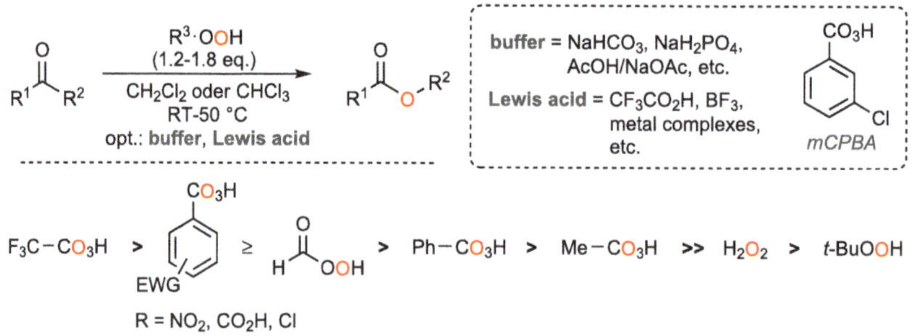

Fig. 4.49 Typical reaction conditions of the Baeyer–Villiger oxidation and relative reactivity of common peracids/peroxides

In the reaction known as *Baeyer–Villiger oxidation*, a wide range of functional groups is tolerated. Isolated alkenes can potentially be epoxidized under the reaction conditions; the chemoselectivity depends on the substrate and the oxidizing agent in these cases [145]. The more electron-rich the olefin and the less its steric hindrance, the more an epoxidation is preferred over the oxidation (see Fig. 4.50).

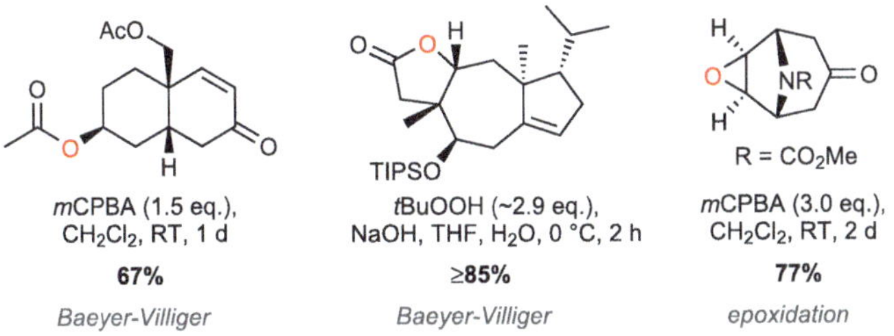

Fig. 4.50 Competition of Baeyer–Villiger oxidations and epoxidations in total syntheses [146–148]

The most common solvents are chloroalkanes, other solvents such as toluene, acetonitrile are also utilized occasionally. Either peracids or peroxides can be used as oxidant. The by far most frequently employed reagent is *meta*-chloroperbenzoic acid (*m*CPBA). The order of reactivity of the various peracids is derived from the strength of the conjugated acid, which is obtained as a by-product alongside the ester. Peroxides are significantly less reactive compared to peracids.

The postulated mechanism consists of two discrete steps. First, the peracid or peroxide attacks the carbonyl carbon and forms the Criegee intermediate **106**. The migrating substituent R^M is transferred to the geminal O-atom in a concerted, cyclic transition state **107**. The hydroxyl proton of the Criegee intermediate **106** can only be transferred in an intramolecular fashion with peracids, which is why an additional catalyst is needed with peroxides. Depending on the reaction conditions (solvent, catalyst) and the substrate, migration of the substituent via **107** usually constitutes the rate-determining step. However, the addition to the Criegee intermediate can also become the slowest step in some reactions (see Fig. 4.51) [142, 143].

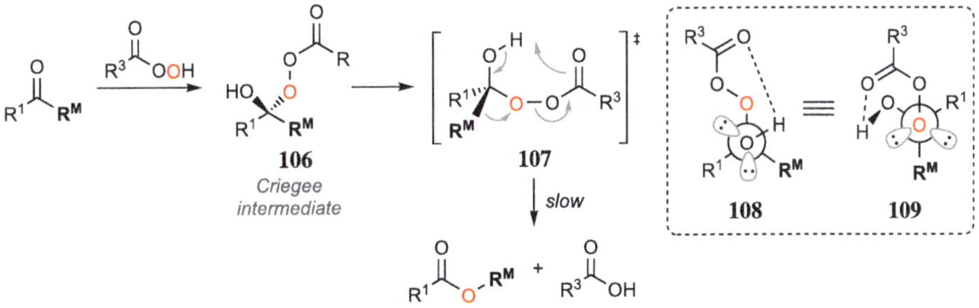

Fig. 4.51 Mechanism of the Baeyer–Villiger oxidation with peracids [149, 150]

The migrating group is aligned in an antiperiplanar manner to the breaking O-O bond of the leaving group as shown in **108, 109**. Considering the valence orbitals, this allows the greatest interaction between the $\sigma_{C\text{-}R^M}$ and $\sigma^*_{O\text{-}O}$ bonds. This is referred to as the *primary* stereoelectronic effect [151]. The antiperiplanar position of the free oxygen electron pair with regard to the R^M group (**108**, *secondary* stereoelectronic effect) facilitates the C-R^M bond breaking and migration through the $n_O \rightarrow \sigma^*_{C\text{-}R^M}$ interaction [149, 150]. Brønsted acids can catalyze the addition step, but not the rearrangement of the Criegee intermediate. According to theoretical studies, the addition should be rate-determining. The presence of traces of catalytically active carboxylic acids in technical grade peroxycarboxylic acids would however explain why the rearrangement of **107** is usually observed to be slower [152].

The regiochemistry depends on the migratory aptitude of the substituents on the carbonyl, [142] which follows the order

tertiary alkyl > secondary alkyl > benzyl > phenyl > primary alkyl > methyl

The migratory aptitude results from a combination of the electronic stabilization of the transition state (primary and secondary stereoelectronic effect) and the steric demand of R^M [153]. In acyclic ketones, the migrating group must usually be a secondary or tertiary alkyl residue. Both the migrating group and the stereochemistry of the product can be well predicted, as the migrating group is transferred while preserving the configuration (see Fig. 4.52).

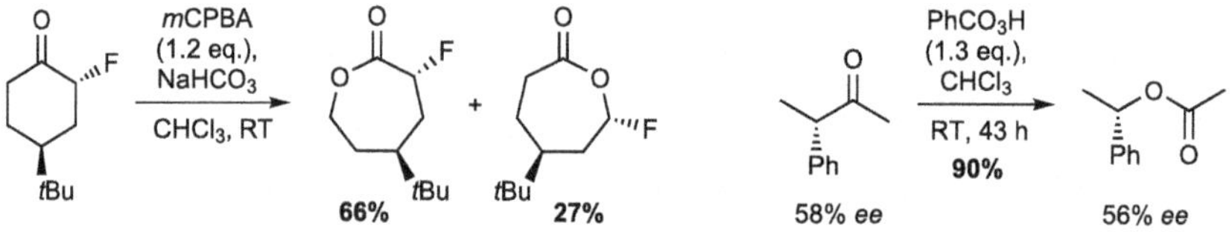

Fig. 4.52 Stereospecificity of the Baeyer–Villiger oxidation [149, 150, 154]

The strongest oxidants, which are also strong acids, should be used in the presence of a buffer to avoid transesterifications. The acids resulting from the peracid oxidants must be separated from the product. While this is easily executed on laboratory scale, purification on scales beyond a certain size poses major problems [18]. Peroxides, especially hydrogen peroxide, are the reagent of choice in these cases, but there are also disadvantages associated with their utilization: H_2O_2 requires the use of H_2O as a cosolvent, which can saponify the formed esters. Furthermore, the sluggishness of the reaction requires the addition of a suitable catalyst.

In recent years, the development of a Baeyer–Villiger oxidation with hydrogen peroxide as a reagent has been intensively pursued, also for improving the sustainability of the transformation [142, 155]. Brønsted acids are suitable as catalysts under these conditions. Transition metal complexes are similarly capable of accelerating the reaction, for example via coordination of the carbonyl group and lowering the LUMOs for an attack by H_2O_2. In addition to an interaction with the substrate, Lewis acids or bases can activate the Criegee intermediate or the peroxide oxidant by increasing the species' electrophilicity or nucleophilicity upon coordination (see Fig. 4.53).

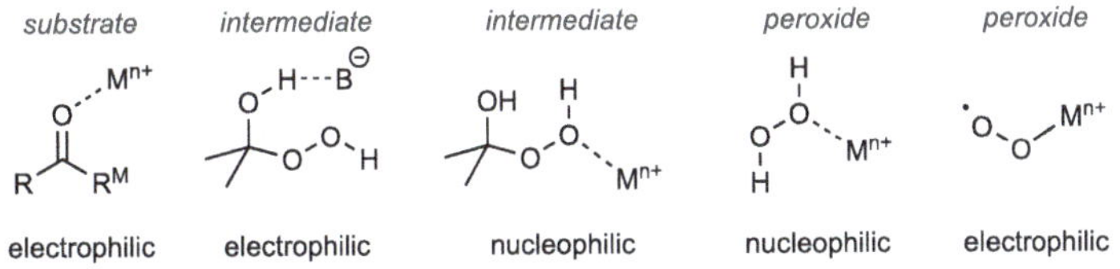

Fig. 4.53 Possible activation modes in catalyzed Baeyer–Villiger oxidations [142]

The charm of utilizing transition metal complexes lies in the possibility of using chiral complexes to enable an asymmetric reaction. High enantioselectivities beyond 90% can so far only be achieved for a few substrates with metal catalysts, but a growing number of enzymatic methods have been published [156, 157]. Since the Baeyer–Villiger oxidation does not introduce a new stereocenter, enantioselective methods are limited to desymmetrizations of prochiral ketones (especially cyclobutanones and cyclohexanones) and kinetic resolutions of racemic mixtures.

A racemic method using H_2O_2 with a relatively broad substrate scope was disclosed by Ishihara and co-workers [158]. The biphasic system allows the selective oxidation of ketones in the presence of CC double bonds by using a combination of a lipophilic Lewis acid and a hydrophilic Brønsted acid. Despite the presence of an aqueous phase, no hydrolysis to the corresponding acid was observed in the selected substrates, although transesterifications did occasionally occur. The postulated mechanism relies on the *in situ* formation of $M[B(C_6F_5)_4]_m \cdot [HO_2CCO_2H]_n$ as the catalytic species, which acts as a highly reactive Brønsted acid and activates the carbonyl group for an attack by the peroxide (see Fig. 4.54).

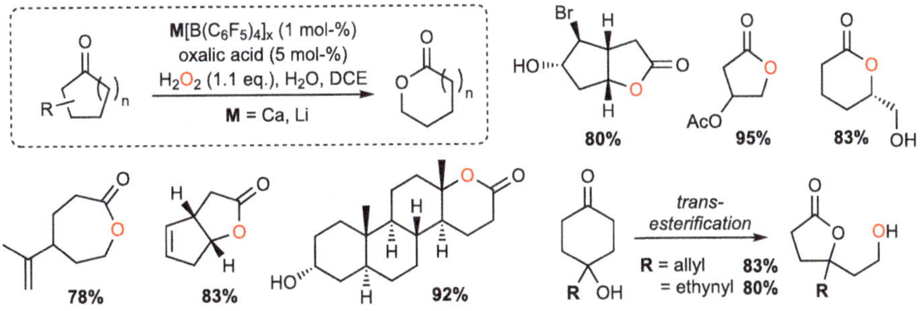

Fig. 4.54 Baeyer–Villiger oxidation using H_2O_2 as oxidant developed by Ishihara [158]

The reaction of aldoximes and ketoximes to amides in acidic medium can be considered the nitrogen counterpart to the Baeyer–Villiger oxidation, despite a different mechanism.

It has been named *Beckmann rearrangement* in honor of its discoverer (see Fig. 4.55) [159].

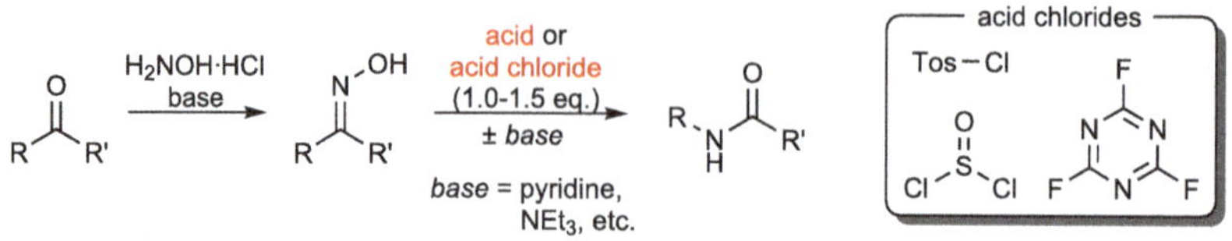

Fig. 4.55 General reaction scheme of the Beckmann rearrangement

The reaction is highly relevant for industrial production: ϵ-caprolactam is synthesized from cyclohexanone on a million-ton scale via Beckmann rearrangement. Hydroxylammonium sulfate is generated *in situ* in an aqueous sulfuric acid-ammonia solution and converted with cyclohexanone to the respective oxime in a weakly acidic solution. This is added as a melt to an aqueous solution of a slight excess of oleum. Due to the strong exothermicity of the rearrangement step, the reaction is conducted in a solution of caprolactam, the dilution limiting the adiabatic temperature increase. In the final step the sulfate is hydrolyzed with diluted ammonia (see Fig. 4.56) [160].

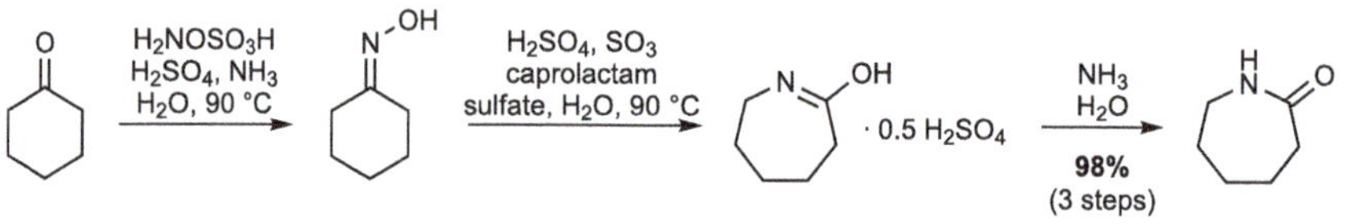

Fig. 4.56 Industrial synthesis of ϵ-caprolactam [160–162]

The postulated mechanism of the Beckmann rearrangement is based on the coordination of an activator A as initial step (**110**). Depending on whether an acid or an acid chloride is employed, different activated oximes result. The migration of the group R in **110** to the nitrogen atom is *trans*-selective with respect to the N-O bond, so depending on the geometry of the oxime, different regioisomers are obtained as product. In the concerted migration step, the cation **112** is formed from the transition state **111** by elimination of the leaving group, to which water adds to afford **113**. The reaction usually proceeds with retention of configuration with respect to the migrating group [163, 164]. After tautomerization, the desired amide is obtained (see Fig. 4.57) [159, 165, 166].

The Beckmann rearrangement is only occasionally encountered in total synthesis. The stereospecificity of the reaction is both a blessing and a curse: the stereochemistry of the oxime dictates the migrating group due to the *trans*-selectivity. Since the geometry of the oxime cannot be consistently controlled, the desired isomer of the oxime is obtained as a

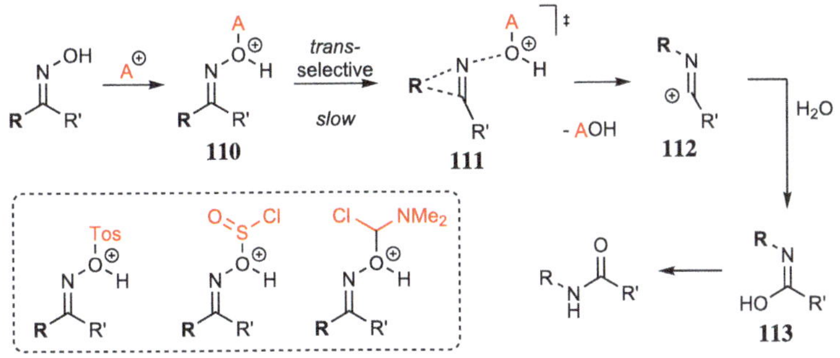

Fig. 4.57 Postulated mechanism of the Beckmann rearrangement and various activated oximes of type **110** [159, 165]

best case scenario. The isomer ratio can be influenced to some extent by varying the reaction conditions [167]. In the less favorable case, the "wrong" oxime isomer can only form the undesired amide analog. The selectivity is not a concern for symmetric substrates, as the resulting product of both oxime isomers is identical (see Fig. 4.58).

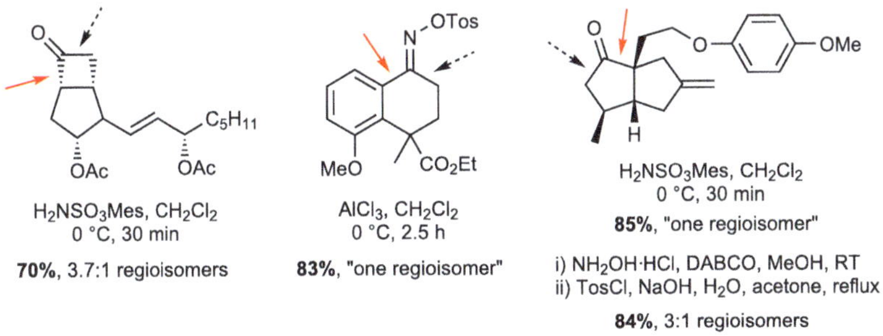

Fig. 4.58 Applications of the Beckmann rearrangement. The preferred migrating bond is indicated by the red arrow [167–169]

Since the initial contributions by Beckmann, numerous variants and subsequent improvements have been reported, most of which rely on the isolation of the oxime or tosyloxime. A preparatively simple and very mild variant was published by Giacomelli and co-workers in the early 2000s. Trichlorotriazine (TCT) serves as an activator, but it does not react directly with the oxime, instead forming the Vilsmeier reagent with the DMF solvent. This can subsequently activate the oxime, analogous to the mechanism shown in Fig. 4.57. Ketoximes are converted to the corresponding amides. As only a minor dependence of the migrating group on the oxime stereoisomer is observed, it is assumed that an interconversion of the *E*- and *Z*-oximes takes place under the reaction conditions. Aldoximes quantitatively afford nitriles (see Fig. 4.59) [170].

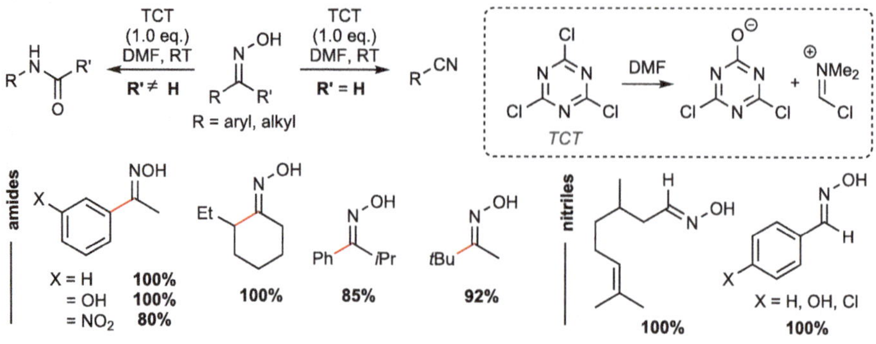

Fig. 4.59 Synthesis of amides and nitriles from oximes according to Giacomelli *et al.*. The migrating bond in ketoximes is marked in red [170]

Due to their polarized CO bond, carbonyl groups allow for selective oxidation to the respective carboxylic acid derivatives or in α-position to the carbonyl group. The increased acidity and thus reactivity of allylic or benzylic CH bonds is another handle that can be exploited for selective functionalizations. Traditionally, oxidations in allylic position were mediated by selenium dioxide with subsequent derivatization. This reaction, occasionally also referred to as *Riley* oxidation, used to be carried out with superstoichiometric amounts of SeO_2 (1.1–2 eq.). Over time the use of a secondary oxidant (*t*BuOOH, PhI(OAc)$_2$, etc.) has been established as the de-facto standard method. The conditions, especially the reaction temperature, should be as mild as possible to avoid the overoxidation of aldehydes to acids or further undesired redox reactions (see Fig. 4.60) [171, 172].

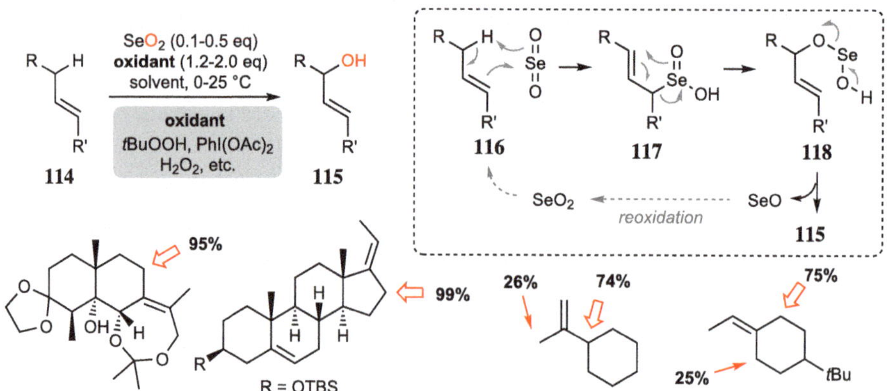

Fig. 4.60 Mechanism of allylic selenium dioxide oxidation and examples of regioselectivity [172–177]

The mechanism consists of a concerted ene reaction between the alkene and SeO_2 to **117**, followed by a 2,3-sigmatropic Wittig rearrangement, with the CC double bond shifting back to its original position. **118** can then be converted to the allylic alcohol **115** upon elimination of SeO. In the presence of a secondary oxidant, SeO can be regenerated from SeO_2 and subsequently reacts with another equivalent of the alkene [172, 173].

Empirical selectivity guidelines have been derived based on the underlying mechanism to differentiate which of the eligible allylic positions are favored for oxidation (see Fig. 4.60) [172]:

1. In trisubstituted olefins, oxidation occurs at the more highly substituted side of the double bond.
2. The reaction rate in trisubstituted olefins decreases in the order $CH_2 > CH_3 > CH$.
3. For cyclic hydrocarbons, oxidation occurs in the ring, if possible, and α to the more substituted carbon atom. If the favored position for attack is tertiary, dienes may result.
4. Disubstituted olefins react in the α-position in the order $CH_2 > CH_3$
5. Vicinal disubstituted cyclic olefins, e.g., cyclohexene, show the trend $CH_2 > CH$. The double bond can possibly rearrange.
6. Terminal olefins yield primary alcohols with concomitant shift of the allylic double bond.

Due to its toxicity and the odor of the by-products, synthetic chemists typically resort to selenium reagents only when alternative methods do not offer comparable selectivity. This mainly applies to allylic and α-carbonyl oxidations, which are still occasionally employed in synthesis. In Guillou's approach to *rac*-codeine, a SeO_2-mediated oxidation was used to introduce the hydroxy group in allylic position (**119**→**120**). However, a subsequent oxidation/reduction sequence had to be used to synthesize the natural product due to the incorrect stereochemistry resulting in the allylic functionalization (see Fig. 4.61) [178].

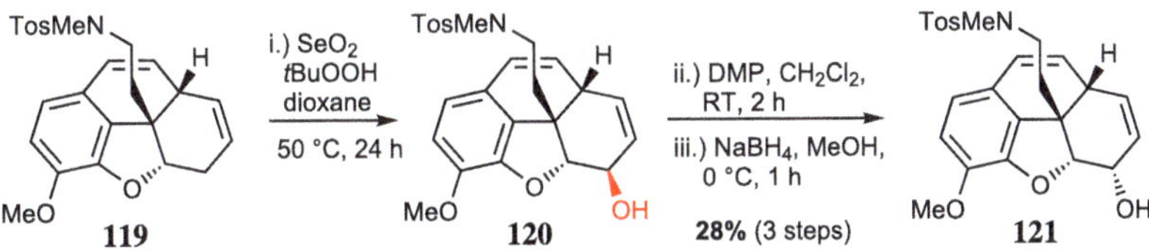

Fig. 4.61 Use of an allylic SeO_2-mediated oxidation in the total synthesis of codeine [178]

Allylic oxidations are a topic of intense research, as many target structures contain allylic alcohols with defined stereochemistry at the OH functionality [179]. An enantioselective, direct CH-oxidation could deliver such structures in one step. Currently, access to this structural motif is typically obtained by a 1, 2-reduction of enones, especially in structurally rigid alkenes such as steroids. The White group developed a method for the selective allylic functionalization of alkenes catalyzed by bis-sulfoxide-complexed Pd salts. After the formation of the allyl-Pd complex **122**, a nucleophile can lead to either the branched or linear product (see Fig. 4.62) [180–182].

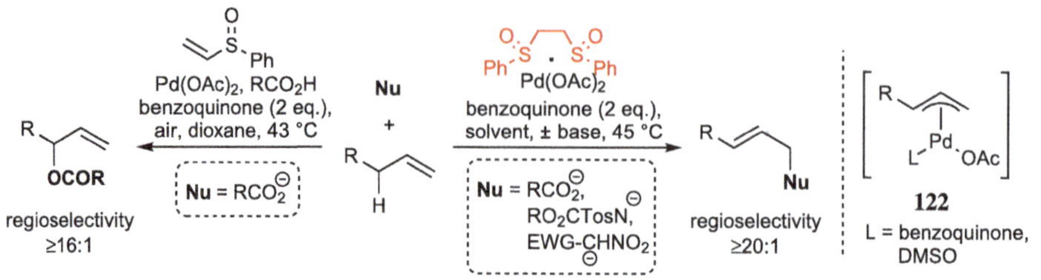

Fig. 4.62 Allylic CH-functionalization according to White *et al.* [180–182]

Instead of benzoquinone, the use of molecular oxygen as oxidant was subsequently introduced as well [183]. The method could later be expanded to forge the allylic C-heteroatom bond in an enantioselective fashion. Hartwig and co-workers reported an enantioselective CH activation by sequential use of Ir and Pd salts. In addition to amines, malonic acid esters, sulfones, and alcohols can be used as nucleophiles (see Fig. 4.63) [184].

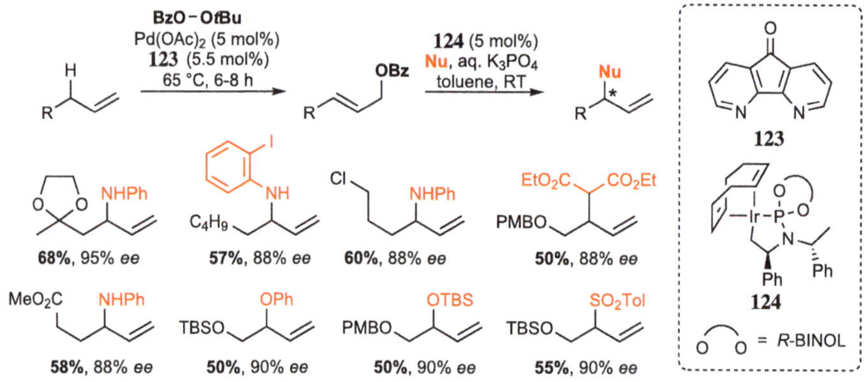

Fig. 4.63 Sequential, enantioselective allylic CH-functionalization [184]

Allylic and benzylic oxidations can also be achieved with stoichiometric use of classic oxidants, foremost with hexavalent chromium reagents and quinones [185]. Cr-based oxidizing agents are, along with selenium dioxide, often utilized to synthesize α,β-unsaturated enones.

Allylic oxidations to enones often yield product mixtures due to lack of regio- and chemoselectivity. The first step of most allylic oxidations is the formation of an allyl radical by breaking the weakest C-H bond. Subsequently, the α,β-unsaturated enone can be quickly formed via a peroxide. Apart from the lower bond strength at the allylic position, the steric environment also plays a role in determining the regioselectivity [185]. Especially reagents based on chromium trioxide, often complexed with nitrogen heterocycles such as 3,5-dimethylpyrazole (3,5-DMP), successfully furnish the oxidation products. Newer methods also use substoichiometric amounts of chromium with a secondary oxidant, for example *t*BuOOH (see Fig. 4.64).

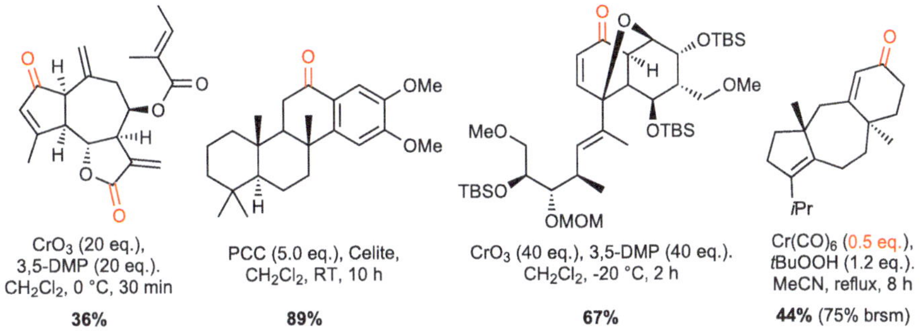
CrO3 (20 eq.), 3,5-DMP (20 eq.). CH2Cl2, 0 °C, 30 min
36%
PCC (5.0 eq.), Celite, CH2Cl2, RT, 10 h
89%
CrO3 (40 eq.), 3,5-DMP (40 eq.). CH2Cl2, -20 °C, 2 h
67%
Cr(CO)6 (0.5 eq.), tBuOOH (1.2 eq.). MeCN, reflux, 8 h
44% (75% brsm)

Fig. 4.64 Synthesis examples of allylic and benzylic oxidations [186–189]

The oxidation of ketones, esters, or amides to the corresponding α, β-unsaturated carbonyls formally constitutes a dehydrogenation, as H_2 is removed [190, 191]. The reaction was traditionally carried out as a two-step process: the introduction of a leaving group followed by its elimination as H-X. The transformation was either mediated by selenium-based reagents or α-positioned C-halogen bonds could be eliminated with non-nucleophilic bases (see Fig. 4.65) [192, 193].

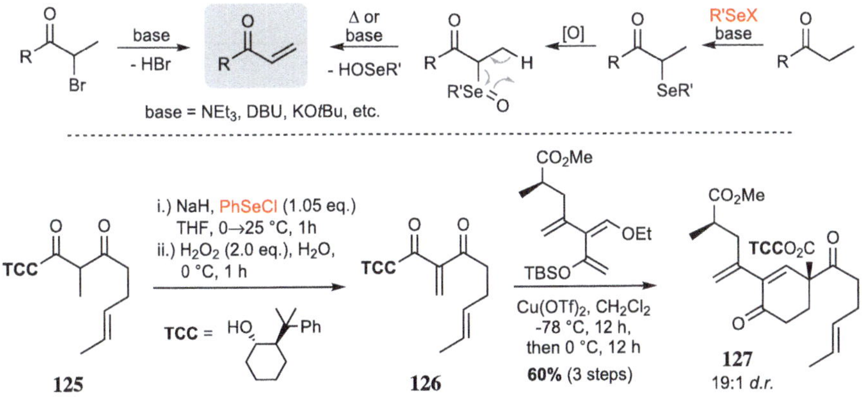
base
- HBr
Δ or base
- HOSeR'
[O]
R'SeX base
base = NEt3, DBU, KOtBu, etc.
i.) NaH, PhSeCl (1.05 eq.) THF, 0→25 °C, 1h
ii.) H2O2 (2.0 eq.), H2O, 0 °C, 1 h
TCC =
Cu(OTf)2, CH2Cl2 -78 °C, 12 h, then 0 °C, 12 h
60% (3 steps)
125
126
127
19:1 d.r.

Fig. 4.65 Oxidation of carbonyl derivatives by elimination of halogen/selenium and application in Shenvi's synthesis of isocyanoterpenes [194]

After the addition of selenium, oxidizing agents such as peroxides, peracids, or ozone are used to form the corresponding selenium oxide. This reactive intermediate undergoes a thermal *syn*-elimination via an intramolecular, cyclic transition state to afford the desired olefin [195, 196]. Aldehydes are stable under the reaction conditions, and a subsequent oxidation to the corresponding carboxylic acid does not occur.

The direct oxidation is simpler from a preparative point of view and usually more efficient in terms of atom economy (see Fig. 4.66). The dehydrogenation of enolates or enolate equivalents can be catalyzed by transition metal complexes. The *Saegusa–Ito oxidation* likely is the best-known variant, in which originally silyl enol ethers were converted into α, β-unsaturated ketones in the presence of stoichiometric amounts of Pd(OAc)$_2$ by dehydrosilylation (method A) [197–199]. Later modifications, especially by the Larock group, saw the introduction of secondary oxidants, enabling a catalytic use of Pd salts (conditions B) [200]. Modern catalytic methods also allow the direct use of the carbonyl as opposed to the silyl enol ether. The mechanism can be rationalized by the initial formation of an α-pallada-carbonyl complex **128**. Under classic Saegusa conditions, TMS-OAc is eliminated. The β-hydride elimination from **128** via **129** yields the desired enone and a palladium hydride complex **130**, which can reductively generate HX and a Pd(0) species. If secondary oxidants are employed, the catalyst can be reoxidized to the active Pd(II) complex [198].

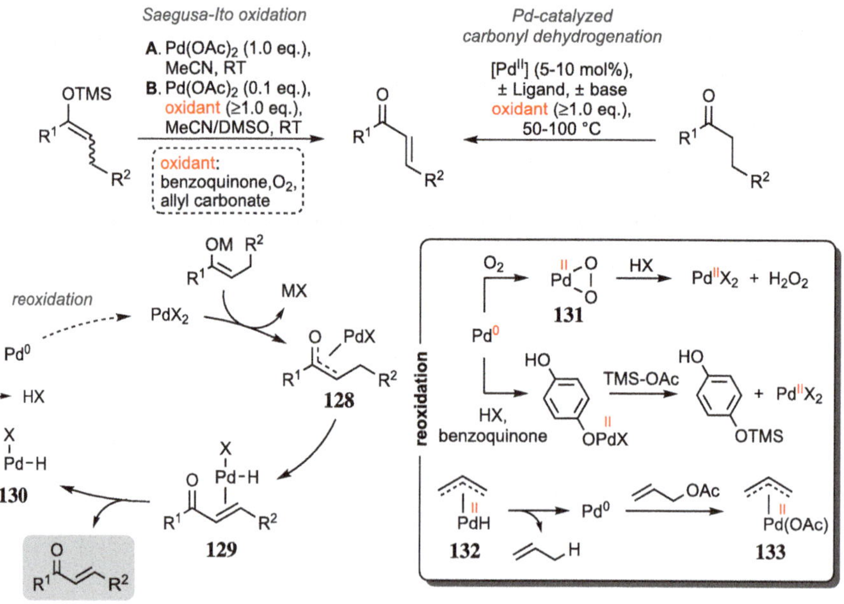

Fig. 4.66 Reaction conditions and mechanism of the dehydrogenation of carbonyl compounds [198, 201, 202]

Depending on the secondary oxidant, the reoxidation proceeds via different routes. Under aerobic conditions, the existence of a Pd-peroxo complex **131** was postulated, which can form PdX$_2$ and hydrogen peroxide in the presence of an acid [202]. Benzoquinone reacts to Pd-hydroquinone and subsequently to the corresponding TMS ether, if the TMS-OAc elimination by-product from the catalytic cycle is also present, such as under Saegusa conditions. With allyl carboxylates and phosphates as oxidants, the reaction instead proceeds via mixed allyl-Pd intermediates (X = Allyl, e.g., **133**) and propene is eliminated instead of AcOH upon reductive elimination from Pd(II) to Pd(0) (**132**) [198].

If multiple regioisomers are possible, the selectivity is guided by the starting material: The silyl enol ether is dehydrated in the β-position to the carbonyl, which is dictated by the intermediates **128** and **129**. The chemo- and regioselectivity can thus be controlled by the formation of the enolate and the corresponding silyl enol ether (kinetic versus thermodynamic enolate, cf. Sect. 2.2). Many functional groups (esters, ketones, nitriles, halides, silyl ethers, epoxides, etc.) are tolerated in the Saegusa–Ito oxidation, but some substrates yield significantly better results when stoichiometric amounts of Pd(OAc)$_2$ are used. Basic functionalities such as nitrogen in isoquinolines can negatively affect the reaction through complexation of the Pd metal, thus necessitating a literature search with similar substrates (see Fig. 4.67).

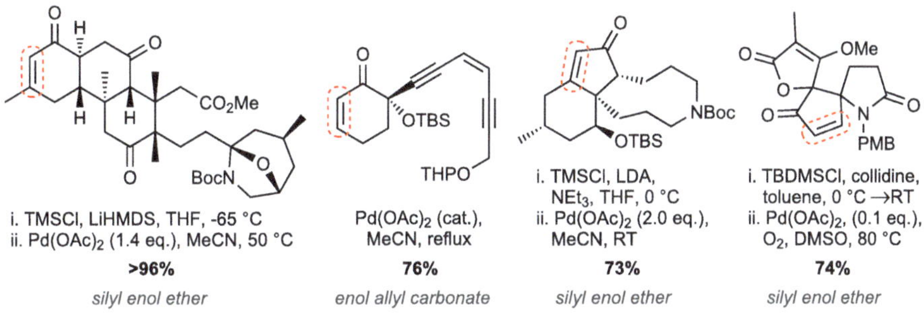

Fig. 4.67 Examples of Saegusa–Ito oxidations in total syntheses [203–206]

The scope of the oxidation has been expanded by the introduction of catalytic methods and the possibility of directly using carbonyl compounds instead of their silyl enol ether derivatives. Particularly the contributions of Newhouse and Stahl have significantly advanced this field: The oxidants used are either allyl derivatives under basic conditions or oxygen in a neutral medium (see Fig. 4.68) [190, 207, 208].

Apart from palladium, the reaction can also be mediated by other transition metals: Methods have been developed for select substrate classes in the presence of Ni, Cu, Pt, and Ir [190].

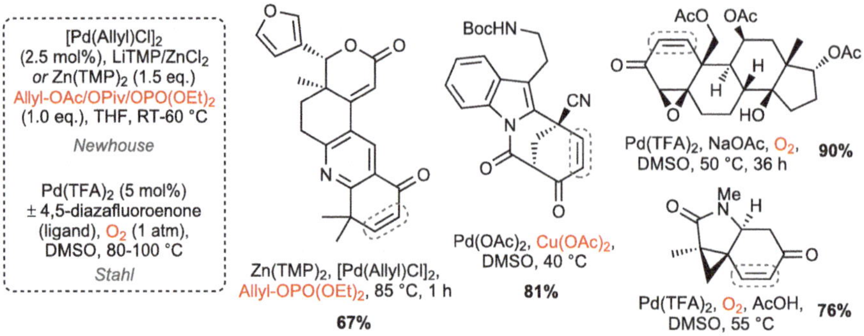

Fig. 4.68 Direct carbonyl dehydrogenations of complex synthesis intermediates [209–212]

4.2   Reduction

The difficulty in choosing a suitable reduction method lies, similar to an oxidation, in achieving high chemoselectivity in the presence of several potentially reactive functional groups (see Fig. 4.69).

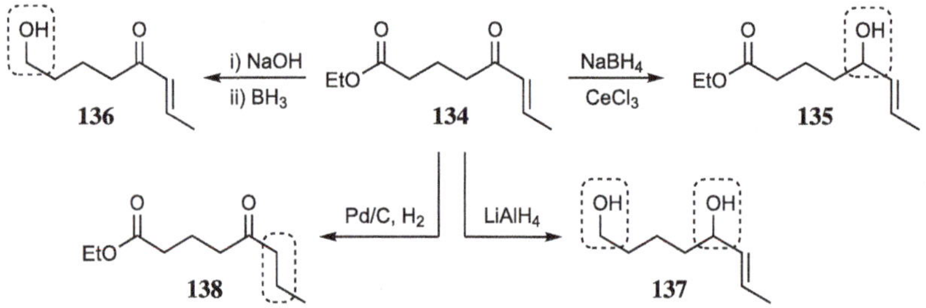

Fig. 4.69 Chemoselective reductions

The highly reactive $LiAlH_4$ converts both carbonyl groups into the respective alcohols, while the milder $NaBH_4$/$CeCl_3$ (*Luche* reduction) selectively reduces only the keto group of the Michael system. The double bond can best be converted to an alkane using elemental hydrogen as a reductant in the presence of heterogeneous catalysts. The ester group similarly requires a nucleophilic reducing agent. However, common methods cannot differentiate between ester and ketone in one step. Following saponification to the carboxylic acid, the chemoselective reaction with BH_3, which is the reductant of choice for carboxylic acids, successfully affords the primary alcohol. In more complex substrates, the diastereoselectivity of a reduction plays an equally crucial role in addition to chemoselectivity, which dictates the individual reducing agent (see Fig. 4.70).

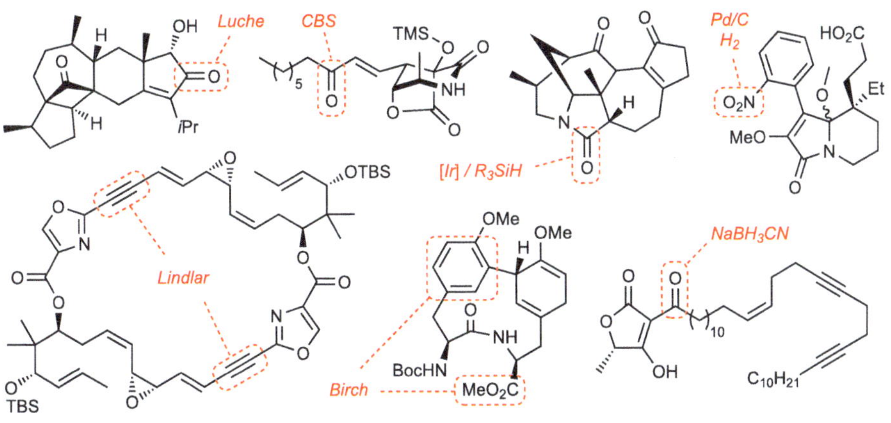

Fig. 4.70 Select examples of various reductions in total syntheses [213–219]

There are four classes of reducing agents that are of particular synthetic importance in this context:

1. Metal hydrides (aluminum hydrides, borohydrides, silanes, stannanes)
2. solvated metals (Li, Na, K, Zn, Sn, etc.)
3. reductive defunctionalization (OH, halogen)
4. hydrogen (under homogeneous or heterogeneous catalysis)

Furthermore, while there are electrochemical, enzymatic, and photochemical methods, these yet remain underutilized to find widespread application in (total) syntheses. Nevertheless, they can undoubtedly outshine the "standard methods" in terms of selectivity for individual substrates.

Once a functionality has been selected for reduction, a suitable reducing agent can be chosen. Carbonyl derivatives are typically reduced by metal hydrides or via homogeneous hydrogenations. Aromatic systems, CC or CN multiple bonds, as well as sulfur derivatives necessitate a much wider range of methods on the other hand (see Table 4.3).

4.2.1   Metal Hydrides

Metal hydrides are by far the most frequently used reducing agents for C-heteroatom single and multiple bonds in an academic setting. The anionic and neutral metal hydrides of aluminum, boron, and silicon are nucleophilic or electrophilic hydride sources (see Fig. 4.71). Reductions with stannanes, on the other hand, normally proceed via a radical mechanism.

The reaction rate roughly correlates with the nucleophilicity of the reducing agent. The selectivity can also be inferred from the nucleophilicity — under the premise of an identical

Table 4.3 Common methods for the reduction of functional groups [220–226]

substrate	product	metal hydride	solv. metal	hydrogenation homogeneous	heterogeneous	others
R–CHO / R(C=O)R'	alcohol	1-7, 10, 12-18	(19, 20)[f]	28-30	Pt, Pd, Ni, (Cu)	–
	alkane	–	–	–	Pd, (Ni)[d]	34-36
R–CO$_2$R'	aldehyde	4[c]	–	–	–	–
	alcohol	2, 3, 5-7, 13	(19, 20)[f]	24-26	Cu, Ru	–
R–CO$_2$H	aldehyde	10[a]	–	–	Pd (33[b,e])	–
	alcohol	2, 10, 11	–	–	Co, Re, (Ru)	–
	alkane	–	–	–	–	39
R–COCl	aldehyde	–	n.d.	–	Pd (33[e])	–
	alcohol	1-7, 14	n.d.	–	–	–
R–OH	alkane	–	–	–	Pd	38
R–CO$_2$NR'$_2$	amine	2, 3, 6, 10, 11	–	24	Ru, Re	–
	aldehyde	5	–	–	–	–
R–CN	amine	2, 11	(19, 20)[f]		Ni, Co, Pt, Pd	–
	aldehyde	3, 4	–		–	–
R‑CH=NR'	amine	8, 9	(19, 20)[f]	–	Ni, Co, Pt, Pd	–
R–NO$_2$	amine	–	(19, 20)[f]	–	Pt, Pd, Ni	40
R–C≡C–R'	Z-alkene	2, 10	–	–	Pd (32[e])	37
	E-alkene	–	19	27	–	–
R‑CH=CH‑R'	alkane	–	(19, 20)[f]	21-23, 31	Pd	37
(arene, X)	cyclohexadiene	–	19	–	–	–
	cyclohexene	–	20	–	–	–
	cyclohexane	–	–	–	Rh, Ru, Ni	–
R(epoxide)R'	alcohol	2, 3, 10, 11	19	possible	Pd	–
R–I/Br/Cl	alkane	2, 3	(19, 20)[f]	possible	Pd	–

methods

metal hydride: 1. NaBH$_4$; 2. LiAlH$_4$; 3. LiBHEt$_3$ (Superhydride); 4. DIBAL; 5. Red-Al; 6. K/L-selectride; 7. LiBH$_4$; 8. NaBH$_3$CN; 9. NaBH(OAc)$_3$; 10. BH$_3$; 11. AlH$_3$; 12. Et$_3$SiH; 13. (MeO)$_3$SiH 14. Bu$_3$SnH; 15. BINAL-H; 16. DIP-Chlorid; 17. Alpine borane; 18. CBS reduction

solv. metals: 19. Li/Na/K, NH$_3$ (*Birch*); 20. Li, EtNH$_2$ (*Benkeser*)

hydrogenation: 21. Rh(PPh$_3$)$_3$Cl (*Wilkinson*); 22. Ir(COD)PPh$_3$Py (*Crabtree*); 23. Rh(COD)(PR$_3$)$_2$ (*Schrock-*
(homogeneous)   Osborn); 24. Ru-MACHO; 25. Saudan cat; 26. Gusev cat; 27. Cp*Ru(COD)Cl (*Fürstner*); 28. Meerwein-Pondorf-Verley reduction; 29. Ru(PR$_3$)$_2$(NR$_3$)$_2$, H$_2$ (*Noyori*); 30. Ru(η^6-Ar)(NR$_3$)$_2$, HCO$_2$H/*i*PrOH (*Noyori transfer hydrogenation*); 31. Ir-PHOX (*Pfaltz*)

(heterogeneous) 32. Pd-Pb(OAc)$_2$/CaCO$_3$ (*Lindlar*); 33. Pd/BaSO$_4$ (*Rosenmund*)

others: 34. Zn/Hg (*Clemmensen*); 35. N$_2$H$_4$, KOH (*Wolf-Kishner*); 36. H$_2$NNHTos, hydride/borane; 37. N$_2$H$_2$ (diimine); 38. Barton deoxygenation; 39. Barton decarboxylation; 40. Fe/Zn, H$^+$

[a] reoxidation; [b] via acid chloride; [c] possibly via Weinreb amide; [d] via thioketone; [e] quinoline as additive; [f] undesired reaction

mechanism. The relative rates of the conversion of octyl bromide to octane with LiBHEt$_3$, LiAlH$_4$ and LiBH$_4$ respectively are 9500, 240 and 1, for example [228].

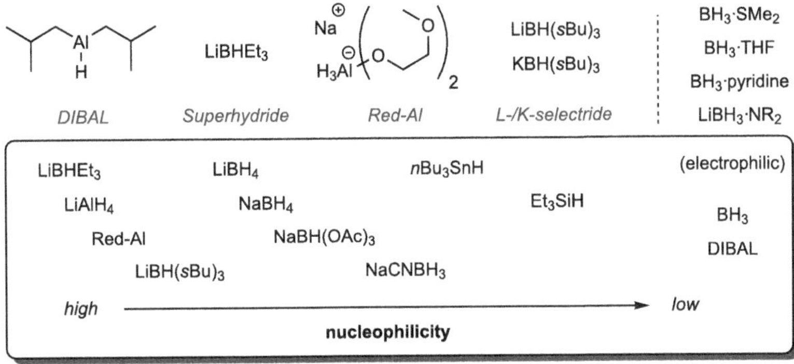

Fig. 4.71 Nucleophilicity of common metal hydrides [221, 227]

The most significant influencing factors on the reactivity and selectivity of metal hydrides include, in addition to temperature, solvent and others:[221]

- *Cations in alanates and borates*: the stronger the Lewis-acidic character of the cation, the easier the polarization of the CO bond can occur. Lithium derivatives are therefore stronger reductants than the corresponding sodium analogs, e.g., $NaBH_4$ versus $LiBH_4$
- *Substitution of the metal center by donors or acceptors*: electron-withdrawing groups (OAc, CN, OR) destabilize the positive partial charge on the metal and lower the reactivity. Electron donors (alkyl residues) increase the reactivity, e.g., $LiBHEt_3$ versus $LiBH_4$ and $NaBH_4$ versus $NaBH(OAc)_3$
- *Steric demand of the substituents*: Sterically shielded hydrides are less reactive than those with easily accessible metal-H bonds, e.g., $LiBHEt_3$ versus $LiBH(sBu)_3$

Table 4.4 presents the reactivities of the most common metal hydrides towards typical functional groups. While this rough overview does not claim to provide reliable relative reaction rates with the respective functional groups for each substrate — after all, steric and electronic properties of the starting materials cannot be taken into account — it is intended to serve as initial reference for the choice of a suitable reducing agent.

The mechanism of reduction with anionic metal hydrides proceeds almost identically for all members of this class of reagents. In the reduction of C-X double bonds (e.g., ketones, esters), a Lewis acid such as the cation of the metal hydride initially coordinates the heteroatom and polarizes the C-X bond, thus lowering the LUMO [239]. The rate-determining step of reduction with nucleophilic metal hydrides is the subsequent C-H bond formation (**139**) [240]. The product of the reduction is an alkoxy metal hydride.

In the case of $LiAlH_4$ and $NaBH_4$, the reagent contains four equivalents of hydride. This infers that the monoalkoxy hydride product of the first reduction can also serve as a reducing agent, as can the di- and trialkoxymetal hydride congeners. However, the reaction rate

Table 4.4 Comparison of the substrate scope of anionic and neutral metal hydrides [220, 222, 229–238]. It is assumed that the substrate is completely reduced to the alcohol, or in the case of amides, nitriles and imines to the amine. *Reactivity:* ++ very fast; + fast; ± slow; (−) very slow; − no reaction

Reagent	aldehyde/ketone	acid halide	lactone	ester	epoxide	carboxylic acid	amide	nitrile	imine	olefin	alkyne
LiBHEt$_3$	+	+	+	+	+	−	+	+[a]	−	−	−
LiAlH$_4$	++	++	++	++	+	+	±	±	+	−	(−)
Red-Al	++	++	+	+	−	+[b]	+[e]	(−)[a]		−	−
K/Li-BH(sBu)$_3$	+	+	+	+	+	−	+		−	−	−
LiBH$_3$NR$_2$	++		+	+	+	−	±	(−)	+	−	
DIBAL	++	+	+[c]	+[d]	+	±[b]	+[e]	+[a]	+	(−)[f]	(−)[f]
AlH$_3$	++	+	++	++	+	+	+	++	−	−	+[g]
9-BBN	++	+	+	(−)	(−)	(−)	+	(−)		+[h]	+[h]
BH$_3$	++	(−)	(−)	(−)	(−)[i]	++	±	±	−[i]	+[h]	+[h]
Li(OtBu)$_3$AlH	++	+[d]	(−)	(−)	(−)	−	−	−	−	−	−
LiBH$_4$	++	++	±	±	±	−	−	−	+	−	−
NaBH$_4$	++	++	(−)	−	−	−	−	−	+	−	−
KBH(OiPr)$_3$	++	±	−	−	−	−	−	−	−	−	−
Na(OAc)$_3$BH	+[j]	−	(−)	(−)	−	−	(−)	−	++	−	−
NaCNBH$_3$	±[k]	−	−	−	−	−	−	−	+	−	−

[a] Reduction to the imine, then hydrolysis to the aldehyde; [b] Sources on reactivity partially contradict each other; [c] Under strictly controlled conditions, reduction to the lactol is possible; [d] Under strictly controlled conditions, reduction to the aldehyde is possible; [e] Reduction can (depending on the substrate) yield both amine and aldehyde; [f] Organoaluminium compounds are formed and can further react or be quenched to the alkane/alkene; [g] Only reactive with propargyl alcohols or propargyl-functionalized substrates, then allene formation; [h] Formation of the organoborane; [i] Significantly increased reactivity in the presence of Lewis acids; [j] Aldehydes can be selectively reduced in the presence of ketones; [k] at pH=3, the reaction of aldehydes/ketones is fast, at pH=6-8 only very slow.

decreases with the alanate as the number of alkoxy groups increases ($k_1 > k_2 > k_3 > k_4$), so that LiAlH$_4$ should remain the dominant reducing agent over the course of the reaction. As a consequence, more than 0.25 equivalents of reductant are usually required. An additional complicating factor is the possibility of *disproportionation* reactions. In this case, each of the reactive species LiAlH$_n$(OR)$_{4-n}$ ($n = 1$–3) can effect the reformation of AlH$_4$ by ligand exchange (OR→H, see Fig. 4.72) [241–243].

These exchange processes were observed with alanates [241, 243]. However, deuteration experiments showed that with borohydrides the corresponding monoalkoxyborohydrides do not disproportionate, but rather react with another equivalent of the substrate to form dialkoxyborohydrides [242]. The mechanistic details for the subsequent steps of the reduction have not yet been clarified. In NaBH$_4$-mediated reductions, the transfer of the first hydride is rate-determining. In contrast, the first step is the fastest in LiAlH$_4$-mediated reactions. As a consequence, a substantial portion of the substrate is reduced by LiAlH$_4$ and not by the intermediate alkoxyalanates in the presence of superstoichiometric amounts of lithium alanate (substrate/LiAlH$_4$ 1:>0.25). In the case of BH$_3$, however, a sequential reac-

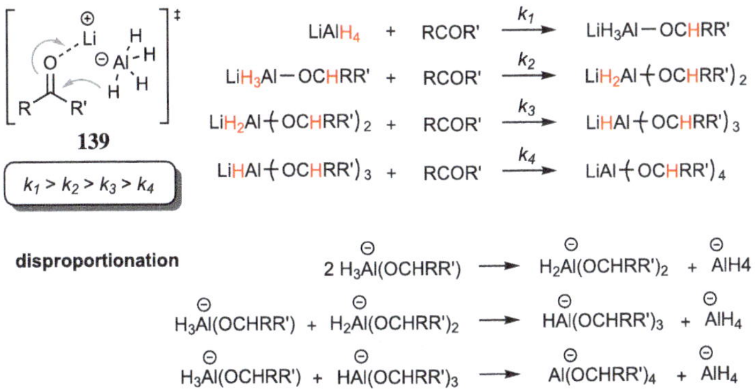

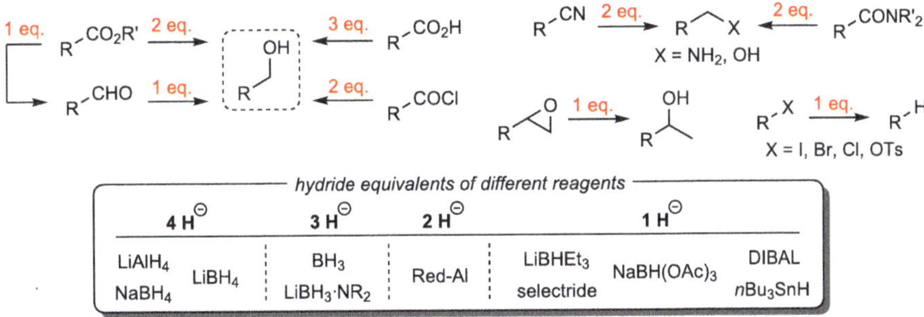

Fig. 4.72 Mechanism of the carbonyl reduction with LiAlH$_4$ [241–243]

tion via the trialkoxy to the tetraalkylborate species is possible due to the slower first step. A rapid disproportionation additionally supports the occurrence of the differently substituted boron analogs over the course of the reduction, since at least the trialkoxyborohydride is stable in ether media.

Depending on the amount of hydrides contained in the respective reagent, different equivalents of the reducing agents are required for the conversion of a substrate (see Fig. 4.73).

Fig. 4.73 Stoichiometry of the reduction of typical functional groups with hydrides [223]

Apart from the disproportionation, aluminum and borohydrides differ in their transition states in certain solvents (**139**, **140** *vs.* **141**). LiAlH$_4$ reacts in alcoholic solution to form tetraalkoxyaluminate, which is why only aprotic solvents can be used in the reaction. Borohydride-mediated reductions in turn can tolerate alcoholic media. In the presence of stoichiometric amounts of alcohol, a dependence of the reaction rate on the protic solvent was observed, but not on the cation. This indicates transition states in which the solvent takes on the role of the Lewis acid and activates of the CO bond by H-bond formation (**141**, see Fig. 4.74) [242, 244, 245]. A recent theoretical study has indicated that the protic solvent faciliates formation of borane-carbonyl intermediates under B–H protonolysis, which undergo an *intra*molecular, 4-membered transition state [246].

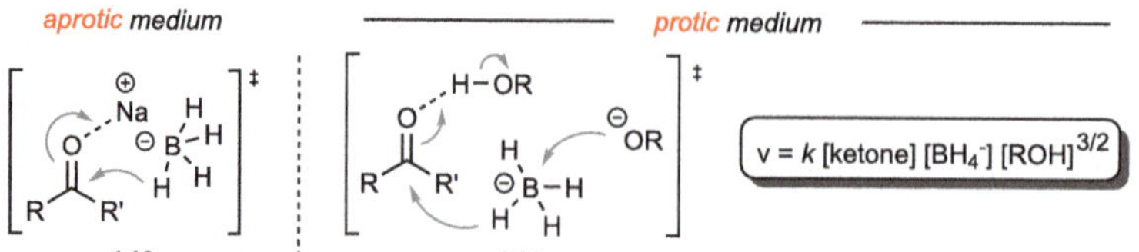

Fig. 4.74 Transition states of the reduction of carbonyl compounds with borohydrides in protic and aprotic solvents [242, 244, 245]

With metal hydrides that carry only one equivalent of hydrogen (LiBHEt$_3$, LiAlH(OtBu)$_3$, NaBH(OAc)$_3$, cf. Fig. 4.73), subsequent reactions and disproportionation do not arise. In the reaction with aluminum hydrides or borohydrides, stoichiometric amounts of aluminum salts or boric acid are produced, respectively. While the work-up with borohydrides usually does not cause major problems, the separation of the product from the aluminum salt is often not trivial, which is why an important focus of the optimization of an alanate reduction also lies on efficient work-up.

The reactivity and selectivity of the metal hydride can be further adjusted by adding rare earth salts. A well-known example of this is the combination of NaBH$_4$ with CeCl$_3$, the so-called *Luche* reduction (see Table 4.5) [247].

Table 4.5 Influence of additives on the reduction of 2-cyclopentenone and mechanism of the Luche reduction [247–249]

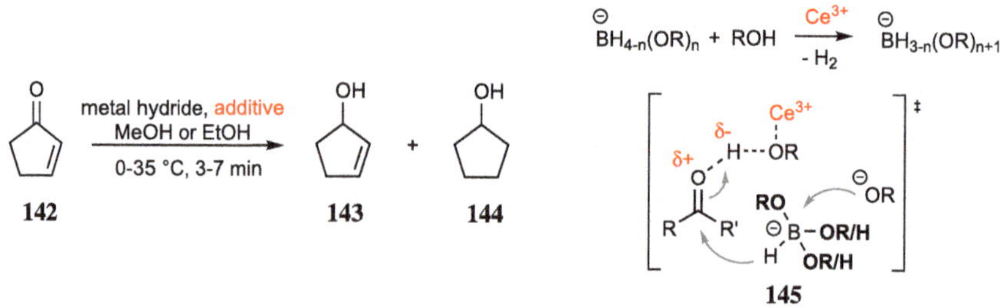

metal hydride	additive	143	144
NaBH$_4$	CeCl$_3$	97	3
NaBH$_4$	SmCl$_3$	94	6
NaBH$_4$	LaCl$_3$	90	10
NaBH$_4$	YCl$_3$	86	14
LiAlH$_4$	CeCl$_3$	64	36
LiAlH$_4$[a,b]	–	14	84[c]
NaBH$_4$[d]	–	0	100

[a] in THF; [b] 17 h at 0 °C; [c] still contains dissolved ketone; [d] 30 min at 78 °C.

Cerium is thought to serve two roles: It catalyzes the formation of the alkoxyborate (mono- or trialkoxyborate), which according to HSAB theory is a harder reductant and therefore should display an improved selectivity for the 1,2-reduction. In addition, it has been postulated that in alcoholic solution the cerium induces an increased acidity of the solvent and thus a stronger polarization of the carbonyl group, similar to **141**, which significantly accelerates the 1, 2-addition due to the lowered LUMO (**145**) [247]. A complexation of the carbonyl group by the Cerium ion was also suggested [249].

Using Luche conditions, the carbonyl group of enones can also be chemoselectively converted into the corresponding allylic alcohol in the presence of other carbonyl functions ($\rightarrow$ Sect. 2.5) [250]. Furthermore, the reagent shows excellent selectivity for ketones over esters, which was impressively demonstrated in several densely functionalized molecules, such as the synthesis of azadirachtin by Ley and co-workers (**147**) (see Fig. 4.75).

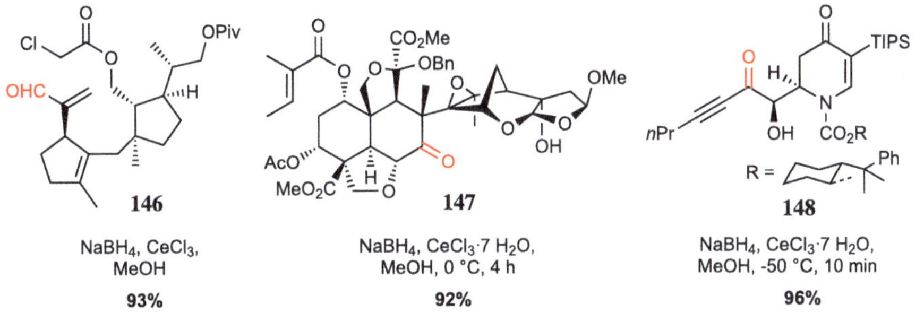

Fig. 4.75 Examples of Luche reductions in total synthesis [251–254]

DIBAL exhibits a similar selectivity scope as $LiAlH_4$ or $NaBH_4$. However, partial reductions of esters, amides, or nitriles to the respective aldehyde are conceivable, given precise control of the reaction conditions (stoichiometry, solvent, temperature) is achieved (see Fig. 4.76).

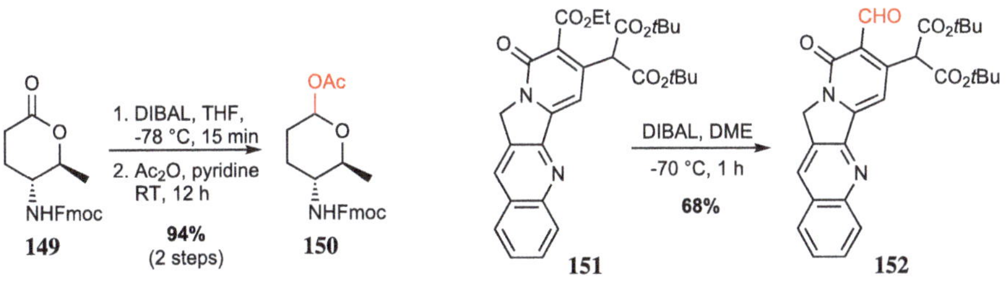

Fig. 4.76 Partial reductions of esters with DIBAL [255, 256]

Mechanistically, the carbonyl heteroatom is initially coordinated by the Lewis acidic metal center (**153**), whereby the alkoxyalanate **155** is obtained via hydride transfer (**154**). **155** decomposes upon aqueous work-up to afford the desired alcohol. With esters (or lactones) in apolar solvents, a strictly stoichiometric use of the reductant and usually low temperatures allow the isolation of the aldehyde (or lactol in the case of lactones) instead of the alcohol. The selectivity is controlled by the stability of the aluminum acetal intermediate **157**. If **157** is stable or decomposes very slowly under the reaction conditions, the subsequent reaction to the alcohol can be suppressed and the aqueous work-up delivers the aldehyde. If in contrast the acetal is reactive enough that the alkoxy group is transferred to the metal center (**158**), the resulting aldehyde can react with another equivalent of DIBAL. The aluminum acetal stability is the deciding factor as the aldehyde reacts much faster than an ester due to the higher electrophilicity of the carbonyl C atom. The factors influencing the stability are unfortunately hardly understood: Apart from low temperatures, nonpolar solvents seem to stabilize the aluminum acetals. The overreduction in polar solvents can be rationalized by a polarization of the metal center in the aluminum acetal through the solvent (**159**, see Fig. 4.77) [233, 257–259].

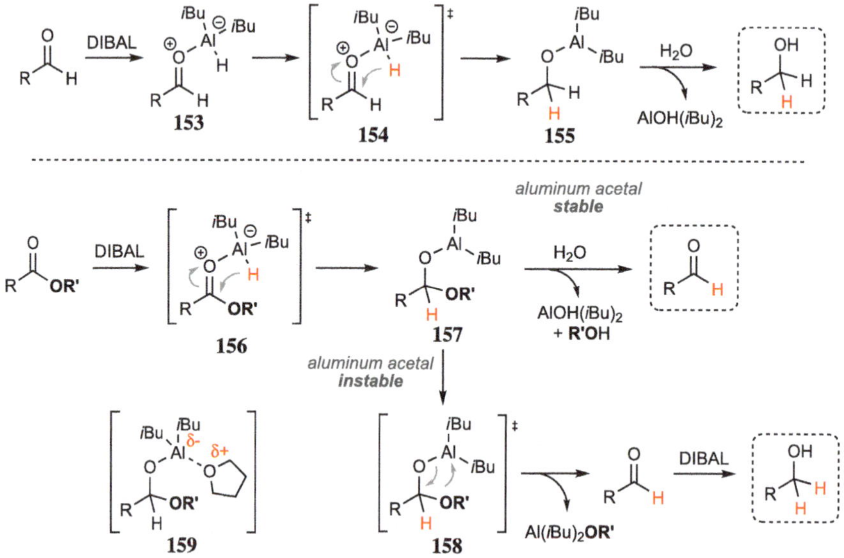

Fig. 4.77 Mechanism of the reaction of aldehydes and esters with DIBAL [233, 257–259]

Optimal reaction conditions for a selective reduction of an ester constitute a precarious balance. Especially the choice of temperature often determines whether the aldehyde or an alcohol is obtained (see Fig. 4.78) [260].

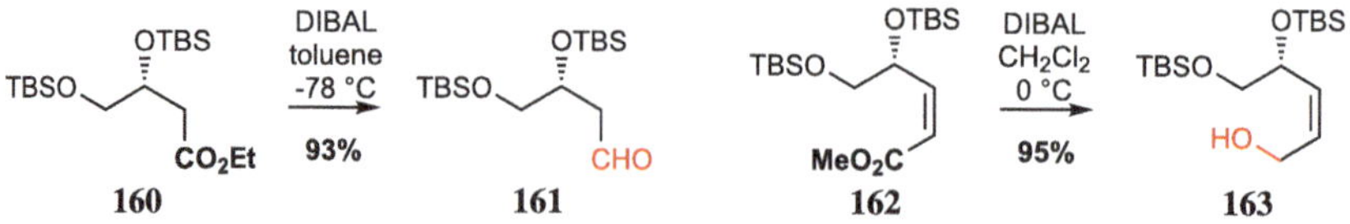

Fig. 4.78 Reduction of an ester to an aldehyde or alcohol in Dias' synthesis of a callystatin A fragment [260]

If a desired partial reduction to the aldehyde is unsuccessful, a two-step sequence (total reduction to the alcohol, oxidation to the aldehyde), or methoxymethylamides (*Weinreb amides*, → Sect. 2.6) are typically adopted as viable alternatives [261, 262]. The charm of Weinreb amides lies in the stabilization of the tetrahedral, anionic hemiacetal intermediate (**165**) received after addition of the nucleophile. This prevents a subsequent conversion with a second equivalent of the nucleophile to give the alcohol (see Fig. 4.79) [263].

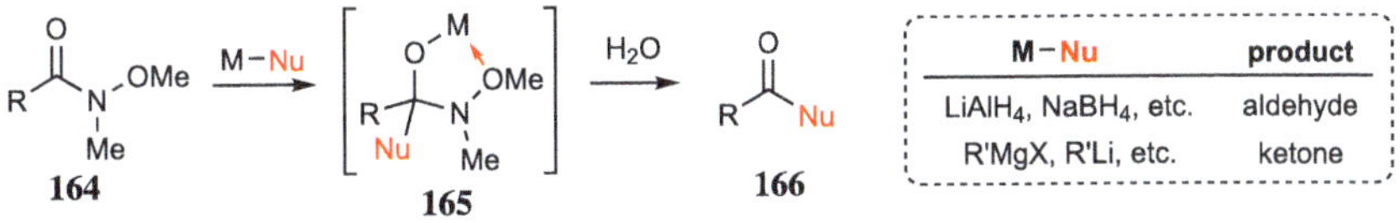

Fig. 4.79 Conversion of Weinreb amides to acyl derivatives (**166**) [263]

In the reduction of nitriles and amides with DIBAL, the aldehyde can be selectively obtained. Mechanistically, this can be understood in the case of the nitrile by stopping the reduction at the stage of the imine (**169**), which releases the aldehyde upon hydrolysis in the work-up (see Fig. 4.80). When amides react with one equivalent of DIBAL, the N,O-acetal **167** can decompose by cleavage of the central C-O bond, affording the iminium ion **168**. In the presence of a second equivalent of DIBAL, either the reduction to the amine or **168** a hydrolysis to the corresponding aldehyde can occur [221]. Interestingly, the reduction of some substrates stalls at the stage of the imine even in the presence of an excess of DIBAL (**170**→**171**) [264].

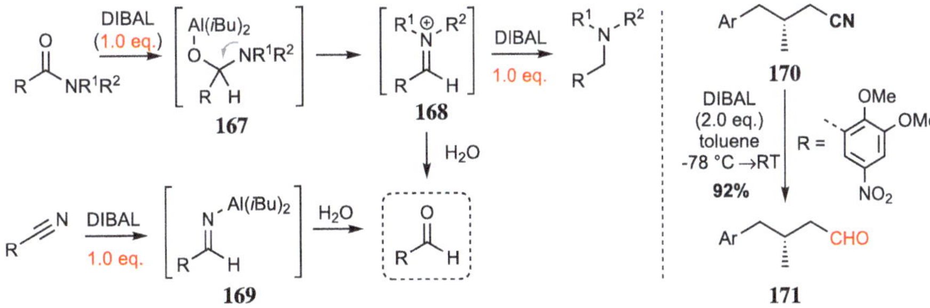

Fig. 4.80 Mechanism of the DIBAL reduction of amides and nitriles and example of a selective nitrile reduction [221, 233, 264, 265]

Esters, lactones, acid halides, and anhydrides necessitate the use of a *nucleophilic* metal hydride ($NaBH_4$, $LiBHEt_3$, etc.). In contrast to these, neutral, *electrophilic* metal hydrides can allow a complementary reactivity. Apart from DIBAL, the most commonly encountered reducing agents of this class are BH_3 and AlH_3. BH_3 displays a preference for the reduction of carboxylic acids in the presence of other reducible functionalities: it leaves most carbonyl derivatives such as esters, lactones, or nitro derivatives untouched, but does react with olefins.

The reactivity of various functional groups towards borane decreases in the order[266].

carboxylic acid > aldehyde > ketone $\gtrsim$ olefin, imine > nitrile, amide, epoxide > ester

These differences in reactivity can be well understood through the respective mechanism (see Fig. 4.81) [266–269].

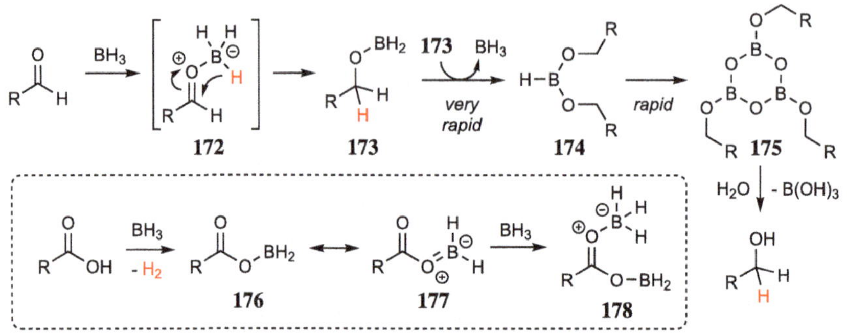

Fig. 4.81 Mechanism of the reaction of BH$_3$ with aldehydes and carboxylic acids [267–269]

The reaction of the electrophilic borane with the carbonyl group largely corresponds to the mechanism of the reduction by alanes. Following initial complexation of the carbonyl group, a hydride transfer via **172** to the borinic ester **173** occurs. This species also constitutes an efficient reducing agent and can react with another equivalent of the aldehyde to form the dialkoxyborane **174**. **174** can be observed as an intermediate in some cases, its reduction potential is however significantly lower than that of BH$_3$ or H$_2$BOR (**173**) [268]. The subsequent conversion to boroxine **175** can be envisioned either by reacting with another equivalent of the substrate or by an equilibration with H$_2$BOR or HB(OR)$_2$ **175**. **175** affords the desired alcohol after aqueous work-up.

A hallmark of borane-mediated reductions is the high selectivity for carboxylic acids, a substrate class that represents one of the least reactive functional groups with other reducing agents. The reason for this selectivity is the formation of the acyloxyborane **176** [266, 269, 270]. During the reaction, this intermediate is formed rapidly with stoichiometric evolution of hydrogen. The carbonyl group in **176** is activated by the strong Lewis-acidic character of boron, evident by its resonance structure **177**, and can be regarded as a mixed anhydride. **176** subsequently affords acid-base complex **178** with another equivalent BH$_3$, which can be quickly converted to trialkoxyboroxine **175**. Whether the initial acyloxyborane is a mono-(**176**), di- or trimeric intermediate seems to depend on the stereoelectronic properties of the substrate. The reactive species has been experimentally confirmed to be **176** [269, 270]. The necessity - in contrast to esters - to use a 1:1 stoichiometry of BH$_3$/carboxylic acid is due to the initial formation of acyloxyborane and the loss of a hydride equivalent as H$_2$.

Carboxylic acids thus require the use of three hydride equivalents, esters/amides two, and aldehydes/ketones can be converted in a substrate/BH$_3$ ratio of 3:1 (see Fig. 4.82).

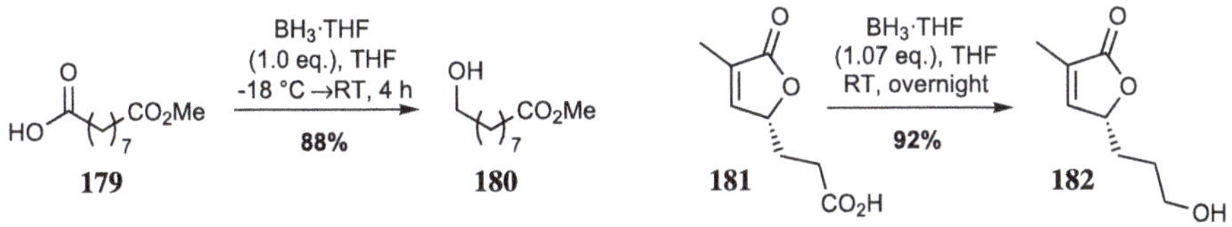

Fig. 4.82 Selective reduction of carboxylic acids with BH$_3$ in total syntheses [271, 272]

If another functional group needs to be reduced with borane in the presence of a carboxylic acid, the carboxylic acid can be protected *in situ* by forming the corresponding carboxylic acid salt. The carboxylate can no longer form an acyloxyborane with BH$_3$ [232].

Borane (BH$_3$) or diborane (B$_2$H$_6$) must either be prepared *in situ* or added complexed with a Lewis base. The strength of the Lewis base determines the reactivity of the respective borane complex. The THF adduct is the most reactive, while aminoboranes are the most stable and react sluggishly even in the presence of proton sources (H$_2$O, ROH, RCO$_2$H). Often, additional Lewis acids are used for activation. The dimethyl sulfide adduct possesses only a slightly diminished reactivity compared to BH$_3$·THF. Unlike the latter, it is stable in solution for extended periods, which usually makes it the reagent of choice [230].

Whether the reduction of acid chlorides with boranes listed in Table 4.4 can be observed depends on the reagent used: While no conversion takes place with diborane, a slow reaction can be observed with BH$_3$·THF [267].

In addition to the reduction of various functionalities such as C-O and C-N multiple bonds, hydrides can cleave C-X single bonds (X = O, halogen, etc.) as nucleophiles in classic substitution reactions. Typical transformations are the opening of epoxides and the formation of alkanes from alkyl halides and alkyl sulfonic acid esters (see Fig. 4.83).

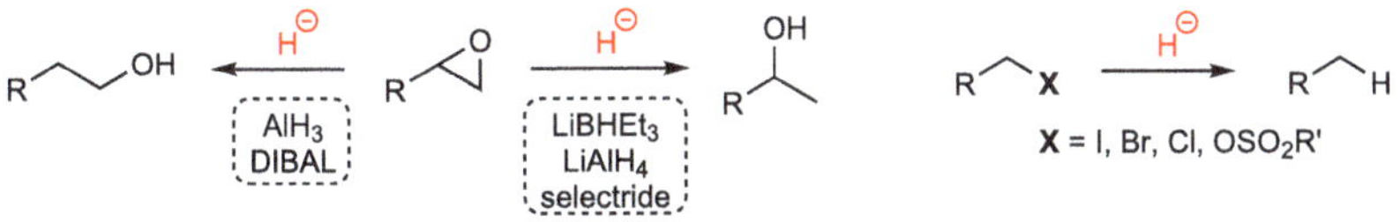

Fig. 4.83 C-X reductions and selectivity of epoxide opening with hydrides [221]

The cleavage of epoxides requires the activation of the C-O bond. This can be achieved via coordination by either a Lewis acidic cation (Li$^+$) or an electrophilic reagent (AlH$_3$, DIBAL). LiBHEt$_3$ is particularly noteworthy as it is the most nucleophilic of the anionic hydrides. The mechanism of ring opening with strongly nucleophilic hydrides (LiBHEt$_3$, LiAlH$_4$, etc.) is a Lewis acid-assisted S$_N$2 attack. The reduction with borohydrides, alkoxyborohydrides, and

Red-Al are very slow. However, the addition of a Lewis acid such as BF_3 or BEt_3 can facilitate an efficient ring opening. The regioselectivity of the epoxide opening of unsymmetrical epoxides is determined by the strength of the Lewis acid-base interaction. Weak electrophiles favor a reaction on the sterically more accessible side of the oxirane — $LiBHEt_3$, $LiAlH_4$ thus provide secondary alcohols as products (see Table 4.6) [221].

Table 4.6 Dependence of the regioselectivity of the reductive opening of epoxides [221, 273–283]

metal hydride	Lewis acid	conditions	183 : 184	185 : 186
$LiAlH(OMe)_3$	–	THF, 0 °C	99 : 1	–
$LiBHEt_3$	–	THF, RT	>98 : <2[a]	100 : 0
$LiAlH_4$	–	THF, 0 °C	96 : 4	100 : 0
$LiAlH(OtBu)_3$	BEt_3	THF, RT	89 : 11	90 : 10
$NaBH_4$-MeOH	–	tBuOH, 82 °C	89 : 11	–
AlH_3	–	THF	74 : 26	91 : 9
$LiBH_4$	–	Et_2O, RT	74 : 26	–
DIBAL	–	toluene, 0 °C	27 : 73	90 : 10
$NaBH_4$	BF_3	THF, RT	27 : 73	–
$NaBH_3CN$	BF_3	THF, RT	3 : 97	3 : 97
$LiAlH_4$	$AlCl_3$	Et_2O, 35 °C	2 : 98	–
BH_3	BF_3	–[a]	0 : 100[b]	–

[a] No further details on the reaction conditions. [b] 2-Phenylethanol was listed as the only product.

In contrast, stronger Lewis acids (AlH_3) yield the primary alcohol instead. They proceed via a S_N1 mechanism, whereby the C-O bond is broken first, thus allowing maximum stabilization of the carbocation [221, 284]. In the case of 2,3-epoxy alcohols, which are easily accessible in enantiomerically pure form by Sharpless epoxidation of allyl alcohols, the hydroxy group can additionally act as a directing group to effect a selective opening of the epoxide [285]. While Red-Al furnishes the 1,3-diol as a product, a reversed regioselectivity can be obtained with DIBAL. Mechanistically, complexation of the hydroxy function by Red-Al is assumed as the initial step [286, 287]. Thomas *et al.* relied on the method developed by Kishi to convert **187** to the corresponding diol **188** upon treatment with Red-Al. The product could subsequently be converted in several steps to the advanced intermediate **189**, which was used in a study on bryostatin analogs (see Fig. 4.84) [288].

Apart from the reductive opening of epoxides, nucleophilic metal hydrides can be used for the substitution of halogens and similar leaving groups (OMs, OTos). The requisite high nucleophilicity of the reagent makes its use in the presence of sensitive functionalities a challenge, which is why these substitutions are typically encountered in early stages of a synthesis, where the substrate does not yet possess a prohibitively high complexity.

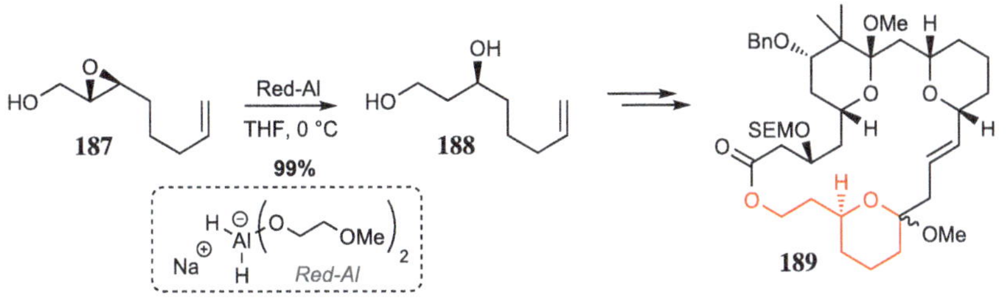

Fig. 4.84 Selective ring opening of the epoxy alcohol **187** in a study on the synthesis of bryostatin analogs by Thomas *et al.* [288]

The leaving group reactivity mirrors the weakness of the C-X bond [289]:

iodide > bromide > chloride   Bn-X $\approx$ allyl-X > RCH$_2$X > R$_2$CHX > R$_3$CX

In the reduction of alkyl halides and tosylates, it is assumed that primary and secondary halides and all tosylates follow a bimolecular nucleophilic substitution, while tertiary halides usually undergo a single electron transfer (*SET*) [290]. Alkyl iodides can even display a SET mechanism in primary or secondary substrates if the steric environment does not allow an S$_N$2 reaction [291].

Use of this approach was reported in the total syntheses of clavosolide A and (–)-nupharolutine (**193**, see Fig. 4.85) [292, 293].

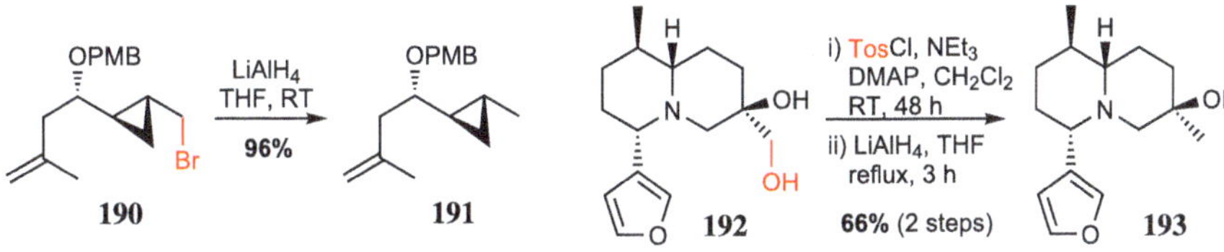

Fig. 4.85 Reductive debromination and detosylation in the synthesis of clavosolide A and (–)-nupharolutine (**193**) [292, 293]

In the presence of stereogenic centers in a molecule, they can impart the preference for the formation of a certain product (substrate induction). The diastereoselectivity can be especially well controlled in rigid, cyclic substrates such as substituted cyclohexanones. The hydride will predominantly approach the sterically less hindered side of a substrate in absence of chelate formation (with metal ions) or coordination of the reagent by the substrate (active reaction control).

For the reaction of *anionic* hydrides (NaBH$_4$, LiAlH$_4$, etc.), the Felkin–Anh model applies for predicting the selectivity in acyclic substrates (**196**, cf. Sect. 2.1). However, tricoordinated, *neutral* reducing agents (BH$_3$, DIBAL) display reversed selectivity. This is rationalized by the conformation of the α-chiral substrate in the 4-membered transition state

(**195**) according to Midland [294]. In **195**, the largest residue or the group with the best C-X acceptor bond is aligned orthogonal to the carbonyl group. The substituent with the least steric demand orients itself in a *synclinal* position with regard to the CO bond to minimize steric repulsion with the reducing agent. The latter approaches the substrate via the Bürgi–Dunitz trajectory from the opposite face of the carbonyl group to the C-X acceptor bond. If the system has a basic functionality in α- or β-position, the cation of the metal hydride can potentially complex it following the previously outlined models by Cram (**194**, 1,2-induction) and Evans (1,3-induction) (see Fig. 4.86) [295, 296]. In the case of a Curtin–Hammett situation, the stereodetermining conformation does not necessarily have to be the energetically favored one, but rather possesses the lowest activation barrier.

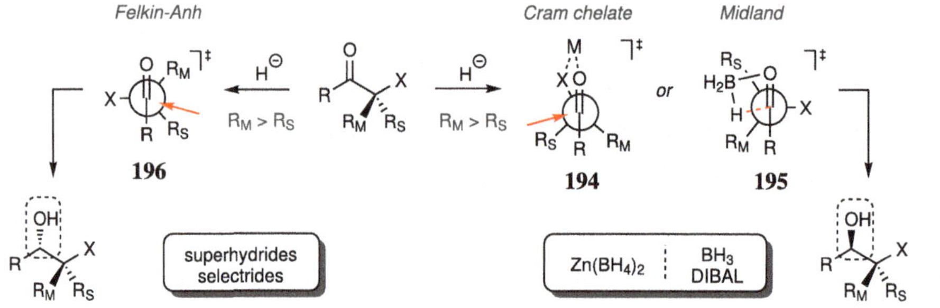

Fig. 4.86 Models for explaining the diastereoselectivity in the reduction of α-chiral carbonyl compounds [294–296]

An example of a change in diastereoselectivity in a carbonyl reduction can be found in Tsuji's contribution on the synthesis of the plant growth promoter brassinolide. The α-chiral ketone **197** can be reduced with DIBAL via **200** to the *anti*-configured allylic alcohol **198**. In the reaction with L-selectride, the opposite diastereomer **199** is obtained by attack of the hydride on the *Si*-face (**201**, see Fig. 4.87) [297].

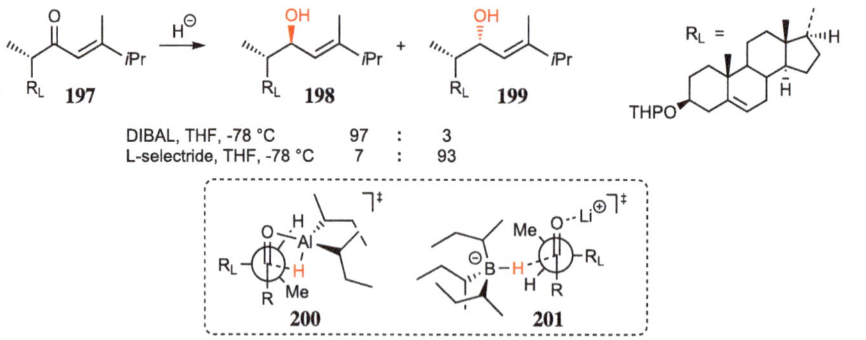

Fig. 4.87 Control of diastereoselectivity in the reduction of **197** [297]

The chosen protecting group of an α-hydroxyketone also influences the diastereoselectivity of the reaction. Sterically less demanding alkyl ethers (OR = OBn, OMOM, OMEM, etc.) can act as basic ligand and complex metal ions, while bulkier silyl ethers (OR = OTBS, OTIPS, OTBDPS) are not capable of this. As a result, a reduction according to the Cram-Chelate model (**194**) is predicted in the former case, whereas latter is expected to follow the Felkin–Anh model (**196**) [221].

In their synthesis of the secondary metabolite pumiliotoxin B, Overman and co-workers were able to selectively convert the α, β-unsaturated ketone **202** to the respective *syn-* (**203**) or *anti*-product (**204**) solely by changing the protecting group of the hydroxy function (see Fig. 4.88). The corresponding transition states **205** and **206** (each with LiAlH$_4$) reliably predict the observed selectivity. The use of the alane iBu$_3$Al with CH$_2$OBn as a protecting group preferentially afforded the *syn*-configured allylic alcohol **203** via a 6-membered [258] transition state (**207**). The large steric demand of the silyl ether further increases the selectivity. The choice of solvent and temperature also significantly impact the ratio of diastereomers [298].

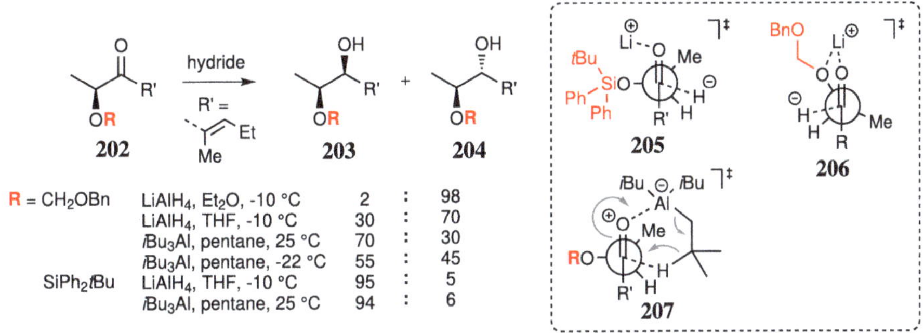

Fig. 4.88 Influence of protecting groups on the selectivity of reduction [298]

The most common reagents for chelate-controlled reductions of ketones are Zn(BH$_4$)$_2$ or Me$_4$NBH(OAc)$_3$ [296, 299, 300]. The latter requires the presence of unprotected hydroxy groups in the substrate, as BH(OAc)$_3^-$ is not sufficiently reactive compared to the intermittent BH(OAc)$_2$OR$^-$ (formed with free OH groups). Me$_4$NBH(OAc)$_3$ is especially suited for use with β-chiral carbonyl compounds, as it enforces an intramolecular hydride transfer. If a Felkin–Anh-controlled addition is desired and chelate formation to be prevented, K-selectride is typically employed. Zn(BH$_4$)$_2$ and the selectrides can thus be used complementarily [301].

As general rule of thumb, one can assume that the larger the steric demand of the reducing agent, the more selective the reaction. The easier a differentiation of the substituents at the α-chiral center due to their steric bulk and the larger the substituent of the carbonyl group, the easier it is to selectively form one product as well (see Table 4.7) [302].

Table 4.7 Dependence of diastereoselectivity on the substituents of the substrate in the reduction of chiral ketones [302]

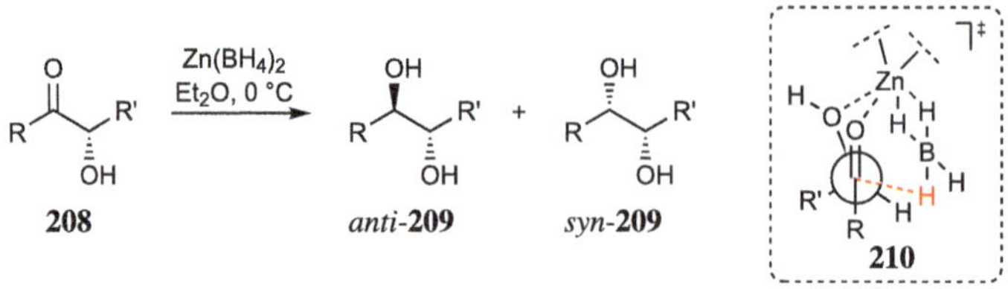

R	R'	*anti*-**209**	*syn*-**209**
*n*Pent	Me	77	23
*n*Bu	Et	89	11
*n*Pr	*n*Pr	>99	<1
Et	*n*Bu	87	13
Me	*n*Pent	85	15
Me	*i*Pr	96	4
*i*Pr	Me	85	15

Novartis conducted studies targeting an industrial production of the complex natural product (+)-discodermolide. Due to the scarcity of natural sources, the secondary metabolite is currently only accessible synthetically. In the 39-step synthesis, the penultimate step was the reduction of a β-hydroxyketone to the corresponding *anti*-1,3-diol using $Me_4NBH(OAc)_3$ (*Evans–Saksena reduction*) [300]. The desired product **212** was isolated in 75% yield as the only diastereomer, which results from the intramolecular reaction of the hydride with ketone **211**, presumably via **213** (see Fig. 4.89) [303].

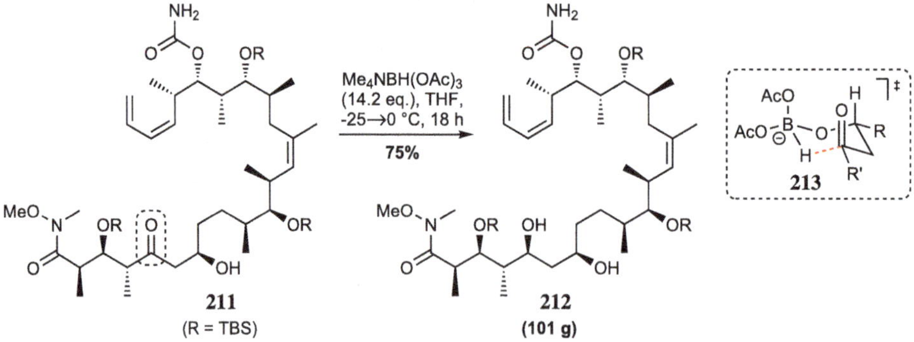

Fig. 4.89 Novartis' synthesis of (+)-discodermolide [303]

Contrary to acyclic systems, the diastereoselectivity in the reduction of cyclohexanone derivatives can be very well controlled in the absence of complexing functionalities solely by the steric demand of the reagent. For example, the reaction of *t*Bu-cyclohexanone with $LiAlH_4$ furnishes the thermodynamically preferred *trans*-product with high yield (see Fig. 4.90).

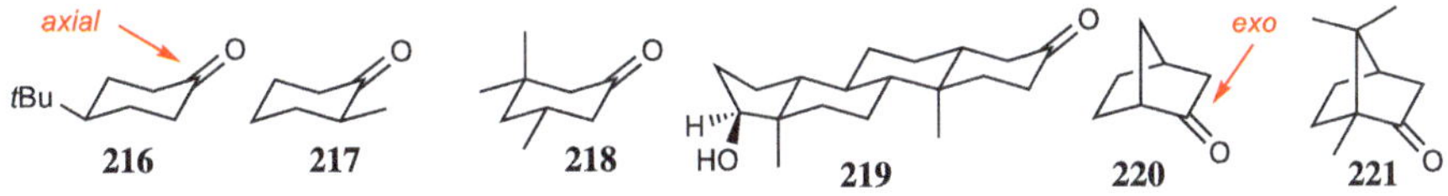

Fig. 4.90 Selectivity in the conversion of 4-tBu-cyclohexanone with $LiAlH_4$

Since the reaction is irreversible, the thermodynamic stability of the products does not determine the selectivity. The inherent challenge with cyclic substrates, especially cyclohexanones, relies on the occurrence of either a steric or torsional interaction during product formation. With acyclic starting materials, both interactions can usually be avoided in the corresponding transition states. The pathway to a transition state is thus determined by which of the two interactions is dominant [304–308]. An axial approach results in a *steric* repulsion of the nucleophile with the hydrogen atoms at the 3,5-position (**214**). In the case of an equatorial attack, a *torsional* strain occurs both from the carbonyl group and the incoming nucleophile with the adjacent methylene groups (**215**). As a consequence, sterically demanding nucleophiles preferentially show an equatorial trajectory and afford the thermodynamically less preferred product, while smaller nucleophiles approach in an axial manner. If there are axial substituents on the ring, this trend is further enhanced by the increase in steric interaction (see Table 4.8).

Table 4.8 Stereoselectivity in the reduction of cyclohexanones [247, 277, 309–323]

reagent	conditions	axial or exo-reduction [%]					
		216	**217**	**218**	**219**	**220**	**221**
DIBAL	Et_2O, 25 °C	97	92	98	-	95	36
$LiAlH_4$	THF, 0 °C	92	76	20	94[a]	92	11
$NaBH_4/CeCl_3$	MeOH, 35 °C	94	69	-	>95	-	68
9-BBN	THF, 25 °C	92	60	-	-	75	9
$LiAlH(OtBu)_3$	THF, 0 °C	91	64	73	-	93	7
$NaBH_4$	MeOH, 20-25 °C	80	66	20	81[b]	-	38
AlH_3	THF, 0 °C	87	79	-	-	93	10
$LiBH_4$	DME, 20 °C	85	71	-	-	-	-
DIBAL	Toluene, 0 °C	61	49	-	-	93	16
L-selectride	THF, 0 °C	7	2	0.2	-	99.6	0.4
K-selectride	THF, 20 °C	-	1	-	<1	-	-

[a] -70 °C; [b] 65 °C.

The reaction of cyclohexanone with aluminum hydrides possesses an early, substrate-like transition state. In contrast, a late transition state is assumed for borohydrides [221]. Although the two transition states for an axial or equatorial approach can be easily derived, the *cause* of this preference has long been controversially discussed. The dominant torsional interaction originally postulated by Felkin now seems to be generally accepted as the rationalization for the axial preference. However, there are alternative explanations provided by the Cieplak model (see Sect. 2.1) or by assuming an asymmetry of the π^*_{CO}-orbitals, whereby the axial orbital lobe has a larger extension than the equatorial one [304, 324]. The preference of small nucleophiles for axial addition, for example, predicts **222** to be the preferred transition structure with LiAlH$_4$. For LiBH(iPr)$_3$, DFT modeling results in **223** as the energetically favored transition state [325]. The different selectivities of norcamphor (**220**) and camphor (**221**) can be attributed to a significantly different steric environment of the exo- and endo-face of the [2.2.1]-bicycloheptane due to the presence of the additional methyl groups. The more sterically demanding a complex metal hydride is, the easier the two faces can be distinguished (see Fig. 4.91).

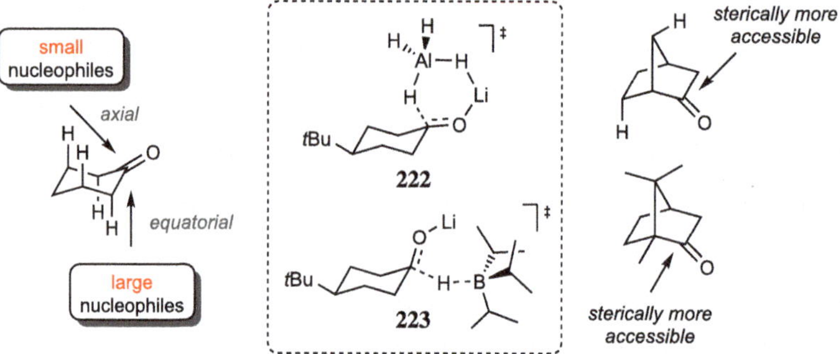

Fig. 4.91 Preferred reaction trajectories and postulated transition states in the reduction of ketones with complex metal hydrides [325]

The molecular strain of the starting material can also be synthetically exploited: Strained cycloalkanones should generally react faster than acyclic substrates, as the rehybridization from sp^2 to sp^3 reduces both the bond strain and torsional strain. As a consequence, acyclic ketones do not always have to be protected in the presence of cyclic ketones to achieve a selective reduction (see Fig. 4.92).

Tin hydrides have the distinguishing property among metal hydrides of being able to participate in radical reactions through homolytic Sn-H bond cleavage, thereby enabling the conversion of a variety of different C-X bonds (X = Cl, Br, I, SePh, SPh, NO$_2$, OC(=S)R) to the corresponding alkanes. They are thus routinely used for radical defunctionalizations. Although formally associated with metal hydrides, tin hydrides will be covered in more detail when introducing the class of defunctionalization reactions (*vide infra*).

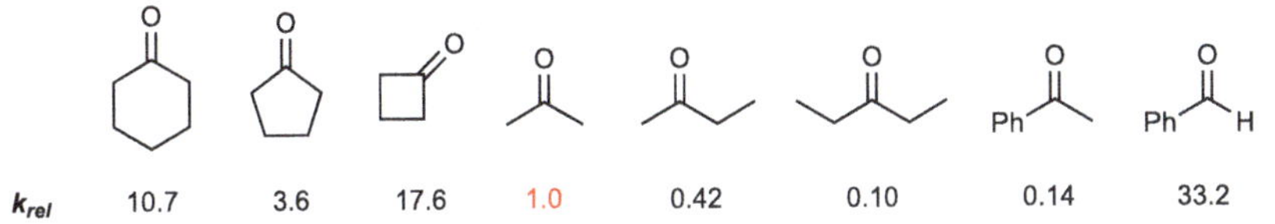

k_{rel} 10.7 3.6 17.6 1.0 0.42 0.10 0.14 33.2

Fig. 4.92 Relative reaction rates of $NaBH_4$ with various carbonyl derivatives at $0\,^\circ C$ in *i*PrOH [326–330]

Silanes can act as a radical H-donor (also for defunctionalizations) or as a hydride donor to facilitate a reduction [331]. The small difference in electronegativity between Si and H polarizes the Si-H bond so that the hydrogen, unlike in alkanes, leans towards a hydridic nature. As a result, silanes are very mild reducing agents, which require activated carboxonium intermediates as reaction partners (**225**, mode A in Fig. 4.94) due to the relatively strong Si-H bond to effect a hydride transfer. The oxidized organosilane by-products are non-toxic and can usually be easily separated from the product (see Fig. 4.93).

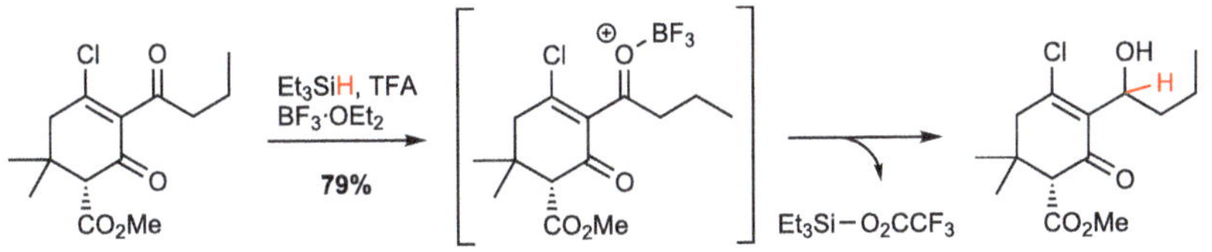

Fig. 4.93 Selective reduction of a ketone with Et_3SiH [332]

In the absence of Lewis or Brønsted acids, carbonyl compounds such as aldehydes or ketones are inert towards silanes [333]. As an alternative activator, additives such as fluorides can be used to form hypervalent hydrosilanides, which are significantly stronger reducing agents than the parent organosilane (mode B, Fig. 4.94) [331].

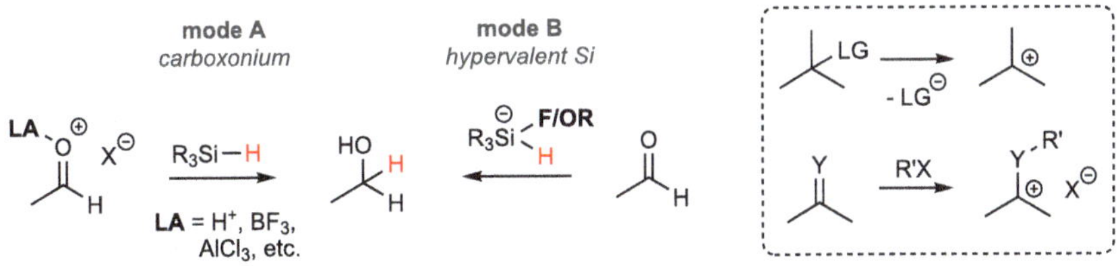

mode A — carboxonium

mode B — hypervalent Si

Fig. 4.94 Reaction modes in the reduction of carbonyl derivatives by organosilanes [331]

Since the addition of a promoter is required for the reaction to proceed and its presence cannot be avoided, the choice of the activator in turn provides an additional handle to tune the selectivity of the conversion. The carboxonium intermediates can be generated either by the cleavage of a leaving group or by the reaction of a CC or CY multiple bond with a Lewis acid. If relying on a leaving group (LG = Cl, Br, I, OH, etc.), reductive defunctionalization reactions of alcohols, alkyl halides, etc. can be realized. The rate-determining step appears in both cases to be the transfer of the hydride.

A mechanistic investigation of the reaction kinetics of various silanes reported the relative reaction rate to be in the order $R_3SiH > R_2SiH_2 > RSiH_3$. Particularly alkyl substituents increase the donor properties, and thus the reactivity, of the organosilane reagents. The donor properties of the silanes follow the order [334]

$$Et_3Si\text{-}H > n\text{Oct}_3Si\text{-}H > Et_2SiH_2 > Ph_2SiH_2 > Ph_3Si\text{-}H > PhSiH_3$$

When choosing the carboxonium approach, the most common additives are Brønsted acids, for example trifluoroacetic acid (TFA), HCl, H_2SO_4 as well as Lewis acids such as BF_3, $B(C_6F_5)_3$, $AlCl_3$, and others. The by far most frequently used reagent combination is Et_3SiH/TFA as a solution in dichloromethane. In contrast, when resorting to hypervalent silanes, alkoxysilanes in combination with nucleophiles such as fluorides are more reactive than organosilanes that only carry alkyl or aryl residues [335]. The combination $(EtO)_3SiH$ with CsF or KF has proven to be particularly advantageous with regard to substrate scope and yields. With trialkoxysilanes, even esters can be converted to the corresponding alcohols, which is rarely successful with Et_3SiH/TFA [331] (see Table 4.9).

Table 4.9 Substrate scope of the reduction with organosilanes [331, 335]. *Reactivity*: ++ very fast; + fast; ± moderately fast to slow; (–) very slow; – no reaction

reagent	aldehyde	ketone	ester	amide	α,β-unsat. carbonyls	olefin	alkyne
Et_3SiH/TFA	++	±	(–)	–	+	±	–
$(MeO)_3SiH$/CsF	++	+	±	–	(–)	(–)	–

Organosilanes allow for the reduction of many functional groups which could also be achieved using aluminum or borohydrides. The advantage is the higher chemoselectivity that is observed in some cases, as the background reaction with other reducible groups can be more easily controlled. However, due to the formation of carbocations, potential side reactions such as rearrangements of the molecular skeleton, fragmentations, and eliminations can occur. The formation of ethers during the reduction of aldehydes may also cause issues. However, the latter can be suppressed by using a combination of H_2O and a non-hydrogen bonding cosolvent (see Fig. 4.95) [336].

Fig. 4.95 Influence of the solvent on the reduction of aldehydes [336, 337]

When utilizing the carboxonium mode, the relative stability of the intermediate is mirrored in the overall reaction rate. Therefore, tri- and tetrasubstituted double bonds are easily reduced, while terminal alkenes can be considered inert for practical purposes.

A nice highlight of said selectivity can be found in Takano's synthesis of the steroidal hormone estrogen. Following a Diels-Alder reaction and a methylation between the C- and D-ring, the reduction of the benzylic double bond of **230** at C9/C11 ensued in the presence of Et$_3$SiH/TFA. Neither the carbonyl group at C17 nor the isolated double bond were affected. The subsequent retro-Diels–Alder reaction of **231** and the final reduction of the double bond in the D-ring finally afforded the desired steroid (see Fig. 4.96) [338].

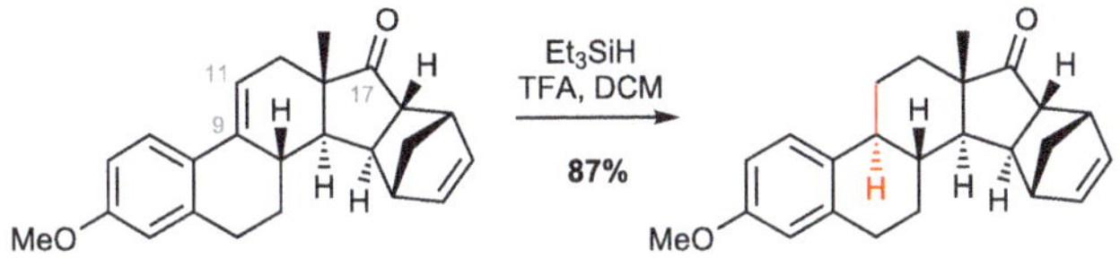

Fig. 4.96 Excerpt of Takano's synthesis of estrogen [338]

Another class of metal hydride reagents are copper hydrides (e.g., the Stryker reagent [(Ph$_3$P)CuH]$_6$), whose main application is the selective 1,4-reduction of Michael systems ($\rightarrow$ Sect. 2.5) [339].

4.2.2   Solvated Metals

The reduction of functional groups by solvated metals comprises a single electron transfer from the metal to the substrate in the presence of a proton source. Electrolytic reductions work analogously (see Sect. 10.1). Metal reductions are typically used for the conversion of polarized multiple bonds, such as CO-, CN-, or conjugated CC-multiple bonds. Isolated alkenes are difficult to reduce - methods utilizing hydrogen as a reductant are best suited in these cases (see Sect. 4.2.4).

Commonly encountered reactions are the partial reduction of arenes (*Birch* reduction), the reduction of carbonyl compounds and alkynes, the cleavage of benzyl ethers, and the reduction of strongly electron-withdrawing functional groups (see Fig. 4.97). The Clemmensen reduction also falls into the category of solvated metal reductions. However, since it is primarily used for the defunctionalization of carbonyl compounds, it will be discussed more in detail in the next subsection.

Predicting the efficiency of a given transformation remains challenging to date: The (thermodynamic) ability to reduce a substrate roughly correlates with the redox potential of the

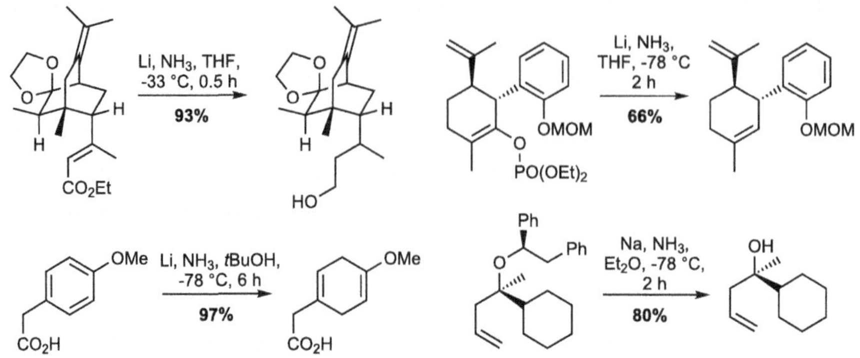

Fig. 4.97 Examples of reductions using solvated metals [178, 340–342]

given metal. The most electronegative metals are the strongest reductants (see Table 4.10) [223]. The observed reaction rate, however, is highly dependent on the reaction conditions. Various studies have so far been unable to show a consistent trend [343, 344]. Furthermore, lithium has the highest molar solubility in ammonia and unwanted side reactions are suppressed by alcoholic cosolvents. Especially on an industrial scale, lithium is therefore the preferred reductant [343]. The stronger reducing agents can also be used to convert CC multiple bonds, while Fe or Sn only reduce highly polarized bonds: while nitro groups are converted, carbonyl and alkene functionalities are not affected.

Table 4.10 Redox potential of various metals [223]

metal	Redox potential [V]
Li	−2.9
K	−2.9
Na	−2.7
Al	−1.34
Zn	−0.76
Fe	−0.44
Sn	−0.14

In addition to an oxidizable metal as an electron transfer agent, a proton source is required. Protic additives such as alcohols (*t*BuOH, EtOH, MeOH), ammonia, amines, ammonium salts, or even water have been used for this purpose. Cosolvents such as Et_2O, THF are normally employed to provide a homogeneous solution at the requisite low reaction temperatures (usually −40 to −78 °C).

The *Birch reduction* of aromatic substrates is conducted in the presence of an excess of a proton source, usually *t*BuOH. Metals dissolved in ammonia display a characteristic dark blue color, which is an excellent indicator for the requisite absence of water. In the reduction of monosubstituted arenes, the substituent can be located either in a vinylic or allylic position in the product. This selectivity is determined by the type of substituent: electron acceptors ultimately occupy saturated, allylic positions, whereas electron donors favor a vinylic position (see Fig. 4.98) [345, 346].

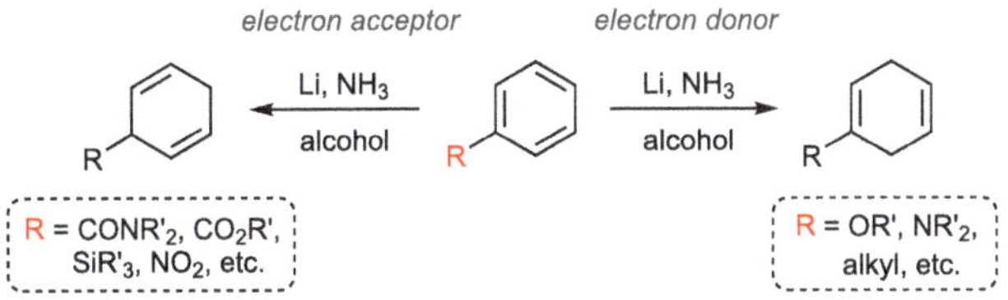

Fig. 4.98 Regioselectivity of the Birch reduction [345, 346]

The localization of the radical and the anionic charge are *para* with regard to each other and are positioned in the carbon framework where they experience maximum stabilization. The regioselectivity of the process is thus substituent-controlled. Electron-withdrawing groups stabilize the negative charge that builds up during the reaction and accelerate the reduction. Donor substituents, on the other hand, generally deactivate the aromatic ring and decrease the rate of reduction [345]. The reaction is first order with respect to aromatic substrate, electrons, metal cations, and proton donor [344].

In the first step, the radical **232** is obtained from the substituted aromatic system. Following the *ortho*-protonation to **233**, the second electron can be transferred and the cyclohexadienyl anion **234** can be formed. The question of *ortho-* vs. *meta*-protonation has been long debated, but recent studies favor protonation in the *ortho*-position relative to functional groups on the aromatic core [346]. After the transfer of the second electron, the non-conjugated diene **235** is generated by protonation of the middle C-atom. The selective protonation of the central atom in pentadienyl systems seems to be a general phenomenon of conjugated linear anions, as the negative charge density at this C-atom is highest [347]. The product with the donor group in the vinylic position thus results from the energetically preferred intermediate **232**. Compared to **236**, the negative charge is destabilized by the +M effect of the molecule's donor group at that position. The rate law of the reaction indicates the protonation of the radical anion **232** to be the rate-determining step [344]. For acceptor-substituted arenes, the stabilities are opposite to donor-substituted substrates: while **237** is destabilized, the electron-withdrawing group has a stabilizing effect on the heightened charge density in the *ipso*-position (**238**). For activated arenes (acceptor-substituted, styrenes, biphenyls, naphthalene, anthracene, etc.), a dianion **239** is afforded in a consecutive second electron transfer [345]. The ultimate protonation of the anion **240** occurs at the *ipso*-position (see Fig. 4.99) [346].

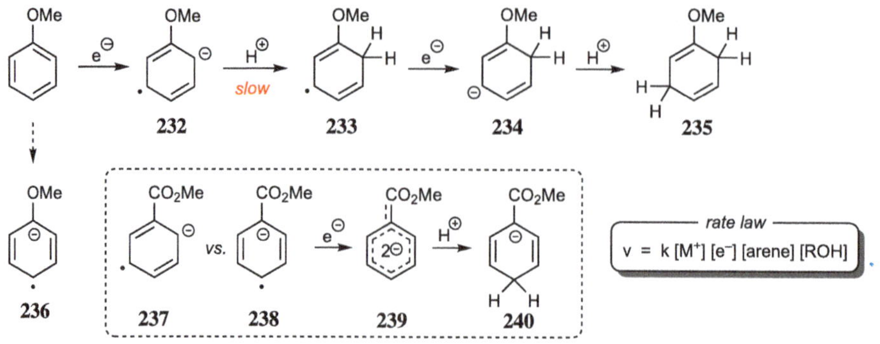

Fig. 4.99 Mechanism of the Birch reduction [344–346]

The reaction usually stops at the stage of the diene, as the isolated double bonds in the product are much more resistant to reduction with single electron transfers than the conjugated aromatic ring. If Birch reductions are carried out on donor-substituted substrates in the absence of an alcohol as a proton source, the formed $LiNH_2$ can, however, lead to double bond isomerization and over-reduction. In the conversion of activated aromatic systems, only a stoichiometric amount of alcohol should be used, as there can also be over-reductions when an excess of alcohol is present [345]. To therefore prevent over-hydrogenation upon quenching of the reaction, it is advised to add water, NH_4Cl, or another acidic salt instead of an alcohol. If primary amines serve as solvents (without an additional proton donor) instead of ammonia/alcohol, aromatic substrates can be selectively reduced to the corresponding cyclic olefin. This stronger reductant combination is also referred to as the *Benkeser* reaction (see Fig. 4.100) [225].

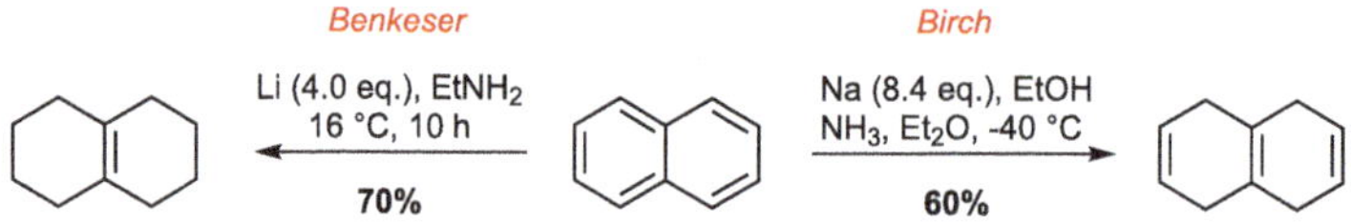

Fig. 4.100 Comparison of the reduction of naphthalene under Birch and Benkeser conditions [348, 349]

The metal-mediated formation of H_2 from ROH or the cleavage of functional groups comprise common competing pathways to the desired reduction. These can often be successfully suppressed by resorting to low temperatures [344, 345].

In the total synthesis of the alkaloid nominine, Peese and Gin accomplished the reduction of the aromatic ring in the advanced intermediate **241** by a Birch reduction with sodium metal and isopropanol as the proton source. The vinyl ether **242** was cleaved to the β, γ-unsaturated ketone by acidic work-up. The final steps to the natural product (**244**) included a pyrrolidine-mediated Diels-Alder reaction, a Wittig reaction to form the *exo*-double bond, and a SeO$_2$-mediated allylic oxidation (see Fig. 4.101) [350].

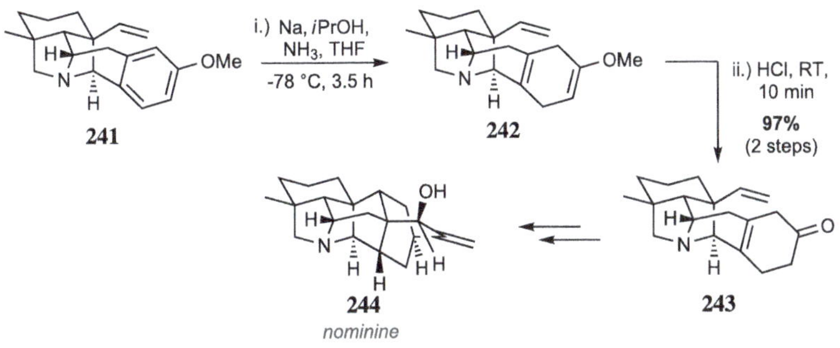

Fig. 4.101 Synthesis of nominine by Peese and Gin [350]

Non-conjugated terminal as well as internal alkenes cannot be reduced under Birch conditions. However, a Benkeser reaction does usually provide the desired products. Allylic or benzylic alcohols can be converted to the respective alkenes or toluene derivatives by cleavage of the C-O bond upon loss of H$_2$O [225].

If only one equivalent of alcohol is used, the transitory anion **245** can be intercepted by alkyl halides. With the use of auxiliary groups or persistent stereogenic centers present in the substrate, a new stereogenic center can be introduced at the cyclohexadiene core **246** through a diastereoselective reaction (see Fig. 4.102) [351].

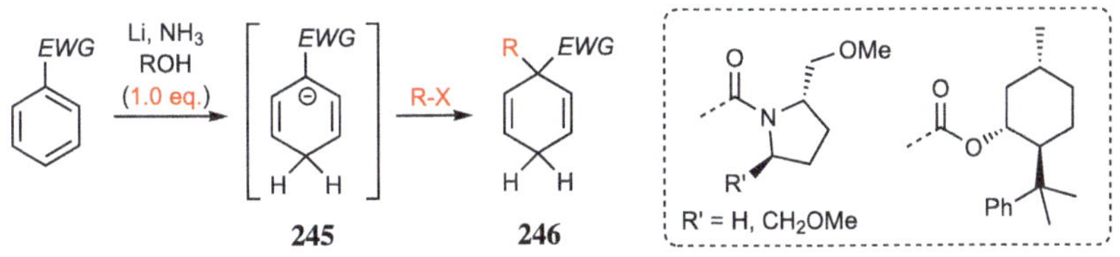

Fig. 4.102 Principle of the alkylating Birch reduction and possible chiral auxiliaries that serve as EWG [351–353]

Schultz and co-workers cleverly exploited this approach to complete the first asymmetric synthesis of the unnatural enantiomer of a Hasubanan alkaloid. The access to (+)-cepharamine (**250**) was achieved by Birch alkylation of the chiral amide **247** with alkyl iodide **248**, selectively introducing the first stereogenic center at C14 in the molecular framework in **249**. Modification of the anisole-derived ring by a reductive cyclization and a Hoffmann rearrangement finally enabled the construction of the tetracyclic structure of the desired alkaloid (see Fig. 4.103) [354].

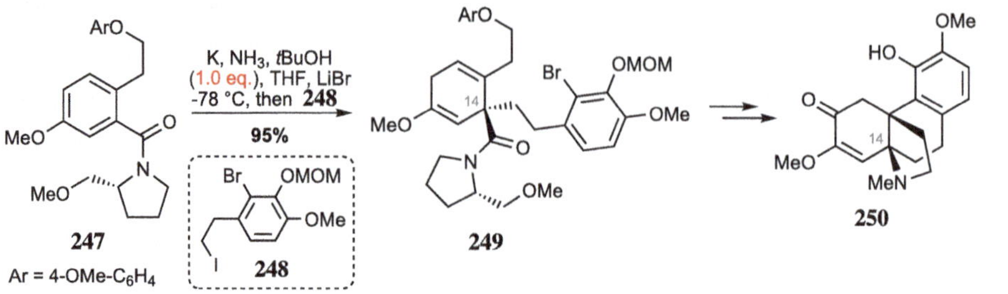

Fig. 4.103 Use of an alkylating Birch reduction in the total synthesis of (+)-cepharamine (**250**) [354]

Birch conditions are unsuitable to fully reduce aromatic systems to the saturated carbo- or heterocycles. However, this conversion is easily possible by heterogeneous hydrogenation, often while retaining the functionalities in the ring periphery (see Sect. 4.2.4). In addition to the CC double bonds in aromatic substrates, triple bonds in alkynes can also be reduced under Birch conditions. Due to the stepwise mechanism of the transformation, the alkynes are usually converted to the corresponding *trans*-alkenes with high selectivity [355].

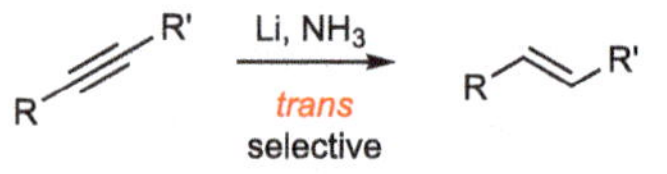

Normally, no additional proton source (ROH, NH_4X) is needed as additive [225]. In case of its use, the proton donor may additionally influence the *cis/trans* ratio of the product, depending on the metal in question [356].

When looking at the alleged mechanism, the first electron transfer to **251** is reversible. The subsequent protonation to the vinyl radical **252** constitutes the rate-determining step of the sequence [357]. Since the inversion barrier between **252** and **254** is relatively low, an equilibration is assumed to take place. The resulting vinyl anions **253**, **255** cannot interconvert. The *cis/trans* ratio of the reduction therefore reflects the energetic ratio of **252** and **254** (see Fig. 4.104).

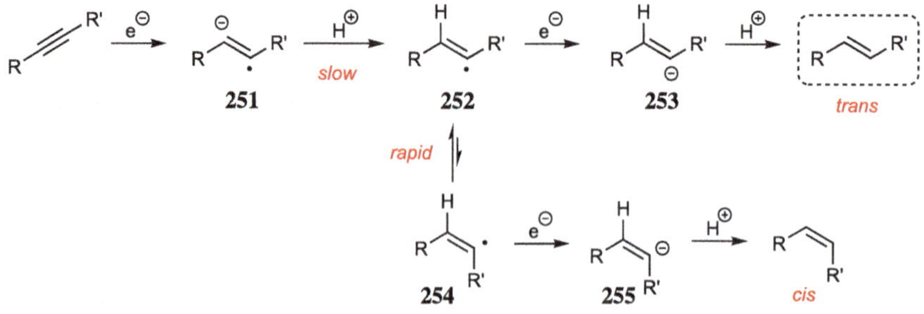

Fig. 4.104 Mechanism of the stepwise reduction of alkynes [355, 357]

Dias used a *trans*-selective reduction in his total synthesis of the bacterial metabolite pironetin (**258**). After substitution of the terminal tosyl group with lithium acetylide and its methylation, the Na-mediated reduction in ammonia afforded the *E*-alkene **257**. Subsequent formation of the dihydropyranone completed the synthesis of the immunosuppressant pironetin (**258**) (see Fig. 4.105) [358].

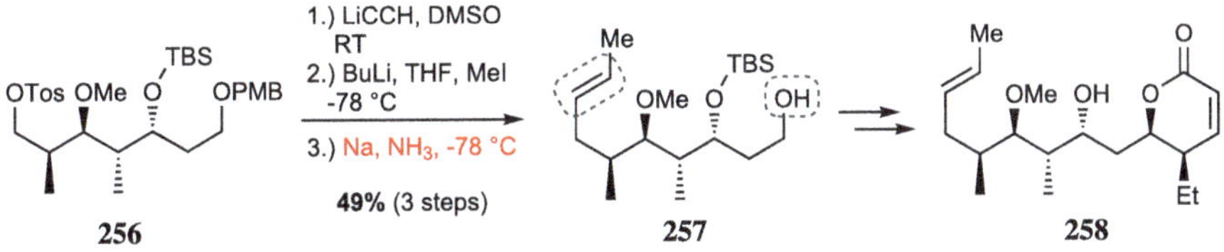

Fig. 4.105 Dias' synthesis of the secondary metabolite pironetin (**258**) [358]

The reduction of carbonyl derivatives, mainly ketones or esters (*Bouveault–Blanc* reaction), by solvated metals is rarely encountered. However, there are occasional examples in syntheses where this reaction can display an advantage over complex metal hydrides or reductions by hydrogen [359]. Even if not intended to occur, the lability of the carbonyl function in the presence of solvated metals should be taken into account under Birch conditions, especially with complex substrates (see Table 4.3).

Birch conditions are perfectly suited for the deprotection of benzyl ethers. Pfizer sought access to the active ingredient sumanirole (**260**), a D2 receptor agonist, which can serve as a probe for investigating neurological processes. To access the secondary amine while simultaneously debenzylating the urea, an ammonia solution of lithium metal was used in the presence of *tert*-amyl alcohol as a proton source. The conversion was stopped by addition of water and subsequently reacted with maleic acid, providing the product as maleate salt in 84% yield and with >99% purity (see Fig. 4.106) [343].

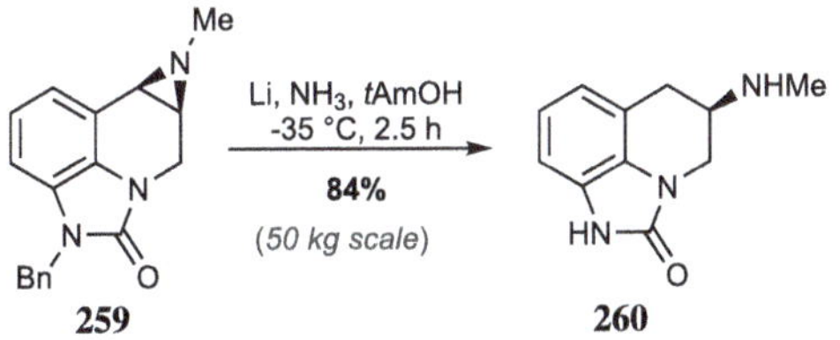

Fig. 4.106 Production of the active ingredient sumanirole (**260**) by Li-mediated reduction on a multi-kg scale [343]

A recent addition to the field of reductions with alkali metals is the use of redox catalysts. The addition of cobalt or iron salts as electron transfer reagents allows the utilization of significantly less redox-active metals such as magnesium as reductants. Thus, reductions of nitro groups, deallylations, pinacol couplings, and desulfonylations are possible (see Fig. 4.107) [360, 361].

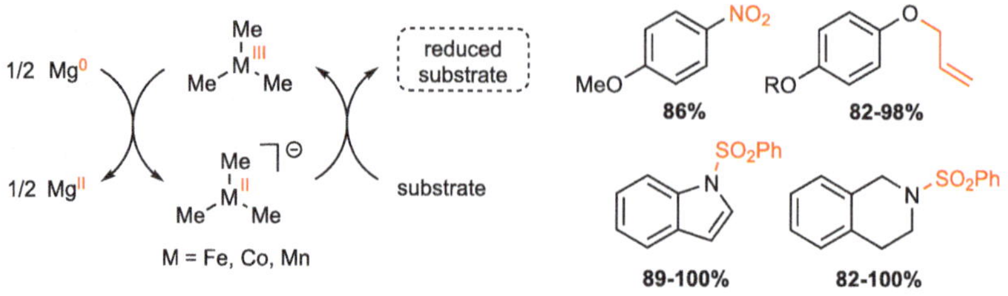

Fig. 4.107 Redox catalysis in reduction with base metals [361]

4.2.3 Reductive Defunctionalization

The defunctionalization of complex molecules is a field of study which has been dominated by the use of organotin compounds to convert alcohols, halides, selenides, or sulfides to the respective alkanes [362]. Beyond tin-based reagents, organosilanes, hydrazones (in combination with tin hydrides), and also metal amalgams can also furnish a selective removal of functional groups. Raney nickel and other heterogeneous catalysts have been employed as well. However, their typical applications focus on the reduction of CC-, CO-, and CN-multiple bonds, which is why they will be excluded for now from the discussion on defunctionalization reactions.

The chemistry of free radicals — in synthesis in general and in defunctionalizations in particular — has been dominated by stannanes (Bu_3SnH, Ph_3SnH) by virtue of their commercial availability, stability, compatibility with a variety of functional groups, the rapid H-atom abstraction, and the ability to maintain radical reactions through a chain mechanism [363, 364].

Radical reactions require a radical initiator, which in the vast majority of cases is either azobisisobutyronitrile (**261**, AIBN) or dibenzoyl peroxide (**262**). The thermally labile

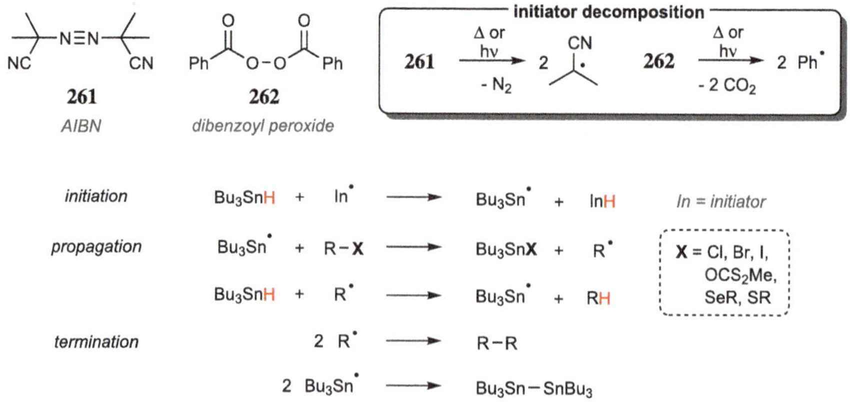

Fig. 4.108 Mechanism of radical defunctionalizations with tin hydrides [365]

initiators irreversibly decompose to the respective primary radicals, which initiate the radical chain reaction with Bu_3SnH (see Fig. 4.108).

The reaction rate is derived from the relative stability of the intermediate radicals, which is why the defunctionalization of tertiary carbon atoms proceeds more rapidly than for secondary or primary carbon atoms.

The presence of CC double bonds can pose a challenge in radical defunctionalizations. The reaction of tin radicals with alkyl chlorides (**263**→**264**) is usually slower than the addition of the radicals to the olefinic bond ($k_1 < k_2$, **263**→**265**). With alkyl bromides, the reaction constants of defunctionalization display a similar order of magnitude compared to the olefin addition ($k_1 \approx k_2$). Only alkyl iodides enable a selective conversion in the presence of CC double bonds (see Fig. 4.109) [365].

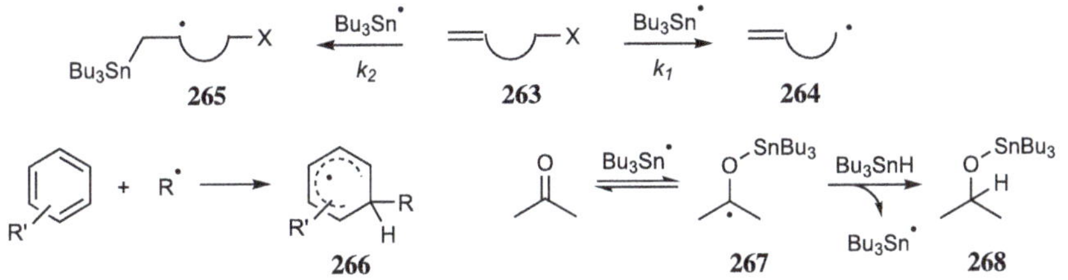

Fig. 4.109 Possible side reactions of various radical defunctionalizations [365, 366]

When employing aromatic, benzylic, and allylic substrates, a recent study has shown that the formation of stable radicals (**266**) results in undesired side reactions of said radicals. Depending on the overall reversibility of their formation, this can slow down or even halt the radical process [366]. Stannyl radicals, on the other hand, are particularly reactive towards carbonyl groups and can provide the radical **267**, which can subsequently consume an equivalent of the reducing agent. In the worst case, **267** is resonance-stabilized by the carbonyl substituents, thereby irreversibly terminating the reaction (see Fig. 4.110).

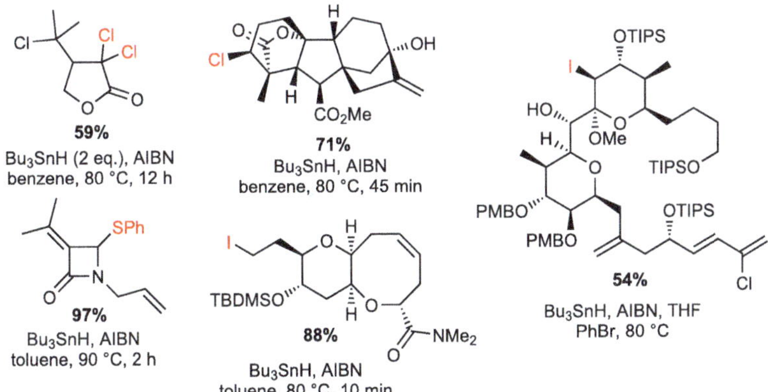

Fig. 4.110 Synthetic examples of radical defunctionalizations [367–371]

In addition to the "classic" defunctionalizations, tin hydrides are an integral part of the *Barton–McCombie deoxygenation*, in which xanthates (C(=S)SR) are used as leaving groups to remove hydroxy functions [372]. In the initial step, a thiocarbonate or xanthate (**269**) is formed by reacting an alcohol with an activated thiocarbonyl derivative (one-step variant). Alternatively, **269** can also be obtained by converting the alcohol with CS_2, followed by methyl iodide addition (two-step one-pot process). The addition of Bu_3SnH and a radical starter furnishes intermediate **270**, which decomposes to the deoxygenated radical **271** and the stannane **272**. **271** affords the desired alkane by reacting with an equivalent of Bu_3SnH. **272** further decomposes to COS and Bu_3SnX (see Fig. 4.111) [372, 373].

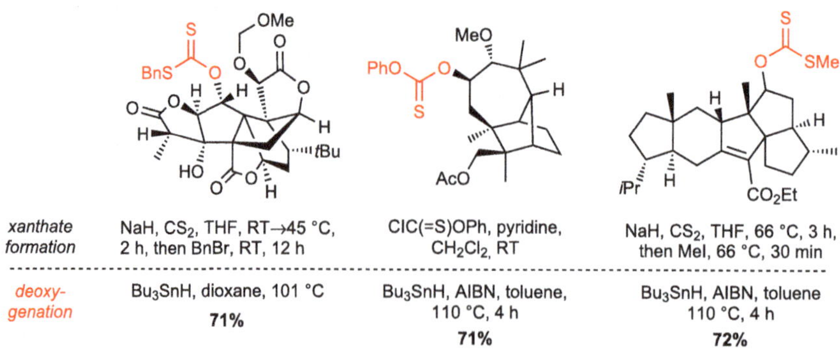

Fig. 4.111 Mechanism of the Barton-McCombie deoxygenation [372, 373]

Typical reaction conditions comprise the use of Bu_3SnH in the presence of catalytic amounts of AIBN in refluxing toluene or benzene. The reaction lends itself to the conversion of secondary alcohols, as tertiary radicals (**271**) form slowly and primary alcohols prefer to abstract hydrogen (**273**) rather than fragment. The concentration of Bu_3SnH is usually kept low to suppress formation of the by-product **273**. This can for example be achieved by the slow, continuous addition of the stannane over the course of the reaction [374]. The Barton-McCombie deoxygenation is no longer used as frequently in modern total syntheses. However, its advantage lies in a very good tolerance towards functional groups (see Fig. 4.112). Only the presence of alkenes in the substrates poses a challenge, for example if

xanthate formation	NaH, CS_2, THF, RT→45 °C, 2 h, then BnBr, RT, 12 h	ClC(=S)OPh, pyridine, CH_2Cl_2, RT	NaH, CS_2, THF, 66 °C, 3 h, then MeI, 66 °C, 30 min
deoxygenation	Bu_3SnH, dioxane, 101 °C **71%**	Bu_3SnH, AIBN, toluene, 110 °C, 4 h **71%**	Bu_3SnH, AIBN, toluene 110 °C, 4 h **72%**

Fig. 4.112 Barton-McCombie deoxygenations in the syntheses of complex natural products [375–377]

their undesired conversion leads to kinetically favored pathways such as the formation of 5- or 6-membered rings.

Tin-mediated defunctionalizations can also be exploited as part of a domino sequence. Ley and co-workers used the radical cleavage of the xanthate in their synthesis of the complex secondary metabolite azadirachtin (**277**) to build the bicyclic-bridged eastern fragment. The reaction of allene **274** led to intermediate **275** under homolytic cleavage of the xanthate. This then provided the desired alkene **276** in a 5-*exo*-cyclization, from which the natural product could be obtained in further steps (see Fig. 4.113) [378].

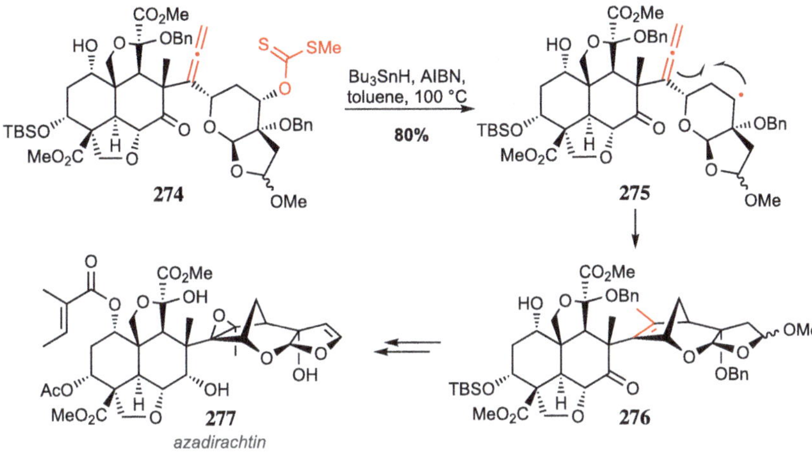

Fig. 4.113 Application of a Barton-McCombie-initiated domino sequence in the synthesis of the natural product azadirachtin [378]

In addition to dehydroxylation, a decarboxylation was analogously developed by Barton and co-workers employing stannanes [379]. For this, a thiohydroxamate ester (**278**, *Barton ester*) is formed from an activated carboxylic acid, for example an acid chloride, which subsequently affords the labile intermediate **279** with tin hydrides. **279** finally decomposes in the presence of hydrogen sources (stannanes, silanes) to the alkane similar to the mechanism of the Barton-McCombie deoxygenation (see Fig. 4.111). By-products are CO_2 and stannylthiopyridine **280** (see Fig. 4.114). If a radical scavenger other than the stannane is utilized under photoirradiation of suitable wavelengths, Barton esters can also be converted into the respective alcohols, alkyl halides or selenides [380].

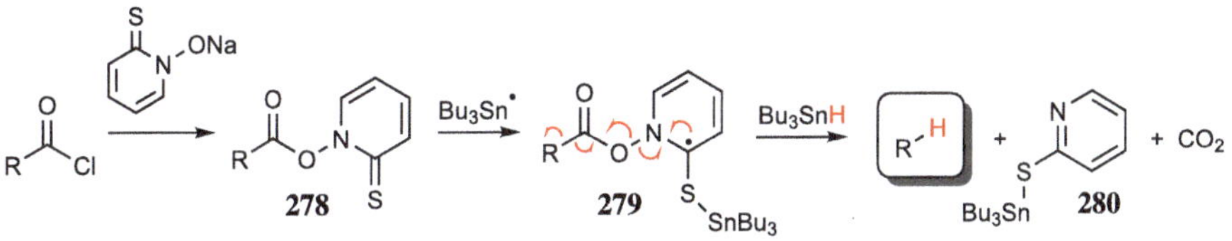

Fig. 4.114 Mechanism of the Barton decarboxylation [379]

Modern variants of radical defunctionalization attempt to only catalytically employ the toxic tin or to completely avoid it, for example by using organosilanes such as (TMS)₃SiH (see below) [381, 382]. Catalytic amounts of an active tin species can be realized by a suitable secondary (stoichiometric) reductant to regenerate the metal hydride. Fu and co-workers developed the first Barton-McCombie deoxygenation using this approach, in which polymethylhydrosiloxane (PMHS) constitutes the secondary reductant. Bu₃SnH is formed *in situ* from BuOH and (Bu₃Sn)₂O, a cheaper and more stable tin source. In addition, the alcohol accelerates the rate-determining recovery of Bu₃SnH from Bu₃Sn-OPh (**283**), the primary by-product of the reduction (see Fig. 4.115) [383].

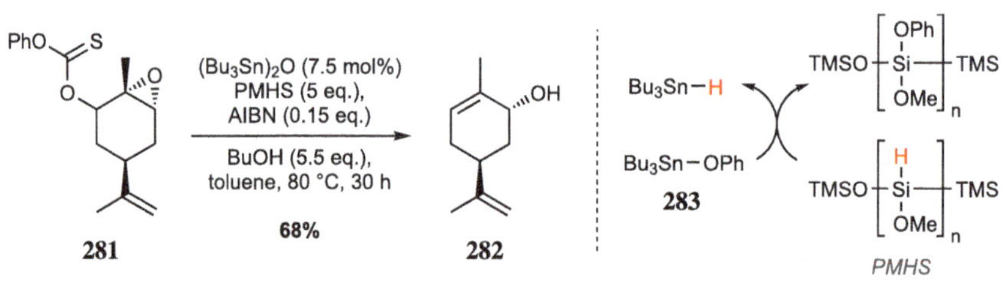

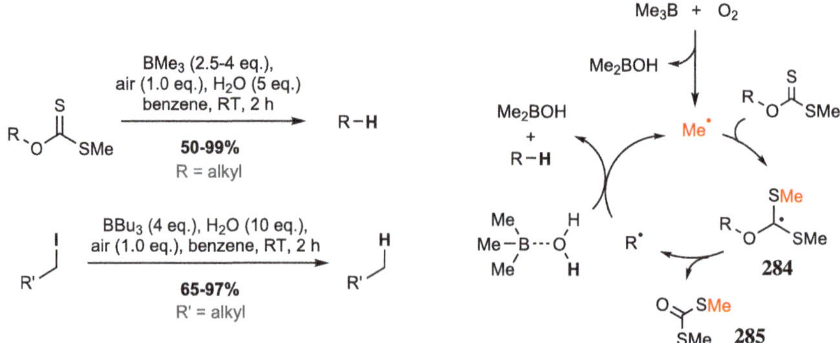

Fig. 4.115 Bu₃SnH-catalyzed Barton-McCombie deoxygenation [383]

Further examples of tin-catalyzed reactions include dehalogenations, the reduction of various functional groups, and cyclization reactions [384]. The Wood group introduced completely tin-free defunctionalizations, which are instead mediated by trialkylboranes [385, 386]. The Barton-McCombie-type defunctionalization uses methyl radicals as radical carriers, which initially form the radical of the respective dithioorthoester **284** from xanthates. Following the elimination of **285**, a H-atom is transferred to the deoxygenated radical R˙ from BMe₃-associated water. The formation of the stable B-O bond of the hydroxide with the borane constitutes the driving force of the transformation (see Fig. 4.116). Subsequent modifications allowed the reaction to be applied to the dehalogenation of alkyl iodides as well. Thus, xanthates and iodides can be selectively removed in the presence of esters, acetals, bromides, and chlorides.

Fig. 4.116 Tin-free defunctionalizations developed by Wood *et al.* [385, 386]

In the Barton decarboxylation, the stannane can often be substituted by a thiol as hydrogen source. Especially *tert*-butylthiol displays similar efficiency but with significantly simplified separation of the respective disulfide analog of by-product **280** [374]. The concept can also be combined with transition metal-catalysis to achieve broadly applicable conditions [387]. Further tin-free variants of dehalogenations rely on photoredox catalysis for the formation of radicals ($\rightarrow$ Chap. 10) [388–390].

Organosilanes can react either in a radical or ionic manner in the removal of halogen or hydroxy groups. Silanes typically display a low tendency for homolytic Si-H bond cleavage, which is why a radical chain reaction cannot be maintained [364]. A notable exception is (TMS)$_3$SiH (*supersilane*), which was developed by Chatgilialoglu in the 1980s and constitutes a viable alternative to Bu$_3$SnH [391, 392]. (TMS)$_3$SiH is on average an order of magnitude slower than Bu$_3$SnH in the reaction with carbon radicals [393]. Iodides, bromides, selenides, isocyanides, acid chlorides, xanthates, and sulfides have been reported as suitable radical precursors. The defunctionalizations also use a radical initiator such as AIBN or dibenzoyl peroxide, and are conducted similarly to conversions with tin hydrides (see Fig. 4.117).

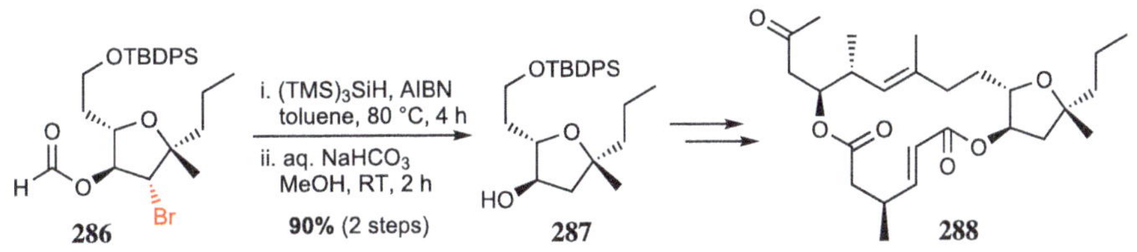

Fig. 4.117 Supersilane-mediated debromination in Fürstner's synthesis of amphidinolide X (**288**) [394]

Under ionic conditions, the carboxonium pathway (mode A, cf. Fig. 4.94) is pursued. An elimination of the functionality targeted for removal only affords high yields if a particularly stable cation is formed. Therefore, primary alkyl halides and alcohols are unsuited for defunctionalization by organosilanes. Tertiary and some secondary alcohols as well as

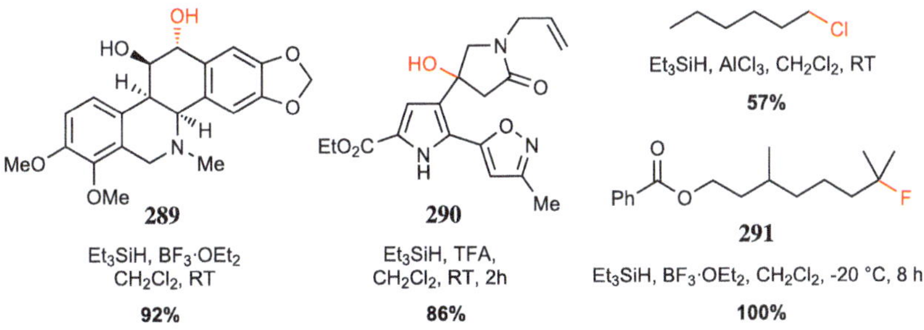

Fig. 4.118 Examples of ionic defunctionalizations with organosilanes [395–398]

benzylic and allylic functionalities are preferentially defunctionalized by Et_3SiH/BF_3, whereas alkyl halides are typically converted using $Et_3SiH/AlCl_3$ (see Fig. 4.118) [331].

The use of frustrated Lewis acids/bases is a more recent advancement of the classical silane/Lewis acid reagent combination. The method reacts silanes with alcohols, carbonyls, halides, amines, or thiols in the presence of $B(C_6F_5)_3$. The activation of the Si-H bond results in a defunctionalization and not a formal addition of H_2, in contrast to the well-known activation of H_2 for the hydrogenation of CC-, CO- and CN-double bonds by frustrated Lewis pairs. The polarization and cleavage of the Si-H bond presumably proceeds via **292** and gives intermediate **293** after reaction with the substrate. As a consequence, primary alcohols are even more reactive than secondary ones, which allows for complementary reactivity to an ionic reaction pathway [399]. The utility of the defunctionalization could be demonstrated in various complex substrates. The performance does portray a strong dependence on the type of borane and silane used, which implies the need for broader optimization studies when resorting to other substrates (see Fig. 4.119) [400].

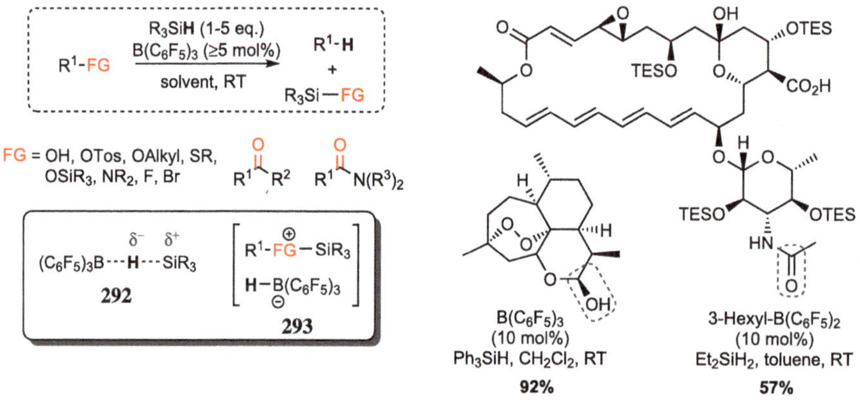

Fig. 4.119 Borane-mediated defunctionalizations to alkanes [399, 400]

The conversion of a ketone or aldehyde to the corresponding alkane can be selectively achieved by various methods. The *Clemmensen* reduction, the *Wolff–Kishner* reaction, and the reduction relying on tosylhydrazones can be considered as the most common approaches (see Table 4.11) [401].

In the Clemmensen reduction, ketones and aldehydes react with amalgamated zinc (Zn/Hg) and hydrochloric acid (as a solution in H_2O or anhydrous in organic solvents). The original conditions used aqueous HCl, whereas modern modifications rely on a solution of ethers (Et_2O, THF) or arenes (benzene, toluene) saturated with gaseous HCl [402]. Diketones and enones as substrates can result in complex product mixtures under Clemmensen conditions. Commonly observed side reactions include pinacol couplings and the introduction of double bonds.

Table 4.11 Comparison of common reduction methods of ketones and aldehydes to hydrocarbons

reduction	conditions	remarks
Clemmensen	Zn/Hg, HCl, Et$_2$O or THF, 0 °C-RT, 1 h	does not reduce aliphatic alcohols, halides; side reactions with α,β-unsaturated and α-functionalized carbonyls
Wolff-Kishner	H$_2$N-NH$_2\cdot$H$_2$O, KOH, HOCH$_2$CH$_2$OH, 190 °C	better suited for complex substrates; side reactions with 1,3- and 1,4-diketones
tosylhydrazone	i.) TosHN-NH$_2$, MeOH, RT; ii.) metal hydrides or catecholborane, solvent, $\pm$H$^+$, 80-100 °C	also used with complex substrates; double bond migration with α,β-unsaturated carbonyls commonly observed

The mechanism of the Clemmensen reduction is not well understood, which is due to the fact that the received products can change when varying the reaction conditions. Since aliphatic alcohols are not reduced to the respective alkanes, their involvement can be ruled out. Allylic and benzylic alcohols, on the other hand, are indeed reactive under the studied conditions. Two reaction mechanisms have been considered, a carbene path (path A) similar to the McMurry reaction and a homogeneous organozinc mechanism (path B) [403, 404]. While the carbene path via **294** and **295** appears very straightforward, [404] there is no general consensus on the involved intermediates of the second path. Both a direct addition of Zn to the carbonyl C-atom (**296**) [403] and a single electron transfer are discussed (**297**) [405–407]. The postulated subsequent intermediates (**298, 299**) could also not be experimentally observed so far (see Fig. 4.120).

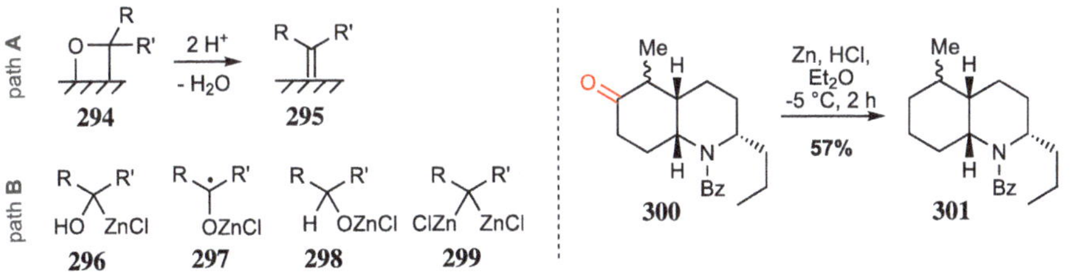

Fig. 4.120 Postulated intermediates of the Clemmensen reduction [403–407] and application in the synthesis of the alkaloid (–)-pumiliotoxin C [408]

The *Wolff–Kishner* reduction is a useful alternative to the Clemmensen reduction or the reductive desulfurization of dithianes with Raney nickel (*vide infra*). By virtue of the absence of any metal, possible side reactions such as the reduction of N-O bonds, imines, hydrazines, azo compounds or other electron-poor functional groups can be prevented [403]. The original conditions involved the addition of preformed hydrazones to a hot melt of KOH or heating of the hydrazone with alcoholic sodium ethanolate. The *Huang–Minlon* modification, developed in the 1940s, is now the standard variant of the process, in which the substrate is converted in the presence of aqueous hydrazine and KOH in hot ethylene glycol.

Mechanistic details of the Wolff-Kishner reduction have not been elucidated, similar to the Clemmenson reduction. A first-order dependence on hydrazone and base was observed with diaryl ketones as substrates [409, 410]. The postulated mechanism comprises the initial formation of the hydrazone **302** and a base-catalyzed tautomerization via **303** to the diazene **304**. After removal of the proton at the terminal nitrogen, N_2 and water are eliminated to presumably afford the carbanion **305**, which can react with a proton source such as water to yield the respective alkane (see Fig. 4.121).

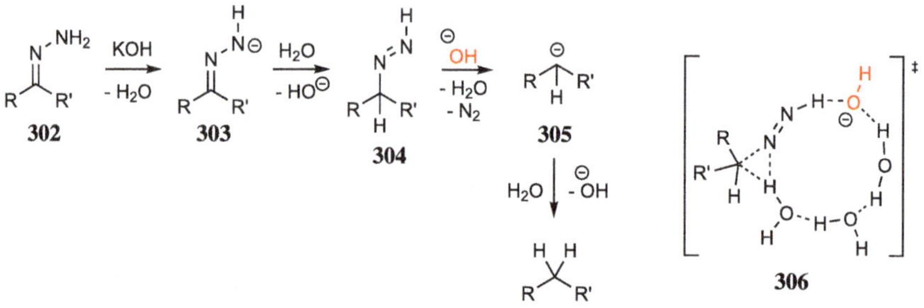

Fig. 4.121 Postulated mechanism of the Wolff-Kishner reduction [409, 411, 412]. The base was highlighted in **306**

The protonation **303**→**304** was postulated to be the rate-determining step of the sequence based on experimental studies with diaryl ketones [409]. However, this has not been further substantiated due to the elusive nature of the intermediates in the subsequent steps starting from **302**. In particular, the occurrence of a carbanion has not been directly observed, although there is circumstantial experimental evidence [411]. A recent theoretical study disputed the existence of **305** in alkyl ketones. Instead, a cyclic, concerted transition state **306** was proposed for the elimination of N_2 under simultaneous protonation of the diazene carbon atom involving three molecules of H_2O [412].

In contrast to the Clemmensen reduction, the Wolff-Kishner reduction is still a frequently encountered method in total syntheses when defunctionalizing ketones to the corresponding alkanes. Merck required kilogram quantities of the substituted imidazole **308** for preclinical studies. The synthesis was achieved using **307** as substrate in a Wolff-Kishner reduction on an industrial scale [413]. Following the *in situ* formation of the hydrazone in diethylene glycol (DEG), the suspension was heated for several hours. The product was then precipitated by adding MeCN/H_2O and, after filtration, isolated in sufficient purity without further work-up (see Fig. 4.122).

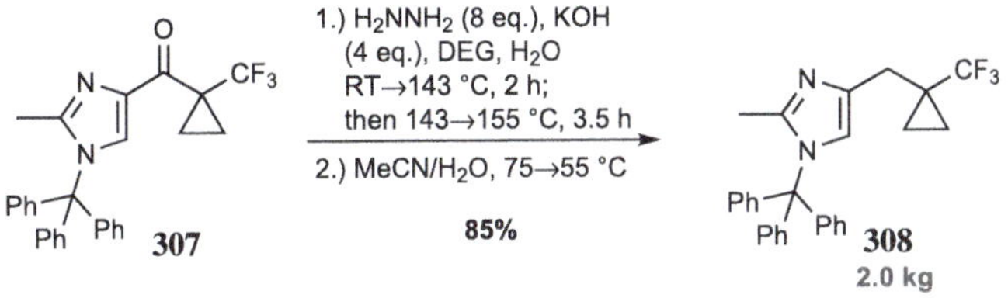

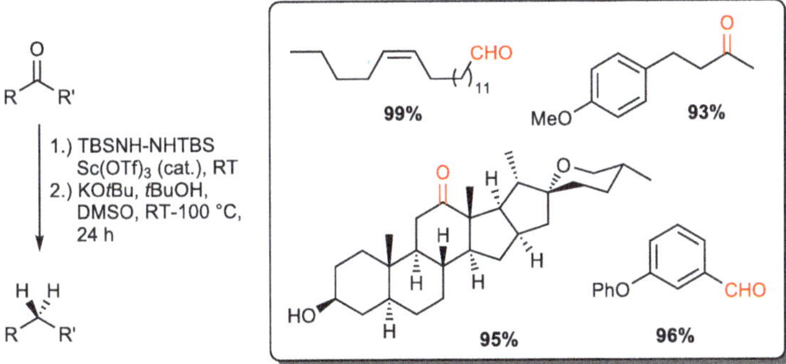

Fig. 4.122 Wolff-Kishner reduction of **307** on a kilogram scale [413]

Since the Huang–Minlon modification also requires temperatures of 190–200 °C for efficient turnover, further "low-temperature variants" were developed that allow the conversion to run at room temperature up to 100 °C [414, 415]. However, these involve either the isolation of the usually unstable hydrazones or require a slow addition of toxic hydrazine over several hours. An elegant improvement was reported by the group of A.G. Myers, in which stable silylhydrazones are converted with KO*t*Bu in *t*BuOH/DMSO to the respective alkanes with high yields at moderate 23–100 °C (see Fig. 4.123) [416].

Fig. 4.123 Wolff-Kishner modification by Myers [416]

A variant of the Wolff-Kishner reduction by Cagliotti and Hutchins involves the use of tosylhydrazones instead of the parent hydrazones [417, 418]. Metal hydrides such as $NaBH_4$, $NaBH_3CN$, or catecholborane are effective reductants. In contrast to the Wolff–Kishner reduction, the tosylhydrazones must be isolated.

If complex metal hydrides are used, an initial addition of the hydride to the carbonyl-C-atom of the hydrazone **309** takes place, followed by the elimination of sulfinic acid from **310** to give **304**. Depending on the reaction conditions, the alcoholic solvent or an added acid can serve as proton source. An omission of the reaction step from **309** to **304** with direct elimination of TosH from **309** is also conceivable. In the reduction of tosylhydrazones or unsubstituted hydrazones, the diazene **304** is the common intermediate. Whether the N_2 loss from the diazene proceeds in a radical or ionic manner has not yet been conclusively elucidated [411, 419]. **309** also provides the desired alkane in the presence of catecholborane.

The substituted hydrazine borane **311** similarly furnishes **304** upon addition of a nucleophile, possibly via decay of the tetravalent borane **312** (see Fig. 4.124) [420].

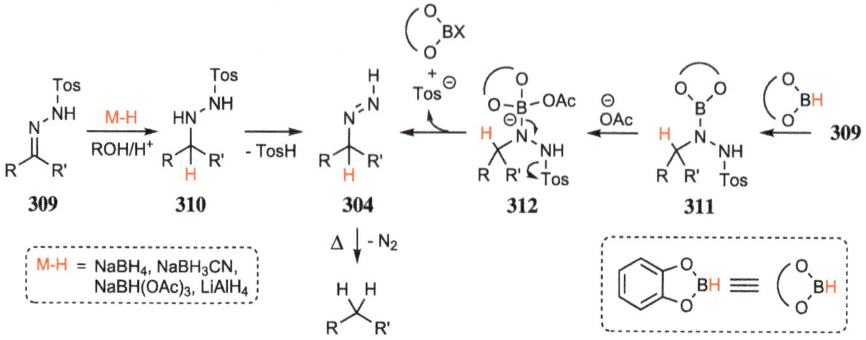

Fig. 4.124 Postulated mechanism of the reduction of tosylhydrazones [411, 419–421]

Enones form the expected alkene as a product, but with transposition of the double bond [422]. Tang and co-workers took advantage of this situation in their synthesis of the diterpenes harringtonolide (**315**) and hainanolide, to access the intermediate **314** starting from **313**. The resultant cyclohexene **314** served in the key step of the synthesis as a dienophile in a Diels–Alder reaction to obtain the 7/5/6/6-fused molecular skeleton of the natural product harringtonolide (**315**) (see Fig. 4.125) [423].

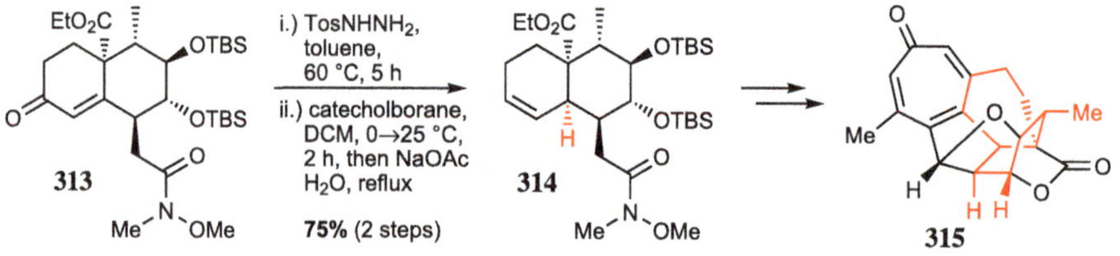

Fig. 4.125 Total synthesis of harringtonolide (**315**) by Tang *et al.* [423] The bicyclodecene framework of **314** is highlighted in **315**

Further methods of deoxygenation include conversion into dithianes or thioketones and subsequent reduction with Raney nickel (see Chap. 4.2.4.2), as well as the formation of an enol triflate followed by Pd-catalyzed reduction. The reduction of the enol triflate can also be achieved by a reductive Stille cross-coupling, but this results in an olefin instead of the alkane. Two elegant examples of this approach can be found in Corey's synthesis of aspidophytin and Keck's synthesis of epothilones B and D (see Fig. 4.126) [424, 425].

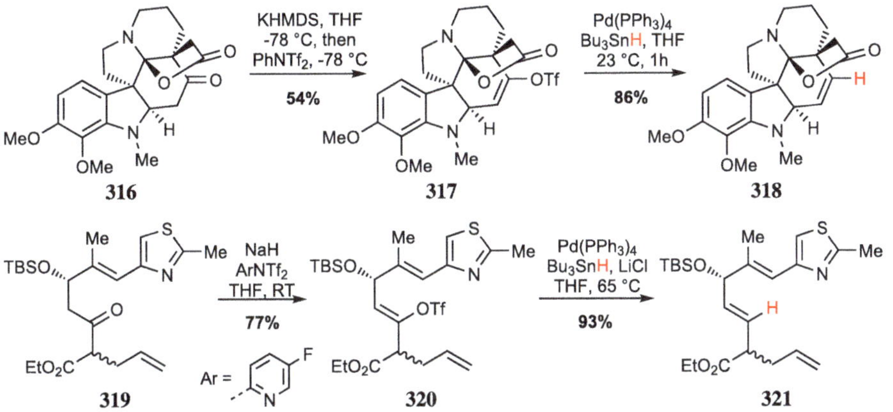

Fig. 4.126 Total syntheses of aspidophytin and epothilone B, D using a reductive Stille cross-coupling [424, 425]

Access to the triflate (**317**, **320**) can be easily obtained by transferring the trifluoromethanesulfonic acid group from an arylbistriflamide (PhNTf$_2$ or *Comins' reagent N,N-bis(trifluoromethanesulfonyl)-2-amino-5-chloropyridine*) to the enolate. In the presence of the Pd precatalyst, the respective alkenes **318**, **321** are provided using Bu$_3$SnH as a hydride source.

4.2.4   Hydrogen as a Reducing Agent

Hydrogen is the most commonly used and by far the cheapest reducing agent in industry — especially when it comes to large-scale reduction, as a quick estimation can show (see Table 4.12).

Table 4.12 Comparison of the costs of different reductants per reduction equivalent. The prices of the metal hydrides were taken from the Sigma-Aldrich catalog (2/2024)

reducing agent	price [\$/mol]
LiAlH$_4$	10.5
NaBH$_4$	4.45
DIBAL	96.8
H$_2$	0.02
H$_2$, 5% Raney Ni	1.40
H$_2$, 5% Pd/C	13.8

The price of the largest pack size was used. LiAlH$_4$: 1111 \$/kg, NaBH$_4$: 471 \$/kg, DIBAL (25% in toluene): 170 \$/kg, H$_2$: 11 \$/kg, Raney-Ni: 110 \$/kg, Pd/C: 1100 \$/kg. M(reactant) is 250 g/mol.

As is evident, hydrogen gas costs at least two orders of magnitude less than other common reducing agents. The cost driver of a hydrogenation on a small scale is the price of the metal catalyst. However, the catalyst can potentially be reused multiple times in hydrogenations. On a large scale, with (almost) complete catalyst recycling, this catalyst cost becomes negligible and the raw material costs of a hydrogenation shrink to the raw material costs of the consumed hydrogen.

Merck developed a commercial route to the diabetes drug sitagliptin (**323**), in which an asymmetric hydrogenation is employed. In the presence of a chiral rhodium-bisphosphine catalyst, the desired amine was obtained in high yield and excellent enantioselectivity, which could be further improved by subsequent crystallization. The detailed, improved route significantly reduced the total amount of waste and completely avoided aqueous waste streams (see Fig. 4.127) [426].

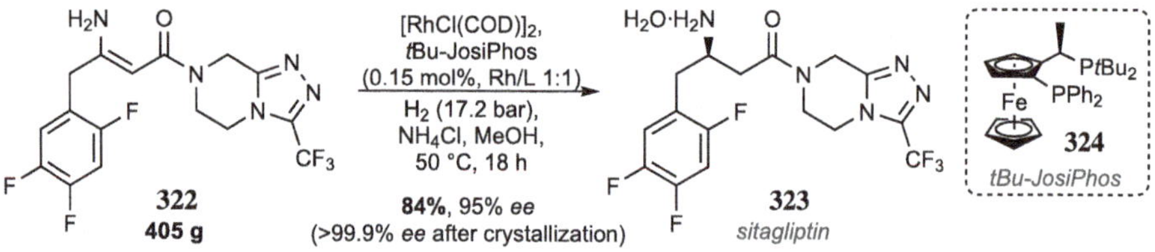

Fig. 4.127 Merck's synthesis of sitagliptin (**323**) [426]

Reductions with hydrogen allow certain transformations that are difficult or impossible to perform selectively with the stoichiometric reductants discussed thus far. Typical examples are debenzylations (Pd/C), hydrogenations of alkynes (Lindlar catalyst), hydrogenations of alkenes (Wilkinson/Crabtree catalyst), hydrodesulfurizations (Raney-Ni), and core hydrogenations of arenes (Ru catalysts).

The extensive material on stereoselective reductions, whether in a diastereo- or enantioselective reaction, is discussed in more detail in Sect. 4.2.5.

4.2.4.1 Homogeneous Hydrogenations

In catalytic hydrogenations, hydrogen is transferred by a catalyst to a substrate. Most reactions use gaseous H_2 as hydrogen source, which is activated by the catalyst. Instead of H_2, hydrogen can also be transferred to the catalyst from alcohols or formic acid in a transfer hydrogenation.

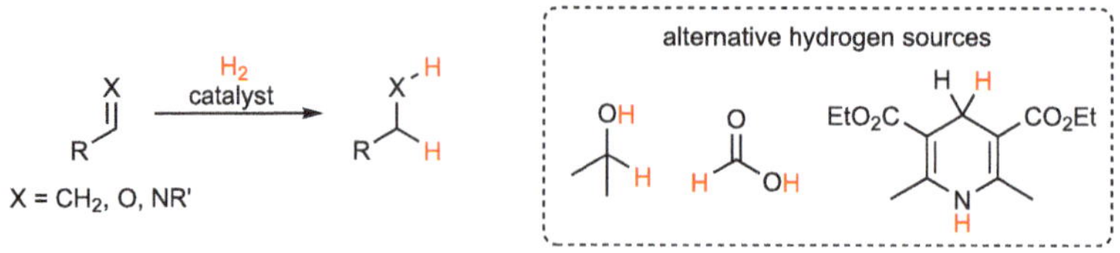

The complexes used as catalysts in homogeneous reactions are usually comprised of late transition metals and phosphine ligands. Particularly Ru-, Rh-, and Ir-based catalysts have emerged as the most widely employed systems [427]. Further catalysts are based on Fe, Co, Pd, and Pt as active metals [428]. The activation and conversion of hydrogen by frustrated Lewis acid/base pairs is a more recent addition, which will also be briefly discussed below and in Chap. 8.

Despite a strongly exothermic reaction of hydrogen with alkenes, the activation barrier in the direct conversion is prohibitively high. Transition metals, however, possess the necessary orbitals to interact directly with and activate hydrogen. The catalytically active species is either a metal monohydride or dihydride, which can subsequently transfer a hydride to the coordinated substrate (see Fig. 4.128) [429].

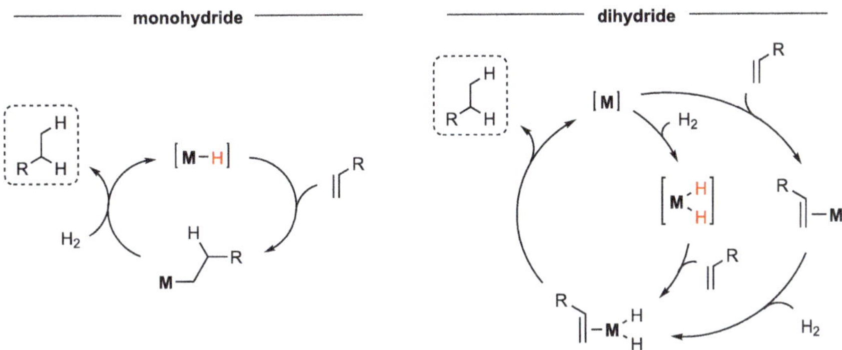

Fig. 4.128 Schematic mechanism with mono- and dihydride metal complexes in the reduction of an alkene [429, 430]

Following the introduction of the first homogeneous catalyst, which displays a reaction rate comparable to heterogeneous catalysts (**325**, *Wilkinson* catalyst), cationic metal complexes were subsequently developed by Crabtree (**326**) and Schrock/Osborn (**327, 328**) (see Fig. 4.129).

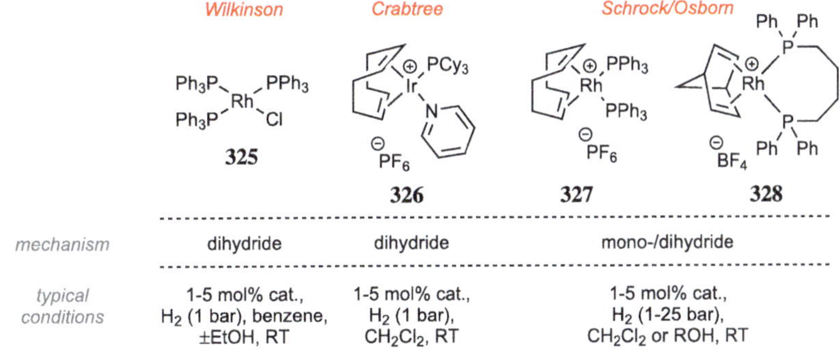

Fig. 4.129 Common homogeneous catalysts for hydrogenations of alkenes [428]

The reactivity of the cationic complexes **326, 327** is significantly higher than the original Wilkinson system. The Schrock-Osborn catalysts can for example be used for the efficient reduction of terminal double bonds in the presence of disubstituted olefins, among other reactions (see Fig. 4.130) [431].

| | Wilkinson | Crabtree | Schrock/Osborn |
	325	**326**	**327**
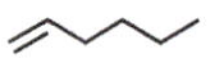	650	6400	4000
	700	4500	10
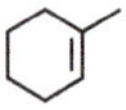	13	3800	–
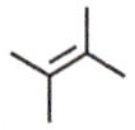	–	4000	–

Fig. 4.130 Reaction rates (in $mol_{substrate}/mol_{cat}\cdot h$) of the hydrogenation of various olefins [431]

As can be easily seen, steric effects play a more significant role than electronic effects. Nevertheless, both do influence the rate of reduction (see Fig. 4.131).

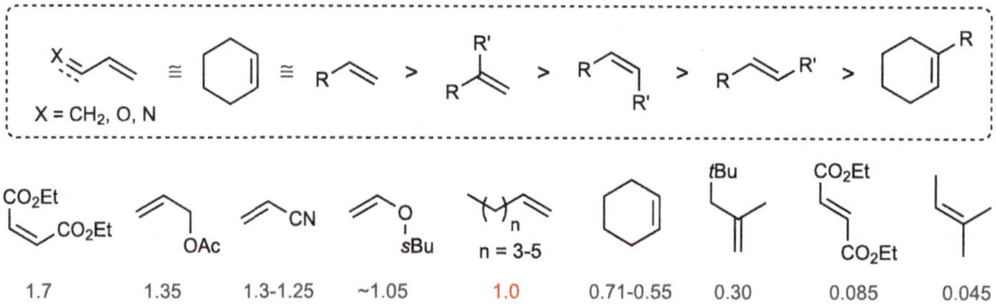

Fig. 4.131 Qualitative order of reactivity in hydrogenations with **325**. [428] The figures indicate the measured relative reaction rates [432]

A hallmark of homogeneous catalysts is their exceptional tolerance towards a variety of functional groups. Since no nucleophilic hydrides are transferred from the catalyst to the substrate, functionalities remain intact that could be affected when utilizing complex metal hydrides as reductants. The Wilkinson catalyst **325**, for example, is compatible with aldehydes, ketones, esters, carboxylic acids, nitriles, or ethers [429]. An elegant application of the Wilkinson catalyst can be found in Fürstner's synthetic approach to the natural products ipomoeassin B and E. The high selectivity of **325** towards disubstituted alkenes was skillfully used to reduce a CC double bond in the presence of two trisubstituted, α,β-unsaturated esters (see Fig. 4.132) [433].

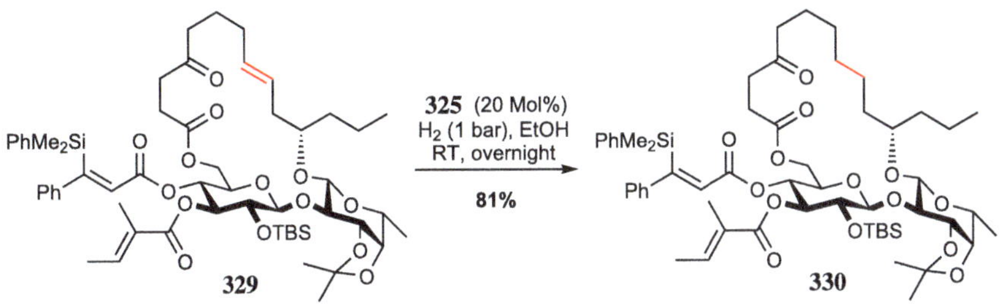

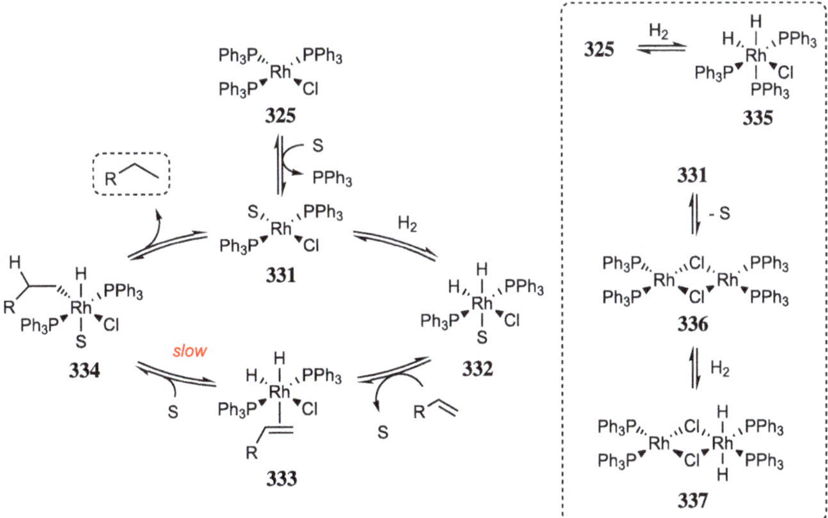

Fig. 4.132 Use of the Wilkinson catalyst in the synthesis of ipomoeassin B and E [433]

The Wilkinson system **325** is the best understood catalyst in the hydrogenation of CC double bonds by virtue of in-depth mechanistic studies of the catalytic cycle, reaction kinetics, and equilibrium reactions of the involved organometallic species. The (pre-)catalyst **325** is initially converted to the catalytically active species **331** upon loss of a phosphine. After the oxidative addition of H_2, the octahedral rhodium(III)dihydride **332** results, which coordinates an alkene (**333**). The subsequent rate-determining step to **334** and the reductive elimination finally regenerate the square-planar Rh(I) complex **331**. The hydrogenation is stereoselective and proceeds via a *syn*-hydrometalation. The regioselectivity usually displays the **primary** alkyl metal complex **334** as the preferred species, in which the metal center is directed to the sterically less hindered terminus of the olefin [434]. If ethanol is used as a cosolvent, the vacant coordination site can temporarily be occupied by a weakly bound molecule of the solvent (see Fig. 4.133) [435]. Several organorhodium species **335-337** have been observed as well, which do not lie on the catalytic cycle and serve as catalyst reservoir.

Fig. 4.133 Catalytic cycle and relevant equilibria in the reduction of olefins with the Wilkinson catalyst **325** [435–437]. S = free coordination site, solvent

Crabtree's cationic iridium complex (**326**) is catalytically much more efficient. Basic mechanistic investigations or the isolation of reactive intermediates have therefore proven unfeasible due to their instability and short lifetime. The postulated activation of the square-planar complex by oxidative addition of hydrogen furnishes the dihydride **338**. Upon addition of further equivalents of hydrogen, cyclooctane is eliminated and an iridium hydride complex is formed whose exact composition has not been fully elucidated (**339–341**) [431]. Vacant coordination sites may additionally be occupied by the solvent. The dissociation of ligands or of strongly coordinating solvents is significantly slower in Ir complexes compared to the respective Rh catalysts, which is why a weakly coordinating solvent is required in the hydrogenation [438]. The structurally related precatalyst **342**, which contains a chiral phosphinooxazoline ligand (*PHOX*), has been investigated much more thoroughly through kinetic and theoretical means than Crabtree catalysts. The PHOX ligand developed by the Pfaltz group is routinely used in the enantioselective hydrogenation of CC double bonds. Based on the mechanistic studies, a M(III)/M(V) catalytic cycle was postulated to be operative. However, this constitutes only one possibility and an analogous M(I)/M(III) mechanism, as present in cationic rhodium complexes, has also been proposed [439]. The intermediate **343**, which is coordinatively saturated by the solvent, is successively converted with alkene and hydrogen via **344** to **345**. Following oxidative addition with concomitant hydride transfer to the ligated alkene, the Ir(V) complex **346** is formed. The hydrogenated product can then be afforded by transferring a hydrogen atom to the alkyl ligand (**347**). After the hydride transfer, the alkane can dissociate from the metal center upon coordination of solvent (see Fig. 4.134) [438, 440, 441].

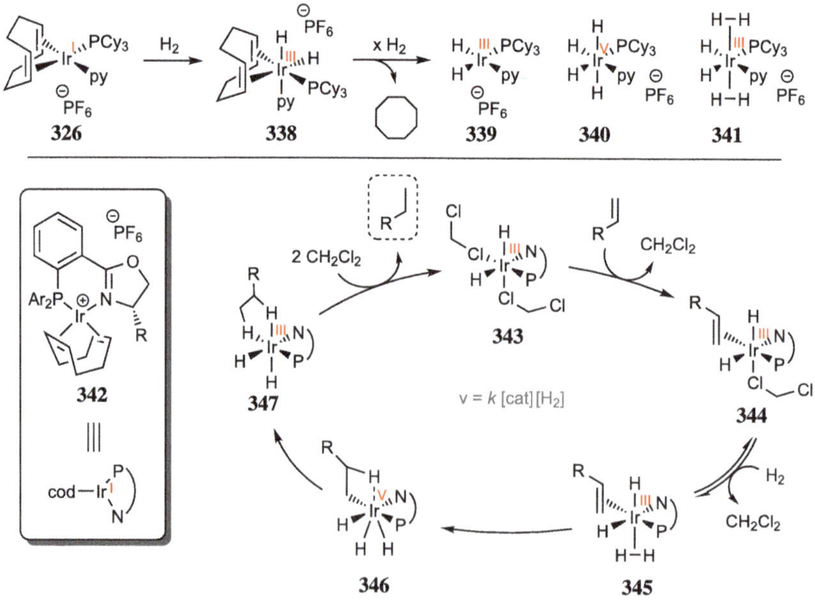

Fig. 4.134 Proposed activation of **326** and postulated catalytic cycle in the hydrogenation with **342** [431, 438, 440, 441]

A.B. Smith and co-workers used the Crabtree catalyst for a diastereoselective reduction in the final stages of their synthesis of the *Daphniphyllum* alkaloid (–)-calyciphylline N. Initial studies on a model substrate indicated that the 1,4-reduction of the conjugated diene could be carried out by Pd(PPh$_3$)$_4$/ZnCl$_2$ with Ph$_2$SiH$_2$ as the reductant. However, when the reaction was conducted with the structurally more complex intermediate **348**, the stoichiometric use of the BArF analog of the Crabtree catalyst (**350**) proved to be the method of choice to obtain the desired product in high yield and acceptable diastereoselectivity (see Fig. 4.135) [442]. **349** was subsequently converted to the natural product by deprotection of the phthalimide and an acid-catalyzed imine formation.

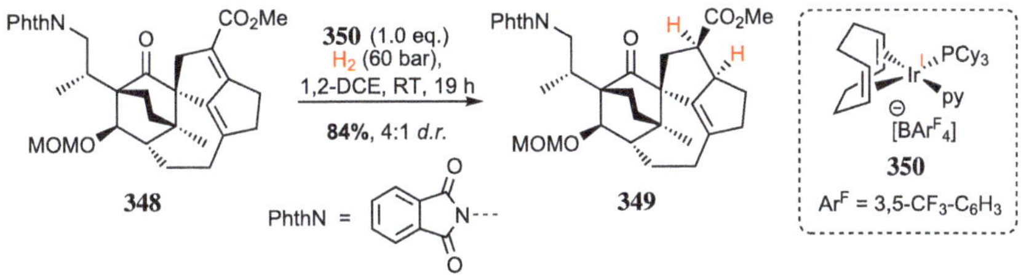

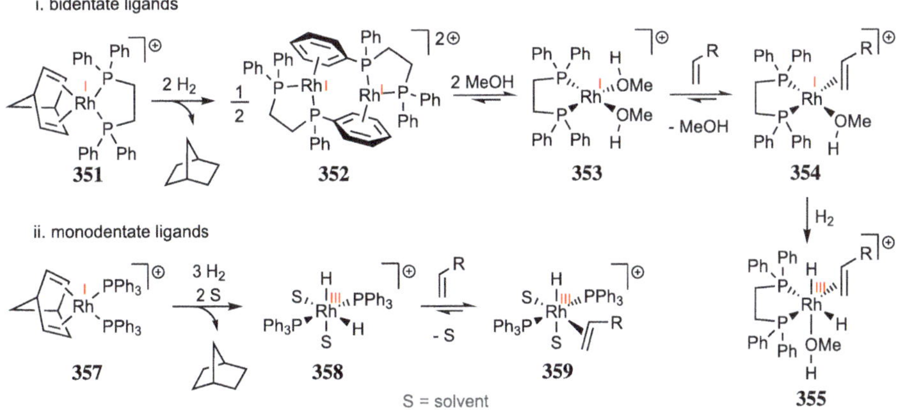

Fig. 4.135 Application of the Crabtree catalyst **350** in the synthesis of calyciphylline N [442]

In the case of the cationic rhodium catalysts **327, 328**, the activation of the catalysts occurs analogously to the Crabtree system. Chelate ligands (**351**) impose the formation of a Rh-bisphosphine complex (**352**, in the presence of MeOH **353**) as the initial intermediate, while monodentate ligands such as PPh$_3$ (**357**) directly lead to a dihydride (**358**). This can be understood by the occurrence of *trans*-positioned H-Rh-phosphine bonds, which would result with chelate ligands following an oxidative addition starting from **353**. In order to minimize this destabilizing *trans*-effect, the addition of hydrogen only takes place after the coordination of the substrate (**354**), which possibly hydrometalates the coordinated alkene in **355** rapidly to furnish a more stable complex (see Fig. 4.136) [443, 444].

Fig. 4.136 Postulated activation of Schrock-Osborn catalysts [443, 444]

Studies by Schrock and Osborn indicated the existence of a mono- and a dihydride-based mechanism, depending on the reaction conditions. The dihydride complex (**358**) is in equilibrium with the corresponding monohydride species (**360**). In the presence of bases (additive, substrate, or free basic ligand), a proton is abstracted from the metal center and a monohydride cycle with neutral rhodium complexes is followed. The reduction in the neutral catalytic cycle is significantly faster than in the case of the cationic complexes (**351–359**), but the tendency for isomerization of double bonds with dienes or alkynes to internal alkenes (via reinsertion/elimination in **360/363**) limits this approach [445]. Even in the absence of base, **360** can be present in small amounts and thus lead to an isomerization of double bonds. The addition of Brønsted acids (e.g., $HClO_4$, HBF_4) pushes the equilibrium towards cationic dihydride complexes and can suppress isomerization (see Fig. 4.137) [445–447]. Many modern asymmetric hydrogenation catalysts use cationic rhodium in combination with chiral ligands. Strongly binding substrates (dienes, α, β-unsaturated carbonyl compounds, etc.) first coordinate to the metal, followed by subsequent uptake of hydrogen (cf. Fig. 4.128) [444].

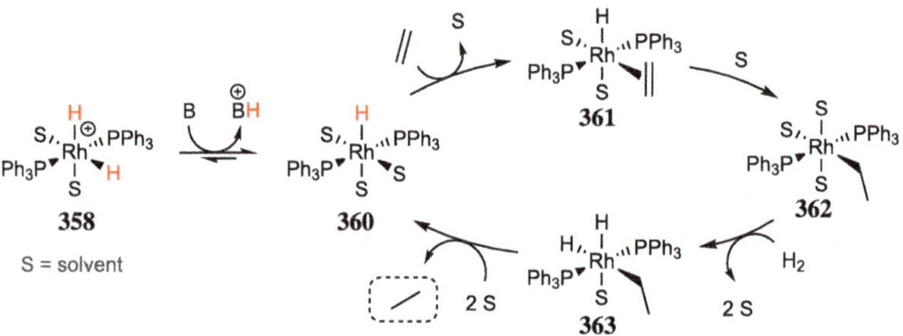

Fig. 4.137 Monohydride mechanism in Schrock-Osborn catalysts [445, 446]

The cationic Ir and Rh catalysts are strongly electrophilic and bind alkenes or other ligands more strongly than the corresponding neutral complexes. This can be exploited in the presence of directing groups to realize diastereoselective reactions [434, 448]. The rigid structure can guarantee high diastereoselectivity especially in cyclic systems, while in acyclic substrates, strong stereoelectronic effects (1,3-allylic strain) or chiral ligands are used in a "*matched*" situation to control the course of the reaction. The directing group coordinates the metal center in the latter, thus effecting a facial differentiation during the transformation. A corresponding intermediate **364** was isolated, in which both the directing hydroxy group and the alkene coordinate the iridium atom in a chelating fashion (see Fig. 4.138) [449]. For cationic iridium complexes, the bond strength of directing functionalities follows the order $OMe > CO_2R > C{=}O > CONH_2$ [449]. Of the two diastereotopic hydride ligands in the intermediate dihydride complex, the one with a H-M-C=C dihedral angle of about 0° should be preferably transferred (**365**), as it displays a stronger interaction with the π^*_{CC} orbital of the substrate [434].

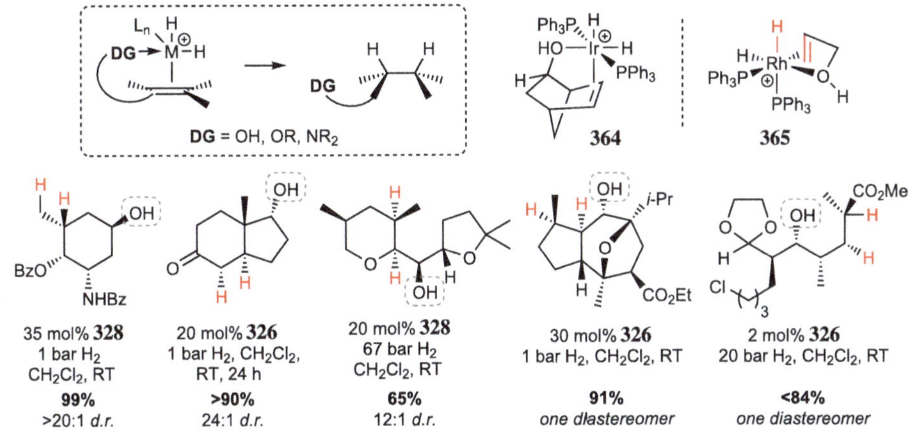

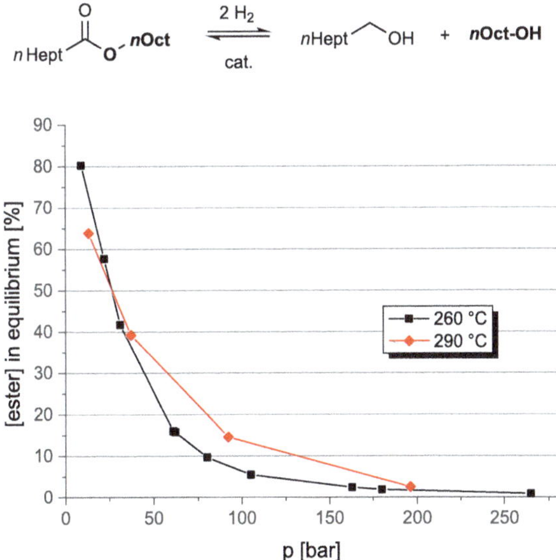

Fig. 4.138 Principle of directed hydrogenation and synthesis examples [434, 449–454]

The reduction of aldehydes, ketones, and imines constitutes an equilibrium reaction, which strongly favors the products. In contrast, the complete conversion of esters or amides is very challenging for thermodynamic reasons, as significant amounts of the substrate can still be present in equilibrium (see Fig. 4.139) [455]. Since the reaction is exothermic, a decrease in the temperature shifts the position of the equilibrium further towards the products.

Fig. 4.139 Position of the reaction equilibrium in the hydrogenation of octanoic acid octyl ester [455]

The conversion of aldehydes in the presence of the Wilkinson catalyst results in the decarbonylation of the substrate and inhibition of the catalyst by CO [456]. The desired hydrogenation of carbonyl derivatives can in turn be successfully achieved under catalytic conditions by resorting to bifunctional ruthenium complexes. The modification of the ligand periphery by the addition of N,N-chelate ligands allows a radical change in selectivity (see Fig. 4.140) [457].

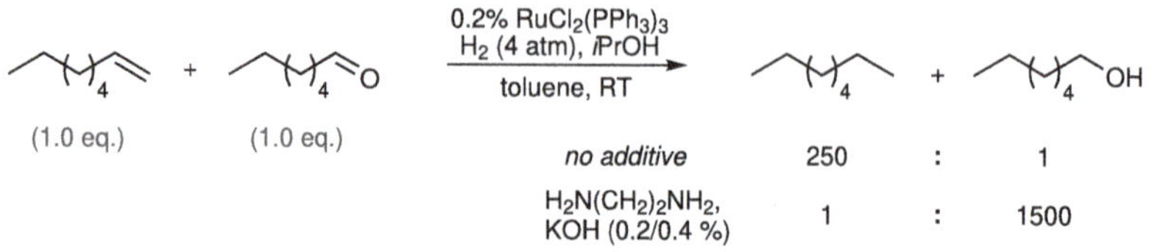

Fig. 4.140 Influence of additives on the relative hydrogenation rate of CC- versus CO-bonds [457]

The majority of catalysts for CO hydrogenations are based on Ru chelate complexes following the same principles. After the development of the first efficient catalyst system by Elsevier employing trisphosphine ligands ($MeC(CH_2PPh_2)_3$, TriPhos), a multitude of further methods with bi- and tridentate P,N-ligands was reported (see Fig. 4.141) [458].

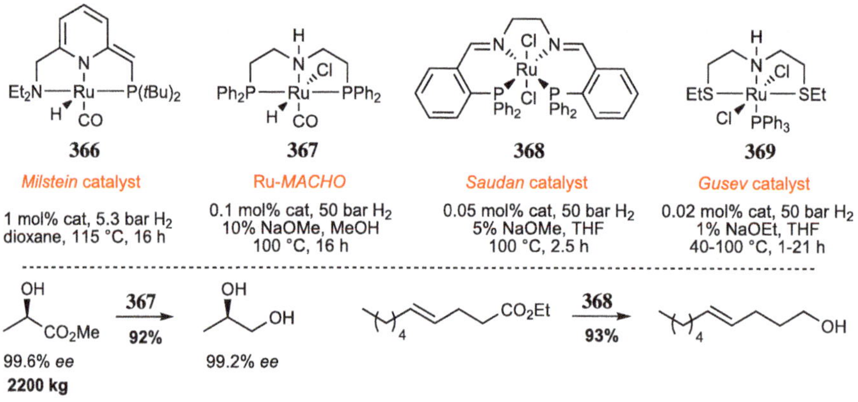

Fig. 4.141 Modern hydrogenation catalysts, typical conditions for ester hydrogenations and selected examples [459–461]

In particular, the methods relying on **367** (Ru-MACHO) and **368** developed by the researchers from Takasago and the Saudan group have significantly expanded the substrate scope while maintaining mild conditions. **367** was specifically developed for industrial applications and even used on a ton scale for the hydrogenation of enantiomerically pure α-chiral esters without suffering significant erosion of enantiomeric excess [460]. The Saudan catalyst, in turn, tolerates CC double bonds in the substrate, which are preserved under the reaction conditions [461]. The Ru-SNS catalysts published by Gusev represent the current state of the art in terms of catalyst loading combined with minimal structural complexity of the ligand [462]. Mechanistic investigations of the TriPhos and Milstein system were able to shed light on the individual steps of the catalytic cycle [463–465]. Additional developments now include Os-, Fe- or Mn-based catalysts that can successfully hydrogenate carboxylic acid derivatives [458, 466, 467].

The homogeneous *trans*-hydrogenation of alkynes to the corresponding *E*-alkenes is a relatively new methodology. Based on earlier work by Trost and Fürstner, [468, 469] the Fürstner group utilized a RuCp*-catalyst (**370**), which *trans*-hydrogenates alkynes with high selectivity while possessing good tolerance towards common functionalities (esters, nitriles, amides, alkenes, nitro groups, thioethers, silyl ethers, carboxylic acids, alcohols) [470, 471]. Small amounts of the alkane are also formed (5–15%). The mechanism presumably proceeds via ruthenacyclopropene **371**, followed by **372** [472, 473]. Only 1,3-dienes or 1,3-enynes are unsuitable substrates, as these functionalities act as catalyst poisons due to the strong π-metal interaction. Further methods for the *trans*-hydrogenation of alkynes have been reported by other research groups. However, these studies are usually not as broadly applicable to various substrates as the Fürstner protocol (see Fig. 4.142) [471, 474].

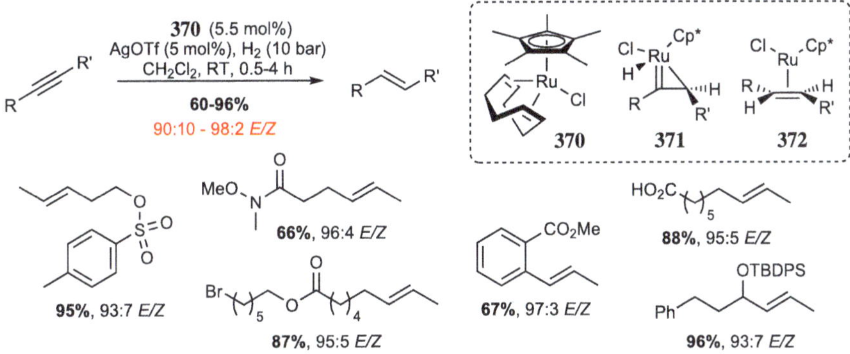

Fig. 4.142 *trans*-Hydrogenation of alkynes by Fürstner *et al.* [470, 472, 473]. The yield represents the cumulative sum of alkene and alkane

In addition to using hydrogen as reductant, organic molecules or salts can also be employed as indirect sources of hydrogen [475–477]. This type of reaction is referred to as **transfer hydrogenation** and is particularly popular in an academic context, as handling of hydrogen is usually more complicated in small scale than working with the respective hydrogen surrogates. Two types of mechanism[III] can be distinguished in transfer hydrogenations: (i) a direct transfer, in which a hydride is transferred from the reductant to the substrate, and (ii) a two-step mechanism, in which the hydride is transferred to the substrate via a metal hydride intermediate. In the direct mechanism, the metal coordinates both reactants, which is why strong Lewis acids are preferred under these conditions (M = Al, Ln). In the two-step mechanism, a

---

[III] Salts of formic acid such as ammonium formate can also thermally decompose to release H_2. Whether this is the operative mechanism or just an unwanted background reaction should be elucidated on a case-by-case basis.

metal hydride is formed as an intermediate. Metals that show a high affinity for hydride ligands thus prefer this second path (M = Ru, Ir, Rh; see Fig. 4.143) [476]. Base metals (Mn, Fe, Co, Ni) have also been successfully used as catalysts in transfer hydrogenations, albeit at the cost of lower activity and typically requiring specialized ligand systems [478].

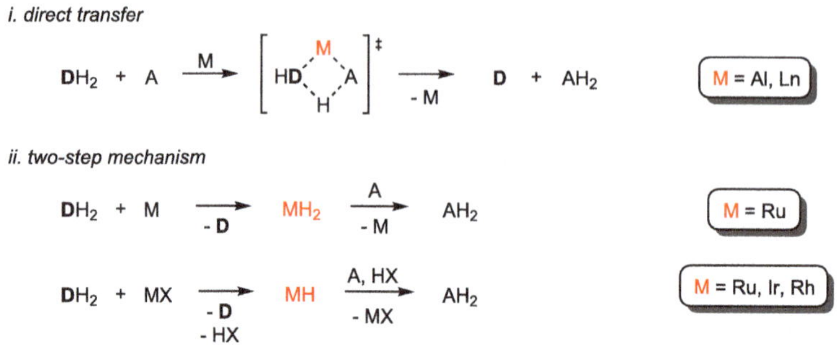

Fig. 4.143 Mechanisms of transfer hydrogenation. D = H$_2$ donor, A = substrate/H$_2$ acceptor, M = metal [476, 479, 480]

The *Meerwein–Ponndorf–Verley* reduction (MPV), the reverse process of the Oppenauer oxidation, can be regarded as the prototype of a transfer hydrogenation [132]. In the reaction, an alcohol (usually isopropanol) is oxidized to the corresponding ketone as the oxidant, while the substrate is reduced to the desired alcohol. Since all steps of the conversion are reversible, the product distribution necessarily reflects the thermodynamic equilibrium. Several techniques are used to achieve complete conversion:

- The reducing agent usually serves as solvent, which shifts the equilibrium towards the products due to its large excess.
- When formic acid is used as a donor, CO_2 is obtained as by-product, which evaporates from the solution and is no longer available for the reverse reaction.
- If the desired product is more volatile than the substrate and reagents, it can be distilled off continuously, thereby shifting the reaction equilibrium towards the products.

The MPV reduction, unlike the Oppenauer oxidation, is still occasionally encountered today, but like its oxidative counterpart, it exhibits some significant disadvantages: The relatively large and often stoichiometric amount of reagents required, side reactions that are difficult to suppress in strongly basic conditions (aldol reactions), and a pronounced sensitivity to moisture often render it prohibitive for a specific application. The MPV reduction mechanistically corresponds to a direct hydride transfer [481]. Since the Oppenauer oxidation and the Meerwein–Ponndorf–Verley reduction are mirror images from a mechanistic point of view, their transition states must be identical. Only the redox potential of the substrates and their concentration determines the position of the thermodynamic reaction equilibrium (see Fig. 4.144, cf. Fig. 4.44).

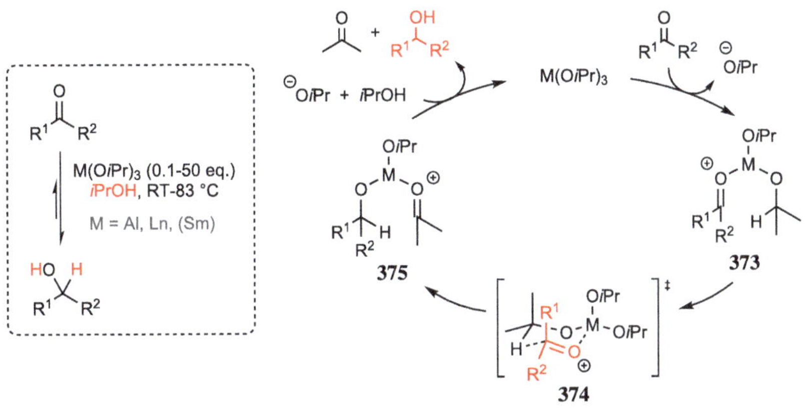

Fig. 4.144 Mechanism of the Meerwein–Ponndorf–Verley reduction [481]

Hale and co-workers utilized a Meerwein–Ponndorf–Verley reduction in their formal total synthesis of the antineoplastic macrolide bryostatin 7. Out of the multitude of surveyed conditions, the MPV reduction proved to be the preferred method, both in terms of diastereoselectivity and yield of the desired alcohol **378**. The reactive Lewis acid Al(O*i*Pr)$_3$ was generated *in situ* from AlMe$_3$. The unwanted diastereomer **379** could be converted back to **377** by reoxidation with TPAP (see Fig. 4.145) [482].

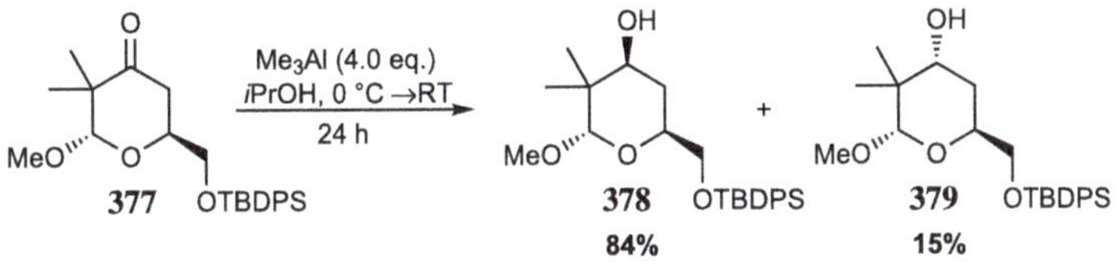

Fig. 4.145 Meerwein–Ponndorf–Verley reduction in Hale's formal total synthesis of bryostatin 7 [482]

Although traditionally a large excess of Al(O*i*Pr)$_3$ is added, substoichiometric amounts are sufficient for an efficient conversion when resorting to modern catalysts. Particularly lanthanide salts and Al complexes with bidentate ligands can be utilized catalytically, as they do not tend to aggregate and thus deactivate to such a large extent as Al(O*i*Pr)$_3$ [476].

In addition to a direct transfer hydrogenation with Ln and Al salts, the use of late transition metal complexes has gained in popularity since the beginning of the 2000s. As the catalytic cycle relies on metal hydrides as intermediates, metals that form stable hydride complexes are particularly suitable for the reaction. Unsurprisingly, metals that have also been employed in classic hydrogenation catalysts enable an efficient transfer hydrogenation [479]. However, most classic hydrogenation catalysts only display moderate to low activity in transfer hydrogenations and vice versa [483]. Based on studies with α-deuterated alcohols, the most widely utilized catalysts derived from Rh, Ir, and Ru complexes proceed via monohydrides as intermediates (see Fig. 4.146, cf. Fig. 4.143) [480]. Other late transition metals such as Fe, Ni, Pd, and Os have also been successfully used in transfer hydrogenations [475, 484].

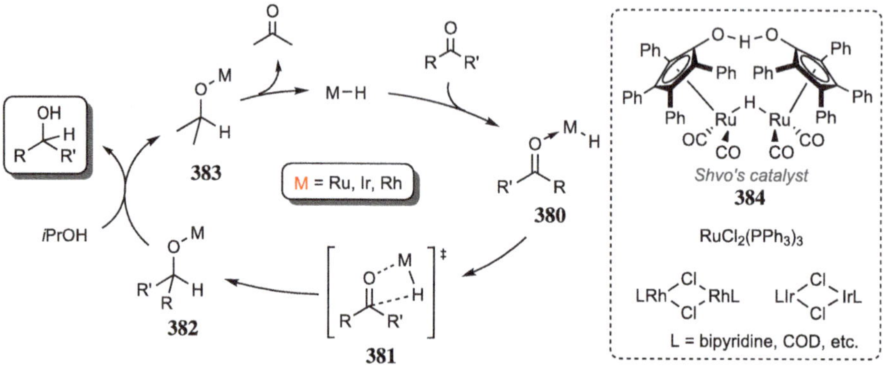

Fig. 4.146 Mechanism of transfer hydrogenation using monohydrides and transition metal complexes that operate via this mechanism [479, 480]

Typical donors of hydrogen are alcohols (*i*PrOH, MeOH, EtOH), formic acid and its salts, Hantzsch esters, cyclohexene, cyclohexadiene, hydrazine, amine boranes, etc. *i*PrOH and formates are usually preferred due to their low cost and easy handling, for example as solvents [475, 484]. Transition metals usually require a base, as (i.) the removal of HX from the metal complex to form the monometallic hydride is strongly endothermic and (ii.) the pH value, which decreases successively during the course of the reaction, is not compatible with many substrates (cf. Fig. 4.143). The HX abstraction can be significantly accelerated by neutralizing the acid by-product with a strong base [485]. An elegant variant is the direct inclusion of a basic functionality in the catalyst itself. The resulting complexes constitute the basis of the Noyori-Ikariya catalysts, which in their variations represent the most successful class of asymmetric transfer hydrogenation catalysts (*vide infra*). The postulated mechanism for the reaction of ketones (**386**) involves an initial transfer of a proton and hydride from isopropanol to **385**. According to theoretical studies of the reaction in solution, a stepwise mechanism is favored over a concerted transfer of H_2 to the substrate (via **390**): After the elimination of acetone, **386** can transfer a hydride to the substrate via transition state **387**. The short-lived intermediate **388** is finally protonated via **389** and the desired product is obtained [486, 487]. With this type of catalysts, the reaction is carried out in an intermolecular manner, without coordination of alcohol or carbonyl functionality (*outer-sphere* mechanism, see Fig. 4.147) [479, 483, 488]. For the transfer of H_2 from *i*PrOH (**385**→**386**), the steps **387**-**389** are followed as well, but in reverse order. With formic acid as H_2 donor, this sequence similarly applies, via **391** [489].

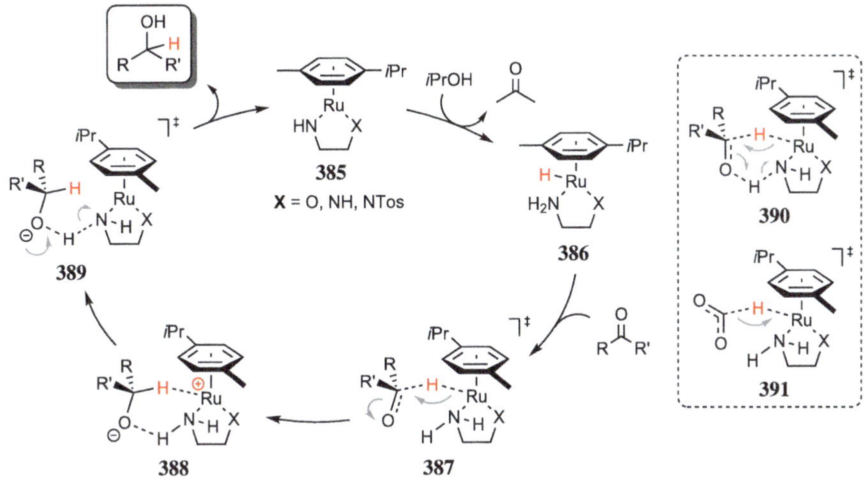

Fig. 4.147 Mechanism of transfer hydrogenation with bifunctional Noyori-Ikarya-type catalysts **385** [486]

Apart from the synthesis of alcohols, transition metal-catalyzed transfer hydrogenations are primarily utilized for the racemization step in dynamic-kinetic resolutions [490]. A prevalent complex for this purpose is Shvo's catalyst **384** (cf. Fig. 11.36, Sect. 11.1) [491]. In their synthesis of the antidepressant norsertraline (**396**), Bäckvall and co-workers used a derivative of Shvo's catalyst (**394**) for the chemoenzymatic resolution of amine **392** [492]. While the enzyme selectively converts the *R*-enantiomer to the respective acetate with *i*PrOAc, **394** racemizes the remaining enantiomer, thus achieving yields beyond 50%. In the further course of the synthesis, **395** was hydrogenated in the presence of the Crabtree catalyst **326**, with the NHAc group serving as a directing element (see Fig. 4.148).

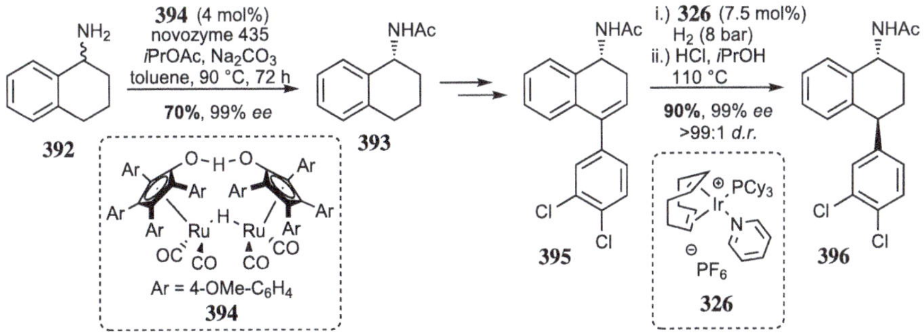

Fig. 4.148 Synthesis of norsertraline (**396**) according to Bäckvall [492]

The Evans–Tishchenko reaction constitutes an intramolecular variant of a transfer hydrogenation for the production of partially protected 1,3-*anti*-diols [493].

Research into **frustrated Lewis pairs** poses a rapidly developing field of homogeneous hydrogenations in recent years ($\rightarrow$ Sect. 8.1) [494–497]. Due to either a sterically demanding environment or a rigid catalyst framework, the Lewis pair cannot form a classic Lewis acid/base adduct, allowing the available Lewis acidity/basicity to separately act as an electron acceptor or donor, respectively. This combination can reversibly cleave the hydrogen bond in a heterolytic fashion (H^+/H^-) in the absence of other metal catalysts (**397**$\rightarrow$**398**) and thus activate it for hydrogenation. An example of these zwitterionic intermediates is **399**. Electron-poor boranes and occasionally phosphonium ions are preferred as Lewis acids. The Lewis base can be represented by phosphines, amines, *N*-heterocyclic carbenes, or even solvents such as ether (see Fig. 4.149) [494–496].

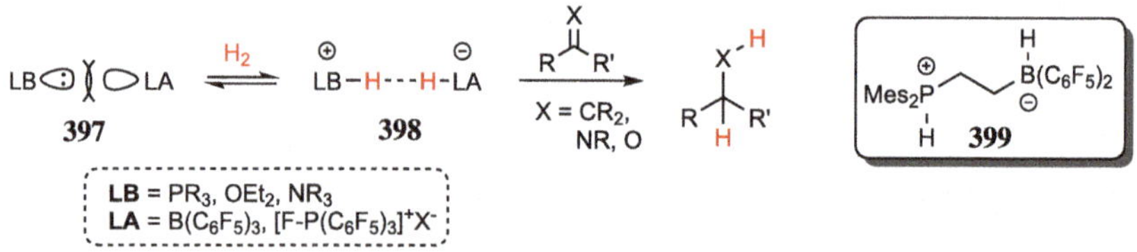

Fig. 4.149 Principle of hydrogenation with frustrated Lewis pairs [494–496]

Substrates include geminal disubstituted alkenes, alkynes, enamines, imines, silyl ethers, aromatic heterocycles, and ketones [494]. In addition to classic hydrogenations, asymmetric reductions can also be carried out with chiral frustrated Lewis pairs [498].

4.2.4.2 Heterogeneous Variants

Hydrogenations are one of the areas of organic synthesis where a significant proportion of applications is mediated by heterogeneous catalysts, even in an academic context. Examples of such reactions include partial hydrogenations of alkynes, benzylic C-O and C-N hydrogenolyses (debenzylations), hydrogenations of arenes to the corresponding cycloalkanes, as well as reductions of nitroarenes. The ubiquitous Pd/C is the catalyst of choice in the academic field, followed by PtO_2 or Raney-Ni (see Fig. 4.150) [499, 500].

Pd/C	Pd/Pb(OAc)$_2$/CaCO$_3$	Pd/BaSO$_4$	PtO$_2$ · H$_2$O	Pd(OH)$_2$
	Lindlar catalyst	*Rosenmund catalyst*	*Adam's catalyst*	*Pearlman's catalyst*

Pt/C	Ni/Al	Rh/C	Rh/Al$_2$O$_3$	(CuO)$_x$·(Cr$_2$O$_3$)$_y$
	Raney nickel			*copper chromite*

Fig. 4.150 Frequently used heterogeneous catalysts

Heterogeneous catalysts can be supported or unsupported. In the case of supported catalysts, the active metal (Pd, Pt, Rh, Cu, Ni, etc.) is situated on an inert carrier material, meaning that the transition metal constitutes only a minor part of the catalyst. A supported metal can be used more efficiently due to the significantly higher surface area of the resultant smaller nanoparticles of the active metal. Additionally, this is usually accompanied by a higher tolerance towards catalyst poisons. With unsupported catalysts, suspensions or slurries of powder catalysts are typically utilized. In addition, metal nanoparticles can also be employed as a colloidal mixture (e.g., Pd black) for more efficient use of the active surface area. The type of active metal and the chosen support contribute significantly to the activity and chemoselectivity of a reaction. Common functional groups are listed in Table 4.13 in descending order of reactivity in hydrogenations. Upon increase of the temperature and especially the pressure, even the most unreactive substrates can be hydrogenated. The challenge of a given hydrogenation therefore usually lies in the selective transformation of a functionality in the presence of other functional groups prone to reduction.

Table 4.13 Typical conditions for the heterogeneous hydrogenation of functional groups [226, 458, 501–504]

reaction	metal	p [bar]	T [°C]	solvent
 $R\!-\!\equiv\!-\!R' \longrightarrow$ (cis-alkene)	Pd (Lindlar)[a]	1-3[b]	5-50	slightly polar
$Ar\!-\!NO_2 \longrightarrow Ar\!-\!NH_2$	Pt, Pd, Ni	1-5	5-50	varies
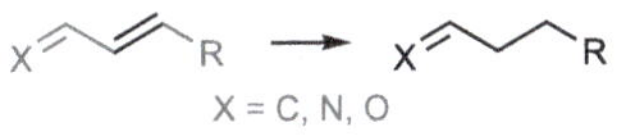 $X\!=\!\!/\!\!\!\!\searrow\!R \longrightarrow X\!=\!\!\searrow\!R$; X = C, N, O	Pd[a]	1-5[b]	5-50	slightly polar
 $R\!\searrow\!\!R' \longrightarrow R\!\searrow\!R'$	Pd, Pt	1-50	5-100	slightly polar
 $R\!-\!CHO \longrightarrow R\!\frown\!OH$	Pt, Pd, Ni Cu	1-10 20-100	20-100 50-150	polar polar
 $R\!-\!CN \longrightarrow R\!\frown\!NH_2$	Ni, Co Pt, Pd	20-100 1-10	20-100 20-100	polar + NH_3 polar + HCl
 $R\!-\!C(\!=\!O)\!-\!R' \longrightarrow R\!-\!CH(OH)\!-\!R'$	Pt, Ni, Ru Cu	5-50 50-100	20-150 100-200	polar polar
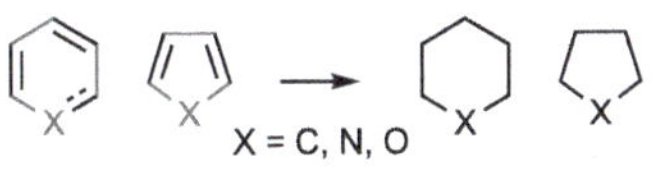 (arene / heteroarene) $\longrightarrow$ (saturated ring); X = C, N, O	Rh Raney Ni Ru	2-20 80-150 80-150	20-80 80-150 100-200	varies varies varies
 $R\!-\!CO_2R' \longrightarrow R\!\frown\!OH$	Cu, Ru	100-300	100-200	varies
 $R\!-\!CO_2H \longrightarrow R\!\frown\!OH$	Co, Re,[c] (Ru)[c]	>150	>150	polar
 $R\!-\!CO_2NR'_2 \longrightarrow R\!\frown\!NR'_2$	Ru, Re[c]	>150	>150	polar

[a] Use of deactivated catalysts; [b] control of H_2-uptake if possible; [c] use of promoters

decreasing reactivity (↓)

Apart from their lower price due to the lower content of (noble) metal, the main advantages of supported catalysts are the higher resistance to deactivation as well as the gain of several additional dimensions for catalyst optimization: The choice of the support, the individual metal loading, the deposition method (impregnation, precipitation, adsorption), post-treatment (often thermal calcination), additives, binders etc. [505]. Due to this flexibility a catalyst can be specifically adapted for a given process [427, 506]. The advantage of this multitude of options is also its greatest drawback: The necessity of an extensive optimization study, in which often purely empirical structure-activity relationships are observed. This is due to the high complexity of heterogeneously catalyzed processes, in which interactions on the catalyst surface are usually detected only indirectly. For example, it took until the 1970s starting from the beginning of the 20th century in its initial development, until Ertl *et al.* elucidated the mechanism of the industrial ammonia synthesis (Haber–Bosch process) [507].

The support must be adapted to the substrates and reaction conditions: For acid-labile substrates, Lewis-acidic carriers such as Al_2O_3 or SiO_2 are unsuitable. In the case of leaching of the active metal or sintering, metal oxide supports are often preferred to achieve stronger metal-support interactions. Common supports comprise carbon (activated carbon, graphite), Al_2O_3, SiO_2, ZrO_2, iron oxides, CeO_2, inorganic carbonates and sulfates, Kieselguhr, zeolites, aluminas, and a variety of other solids [499].

While in the academic environment elemental metals and their oxides or sulfides are often used as unsupported catalysts, which in many cases are not recycled after a reaction, industrial syntheses often rely on supported metal catalysts — with Raney nickel as a notable exception (see Table 4.14). Apart from platinum group metals, the active metals usually need to be pre-reduced and thus activated for a hydrogenation. Depending on the redox potential, this can also be achieved directly *in situ* under the reaction conditions.

Table 4.14 Active metals and their characteristics in supported catalysts [501]

metal	characteristics
Pd	metal of choice for hydrogenolysis (C-halogen, benzylic/allylic CO and CN bonds) and CC double bonds; strong tendency for double bond isomerization
Pt	very active for reduction of CC, CN and CO double bonds as well as $Ar-NO_2$. No double bond isomerizations. C-halogen and allylic/benzylic CO bonds usually remain intact
(Raney) Ni	reduction of arenes, $R-CO_2R'$, R-CN, sulfur derivatives, ketones
Ru	classic catalyst for the reduction of arenes, as well as ketones, esters
Rh	hydrogenation of arenes under mild conditions
Cu	reduction of esters, lactones, anhydrides
Co	catalyst for the reduction of R-CN, $R-CO_2H$, very acid-resistant
Re	reduction of carboxylic acids, amides, very acid-resistant

The activity and selectivity of a catalyst in a reaction is usually influenced by the following variables in descending order [501]:

active metal > additives, doping > reaction medium, catalyst support > catalyst structure (morphology, particle size, porosity) > reaction conditions

The active metal undoubtedly imparts the greatest influence. However, the selective addition of additives or doping of the catalyst can in many cases significantly change the observed chemoselectivity. The solvent has been shown to impact not only the chemoselectivity but also the regioselectivity of a given transformation (see Fig. 4.151).

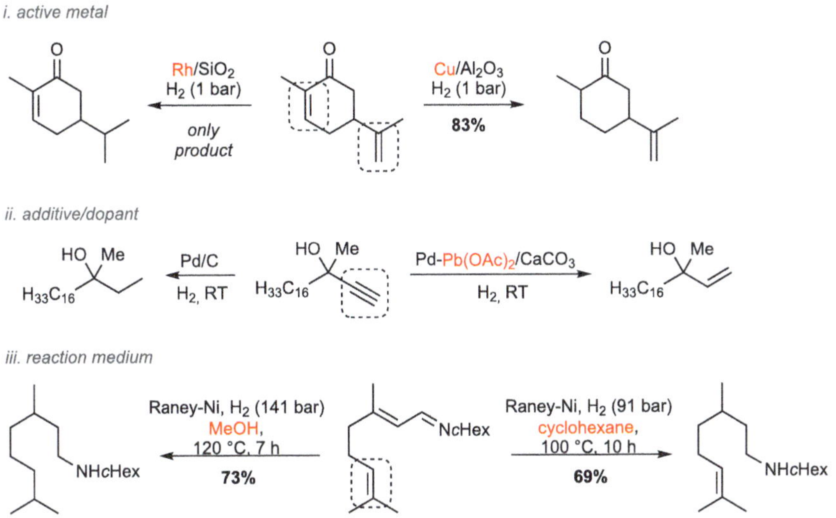

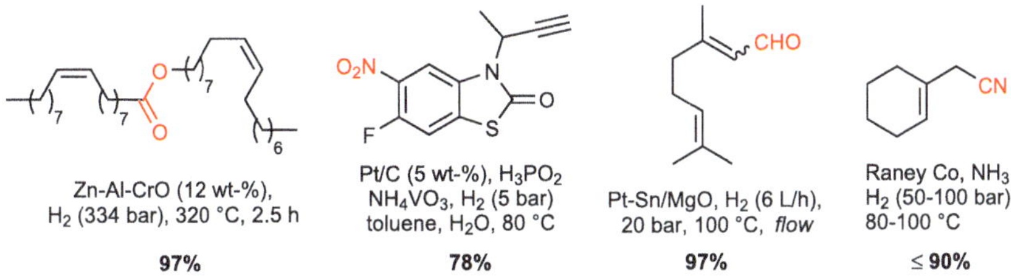

Fig. 4.151 Influence of various factors on the chemoselectivity of hydrogenations [508–510]

While acidic and basic solvents usually have a predictable effect, for example the selective activation of basic functionalities towards hydrogenation or hydrogenolysis in an acidic medium, the impact of a neutral medium is more challenging to predict. For polar substrates, polar media should be used if possible, while nonpolar substrates often benefit when a similarly nonpolar medium is chosen. The substrate does not need to have high solubility in the reaction medium. However, good solubility of the product is essential so that no solid can cover the catalyst surface and permanently deactivate it. Suitable solvents are non-reducible or show a great reluctance to do so, such as water, acetic acid, methanol/ethanol, alkanes, sterically demanding esters, acetone. However, under high H_2 pressure, esters and ethers are also susceptible to hydrogenation and hydrogenolysis.

Many synthetic problems can be addressed with the wide range of available standard catalysts. However, if a substrate has two functionalities with similar reactivity (e.g., alkyne and nitro groups) or if one functionality is to be reduced in the presence of a more reactive one, these challenging reactions often demand tailor-made catalysts (see Fig. 4.152) [427].

Fig. 4.152 Examples of challenging chemoselective reductions [427, 511–513]. Hydrogenation of the highlighted functionality to the respective alcohol or primary amine

The mechanism of heterogeneous hydrogenation has been thoroughly investigated and is largely understood, even though some details have not been finally elucidated due to the presence of up to three phases (solid catalyst, dissolved substrate/product, gaseous hydrogen) [514]. The *Horiuti–Polanyi* mechanism[515] consists of several steps:

- adsorption and activation of H_2
- adsorption of the substrate
- transfer of an H-atom, formation of a metal alkyl species
- transfer of the second H-atom and desorption of the product.

The dissociative adsorption of H_2 activates hydrogen (**401**). Following the adsorption of the substrate, the alkene forms a π-bond with the metal surface and is thus likewise activated (**402**). After the transfer of the first hydrogen atom, a metal alkyl intermediate **403** is generated, which affords the hydrogenated product upon the second H-transfer. All steps of the catalytic cycle are usually reversible. However, at high hydrogen pressure and low temperature, it can be assumed that the transfer of the last H-atom and the desorption of the product can be considered irreversible for practical purposes (see Fig. 4.153) [516].

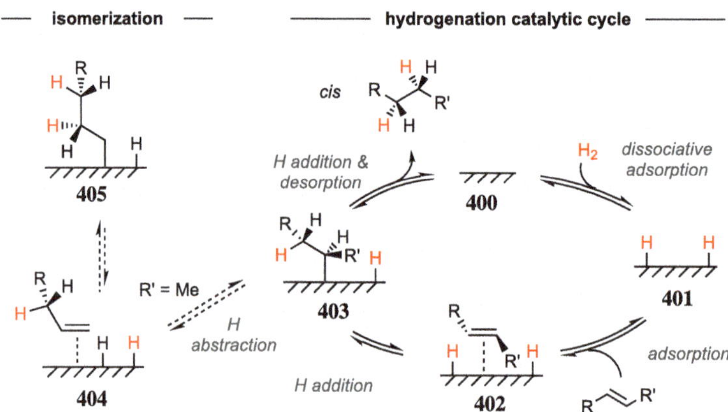

Fig. 4.153 Horiuti–Polanyi mechanism of alkene hydrogenation [514]

Almost all heterogeneous catalysts tend to isomerize double bonds. Due to the reversibility of the steps in the Horiuti–Polanyi mechanism, the alkyl metal species **403** can transfer back an H-atom from a different position in the molecule to the catalyst and thus yields an isomerized alkene (**404**). The extent of isomerization by different metal catalysts decreases in the order [499]

$$Pd > Ni > Rh > Ru > Os \approx Ir \approx Pt$$

Heterogeneous reductions of double bonds mechanistically proceed via a *cis*-hydrogenation. The alkene in **404** can also take up another H-atom via **405**, which affords products with *trans*-aligned H-atoms. The isomerization over Pd and Ni catalysts largely proceeds via π-allyl intermediates and does not follow the classic Horiuti–Polanyi mechanism [226, 516].

In an irreversible reaction and without mass transport limitation, the reaction rate of the hydrogenation

$$A + H_2 \xrightarrow{cat.} P$$

depends on the surface coverage with H_2 and the substrate[IV]:

$$v = k\,[cat]\,\Theta_{H_2}\Theta_A \tag{4.1}$$

$$\approx k\,[cat]\,[H_2]^n\,\Theta_A \tag{4.2}$$

$$\approx k\,[cat]\,\frac{[H_2]^n\,K_A\,[A]}{1 + K_A[A] + K_{H_2}[H_2] + K_P[P] + K_{BP}[BP] + K_S[S] + K_M[M]} \tag{4.3}$$

Θ	surface coverage of the catalyst with substrate A and hydrogen
K	adsorption constants of A, P, by-products (BP), solvent (S) as well as modifiers and additives (M)
$[A],[H_2],[P],$ $[BP], [S], [M]$	concentration of the components in solution

In practice, the surface coverage of the hydrogen can be approximated with its reaction order (Eqs. 4.2 and 4.3). In contrast to homogeneously catalyzed, irreversible reactions, the coverage of the catalyst surface with the product, by-products, solvent, and other modifiers also plays a role. If the product adsorption is very weak, the reaction remains zeroth order in A until almost complete conversion, while a homogeneous reaction shows a linear dependence

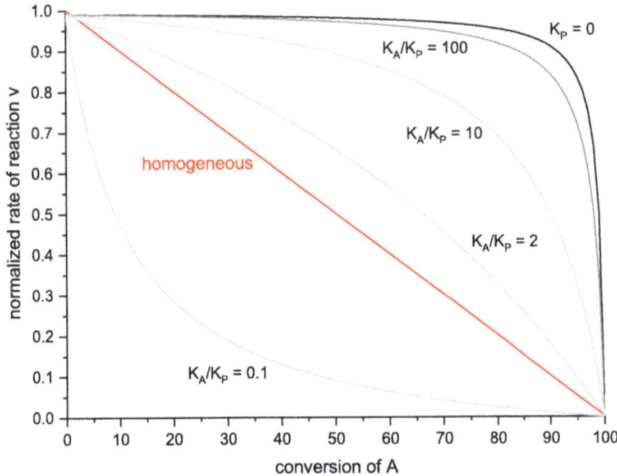

Fig. 4.154 Comparison of the reaction rate of homogeneously ($v = k[cat][A][H_2]$) or heterogeneously catalyzed reactions ($v = k[cat][H_2]\Theta_A$) with varying strength of product adsorption, when $[H_2]$ = constant

---

[IV] The equations are based on a Langmuir–Hinshelwood–Hougen–Watson kinetic, in which the reaction of A with H_2 on the catalyst surface is considered rate-determining.

on A. The stronger the product adsorption, the slower the reaction in the presence of the heterogeneous catalyst due to product inhibition (see Fig. 4.154) [226].

In the absence of directing, Lewis-basic groups, heterogeneous catalysts direct the hydrogenation to occur on the sterically least hindered side. If coordinating groups are present in the substrate, the use of alcoholic solvents negates their effect and thus only the steric environment of the starting material determines the hydrogenation selectivity [517]. Heterogeneous catalysts can even be more sensitive than homogeneous variants to the steric environment of the functional group to be reduced. This is due to the strong steric interaction of the substrate with the heterogeneous catalyst as it must be adsorbed directly on the catalyst surface for H_2 to be transferred.

In the case of alkenes, the reactivity (and thus also the chemoselectivity) with respect to the steric environment follows a similar order under heterogeneous catalysis as displayed by the Wilkinson catalyst (see Fig. 4.155, cf. Fig. 4.131) [501].

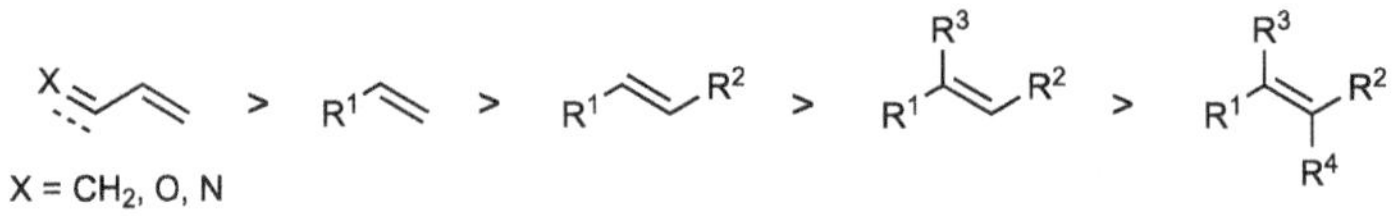

Fig. 4.155 Order of reactivity of alkenes in heterogeneous hydrogenations [501]

The more substituted the alkene is, the slower the hydrogenation. However, the rate can vary from catalyst to catalyst for di- or trisubstituted double bonds. In principle, isolated double bonds in polyenes can be selectively hydrogenated as long as the degree of substitution of the different alkene moieties differs. Overman and co-workers exploited this effect in their synthesis of the tetracyclic diterpene scopadulcic acid B. After reducing the aromatic ring (**406**) by a Birch reduction and subsequent methylation, the disubstituted double bond in **407** could be selectively hydrogenated to give **408** (see Fig. 4.156) [518].

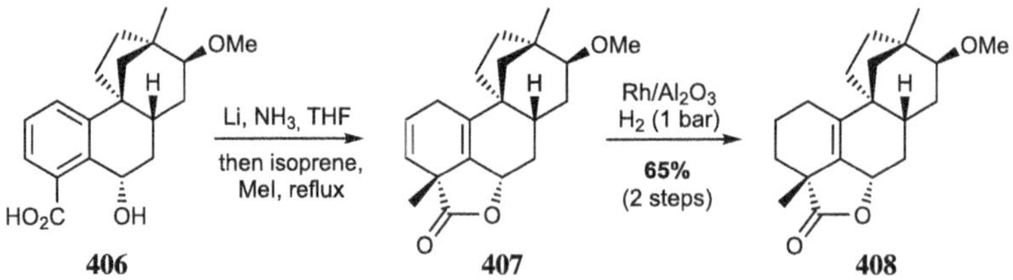

Fig. 4.156 Selective reduction in the synthesis of scopadulcic acid B [518]

For alkenes, the reactivity of the double bond towards reduction by various metal catalysts decreases in the order [226]

$$Pd \geq Rh > Pt \geq Ni \gg Ru$$

For simple CC hydrogenations, Pd, Pt, or Rh catalysts are therefore preferably used, which often allows the reductions to be carried out under neutral conditions at room temperature (see Fig. 4.157).

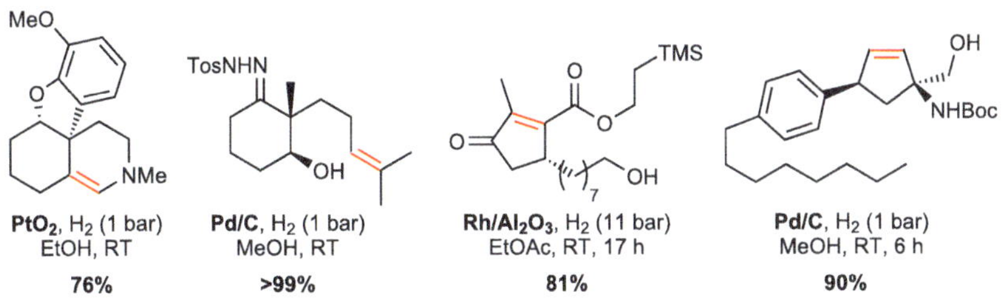

Fig. 4.157 Application of heterogeneous hydrogenations in the total syntheses of natural products [519–522]

Normally, 1–10 wt.% catalyst relative to the substrate is employed. As observed for homogeneous complexes, Lewis bases can similarly inhibit heterogeneous catalysts. These are primarily sulfur derivatives, basic nitrogen compounds, phosphorus compounds, or halides (especially iodides). The higher the oxidation state of the inhibitor, the less problematic the compounds are. For example, thiols or thioethers are usually strong inhibitors, while sulfates do not pose a major problem [107].

The "negative" properties of these catalyst poisons can, however, be deliberately exploited to modify the activity and also selectivity of a catalyst. A prominent example of this strategy is the Lindlar catalyst (Pd-**Pb(OAc)**$_2$/ CaCO$_3$), which mediates the reduction of alkynes to *cis*-olefins without affording a significant over-reduction of the alkene to alkane (see Fig. 4.158) [523].

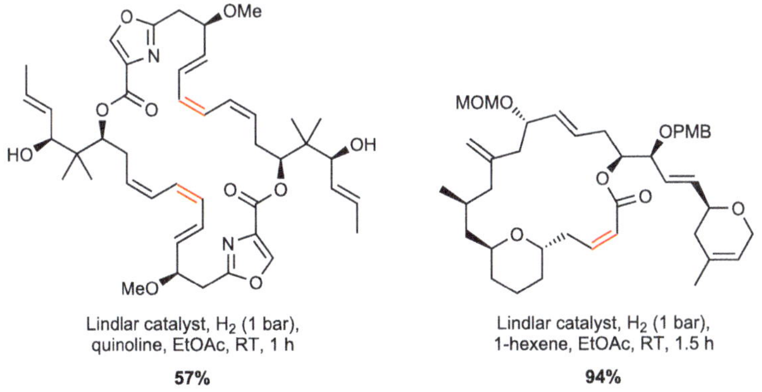

Fig. 4.158 Selective reductions of alkynes in the syntheses of disorazol C$_1$ and laulimalide [524, 525]

The modification of reactivity for a selective hydrogenation of triple bonds by the Lindlar system and other catalysts is related to the nature of the active metal centers on the catalyst surface. For supported Pd, Rh, and Ir, there are two types of catalytically active centers, one of which can hydrogenate alkenes and alkynes, while the second only reduces alkenes [526]. This second metal center can be inhibited by carbon monoxide, so that the selectivity for the reduction of alkynes to alkenes is increased at the cost of decreasing the overall reaction rate. The Pb(OAc)$_2$ normally used for Lindlar catalysts can probably increase the selectivity via the same mechanism, [226, 527] while quinoline shows a higher affinity than alkenes for the binding sites of the active center and can displace them [528].

In direct competition, substrates with CC-triple bonds bind much more strongly to the catalyst surface than double bonds [529]. Due to the stronger chemisorption, alkynes can displace intermediate alkenes from the surface. Furthermore, the reaction conditions should allow for easy exchange of intermediate alkenes by the starting material: A low hydrogen partial pressure and a small amount of catalyst increase the chemoselectivity by depleting the catalyst of hydrogen. For particularly challenging substrates, the reaction can also be stopped after the consumption of one equivalent of hydrogen, as the hydrogen uptake slows down noticeably after the reduction of the triple bond, but usually does not come to a complete halt (see Fig. 4.159) [530].

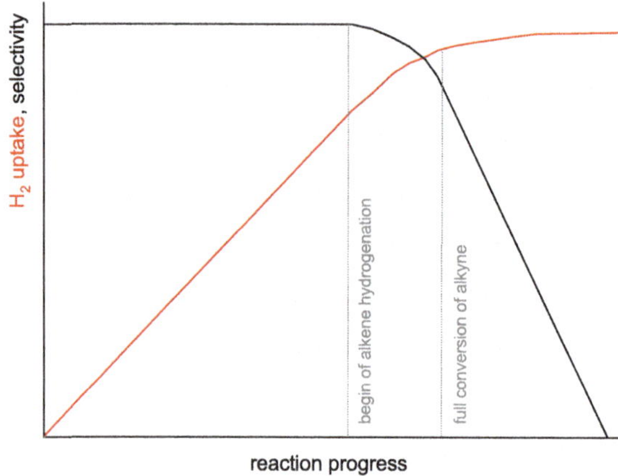

Fig. 4.159 Schematic H$_2$ uptake and selectivity of the semihydrogenation of alkynes

With a high difference in adsorption affinity, only the alkyne is hydrogenated as long as there is enough starting material to occupy all metal centers of the catalyst. As soon as there are fewer substrate molecules than reactive centers, a hydrogenation of the alkene will inevitably occur. The selectivity therefore decreases with increasing conversion. If a large excess of catalyst is used, this leads to a higher reaction rate, but the selectivity will decrease as more metal centers are available to reduce the alkene. An elegant trick is the addition of terminal alkenes as sacrificial reagents (cf. Fig. 4.158), which are preferentially reduced after reaction of the triple bond, so that the slightly less reactive double bond of the desired product is not affected [531].

The total synthesis of the tricyclic polyether ascospiroacetal A (**411**) by the Britton group impressively demonstrated the outstanding selectivity of the Lindlar system [532]. Following the Sonogashira coupling of the terminal alkyne **409** with vinyl iodide **410** to give the enyne, the crude product was directly converted to the *cis*-alkene **411** under reductive conditions. The ester, acid, and ketal groups remained intact, while the reduction of the enyne stopped at the stage of the conjugated diene (see Fig. 4.160).

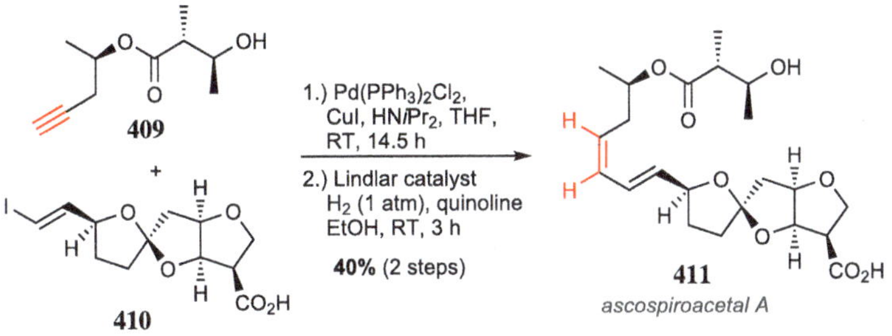

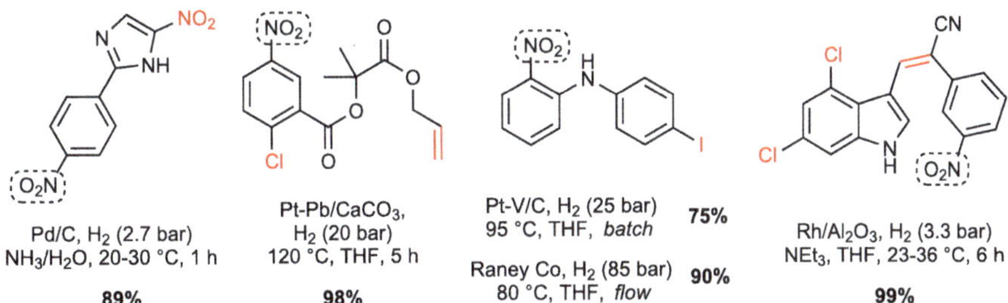

Fig. 4.160 Use of an alkyne semihydrogenation in the synthesis of ascospiroacetal A (**411**) by Britton *et al.* [532]

Since substituted anilines are a common structural motif of bioactive compounds, the reduction of nitroarenes is one of the most important reductions for pharmaceutical and agrochemical applications. The selective hydrogenation of the nitro group in highly function-alized substrates is challenging, as many catalysts can also reduce other functional groups, such as CC multiple bonds, halogens, nitriles, heterocycles, or benzyl groups. Therefore, the choice of the appropriate catalyst and the optimal reaction conditions often is the subject of extensive optimization studies [226, 512, 533]. The classic catalysts rely on Pt and Pd as active metals, usually supported on carbon. To achieve high selectivity, modifiers such as additional metals, organic bases, or acids are often used. Particularly halogenated nitroarenes represent important intermediates in pharmaceutical applications, with the extent of dehalo-genation decreasing when moving to the lighter homologues (Ar-I > Ar-Br > Ar-Cl > Ar-F). Alkynes and to some degree alkenes are also affected particularly often (see Fig. 4.161).

Fig. 4.161 Selective hydrogenations of nitroarenes to the corresponding aniline in the presence of sensitive functionalities (red) that remain intact under reaction conditions [534–537]

The last major class of heterogeneous hydrogenations in organic synthesis is the hydrogenolysis of C-heteroatom single bonds. Usually, this involves the cleavage of benzyl or allyl ethers or amines, which often serve as protecting groups during a synthesis. Of the common active metals, Pd displays a particularly high tendency for hydrogenolysis. Standard conditions consist of Pd/C and an H_2 atmosphere (1 bar) in alcoholic solvents, sometimes in the presence of catalytic amounts of Brønsted acids. Most functional groups remain intact under these conditions. However, the reduction of CC multiple bonds or Ar-NO$_2$ groups usually cannot be fully prevented under these circumstances (see Fig. 4.162) [501]. In turn, if a debenzylation is to be suppressed during the reduction of another functional group, Pd catalysts should be avoided in favor of other metals (see Table 4.14).

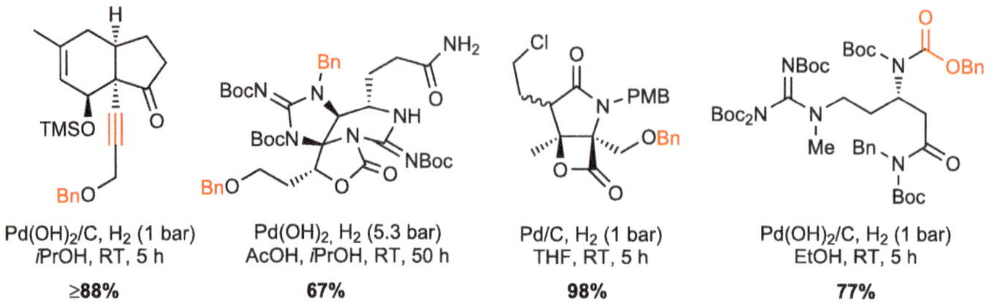
Pd(OH)2/C, H2 (1 bar)
iPrOH, RT, 5 h
≥88%
Pd(OH)2, H2 (5.3 bar)
AcOH, iPrOH, RT, 50 h
67%
Pd/C, H2 (1 bar)
THF, RT, 5 h
98%
Pd(OH)2/C, H2 (1 bar)
EtOH, RT, 5 h
77%

Fig. 4.162 Examples of debenzylation reactions. The highlighted groups are reduced under the reaction conditions [538–541]

An alternative method for the deoxygenation of carbonyl derivatives to the corresponding alkanes is a two-step process in which a thioacetal or thioketone is generated first and then chemoselectively reduced [542]. The mechanism has not been fully elucidated yet, but it is assumed to involve radical intermediates [543]. It was postulated that Raney nickel is converted to nickel sulfide, so it acts as a reagent and is consumed during the reaction. An alternative explanation rationalizes the outcome on the basis of protic solvents serving as hydrogen equivalents or residual hydrogen still adsorbed on the catalyst from its preparation, without resorting to stoichiometric NiS formation. A limitation of the methodology is the incompatibility with CC multiple bonds [542]. Wipf and co-workers first used Lawesson's reagent (**414**) to convert lactam **412** into thiolactam **413**. The subsequent reduction afforded the *Stemona* alkaloid (–)-stenine (**415**) in 78% yield (see Fig. 4.163) [544].

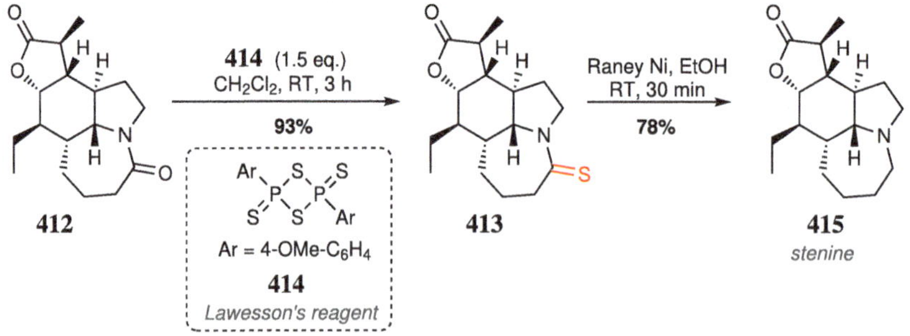
414 (1.5 eq.)
CH2Cl2, RT, 3 h
93%
Ar = 4-OMe-C6H4
414
Lawesson's reagent
Raney Ni, EtOH
RT, 30 min
78%
412
413
415
stenine

Fig. 4.163 Total synthesis of (–)-stenine (**415**) by Wipf *et al.* [544]

A method for the preparation of aldehydes that is rarely used today is the reduction of the respective acid chlorides. The so-called *Rosenmund* reduction uses specifically poisoned Pd catalysts in conjunction with certain additives, similar to the Lindlar system [545, 546]. Either

Pd/BaSO$_4$ (Rosenmund catalyst) or Pd/C with addition of quinoline or sulfur derivatives can be employed. Tertiary amines can be added to sequester the HCl by-product formed *in situ* in the case of acid-sensitive substrates.

The core hydrogenation of arenes to saturated hydrocarbons is of crucial importance for the production of basic and fine chemicals. However, since the necessary conditions usually also reduce many synthetically relevant functionalities, this reaction plays a niche role in synthetic approaches to complex natural substances [427].

4.2.5   Asymmetric Reductions

As already discussed in the chapter on carbonyl chemistry, there are several handles which can be exploited for the induction of stereoinformation. Stereoselective reactions can be achieved either by exploiting stereogenic centers in the substrate or by employing chiral reagents/catalysts. While substrate control in chiral carbonyl derivatives (for example in the addition of metal hydrides) can be described by models for 1,2- (Felkin–Anh) and 1,3-induction (Evans), there exists a wide variety of enantioselective, catalyst-induced methods for the asymmetric reduction of prochiral carbonyl compounds. Such a reduction of prochiral starting materials must be kinetically controlled, as the enantiomeric products have identical thermodynamic properties and are therefore isoenergetic (see Fig. 4.164).

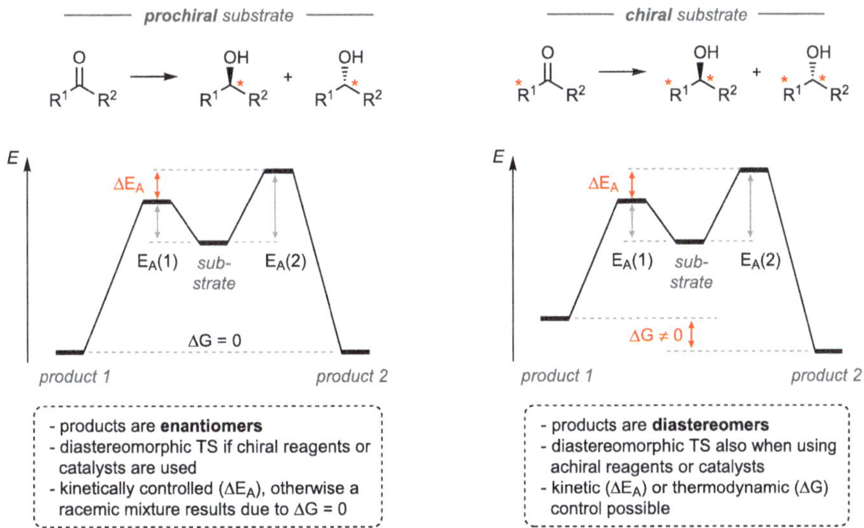

Fig. 4.164 Energy diagrams for the reduction of prochiral and chiral substrates

In addition to the reduction of CO double bonds, the second major class of asymmetric reductions is the conversion of olefins to chiral alkanes. The reduction of alkenes is now dominated by hydrogenations in the presence of chiral transition metal catalysts. The synthesis of secondary alcohols from the corresponding prochiral ketones, on the other hand, can be effected by the action of BH_3 or other methods in addition to a reduction with hydrogen [547, 548].

4.2.5.1 Enantioselective Reductions of Carbonyl Derivatives

Initial studies primarily relied on stoichiometric amounts of chiral reagents to convert prochiral ketones to secondary alcohols with moderate to good enantiomeric excess. While BINAL-H (**416**) constitutes a complex metal hydride with the corresponding reactivity scope, the electrophilic dialkyl chlorides (**417**) and trialkyl boranes (**418**), both derived from α-pinene, have established themselves in parallel (Fig. 4.165) [549].

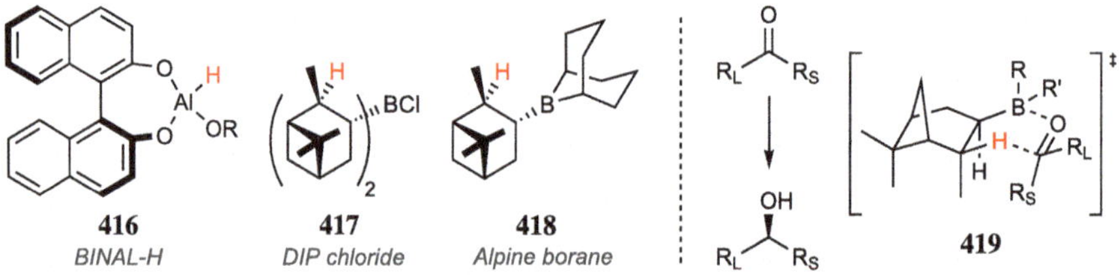

Fig. 4.165 Stoichiometrically used reductants and transition state of the reduction with **417**, **418** [549]

The reduction with BINAL-H assumes a six-membered, chair-like transition state [550]. The reductant in the reaction with the boranes **417** or **418** is not the borane itself. The hydride is instead transferred from the isopinocampheyl moiety to the carbonyl group with concomitant elimination of pinene (**419**). A comparison of the substrate scope of these stoichiometric reducing agents shows the degree of asymmetric induction that can be achieved with different substrates (see Table 4.15).

Table 4.15 Enantioselectivity in the reduction of prochiral ketones with chiral reducing agents [549, 550]

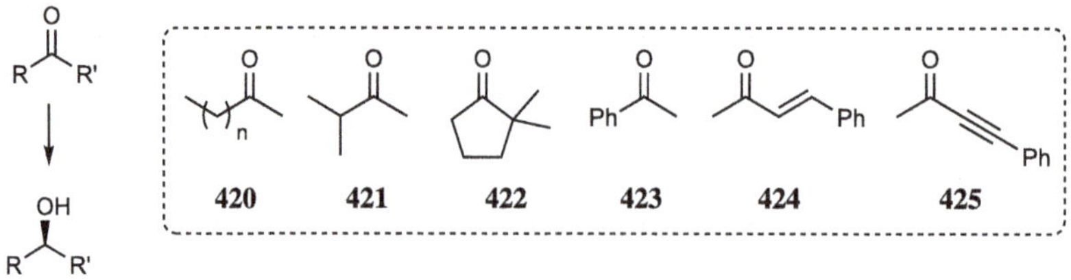

reagent	enantiomeric excess [%]					
	420	**421**	**422**	**423**	**424**	**425**
BINAL-H (**416**)	–	78	11	95	70	89
DIP-Cl (**417**)	4 (n=1)	32	98	98	12	21
Alpine borane (**418**)	48 (n=7)	62	20	87	56	78

Brown and co-workers were able to demonstrate the applicability of DIP-chloride (**417**) in the synthesis of a number of bioactive compounds. The antidepressant fluoxetine (**428**) could be obtained by reducing the ketone **426** to the corresponding chloroalcohol **427** with an excellent enantiomeric excess of 96%. After a subsequent Mitsunobu reaction and introduction of the amino functionality, this route was used to synthesize not only the commercial active ingredient but also other bioactive compounds used as antidepressants (see Fig. 4.166) [551].

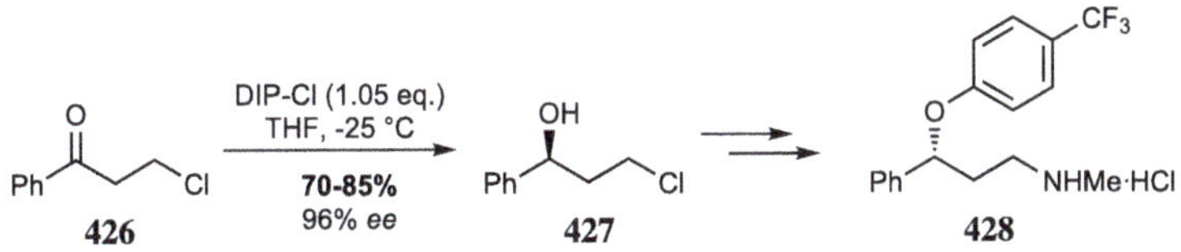

Fig. 4.166 Synthesis of fluoxetine (**428**) by Brown *et al.* [551]

Based on the work of Itsuno, oxazaborolidines were introduced by Corey, Bakshi, and Shibata in the 1980s as efficient catalysts [552]. The *CBS reduction*, named after the authors, is one of the most frequently used reactions for the enantioselective reduction of prochiral ketones by virtue of its simplicity and high enantioselectivity (see Fig. 4.167) [553].

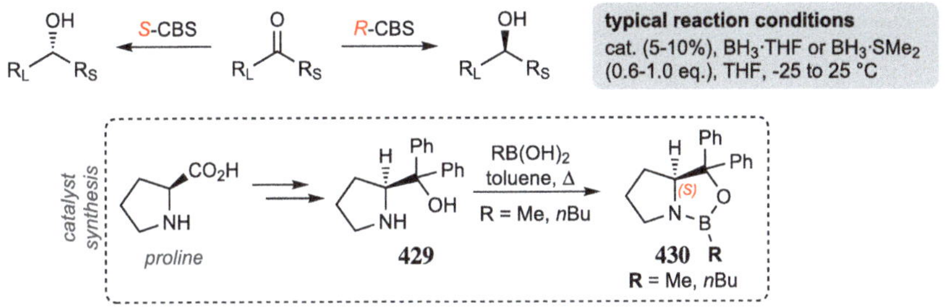

Fig. 4.167 Control of the enantioselectivity in the CBS reduction, synthesis of the oxazaborolidine *S*-**430**, and typical reaction conditions [553]

Mechanistically, the high selectivity of the catalyst relies on the presence of Lewis acidic and basic centers, which can activate both the carbonyl group and BH_3. The coordination of the electrophilic borane to the oxazaborolidine increases its nucleophilicity as a hydride donor and in turn the Lewis acidity of the endocyclic boron atom. The required activation of the reductant ensures that no undesired background reaction occurs, which would lead to a decreased stereoselectivity [552].

Following the reaction of the catalyst **430** with BH_3, complex **431** is formed, which is subsequently coordinated by the ketone. In the transition state **432**, the ketone can coordinate the boron in a syn- or antiperiplanar arrangement with regard to the large substituent R_L. Due to a combination of steric repulsion of the substituent at boron (R = Me, Bu) with R_L and stabilizing van der Waals attraction, the large substituent points away from the boron as in **432** to minimize steric repulsion [554]. The resulting intermediate **433** can either directly collapse to give **434** or via **435** by uptake of a second borane to furnish **431**. After acidic work-up, the desired alcohol is isolated from **434** by protodeboronation (see Fig. 4.168) [552].

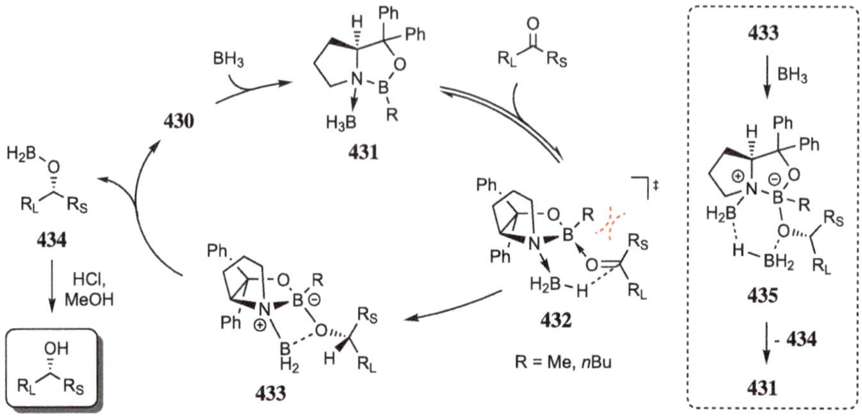

Fig. 4.168 Mechanism of the CBS reduction [552]

The CBS reduction has been used to great effect in a multitude of total syntheses. Esters, olefins, alkynes, vinyl-/aryl halides, thioethers, (thio-)acetals, amides, silyl ethers, and silanes are tolerated under the reaction conditions. Amines, however, must be protected [552, 553]. Particularly impressive examples in advanced stages are provided by the total syntheses of okadaic acid, bryostatin 2, desmethyllaulimalide, and archazolid A, in addition to the examples shown in Fig. 4.169 [555–558]. Instead of BH_3, alternative boranes can also be employed, with catecholborane proving to be very advantageous due to its good reactivity.

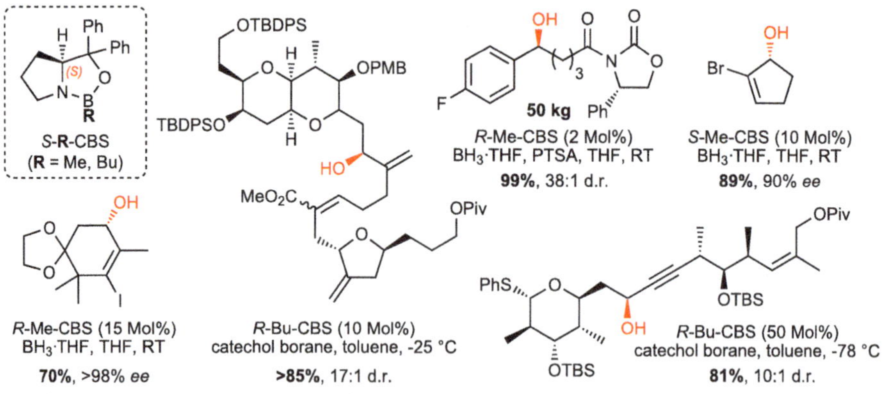

Fig. 4.169 Applications of CBS reduction in the presence of representative functional groups [559–563]

The reduction of ketimines in the presence of oxazaborolidines is only moderately effective. The lower electrophilicity of the imine carbon atom and the equilibrium between *E*- and *Z*-imines reduce both the yield as well as the enantioselectivity of the reaction [553].

The switch from main group-based, Lewis-acidic catalysts to transition metal complexes, on the other hand, also allows the selective isolation of products that are not efficiently accessible by a CBS reduction. Noyori's work on Ru-BINAP complexes represented a breakthrough in the field of enantioselective reduction of carbonyls, where these metal complexes proved to be competent asymmetric hydrogenation catalysts (see Fig. 4.170) [564, 565].

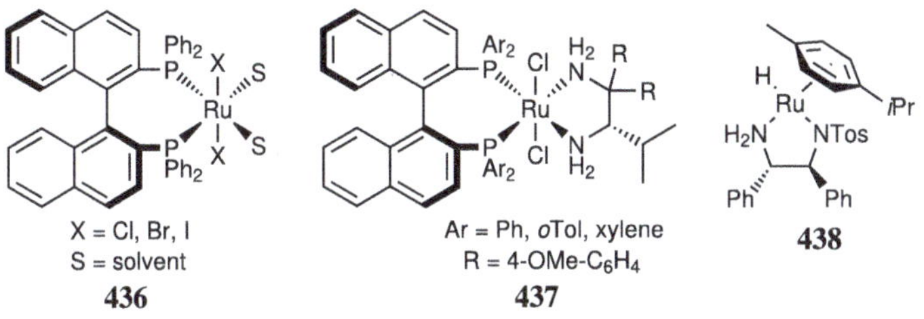

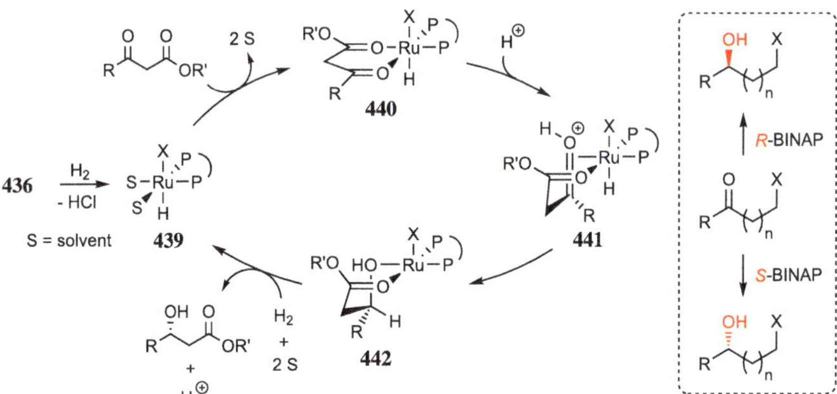

Fig. 4.170 Noyori catalysts for the enantioselective reduction of ketones

The first systems (**436**) necessitated the presence of a Lewis-basic functional group (ester, alcohol, etc.) in the β-position of the substrate for enforcing a structurally rigid transition state to afford high enantioselectivity. The polymeric precatalyst **436** can furnish the catalytically active complex **439** in the presence of hydrogen. Following the coordination of the substrate (**440**), its keto group in **441** is activated by protonation. The hydridic hydrogen at the metal center is transferred to the carbonyl group. Subsequently, the product is liberated from **442** by displacement with a solvent molecule. The catalytically active species **439** is finally regenerated by reaction with H$_2$ (see Fig. 4.171) [564].

Fig. 4.171 Mechanism of hydrogenation with **436** [564]

For the catalysts of type **436**, it was unexpectedly observed that an addition of diamines resulted in an increase in the rate of reaction by two orders of magnitude [566]. This led to the development of metal complexes bearing both P,P- and N,N-chelate ligands (**437**). The presence of a second basic functionality (ester, alcohol) in the substrate is critical to attain a high activity and selectivity with **436**. The Ru complexes subsequently introduced by Noyori (**437**), on the other hand, enable the asymmetric hydrogenation of unfunctionalized ketones. If both chelate ligands are chiral, a strong stereochemical induction can be achieved in a *matched* situation for the ligands, which can even override the inherent directing effect of stereogenic centers in the substrate. The currently assumed mechanism (see Fig. 4.172) proceeds via various intermediates which are still controversially discussed [567–570].

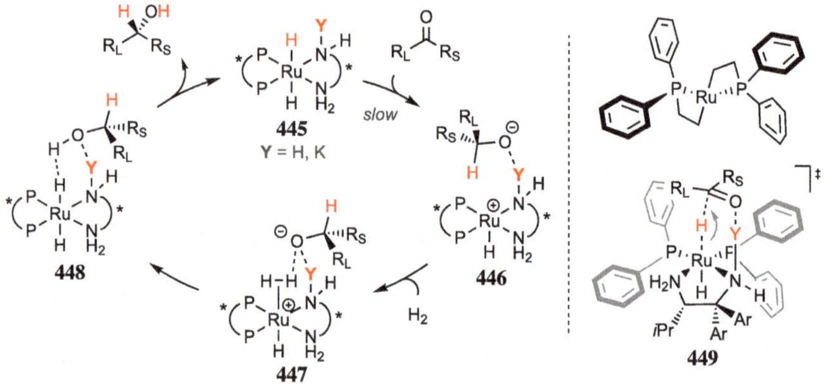

Fig. 4.172 Postulated mechanism for the reduction with **437**, stereochemical model and transition state of hydrogenation (**449**). The naphthyl residues in BINAP are omitted for clarity [567–569, 571]

Starting from the reactive species **445**, the enantiomerically enriched product **446** is formed in the rate-determining step, which also constitutes the stereochemically-inducing step. In the presence of an excess of base, a stabilizing interaction was postulated via a metal ion in **446** and the transition state **449** (Y = K). After reorganization and coordination of H_2, **447** can heterolytically split the hydrogen reductant and transfer a proton to the alkoxide, which dissociates from **448**. The regenerated Ru-hydride species **445** then reenters the catalytic cycle. The stereochemical model of the bound bisphosphine shows a structure in which the P-aryl substituents alternately point towards or away from the substrate in the four quadrants. By choosing the right diamine, the steric interactions and thus the asymmetry at the metal center are further enhanced. The postulated six-membered transition state **449** proceeds via an *outer-sphere* mechanism in the H_2 transfer, in which the ketone is oriented in such a way that a repulsive interaction of the R_L substituent with the bisphosphine and the α-positioned iPr-group of the diamine backbone is minimized.

The methodology is chemoselective for C=O- over C=C- and C≡C-bonds and further tolerates many functionalities such as NO_2, halogens, acetals, esters, amides, and amines. An illustrative example by the Merck process development department shows that the hydrogenation of a complex heterocyclic ketone in the presence of only minor amounts of catalyst can afford the desired product with quantitative yield and excellent selectivity (see Fig. 4.173) [572].

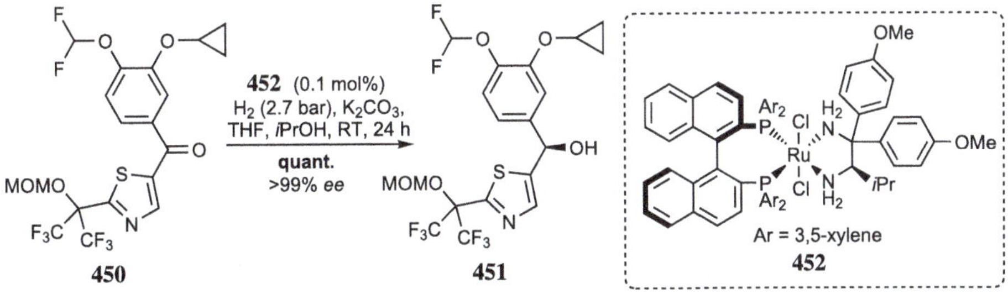

Fig. 4.173 Hydrogenation of a ketone in the synthesis of a PDE-IV inhibitor [572]

In addition to the BINAP-based phosphines introduced by Noyori, a plethora of further chiral ligands is suitable for enantioinduction. Other metal complexes, especially those based on iridium, also enable an efficient and selective conversion, not only for CO- but also for CN-bond reductions [573–576].

The commercial synthesis of the herbicide *S*-metolachlor by Ciba-Geigy is one of the largest industrial applications of asymmetric hydrogenations. It relies on the Ir-catalyzed reduction of imine **453**, for which the class of ferrocene-based Josiphos ligands was specifically developed at the time. 10 tons of the desired amine **454** could be produced in a single batch in 79% *ee*, requiring only 34 g of the Ir complex, 68 g of ligand (**456**), and a few grams of H_2SO_4 and NaI as additive. The process is also solvent-free, which obviates the need for a cumbersome recycling of the solvent, further improving the economic aspects of the synthesis (see Fig. 4.165) [577] (Fig. 4.174).

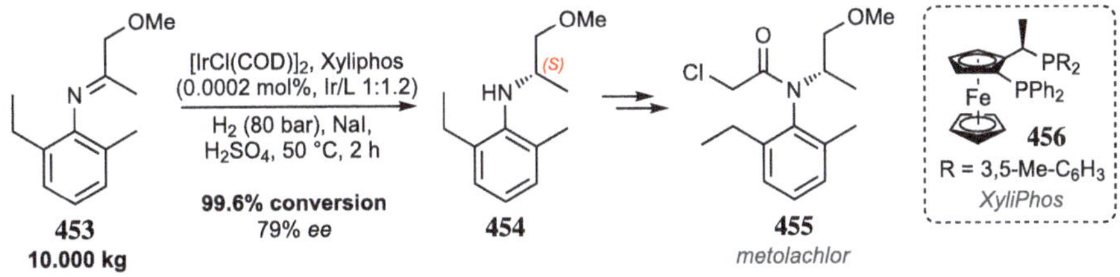

Fig. 4.174 Industrial synthesis of the herbicide metolachlor (**455**) [577]

As already mentioned on Sect. 4.2.4.2, diamine-ligated ruthenium-arene complexes such as **438** are extremely efficient catalysts, particularly for asymmetric transfer hydrogenations. The **Noyori–Ikariya catalyst** (not to be confused with the other Noyori catalysts **436**, **437**) can selectively convert a wide range of ketones to the corresponding secondary alcohols in the presence of redox-sensitive functional groups (see Fig. 4.175) [578, 579]. The reaction can be carried out under slightly basic conditions at room temperature, converting prochiral ketones with high enantioselectivities. If secondary alcohols are used as reductants the reaction becomes reversible, which is why they are often present in large excess, for example as solvents. Extended reaction times can also lead to an erosion of enantioselectivity [547].

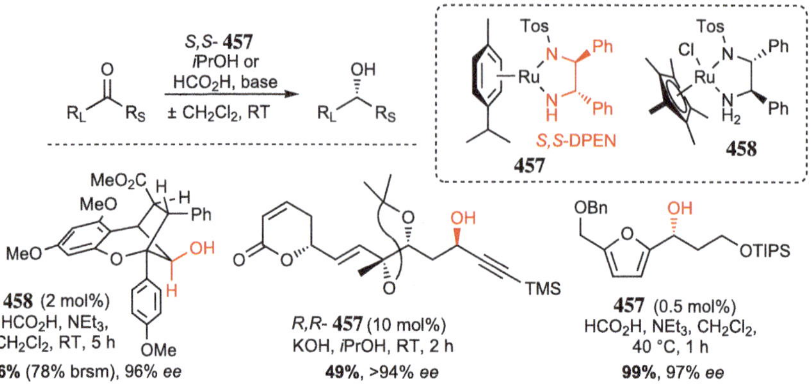

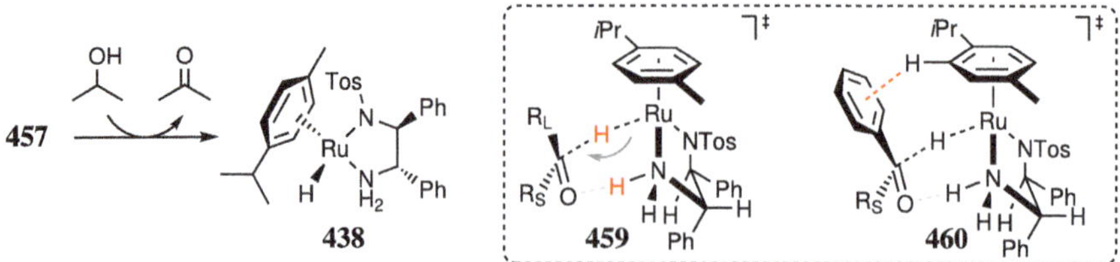

Fig. 4.175 Typical reaction conditions and applications of the asymmetric transfer hydrogenation according to Noyori [580–582]

The catalytically active complex **457** consists of the η^6-bound arene (here: cymene) and tosylated diphenylethylenediamine (DPEN). In analogy to the general mechanism of transfer hydrogenations with Noyori-type catalysts (see Fig. 4.147), the catalytic cycle is also initiated by the transfer of H_2 from iPrOH or formic acid to the metal complex. The enantiodiscriminating step is the *outer-sphere* reduction of the carbonyl group via **459**, in which the sterically more demanding substitutent on the carbonyl is oriented in a β-position [583]. Instead of a concerted transition state with concomitant transfer of both H-atoms to the reactant, a stepwise conversion seems to be rather indicated based on recent mechanistic investigations [486]. The source of the proton (as opposed to the hydride) that is transferred to the reduced product in the second step can either be the nitrogen-bound hydrogen of the diamine ligand (inner-sphere mechanism in iPrOH as solvent) or it may stem from the solvent/reagent (outer-sphere transfer from HCO_2H) [584]. In the case of aryl ketones (R_L = Ph) or α,β-unsaturated ketones, an additional stabilizing π-CH interaction of a proton on the cymene ligand with R_L has been postulated for **460** [483]. The conversion of aryl ketones therefore usually yields excellent *ee* values (see Fig. 4.176).

Fig. 4.176 Intermediates and transition states of the reduction with **457**. [479, 483, 486, 583, 585]

Marshall and Ellis used the Noyori transfer hydrogenation to access a key intermediate en route to the polyketide cytostatin. The ynone **461** was quantitatively converted in isopropanol as solvent to the respective propargylic alcohol **462**, which was isolated diastereomerically pure. The subsequent steps to the C3-C13 fragment of the phosphatase inhibitor included the hydrogenation of the triple bond, a debenzylation, and a carbon chain homologation following an adjustment of the oxidation state (see Fig. 4.177) [586].

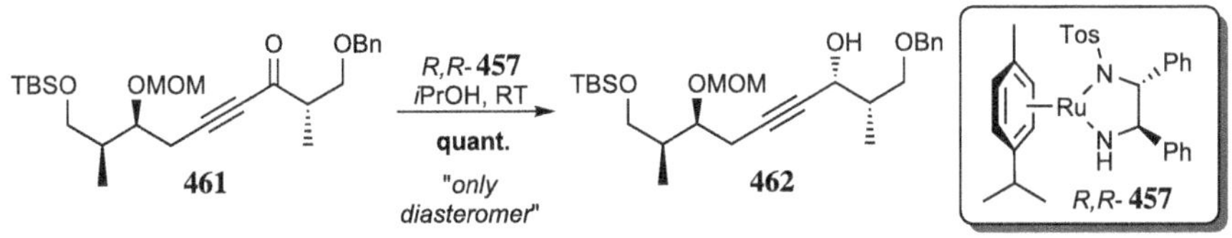

Fig. 4.177 Application of the Noyori transfer hydrogenation by Marshall [586]

4.2.5.2 Enantioselective Reductions of Alkenes

The asymmetric hydrogenation of alkenes was already studied extensively before the advent of enantioselective carbonyl hydrogenations. The reaction usually displays excellent chemo-, regio-, and enantioselectivity. The reduction of alkenes, which bear coordinating groups such as amides or carboxylic acids in proximity to the double bond, is typically carried out in the presence of Rh(I)- and Ru(II) complexes, which require a coordinative fixation of the substrate. Complementary to this, derivatives of the Ir-based Crabtree catalyst can also reduce di-, tri-, and tetrasubstituted alkenes without the need for coordinating groups (see Fig. 4.178). Although some homogeneous catalysts do not contain chiral phosphines as a stereodirecting element, they are the dominant ligands in asymmetric hydrogenations [427, 439, 575, 587–589].

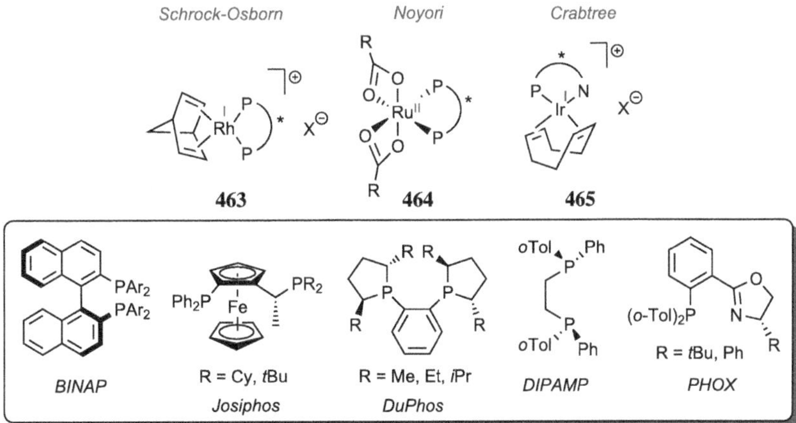

Fig. 4.178 Commonly employed classes of asymmetric catalysts and chiral ligands [439, 575, 587, 588]

The mechanism of enantioselective hydrogenation was intensively studied with the chiral Schrock–Osborn-type catalyst **463**. The activation of the cationic bisphosphine-Rh(I) complexes proceeds as shown in Fig. 4.136 [443]. Based on fundamental work by Halpern on the reduction of prochiral enamines, a mechanism for the hydrogenation of substrates with coordinating or directing groups (DG) can be formulated (see Fig. 4.179) [590].

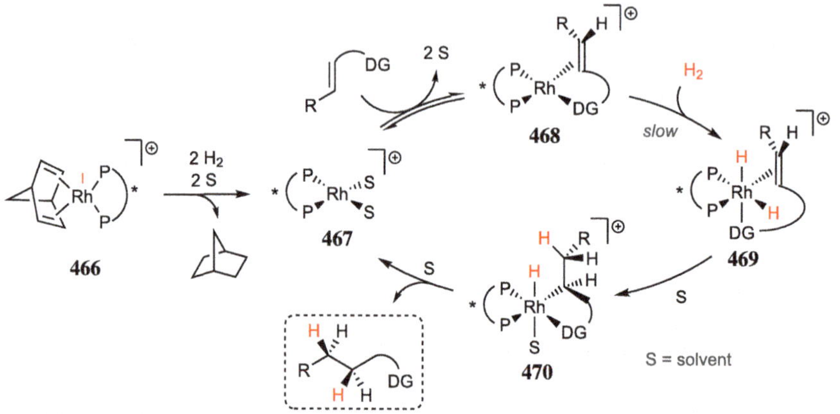

Fig. 4.179 Mechanism of asymmetric hydrogenation of alkenes with chiral Schrock–Osborn-type catalysts [590]

Following the reversible coordination of the substrate by complex **467** with displacement of two molecules of solvent, hydrogen is oxidatively added via **468** as the rate-determining step. Depending on the chosen ligands, the enantiodifferentiating coordination of the substrate controls which enantiomer of the product is ultimately obtained. The two possible adducts **468** of the substrate with the metal center (coordination of the *Re*- or *Si*-face of the alkene) yield two diastereomeric complexes (only one shown). The energetically less favorable and thus less populated adduct reacts significantly faster with H_2. The enantiomeric excess thus results from the difference in activation energies $\Delta\Delta G^{\ddagger}$ according to the Curtin–Hammett principle. Both the *cis*-addition of H_2 to the substrate (**469**→**470**) and the reductive elimination (**470**→**467**) are stereospecific. The enantiomeric ratio of the product thus mirrors the energetic situation of the rate-determining step. A high partial pressure of hydrogen counteracts the reversibility in the coordination of the substrate (**467**→**468**). This results in a diminished enantiomeric excess and sometimes even reversal of stereoselectivity, as the less reactive but more populated conformer of **468** can also provide the product under these conditions (see Fig. 4.180) [590].

The first large-scale application of an asymmetric hydrogenation was the commercial production of L-DOPA (**474**), developed by Monsanto on a metric ton scale. The key step is the enantioselective hydrogenation of the prochiral enamide **473** in the presence of the air-stable Rh complex **475**, which contains the bisphosphine DIPAMP as chiral ligand. The

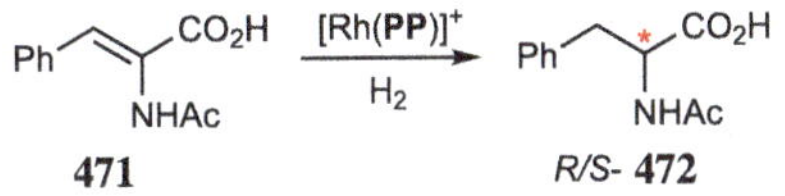

H_2 [bar]	ee [%] (configuration) *R,R*-DIOP	ee [%] (configuration) *R,R*-DIPAMP
1	55.2 (*R*)	
5	8.4 (*R*)	63.6 (*S*)
20	0.5 (*S*)	29.9 (*S*)
50	4.9 (*S*)	

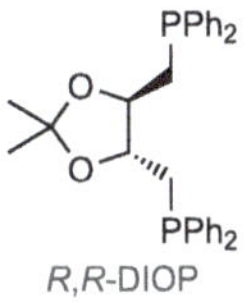

Fig. 4.180 Pressure dependence of the enantiomeric excess in the hydrogenation of prochiral enamines [590]

product could be isolated in almost quantitative yield with excellent enantiomeric excess (95%) and was subsequently further enriched by recrystallization. The amino group was then deprotected under acidic conditions (see Fig. 4.181) [591].

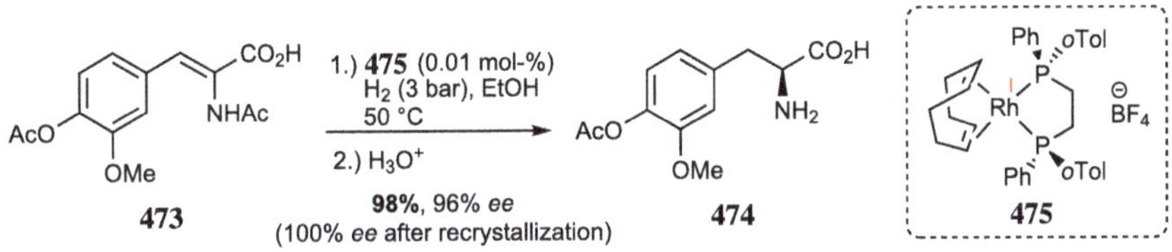

Fig. 4.181 Key step in Monsanto's L-DOPA process [591]

In addition to the use of Schrock-Osborn complexes, Noyori systems based on Ru can also be used for substrates bearing directing groups. The mechanism is similar to that of Rh catalysts (see Fig. 4.182) [567, 592]. Following the activation of the precatalyst **476** by treatment with H_2, the square-planar Ru complex **477** is obtained. The catalytic cycle is characterized by the formation of a monohydride complex (**477**) preceding the coordination of the olefinic substrate. The alkene subsequently reversibly inserts into the metal-H bond of the short-lived intermediate **478**. The Ru-C bond in **479** is cleaved by H_2 to form the enantiomerically enriched product and regenerate the monohydride species **477**, which can reenter the catalytic cycle. Alternatively, the product can also be displaced from **479** upon substitution by the solvent (**480**), with **477** subsequently being regenerated under reductive conditions. The stereochemistry of the product is determined by **478**, in which the olefin can be coordinated to the metal center either via its *Re*- or *Si*-face. Since the metal complex already contains a hydride, a comparable rate of migratory insertion into the Ru-C bond for both diastereomeric complexes (**478**$_{Re}$, **478**$_{Si}$) was postulated. The relative energetic stability of these complexes is thus reflected in the enantiomeric excess of the product. The reductive elimination of the alkane (**479**→**477**) proceeds stereospecifically with retention of configuration [567, 592].

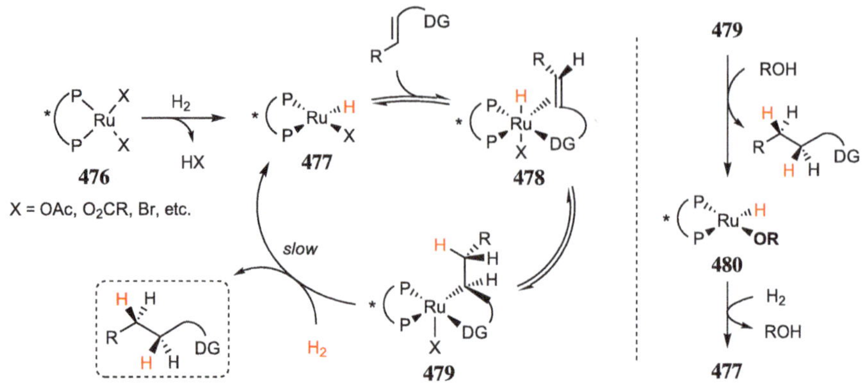

Fig. 4.182 Mechanism of the hydrogenation of alkenes with chiral Noyori-type Ru catalysts [567, 592]

Merck relied on Noyori's BINAP-complexed Ru catalyst to synthesize the $PPAR_\gamma$ agonist (peroxisome proliferator-activated receptor) **483** on a kg scale, which can be used for the treatment of *diabetes mellitus* type II. The chosen route via an asymmetric hydrogenation was able to provide over 3 kg of the pharmaceutically active ingredient with a total yield of 50% and in 99.5% *ee*, enabling toxicological and initial clinical studies to be conducted (see Fig. 4.183) [593].

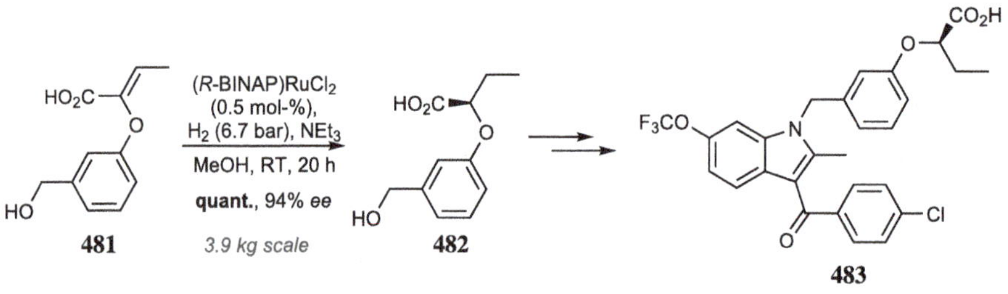

Fig. 4.183 Merck's synthesis of a $PPAR_\gamma$ agonist (**483**) [593]

Unfunctionalized olefins can be converted selectively to the enantiomerically enriched alkanes by resorting to iridium catalysts. The Crabtree-type catalysts **484** used in asymmetric reductions usually contain a chiral N,P-chelate ligand and tetrakis(3,5-bis(trifluoromethyl)phenyl)borate (*BARF*) as non-coordinating counterion [439, 594, 595]. Seminal contributions were carried out in the Pfaltz laboratory, which introduced the class of phosphinooxazolines (*PHOX*) as particularly efficient ligands. While the number of available ligands for **484** has steadily increased, most published ligands provide high enantioselectivity only for a limited substrate scope, which may be why a broader application in an industrial environment is still pending (Fig. 4.184).

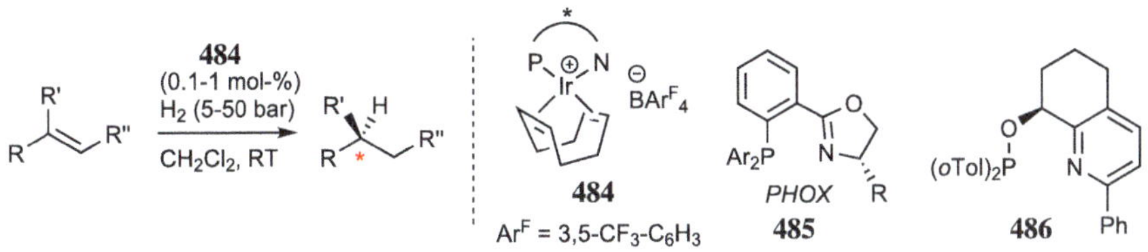

Fig. 4.184 Typical conditions of asymmetric hydrogenations with Crabtree catalysts [439]

The mechanism of hydrogenation using the Ir(III)/Ir(V) species has been investigated by several groups (cf. Fig. 4.134) [438, 440, 596]. Although the available experimental data is still lacking for a complete understanding of the asymmetric reaction, the broad strokes of the catalytic cycle have been postulated. The facial selectivity in the coordination of the alkene by **487** favors one of the two possible complexes, **488** or **490**. After the insertion of the alkene into the Ir-H bond, the diastereomeric complexes **489** and **491** result. While the coordination of one side of the alkene leads to enantiomer A, the diastereomeric complex **489** delivers the respective mirror image (see Fig. 4.185) [439, 596].

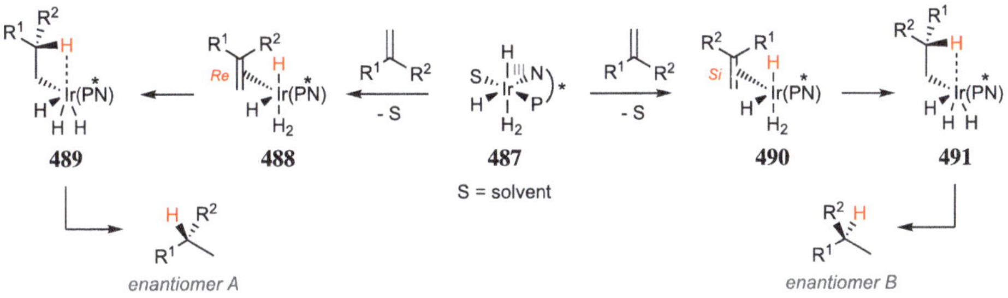

Fig. 4.185 Important intermediates in the hydrogenation of alkenes with chiral Crabtree catalysts [439, 596]

Especially Rh-catalyzed hydrogenations are commonly encountered on an industrial scale, even if the high price of the precious metal only allows for very efficient reductions in a commercial synthesis [597]. Some particularly impressive examples, several of which were carried out on an industrial scale, are shown in Fig. 4.186.

In addition to the reduction of isolated CC double bonds, either the CC or the CO double bond can be selectively reduced in α, β-unsaturated carbonyl derivatives. The method of choice used for a 1,4-selective reduction is mediated by copper hydrides, which were already discussed in Sect. 2.5.

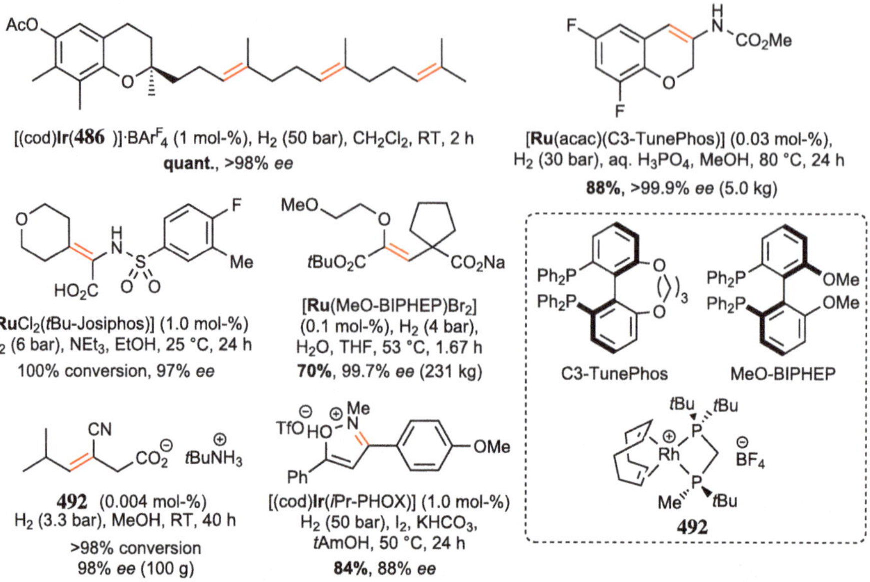

Fig. 4.186 Asymmetric hydrogenations in the synthesis of intermediates of pharmaceutically active compounds [598–603]

References

1. W. Adam, C. R. Saha-Möller, P. A. Ganeshpure, *Chem. Rev.* **2001**, *101*, 3499–3548.
2. E. Wojaczynska, J. Wojaczynski, *Chem. Rev.* **2010**, *110*, 4303–4356.
3. K. A. Stingl, S. B. Tsogoeva, *Tetrahedron: Asymm.* **2010**, *21*, 1055–1074.
4. S. Youssif, *Arkivoc* **2001**, 242–268.
5. S. Mahapatra, R. G. Carter, *J. Am. Chem. Soc.* **2013**, *135*, 10792–10803.
6. S. Hanessian, R. R. Vakiti, S. Dorich, S. Banerjee, F. Lecomte, J. R. DelValle, J. Zhang, B. Deschenes-Simard, *Angew. Chem. Int. Ed.* **2011**, *50*, 3497–3500.
7. O. O. Fadeyi, C. W. Lindsley, *Org. Lett.* **2009**, *11*, 3950–3952.
8. S. Hanessian, T. Focken, X. Mi, R. Oza, B. Chen, D. Ritson, R. Beaudegnies, *J. Org. Chem.* **2010**, *75*, 5601–5618.
9. B. D. Williams, A. B. Smith, III, *J. Org. Chem.* **2014**, *79*, 9284–9296.
10. A. B. Smith, III, T. Bosanac, K. Basu, *J. Am. Chem. Soc.* **2009**, *131*, 2348–2358.
11. O. Iwamoto, H. Koshino, D. Hashizume, K. Nagasawa, *Angew. Chem. Int. Ed.* **2007**, *46*, 8625–8628.
12. D. J. Mergott, S. A. Frank, W. R. Roush, *Proc. Nat. Acad. Sci.* **2004**, *101*, 11955–11959.
13. G. Tojo, M. Fernández, *Oxidation of Alcohols to Carboxylic Acids*, Springer Science + Business Media, New York, 1st ed., **2007**.
14. G. Tojo, M. Fernández, *Oxidation of Alcohols to Aldehydes and Ketones*, Springer Science + Business Media, New York, 1st ed., **2006**.
15. F. A. Luzzio, *Org. React.* **1998**, *53*, 1–221.
16. J. M. Bobbitt, C. Brückner, N. Merbouh, *Org. React.* **2010**, *2*, 103–424.

17. T. T. Tidwell, *Org. React.* **1990**, *39*, 297–555.

18. S. Caron, R. W. Dugger, S. G. Ruggeri, J. A. Ragan, D. H. B. Ripin, *Chem. Rev.* **2006**, *106*, 2943–2989.

19. F. H. Westheimer, N. Nicolaides, *J. Am. Chem. Soc.* **1949**, *71*, 25–28.

20. S. L. Scott, A. Bakac, J. H. Espenson, *J. Am. Chem. Soc.* **1992**, *114*, 4205–4213.

21. J. F. Perez-Benito, C. Arias, *Can. J. Chem.* **1993**, *71*, 649–655.

22. J. Herscovici, M.-J. Egron, K. Antonakis, *J. Chem. Soc. Perkin Trans. 1* **1982**, 1967–1973.

23. S. Czernecki, C. Georgoulis, C. L. Stevens, K. Vijayakumaran, *Tetrahedron Lett.* **1985**, *26*, 1699–1702.

24. A. Bélanger, D. J. F. Berney, H.-J. Borschberg, R. Brousseau, A. Doutheau, R. Durand, H. Katayama, R. Lapalme, D. M. Leturc, C.-C. Liao, F. N. MacLachlan, J.-P. Maffrand, F. Marazza, R. Martino, C. Moreau, L. Saint-Laurent, R. Saintonge, P. Soucy, L. Ruest, P. Deslongchamps, *Can. J. Chem.* **1979**, *57*, 3348–3354.

25. H. Kuwajima, T. Tanahashi, K. Inoue, H. Inouye, *Chem. Pharm. Bull.* **1999**, *47*, 1634–1637.

26. D. S. Tan, G. B. Dudley, S. J. Danishefsky, *Angew. Chem. Int. Ed.* **2002**, *41*, 2185–2188.

27. E. J. Corey, G. Schmidt, *Tetrahedron Lett.* **1979**, *20*, 399–402.

28. M. Zhao, J. Li, Z. Song, R. Desmond, D. M. Tschaen, E. J. J. Grabowski, P. J. Reider, *Tetrahedron Lett.* **1998**, *39*, 5323–5326.

29. T. T. Tidwell, *Synthesis* **1990**, 857–870.

30. M. Seki, Y. Mori, M. Hatsuda, S. Yamada, *J. Org. Chem.* **2002**, *67*, 5527–5536.

31. T. T. Tidwell, *Org. React.* **1990**, *39*, 297–572.

32. A. J. Mancuso, D. Swern, *Synthesis* **1981**, 165–185.

33. K. Omura, D. Swern, *Tetrahedron* **1978**, *34*, 1651–1660.

34. S. K. Bur, A. Padwa, *Chem. Rev.* **2004**, *104*, 2401–2432.

35. D. A. Longbottom, A. J. Morrison, D. J. Dixon, S. V. Ley, *Angew. Chem. Int. Ed.* **2002**, *41*, 2786–2790.

36. D. T. Hog, F. M. E. Huber, P. Mayer, D. Trauner, *Angew. Chem. Int. Ed.* **2014**, *53*, 8513–8517.

37. H. Tohma, Y. Kita, *Adv. Synth. Catal.* **2004**, *346*, 111–124.

38. M. Uyanik, K. Ishihara, *Chem. Commun.* **2009**, 2086–2099.

39. A. Yoshimura, V. V. Zhdankin, *Chem. Rev.* **2016**, *116*, 3328–3435.

40. U. Ladziata, V. V. Zhdankin, *Arkivoc* **2006**, 26–58.

41. A. Chipman, *Asian J. Org. Chem.* **2022**, *11*, e202100522.

42. A. Duschek, S. F. Kirsch, *Angew. Chem. Int. Ed.* **2011**, *50*, 1524–1552.

43. J. D. More, N. S. Finney, *Org. Lett.* **2002**, *4*, 3001–3003.

44. M. Ocejo, J. L. Vicario, D. Badía, L. Carrillo, E. Reyes, *Synlett* **2005**, 2110–2112.

45. M. Frigerio, M. Santagostino, S. Sputore, G. Palmisano, *J. Org. Chem.* **1995**, *60*, 7272–7276.

46. M. Frigerio, M. Santagostino, S. Sputore, *J. Org. Chem.* **1999**, *64*, 4537–4538.

47. R. E. Ireland, L. Liu, *J. Org. Chem.* **1993**, *58*, 2899.

48. L. McDermott, D. Aljovic, Z. G. Walters, F. Peng, R. Zhao, K. R. Campos, N. K. Garg, *Org. Lett.* **2025**, *27*, 13165–13169.

49. D. B. Dess, J. C. Martin, *J. Am. Chem. Soc.* **1991**, *113*, 7277–7287.

50. A. Ozanne, L. Pouységu, D. Depernet, B. Francois, S. Quideau, *Org. Lett.* **2003**, *5*, 2903–2906.

51. D. B. Dess, J. C. Martin, *J. Org. Chem.* **1983**, *48*, 4156–4158.

52. H. Jiang, T.-Y. Sun, X. Wang, Y. Xie, X. Zhang, Y.-D. Wu, H. F. Schaefer, III, *Org. Lett.* **2017**, *19*, 6502–6505.

53. J. T. Su, W. A. Goddard III, *J. Am. Chem. Soc.* **2005**, *127*, 14146–14147.

54. S. D. Meyer, S. L. Schreiber, *J. Org. Chem.* **1994**, *59*, 7549–7552.

55. K. C. Nicolaou, P. S. Baran, R. Kranich, Y.-L. Zhong, K. Sugita, N. Zou, *Angew. Chem. Int. Ed.* **2001**, *40*, 202–206.

56. P. J. Stevenson, A. B. Treacy, M. Nieuwenhuyzen, *J. Chem. Soc. Perkin Trans. 2* **1997**, 589–591.

57. B. Chandrasekhar, S. Athe, P. P. Reddy, S. Ghosh, *Org. Biomol. Chem.* **2015**, *13*, 115–124.

58. E. Corey, A. Palani, *Tetrahedron Lett.* **1995**, *36*, 3485–3488.

59. T. Zoller, P. Breuilles, D. Uguen, A. D. Cian, J. Fischer, *Tetrahedron Lett.* **1999**, *40*, 6253–6256.

60. K. C. Nicolaou, Y.-L. Zhong, P. S. Baran, *J. Am. Chem. Soc.* **2000**, *122*, 7596–7597.

61. C. H. Kim, H. J. An, W. K. Shin, W. Yu, S. K. Woo, S. K. Jung, E. Lee, *Angew. Chem. Int. Ed.* **2006**, *45*, 8019–8021.

62. M. Uyanik, M. Akakura, K. Ishihara, *J. Am. Chem. Soc.* **2009**, *131*, 251–262.

63. P. L. Anelli, C. Biffi, F. Montanari, S. Quici, *J. Org. Chem.* **1987**, *52*, 2559–2562.

64. A. E. de Nooy, A. C. Besemer, H. van Bekkum, *Synthesis* **1996**, 1153–1174.

65. T. Vogler, A. Studer, *Synthesis* **2008**, 1979–1993.

66. R. Ciriminna, M. Pagliaro, *Org. Process Res. Dev.* **2010**, *14*, 245–251.

67. L. Tebben, A. Studer, *Angew. Chem. Int. Ed.* **2011**, *50*, 5034–5068.

68. D. Leifert, A. Studer, *Chem. Rev.* **2023**, *123*, 10302–10380.

69. A. D. Mico, R. Margarita, L. Parlanti, A. Vescovi, G. Piancatelli, *J. Org. Chem.* **1997**, *62*, 6974–6977.

70. J. B. Epp, T. S. Widlanski, *J. Org. Chem.* **1999**, *64*, 293–295.

71. M. Zhao, J. Li, E. Mano, Z. Song, D. M. Tschaen, E. J. J. Grabowski, P. J. Reider, *J. Org. Chem.* **1999**, *64*, 2564–2566.

72. A. de Nooy, A. Besemer, H. van Bekkum, *Recl. Trav. Chim. Pays-Bas* **1994**, *113*, 165–166.

73. W. F. Bailey, J. M. Bobbitt, K. B. Wiberg, *J. Org. Chem.* **2007**, *72*, 4504–4509.

74. M. F. Semmelhack, C. R. Schmid, D. A. Cortés, *Tetrahedron Lett.* **1986**, *27*, 1119–1122.

75. T. A. Hamlin, C. B. Kelly, J. M. Ovian, R. J. Wiles, L. J. Tilley, N. E. Leadbeater, *J. Org. Chem.* **2015**, *80*, 8150–8167.

76. S. A. Miller, J. Nandi, N. E. Leadbeater, N. A. Eddy, *Eur. J. Org. Chem.* **2020**, 108–112.

77. E. Tyrrell, G. A. Skinner, J. Janes, G. Milsom, *Synlett* **2002**, 1073–1076.

78. P. L. Anelli, S. Banfi, F. Montanari, S. Quici, *J. Org. Chem.* **1989**, *54*, 2970–2972.

79. J. Einhorn, C. Einhorn, F. Ratajczak, J.-L. Pierre, *J. Org. Chem.* **1997**, *61*, 7452–7454.

80. A. B. Benowitz, S. Fidanze, P. L. C. Small, Y. Kishi, *J. Am. Chem. Soc.* **2001**, *123*, 5128–5129.

81. M. A. Iron, A. M. Szpilman, *Chem. Eur. J.* **2017**, *23*, 1368–1378.

82. J. M. Hoover, S. S. Stahl, *J. Am. Chem. Soc.* **2011**, *133*, 16901–16910.

83. R. C. Walroth, K. C. Miles, J. T. Lukens, S. N. MacMillan, S. S. Stahl, K. M. Lancaster, *J. Am. Chem. Soc.* **2017**, *139*, 13507–13517.

84. M. C. Ryan, L. D. Whitmire, S. D. McCann, S. S. Stahl, *Inorg. Chem.* **2019**, *58*, 10194–10200.

85. A. N. Lowell, M. D. DeMars, S. T. Slocum, F. Yu, K. Anand, J. A. Chemler, N. Korakavi, J. K. Priessnitz, S. R. Park, A. A. Koch, P. J. Schultz, D. H. Sherman, *J. Am. Chem. Soc.* **2017**, *139*, 7913–7920.

86. V. von Kiedrowski, F. Quentin, M. Hiersemann, *Org. Lett.* **2017**, *19*, 4391–4394.

87. S. V. Ley, J. Norman, W. P. Griffith, S. P. Marsden, *Synthesis* **1994**, 639–666.

88. A.-K. C. Schmidt, C. B. W. Stark, *Org. Lett.* **2011**, *13*, 4164–4167.

89. W. P. Griffith, *Chem. Soc. Rev.* **1992**, *21*, 179–185.

90. T. J. Zerk, P. W. Moore, J. S. Harbort, S. Chow, L. Byrne, G. A. Koutsantonis, J. R. Harmer, M. Martínez, C. M. Williams, P. V. Bernhardt, *Chem. Sci.* **2017**, *8*, 8435–8442.

91. W. D. Chandler, Z. Wang, D. G. Lee, *Can. J. Chem.* **2005**, *83*, 1212–1221.

92. D. G. Lee, Z. Wang, W. D. Chandler, *J. Org. Chem.* **1992**, *57*, 3276–3277.

93. T. J. Zerk, P. W. Moore, C. M. Williams, P. V. Bernhardt, *Chem. Commun.* **2016**, *52*, 10301–10304.

94. J. Chan, T. F. Jamison, *J. Am. Chem. Soc.* **2003**, *125*, 11514–11515.

95. M. A. Kienzler, S. Suseno, D. Trauner, *J. Am. Chem. Soc.* **2008**, *130*, 8604–8605.

96. D. M. Mans, W. H. Pearson, *Org. Lett.* **2004**, *6*, 3305–3308.

97. S. Iimura, L. E. Overman, R. Paulini, A. Zakarian, *J. Am. Chem. Soc.* **2006**, *128*, 13095–13101.

98. K. C. Nicolaou, H. Ueno, J.-J. Liu, P. G. Nantermet, Z. Yang, J. Renaud, K. Paulvannan, R. Chadha, *J. Am. Chem. Soc.* **1995**, *117*, 653–659.

99. P. H. J. Carlsen, T. Katsuki, V. S. Martin, K. B. Sharpless, *J. Org. Chem.* **1981**, *46*, 3936–3938.

100. A. E. Boelrijk, J. Reedijk, *J. Mol. Catal.* **1994**, *89*, 63–76.

101. W. C. Still, H. Ohmizu, *J. Org. Chem.* **1981**, *46*, 5244–5246.

102. M. Besson, P. Gallezot, *Catal. Today* **2000**, *57*, 127–141.

103. P. Gallezot, *Catal. Today* **1997**, *37*, 405–418.

104. T. Mallat, A. Baiker, *Catal. Today* **1994**, *19*, 247–284.

105. P. Liu, E. N. Jacobsen, *J. Am. Chem. Soc.* **2001**, *123*, 10772–10773.

106. M. Rottenberg, P. Baertschi, *Helv. Chim. Acta* **1956**, *39*, 1973–1975.

107. M. D. Argyle, C. H. Bartholomew, *Catalysts* **2015**, *5*, 145–269.

108. T. Mallat, A. Baiker, *Chem. Rev.* **2004**, *104*, 3037–3058.

109. M. Bols, *J. Org. Chem.* **1991**, *56*, 5943–5945.

110. J. Fried, J. C. Sib, *Tetrahedron Lett.* **1973**, *40*, 3899–3902.

111. L. Johnson, D. L. Verraest, J. van Haveren, K. Hakala, J. A. Peters, H. van Bekkum, *Tetrahedron: Asymm.* **1994**, *5*, 2475–2484.

112. A. J. Fatiadi, *Synthesis* **1987**, 85–127.

113. R. Stewart, *J. Am. Chem. Soc.* **1957**, *79*, 3057–3061.

114. M. Kordes, H. Winsel, A. de Meijere, *Eur. J. Org. Chem.* **2000**, 3235–3245.

115. M. J. Begley, L. Crombie, R. C. F. Jones, C. J. Palmer, *J. Chem. Soc. Perkin Trans. 1* **1987**, 353–357.

116. A. Raach, O. Reiser, *J. Prakt. Chem.* **2000**, *342*, 605–608.

117. C. R. Chinake, O. Olojo, R. H. Simoyi, *J. Phys. Chem. A* **1998**, *102*, 606–611.

118. E. Dacanale, F. Montanari, *J. Org. Chem.* **1986**, *51*, 567–569.

119. H. S. Isbell, L. T. Sniegoski, *J. Res. Natl. Bur. Stand.* **1964**, *68A*, 301–304.

120. A. A. Hussein, A. A. M. Al-Hadedi, A. J. Mahrath, G. A. I. Moustafa, F. A. Almalki, A. Alqahtani, S. Shityakov, M. E. Algazally, *R. Soc. Open Sci.* **2020**, *7*, 191568.

121. K. S. Goh, C.-H. Tan, *RSC Adv.* **2012**, *2*, 5536–5538.

122. J. T. Arcari, C. Buckley, L. M. Buzon, Y. Dan, T. Houck, Z. Jin, C. Li, P. Li, K. Liu, Y. Liu, J. Magano, L. A. Martinez-Alsina, B. Nguyen, R. Pearson, J. L. Piper, J. A. Ragan, M. R. Reese, W. Tian, J. Van Haitsma, M. G. Vetelino, Y. Zhang, *Org. Process Res. Dev.* **2024**, *28*, 2935–2944.

123. R. K. Boeckman, M. del Rosario Rico Ferreira, L. H. Mitchell, P. Shao, *J. Am. Chem. Soc.* **2002**, *124*, 190–191.

124. M. Kobayashi, W. Wang, Y. Tsutsui, M. Sugimoto, N. Murakami, *Tetrahedron Lett.* **1998**, *39*, 8291–8294.

125. L. F. Tietze, S.-C. Duefert, J. Clerc, M. Bischoff, C. Maaß, D. Stalke, *Angew. Chem. Int. Ed.* **2013**, *52*, 3191–3194.

126. J. Einhorn, C. Einhorn, F. Ratajczak, J.-L. Pierre, *J. Org. Chem.* **1996**, *61*, 7452–7454.

127. M. Kaspar, E. Kudova, *J. Org. Chem.* **2022**, *87*, 9157–9170.

128. F. J. Kakis, M. Fetizon, N. Douchkine, M. Golfier, P. Mourgues, T. Prange, *J. Org. Chem.* **1974**, *39*, 523–533.

129. H. Adkins, R. M. Elofson, A. G. Rossow, C. C. Robinson, *J. Am. Chem. Soc.* **1949**, *71*, 3622–3629.

130. R. V. Stevens, K. T. Chapman, C. A. Stubbs, W. W. Tam, K. F. Albizati, *Tetrahedron Lett.* **1982**, *23*, 4647–4650.

131. J. B. Arterburn, *Tetrahedron* **2001**, *57*, 9765–9788.

132. C. F. de Graauw, J. A. Peters, H. van Bekkum, J. Huskens, *Synthesis* **1994**, 1007–1017.
133. C. R. Graves, E. J. Campbell, S. T. Nguyen, *Tetrahedron: Asymm.* **2005**, *16*, 3460–3468.
134. K. Ishihara, H. Kurihara, H. Yamamoto, *J. Org. Chem.* **1997**, *62*, 5664–5665.
135. T. Suzuki, K. Morita, M. Tsuchida, K. Hiroi, *JOC* **2003**, *68*, 1601–1602.
136. M. G. Coleman, A. N. Brown, B. A. Bolton, H. Guan, *Adv. Synth. Catal.* **2010**, *352*, 967–970.
137. A. J. Fatiadi, *Synthesis* **1976**, 65–104.
138. A. J. Fatiadi, *Synthesis* **1976**, 133–167.
139. J. C. Chabala, A. Rosegay, M. A. R. Walsh, *J. Agric. Food Chem.* **1981**, *29*, 881–884.
140. H. Kwart, T. J. George, *J. Org. Chem.* **1979**, *44*, 162–164.
141. A. Ohki, T. Ishiguchi, K. Kuzumi, *Tetrahedron* **1979**, *35*, 1737–1743.
142. G.-J. ten Brink, I. W. C. E. Arends, R. A. Sheldon, *Chem. Rev.* **2004**, *104*, 4105–4123.
143. M. Renz, B. Meunier, *Eur. J. Org. Chem.* **1999**, 737–750.
144. G. Strukul, *Angew. Chem. Int. Ed.* **1998**, *37*, 1198–1209.
145. L. Zhou, L. Lin, X. Liu, X. Feng in *Molecular Rearrangements in Organic Synthesis* (Ed.: C. M. Rojas), John Wiley & Sons, **2016**, Chapter 2, pp. 35–57.
146. K. Mukai, D. Urabe, S. Kasuya, N. Aoki, M. Inoue, *Angew. Chem. Int. Ed.* **2013**, *52*, 5300–5304.
147. M. E. Jung, R. M. Lui, *J. Org. Chem.* **2010**, *75*, 7146–7158.
148. O. Affolter, A. Baro, W. Frey, S. Laschat, *Tetrahedron* **2009**, *65*, 6626–6634.
149. C. M. Crudden, A. C. Chen, L. A. Calhoun, *Angew. Chem. Int. Ed.* **2000**, *39*, 2851–2855.
150. F. Grein, A. C. Chen, D. Edwards, C. M. Crudden, *J. Org. Chem.* **2006**, *71*, 861–872.
151. R. M. Goodman, Y. Kishi, *J. Am. Chem. Soc.* **1998**, *120*, 9392–9393.
152. R. D. Bach, *J. Org. Chem.* **2012**, *77*, 6801–6815.
153. Y. Itoh, M. Yamanaka, K. Mikami, *J. Org. Chem.* **2013**, *78*, 146–153.
154. K. Mislow, J. Brenner, *J. Am. Chem. Soc.* **1953**, *75*, 2318–2322.
155. M. Uyanik, K. Ishihara, *ACS Catal.* **2013**, *3*, 513–520.
156. H. Leisch, K. Morley, P. C. K. Lau, *Chem. Rev.* **2011**, *111*, 4165–4222.
157. G. de Gonzalo, M. D. Mihovilovic, M. W. Fraaije, *ChemBioChem* **2010**, *11*, 2208–2231.
158. M. Uyanik, D. Nakashima, K. Ishihara, *Angew. Chem. Int. Ed.* **2012**, *51*, 9093–9096.
159. R. E. Gawley, *Org. React.* **1988**, *35*, 1–420.
160. J. Ritz, H. Fuchs, H. Kieczka, W. C. Moran in *Caprolactam in Ullmann's Encyclopedia of Industrial Chemistry*, **2012**.
161. J. Tinge, M. Groothaert, H. op het Veld, J. Ritz, H. Fuchs, H. Kieczka, W. C. Moran in *Caprolactam in Ullmann's Encyclopedia of Industrial Chemistry*, **2018**.
162. K. K. Kelly, J. S. Matthews, *J. Org. Chem.* **1971**, *36*, 2159–2161.
163. R. K. Hill, O. T. Chortyk, *J. Am. Chem. Soc.* **1962**, *84*, 1064–1065.
164. J. Kenyon, D. P. Young, *J. Chem. Soc.* **1941**, 263–267.
165. S. Yamabe, N. Tsuchida, S. Yamazaki, *J. Org. Chem.* **2005**, *70*, 10638–10644.
166. S. P. Verevkin, V. N. Emel'yanenko, A. V. Toktonov, P. Goodrich, C. Hardacre, *Ind. Eng. Chem. Res.* **2009**, *48*, 9809–9816.
167. S. Xu, D. Unabara, D. Uemura, H. Arimoto, *Chem. Asian J.* **2014**, *9*, 367–375.
168. W. Bartmann, G. Beck, J. Knolle, R. H. Rupp, *Tetrahedron Lett.* **1982**, *23*, 3647–3650.
169. C. C. Nawrat, R. R. A. Kitson, C. J. Moody, *Org. Lett.* **2014**, *16*, 1896–1899.
170. L. De Luca, G. Giacomelli, A. Porcheddu, *J. Org. Chem.* **2002**, *67*, 6272–6274.
171. J. Mlochowski, H. Wójtowicz-Mlochowska, *Molecules* **2015**, *20*, 10205–10243.
172. N. Rabjohn, *Org. React.* **1976**, *24*, 261–415.
173. D. A. Singleton, C. Hang, *J. Org. Chem.* **2000**, *65*, 7554–7560.
174. W. J. Xia, D. R. Li, L. Shi, Y. Q. Tu, *Tetrahedron Lett.* **2002**, *43*, 627–630.
175. W. Yu, Z. Jin, *J. Am. Chem. Soc.* **2001**, *123*, 3369–3370.
176. H. Rapoport, U. T. Bhalerao, *J. Am. Chem. Soc.* **1971**, *93*, 4835–4840.

177. G. Park, J. Hwang, W. S. Jung, C. S. Ra, *Bull. Korean Chem. Soc.* **2005**, *26*, 1856–1860.

178. M. Varin, E. Barré, B. Iorga, C. Guillou, *Chem. Eur. J.* **2008**, *14*, 6606–6608.

179. C. J. Engelin, P. Fristrup, *Molecules* **2011**, *16*, 951–969.

180. M. S. Chen, N. Prabagaran, N. A. Labenz, M. C. White, *J. Am. Chem. Soc.* **2005**, *127*, 6970–6971.

181. S. A. Reed, A. R. Mazzotti, M. C. White, *J. Am. Chem. Soc.* **2009**, *131*, 11701–11706.

182. A. J. Young, M. C. White, *J. Am. Chem. Soc.* **2008**, *130*, 14090–14091.

183. C. C. Pattillo, I. I. Strambeanu, P. Calleja, N. A. Vermeulen, T. Mizuno, M. C. White, *J. Am. Chem. Soc.* **2016**, *138*, 1265–1272.

184. A. Sharma, J. F. Hartwig, *J. Am. Chem. Soc.* **2013**, *135*, 17983–17989.

185. V. Weidmann, W. Maison, *Synthesis* **2013**, *45*, 2201–2221.

186. T. C. Johnson, M. R. Chin, T. Han, J. P. Shen, T. Rana, D. Siegel, *J. Am. Chem. Soc.* **2016**, *138*, 6068–6073.

187. K. Nagaraju, R. Chegondi, S. Chandrasekhar, *Org. Lett.* **2016**, *18*, 2684–2687.

188. S. Marchart, A. Gromov, J. Mulzer, *Angew. Chem. Int. Ed.* **2010**, *49*, 2050–2053.

189. G. Majetich, J. S. Song, C. Ringold, G. A. Nemeth, M. G. Newton, *J. Org. Chem.* **1991**, *56*, 3973–3988.

190. S. Gnaim, J. C. Vantourout, F. Serpier, P.-G. Echeverria, P. S. Baran, *ACS Catal.* **2021**, *11*, 883–892.

191. S. S. Stahl, T. Diao in *Comp. Org. Synth. II* (Ed.: P. Knochel), Elsevier, **2014**, Chapter 7.06, pp. 178–212.

192. H. J. Reich, S. Wollowitz, *Org. React.* **1993**, *44*, 1–296.

193. H. J. Reich, *Acc. Chem. Res.* **1979**, *12*, 22–30.

194. H.-H. Lu, S. V. Pronin, Y. Antonova-Koch, S. Meister, E. A. Winzeler, R. A. Shenvi, *J. Am. Chem. Soc.* **2016**, *138*, 7268–7271.

195. H. J. Reich, S. Wollowitz, J. E. Trend, F. Chow, D. F. Wendelborn, *J. Org. Chem.* **1978**, *43*, 1697–1705.

196. S. Uemura, Y. Hirai, K. Ohe, N. Sugita, *J. Chem. Soc. Chem. Commun.* **1985**, 1037–1038.

197. Y. Ito, T. Hirao, T. Saegusa, *J. Org. Chem.* **1978**, *43*, 1011–1013.

198. J. Le Bras, J. Muzart, *Org. React.* **2019**, *98*, 1–172.

199. Y. Ito, M. Suginome in *Handbook of Organopalladium Chemistry for Organic Synthesis* (Ed.: E. Negishi), John Wiley & Sons, **2002**, Chapter VIII.3.1, pp. 2873–2879.

200. R. C. Larock, T. R. Hightower, G. A. Kraus, P. Hahn, D. Zheng, *Tetrahedron Lett.* **1995**, *36*, 2423–2426.

201. S. Porth, J. W. Bats, D. Trauner, G. Giester, J. Mulzer, *Angew. Chem. Int. Ed.* **1999**, *38*, 2015–2016.

202. D. Wang, A. B. Weinstein, P. B. White, S. S. Stahl, *Chem. Rev.* **2018**, *118*, 2636–2679.

203. F. Yoshimura, M. Sasaki, I. Hattori, K. Komatsu, M. Sakai, K. Tanino, M. Miyashita, *Chem. Eur. J.* **2009**, *15*, 6626–6644.

204. P. Magnus, G. F. Miknis, N. J. Press, D. Grandjean, G. M. Taylor, J. Harling, *J. Am. Chem. Soc.* **1997**, *119*, 6739–6748.

205. A. Nakayama, N. Kogure, M. Kitajima, H. Takayama, *Org. Lett.* **2009**, *11*, 5554–5557.

206. A. S. Kende, J. I. M. Hernando, J. B. J. Milbank, *Org. Lett.* **2001**, *3*, 2505–2508.

207. T. Hirao, *J. Org. Chem.* **2019**, *84*, 1687–1692.

208. D. Huang, T. R. Newhouse, *Acc. Chem. Res.* **2021**, *54*, 118–1130.

209. A. W. Schuppe, Y. Zhao, Y. Liu, T. R. Newhouse, *J. Am. Chem. Soc.* **2019**, *141*, 9191–9196.

210. J. Zhou, D.-X. Tan, F.-S. Han, *Angew. Chem. Int. Ed.* **2020**, *59*, 18731–18740.

211. H. R. Khatri, B. Bhattarai, W. Kaplan, Z. Li, M. J. C. Long, Y. Aye, P. Nagorny, *J. Am. Chem. Soc.* **2019**, *141*, 4849–4860.

212. S. R. McCabe, P. Wipf, *Angew. Chem. Int. Ed.* **2017**, *56*, 324–327.
213. N. Zhao, S. Yin, S. Xie, H. Yan, P. Ren, G. Chen, F. Chen, J. Xu, *Angew. Chem. Int. Ed.* **2018**, *57*, 3386–3390.
214. M. Wohlfahrt, K. Harms, U. Koert, *Angew. Chem. Int. Ed.* **2011**, *50*, 8404–8406.
215. H. Shi, I. N. Michaelides, B. Darses, P. Jakubec, Q. N. N. Nguyen, R. S. Paton, D. J. Dixon, *J. Am. Chem. Soc.* **2017**, *139*, 17755–17758.
216. D. Dagoneau, Z. Xu, Q. Wang, J. Zhu, *Angew. Chem. Int. Ed.* **2016**, *55*, 760–763.
217. K. C. Nicolaou, J. Krieger, G. M. Murhade, P. Subramanian, B. D. Dherange, D. Vourloumis, S. Munneke, B. Lin, C. Gu, H. Sarvaiya, J. Sandoval, Z. Zhang, M. Aujay, J. W. Purcell, J. Gavrilyuk, *J. Am. Chem. Soc.* **2020**, *142*, 15476–15487.
218. J. B. Cox, A. Kimishima, J. L. Wood, *J. Am. Chem. Soc.* **2019**, *141*, 25–28.
219. J. Adrian, C. B. W. Stark, *J. Org. Chem.* **2016**, *81*, 8175–8186.
220. H. C. Brown, S. Krishnamurthy, *Tetrahedron* **1979**, *35*, 567–607.
221. J. Seyden-Penne, *Reductions by the Alumino- and Borohydrides in Organic Synthesis*, Wiley-VCH, 2nd ed., **1997**.
222. B. Kammermeier in *Reduction in Ullmann's Encyclopedia of Industrial Chemistry*, **2000**.
223. M. Hudlicky, *Reductions in Organic Synthesis*, 2nd ed., **1996**.
224. J. Magano, J. R. Dunetz, *Org. Process Res. Dev.* **2012**, *16*, 1156–1184.
225. E. M. Kaiser, *Synthesis* **1972**, 391–415.
226. R. L. Augustine, *Heterogeneous Catalysis for the Synthetic Chemist*, Marcel Dekker, INC, **1996**.
227. D. Richter, H. Mayr, *Angew. Chem. Int. Ed.* **2009**, *48*, 1958–1961.
228. H. C. Brown, S. Krishnamurthy, *J. Am. Chem. Soc.* **1973**, *95*, 1669–1671.
229. G. D. Paderes, P. Metivier, W. L. Jorgensen, *J. Org. Chem.* **1991**, *56*, 4718–4733.
230. E. R. Burkhardt, K. Matos, *Chem. Rev.* **2006**, *106*, 2617–2650.
231. J. S. Cha, *Bull. Korean Chem. Soc.* **2011**, *32*, 1808–1846.
232. E. R. H. Walker, *Chem. Soc. Rev.* **1976**, *5*, 23–50.
233. E. Winterfeldt, *Synthesis* **1975**, 617–630.
234. A. F. Abdel-Magid, S. J. Mehrman, *Org. Process Res. Dev.* **2006**, *10*, 971–1031.
235. C. F. Lane, *Aldrichimica Acta* **1975**, *8*, 3–10.
236. J. S. Cha, M. K. Jeong, O. O. Kwon, K. D. Lee, H. S. Lee, *Bull. Korean Chem. Soc.* **1994**, *15*, 873–881.
237. V. Bazant, M. Capka, M. Cerný, V. Chvalovský, K. Kochloefl, M. Kraus, J. Málek, *Tetrahedron Lett.* **1968**, *9*, 3303–3306.
238. L. Pasumansky, C. T. Goralski, B. Singaram, *Org. Process Res. Dev.* **2006**, *10*, 959–970.
239. J.-M. Lefour, A. Loupy, *Tetrahedron* **1978**, *34*, 2597–2605.
240. J. J. Gajewski, W. Bocian, N. J. Harris, L. P. Olson, J. P. Gajewski, *J. Am. Chem. Soc.* **1999**, *121*, 326–334.
241. E. C. Ashby, J. R. Boone, *J. Am. Chem. Soc.* **1976**, *98*, 5524–5531.
242. D. C. Wigfield, *Tetrahedron* **1979**, *35*, 449–462.
243. H. Haubenstock, E. L. Eliel, *J. Am. Chem. Soc.* **1962**, *84*, 2363–2368.
244. D. C. Wigfield, F. W. Gowland, *J. Org. Chem.* **1977**, *42*, 1108–1109.
245. O. Eisenstein, H. B. Schlegel, M. M. Kayser, *J. Org. Chem.* **1982**, *47*, 2886–2891.
246. S. Pal, *Organometallics* **2023**, *42*, 3099–3108.
247. A. L. Gemal, J.-L. Luche, *J. Am. Chem. Soc.* **1981**, *103*, 5454–5459.
248. H. C. Brown, H. M. Hess, *J. Org. Chem.* **1969**, *34*, 2206–2209.
249. S. Fukuzawa, T. Fujinami, S. Yamauchi, S. Sakai, *J. Chem. Soc. Perkin Trans. 1* **1986**, 1929–1932.
250. G. A. Molander, *Chem. Rev.* **1992**, *92*, 29–68.

251. N. Kato, H. Kataoka, S. Ohbuchi, S. Tanaka, H. Takeshita, *J. Chem. Soc. Chem. Commun.* **1988**, 354–356.

252. G. E. Veitch, E. Beckmann, B. J. Burke, A. Boyer, C. Ayats, S. V. Ley, *Angew. Chem. Int. Ed.* **2007**, *46*, 7633–7635.

253. A. A. Denholm, L. Jennens, S. V. Ley, A. Wood, *Tetrahedron* **1995**, *51*, 6591–6604.

254. J. T. Kuethe, D. L. Comins, *J. Org. Chem.* **2004**, *69*, 5219–5231.

255. D. A. Evans, W. C. Black, *J. Am. Chem. Soc.* **1993**, *115*, 4497–4513.

256. M. Boch, T. Korth, J. M. Nelke, D. Pike, H. Radunz, E. Winterfeldt, *Chem. Ber.* **1972**, *105*, 2126–2142.

257. A. Boussonnière, R. Bénéteau, J. Lebreton, F. Dénès, *Eur. J. Org. Chem.* **2013**, 7853–7866.

258. E. C. Ashby, S. H. Yu, *J. Org. Chem.* **1970**, *35*, 1034–1040.

259. B. Mudryk, C. A. Shook, T. Cohen, *J. Am. Chem. Soc.* **1990**, *112*, 6389–6391.

260. L. C. Dias, P. R. R. Meira, *Tetrahedron Lett.* **2002**, *43*, 8883–8885.

261. M. Mentzel, P. H. M. R. Hoffmann, *J. Prakt. Chem.* **1997**, *339*, 517–524.

262. J. Singh, N. Satyamurthi, I. Singh Aidhen, *J. Prakt. Chem.* **2000**, *342*, 340–347.

263. S. Nahm, S. M. Weinreb, *Tetrahedron Lett.* **1981**, *22*, 3815–3818.

264. M. B. Andrus, E. L. Meredith, E. J. Hicken, B. L. Simmons, R. R. Glancey, W. Ma, *J. Org. Chem.* **2003**, *68*, 8162–8169.

265. Akzo Nobel, *Technical Bulletin OMS 06.388.01*, **2006**.

266. N. M. Yoon, C. S. Pak, H. C. Brown, S. Krishnamurthy, T. P. Stocky, *J. Org. Chem.* **1973**, *38*, 2786–2792.

267. C. F. Lane, *Chem. Rev.* **1976**, *76*, 773–799.

268. M. DiMare, *J. Org. Chem.* **1996**, *61*, 8378–8385.

269. H. C. Brown, T. P. Stocky, *J. Am. Chem. Soc.* **1977**, *99*, 8218–8226.

270. P. C. Lobben, S. S.-W. Leung, S. Tummala, *Org. Process Res. Dev.* **2004**, *8*, 1072–1075.

271. B. W. Gung, H. Dickson, *Org. Lett.* **2002**, *4*, 2517–2519.

272. S.-K. Khim, A. G. Schultz, *J. Org. Chem.* **2004**, *69*, 7734–7736.

273. S. Krishnamurthy, R. M. Schubert, H. C. Brown, *J. Am. Chem. Soc.* **1973**, *95*, 8486–8487.

274. H. C. Brown, N. M. Yoon, *J. Am. Chem. Soc.* **1966**, *88*, 1464–1472.

275. A. Ookawa, H. Hiratsuka, K. Soai, *Bull. Chem. Soc. Japan* **1987**, *60*, 1813–1817.

276. S. Krishnamurthy, H. C. Brown, *J. Org. Chem.* **1979**, *44*, 3678–3682.

277. N. M. Yoon, Y. S. Gyoung, *J. Org. Chem.* **1985**, *50*, 2443–2450.

278. R. Fuchs, *J. Am. Chem. Soc.* **1956**, *78*, 5612–5613.

279. K. Soai, A. Ookawa, *J. Org. Chem.* **1986**, *51*, 4000–4005.

280. H. C. Brown, B. C. S. Rao, *J. Am. Chem. Soc.* **1960**, *82*, 681–686.

281. J. J. Eisch, Z. R. Liu, M. Singh, *J. Org. Chem.* **1992**, *57*, 1618–1621.

282. R. O. Hutchins, I. M. Taffer, W. Burgoyne, *J. Org. Chem.* **1981**, *46*, 5214–5215.

283. E. L. Eliel, D. W. Delmonte, *J. Am. Chem. Soc.* **1958**, *80*, 1744–1752.

284. T. Hansen, P. Vermeeren, A. Haim, M. J. H. van Dorp, J. D. C. Codée, F. M. Bickelhaupt, T. A. Hamlin, *Eur. J. Org. Chem.* **2020**, 3822–3828.

285. A. Riera, M. Moreno, *Molecules* **2010**, *15*, 1041–1073.

286. J. M. Finan, Y. Kishi, *Tetrahedron Lett.* **1982**, *23*, 2719–2722.

287. D. Tanner, T. Groth, *Tetrahedron* **1997**, *53*, 16139–1614.

288. M. Ball, B. J. Bradshaw, R. Dumeunier, T. J. Gregson, S. MacCormick, H. Omori, E. J. Thomas, *Tetrahedron Lett.* **2006**, *47*, 2223–2227.

289. S. Krishnamurthy, H. C. Brown, *J. Org. Chem.* **1982**, *47*, 276–280.

290. E. C. Ashby, R. N. DePriest, A. B. Goel, B. Wenderoth, T. N. Pham, *J. Org. Chem.* **1984**, *49*, 3545–3556.

291. E. C. Ashby, C. O. Welder, *J. Org. Chem.* **1997**, *62*, 3542–3551.

292. J. B. Son, S. N. Kim, N. Y. Kim, D. H. Lee, *Org. Lett.* **2006**, *8*, 661–664.

293. K. M. Goodenough, W. J. Moran, P. Raubo, J. P. A. Harrity, *J. Org. Chem.* **2005**, *70*, 207–213.

294. M. M. Midland, Y. C. Kwon, *J. Am. Chem. Soc.* **1983**, *105*, 3725–3727.

295. M. T. Reetz, *Acc. Chem. Res.* **1993**, *26*, 462–468.

296. T. Oishi, T. Nakata, *Acc. Chem. Res.* **1984**, *17*, 338–344.

297. T. Takahashi, A. Ootake, H. Yamada, J. Tsuji, *Tetrahedron Lett.* **1985**, *26*, 69–72.

298. L. E. Overman, R. J. McCready, *Tetrahedron Lett.* **1982**, *23*, 2355–2358.

299. B. C. Ranu, *Synlett* **1993**, 885–892.

300. D. A. Evans, K. T. Chapman, E. M. Carreira, *J. Am. Chem. Soc.* **1988**, *110*, 3560–3578.

301. T. Takahashi, M. Miyazawa, J. Tsuji, *Tetrahedron Lett.* **1985**, *26*, 5139–5142.

302. T. Nakata, T. Tanaka, T. Oishi, *Tetrahedron Lett.* **1983**, *24*, 2653–2656.

303. S. J. Mickel, D. Niederer, R. Daeffler, A. Osmani, E. Kuesters, E. Schmid, K. Schaer, R. Gamboni, W. Chen, E. Loeser, F. R. Kinder, K. Konigsberger, K. Prasad, T. M. Ramsey, O. Repic, R.-M. Wang, G. Florence, I. Lyothier, I. Paterson, *Org. Process Res. Dev.* **2004**, *8*, 122–130.

304. B. W. Gung, *Tetrahedron* **1996**, *52*, 5263–5301.

305. E. C. Ashby, J. T. Laemmle, *Chem. Rev.* **1975**, *75*, 521–546.

306. M. Cherest, *Tetrahedron* **1980**, *36*, 1593–1598.

307. B. W. Gung, *Chem. Rev.* **1999**, *99*, 1377–1386.

308. M. Cherest, H. Felkin, *Tetrahedron Lett.* **1968**, *9*, 2205–2209.

309. J. S. Cha, O. O. Kwon, J. M. Kim, S. D. Cho, *Synlett* **1997**, 1465–1466.

310. E. C. Ashby, J. R. Boone, *J. Org. Chem.* **1976**, *41*, 2890–2903.

311. P. T. Lansbury, R. E. MacLeay, *J. Org. Chem.* **1963**, *28*, 1940–1941.

312. D. C. Ayres, D. N. Kirk, R. Sawdaye, *J. Chem. Soc. B* **1970**, 505–510.

313. H. C. Brown, W. C. Dickason, *J. Am. Chem. Soc.* **1970**, *92*, 709–710.

314. H. C. Brown, S. Krishnamurthy, N. M. Yoon, *J. Org. Chem.* **1976**, *41*, 1778–1791.

315. J. Klein, E. Dunkelblum, E. Eliel, Y. Senda, *Tetrahedron Lett.* **1968**, *9*, 6127–6130.

316. E. C. Ashby, J. P. Sevenair, F. R. Dobbs, *J. Org. Chem.* **1971**, *36*, 197–199.

317. R. J. McMahon, K. E. Wiegers, S. G. Smith, *J. Org. Chem.* **1981**, *46*, 99–101.

318. N. M. Yoon, H. C. Brown, *J. Am. Chem. Soc.* **1968**, *90*, 2927–2938.

319. M. C. Barden, J. Schwartz, *J. Org. Chem.* **1995**, *60*, 5963–5965.

320. H. Handel, J.-L. Pierre, *Tetrahedron Lett.* **1976**, *17*, 2029–2032.

321. H. C. Brown, S. Krishnamurthy, *J. Am. Chem. Soc.* **1972**, *94*, 7159–7161.

322. C. A. Brown, *J. Am. Chem. Soc.* **1973**, *95*, 4100–4102.

323. I. Gavrilovic, K. Mitchell, A. D. Brailsford, D. A. Cowan, A. T. Kicman, R. J. Ansell, *Steroids* **2011**, *76*, 478–483.

324. Y.-D. Wu, K. N. Houk, M. N. Paddon-Row, *Angew. Chem. Int. Ed. Engl.* **1992**, *31*, 1019–1021.

325. S. R. N. A. G. Jimenez-Oses, D. L. Comins, K. N. Houk, *J. Org. Chem.* **2014**, *79*, 11609–11618.

326. H. C. Brown, K. Ichikawa, *Tetrahedron* **1957**, *1*, 221–230.

327. H. C. Brown, O. H. Wheeler, K. Ichikawa, *Tetrahedron* **1957**, *1*, 214–220.

328. H. C. Brown, K. Ichikawa, *J. Am. Chem. Soc.* **1962**, *84*, 373–376.

329. D. C. Wigfield, D. J. Phelps, *J. Chem. Soc. Perkin Trans. 1* **1972**, 680–683.

330. D. C. Wigfield, D. J. Phelps, *J. Am. Chem. Soc.* **1974**, *96*, 543–549.

331. G. L. Larson, J. L. Fry, *Ionic and Organometallic-Catalyzed Organosilane Reductions*, John Wiley & Sons, **2010**.

332. F. A. Lakhvich, L. Lich, D. B. Rubinov, I. L. Rubinova, A. A. Akhren, *J Org. Chem. USSR (Engl. Transl.)* **1990**, *25*, 1493.

333. D. N. Kursanov, Z. N. Parnes, N. M. Loin, *Synthesis* **1975**, 633–651.

334. F. A. Carey, H. S. Tremper, *J. Am. Chem. Soc.* **1968**, *90*, 2578–2583.

335. R. J. P. Corriu, R. Perz, C. Reye, *Tetrahedron* **1983**, *39*, 999–1009.

336. M. P. Doyle, D. J. DeBruyn, S. J. Donnelly, D. A. Kooistra, A. A. Odubela, C. T. West, S. M. Zonnebelt, *J. Org. Chem.* **1974**, *39*, 2740–2747.

337. M. P. Doyle, C. T. West, S. J. Donnelly, C. C. McOsker, *J. Organomet. Chem.* **1976**, *117*, 129–140.

338. S. Takano, M. Moriya, K. Ogasawara, *Tetrahedron Lett.* **1992**, *33*, 1909–1910.

339. C. Deutsch, N. Krause, B. H. Lipshutz, *Chem. Rev.* **2008**, *108*, 2916–2927.

340. A. Srikrishna, G. Ravi, G. Satyanarayana, *Tetrahedron Lett.* **2007**, *48*, 73–76.

341. M. R. Agharahimi, N. A. LeBel, *J. Org. Chem.* **1995**, *60*, 1856–1863.

342. L. F. Tietze, T. Kinzel, T. Wolfram, *Chem. Eur. J.* **2009**, *15*, 6199–6210.

343. D. K. Joshi, J. W. Sutton, S. Carver, J. P. Blanchard, *Org. Process Res. Dev.* **2005**, *9*, 997–1002.

344. A. Greenfield, U. Schindewolf, *Ber. Bunsenges. Phys. Chem.* **1998**, *102*, 1808–1814.

345. P. W. Rabideau, Z. Marcinow, *Org. React.* **1992**, *42*, 1–334.

346. H. E. Zimmerman, *Acc. Chem. Res.* **2012**, *45*, 164–170.

347. M. J. S. Dewar, M. A. Fox, D. J. Nelson, *J. Organomet. Chem.* **1980**, *185*, 157–181.

348. R. A. Benkeser, R. E. Robinson, D. M. Sauve, O. H. Thomas, *J. Am. Chem. Soc.* **1955**, *77*, 3230–3233.

349. A. J. Birch, A. R. Murray, H. Smith, *J. Chem. Soc.* **1951**, 1945–1950.

350. K. M. Peese, D. Y. Gin, *J. Am. Chem. Soc.* **2006**, *128*, 8734–8735.

351. A. G. Schultz, *Chem. Commun.* **1999**, 1263–1271.

352. T. J. Donohoe, M. Helliwell, C. A. Stevenson, T. Ladduwahetty, *Tetrahedron Lett.* **1998**, *39*, 3071–3074.

353. T. J. Donohoe, P. M. Guyo, M. Helliwell, *Tetrahedron Lett.* **1999**, *40*, 435–438.

354. A. G. Schultz, A. Wang, *J. Am. Chem. Soc.* **1998**, *120*, 8259–8260.

355. R. Damrauer, *J. Org. Chem.* **2006**, *71*, 9165–9171.

356. D. Kaufman, E. Johnson, M. D. Mosher, *Tetrahedron Lett.* **2005**, *46*, 5613–5615.

357. R. R. Dewald, C. J. Ekstein, W. M. Song, *J. Am. Chem. Soc.* **1987**, *109*, 6921–6922.

358. L. C. Dias, L. G. de Oliveira, , M. A. de Sousa, *Org. Lett.* **2003**, *5*, 265–268.

359. S. K. Pradhan, *Tetrahedron* **1986**, *42*, 6351–6388.

360. R. W. Hoffmann, *Angew. Chem. Int. Ed.* **2005**, *44*, 6277–6279.

361. M. Uchiyama, Y. Matsumoto, S. Nakamura, T. Ohwada, N. Kobayashi, N. Yamashita, A. Matsumiya, T. Sakamoto, *J. Am. Chem. Soc.* **2004**, *126*, 8755–8759.

362. W. P. Neumann, *Synthesis* **1987**, 665–683.

363. D. Crich, S. Sun, *J. Org. Chem.* **1996**, *61*, 7200–7201.

364. C. Y. Lin, J. Peh, M. L. Coote, *J. Org. Chem.* **2011**, *76*, 1715–1726.

365. W. B. Motherwell, D. Crich, *Free Radical Chain Reactions in Organic Synthesis (Best Synthetic Methods)*, Academic Press, **1992**.

366. K. U. Ingold, V. W. Bowry, *J. Org. Chem.* **2015**, *80*, 1321–1331.

367. S. Takano, S. Nishizawa, M. Akiyama, K. Ogasawara, *Synthesis* **1984**, 949–950.

368. J. D. Buynak, M. N. Rao, H. Pajouhesh, R. Y. Chandrasekaran, K. Finn, P. D. Meester, S. C. Chu, *J. Org. Chem.* **1985**, *50*, 4245–4252.

369. Z. J. Duri, B. M. Fraga, J. R. Hanson, *J. Chem. Soc. Perkin Trans. 1* **1981**, 161–164.

370. T. Sohn, D. Kim, R. S. Paton, *Chem. Eur. J.* **2015**, 15988–15997.

371. M. M. Hayward, R. M. Roth, K. J. Duffy, P. I. Dalko, K. L. Stevens, J. Guo, Y. Kishi, *Angew. Chem. Int. Ed.* **1998**, *37*, 192–196.

372. D. Crich, L. Quintero, *Chem. Rev.* **1989**, *89*, 1413–1432.

373. S. W. McCombie, B. Quiclet-Sire, S. Z. Zard, *Tetrahedron* **2018**, *74*, 4969–4979.

374. D. P. Curran, *Synthesis* **1988**, 417–439.

375. E. J. Corey, A. K. Ghosh, *Tetrahedron Lett.* **1988**, *29*, 3205–3206.

376. B. Lei, A. G. Fallis, *J. Am. Chem. Soc.* **1990**, *112*, 4609–4610.

377. T. Hudlicky, L. Radesca-Kwart, L.-Q. Li, T. Bryant, *Tetrahedron Lett.* **1988**, *29*, 3283–3286.
378. G. E. Veitch, E. Beckmann, B. J. Burke, A. Boyer, S. L. Maslen, S. V. Ley, *Angew. Chem. Int. Ed.* **2007**, *46*, 7629–7632.
379. M. F. Saraiva, M. R. Couri, M. L. Hyaric, M. V. de Almeida, *Tetrahedron* **2009**, *65*, 3563–3572.
380. J. Zhu, A. J. Klunder, B. Zwanenburg, *Tetrahedron* **1995**, *51*, 5099–5116.
381. A. Studer, S. Amrein, *Synthesis* **2002**, 835–849.
382. P. A. Baguley, J. C. Walton, *Angew. Chem. Int. Ed.* **1998**, *37*, 3072–3082.
383. R. M. Lopez, D. S. Hays, G. C. Fu, *J. Am. Chem. Soc.* **1997**, *119*, 6949–6950.
384. E. Le Grognec, J.-M. Chretien, F. Zammattio, J.-P. Quintard, *Chem. Rev.* **2015**, *115*, 10207–10260.
385. D. A. Spiegel, K. B. Wiberg, L. N. Schacherer, M. R. Medeiros, J. L. Wood, *J. Am. Chem. Soc.* **2005**, *127*, 12513–12515.
386. M. R. Medeiros, L. N. Schacherer, D. A. Spiegel, J. L. Wood, *Org. Lett.* **2007**, *9*, 4427–4429.
387. T. Qin, L. R. Malins, J. T. Edwards, R. R. Merchant, A. J. E. Novak, J. Z. Zhong, R. B. Mills, M. Yan, C. Yuan, M. D. Eastgate, P. S. Baran, *Angew. Chem. Int. Ed.* **2017**, *56*, 260–265.
388. J. M. R. Narayanam, J. W. Tucker, C. R. J. Stephenson, *J. Am. Chem. Soc.* **2009**, *131*, 8756–8757.
389. J. H. Lee, S. Mho, *J. Org. Chem.* **2015**, *80*, 3309–3314.
390. H. Yamamoto, K. Yamaoka, A. Shinohara, K. Shibata, K. Takao, A. Ogura, *Chem. Sci.* **2023**, *14*, 11243–11250.
391. C. Chatgilialoglu, *Chem. Eur. J.* **2008**, *14*, 2310–2320.
392. C. Chatgilialoglu, *Acc. Chem. Res.* **1992**, *25*, 188–194.
393. C. Chatgilialoglu, J. Dickhaut, B. Giese, *J. Org. Chem.* **1991**, *56*, 6399–6403.
394. A. Fürstner, E. Kattnig, O. Lepage, *J. Am. Chem. Soc.* **2006**, *128*, 9194–9204.
395. M. Hanaoka, S. Yoshida, C. Mukai, *Tetrahedron Lett.* **1985**, *26*, 5163–5166.
396. R. J. Sundberg, G. S. Hamilton, J. P. Laurino, *J. Org. Chem.* **1988**, *53*, 976–983.
397. F. C. Whitmore, E. W. Pietrusza, L. H. Sommer, *J. Am. Chem. Soc.* **1947**, *69*, 2108–2110.
398. K. Hirano, K. Fujita, H. Yorimitsu, H. Shinokubo, K. Oshima, *Tetrahedron Lett.* **2004**, *45*, 2555–2557.
399. H. Fang, M. Oestreich, *Chem. Sci.* **2020**, *11*, 12604–12615.
400. T. A. Bender, P. R. Payne, M. R. Gagné, *Nat. Chem.* **2017**, *10*, 85–90.
401. J. W. Burton in *Comp. Org. Synth. II* (Ed.: J. Clayden), Elsevier, **2014**, Chapter 8.12, pp. 446–478.
402. M. Toda, M. Hayashi, Y. Hirata, S. Yamamura, *Bull. Chem. Soc. Japan* **1972**, *45*, 264–266.
403. E. Vedejs, *Org. React.* **1975**, *22*, 401–422.
404. J. Burdon, R. C. Price, *J. Chem. Soc. Chem. Commun.* **1986**, 893–894.
405. M. L. Di Vona, V. Rosnati, *J. Org. Chem.* **1991**, *56*, 4269–4273.
406. M. L. Di Vona, B. Floris, L. Luchetti, V. Rosnati, *Tetrahedron Lett.* **1990**, *31*, 6081–6084.
407. C. Villiers, M. Ephritikhine, *Chem. Eur. J.* **2001**, *7*, 3043–3051.
408. M. Naruse, S. Aoyagi, C. Kibayashi, *J. Chem. Soc. Perkin Trans. 1* **1986**, 1113–1124.
409. H. H. Szmant, C. E. Alciaturi, *J. Solution Chem.* **1978**, *7*, 269–281.
410. H. H. Szmant, H. F. Harnsberger, T. J. Butler, W. P. Barie, *J. Am. Chem. Soc.* **1952**, *74*, 2724–2728.
411. D. F. Taber, S. J. Stachel, *Tetrahedron Lett.* **1992**, *33*, 903–906.
412. S. Yamabe, G. Zeng, W. Guan, S. Sakaki, *Beilstein J. Org. Chem.* **2014**, *10*, 259–270.
413. J. T. Kuethe, K. G. Childers, Z. Peng, M. Journet, G. R. Humphrey, *Org. Process Res. Dev.* **2009**, *13*, 576–580.
414. D. J. Cram, M. R. V. Sahyun, *J. Am. Chem. Soc.* **1962**, *84*, 1734–1735.
415. M. F. Grundon, H. B. Henbest, M. D. Scott, *J. Chem. Soc.* **1963**, 1855–1858.
416. M. E. Furrow, A. G. Myers, *J. Am. Chem. Soc.* **2004**, *126*, 5436–5445.

417. L. Cagliotti, *Tetrahedron* **1966**, *22*, 487–493.
418. R. O. Hutchins, B. Maryanoff, C. Milewski, *J. Am. Chem. Soc.* **1971**, *93*, 1793–1794.
419. D. F. Taber, Y. Wang, S. J. Stachel, *Tetrahedron Lett.* **1993**, *34*, 6209–6210.
420. G. W. Kabalka, D. T. C. Yang, J. D. Baker, *J. Org. Chem.* **1976**, *41*, 574–575.
421. V. P. Miller, D. Yang, T. M. Weigel, O. Han, H. Liu, *J. Org. Chem.* **1989**, *54*, 4175–4188.
422. W. Qi, M. C. McIntosh, *Org. Lett.* **2008**, *10*, 357–359.
423. M. Zhang, N. Liu, W. Tang, *J. Am. Chem. Soc.* **2013**, *135*, 12434–12438.
424. F. He, Y. Bo, J. D. Altom, E. J. Corey, *J. Am. Chem. Soc.* **1999**, *121*, 6771–6772.
425. G. E. Keck, R. L. Giles, V. J. Cee, C. A. Wager, T. Yu, M. B. Kraft, *J. Org. Chem.* **2008**, *73*, 9675–9691.
426. K. B. Hansen, Y. Hsiao, F. Xu, N. Rivera, A. Clausen, M. Kubryk, S. Krska, T. Rosner, B. Simmons, J. Balsells, N. Ikemoto, Y. Sun, F. Spindler, C. Malan, E. J. J. Grabowski, J. D. Armstrong III., *J. Am. Chem. Soc.* **2009**, *131*, 8798–8804.
427. H.-U. Blaser, C. Malan, B. Pugin, F. Spindler, H. Steiner, M. Studer, *Adv. Synth. Catal.* **2003**, *345*, 103–151.
428. *Handbook of Homogenous Hydrogenation, Vol. 1-3*, (Eds.: J. G. de Vries, C. J. Elsevier), Wiley-VCH, **2007**.
429. R. E. Harmon, S. K. Gupta, D. J. Brown, *Chem. Rev.* **1973**, *73*, 21–52.
430. L. A. Oro, D. Carmona in *Handbook of Homogenous Hydrogenation* (Eds.: J. G. de Vries, C. J. Elsevier), Wiley-VCH, **2007**, Chapter 1.
431. R. Crabtree, *Acc. Chem. Res.* **1979**, *12*, 331–337.
432. J. P. Candlin, A. R. Oldham, *Discuss. Faraday Soc.* **1968**, *46*, 60–71.
433. A. Fürstner, T. Nagano, *J. Am. Chem. Soc.* **2007**, *129*, 1906–1907.
434. A. H. Hoveyda, D. A. Evans, G. C. Fu, *Chem. Rev.* **1993**, *93*, 1307–1370.
435. J. Halpern, *Inorg. Chim. Acta* **1981**, *50*, 11–19.
436. S. B. Duckett, C. L. Newell, R. Eisenberg, *J. Am. Chem. Soc.* **1994**, *116*, 10548–10556.
437. S. B. Duckett, C. L. Newell, R. Eisenberg, *J. Am. Chem. Soc.* **1997**, *119*, 2068.
438. R. H. Crabtree in *Handbook of Homogenous Hydrogenation* (Eds.: J. G. de Vries, C. J. Elsevier), Wiley-VCH, **2007**, Chapter 2.
439. J. J. Verendel, O. Pàmies, M. Diéguez, P. G. Andersson, *Chem. Rev.* **2014**, *114*, 2130–2169.
440. P. Brandt, C. Hedberg, P. G. Andersson, *Chem. Eur. J.* **2003**, *9*, 339–347.
441. G. Helmchen, *Chem. Eur. J.* **2023**, *29*, e202301488.
442. A. Shvartsbart, A. B. Smith, *J. Am. Chem. Soc.* **2014**, *136*, 870–873.
443. J. Halpern, D. P. Riley, A. S. Chand, J. J. Pluth, *J. Am. Chem. Soc.* **1977**, *99*, 8055–8057.
444. C. R. Landis, J. Halpern, *J. Am. Chem. Soc.* **1987**, *109*, 1746–1754.
445. R. R. Schrock, J. A. Osborn, *J. Am. Chem. Soc.* **1976**, *98*, 2134–2143.
446. R. R. Schrock, J. A. Osborn, *J. Am. Chem. Soc.* **1976**, *98*, 2143–2147.
447. R. R. Schrock, J. A. Osborn, *J. Am. Chem. Soc.* **1976**, *98*, 4450–4455.
448. J. M. Brown, *Angew. Chem. Int. Ed. Engl.* **1987**, *26*, 190–203.
449. R. H. Crabtree, M. W. Davis, *J. Org. Chem.* **1986**, *51*, 2655–2661.
450. G. Stork, D. E. Kahne, *J. Am. Chem. Soc.* **1983**, *105*, 1072–1073.
451. A. S. Machado, A. Olesker, S. Castillon, G. Lukacs, *J. Chem. Soc. Chem. Commun.* **1985**, 330–332.
452. D. A. Evans, M. M. Morrissey, *J. Am. Chem. Soc.* **1984**, *106*, 3866–3868.
453. K. C. Nicolaou, Q. Kang, S. Y. Ng, D. Y.-K. Chen, *J. Am. Chem. Soc.* **2010**, *132*, 8219–8222.
454. S. Ho, D. L. Sackett, J. L. Leighton, *J. Am. Chem. Soc.* **2015**, *137*, 14047–14050.
455. H. Adkins, R. E. Burks, *J. Am. Chem. Soc.* **1948**, *70*, 4174–4177.
456. J. A. Osborn, F. H. Jardine, J. F. Young, G. Wilkinson, *J. Chem. Soc. A* **1966**, 1711–1732.
457. T. Ohkuma, H. Ooka, T. Ikariya, R. Noyori, *J. Am. Chem. Soc.* **1995**, *117*, 10417–10418.

458. J. Pritchard, G. A. Filonenko, R. van Putten, E. J. M. Hensen, E. A. Pidko, *Chem. Soc. Rev.* **2015**, *44*, 3808–3833.

459. J. Zhang, G. Leitus, Y. Ben-David, D. Milstein, *Angew. Chem. Int. Ed.* **2006**, *45*, 1113–1115.

460. W. Kuriyama, T. Matsumoto, O. Ogata, Y. Ino, K. Aoki, S. Tanaka, K. Ishida, T. Kobayashi, N. Sayo, T. Saito, *Org. Process Res. Dev.* **2012**, *16*, 166–171.

461. L. A. Saudan, C. M. Saudan, C. Debieux, P. Wyss, *Angew. Chem. Int. Ed.* **2007**, *46*, 7473–7476.

462. A. Zanotti-Gerosa, D. Grainger, L. Todd, G. Grasa, L. Milner, E. Boddie, L. Browne, I. Egerton, L. Wong, *Chim. Oggi* **2019**, *37*, 8–11.

463. T. vom Stein, M. Meuresch, D. Limper, M. Schmitz, M. Hölscher, J. Coetzee, D. J. Cole-Hamilton, J. Klankermayer, W. Leitner, *J. Am. Chem. Soc.* **2014**, *136*, 13217–13225.

464. S. Qu, H. Dai, Y. Dang, C. Song, Z.-X. Wang, H. Guan, *ACS Catal.* **2014**, *4*, 4377–4388.

465. D. G. Gusev, *Organometallics* **2020**, *39*, 258–270.

466. G. Chelucci, S. Baldino, W. Baratta, *Acc. Chem. Res.* **2015**, *48*, 363–379.

467. M. K. Pandey, J. Choudhury, *ACS Omega* **2020**, *5*, 30775–30786.

468. B. M. Trost, Z. T. Ball, T. Jöge, *J. Am. Chem. Soc.* **2002**, *124*, 7922–7923.

469. A. Fürstner, K. Radkowski, *Chem. Commun.* **2002**, 2182–2183.

470. K. Radkowski, B. Sundararaju, A. Fürstner, *Angew. Chem. Int. Ed.* **2013**, *52*, 355–360.

471. A. Fürstner, *J. Am. Chem. Soc.* **2019**, *141*, 11–24.

472. M. Leutzsch, L. M. Wolf, P. Gupta, M. Fuchs, W. Thiel, C. Farès, A. Fürstner, *Angew. Chem. Int. Ed.* **2015**, *54*, 12431–12436.

473. A. Guthertz, M. Leutzsch, L. M. Wolf, P. Gupta, S. M. Rummelt, R. Goddard, C. Farès, W. Thiel, A. Fürstner, *J. Am. Chem. Soc.* **2018**, *140*, 3156–3169.

474. K. C. K. Swamy, A. S. Reddy, K. Sandeep, A. Kalyani, *Tetrahedron Lett.* **2018**, *59*, 419–429.

475. D. Wan, D. Astruc, *Chem. Rev.* **2015**, *115*, 6621–6686.

476. D. Klomp, U. Hanefeld, J. A. Peters in *Handbook of Homogenous Hydrogenation* (Eds.: J. G. de Vries, C. J. Elsevier), Wiley-VCH, **2007**, Chapter 20.

477. S. G. Ouellet, A. M. Walji, D. W. C. MacMillan, *Acc. Chem. Res.* **2007**, *40*, 1327–1339.

478. D. Baidilov, D. Hayrapetyan, A. Y. Khalimon, *Tetrahedron* **2021**, *98*, 132435.

479. J. S. M. Samec, J.-E. Bäckvall, P. G. Andersson, P. Brandt, *Chem. Soc. Rev.* **2006**, *35*, 237–248.

480. O. Pàmies, J.-E. Bäckvall, *Chem. Eur. J.* **2001**, *7*, 5052–5058.

481. D. Klomp, T. Maschmeyer, U. Hanefeld, J. A. Peters, *Chem. Eur. J.* **2004**, *10*, 2088–2093.

482. S. Manaviazar, M. Frigerio, G. S. Bhatia, M. G. Hummersone, A. E. Aliev, K. J. Hale, *Org. Lett.* **2006**, *8*, 4477–4480.

483. C. A. Sandoval, T. Ohkuma, N. Utsumi, K. Tsutsumi, K. Murata, R. Noyori, *Chem. Asian J.* **2006**, *1-2*, 102–110.

484. F. Foubelo, C. Nájera, M. Yus, *Tet. Asymm.* **2015**, *26*, 769–790.

485. M. Yamakawa, H. Ito, R. Noyori, *J. Am. Chem. Soc.* **2000**, *122*, 1466–1478.

486. P. A. Dub, J. C. Gordon, *Dalton Trans.* **2016**, *45*, 6756–6781.

487. P. A. Dub, T. Ikariya, *J. Am. Chem. Soc.* **2013**, *135*, 2604–2619.

488. J.-W. Handgraaf, E. J. Meijer, *J. Am. Chem. Soc.* **2007**, *129*, 3099–3103.

489. T. Koike, T. Ikariya, *Adv. Synth. Catal.* **2004**, *346*, 37–41.

490. J.-E. Bäckvall, *J. Organomet. Chem.* **2002**, *652*, 105–111.

491. B. L. Conley, M. K. Pennington-Boggio, E. Boz, T. J. Williams, *Chem. Rev.* **2010**, *110*, 2294–2312.

492. L. K. Thalén, D. Zhao, J.-B. Sortais, J. Paetzold, C. Hoben, J.-E. Bäckvall, *Chem. Eur. J.* **2009**, *15*, 3403–3410.

493. K. J. Ralston, A. N. Hulme, *Synthesis* **2012**, *44*, 2310–2324.

494. D. W. Stephan, *J. Am. Chem. Soc.* **2015**, *137*, 10018–10032.

495. D. W. Stephan, G. Erker, *Angew. Chem. Int. Ed.* **2015**, *54*, 6400–6441.

496. J. Lam, K. M. Szkop, E. Mosaferi, D. W. Stephan, *Chem. Soc. Rev.* **2019**, *48*, 3592–3612.

497. D. W. Stephan, *J. Am. Chem. Soc.* **2021**, *143*, 20002–20014.

498. L. Shi, Y.-G. Zhou, *ChemCatChem* **2015**, *7*, 54–56.

499. S. Nishimura, *Handbook of Heterogenous Catalytic Hydrogenation for Organic Synthesis*, John Wiley & Sons, **2001**.

500. P. N. Rylander, *Hydrogenation Methods*, Academic Press, **1985**.

501. H.-U. Blaser, A. Schnyder, H. Steiner, F. Rössler, P. Baumeister in *Handbook of Heterogeneous Catalysis* (Eds.: G. Ertl, H. Knözinger, F. Schüth, J. Weitkamp), Wiley-VCH, **2008**, Chapter 14.10.2.

502. A. M. Smith, R. Whyman, *Chem. Rev.* **2014**, *114*, 5477–5510.

503. M. Tamura, Y. Nakagawa, K. Tomishige, *Asian J. Org. Chem.* **2020**, *9*, 126–143.

504. R. Qu, K. June, M. Beller, *Chem. Rev.* **2023**, *123*, 1103–1165.

505. O. Deutschmann, H. Knözinger, K. Kochloefl, T. Turek, *Ullmann's Encyclopedia of Industrial Chemistry* **2011**.

506. J. Pérez-Ramírez, C. H. Christensen, K. Egeblad, C. H. Christensen, J. C. Groen, *Chem. Soc. Rev.* **2013**, *42*, 6094–6112.

507. G. Ertl, *Angew. Chem. Int. Ed. Engl.* **1990**, *29*, 1219–1227.

508. N. Ravasio, M. Antenori, M. Gargano, M. Rossi, *J. Mol. Catal.* **1992**, *74*, 267–274.

509. H. Lindlar, *Helv. Chim. Acta* **1952**, *35*, 446–450.

510. A. G. Caldwell, E. R. H. Jones, *J. Chem. Soc.* **1946**, 597–599.

511. H. Boerma, *Stud. Surf. Sci. Catal.* **1976**, *1*, 105–118.

512. H.-U. Blaser, H. Steiner, M. Studer, *ChemCatChem* **2009**, *1*, 210–221.

513. S. Recchia, C. Dossi, N. Poli, A. Fusi, L. Sordelli, R. Psaro, *J. Catal.* **1999**, *184*, 1–4.

514. F. Zaera, *Phys.Chem. Chem. Phys.* **2013**, *15*, 11988–12003.

515. I. Horiuti, M. Polanyi, *Trans. Faraday Soc.* **1934**, *30*, 1164–1172.

516. R. L. Augustine, F. Yaghmaie, J. F. V. Peppen, *J. Org. Chem.* **1984**, *49*, 1865–1870.

517. L. Cerveny, V. Ruzicka, *Catal. Rev. – Sci. Eng.* **1982**, *24*, 503–566.

518. L. E. Overman, D. J. Ricca, V. D. Tran, *J. Am. Chem. Soc.* **1993**, *115*, 2042–2044.

519. W. Hsin, L.-T. Chang, H.-L. Liou, *Synlett* **2008**, 2299–2302.

520. R. J. Sharpe, J. S. Johnson, *J. Org. Chem.* **2015**, *80*, 9740–9766.

521. A. Kondoh, A. Arlt, B. Gabor, A. Fürstner, *Chem. Eur. J.* **2013**, *19*, 7731–7738.

522. C. C. Oliveira, E. A. F. dos Santos, J. H. B. Nunes, C. R. D. Correia, *J. Org. Chem.* **2012**, *77*, 8182–8190.

523. C. Oger, L. Balas, T. Durand, J.-M. Galano, *Chem. Rev.* **2013**, *113*, 1313–1350.

524. P. Wipf, T. H. Graham, *J. Am. Chem. Soc.* **2004**, *126*, 15346–15347.

525. A. K. Ghosh, Y. Wang, J. T. Kim, *J. Org. Chem.* **2001**, *66*, 8973–8982.

526. A. S. Al-Ammar, S. J. Thomson, G. Webb, *J. Chem. Soc. Chem. Commun.* **1977**, 323–325.

527. R. Schlögl, K. Noack, H. Zbinden, A. Reller, *Helv. Chim. Acta* **1987**, *70*, 627–679.

528. A. Steenhoek, B. H. van Wijngaarden, H. J. J. Pabon, *Recl. Trav. Chim. Pays-Bas* **1971**, *90*, 961–973.

529. A. Farkas, *Trans. Faraday Soc.* **1939**, *35*, 906–917.

530. E. N. Marvell, T. Li, *Synthesis* **1973**, 457–468.

531. T.-L. Ho, S.-H. Liu, *Synth. Commun.* **1987**, *17*, 969–973.

532. S. Chang, S. Hur, R. Britton, *Chem. Eur. J.* **2015**, *21*, 16646–16653.

533. M. Hoogenraad, J. B. van der Linden, A. A. Smith, B. Hughes, A. M. Derrick, L. J. Harris, P. D. Higginson, A. J. Pettman, *Org. Process Res. Dev.* **2004**, *8*, 469–476.

534. S. H. P. William H. Jones, M. Sletzinger, *Ann. N. Y. Acad. Sci.* **1973**, *214*, 150–157.

535. U. Siegrist, P. Baumeister, H.-U. Blaser, M. Studer, *Catalysis of Organic Reactions (Chemical Industries)*, **1998**, pp. 207–219.

536. P. Loos, H. Alex, J. Hassfeld, K. Lovis, J. Platzek, N. Steinfeldt, S. Hübner, *Org. Process Res. Dev.* **2016**, *20*, 452–464.

537. T. J. N. Watson, S. W. Horgan, R. S. Shah, R. A. Farr, R. A. Schnettler, C. R. Nevill, F. J. Weiberth, E. W. Huber, B. M. Baron, M. E. Webster, R. K. Mishra, B. L. Harrison, P. L. Nyce, C. L. Rand, C. T. Goralski, *Org. Process Res. Dev.* **2000**, *4*, 477–487.

538. H. Zaimoku, H. Nishide, A. Nishibata, N. Goto, T. Taniguchi, H. Ishibashi, *Org. Lett.* **2013**, *15*, 2140–2143.

539. V. R. Bhonde, R. E. Looper, *J. Am. Chem. Soc.* **2011**, *133*, 20172–20174.

540. G. Ma, H. Nguyen, D. Romo, *Org. Lett.* **2007**, *9*, 2143–2146.

541. Y. Ichikawa, K. Hirata, M. Ohbayashi, M. Isobe, *Chem. Eur. J.* **2004**, *10*, 3241–3251.

542. G. R. Pettit, E. E. van Tamelen, *Org. React.* **1962**, *12*, 356–529.

543. W. A. Bonner, R. A. Grimm in *The Chemistry of Organic Sulfur Compounds, Vol. 2* (Eds.: N. Kharasch, C. Y. Meyers), Pergamon Press, **1966**, Chapter 2, pp. 35–71.

544. P. Wipf, Y. Kim, D. M. Goldstein, *J. Am. Chem. Soc.* **1995**, *117*, 11106–11112.

545. A. W. Burgstahler, L. O. Weigel, C. G. Shaefer, *Synthesis* **1976**, 767–768.

546. V. G. Yadav, S. B. Chandalia, *Org. Process Res. Dev.* **1997**, *1*, 226–232.

547. D. J. Ager, A. H. M. de Vries, J. G. de Vries, *Chem. Soc. Rev.* **2012**, *41*, 3340–3380.

548. A. Zanotti-Gerosa, W. Hems, M. Groarke, F. Hancock, *Platinum Metals Rev.* **2005**, *49*, 158–165.

549. M. M. Midland, *Chem. Rev.* **1989**, *89*, 1553–1561.

550. R. Noyori, I. Tomino, M. N. Y. Tanimoto, *J. Am. Chem. Soc.* **1984**, *106*, 6709–6716.

551. M. Srebnik, P. V. Ramachandran, H. C. Brown, *J. Org. Chem.* **1988**, *53*, 2916–2920.

552. E. J. Corey, C. J. Helal, *Angew. Chem. Int. Ed.* **1998**, *37*, 1986–2012.

553. B. T. Cho, *Tetrahedron* **2006**, *62*, 7621–7643.

554. C. Eschmann, L. Song, P. R. Schreiner, *Angew. Chem. Int. Ed.* **2021**, *60*, 4823–4832.

555. S. F. Sabes, R. A. Urbanek, C. J. Forsyth, *J. Am. Chem. Soc.* **1998**, *120*, 2534–2542.

556. D. A. Evans, P. H. Carter, E. M. Carreira, A. B. Charette, J. A. Prunet, M. Lautens, *J. Am. Chem. Soc.* **1999**, *121*, 7540–7552.

557. I. Paterson, H. Bergmann, D. Menche, A. Berkessel, *Org. Lett.* **2004**, *6*, 1293–1295.

558. D. Menche, J. Hassfeld, J. Li, K. Mayer, S. Rudolph, *J. Org. Chem.* **2009**, *74*, 7220–7229.

559. M. J. Di Grandi, D. K. Jung, W. J. Krol, S. J. Danishefsky, *J. Org. Chem.* **1993**, *58*, 4989–4992.

560. D. P. Stamos, S. S. Chen, Y. Kishi, *J. Org. Chem.* **1997**, *62*, 7552–7553.

561. X. Fu, T. L. McAllister, T. K. Thiruvengadam, C.-H. Tann, D. Su, *Tetrahedron Lett.* **2003**, *44*, 801–804.

562. E. J. Corey, K. Rao, *Tetrahedron Lett.* **1991**, *32*, 4623–4626.

563. D. T. Hung, J. B. Nerenberg, S. L. Schreiber, *J. Am. Chem. Soc.* **1996**, *118*, 11054–11080.

564. R. Noyori, T. Ohkuma, *Angew. Chem. Int. Ed.* **2001**, *40*, 40–73.

565. H. Shimizu, I. Nagasaki, K. Matsumura, N. Sayo, T. Saito, *Acc. Chem. Res.* **2007**, *40*, 1385–1393.

566. T. Ohkuma, H. Ooka, S. Hashiguchi, T. Ikariya, R. Noyori, *J. Am. Chem. Soc.* **1995**, *117*, 2675–2676.

567. M. Kitamura, H. Nakatsuka, *Chem. Commun.* **2011**, *47*, 842–846.

568. F. Hasanayn, R. H. Morris, *Inorg. Chem.* **2012**, *51*, 10808–10818.

569. P. A. Dub, N. J. H. N. R. L. Martin, J. C. Gordon, *J. Am. Chem. Soc.* **2014**, *136*, 3505–3521.

570. R. J. Hamilton, C. G. Leong, G. Bigam, M. Miskolzie, S. H. Bergens, *J. Am. Chem. Soc.* **2005**, *127*, 4152–4153.

571. K. Abdur-Rashid, M. Faatz, A. J. Lough, R. H. Morris, *J. Am. Chem. Soc.* **2001**, *123*, 7473–7474.

572. C. Chen, R. A. Reamer, J. R. Chilenski, C. J. McWilliams, *Org. Lett.* **2003**, *5*, 5039–5042.

573. J.-P. Genet, P. Phansavath, V. Ratovelomanana-Vidal, *Isr. J. Chem.* **2021**, *61*, 409–426.

574. J.-H. Xie, D.-H. Bao, Q.-L. Zhou, *Synthesis* **2015**, *47*, 460–471.

575. W. Tang, X. Zhang, *Chem. Rev.* **2003**, *103*, 3029–3069.

576. N. U. D. Reshi, V. B. Saptal, M. Beller, J. K. Bera, *ACS Catal.* **2021**, *11*, 13809–13837.

577. H.-U. Blaser, *Adv. Synth. Catal.* **2002**, *344*, 17–31.

578. A. E. Cotman, *Chem. Eur. J.* **2021**, *27*, 39–53.

579. J. Václavík, P. Sot, B. Vilhanová, J. Pechácek, M. Kuzma, P. Kacer, *Molecules* **2013**, *18*, 6804–6828.

580. S. D. Stone, N. J. Lajkiewicz, L. Whitesell, A. Hilmy, J. A. Porco, *J. Am. Chem. Soc.* **2015**, *137*, 525–530.

581. K. Fujii, K. Maki, M. Kanai, M. Shibasaki, *Org. Lett.* **2003**, *5*, 733–736.

582. L. Zhu, Y. Liu, R. Ma, R. Tong, *Angew. Chem. Int. Ed.* **2015**, *54*, 627–632.

583. A. M. R. Hall, D. B. G. Berry, J. N. Crossley, A. Codina, I. Clegg, J. P. Lowe, A. Buchard, U. Hintermair, *ACS Catal.* **2021**, *11*, 13649–13659.

584. N. V. Tkachenko, P. Rublev, P. A. Dub, *ACS Catal.* **2022**, *12*, 13149–13157.

585. T. Ikariya, A. J. Blacker, *Acc. Chem. Res.* **2007**, *40*, 1300–1308.

586. J. A. Marshall, K. Ellis, *Tetrahedron Lett.* **2004**, *45*, 1351–1353.

587. C. S. G. Seo, R. H. Morris, *Organometallics* **2019**, *38*, 47–65.

588. P. Etayo, A. Vidal-Ferran, *Chem. Soc. Rev.* **2013**, *42*, 728–754.

589. J.-P. Genet, *Acc. Chem. Res.* **2003**, *36*, 908–918.

590. J. Halpern, *Science* **1982**, *217*, 401–407.

591. W. S. Knowles, *Acc. Chem. Res.* **1983**, *16*, 106–112.

592. M. Kitamura, M. Tsukamoto, Y. Bessho, M. Yoshimura, U. Kobs, M. Widhalm, R. Noyori, *J. Am. Chem. Soc.* **2002**, *124*, 6649–6667.

593. P. E. Maligres, G. R. Humphrey, J.-F. Marcoux, M. C. Hillier, D. Zhao, S. Krska, E. J. J. Grabowski, *Org. Process Res. Dev.* **2009**, *13*, 525–534.

594. S. J. Roseblade, A. Pfaltz, *Acc. Chem. Res.* **2007**, *40*, 1402–1411.

595. A. Pfaltz, J. Blankenstein, R. Hilgraf, E. Hörmann, S. McIntyre, F. Menges, M. Schönleber, S. P. Smidt, B. Wüstenberg, N. Zimmermann, *Adv. Synth. Catal.* **2003**, *345*, 33–43.

596. B. B. C. Peters, P. G. Andersson, *J. Am. Chem. Soc.* **2022**, *144*, 16252–16261.

597. N. B. Johnson, I. C. Lennon, P. H. Moran, J. A. Ramsden, *Acc. Chem. Res.* **2007**, *40*, 1291–1299.

598. S. Bell, B. Wüstenberg, S. Kaiser, F. Menges, T. Netscher, A. Pfaltz, *Science* **2006**, *311*, 642–644.

599. A. Beliaev, *Org. Process Res. Dev.* **2016**, *20*, 724–732.

600. C. S. Shultz, S. D. Dreher, N. Ikemoto, J. M. Williams, E. J. J. Grabowski, S. W. Krska, Y. Sun, P. G. Dormer, L. DiMichele, *Org. Lett.* **2005**, *7*, 3405–3408.

601. M. Bulliard, B. Laboue, J. Lastennet, S. Roussiasse, *Org. Process Res. Dev.* **2001**, *5*, 438–441.

602. G. Hoge, H.-P. Wu, W. S. Kissel, D. A. Pflum, D. J. Greene, J. Bao, *J. Am. Chem. Soc.* **2004**, *126*, 5966–5967.

603. R. Ikeda, R. Kuwano, *Chem. Eur. J.* **2016**, *22*, 8610–8618.

Pericyclic Reactions

Diels-Alder cycloadditions occupy a prominent role in the realm of pericyclic reactions, being featured as the key bond-forming step in a multitude of highly complex natural product syntheses. The atom economy of pericyclic reactions is considered to be ideal: Given that often no additives are required for the activation of substrates and several bonds can be forged stereospecifically in one step without the elimination of a leaving group, it is judged particularly efficient by sustainability standards. The Roush route to synthesize the polyketide (–)-spinosyn A in a domino olefination-transannular cycloaddition sequence via the advanced intermediate **2** represents just one of the impressive applications of a pericyclic reaction (see Fig. 5.1) [1].

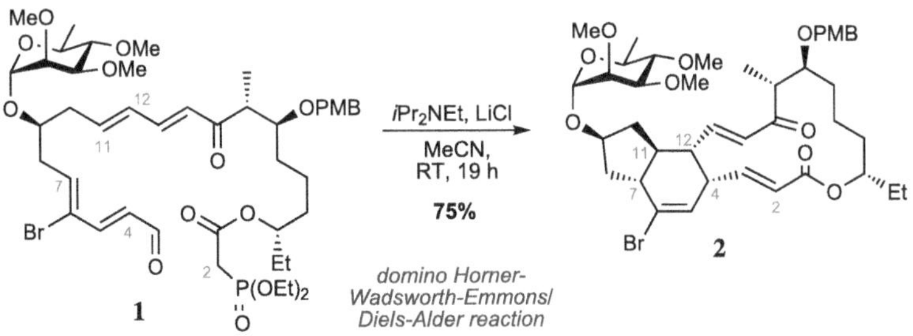

Fig. 5.1 Synthesis of (–)-spinosyn A by Roush *et al.* [1]

Discovered in the first half of the twentieth century, so-called thermo-reorganization processes (electrocyclic reactions) were referred to as "no-mechanism" reactions [2], as their mechanism could not be elucidated by the available analytical approaches. The mutual

© The Author(s), under exclusive license to Springer-Verlag GmbH, DE, part of Springer Nature 2026
A. Düfert, *Methods of Organic Synthesis*,
https://doi.org/10.1007/978-3-662-70963-4_5

relationship of these reactions was obvious at first glance, yet a comprehensive explanation of their reactivity and stereospecificity was lacking. Woodward, who together with Hoffmann first recognized these common features, referred to them before their discovery as the "four mysterious reactions" [3], which prompted him to investigate the topic more in-depth and whose mechanism became clear in light of the newly gained understanding (see Fig. 5.2).

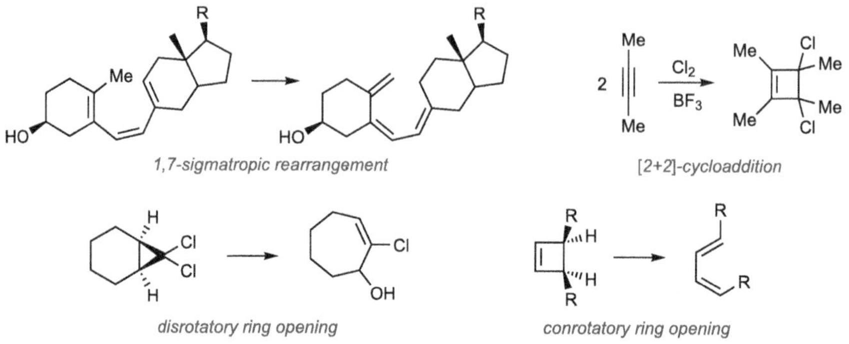

Fig. 5.2 Woodward's "four mysterious reactions" [3]

Pericyclic reactions differ mechanistically from classic organic reactions, as their dominant interaction is based neither on an acid/base nor an electrophile/nucleophile interaction. They are instead *orbitally controlled*. These are processes in which all bond changes occur simultaneously (*concerted*), without the occurrence of intermediates and a cyclic arrangement of continuously interacting atoms during the transition state (see Fig. 5.3) [4]. Although bond formations and cleavages occur simultaneously, they do not have to be equally advanced at all involved atoms. A concerted reaction is therefore a simultaneous, but not necessarily a synchronous bond formation/cleavage. Especially reactions involving heteroatoms can proceed in an asynchronous manner [5–7].

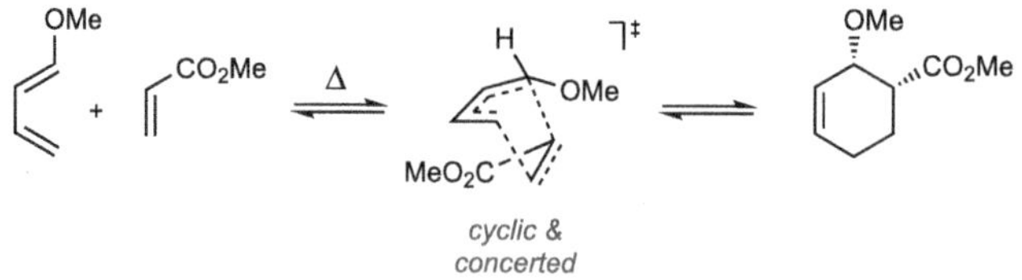

Fig. 5.3 Thermal [4+2]-cycloaddition

Indeed, the consideration of orbital interactions of molecules and during reactions was a field of study that was not associated with organic chemistry at the time. The conclusion drawn by the involved research groups (Oosterhoff, Havinga, Woodward, Hoffmann, Fukui)

[8–13], that pericyclic reaction not only proceed orbitally controlled, but that this implies the necessity of a preservation of orbital symmetry throughout its course, is thus all the more significant. It was for this recognition that Hoffmann and Fukui were ultimately awarded the Nobel Prize in Chemistry in 1981.

Notable pericyclic reactions that are regarded part of the standard canon of synthetic methods include the Diels-Alder reaction, Cope rearrangement, Claisen rearrangement, Huisgen cycloaddition, ozonolysis, Nazarov cyclization, ene reaction, Staudinger reaction, and the Favorskii reaction, among others.

5.1 Theoretical and Mechanistic Principles

Pericyclic reactions are categorized according to their reaction conditions as thermal or photochemical processes, which are considered to afford complementary products with regard to their stereochemical course. If a reaction is not allowed under thermal conditions for reasons of orbital symmetry, it can usually be realized photochemically. The various classes of pericyclic reactions comprise cycloadditions (including cheletropic reactions), electrocyclic reactions, sigmatropic rearrangements, and group transfer reactions, including ene reactions (see Fig. 5.4) [4].

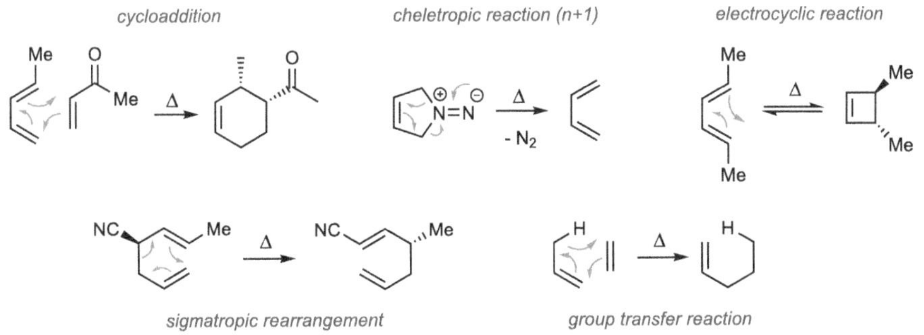

Fig. 5.4 Classes of pericyclic reactions [4]

In a *cycloaddition*, two σ bonds are formed between two conjugated π systems and two π bonds are broken. The coupling of three components, exchanging three π bonds for three σ bonds, is also occasionally encountered. As a further subcategory, 1,3-dipolar cycloadditions will be discussed in detail as well (*vide infra*). *Cheletropic reactions* are a special case of cycloadditions, where both σ bonds are formed or broken at the same atom of one component. It usually constitutes a retro-cycloaddition with concomitant release of gaseous products

(N_2, CO, SO_2, etc.). In an *electrocyclic reaction*, the ends of a π system are connected. This involves the tautomerism of the involved double bonds, breaking an existing double bond and forming a new single bond. A *sigmatropic rearrangement* entails the migration of a conjugated π system, whereby a σ bond is newly formed while simultaneously breaking another σ bond. In *group transfer reactions*, groups or atoms are transferred onto a π system.

The driving force of the reactions lies in the energy gain resulting from the difference between the formed σ bonds compared to the broken π bonds. Rearrangements occur due to the increased stability of the constitutional isomers of the products compared to the substrates. A normal Diels-Alder reaction, for example, is approximately exothermic by 150 kJ/mol (see Fig. 5.5).

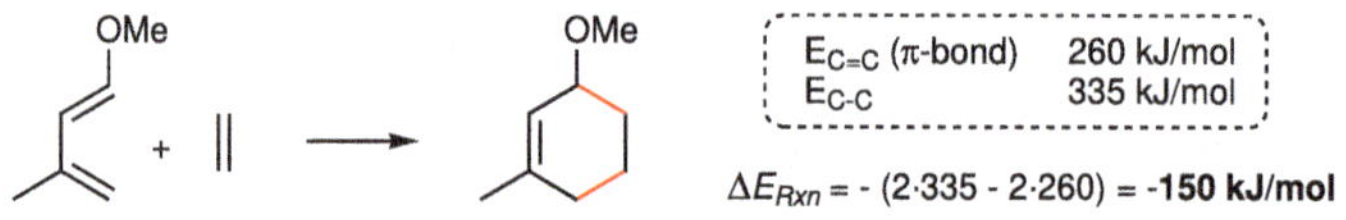

Fig. 5.5 Driving force of the Diels-Alder reaction

Unlike two-center reactions, concerted transformations necessitate a highly ordered transition state. The substrates must be precisely aligned to effect maximum interaction and thus stabilization of the transition state. This interplay of entropy and enthalpy governs the rate-determining activation process. Concerted reactions therefore generally exhibit large negative entropies of activation ($\Delta S^{\ddagger}$) while possessing only moderate activation enthalpy contributions [14]. Many pericyclic reactions are reversible. Depending on the steric and electronic stabilization of the products (e.g., a high ring strain of the substrate), their relative energetic position, and entropic factors (e.g., the release of N_2 in the above-mentioned cheletropic reaction), a reverse reaction may require an unsurmountable activation energy for practical purposes.

In pericyclic reactions, it is specified for each component whether it reacts on opposite faces of the molecule or on the same face, which usually considers only the reacting π systems. Reactions on the same face are referred to as *suprafacial*. Reactions on different sides are termed *antarafacial*, whereby these two reactions pathways display a complementary stereospecificity [9]. For example, if the conversion of **3** proceeds under a suprafacial or antarafacial topology in an electrocyclic reaction, different diastereomers result as a product due to the deviating transition states **4** and **5**, respectively (see Fig. 5.6).

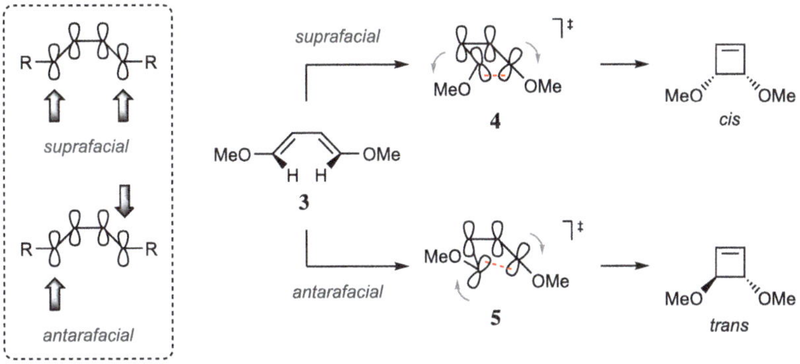

Fig. 5.6 Suprafacial and antarafacial reaction topology of an electrocyclization

Since the ring closure occurs under thermal conditions or photochemical conditions, the stereoselectivity can be controlled by choosing the reaction conditions, as implied in the depicted example. In addition to the orbital interactions of the components with respect to the molecular axis, the relative sense of rotation at the termini of the π-system needs to also be considered in electrocyclic reactions. If both molecular termini rotate in the same direction, this is called a *conrotatory* reaction pathway. If they rotate in different directions, this is referred to as *disrotatory* course of reaction (see Fig. 5.7) [15].

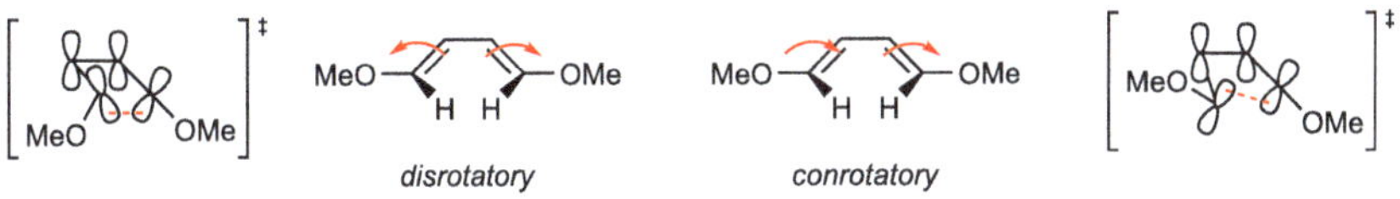

Fig. 5.7 Disrotatory *versus* conrotatory sense of rotation in electrocyclic reactions [15]

A comparison of Figs. 5.6 and 5.7 implies that for a conrotatory rotation, the π-system must undergo an antarafacial interaction. A suprafacial interaction results from a disrotatory course.

Apart from the facial approach of the individual components and the type of interacting bonds at these components (π, σ, and **n** for free electrons), the number of involved electrons is also indicated in the classification of pericyclic reactions. This number often coincides with the number of involved atoms. In components with an odd number of involved atoms, such as 1,3-dipolar cycloaddition and electrocyclic reactions of cyclopropyl cations, the number of electrons and atoms is dissimilar. The notation follows the scheme [9]:

[bond type #electrons facial course]

The facial reaction path is abbreviated with the letter s or a, so that one arrives at the description of components such as $[\pi 4a]$ and $[\sigma 2s]$. If one considers the electrocyclic reactions shown in Fig. 5.6 as well as a representative selection of other pericyclic processes, one obtains the description of the reaction components shown in Fig. 5.8. [3,3]-sigmatropic rearrangements can be broken down into their individual components in various ways: either one considers three two-electron systems or an allyl anion and cation each (see Fig. 5.20).

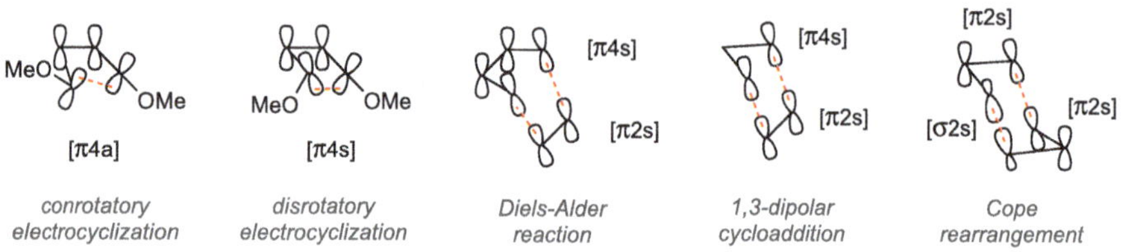

MeO — OMe	$[\pi 4a]$	conrotatory electrocyclization

MeO — OMe	$[\pi 4s]$	disrotatory electrocyclization

$[\pi 4s]$	$[\pi 2s]$	Diels-Alder reaction

$[\pi 4s]$	$[\pi 2s]$	1,3-dipolar cycloaddition

$[\pi 2s]$	$[\sigma 2s]$	$[\pi 2s]$	Cope rearrangement

Fig. 5.8 Classification of the reacting components of various pericyclic reactions

5.1.1 Conservation of Symmetry and Correlation Diagrams

Pericyclic reactions proceed under orbital control. These can only interact if they have the same symmetry (symmetric or antisymmetric) with respect to the components' inherent symmetry elements (see below). This implies that for a **continuous** interaction of the atoms in the cyclic transition state, there must be no change in orbital symmetry over the course of the reaction starting from the reactants and arriving at the products [16]. This is based on the symmetry properties of the interacting wave functions and the Hamiltonian, which can only interact as long as they are not orthogonal. Woodward and Hoffmann initially studied and reported symmetry-based selection rules for electrocyclic reactions and subsequently expanded the concept to cover all pericyclic processes with the aid of orbital correlation diagrams. These correlation diagrams enable a tracking of which orbitals of the substrates are transformed into which orbitals of the products along the reaction coordinate. A reaction only proceeds if all electrons of the reactants can transition to the products without a symmetry-induced barrier. Therefore, reactions are classified according to whether they are *symmetry-allowed* and *symmetry-forbidden*. Just because a reaction is symmetry-allowed, however, does not imply it automatically proceeds, as, for example, steric effects could prevent its occurrence. It merely indicates that no additional electronic barrier exists.

Very seldom a symmetry-forbidden path may be preferred in lieu of the expected allowed trajectory. It is possible to design features in molecules and transition states that preclude or disfavor geometrical achievement of an allowed transition state. It is also possible to endow a "forbidden" transition state with substantial diradical or ionic character (electronic asymmetry, strong donor/acceptor polarization) at the bond-forming loci. Where such strong electronic asymmetries occur, violations to the expected reaction pathway as delineated by the Woodward–Hoffmann rules may ensue, often on the grounds of a stepwise as opposed to a concerted mechanism (*vide infra*) [17].

The classification into symmetric (S) and antisymmetric behavior (A) of orbitals and states (see below) depicted in correlation diagrams refers to the symmetry elements maintained throughout the entire course of the reaction (substrates, transition state, products) [9]. For example, in the cycloaddition of butadiene to ethene, only the mirror plane σ is preserved, while in the conrotatory electrocyclization of butadiene, a C_2 rotation axis is retained (see Fig. 5.9).

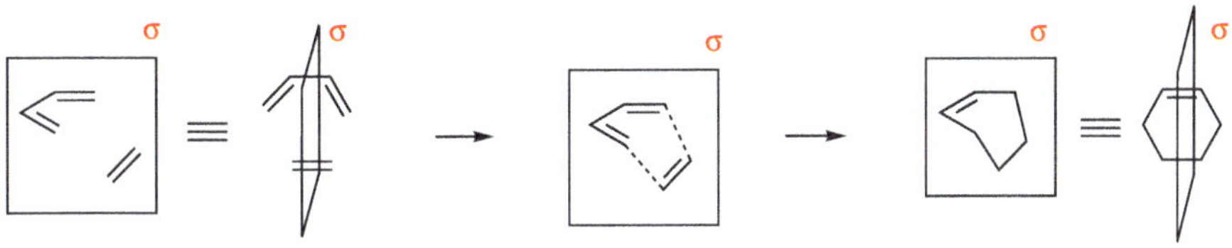

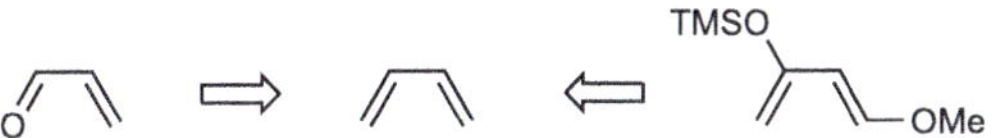

Fig. 5.9 Preservation of symmetry elements over the course of the reaction

The reaction of butadiene with ethene is symmetry-allowed, but given the energetic position of the involved orbitals, it still possesses a high activation barrier. Even though the introduction of substituents to the components can positively influence this outcome, the preservation of any symmetry element would in principle cease upon the introduction of these substituents. The disturbance or discontinuation of symmetry can be considered negligible as long as the influence of the substituent has little effect on the involved orbitals. Thus, only the symmetry of the reactive orbitals should be considered as a reasonable approximation. In this way, isoelectronic systems can be approximated as being symmetry-equivalent (see Fig. 5.10). It has also been shown by theoretical means that the preservation of a minimal degree of symmetry is still sufficient for the progression of a concerted process. However, cycloadditions display a decreasing reaction rate due to an increase in the activation enthalpy, the further they deviate from an ideal and unperturbed symmetry [18]. The asynchronicity of a transition state that typically accompanies asymmetric substrates is founded on the minimization of the closed-shell repulsion. Whereas following a more asynchronous reaction mode costs favorable HOMO-LUMO orbital overlap, it also lowers repulsive occupied–occupied orbital overlap. The extent of asynchronicity is thus governed by the balance between these stabilizing (electron repulsion minimization) and destabilizing (HOMO-LUMO overlap) interactions [7].

Fig. 5.10 Reduction to the highest inherent symmetry of the orbitals involved in the reaction

The molecular orbital correlation diagram for the prototypical cycloaddition of butadiene and ethene is shown in Fig. 5.11. The orbitals of reactants and their behavior with respect to the symmetry plane σ were included in energetically ascending order. After populating the molecular orbitals with electrons, the orbitals of the same symmetry of the reactants were connected with those of the product, again in the order of ascending energy. It shows that the substrate orbitals occupied in the ground state are all bonding and can only be converted into bonding orbitals of the product. Thus, there is no symmetry-derived barrier for the reaction. This is referred to as a *symmetry-allowed* pericyclic reaction. Of course, there is still an activation energy for the depicted [4+2]-cycloaddition, which cannot be derived from the correlation diagram. It is about 85 kJ/mol and is caused by energy changes during rehybridization of orbitals, the stretching and compression of bonds, and the distortion of bond angles.

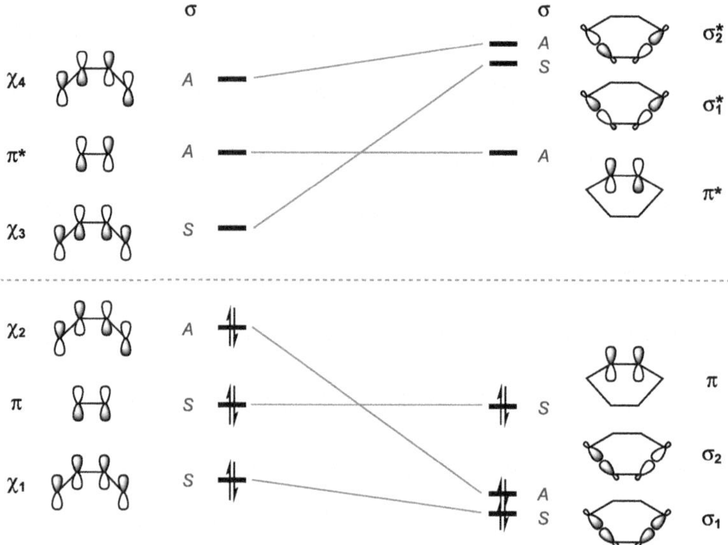

Fig. 5.11 Orbital correlation diagram of the cycloaddition of butadiene and ethene and symmetry of the involved orbitals. The dashed line indicates the energetic boundary between bonding and antibonding orbitals

The second prototypical pericyclic reaction is the [2+2]-cycloaddition of two ethene molecules. If one proceeds in an analogous fashion, the correlation diagram shown in Fig. 5.12 is obtained.

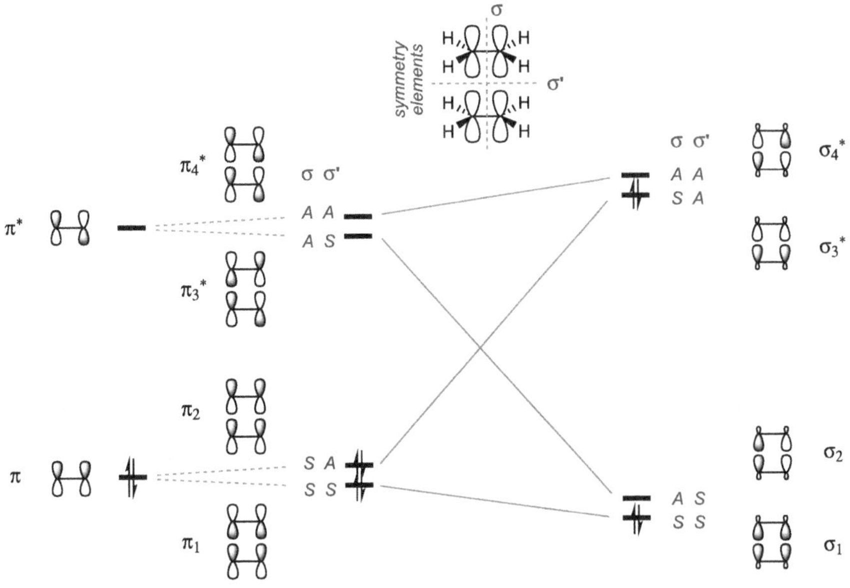

Fig. 5.12 Orbital correlation diagram of the thermal dimerization of ethene to cyclobutane and symmetry of the involved orbitals

Based on the depicted thermal reaction, the relevant symmetry elements obtained are two mirror planes σ and σ'. The orbitals π and π^* are degenerate at a large distance between the two ethene molecules (shown far left), but as they approach each other, the degeneracy is lifted and the orbitals π_1 to π_4^* result. After determining the symmetry properties of the MOs of the reactants and the product, the MOs of the reactants are populated with the interacting electrons (here a total of 4). After correlating the orbitals of reactants and the product, it can be seen that two electrons of the reactants are transferred from the bonding π_2 orbital to the antibonding σ_3^* orbital in the resulting cyclobutane. The product, for symmetry reasons, is in a doubly excited state, as two electrons are occupying an excited orbital. This means that the reactants cannot readily react to give the product, as there is an electronic barrier. The reaction is therefore *symmetry-forbidden* and is not observed to occur. The height of the symmetry-imposed barrier can be estimated by taking into account the energy required to lift two bonding electrons from an occupied, bonding level to an antibonding level. It is about 5 eV or 480 kJ/mol [9].

Under photochemical reaction conditions, i.e., by irradiation of light, an electron can be lifted into the first excited state. If the correlation diagram is modified accordingly, the reaction pathway changes (see Fig. 5.13).

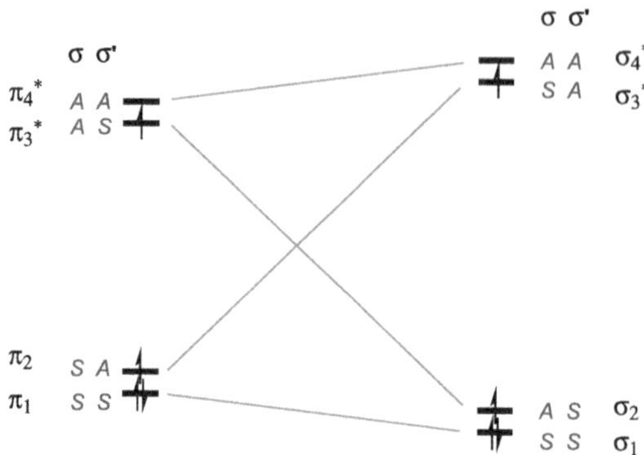

Fig. 5.13 Orbital correlation diagram of the photochemical [2+2]-cycloaddition of ethene. See Fig. 5.12 for details of the involved orbitals

As is evident in the orbital correlation diagram, in a photochemical [2+2]-cycloaddition, the number of electrons in bonding orbitals and thus the bond order of the reactants and the product is the same. Both the reactant and the product are in the 1st excited state. As a result, there is no symmetry-imposed barrier, the reaction is *symmetry-allowed*.

The orbital correlation diagrams are intuitively comprehensible and have found wide applications. However, they cannot provide quantitative information. In contrast to orbitals, which result from approximations and assumptions about the involved systems, quantum mechanical states represent real solutions to the Schrödinger equation. A state can be approximately derived from the electron configuration of the reactants and products,[1] allowing the creation of *state* correlation diagrams [19]. They enable at least a quantitative description of reactants and products. The symmetry of the states results from the multiplication of the orbital symmetries ($A \times A = S$; $S \times S = S$; $S \times A = A$). From them, it becomes even clearer why the [4+2]-cycloaddition is thermally allowed and the [2+2]-cycloaddition is photochemically allowed. In the Diels-Alder reaction, the ground state of the reactants transitions to the ground state of the product. However, the 1st excited state of the reactants correlates with higher excited states of the product, which is why this course is energetically very unfavorable and thus symmetry-forbidden. In the dimerization of ethene, the ground states of the reactants and the product do not correlate with each other, but rather with higher excited states (see Fig. 5.12). Since energetic configurations of the same symmetry are not allowed to cross, an *avoided crossing* is observed at this point instead. It results from the fact that the states interact more strongly the closer they come to each other energetically. The

---

[1] Here, the interaction or mixing of different electron configurations is neglected, which leads to a slight change in the state energies. However, since one configuration will dominate a state, this approach is suitable for an initial approximation. Thus, orbital correlation diagrams can be considered as approximately equivalent to state correlation diagrams.

energetically lower state is consequently lowered further, the higher one is energetically raised accordingly. At the isoenergetic point, the interactions and the effects on the state energies are maximized, thereby avoiding the crossing of these states. Nevertheless, due to the strong interactions occurring in this area, the states can mix and partially lose their identity, which is referred to as *configuration interaction* [20]. The same avoided crossing is also observed in the case of the excited states of the Diels-Alder reaction (not depicted). The high energy barrier of the thermal [2+2]-cycloaddition illustrates why this reaction is symmetry-forbidden. The first excited state of the two ethenes, which can be reached under photochemical irradiation, correlates with the first excited state of the product. This reaction course is symmetry-allowed (see Fig. 5.14).

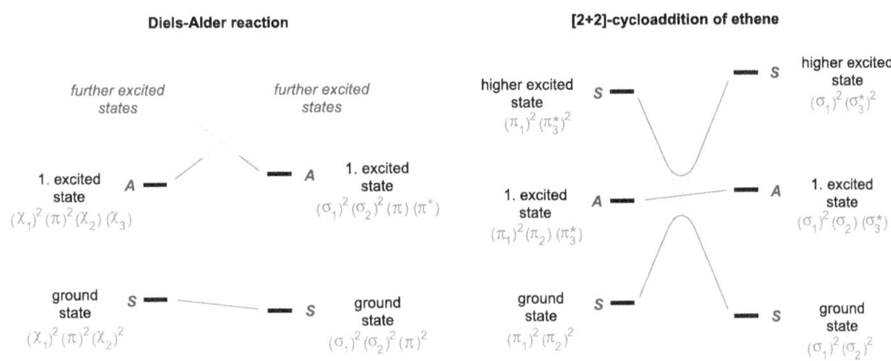

Fig. 5.14 State correlation diagram of the [4+2] and [2+2] cycloaddition [9]

Finally, the electrocyclic ring closure of butadiene to cyclobutene under both thermal and photochemical conditions will be analyzed. The correlation diagram makes it clear why, depending on the reaction conditions, either the *syn-* or the *anti*-configured product is obtained in an electrocyclization. In a conrotatory reaction, which leads to the *anti*-product, the reaction is only allowed if the butadiene is in the ground state. Therefore, only this product is isolated under thermal conditions. In a disrotatory ring closure, the conserved symmetry element changes and thus the symmetry behavior of the involved orbitals is reversed. As a consequence, the first excited state must react to form the *syn*-product to avoid a symmetry-imposed reaction barrier. Therefore, under photochemical conditions, the selective formation of the *syn*-product is observed (see Fig. 5.15) [15, 19].

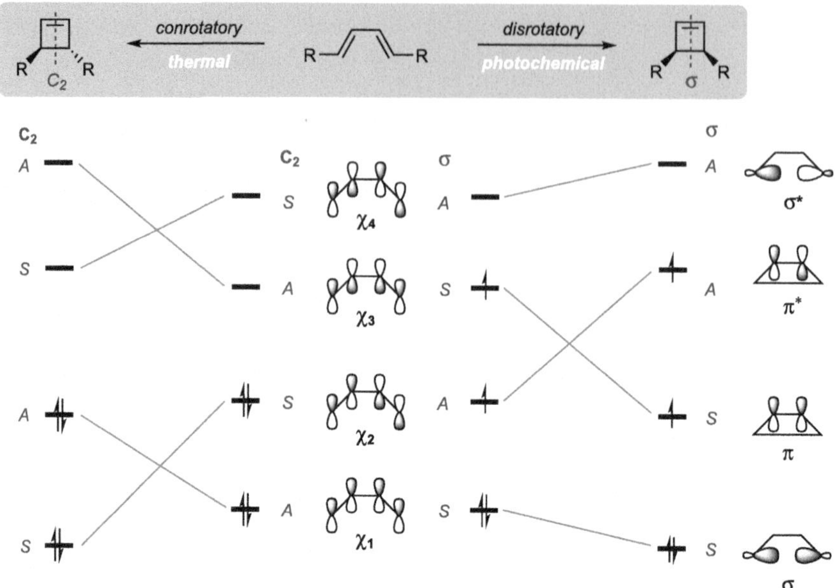

Fig. 5.15 Schematic orbital correlation diagram of the electrocyclization of substituted butadienes [9, 15, 19]

5.1.2 Perturbation Theory and Frontier Orbital Interactions

Correlation diagrams unfortunately possess some disadvantages, even though their approach is temptingly simple and the principle of symmetry conservation discovered by Woodward and Hoffmann can be most clearly understood with them: First of all, they reach their limits when the substrates are intrinsically asymmetric, for example, in the case of reactants bearing strongly polar substituents. State correlation diagrams still reliably lead to the correct result, but their construction is no longer a simple matter. Moreover, for many cases where a reaction contains a component with an antarafacial pathway, a correlation diagram cannot be constructed anymore. Second of all, the reaction of (photochemically) excited states can entail some complications and ultimately result in a false prediction of whether the process is symmetry-allowed or not. This deviation can occur when the reacting excited state differs from the one reached in the initial excitation. Especially singlet-triplet splittings of different excited states can deviate to such a degree that the symmetries of the lowest singlet and the lowest triplet state differ rather significantly. In addition, radiationless relaxation processes can occur, so that the processes taking place following the irradiation occur from the vibrationally excited ground state [9]. Third of all, correlation diagrams do not make any statements about the experimentally observed pronounced regioselectivity of cycloadditions. For this reason, the consideration of the frontier orbitals of the reactants is often chosen instead, using perturbation mechanics, which allows a (simple) quantitative assessment of pericyclic processes.[II]

The interaction of two components is described in perturbation theory by the Klopman–Salem equation [21–24].

$$\Delta E = -\sum_{a,b}(q_a + q_b)\beta_{ab}S_{ab} + \sum_{k,l}\frac{Q_k Q_l}{\epsilon R_{kl}}$$

$$+ 2\sum_r^{occ}\sum_s^{virt} - \sum_r^{virt}\sum_s^{occ}\frac{(\sum_{rs}c_{ra}c_{sb}\beta_{ab})^2}{E_r - E_s} \tag{5.1}$$

q_a electron population in atomic orbital a
β_{ab} interorbital interaction of atomic orbitals a, b (resonance integral)
S_{ab} overlap integral of atomic orbitals a, b
Q_k electron density (charge) at atom k
R_{kl} distance between atoms k and l
c_{ra} coefficient of atomic orbital a in molecular orbital r, the indices r and s each cover the MOs of a reactant
E_r energy of the molecular orbital r

ΔE represents the change in energy of the two components when they come into sufficient proximity to interact. A stabilization indicates a significant decrease of the energy (cf. Figure 5.17). Therefore, terms with a negative sign are stabilizing, while terms with a positive sign raise the components' energy. This perturbation approach is based on several simplifications and does not take into account some interactions,[III] so that it is now used mainly for a qualitative understanding since the advent/availability of more accurate quantum mechanical methods.

The first term refers to the repulsive interaction of occupied orbitals (*closed-shell repulsion*). The resonance integral results in a change in sign, turning the entire term positive and thus attributing it a destabilizing effect. Despite the interorbital interaction of the first order[IV] between two occupied orbitals being very small, the sum of all interactions over

---

[II] A third approach considers the topology of the involved orbitals in the transition state and classifies them as aromatic and antiaromatic. This approach will not be discussed further here, please refer to: Dewar, *Angew. Chem. Int. Ed. Engl.* **1971**, *10*, 761–776; Zimmermann, *Acc. Chem. Res.* **1971**, *4*, 272–280; Rzepa, *J. Chem. Ed.* **2007**, *84*, 1535–1540.

[III] Perturbation theory generally only assumes small interactions. Furthermore, some effects such as nucleus–electron interactions are not taken into account in perturbation theory, among others.

[IV] The order refers to the power of the terms, which is why the closed-shell interaction is referred to as first-order interaction, while the third term is referred to as the second-order interaction.

all orbitals leads to a strong destabilization. The closed-shell repulsion thus represents the largest contributor to the activation enthalpy of a reaction. Since the first term is approximately equivalent for different reaction paths, it can be neglected hereafter to account for the selectivity of a reaction. The formation of regioisomeric products, for example, can be rationalized in terms of energy differences of the competing pathways, which can be approximated to be equivalent in terms of the closed-shell repulsion. This assumption can sometimes lead to contradicting results between theory and experiment, but in most cases there is a reasonable agreement.

The second term takes into account the attraction or repulsion of charged or polarized components (Coulomb or *electrostatic interaction*). In analogy to the classic electromagnetic force, like-charged particles repel each other and oppositely charged particles attract each other. Depending on the individual signs of Q_k and Q_l, the positive sign in front of the expression thus leads to a stabilization (negative term) or destabilization (positive term). For interactions of ions, integer values can be used, but polarized atoms or molecular fragments usually lead to a partial polarization with non-integer charges. The second term usually plays no or only a minor role in pericyclic reactions. An important exception, however, are hetero-Diels-Alder reactions, in which an asynchronous reaction path with partial polarization of the components can be observed (see Fig. 5.16).

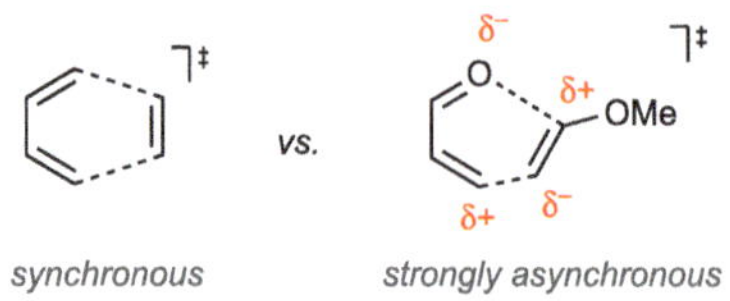

Fig. 5.16 Occurrence of asynchronous reaction pathways in hetero-Diels-Alder reactions

The third term represents the interaction of empty with filled orbitals between both components (*second-order perturbation*). The strength of the interaction is inversely proportional to the energy difference of the interacting orbitals, which is why mathematically a negative sign and thus a net stabilization of the reaction follows. It can be concluded from the third term that the orbital interaction of the reactants becomes stronger the more similar their energetic level is. Fukui concluded from this that for a qualitative description of pericyclic reactions, the consideration of the frontier orbitals is sufficient, as their interaction will dominate the entire process, despite all molecular orbitals interacting with each other of course [10]. The frontier orbitals are the highest occupied (HOMO) and the lowest unoccupied (LUMO) molecular orbitals. The interaction or mixing of orbitals is schematically shown in Fig. 5.17 and also forms the basis for avoided crossings in correlation diagrams.

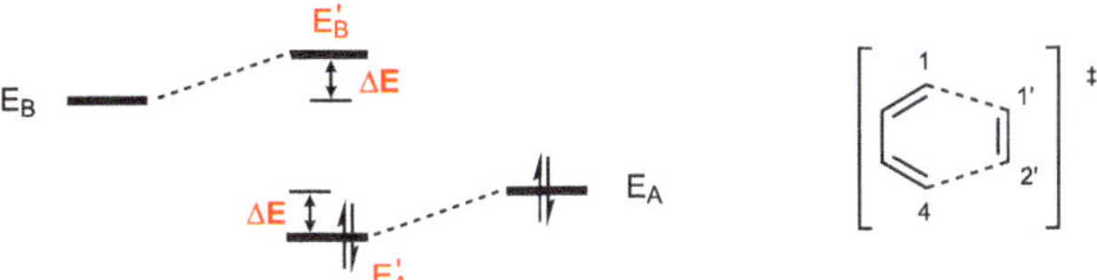

Fig. 5.17 Schematic representation of a stabilizing orbital interaction

Since the first term always causes a repulsion and polarization effects usually play a minor role in most pericyclic reactions, the orbital interactions largely determine the course of the reaction. This assumption appears reasonable: when considering different competing reaction paths, the first term should be almost the same size for all paths. Although the effect of the closed-shell repulsion thus contributes the lion's share to the activation enthalpy, the orbital interaction leads to the choice of the *dominant* reaction path. The regioselectivity of cycloadditions can thus be rationalized and predicted to a certain degree: For example when analyzing a Diels-Alder reaction (cf. Fig. 5.17), if one only considers the interaction between the terminal atoms 1 of the diene and 1' of the dienophile, as well as between atoms 4 of the diene and 2' of the dienophile, we arrive at the following estimate for the interaction of the components, neglecting further effects:

$$\Delta E \cong \frac{(c_{HO,1}\, c_{LU,1'}\, \beta_{1,1'} + c_{HO,4}\, c_{LU,2'}\, \beta_{4,2'})^2}{E_{LU,Dienophile} - E_{HO,Diene}}$$
$$+ \frac{(c_{LU,1}\, c_{HO,1'}\, \beta_{1,1'} + c_{LU,4}\, c_{HO,2'}\, \beta_{4,2'})^2}{E_{LU,Diene} - E_{HO,Dienophile}} \tag{5.2}$$

The HOMO and LUMO energies, unlike the transition structures, are easy to calculate and are also accessible from physical measurements. Thus, the HOMO energy roughly corresponds to the negative value of the first ionization energy (ionization potential) [25, 26], while the LUMO energy is accessible from UV-spectroscopic measurements and approximately equals the electron affinity [27, 28].[V]

Houk and co-workers applied the Klopman–Salem equation to confirm that a perturbation approach allows a good qualitative description and prediction of reactivities and selectivities in pericyclic reactions [29]. Despite the fact that those reactivities/selectivities of many reactions can be modeled by quantitative frontier orbital calculations, this approach has some weaknesses.[VI] Alternative approaches (conceptual DFT, HSAB, configuration mixing)

[V] The assumption is valid according to Koopmans' theorem only if reorganization effects are neglected.

[VI] Secondary orbital interactions constitute interactions of orbitals that are not directly involved in bond formation. They are not taken into account in frontier orbital theory, but can be more dominant than primary orbital interactions in some reactions. The mixing of electronic states in the transition state (configuration interaction, see Sect. 5.5.1.1) can also lead to a wrong prediction of reactivity, as have been pursued, but they are usually also based on a consideration of the frontier orbitals and their energetic interaction. Their advantage lies in the use of other, sometimes more accurate models. However, these are not as intuitively comprehensible and require a quantum mechanical calculation [30–33]. The frontier orbital approach is nowadays typically used for the qualitative description of pericyclic processes. The Klopman–Salem equation will be revisited in its simplified form in the subsequent chapter to explain the regioselectivity of Diels-Alder reactions (*vide infra*).

While correlation diagrams are difficult or even impossible to construct in some cases and the orbital coefficients must be available for an assessment via perturbation theory, a simple — purely qualitative — model is the examination of the frontier orbitals introduced by Fukui. This is, as already explained using the Klopman–Salem equation, a quite plausible approach. The interaction of the HOMO of one component with the LUMO of the other is considered and vice versa. A reaction is then regarded as symmetry-allowed if both interactions are bonding (see Fig. 5.18).

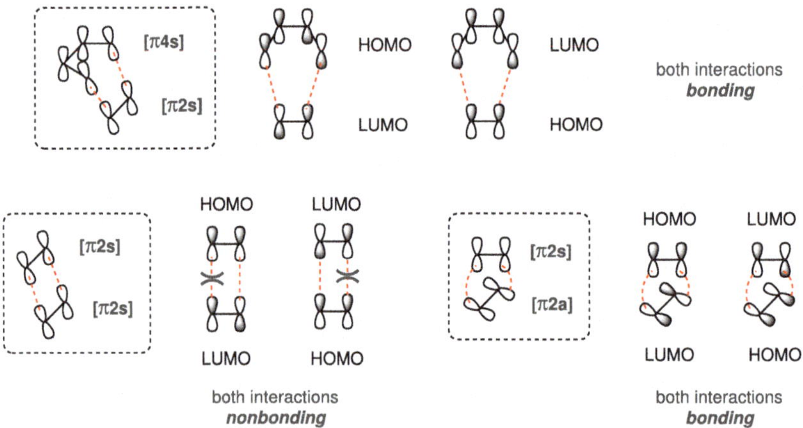

Fig. 5.18 Frontier orbital interactions in the thermal Diels-Alder reaction and [2+2]-cycloaddition

From the analysis of the frontier orbitals, it is concluded that the [4+2]-cycloaddition is allowed and the [2+2]-cycloaddition is forbidden, which had already been deduced from the correlation diagrams. Figure 5.18 only covers one variant of the orbital interactions. This is valid, as all permutations must always lead to the same result. If one of the two components does not react supra-, but antarafacially in the [2+2]-cycloaddition, the reaction becomes symmetry-allowed. In fact, there are examples of symmetry-allowed, thermal [2+2]-cycloadditions (e.g., when one of the components is a ketene). Two normal olefins cannot follow this path for steric reasons, but, in principle, this would be allowed on the

the frontier orbital theory relies on ground state configurations to describe reactivity. The reactivity pattern of cycloadditions with inverse electron demand is occasionally also miscalculated [30].

grounds of symmetry. A somewhat different picture emerges for photochemical reactions. Due to the excitation of a component from the HOMO into the SOMO, the SOMO/LUMO interaction bears a closer examination. In photochemical reactions, the interaction of the SOMO with the LUMO is considered and vice versa. Depending on the attack (suprafacial vs. antarafacial), [4+2]- and [2+2]-cycloadditions can be considered as representative (see Fig. 5.19).

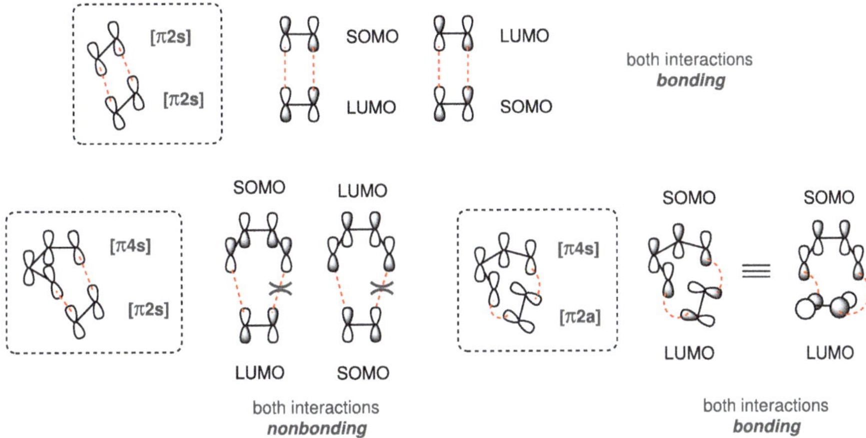

Fig. 5.19 Frontier orbital interactions in the photochemical Diels-Alder reaction and [2+2]-cycloaddition

As expected, the [2+2]-cycloaddition is symmetry-allowed under photochemical conditions. Since the SOMO of one component is identical to the LUMO of the other component (including its symmetry behavior), the frontier orbital analysis is very simple. In contrast, the Diels-Alder reaction is symmetry-forbidden under photochemical conditions when both components react suprafacially (see Fig. 5.11 for the MOs). However, if one of the components were allowed to react in an antarafacial manner, the photochemical [4+2]-cycloaddition again becomes symmetry-allowed. Even though this information could not be obtained from correlation diagrams, it is experimentally observed. As a competing reaction, a [2+2]-cycloaddition can occur instead of the desired Diels-Alder reaction.

Finally, electrocyclic reactions and sigmatropic rearrangements shall be further analyzed (see Fig. 5.20).

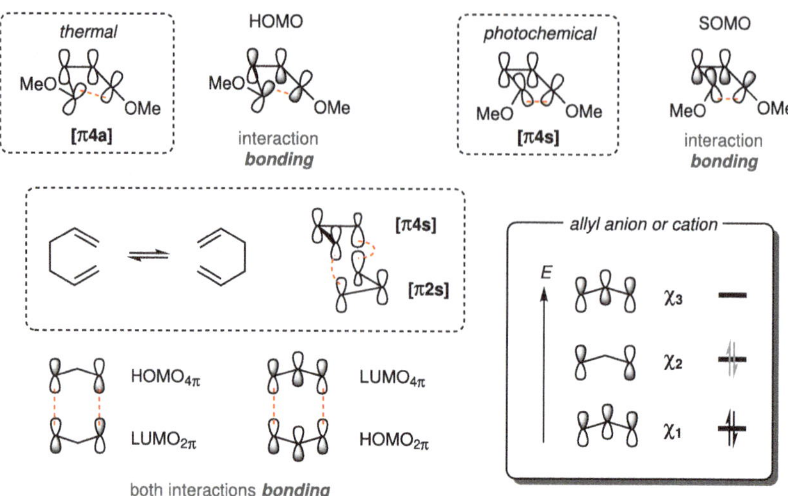

Fig. 5.20 Frontier orbital interactions in electrocyclic reactions and the Cope rearrangement

The electrocyclic transformations react from the HOMO or SOMO under photochemical conditions. It is clearly evident that a cyclization under thermal conditions is only allowed if proceeding antarafacially, thus resulting in a conrotatory ring closure. Under photochemical conditions, the reaction may only proceed suprafacially, leading to a disrotatory ring closure. For the previously discussed Cope rearrangement, it appears to be more complicated, as no conjugated π-system exists. The transformation is therefore approximated by assuming the reactant to consist of two interacting allyl components. If the electrons of the σ-bonds are ascribed to the two π-systems (the σ-bond also participates in the reaction), a total of six electrons are received to participate in the interaction. These are distributed to the two allyl systems, resulting in an allyl cation and an allyl anion equivalent. If the depicted MO scheme is applied in the previously outlined way, the [3,3]-sigmatropic rearrangement can be classified as symmetry-allowed.

To further simplify this procedure, Woodward and Hoffmann established general selection rules (the so-called *general Woodward–Hoffmann rules*) that signify a coherent mechanistic picture, which is underlying for all pericyclic reactions [9]. For this, only the number of involved components, the interacting electrons and the reactant's facial selectivity need to be considered.

*Thermal reactions are symmetry-allowed when the sum of the $(4n+2)_s$ and $(4m)_a$ components is **odd**.*

*Photochemical reactions are symmetry-allowed when the sum of the $(4n+2)_s$ and $(4m)_a$ components is **even**.*

For these rules, zero can be substituted for *n*, and it is also a valid total sum, whereby it is defined as an even number.

These rules may seem confounding at first glance, but they greatly facilitate a simple prediction of reactions once you have internalized them. Considering the reactions already discussed using correlation diagrams and through frontier orbital theory, these selection rules can be understood in more detail (see Fig. 5.21).

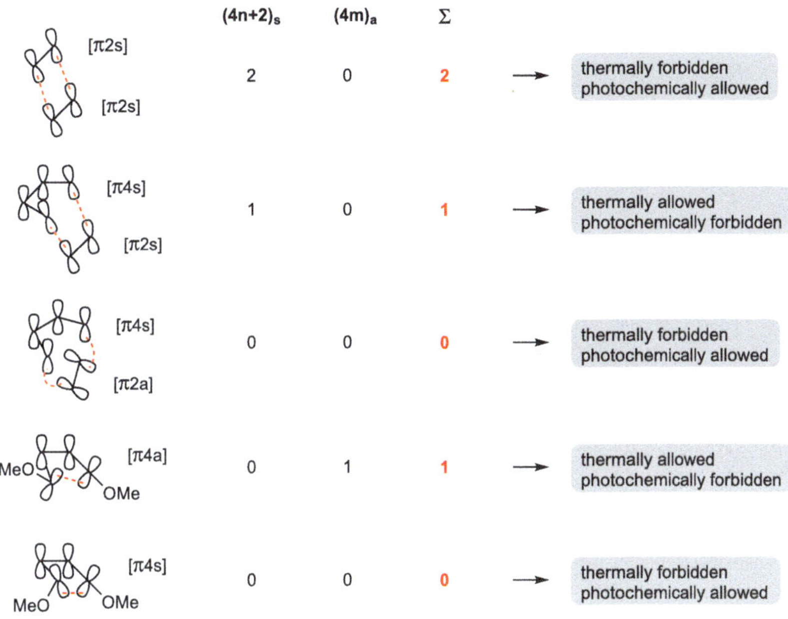

Fig. 5.21 Examples of the general Woodward–Hoffmann rules

The [2+2]-cycloaddition is the simplest case. Both components react suprafacially and have two electrons involved in the transition state. Thus, we get two components of the $(4n + 2)_s$ category, as zero can also be substituted for *n*. The sum is therefore even, and the reaction is only allowed when proceeding photochemically. The Diels-Alder reaction also contains two suprafacially reacting components, one of which contributes four electrons and the other two. The two-electron component thus belongs to the $(4n + 2)_s$ category, but the four-electron component does not, as only components with 2, 6, 10, etc. electrons are included. Since it also does not fit into the $(4m)_a$ category, it is not counted at all. Thus, we get a sum of one, and the reaction is decreed thermally allowed. If one of the two components reacts antarafacially (which one is irrelevant), the reaction suddenly becomes photochemically allowed and thermally forbidden. In the depicted example, neither of the two components fits into the categories anymore, but since zero is considered an even number, we arrive at the expected result. For the electrocyclic reactions, it was already derived that a conrotatory ring closure is associated with an antarafacial interaction and it is only allowed when occurring thermally (the sum is one). The disrotatory ring closure, on the other hand, is only photochemically feasible and, like the "photochemical" Diels-Alder reaction, yields a sum of zero.

The further applicability of the frontier orbital analysis and the general Woodward–Hoffmann rules will be examined more closely in the following chapters, covering various variants of pericyclic reactions of both a diastereo- and enantioselective nature.

5.2 Cycloadditions

Cycloaddition reactions represent the largest class of pericyclic reactions, with the Diels-Alder reaction constituting the prototype of a [4+2]-cycloaddition. Cycloadditions are stereospecific due to the necessary preservation of orbital symmetry and usually display a high regioselectivity. Their contribution as an elegant tool for CC bond construction quickly became apparent, and they have since proven indispensable for the formation of cyclic structures in both academic and industrial syntheses (see Fig. 5.22) [34–38].

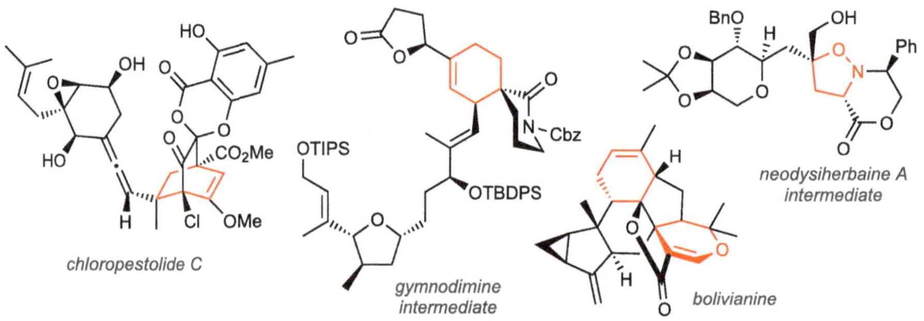

Fig. 5.22 Complex applications of the Diels-Alder and 1,3-dipolar cycloaddition in natural product syntheses [39–42]

Apart from [4+2]-cycloadditions, further subcategories include 1,3-dipolar cycloadditions and [2+2]-cycloadditions. Beyond these standard transformations, examples of [3+3]-, [4+3]-, [2+2+2]-cycloadditions and reactions of higher order have also been reported [43–45].

The somewhat less frequently encountered yet nonetheless immensely important 1,3-dipolar cycloadditions entail the reaction of a zwitterionic, usually electron-poor component (termed 1,3-dipole) with alkenes or alkynes (called dipolarophile). Cycloadditions find such a wide application because they can selectively and predictably introduce up to four stereogenic centers in one step. Especially in the 1950s to 1970s, when the highly stereoselective synthesis of acyclic systems was nigh impossible due to a lack of suitable methods, cyclic intermediates were often used to access the desired natural products with the correct stereoinformation (see Fig. 5.23) [46].

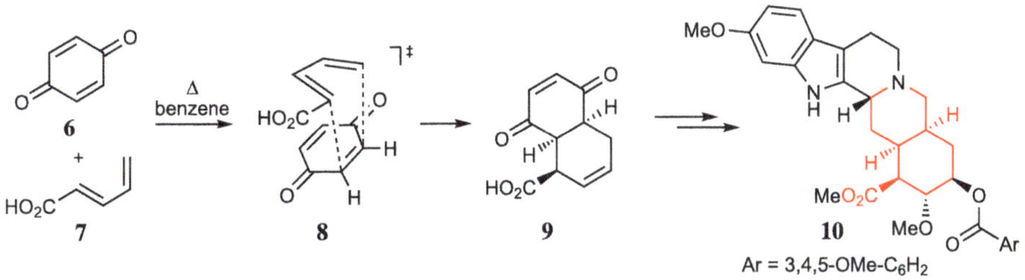

Fig. 5.23 Woodward's synthesis of reserpine (**10**) using a Diels-Alder reaction as key step

5.2.1 Reactivity and Selectivity

5.2.1.1 Diastereoselectivity — Concerted versus Stepwise Cycloadditions

All cycloadditions are stereospecific, regardless of whether they proceed suprafacially or antarafacially with regard to the individual components. Therefore only one diastereomer should be obtained due to the symmetry-related activation barrier. In the absence of stereogenic centers or chiral catalysts to effect a facial differentiation, a racemic mixture is always obtained (see Fig. 5.24).

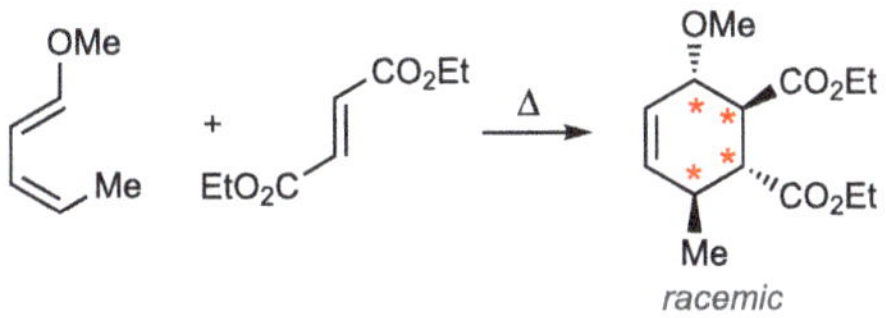

Fig. 5.24 Stereoselectivity of the Diels-Alder reaction

In the case of the Diels-Alder reaction shown in Fig. 5.24, the double bond geometry of both components is decisive as it is transferred to the product. An *E*-configured double bond of the dienophile leads to the *trans*-arrangement of the substituents in the product, whereas a *Z*-double bond results in a *syn*-compound. In the diene, *E*-configured substituents are rotated towards the approaching dienophile. If the diene is approached from above, these groups will also be positioned alike in the product. *Z*-positioned substituents are rotated in the opposite direction (see Fig. 5.25) [6].

There are examples where a mixture of different diastereomers is obtained. This suggests a stepwise rather than a concerted mechanism. This preferentially occurs when heteroatoms are involved in the cycloaddition, e.g., in hetero-Diels-Alder reactions, in the Staudinger

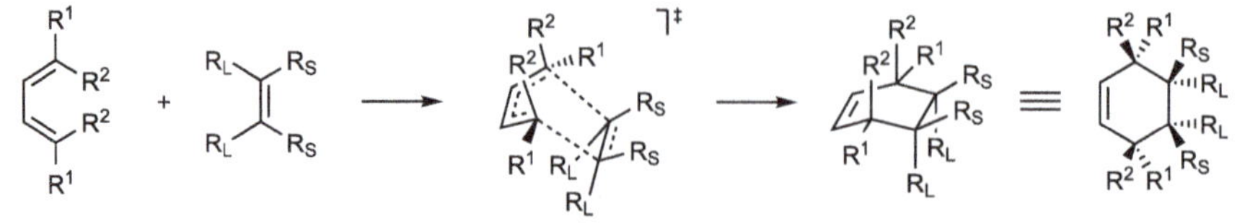

Fig. 5.25 Stereospecificity of cycloadditions

reaction, and similar transformations. In these cases, radical or zwitterionic intermediates (**12** and **13**) can play a role, in which a rotation around the terminal CC single bond occurs before the final ring closure, resulting in mixtures of diastereomers (see Fig. 5.26) [6].

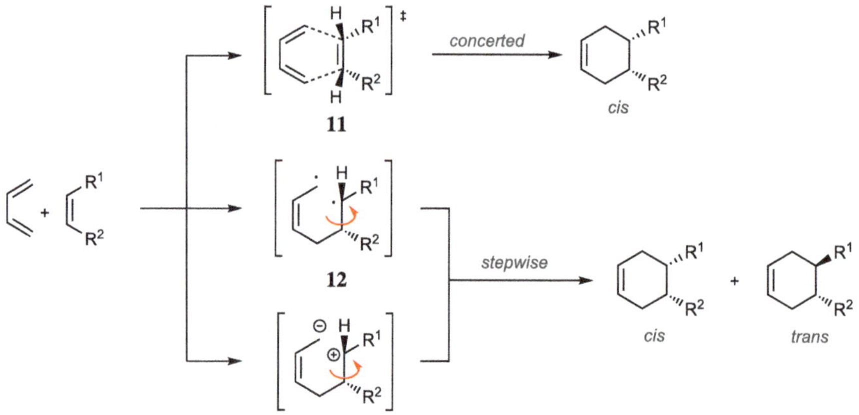

Fig. 5.26 Concerted versus stepwise mechanism of cycloaddition reactions [6]

A large HOMO-LUMO difference of both components (cf. Eq. 5.1, Sect. 5.1.2) and sterically demanding substrates decelerate the synchronous, concerted pathway and alternative, two-step trajectories become possible. Substituents that stabilize ionic or radical intermediates can additionally support a strongly asynchronous or stepwise reaction. The loss in stereoselectivity correlates with the lifetime of the intermediate, the rotation barrier, and, if an equilibrium between isomeric intermediates exists, the relative energies of these rotational isomers [6]. In the reaction of hexachloropentacyclodiene with various E-olefins, an increasing proportion of the *cis*-isomer was observed in the presence of substituents that help stabilize biradical intermediates (see Fig. 5.27).

Even if a reaction affords the "regular" diastereomer with complete stereoselectivity, this does not necessarily imply a fully concerted pathway. A CC bond rotation in the stepwise mechanism may simply exhibit too large a barrier, or the lifetime of the intermediates may be too short to result in a significant loss of selectivity. Depending on the relative energetic position of the non-concerted transition state, both reaction paths can thus occur.

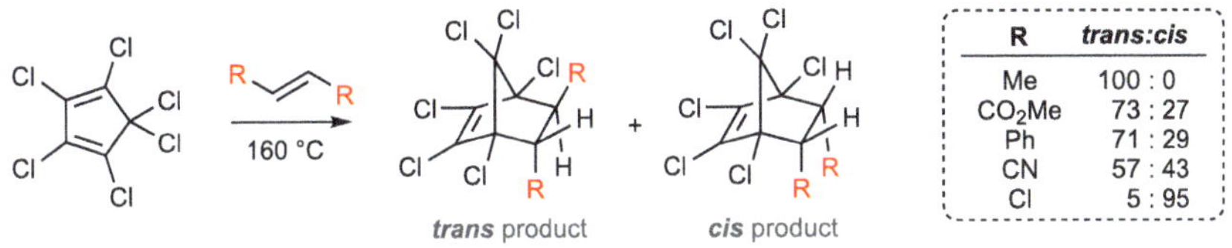

R	trans:cis
Me	100 : 0
CO$_2$Me	73 : 27
Ph	71 : 29
CN	57 : 43
Cl	5 : 95

Fig. 5.27 Dependence of the stereoerosion of non-concerted [4+2]-cycloadditions on the substitution pattern [47]

5.2.1.2 Reaction Rates and Substituent Effects

For a cycloaddition of a substrate bearing an extended π-system, the electrons in the conjugated π-system must be able to sustain a continuous interaction over the course of the reaction. The double bonds of acyclic polyenes must therefore be able to assume a synperiplanar conformation, which is also referred to as s-cis or cisoid (see Fig. 5.28) [48, 49].

Fig. 5.28 Possible conformations of butadienes

For unsubstituted polyenes or those bearing small substituents, this is a rather trivial matter. However, the more sterically hindered the substituents are, the more challenging the isomerization of the antiperiplanar or s-trans into the synperiplanar/s-cis conformation may become.

In the case of *tert*-butylbutadiene, a transoid conformation can hardly be adopted, which is why it is converted much more rapidly in the Diels-Alder reaction with maleic anhydride compared to the unsubstituted butadiene. On the other hand, 2,3-di-*tert*-butylbutadiene cannot assume a planar structure due to the strong steric repulsion, thus precluding any formation of the cyclohexene product even at significantly elevated temperatures. While large substituents at C-2 favor a cycloaddition, 1-substituted Z-dienes show a greatly reduced reactivity by virtue of the preferred adoption of the s-trans conformer (see Fig. 5.29) [48].

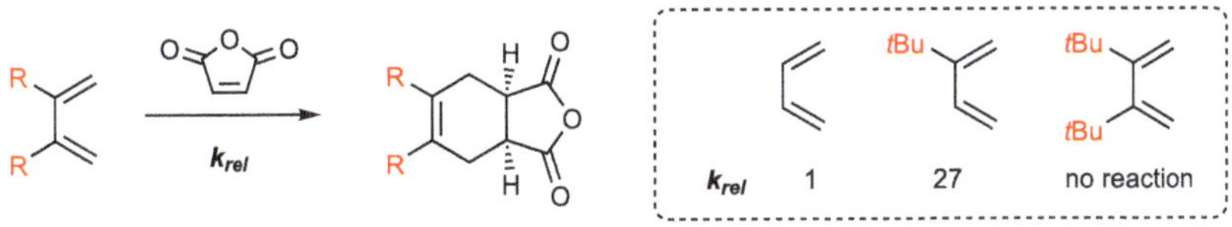

Fig. 5.29 Steric hindrance in the formation of s-cis dienes and influence on the reaction rate of the [4+2]-cycloaddition [48]

Polyenes that are forced into a s-cis conformation (*conformationally locked*), whether through steric or electronic effects, possess outstanding reactivity. Cyclopentadiene and furan particularly lend themselves for use in cycloadditions, especially when the conversion of less reactive dienophiles is desired. Furthermore, cyclic substrates possessing a strained polyene system result in a decreased activation energy by raising the ground state energies of the reactant if this ameliorates the ring strain (see Fig. 5.30) [50].

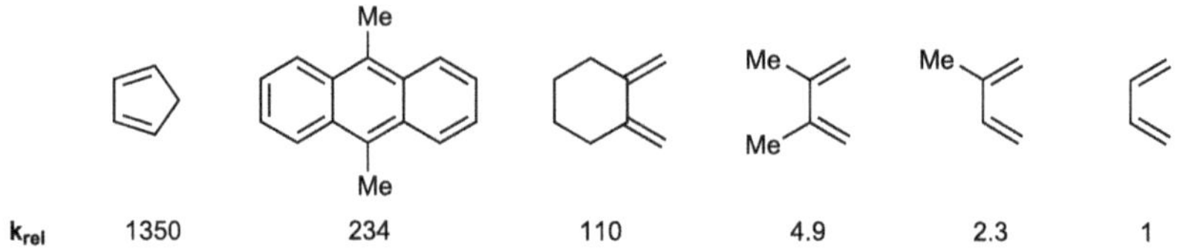

Fig. 5.30 Relative reactivity of various dienes in the cycloaddition with maleic anhydride [50]

While steric effects can have a significant impact on the reaction rate and stereoselectivity, electronic effects are decidedly more crucial. The rate of cycloadditions can be immensely increased by the selective introduction of various substituents. Generally, the reaction of two components with *opposite electronic character* (electron-poor and electron-rich) is particularly favorable. This can be rationalized by the intensifying interaction of the reactants as the interacting orbitals come into proximity and the overlap increases. The effect of the electronic character, i.e., the substitution pattern, of a component changes the energetic position of the molecular orbitals and thus influences the reactivity in a cycloaddition. The perturbation theory, as expressed by the Klopman–Salem equation, provides a demonstrative representation of this principle when considering the second-order perturbation term according to equation (5.2).

$$\Delta E \cong \frac{(c_{HO,1}\, c_{LU,1'}\, \beta_{1,1'} + c_{HO,4}\, c_{LU,2'}\, \beta_{4,2'})^2}{E_{LU,Dienophile} - E_{HO,Diene}}$$

$$+ \frac{(c_{LU,1}\, c_{HO,1'}\, \beta_{1,1'} + c_{LU,4}\, c_{HO,2'}\, \beta_{4,2'})^2}{E_{LU,Diene} - E_{HO,Dienophile}} \tag{5.2}$$

The activation barrier and reaction rate are directly derived from the components' energy gap between HOMO and LUMO, which determines the strength of their interaction. According to Houk, substituents on the reacting systems can be classified according to their influence on the frontier orbitals [51]:

- **X**-substituents are electron donors, whose non-bonding orbitals can overlap with the respective π-system. They raise both the HOMO and the LUMO energies. Typical X-substituents are –OR, –NR$_2$, –SR, Br, etc.

- **Z**-substituents are electron acceptors and typically conjugating; they result in a decrease of the LUMO and HOMO energy. Typical Z-substituents are $-CN$, $-CHO$, CO_2R, $-NO_2$, etc.
- **C**-substituents bear a π-system that is in conjugation with the double bonds of the reactants. They lower the LUMO energy and raise the HOMO energy, thus compressing frontier orbital separation. Typical C-substituents are phenyl or vinyl groups.

Depending on which component has which electronic properties, cycloadditions can be roughly divided into three classes, which were originally devised by Sustmann for Diels-Alder reactions [6]. This categorization is based on the consideration of the frontier orbital interactions (HOMO/LUMO) of both components and the identification of the dominant interaction (see Fig. 5.31).

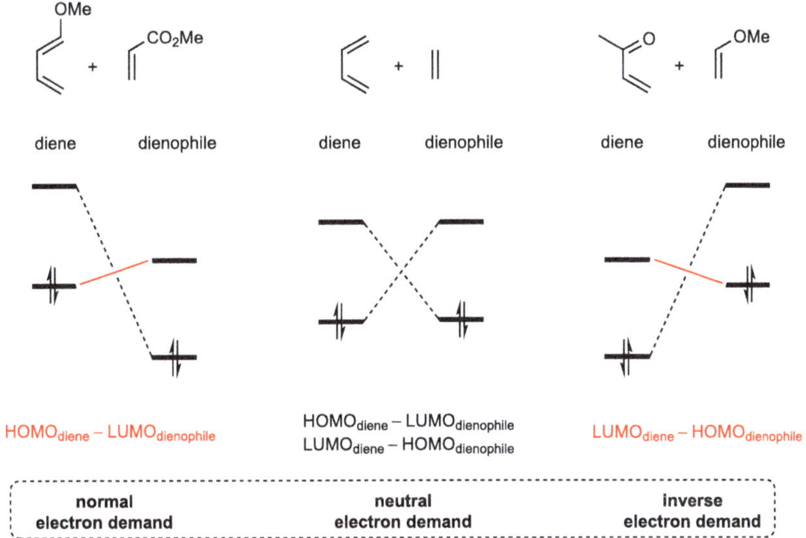

Fig. 5.31 Classifications of the Diels-Alder reaction, example reactions and dominant orbital interactions

There are cycloadditions with

- *normal electron demand* — The dominant interaction is $HOMO_{diene}$–$LUMO_{dienophile}$. Most of the Diels-Alder reactions fall into this category. It occurs with electron-rich dienes and electron-poor dienophiles.
- *neutral electron demand* — Both HOMO/LUMO interactions must be considered. If both components have similar electronic properties, they fall into this category. These components react only very slowly or not at all.

- *inverse electron demand* — The dominant interaction is $LUMO_{diene}$– $HOMO_{dienophile}$. Hetero-Diels-Alder reactions of oxa- and azabutadienes, the reaction of electron-poor dienes with electron-rich dienophiles, and 1,3-dipolar cycloadditions fall into this category. In the latter case, the dipolar component is electron-poor and the dipolarophile is electron-rich (see Sect. 5.2.3).

C-substituents always have an accelerating influence, as they reduce the HOMO/LUMO gap. In cycloadditions with normal electron demand (especially Diels-Alder reactions), X-substituents on the diene and Z-substituents on the dienophile increase the reaction rate (see Fig. 5.32) [52]. Cycloadditions with inverse electron demand (hetero-Diels-Alder reactions and 1,3-dipolar cycloadditions) can therefore be accelerated by the reversed substitution pattern. In general, 1-substituted dienes possess higher reactivity compared to 2-substituted systems [49].

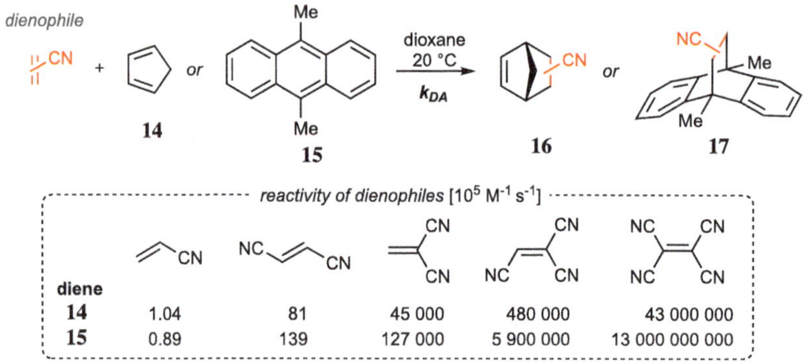

Fig. 5.32 Comparison of the reactivity of differently substituted cyanoethylenes [52]

Since a narrow energy gap of two reacting components leads to a strong interaction and thus lowering of the activation energy,[VII] it can be explained why, in contrast to normal and inverse Diels-Alder reactions, the reactions with neutral electron demand are very sluggish or do not proceed at all: Due to the similar electronic character of both reactants, the HOMO/LUMO gap is very wide, the interaction weak, and the activation barrier high.

5.2.1.3 Regioselectivity and Periselectivity

Cycloadditions often exhibit a clear regioselectivity (see Fig. 5.33). Initially, the rationalization of this phenomenon was based on a combination of steric and electronic effects, yet a satisfactory and reliable explanation remained elusive [6]. However, several quantitative approaches have become available over the years, of which perturbation theory in particular allows a quick and in most instances correct prediction (see Sect. 5.1.2).

[VII] Provided the interaction is of a stabilizing nature. While this can also be traced back to the Klopman–Salem equation (3rd term), this constitutes a general principle that is not limited to perturbation theory.

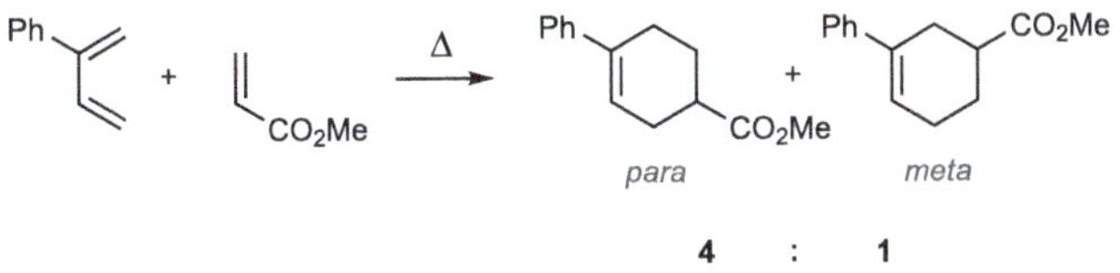

Fig. 5.33 Regioselectivity of the Diels-Alder reaction [29]

If we consider the simplified Klopman–Salem equation (5.2), it is clear from the above explanations why the reaction shown in Fig. 5.33 proceeds. The dominant orbital interaction in the Diels-Alder reaction with normal electron demand is $HOMO_{diene}$–$LUMO_{dienophile}$. While this energy gap determines the general rate of reaction, it is identical for the pathways to the different regioisomers. The regioselectivity of a cycloaddition, on the other hand, is determined by the numerator in the third term of the Klopman–Salem equation.

$$(c_{HO,1}\, c_{LU,1'}\, \beta_{1,1'} + c_{HO,4}\, c_{LU,2'}\, \beta_{4,2'})^2 \tag{5.3}$$

By knowing the relative size of the eigenvector coefficients $c_{HO/LU}$ of the dominant HOMO/LUMO interaction, the favored orientation in the transition state can be predicted.[VIII] In general, the interaction is approximately greatest when the **largest** orbital coefficients at the two components can interact with each other, as well as the smallest (see Fig. 5.34).

Fig. 5.34 Orbital interactions as the cause of regioselectivity

The effects of the X-, Z- and C-substituents on the energetic position of the orbitals and their coefficients can be understood from Figs. 5.35, 5.36, and 5.37 [51]. The relative size of the orbital coefficients and the energetic position of HOMO and LUMO (in eV) have been averaged over many systems. The phases of the MO coefficients (shown here as dark and light) clarify whether the two ends of a component have the same or opposite signs in their molecular orbitals.

[VIII] As mentioned in Sect. 5.1, frontier molecular orbital theory uses properties and thus coefficients of the ground state of the reactants for the discussion of the reactivity. This shortcoming usually results in sufficient accuracy of prediction for thermal reaction but becomes less reliable for photochemical reactions that proceed in excited states.

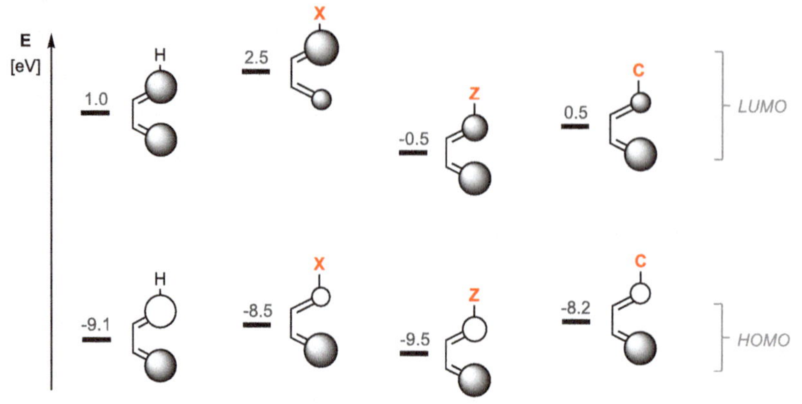

Fig. 5.35 Orbital energies and coefficients of 1-substituted dienes [51]

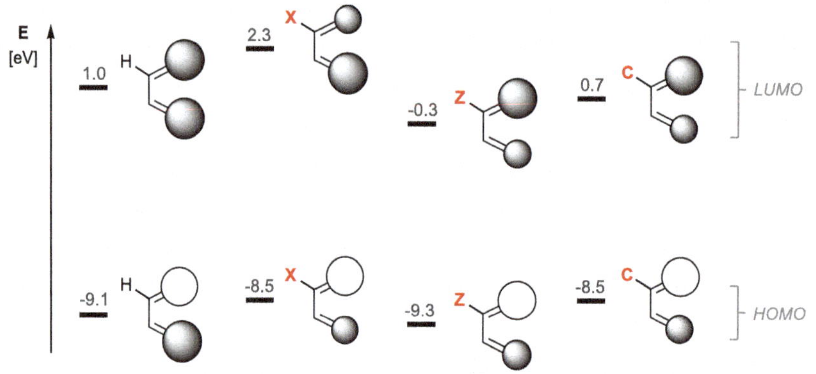

Fig. 5.36 Orbital energies and coefficients of 2-substituted dienes [51]

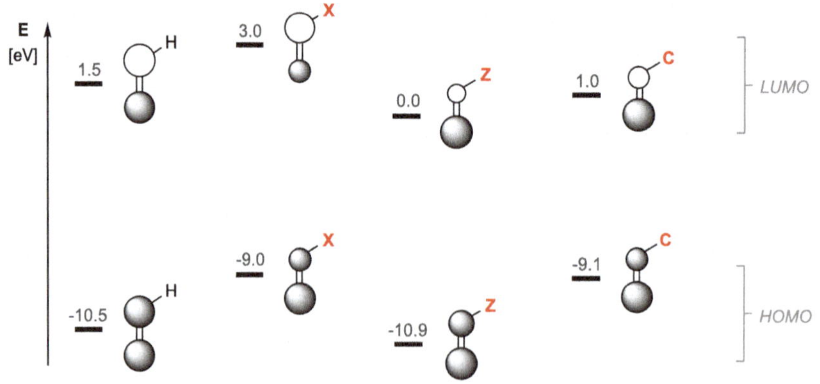

Fig. 5.37 Orbital energies and coefficients of monosubstituted dienophiles [51]

An appropriate substitution pattern can thus not only accelerate the reaction by narrowing the HOMO/LUMO gap (see Fig. 5.32), but the regioselectivity can be furthermore improved (see Fig. 5.38).

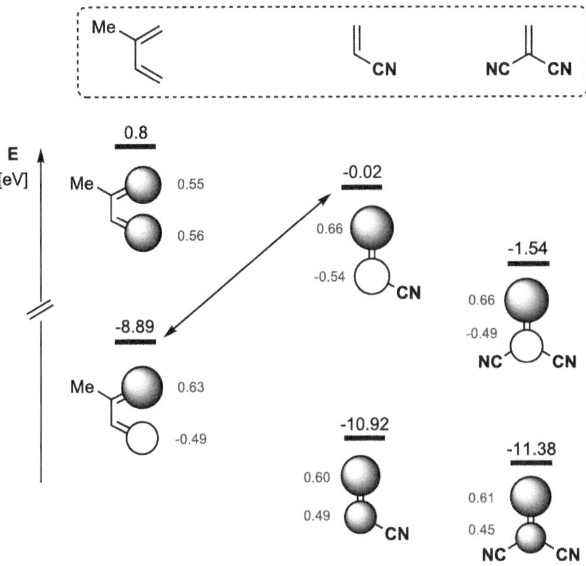

Fig. 5.38 Orbital energies and coefficients of isoprene as well as mono- and geminally disubstituted cyanoethylene. The size of the eigenvector coefficients is also given [53]

According to Fig. 5.38, a geminally disubstituted dienophile should not only display a drastically increased reactivity (cf. Fig. 5.32) but also a significantly improved regioselectivity due to the changed orbital coefficients. Cycloadditions are thus one of the few exceptions where increases in reactivity and selectivity are not mutually exclusive but can be jointly achieved. In addition to the substitution pattern (internal factor), other methods (addition of Lewis acids, high-pressure reactions; external factors) can be used to further positively modify the selectivity and accelerate the reaction (*vide infra*).

According to the principle of opposite electronic character, electron-rich dienes (X- and C-substituted) usually react with electron-poor dienophiles (Z-substituted) and electron-poor dienes (Z-substituted) with electron-rich dienophiles (X- and C-substituted). In the dominant orbital interaction, electron-rich dienes and dienophiles react from the HOMO and electron-poor dienophiles and dienes from the LUMO. This results in the regioselectivities shown in Fig. 5.39 according to the guidelines established by Houk [51].

For [2+2]-cycloadditions, Fig. 5.39 can also be referred to. However, since the photochemically excited state can differ somewhat from the LUMO (cf. Sect. 5.1), deviations from the predicted selectivities are more common in these cases.

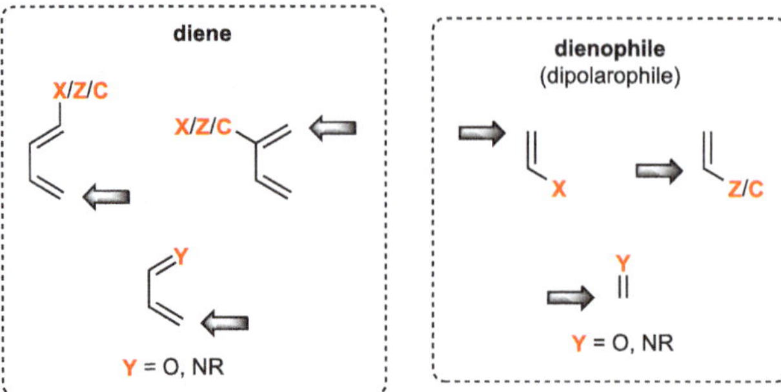

Fig. 5.39 Regioselectivity of Diels-Alder reactions. The arrow indicates the terminus with the largest orbital coefficient [51]

The same principles apply to 1,3-dipolar cycloadditions [54, 55]. The majority of reactions of 1,3-dipoles with electron-rich, X-substituted dipolarophiles are LUMO-controlled. Substrates for which this is not the case are too unreactive for synthetically relevant applications. However, when reacting with C- and Z-substituted dipolarophiles, both interactions must be considered and deviating regioselectivities can occur. LUMO-controlled reactions lead to the formation of the addition product bearing the substituent of the dipolarophile at the anionic terminus (see Fig. 5.40).

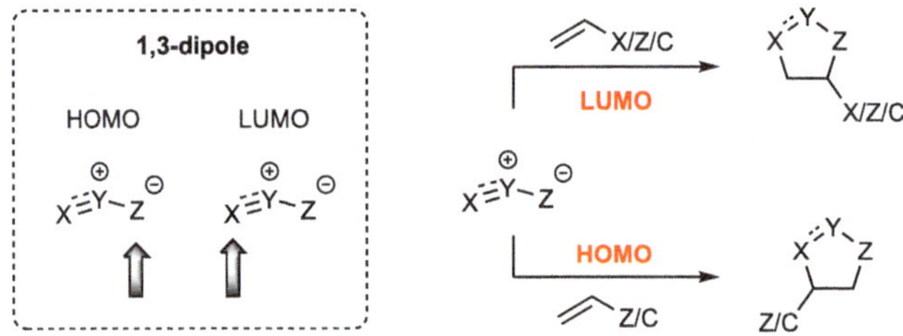

Fig. 5.40 Regioselectivity of 1,3-dipolar cycloadditions. The arrow indicates the terminus with the largest orbital coefficient [54]

The frontier orbitals of the dipoles can be derived from the parent allyl anion. Since the HOMO possesses a nodal plane at the central atom, a substitution at the central atom should have no significant influence as an approximation, while a terminal substitution should result in a pronounced effect. The opposite applies to the LUMO, as there is a higher electron density at the central atom compared to the anionic terminus. In case of unsymmetric dipoles, the electron density is likewise somewhat skewed compared to the parent allyl anion, but general trends still hold true (see Fig. 5.41).

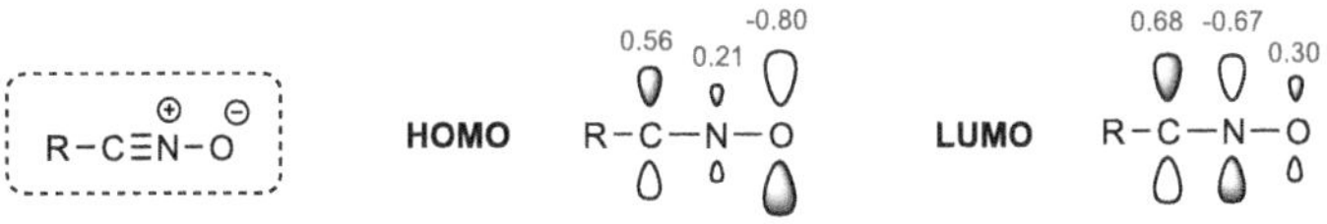

Fig. 5.41 Frontier orbitals of nitrile oxide. The numbers indicate the eigenvectors of the π-orbitals at the three atoms [56]

The trends in the frontier orbitals of dipolarophiles were systematically investigated by Houk via quantum mechanical calculations and by comparison with experimental data [54, 56]. Regardless of the type of substituents (X/Z/C), their influence on the size of the eigenvectors at the termini and the energetic position of the MOs will in most cases proceed in the following order:

- HOMO: anionic terminus > neutral terminus $\gg$ central atom
- LUMO: neutral terminus $\cong$ central atom > anionic terminus

In general, when a dipole is substituted, it can be assumed that the substituents' influence on the HOMO/LUMO level will only be about half as large as with a corresponding substitution of the dipolarophile [55]. Despite the fact that non-HOMO/LUMO orbital interactions, closed-shell repulsion, and Coulomb interactions contribute to the overall interaction, explanations of reactivity and regioselectivity according to Fukui's frontier orbital analysis are remarkably successful. The biggest challenge of these molecular orbital analysis is the steric hindrance of the reactants. Thus, a path strongly favored by electronic criteria can become impossible if steric repulsion is dominant. Especially 1,3-dipolar cycloadditions, which in many cases require the inclusion of both the HOMO and the LUMO interaction of the dipole, are particularly susceptible to this.

In polyenes, various addition paths to different constitutional isomers may be possible. The differentiation between two symmetry-allowed pericyclic processes under identical reaction conditions is referred to as *periselectivity* (see Fig. 5.42) [4].

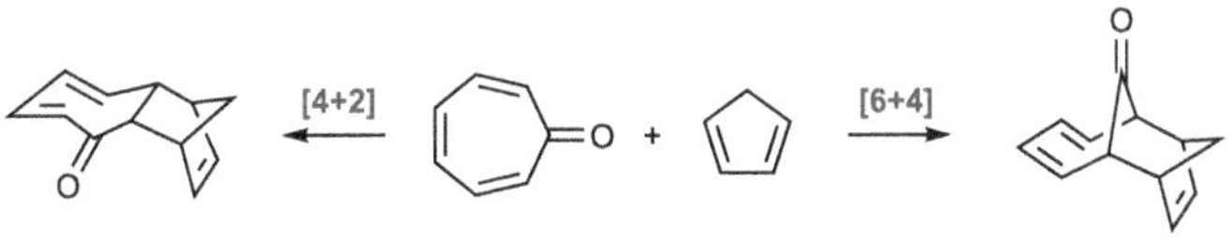

Fig. 5.42 Competition of [4+2]- and [6+4]-cycloadditions in the reaction of cyclopentadiene and tropone

In analogy to questions of regioselectivity, the favored path is derived from the dominant orbital interactions, specifically from the orbital coefficients [57]. The formation of regioisomeric products can thus be subsumed under periselectivity. Unfortunately, it is currently not possible to predict the periselectivity without quantum mechanical calculations (other than the regioselectivity). The larger the involved π-system, the more possible reaction paths are conceivable. The above reaction of tropone and cyclopentadiene consists of eight competing symmetry-allowed reaction paths (as well as ten further symmetry-forbidden ones). The dominant reaction path is the [6+4]-cycloaddition displayed on the right. As a guiding principle, one reaction path dominates in lower order cycloadditions due to the limited number of alternative combinations, as can be seen in the comparison of the afforded and possible products in several Diels-Alder reactions (see Fig. 5.43).

Fig. 5.43 Competition of the Diels-Alder reaction of two different dienes [58, 59]

The influence of substituents on the potential switch of the reaction mode is a rather complex problem. In general, the influence of monosubstitution of one component has no to very little impact on the periselectivity, assuming that steric effects do not play a major role [57]. The dominant reaction path persists. If substituents are introduced on each component, the mode is most likely to change when both substituents have different electronic properties (e.g., an ester and an ether functionality). As soon as two or more substituents of the same category are attached to one component, an alternative reaction path can often be followed (see Fig. 5.44).

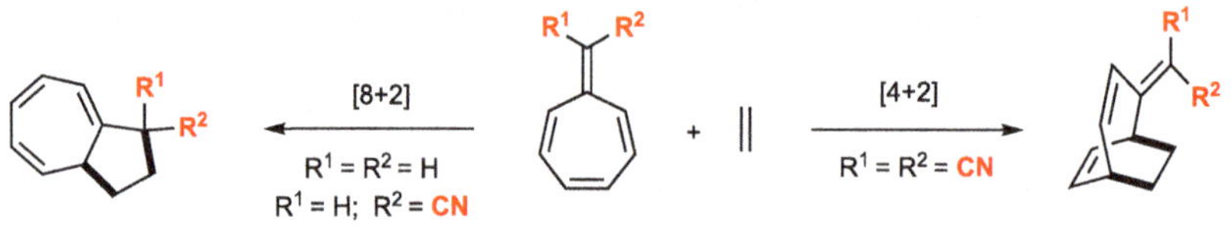

Fig. 5.44 Change of the reaction mode by successive introduction of substituents. The newly formed bonds have been highlighted [57]

Theoretical modeling can often predict the correct reaction path. However, sometimes even a quantum mechanical calculation of the competing reaction modes can fail, as a bifurcation of the reaction occurs after a shared, ambimodal transition state on the potential energy surface [60, 61].

5.2.2 Diels-Alder Reactions

Diels-Alder reactions constitute a significant proportion of the published pericyclic reactions and, along with Aldol reactions, cross-couplings, and other transformations, are a frequently employed reaction classes in the syntheses of complex target structures (see Fig. 5.45) [34, 62–65].

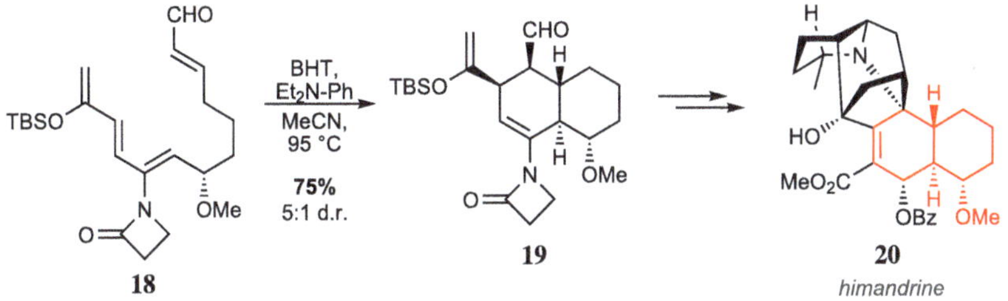

Fig. 5.45 Diels-Alder reaction in Movassaghi's synthesis of (–)-himandrine (**20**) [66]

Due to the vast amount of published material[67–74] and the extensive studies of the theoretical foundations, the reactivities in Diels-Alder reactions can be easily predicted without computational effort, which further enhances their attractiveness for use in synthetic endeavors. This is complemented by the stereospecificity of the reactions and the possibility of asymmetric variants for select types of reactants.

Based on the discussed theoretical principles, typical substrates for Diels-Alder reactions comprise the combination of electron-rich dienes with electron-poor dienophiles as well as electron-poor dienes that react with electron-rich dienophiles. In addition to cyclopentadiene and furan, the silyloxy-substituted methoxybutadiene introduced by Danishefsky (*Danishefsky's diene*) has emerged as key component of many syntheses by virtue of its excellent reactivity and regioselectivity and the presence of two additional functionalities for derivatizations [75–77]. α,β-unsaturated aldehydes are the classic substrates among the dienophiles, along with cyanoethylene and maleic anhydride (see Fig. 5.46).

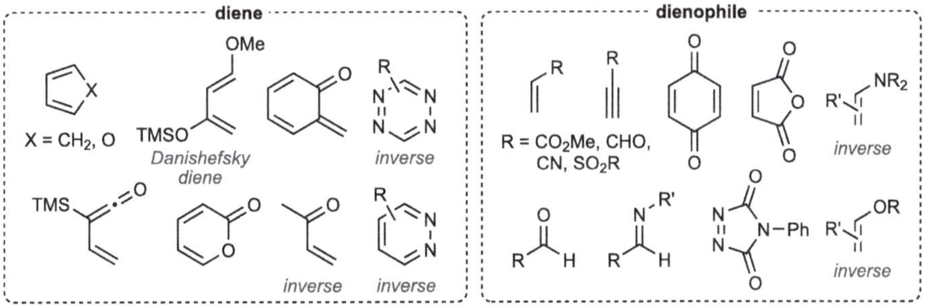

Fig. 5.46 Common substrates of Diels-Alder reactions with normal and inverse electron demand

Ketenes are counterintuitively poorly suited as a dienophile component for affording to the desired cyclohexenones. Therefore, masked substrates (ketene equivalents) are employed to introduce these functionalities, which can be converted to the corresponding product in one or two further steps [78].

5.2.2.1 *Endo-/exo*-Selectivity

In the Diels-Alder reaction, the dienophile can approach in such a way that the substituent is oriented towards the diene due to electronic or steric interactions, which is referred to as an *endo*-attack. The reverse orientation, in which the substituent of the dienophile is pointing away from the diene, is termed an *exo*-attack. Depending on the transition state, different diastereomers result as products (see Fig. 5.47).

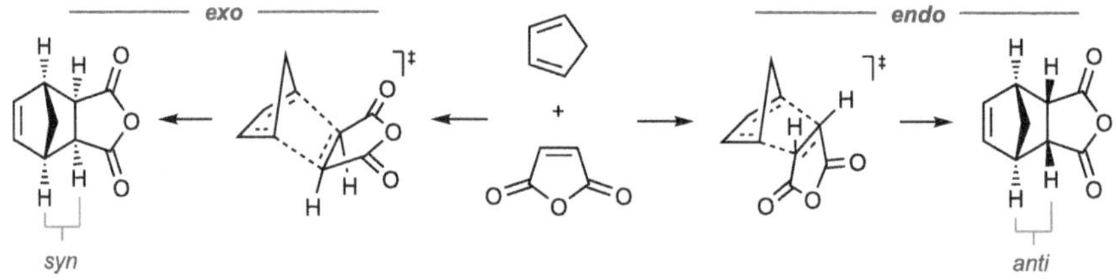

Fig. 5.47 Formation of the *endo-* and *exo*-cycloaddition products

Especially in the presence of unsaturated functional groups on the dienophile, the *endo*-addition is preferred, as was observed early on by Alder (so-called *Alder endo-rule*). The *endo/exo*-ratio is dependent on both the solvent and temperature [79, 80]. The empirical *endo*-rule can successfully predict the dominant stereoselectivity in kinetically controlled, thermal Diels-Alder reactions with very rigid and reactive dienophiles such as maleic anhydride and benzoquinone. However, in acyclic systems, the reverse preference is observed in some cases depending on the substitution pattern, which is attributed to a destabilization of the *endo*-transition state by steric repulsion [81].

The potential energy surface of most Diels-Alder reactions shows a lower activation barrier for the formation of the *endo*-product. In contrast, the *exo*-product typically is thermodynamically more stable and the product ratio can be shifted towards the *exo*-product under equilibrium conditions at elevated temperatures (see Figs. 5.48 and 5.49). The Diels-Alder reaction is thus *reversible*.

Originally proposed by Woodward and Hoffmann, so-called *secondary orbital interactions (SOI)* were postulated to account for the regioselectivity. These can occur when the dienophile contains additional substituents possessing π-orbitals in proximity to the reacting double bond. These π-orbitals enable an additional π-π-interaction with the diene reactant. Secondary orbital interactions participate in the reorganization of the π-system's electron density over the course of the reaction, but do not contribute to the σ-bond formation or scis-

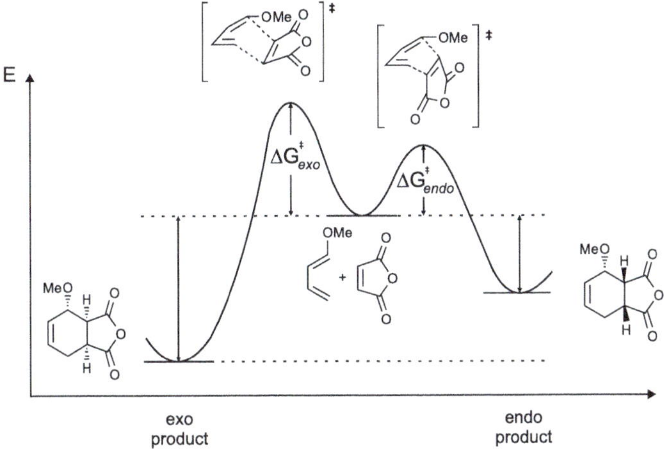

Fig. 5.48 Energy profile of the Diels-Alder reaction. Competition of *exo*- and *endo*-product formation

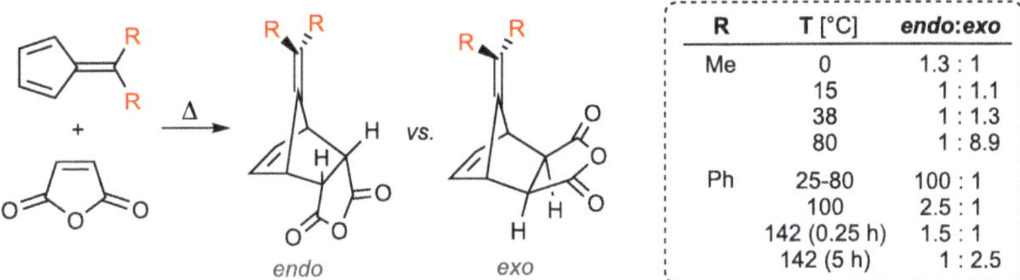

R	T [°C]	endo:exo
Me	0	1.3 : 1
	15	1 : 1.1
	38	1 : 1.3
	80	1 : 8.9
Ph	25–80	100 : 1
	100	2.5 : 1
	142 (0.25 h)	1.5 : 1
	142 (5 h)	1 : 2.5

Fig. 5.49 Dependence of the *endo/exo*-ratio on the reaction temperature and time [79]

sion. Due to the stabilizing interaction (the interacting orbitals on diene and dienophile have the same phase), the formation of the *endo*-product is facilitated, as no secondary orbital interaction can occur in the *exo*-transition state (see Fig. 5.50).

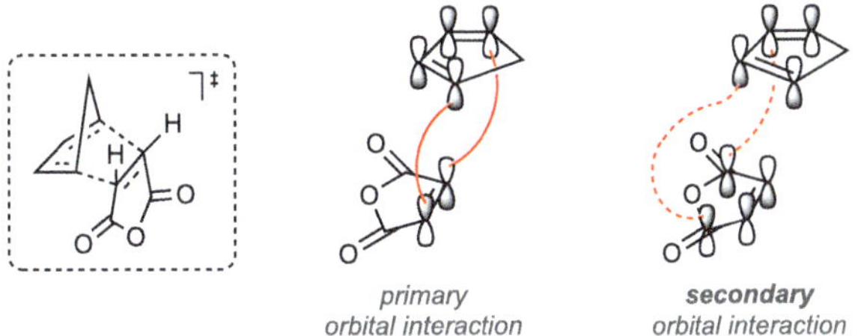

Fig. 5.50 Secondary orbital interactions in the *endo*-transition state of the Diels-Alder reaction

The existence of secondary orbital interactions is still controversially debated [82–88]. In addition to secondary orbital interactions, van der Waals forces, partial charge transfers, and activation volumes have also been cited as rationalization of the *endo*-preference. A general consensus on the underlying principle is still pending in the scientific community.

Publications often rely on the citation of the relative product configuration and the substrate's double bond geometry in addition to the prefixes *exo/endo* to precisely classify the transition states. For higher substituted dienophiles, the designation is based on the orientation of the substituent with the highest priority according to the Cahn–Ingold–Prelog rules. This results in the four possible transition states shown in Fig. 5.51.

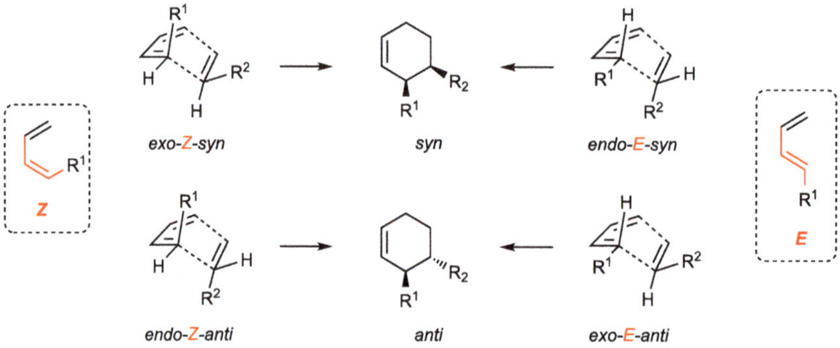

Fig. 5.51 Transition structures of intermolecular Diels-Alder reactions

Since the *syn*-product is obtained as a racemic mixture, *exo-Z-syn* and *endo-E-syn* transition states afford the identical diastereomer. The same also applies to the *anti*-configured addition product. In intramolecular reactions, the designation *exo* or *endo* depends on the orientation of the (carbon) chain through which the diene and dienophile are connected (see Fig. 5.52). Some of the aforementioned transition states may not be feasible due to an unfavorable ring strain, steric repulsion, or other (stereo)electronic effects [73, 89].

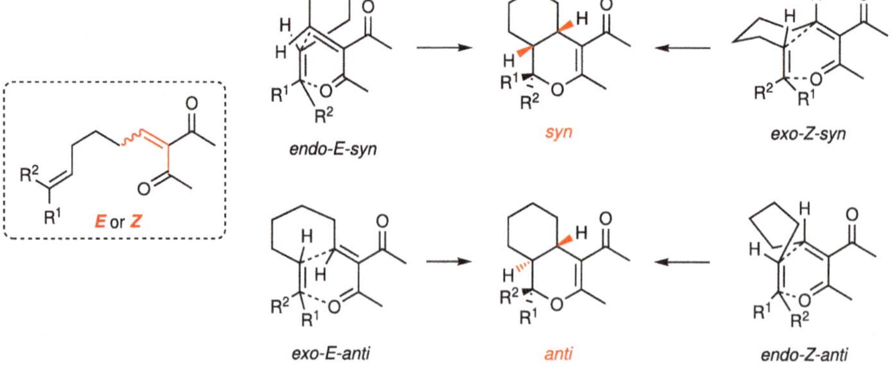

Fig. 5.52 Transition structures of intramolecular Diels-Alder reactions [89]

5.2.2.2 Hetero-Diels-Alder Reactions

In hetero-Diels-Alder reactions, a reactant contains at least one heteroatom that is transferred to the six-membered ring of the product (see Fig. 5.53) [90].

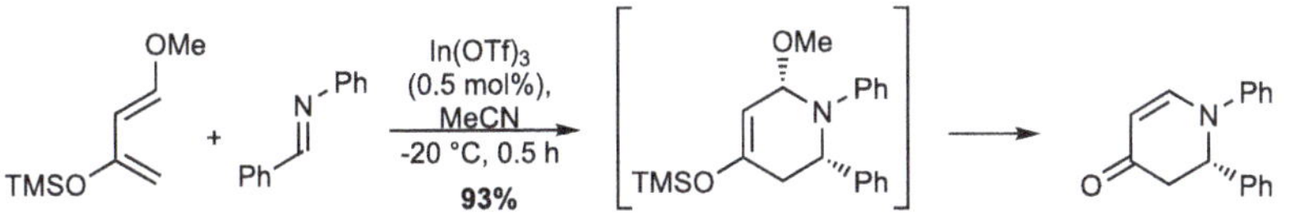

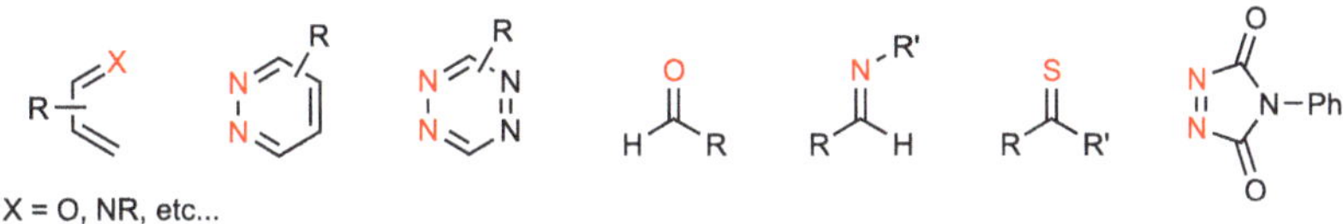

Fig. 5.53 Example of a hetero-Diels-Alder reaction [91]

Most often, a diene is converted with a carbonyl derivative as dienophile such as an aldehyde, furnishing a dihydropyran. The reverse example is a Diels-Alder reaction with inverse electron demand, where the heteroatom is present in the diene. α, β-Unsaturated aldehydes, as well as tetrazines, are typical representatives of electron-poor dienes. The most commonly involved heteroatoms are nitrogen or oxygen (aza- or oxo-Diels-Alder reaction), [92–97] but there are also substrates where sulfur or even phosphorus are incorporated (see Fig. 5.54) [98, 99].

Fig. 5.54 Common heteroatom-containing dienes and dienophiles

The introduction of heteroatoms lowers the HOMO, as well as the LUMO in most instances, and changes the orbital coefficients at the reacting termini. Many heteroatom-containing substrates preferentially react with electron-rich coupling partners, which is why a cycloaddition with inverse electron demand is the common reaction mode for heterodienes (see Fig. 5.55). Beyond the inherent influence of the heteroatom on the HOMO/LUMO levels, the added possibility to drastically reduce the electron density in the π-system by means of Lewis acid coordination at the heteroatoms is a crucial advantage. As a result, HOMO and especially LUMO are lowered (cf. Fig. 5.65, Sect. 5.2.2.4) [100]. In addition, the effects on the orbital energy and coefficients can be further augmented in Aza-Diels-Alder reactions by attaching electron-withdrawing substituents to the nitrogen. The resulting enhanced reactivity signifies that hetero-Diels-Alder reactions can often proceed under milder conditions than the formation of the corresponding carbocycles [92].

Another advantage of hetero-Diels-Alder reactions is the option to exploit the inherent asymmetry of the diene or dienophile. Due to the different orbital coefficients at the reacting termini, a good to excellent regioselectivity typically ensues. However, given that a significant number of hetero-Diels-Alder reactions follows a stepwise mechanism as the preferred

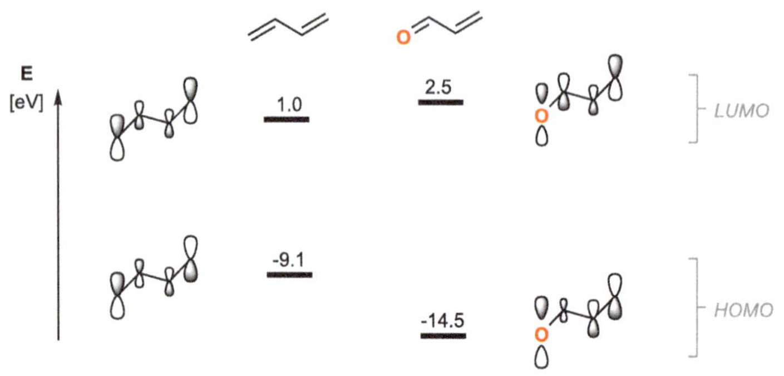

Fig. 5.55 Comparison of the orbital position and coefficients of butadiene and acrolein [92]

path, the diastereoselectivity of these processes cannot be predicted with the same certainty as in the corresponding "classical" [4+2]-cycloaddition [90].

Since almost all natural products and most pharmaceuticals contain heterocyclic moieties, hetero-Diels-Alder reactions are particularly suitable for accessing these compound classes. Heathcock and co-workers utilized a biomimetic domino reaction to construct the secodaphniphyllate **26**. The first step of the reaction sequence consists of an aza-Diels-Alder reaction to **23**, which affords the target molecular skeleton in **24** with the subsequent aza-Prins reaction (see Fig. 5.36) [101, 102] (Fig. 5.56).

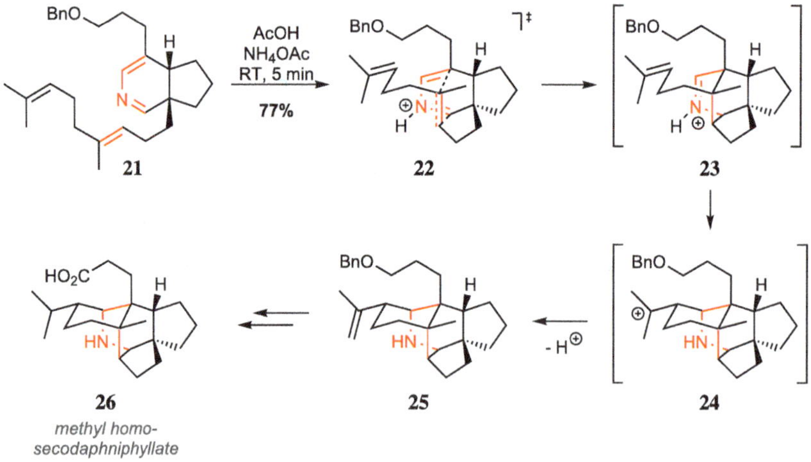

Fig. 5.56 Heathcock's biomimetic synthesis of methyl homosecodaphniphyllate [101, 102]

5.2.2.3 Retro-Diels-Alder Reactions

The Diels-Alder reaction is a reversible process. If two reactants cyclize in an intermolecular fashion, the substrates possess a higher entropy than the product as two components react to afford one carbo- or heterocycle. In principle, all Diels-Alder reactions can be reversed to give their substrates at sufficiently high temperatures due to the linear Gibbs energy relation:

$$\Delta G = \Delta H - T\,\Delta S \tag{5.4}$$

As previously indicated, this is one mechanism for the loss of *endo*-selectivity in some cycloadditions. Aromatic dienes (furan, anthracene) undergo cycloreversion comparatively effortlessly due to the restoration of aromaticity.

The main application of the retro-Diels-Alder reaction lies in the intermittent formation of reactive olefins and other π-bonds, which can subsequently react with a second substrate in an inter- or intramolecular manner as part of a domino or one-pot reaction [103–105]. Anthracene, 9,10-dimethylanthracene, cyclopentadiene, dimethylfulvene, and tetrazines are particularly suited as dienes for a cycloreversion. As dienophiles, CO_2, N_2, and R-CN are commonly employed, given that they are removed from the reaction mixture as gases, thus constituting the driving force of the fragmentation (see Fig. 5.57).

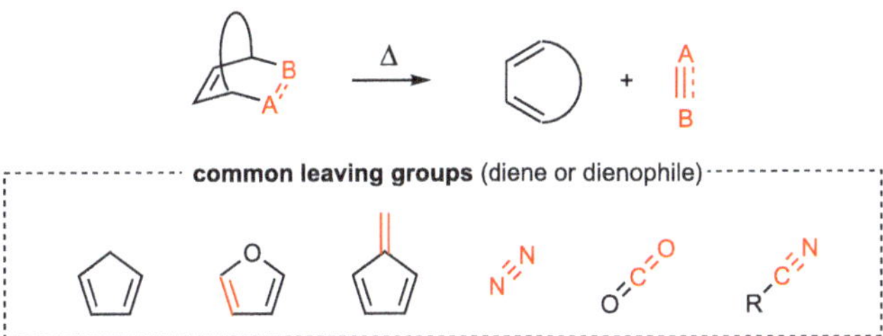

Fig. 5.57 General scheme and common components of retro-Diels-Alder reactions

Mander and Thomson used a domino cycloreversion/cycloaddition reaction as the key step in their synthesis of the natural product sordaricin (**31**). Following the retro-Diels-Alder reaction with concomitant release of cyclopentadiene, the resulting geminally disubstituted cyclopentadiene **28** was converted in a concluding intramolecular [4+2] addition favoring the *exo-Z-syn*-transition state **29** to afford **30** (see Fig. 5.58) [106].

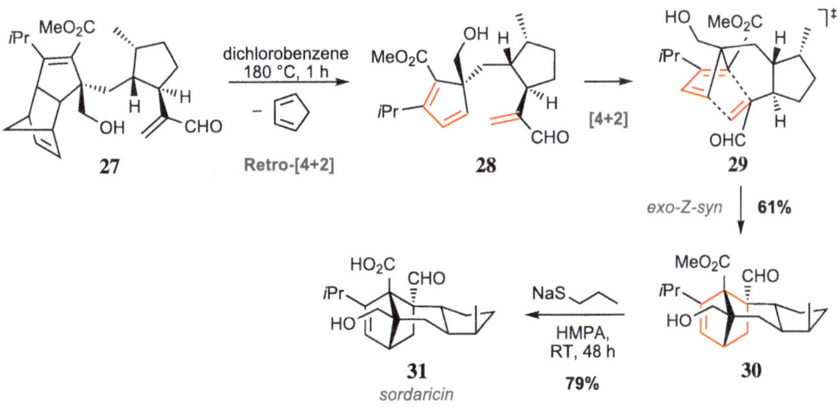

Fig. 5.58 Synthesis of sordaricin (**31**) by Mander *et al.* [106]

5.2.2.4 Modern Variants of the Diels-Alder Cycloaddition and Use in Natural Product Synthesis

By virtue of the expedient properties of the Diels-Alder reaction discussed at the beginning of this chapter (diastereoselectivity, yield, substrate scope, predictability/reliability), this cycloaddition is an important and frequently encountered method in the synthesis of complex natural products. Needless to say, only a very limited selection of synthetic applications can be covered herein. For a more comprehensive overview, the interested reader is referred to the relevant review articles [34, 62–65].

The racemic synthesis of the antineoplastic diterpenoid myrocin C, which was carried out in the Danishefsky group, used two Diels-Alder reactions for the efficient construction of the [6,6,6]-fused molecular framework. The silyloxy diene **32** was converted with benzoquinone in an initial intermolecular cycloaddition to afford **33**. Subsequent manipulation of the bridging alkene moiety comprising an oxidative cleavage, cyclopropanation, and linkage with a fumaric acid derivative furnished the precursor of the second Diels-Alder reaction. **34** was converted intramolecularly to afford the [6,6,6]-fused skeleton of myrocin C (**35**) in refluxing benzene. The carbonyl group was then removed oxidatively and the missing hydroxy functionalities were introduced via epoxidation and O_2-mediated oxidation. The natural product myrocin C (**36**) could finally be obtained as a racemic mixture and its structure was confirmed by X-ray crystallography (see Fig. 5.59) [107].

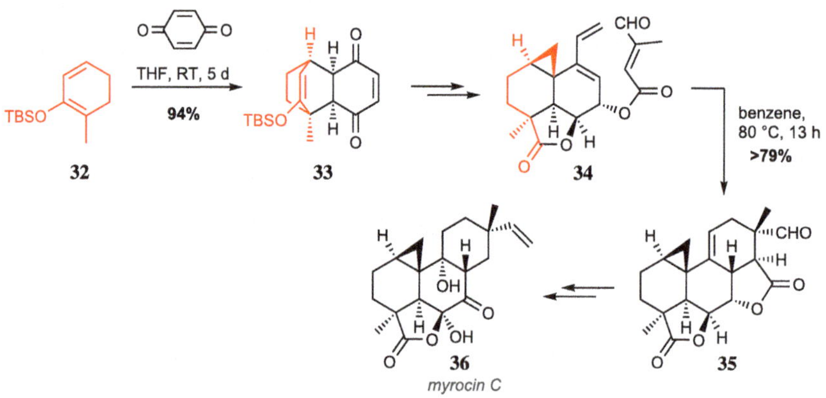

Fig. 5.59 Danishefsky's synthesis of myrocin C (**36**) [107]

In addition to classic [4+2]-cycloadditions, hetero-Diels-Alder reactions are also regularly used in the synthesis of natural products or derivatives thereof. Jacobi's synthesis of the alkaloid (–)-norsecurinine (**41**) relied on a hetero-Diels-Alder/retro-Diels-Alder sequence as key step. Starting from oxazole **37**, a Michael addition to the silyl-protected enynone to **38** initiated the domino reaction, after which the intramolecular Diels-Alder reaction proceeded to afford **39**. Upon release of acetonitrile from **39**, the substituted furan **40** was obtained in the concluding cycloreversion. Subsequent release of the masked lactone and construction of the bridged tetracyclic ring system provided the natural product (−)-norsecurinine (**41**) in 11 steps with a total yield of 5% (see Fig. 5.60) [108].

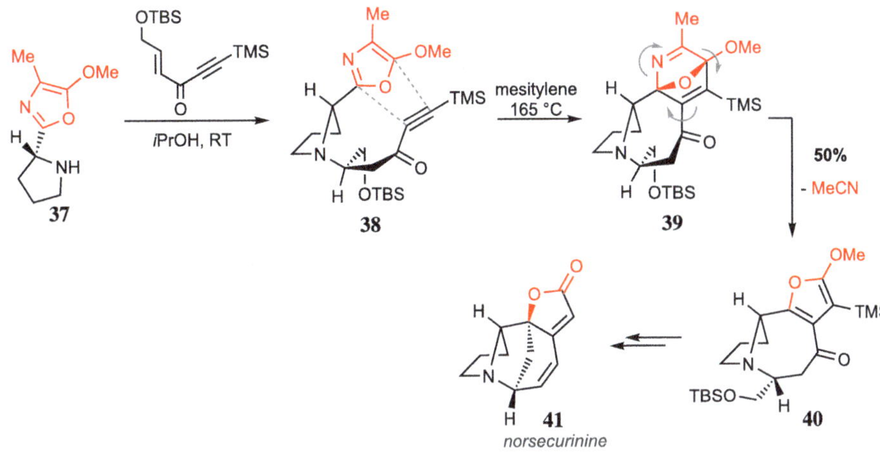

Fig. 5.60 Synthesis of (−)-norsecurinine (**41**) reported by Jacobi [108]

While simple heating affords good yields in many cases with reactive substrates under thermal conditions, a significant share of less reactive substrates requires further activation to enable their conversion in an acceptable timeframe. In addition to the well-established use of Lewis acids previously mentioned for hetero-Diels-Alder reactions, the application of a high reaction pressure is often observed to facilitate an increase in the rate of reaction (see Fig. 5.61) [109, 110].

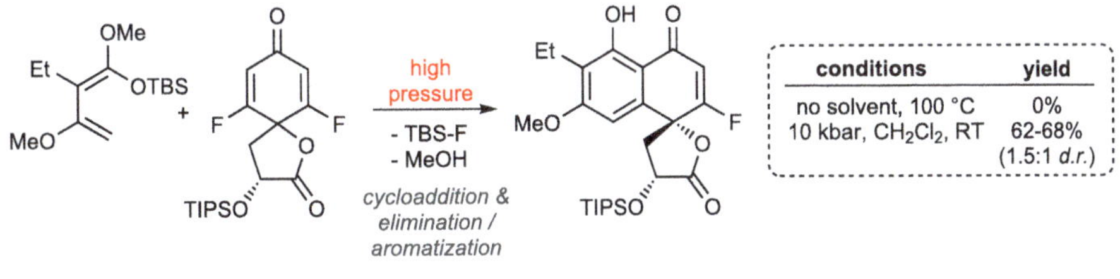

Fig. 5.61 Application of high-pressure conditions in Suzuki's approach to perenniporides A-C [111]

The *activation volume* $\Delta V^{\ddagger}$ describes the change in the rate constant k of a reaction $A + B \rightarrow [TS]^{\ddagger} \rightarrow P$ as a function of pressure under isothermal conditions [110].

$$\Delta V^{\ddagger} = -RT \left(\frac{\partial lnk}{\partial P} \right)_T \tag{5.5}$$

$\Delta V^{\ddagger}$ is composed of the partial molar volumes of the reactants and the transition state.

$$\Delta V^{\ddagger} = \overline{V}^{\ddagger} - \overline{V}_A - \overline{V}_B \tag{5.6}$$

At pressures above 10 kbar, the linear dependence of the rate of reaction on the pressure flattens out, which is why a further increase in pressure no longer displays an added beneficial effect [109].

In addition to a general acceleration of Diels-Alder reactions, an improvement in diastereoselectivity (*endo/exo*-ratio) is often observed alongside. This phenomenon is normally attributed to the different activation volumes of the respective *endo*- and *exo*-transition states. Assuming that the *endo*-transition state occupies a smaller volume $\overline{V}^{\ddagger}_{endo}$, it would be progressively favored as pressure increases (see Fig. 5.62).

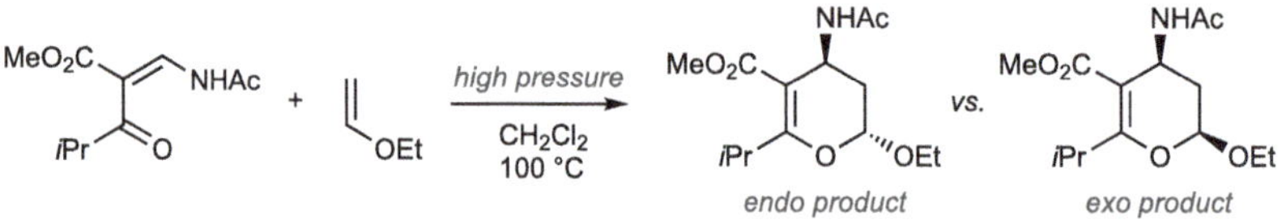

p [kbar]	k_{rel}	endo:exo
0.5	1.00	1 : 1.29
0.75	1.80	1 : 1.24
1.0	1.84	1 : 1.16
1.5	3.16	1 : 1.11
2.0	5.55	1 : 1.06
3.5	-	1.07 : 1
5.0	-	1.14 : 1

Fig. 5.62 Dependence of reaction rate and the *endo/exo*-ratio on pressure [112]

Despite the fact that high-pressure conditions might offer a benefit for various transformations beyond pericyclic reactions, the vast majority of its applications cover cycloadditions (see Fig. 5.63) [113]. Regardless of its promise (and success) particularly with labile substrates, the technique is only applied sporadically, which may probably be due to the limited availability of high-pressure equipment in most synthetic laboratories.

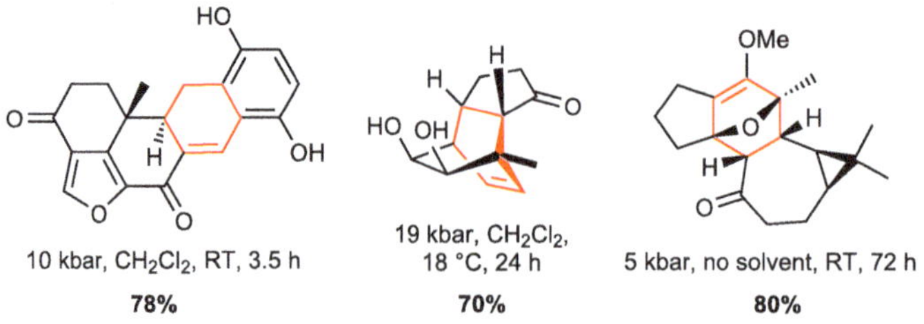

Fig. 5.63 High-pressure cycloadditions in natural product syntheses [114–116]

The second and nowadays "standard" approach is the addition of Lewis acids to effect an acceleration of the reaction. Upon complexation of the dienophile (or heterodiene in hetero-Diels-Alder reactions) by a Brønsted or Lewis acid, a dramatic influence on the reactivity and selectivity of the Diels-Alder reaction can be observed. The reaction rate is raised by several orders of magnitude, the *endo*-selectivity increases and an enhanced regioselectivity is obtained [100]. The acid-catalyzed Diels-Alder reaction thus represents one of the few exceptions to the reactivity/selectivity paradigm, wherein an increase in one implies a decrease in the other (see Fig. 5.64) [117].

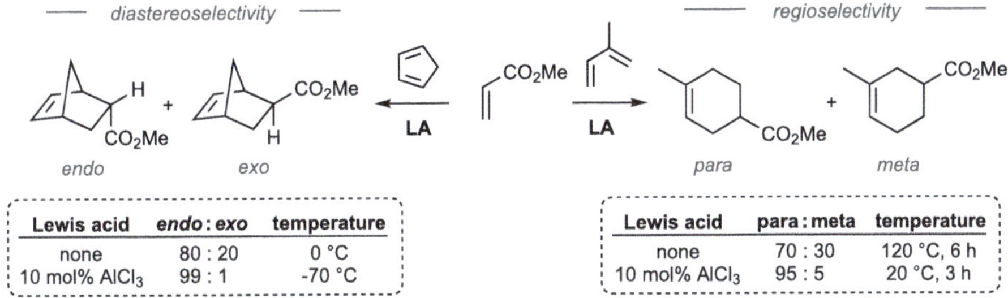

Lewis acid	*endo* : *exo*	temperature
none	80 : 20	0 °C
10 mol% AlCl₃	99 : 1	-70 °C

Lewis acid	*para* : *meta*	temperature
none	70 : 30	120 °C, 6 h
10 mol% AlCl₃	95 : 5	20 °C, 3 h

Fig. 5.64 Impact of Lewis acids on the diastereo- and regioselectivity [48, 118, 119]

The observed effects are a consequence of the reduced level of the LUMO in the dienophile through a decrease in electron density and the resulting change in the orbital coefficients. This induces an amplification of the differences in the orbital coefficients at the reacting termini of the dienophile and thus an increased regio-, stereo-, and periselectivity (see Fig. 5.65). Also, a reduction of electron–electron repulsion was postulated as an alternative rationale [120]. These orbital models imply an increased asynchronicity during bond formation. Taking into account electrostatic effects (first-order interactions), a stepwise mechanism with cationic intermediates can occur as a consequence in some cases [100].

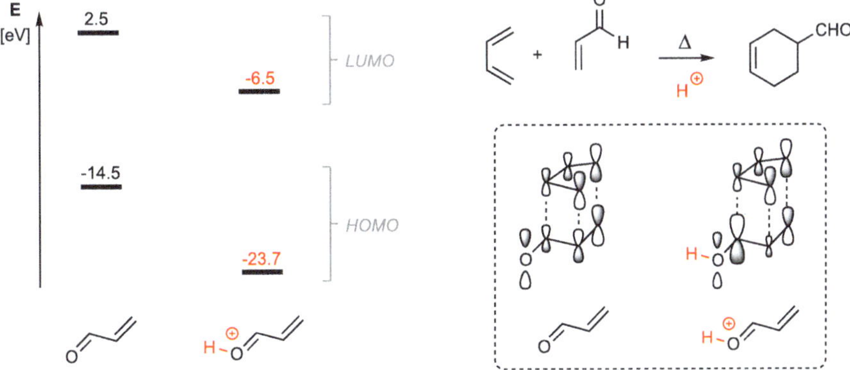

Fig. 5.65 Change in the orbital energies and coefficients of acrolein upon protonation [100]

The choice of the Lewis acid can have an influence on the *endo/exo*-selectivity of the cycloaddition. These effects are largely attributed to steric repulsion. Thus, the use of a sterically highly hindered Lewis acid can favor the formation of the *exo*-product (contrary to the electronic preference), while smaller Lewis acids form the electronically preferred product (see Fig. 5.66) [121].

Lewis acid	yield [%]	*endo* : *exo*
AlCl$_3$	71	98 : 2
Eu(fod)$_3$	33	65 : 35
TMSOTf	70	30 : 70
TBSOTf	84	2 : 98

Fig. 5.66 Influence of the Lewis acid on the *endo/exo*-ratio of the [4+2]-cycloaddition [121]

The Diels-Alder reaction has been studied in the presence of a plethora of Lewis acids, from strongly Lewis acidic compounds such as AlCl$_3$, TiCl$_4$, SnCl$_4$, ZnBr$_2$, ZnCl$_2$, lanthanide-based additives to metal-free (e.g., BF$_3$, TMSOTf) and chiral Lewis acids (*vide infra*). The use of main group element-based Lewis acids carries certain disadvantages. First of all, they bind too strongly to the substrate and, due to the unfavorable coordination equilibrium, must be used in large excess to positively impact the reaction. Second of all, they are usually highly reactive and extremely hydrophobic. In recent years, lanthanide salts such as Sc(OTf)$_3$, In(OTf)$_3$, and Yb(OTf)$_3$ as well as silyl-based organocatalysts have come into focus, as they are able to accelerate the cycloaddition very effectively even with small amounts of catalyst and are largely insensitive to small traces of water in the reaction mixture [96]. Apart from these, weakly acidic, non-nucleophilic phenols afford the desired DA adducts under mild conditions and are encountered particularly often in the total synthesis of natural products.

Tadano and co-workers employed an intramolecular Diels-Alder reaction in their synthesis of the macrocyclic antibiotic (+)-tubelactomycin A (**45**). The preferred *exo-E-anti* arrangement of the boat transition state **43** is attributed to the evasion of a diaxial interaction of the two methyl groups. The sterically highly hindered and weakly acidic phenol BHT was used as a Lewis acid to obtain the desired intermediate **44** in 93% yield with an *exo:endo*-selectivity of 8:1. Whether the aldehyde in **43** is merely protonated or interacts via an H-bond with BHT was not further investigated. **44** could subsequently be converted to the natural product **45** in 13 further steps (see Fig. 5.67) [122, 123].

Taking a cue from analogous approaches in aldol reactions, **asymmetric** cycloadditions can be realized either by an enantioinduction from stereoinformation in the reactants (persistent stereogenic centers or auxiliaries) [124] as well as by external stimuli (chiral reagents or catalysts) [125–127].

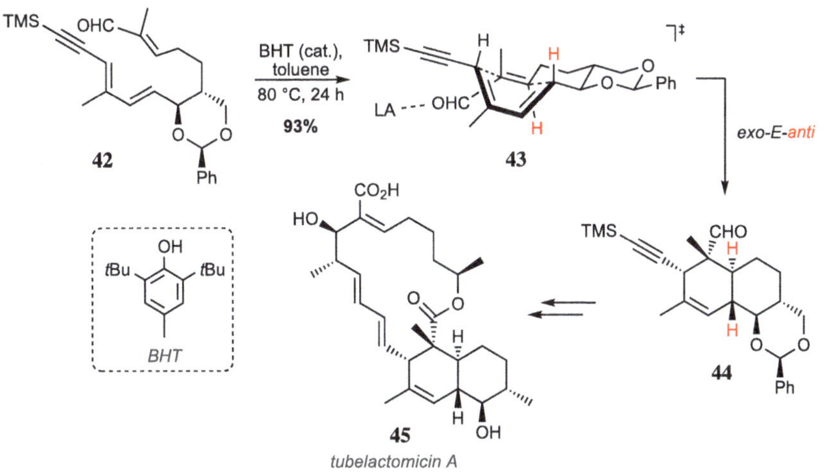

Fig. 5.67 Synthesis of (+)-tubelactomycin A (45) by Tadano *et al.* [122, 123]. The *anti*-arrangement refers to the highlighted H-atoms

Substrate-controlled reactions are almost exclusively used in intramolecular Diels-Alder reactions, in which typically all possible transition states except one are disfavored for steric reasons. This may be achieved through various means: by locking the conformation via temporary silyl groups, by preventing certain reaction paths through varying chain lengths between the reactive ends, or by the introduction of sterically shielding substituents. Despite its success, this approach is not generally applicable but rather highly substrate dependent (cf. Fig. 5.52) and can therefore only be illustrated by referring to some representative examples.

Miyashita and co-workers employed an intramolecular Diels-Alder reaction for the selective construction of the B- and C-rings in the marine alkaloid norzoanthamine. The configuration of the newly formed stereocenters in 48 was controlled by the stereogenic elements present in the substrate 46; these dictated the cycloaddition to proceed via a rigid *exo-E-anti*-transition state 47. The *endo*-isomer was also obtained as minor product via an *endo-E-syn* path, which may be a result of the requisite high reaction temperature. It could afterwards be separated from the desired stereoisomer by crystallization. The *trans,anti,trans*-linked perhydrophenanthrene framework 48 was subsequently converted to the complex natural product in a further 28 steps (see Fig. 5.68) [128].

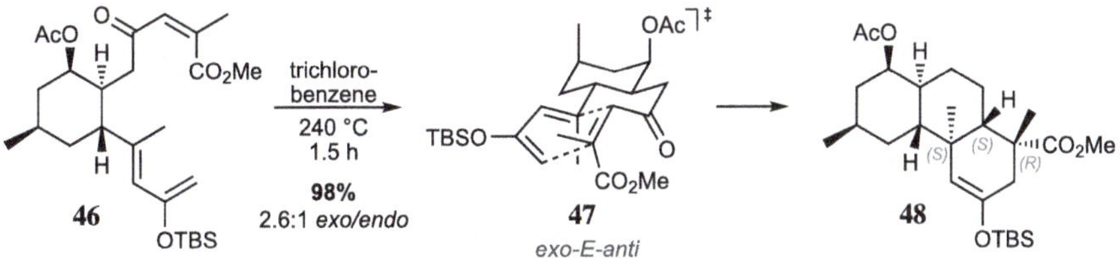

Fig. 5.68 Synthesis of norzoanthamine according to Miyashita *et al.* [128]

The attachment of *auxiliaries* is largely realized by appending them to the dienophile via an ester or amide linkage, typically relying on acrylic acid derivatives or related structures [124]. The challenge of intermolecular variants lies in selectively favoring only one reactive conformation. The most commonly employed auxiliary-augmented dienophiles often use menthol/phenylmenthol, the Oppolzer auxiliary derived from camphorsulfonic acid, the Evans auxiliary synthesized from amino acids, as well as lactate esters as stereoinductors (see Fig. 5.69).

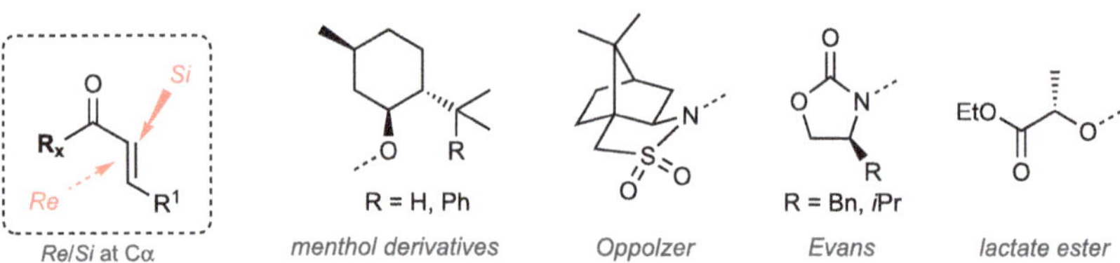

Fig. 5.69 Common dienophiles in auxiliary-controlled asymmetric Diels-Alder reactions [124] R_x signifies the auxiliary group

Apart from menthol esters, which impart the desired reactive conformation through steric shielding, both carbonyl and sulfonyl groups in the other auxiliaries shown in Fig. 5.69 are locked into position by addition of a coordinating Lewis acid [124]. With non-chelated α, β-unsaturated esters, the s-*trans*-orientation of the substrate typically constitutes the reactive conformation [129]. In **49**, this is achieved by steric repulsion with $AlCl_3$, while the phenyl group can shield the back of the double bond through an alignment via a π-interaction. In contrast, α, β-unsaturated amides prefer the s-*cis*-conformation. While the ethyl group on the metal ion shields the upper *Re*-face in **50**, shielding of the rear *Si*-face occurs in **51**. With benzyl-substituted Evans auxiliaries, there is also a stabilizing π-interaction of the aryl group with the CC double bond [130]. In the case of lactate esters (**52**), the facial selectivity is based on the shielding of one face of the dienophile by the bulky, octahedral $TiCl_4$, which forms a rigid chelate with both carbonyl groups of the diester (see Fig. 5.70) [131].

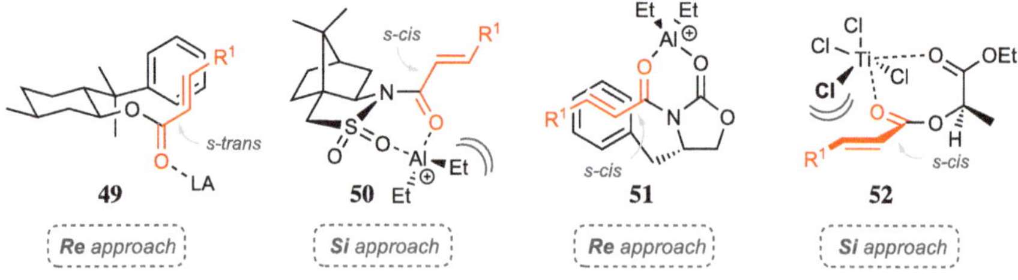

Fig. 5.70 Reactive conformations of common auxiliary-bearing dienophiles [124, 130–132]

Evans and Black resorted to an intramolecular Diels-Alder reaction in their synthesis of the tetracyclic macrolide (+)-lepicidin A (**56**). The diene subunit in **53** was assembled by a Stille coupling of the macrocyclic vinyl stannane with a suitably functionalized vinyl iodide of the side chain. The good diastereoselectivity of 10:1 in the Lewis acid-catalyzed intramolecular [4+2]-cycloaddition was accomplished by the introduction of the Evans auxiliary. The addition product **55** (a *trans*-hydrindene) was obtained presumably via the transition state **54**. Refunctionalization and an intramolecular aldol condensation afforded the differentially protected aglycone moiety. Successive glycosidations with 2,3,4-tri-*O*-methyl-D-rhamnose and L-forosamine followed by deprotection and methylation furnished the natural product (+)-lepicidin A (**56**) and confirmed its absolute configuration (see Fig. 5.71) [133].

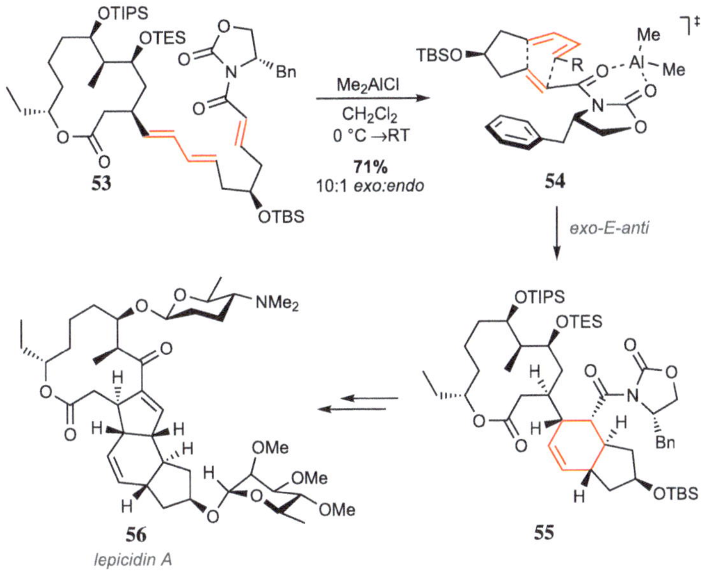

Fig. 5.71 Evans' synthesis of (+)-lepicidin A (**56**) [133]

The discovery that Lewis acids can effect a significant rate enhancement in cycloadditions logically led to the use of *chiral Lewis acids* for the induction of a desired asymmetric reaction topology [126]. This precludes the need for an introduction and subsequent cleavage of auxiliary groups while still enabling an enantioselective transformation. To avoid multiple reactive conformations enabled by too many degrees of rotational freedom, a coordination of the Lewis acid to two points of the molecule is usually resorted to — either through a double Lewis base/acid interaction or an additional π-interaction. Although the first example of the use of a chiral Lewis acid was published as early as 1979 [134], the seminal studies in this field were carried out from the late 1980s to the mid-2000s (see Fig. 5.72) [135–145].

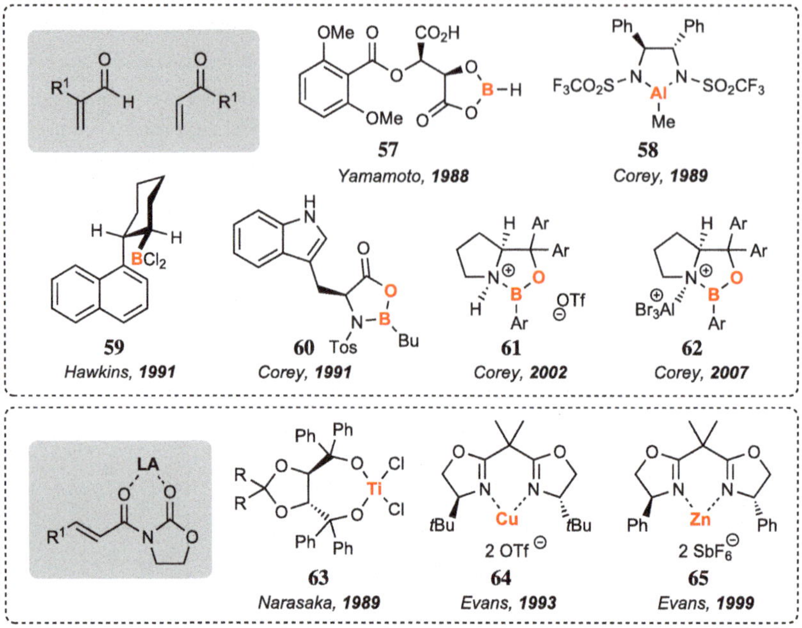

Fig. 5.72 Common chiral Lewis acids in enantioselective Diels-Alder reactions. The complexing atoms of the Lewis acids are highlighted

The research groups of Corey (**58, 60, 61**) and Evans (**64, 65**) investigated different catalyst systems over an extended period of time and intensively analyzed their mechanisms and substrate scope. They can be considered to be among the most efficient systems. The Lewis acids introduced by Yamamoto, Narasaka, and Hawkins (**57, 63, 59**) are no longer in use today, as the complementary methods of Evans and Corey have superseded them due to their broad substrate scope. They nevertheless represent inspiring examples of rational catalyst design and provide good to excellent enantiomeric excess [135, 136, 138]. **58** and **60** were the first chiral Lewis acids introduced by Corey and did not possess as wide an applicability as the subsequently published catalysts **61** and **62**. **58** relies on the respective amide of the oxazolidinone to ensure a conformational locking of the dienophile in the transition state **66** (see Fig. 5.73) [137]. By shielding the *Si*-face in **66** through one of the phenyl residues, only the addition from the direction facing away from the viewer is favored (*Re*-face) [146]. **60** catalyzes the conversion of α, β-unsaturated aldehydes as substrates. In addition, the Lewis acid necessitates the presence of an α-substituent on the aldehyde component to allow for good facial discrimination and thus high enantioinduction [139]. The catalyst is directly attached to the aldehyde functionality via a dative O-B bond and a long-range hydrogen bond. The substrate is thus fixed at three points through an additional π-interaction of the indole moiety in the transition state **67**. The approach of the diene occurs from the back (*Re*-face) [147].

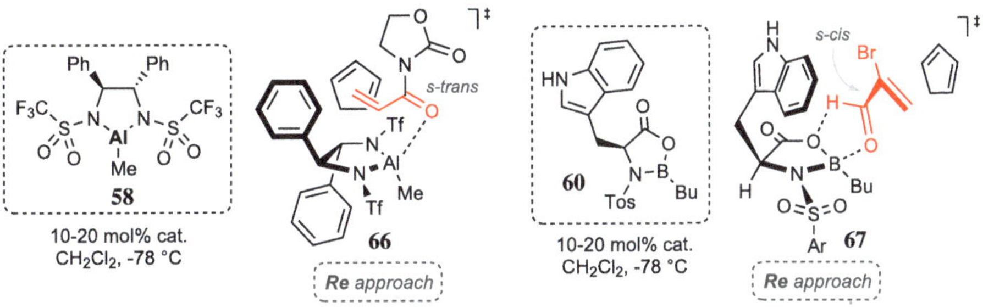

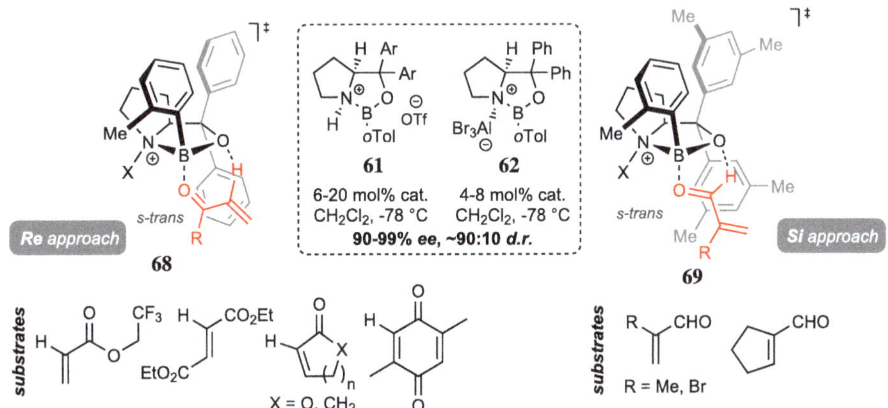

Fig. 5.73 Transition states of the cycloadditions in the presence of **58** and **60** [137, 139, 146, 147]

The proline-derived oxazaborolidine **61** represented a significant advancement in terms of its substrate scope and general applicability compared to the previously published chiral Lewis acids **58** and **60**. In addition to the enantioselectivity of the transformation, the *endo/exo*-ratio can additionally be controlled, depending on whether α-substituted acroleins or α, β-unsaturated ketones, esters or lactones are used. The *exo/endo*-selectivity was observed to be highly substrate dependent, as α-substituted acroleins generally prefer to form the *exo*-product. This *endo/exo*-differentiation is attributed to the formation of a B,O-chelate in the transition states **68** and **69**. Depending on which hydrogen atom is available, either a five-membered (**69**) or six-membered chelate (**68**) is formed. The aryl groups shield the back face. Depending on the substrate, an additional π-interaction of the aryl moiety with the dienophile may occur. The diene approaches from the direction of the easily accessible front face. Interestingly, the attack of the diene in **68** shows the opposite facial selectivity compared to **69**. However, since the *endo*-product is formed via **68** and the *exo*-product via **69**, the same enantiomer results (see Fig. 5.74) [126]. The subsequently introduced oxazaborolidines of the type **62** and related systems operate along the same principles and require a lower catalyst load in most cases (see Figs. 5.74 and 5.75) [148].

Fig. 5.74 Enantioselective Diels-Alder reactions in the presence of **61**, **62** [126, 148]

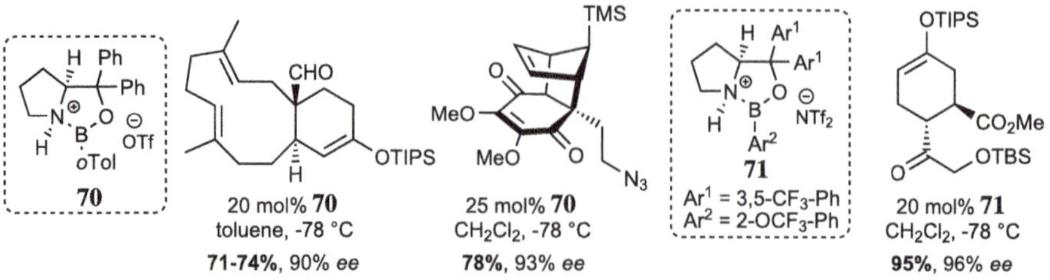

Fig. 5.75 Examples of asymmetric Diels-Alder reactions in total syntheses with chiral oxazaboro-lidines as enantioinductors [149–151]

The bisoxazoline (box)-based chiral Lewis acids introduced by Evans (**64, 65**) have been applied for a variety of other reactions in addition to enantioselective Diels-Alder cycloadditions (Aldol reaction, allylation, cyclopropanation, Michael addition, Mannich reaction). They rely on a bidentate coordination of the central metal atom by the dienophile. This necessitates the presence of a second carbonyl group in the substrate to allow for an efficient facial discrimination. Most conversions employ acrylate amides with oxazolidinones as the amine component for this purpose (cf Fig. 5.72) [141, 142].

Apart from their counterion and the shielding substituents, the chiral Lewis acids **64** and **65** make use of the identical bisoxazoline backbone but differ in the choice of the coordinated metal. Upon switch of the central metal ion, the opposite enantiomer of the product can be obtained. This is attributed to the metals' bonding mode: copper(II) prefers a distorted square-planar coordination geometry, whereas Zn(II) favors a distorted tetrahedral ligand sphere. As a consequence, the coordination geometry of the metal in **73** or **74** facilitates an approach of the diene from either the *Re-* or *Si-*face of the dienophile, thus enabling the access to either enantiomer (see Fig. 5.76) [141].

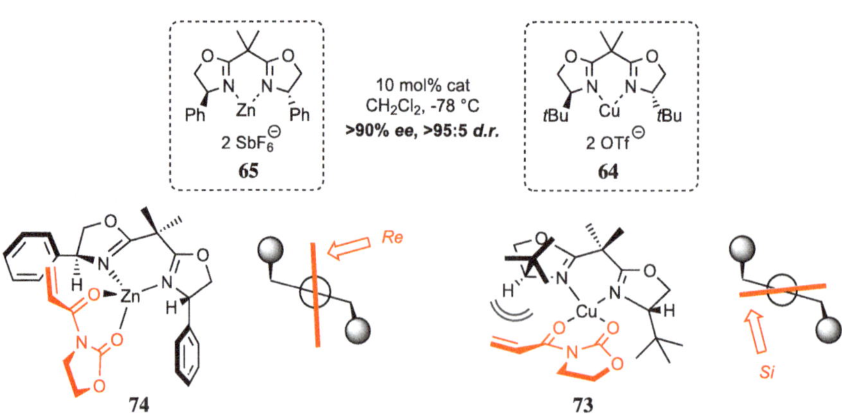

Fig. 5.76 Enantioselective Diels-Alder reactions in the presence of **64** and **65** [141]

Romo and co-workers made use of the method developed by Evans for their synthesis of the marine toxin (–)-gymnodimine (**77**). The protected amide **75** served as dienophile and its exocyclic olefin afforded the spiro-lactam **76** with excellent enantio- and diastereoselectivity. **76** comprised a building block of the right-hand fragment, which was subsequently attached to the left tetrahydrofuran segment via Nozaki–Hiyama–Kishi coupling. The extremely sensitive butenolide could be appended in the final steps of the sequence by a vinylogous Mukaiyama aldol reaction to circumvent complications arising from a potential opening of the lactone (see Fig. 5.77) [152].

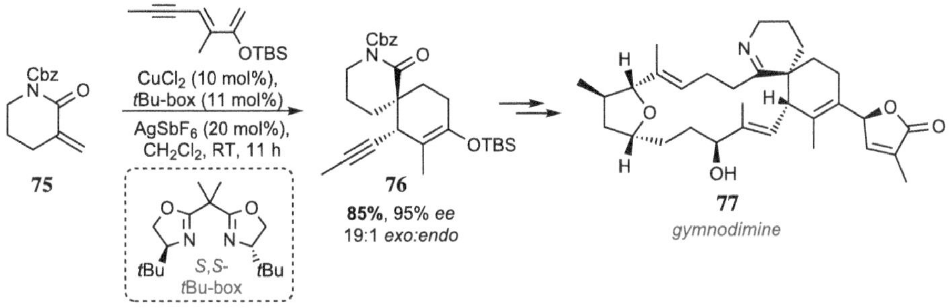

Fig. 5.77 Romo's approach to (–)-gymnodimine (**77**) [152]

Facial differentiation in hetero-Diels-Alder reactions can also be achieved by adding a chiral Lewis acid [153]. Typically, dienes are reacted with carbonyls or imines as dienophiles for this purpose (normal electron demand), but occasional examples have been reported of hetero-Diels-Alder reactions with inverse electron demand, where the heteroatom has been incorporated into the diene component [154].

Jacobsen and co-workers disclosed probably the most versatile method to date for the conversion of electron-rich dienes with activated and unactivated aldehydes [155, 156]. The chromium-based Lewis acid **78** used for this purpose relies on an aminoindanol backbone to induce chirality. Due to the availability of both enantiomers of the ligand via the asymmetric Sharpless aminohydroxylation (see Sect. 3.3), both enantiomers of the cycloaddition product can thus be selectively accessed. The postulated transition state **79** was derived from the crystal structure of **78**, which exists as a water-bridged dimer to saturate the octahedral coordination sphere of the chromium. Therein, the dienophile (or heterodiene) coordinates to the metal center of the Lewis acid, whose dimeric composition is preserved throughout the reaction. Due to the shielding of the lower side, the approach of the diene can only occur from the top face. The addition of molecular sieves is critical for the efficient turnover of the transformation. Their purported role can be traced back to the abstraction of H_2O from the metal center, thereby creating an open coordination site for the substrate (see Fig. 5.78) [156].

Jacobsen and co-workers reported the synthesis of the highly oxygenated bacterial natural product FR-901464 (**83**), which exhibits an impressive cytotoxicity of less than 1 nmol/L.

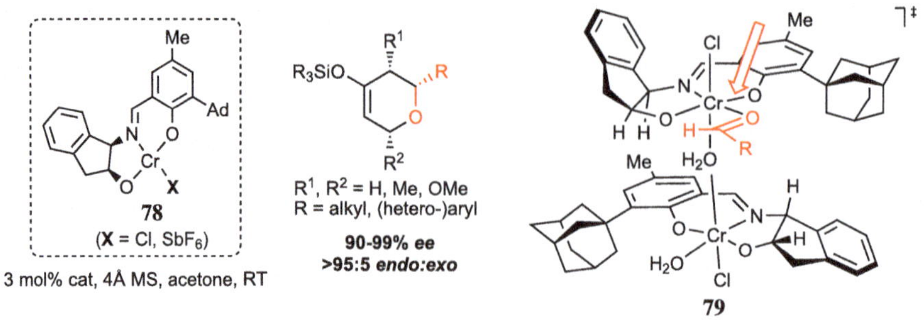

Fig. 5.78 Asymmetric hetero-Diels-Alder reactions by Jacobsen and postulated transition state with dimeric, water-bridged chromium salen complex [155, 156]

The central tetrahydropyran unit **82** was synthesized in an asymmetric hetero-Diels-Alder reaction from **80** and **81** with very good yield and excellent enantiomeric excess of 95% in the presence of the chlorine-ligated Cr catalyst **78**. The concluding coupling with the right and left parts of the molecule yielded the desired natural product FR-901464 (**83**) and enabled the synthesis of two further structural analogs (see Fig. 5.79) [157].

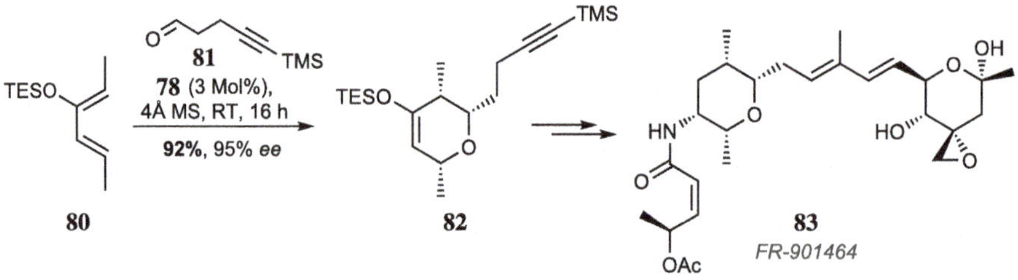

Fig. 5.79 Jacobsen's synthesis of FR-901464 (**83**) [157]

The innate potential of the asymmetric method introduced by the Jacobsen group was further demonstrated by its application in the access to a variety of natural products, for example, in the total syntheses of fostriecin, (+)-ambruticin, several iridoid alkaloids, (−)-dactylolide and gambierol [158–162].

List disclosed a new approach to asymmetric cycloadditions by separating the catalyst into two active components. The method uses silylium cations as achiral Lewis acids for substrate activation (**85**) and a chiral counterion (IDPi⁻) to effect facial differentiation. The sterically very demanding anionic catalyst IDPi mimics peptides by creating a chiral pocket in which the reaction takes place. Due to its flexible structure in the periphery, different substrates can be accommodated (key/lock principle). The catalytic cycle is initiated by transferring the SiEt₃ group from the silyl donor to the precatalyst, thus generating the catalytically active species **84**. This silyl-activated organocatalyst can subsequently react with the substrate and afford silylium cation **85**. The chiral anion of the iminodiphosphorane shields one side of the substrate and allows high enantioselectivity, despite the interaction with the reactant being only electrostatic and not covalent in nature (see Fig. 5.80) [163].

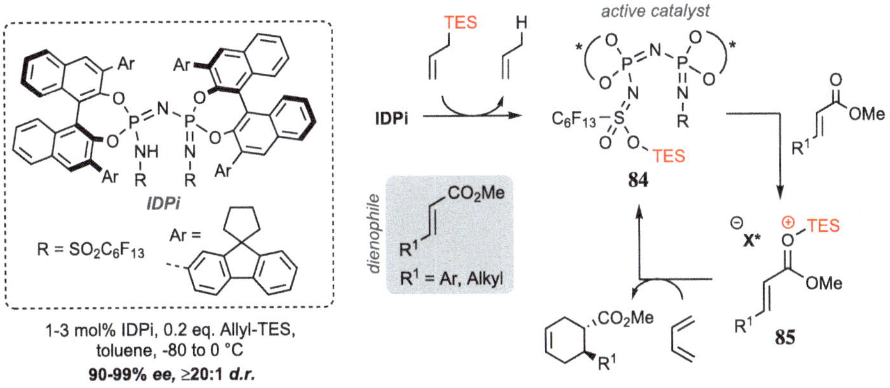

Fig. 5.80 Chiral counterion-directed, asymmetric Diels-Alder reaction reported by List *et al.* [163]

5.2.3 1,3-Dipolar Cycloadditions

Apart from Diels-Alder reactions, 1,3-dipolar cycloadditions constitute the second most frequently used class of cycloadditions in the synthesis of natural products or pharmaceuticals [44, 164–168]. Unlike the former, which only offer limited handles for derivatization, a variety of different products can be formed in a few steps following 1,3-dipolar cycloadditions due to the involvement of (several) heteroatoms. Both types of cycloadditions can often be applied to a complementing substrate scope. Dipolar cycloadditions are more frequently used in an academic than in an industrial context. The illustrated commercial applications, however, demonstrate the high efficiency of the reaction, which could not be achieved via alternative routes to the same extent (see Fig. 5.81).

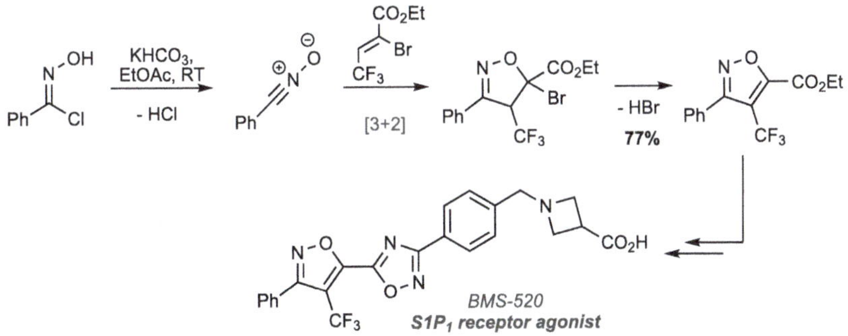

Fig. 5.81 Synthesis of the drug candidate BMS-520 for preclinical studies [169]

Depending on the reactant, the substrates of the cycloaddition can be divided into two categories: dipoles of the allyl and propargyl type (see Fig. 5.82) [170]. The coupling partner of 1,3-dipoles is referred to as a dipolarophile.

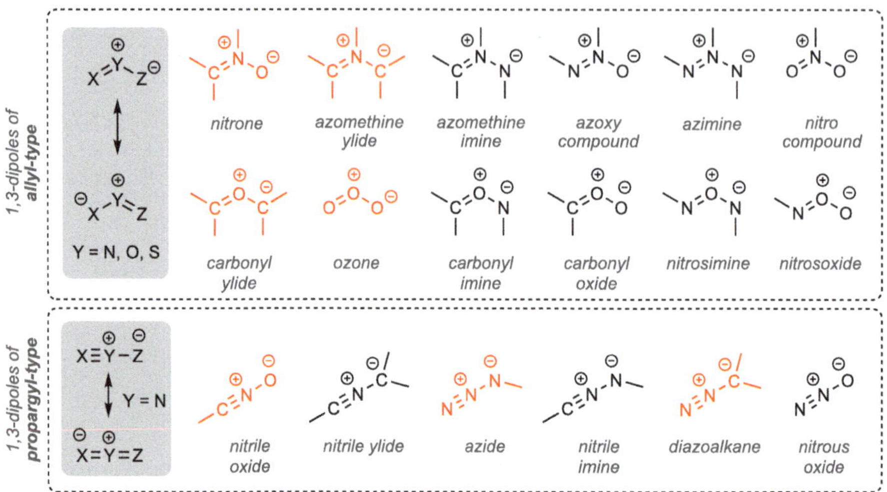

Fig. 5.82 Common 1,3-dipoles. Important species have been highlighted in red

Of these two substrate classes, nitrones, azomethine ylides, carbonyl ylides, and ozone (allyl type) as well as nitrile oxides and diazoalkanes (propargyl type) play a particularly important role. Azides themselves are rarely encountered in a classic 1,3-dipolar cycloaddition, unless the dipolarophile is a highly strained alkyne. In the context of click reactions, however, do they afford the cycloaddition products — albeit via a non-concerted mechanism in the presence of Cu(I) and Ru(I) compounds — and are indispensable in many areas of catalysis, biochemistry, and synthetic chemistry (see Chap. 9).

5.2.3.1 Mechanism of 1,3-Dipolar Cycloadditions

1,3-dipolar cycloadditions are based on the mechanistic principles discussed at the beginning of the chapter, which apply to all pericyclic processes. Similar to other cycloadditions, they can be divided into three categories according to the simplified Klopman–Salem equation, which are distinguished by the dominant orbital interaction of the process: reactions with normal, neutral, or inverse electron demand. Some dipoles can react from both the HOMO and the LUMO, this depends on whether the dipolarophile is electron-poor or electron-rich (see Fig. 5.83) [14, 54, 55].

The regioselectivity of 1,3-dipolar cycloadditions and the influence of substituents on the 1,3-dipoles have already been discussed (see Sect. 5.2.2). In conjunction with the above

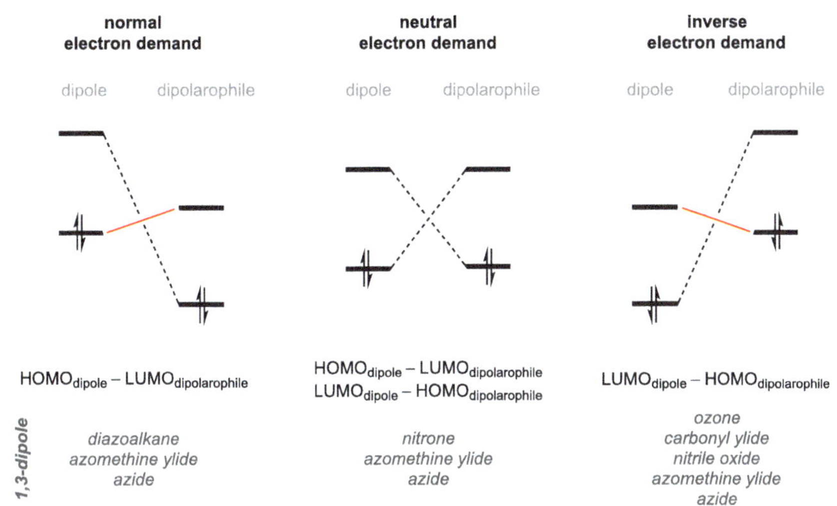

Fig. 5.83 Classifications of 1,3-dipolar cycloadditions, dominant orbital interactions, and distribution of the most important dipoles [54]

classification of the dipoles according to their dominant orbital interactions, the main product of the cycloaddition can thus be predicted.

For *nitrones*, both types of interactions must be taken into account. Even in a reaction with electron-poor dipolarophiles the addition is mainly HOMO-controlled. Given that nitrones possess similar HOMO coefficients at both termini and thus should not display a pronounced preference for either isomer, the regioselectivity partly results from the LUMO coefficients. The preferred regioisomer is usually the 5-substituted isoxazoli(di)ne when reacting with X-/Z-/C-substituted alkenes or alkynes (see Fig. 5.84). The stronger the influence of the HOMO becomes (e.g., when reacting with acrylonitrile or nitroethene), the higher the share of the minor isomer in the product mixture is [54].

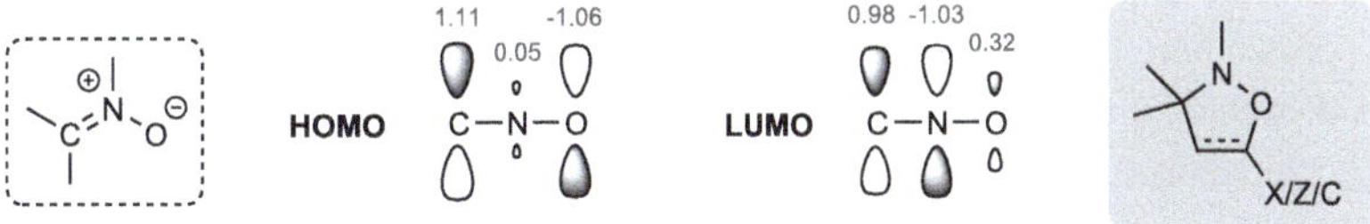

Fig. 5.84 Frontier orbitals of nitrones and preferred regioisomer of the cycloaddition. The cited numbers are the orbital coefficients [54]

Azomethine ylides are C_2-symmetric and therefore consequently do not show any regioselectivity. Due to the small HOMO/LUMO difference, reactions with X-, Z- and C-substituted dipolarophiles all proceed, with the dominant orbital interaction varying accordingly [165]. As a result, they undergo cycloadditions with normal, neutral, and inverse electron demand.

Carbonyl ylides usually react from the LUMO. *Nitrile oxides* react LUMO-, *diazoalkanes* mainly HOMO-controlled. The reaction with heterodipolarophiles (aldehydes, imines, ketones, etc.) leads, with the exception of nitrile ylides, preferably to the regioisomer shown in Fig. 5.85. For a deeper understanding and further examples, the interested reader is recommended to consult Houk's systematic investigation of frontier orbital interactions in 1,3-dipolar cycloadditions [54].

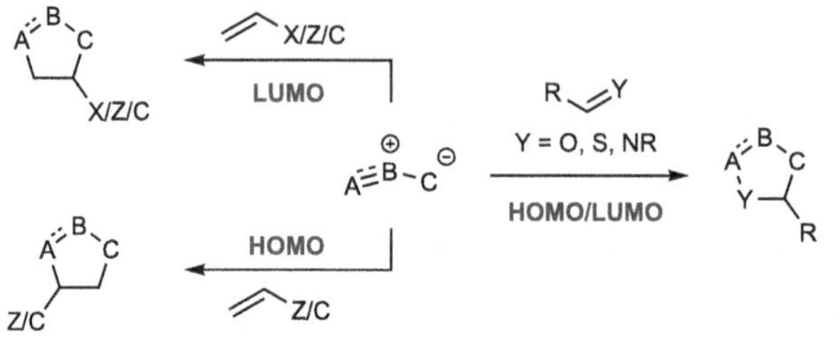

Fig. 5.85 Regioselectivity of 1,3-dipolar cycloadditions

HOMO-controlled reactions can be accelerated by introducing Z-type substituents to the dipolarophile. In LUMO-controlled reactions, X-substituents on the dipolarophile lead to an increase in rate. C-substituents generally have a positive effect on the transformation. On the other hand, the substitution pattern of the dipole can also have a positive or negative effect on the reaction rate depending on the operative dominant interaction. Cycloadditions of diazoalkanes, which react almost exclusively from the HOMO, are increasingly slowed down by successive substitution with electron-withdrawing groups and a resulting lowering of the HOMO. For example, the introduction of two carbonyl groups in the α-position of the C-terminus inhibits a reaction with highly reactive norbornene, which rapidly takes place with the unsubstituted dipole [14].

In addition, steric effects play a much larger role compared to classical [4+2]-cycloadditions and can result in the regioisomer predicted as less favorable to become the major product [14]. 1,3-dipoles can adopt a variety of diastereomeric geometries and conformers (shown here using the example of azomethine ylides), which can result in product mixtures of stereoisomers (see Fig. 5.86) [165].

Many cycloadditions are asynchronous due to the asymmetric nature of the 1,3-dipole. Modern theoretical treatise has shown that due to the asymmetric electron density at the 1,3-dipole, a pseudodiradical, pseudoradical, zwitterionic, or carbenoid electronic structure results, which imply the possibility of highly asynchronous or non-concerted transition states [171]. Unfortunately, no clear rule of thumb can be provided, unlike with Diels-Alder or hetero-Diels-Alder reactions. The degree of asynchronicity must therefore be investigated on a case-by-case basis.

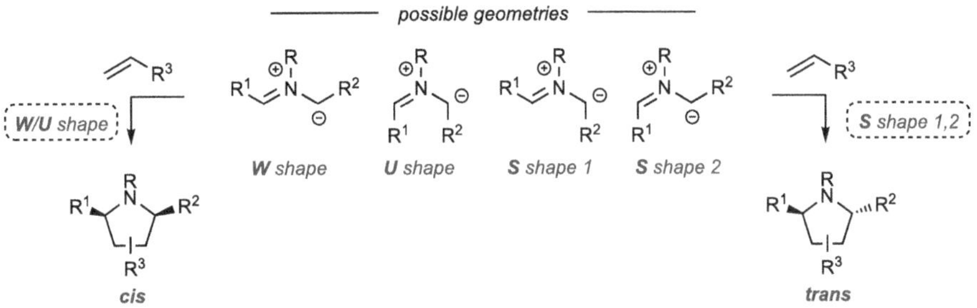

Fig. 5.86 W-, U- and S-shape of azomethine ylides and resulting diastereomers [165]

5.2.3.2 Preparation of 1,3-Dipoles and Derivatization of the Products

Given their highly reactive nature, a significant number of dipoles is not isolable or indefinitely stable. For this reason, one of the most common approaches is to prepare them *in situ* and subsequently convert them in a one-pot reaction. The standard methods for preparing the most common dipoles are summarized in Figs. 5.87 and 5.88 [164–168].

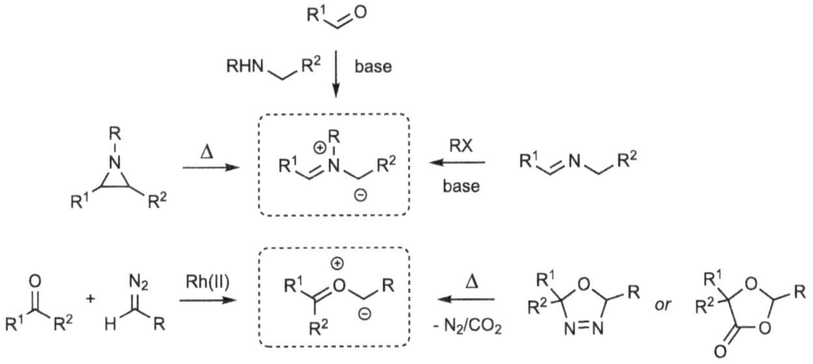

Fig. 5.87 Preparation of azomethine and carbonyl ylides

Azomethine ylides are primarily accessed by one of two ways: Either starting from aldehydes and secondary amines in a basic medium or by conversion of imines with electrophiles. The preferred and almost exclusively encountered method for generating *carbonyl ylides* is the reaction of a carbonyl component with a diazo compound in the presence of catalytic amounts of Rh(II) salts, obtaining rhodium carbenes as intermediates. This variant is particularly often used in intramolecular cycloadditions.

Nitrones are either accessible from carbonyl compounds by reaction with hydroxylamines or from oximes and alkyl bromides. *Nitrile oxides* can be prepared from oximes with NCS and a base or from nitro compounds and isocyanates in a basic medium. To form *diazoalkanes*, two methods are commonly employed: either a carbamate can be converted to the diazo compound via the corresponding nitroso carbamate or an N$_2$ unit can be transferred from

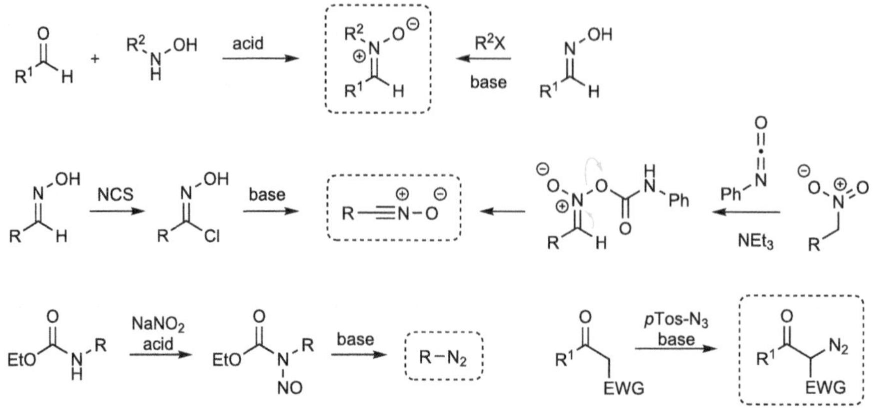

Fig. 5.88 Preparation of nitrones, nitrile oxides, and diazo compounds

tosyl azide to α-substituted carbonyl compounds [172, 173]. The multitude of possible derivatizations of the addition products cannot be covered at this point. This range will be illustrated resorting to an excerpt of a few 1,3-dipoles. Isoxazoles, isoxazolines, and isoxazolidines (cycloaddition products of nitrones or nitrile oxides with alkenes or alkynes) can easily be converted to different acyclic downstream products by virtue of their N-O bond lability (see Fig. 5.89) [167].

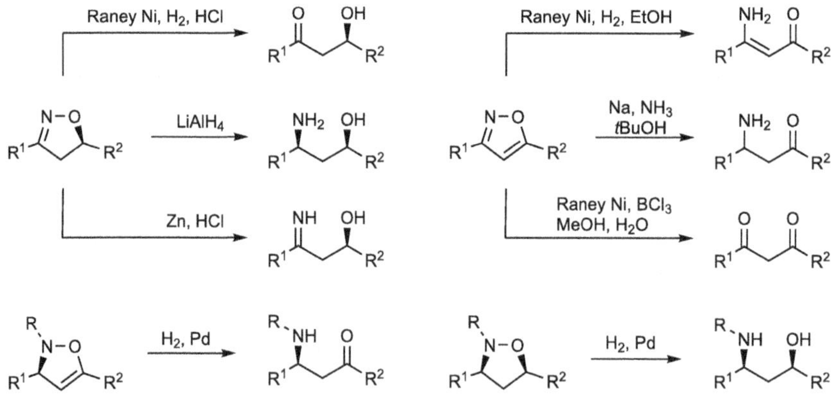

Fig. 5.89 Derivatization of isoxazoles, isoxazolines, and isoxazolidines [167]

The depicted products received after only one step can subsequently be further transformed almost at will with various oxidation/reduction methods, reactions with electrophiles, nucleophilic additions, and the like.

5.2.3.3 Advanced Applications of 1,3-Dipolar Cycloadditions

One of the challenges of 1,3-dipolar cycloadditions is rooted in the susceptibility of their regioselectivity to electronic and especially steric influences. One possibility to circumvent this conundrum was intensively studied in the research group of Kanemasa. In their approach, metal ions are exploited as a coordinative bridge between the two reactants to ensure an intramolecular reaction, limit the degrees of freedom, and assert a high stereocontrol. Thus, highly selective conversions of nitrile oxides, nitrones, and diazo compounds could be realized (see Fig. 5.90) [174].

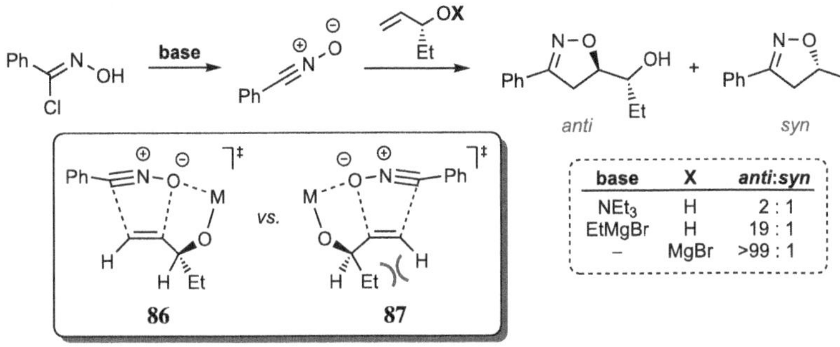

Fig. 5.90 MgBr-controlled regio- and stereoselective nitrile oxide cycloadditions by Kanemasa *et al.* [174]

The diastereo- and regioselectivity of the magnesium-mediated cycloaddition of nitrile oxides with substituted allylic alcohols can be rationalized by the bridged transition states **86** and **87**. In the absence of chelating metal ions, little to no diastereoselectivity is observed. The mixture of diastereomeric regioisomers (alternative regioisomer not shown) is only formed in 40%, which can be attributed to a diminished rate of conversion. The presence of the weakly acidic magnesium cation, which stems either from the requisite base in the nitrile oxide formation or is introduced via the corresponding magnesium alcoholate dipolarophile, leads to the coordination of both the allylic alcohol and the nitrile oxide. The henceforth intramolecular reaction exclusively affords the desired regioisomer in up to 99% yield. The diastereomeric *anti*- and *syn*-configured addition products are obtained via the competing transition states **86** and **87**. Due to the occurrence of a 1,3-allylic strain in **87**, the formation of the *anti*-product is favored.

Especially azomethine ylides and carbonyl ylides are frequently employed in natural product syntheses. Lee and co-workers disclosed the formal total synthesis of the antibiotic platensimycin (**92**), in which a 1,3-dipolar cycloaddition of an *in situ*-formed carbonyl ylide constitutes the key step to build the doubly bridged, cyclic ether framework. **89** was initially formed from **88** in the presence of dirhodium tetraacetate and converted in an intramolecular reaction to **90**. The doubly bridged bicycle **91**, which is an advanced intermediate in Nicolaou's total synthesis, was subsequently prepared in five further steps from **90** (see Fig. 5.91) [175, 176].

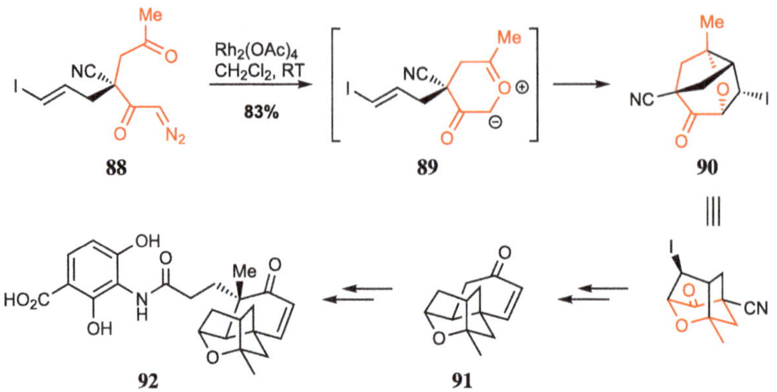

Fig. 5.91 Lee's formal total synthesis of platensimycin (**92**) [175]

In Denmark's syntheses of the glycosidase inhibitors (+)-castanospermine, (+)-6-epicastanospermine, (+)-australine (**97**), and (+)-3-epiaustralin, the 5/5 or 5/6-fused alkaloids were all derived from a common precursor (**96**). Its synthesis was realized by way of a Diels-Alder/1,3-dipolar cycloaddition sequence as key step. The enantioselective Diels-Alder reaction was achieved by the use of a chiral enol ether in the presence of an Al-based Lewis acid, exclusively affording the *exo*-product **94** with an excellent facial selectivity of 44:1 (*Si/Re* ratio). Following the isolation of **94**, the intramolecular dipolar cycloaddition with the diene side chain **96** ensued, whereby the exclusive *exo*-selectivity could be attributed to the half-chair transition state **95**. The chiral alkoxy group therein adopts an axial position, presumably due to the anomeric stabilization of the depicted conformer. The *exo/endo*-selectivity results from the conformational arrangement dictated by the O-Si-O bridge, which for steric reasons only allows the approach shown in transition state **95** (see Fig. 5.92) [177].

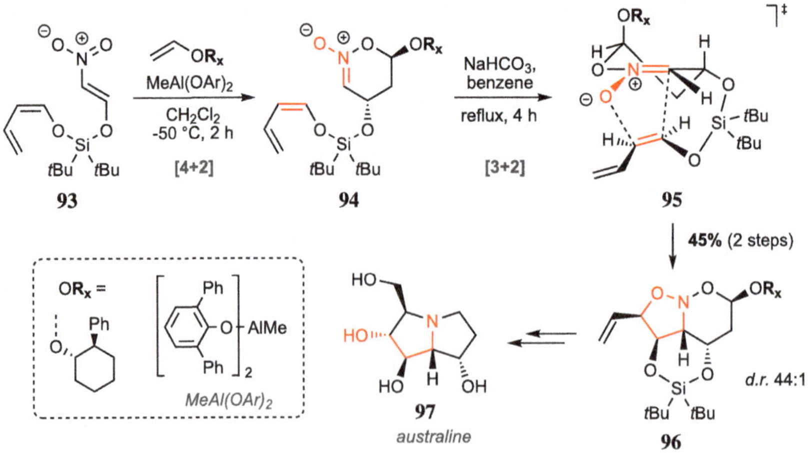

Fig. 5.92 Denmark's synthesis of (+)-australine (**97**) [177]

In analogy to the reversibility of [4+2]-cycloadditions, 1,3-dipolar cycloreversions can similarly be realized [178]. Other than a dedicated synthesis and (possibly unsuccessful) isolation, this represents a convenient approach to generate a reactive dipolarophile *in situ* (see Fig. 5.87). Boger and co-workers reported the synthesis of the bisindole *vinca* alkaloids vindoline (**102**), vindorosine (**103**), and several vinblastine analogs, which are important drugs for the treatment of various types of cancer, using a dipolar cycloreversion to generate a carbonyl ylide (see Fig. 5.93) [179].

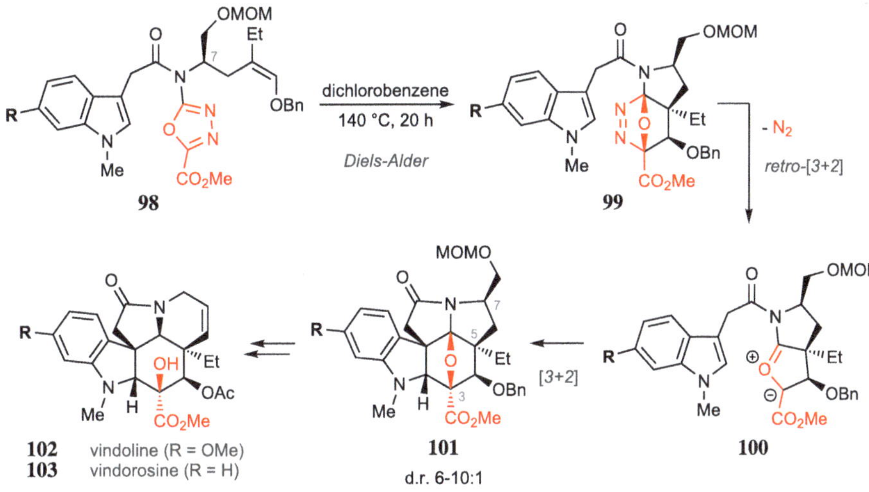

Fig. 5.93 Syntheses of vindoline (**102**) and vindorosine (**103**) by Boger *et al.* [179]

The domino Diels-Alder/cycloreversion/1,3-dipolar cycloaddition reaction, in which four C-C bonds, three rings, and six stereogenic centers are formed, presumably proceeds via the purported intermediates **99** and **100**. The stereocenter on the side chain appended to the dienophile controls the facial selectivity of the initial [4+2]-cycloaddition by steric shielding. The protected hydroxymethyl group at C-7 and the ethyl residue at C-5 in the newly formed five-membered ring in **99** are *trans*-configured. Following loss of dinitrogen in the cycloreversion, the carbonyl ylide **100** reacts *endo*-selectively to give **101**. The final stereochemistry could thus be effectively controlled by the C-7 stereocenter already present at the outset of the transformation. The same approach was used by the Boger group for the subsequent syntheses of the alkaloids aspidoalbidine and 1-acetylaspidoalbidine [180].

The principles of an asymmetric induction are identical to those discussed for Diels-Alder reactions: the use of stereoinformation present in the substrate (persistent stereogenic centers, auxiliaries) as well as chiral catalysts, typically Lewis acids. Due to the different electron distribution of dipoles compared to dienophiles and often reversed electron demand of the components (inverse instead of normal electron demand), 3+2 cycloadditions require different enantioinductors than those presented in the previous section. Here, the reader is referred to the relevant literature as a starting point into the topic [181].

The organic synthesis department of AbbVie was required to find an efficient entry for the production of the clinical candidate ABBV-3221. The active pharmaceutical ingredient (API) targets the cystic fibrosis trans-membrane receptor and is intended for use in a multi-drug therapy, as typical for the treatment of this fatal genetic disease. Since the API advanced to the clinical stage, multi-kilogram quantities of the drug candidate were required. An asymmetric, Cu-catalyzed [3+2]-cycloaddition was identified as key step for the construction of the central pyrrolidine. A ferrocene-based isoxazole-substituted phosphine ligand was selected following an extensive screening and provided the cycloaddition product in acceptable yield and excellent enantiomeric excess. Following the subsequent optimization, the desired pyrrolidine could be synthesized on a 19 kg scale in a pilot campaign (see Fig. 5.94) [182].

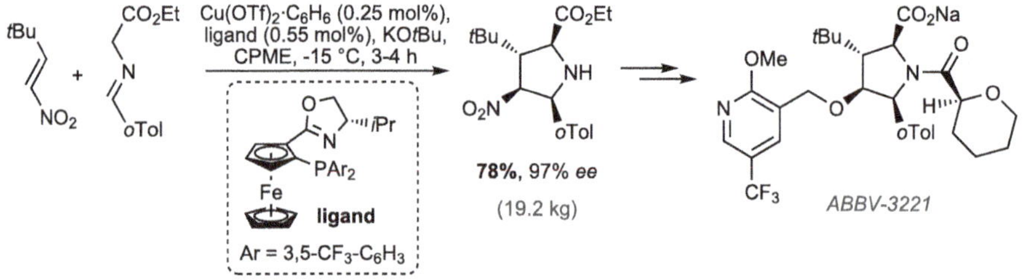

Fig. 5.94 AbbVie's approach to ABBV-3221 on a kg scale [182]

5.2.4 [2+2]-Cycloadditions

[2+2]-Cycloadditions represent the most common type of photochemical reactions in pericyclic processes. They allow for a straightforward access to cyclobutanes, oxetanes, and β-lactones and are thus the preferred synthetic approach for these product classes [38, 43, 183–185].

The group of Stoltz used a photochemical [2+2]-cycloaddition to construct the 5/6/4/5-fused intermediate **105** in their synthesis of the norcembranoid diterpene scabrolide A (**106**). The strained cyclobutane was subsequently converted to the desired cycloheptyl motif in a fragmentation, thereby forming the 5/6/7-fused carbon skeleton of the natural product and completing the total synthesis (see Fig. 5.95) [186].

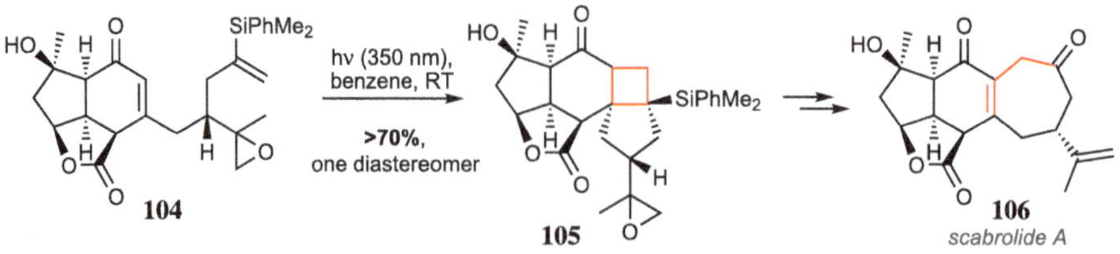

Fig. 5.95 Stoltz' synthesis of scabrolide A (**106**) [186]

A variety of different components are suitable as substrates of a [2+2]-cycloaddition, ranging from olefins to carbonyl compounds, ketenes or allenes (see Fig. 5.96) [187, 188].

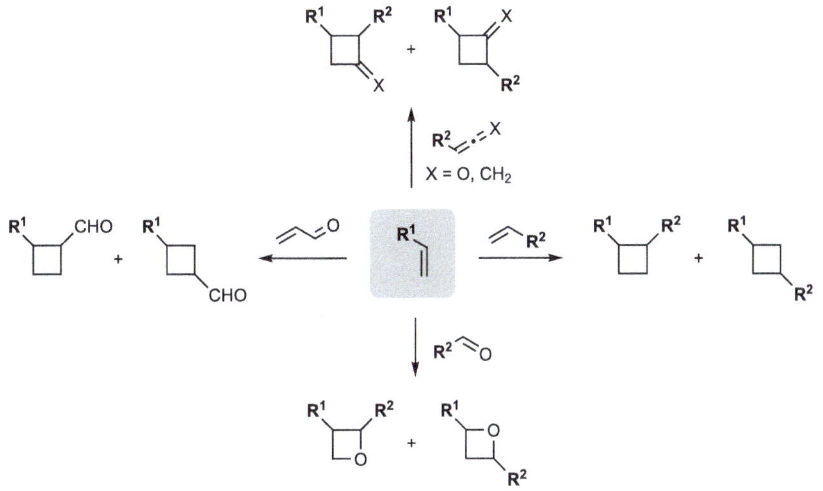

Fig. 5.96 Schematic overview of select photochemical and thermal [2+2]-cycloadditions

The reaction of ketenes with imines (*Staudinger* reaction) proceeds via ionic intermediates, with the ring closure comprising an electrocyclization. This reaction is therefore not part of the [2+2]-cycloaddition canon and will be discussed in more detail in Sect. 6.3.

5.2.4.1 Mechanism of Photochemical and Thermal Reactions

[2+2]-cycloadditions can be conducted under both thermal and photochemical conditions. Under photochemical control, both components react in suprafacial manner. In a thermal cycloaddition, one of the components must react antarafacially to satisfy the symmetry-derived selection rules (see Fig. 5.97). An antarafacial pathway is only possible with ketenes or allenes as reactants for steric reasons, as the central carbon atom does not bear any substituents. The reaction conditions are comparatively harsh and often necessitate elevated temperatures of up to 200°C.

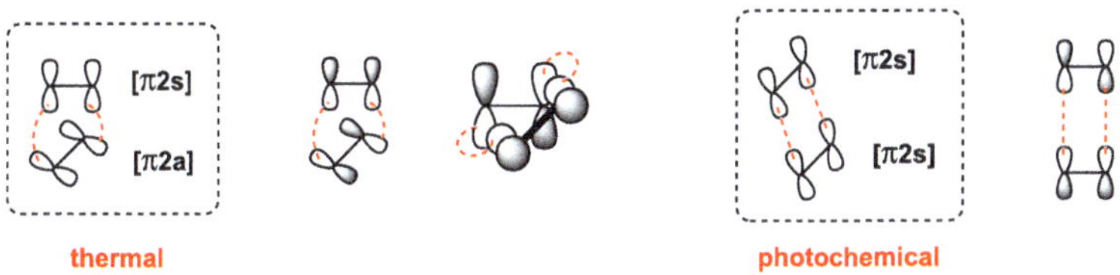

Fig. 5.97 Facial topology of [2+2]-cycloadditions

The mechanism of the photochemical variant merits a more detailed examination: The direct excitation of an alkene leads to the occupation of the lowest excited singlet state (S_1). Aryl-substituted alkenes react via a singlet reaction path by means of a $\pi \rightarrow \pi^*$ excitation. α,β-Unsaturated carbonyl compounds proceed via a n$\rightarrow \pi^*$ transition to the S_1 state. The transition to the triplet state (intersystem crossing, *ISC*) is fast in α,β-unsaturated carbonyl compounds. For these substrates, product formation likely proceeds from the T_1 state, whereby 1,4-biradical intermediates have been postulated to occur in some instances. These intermediates have been detected successfully and could be further studied spectroscopically [189]. Apart from the reaction trajectory leading to the product, the biradical can also decompose and revert to the reactants. Barring a direct excitation and subsequent singlet and triplet pathways, the third possibility lies in the use of a promoter (sensitizer). It is photochemically excited in place of the alkene ($S_0 \rightarrow S_1 \rightarrow T_1$) and then transfers the absorbed energy to the substrate via a radiationless process to similarly arrive at the triplet state. The energy transfer from the promoter to the alkene occurs quickly if the triplet energy of the alkene acceptor is lower than that of the promoter (cf. Sect. 10.2.1, Fig. 10.14). Furthermore, the photochemical isomerization of the double bond is an additional competing reaction for vicinal disubstituted olefins (see Fig. 5.98) [43, 183, 190].

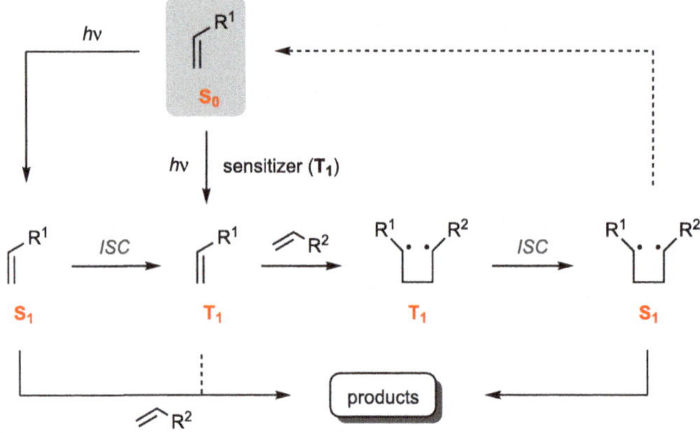

Fig. 5.98 Mechanism of the photochemical [2+2]-cycloaddition [43, 183, 190]

The *Paternò–Büchi reaction*, in which an olefin is converted under photochemical stimulation with a carbonyl compound, proceeds via a similar mechanism [38, 183, 191]. In cycloadditions of carbonyl derivatives (aldehydes, α, β-unsaturated aldehydes/ketones/amides/esters), the carbonyl component is most commonly excited and reacts from the respective triplet state following an ISC process (substrate dependent: n$\rightarrow \pi^*$ or $\pi \rightarrow \pi^*$ triplet). This excited reactant can subsequently couple with the other substrate (olefin, alkyne, etc.) to form a 1,4-biradical. The Paternò–Büchi reaction therefore generally does not follow a concerted reaction pathway. While intermediate exciplexes, which could be detected in some cases [192], can emerge en route to a biradical, their influence on the regio- and stereochemistry of the products is not yet fully understood. In the reaction of electron-rich alkenes with electron-poor carbonyl compounds, a complete electron transfer has even been confirmed as a mechanism. The number of natural products containing an oxetane moiety is rather limited, which is why natural product syntheses incorporating a Paternò–Büchi reaction usually comprise a subsequent cleavage of the respective oxetane [38].

The regioselectivity of photochemical cycloadditions can be well explained on a qualitative level by either the Klopman–Salem equation or a frontier orbital (SOMO/LUMO) interaction of the components. With unsymmetric components, two possible trajectories can be observed: A head-to-head or a head-to-tail coupling. In monosubstituted components, a head-to-head connection occurs when both reactants are either electron-rich or electron-poor. A homodimerization of olefins therefore almost exclusively furnishes the head-to-head dimers. If two components have different electronic properties, a head-to-tail coupling is favored (see Fig. 5.99) [183]. In addition, the occurrence of hydrogen bonds can influence the regioselectivity, for example, in the reaction of α, β-unsaturated carbonyls with allylic alcohols.

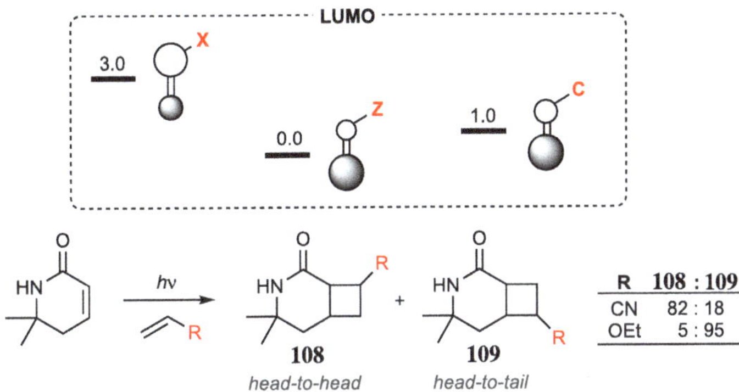

Fig. 5.99 Orbital energies (in eV) and coefficients of the LUMOs of monosubstituted olefins [51]. Dependence of the regioselectivity on the substitution pattern [193]

While some of the thermal [2+2]-cycloadditions proceed in a concerted fashion, the product formation of many thermal reactions has been envisaged to occur via a stepwise, radical mechanism. In these cases, bond formation is asynchronous. In the conversion of enones with electron-rich alkenes, the first bond in the formation of the radical intermediates usually develops between the more electrophilic α-C-atom and the less substituted terminus of the alkene. In contrast, the β-C-atom of the enone initially reacts with the less substituted terminus of the alkene with electron-poor olefins as substrates. While the regioselectivity of the products thus roughly mirrors the radical stabilities of the corresponding primary intermediates, it in fact is derived from the reactivity of the excited $^3(\pi\text{-}\pi^*)$-state (see Fig. 5.100) [194].

The synthetic utility of intermolecular cycloadditions is limited by the occasionally challenging regioselectivity. The diastereoselectivity *endo* versus *exo* is also not as pronounced and predictable as in the related [4+2]-cycloadditions. However, the *exo*-product is typi-

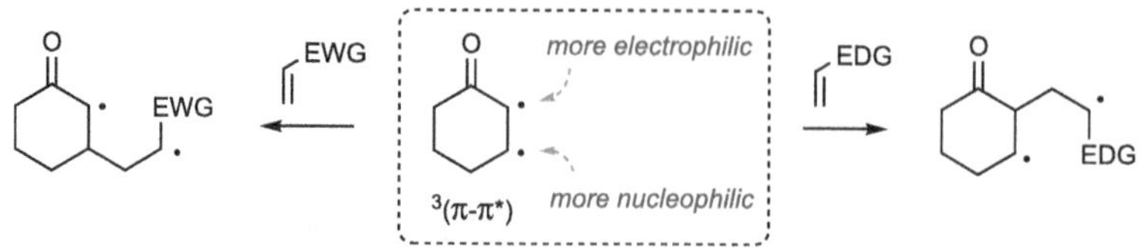

Fig. 5.100 Reactivity of the excited state in the photochemical cycloaddition of cyclohexenone with olefins [194]

cally slightly favored. This preference can usually be attributed to purely steric reasons, as stabilizing secondary orbital interactions (should they be the cause of the *endo*-selectivity) cannot occur with the employed olefins. The regioselectivity issue can furthermore be usually circumvented by an ensuring an intramolecular course of reaction [43].

For 1,2-disubstituted alkenes, the stereochemical integrity of the double bond is a question of the operative mechanism [190]. Concerted reactions proceed stereospecifically, while a stepwise mechanism can result in the isomerization of the initial double bond geometry. Cyclic alkenes or alkenones cannot usually isomerize in normal rings (smaller than eight-membered), which is why they are preferred as substrates. Given the intermediacy of radical species whose lifetime would allow a rotation of single bonds, the original E or Z geometry may not be completely preserved. The extent of stereoloss depends on the lifetime of the intermediate [189] and the magnitude of the rotational barrier, with the thermodynamically more stable *trans* product being preferentially obtained. Thus, thermodynamics dictate the product distribution, which may further be supported by the reversibility of the process (decomposition of the biradicals to the reactants), resulting in stereoconvergent reactions on occasion (see Fig. 5.101) [191].

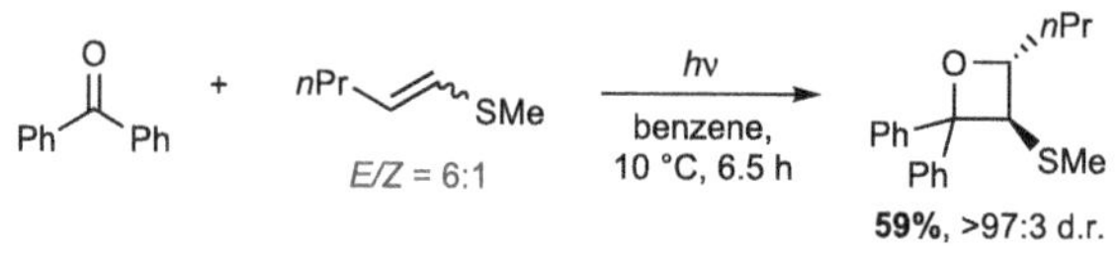

Fig. 5.101 Stereoconvergence in the Paternò–Büchi reaction [195]

Given the presence of a triplet mechanism, the preferred formation of the thermodynamically less stable *cis*-cycloadducts cannot be sufficiently rationalized by the above reasoning. In triplet reactions, the properties of the intermittent biradicals determine the formation of a respective diastereomer. Based on theoretical considerations, it was suggested that the molecular geometry prior to the ISC process from the biradicals en route to the singlet energy hypersurface is responsible for the observed product selectivities via a kind of "memory effect" ($T_1 \rightarrow S_1 \rightarrow$ products) [196]. This may thus serve as possible explanation for the preferred formation of thermodynamically disfavored products.

When employing cyclohexenones, the formation of both *cis*- and *trans*-fused addition products can occasionally be observed (see Fig. 5.102) [197]. The reason for this is not clear, although several explanations have been postulated [190, 198]. It is now generally accepted that both stereoisomers originate from the same excited enone, for which there is much experimental evidence. One theory was able to explain the formation of both stereoisomers by attributing their origin to two differently distorted excited states [198].

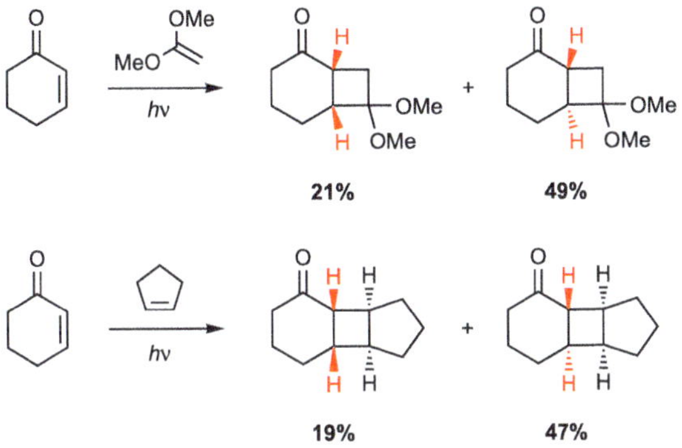

21%   **49%**

19%   **47%**

Fig. 5.102 Stereointegrity in the cycloaddition of cyclohexenones [197]

In contrast to photochemical [2+2]-cycloadditions, which allows the conversion of all compounds containing C-C or C-heteroatom double bonds, only a few compounds successfully participate in a thermal [2+2]-cycloaddition. The most commonly encountered reactants are ketenes and allenes [187]. According to the Woodward-Hoffmann rules, a $[\pi 2s + \pi 2s + \pi 2s]$-cycloaddition may also serve as rationalization to account for a symmetry-allowed, thermal reaction in addition to a $[\pi 2s + \pi 2a]$-topology. Based on the HOMO and LUMO of the ketene, the transition state geometry **110** can be derived either by the necessity for an antarafacial topology $(\pi 2s + \pi 2a)$ or by the involvement of two orthogonal π systems jointly participating in the reaction $(\pi 2s + \pi 2s + \pi 2s)$ (see Fig. 5.103) [199].

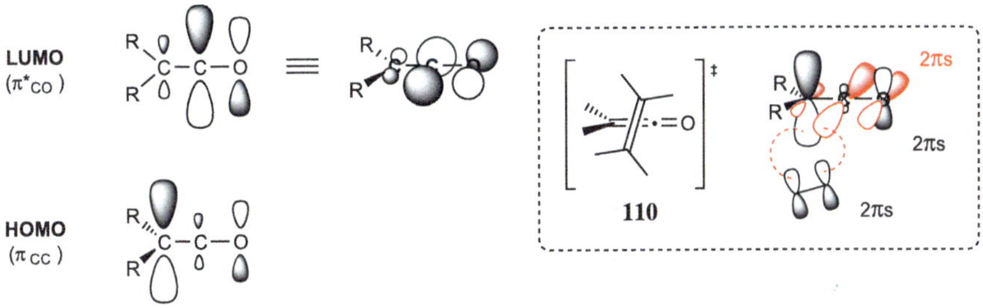

LUMO
$(\pi^{*}{}_{CO})$

HOMO
(π_{CC})

110

Fig. 5.103 Frontier orbitals and transition state geometry in the [2+2]-cycloaddition of ketenes [199]

The question of whether the thermal [2+2]-cycloaddition of allenes and ketenes proceeds stepwise or in a concerted manner remains controversial [200–203]. While *E*-configured olefins yield a mixture of *cis*- and *trans*-products, the corresponding *Z*-olefins add to ketenes stereoselectively. Studies of kinetic isotope effects support a stepwise mechanism and theoretical investigations indicate highly asynchronous transition states, which however does not automatically translate to a stepwise mechanism (cf. hetero-Diels-Alder reactions). Nevertheless, the majority of published studies now seem to favor a radical mechanism with discrete intermediates over a concerted process.

5.2.4.2 Applications and Modern Variants

Ketenes, allenes, olefins, alkynes, and carbonyls are the compounds normally employed as substrates in [2+2]-cycloadditions, and this range of reactants enables the synthesis a wide variety of different products (see Fig. 5.104) [38, 43, 185, 204].

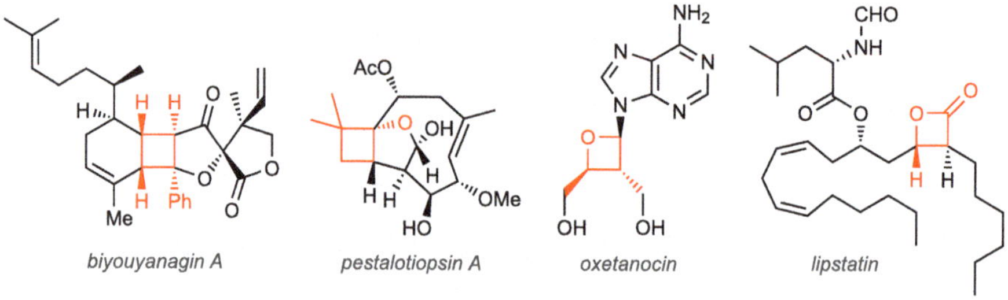

Fig. 5.104 Select natural products synthesized via [2+2]-cycloadditions [205–208]

The [2+2]-cycloaddition is typically used in two ways: Either a four-membered ring itself comprises the target structure, for whose synthesis there are not many effective and reliable methods available. On the other hand, these small rings can be cleaved due to their high energy content (ring strain) and thus serve as intermediates in a complex synthesis (see Fig. 5.105) [209].

The cycloreversion (bond cleavage of **A** and **B**) can afford products that formally correspond to a bond metathesis. Path **A** is the most frequently pursued cleavage mode. Especially when using *β*-hydroxy or *β*-alkoxyalkenones, the subsequent retro-Aldol reaction has proven to be a versatile tool in synthesis. The domino photo-cycloaddition/retro-Aldol reaction is also referred to as the *de-Mayo* reaction. Büchi and co-workers used the de-Mayo reaction to construct the molecular skeleton of the iridoid alkaloid loganine (**120**). Following the ring closure to the primary cycloaddition product **117**, a retro-Aldol reaction (bond breaking **A**) afforded the dialdehyde **118**. The subsequent hemiacetal formation yielded the intermediate **119** containing the 5/6-fused loganine framework, which could subsequently be converted to the secondary metabolite **120** (see Fig. 5.106) [210].

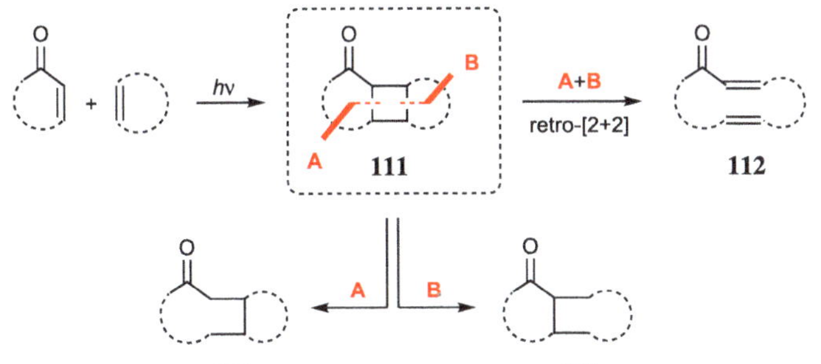

Fig. 5.105 Common fragmentations of [2+2]-cycloaddition products

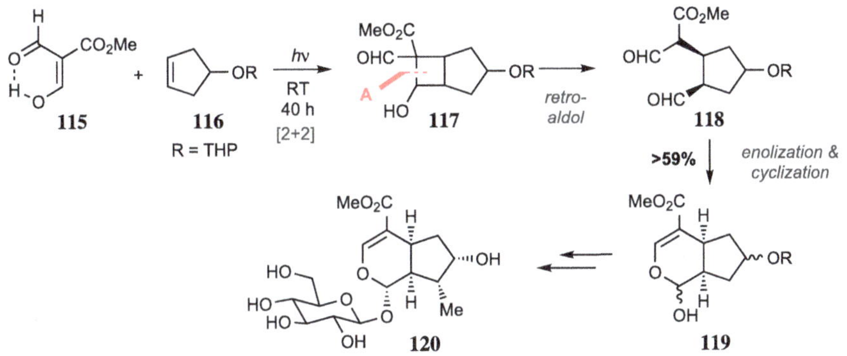

Fig. 5.106 Synthesis of loganine reported by Büchi *et al.* [210]

A fragmentation does not always have to occur in the context of a domino reaction or even directly following the cycloaddition step, but can furnish the desired products at a later stage of a given synthesis. Two representative examples can be found in the syntheses of the terpenes aphanamol I (**123**) and (±)-ingenol (**128**, see Fig. 5.107) [211, 212].

In the synthesis of aphanamol I (**123**), the primary cycloaddition product **121** was first converted to the epoxide **122**, which then delivered the natural product **123** upon saponification of the ester with concomitant opening of the epoxide. Winkler's racemic synthesis of the anti-HIV metabolite ingenol (**128**) relied on an intramolecular photochemical [2+2]-cycloaddition to afford the tricycle **125** from **124**. Following saponification of the ester and acetal cleavage in **125**, a retro-Aldol reaction furnished the seven-membered bridgehead ring of **127**. The highly oxygenated natural product could subsequently be completed in further 32 steps to achieve the first racemic total synthesis.

While the majority of reactants in photochemical cycloadditions constitutes stable compounds, ketenes can generally be considered unstable and must be generated *in situ* prior to their use in thermal reactions [213]. The most common methods draw on carboxylic

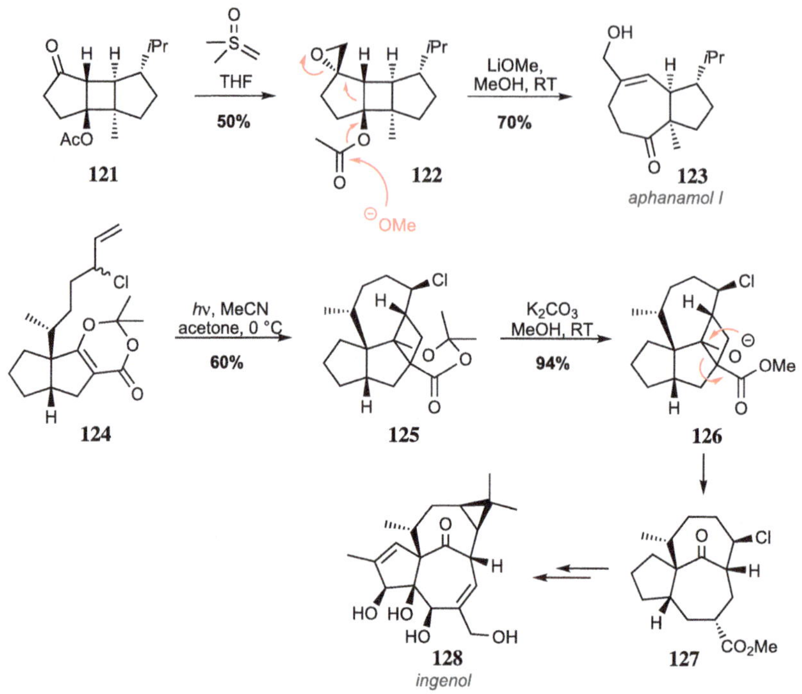

Fig. 5.107 Syntheses of aphanamol I (**123**) and (±)-ingenol (**128**) [211, 212]

acid derivatives as precursors of the respective ketenes. In addition, the rearrangement of
α-diazoketones can also be utilized for ketene formation (see Fig. 5.108).

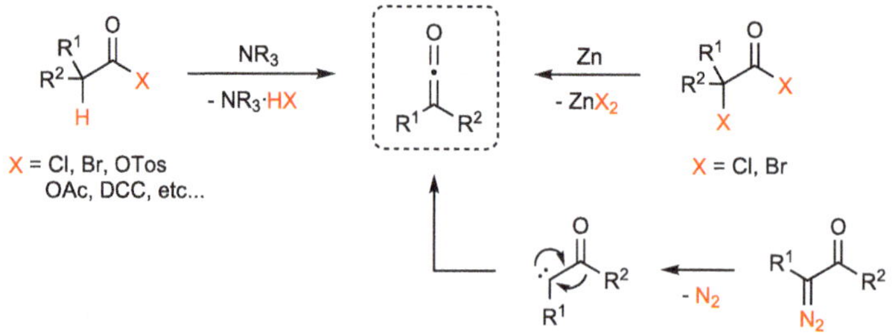

Fig. 5.108 Preparation of ketenes [213]

Danishefsky and co-workers employed the thermal [2+2]-cycloaddition of an olefin with
a ketene for their synthesis and structural revision of the neurotrophically active diterpene
(±)-tricholomalide A (**131**). The 5/7/4-fused ring system in **130** could be constructed from
dichloroketene, which was generated *in situ* from trichloroacetyl chloride in the presence

of elemental zinc. The target natural product **131** was afterwards obtained in a further 12 steps, which comprised an oxidative ring expansion of the cyclobutanone, the introduction of the keto function in the eastern hemisphere, and the concluding ring closure to the cyclic allylic ether in a Michael addition (see Fig. 5.109) [214].

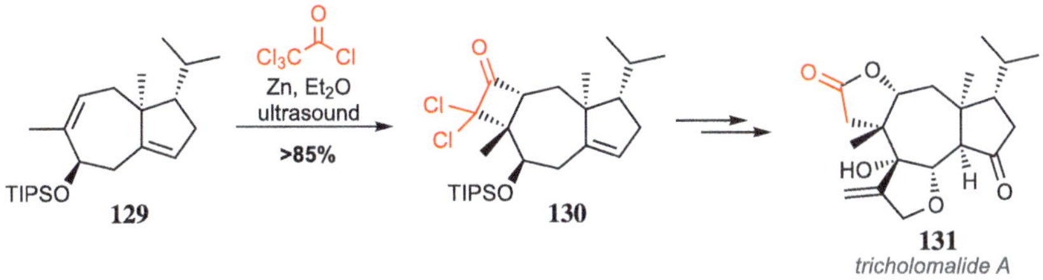

Fig. 5.109 Danishefsky's synthesis of (±)-Tricholomalid A (**131**) [214]

While [2+2]-cycloadditions similarly offer the possibility of an enantioselective reaction, no wealth of asymmetric methodologies exists, which stands in stark contrast to the Diels-Alder reaction. In addition to the obvious approach of using auxiliaries for a diastereoselective reaction, the use of catalytic methods naturally constitutes the most preferred avenue. The Bach group developed **137** as a catalyst for this purpose, which fixates the substrate by means of non-covalent hydrogen bonds and enables the shielding of one face of the reactant by the large aryl substituent [215]. The synthesis of the alkaloid (+)-meloscine (**136**) is based on an enantioselective [2+2]-cycloaddition of **132** and **133** in the presence of the organocatalyst **137** (see Fig. 5.110) [216]. After the ring expansion to **135**, the natural product (**136**) was obtained in a total of 15 steps with an overall yield of 7%.

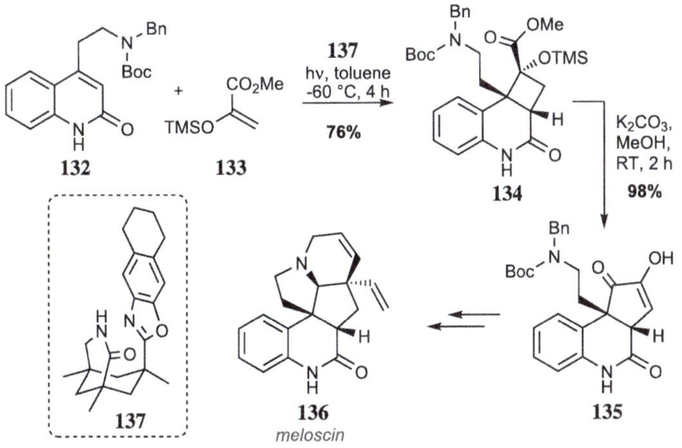

Fig. 5.110 Synthesis of (+)-meloscine (**136**) by the Bach group [216]

Apart from the classic photochemical [2+2]-cycloaddition of two olefins and the Paternò–Büchi reaction, there exists a third common variant of photocycloadditions. For this, metal salts are employed as homogeneous metal-based catalysts [217]. These either act as normal Lewis acids or as photosensitizers. For example, a copper-alkene complex can mediate an elevation of the olefin to its S_1 state following its excitation at 250 nm. These complexes enable an asymmetric reaction by facial discrimination induced through chiral ligands [218–221].

Ito and Iguchi utilized this principle in their synthesis of the marine prostanoid (+)-tricycloclavulone (**139**), which possesses antiproliferative properties. Although the asymmetric cycloaddition proceeds with only moderate enantiomeric excess of 73%, an advanced intermediate could be recrystallized enantiomerically pure. The natural product (+)-tricycloclavulone (**139**) could thus be synthesized in 21 steps (see Fig. 5.111) [222].

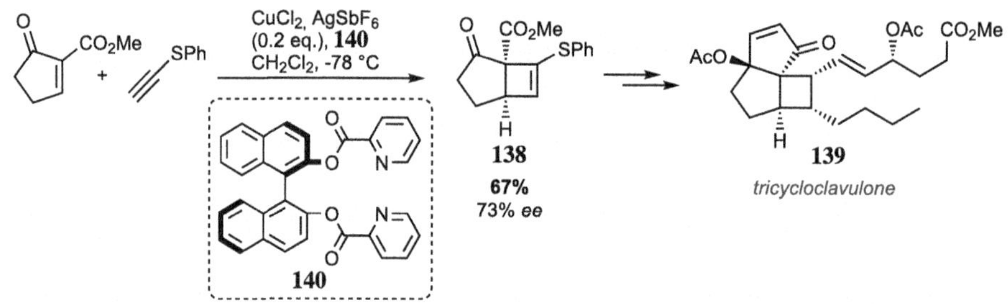

Fig. 5.111 Synthesis of (+)-tricycloclavulone (**139**) disclosed by Ito and Iguchi [222]

5.2.5 Cheletropic Reactions

Cheletropic reactions entail the addition (or extrusion) of $[2\pi]$-electrophiles to a polyene, usually an alkene or diene, in which two σ-bonds terminating at a single atom are formed or cleaved in concert. The reaction mode is categorized based on the involved *atoms* as [2+1], [4+1], etc.[IX] Cheletropic reactions thus represent a special subcategory of cycloadditions. The driving force of the reaction is, as with other cycloadditions, the conversion of π- into σ-bonds or, if a reverse cycloaddition takes place, the release of gaseous products (N_2, CO, SO_2, etc.) and the concomitant gain in entropy in addition to the reaction enthalpy. Well-

---

[IX] In the literature, square brackets are occasionally used to denote the electrons involved in the process, while the number of associated atoms is indicated in round brackets. The Simmons–Smith cyclopropanation would thus be a (2+1)- and [2+0]-cycloaddition. Since writing everything in square brackets constitutes the more common variant, it will be used henceforth. However, the reader should be able to distinguish between the number of atoms and electrons involved to avoid any misconception.

known examples of cheletropic reactions comprise the Simmons–Smith cyclopropanation and the Ramberg–Bäcklund rearrangement (see Fig. 5.112) [223–225].

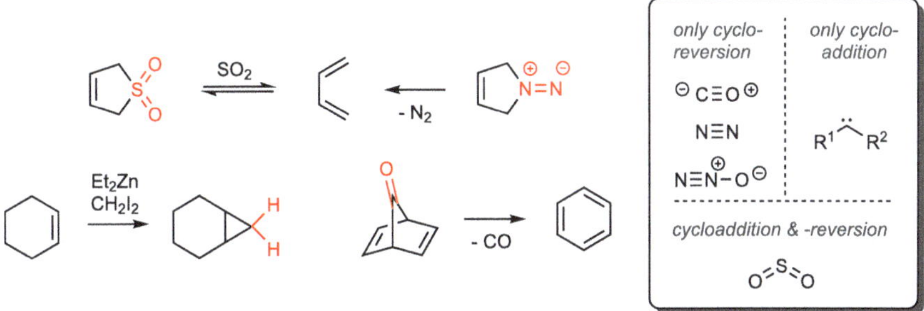

Fig. 5.112 Examples of cheletropic reactions

5.2.5.1 Mechanism and Variants

Cheletropic reactions proceed stereospecifically with respect to both components [226]. According to Woodward and Hoffmann, the transformation is classified, based on the atomic component, as possessing a linear or nonlinear reaction topology. In the linear trajectory, the atomic component reacts in an antarafacial manner with respect to its nodal orbital surface [9]. This topology at the atomic component concomitantly determines whether the reaction proceeds supra- or antarafacially at the polyene reactant. For atomic components, the HOMO is a sp^2-hybrid orbital, while the LUMO corresponds to a p-orbital. In cyclopropanations, a nonlinear course is followed for reasons of symmetry. In the [4+1]-cycloaddition of SO_2 to a diene, on the other hand, a linear approach occurs, with the diene undergoing a disrotatory ring closure during bond formation (see Fig. 5.113). An orbital analysis revealed that under standard conditions and with most substrates, the polyene typically reacts from the HOMO (χ_n) and the atomic component from the LUMO (ω_0). Depending on the orbital positions derived from the substitution pattern and the identity of the components, the reverse pairing ($\text{LUMO}_{polyene}$–HOMO_{atom}) can also make a significant contribution during the transition state [227–231]. Very electron-rich carbenes, such as $C(OMe)_2$ or $PhCNMe_2$, rather react from the HOMO in an almost linear fashion [228, 229].

In principle, a nonlinear attack can also occur in the addition of SO_2 to a diene if the ring closure proceeds in a conrotatory manner from an excited state. While the corresponding photochemical process does not proceed with fully conserved stereospecificity, it does display a clear preference for the conrotatory path, as the excited diene must participate in an antarafacial fashion with respect to the nodal plane to conserve the orbital symmetry [9]. As outlined above in Sect. 5.1, the facial selectivity changes with the increase/decrease of the number of electrons by two according to the general Woodward–Hoffmann rules. There-

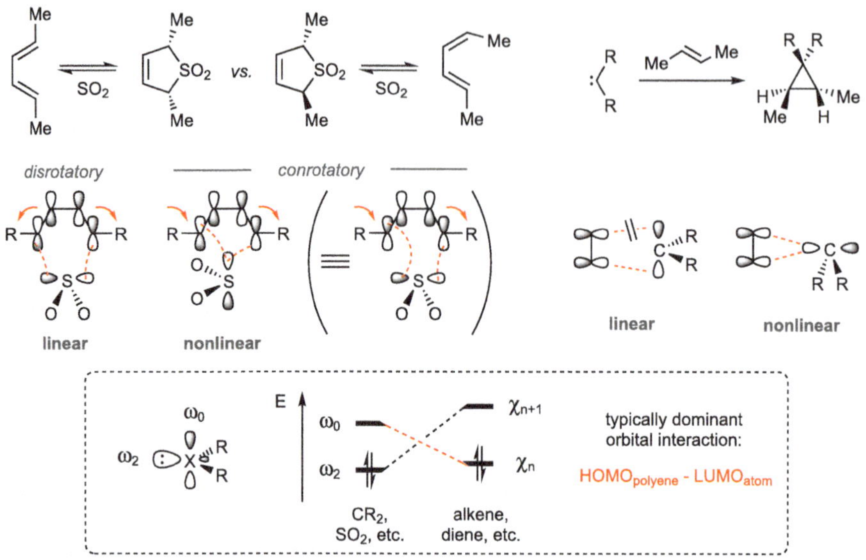

Fig. 5.113 Examples of cheletropic reactions and relevant orbital interactions in thermal reactions [232]

fore, when changing from an alkene to a diene, a change from a supra- to an antarafacial topology must occur and thus a linear approach will result instead of a nonlinear attack. While experimental evidence indicated the occurrence of a concerted [4+1]-cycloaddition of carbenes, a change of the addition topology for symmetry reasons upon proceeding from an alkene to a diene as the π-bearing reactant was not part of the study [233]. Even though non-stabilized carbenes are suitable substrates for cyclopropanations, the innate instability of the carbenes limits direct access to many target products. Barring the Simmons–Smith reaction, alternative cyclopropanation methods (e.g., under Cu- or Rh-catalysis) are often resorted to for this reason (*vide infra*). Dihalogen carbenes are exempt from this rule of thumb, as they can easily be employed as substrates in the addition to alkenes due to their stability [234].

The formation of cyclopropanes by reaction of singlet carbenes or metal carbenoids with alkenes constitutes the synthetically most important class of cheletropic reactions. However, new approaches relying on photocatalysis, electrosynthesis, intermediate anions, or metal catalysis are increasingly gaining traction to assemble cyclopropanes [235]. In the *Simmons–Smith cyclopropanation*, zinc or samarium carbenoids add to the double bond in a cycloaddition-like [2+1]-mechanism (see Fig. 5.114) [223, 224].

The most commonly encountered reaction conditions use Zn/Cu couple to generate the Zn carbenoids. The Furukawa modification, which instead relies on diethylzinc as organometallic precursor, is preferred when electron-poor and thus less nucleophilic alkenes need to be

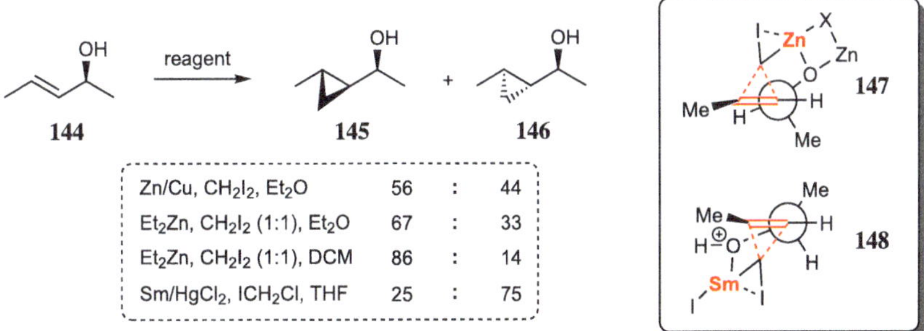

Fig. 5.114 Postulated mechanism of the Simmons–Smith cyclopropanation and typical reaction conditions [236]

converted to their corresponding cyclopropanes. Since the requisite reagents are commercially available and do not necessitate additional purification, this method is considered very reliable. Furthermore, the addition of trifluoroacetic acid, substituted phenols, and phosphates can have a beneficial effect on the reaction rate [237, 238].

Exploiting their Lewis-basic properties in the substrate for a directed transfer of the cyclopropanation reagent, coordinating groups such as hydroxy, alkoxy, amino and related functionalities allow the conduction of highly regio- and diastereoselective reactions. Cyclic allylic alcohols usually display excellent diastereoselectivities. Uniformly high diastereoselectivities ($>100:1$) can also be achieved in the case of acyclic, Z-configured disubstituted alkenes due to the occurrence of a 1,3-allylic strain. In contrast, in cyclopropanations involving the corresponding E-configured alkenes, only a minor preference for one product ($<2:1$) is observed under identical conditions. The Sm carbenoids introduced by Molander afford the opposite stereoselectivity (*anti* instead of *syn*) compared to the Zn homologues for sterically less demanding substituents at the alkene. This is rationalized by the different coordination modes and requisite bond angles in the transition states (see Fig. 5.115) [223].

Zn/Cu, CH$_2$I$_2$, Et$_2$O	56	:	44	
Et$_2$Zn, CH$_2$I$_2$ (1:1), Et$_2$O	67	:	33	
Et$_2$Zn, CH$_2$I$_2$ (1:1), DCM	86	:	14	
Sm/HgCl$_2$, ICH$_2$Cl, THF	25	:	75	

Fig. 5.115 Diastereoselectivity in the cyclopropanation of pent-3-en-2-ol and postulated transition states [223, 239]. Further substituents on the second Zn atom in **147** are not shown

In the directed variant, the organozinc reagent reacts with allylic alcohols to form a zinc alkoxide, which affords the expected *syn*-addition product via **147**. The O-Zn group is positioned in an anticlinal fashion with respect to the double bond. A theoretical treatise implied the presence of ZnX_2 or a second zinc alkoxide, to which the resultant zinc alkoxides are probably still coordinated to [236]. In the case of Sm carbenoids, given no deprotonation of the OH group takes place through the weakly basic reagents, a positive charge would be localized at the oxygen atom. The synclinal oxonium moiety would subsequently reduce the electron density and thus reactivity of the alkene via a $\pi \rightarrow \sigma^*$ interaction. To circumvent this destabilization, a synperiplanar conformation is adopted between the hydroxy group and the alkene double bond, whereby the samarium organyls show a reversed diastereoselectivity via the resulting transition state **148**. If steric interactions become dominant, for example, through the emergence of a 1,3-allylic strain when using Z-alkenes or in the presence of sterically demanding substituents in E-alkenes, the identical diastereomer is obtained via **147**, regardless of whether zinc- or samarium-based organometallic reagents are used.

The selectivity is intricately liked to the choice and stoichiometry of the reagents used ($EtZnCH_2I$ *vs.* $Zn(CH_2I)_2$ *vs.* $IZnCH_2I$). Especially the intrinsic properties of the solvent (coordinating *vs.* non-coordinating) determine whether the intramolecular coordination allows an efficient stereoinduction [240]. Apart from the nucleophilicity of the alkene, which largely impacts the observed regioselectivity, the directing effect of coordinating groups can also be exploited to support the conversion at the desired reaction site if various alkene moieties are viable. For example, iBu$_3$Al selectively leads to the addition at the isolated double bond in contrast to the reaction with the Simmons–Smith reagent in the cyclopropanation of geraniol (see Fig. 5.116).

	150		151
Et$_2$Zn, CH$_2$I$_2$	74	:	2
Sm/HgCl$_2$, ICH$_2$Cl	98	:	0
iBu$_3$Al, CH$_2$I$_2$	1	:	76

Fig. 5.116 Dependence of the regioselectivity in the Simmons–Smith cyclopropanation on the choice of reagents [241]

The synthesis of alkenes by extrusion of SO_2 from α-halosulfones (**152**) is referred to as *Ramberg–Bäcklund rearrangement* [225]. The two-step process proceeds via episulfone intermediates (**154**), which are formed following the deprotonation of **152**. While several mechanistic studies led to the postulate of a S_N2-type mechanism for the formation of the episulfones, the close structural relationship of the reaction with the Favorskii rearrangement renders a disrotatory, electrocyclic process as conceivable as well (see Fig. 5.163). The final cleavage of SO_2 proceeds stereoselectively to give the corresponding alkene, resulting in a net inversion at both stereogenic centers from **152** to **155** [242]. Since a linear extrusion is symmetry-forbidden, a concerted process would proceed following a nonlinear topology, similar to [2+1]-cycloadditions of carbenes [9]. A pathway via radical intermediates has also been postulated [243]. More recent theoretical studies rather support a concerted reaction [244]. In the presence of a base, an acceleration of the product formation from **154** has been observed. The underlying rationale governing the mechanism is still controversially discussed (see Fig. 5.117) [243, 245, 246].

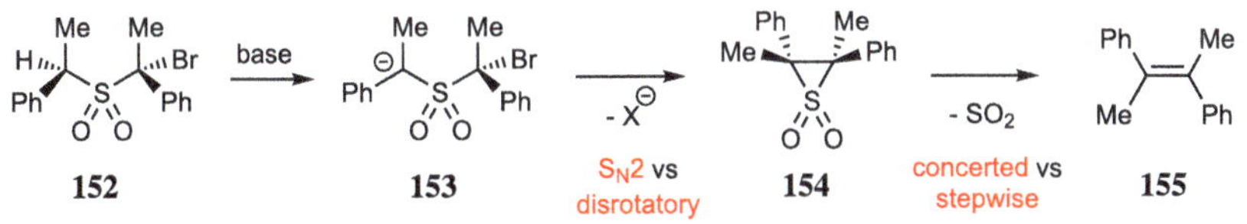

Fig. 5.117 Postulated mechanism of the Ramberg–Bäcklund rearrangement [242]

Ramberg–Bäcklund rearrangements are often used for the intramolecular construction of di-, tri-, or tetrasubstituted olefins. In aqueous medium, Z-alkenes are obtained as the major product, whereas in polar aprotic solvents, E-alkenes can also be selectively accessed in the presence of strong bases. Furthermore, sulfones can be directly converted to alkenes if a halogenation takes place *in situ* prior to the episulfone formation by adding CCl_4 or CF_2Br_2 (Meyers modification) [247].

5.2.5.2 Synthetic Applications

The illustrated cheletropic reactions have only been sparsely utilized in total syntheses. The extrusion of SO_2 to afford dienes is a common approach to generate highly reactive diene precursors for cycloadditions.

(+)-Rhishirilide B (**161**) is a glutathione-S-transferase inhibitor and was isolated from bacteria of the strain *streptomyces rishiriensis* OFR-1056. The natural product is particularly interesting due to its bioactivity profile, as the target enzyme governs the resistance to cancer therapeutics [248]. In Pettus' enantioselective synthesis, a domino-cheletropic/Diels-Alder reaction constitutes the key step of the group's approach (see Fig. 5.118). Following the elimination of SO_2 in the linear cheletropic extrusion, the diene **157** reacted with the enantiomerically pure benzoquinone derivative **158** to furnish the primary cycloaddition product. Intermediate **159** then afforded the desired α, β-unsaturated ketone in a β-elimination. A subsequent oxidative aromatization with DDQ completed the tricyclic core of the natural product, from which the intricate structure of (+)-rhishirilide B (**161**) was accessible in further eight steps.

In the enantioselective synthesis of the bacterial antibiotic (+)-ambruticine, Liu and Jacobsen relied on the Simmons–Smith reaction to introduce the central trisubstituted cyclopropyl moiety [159]. The allylic alcohol **162** was reacted with the *in situ* generated zinc carbenoid, whereby the order of addition and the exact stoichiometry was observed to have a decisive influence. The cyclopropane **164** could be obtained in 86% yield, presumably via transition

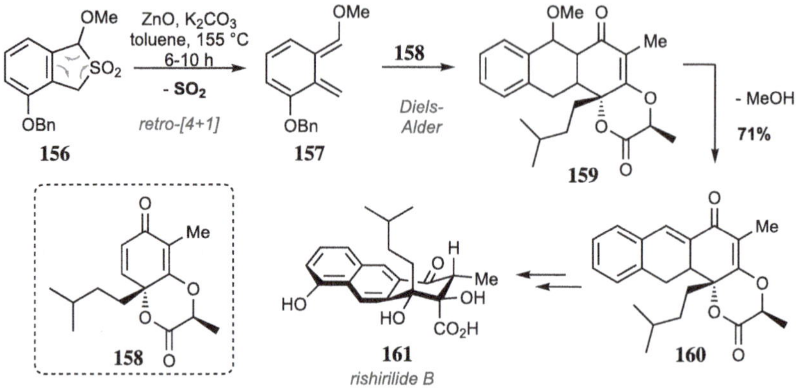

Fig. 5.118 Pettus' synthesis of (+)-rhishirilide B (**161**)

state **166**. In the assumed transition geometry, the zinc organyl is coordinated by one of the amide groups and heterocyle of the dioxaborolane reagent, thus selectively directing the approach of the carbene to occur on the β-face [249]. Other notable steps in the total synthesis entail an enantioselective hetero-Diels-Alder reaction using the chiral Lewis acid **78** introduced by Jacobsen (see Fig. 5.78) and an asymmetric hydroformylation (see Fig. 5.119) [159].

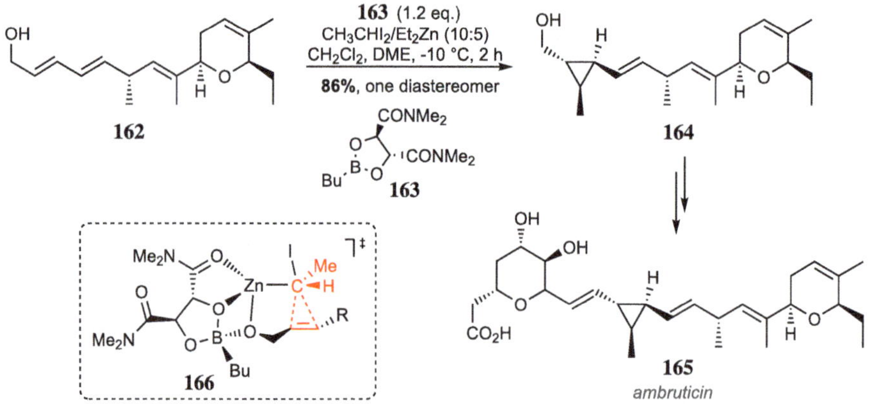

Fig. 5.119 Synthesis of (+)-ambruticine reported by Jacobsen *et al.* [159]

The stoichiometric dioxaborolane **163** has seen the largest range of applications for the asymmetric access to cyclopropanes. To this day it constitutes the reagent of choice in the presence of many functional groups [240]. Apart from its use in a large number of total syntheses (e.g., curacin A, FR-900848, U-106305, doliculide, calipeltoside A) [250–254], it was also successfully employed on an industrial scale [255].

The transition metal-mediated carbene transfer from diazo compounds (see Sect. 6.5) has over time outdistanced the Simmons-Smith reaction and is now the most common method for constructing cyclopropanes. The mechanism is not yet fully understood in its entirety. The reaction was observed to proceed at least for some of the studied systems in a concerted fashion, but strongly asynchronously [223, 256–259]. Alternatively, a multi-step process can occur as well, forming metalacyclobutanes as the primary product [260–262]. Particularly chiral Cu-, Rh-, Ru- and Co-based catalysts permitted the development of asymmetric variants [263, 264].

The Uyeda group recently disclosed an asymmetric, co-catalyzed cyclopropanation using non-stabilized carbenes generated from dichloroalkanes, which forego the need for resorting to unstable diazoalkanes. The Zn^0 is assumed to abstract Cl and facilitates formation of the cationic metal-carbenoid intermediate, which would transfer the C1-fragment via transition state **169** [265]. A modification of this method was used by Pfizer in their commercial synthesis of the SARS-CoV-2 drug nirmatrelvir, the active pharmaceutical ingredient of paxlovid (see Fig. 5.120) [266].

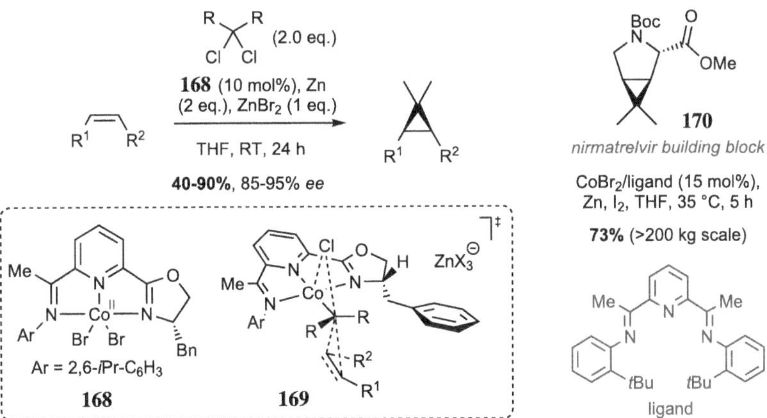

Fig. 5.120 Co-catalyzed cyclopropanation by Uyeda *et al.* and application on the industrial-scale manufacture of nirmatrelvir building block **170** [265, 266]

The secondary metabolite hirsutellone B (**175**), produced by fungi of the class *hirsutella nivea* BCC 2594, possesses potent activity against *mycobacterium tuberculosis*. The Nicolaou group effected the macrocyclization via a Ramberg–Bäcklund rearrangement and thus provided the handle to subsequently introduce the pyrrolidinone moiety. The treatment of thioacetate **171** with NaOMe in methanolic solution and successive oxidation with H_2O_2 and Na_2WO_4 afforded the macrocyclic sulfone **172** with 79% yield. A modification of the Meyers variant of the Ramberg–Bäcklund rearrangement led, via the *in situ* bromination with CF_2Br_2, to the formation of the episulfone, which selectively provided the Z-alkene **173** following the extrusion of SO_2 [267]. After subsequent functionalization with methylcyanoformate, the α-ketoester **174** could be isolated in 61% yield and was ultimately converted to the natural product **175** in a few steps (see Fig. 5.121) [268].

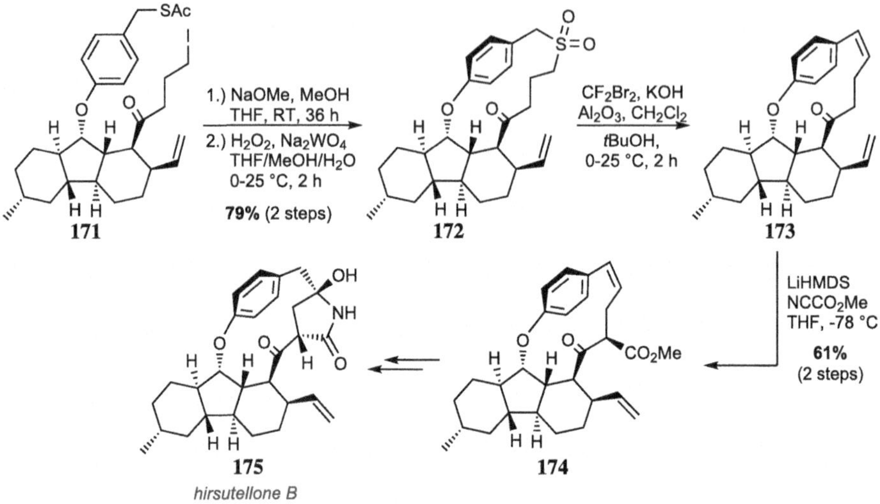

Fig. 5.121 Intramolecular Ramberg–Bäcklund rearrangement in the synthesis of hirsutellone B [268]

5.2.6 Further Cycloadditions

Apart from the cycloadditions discussed so far, there are a number of other important pericyclic additions that are not encountered as frequently as the [4+2]-, [2+2]-, and 1,3-dipolar cycloadditions. These include [2+2+2]-cycloadditions for the construction of substituted benzenes or pyridines, photochemical *meta*-cycloadditions of aromatic systems, as well as [σ, π]-cycloadditions (see Fig. 5.122).

These cycloadditions similarly follow the general Woodward–Hoffmann rules [9]. While this is relatively easy to understand for some cycloadditions, there are more complex reactions where the correct stereochemical topology only becomes apparent upon close examination. A detailed discussion of these variants of pericyclic reactions must be omitted due to lack of space, but there exists a wealth of literature covering this topic [9, 45, 269–279]. Among these less frequently perused variants, the [2+2+2]-cycloaddition still plays a significant role [280–287].

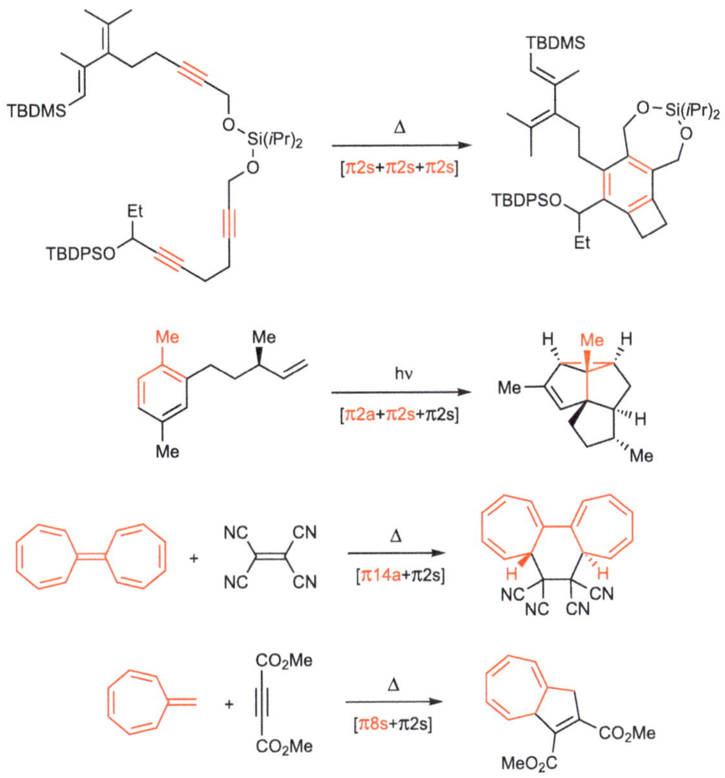

Fig. 5.122 Examples of further cycloadditions

5.3   Sigmatropic Rearrangements

Sigmatropic rearrangements constitute the second major class of pericyclic reactions governed by the conservation of orbital symmetry. They involve a reorganization of the bond structure, in which a σ-bond of a substitutent either migrates along a conjugated π-system, or a new σ-bond is formed at the end of two non-conjugated π-systems with concomitant cleavage of a second σ-bond. Overall, the number of σ- and π-bonds is thus preserved while the connectivity of the migrating group is changed. Various name reactions fall into this category, of which the most well-known encompass the Claisen, Cope, and Wagner–Meerwein rearrangement (see Fig. 5.123).

The nomenclature of [m,n]-sigmatropic rearrangements refers to the migration of a typically allylic σ-bond to a new position which is m-1 and n-1 atoms removed from the originally bonded position [9]. The rearrangement of hydrogen atoms (**176**→**177**) is therefore alluded to as [1,n]-rearrangements, as the position of one of the newly formed bonds will automatically be at the H-atom itself. If the second component is not a single atom, but,

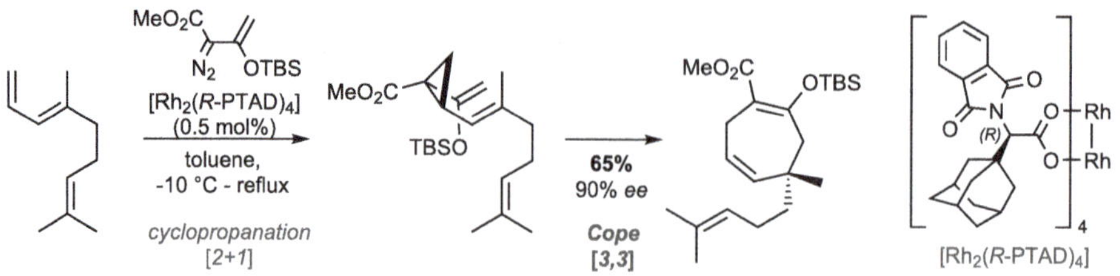

Fig. 5.123 Rearrangement key step in Stoltz and Williams' synthesis of 5-*epi*-vibsanin E [288]

for example, an allyl group (**178**), a topological distinction must be made with respect to both π-systems. Thus, the Cope rearrangement of **178** is a sigmatropic reaction of the order [3,3], as the newly forming σ-bond is two atoms away from the initial bonding position with regard to both components (see Fig. 5.124).

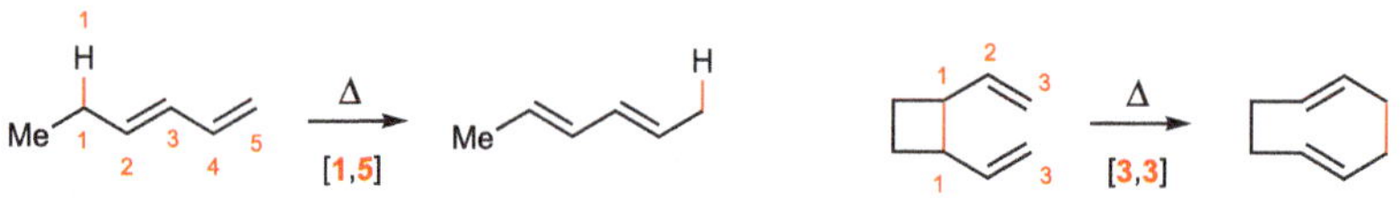

Fig. 5.124 Classification of sigmatropic rearrangements

5.3.1 Mechanism and Selectivity

In contrast to cycloadditions, the driving force of the reaction is not founded on the exchange of σ- for π-bonds, but rather on the formation of a more stable constitutional isomer.

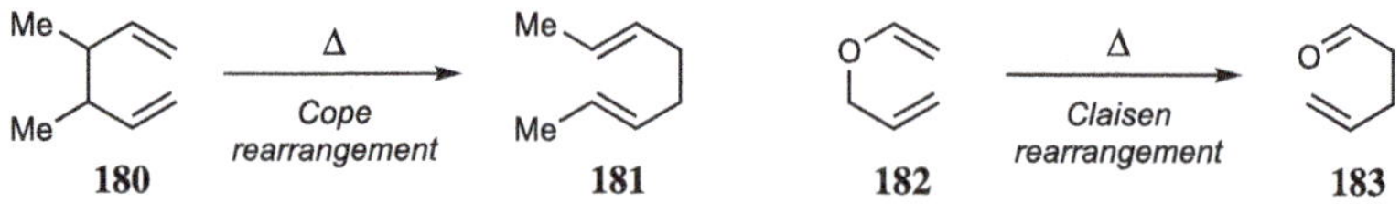

In the Cope rearrangement of the 1,5-diene **180**, the resulting diene **181** is energetically more stable due to its more highly substituted double bonds. The driving force of the Claisen rearrangement is based on the exchange of a CC- for a CO-double bond (**182** *vs.* **183**). The reduction of ring strain by the cleavage of small rings (3- or 4-membered) may also constitute a possible driving force, depending on the substrate (see Fig. 5.123).

In terms of the interacting orbitals, the concept of a supra- or antarafacial topology with regard to the nodal plane delineated in Fig. 5.8 is retained. Both pathways are possible, depending on the individual reaction conditions and steric prerequisites of the substrates (see Fig. 5.125).

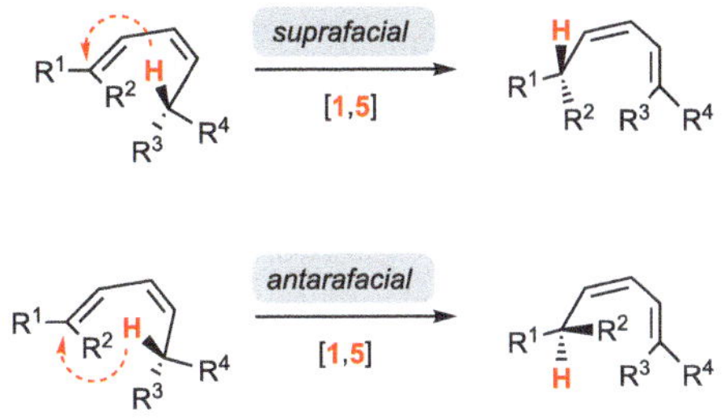

Fig. 5.125 Topology of sigmatropic rearrangements

Apart from hydrogen shift reactions, all rearrangements necessitate a classification of the topology of both components according to the general Woodward–Hoffmann rules in order to judge whether this type of reaction is symmetry-allowed under the chosen thermal or photochemical conditions. Barring the frontier orbital approach, an investigation of the conservation of orbital symmetry relying on correlation diagrams is similarly viable. By virtue of its clarity and quick applicability for estimating whether a planned reaction is allowed, the frontier orbital treatise may constitute the more simple and low-effort avenue. Woodward and Hoffmann provided a simple mnemonic for this purpose (see Table 5.1) [9].

Table 5.1 Selection rules of sigmatropic reactions of the order [m,n]

number of electrons	thermal	photochemical
4q	antara/supra	supra/supra, antara/antara
4q+2	supra/supra, antara/antara	antara/supra

In the reactions shown in Fig. 5.125, for example, a suprafacial participation is only allowed under thermal conditions. There are six electrons involved in the process and the H-atom reacts in a suprafacial manner by necessity. Therefore, the second component is required to react suprafacially to avoid the process becoming symmetry-forbidden. In the Cope rearrangement of **180**, both components (two allyl systems with a total of six electrons) react in a suprafacial fashion under thermal conditions. The alternative antara-/antarafacial pathway is inaccessible due to the occurrence of steric strain in the transition state.

For [1,n]-sigmatropic rearrangements of H-atoms, the scheme can be further simplified, as the H-atom always participates in suprafacial manner. It is therefore only necessary to classify the topology of the reacting π-system (Table 5.2).

Table 5.2 Selection rules of [1,n]-sigmatropic H-rearrangements

number of electrons	thermal	photochemical
4q	antarafacial	suprafacial
4q+2	suprafacial	antarafacial

A more detailed analysis of the involved orbitals and the transition state geometries will be carried out subsequently, divided according to the respective order of rearrangement. The advanced applications of all types of sigmatropic rearrangements will then be outlined by select examples.

5.3.1.1 [1,*n*]-Sigmatropic Rearrangements

[1,*n*]-sigmatropic rearrangements entail, in addition to the "classic" hydrogen shift reactions, rearrangements of methyl groups along a π-system, among others. Furthermore, Wagner–Meerwein rearrangements can also be identified as belonging to the canon of sigmatropic rearrangements by analysis of the involved orbitals. The most frequently encountered [1,*n*]-sigmatropic rearrangements and the involved orbitals are shown in Fig. 5.126.

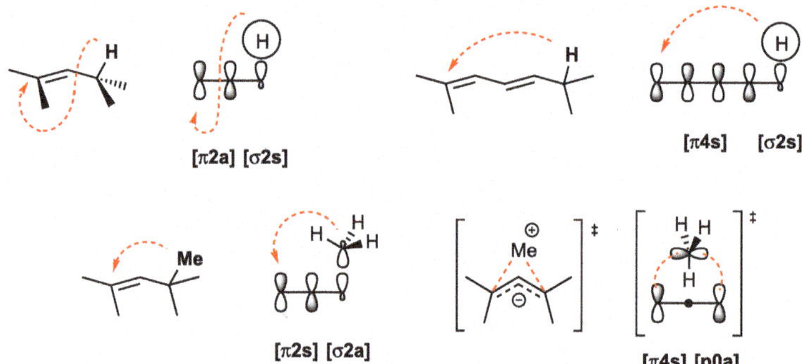

Fig. 5.126 Common [1,*n*]-sigmatropic rearrangements

The [1,3]-hydrogen shift allowed under thermal conditions proceeds suprafacially, while in the [1,5]-rearrangement the π-system participates with antarafacial topology. Assuming that a proton always follows a suprafacial pathway (due to the involved isotropic s-orbital), the dependence of the observed stereochemical product configuration can be easily rationalized: upon photochemical excitation, the stereochemistry of the product is reversed. The reactions follow the scheme described in Table 5.2. In practice, a thermal [1,3]-sigmatropic hydrogen shift is seldom observed, as the system is too strained to allow the requisite antarafacial topology. Nevertheless, the reaction is allowed on the basis of orbital symmetry.

If the involved frontier orbitals as depicted in Fig. 5.126 are tabulated according to the general Woodward–Hoffmann rules, the result is an odd sum, which corresponds to a thermally allowed process (cf. Sect. 5.1).

Intriguingly, suprafacial [1,3]-sigmatropic rearrangements of methyl groups are experimentally observed under *thermal* conditions. While this may appear to violate the Woodward–Hoffmann rules, it can be rationalized upon closer examination of the frontier orbitals. Assuming that alkyl groups or other carbon-bonded groups react antarafacially and thus with inversion, a suprafacial course with respect to the π-system becomes symmetry-allowed (see Fig. 5.127).

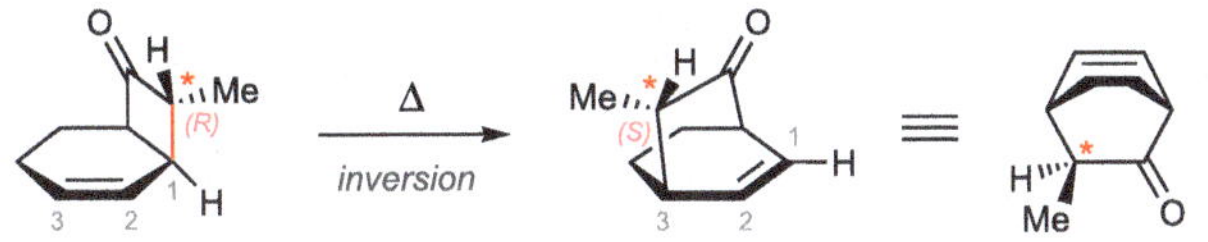

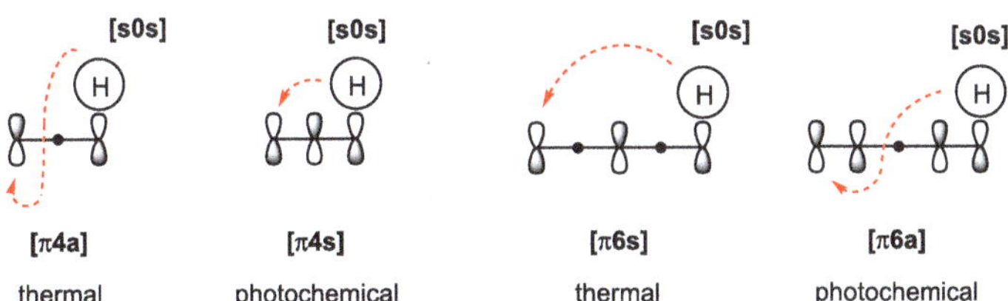

Fig. 5.127 Suprafacial [1,3]C-sigmatropic rearrangement under thermal conditions [289]

Another way to examine the reactions via an orbital analysis is to consider the transition states (as opposed to the previously considered ground states of the reactants), which was already depicted for the last rearrangement shown in Fig. 5.126. The π-system is considered an allyl anion and the methyl group is described as a cation as an approximation. This approach carries two advantages: First of all, the frontier orbitals no longer need to be constructed by linear combination of the π- and σ-systems. The π-system in [1,n]-sigmatropic rearrangements is approximated as an anion and subsequently composed in such a fashion that all odd atoms receive a p-orbital with alternating sign. This also simplifies the analysis of photochemical [1,n]-sigmatropic rearrangements (see Fig. 5.128).

Fig. 5.128 Facial course of [1,n]-sigmatropic rearrangements

Second of all, one component is eliminated in [3,3]- and [2,3]-rearrangements (the σ-component is integrated into one of the π-systems), which slightly simplifies the equation in the general Woodward–Hoffmann rules. Although the classification of a component with zero electrons is not actually permissible according to the Woodward–Hoffmann rules, this works perfectly well within the approximation as a transition state.

Wagner–Meerwein rearrangements can also be understood as sigmatropic rearrangements, which accounts for the absence of an "anionic" equivalent. If an α-substituted cation is defined as a 2π-system, a [1,2]-rearrangement must proceed suprafacially and with retention at the migrating alkyl group. In a theoretical anionic variant, one of the two components would have to react antarafacially, which would become even more sterically prohibitive in a [1,2]-shift compared to the already scarcely observed [1,3]-shift (see Fig. 5.129).

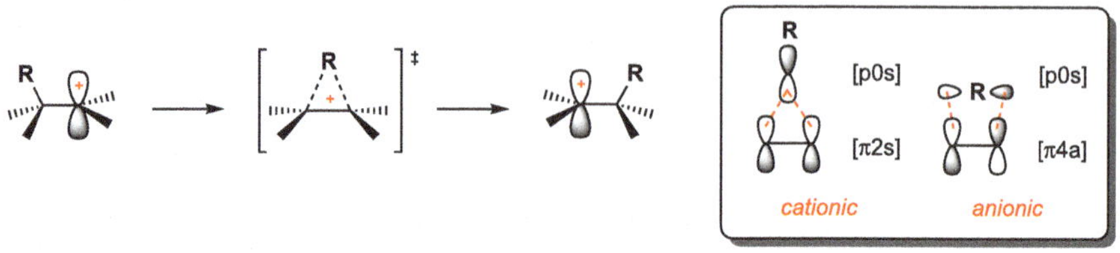

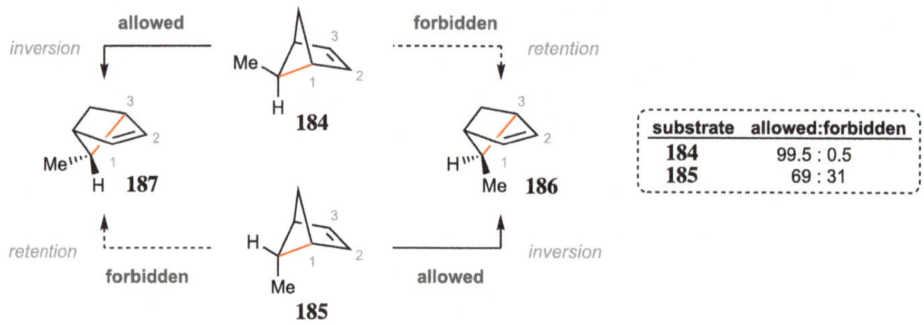

Fig. 5.129 Orbital analysis of cationic and anionic Wagner–Meerwein rearrangements

A less frequently observed variant of [1,3]-sigmatropic rearrangements is the breaking of a C-O- rather than a C-C-σ bond. However, it can be deliberately used for effecting a desired reaction under certain conditions [290]. Apart from the cases discussed thus far, some transformations have been observed whose product distribution can only be achieved by involving formally symmetry-forbidden transition states (see Fig. 5.130) [289, 291, 292].

Fig. 5.130 Stereochemical course of [1,3]C-sigmatropic rearrangements [291]

For the substrates used in the study, a suprafacial process with retention (symmetry-forbidden) could also be observed in addition to a suprafacial rearrangement with inversion of the methyl-substituted stereocenter (symmetry-allowed). Depending on the steric environment of the substrate, the symmetry-forbidden rearrangement can dominate in some cases. The cause of these seemingly symmetry-forbidden processes was long controversially discussed [289], but it is now assumed that the majority of these rearrangements proceeds via diradical intermediates [291].

5.3.1.2 [3,3]-Sigmatropic Rearrangements

The synthetically most important sigmatropic rearrangements are [3,3]-processes, which entail either a rearrangement limited to the carbon skeleton (Cope rearrangement) or encompass the cleavage of a C-heteroatom-σ-bond under concomitant formation of a new π-bond (Claisen rearrangement, see Fig. 5.131)) [293–296]. Barring oxy-Cope rearrangements, these sigmatropic rearrangements constitute equilibrium processes and the product composition depends on the relative stability of reactant and product.

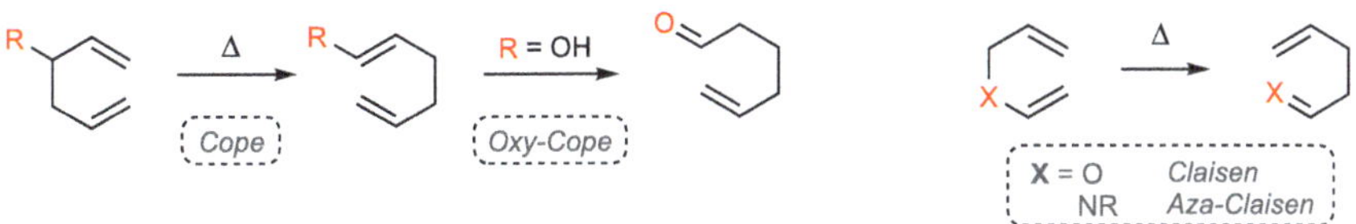

Fig. 5.131 Important sigmatropic name reactions

The *Cope* rearrangement is the thermal isomerization of a 1,5-diene to a regioisomeric 1,5-diene (see Fig. 5.132). The driving force of the reaction originates from the formation of the thermodynamically more stable constitutional isomer. In the *oxy-Cope* rearrangement, a hydroxy group is attached to the rearranging 1,5-diene, which tautomerizes to a carbonyl group following the Cope rearrangement. This subsequent carbonyl formation removes the rearrangement product from the equilibrium and drives the reaction to completion. In the *Claisen* rearrangement, allyl vinyl ethers are converted to γ, δ-unsaturated carbonyl compounds. The substrate of the *aza-Claisen* rearrangement is an allyl vinyl amine and the obtained product is the corresponding imine.

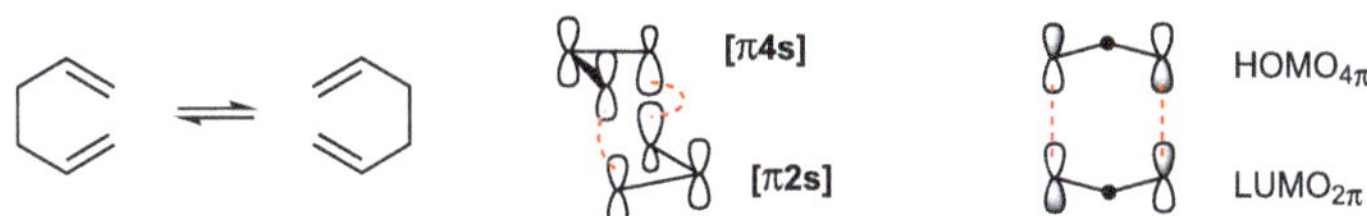

Fig. 5.132 Orbitals involved in the [3,3]-sigmatropic rearrangement as exemplified by the Cope rearrangement

The dominant transition state of concerted [3,3]-sigmatropic rearrangements is chair-like (**189**). Even though a boat conformation can also occur (**188**), it is generally considered energetically less favorable, unless forced by steric constraints of the substrate. Not all Cope rearrangements proceed in a concerted fashion: The presence of radical-stabilizing groups can favor a stepwise mechanism in which the bond breaking precedes the bond making [297]. In this case, diradical intermediates are postulated to occur. The opposite case has been theoretically analyzed as well with prior bond making, which renders the transition state "tight" due to the involved bond distances in **190** [298]. **188** and **189** are considered "loose" transition states [298]. Claisen rearrangements can theoretically proceed both via concerted and diradical pathways [299]. This has been studied experimentally far less extensively and yet awaits an in-depth treatise by modern quantum mechanical methods (see Fig. 5.133).

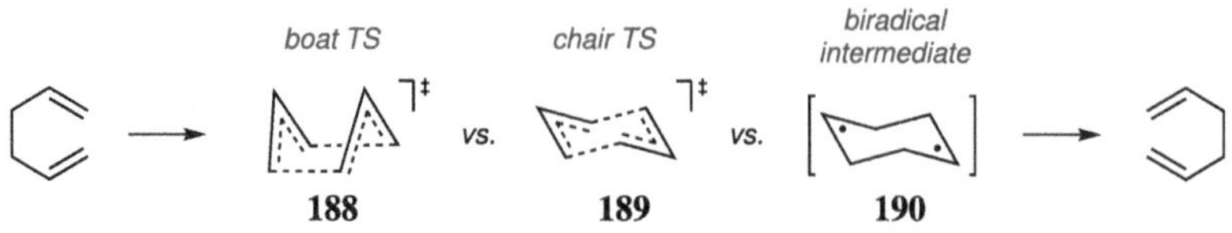

Fig. 5.133 Possible reaction paths of the Cope rearrangement [298]

Their popularity in total syntheses may be rooted to large degree in its high stereochemical predictability: Almost all [3,3]-sigmatropic rearrangements proceed through the chair-like transition state **189**. Depending on the position of substituents or the configurations of the two double bonds, a stereospecific transfer of this stereoinformation from reactant to the product takes place (see Fig. 5.134).

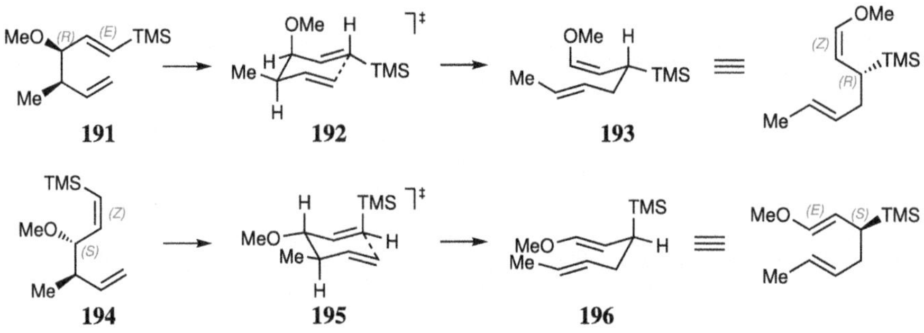

Fig. 5.134 Stereochemistry of [3,3]-sigmatropic rearrangements via chair transition states [294]

The Cope rearrangements of the dienes **191** and **194** proceed via the transition states **192** and **195**, respectively. When attempting a prediction of the resultant stereochemistry in the corresponding products, it is advisable to first convert the reactants to the operative chair-like conformer. From this, access to the products **193** and **196** via the respective transition structures is easily comprehensible if the most stable conformer is also the reactive one (cf. A-values, Chap. 1). Under certain circumstances, an inversion of the chair may still need to be performed to obtain the correct, dominant transition structure. Doering and Roth insightfully concluded from the Cope rearrangement of 3,4-dimethylhexadiene that the stereochemistry of the products allows the determination of the relative energies of the involved transition states (see Fig. 5.135) [2]. The boat transition states play only a minor role for the studied diastereomeric substrates. The high proportion of the product resulting from **198** is initially surprising, as its share does not reflect the ratio derived from the A-values . This observation can be attributed to the avoidance of a 1,2-*gauche* interaction of the two methyl groups in the equatorial position (**197**).

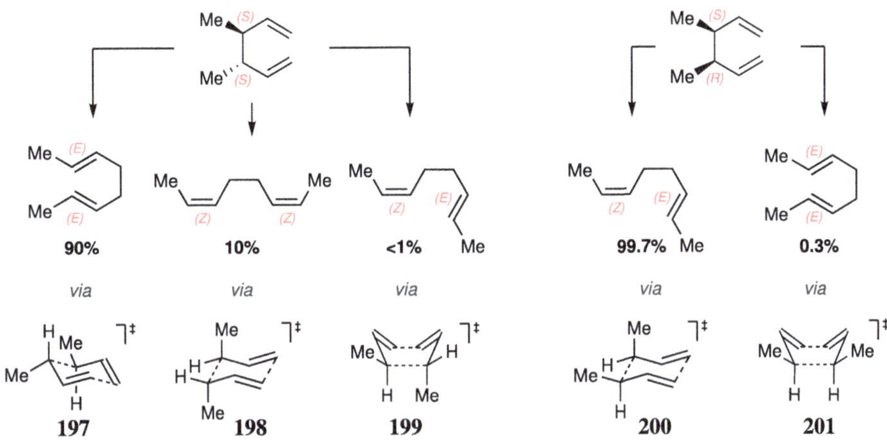

Fig. 5.135 Dependence of product distribution on the type of transition state [2]

The Claisen rearrangement also typically entails a chair-like transition state, where the structures shown in Fig. 5.134 only need to be slightly modified to arrive at the Claisen analogs. Figure 5.136 vividly illustrates the possibility of two competing chair-like transition states (**204** *vs.* **207**), whose substrates are in a rapid conformational equilibrium (**203** *vs.* **206**, Curtin–Hammett situation). In both conformers, the methyl substituent assumes a pseudoaxial position and the silyl group is resting in a pseudo-equatorial position as imposed by the double bond configuration. This results in the methoxy group adopting an axial position in **203** or **204**, whereas it is equatorial in **206** and **207**. As a result, **208** is obtained as the major product (see Fig. 5.136) [294].

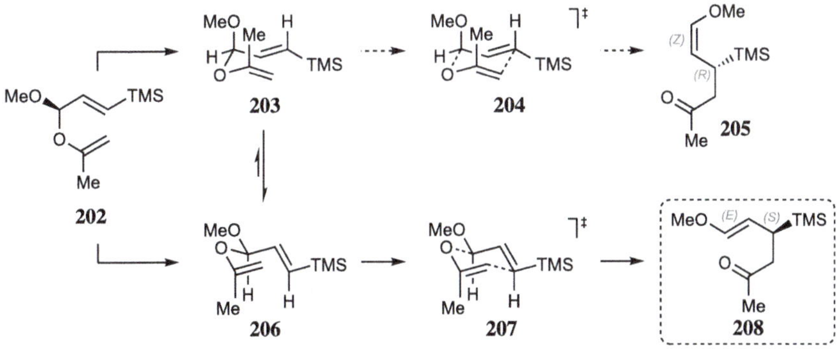

Fig. 5.136 Competing pathways of a Claisen rearrangements possessing a conformational pre-equilibrium [294]

Boat transition states are mainly encountered in cyclic substrates, which grants the double bonds only a limited degree of flexibility due to the rigid ring structure [295]. In the synthesis of the glutamic acid mimetic (–)-kainic acid (**212**), Knight and co-workers employed an Ireland–Claisen rearrangement of **209**, which is forced by the macrolide ring to proceed via the boat transition state **210**. **211** was subsequently converted to the natural product **212** in four further steps (see Fig. 5.137) [300].

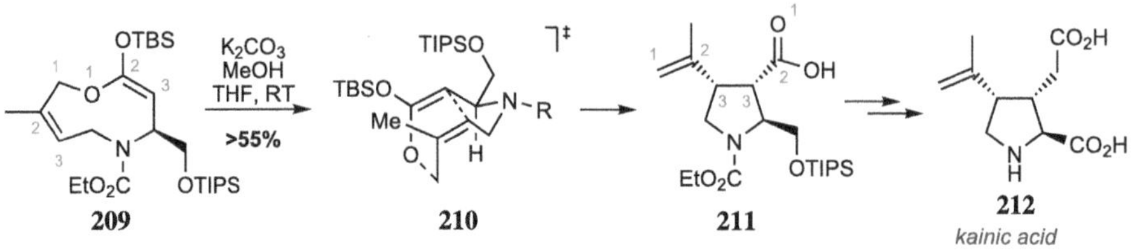

Fig. 5.137 Knight's synthesis of (–)-kainic acid (**212**) [300]

Trisubstituted cyclohexenes of the type **213** mainly afford the axial product (**215**) in a Claisen rearrangement. This can be rationalized by the favored chair-like transition state **214** over a twist-like transition state (not depicted) as predicted by the Fürst–Plattner rule (see Fig. 5.138 and confer Sect. 1.1.3) [301]. Exocyclic olefins, on the other hand, display no facial selectivity [302].

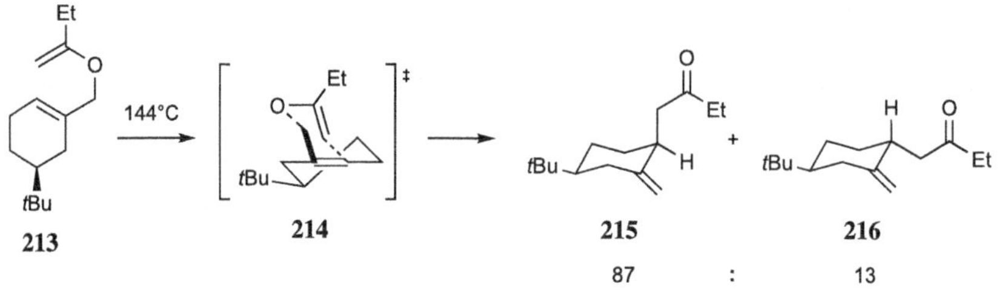

Fig. 5.138 Axial preference of Claisen rearrangements in endocyclic cyclohexenes [301]

The research group led by Steven Ley has been particularly dedicated to the synthesis of complex natural products. The synthesis of the natural insecticide and growth disruptor azadirachtin constitutes one of their most outstanding achievements (see Fig. 5.139). The coupling of the western decalin moiety with the eastern pyran fragment represented the key step in their approach to this highly complex secondary metabolite, which comprises

16 contiguous stereocenters (seven of them quaternary) as well as seven fused or bridged rings. Since a direct connection of the two sterically very demanding fragments was not possible, this necessitated a detour through an ingenious relay strategy via a thermal Claisen rearrangement of the advanced intermediate **226**. The desired reaction proceeded both under thermal microwave conditions and in the presence of an activating Lewis acid. Due to the steric repulsion of the propargyl group and the β-facing, fused tetrahydrofuranyl moiety, the approach of the alkyne occurred from the α-face. To account for the observed product stereochemistry, the reaction presumably proceeds via a twist transition state as an inversion of the *trans*-decalin backbone is not possible (contrary to the expected Fürst–Plattner selectivity). The [3,3]-sigmatropic rearrangement accordingly furnished the quaternary stereocenter with the desired configuration and the requisite allene functionality, which served as handle for a subsequent radical cyclization (cf. Figure 10.1, Chap. 10) [303].

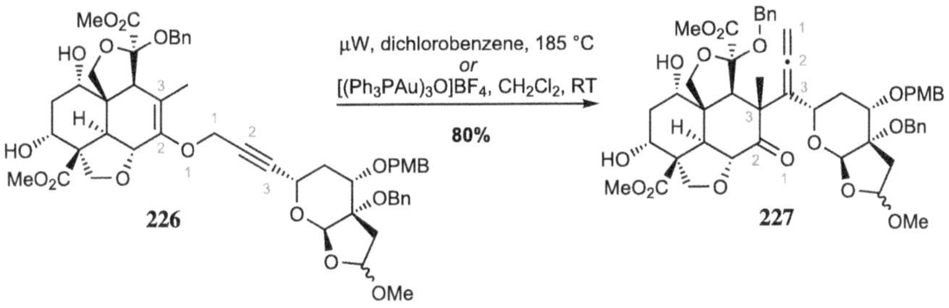

Fig. 5.139 Ley's approach to azadirachtin [303]

The classic Cope and Claisen rearrangements possess a number of different subtypes, which can be classified based on their substrates. In addition to the oxy- and aza-Cope rearrangement, there additionally exists the Ireland–Claisen rearrangement shown in Fig. 5.137 (see also Fig. 5.142).

The oxy-Cope rearrangement is commonly conducted in the presence of a Brønsted base. This variant is referred to as *anionic oxy-Cope rearrangement* [304, 305]. The transformation proceeds *significantly* faster with a deprotonated alcohol compared to a reaction under neutral conditions, many reactions can even be carried out at room temperature. The tremendous acceleration of the reaction rate, which becomes up to 10^{17} times more rapid upon addition of a base [306], is attributed to two synergistic effects. The substrate partially delocalizes the oxygen-centered charge. First of all, the energy level of the reactant is raised as a consequence and leads to a net decrease of the activation energy, as the transition state is less affected by the negative charge. Second of all, there exists a strong $n_O{\rightarrow}\sigma^*_{CC}$ interaction, which significantly weakens the σ bond to be cleaved and further lowers the reaction barrier (see Fig. 5.140).

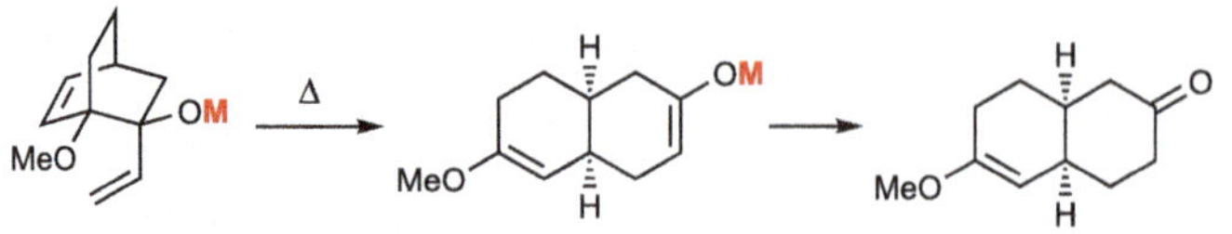

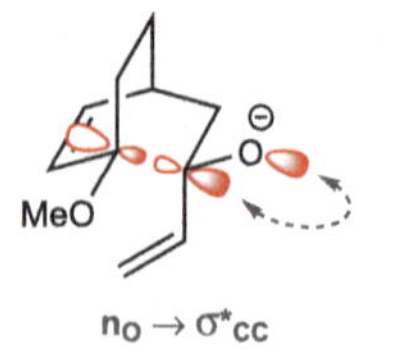

M	E_A [kcal/mol]	T [°C]
H	35.9	175-215
K	19.4	10-55
K, 18-crown-6	18.2	-20-5

Fig. 5.140 Anionic oxy-Cope reaction, activation energies, and resulting requisite reaction temperatures [306]

The judicious choice of the counterion is particularly critical. While the Li or MgBr alcoholate did not result in such a pronounced acceleration for the reaction depicted in Fig. 5.140, the corresponding potassium alcoholate increased the reactivity by a factor of 10^{10} at room temperature. By adding 18-crown-6, a further increase in rate by two orders of magnitude could be achieved, from which it was concluded that an efficient dissociation of the ion pair is pivotal for effecting a strong acceleration [306]. This results in an approximate reactivity scale of the alcoholates in the order

$$K^+/18\text{-crown-6} \approx K^+/\text{HMPA} > K^+ \gg Na^+ > Li^+, MgBr^+ \ggg H$$

An example of a particularly complex anionic oxy-Cope rearrangement is found in Boeckmann's synthesis of the cyclooctanoid natural product pleuromutilin (**234**), in which the pericyclic reaction was used to construct the challenging eight-membered ring structure (see Fig. 5.141) [307].

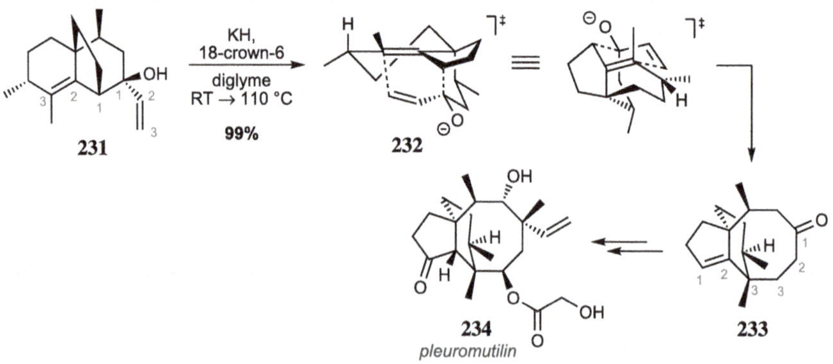

Fig. 5.141 Boeckmann's synthesis of (±)-pleuromutilin (**234**) [307]

The bridged bicycle **231** comprising a tetrasubstituted, endocyclic double bond allows only one degree of freedom for rotation of the acyclic alkene appendage because of its highly rigid framework. In the chair-like transition state **232**, the two fused six-membered rings of the substrate adopt a chair and a half-chair conformation, while the five-membered ring assumes an envelope-like structure. Following the concluding tautomerization, the 5/6/8-fused product **233** is obtained. This could subsequently be converted to the natural product (±)-pleuromutilin (**234**) in several further steps.

The Ireland–Claisen rearrangement is a mild variant of the Claisen rearrangement using a carboxylic acid allyl ester, which is converted to the corresponding silyl-stabilized enolate. The resultant O-Si bond is cleaved in the reaction, affording carboxylic acids as products after the work-up (see Fig. 5.142) [308, 309].

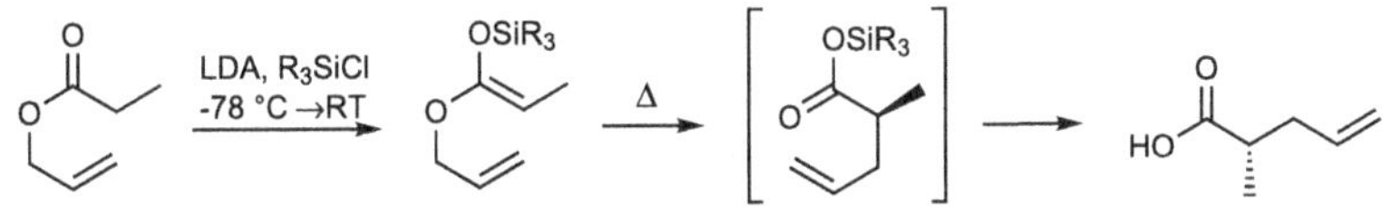

Fig. 5.142 General scheme of the Ireland–Claisen rearrangement

An advantage of the Ireland–Claisen rearrangement lies in the ability to determine the enolate geometry (*E* or *Z*) by skillful choice the reaction conditions (solvent, base; see Sect. 2.2). In addition, the reaction proceeds at relatively moderate temperatures ($<100\,^{\circ}$C). For these reasons, the Ireland–Claisen rearrangement has been successfully incorporated into a variety of natural product syntheses [294].

Deslongchamps employed an Ireland–Claisen rearrangement in his formal total synthesis of the macrocyclic antibiotic erythromycin A. The formation of the silyl enol ether starting from **235** afforded the rearrangement product **236** with moderate yield. The minor C-4 epimer presumably stems from the proportionally obtained *E*-configured enolate precursor, if the transition state **237** can be assumed to be dominant (see Fig. 5.143) [310].

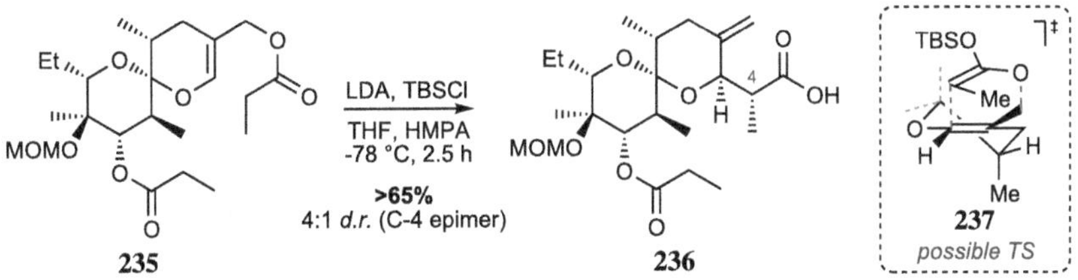

Fig. 5.143 Formal synthesis of erythromycin A according to Deslongchamps *et al.* [310]

Apart from the subtypes mentioned above, further variants have been reported [294], such as the Johnson–Claisen orthoester rearrangement and the Overman rearrangement, as well as the Eschenmoser and Bellus–Claisen rearrangements [311–316].

5.3.1.3 [2,3]-Sigmatropic Rearrangements

In the [2,3]-sigmatropic rearrangement, olefins bearing heteroatom substituents in the allylic position rearrange to give the corresponding thermodynamically more stable constitutional isomers. The thermally allowed reaction proceeds through a five-membered, cyclic transition state. Depending on the type of heteroatoms X, Y and the electronic state of Y (anion, lone pair, ylide), different classes of [2,3]-sigmatropic rearrangements can be distinguished (see Fig. 5.144) [317–321].

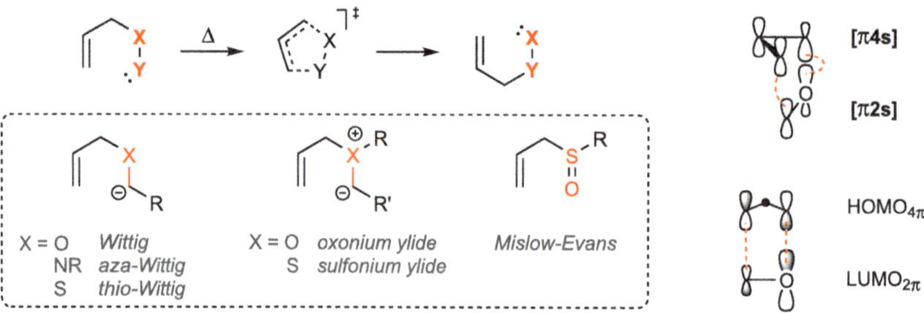

Fig. 5.144 General reaction scheme, common variants of [2,3]-sigmatropic rearrangements, and relevant orbital interactions of the Wittig rearrangement [317]

The most commonly encountered variant is the Wittig rearrangement of anionic allyl ethers. Related reactions such as the Mislow–Evans rearrangement of allyl sulfoxides have been introduced as well. The six electrons involved in the Wittig rearrangement can be formally divided between an allyl anion and a carbanion. An analysis of the involved frontier orbitals and their topology yields a $[\pi 4 s]$- (allyl system) and a $[\pi 2 s]$-component (carbanion). Given that the reaction is accelerated by decreasing the frontier orbital gap (cf. Sect. 5.1), a substitution pattern to raise the HOMO and lower the LUMO would thus be highly beneficial. As the postulated dominant orbital interaction is HOMO$_{\text{anion}}$-LUMO$_{\text{allyl}}$, the larger a stabilization of the allyl system and destabilization of the carbanion by electron-withdrawing and electron-donating substituents, respectively, the more readily the rearrangement should occur (cf. Sect. 5.2.1.2). While no quantitative evidence for this prediction has yet been reported, a semiquantitative study for thio-Wittig rearrangements was undertaken and showed that the introduction of substituents lowering the HOMO$_{\text{anion}}$ and raising the LUMO$_{\text{allyl}}$ did slow down the reaction (see Fig. 5.145) [317].

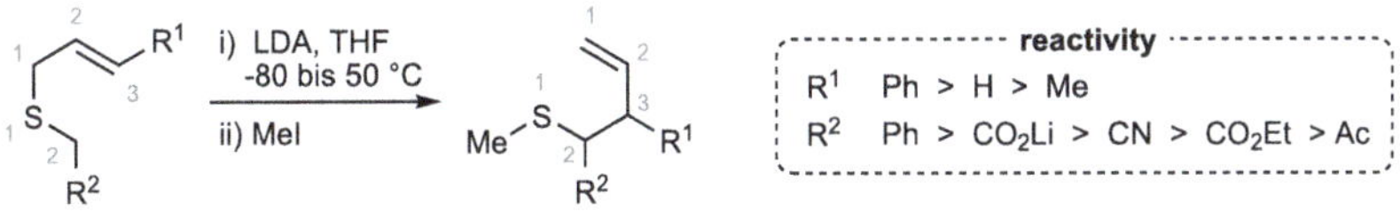

Fig. 5.145 Influence of substituents on the reaction speed of the thio-Wittig rearrangement [317]

Since the nature of the substrate in [2,3]-sigmatropic rearrangements permits a competing radical [1,2]-rearrangement, these pericyclic reactions should be carried out at low temperatures to minimize this alternative pathway. The extent of product contamination by competing processes thus depends on the substrates' structural environment, the reaction temperature, and in some cases even on the base used (e.g., in Wittig rearrangements) (see Fig. 5.146) [322].

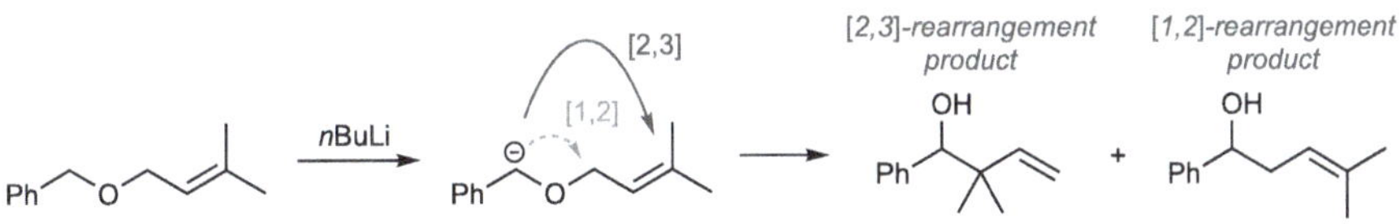

Fig. 5.146 Influence of temperature on the product distribution of the Wittig rearrangement [322]

The diastereoselectivity of the [2,3]-sigmatropic rearrangement can be derived from a putative rigid five-membered transition states [323, 324]. The preferred pseudo-equatorial arrangement of the substituents in the envelope transition state explains why products with E-configured double bonds are preferentially formed. The reaction is thus often diastereoconvergent (see Fig. 5.147).

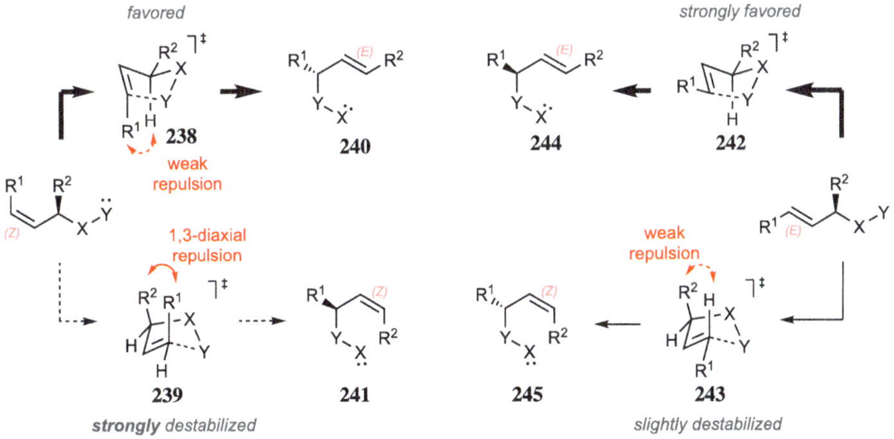

Fig. 5.147 Transition states and products of E- and Z-configured reactants in the [2,3]-sigmatropic rearrangement [323–325]

In all competing transition states, 1,3-diaxial interactions determine the energetically most favorable reaction path, from which one of the diastereomeric products **240**, **241** or **244**, **245** results. For the *Z*-configured substrate, **239** is additionally strongly destabilized by the occurrence of a 1,3-allylic strain in the transition state, which is why *E*-configured alkenes typically yield better diastereoselectivity [325].

Piers relied on a Wittig rearrangement in his total synthesis of the diterpene (±)-sarcodonin G to introduce the homoallyl side chain on the A-ring of the natural product [326]. Following preparation of the tin precursor **246**, the carbanionic reactant **247** could be generated *in situ* by a Sn/Li exchange. By virtue of the highly rigid *trans*-decalin skeleton, the equally constrained transition state **248** exclusively afforded **249** as the only diastereomer in 88% yield. Subsequent transformations then furnished the antifungal secondary metabolite sarcodonin G as racemic mixture (**250**, Fig. 5.148) [326].

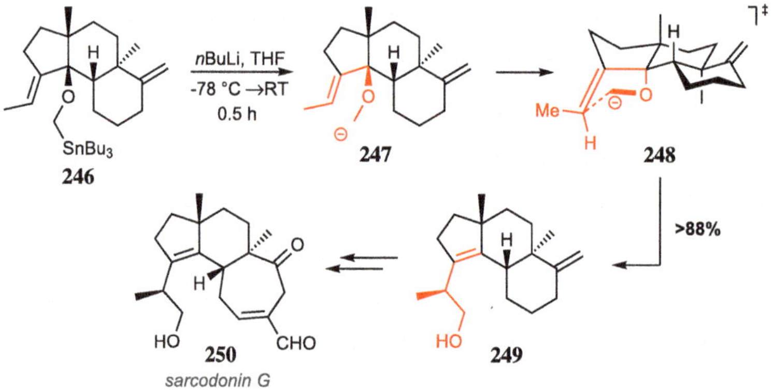

Fig. 5.148 Synthesis of (±)-sarcodonin G (**250**) according to Piers *et al.* [326]

Vigulariol (**255**) originates from the marine *cladiellin*-cembranoid family, which displays a wide range of biological activity such as cytotoxicity against various tumor cell lines. The first synthesis of the racemic natural product by Clark and co-workers was based on a Cu-catalyzed conversion of the diazo intermediate **251**, which generated the oxonium ylide **252** *in situ* and subsequently underwent a Wittig rearrangement to construct the B/C ring framework. The substrate afforded the *Z*-isomer of the bicyclic product **254** as major product, which can be rationalized by energetically dominant trajectory passing through transition state **253**. The oxepin **254** was subsequently converted to the natural product vigulariol (**255**) in 12 further steps (see Fig. 5.149) [327].

5.3.2 Modern Variants and Synthetic Applications

Sigmatropic reactions have frequently been drawn on to efficiently synthesize complex molecular architectures [328, 329]. In contrast to cycloadditions, which have seen the intro-

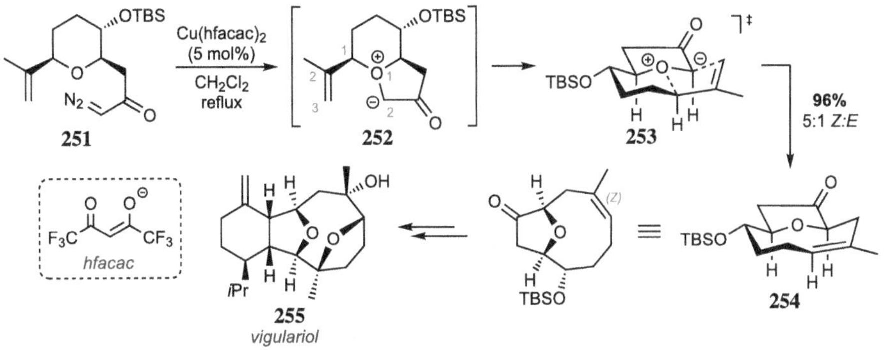

Fig. 5.149 Clark's total synthesis of (±)-Vigulariol (**255**) [327]

duction of many methodologies for an enantioselective transformation, the corresponding asymmetric sigmatropic rearrangements have been studied much less intensively, if the number of publications qualifies as a suitable indicator. While in Diels-Alder reactions two components can serve as handles to influence the reactivity and selectivity, all modifications in sigmatropic rearrangements have to be accommodated in *one* substrate only. Apart from the challenge of compatibility of all desired functionalities, the synthetic effort to access the requisite substrate can prevent an efficient use of a methodology. Nevertheless, several examples have been developed for [3,3]-sigmatropic rearrangements that have been rendered asymmetric through the use of chiral Lewis acids [330–332]. In Corey's synthesis of the anti-inflammatory natural product (+)-fuscol (**260**), an enantioselective Ireland–Claisen rearrangement was cleverly incorporated to set two of the three required stereocenters. In the presence of the chiral diazaborolidine **258**, which had previously been disclosed by Corey for use in enantioselective aldol reactions, the desired enantiomer **259** could be selectively obtained from the *E*-enolate of **256** via the transition state **257**. In addition to the major product, another diastereomer was isolated in 19%, which could result from the participation of a boat transition state or the partial formation of the *Z*-enolate prior to the pericyclic reaction. **259** could afterwards be converted to the natural product in 11 more steps (see Fig. 5.150) [333].

Despite their high stereoinduction, the Lewis acids often bind more strongly to the product compared to the reactant (e.g., vinyl allyl ether *vs.* aldehydes) and tie up the stereoinductor. The high, sometimes even stoichiometric, amount of chiral Lewis acids thus constitutes the main challenge preventing their more widespread use.

A recent work on enantioselective, catalytic Claisen rearrangements by Yamamoto relied on enol phosphonates as substrates. They were converted to the corresponding α-ketophosphonates in the presence of a copper catalyst, which could then be further derivatized (see Fig. 5.151) [334].

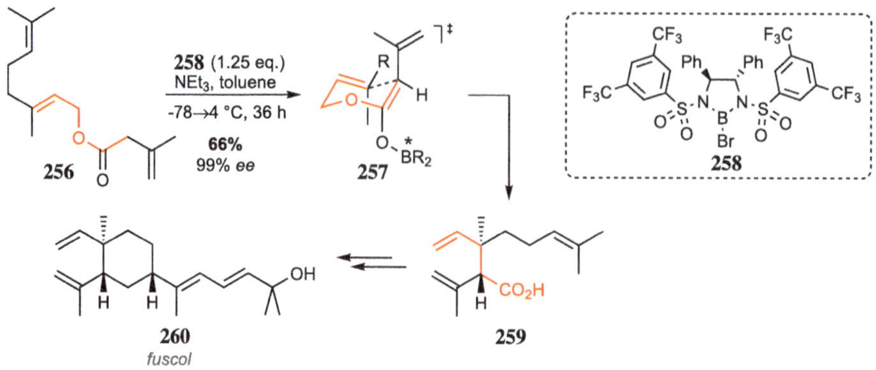

Fig. 5.150 Corey's synthesis of (+)-fuscol (**260**) using an enantioselective Ireland–Claisen rearrangement [333]

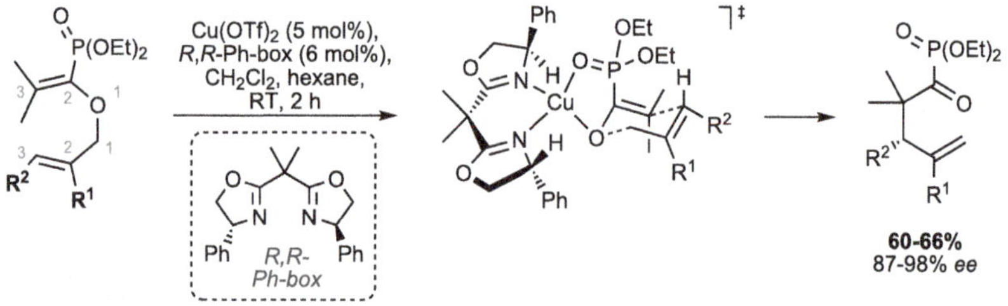

Fig. 5.151 Asymmetric Claisen rearrangement by Yamamoto [334]

The enantioinduction was attributed to the interaction of the phenyl moieties of the ligand with the terminal methyl groups, which is why a geminal disubstitution at C3 is essential for high *ee* values. The substitution pattern additionally invokes an acceleration of the reaction by the Thorpe–Ingold effect, without which the rearrangement would require elevated temperatures and not occur at room temperature.

White and co-workers skillfully utilized a domino Petasis olefination/Claisen rearrangement in their synthesis of the oxylipin secondary metabolites solandelactone A, B, E, and F. In the presence of the Petasis reagent Cp$_2$TiMe$_2$, the diene **262** was generated *in situ* as a diastereomeric mixture, from which the eight-membered, unsaturated lactone **264** was obtained in a Claisen rearrangement. The transition state **263** — shown here only for one diastereomer of **262** — explains why, despite diastereomeric starting materials, only one product was obtained, as no new stereocenter is introduced. Following the construction of the eight-membered ring, subsequent steps comprised a desilylation, oxidation of the terminal hydroxy group, and a Nozaki–Hiyama–Kishi reaction with vinyl iodide **266** to afford the natural product solandelactone E (**265**) as a 3.5:1 mixture with its epimer solandelactone F (see Fig. 5.152) [335].

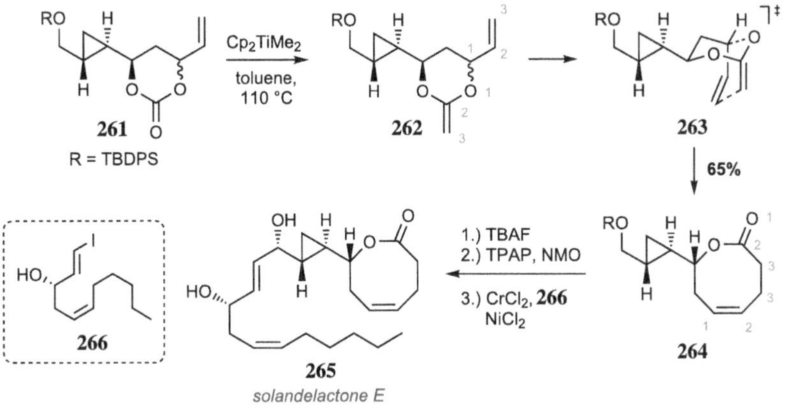

Fig. 5.152 White's synthesis of Solandelactone E [335]

Despite intensive efforts, there are hardly any examples of enantioselective, catalytic Cope rearrangements known to date. Not for lack of trying, the absence of coordinating functionalities in the 1,5-diene substrates required to ensure a strong interaction with the chiral catalysts simply constitutes a nearly unsurmountable challenge [336]. Furthermore, Cope rearrangements are potentially reversible, as the energy difference between starting material and product is significantly lower than in Claisen rearrangements. For this reason, asymmetric Cope rearrangements are exclusively encountered as part of a domino reaction in complex synthetic settings, in which a stereocenter is introduced enantioselectively prior to the sigmatropic rearrangement, thus enabling a *diastereoselective* reaction under substrate control.

Asymmetric Wittig rearrangements are almost entirely based on the use of stereogenic centers present in the substrate, either in the form of auxiliaries or as persistent stereogenic centers [325, 337]. A more recent work by Maezaki shows that at least for certain substrates, high enantioselectivities are possible under catalytic conditions using chiral ligands (see Fig. 5.153) [338].

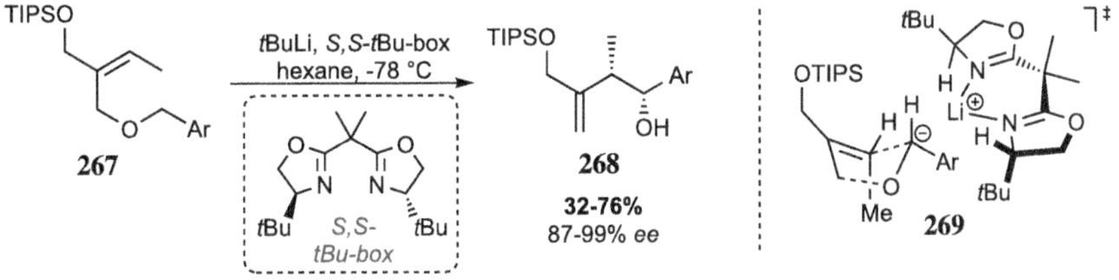

Fig. 5.153 Asymmetric [2,3]-Wittig rearrangement according to Maezaki and possible transition state **269** [338]

As early as 1999, Nakai reported that bisoxazoline (box) ligands allow a moderate enantiomeric excess of 65% [339]. Although no transition state was postulated, a presumed trajectory passing through **269** would predict the observed enantiomer **268** as the main product. The repulsion of the *t*Bu groups of the ligand with the methyl group of the alkene would be expected to constitute the dominant interaction in a purported transition state **269** enabling an enantioinduction.

5.4   Electrocyclic Reactions

An electrocyclic reaction is pericyclic rearrangement where the net result is one π-bond being converted into one σ-bond or *vice versa*. This typically refers to a cyclization of polyenes; however, ring-opening reactions of cyclopropane cations also fall under the regime of these pericyclic processes. The driving force usually relies on the energy difference of the breaking of a π-bond and its conversion into a σ-bond. For the cleavage of small rings, the release of ring strain can be similarly exploited. The term electrocyclization is often synonymously used for both the cyclization as well as the cleavage of a (hetero)cyclic substrate or intermediate; this reverse process is more correctly referred to as an electrocyclic ring opening. Electrocyclic reactions are classified based on the total number of involved π-electrons (see Fig. 5.154) [9].

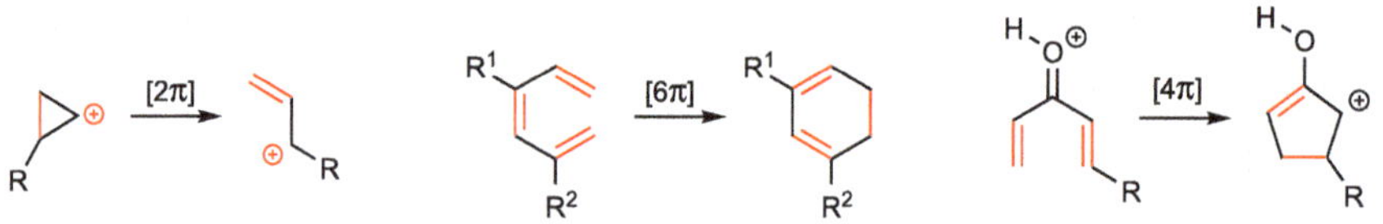

Fig. 5.154 Classification of electrocyclic reactions [9]

The processes are reversible, which accounts for the preferential formation of the thermodynamically most favorable product. As depicted in Fig. 5.154, ring openings of three-membered rings or the cyclization to cyclopentenes also fall under these pericyclic processes and are similarly governed by the conservation of their orbital symmetry. Well-known name reactions comprise the Nazarov cyclization, the Staudinger and the Favorskii reaction, which have been used to great effect in the syntheses of complex intermediates (see Fig. 5.155). Higher order electrocyclizations with involving 10π-atoms or more have also been reported [340].

5.4.1   Mechanism and Selectivity

As previously illustrated in Sect. 5.1, particularly ring-opening reactions of cyclobutenes constitute the prototypical electrocyclic reaction. These transformations proceed

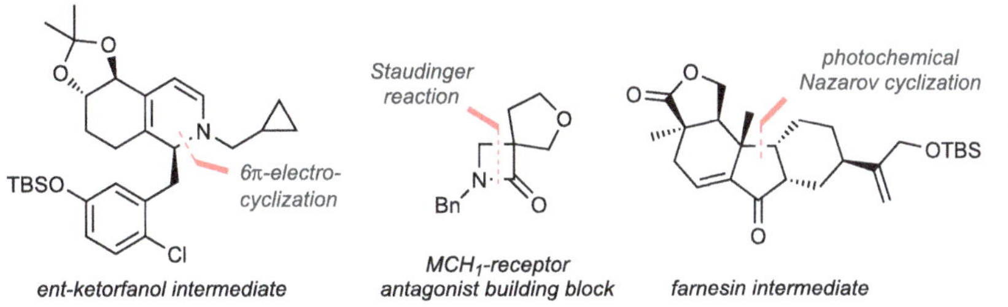

Fig. 5.155 Products of electrocyclic reactions in the synthesis of bioactive compounds [341–343]

stereospecifically and yield, depending on whether *cis-* or *trans*-substituted cyclobutenes are used as substrates, selectively the *Z,E-*, *Z,Z-*, or *E,E*-diene. An analysis of the relevant orbitals of the involved σ- and π-bond classifies this process as a 4π-ring opening (see Fig. 5.15). Apart from the energy gain of the σ/π-bond interconversion, the driving force is further bolstered by the release of ring strain upon opening of the rigid cyclobutene (see Fig. 5.156).

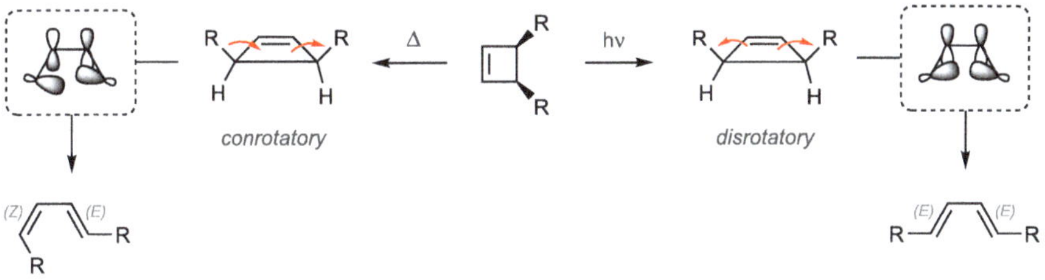

Fig. 5.156 Dependence of the ring-opening modes of cyclobutenes on the reaction conditions. See Fig. 5.15 for the involved orbitals

The mode of ring opening describes the sense of rotation at the reacting termini, which can either turn in the same (*conrotatory*) or opposite direction (*disrotatory*). It is determined by the individual orbital occupancy, which directly results from the employed reaction conditions (thermal *vs.* photochemical). Based on this principle, the sense of rotation can also be derived for other ring systems. As with pericyclic reactions in general, the number of valence electrons rather than the ring size constitutes the decisive factor governing the reaction topology and thus sense of rotation (see Fig. 5.157) [10].

The 6π-cyclization occurs under thermal conditions with suprafacial, under photochemical conditions with antarafacial topology, which is why **270** proceeds in a disrotatory and **272** in a conrotatory fashion to afford **271** or **273**, respectively. The electrocyclization of ketodienes (Nazarov cyclization) catalyzed by Lewis acids is a concerted [π4a] process under thermal conditions and shows a conrotatory sense of rotation in the formation of **275** from **274**. Apart from a ring-closing alkene metathesis, the cyclization of conjugated

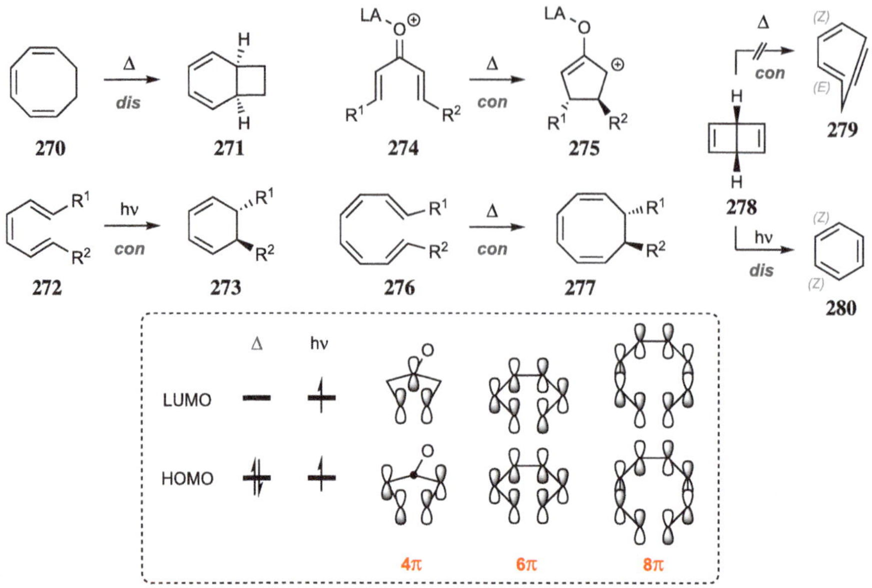

Fig. 5.157 Examples of electrocyclic reactions under photochemical and thermal reaction conditions and corresponding orbital diagrams

tetraenes is a common motif to access eight-membered polyenes in total synthesis. **276** thus affords the *anti*-substituted cyclotriene **277** under thermal conditions.

Even if the orbital symmetry in a given reaction would be conserved, this does not automatically result in the formation of the product, it merely refers to the occurrence of an *additional* symmetry-derived barrier for the transformation. The conversion of Dewar benzene (**278**) is a classic example: Under thermal conditions, a conrotatory ring opening is expected while a disrotatory sense of rotation would result under photochemical control. While benzene is quickly generated via photochemical irradiation, thermal conditions would afford **279** containing an *E*-double bond. The preferred thermal mode of ring opening cannot operate, as the resultant **279** is extraordinarily strained — despite the symmetry-allowed nature of the transformation. Instead, **278** shows an unexpectedly high thermal stability with a half-life of 2 days at room temperature [344]. The thermal reaction therefore proceeds via a non-concerted mechanism, possibly involving radicals [345].

5.4.2 Torquoselectivity

In a con- or disrotatory electrocyclic reaction, two pathways are possible: The direction of twisting at the reacting termini can either result in an inward or outward rotation of the substituents. In the ring opening of cyclobutenes, this would either afford an *E*- or *Z*-configured diene, respectively. Depending on the stereoelectronic properties of the functionalities located on the involved ring, a preferential direction of twisting in the breaking of σ-bonds in the transition state can be deduced from the isolated products (inwards vs. outwards, shortened to *in/out*). This phenomenon is referred to as *torquoselectivity* (see Fig. 5.158) [346].

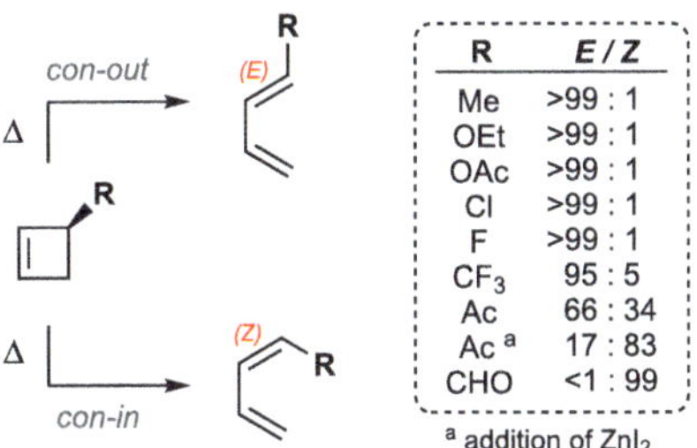

Fig. 5.158 Torquoselectivity of electrocyclic ring openings of monosubstituted cyclobutenes [346, 347]

Based on the observed preferences of different substituents, it can be easily concluded that electron-donating functional groups favor a *con-out* mode, while electron-withdrawing functionalities afford the *Z*-diene as product via a *con-in* transition state. The sense of rotation can be derived based on the dominant orbital interactions. A stabilization of the transition states can be achieved either by the introduction of electron donors or acceptors: Upon an inward rotation of the terminal C-R bond (R = acceptor) to form a *Z*-diene, a delocalization of electron density from the σ-bond of the cyclobutene to the substituent R occurs ($\sigma \rightarrow \pi_A^*/\sigma_A^*$), thus partially depleting the σ-bond and lowering the bond order. In an outward rotation, this interaction would not take place quite as efficiently, as only an interaction with the C-atom on which the R group is located is possible — the other terminal C-atom can hardly contribute to the interaction. In donor-substituted systems, an outward rotation is preferred, in which electron density from the substituent can populate the σ^*-orbital of the cyclobutene ($\pi_D/n_D/\sigma_D \rightarrow \sigma^*$), also effectively decreasing the bond order (see Fig. 5.159) [346]. An intensification of the electron-withdrawing properties of substituents, e.g., by addition of Lewis acids that coordinate to R, can additionally shift the product ratio in favor of an inward rotation of the substituent [347].

While this mnemonic describing the direction of rotation is relatively universal for all donors, an unfavorable steric repulsion during an inward rotation can energetically discourage this reaction path for weak acceptors. Thus, in the absence of further donors/acceptors, CF_3 substituents located at C1 preferentially form *E*-dienes. For monosubstituted reactants, almost all common acceptor substituents prefer to rotate outward, except for CHO and NO functionalities. In substrates bearing multiple substituents, these effects can either be oppos-

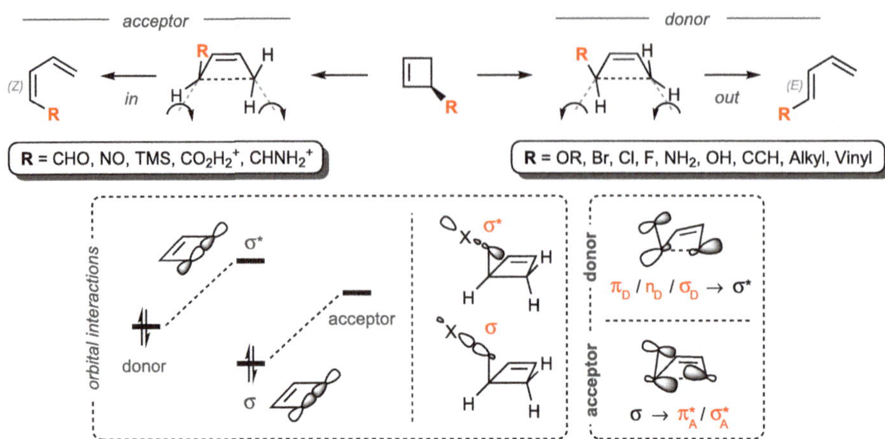

Fig. 5.159 Mechanism of torquoselectivity based on the dominant orbital interactions of 3-substituted cyclobutenes [346]

ing or reinforcing, whereby even the sterically highly hindered diene may become the major product (see Fig. 5.160) [346].

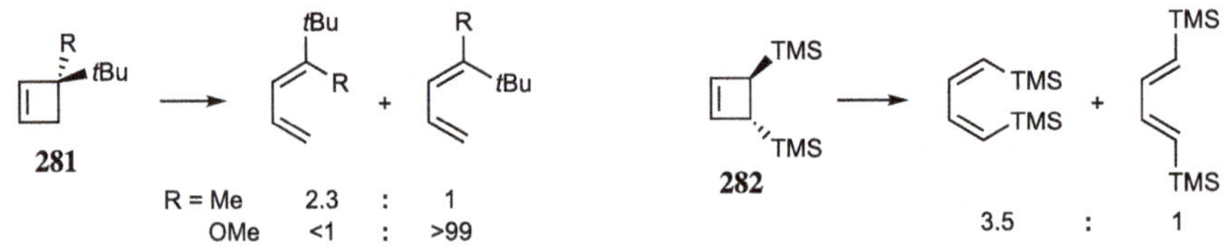

Fig. 5.160 Formation of sterically strained dienes from substituted cyclobutenes via electrocyclic ring openings [346, 348]

Alkoxy groups possess a high-lying HOMO that, due to their lone pairs, strongly favors an outward rotation through interaction with the σ^*-orbital. They are therefore good donors, which accounts for the inward rotation of the sterically demanding but only weakly donating *t*Butyl substituent in the geminally disubstituted cyclobutene **281**. Alkyl-bearing substrates, the functionality being only a weak σ-donor, often are governed by steric rather than electronic interactions, resulting in the expected selectivity of the methyl congener of **281** with a preferred *con-out* mode. The disilylcyclobutene **282** studied by Murakami even predominantly forms the *E,E*-diene. The TMS group is a good σ^*-acceptor with a low-lying LUMO due to the polarizable C-Si bond. As a consequence, electron density is delocalized from the σ-orbital of the C-C bond into the antibonding $\sigma^*_{C\text{-}Si}$-orbital and the transition state of a *con-in* mode is stabilized [348, 349].

Apart from cyclobutenes, other electrocyclic reactions also show torquoselectivity. In the cationic ring opening of cyclopropanes, a dependence between the relative reaction rate and the type of introduced substituents was established early on. It can be traced back to the stereoelectronic interactions in the respective transition states (see Fig. 5.161) [350].

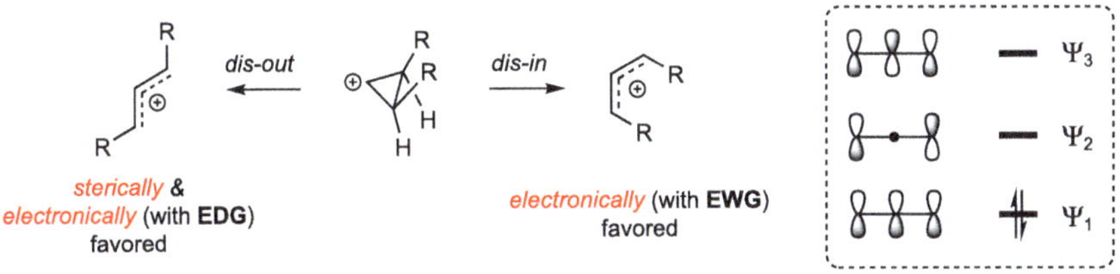

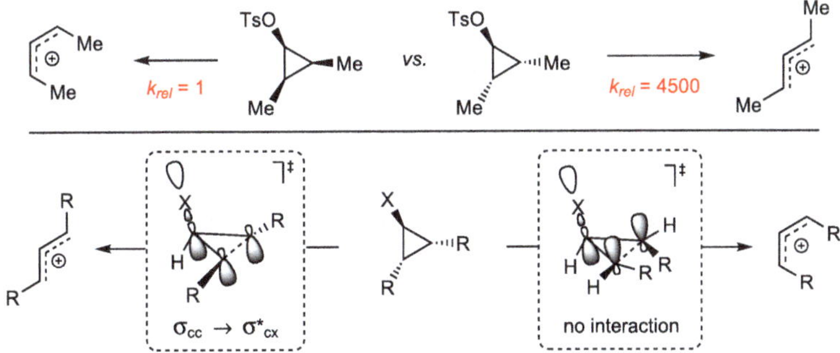

Fig. 5.161 Torquoselectivity in substituted cyclopropyl cations and orbitals of the allyl cation

Based on the orbitals of the parent allyl cation, cyclopropyl cations participate in a disrotatory ring opening under thermal conditions ($\pi 2\,$s). The two possible modes (*in* vs. *out*) would be oppositely stabilized by donors and acceptors according to the guidelines delineated above. For non-fused substrates, the *dis-out* mode would be preferred based on the steric repulsion of the substitutents during the transition states. If the two substituents are part of a ring fused with the cyclopropane, only the corresponding product resulting from a *dis-in* mode is observed, as the opposite mode would induce a prohibitively high ring strain.

The above assumptions only apply if the extrusion of the leaving group is concluded prior to the initiation of the cyclization. In fact, a significant influence of the relative stereochemistry of the leaving group on the reactivity of the substrates was observed. This indicates that its cleavage and the ring opening would proceed simultaneously rather than the extrusion preceding the electrocyclic reaction (see Fig. 5.162) [350].

Fig. 5.162 Relative reactivity of the opening of *cis,cis-* and *trans,trans-*tosylcyclopropanes and origin of the torquoselectivity of *trans,trans-*tosylcyclopropanes [350]

The electronic properties of the alkyl groups play only a subordinate role with regard to torquoselectivity in the depicted substrate (cf. Fig. 5.159). Instead, depending on the direction of disrotatory rotation, a $\sigma_{CC} \rightarrow \sigma^*_{CX}$ interaction can occur via the delocalization of electron density from the breaking CC bond into the leaving group. This would weaken the bond of the leaving group and consequently effect a faster bond breaking by stabilizing the respective transition state. With *trans,trans*-tosylcyclopropane, the sterically preferred *trans,trans*-allyl cation would need to be formed to experience such an acceleration. However, for *cis,cis*-tosylcyclopropanes, it is the corresponding *cis,cis*-product, which leads to a significant retardation of the transformation by destabilizing the transition state on steric grounds.

The reverse reaction, i.e., the formation of cyclopropanes from allyl derivatives, follows the same guidelines as the ring opening (cf. Fig. 5.161). In the *Favorskii rearrangement*, an α-haloketone is deprotonated and subsequently converted into a cyclopropanone with inversion of the configuration at the C-Cl center. Although the mechanism is often attributed to a S_N2 pathway, it was recently shown that the rearrangement proceeds in an electrocyclic fashion (see Fig. 5.163) [351].

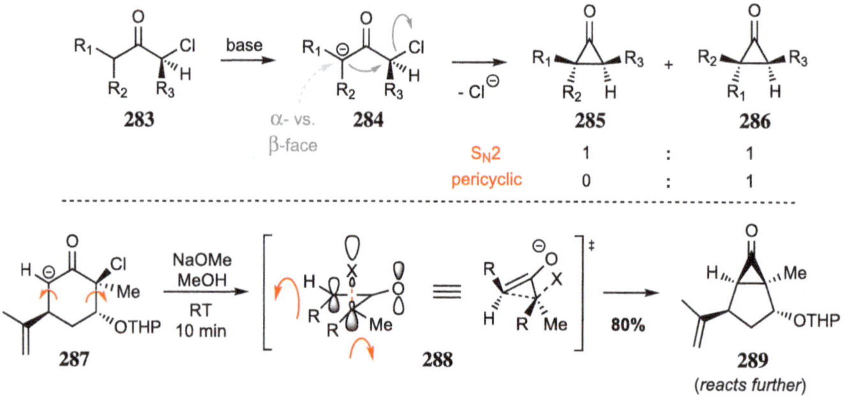

Fig. 5.163 General mechanism of the Favorskii rearrangement and expected product distribution in S_N2 versus electrocyclic reaction course. Favorskii rearrangement of **287** in the synthesis of (+)-iridomyrmecin by Lee [351, 352]

Following deprotonation of the substrate **283**, the ring closure can occur either via the α- or β-face of the enolate **284**. In a nucleophilic substitution, a product ratio (**285** *vs.* **286**) of 1:1 would result. In contrast, an electrocyclic transformation would only yield **286** via a disrotatory ring closure, as for symmetry reasons the α-side of the enolate must interact with the σ^*_{CCl} bond. In fact, only **286** was observed, strongly supporting the notion that the mechanism is pericyclic in nature [351].

The Lee group used the carvone-derived ketone **287** as substrate in their synthesis of (+)-iridomyrmecin, which exclusively generated **289** under basic conditions. The *dis-out*-mode via transition state **288** resulted in the formation of the fused 5/3-ring system based on the minimization of steric interactions. **289** was unstable under the reaction conditions and the cyclopropenone moiety was successively cleaved to afford the corresponding ester in 80% yield [352].

5.4.3 Variations of Electrocyclic Reactions

Electrocyclic reactions are primarily employed to selectively synthesize carbocycles bearing a complex substitution pattern. While this seems like a wondrous discovery by modern age chemistry, nature has been relying on this methodology for millions of years. Without a doubt, the most frequently performed variant of this class of pericyclic reactions is observed in the biochemical generation of vitamin D_3 proceeding in human skin cell tissue (Fig. 5.164) [353, 354].

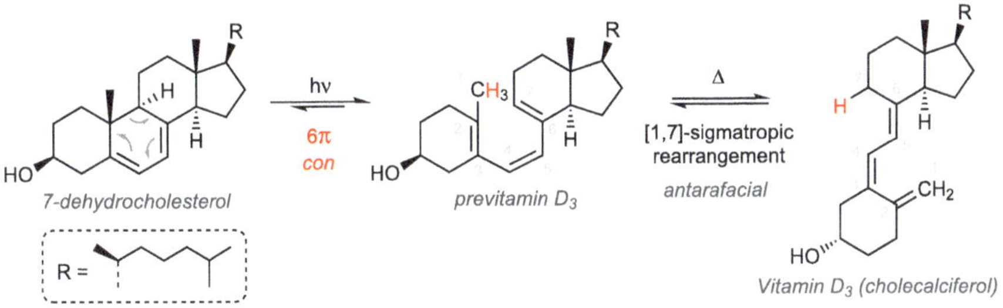

Fig. 5.164 Biosynthesis of vitamin D_3 through photochemically initiated electrocyclic ring opening and thermal [1,7]-sigmatropic hydrogen shift [353, 354]

In addition to the electrocyclic reactions discussed so far, such as the ring opening of cyclobutenes, there are a number of "non-classical" variants that may not appear to be located in the realm of pericyclic reactions at first glance. These include transformations such as the Staudinger-β-lactam synthesis or the Nazarov cyclization, apart from the already discussed Favorskii rearrangement and ring opening of substituted cyclopropanes.

In the *Nazarov cyclization*, pentadienyl cations are converted to cyclopentenones via oxoallyl intermediates. The 4π-electrocyclization proceeds with conrotatory sense of rotation under thermal conditions, the photochemical reaction carries a disrotatory topology. The reaction can be mechanistically divided into several discrete steps: initial generation of the pentadienyl cation **291**, conrotatory ring closure (under thermal conditions) to the oxoallyl cation **292**, formation of the enolate **293**, and concluding protonation of **293** to afford the cyclopentenone **294** (see Fig. 5.165) [355–357].

The generation of the pentadienyl cation **291** starting from **290** is not the only approach, there are several methods relying on alternative substrates [358]. The above sequence can

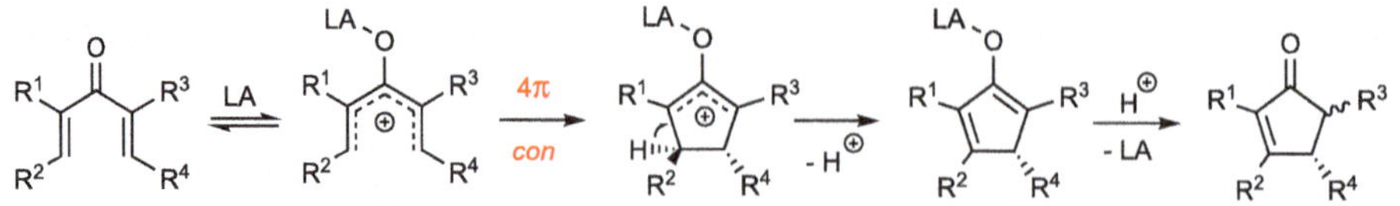

Fig. 5.165 Detailed mechanism of the Nazarov cyclization [355]

furthermore be incorporated in the framework of a domino reaction, either terminating with a Nazarov cyclization or the electrocyclization constituting the initiation of a cascade, usually starting from **292** (*vide infra*). The reaction's synthetic significance is limited by several challenges under classical conditions: (1.) the necessity of using strong Lewis acids, which curtails the tolerance towards many functional groups; in addition, stoichiometric amounts of Lewis acid are typically required; (2.) the elimination of the proton in **292** is not regioselective; and (3.) the protonation of the enolate is rarely diastereoselective. Many of these problems, however, have been (partially) solved or successfully circumvented in subsequent studies, with the regioselective elimination in **292** receiving a particular focus [355–357].

Based on the substitution pattern of the substrate, a stabilization or destabilization of the positive charge in **292** can lead to the preferential formation of a regioisomer of **293**. The introduction of silicon or fluorine substituents has been used to deliberately tread this alternative path. Their influence can ultimately be rendered "invisible" by their subsequent removal from the substrate in the final step through a potential elimination [359, 360]. Alternatively, selective eliminations can also be achieved through polarization of the substrate by substituents with opposing electronic properties and a chelating Lewis acid (see Fig. 5.166) [361].

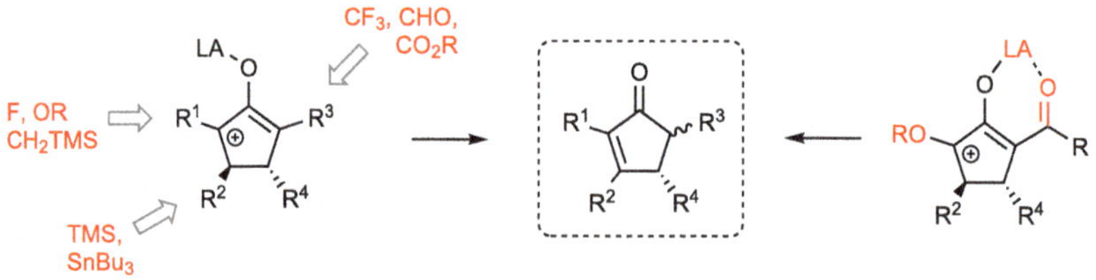

Fig. 5.166 Silyl- or fluorine-directed and polarized Nazarov cyclization. The color-marked substituents in the indicated position each stabilize the depicted regioisomer of the oxoallyl cation [359–361]

Aside from the usual straightforward cyclization, individual intermediates of the transformation can be intercepted within the framework of a domino reaction (see Fig. 5.167) [355–357, 362, 363]. The oxyallyl cation **292** is often further functionalized by a cycloaddition (dipolar [3+2]-cycloaddition in **295**), Friedel–Crafts reaction (**297**), or alkylation with an alkene (6-*endo-trig* with **298** or 5-*exo-trig* with **299**). Additionally, an *O*-alkylation can be realized, for example, as a concluding step of an alkene functionalization as in **297** (not shown).

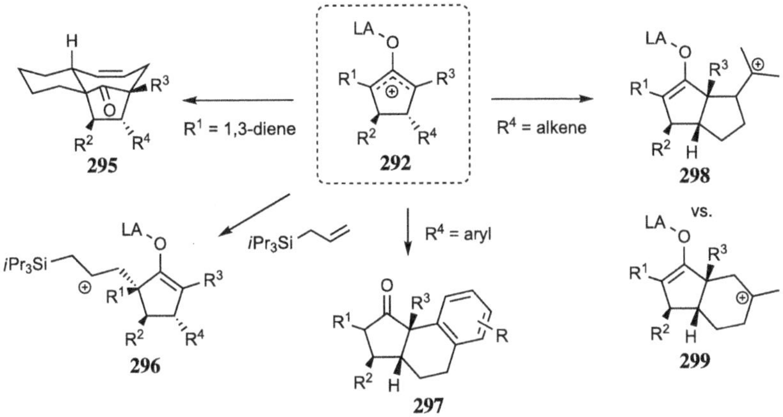

Fig. 5.167 Examples of interrupted Nazarov cyclizations [355]

Based on extensive work by the Denmark group, it was shown that substituents in the α-position (R^1, R^3) generally accelerate the reaction when they are electron-donating. Conversely, electron acceptors in β-position (R^2, R^4) increase the reaction rate. The necessity of using stoichiometric amounts of strong Lewis acids (e.g., BF_3, $SnCl_4$, $TiCl_4$, $AlCl_3$, etc.) is due to the high stability of **293**, which prevents efficient protonation under concomitant cleavage of the Lewis acid and thus the realization of a catalytic cycle. The use of highly reactive substrates could enable catalytic variants through careful choice of the α- and β-substituents. The spectrum of active catalysts has additionally been expanded significantly since the turn of the 21-st century [364].

All asymmetric methods of the Nazarov cyclization are based on controlling the direction of rotation during the thermal conrotatory ring closure. This is achieved either by the transfer or induction of a stereogenic center in the substrate, as well as the use of auxiliaries or chiral Lewis acids as stereoinductors. Despite some new promising approaches, such as the use of BINOL-derived organocatalysts, the most common method remains a reaction with Cu- or Sc-pybox/box salts [355, 356]. Unfortunately, to date there still exists no universally applicable, asymmetric method, which emphasizes the multitude challenges of this transformation [364].

Due to the ubiquitous occurrence of cyclopentanes in secondary metabolites, the Nazarov cyclization has been used as a key step in several total syntheses (see Fig. 5.168). Their incorporation was mainly facilitated by the development of highly regio- and stereoselective methods using milder Lewis acids [355, 364].

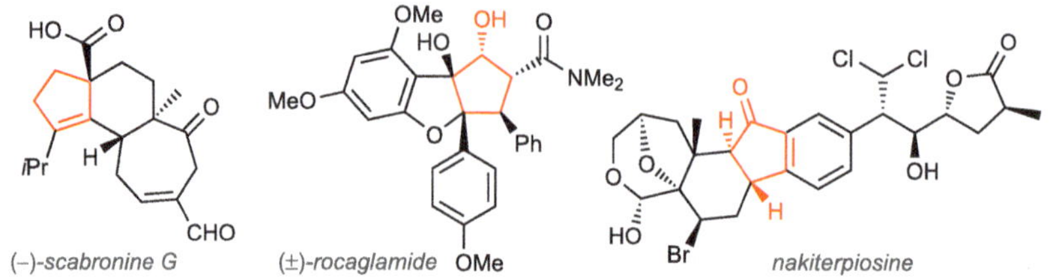

Fig. 5.168 Selection of natural products synthesized using a Nazarov cyclization [365–367]

The synthesis of β-lactams from ketenes and imines via a *Staudinger reaction* formally corresponds to a [2+2]-cycloaddition. However, it mechanistically proceeds via zwitterionic intermediates, which afford the corresponding lactam in an electrocyclization. The heterocyclic products are commonly isolated as a mixture of the 3/4-*syn*- or *anti*-product **303** (see Fig. 5.169) [368, 369].

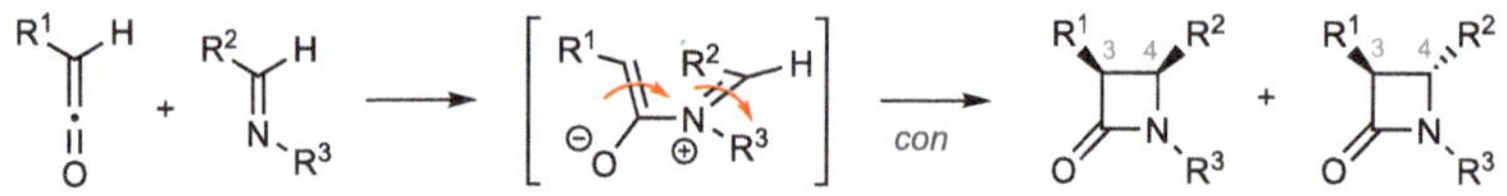

Fig. 5.169 Mechanism of the synthesis of β-lactams from ketenes and imines [368]

An experimental mechanistic investigation through direct detection of the intermediates is complicated by the fact that ketenes are usually only generated *in situ* and that reactions are very rapid by virtue of their high reactivity. Despite these challenges, it is now considered scientific consensus that the majority of the reactants follow the mechanism outlined in Fig. 5.169. Acyclic imines of the structure **301** usually afford the *syn*-isomer as major product [369].

The dependence of the stereoselectivity on the structure of the reactants was systematically studied [370, 371]. Based on these investigations, *trans*-imines (**301**) are expected to stereospecifically yield *syn*-configured lactams whereas *cis*-imines (e.g., cyclic imines) yield the respective *anti*-products (see Fig. 5.170) [370].

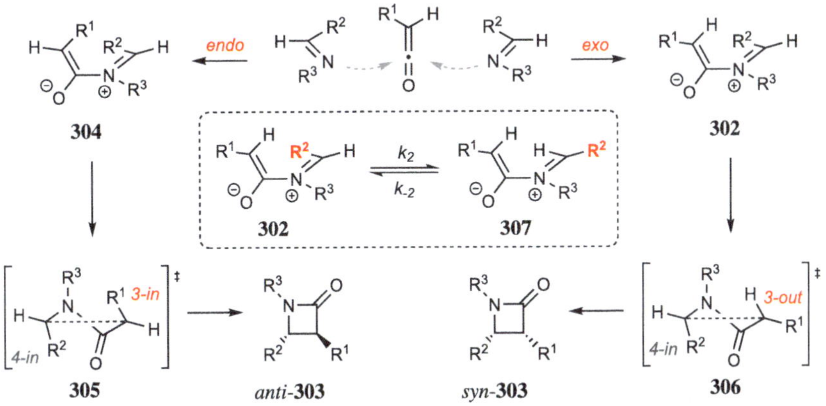

Fig. 5.170 Origin of the stereoselectivity in the Staudinger synthesis [370, 371]

The two possible stereoisomers can be rationalized by the approach of the imine either from the direction of the substituent R^1 (*endo*-attack) or from the opposite face (*exo*-attack). The two isomers **304** and **302** undergo a conrotatory ring closure to the corresponding products *anti*-**303** and *syn*-**303**. Apart from the sterically preferred *exo*-attack, electronic reasons similarly favor *syn*-β-lactams: Based on torquoselectivity, donor substituents on the ketene promote a *3-out*-mode (**306**). This can be deduced from the observation that the selectivity reverses from *syn* to *anti* upon change of the electronic nature of the substrates from electron-rich to electron-poor. Furthermore there are additional factors facilitating a deviation from the expected *syn*-selectivity: It is assumed that after an *exo*-attack, **302** can isomerize to the sterically preferred intermediate **307**, which affords *anti*-**303**, respectively. The *syn*-selectivity usually decreases upon raising of the reaction temperature, which can be attributed to an increased isomerization rate of the zwitterionic intermediate **302**, **307**. Electron-withdrawing substituents on the ketene and electron-donating functionalities on the imine increase the lifetime of the zwitterionic intermediate and thus promote formation of the thermodynamically more stable *anti*-**303** via isomerization. In addition, if the imine substituent R^3 only possesses small steric demand, the thermodynamic equilibration is further accelerated, while large groups both slow down the isomerization and furnish a sterically much less stable intermediate **307**. Overall, the selectivity can thus be regulated by the handles shown in Fig. 5.171 [370].

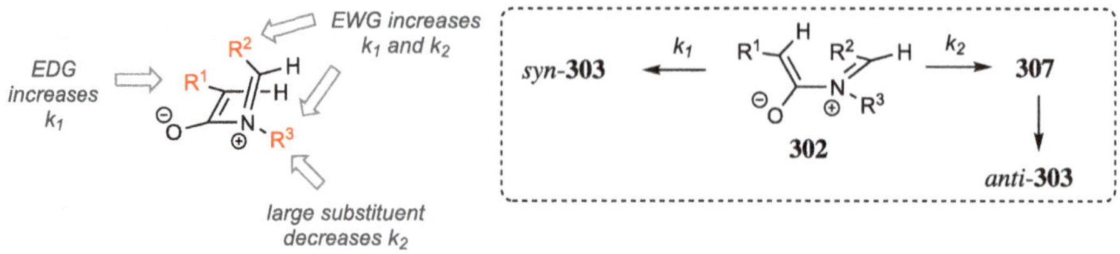

Fig. 5.171 Control factors of diastereoselectivity in Staudinger synthesis [370]

Controlling the direction of rotation during ring closure does not determine the relative, but the absolute configuration of R^1 and R^2 in the final product. The torquoselectivity can be used in conjunction with chiral substrates (ketenes, imines) to obtain only one enantiomer or diastereomer [369, 372]. The first catalytic enantioselective synthesis of lactams was disclosed by Lectka in 2000, wherein ketenes and imines were reacted in the presence of chiral amines [373]. Further catalysts, both in chiral and achiral form, have been developed and applied successfully in the Staudinger reaction. Interestingly, the enantioselective, catalytic variant does not rely on a torquoselective step, unlike other diastereoselective or enantioselective electrocyclizations. Instead, an alternative, non-concerted path is pursued, which can proceed via cycle A or B depending on the nucleophile in question. The general catalytic cycle shown in Fig. 5.172 also applies to the transformation in the presence of an achiral, nucleophilic catalyst [368, 369, 374].

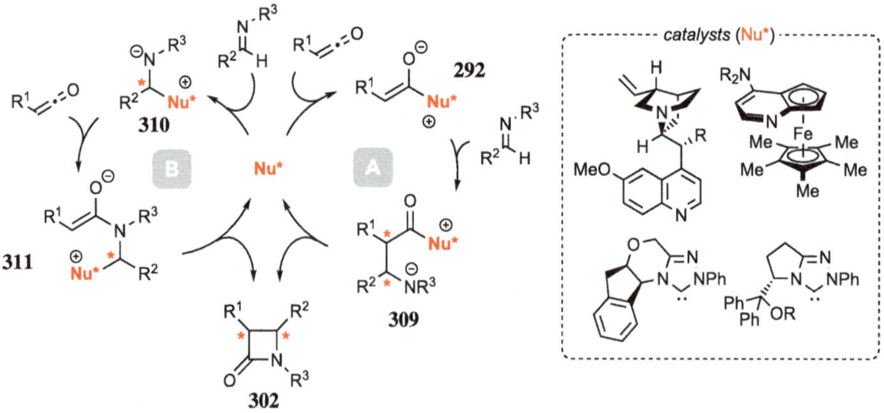

Fig. 5.172 Postulated mechanisms of the enantioselective, catalytic Staudinger reaction and common catalysts [368, 369, 374]

Electrocyclic reactions are particularly well suited for the formation or opening of cyclic (poly-)enes when the driving force is aided either by the release of ring strain or by the removal of the respective product of a ring opening through a subsequent reaction. Accordingly, this qualifies pericyclic processes for incorporation into a domino reaction. In Boger's synthesis of selected tropoisoquinolines, a succession of several pericyclic reactions was used to construct the fused tropone core of granditropone, grandirubrine, imerubrine, and isoimerubrine. Following a [4+2] cycloaddition of **312** with the protected propenone, **313** was first obtained. Subsequent addition of HCl afforded diene **314** in a retro-Diels-Alder reaction with concomitant loss of CO_2. The concluding disrotatory ring opening yielded the tropolone **315**. The secondary metabolite grandirubrine was then isolated after two further steps (see Fig. 5.173) [375].

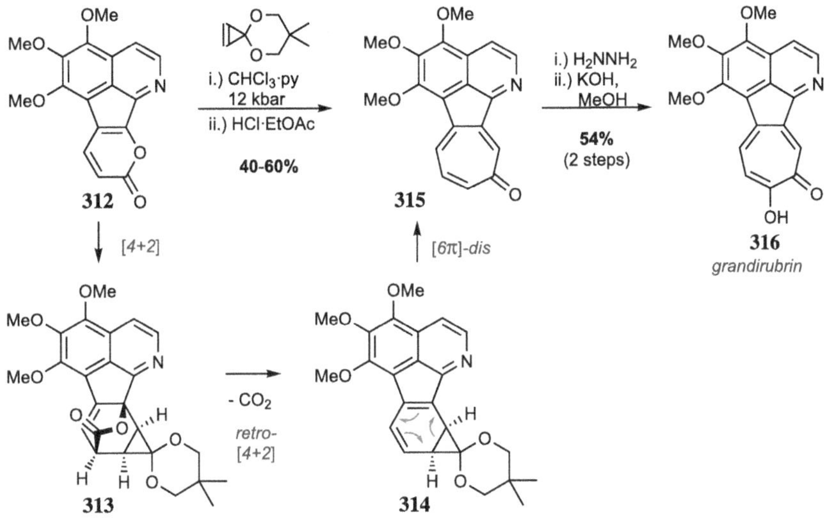

Fig. 5.173 Boger's synthesis of grandirubrine [375]

Intimate knowledge of a biosynthetic route to a natural product can occasionally be used to great effect by employing a biomimetic reaction sequence for the synthesis of the desired molecule. Thus, Nicolaou and co-workers relied on the biosynthetic mechanism of endiandric acid, which had previously been postulated by Black [376], to rapidly assemble the corresponding (±)-methyl ester. Starting from the acyclic precursor **317**, **318** was received after an initial Lindlar reduction to the polyene **319**, which was followed by an impressive, double electrocyclization/Diels-Alder sequence. The commencing reduction of both alkyne units exposed the skipped hexaene backbone. The first 8π-conrotatory electrocyclization ensued, wherein the purported transition state **320** places the two appendant residues in *anti*-position to yield **321**. This was followed by the second (6π-disrotatory) electrocyclization to forge the 6/4-fused ring framework (**323**). The concluding Diels-Alder reaction can proceed either via an *exo-E-anti-* (**324**) or *endo-E-syn*-transition state (**325**). However, due to the short CH_2 bridge to the diene only **324** could be realized, which resulted in the exclusive formation of the desired product **318** (see Fig. 5.174) [377].

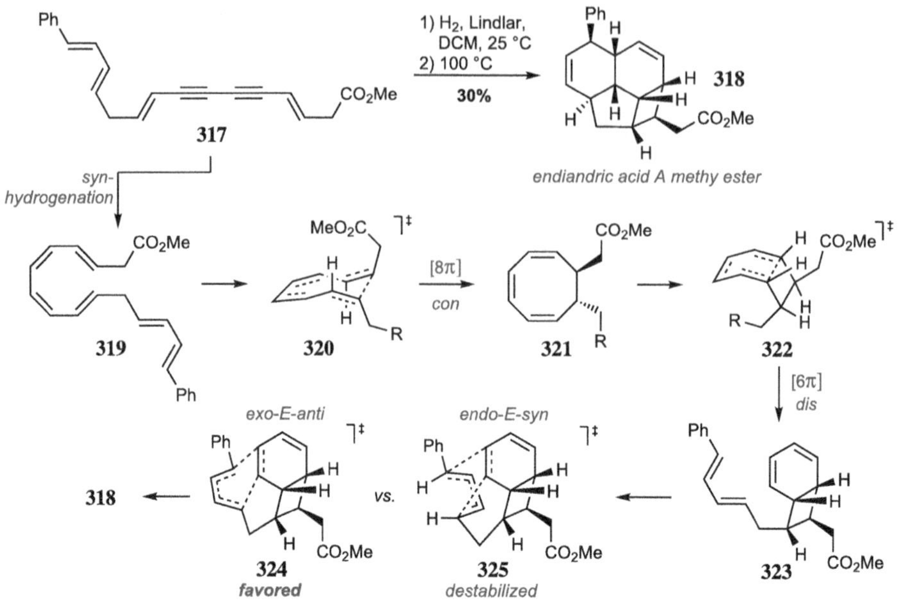

Fig. 5.174 Synthesis of the ($\pm$)-endiandric acid A methyl ester by Nicolaou *et al.* [377]

A similar sequence was exploited by the Trauner group in their approach to the fungal lipid PF-1018. Unlike the racemic endiandric acids, PF-1018 was isolated as an enantiomerically pure compound. As the single remote stereocenter in the proposed biosynthetic precursor is unlikely to govern the diastereoselectivity of the initial 8π electrocyclization, this pericyclic reaction may biosynthetically proceed as a separate first step, wherein a chiral enzyme environment may effect the enantioinduction. Synthetically, the sequence was initiated by a Cu-cocatalyzed Stille cross-coupling of **326** and **327**, after which the tetraene **328** first underwent an 8π-conrotatory cyclization to the cyclooctatriene **329**. The periselectivity of the conjugated system favored a final Diels-Alder cycloaddition over an alternative 6π-electrocyclization, whereby the desired advanced intermediate **330** could be isolated in 42% yield. A theoretical investigation of the sequence revealed that the sterically demanding TBS protecting group at C2 destabilized the transition state of the competing 6π-cyclization relative to the [4+2]-cycloaddition. The preceding, reversible 8π-cyclization afforded several products (not shown) that could interconvert and thus constitute a pre-equilibrium for the concluding cycloaddition (Curtin–Hammett control). The overall enantioinduction was provided by the configuration of the stereocenter at C3 (see Fig. 5.175) [378].

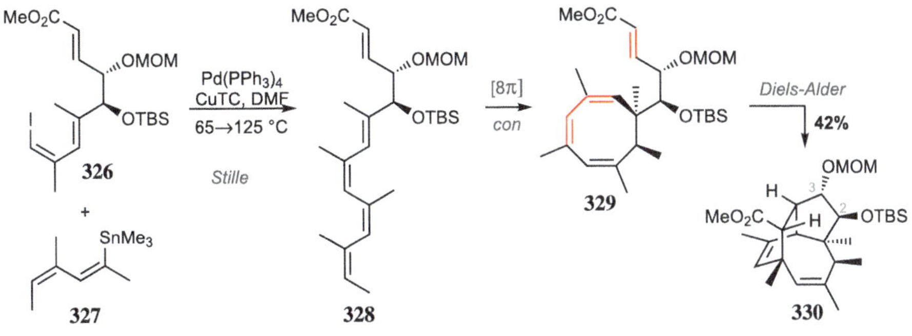

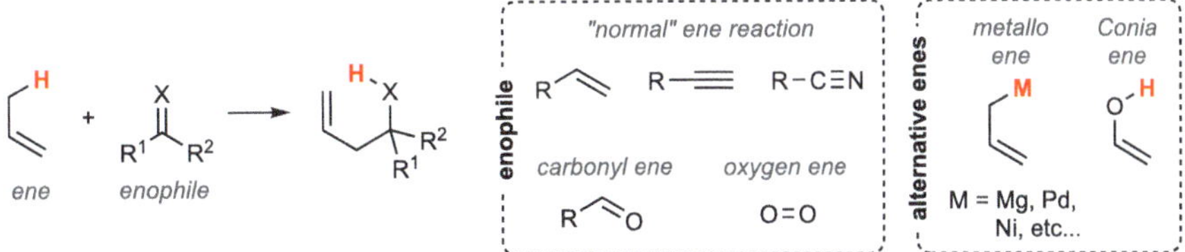

Fig. 5.175 Key sequence in Trauner's approach to PF-1018 [378]

5.5 Group Transfer Reactions

The final category of pericyclic reactions covered herein are group transfer reactions. They encompass the reaction of a (poly-)ene with a second substrate, in the course of which, however, no carbo- or heterocycle is formed. The most commonly used group transfer reactions are ene reactions and the reductions of alkenes/alkynes by diimide.

5.5.1 Ene Reactions

Ene reactions involve the mostly intermolecular conversion of allyl derivatives (so-called *enes*) with C-X multiple bonds (*enophiles*), such as alkenes, carbonyl compounds, and singlet oxygen (see Fig. 5.176) [379].

Fig. 5.176 General reaction scheme of ene reactions and classification of subcategories

Subclasses of the ene reaction comprise carbonyl ene [380, 381], metallo-ene [382, 383], Conia-ene, oxygen-ene[384], as well as retro-ene reactions [385], among others. Especially the carbonyl-ene reaction represents the most frequently used variant as it lends itself for an atom-economical access to homoallylic alcohols.

5.5.1.1 Mechanism and Selectivity

The ene reaction employs the same substrates as enophiles that similarly display a high reactivity in cycloadditions. The low activation barrier in both processes accounts for the ene reaction being also a common side reaction of Diels-Alder reactions. In analogy to [4+2]-cycloadditions, ene reactions proceed via a concerted, six-membered transition state. The comparatively higher activation barrier, however, necessitates significantly increased reaction temperatures, which limits the overall substrate scope [386]. As with other pericyclic reactions, a modification of the orbital energy levels can yield a substantial acceleration of the process, e.g., by using activated substrates or by employing Lewis acids as additives.

The $HOMO_{ene}$-$LUMO_{enophile}$ interaction constitutes the dominant orbital interaction. It accounts for the observed decrease in activation energy if electron-withdrawing substituents are introduced on the enophile and electron donors on the ene. An intramolecular reaction is also favored for entropic reasons due to the proximity of their reactive units, which is why especially intramolecular carbonyl-ene reactions of enals are a common motif (see Fig. 5.177) [380, 381].

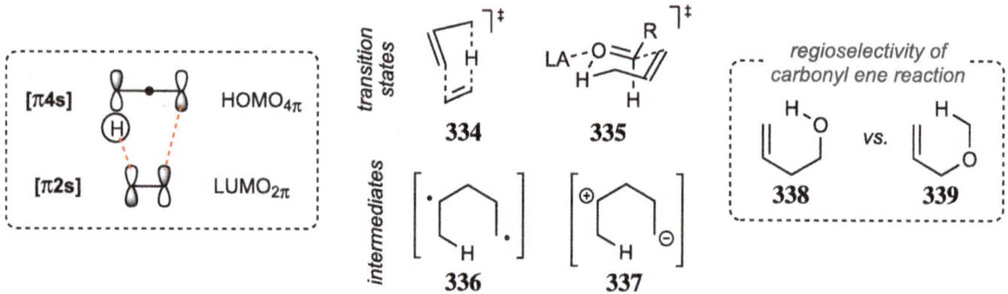

Fig. 5.177 Dominant orbital interactions, transition states of the ene reaction, and intermediates of a stepwise mechanism [387]

In the presence of Lewis acids, a stepwise mechanism with radical (**336**) or even zwitterionic intermediates (**337**) can be preferred instead of a concerted pathway. Depending on the chosen reaction conditions, a smooth transition between both cases can occur, which also has repercussions on the regio- and stereoselectivity of the rearrangement. Thermal ene reactions are often concerted and show early, envelope-like transition states (**334**). When Lewis acid additives are employed, late, chair-like transition states often occur (e.g., **335**), in which discrete intermediates with localized charges can be encountered. Radical intermediates are a recurring theme in thermal processes that do not allow a concerted pathway for steric reasons, for example, in strained cyclic substrates [387].

1,1-disubstituted, electron-rich allylic systems are the most reactive ene components on electronic grounds, followed by mono- or 1,2-disubstituted alkenes. In Lewis acid-assisted ene reactions, the steric environment of the coordinated additive also plays a major role. The enophile should be electrophilic, either by resorting to intrinsically highly polarized substrates (e.g., carbonyls, nitriles) or by introduction of electron-withdrawing substitutents. Olefins are relatively inert as enophiles, whereas alkyne derivatives are significantly more reactive. Under increased pressure, alkynes react with alkenes to form 1,4-dienes. Unsymmetrical substrates can afford different regioisomers. If a carbonyl compound is used as an enophile, the corresponding homoallylic alcohol **338** constitutes the major product instead of the respective ether **339**. The carbonyl O-atom preferentially reacts with the allylic hydrogen atom.

Common Lewis acids are depicted in Fig. 5.178. Initially, aluminum salts were routinely chosen as Lewis acids, yet often required stoichiometric amounts. They have been replaced by more efficient transition metal catalysts over time, with the subsequent establishment of organocatalytic catalysts as well. The use of a Lewis acid requires the presence of basic coordination sites to effect a covalent interaction, which renders heteroatom-containing substrates particularly suitable [380, 388].

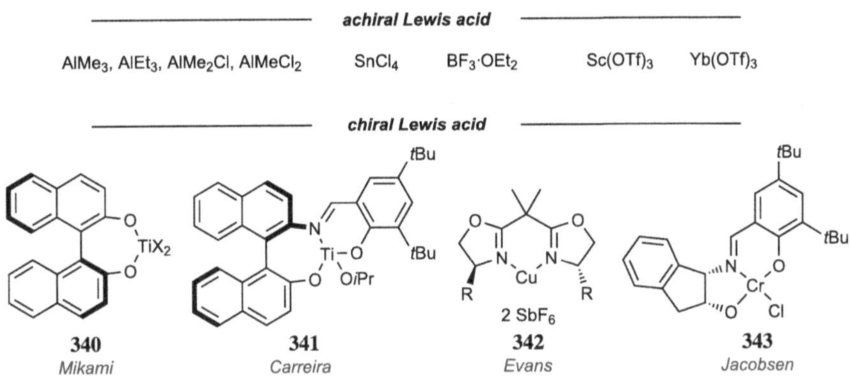

Fig. 5.178 Common Lewis acids used in the carbonyl-ene reaction [380, 388]

Under thermal conditions, the steric accessibility of the multiple bond and the allylic H-atom of the ene is a critical parameter [387]: As a consequence, the reaction rate usually decreases in the order $H_{primary} > H_{secondary} > H_{tertiary}$, irrespective of the thermodynamic stability of the product. Depending on the ene/enophile combination in question, the more highly substituted product can also be obtained, for example, with azodicarboxylate as an enophile. Lewis acid-catalyzed reactions typically afford the more highly substituted homoallylic alcohol as major product. This can be understood by the build-up of a partial charge in the transition state or the occurrence of ionic intermediates, as these charges are better stabilized in the respective regioisomeric transition states/intermediates (see Fig. 5.179).

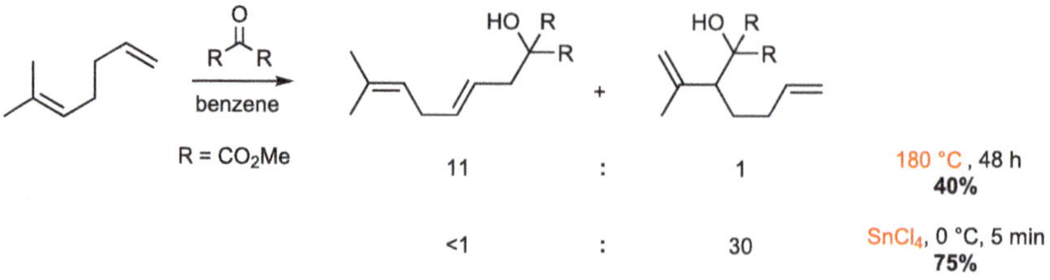

Fig. 5.179 Regioselectivity in the thermal *versus* Lewis acid-catalyzed ene reaction [389]

Enantioselective ene reactions were first comprehensively studied by Mikami, whose titanium-BINOL system **340** still remains to this day one of the most common reagents for asymmetric stereoinduction in carbonyl-ene reactions. Newer methods based on the transition metal complexes **341-343** shown in Fig. 5.178 were introduced by the research groups of Carreira, Evans, and Jacobsen. They are characterized by an increased Lewis acidity compared to the Mikami system, which is why less reactive substrates, such as aldehydes or ketones like methyl pyruvate, can be converted to the desired products under mild reaction conditions with high yields [380, 388].

5.5.1.2 Applications of the Ene Reaction

Laulimalide is a marine natural product possessing antimitotic properties that are conveyed via microtubule stabilization of the mitotic spindle apparatus. Mulzer's improved total synthesis of the macrolide relied on an asymmetric ene reaction to construct the C15 stereocenter on the right-hand fragment (see Fig. 5.180) [390, 391].

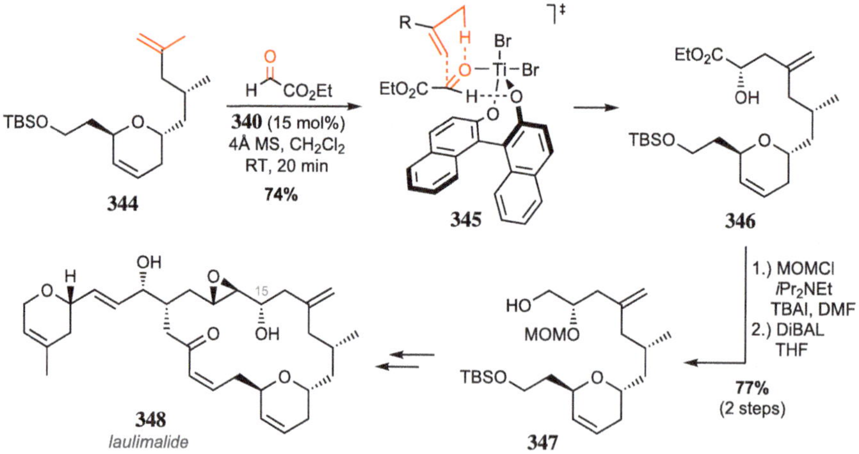

Fig. 5.180 Formal total synthesis of laulimalide by Mulzer [390, 391]

The reaction of the terminal alkene with ethyl glyoxylate in the presence of catalytic amounts of *R*-**340** afforded the desired alcohol **346** with excellent diastereoselectivity of

>95% ds. The purported envelope transition state **345** was postulated by Corey for the carbonyl-ene reaction using the Mikami system **340** [392]. Intermediate **347** is a key fragment of the first total synthesis by the Fürstner group, which can be converted into the natural product in 10 further steps, thus completing the formal total synthesis.

L-Menthol is one of the most important flavor compounds with an estimated global demand of >35 kt/a. BASF developed a large-scale synthesis starting from neral, the Z-configured citral isomer [393–395]. The key steps encompass the asymmetric hydrogenation of neral (**349**), followed by an ene reaction of citronellal (**350**) to form isopulegol (**353**), which can finally be reduced to menthol. The asymmetric reduction of neral (**349**) using a Rh/Chiraphos catalyst described in the patent was continuously operated for 19 days and furnished the desired product. Despite the multitude of catalysts and ligands screened, the stereoinduction could not be brought to exceed 87–89% *ee* [393]. In the second step, L-isopulegol (**353**) was produced in a diastereoselective ene reaction from citronellal (**350**). The sterically demanding bisphenol **351** forms the presumably di- or oligomeric Lewis acid as active catalyst with AlEt3. The addition of trifluoroacetone suppressed the formation of unwanted by-products, especially the dimerized ester citronellyl citronellate. The relative stereochemistry in **353** follows from the arrangement of the substituents in the transition state **352**, via which L-isopulegol (**353**) was obtained with high selectivity in >96% yield and >98% *de* (see Fig. 5.181).

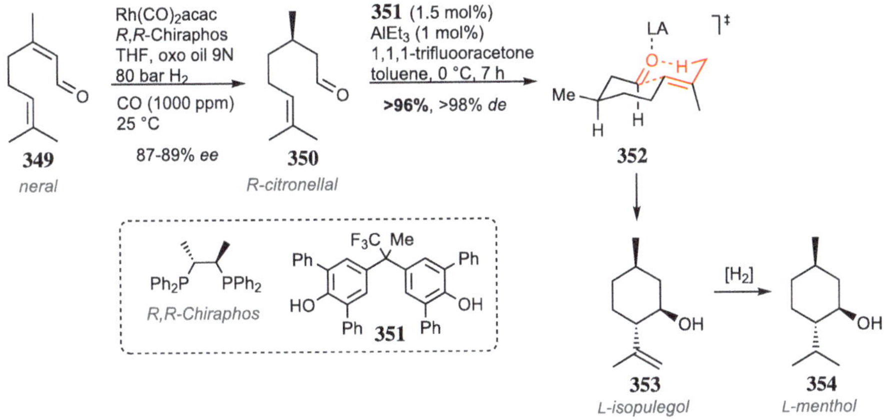

Fig. 5.181 BASF's menthol synthesis starting from neral (**349**) [393–395]

In their synthetic approach to the class of spongian diterpenes, the common intermediate **357** was exploited by Liang and co-workers to obtain the secondary metabolites cheloviolene C, seconorrisolide B, and seconorrisolide C. Starting from **355**, the domino sequence first saw the extrusion of cyclopentadiene in a retro-Diels-Alder reaction to expose the reactive enophile in **356**. Following the stereospecific ene reaction, the common intermediate **357** could be isolated in 84% yield, and was subsequently used as a platform in their approach to various diterpenes (see Fig. 5.182) [396].

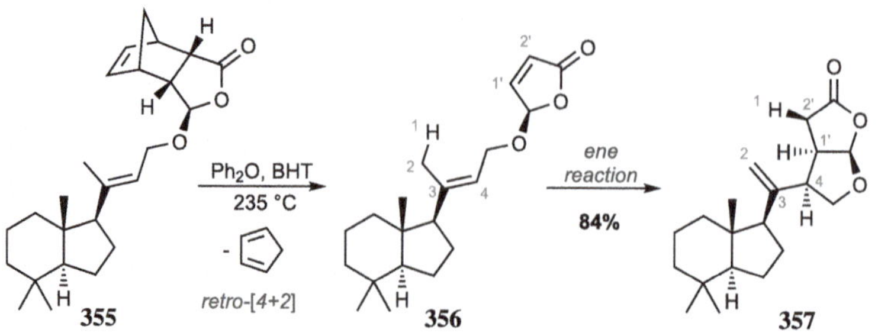

Fig. 5.182 Synthesis of spongian diterpenes according to Liang *et al.* [396]

5.5.2 Diimide Reductions

Diimides are no longer encountered as frequently in the current synthetic literature for the reduction of CC-multiple bonds. This can be largely attributed to the highly labile nature of the diimide and the availability of alternative, similarly mild reduction methods. With certain motifs or functionalities, it is however still in use, such as in the preparation of Z-vinyl halides [397]. With diimide, strongly polarized bonds (e.g., nitrile, nitro, imine, disulfide, sulfoxide functionalities, C-halogen bonds) remain fully intact. Isomerizations of the double bond or a hydrogenolysis also do not occur due to the pericyclic mechanism. Diimide is extremely unstable and is generated *in situ* from suitable precursors (see Fig. 5.183) [398–400].

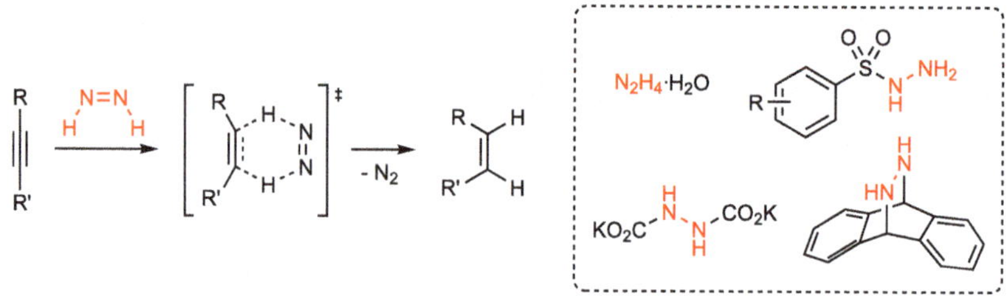

Fig. 5.183 Mechanism of diimide reduction and common diimide precursors [398–400]

From the large body of published reactions, the following guidelines could be derived [399]: *trans*-olefins are hydrogenated about 3–10 times faster than corresponding *cis*-olefins. The reaction is very sensitive to the steric environment of the starting material. Thus, the reaction rate decreases rapidly with an increasing degree of substitution of an olefin. The main side reaction is the disproportionation of the diimide into dinitrogen and hydrazine, which is why superstoichiometric amounts of diimide are required. The intermittent alkenes can usually be isolated only in low yield, as the reduction of alkynes and alkenes shows comparable reaction rates. An exception to this rule of thumb is the reduction of 1-haloalkynes, which can be selectively halted at the olefinic stage.

The Trost group used a diimide reduction in their synthesis of fostriecin (**360**) to construct the respective Z-alkene from a silyl alkyne. The cytotoxic natural product features a *cis*-double bond, the construction of which proved surprisingly resilient to realize. Several methods were attempted to convert the silylated alkyne **358** to the Z-alkene. All other screened approaches led to over-reduction or very low turnovers. Diimide, which was generated from arylsulfonyl hydrazine **361** in a thermal decomposition, was able to provide the desired vinyl silane **359** with moderate yield, without affecting the α, β-unsaturated lactone, the isolated double bond, or the silyl protecting groups under the slightly basic conditions. The endgame included a Hiyama cross-coupling of the vinyl silane and the phosphorylation of the secondary hydroxy function (see Fig. 5.184) [401].

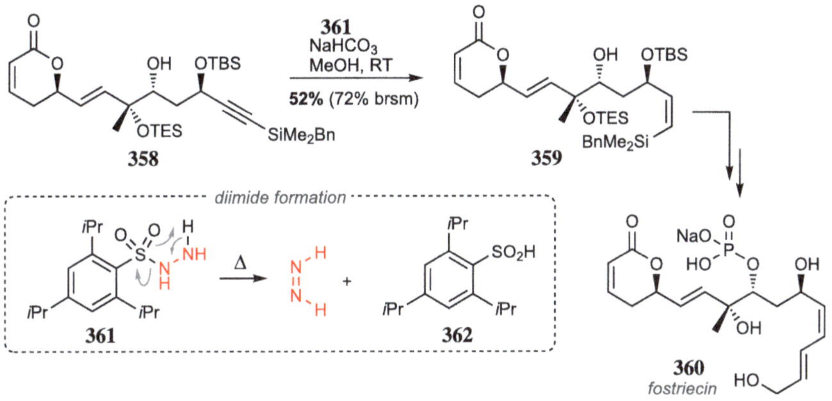

Fig. 5.184 Trost's synthesis of fostriecin [401]

One advantage of the reduction utilizing hydrazine lies in its atom economy, as the only by-product is N_2, making it potentially suitable for a large-scale process. This similarly constitutes the greatest challenge of a thermal runaway reaction with concomitant gas release that can be very difficult to control at scale. A successful example of an industrial application can be found in the synthesis of the antimalarial endoperoxide drug artemisinine (see Fig. 5.185) [402].

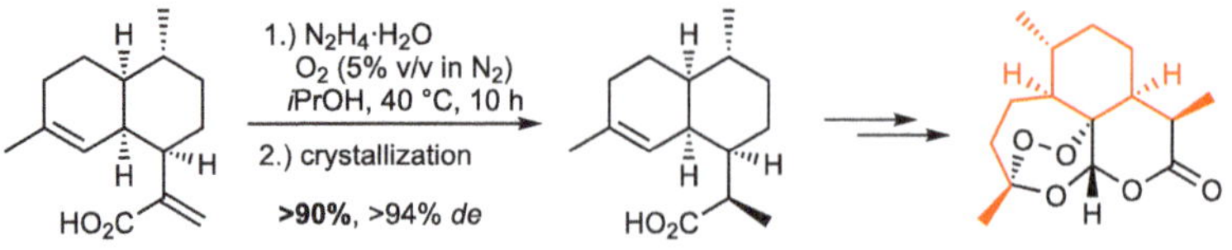

Fig. 5.185 Sanofi's synthesis of artemisinine [402]

Diimide was generated through oxidation of hydrazine with dilute oxygen and the reaction could be monitored by *in situ* infrared spectroscopy. After the highly diastereoselective reduction of artemisinic acid **363** to dihydroartemisinic acid **364**, the product of the multi-kilogram batch could be isolated in high purity by crystallization at pH = 6-7. **364** was subsequently converted to the active ingredient **365** in three further steps.

References

1. D. J. Mergott, S. A. Frank, W. R. Roush, *Proc. Nat. Acad. Sci.* **2004**, *101*, 11955–11959
2. W. von E. Doering, W. R. Roth, *Tetrahedron* **1962**, *18*, 67–74
3. J. I. Seeman, *J. Org. Chem.* **2015**, *80*, 11632–11671
4. A. D. McNaught, A. Wilkinson (Eds.), *IUPAC. Compendium of Chemical Terminology*, 2. Edition ("Gold Book"), Blackwell Scientific Publications, Oxford, **1997**.
5. K. N. Houk, Y. Li, J. D. Evanseck, *Angew. Chem. Int. Ed. Engl.* **1992**, *31*, 682–708
6. J. Sauer, R. Sustmann, *Angew. Chem. Int. Ed. Engl.* **1980**, *19*, 779–807
7. P. Vermeeren, T. A. Hamlin, F. M. Bickelhaupt, *Phys. Chem. Chem. Phys.* **2021**, *23*, 20095–20106
8. E. Havinga, J. L. M. A. Schlatmann, *Tetrahedron* **1961**, *16*, 146–152
9. R. B. Woodward, R. Hoffmann, *Angew. Chem. Int. Ed. Engl.* **1969**, *8*, 781–853
10. K. Fukui, *Acc. Chem. Res.* **1971**, *4*, 57–64
11. R. Hoffmann, *Angew. Chem. Int. Ed.* **2004**, *43*, 6586–6590
12. E. J. Corey, *J. Org. Chem.* **2004**, *69*, 2917–2919
13. K. N. Houk, F. Liu, Z. Yang, J. I. Seeman, *Angew. Chem. Int. Ed.* **2021**, *60*, 12660–12681
14. R. Huisgen, *Angew. Chem. Int. Ed. Engl.* **1963**, *2*, 633–645
15. R. B. Woodward, R. Hoffmann, *J. Am. Chem. Soc.* **1965**, *87*, 395–397
16. T. Sekikawa, N. Saito, Y. Kurimoto, N. Ishii, T. Mizuno, T. Kanai, J. Itatani, K. Saita, T. Taketsugu, *Phys. Chem. Chem. Phys.* **2023**, *25*, 8497–8506
17. Q. Zhou, G. Kukier, I. Gordiy, R. Hoffmann, J. I. Seeman, K. N. Houk, *J. Org. Chem.* **2024**, *89*, 1018–1034
18. I. Tuvi-Arad, D. Avnir, *Chem. Eur. J.* **2012**, *18*, 10014–10020
19. H. C. Longuet-Higgins, E. W. Abrahamson, *J. Am. Chem. Soc.* **1965**, *87*, 2045–2046
20. W. Kauzmann, *Quantum Chemistry*, Academic Press, **1957**
21. G. Klopman, R. F. Hudson, *Theoret. Chim. Acta* **1967**, *8*, 165–174
22. G. Klopman, *J. Am. Chem. Soc.* **1968**, *90*, 223–235
23. L. Salem, *J. Am. Chem. Soc.* **1968**, *90*, 543–552
24. L. Salem, *J. Am. Chem. Soc.* **1968**, *90*, 553–566

25. R. Sustmann, R. Schubert, *Tetrahedron Lett.* **1972**, *13*, 2739–2742

26. R. Sustmann, H. Trill, *Tetrahedron Lett.* **1972**, *13*, 4271–4274

27. G. Desimoni, P. P. Righetti, E. Selva, G. Tacconi, V. Riganti, M. Specchiarello, *Tetrahedron* **1977**, *33*, 2829–2836

28. N. D. Epiotis, *J. Am. Chem. Soc.* **1972**, *94*, 1924–1934

29. K. N. Houk, *Acc. Chem. Res.* **1975**, *8*, 361–369

30. D. H. Ess, G. O. Jones, K. N. Houk, *Adv. Synth. Catal.* **2006**, *348*, 2337–2361

31. N. D. Epiotis, *Angew. Chem. Int. Ed. Engl.* **1974**, *13*, 751–780

32. P. W. Ayers, C. Morell, F. D. Proft, P. Geerlings, *Chem. Eur. J.* **2007**, *13*, 8240–8247

33. P. Geerlings, P. W. Ayers, A. Toro-Labbé, P. K. Chattaraj, F. D. Proft, *Acc. Chem. Res.* **2012**, *45*, 683–695

34. K. C. Nicolaou, S. A. Snyder, T. Montagnon, G. Vassilikogiannakis, *Angew. Chem. Int. Ed.* **2002**, *41*, 1668–1698

35. M. Juhl, D. Tanner, *Chem. Soc. Rev.* **2009**, *38*, 2983–2992

36. M. M. Heravi, V. F. Vavsari, *RSC Adv.* **2015**, *5*, 50890–50912

37. J.-A. Funel, S. Abele, *Angew. Chem. Int. Ed.* **2013**, *52*, 3822–3863

38. T. Bach, J. P. Hehn, *Angew. Chem. Int. Ed.* **2011**, *50*, 1005–1045

39. T. Suzuki, S. Watanabe, W. Ikeda, S. Kobayashi, K. Tanino, *J. Org. Chem.* **2021**, *86*, 15597–15605

40. T. Tsujimoto, J. Ishihara, M. Horie, A. Murai, *Synlett* **2002**, 399–402

41. T. Hirai, K. Shibata, Y. Niwano, M. Shiozaki, Y. Hashimoto, N. Morita, S. Ban, O. Tamura, *Org. Lett.* **2017**, *19*, 6320–6323

42. C. Yuan, B. Du, L. Yang, B. Liu, *J. Am. Chem. Soc.* **2013**, *135*, 9291–9294

43. S. Poplata, A. Tröster, Y.-Q. Zou, T. Bach, *Chem. Rev.* **2016**, *116*, 9748–9815

44. M. Breugst, H.-U. Reissig, *Angew. Chem. Int. Ed.* **2020**, *59*, 12293–12307

45. D. McLeod, M. K. Thøgersen, N. I. Jessen, K. A. Jørgensen, C. S. Jamieson, X.-S. Xue, K. N. Houk, F. Liu, R. Hoffmann, *Acc. Chem. Res.* **2019**, *52*, 3488–3501

46. R. B. Woodward, F. E. Bader, H. Bickel, A. J. Frey, R. W. Kierstead, *Tetrahedron* **1958**, *2*, 1–57

47. V. Mark, *J. Org. Chem.* **1974**, *39*, 3179–3181

48. J. Sauer, *Angew. Chem. Int. Ed.* **1967**, *6*, 16–33

49. D. Craigh, J. J. Shipman, R. B. Fowler, *J. Am. Chem. Soc.* **1961**, *83*, 2885–2891

50. J. Sauer, D. Lang, A. Mielert, *Angew. Chem. Int. Ed.* **1962**, *1*, 268–269

51. K. N. Houk, *J. Am. Chem. Soc.* **1973**, *95*, 4092–4094

52. R. Huisgen, R. Schug, *J. Am. Chem. Soc.* **1976**, *98*, 7819–7821

53. K. N. Houk, L. L. Munchausen, *J. Am. Chem. Soc.* **1976**, *98*, 937–946

54. K. N. Houk, J. Sims, C. R. Watts, L. J. Luskus, *J. Am. Chem. Soc.* **1973**, *95*, 7301–7315

55. R. Sustmann, *Tetrahedron Lett.* **1971**, *12*, 2717–2720

56. K. N. Houk, J. Sims, R. E. Duke, R. W. Strozier, J. K. George, *J. Am. Chem. Soc.* **1973**, *95*, 7287–7301

57. M. N. Paddon-Row, *Aust. J. Chem.* **1974**, *27*, 299–313

58. J. B. Thomas, J. R. Waas, M. Harmata, D. A. Singleton, *J. Am. Chem. Soc.* **2008**, *130*, 14544–14555

59. W. K. Johnson, *J. Org. Chem.* **1959**, *24*, 864–865

60. B. R. Ussing, C. Hang, D. A. Singleton, *J. Am. Chem. Soc.* **2006**, *128*, 7594–7607

61. Z. Yang, X. Dong, Y. Yu, P. Yu, Y. Li, C. Jamieson, K. N. Houk, *J. Am. Chem. Soc.* **2018**, *140*, 3061–3067

62. E. Marsault, A. Toró, P. Nowak, P. Deslongchamps, *Tetrahedron* **2001**, *57*, 4243–4260

63. K. Takao, R. Munakata, K. Tadano, *Chem. Rev.* **2005**, *105*, 4779–4807

64. T. Kametani, S. Hibino, *Adv- Heterocycl. Chem.* **1987**, *42*, 245–333

65. R. R. Schmidt, *Acc. Chem. Res.* **1986**, *19*, 250–259

66. M. Movassaghi, M. Tjandra, J. Qi, *J. Am. Chem. Soc.* **2009**, *131*, 9648–9650

67. B. R. Bear, S. M. Sparks, K. J. Shea, *Angew. Chem. Int. Ed.* **2001**, *40*, 820–849

68. A. R. Katritzky, N. Dennis, *Chem. Rev.* **1989**, *89*, 827–861

69. K. Afarinkia, V. Vinader, T. D. Nelson, G. H. Posner, *Tetrahedron* **1992**, *48*, 9111–9171

70. S. Cossu, F. Fabris, O. De Lucchi, *Synlett* **1997**, 1327–1334

71. C. O. Kappe, S. S. Murphree, A. Padwa, *Tetrahedron* **1997**, *53*, 14179–14233

72. A. Kumar, *Chem. Rev.* **2001**, *101*, 1–19

73. D. Craig, *Chem. Soc. Rev.* **1987**, *16*, 187–239

74. G. Mehta, R. Uma, *Acc. Chem. Res.* **2000**, *33*, 278–286

75. S. Harada, A. Nishida, *Asian J. Org. Chem.* **2019**, *8*, 732–745

76. S. Danishefsky, *Acc. Chem. Res.* **1981**, *14*, 400–406

77. M. Petrzilka, J. I. Grayson, *Synthesis* **1981**, 753–786

78. V. K. Aggarwal, A. Ali, M. P. Coogan, *Tetrahedron* **1999**, *55*, 293–312

79. J. G. Martin, J. K. Hill, *Chem. Rev.* **1961**, *61*, 537–562

80. J. R. L. Smith, R. O. C. Norman, M. R. Stillings, *Tetrahedron* **1978**, *34*, 1381–1383

81. W. J. Lording, T. Fallon, M. S. Sherburn, M. N. Paddon-Row, *Chem. Sci.* **2020**, *11*, 11915–11926

82. J. I. García, J. A. Mayoral, L. Salvatella, *Acc. Chem. Res.* **2000**, *33*, 658–664

83. C. S. Wannere, A. Paul, R. Herges, K. N. Houk, H. F. Schaeffer III., P. v. R. Schleyer, *J. Comp. Chem.* **2007**, *28*, 344–361

84. A. Arrieta, F. P. Cossío, B. Lecea, *J. Org. Chem.* **2001**, *66*, 6178–6180

85. I. Fernández, F. M. Bickelhaupt, *J. Comp. Chem.* **2013**, *35*, 371–376

86. N. H. Werstiuk, W. Sokol, *Can. J. Chem.* **2008**, *86*, 737–744

87. J. I. García, J. A. Mayoral, L. Salvatella, *Eur. J. Org. Chem.* **2005**, 85–90

88. M. Ramirez, D. Svatunek, F. Liu, N. K. Garg, K. N. Houk, *Angew. Chem. Int. Ed.* **2021**, *60*, 14989–14997

89. L. F. Tietze, H. Geissler, J. Fennen, T. Brumby, S. Brand, G. Schulz, *J. Org. Chem.* **1994**, *59*, 182–191

90. S. M. Weinreb, R. R. Staib, *Tetrahedron* **1982**, *38*, 3087–3128

91. T. Ali, K. K. Chauhan, C. G. Frost, *Tetrahedron Lett.* **1999**, *40*, 5621–5624

92. G. Desimoni, G. Tacconi, *Chem. Rev.* **1975**, *75*, 651–692

93. D. L. Boger, *Tetrahedron* **1983**, *39*, 2869–2939

94. S. M. Weinreb, P. M. Scola, *Chem. Rev.* **1989**, *89*, 1525–1534

95. M. Behforouz, M. Ahmadian, *Tetrahedron* **2000**, *56*, 5259–5288

96. P. Buonora, J.-C. Olsen, T. Oh, *Tetrahedron* **2001**, *57*, 6099–6138

97. N. Saracoglu, *Tetrahedron* **2007**, *63*, 4199–4236

98. J. Rapp, R. Huisgen, *Tetrahedron* **1997**, *53*, 961–970

99. R. K. Bansal, K. Karaghiosoff, N. Gupta, N. Gandhia, S. K. Kumawat, *Tetrahedron* **2005**, *61*, 10521–10528

100. K. N. Houk, R. W. Strozier, *J. Am. Chem. Soc.* **1973**, *95*, 4094–4096

101. R. B. Ruggeri, M. M. Hansen, C. H. Heathcock, *J. Am. Chem. Soc.* **1988**, *110*, 8734–8736

102. C. H. Heathcock, M. M. Hansen, R. B. Ruggeri, J. C. Kath, *J. Org. Chem.* **1992**, *57*, 2544–2553

103. A. Ichihara, *Synthesis* **1987**, 207–222

104. M.-C. Lasne, J.-L. Ripoll, *Synthesis* **1985**, 121–143

105. A. J. H. Klunder, J. Zhu, B. Zwanenburg, *Chem. Rev.* **1999**, *99*, 1163–1190

106. L. N. Mander, R. J. Thomson, *Org. Lett.* **2003**, *5*, 1321–1324

107. M. Y. Chu-Moyer, S. J. Danishefsky, *J. Am. Chem. Soc.* **1992**, *114*, 8333–8334

108. P. A. Jacobi, C. A. Blum, R. W. DeSimone, U. E. S. Udodong, *Tetrahedron Lett.* **1989**, *30*, 7173–7176

109. K. Matsumoto, H. Hamana, H. Iida, *Helv. Chim. Acta* **2005**, *88*, 2033–2234
110. J. R. McCabe, C. A. Eckert, *Acc. Chem. Res.* **1974**, *7*, 251–257
111. M. Morita, K. Ohmori, K. Suzuki, *Org. Lett.* **2015**, *17*, 5634–5637
112. L. F. Tietze, T. Hübsch, C. Ott, G. Kuchta, M. Buback, *Liebigs Ann.* **1995**, 1–7
113. C. L. Hugelshofer, T. Magauer, *Synthesis* **2014**, *46*, 1279–1296
114. M. A. Kienzler, S. Suseno, D. Trauner, *J. Am. Chem. Soc.* **2008**, *130*, 8604–8605
115. M. G. Banwell, A. J. Edwards, G. J. Harfoot, K. A. Jolliffe, *Tetrahedron* **2004**, *60*, 535–547
116. A. B. Smith, III, N. J. Liverton, N. J. Hrib, H. Sivaramakrishnan, K. Winzenberg, *J. Am. Chem. Soc.* **1986**, *108*, 3040–3048
117. U. Pindur, G. Lutz, C. Otto, *Chem. Rev.* **1993**, *93*, 741–761
118. J. Sauer, J. Kredel, *Angew. Chem.* **1965**, *77*, 1037
119. T. Inukai, T. Kojima, *J. Org. Chem.* **1966**, *31*, 1121–1123
120. P. Vermeeren, T. A. Hamlin, I. Fernndez, F. M. Bickelhaupt, *Angew. Chem. Int. Ed.* **2020**, *59*, 6201–6206
121. D. Nogue, R. Paugam, L. Wartski, *Tetrahedron Lett.* **1992**, *35*, 1265–1258
122. T. Motozaki, K. Sawamura, A. Suzuki, K. Yoshida, T. Ueki, A. Ohara, R. Munakata, K. Takao, K. Tadano, *Org. Lett.* **2005**, *7*, 2261–2264
123. T. Motozaki, K. Sawamura, A. Suzuki, K. Yoshida, T. Ueki, A. Ohara, R. Munakata, K. Takao, K. Tadano, *Org. Lett.* **2005**, *7*, 2265–2267
124. W. Oppolzer, *Angew. Chem. Int. Ed. Engl.* **1984**, *23*, 876–889
125. K. Narasaka, *Synthesis* **1991**, 1–11
126. E. J. Corey, *Angew. Chem. Int. Ed.* **2002**, *41*, 1650–1667
127. J. S. Johnson, D. A. Evans, *Acc. Chem. Res.* **2000**, *33*, 325–335
128. M. Miyashita, M. Sasaki, I. Hattori, M. Sakai, K. Tanino, *Science* **2004**, *305*, 495–499
129. S. Shambayati, W. E. Crowe, S. L. Schreiber, *Angew. Chem. Int. Ed. Engl.* **1990**, *29*, 256–272
130. D. A. Evans, K. T. Chapman, D. T. Hung, A. T. Kawaguchi, *Angew. Chem. Int. Ed. Engl.* **1987**, *26*, 1184–1186
131. T. Poll, J. O. Metter, G. Helmchen, *Angew. Chem. Int. Ed. Engl.* **1985**, *24*, 112–113
132. E. J. Corey, H. E. Ensley, *J. Am. Chem. Soc.* **1975**, *97*, 6908–6909
133. D. A. Evans, W. C. Black, *J. Am. Chem. Soc.* **1993**, *115*, 4497–4513
134. S. Hashimoto, N. Komeshima, K. Koga, *J. Chem. Soc. Chem. Commun.* **1979**, 437–438
135. K. Furuta, Y. Miwa, K. Iwanaga, H. Yamamoto, *J. Am. Chem. Soc.* **1988**, *110*, 6254–6255
136. K. Narasaka, N. Iwasawa, M. Inoue, T. Yamada, M. Nakashima, J. Sugimori, *J. Am. Chem. Soc.* **1989**, *111*, 5340–5345
137. E. J. Corey, R. Imwinkelried, S. Pikul, Y. B. Xiang, *J. Am. Chem. Soc.* **1989**, *111*, 5493–5495
138. J. M. Hawkins, S. Loren, *J. Am. Chem. Soc.* **1991**, *113*, 7794–7795
139. E. J. Corey, T. P. Loh, *J. Am. Chem. Soc.* **1991**, *113*, 8966–8967
140. D. A. Evans, S. J. Miller, T. Lectka, *J. Am. Chem. Soc.* **1993**, *115*, 6460–6461
141. D. A. Evans, S. J. Miller, T. Lectka, P. von Matt, *J. Am. Chem. Soc.* **1999**, *121*, 7559–7573
142. D. A. Evans, D. M. Barnes, J. S. Johnson, T. Lectka, P. von Matt, S. J. Miller, J. A. Murry, R. D. Norcross, E. A. Shaughnessy, K. R. Campos, *J. Am. Chem. Soc.* **1999**, *121*, 7582–7594
143. E. J. Corey, T. Shibata, T. W. Lee, *J. Am. Chem. Soc.* **2002**, *124*, 3808–3809
144. D. H. Ryu, T. W. Lee, E. J. Corey, *J. Am. Chem. Soc.* **2002**, *124*, 9992–9993
145. D. Liu, E. Canales, E. J. Corey, *J. Am. Chem. Soc.* **2007**, *129*, 1498–1499
146. E. J. Corey, S. Sarshar, J. Bordner, *J. Am. Chem. Soc.* **1992**, *114*, 7938–7939
147. E. J. Corey, T.-P. Loh, T. D. Roper, M. D. Azimioara, M. C. Noe, *J. Am. Chem. Soc.* **1992**, *114*, 8290–8291
148. E. J. Corey, *Angew. Chem. Int. Ed.* **2009**, *48*, 2100–2117
149. S. A. Snyder, E. J. Corey, *J. Am. Chem. Soc.* **2006**, *128*, 740–742

150. S. B. Herzon, N. A. Calandra, S. M. King, *Angew. Chem. Int. Ed.* **2011**, *50*, 8863–8866

151. K.-Y. Wang, D.-D. Liu, T.-W. Sun, Y. Lu, S.-L. Zhang, Y.-H. Li, Y.-X. Han, H.-Y. Liu, C. Peng, Q.-Y. Wang, J.-H. Chen, Z. Yang, *J. Org. Chem.* **2018**, *83*, 6907–6923

152. K. Kong, Z. Moussa, C. Lee, D. Romo, *J. Am. Chem. Soc.* **2011**, *133*, 19844–19856

153. H. Pellissier, *Tetrahedron* **2009**, *65*, 2839–2877

154. K. Maruoka, T. Itoh, T. Shirasaka, H. Yamamoto, *J. Am. Chem. Soc.* **1988**, *110*, 310–312

155. A. G. Dossetter, T. F. Jamison, E. N. Jacobsen, *Angew. Chem. Int. Ed.* **1999**, *38*, 2398–2400

156. K. Gademann, D. E. Chavez, E. N. Jacobsen, *Angew. Chem. Int. Ed.* **2002**, *41*, 3059–3061

157. C. F. Thompson, T. F. Jamison, E. N. Jacobsen, *J. Am. Chem. Soc.* **2001**, *123*, 9975–9983

158. D. E. Chavez, E. N. Jacobsen, *Angew. Chem. Int. Ed.* **2001**, *40*, 3667–3670

159. P. Liu, E. N. Jacobsen, *J. Am. Chem. Soc.* **2001**, *123*, 10772–10773

160. M. S. Taylor, E. N. Jacobsen, *Proc. Nat. Acad. Sci.* **2004**, *101*, 5368–5373

161. I. Louis, N. L. Hungerford, E. J. Humphries, M. D. McLeod, *Org. Lett.* **2006**, *8*, 1117–1120

162. U. Majumder, J. M. Cox, H. W. B. Johnson, J. D. Rainier, *Chem. Eur. J.* **2006**, *12*, 1747–1753

163. T. Gatzenmeier, M. Turberg, D. Yepes, Y. Xie, F. Neese, G. Bistoni, B. List, *J. Am. Chem. Soc.* **2018**, *140*, 12671–12676

164. A. Padwa, *Tetrahedron* **2011**, *67*, 8057–8072

165. I. Coldham, R. Hufton, *Chem. Rev.* **2005**, *105*, 2765–2809

166. K. V. Gothelf, K. A. Jørgensen, *Chem. Rev.* **1998**, *98*, 863–909

167. A. I. Kotyatkina, V. N. Zhabinsky, V. A. Khripach, *Russ. Chem. Rev.* **2001**, *70*, 641–753

168. V. Naira, T. D. Suja, *Tetrahedron* **2007**, *63*, 12247–12275

169. X. Hou, J. Zhu, B.-C. Chen, S. H. Watterson, W. J. Pitts, A. J. Dyckman, P. H. Carter, A. Mathur, H. Zhang, *Org. Process Res. Dev.* **2016**, *20*, 989–995

170. R. Huisgen, *Angew. Chem. Int. Ed. Engl.* **1963**, *2*, 565–598

171. M. Ríos-Gutiérrez, L. R. Domingo, *Eur. J. Org. Chem.* **2019**, 267–282

172. M. Regitz, *Synthesis* **1972**, 351–373

173. T. Ye, M. A. McKervey, *Chem. Rev.* **1994**, *94*, 1091–1160

174. S. Kanemasa, *Synlett* **2002**, 1371–1387

175. C. H. Kim, K. P. Jang, S. Y. Choi, Y. K. Chung, E. Lee, *Angew. Chem. Int. Ed.* **2008**, *47*, 4009–4011

176. K. C. Nicolaou, A. Li, D. J. Edmonds, *Angew. Chem. Int. Ed.* **2006**, *45*, 7086–7090

177. S. E. Denmark, E. A. Martinborough, *J. Am. Chem. Soc.* **1999**, *121*, 3046–3056

178. G. Bianchi, C. D. Micheli, R. Gandolfi, *Angew. Chem. Int. Ed. Engl.* **1979**, *18*, 721–738

179. H. Ishikawa, D. A. Colby, S. Seto, P. Va, A. Tam, H. Kakei, T. J. Rayl, I. Hwang, D. L. Boger, *J. Am. Chem. Soc.* **2009**, *131*, 4904–4916

180. E. L. Campbell, A. M. Zuhl, C. M. Liu, D. L. Boger, *J. Am. Chem. Soc.* **2010**, *132*, 3009–3012

181. T. Hashimoto, K. Maruoka, *Chem. Rev.* **2015**, *115*, 5366–5412

182. J. Hartung, S. N. Greszler, R. C. Klix, J. M. Kallemeyn, *Org. Process Res. Dev.* **2019**, *23*, 2532–2537

183. T. Bach, *Synthesis* **1998**, 683–703

184. M. T. Crimmins, *Chem. Rev.* **1988**, *88*, 1453–1473

185. N. Hoffmann, *Chem. Rev.* **2008**, *108*, 1052–1103

186. N. J. Hafeman, S. A. Loskot, C. E. Reimann, B. P. Pritchett, S. C. Virgil, B. M. Stoltz, *J. Am. Chem. Soc.* **2020**, *142*, 8585–8590

187. B. Alcaide, P. Almendros, C. Aragoncillo, *Chem. Soc. Rev.* **2010**, *39*, 783–816

188. E. Lee-Ruff, G. Mladenova, *Chem. Rev.* **2003**, *103*, 1449–1483

189. N. A. Kaprinidis, G. Lem, S. H. Courtney, D. I. Schuster, *J. Am. Chem. Soc.* **1993**, *115*, 3324–3325

190. D. I. Schuster, G. Lem, N. A. Kaprinidis, *Chem. Rev.* **1993**, *93*, 3–22

191. M. D'Auria, *Photochem. Photobiol. Sci.* **2019**, *18*, 2297–2362

192. K. A. Schnapp, R. M. Wilson, D. M. Ho, R. A. Caldwell, D. Creed, *J. Am. Chem. Soc.* **1990**, *112*, 3700–3702

193. T. Suishu, T. Shimo, K. Somekawa, *Tetrahedron* **1997**, *53*, 3545–3556

194. J. L. Broeker, J. E. Eksterowicz, A. J. Belk, K. N. Houk, *J. Am. Chem. Soc.* **1995**, *117*, 1847–1848

195. T. H. Morris, E. H. Smith, R. Walsh, *J. Chem. Soc. Chem. Commun.* **1987**, 964–965

196. A. G. Griesbeck, H. Mauder, S. Stadtmüller, *Acc. Chem. Res.* **1994**, *27*, 70–75

197. E. J. Corey, J. D. Bass, R. LeMahieu, R. B. Mitra, *J. Am. Chem. Soc.* **1964**, *86*, 5570–5583

198. R. Shen, E. J. Corey, *Org. Lett.* **2007**, *9*, 1057–1059

199. S. Yamabe, K. Kuwata, T. Minato, *Theor. Chem. Acc.* **1999**, *102*, 139–146

200. M. R. Siebert, J. M. Osbourn, K. M. Brummond, D. J. Tantillo, *J. Am. Chem. Soc.* **2010**, *132*, 11952–11966

201. D. Siri, A. Gaudel-Siri, J.-M. Pons, D. Liotard, M. Rajzmann, *J. Mol. Struct. (Theochem)* **2002**, *588*, 71–78

202. F. Bernardi, A. Bottoni, M. Olivucci, A. Venturini, M. A. Robb, *J. Chem. Soc. Faraday Trans.* **1994**, *90*, 1617–1630

203. K. N. Houk, Y. Li, J. Storer, L. Raimondi, B. Beno, *J. Chem. Soc. Faraday Trans.* **1994**, *90*, 1599–1604

204. J. Iriondo-Alberdi, M. F. Greaney, *Eur. J. Org. Chem.* **2007**, 4801–4815

205. K. C. Nicolaou, D. Sarlah, D. M. Shaw, *Angew. Chem. Int. Ed.* **2007**, *46*, 4708–4711

206. K. Takao, N. Hayakawa, R. Yamada, T. Yamaguchi, H. Saegusa, M. Uchida, S. Samejima, K. Tadano, *J. Org. Chem.* **2009**, *74*, 6452–6461

207. R. Hambalek, G. Just, *Tetrahedron Lett.* **1990**, *31*, 5445–5448

208. A. Pommier, J.-M. Pons, P. J. Kocienski, *J. Org. Chem.* **1995**, *60*, 7334–7339

209. J. D. Winkler, C. M. Bowen, F. Liotta, *Chem. Rev.* **1995**, *95*, 2003–2020

210. G. Büchi, J. A. Carlson, J. E. Powell, L. F. Tietze, *J. Am. Chem. Soc.* **1973**, *95*, 540–545

211. T. Hansson, B. Wickberg, *J. Org. Chem.* **1992**, *57*, 5370–5376

212. J. D. Winkler, M. B. Rouse, M. F. Greaney, S. J. Harrison, Y. T. Jeon, *J. Am. Chem. Soc.* **2002**, *124*, 9726–9728

213. A. D. Allen, T. T. Tidwell, *Chem. Rev.* **2013**, *113*, 7287–7342

214. Z. Wang, S.-J. Min, S. J. Danishefsky, *J. Am. Chem. Soc.* **2009**, *131*, 10848–10849

215. T. Bach, H. Bergmann, *J. Am. Chem. Soc.* **2000**, *122*, 11525–11526

216. P. Selig, T. Bach, *Angew. Chem. Int. Ed.* **2008**, *47*, 5082–5084

217. R. G. Salomon, *Tetrahedron* **1983**, *39*, 485–575

218. K. Narasaka, Y. Hayashi, H. Shimadzu, S. Niihata, *J. Am. Chem. Soc.* **1992**, *114*, 8869–8885

219. H. Butenschön, *Angew. Chem. Int. Ed.* **2008**, *47*, 3492–3495

220. H. Teller, S. Flügge, R. Goddard, A. Fürstner, *Angew. Chem. Int. Ed.* **2010**, *49*, 1949–1953

221. K. Aikawa, Y. Hioki, N. Shimizu, K. Mikami, *J. Am. Chem. Soc.* **2011**, *133*, 20092–20095

222. H. Ito, M. Hasegawa, Y. Takenaka, T. Kobayashi, K. Iguchi, *J. Am. Chem. Soc.* **2004**, *126*, 4520–4521

223. H. Lebel, J.-F. Marcoux, C. Molinaro, A. B. Charette, *Chem. Rev.* **2003**, *103*, 977–1050

224. J. M. Concellón, H. Rodríguez-Solla, C. Concellón, V. del Amo, *Chem. Soc. Rev.* **2010**, *39*, 4103–4113

225. R. J. K. Taylor, G. Casy, *Org. React.* **2004**, 359–475

226. S. D. McGregor, D. M. Lemal, *J. Am. Chem. Soc.* **1966**, *88*, 2858–2859

227. N. G. Rondan, K. N. Houk, R. A. Moss, *J. Am. Chem. Soc.* **1980**, *102*, 1770–1776

228. R. A. Moss, *Acc. Chem. Res.* **1980**, *13*, 58–64

229. R. A. Moss, *Acc. Chem. Res.* **1989**, *22*, 15–21

230. D. Suarez, T. L. Sordo, J. A. Sordo, *J. Org. Chem.* **1995**, *60*, 2848–2852

231. K. N. Houk, N. G. Rondau, J. Mareda, *Tetrahedron* **1985**, *41*, 1555–1563
232. I. Fleming, *Frontier Orbitals and Organic Chemical Reactions*, John Wiley & Sons, **1976**
233. L. Boisvert, F. Beaumier, C. Spino, *Org. Lett.* **2007**, *9*, 5361–5363
234. M. Fedorynski, *Chem. Rev.* **2003**, *103*, 1099–1132
235. C. B. Kelly, L. Thai-Savard, J. Hu, T. B. Marder, G. A. Molander, A. B. Charette, *ChemCatChem* **2024**, *16*, e202400110
236. M. Nakamura, A. Hirai, E. Nakamura, *J. Am. Chem. Soc.* **2003**, *125*, 2341–2350
237. J. C. Lorenz, J. Long, Z. Yang, S. Xue, Y. Xie, Y. Shi, *J. Org. Chem.* **2004**, *69*, 327–334
238. A. Voituriez, L. E. Zimmer, A. B. Charette, *J. Org. Chem.* **2010**, *75*, 1244–1250
239. A. B. Charette, H. Lebel, *J. Org. Chem.* **1995**, *60*, 2966–2967
240. A. B. Charette, J.-F. Marcoux, *Synlett* **1995**, 1197–1207
241. G. A. Molander, L. S. Harring, *J. Org. Chem.* **1989**, *54*, 3525–3532
242. F. G. Bordwell, E. Doomes, P. W. R. Corfield, *J. Am. Chem. Soc.* **1970**, *92*, 2581–2583
243. F. G. Bordwell, J. M. W. Jr., E. B. H. Jr., B. B. Jarvis, *J. Am. Chem. Soc.* **1968**, *90*, 429–435
244. D. Suárez, J. A. Sordo, T. L. Sordo, *J. Phys. Chem.* **1996**, *100*, 13462–13465
245. S. Matsumura, T. Nagai, N. Tokura, *Bull. Chem. Soc. Jpn.* **1968**, *41*, 2672–2675
246. J. F. King, M. S. Gill, D. F. Klassen, *Pure Appl. Chem.* **1996**, *68*, 825–830
247. C. Y. Meyers, A. M. Malte, W. S. Matthews, *J. Am. Chem. Soc.* **1969**, *91*, 7510–7512
248. L. H. Mejorado, T. R. R. Pettus, *J. Am. Chem. Soc.* **2006**, *128*, 15625–15631
249. T. Wang, Y. Liang, Z.-X. Yu, *J. Am. Chem. Soc.* **2011**, *133*, 9343–9353
250. J. D. White, T.-S. Kim, M. Nambu, *J. Am. Chem. Soc.* **1995**, *117*, 5612–5613
251. A. G. M. Barrett, D. Hamprecht, A. J. P. White, D. J. Williams, *J. Am. Chem. Soc.* **1996**, *118*, 7863–7864
252. A. G. M. Barrett, K. Kasdorf, *J. Am. Chem. Soc.* **1996**, *118*, 11030–11037
253. A. K. Ghosh, C. Liu, *Org. Lett.* **2001**, *3*, 635–638
254. H. Huang, J. S. Panek, *Org. Lett.* **2004**, *6*, 4384–4385
255. R. Anthes, S. Benoit, C.-K. Chen, E. A. Corbett, R. M. Corbett, A. J. DelMonte, S. Gingras, R. C. Livingston, Y. Pendri, J. Sausker, M. Soumeillant, *Org. Process Res. Dev.* **2008**, *12*, 178–182
256. D. T. Nowlan, T. M. Gregg, H. M. L. Davies, D. A. Singleton, *J. Am. Chem. Soc.* **2003**, *125*, 15902–15911
257. J. M. Fraile, J. I. García, V. Martínez-Merino, J. A. Mayoral, L. Salvatella, *J. Am. Chem. Soc.* **2001**, *123*, 7616–7625
258. F. Planas, M. Costantini, M. Montesinos-Magraner, F. Himo, A. Mendoza, *ACS Catal.* **2021**, *11*, 10950–10963
259. Y.-S. Xue, Y.-P. Caia, Z.-X. Chen, *RSC Adv.* **2015**, *5*, 57781–57791
260. T. Rasmussen, J. F. Jensen, N. Østergaard, D. Tanner, T. Ziegler, P.-O. Norrby, *Chem. Eur. J.* **2002**, *8*, 177–184
261. B. F. Straub, I. Gruber, F. Rominger, P. Hofmann, *J. Organomet. Chem.* **2003**, *684*, 124–143
262. C. Özen, N. S. Tüzün, *Organometallics* **2008**, *27*, 4600–4610
263. H. Pellissier, *Tetrahedron* **2008**, *64*, 7041–7095
264. R. Dalpozzo in *Asymmetric Synthesis of Three-Membered Rings*, Wiley-VCH, **2017**, Chapter 1, pp. 1–204
265. K. E. Berger, R. J. Martinez, J. Zhou, C. Uyeda, *J. Am. Chem. Soc.* **2023**, *145*, 9441–9447
266. R. F. Algera, C. Allais, A. F. Baldwin, T. Busch, F. Colombo, M. Colombo, C. Depretz, Y. R. Dumond, A. R. F. Quintero, M. Heredia, J. Jung, A. Lall, T. Lee, Y. Liu, S. Mandelli, M. Mantel, R. Morris, J. Mustakis, B. Nguyen, R. Pearson, J. L. Piper, J. A. Ragan, B. Ruffin, C. Talicska, S. Tcyrulnikov, C. Uyeda, R. M. Weekly, M. Zeng, *Org. Process Res. Dev.* **2023**, *27*, 2260–2270
267. T.-L. Chan, S. Fong, Y. Li, T.-O. Man, C.-D. Poon, *J. Chem. Soc. Chem. Commun.* **1994**, 1771–1772

268. K. C. Nicolaou, D. Sarlah, T. R. Wu, W. Zhan, *Angew. Chem. Int. Ed.* **2009**, *48*, 6870–6874

269. M. Lautens, W. Klute, W. Tam, *Chem. Rev.* **1996**, *96*, 49–92

270. J. Cornelisse, *Chem. Rev.* **1993**, *93*, 615–669

271. P. A. Wender, R. Ternansky, M. deLong, S. Singh, A. Olivero, K. Rice, *Pure Appl. Chem.* **1990**, *62*, 1597–1602

272. P. A. Inglesby, P. A. Evans, *Chem. Soc. Rev.* **2010**, *39*, 2791–2805

273. M. Harmata, *Adv. Synth. Catal.* **2006**, *348*, 2297–2306

274. M. Harmata, *Acc. Chem. Res.* **2001**, *34*, 595–605

275. V. Nair, K. G. Abhilash, *Synlett* **2008**, 301–312

276. H. Pellissier, *Adv. Synth. Catal.* **2011**, *353*, 189–218

277. J. H. Rigby, *Tetrahedron* **1999**, *55*, 4521–4538

278. J. H. Rigby, *Acc. Chem. Res.* **1993**, *26*, 579–585

279. T. Shibata, *Adv. Synth. Catal.* **2006**, *348*, 2328–2336

280. G. Domínguez, J. Pérez-Castells, *Chem. Soc. Rev.* **2011**, *40*, 3430–3444

281. S. Kotha, E. Brahmachary, K. Lahiri, *Eur. J. Org. Chem.* **2005**, 4741–4767

282. T. Shibata, K. Tsuchikama, *Org. Biomol. Chem.* **2008**, *6*, 1317–1323

283. P. R. Chopade, J. Louie, *Adv. Synth. Catal.* **2006**, *348*, 2307–2327

284. B. Heller, M. Hapke, *Chem. Soc. Rev.* **2007**, *36*, 1085–1094

285. J. A. Varela, C. Saá, *Synlett* **2008**, 2571–2578

286. J. A. Varela, C. Saá, *Chem. Rev.* **2003**, *103*, 3787–3801

287. K. Tanaka, *Synlett* **2007**, 1977–1994

288. B. D. Schwartz, J. R. Denton, Y. Lian, H. M. L. Davies, C. M. Williams, *J. Am. Chem. Soc.* **2009**, *131*, 8329–8332

289. J. A. Berson, *Acc. Chem. Res.* **1972**, *5*, 406–414

290. C. G. Nasveschuk, T. Rovis, *Org. Biomol. Chem.* **2008**, *6*, 240–254

291. P. A. Leber, J. E. Baldwin, *Acc. Chem. Res.* **2002**, *35*, 279–287

292. X. S. Bogle, P. A. Leber, L. A. McCullough, D. C. Powers, *J. Org. Chem.* **2005**, *70*, 8913–8918

293. N. Graulich, *WIREs Comput. Mol. Sci.* **2011**, *1*, 172–190

294. A. M. M. Castro, *Chem. Rev.* **2004**, *104*, 2939–3002

295. F. E. Ziegler, *Chem. Rev.* **1988**, *88*, 1423–1452

296. R. P. Lutz, *Chem. Rev.* **1984**, *84*, 205–247

297. W. R. Roth, F. Hunold, *Liebigs Ann.* **1996**, 1917–1928

298. S. Sakai, *Int. J. Quantum Chem.* **2000**, *80*, 1099–1106

299. M. J. S. Dewar, C. Jie, *J. Am. Chem. Soc.* **1989**, *111*, 511–519

300. J. Cooper, D. W. Knight, P. T. Gallagher, *J. Chem. Soc. Perkin Trans. 1* **1992**, 553–559

301. R. E. Ireland, M. D. Varney, *J. Org. Chem.* **1983**, *48*, 829–1833

302. H. O. House, J. Lubinkowski, J. J. Good, *J. Org. Chem.* **1975**, *40*, 86–92

303. G. E. Veitch, E. Beckmann, B. J. Burke, A. Boyer, S. L. Maslen, S. V. Ley, *Angew. Chem. Int. Ed.* **2007**, *46*, 7629–7632

304. L. A. Paquette, *Tetrahedron* **1997**, *53*, 13971–14020

305. L. A. Paquette, *Angew. Chem. Int. Ed. Engl.* **1990**, *29*, 609–626

306. D. A. Evans, A. M. Golob, *J. Am. Chem. Soc.* **1975**, *97*, 4765–4766

307. R. K. Boeckman, D. M. Springer, T. R. Alessi, *J. Am. Chem. Soc.* **1989**, *111*, 8284–8286

308. Y. Chai, S. Chong, H. A. Lindsay, C. McFarland, M. C. McIntosh, *Tetrahedron* **2002**, *58*, 2905–2928

309. S. Pereira, M. Srebnik, *Aldrichim. Acta* **1993**, *26*, 17–29

310. B. Bernet, P. M. Bishop, M. Caron, T. Kawamata, B. L. Roy, L. Ruest, G. Sauvé, P. Soucy, P. Deslongchamps, *Can. J. Chem.* **1985**, *63*, 2810–2814

311. C. Schneider, *Synlett* **2001**, 1079–1090

312. W. S. Johnson, L. Werthemann, W. R. Bartlett, T. J. Brocksom, T.-T. Li, D. J. Faulkner, M. R. Peterson, *J. Am. Chem. Soc.* **1970**, *92*, 741–743

313. L. E. Overman, *Acc. Chem. Res.* **1980**, *13*, 218–224

314. L. E. Overman, *Angew. Chem. Int. Ed. Engl.* **1984**, *23*, 579–586

315. A. E. Wick, D. Felix, K. Steen, A. Eschenmoser, *Helv. Chim. Acta* **1964**, *47*, 2425–2429

316. J. Gonda, *Angew. Chem. Int. Ed.* **2004**, *43*, 3516–3524

317. T. Nakai, K. Mikami, *Chem. Rev.* **1986**, *86*, 885–902

318. K. Mikami, T. Nakai, *Synthesis* **1991**, 594–604

319. Y. Zhang, J. Wang, *Coord. Chem. Rev.* **2010**, *254*, 941–953

320. P. Somfai, O. Panknin, *Synlett* **2007**, 1190–1202

321. J. B. Sweeney, *Chem. Soc. Rev.* **2009**, *38*, 1027–1038

322. V. Rautenstrauch, *J. Chem. Soc. D* **1970**, 4–6

323. Y.-D. Wu, K. N. Houk, J. A. Marshall, *J. Org. Chem.* **1990**, *55*, 1421–1423

324. Y.-D. Wu, K. N. Houk, *J. Org. Chem.* **1991**, *56*, 5657–5661

325. T. H. West, S. S. M. Spoehrle, K. Kasten, J. E. Taylor, A. D. Smith, *ACS Catal.* **2015**, *5*, 7446–7479

326. E. Piers, M. Gilbert, K. L. Cook, *Org. Lett.* **2000**, *2*, 1407–1410

327. J. S. Clark, S. T. Hayes, C. Wilson, L. Gobbi, *Angew. Chem. Int. Ed.* **2007**, *46*, 437–440

328. E. A. Ilardi, C. E. Stivala, A. Zakarian, *Chem. Soc. Rev.* **2009**, *38*, 3133–3148

329. A. C. Jones, J. A. May, R. Sarpong, B. M. Stoltz, *Angew. Chem. Int. Ed.* **2014**, *53*, 2556–2591

330. U. Nubbemeyer, *Synthesis* **2003**, 961–1008

331. H. Ito, T. Taguchi, *Chem. Soc. Rev.* **1999**, *28*, 43–50

332. D. Enders, M. Knopp, R. Schiffers, *Tetrahedron: Asymmetry* **1996**, *7*, 1847–1882

333. E. J. Corey, B. E. Roberts, B. R. Dixon, *J. Am. Chem. Soc.* **1995**, *117*, 193–196

334. J. Tan, C.-H. Cheon, H. Yamamoto, *Angew. Chem. Int. Ed.* **2012**, *51*, 8264–8267

335. J. D. White, C. M. Lincoln, J. Yang, W. H. C. Martin, D. B. Chan, *J. Org. Chem.* **2008**, *73*, 4139–4150

336. D. Kaldre, J. L. Gleason, *Angew. Chem. Int. Ed.* **2016**, *55*, 11557–11561

337. T. Nakai, K. Tomooka, *Pure Appl. Chem.* **1997**, *69*, 595–600

338. Y. Hirokawa, M. Kitamura, N. Maezaki, *Tetrahedron: Asymm.* **2008**, *19*, 1167–1170

339. K. Tomooka, K. Yamamoto, T. Nakai, *Angew. Chem. Int. Ed.* **1999**, *38*, 3741–3742

340. A. de Cózar, A. Arrieta, I. Arrastia, F. P. Cossío, *ChemPlusChem* **2023**, *88*, e202300482

341. E. M. Phillips, T. Mesganaw, A. Patel, S. Duttwyler, B. Q. Mercado, K. N. Houk, J. A. Ellman, *Angew. Chem. Int. Ed.* **2015**, *54*, 12044–12048

342. S. Karlsson, R. Bergman, C. Löfberg, P. R. Moore, F. Pontén, J. Tholander, H. Sörensen, *Org. Process Res. Dev.* **2015**, *19*, 2067–2074

343. Y. Que, H. Shao, H. He, S. Gao, *Angew. Chem. Int. Ed.* **2020**, *59*, 7444–7449

344. E. E. van Tamelen, S. P. Pappas, K. L. Kirk, *J. Am. Chem. Soc.* **1971**, *93*, 6092–6101

345. M. J. Goldstein, R. S. Leight, *J. Am. Chem. Soc.* **1977**, *91*, 8112–8114

346. W. R. Dolbier, H. Koroniak, K. N. Houk, C. Sheu, *Acc. Chem. Res.* **1996**, *29*, 471–477

347. S. Niwayama, *J. Org. Chem.* **1996**, *61*, 640–646

348. M. Murakami, M. Hasegawa, *Angew. Chem. Int. Ed.* **2004**, *43*, 4874–4876

349. P. S. Lee, X. Zhang, K. N. Houk, *J. Am. Chem. Soc.* **2003**, *125*, 5072–5079

350. C. H. DePuy, *Acc. Chem. Res.* **1968**, *1*, 33–41

351. G. D. Hamblin, R. P. Jimenez, T. S. Sorensen, *J. Org. Chem.* **2007**, *72*, 8033–8045

352. E. Lee, C. H. Yoon, *J. Chem. Soc. Chem. Commun.* **1994**, 479–481

353. X. Q. Tian, T. C. Chen, L. Y. Matsuoka, J. Wortsman, M. F. Holick, *J. Biol. Chem.* **1993**, *268*, 14888–14892

354. M. F. Holick, *Am. J. Clin. Nutr.* **1995**, *61*, 638S–645S

355. A. J. Frontier, C. Collison, *Tetrahedron* **2005**, *61*, 7577–7606

356. H. Pellissier, *Tetrahedron* **2005**, *61*, 6479–6517

357. M. A. Tius, *Eur. J. Org. Chem.* **2005**, 2193–2206

358. W. T. Spencer, T. Vaidya, A. J. Frontier, *Eur. J. Org. Chem.* **2013**, 3621–3633

359. J. Ichikawa, *Pure Appl. Chem.* **2000**, *72*, 1685–1689

360. S. E. Denmark, K. L. Habermas, G. A. Hite, *Helv. Chim. Acta* **1988**, *71*, 168–194

361. W. He, I. R. Herrick, T. A. Atesin, P. A. Caruana, C. A. Kellenberger, A. J. Frontier, *J. Am. Chem. Soc.* **2008**, *130*, 1003–1011

362. J. A. Bender, A. E. Blize, C. C. Browder, S. Giese, F. G. West, *J. Org. Chem.* **1998**, *63*, 2430–2431

363. A. V. Yadykov, V. Z. Shirinian, *Adv. Synth. Catal.* **2020**, *362*, 702–723

364. T. Vaidya, R. Eisenberg, A. J. Frontier, *ChemCatChem* **2011**, *3*, 1531–1548

365. S. P. Waters, Y. Tian, Y.-M. Li, S. J. Danishefsky, *J. Am. Chem. Soc.* **2005**, *127*, 13514–13515

366. J. A. Malona, K. Cariou, A. J. Frontier, *J. Am. Chem. Soc.* **2009**, *131*, 7560–7561

367. S. Gao, Q. Wang, C. Chen, *J. Am. Chem. Soc.* **2009**, *131*, 1410–1412

368. F. P. Cossio, A. Arrieta, M. A. Sierra, *Acc. Chem. Res.* **2008**, *41*, 925–936

369. N. Fu, T. T. Tidwell, *Tetrahedron* **2008**, *64*, 10465–10496

370. J. Xu, *Arkivoc* **2008**, *ix*, 21–44

371. F. P. Cossío, A. de Cózar, M. A. Sierra, L. Casarrubios, J. G. Muntaner, B. K. Banik, D. Bandyopadhyay, *RSC Adv.* **2022**, *12*, 104–117

372. A. Landa, A. Mielgo, M. Oiarbide, C. Palomo, *Org. React.* **2018**, *95*, 423–594

373. S. France, A. Weatherwax, A. E. Taggi, T. Lectka, *Acc. Chem. Res.* **2004**, *37*, 592–600

374. C. R. Pitts, T. Lectka, *Chem. Rev.* **2014**, *114*, 7930–7953

375. D. L. Boger, K. Takahashi, *J. Am. Chem. Soc.* **1995**, *117*, 12452–12459

376. W. M. Bandaranayake, J. E. Banfield, D. S. C. Black, *J. Chem. Soc. Chem. Commun.* **1980**, 902–903

377. K. C. Nicolaou, N. A. Petasis, R. E. Zipkin, *J. Am. Chem. Soc.* **1982**, *104*, 5560–5562

378. H. Quintela-Varela, C. S. Jamieson, Q. Shao, K. N. Houk, D. Trauner, *Angew. Chem. Int. Ed.* **2020**, *59*, 5263–5267

379. H. M. R. Hoffmann, *Angew. Chem. Int. Ed. Engl.* **1969**, *8*, 556–577

380. M. L. Clarke, M. B. France, *Tetrahedron* **2008**, *64*, 9003–9031

381. B. B. Snider in *Ene Reactions with Alkenes as Enophiles, Vol. Vol. 5* (Ed.: B. Trost), Pergamon Press, Oxford, **1991**, pp. 1–27

382. W. Oppolzer, *Organomet. Reagents Org. Synth.* **1994**, 161–183

383. W. Oppolzer, *Angew. Chem. Int. Ed. Engl.* **1989**, *23*, 38–52

384. M. N. Alberti, M. Orfanopoulos, *Synlett* **2010**, *7*, 999–1026

385. J.-L. Ripoll, Y. Vallée, *Synthesis* **1993**, 659–677

386. B. B. Snider, *Acc. Chem. Res.* **1980**, *13*, 426–432

387. K. Mikami, M. Shimizu, *Chem. Rev.* **1992**, *92*, 1021–1050

388. A. Bakhtiari, J. Safaei-Ghomi, *Synlett* **2019**, *30*, 1738–1764

389. M. F. Salomon, S. N. Pardo, R. G. Salomon, *J. Org. Chem.* **1984**, *49*, 2446–2454

390. M. R. Pitts, J. Mulzer, *Tetrahedron Lett.* **2002**, *43*, 8471–8473

391. J. Mulzer, E. Öhler, *Angew. Chem. Int. Ed.* **2001**, *40*, 3842–3846

392. E. J. Corey, D. Barnes-Seeman, T. W. Lee, S. N. Goodman, *Tetrahedron Lett.* **1997**, *38*, 6513–6516

393. C. Jäkel, R. Paciello, US-Patent 7,534,921 B1, **2009**

394. M. Friedrich, K. Ebel, N. Götz, US-Patent 7,550,633 B2, **2009**

395. M. Friedrich, K. Ebel, N. Götz, W. Krause, C. Zahm, US-Patent 7,608,742 B2, **2009**

396. T. Qiao, Y. Wang, S. Zheng, H. Kang, G. Liang, *Angew. Chem. Int. Ed.* **2020**, *59*, 14111–14114

397. C. Oger, L. Balas, T. Durand, J.-M. Galano, *Chem. Rev.* **2013**, *113*, 1313–1350
398. C. E. Miller, *J. Chem. Ed.* **1965**, *42*, 254–259
399. S. Hünig, H. R. Müller, W. Thier, *Angew. Chem. Int. Ed. Engl.* **1965**, *4*, 271–280
400. D. J. Pasto, R. T. Taylor, *Org. React.* **1991**, *40*, 91–155
401. B. M. Trost, M. U. Frederiksen, J. P. N. Papillon, P. E. Harrington, S. Shin, B. T. Shireman, *J. Am. Chem. Soc.* **2005**, *127*, 3666–3667
402. M. P. Feth, K. Rossen, A. Burgard, *Org. Process Res. Dev.* **2013**, *17*, 282–293

Transition Metal-Catalyzed Coupling Reactions 6

Transition metal-catalyzed coupling reactions are among the most important transformations in organic chemistry for the construction of highly functionalized molecules [1, 2]. This significance has now also extended to their routine use in medicinal chemistry routes as well as in industrial syntheses of pharmaceuticals and agrochemicals (see Fig. 6.1) [3–6].

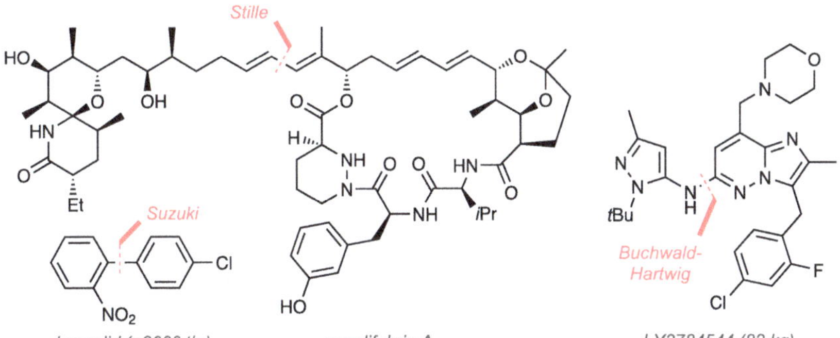

Fig. 6.1 Examples of coupling reactions in the synthesis of bioactive substances [7–9]

In addition to CC bond formation, the forging of C-heteroatom bonds (C-N, C-O, C-S, C-F) has been particularly extended by the seminal studies of Buchwald and Hartwig, and continues to increasingly occupy a growing extent in the standard canon of synthetic chemistry. An analysis of present synthetic methodologies in medicinal chemistry syntheses showed that the Suzuki, Sonogashira, and Buchwald–Hartwig coupling are among the twenty most frequently encountered reactions in exploratory drug research [10].

A. Düfert, *Methods of Organic Synthesis*,
https://doi.org/10.1007/978-3-662-70963-4_6

Classic couplings are categorized by the choice of the nucleophile used, with a halide, triflate or tosylate typically serving as the electrophile (see Fig. 6.2).

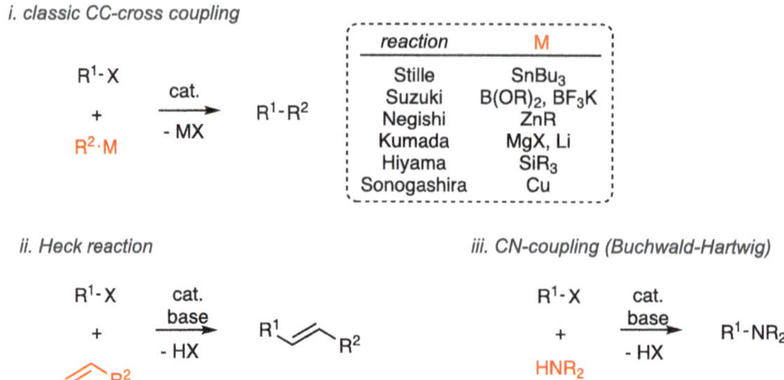

Fig. 6.2 General scheme of coupling reactions

The field has traditionally been dominated by palladium catalysts [11], but newer variants also rely on the use of Ni, Cu, and Fe complexes. While nickel and palladium catalysts primarily cover traditional CC cross-couplings, copper complexes are notably suited for CN bond formation [12–14]. Ni- and Pd-catalyzed cross-couplings display an almost complementary substrate scope with regard to the utilized electrophiles: While Pd complexes are particularly adept at converting halides, triflates and occasionally tosylates, the transformation of phosphates, sulfamates, esters, and even aryl ethers via C-O bond cleavage can be selectively effected by Ni catalysis (see Fig. 6.3) [15–18]. Iron catalysts specifically lend themselves for Kumada cross-couplings, except for select substrate classes, where tailor-made Fe complexes also allow the combination with other organometallic reagents (Negishi, Suzuki). Despite intensive research, catalysts based on first-row transition metals generally exhibit lower reactivities than Pd catalysts, require higher amounts of catalyst, and thus result in harsher conditions [14, 19].

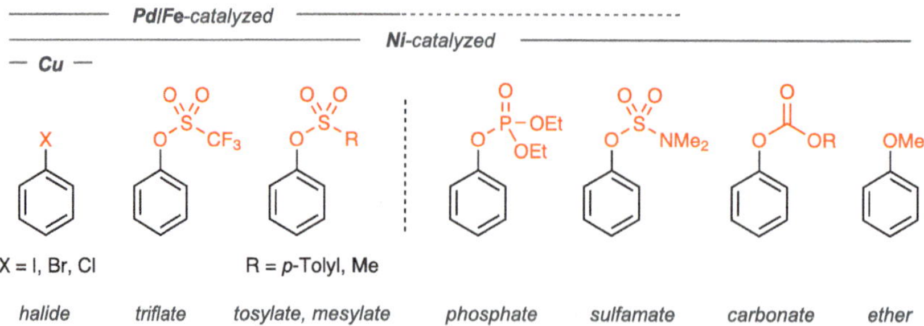

Fig. 6.3 Scope of commonly used electrophiles

The choice of the organometallic species largely determines the substrate scope, the tolerance towards functional groups, the applicability of a given catalyst and the reactivity (see Table 6.1). While initially only aryl halides could be used as electrophiles, the spectrum of coupling partners was subsequently expanded to include alkenyl and alkyl groups [20, 21]. The Sonogashira coupling relies on *in situ* generated copper acetylides as nucleophiles. However, Cu-free methods have also been developed.

Table 6.1 Overview of the different coupling reactions under Pd catalysis [3, 24, 25]

$$R^1\text{-}X \;+\; R^2\text{-}M \;\xrightarrow{\;PdL_n\;}\; R^1\text{-}R^2$$

reactivity of nucleophiles (*classic cross couplings*)

$$\text{Li, MgX} \gg \text{ZnR} > \text{B(OR)}_2 \geq \text{SnBu}_3 > \text{SiR}_3$$

reaction	R^2-M	conditions	activator	FG incompatibility
Stille	R^2-SnBu$_3$	neutral	F$^-$ (optional)	(acid)[c]
Suzuki	R^2-B(OR)$_2$/-BF$_3$K	basic	RO$^-$ / OH$^-$	(1°/2° amine)[c]
Negishi	R^2-ZnR	slightly basic [b]	–	alcohol, acid, 1°/2° amine
Kumada	R^2-MgX/-Li [a]	basic [b]	–	alcohol, acid, 1°/2° amine, R-NO$_2$ aldehyde, (ketone, ester, amide)[c]
Hiyama	R^2-SiR$_3$	neutral - basic	F$^-$ / ring strain	(silyl ether)[d]
Sonogashira	R^2-Cu	basic	–	(aldehyde, especially α,β-unsaturated)[c]
Heck	R^2-vinyl	basic	RO$^-$ / OH$^-$	(1°/2° amine)[c]
Buchwald-Hartwig	HNR$_2$	basic	RO$^-$ / OH$^-$	alcohol, acid, aldehyde, (1°/2° amine)[c]

[a] high homocoupling share possible; [b] basicity of organometal reagent; [c] substrate-/condition-dependent; [d] labile under F$^-$-use

The reactivity of the organometallic reagent in the transmetalation roughly correlates with their nucleophilicity: Only boronic acid derivatives are more reactive than their relative nucleophilicity suggests. This can be understood by their acid/base behavior under the employed conditions: in the basic, mixed alcoholic/aqueous medium, the transmetalation proceeds via a tetracoordinated anionic Pd-boronate complex and not via a tricoordinated, neutral boron, which greatly enhances the reactivity (see Sect. 6.1) [22, 23].

The Suzuki–Miyaura and Stille cross-couplings are by far the most versatile CC cross-couplings by virtue of the range of substrates that have been successfully employed. As counterbalance, the synthesis of boronic acids is usually an onerous and costly process via the corresponding Li- or Mg-intermediates, which is why the latters' direct use — if possible — is highly preferable. Stannanes, on the other hand, are extremely toxic and removing Sn traces from the product can be very laborious, which greatly limits their application on large scale, especially in the preparation of drugs. Each method therefore has its justification, also against the backdrop of their reactivity differences, wherein less reactive coupling partners necessitate the use of more nucleophilic Zn- or Mg-based nucleophiles [25].

The incompatibility with certain functional groups is either dictated by the requisite conditions or the nucleophile. For example, the presence of protic functionalities in Negishi, Kumada, and Buchwald–Hartwig couplings requires an additional equivalent of base or of the organometallic reagent. Nitro groups are redox-sensitive and can cause the reaction to stall or the electrophile to decompose with many substrates, especially in combination with Li-/Mg-reagents. Alkynes are usually tolerated in the coupling reaction. Depending on the substrate, a carbopalladation of the triple bond may occur after oxidative oxidation instead of a transmetalation in a few cases. Alkenes remain intact in almost all cross-couplings. The presence of multiple alkenes may however lead to competing reactions in the Heck reaction. Furthermore, additions to triple bonds are usually faster than additions to double bonds.

Under Pd catalysis, usually 0.1–1 mol% of catalyst is added, while Ni, Cu, and Fe catalysts normally require 5–10 mol% of the transition metal complexes. The nucleophile is used in a slight excess (1.1–1.3 eq.). As solvents, nonpolar to polar-aprotic solvents such as THF, dioxane, DME, or toluene are typically employed. Suzuki cross-couplings require Lewis bases as activators of the boronic acid, which is why water or alkanols are added as cosolvents. As precatalysts, metal salts or complexes based on Pd(II), Ni(II), Cu(I), and Fe(III) are typically used. **1–4** are complexes from which L-Pd^0 can be eliminated and thus activated with high efficiency (see Fig. 6.4).

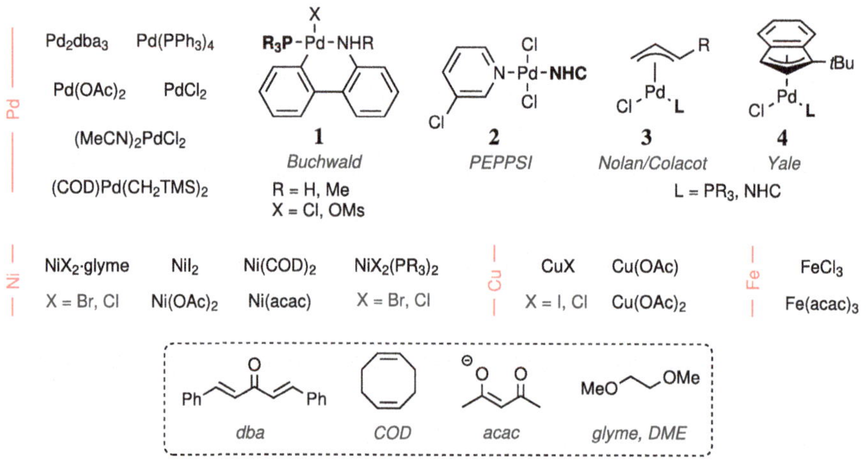

Fig. 6.4 Frequently encountered transition metal precatalysts

While originally only PPh$_3$ or chelate analogs thereof were resorted to as ligands, a variety of different ligands has been established over the years to continuously expand the substrate scope. In particular, electron-rich, sterically demanding phosphines and *N*-heterocyclic carbenes (*NHC*) have revolutionized the field of Pd-catalyzed coupling reactions. For Ni and Cu catalysts, N,N- and N,O-chelate ligands are often used [12, 13, 15, 16]. Fe catalysts typically require complex pincer ligands for an efficient transformation and to suppress catalyst decomposition due to ligand loss. In addition, numerous examples of simple ligands (acac, NMP) or classic bisphosphines as well as NHCs have been reported (see Fig. 6.5) [26, 27].

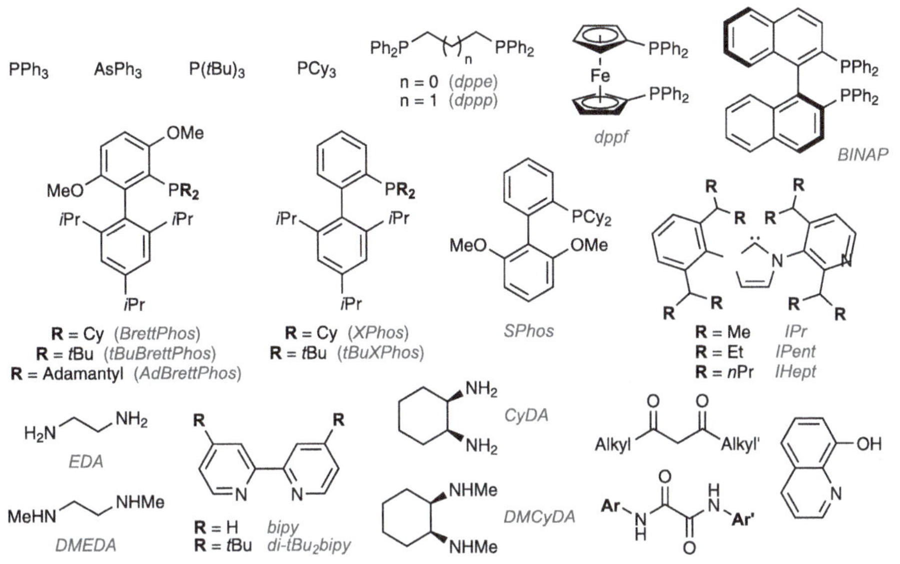

Fig. 6.5 Common ligands for Pd/Ni/Cu/Fe-catalyzed coupling reactions

Beyond the choice of Pd source and ligand, very different results can still be observed depending on the reaction conditions, also including the tolerance towards a range of functional groups. Researchers at Merck investigated the synthesis of the $GABA_A$ $\alpha_{2/3}$-selective agonist **6**, which was planned to include a Suzuki–Miyaura cross-coupling in the concluding step. Anhydrous conditions, which had previously been developed as an efficient system with other substrates, effected no conversion with 3-pyridylboronic acid. The use of alternative bases and solvents also did not prove successful. The isolation of the desired product **6** could however be finally achieved upon addition of water. Depending on the choice of an organic cosolvent, a significant proportion of the respective rearrangement product **7** was observed alongside. While dimethylacetamide, acetonitrile, and ethanol favored the unwanted Dimroth rearrangement of **6** to **7**, it was largely suppressed in dioxane, THF, and 2-Me-THF. The desired product was thus afforded in 95.6% yield when subjected to the optimized reaction conditions, with the ratio **6:7** being 102:1 (see Fig. 6.6) [28].

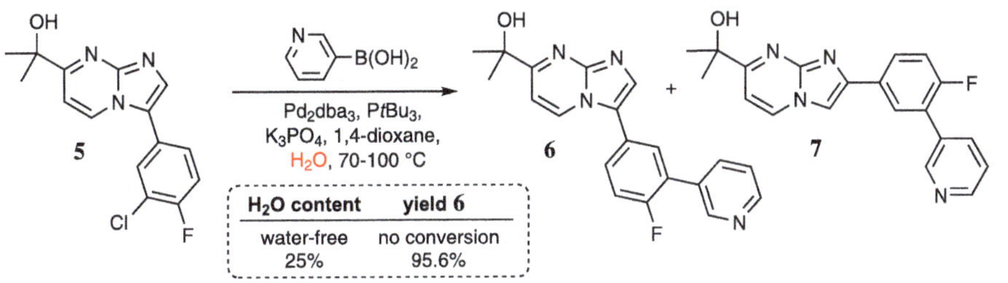

Fig. 6.6 Influence of water content on the turnover of a Suzuki cross-coupling [28]

Apart from the Sonogashira reaction, all the processes listed above catalyze a C_{sp^2}-C_{sp^2} bond formation particularly efficiently. The coupling of alkyl groups is very challenging with alkyl halides or triflates as electrophiles due to a β-H elimination of the alkyl-Pd complexes. Alternatively, nucleophiles with metal-alkyl bonds can be used: magnesium-, boron-, tin-, and silicon-based reagents allow the formation of stable R'_{sp^3} organometallic species. Beyond isolated examples, there is currently no universally valid protocol for alkyl-alkyl cross-couplings; however, good progress has been made so far with Stille, Negishi, and Suzuki reactions.

In addition to the expansion of the substrate scope of classic CC-/CN-/CO-coupling, i.e., the conversion of an electrophile (with elimination of a leaving group) with a nucleophile, a parallel development of alternative reaction modes has taken place since the 2000s. The activation of a CH bond instead of CX in the electrophile as well as the coupling of two electrophiles under oxidative conditions has experienced a steady rise in popularity (see Fig. 6.7) [29–34].

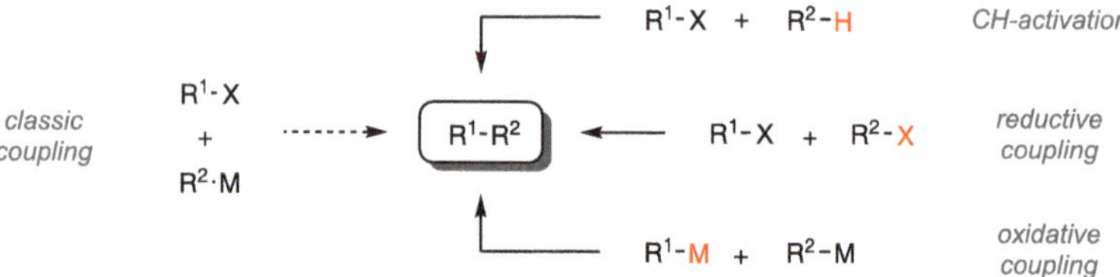

Fig. 6.7 General scheme of alternative coupling modes

While oxidative and reductive couplings continue to expand the canon of possible reactions, CH activation has meanwhile developed into a very mature field of research. Considering the amount and breadth of reported methodological studies, metal catalysts based on 3d elements are becoming dominant, and their application in synthetic endeavors towards secondary metabolites is steadily increasing (see Sect. 6.5) [31].

6.1   Mechanistic Aspects

Given the dominance of Pd catalysts, the mechanism of coupling reactions has been comprehensively studied for this type of metal complexes. The chemistry of Ni, Cu, and Fe catalysts allows for largely complementary reactivity, which can be attributed to the differences of these active metals to Pd. However, in-depth studies of the mechanistic details of alternative transition metals are often hindered by the lower stability of these complexes, a wide range of stable oxidation states, as well as a strong dependence of the mechanism on the individual reaction conditions. Therefore, the mechanistic aspects of Pd-catalyzed processes will be discussed first in-depth and the differences to other transition metal catalysts will be covered subsequently.

J. K. Stille published a general mechanism of cross-coupling reactions with stannanes in 1986, which is universally accepted today and can be applied to all cross-couplings [35]. All palladium-catalyzed reactions follow an identical, universal scheme: Following the initial formation of the catalytically active species from the precatalyst, the oxidative addition of the electrophile to Pd^0, a transmetalation of the nucleophile, and finally a reductive elimination of the coupling product occur; in addition, isomerizations may sometimes arise (see Fig. 6.8).

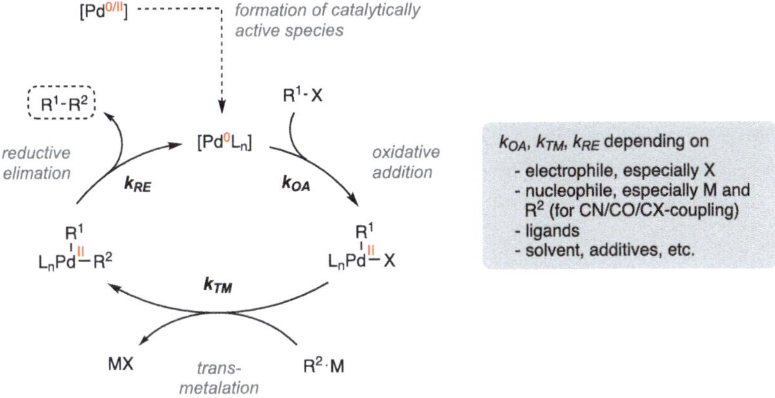

Fig. 6.8 General mechanism of cross-coupling reactions [35]

The relative rate of reaction and thus the identity of the rate-determining step of the catalytic cycle depend to a substantial degree on the coupling partners and the individual Pd catalyst. With the steadily improving understanding of these influencing factors, transformations originally perceived as "impossible" were successively realized by systematic optimization, most notably of the employed ligands [19].

Reactive Pd^0 complexes (Pd^0L_2, Pd^0L_1) are coordinatively and electronically unsaturated at the d^{10} metal center. While an *oxidative addition* involves the desired completion of the coordinative shell, the palladium in turn experiences a loss of electron density in order to insert into the R^1-X bond. Particularly electron-rich ligands facilitate the donation of electrons and accelerate the oxidative addition. The substrate can be oxidatively added more easily, the weaker the CX bond is, which is why the order of reactivity for the oxidative addition follows the trend ArI > ArBr > ArCl in aryl halides. The steric demand of the ancillary (=spectator) ligand plays a special role: beyond a certain size the monoligated species PdL_1 becomes dominant instead of PdL_2 complexes due to steric repulsion in the ligand sphere of the metal. The barrier of the oxidative addition is significantly lowered, so that even less reactive aryl chlorides can be converted [36, 37].

Given the usually transient nature of the involved intermediates, the *transmetalation* remains the least understood elemental step in many reactions. The driving force is based on the difference in bond energies between the more electronegative Pd and the more electropositive metals of the coupling partners (Pauling electronegativities: Pd = 2.20, B = 2.04, Sn = 1.96, Si = 1.90, Zn = 1.65, Mg = 1.31). The reactive ligand X at the Pd will prefer to form an MX bond with the more electropositive metal of the nucleophilic reagent. The less electron-rich the metal center of the catalyst is, the larger becomes the driving force [25, 38]. These thermodynamic trends have also been confirmed for the kinetics of transmetalation, which follow the reverse order of oxidative addition: ArOTf > ArCl > ArBr > ArI. In the case of triflate, OTf^- is usually fully dissociated from the metal center, resulting in a formally cationic complex that possess an available coordination site for the approach of the nucleophile. A strong Pd-ligand bond (with good donors) in turn makes the requisite dissociation of L in PdL_2 complexes more difficult, and the *trans*-effect further destabilizes the transmetalation product. Steric effects also have a significant influence on transmetalation: d^{18} complexes like $ArXPd(II)L_2$ usually undergo a dissociative mechanism. A sterically demanding ancillary ligand increases the rate of ligand dissociation by stabilizing coordinatively unsaturated $L_1Pd(II)$ complexes and thus has a positive effect on transmetalation [36, 37].

The *reductive elimination* as a product-forming step can be influenced by both steric and electronic effects: The more sterically demanding a ligand, the easier the elimination of the coupling product becomes. The two groups to be eliminated are brought into closer proximity by the steric repulsion of the non-reactive, ancillary ligands, thereby lowering the reaction barrier of the elimination step. In addition, monoligated $L_1Pd(II)$ species eliminate the product more rapidly than tetracoordinated $L_2Pd(II)$ complexes.[I] The regeneration of the Pd(0) species via monoligated Pd intermediates is also supported by sterically demanding ligands. In addition, bidentate ligands can favor this step, as they only allow *cis*-configured complexes (*vide infra*). From an electronic point of view, weak electron donors as ancillary ligands support a reductive elimination, as they render the metal center less electron-rich than stronger σ donors. The nature of the reactive ligands R^1 and R^2 likewise influences the reductive elimination to a significant degree. Aryl groups — in CC as well as CX couplings — are eliminated faster than alkynyl or alkyl groups. CC couplings are the easiest in terms of their activation barrier, C-heteroatom couplings follow the trend C-P > C-S > C-N > C-O $\gg$ C-F. Less electron-rich groups R^2 eliminate more slowly, as a more polarized R^1/R^2-Pd bond increases the bond strength. Thus, the reductive elimination is the rate-determining step in many CX couplings (see below). Electron-rich aryl groups accordingly result in faster product formation, as an interaction with the π system lowers the activation energy (mesomeric instead of inductive effect) [37, 40].

As a consequence, strongly donating ancillary ligands accelerate the oxidative addition, but slow down the transmetalation and reductive elimination. A large steric demand of the non-reactive ligand seems to generally have a beneficial effect on all steps, as long as monoligated complexes result as the dominant intermediate species [39]. (see Fig. 6.9).

[I] In the published literature, the term "monoligated" has become established describing metal complexes in which the metal center is coordinated by only one *ancillary* or spectator ligand.

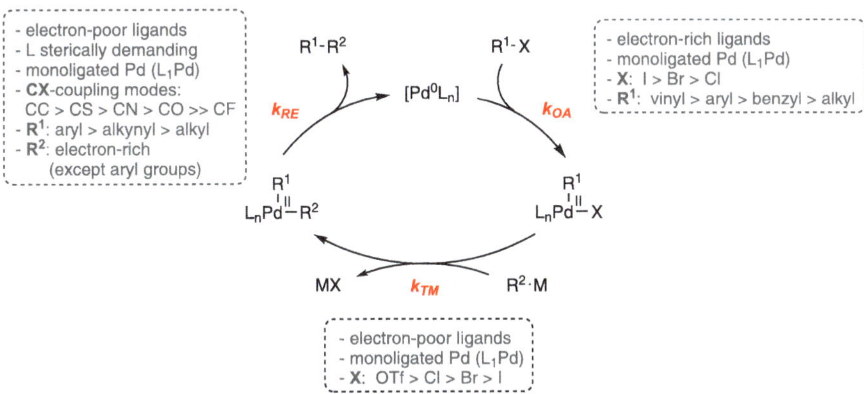

Fig. 6.9 Handles for raising the rate of the individual elementary steps of the catalytic cycle under Pd catalysis [36, 37, 40]

Early Pd-catalyzed CC cross-couplings utilizing first-generation ligand systems (PPh_3) were already limited by their stereoelectronic properties in the oxidative addition, which is why aryl iodides in particular constituted optimal substrates. With the advent of electron-rich and sterically demanding ligands (biaryl phosphines SPhos, XPhos, BrettPhos as well as NHCs), the field of CC cross-couplings was successively expanded in substrate scope, ultimately also enabling CX bond formations that had mainly been limited by their slow reductive elimination (CN-, CO-, CF-couplings) [19].

In order to carry out a systematic optimization of a reaction through guided modification of the coordination environment, quantitative, comparable measures of the steric and electronic properties of different ligands are required. The donor strength of a ligand can be quantified using several methods. An approach proposed by Tolman for phosphines and now also incorporated for other ligands is the experimental determination of the carbonyl vibration in $Ni(CO)_3L$ complexes (*Tolman electronic parameter, TEP*) [41]. The wavenumber of the highest CO stretching vibration is inversely proportional to the donor strength of a ligand (see Fig. 6.10). In general, the strength of the σ bond can be enhanced by successively changing from aryl to alkyl ligands. Many *N*-heterocyclic carbenes are extremely strong donors, while nitrogen-based ligands transfer less electron density to the metal center. sp^2-hybridized nitrogen atoms (imines as well as 5- and 6-membered heterocycles) however, are good π acceptors and can increase the electrophilicity of a metal center. For phosphines, the pK_a values of the conjugate acid can alternatively be used as a measure of the donor strength. To capture their steric demand, Tolman also proposed for phosphines the cone angle of metal and ligand in $Ni(CO)_3L$ complexes at a Ni-P distance of 2.28 Å [41]. Since few ligands have a cone symmetry and conformational changes cannot be captured in this way, the groups of Nolan and Cavallo developed the concept of the buried volume. If a sphere is constructed around the metal center (r = 3.5 Å), a ligand occupies a portion of the volume of this sphere, which reflects the steric demand even with highly asymmetric ligands. Based on crystal structures of L-AuCl complexes, the occupied portion of the coordination sphere was obtained at a distance of Au-L = 2.28 Å for many common ligands [42].

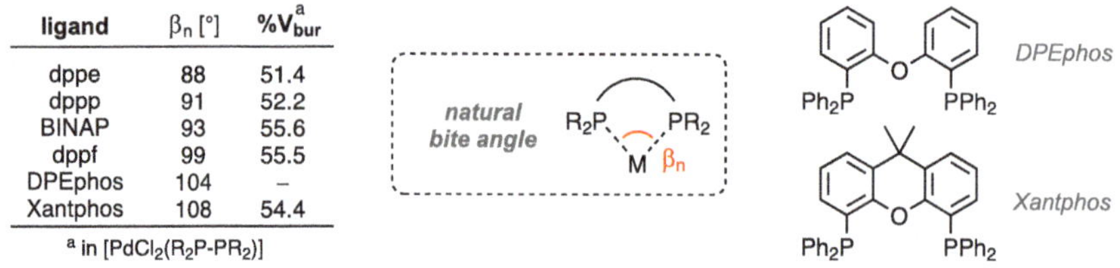

ligand	TEP [cm^{-1}]	θ [°]	%V_{bur}
P(C$_6$F$_5$)$_3$	2090	184	37.3
P(OPh)$_3$	2085	128	30.7
P(4-F-C$_6$H$_4$)$_3$	2071	145	29.8
PPh$_3$	2069	145	29.9
P(pTol)$_3$	2067	145	29.3
P(Ph)$_2$Me	2067	136	–
Ph(Me$_2$)Ph	2065	122	26.2
PtBu$_3$	2056	182	38.1
PCy$_3$	2056	170	33.4
SPhos	2054	208	49.7
XPhos	2053	221	53.1
IPr	2050	–	39.0

Fig. 6.10 Comparison of the donor strength and steric demand of ligands commonly encountered in Pd-mediated coupling reactions [42, 43]

In addition, chelate ligands such as bisphosphines, where the concept of the cone angle obviously fails, are classified according to their L-M-L bond angle (see Fig. 6.11). A wide bond angle can exert two effects: it increases the effective steric bulk of the bidentate ligand and furthermore favors or impedes certain geometries in transition metal complexes. Square planar complexes of divalent metals of the Ni group (Ni, Pd, Pt) are stabilized by angles around 90°, while wider angles support both the oxidation state 0 and trigonal or tetrahedral geometries. The corresponding effect on the energetic sequence of the orbitals thus also potentially results in an altered reactivity. Wide bond angles support reductive eliminations, but the effect of monoligation (T-shaped tricoordinated complexes) is more pronounced than the use of bisphosphines with a large bond angle [44]. In addition, the donor strength of the phosphines can be varied accordingly to further modulate the reactivity.

ligand	β_n [°]	%V_{bur}^a
dppe	88	51.4
dppp	91	52.2
BINAP	93	55.6
dppf	99	55.5
DPEphos	104	–
Xantphos	108	54.4

^{a} in [PdCl$_2$(R$_2$P-PR$_2$)]

Fig. 6.11 Natural bite angle and buried volume of bisphosphines [42, 44]

Initial, in-depth mechanistic studies were conducted in the late 1980s by J. K. Stille on the elemental steps of the cross-coupling with organotin reagents. To this day, the Stille reaction likely is the most studied and best understood Pd-catalyzed CC coupling reaction, closely followed by the Suzuki cross-coupling. The differences in the reactions are naturally found in the transmetalation itself as well as in the preceding steps for generation of the nucleophile. In addition, additional equilibria may occur within or outside the operative catalytic cycle (*off-cycle* species) [36, 45, 46].

The Heck reaction, which also constitutes a Pd(0)-catalyzed process, mechanistically differs from cross-coupling reactions by the presence of the insertion step instead of a transmetalation and the requisite coordination dynamics of the reactant and product. Pd(II)-catalyzed reactions such as the Wacker oxidation also proceed through different species in their catalytic cycles, even if many intermediates are shared. The exact mechanisms of these reactions are explained in more detail in the respective subchapters.

Comparison of Palladium with Alternative Metal Catalysts

Ni, Cu, and Fe catalysts differ significantly in their corresponding mechanism when compared to Pd-catalyzed reactions. The narrower understanding of catalytic cycles mediated by alternative metals also limits a comparable rational approach to optimizing a coupling reaction. Regardless, a few general principles can be derived (see Table 6.2).[II]

Table 6.2 Comparison of different metal catalysts

	Pd	*Ni*	*Cu*	*Fe*
electrophiles	I, Br, Cl, OTf, OTs	I, Br, Cl, OTf, OTs, $OPO(OR)_2$, OSO_2NR_2, OCO_2R, OR	I, (Br)	I, Br, Cl, OTf, OTs
coupling modes	CC, CN, CO, CS, CF	CC, CN, CO, CS	CN, CO	CC
ligands	PR_3, NHC	N,N-/N,O-chelates, PR_3, NHC	"ligand-free", N,N-/ N,O-chelates	"ligand-free", N,N-/N,O-chelates, seldom NHCs, bisphosphines
CC-nucleophiles	Sn, B, Zn, Mg, Si, Cu	B, Zn, Mg	–	Mg, seldom Zn, B
catalytic cycle	Pd^0/Pd^{II}	Ni^0/Ni^{II} possibly Ni^I/Ni^{III}	Cu^I/Cu^{III}	*unknown* (likely Fe^{-II}/Fe^0 or Fe^I/Fe^{III})

Nickel is capable of acting as a two-electron donor similar to palladium, but possesses a $4s^2 3d^8$ electron configuration instead of $4d^{10}$. In principle, oxidative additions and reductive eliminations can therefore be carried out in an analogous fashion. Unlike palladium, however, these can occur via both Ni(0)/Ni(II) and Ni(I)/Ni(III) catalytic cycles. The smaller size makes it more nucleophilic on the one hand, making Ni complexes more reactive towards substrates considered quasi-inert under Pd catalysis [17]. The higher oxophilicity of Ni also enables the activation of tosylates, phosphates, and esters for a C-O bond cleavage by coordination of the metal center to the oxygen [14, 47]. On the other hand, the coordination

---

[II] Additional variations beyond the listed range have been reported but typically require highly specialized catalysts or only possess a focused substrate scope.

sphere experiences a higher steric repulsion when using sterically demanding ancillary ligands. Thus, while modern phosphines and NHCs can be employed as ligands to some extent, challenging reactions can often also be carried out with less complex ligands or even under "ligand-free" conditions [17]. The rate of oxidative addition with various leaving groups roughly corresponds to the trend in the presence of Pd catalysts: Ar-I > Ar-Br > Ar-Cl > OTs > OTf. Nevertheless, the relative reactivity of aryl tosylates, triflates, phosphates, etc. largely depends on the choice of Ni catalyst [48]. The oxidative addition of aryl halides may proceed via a radical halogen abstraction in lieu of the concerted, two-electron mechanism [49]. As with the other Pd alternatives, the disadvantage of Ni-catalyzed reactions is rooted in its need for harsher reaction conditions (temperature, amount of catalyst), which limits the usefulness for more complex substrates. However, Ni(II)-Alkyl intermediates are more reluctant than the respective Pd complexes to undergo β-hydride eliminations, thus offering a potential for C_{sp^3}-coupling reactions [14].

The role of **copper** catalysts has been specifically studied in the context of CN-coupling reactions (Buchwald–Hartwig aminations) [13]. Cu may proceed via two-electron processes analogous to Pd, but rather through the oxidation states Cu(I)/Cu(III). For alkyl halides, a stepwise, radical mechanism involving Cu(I→II→III) has also been proposed [50]. Unlike Pd(0), Cu(I) is stable even in the absence of added stabilizing ligands. Coordinating ligands are in contrast required to dissolve the Cu precatalysts (e.g., CuI). Cu(I) is isoelectronic to Ni(0), not Pd(0). Accordingly, certain similarities are observed in their reactivity and coordination behavior: The smaller atomic radius leads to shorter bonds and, due to the positive charge of Cu(I), a higher Lewis acidity with an increased affinity for O- and N-based ligands. In contrast to Ni, however, it is less nucleophilic than Pd. The even smaller coordination sphere compared to Ni results in the fact that a fine-tuning of ligands, for example the addition of electron-rich, bulky phosphines, cannot be exploited as a lever, as the steric demand of these phosphines is too large for Cu. Given the lack of stereoelectronic modulation of the metal center by the ligands, only aryl iodides and activated aryl bromides as substrates can be efficiently converted. Since Cu(III) is rather unstable compared to Pd(II) and Ni(II), it is assumed that all Cu(III) species should be intrinsically short-lived. Oxidative additions are therefore reversible. As a consequence, it was postulated that a changed sequence of elementary steps occurs to take this instability into account: While Pd experiences the oxidative addition prior to the transmetalation, under Cu catalysis the coupling partner coordinates first ($L_n Cu^I(NHR)_1$) before the electrophile is added and would then quickly afford the coupling product under reductive elimination. The central function of the ancillary ligands primarily culminates in the circumvention of a double transmetalation to give the unreactive complex $L_n Cu^I(NHR)_2$. In order to facilitate the rate-determining oxidative addition despite the lower nucleophilicity of Cu(I), a stabilizing interaction with the leaving group X of the electrophile was postulated. This requisite interaction makes the oxidative addition more dependent on the electronic properties of the leaving group compared to the effect of the R^1 moiety. For this reason, there are no broadly applicable methods for the coupling of aryl chlorides, -triflates, and tosylates. These distinct characteristics lead to the fact that the substrate scope for both the electrophile and the amine component can be considered complementary for Pd and Cu, even if the use of Cu catalysts typically necessitates higher amounts of the respective transition metal complexes [13].

Iron can span the broadest range of formal oxidation states of the metals discussed herein (–II to +VI). This imparts an electronic ambivalence strongly correlated with the oxidation state, so that it can act as a nucleophile in low oxidation states and as a Lewis acid in higher ones, for example $FeCl_3$ serving as Lewis acid in Friedel–Crafts acylations. In addition, like nickel, it can enable both single and two-electron transfer processes, with single electron transfers and thus radical intermediates often constituting competitive or even dominant pathways. Different spin states can furthermore be assumed, which can also significantly influence the reactivity of the metal center. Fe-catalyzed reactions are therefore particularly susceptible to unwanted side reactions if not tightly controlled, especially homocouplings. The modulation of the energetic sequence of the d-orbitals and thus the redox properties can be supported by using redox-active chelate ligands (*non-innocent*) or metallic co-catalysts. One rationalization for the successful use in Kumada cross-couplings — in contrast to much less frequent combination with other organometallic reagents — may lie in the purported interaction of the iron center with Mg^{2+} cations [51]. The reactivity towards different electrophiles is complementary to analogous Pd or Ni complexes: While under Pd/Ni catalysis, aryl iodides or bromides oxidatively add fastest, Fe catalysts are best suited for the corresponding chloride, triflate, or tosylate substrates [27]. In Kumada cross-couplings, a catalytic cycle encompassing initial oxidative addition, transmetalation, and reductive elimination is assumed, in analogy to palladium-mediated reactions [27]. Similar to Cu, Fe has a high affinity for O- and N-based ligands, but in general, the dissociation of ligands from the metal center is several orders of magnitude faster than with the nobler homologues. A strong metal-ligand interaction to ensure ligand attachment is therefore crucial for controlling selectivity, which is why chelate or strong π-acceptor ligands are resorted to comparatively often [51].

The role of metals is not always clear and can lead to misinterpretations without careful control experiments. Based on the lack of reproducibility of their iron-catalyzed CN cross-coupling by other groups, Bolm and co-workers demonstrated that the reaction is in fact a Cu-catalyzed process in cooperation with the Buchwald group. When using high-purity $FeCl_3$ (>99.99%), the arylation of pyrazole, benzamide, phenol, and thiophenol was only rendered viable through the addition of Cu_2O traces (5–10 ppm), thus matching the previously reported yields (see Fig. 6.12) [52].

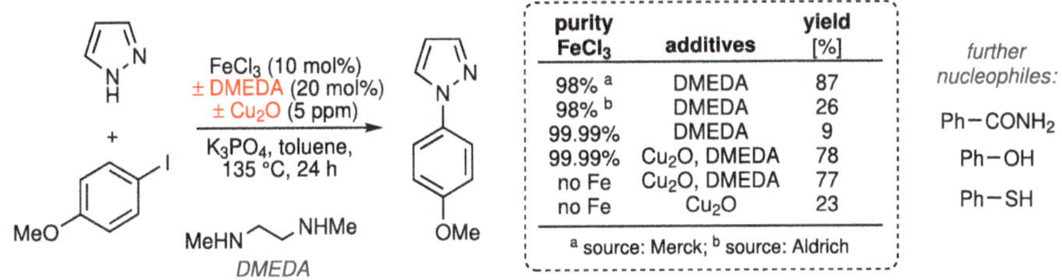

purity FeCl$_3$	additives	yield [%]
98% [a]	DMEDA	87
98% [b]	DMEDA	26
99.99%	DMEDA	9
99.99%	Cu$_2$O, DMEDA	78
no Fe	Cu$_2$O, DMEDA	77
no Fe	Cu$_2$O	23

[a] source: Merck; [b] source: Aldrich

Fig. 6.12 Influence of Cu traces on supposedly $FeCl_3$-catalyzed CN-, CO-, CS-couplings [52]

This accordingly might skew the view on all iron-catalyzed reactions in which no strictly Cu-free iron catalysts were used and also makes some "metal-free" reactions appear questionable if proper control experiments are omitted. Nevertheless, even with carefully conducted checks, the true nature of the catalyst may initially resist elucidation and remain obscure. As a recent case of the alleged Pd-free Suzuki cross-coupling demonstrates, standard test methods did not show significant traces of transition metals in the reaction mixture or the amine catalyst. Only the combination of various control experiments allowed the detection of "hidden" Pd reservoirs and clarified the operative mechanism [53–56].

6.1.1 Formation of the Catalytically Active Species

The generation of the active catalyst from the employed precatalysts, while officially not part of the catalytic cycle, still has a decisive influence on the product yield yet often remains the most poorly understood step of the entire process. The effect of this initiation step is obvious when considering the typical screening of different metal sources for optimizing a conversion in a methodological development. The complexity of the activation step in the reaction mixture often results in its omission from detailed, mechanistic studies.

Pd(0) is unstable and sensitive to air. The same applies to most low-valent Ni(0) and Fe species, which require handling under an inert atmosphere, preferably in a glove box [57]. Despite it being most efficient in some instances to directly use a Pd(0) source, most methods routinely rely on stable Pd(II) compounds as catalyst precursors, which subsequently have to be reduced *in situ*. For most Pd-ligand combinations, a 14-electron palladium(0) complex of the type [PdL$_2$] is assumed to be the catalytically active species, which is coordinatively unsaturated and therefore equally reactive. Its existence, although long postulated [58], could ultimately be confirmed experimentally as well [59]. For sterically very demanding ligands such as PtBu$_3$, biaryl phosphines (SPhos, XPhos, etc.) as well as many NHCs, however, the monosubstituted complex [PdL] constitutes the active species, a highly reactive 12-electron complex [39, 60]. For Pd(OAc)$_2$, the formation of a catalytically active anionic complex [Pd(OAc)(PPh$_3$)$_2$]$^-$ was also observed under certain reaction conditions [61]. The research

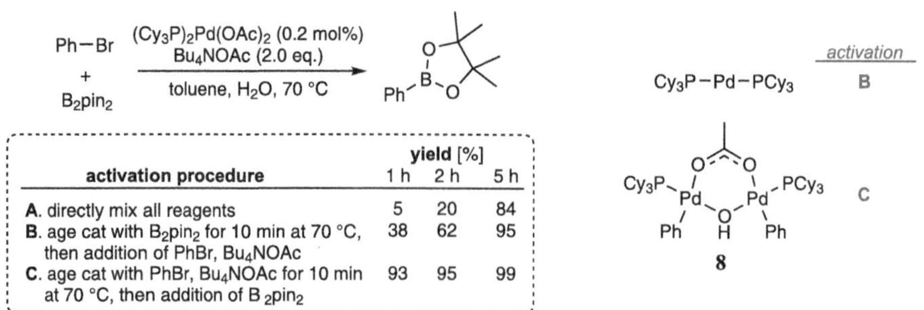

activation procedure	yield [%]		
	1 h	2 h	5 h
A. directly mix all reagents	5	20	84
B. age cat with B$_2$pin$_2$ for 10 min at 70 °C, then addition of PhBr, Bu$_4$NOAc	38	62	95
C. age cat with PhBr, Bu$_4$NOAc for 10 min at 70 °C, then addition of B$_2$pin$_2$	93	95	99

Fig. 6.13 Influence of catalyst activation on a C–B cross-coupling [62]

department at Bristol-Myers Squibb (BMS) investigated the influence of the activation protocol under typical Miyaura borylation conditions and found that different Pd species were formed depending on the activation procedure (see Fig. 6.13) [62].

Aging the catalyst with B_2pin_2 yielded the expected species $Pd^0(PCy_3)_2$ (method B), but the base-mediated reduction afforded the unstable complex $Pd^0(PCy_3)_1$, which oxidatively added PhBr and formed the dimeric Pd species **8** (method C). **8** was competent as a catalyst and displayed a fourfold higher reactivity than $Pd^0(PCy_3)_2$ due to the elimination of the induction period. Without careful control of the activation (method A), the reaction was observed to become slower. The resultant Pd species was not further determined under these conditions.

As is evident by the work of BMS and others, various ways to generate the catalytically active species exist depending on the reaction conditions [62, 63]. The Pd^0 complex can either be accessed by the direct addition of a palladium(0) precursor (e.g., $Pd(PPh_3)_4$, Pd_2dba_3) with subsequent ligand dissociation or by an *in situ* reduction of a suitable palladium(II) compound (see Fig. 6.14) [64].

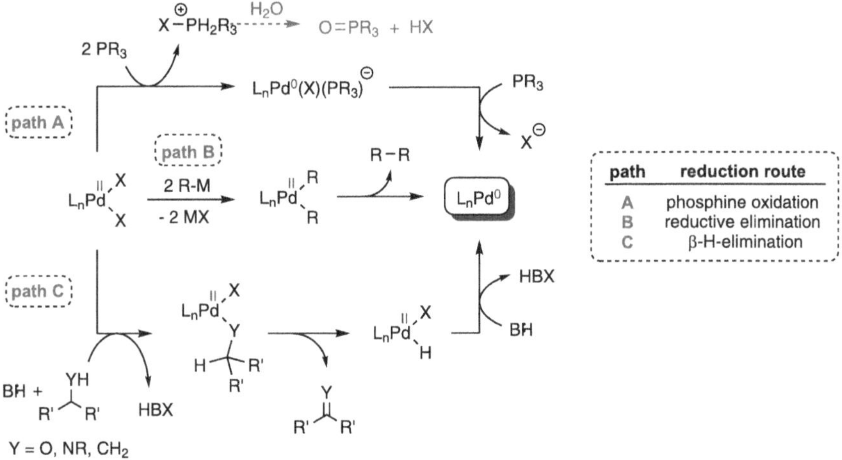

Fig. 6.14 Possible mechanisms for generating the reactive Pd^0 species [65]

In the presence of phosphine ligands, the Pd^0 species is typically generated via oxidation of the phosphine, with $AcO\text{-}PR_3^+$ being formed as by-product in the absence of H_2O and $O=PR_3$ in aqueous environment. The addition of catalytic amounts of H_2O for the activation of the precatalyst can even lead to a significant improvement in yields [66]. If a precatalyst other than $Pd(OAc)_2$ is used, hard nucleophiles (e.g., OH^-, OR^-, H_2O, etc.) are additionally required. In phosphine-free catalyst systems, the active species is obtained either by double transmetalation with R-M and reductive elimination of R-R, or by β-hydride elimination of alcohols or amines and subsequent reductive elimination of HX. Pd^0 complexes can dissociate one or more ligands to form the 14-electron species, depending on the respective ligands (monodentate or chelating ligands).

Two Pd sources, Pd_2dba_3 and $(COD)Pd(CH_2TMS)_2$ merit a further comment: Pd_2dba_3 is the most frequently encountered Pd(0) precatalyst by virtue of its high stability even in the absence of an inert atmosphere. Many methods use Pd_2dba_3 as no further reduction of the metal is necessary, which could potentially consume an equivalent of phosphine or nucleophile. However, its innate stability caused by the high bond strength of the dba also results in a low reactivity towards a successive ligand exchange [67, 68]. In practice, it can therefore be helpful to heat a Pd_2dba_3 solution in the presence of the desired ligand for a few minutes at $120\,°C$ before adding the other reagents and initiating the reaction [69].

$(COD)Pd(CH_2TMS)_2$ is a precatalyst, which extremely efficiently generates $(COD)Pd(0)$ via reductive elimination of $TMSCH_2CH_2TMS$. This species is highly reactive and can form the active catalyst without any induction period through rapid coordination of further ligands or substitution of COD. Therefore, it is excellent for excluding the effect of the Pd source in a ligand screening or for preparing stoichiometric amounts of certain Pd complexes (see Fig. 6.15).

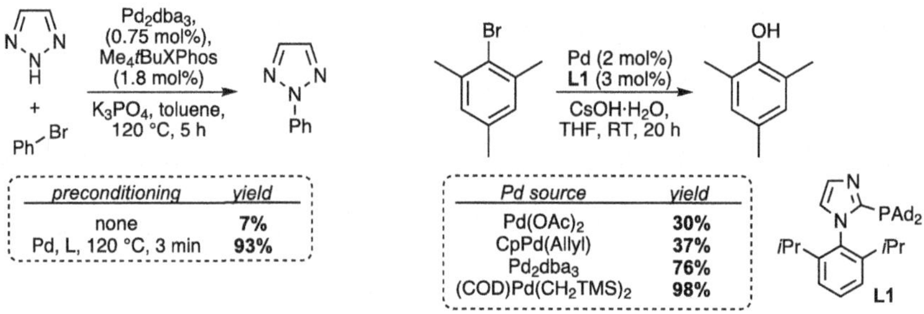

Fig. 6.15 Influence of catalyst preactivation and the Pd source [69, 70]

As can be seen from the cited examples, the efficient generation of the catalytically active species constitutes an often underestimated element of method optimization. In order to minimize the impact of a potential induction period, a series of Pd(II) compounds has been introduced (**1–4**), which can generate Pd(0) particularly quickly and comprise a clearly defined Pd/ligand stoichiometry (see Fig. 6.16) [65].

The aminobiphenyl palladacycles introduced by Buchwald can release the desired catalyst under base-mediated elimination of carbazole. They were specifically developed for use with phosphines. The successive modifications from the 2nd to the 4th generation (G2–G4) have expanded the usable ligand scope to include sterically highly demanding phosphines, increased the solution stability, and prevented the stalling of a reaction by the generated carbazole by introducing the *N*-methyl derivative [71]. Pd-NHC complexes can instead be efficiently formed with PEPPSI precatalysts (*pyridine-enhanced precatalyst preparation, stabilization, and initiation*) developed by Organ *et al.* [72]. Alternatives to the PEPPSI and Buchwald precatalysts are systems based on allyl, crotyl, cinnamyl (**3**), or indenyl complexes (**4**). Originally designed for NHCs as ligands, they have subsequently been expanded for combination with electron-rich, sterically demanding phosphine ligands such as XPhos [73].

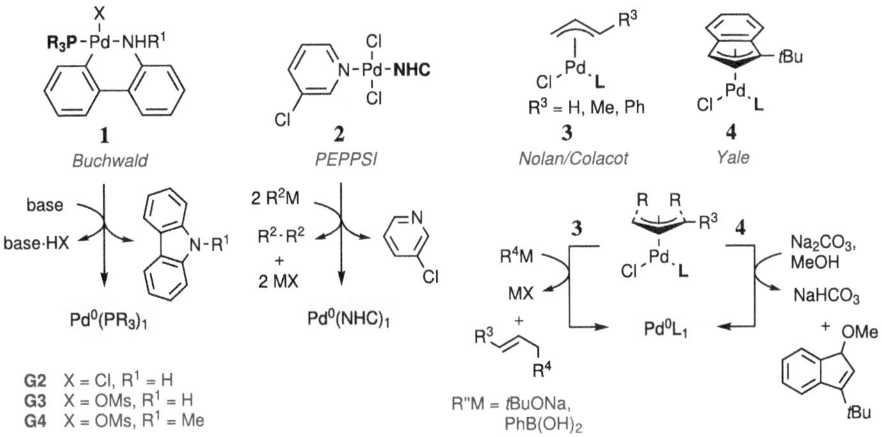

Fig. 6.16 Activation paths of the Pd(II) precatalysts **1–4** [65]

Particularly the complex derived from cinnamic acid shows the fastest activation (cinnamyl > crotyl > allyl) and can most efficiently suppress the formation of unreactive dimers resulting from comproportionation [65]. The indenyl complex **4** (*Yale* precatalyst) devised by Hazari and co-workers can fully suppress dimer formation due to repulsion by the sterically demanding *t*Bu substituent. Beyond these relatively frequently encountered precatalysts, a multitude of alternative systems has been devised. They typically find niche applications but often either lack broad applicability or are too costly and laborious to prepare [65].

6.1.2 Oxidative Addition

Apart from the identity of the ligand and the transition metal, the rate of oxidative additions is to a large degree influenced by the substrate and its electronic and steric properties [74]. Low-valent palladium(0) is relatively electron-rich and consequently nucleophilic, and usually inserts irreversibly into polarized C-X bonds, in which the carbon atom is rather electron-poor [75]. The reactivity of the electrophile mirrors the quality of X as a leaving group and decreases with increasing strength of the C_{sp^2}-X bond in the order

$$I \quad > \quad OTf \; \simeq \; Br \quad \gg \quad Cl$$

Vinyl electrophiles are more reactive compared to the corresponding arene substrates, which are in turn converted more rapidly than benzyl or allyl derivatives. Alkyl electrophiles possess the highest reluctance to oxidatively add to Pd(0). The addition modes for C_{sp^2}- and C_{sp^3}-hybridized substrates differ fundamentally, with the conversion of alkyl halides or triflates being a development that only became widely accessible upon introduction of modern ligand systems.

Based on extensive theoretical and experimental investigations, a concerted interaction between the reactive $[L_nPd]$ species and the electrophile in a three-center transition state (**11**) was postulated for the oxidative addition of **aryl** halides. A recent study employing computational and experimental techniques indicated that the concerted three-center transition state **11** might only apply for some substrate/catalyst combinations and that a Pd-arene interaction allows for an alternative four-center displacement transition state (**12**). Triflate electrophiles always seem to follow **12**. Whereas monoligated complexes tend to prefer the concerted transition state **11**, most bidentate chelate ligands induce a preference for the displacement pathway **12** [76]. Analysis of aryl halides bearing a variety of different substituents yielded a Hammett correlation indicating that during oxidative addition a positive charge forms at the metal center and a negative charge in the aromatic group (**13**, see Fig. 6.17).

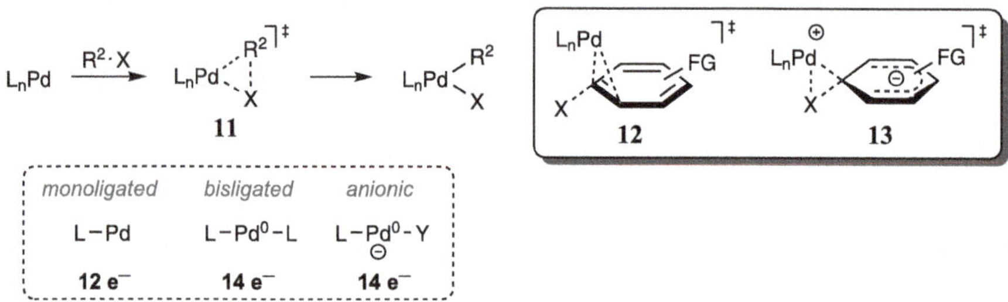

Fig. 6.17 General scheme of oxidative addition

The elemental step is accordingly accelerated by electron-withdrawing substituents, which can stabilize the charge build-up in the transition state (see Fig. 6.18) [77]. This reactivity correlation can be roughly applied to all coupling reactions that have an oxidative addition as the rate-determining step.

FG	k_{rel}	$\sigma_p^-,\ \sigma_m$
4-CN	326	1.00
4-CO$_2$Et	66	0.64
4-CF$_3$	48	0.65
3-CN	47	0.56
3-Cl	11	0.37
3-CO$_2$Et	7.4	0.37
4-Cl	5.7	0.19
3-F	5.4	0.34
3-OMe	1.5	0.12
H	1.0	0.00
4-F	0.93	-0.03
4-OMe	0.26	-0.26

Fig. 6.18 Relative reactivity of various donor- and acceptor-substituted aryl bromides in Negishi cross-coupling reactions [77, 78]

The use of electron-rich ancillary ligands enhances the electron density at the palladium and accelerates the oxidative addition. If they are also sterically demanding, they raise the ground state energy level of the palladium complex and thus enable a net lowering of the activation barrier for an oxidative addition as the transition state is less affected by the steric influence. The coordination environment of the metal center seems to depend on the reaction conditions and thus has not yet been elucidated for all major catalyst/substrate combinations. Electron-rich, sterically highly demanding ligands (biaryl phosphines, NHCs, PtBu$_3$) promote formation of a monoligated 12-electron species based on the steric repulsion in PdL$_2$ complexes. PdL$_1$ is thus assumed to be the operative intermediate for all electrophiles with this ligand class. Recent studies have indicated that monoligated species may be competitive to dominant for oxidative addition even with less bulky ligands such as PPh$_3$, but more work is required on that front to arrive at a holistic picture [79].[III] Aryl iodides, bromides, and chlorides can undergo slightly different mechanisms with regard to the ligand exchange reactions at the metal center (associative versus dissociative) [80]. Chelate ligands also promote oxidative addition, as the angled structure bends the ligand backbone away from the substrate, thus facilitating an uncontested approach to the complex [81]. With monodentate phosphines having lower steric demand such as PPh$_3$, the coordination sphere has been observed to vary with the identity of the leaving group and the chosen reaction conditions. Different studies suggest a PdL$_2$ complex as the reactive species for aryl iodides. For aryl bromides and chlorides, both mono- and bisligated complexes seem plausible. Both monoligated and anionic Pd complexes have been postulated as being operative in the oxidative addition of aryl triflates, depending on the reaction conditions [82–84]. The more electron-rich the metal center is, the less anionic ate-complexes are preferred. [84]

Alkyl halides or triflates have long posed particularly challenging substrates, as the insertion into the C$_{sp^3}$-X bond is significantly slower than with the respective aryl or vinyl analogues. Based on theoretical investigations, this was attributed to the involvement of the π^* orbital in the d$_{Pd}$→σ^*_{ArX} interaction with aryl groups, which is absent when using alkyl halides [85]. Second of all, they tend to undergo β-hydride elimination if the transmetalation of the nucleophile does not proceed sufficiently fast (see Fig. 6.19) [19].

Various experimental studies suggested the involvement of an associative bimolecular process (S$_N$2 reaction) with concomitant extrusion of the leaving group instead of a concerted insertion for alkyl- or allyl-based electrophiles. The addition product is then obtained by subsequent coordination of the anion X$^-$ to the metal center. With allyl chlorides, a pronounced solvent dependence of the stereochemical outcome was observed in various cross-coupling reactions: The configuration of the product is a result of the oxidative addition, which proceeds with complete or nearly complete retention of configuration in weakly coordinating solvents, as would be expected in a concerted transition state. In contrast, the

---

[III] Even though the participation of monoligated complexes during oxidative addition (and to some extent transmetalation) may play a role for **all** classes of ancillary ligands, their inclusion in the discussion and figures remains limited to cases where a scientific consensus has already been reached. See Colacot, *Chem. Rev.* **2022**, *122*, 16911–17240 for further details.

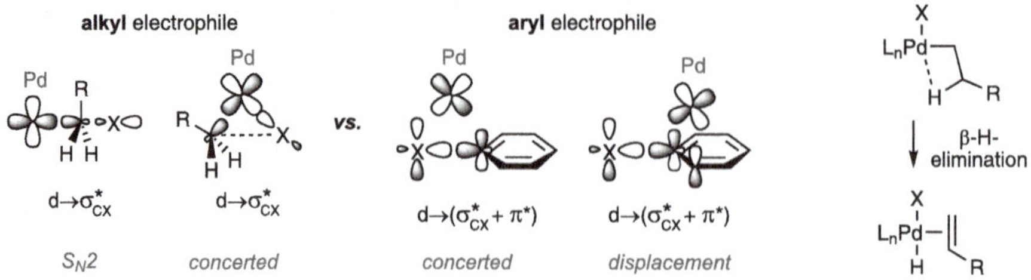

Fig. 6.19 Orbital interactions in the transition state of oxidative addition and possible β-hydride elimination [19, 76, 85]

use of polar coordinating solvents such as MeCN or DMSO yields a complete or nearly complete inversion of configuration (see Fig. 6.20) [86–88]. In the reaction of non-activated alkyl halides or tosylates, inversion was observed as the preferred stereochemical course of reaction even in weakly-coordinating THF, from which the presence of a S_N2-like mechanism was concluded. The reaction of α-sulfonyl bromides is even completely stereospecific and proceeds with full inversion of configuration. As a consequence, the oxidative addition becomes slower with decreasing nucleofugality of the leaving group. Also, sterically more hindered substrates and secondary alkyl halides are converted less rapidly.

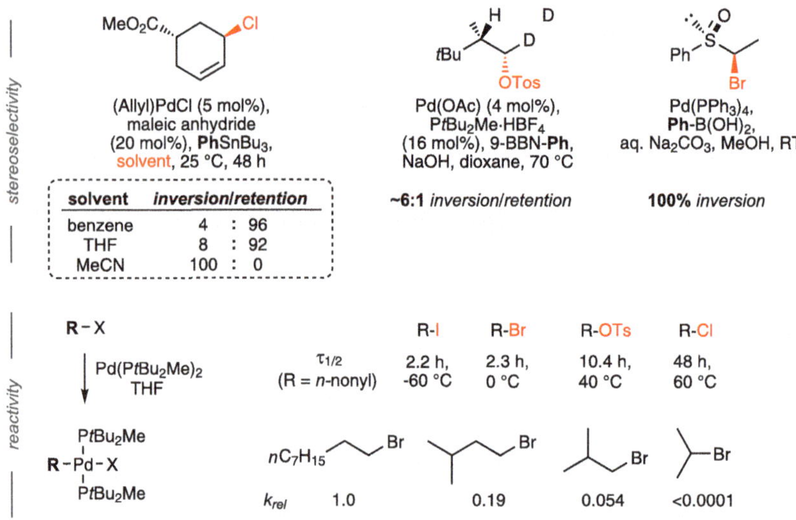

Fig. 6.20 Indications of an S_N2-like mechanism with alkyl electrophiles [87, 89–91]

Cu catalysts also proceed via an S_N2 mechanism in the oxidative addition [92, 93], while various mechanistic variants have been postulated for Fe complexes [94]. In comparison, nickel tends towards single electron transfer processes (*SET*) by virtue of its lower ligand field splitting of the d-orbital energy levels. Therefore, an electron can be more easily added to the $d_{x^2-y^2}$ orbital in square-planar Ni complexes or the coordination environment can move to a tetrahedral geometry. As a consequence, Ni catalysts more readily undergo a radical pathway than analogous Pd systems during oxidative addition, especially with alkyl electrophiles [19, 95]. Ni complexes therefore particularly lend themselves for secondary or even tertiary alkyl halides as substrates, which are hardly or not at all reactive under Pd catalysis due to the operative mechanism (see Fig. 6.21).

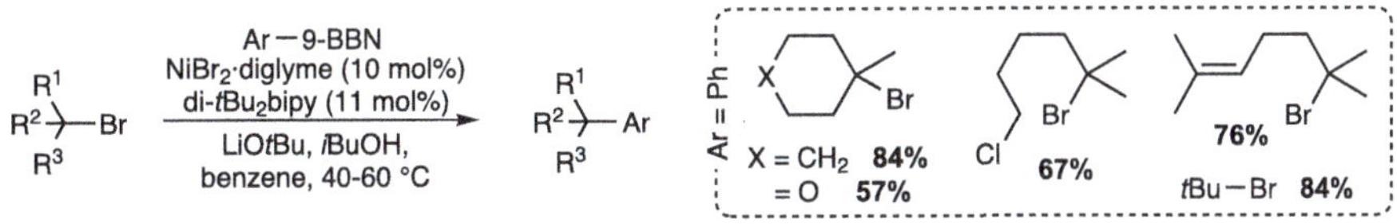

Fig. 6.21 Ni-catalyzed Suzuki coupling of *tert*-alkyl bromides reported by Fu *et al.* [96]

The formation of alkyl radicals via SET from the corresponding halides often leads to a significant amount of homocoupling, dehalogenation, or dehydrohalogenation products. However, with tailor-made catalysts, the recombination with the Ni halide complexes can proceed very efficiently, resulting in high yields of the desired coupling products [95].

In the presence of monodentate ancillary ligands, the *cis-trans* isomerization to the thermodynamically more stable complex constitutes the final step of oxidative addition. Two competing reaction paths were found in a model study for the *cis-trans* isomerization of the primary products resulting from oxidative addition (*cis*-complex **14**). These encompass either a direct or a solvent-assisted associative exchange of PPh$_3$ for a halide ligand coordinated to a second palladium complex. This results in an iodine-bridged intermediate **17**, which rearranges to give the corresponding *trans*-configured complexes **18** and **19**, respectively (see Fig. 6.22) [97].

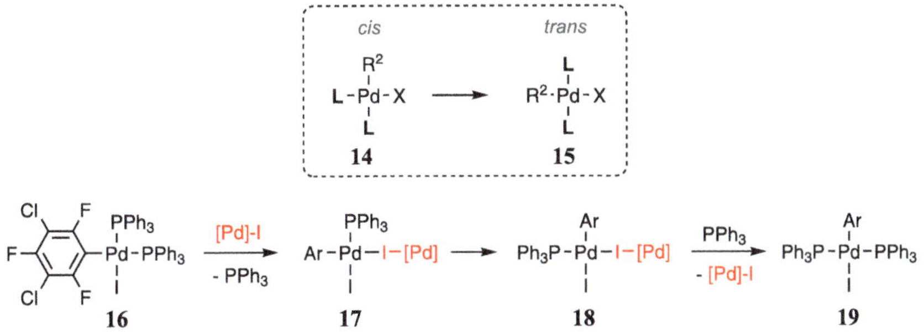

Fig. 6.22 Model reaction of the *cis-trans* isomerization [97]

For electrophiles bearing several reactive CX bonds, either one bond is intrinsically favored or the oxidative addition can be controlled by careful choice of the reaction conditions [98, 99]. The simplest case occurs when different leaving groups are present: Here, the relative reactivity of the leaving groups towards oxidative addition determines the site-selectivity of the coupling reaction, provided the step proceeds irreversibly. In addition, the reaction conditions (metal, ligand, solvent) can support or override the inherent selectivity (see Fig. 6.23).

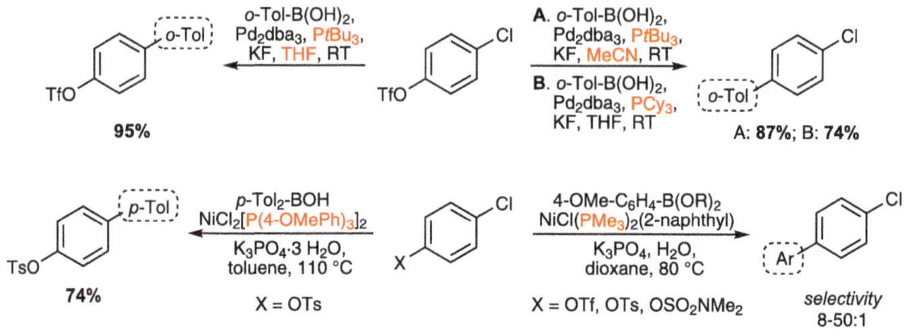

Fig. 6.23 Examples of Pd- and Ni-catalyzed methods for chemoselective cross-couplings [83, 100–102]

If two identical leaving groups are present in a substrate, a selective reaction may still be possible (see Fig. 6.24). Either the two leaving groups possess a difference in electrophilicity imposed by the intrinsic polarities of the ring carbon atoms or the substrate bears substituents that enable a control of the oxidative addition by steric shielding or can act as a directing group (passive versus active control) [76, 99, 103, 104].

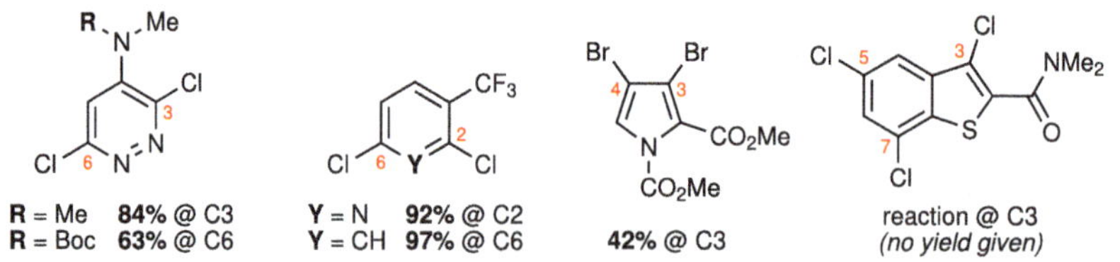

Fig. 6.24 Site-selective Suzuki cross-couplings of polyhalogenated heterocycles [99]

6.1.3 Transmetalation

The transmetalation constitutes the one step of the catalytic cycle in which the various cross-coupling methods differ. CN cross-couplings and Heck reactions contain other elementary steps or the same steps in a different order, which is why these mechanistic details will be covered in the respective sections of these methods (see Sects. 6.3, 6.4). In the simplest terms, transmetalation is a ligand exchange between two metal centers, the driving force of which is based on a gain in stability. Square-planar d^{16} or d^{14} complexes (Ni(II), Pd(II), Cu(III)) usually proceed via an associative path, i.e., the formation of pentacoordinated 18-electron species as intermediates (see Fig. 6.25).

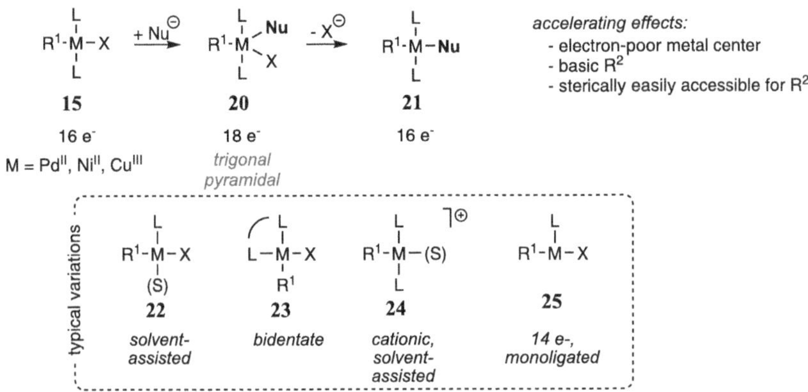

Fig. 6.25 General scheme of associative transmetalation

A multitude of variations is possible: The *trans*-configured complex **15** can initially exchange an ancillary ligand for a solvent molecule (**22**). In the case of bidentate ligands, **23** should be present as the reactive species. Triflates, tosylates, or similar counterions are only weakly coordinated to the metal center or not at all. In these cases the mechanism can transition either via the cationic complex **24** with solvent assistance or even via a monoligated 14-electron species. For sterically very demanding ligands, even with halides as the leaving group, the T-shaped complex **25** has been postulated to be the reactive intermediate. Depending on the reaction conditions, the metal center and the ligands, a broad range of possible pre- and post-transmetalation equilibria may exist.

In general, an electron-poor metal center and electron-rich nucleophiles would decrease the activation barrier of transmetalation. However, the exact elementary steps including all occurring equilibria and off-cycle species need to be taken into account, which greatly complicates the picture and makes comparative statements about different classes of organometal donors (almost) impossible. The polarity of the carbon-metal bond and thus the nucleophilicity for organostannanes, -silanes, and -boranes is rather low, which is why basic additives are required for activation of the nucleophile in Hiyama and Suzuki cross-coupling reactions (see Table 6.1). These shared traits signify a certain kinship of these three methods with regard to their mechanistic characteristics. Zinc- and lithium/magnesium-based reagents, on the other hand, are significantly more reactive and, due to their polar C-M bonds, also possess a pronounced aggregation behavior. If we turn back to enolate chemistry, where the aggregation behavior also played a significant role as illustrated in a preceding chapter, certain observations from cross-coupling chemistry may be put into perspective: a decreased reactivity was observed for lithium species with a higher degree of aggregation based on steric interactions, i.e., a low accessibility of the electrophile to the nucleophilic lithium enolate ($\rightarrow$ Sect. 2.2.1). These effects do not correlate with the electronic properties of the nucleophiles, but still might potentially influence the reactivity in transmetalation in a similar manner. A case-by-case analysis is therefore almost inevitable in most cases.

The Stille reaction in particular has been extensively studied due to the stability of the nucleophiles [45, 105]. Many of the fundamental insights about cross-couplings are therefore derived from stannane-mediated transformations. While the effect of nucleophiles with different electron-rich migrating groups has been thoroughly examined for certain catalysts and organometal donors, the comparison of transmetalation rates for different substituent classes has so far only been carried out for the Stille reaction in competition experiments. It was observed that alkynyl and vinyl groups transmetalate more rapidly than the respective aryl groups in the reaction with acyl chlorides, while alkyl groups are transferred very slowly (see Fig. 6.26) [106].

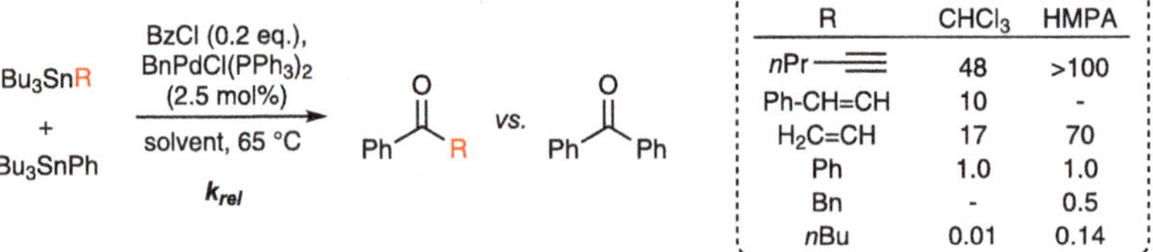

R	CHCl$_3$	HMPA
nPr—≡	48	>100
Ph-CH=CH	10	-
H$_2$C=CH	17	70
Ph	1.0	1.0
Bn	-	0.5
nBu	0.01	0.14

Fig. 6.26 Relative reactivity of different stannanes in the cross-coupling with benzoyl chloride [106]

The order of transmetalation rates can be assumed to be analogous for silicon and boron organyls [35]:

$$\text{alkynyl} \quad > \quad \text{vinyl} \quad > \quad \text{aryl} \quad > \quad \text{allyl} \quad \backsimeq \quad \text{benzyl} \quad \ggg \quad \text{alkyl}$$

For the transfer of a migrating group R, two mechanisms are generally considered as viable: an open or a cyclic transition state. The cyclic mechanism necessitates the presence of a bridging anionic ligand, typically a halide, as well as a neutral ligand that facilitates the attachment of the stannane in the coordination periphery of the metal center. This can be achieved either by weak donors such as electron-poor phosphines or sterically very demanding ligands, which prefer monoligated complexes and thus ensure an open coordination site is retained. In the case of weakly coordinating leaving groups X (e.g., triflate) as well as in highly polar and/or nucleophilic solvents, an open transition state has been postulated to remain the operative pathway, presumably via intermediate solvent complexes (see Fig. 6.27) [105, 107].

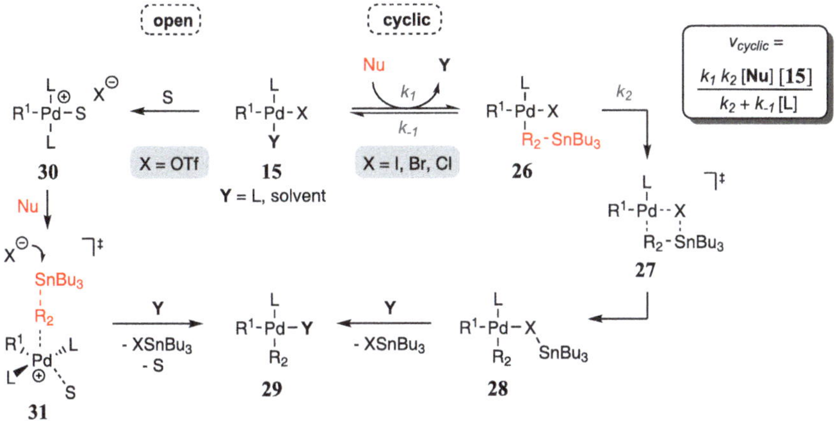

Fig. 6.27 Competition between open and closed transition state in the transmetalation of stannanes [45, 105, 107]

Apart from phosphines, the fourth coordination site (ligand Y) can also be occupied by a solvent molecule in the *trans*-configured oxidative addition complex **15**, or a T-shaped, monoligated 14 e⁻ species might be preferred instead (corresponding to **22** or **25**, see Fig. 6.25). The involvement of intermittent Pd-F complexes has also been hypothesized [108]. Following an approach of the stannane, theoretical studies have postulated the existence of a short-lived π-complex **26** for vinyl and aryl stannanes, which affords the bridged Pd-X-Sn species **28** via the cyclic transition state **27**. The proposed mechanism is consistent with the experimentally determined rate law, which also takes into account the observed decelerating effect of excess ligand [107]. In the concluding step, a phosphine or solvent molecule can be ligated again to obtain a square-planar 16-electron species, or the subsequent intermediate corresponds to a coordinatively unsaturated, T-shaped 14-electron complex. These details are highly dependent on the respective system (ligand, solvent). Theoretical investigations support the outlined, postulated elemental steps [109, 110].

The use of triflates promotes the preferred formation of a cationic solvent complex of type **30** as initial step, which can transmetalate very rapidly by virtue of its strongly electrophilic metal center. The conversion of the nucleophile is probably further supported by attack of X⁻ on the stannane in the transition state [45]. The concluding reductive elimination ensues from **29**. Contrary to the simplified scheme in Fig. 6.27, both *cis*-complexes (**29**) and the corresponding *trans*-species (not shown) can result from the transmetalation and may incur additional isomerization steps.

The viability of an open mechanism cannot be dismissed as the potentially operative pathway in halide complexes, provided the rate of such a process becomes competitive by the intermediacy a sufficiently electrophilic complex (e.g., with a strongly coordinating solvent as ligand) [105]. Also the lack of a good bridging ligand through sequestration/abstraction of the halide may force an open mechanism.

When using chiral stannanes, both an inversion and a retention with respect to the transferred alkyl group were observed depending on the reaction conditions. The stereochemical course of the transmetalation is consistent with the mechanism shown in Fig. 6.27: The open transition state proceeds with inversion at the α-carbon atom directly bound to the tin (via **34**), the cyclic path proceeds with retention of the configuration via transition state **32**. In the transfer of aryl groups, a cyclic pathway would yield **33** as transition state (see Fig. 6.28) [105].

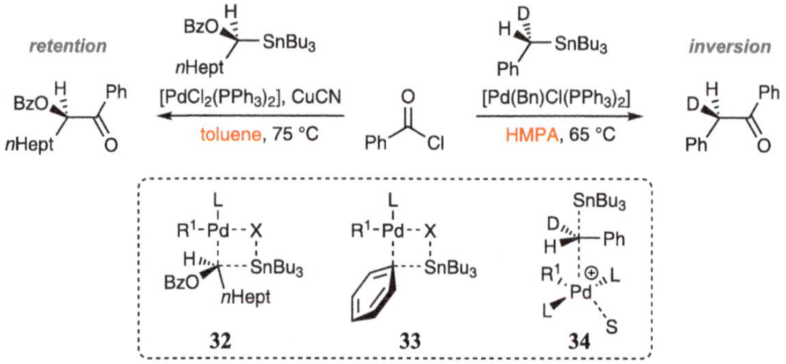

Fig. 6.28 Stereochemical course of the Stille reaction [106, 111]

Given its prevalence also in an industrial context, the elucidation of the mechanistic details in the Suzuki cross-coupling has likewise received significant attention, especially in recent years. As with the Stille reaction (and the Hiyama cross-coupling, see below), the dominant path seems to proceed through a four-membered, cyclic transition state **38** [112]. Current studies with chiral alkylboranes have additionally indicated the possibility of open transition structures such as **39**, which would rationalize the observed inversion of configuration at the nucleophile in rare cases [113]. In the Suzuki cross-coupling, the organoboron reagent cannot directly transfer R^2 due to its low nucleophilicity. The addition of Lewis bases allows the formation of a Pd-O-B intermediate **36** from the oxidative addition complex **15** and the boron nucleophile, in which R^2 is then transferred to Pd via the cyclic transition state **38** [114, 115]. Currently, two alternative processes (path A and B) have been proposed to initiate the transmetalation event, which differ in the precise role of the requisite base. Its purpose can accordingly be assigned to facilitate either the formation of a boronate from the corresponding boronic acid or of a Pd-hydroxo species **35** with elimination of X^- from **15** [116] (Fig. 6.29).

The reactions from **15** to **36** are equilibria and thus reversible. Apart from kinetic factors, the position of the equilibria likewise may influence the relative importance and thus proportion of molecules traversing path A or B. Depending on the base, the amount of H_2O, the identity of X, the type and stoichiometry of ligands, and the presence of other additives, the preferred reaction path can change [114, 117–120, 120–123]. In the presence of an excess

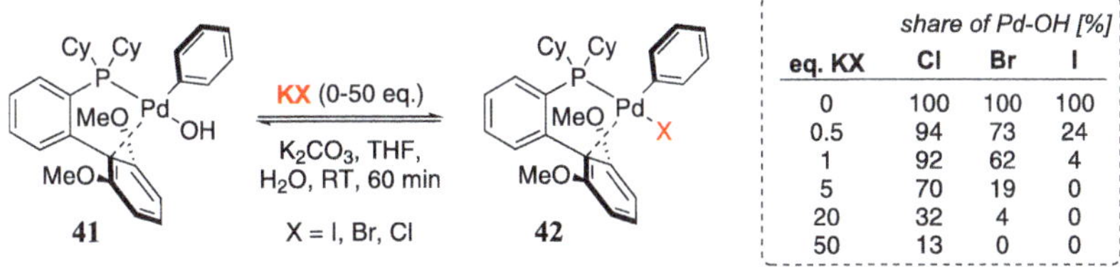

Fig. 6.29 Bifurcation in the reaction pathways of Suzuki cross-couplings [114, 116]

of ligand, **40** was also detected as a competitive intermediate under the employed conditions [114]. The counterion X likewise possesses a dual purpose: Since a multifold excess of X^- is incurred over the course of the reaction under catalytic conditions, the possibility of a change in the dominant mechanism during the course of the reaction has been postulated. When using biaryl phosphine ligands (L = SPhos), the equilibrium increasingly shifts from **41** to **42** as the excess of halide accumulates (see Fig. 6.30) [119]. Whether this also occurs with other classes of ligands or can be transferred to further cross-couplings has not yet been systematically investigated [121].

	share of Pd-OH [%]		
eq. KX	Cl	Br	I
0	100	100	100
0.5	94	73	24
1	92	62	4
5	70	19	0
20	32	4	0
50	13	0	0

Fig. 6.30 Dependence of the equilibrium position **41/42** on the stoichiometry and the identity of the halide [119]

Anhydrous conditions also require the addition of a Lewis base, which typically consist of alkoxylates or salts such as KF. The detailed mechanism of the transmetalation has not been elucidated under these conditions. Whether similarly two competing reaction paths exist, via a Pd-alkoxy complex or an alkoxy boronate, has not yet been studied.

As with Suzuki cross-couplings, the nucleophilicity of organosilicon reagents is insufficient to ensure an effective conversion. Hiyama cross-couplings therefore also require activation of the silicon organyl, either through exogenous additives (e.g., the addition of F^-) or through endogenous modification, e.g., by resorting to intrinsically more reactive homologues such as silanolates. Acyclic silanes (e.g., trimethylvinylsilane) do not undergo CC coupling in the absence of fluorides. Given the analogy of Si and Sn, pentacoordinated intermediates are assumed to be the reactive species and their relevance has been confirmed in theoretical studies [124]. In the transmetalation step, both inversion and retention of the stereochemistry at the transferred group at the organosilicon reagent were observed depending on the temperature and solvent. As with the Stille and Suzuki cross-coupling, this outcome likewise indicates the kinship of Si, Sn, and B by possessing competing open and cyclic transition states (see Fig. 6.31) [125].

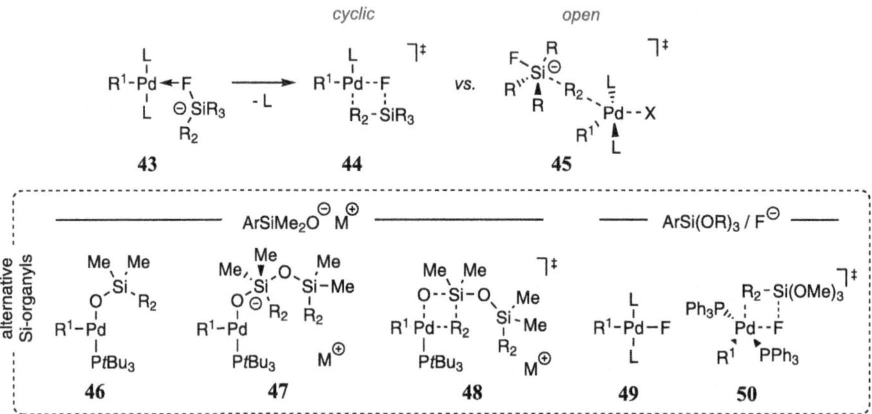

Fig. 6.31 Stereochemical course of the Hiyama reaction and postulated transition states [126, 127]

Upon activation of the silane by fluoride, the formation of an intermittent, fluorine-bridged complex **43** was postulated, which can transfer R^2 via the cyclic transition state **44**. Strongly coordinating solvents and high temperatures destabilize the weak Pd-F-Si bridge and instead favor the acyclic transition state **45**, which results in an inversion of the configuration [125]. Whether the transfer of the aryl group is preceded by either an activation of the silane by forming the pentacoordinated species R^2SiR_3F or an X→F exchange at the Pd center to $[PdR^1L_2F]$ (**49**) has not yet been fully clarified. Theoretical and experimental studies rather indicate an X→F substitution at the Pd prior to coordination of the silane as operative pathway [124, 126]. The mechanism of transmetalation with silanolates as nucleophiles was studied in $Pd(PtBu_3)_2$ complexes. Analogous to Suzuki cross-couplings, the intermediacy of an oxo-bridged complex **46** was postulated, which activates the Si center by adopting a pentavalent, anionic structure with a second equivalent of silanolate (**47**). In the cyclic transition state **48**, R^2 is then transferred to the Pd center [127]. For fluorine-activated trialkoxysilanes, the role of fluoride anions was elucidated employing electrochemical analytical techniques. $R^2Si(OR)_3F$ lacks the requisite reactivity for a direct transmetalation (cyclic or open), which is why initial formation of fluorine complex **49** is assumed to occur. The associative, cyclic transition state **50** subsequently affords to the desired coupling product [126].

The high basicity and polarity of the C-M bond indicate that the conclusions drawn from the mechanistic studies on Stille/Suzuki/Hiyama cross-coupling are not directly transferable to the more reactive Zn/Mg/Li homologues. Given the considerable reactivity of the organometallic reagents, the transmetalation step of the Negishi and Kumada cross-coupling are less well-understood. In particular, the stoichiometric amounts of salts and the multitude of possible metal-metal transmetalations and Schlenk equilibria contribute to this significant complexity [128]. While general statements consequently are difficult to arrive at, a few trends can be derived.

The reaction with organozinc reagents produces ZnX_2 over the course of the reaction. As various theoretical and experimental studies have shown, zinc salts can inhibit several steps of the transformation by forming Pd-Zn clusters that can act as catalyst sink (see below) [129–131]. The presence of ammonium, alkali or alkaline earth halides furnishes $MX \cdot ZnX_2$ complexes, thus counteracting the generation of the heteroleptic Pd-Zn complexes. LiBr has been observed to be the most efficient additive in this regard [132]. Li salts can additionally break up organozinc aggregates and form higher order zincates $(R^2ZnX_3M_2)$ [132]. Since zincates have also been ruled out as transmetalating species under certain conditions, their exact role has not yet been conclusively elucidated [133]. The presence of salt additives increases the polarity of the solvent and thus may also facilitate the transmetalation of ArZnX, which is less reactive than Ar_2Zn (see Fig. 6.32) [134].

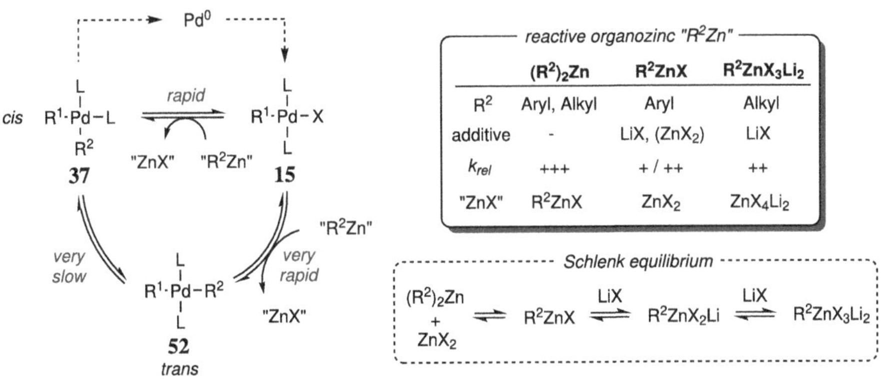

Fig. 6.32 Relevant equilibria of transmetalation in Negishi cross-coupling and possible reactive nucleophiles [133–135]

The formation of the *cis*-configured complex **37** required for reductive elimination proceeds rapidly from **15**. However, the unproductive *trans*-complex **52** is generated even faster as a competitive reaction. Given the reluctance of a *cis-trans* isomerization in the form of a direct conversion from **52** to **37**, a detour via a retro-transmetalation and formation of **15** en route to **37** seems prudent [135]. Prior to transmetalation, a μ-X-complex **53** initially enables the incorporation of the nucleophile according to theoretical studies. The associative transmetalation via **54** transfers R^2. The postulated transmetalation product **55**, in which

$ZnX_2(S)$ remains partially attached through R^2, finally eliminates the Zn salt and affords **52** or the *cis*-configured intermediate **37**. Isomerizations and/or reductive elimination may ensue [130, 135]. Chiral organozinc reagent can be synthesized in principle to further probe the nature of the transmetalation as was accomplished in the Suzuki/Stille/Hiyama cross-coupling. Given the difficulty in their handling and limitedly configurational stability this presented an insurmountable hurdle for the longest time [136, 137]. In a recent study, chiral Zn nucleophiles could be employed for the first time in Pd-catalyzed cross-couplings. They delivered the corresponding CC coupling products while retaining the configuration of the transferred moiety, which would strongly indicate the occurrence of a cyclic transition state like **54** [138].

Secondary transmetalations can additionally be observed with unwanted back-transfer of R^1 to the zinc organyl both with aryl and alkyl nucleophiles, as well as β-H eliminations when alkyl groups are directly ligated to Pd [139, 140]. The detection of the homocoupling product as well as alkenes or ArH can therefore be attributed to these competing side reactions. The role of the nucleophilic organozinc reagent becomes even more convoluted if the secondary equilibria are considered as well. An extensive network of transition metal-zinc complexes has been postulated especially for low-valent Pd(0) complexes and to a limited extent for Pd(II) as well as at Ni(II) complexes (e.g., **57–59**), of which only some furnish the desired product. A significant proportion are inhibitive or indifferent with regard to product formation (see Fig. 6.33) [131].

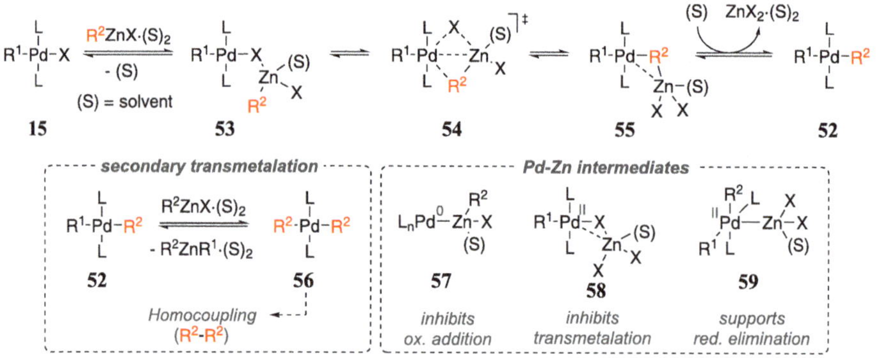

Fig. 6.33 Details of the product-forming transmetalation, the secondary transmetalation and postulated Pd-Zn intermediates [130, 131, 139, 140]. The respective *cis*-complexes were omitted for clarity.

The reactivity of organomagnesium and -lithium compounds is superior to the corresponding zinc organyls due to the more polar C-M bond. Both classes of nucleophiles nevertheless share a similar mechanism. Given the enhanced reactivity, the synthetic breadth of the Kumada cross-coupling is narrower compared to the corresponding zinc-based reagents, which can be attributed to a more limited spectrum of tolerated functional groups. The inherent reactivity, coupled with the extensive equilibria of Grignard reagents in solution, has so far precluded a systematic study on the mechanism of the Kumada cross-coupling. The detailed mechanistic work on the Negishi coupling therefore allows conclusions to be drawn for the Kumada cross-coupling to a certain extent by virtue of their apparent analogous reactivity. The Schlenk equilibrium illustrated herein for organozinc reagents was originally reported for Grignard reagents, which is why a similar scope of reactive Mg nucleophiles can be assumed [141].

Hoffmann and co-workers successfully investigated the stereospecificity of the transmetalation step in the Kumada cross-coupling using an enantiomerically enriched secondary alkyl Grignard reagent **60** at $-78\,°C$ as probe. Almost complete retention was observed, indicating a cyclic transition state (**62**). However, given the tendency of magnesium organyls for single electron transfer (SET), a change in the mechanism at temperatures above $-78\,°C$ cannot be ruled out. The Ni-catalyzed variant of the Kumada cross-coupling similarly proceeded with full retention of the configuration from **60** to **61**, while Fe and Co catalysts displayed a partial enantioerosion (see Fig. 6.34) [142].

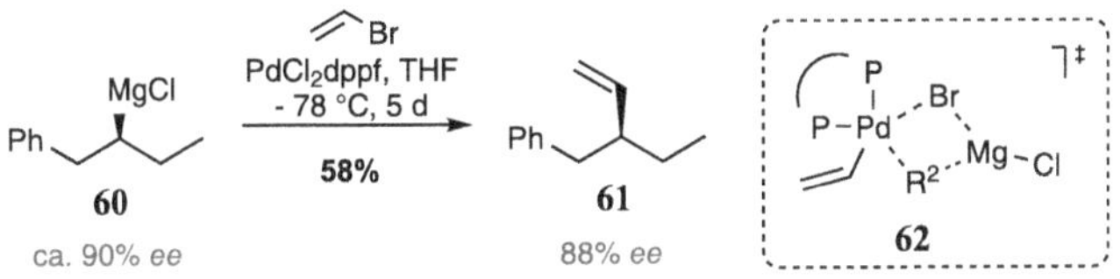

Fig. 6.34 Stereochemical course of the Kumada cross-coupling and possible transition state of transmetalation [142]

The nature of Pd-Mg interactions (whether inhibitory or activating) has not been studied in detail yet. Nevertheless, the high yields observed for Fe-catalyzed cross-couplings with Grignard reagents indicate that transition metal-Mg interactions might play a significant role. Mechanistic investigations have been conducted especially on the Ni-catalyzed Kumada coupling. In addition to the routinely assumed monomeric species, the involvement of homoleptic dinickel complexes was also observed with certain Ni-ligand combinations [143]. Many Ni-catalyzed processes furthermore proceed via a radical mechanism instead of a classic two-electron process.

The *in situ* formation of the organometallic species sets the Sonogashira reaction apart from the other cross-coupling reactions. The reaction most often relies on Cu-based nucleophiles, which are generated from copper(I) salts. Nevertheless, copper-free coupling methods have additionally been developed, and Zn- or Al-acetylides have also been successfully used as nucleophiles [144]. It is assumed that the terminal hydrogen atom of the alkyne is acidified by coordination to the copper(I) salt (**64**) and can thus be deprotonated by the added base, after which the organocopper species **65** is obtained. **65** can subsequently transfer the alkyne to the Pd complex (see Fig. 6.35).

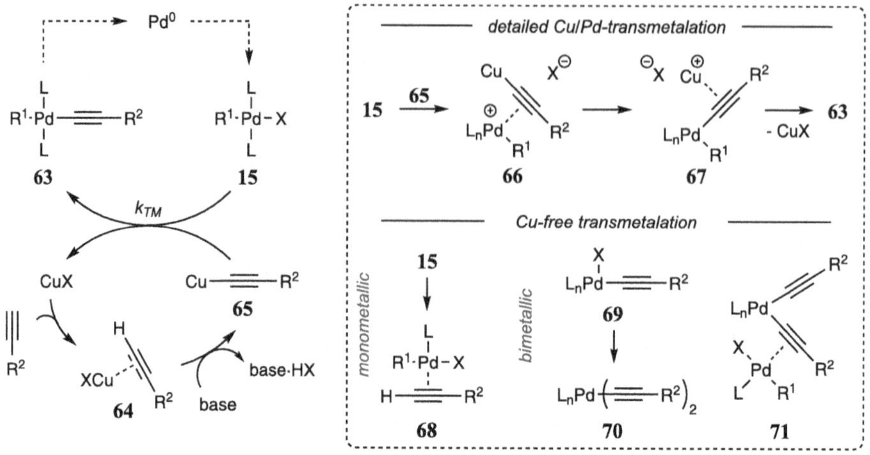

Fig. 6.35 Postulated mechanisms of Cu-cocatalyzed and Cu-free transmetalations in the Sonogashira reaction. The remaining (solvent) ligands on Cu are omitted for clarity [145–148]

Since the amines typically employed as base do not possess the requisite basicity for efficient deprotonation of the alkyne, an additional activation of the terminal CH bond is necessary. The purported Cu-alkyne complex **64** [149] should display sufficient acidity. Since the corresponding Ag-alkyne homologues have been successfully identified by NMR spectroscopy in Ag-cocatalyzed Sonogashira reactions [150], its existence can be assumed by analogy. Further details of the secondary Cu cycle have only been elucidated to a limited degree [145]. The transmetalation step was experimentally determined to be rate-determining with aryl iodides and PPh₃ [151]. Compared to other transmetalation variants in which Pd iodides were observed to transmetalate very slowly, the observation of the transmetalation being rate-determining should not automatically be transferable to other electrophiles and catalysts. The first in-depth study on the elementary steps of transmetalation was reported by Chen *et al.* The authors concluded a slow associative ligand exchange of **15** and **65**, which should afford the heteroleptic species **66**. Following a fast σ/π rearrangement to **67**, CuX can then dissociate [146].

The published methods using Cu-free conditions obviously do not proceed via the generic mechanism relying on **64/65**. A direct activation of the alkyne by coordination to the Pd center was assumed (**68**) to initiate the catalytic cycle, which would then provide **63** in an intramolecular reaction mediated by the base [152, 153]. Recent studies suggest an alternative bimetallic mechanism involving a synergistic interaction of two Pd complexes (via **15→69→70→71→63+69**), in which a Pd-Pd transmetalation was postulated [148].

6.1.4 Reductive Elimination

The reductive elimination constitutes the concluding, product-forming step of the catalytic cycle. The reaction involves the covalent coupling of two ligands with concomitant reduction of the metal's oxidation state by two. Barring alkyl-alkyl eliminations, the CC coupling possesses a low activation barrier and proceeds very rapidly. In contrast, the Pd-catalyzed formation of CX bonds (X = N, O, S, F) displays a significantly higher activation energy. These coupling reactions have long been unattainable due to their slow reductive elimination and only became accessible with the advent of more efficient ligands to accelerate the limiting step (Fig. 6.36) [40, 154].

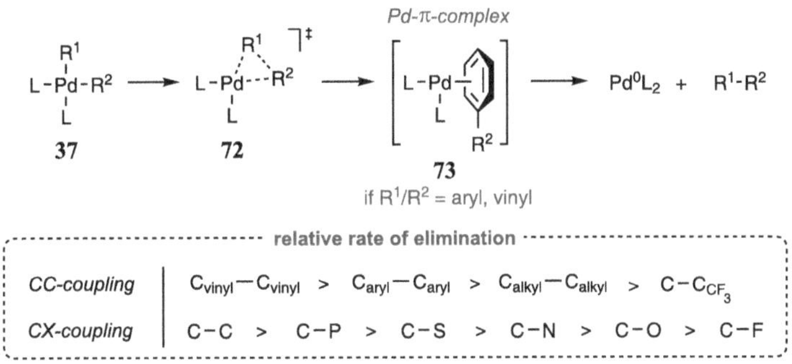

Fig. 6.36 Reaction scheme of the reductive elimination [155]

As the counterpart to oxidative addition, the reductive elimination typically is an irreversible, concerted process starting from the *cis*-configured complex **37**. The involvement of the π-system in the orbital interactions of reductive eliminations allows vinyl and aryl groups to eliminate significantly faster than reactive ligands lacking a π-system [36]. Following transition state **72**, a transient π-complex **73** can occur by coordination to the metal center prior to the dissociation of the product in the case of products bearing vinyl, aryl, or alkynyl groups. In CC and CX couplings, the extent of the driving force for the generation of R^1-R^2 is determined by the relative strength of the Pd-R^1/R^2 bonds. The reductive elimination of alkanes is significantly slower due to the absence of a π-system that might stabilize the transition state. Only C-CF$_3$ coupling reactions display a more pronounced reluctance at bond formation by virtue of their strong Pd-CF$_3$ bond with partial ionic character [155, 156]. Fluorinated alkyl groups can display additional secondary orbital interactions with the metal that further modulate the barrier for reductive elimination [157]. The more electron-poor the metal center, the greater its tendency to increase the electron density via a reductive elimination of ligands. Electron-poor ancillary ligands accordingly raise the rate of reductive elimination. This contrasts with the described effect of the high Pd-CF$_3$ bond strength, as this directly influences the driving force of the reaction and presumably also impacts the transition states. The more electron-rich the groups to be eliminated are, the easier the Pd-R^1/R^2 bonds can be accordingly cleaved in a heterolytic fashion.[IV] Electron-poor phosphines and electron-rich groups R^1/R^2 therefore constitute an optimal combination to facilitate reductive elimination (see Fig. 6.37) [40].

R	k_{rel}
Me	>600
Bn	>250
CH₂COPh	31
CH₂CF₃	1.7
CH₂CN	1
CF₃	– a

a no reaction

X	k_{rel}
Cl	0.30
H	1
Me	1.5
OMe	1.7

Fig. 6.37 Influence of the electronic properties of the reactive ligands on reductive elimination [158, 159]

The influence of the electron density in the reactive ligands increases with the hardness of the coordinated groups. C-substituents are thus particularly malleable, also CN bond formations can be selectively promoted. CS or CP bond formations display only a limited sensitivity to the electronic properties of the moieties bound to the more polarizable S or P atom [40].

Interestingly, the reductive elimination of two reactive ligands with good σ-donor ability is slower than the combination of an electron-rich with an electron-poor residue. This seems to suggest a synergistic effect, in which the groups R^1/R^2 are cleaved fastest when they have more strongly opposing electronic properties. The cause of this counterintuitive finding is not yet fully understood, although there are several explanatory approaches [40].

Sterically demanding ancillary ligands such as PtBu₃ facilitate the dissociation of a ligand to a coordinatively unsaturated T-shaped complex with 14 valence electrons. T-shaped mono-ligated Pd(II) complexes undergo reductive elimination much faster than the corresponding square-planar bisligated species [155]. Sterically demanding, electron-rich phosphines or NHCs thus favor oxidative addition and reductive elimination and accelerate cross-coupling, unless transmetalation is the rate-determining step. In the case of bisphosphine chelate ligands, the elimination of the coupling product is particularly favorable if ligands with wide bond angles are coordinated. This is attributed to a destabilizing steric repulsion with a concomitant reduction of the R^1-Pd-R^2 bonding angle, which facilitate a reduction of the coordination number at the Pd center [44, 160]. Alternatively, the dissociation of one of the bisphosphine arms can occur to form the coordinatively unsaturated, transient 14-electron

---

[IV] This is not entirely correct for aryl ligands. For monoaryl complexes, the rate is faster with more electron-poor aryl groups due to resonance effects dominating over inductive effects. See Hartwig, *Inorg. Chem.* **2007**, *46*, 1936 for details.

complex [161]. The origin of the increased reactivity, which is also observed in oxidative addition, can be understood using correlation diagrams (see Fig. 6.38).[V]

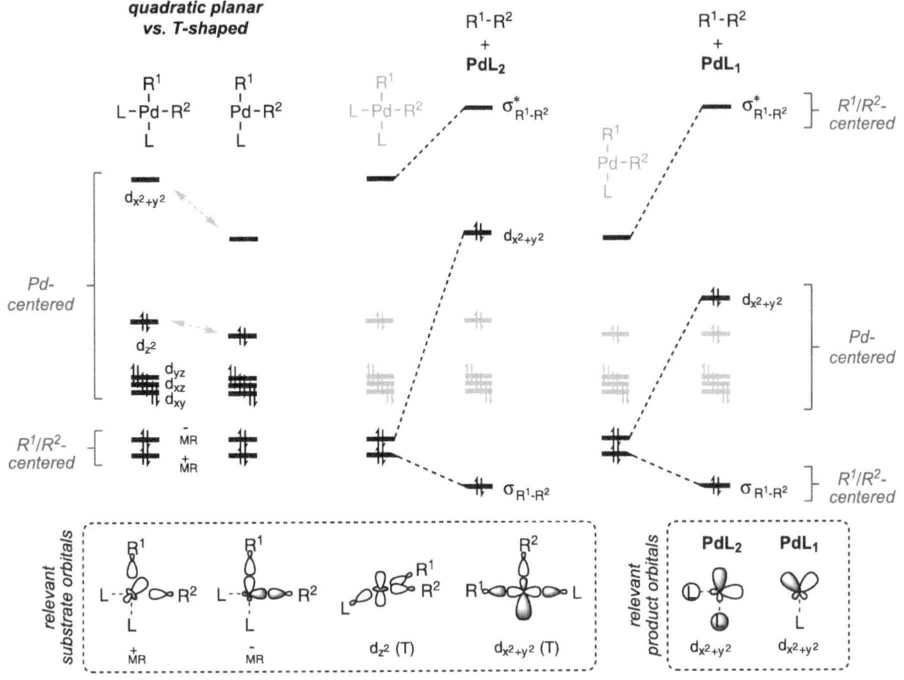

Fig. 6.38 Correlation diagrams of reductive elimination from square-planar and T-shaped Pd complexes and relevant MOs for reactants and products [162–164]

In a direct comparison of the $PdL_2R^1R^2$ and $PdL_1R^1R^2$ species, the position of the d_{z^2} and $d_{x^2+y^2}$ orbitals changes by removing a ligand under rehybridization. The Pd-R^1/R^2 orbitals σ_{MR}^+ and σ_{MR}^- are largely ligand-centered (in total 2×2 electrons each), while the MOs with substantial d character are metal-centered. This results in the d^8 configuration of Pd with the oxidation state +2. The degenerate orbitals d_{xy}, d_{xz} and d_{yz} do not change significantly when altering the coordination periphery and during the R^1/R^2 elimination — they can therefore be excluded from the analysis. Maintaining the overall symmetry, the symmetric σ_{MR}^+ MO is converted into the $\sigma_{R^1-R^2}$ orbital during reductive elimination, while the antisymmetric σ_{MR}^- MO affords the new $d_{x^2+y^2}$ orbital. This transfers an electron pair from a ligand-centered orbital to the metal center and changes the oxidation state from II→0. The new $d_{x^2+y^2}$ orbital is antibonding in a PdL_2 fragment, while it is nonbonding in a PdL_1 species (L lies in the nodal plane) and therefore has a lower energy [162, 163].

[V] The orbital designations were taken for easier understanding from the nomenclature of ligand field theory for pure metal fragments. The correct designations can be found in the cited sources.

An analysis limited to the ground state energies in reactants and products would be misleading: Since the difference in reaction rate occurs in both reductive elimination and oxidative addition, a net energy gain in elimination would lead to a **higher** energy barrier during oxidative addition. In fact, the opposite is experimentally observed. The origin of the decreased activation energy upon the transition of the geometry to T-shaped complexes is therefore mainly rooted in the energetic level of the transition states and reactive intermediates [165].

First-row transition metals (Ni, Cu, Fe) have a lower activation barrier for reductive elimination compared to Pd. This can also be rationalized by the change in the energetic levels of the relevant metal-centered orbitals, as the energetic rearrangement of the MOs by rehybridization requires more energy in Pd [162]. Accordingly, reductive elimination in Pd complexes typically constitutes the rate-determining step in C-heteroatom coupling reactions, whereas Cu-, Ni-, or Fe-based catalysts it is less of a hurdle (see Fig. 6.39).

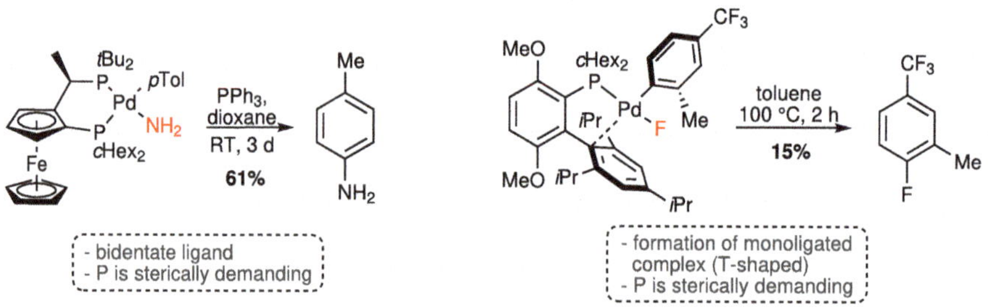

Fig. 6.39 Examples of challenging reductive CX eliminations and positive influence of the ancillary ligand on the rate [160, 166]

The reductive elimination is stereospecific in a concerted mechanism and proceeds with retention of configuration [167]. In rare cases, however, a stepwise, ionic mechanism for CX couplings can also be observed in Pd complexes [168, 169].

6.1.5 Off-Cycle Species and Inhibition of the Catalytic Cycle

While the handles for influencing the catalytic cycle of the various cross-couplings depicted in Fig. 6.8 are well understood, there are a multitude of reactions that lead to the generation of intermediates lying outside the catalytic cycle (*off-cycle* species). In the best case, they can serve as a metal reservoir if their formation is reversible. However, often a decomposition of the catalyst is caused by irreversible ligand loss and subsequent agglomeration. For example, unreactive, colloidal *palladium black* can form when soluble Pd clusters exceed a certain size [170–172]. An understanding of these processes is therefore advisable or even necessary during the optimization of a reaction beyond a laborious "trial and error" approach [173] (Fig. 6.40).

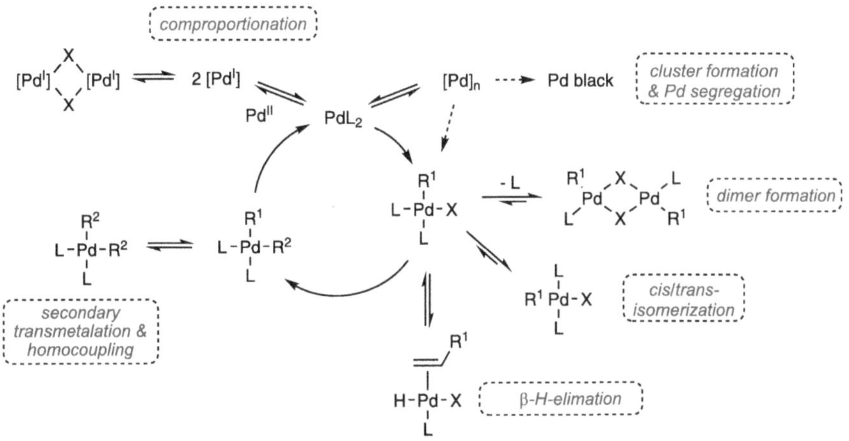

Fig. 6.40 Select possible off-cycle species and decomposition pathways in Pd-catalyzed cross-couplings

The identification of the resting state of transition metal-catalyzed couplings is relatively straightforward when relying on phosphines as ligands, as the reaction can be studied without interference using ^{31}P-NMR spectroscopy. Here, the most abundant species correlates with a slow subsequent conversion. If, for example, the transmetalation product dominates, the reductive elimination should be the slowest step of the catalytic cycle. As a result, the choice of a more suitable ligand system should result in an acceleration of the rate-determining elementary step (cf. Fig. 6.9). The groups of Buchwald and Hartwig successfully exploited this approach to gain an understanding of the limiting step and subsequently could develop a solution in many challenging coupling reactions. For example, it was observed that a Suzuki cross-coupling stalled when employing unprotected, nitrogen-rich heterocycles as electrophiles if acidic $N_{sp^2}H$ functionalities were present in the substrate. Theoretical investigations were able to rule out an oxidative addition or reductive elimination as the rate-determining step. It was noted that the inhibition correlates with the pKa value of the heterocycles (see Fig. 1.43, Sect. 1.1), which is why the formation of dimeric, nitrogen-bridged Pd complexes was indicated as a thermodynamic sink. In addition to increasing the reaction temperature to overcome the barrier of the reverse reaction from the dimer **74**, a change of the organometallic reagent to a more nucleophilic species was also identified as a viable approach for these challenging substrates (see Fig. 6.41) [174].

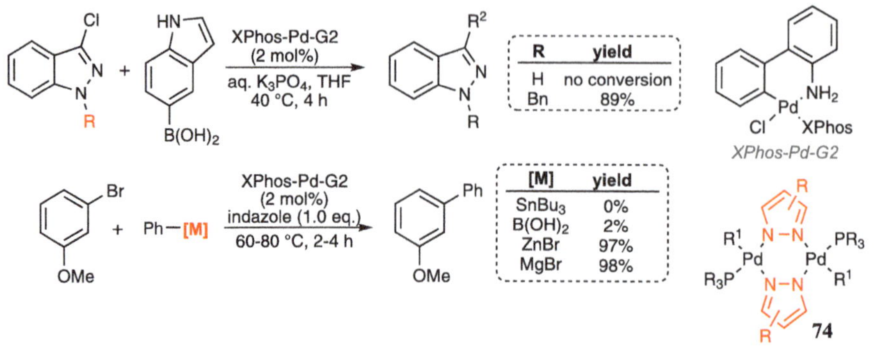

Fig. 6.41 Formation of off-cycle dimers of the type **74** in the Suzuki coupling of unprotected, N-containing heterocycles [174]

6.2   CC Cross-Couplings

As became evident in Fig. 6.2, classic cross-couplings for the construction of C-C single bonds are related by a common mechanism. Only the identity of the nucleophile and the addition of certain activators and additives distinguish these methods. Although the Suzuki cross-coupling is the most commonly encountered method for Pd-catalyzed CC couplings [10], all methods are characterized by inherent advantages and disadvantages, which in many cases justify their use over an alternative cross-coupling (see Table 6.3).

Table 6.3 Advantages and disadvantages of different CC cross-couplings [3, 24, 25]

	R^2-M	advantages	disadvantages
Stille	SnBu$_3$	- excellent FG tolerance - neutral conditions - tolerates H$_2$O	- high toxicity of stannanes - difficult removal of Sn residues from crude product
Suzuki	B(OR)$_2$ BF$_3$K	- high FG tolerance - many nucleophiles commercially available - many variants of boron derivatives viable - H$_2$O often cosolvent	- basic conditions - purification of boron reagents very challenging or not possible - high cost of nucleophiles
Negishi	ZnR ZnX	- medium to high FG tolerance - many nucleophiles commercially available - neutral to slightly basic conditions	- basic conditions - sensitive to H$_2$O - (partial) homocoupling possible - nucleophiles can hardly be stored
Kumada	MgX, Li	- conversion of unreactive substrates possible - Mg/Li-organyls often cheap	- very basic conditions - sensitive towards H$_2$O - mediochre FG tolerance - high share of homocoupling possible
Hiyama	SiR$_3$	- high FG tolerance - neutral to slightly basic conditions	- unreactive silanes require strong activation (typically additives) - laborious synthesis of complex silanes
Sonogashira	Cu	- high FG tolerance - alkynes can be used directly	- basic conditions - electron-poor alkynes not reactive

All cross-couplings can convert aryl or vinyl iodides, bromides, and triflates to the corresponding products with high yield using aryl-based nucleophiles. The choice of a method for these substrates is therefore not primarily dependent on the question of yield, but on secondary factors: The complexity of purification (removal of by-products such as stannanes in the Stille cross-coupling) as well as the individual experience with a given method. On an industrial scale, the availability and price of the starting materials and/or special ligands as well as the usable range of solvents (taking into account safety and sustainability aspects) play an equally important role. The compatibility of functional groups is usually not a critical issue, as the yields generally tend to be high in aryl-aryl couplings. However, when moving to heterocyclic or alkyl-derived electrophiles or nucleophiles, the canon of "standard" methods may quickly shrink: With these more challenging substrates, the kinetics of undesired side reactions often determine which cross-coupling possesses the optimal reactivity profile in a given transformation. Thus, the rate of transmetalation must be higher than competing processes (see Fig. 6.42).

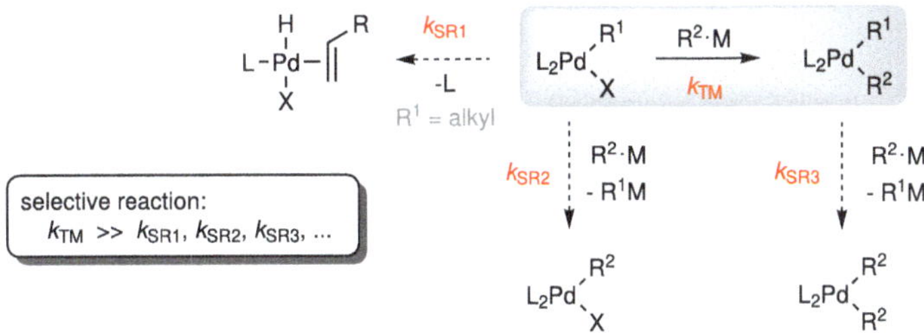

Fig. 6.42 Competition of various undesirable side reactions

Alkyl electrophiles can now be considered established and reliable as substrates following the progress of the last decades. The problem of β-H elimination can be circumvented as long as the transmetalation is fast enough ($k_{TM} > k_{SR1}$). A hallmark of the more polar nucleophiles such as organozinc (Negishi) or organomagnesium reagents (Kumada) is their enhanced reactivity compared to silanes or stannanes by virtue of their higher nucleophilicity. These coupling reagents are therefore particularly suitable for the conversion of the less reactive alkyl halides, which is why the Negishi cross-coupling in particular is considered the method of choice for the conversion of these electrophiles. With vinyl groups as coupling partners, *E/Z*-isomerization can sometimes be observed following oxidative addition via Pd-Pd transmetalations [175].

Regarding the nucleophiles, the availability of preparative methods for the synthesis of a desired organometallic reagent determines whether certain cross-couplings are feasible for a given substrate class at all: With alkyl nucleophiles, organozinc reagents dominate whereas vinyl nucleophiles commonly are based on boron and tin, as the access to these reagents with these metals either tolerates a wide range of functional groups or the preparation is operatively simple. In addition, the stability of the organometallic compounds also plays a role to some extent as a preparation on demand is not always feasible or desirable.

6.2.1 Synthesis of the Organometallic Reagents

All Sn/B/Zn/Si organometallics can be prepared via transmetalation from the corresponding Li- or Mg-organyls. Since organolithium compounds generally display high reactivity and only select electrophilic functionalities are tolerated (ester, ketone, etc.), this accordingly limits the scope of available Sn/B/Zn/Si reagents accessed via this route. This approach is typically employed for aryl and vinyl nucleophiles (see Fig. 6.43).

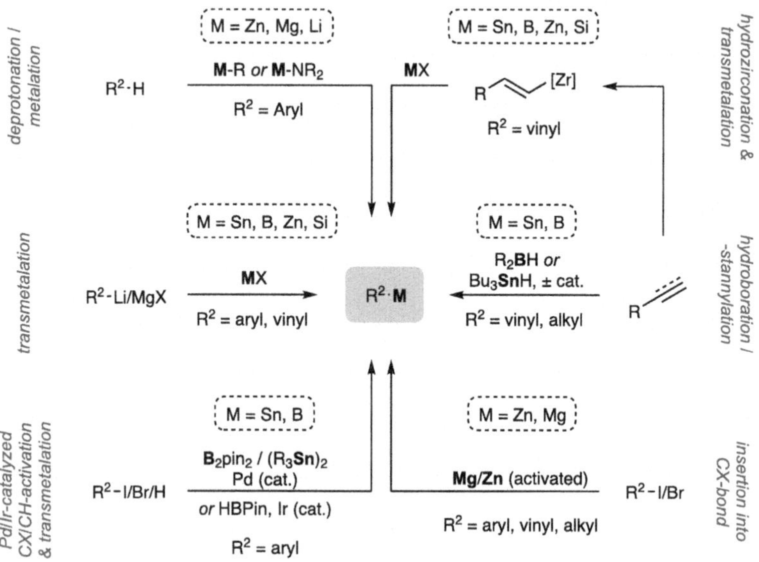

Fig. 6.43 Overview of synthetic routes to organometallic reagents [35, 176–178]

For direct transmetalation, metal salts or, in the case of boranes, trialkoxyborates are commonly used. Since the reaction can be carried out at very low temperatures, typically at −78 °C, hardly any side reactions occur — provided the functionalities in the starting material are stable when forming the Li or Mg organyls (see Fig. 6.44) [179].

The synthesis of highly functionalized organomagnesium reagents was significantly advanced by the Knochel group [179, 184–186]. The insertion into C-I bonds proceeds more rapidly than into C-Br bonds. The more electron-poor the electrophiles are, the faster the desired Grignard reagent is formed. The presence of Li salts catalyzes the halogen-metal exchange, as well as dramatically enhancing the reactivity and chemoselectivity of

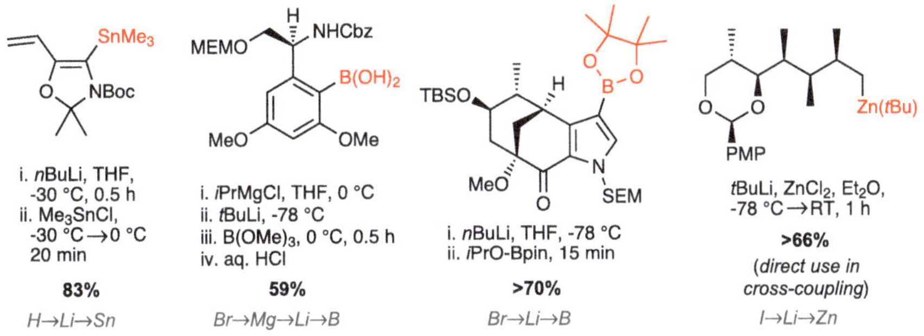

Fig. 6.44 Transmetalations from organolithium compounds [180–183]

the resultant organomagnesium reagents. Thus, aryl nucleophiles bearing nitrile or ester functionalities can be prepared by reacting the corresponding aryl iodides and bromides with iPrMgCl·LiCl (so-called *Turbo-Grignard* reagent) via transmetalation [187, 188]. The addition of LiCl presumably shifts the Schlenk equilibrium towards the formation of diorgano compound iPr(THF)MgAr·Li(THF)$_3$Cl·iPrX, which was postulated to be the operative reagent in a recent study [189]. Grignard reagents of the type sBu$_2$Mg·LiOR are even more reactive and enable a Cl-Mg exchange of very electron-rich aryl halides [190]. The resulting diorganozinc compounds similarly possess an enhanced reactivity compared to organozinc halides and can perform the rate-determining transmetalation more rapidly [191]. Alternatively, substituted (hetero)arenes can be synthesized via a directed ortho-metalation. In addition to Mg amides, the respective Zn amides can also be utilized to access the desired nucleophiles, both of which are similarly improved by addition of LiCl (see Fig. 6.45) [192–194].

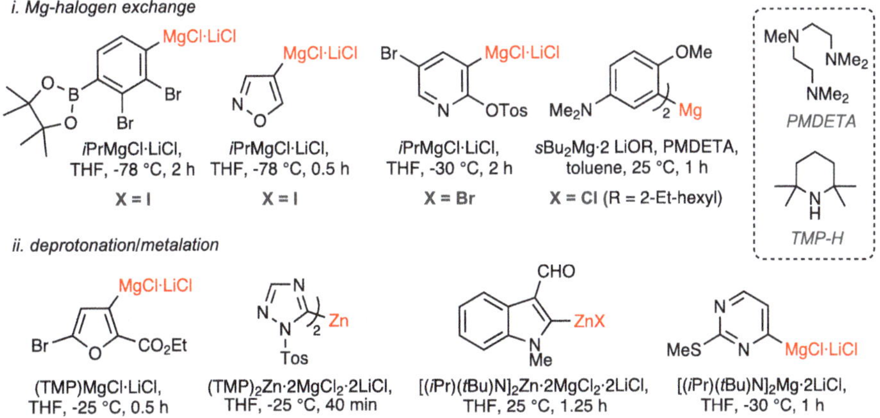

Fig. 6.45 Mg-X exchange and deprotonation/metallation [190, 192–197]

Apart from aryl nucleophiles, the respective vinyl compounds are frequently employed as well in CC cross-couplings. Their synthesis usually relies on the hydrostannylation or -boration of the corresponding alkynes. In addition, alkenes can be converted via the same route to metal alkylates. Since the hydroboration of alkenes is typically faster than the conversion of alkynes, a hydrostannylation is preferred over a hydroboration for alkynes (see Sect. 3.3). The addition of boranes is highly chemo-, regio-, and diastereoselective, with the terminal regioisomer of the alkenyl-/alkyl-borane being preferentially formed (*anti*-Markovnikov addition). Catecholborane and 9-BBN are used most often as boron source. The rate of hydroboration is strongly influenced by both electronic and steric factors, with electron-rich, sterically unhindered substrates adding the borane most rapidly (see Fig. 6.46) [198].

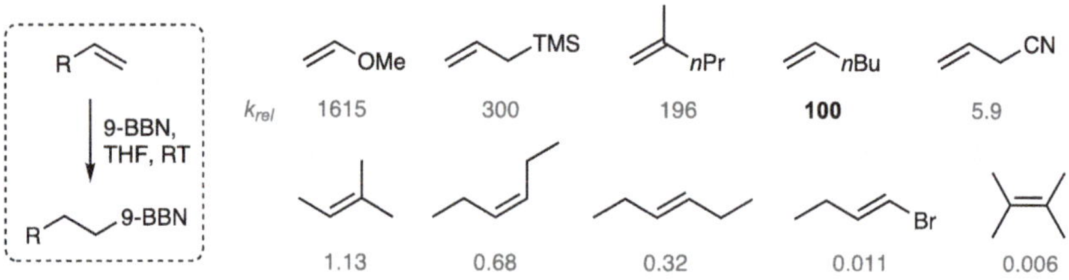

Fig. 6.46 Relative reactivities in the hydroboration of olefins [198]

When using Bu_3SnH, the addition of a radical initiator such as AIBN or Et_3B/air can result in a radical pathway, which may afford an *E/Z* mixture if alkynes are used as substrates [199]. Alternatively, the reaction can be carried out in the presence of a transition metal catalyst with boranes as well as stannanes (see Fig. 6.47). In hydrostannylation, Pd catalysts facilitate a C-Sn bond formation at the terminal carbon atom of the double or triple bond, while Mo catalysts favor the internal product [200]. A second variant is the use of a hydrozirconation, followed by a transmetalation of the intermittent vinylzirconocene to the desired metal reagent.

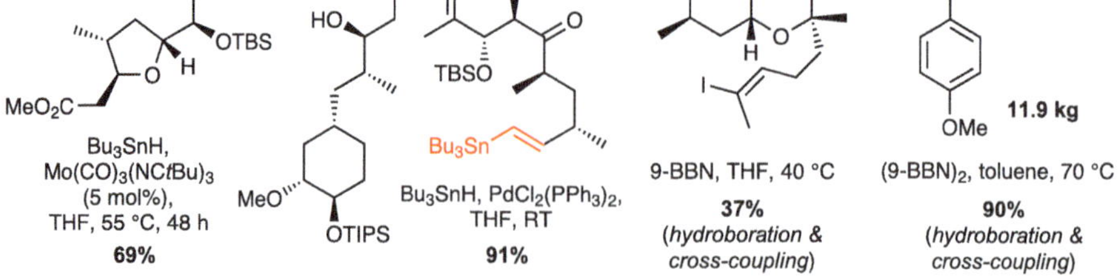

Fig. 6.47 Hydrometalation of complex alkynes or alkenes [201–204]

The hydrometalation approach is highly suited to selectively afford densely functionalized vinyl derivatives. Smith and co-workers resorted to a late-stage hydrostannylation of an alkynyl bromide in their synthesis of phorboxazole A to assemble the vinyl bromide side chain of the natural product. Following the hydrostannylation of **75**, Bu_3SnBr is presumably eliminated in a geminal fashion [200]. The resulting carbene can successively react with a second equivalent of Bu_3SnH to furnish the vinyl stannane. A subsequent Sn-Br exchange and the global deprotection finally yielded the desired macrocyclic natural product **76** (see Fig. 6.48) [205].

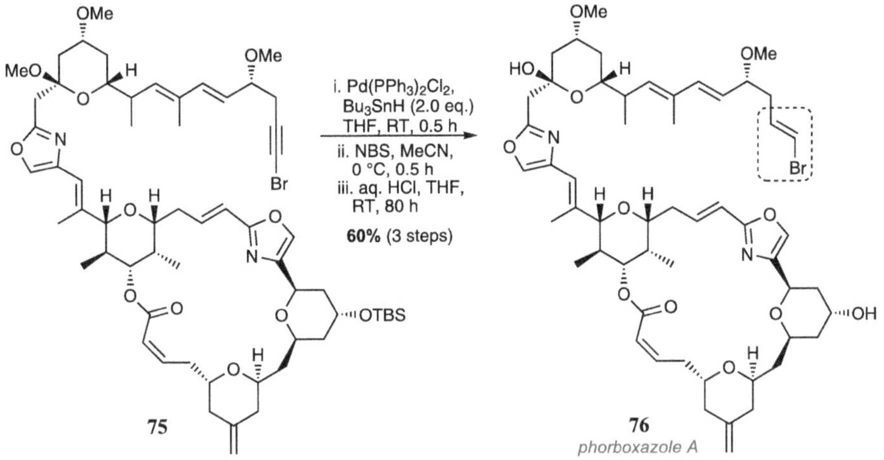

Fig. 6.48 Synthesis of phorboxazole A (**76**) disclosed by Smith *et al.* [205]

Tin and boron functionalities can also be introduced from the corresponding aryl halides via Pd catalysis. The transformation proceeds in a similar fashion as classic cross-couplings, but instead of a CC bond, a C-Sn or C-B bond is formed. Diboranes (*Miyaura borylation*) or distannanes (*Stille–Kelly coupling*) serve as metal source. The resulting aryl boranes and aryl stannanes can also be directly converted to the CC coupling products in a one-pot approach as part of a domino reaction. The Miyaura borylation can be carried out with alkenyl halides and triflates as well. Although the original conditions required KOAc as base and DMSO or dioxane as solvent (conditions A), significantly milder conditions were subsequently introduced [206]. It was observed that KOAc leads to the formation of the inhibiting off-cycle species $PdR^1L(OAc)_2$ in a recent study. The use of ethylhexanoate minimizes concentration of the latter and, in combination with a more active precatalyst and ligands, allowed for successful isolation of the borylation product (conditions B) [207]. In addition, a rhodium- or iridium-catalyzed C-H activation can effect an H-B exchange (see Fig. 6.49) [208, 209].

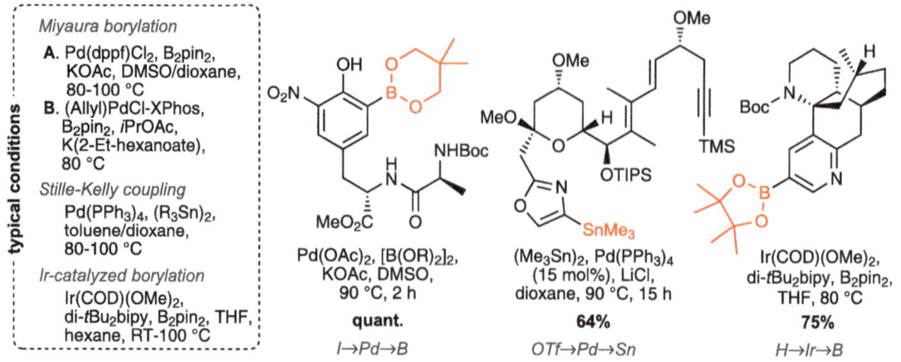

Fig. 6.49 Typical conditions and applications of metalations via C-M cross-coupling or CH activation [210–212]

Stannanes and silanes are extremely stable and can also be further derivatized subsequently, as can some boron derivatives. Accordingly, further functionalities can be introduced into these reagents following the metalation (see Fig. 6.50).

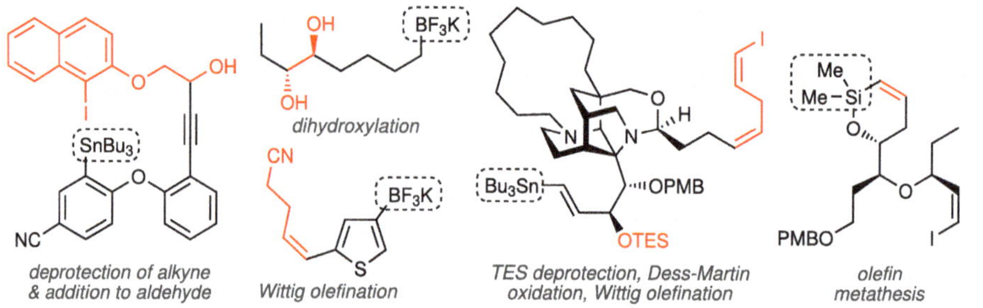

Fig. 6.50 Derivatization of stannanes, silanes, and trifluoroborates [213–217]. The changes after metalation are highlighted in red

Magnesium or zinc metal can also directly insert into the carbon-halogen bond. The reaction of aryl iodides with Mg (often in the presence of I₂ as an activator) constitutes part of the standard canon of organic transformations and proceeds easily. The formation of Grignard reagents from aryl bromides, on the other hand, necessitates the use of activated magnesium [218]. In the few reported cases of aryl chloride substrates, strongly forcing conditions were additionally required. The most common activation encompasses the reduction of metal halides (Mg, Zn, Cu, etc.) by elemental Li, K or Li-naphthalide (*Rieke metals*) [219, 220]. In contrast to Grignard reagents, the reaction of aryl iodides does not proceed in the presence of elemental zinc but requires activation of the metal, for example, Rieke-Zn. Alternatively, addition of LiCl/TMSCl or I₂ also allow access to the corresponding Zn reagents [221, 222].

The reaction with Zn follows a radical pathway, which precludes the stereoselective synthesis of secondary alkyl zinc compounds via direct insertion (see Fig. 6.51) [137].

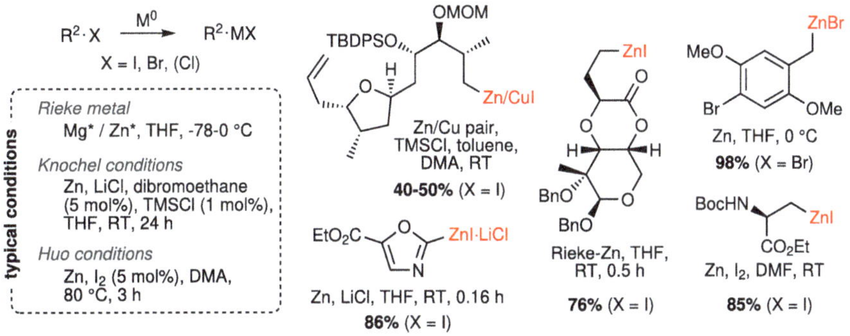

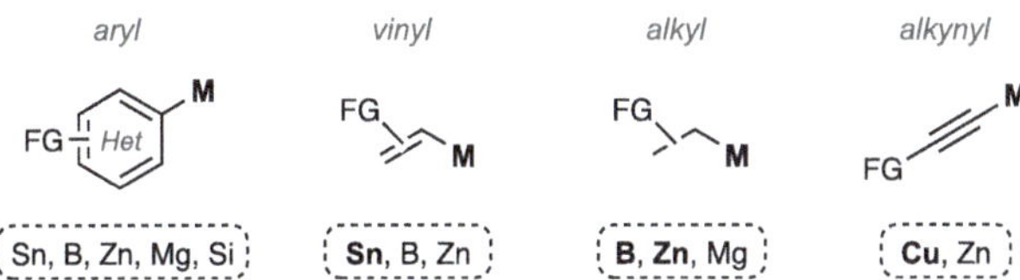

Fig. 6.51 Synthesis of organometallic reagents by direct insertion. The cited yields include the metalation and subsequent cross-coupling [221–227]

Considering the synthetic accessibility of the various organometallic reagents, a preference for certain CC couplings with complex intermediates can be understood from this vantage point: While all methods are well suited for (hetero-)biaryl synthesis, vinyl nucleophiles routinely resort to Stille and to a lesser extent Negishi or Suzuki conditions. Alkyl-based metal organyls, on the other hand, are reliably afforded via hydroboration from alkenes and by direct insertion of activated Zn into alkyl halides. For these substrate classes, Suzuki and Negishi cross-couplings dominate accordingly. Alkynes are easily deprotonated due to their acidic CH bond. If they are not directly converted to Cu acetylides *in situ* in a Sonogashira cross-coupling, the corresponding organozinc derivatives are usually used (see Fig. 6.52).

Fig. 6.52 Common organometallic reagents for the cross-coupling of different substrate classes [35, 176–178]

6.2.2   Characteristics of the Various Cross-Coupling Methods

While the combinations for nucleophiles shown in Fig. 6.52 indicate a certain preference for individual coupling methods in synthetic ventures, the various cross-couplings are distinguished by additional characteristics that permit further optimization. Hereinafter, the advantages and disadvantages as well as the role of certain variations or additives of each CC cross-coupling method will be delineated individually [228].

The neutral reaction conditions of the **Stille–Migita** cross-coupling and the increased nucleophilicity of stannanes compared to boron derivatives usually afford very high yields, even in the presence of functional groups regularly observed to be detrimental for conversion or yield (carboxylic acids, amides, esters, nitro groups, ethers, amines, alcohols, ketones). Furthermore, the nucleophiles are neither susceptible to air nor moisture, which could pose challenges especially on a small scale. In contrast, the toxicity and lipophilicity of the tin components are prohibitive for their application on an industrial scale [228].

Additives can significantly accelerate the reaction. The addition of Cu(I) salts can lead to an increase in rate by up to two orders of magnitude in individual cases [229, 230]. Transmetalations of bisligated complexes (R^1XPdL_2) that comprise a pre-equilibrium, in which a ligand is substituted by a solvent molecule, dissociate an equivalent of the ancillary ligand. The higher the concentration of free ligand in solution, the more the position of the equilibrium is shifted towards the starting materials. This deceleration due to ligand inhibition is strongly pronounced when the ligand is a good donor and is added in large excess. The addition of Cu(I) salts can support the transmetalation by sequestering the free ligand. The Cu effect accordingly diminishes with decreasing donor ability of the ligand. For the ubiquitous PPh_3, the distinct self-inhibition is effectively prevented upon addition of CuI scavengers. With weakly coordinating $AsPh_3$, the self-inhibition is weaker and the addition of CuI shows hardly any effect (see Fig. 6.53) [231, 232].

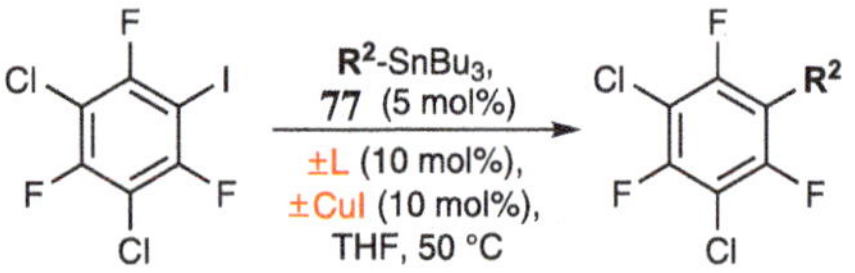

L	additive	k_{rel} R^2 = vinyl	Ph
PPh_3	–	1.0	0.05
	L	~0	~0
	L, CuI	0.31	0.02
$AsPh_3$	–	34.0	1.96
	L	1.13	0.12
	L, CuI	1.62	0.12

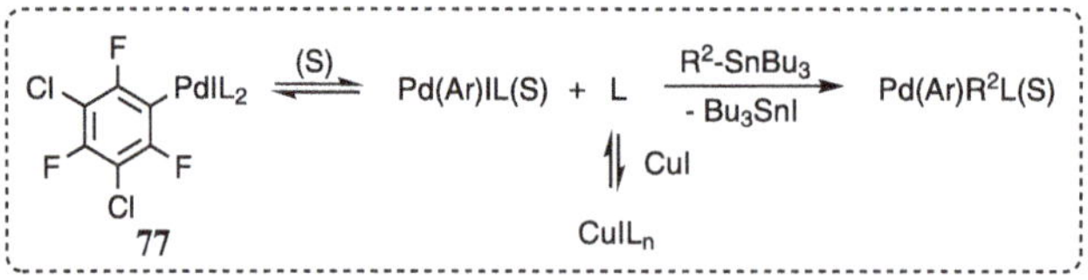

Fig. 6.53 Self-inhibition of the Stille reaction by the presence of free ligand and effect of Cu additives [229, 230]

Modern catalyst systems with electron-rich, sterically demanding ligands usually proceed via monoligated species that do not dissociate neutral ligands during the catalytic cycle or prevent the formation of bisligated species on steric grounds. The use of Cu salt additives should therefore not yield any benefit if the transmetalation either already proceeds sufficiently fast or no significant self-inhibition is observed [233].

In highly polar solvents like NMP or DMF, the latter effect occurs: The aryl or alkenyl stannane reacts in the presence of soft, weak donor ligands (such as $AsPh_3$) with CuI to form an organocopper species via Sn/Cu transmetalation. This can carry out the transmetalation step of R^2 more effectively than the parent stannane [232, 234].

The Stille cross-coupling can transpire solely Cu-mediated for very reactive electrophiles. Mechanistically, the reaction deviates in the order of the elementary steps: presumably an initial transmetalation to the (solvent)Cu-R^2 species occurs, followed by an oxidative addition of the electrophile and the concluding reductive elimination [235], which is reminiscent of the postulated catalytic cycle for Cu-mediated CN coupling reactions. While the conditions initially disclosed by Liebeskind necessitated stoichiometric amounts of Cu-2-thiophene carboxylate (CuTC), subsequently developed, improved variants rendered the reaction catalytic with regard to the requisite Cu content (5–10 mol%) [236, 237]. Given the diminished efficiency of transition metals other than palladium (cf. Chap 6.1), the Cu-catalyzed reaction often requires elevated temperatures. Nicolaou and co-workers employed the Pd-free Stille cross-coupling in their synthesis of disorazoles A_1 and B_1, to evade potential side reactions such as isomerizations caused by the multiple olefinic bonds present in the substrates (**78, 79**) and product (**80**) under Pd catalysis (see Fig. 6.54) [238].

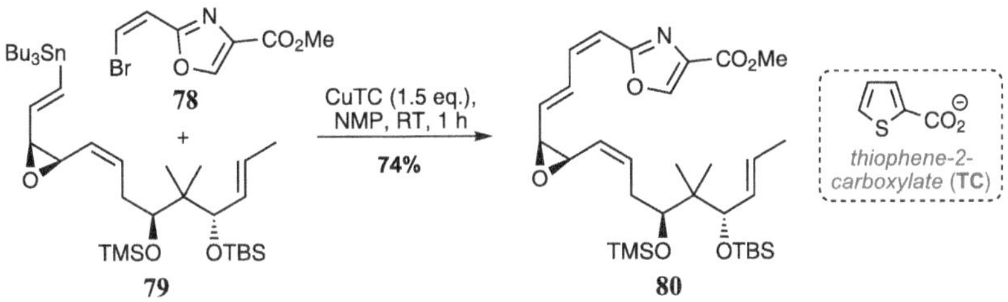

Fig. 6.54 Pd-free Stille coupling in Nicolaou's syntheses of disorazoles A_1, B_1 [238]

Since the Stille reaction exhibits remarkable tolerance towards many functional groups, it is predominantly used in the synthesis of densely functionalized natural products, often as a late-stage coupling step [1]. The Danishefsky group resorted to the Stille–Migita cross-coupling in their total synthesis of the enediyne cytostatic *rac*-dynemicin A (**83**) for the intramolecular stitching cyclization to assemble the bridging 10-membered ring. The bisiodoalkyne **81** was reacted with Z-1,2-bisstannylethene to obtain the highly strained enediyne intermediate **82**. The tetracyclic epoxide **82** was subsequently converted to the target metabolite **83** in further steps (see Fig. 6.55) [239].

The **Suzuki–Miyaura** reaction arguably constitutes the most versatile cross-coupling method and is routinely used on an industrial scale for the synthesis of biaryl motifs [3, 228, 240–242]. The mild reaction conditions, high tolerance towards functional groups, and commercial availability of many boron derivatives enable the facile construction of complex target molecules. The transformation is sensitive to the presence of oxygen and

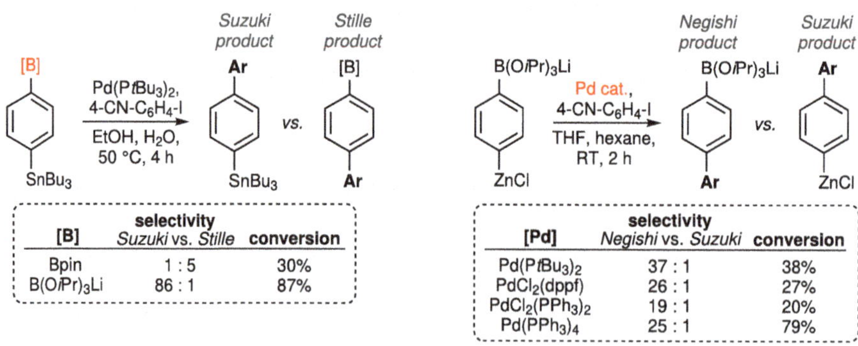

Fig. 6.55 Synthesis of dynemicin A (**83**) by Danishefsky *et al.* [239]

the conversion/yield can thus be impeded. Degassing the solvents accordingly helps to minimize the formation of homocoupling and protodeborylation products. Since the boron reagents exhibit only low nucleophilicity, bases are needed as additives to enable an efficient turnover (see Fig. 6.29 for mechanistic details). Inorganic salts (K_3PO_4, K_2CO_3, KOH) usually serve as base in combination with water as a cosolvent or additive. In the absence of base, no reaction occurs, as the nucleophilicity of the free boronic acids is even lower than that of silanes. For example, B/Sn-bimetallic nucleophiles as reactivity probes can be preferably derivatized at the tin functionality with boronate esters, while the introduction of trialkoxyborates switches the dominant path towards a Suzuki–Miyaura coupling. The Zn-C bond is, in line with the order of nucleophilicity, even more reactive than the respective B-C bond in Zn/B-bimetallic reagents, irrespective of the employed catalyst (see Fig. 6.56) [243, 244].

[B]	selectivity Suzuki vs. Stille	conversion
Bpin	1 : 5	30%
B(O*i*Pr)₃Li	86 : 1	87%

[Pd]	selectivity Negishi vs. Suzuki	conversion
Pd(P*t*Bu₃)₂	37 : 1	38%
PdCl₂(dppf)	26 : 1	27%
PdCl₂(PPh₃)₂	19 : 1	20%
Pd(PPh₃)₄	25 : 1	79%

Fig. 6.56 Competitive reactions of bimetallic nucleophiles [243]

In aqueous media, the choice of a particular base has little impact on the overall reaction. Under these conditions, this can be accounted for by OH^- being the operative species, independent of the identity of the employed base, and the commonly large excess thereof. In anhydrous systems, this assumption is not correct and the solubility of the base in the reaction medium and the extent of dissociation play a crucial role. The stability constants of the hydroxide salts, for example, decrease with the atomic number ($Cs^+ < K^+ < Na^+ < Li^+$), so it can be assumed that cesium bases increase the concentration of OH^- compared to bases of the lower alkali homologues [245, 246].

The variability of boron reagents allows for optimization of the cross-coupling not only in terms of the catalyst system, but also the corresponding nucleophile (see Fig. 6.57) [176].

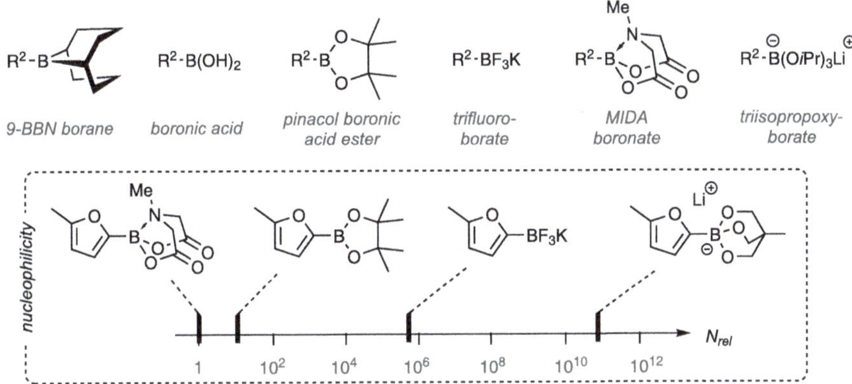

Fig. 6.57 Commonly used boron reagents and their relative nucleophilicity [22, 176]

Boronic acids are unstable towards a variety of organic reagents, which is why they are rarely modified over several steps prior to the coupling but instead commercially available substrates are directly employed. Unlike the parent boronic acids, which can also exist as di- or trimers, boronic acid esters can be considered as "protected" boronic acids and remain solely monomeric. Pinacol boronic acid esters are significantly more resilient than free boronic acid towards *protodeboronation*, the hydrolysis of aryl boron reagents with concomitant liberation of $B(OH)_3$ [247, 248]. The formation of trialkoxyborates, which are the most nucleophilic boron reagents, can even enable the conversion of very unreactive nucleophiles [249–251] (Fig. 6.58).

	conversion [%]	yield [%]
[B]		
$B(OH)_2$	36	8
Bpin	73	49
BF_3K	37	10
$B(OiPr)_3Li$	100	68

Fig. 6.58 Reactivity of different 2-pyridyl organoboron reagents [250]

The tetrahedral *N*-methyliminodiacetic acid boronates (*MIDA* boronates) were introduced by Burke in order to deactivate the boron species so strongly that cross-coupling can be fully suppressed in the absence of water [252, 253]. The MIDA boronates can be easily reverted back to the respective boronic acids by the action of aqueous base [254]. In this way, nucleophilic functionalities can be temporarily protected, reactivated, and exploited for iterative couplings. The complex polyene **87** constitutes the backbone of the secondary metabolite vacidin A. It was assembled by selective Suzuki–Miyaura coupling of pentene-1-boronic acid to give **84**, followed by hydrolysis of the MIDA boronate and renewed cross-coupling to **86**. The concluding domino hydrolysis/cross-coupling reaction could afford the desired product **87** in 32% yield over the 5-step sequence (see Fig. 6.59) [255].

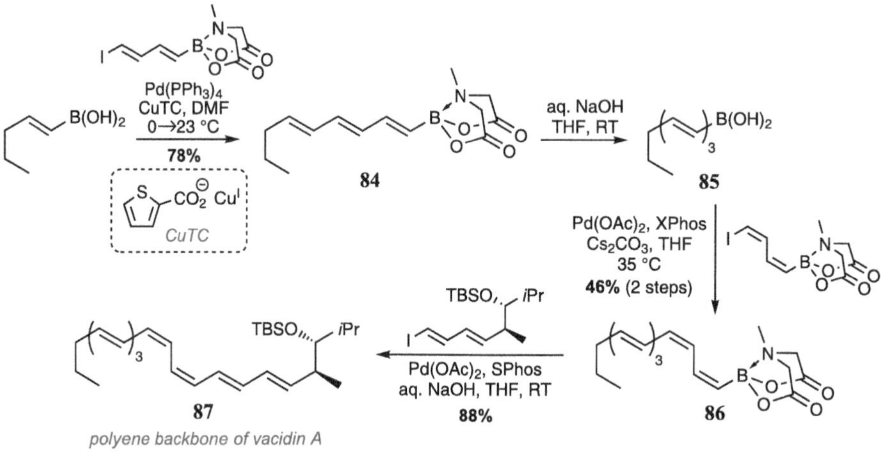

Fig. 6.59 Iterative Suzuki–Miyaura coupling in the synthesis of a vacidin A fragment [255]

Beyond boronic acids and boronate esters commonly derived from catechol or pinacol, the trifluoroborates introduced by Molander offer additional merit in some instances [256, 257]. K^+ salts are insoluble in most organic solvents and require polar solvents, which is why the corresponding Bu_4N salts can provide improved yields due to their enhanced solubility. Alternatively, Bu_4NI can be added stoichiometrically in the reaction of K^+ salts. $[RBF_3]K$ is not the operative species in the transmetalation, as no conversion occurs in the absence of both base and water. Originally, $[RBF_2(OH)]^-$ or $[RBF(OH)_2]^-$ were proposed as reactive intermediates [258], but a detailed study unveiled strong indications that a slow hydrolysis to boronic acid and its subsequent reaction to be the operative mechanism [259–261]. A hallmark of trifluoroborates compared to other boron nucleophiles is their dramatically enhanced stability. Alkyl trifluoroborates are easily accessible and allow the coupling to proceed with only a modicum of side reactions such as β-H eliminations, protodeboronations, or olefin reinsertions [262]. Even secondary alkyl trifluoroborates can be converted under optimized conditions without significant isomerization to the branched product (see Fig. 6.60).

The **Negishi** cross-coupling has the advantage that the employed organozinc reagents manage the precarious balance of high reactivity while tolerating many functional groups. The synthetic accessibility of many organozinc compounds, especially alkyl organozinc compounds, endorses the Negishi cross-coupling for alkyl coupling reactions. Magauer and co-workers resorted to this method in their synthesis of the antiviral meroterpenoid (+)-

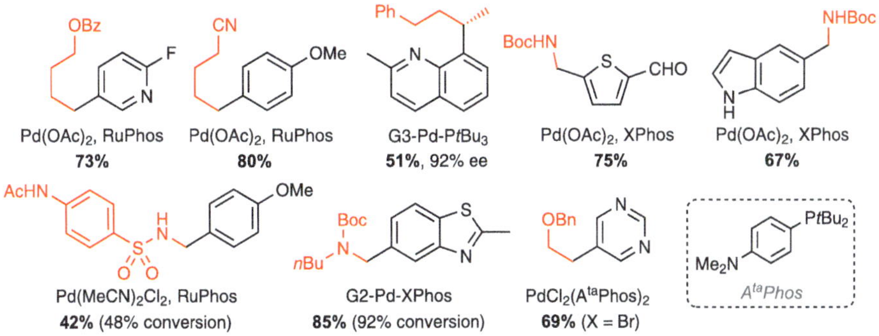

Fig. 6.60 Substrate scope in the coupling of alkyl trifluoroborates with heteroaryl chlorides. The BF_3K-derived moiety is highlighted in red [263–268]

stachyflin to construct the central benzylic bond in **90**. The nucleophile **89** was obtained by Li/Zn exchange starting from the corresponding alkyl iodide (see Fig. 6.61) [269].

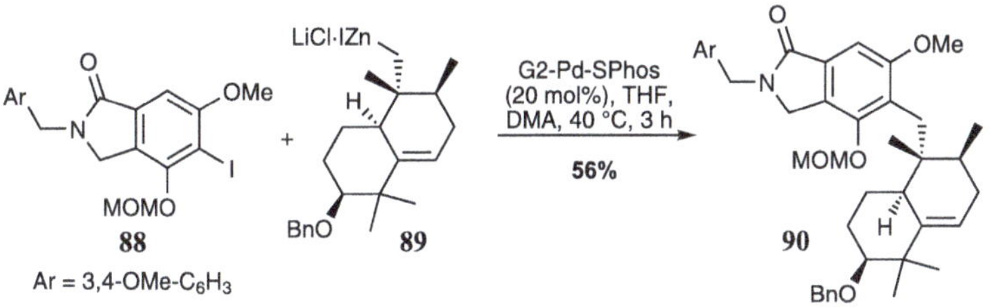

Fig. 6.61 Magauer's synthesis of (+)-stachyflin [269].

The reaction can also be carried out under Ni catalysis [270]. Especially biaryl building blocks can be obtained efficiently, but applications in total syntheses have so far only been reported sporadically (see Fig. 6.62) [271].

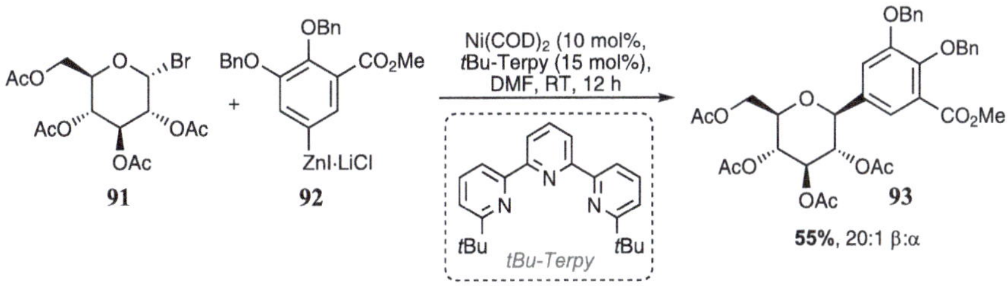

Fig. 6.62 Construction of the C-glycoside salmochelin SX by Gagné *et al.* [272]

The cross-coupling of Grignard reagents and organolithium compounds is referred to as **Kumada–Corriu** coupling. In recent publications, the cross-coupling of Li- based nucleophiles has also been termed the **Murahashi** coupling [273]. The general applicability of the reaction is hampered by the high basicity of the organometallic compounds. Modern ligand systems can partially compensate for this, but the tolerance towards electrophilic functional groups remains the main challenge of the method [274, 275].

Jacobsen and co-workers relied on a Kumada cross-coupling for the C_2-homologation of the right fragment in their synthesis of (+)-ambruticin (**96**). The antifungal secondary metabolite was then constructed via an asymmetric Rh-catalyzed hydroformylation and subsequent enantioselective cyclopropanation, followed by a concluding Julia-Kocienski olefination. The two tetrahydropyran rings could be obtained through asymmetric hetero-Diels–Alder reactions (see Fig. 6.63) [276].

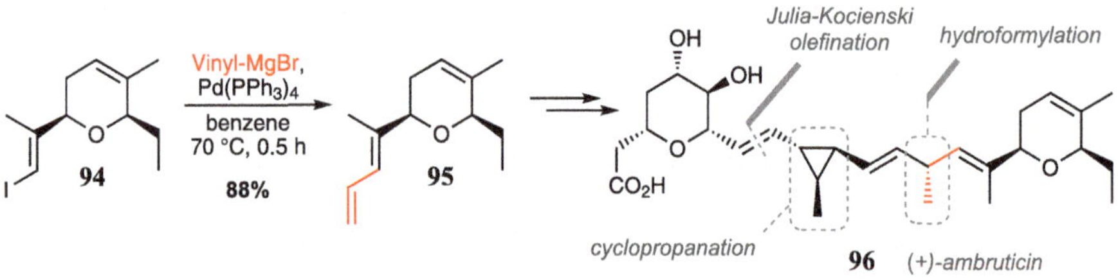

Fig. 6.63 Kumada cross-coupling in the total synthesis of (+)-ambruticin (**96**) [276]

The development of the Ni-catalyzed cross-coupling of Grignard reagents was even disclosed prior to the introduction of the Pd-mediated process by the groups of Kumada and Corriu, building on earlier work by Kochi [277–279]. Accordingly, there are many nickel-catalyzed examples of this transformation, yet the conversion of organolithium compounds has so far been largely limited to Pd catalysis [280]. The advantage of the Kumada reaction over other cross-couplings is the possibility of directly using lithium or magnesium organyls, without further transmetalation to the corresponding zinc or boron derivatives. This makes the Kumada reaction interesting on an industrial scale due to its atom economy, especially for reactions of simple substrates lacking unstable functional groups (see Fig. 6.64) [3].

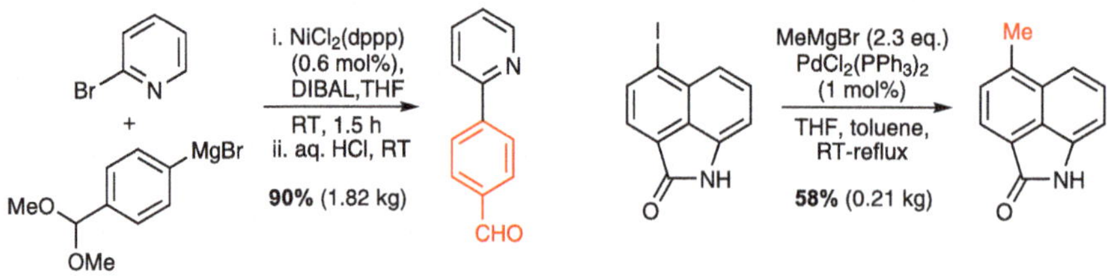

Fig. 6.64 Examples of industrial applications of the Kumada–Corriu cross-coupling [281, 282]

PPh_3 usually endows sufficient reactivity as a ligand, although in many cases the utilization of bidentate ligands leads to an enhanced selectivity. Their use reduces the proportion of homocouplings, isomerization reactions, and β-H eliminations for both Pd- and Ni-based catalysts. In nickel-catalyzed reactions, this effect of bidentate ligands is mainly attributed to the absence of *trans*-complexes, thus avoiding unproductive isomerizations, which results in the relative increase of rate of the catalytic cycle [228].

The presence of *i*PrI can significantly accelerate the Kumada coupling reaction. While it is liberated as a by-product in the Li/Mg exchange from the respective iodides, it can also be purposely introduced as an additive. The acceleration is presumably caused by an alleged change from an ionic to a radical mechanism (see Fig. 6.65) [283].

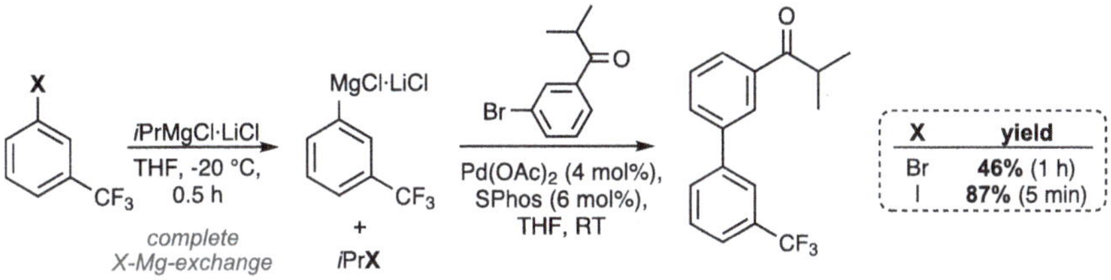

Fig. 6.65 Effect of *i*PrI on the Kumada cross-coupling [283]

In addition to Ni catalysts, Fe-based systems also furnish the coupling products in high yields [26, 27]. A stabilizing interaction of the iron center with Mg^{2+} cations has been postulated to rationalize the high selectivity observed in the reaction of Grignard reagents compared to other nucleophiles [51, 284]. While Fe(acac)$_3$ as precatalyst often provides excellent results even in the absence of additional ligands, the use of additives such as NMP as a cosolvent can significantly enhance the reaction rate and selectivity [285–287]. Although NMP increases the yield of coupling product particularly with vinyl halides, it can inhibit product formation with some substrates and precatalyst species [288]. In addition to vinyl and aryl halides as well as triflates or tosylates, acyl halides can also be converted; they often even react significantly faster than aryl halides (see Fig. 6.66).

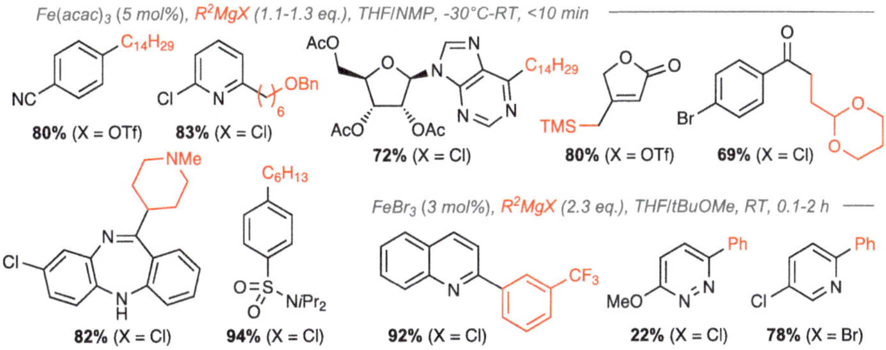

Fig. 6.66 Substrate scope of Fe-catalyzed Kumada cross-couplings [286–289]

Electron-rich aryl halides and alkyl halides as electrophiles as well as the conversion of secondary alkyl organomagnesium nucleophiles may however necessitate the use of more specialized catalyst systems such as FeF_3/SIPr, $FeCl_3$/IPr, or an Fe-salen complex [27].

The mechanism and even the oxidation state of the metal in the iron-catalyzed Kumada cross-coupling is controversially debated [51, 290]. Based on the disclosed studies, a dependence of the mechanism on the individual catalyst system seems possible [27]. In the case of monodentate or chelate ligands, added NMP is probably only weakly attached to the metal center and prone to rapid displacement [291]. The reaction is only encountered to a limited extent in natural product syntheses due to the partially limited tolerance towards functional groups. Fürstner and co-workers used two Fe-catalyzed couplings of alkyl Grignard reagents in their approach to the marine macrolide latrunculin B (**99**). The key fragments **97** and **98** were isolated in high yield and subsequently converted to the natural product (see Fig. 6.67) [292].

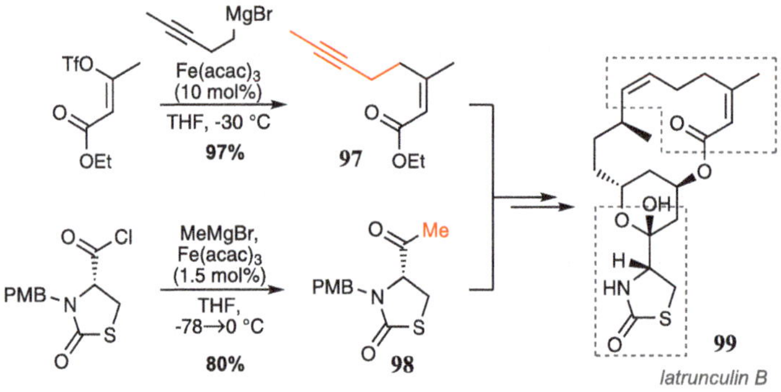

Fig. 6.67 Use of $Fe(acac)_3$-catalyzed cross-couplings in Fürstner's synthesis of latrunculin B (**99**) [292]

The **Hiyama** cross-coupling utilizes organosilicon nucleophiles. It resembles the Suzuki–Miyaura coupling in terms of the range of available Si-organyls and the necessity of activation. However, the weakly polar to nonpolar C-metal bond similarly demonstrates a certain kinship with the Stille–Migita coupling, as the organosilicon reagents usually exhibit very high stability in acidic and basic media as well as under oxidative or reductive conditions. The tetrasubstituted organosilicon compounds lack the requisite reactivity for cross-coupling of acyclic silanes, which is why stoichiometric fluoride additives were initially introduced as activators, exploiting the high fluorophilicity of silicon ($\Delta G_{Si-F} = 159\,kcal\,mol^{-1}$) [293]. Barring the aspect of atom economy (consumption of 1 equivalent of activator), the need for activation by fluoride constitutes the most notable disadvantage of the method: both glass and steel vessels are susceptible to corrosion by fluoride, which complicates the handling of the reaction. Furthermore, silyl protecting groups ubiquitously present in natural product syntheses are usually labile under the reaction conditions (see Fig. 6.68) [294–296].

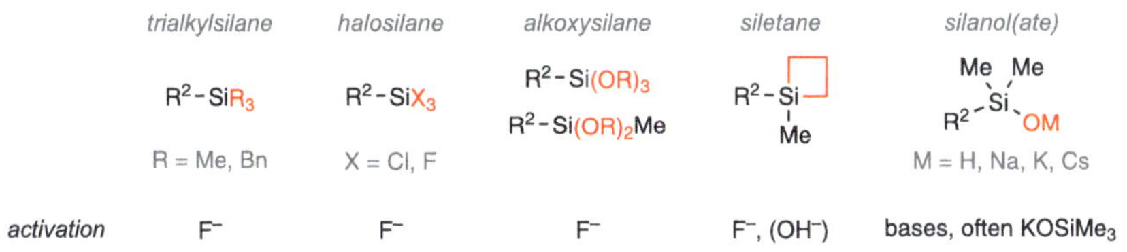

Fig. 6.68 Common silyl-based nucleophiles in Hiyama cross-couplings [295]

Halosilanes possess enhanced reactivity but tend to hydrolyze easily and lack the innate advantageous stability of other silanes required for derivatization prior to or after a coupling; accordingly, they did not find widespread application. Alkoxysilanes, silanols/silanolates, and siletanes, on the other hand, constitute useful extensions of acyclic trialkylsilanes and are based on exogenous or endogenous approaches for activating the Si-C bond. In siletanes, a reduction in ring strain upon transitioning to a pentacoordinated intermediate **100** serves as the driving force. This concept is also referred to as "*strain release*" Lewis acidity. The mechanistic pathways of siletanes and silanols are hypothesized to subsequently converge (see Fig. 6.31), as silanol **101** and disiloxane **102** could be detected when siletanes were activated by a fluoride (see Fig. 6.69) [297, 298].

Fig. 6.69 Postulated mechanism in the reaction of siletanes [297, 298]

Silanols and silanolates enable a fluoride-free cross-coupling, thus compensating a major disadvantage of the Hiyama coupling reaction [299]. While silanols require a Lewis base as an activator, silanolates are suffciently reactive as an "active form" even in the absence of further additives. The advantages of innately pre-activated silanolate salts comprise their simple synthesis, high stability, and insensitivity to water and oxygen. They furthermore circumvent the formation of unreactive disiloxane by-products like **102**, which is an undesired pathway in the Hiyama coupling of silanols [299].

Although fluoride-free variants with silanolates as nucleophiles have now become established, a fluoride additive is still utilized in a significant part of the applications, usually in the form of highly soluble salts such as ammonium (TBAF, TASF) or cesium fluoride. Even "ligand-free" conditions often afford the coupling product in acceptable yields, although the scope of reactive electrophiles in these cases remains limited to alkenyl and aryl iodides/bromides. With careful use of TBAF, silyl protecting groups in the substrate may remain intact during the reaction (see Fig. 6.70) [300].

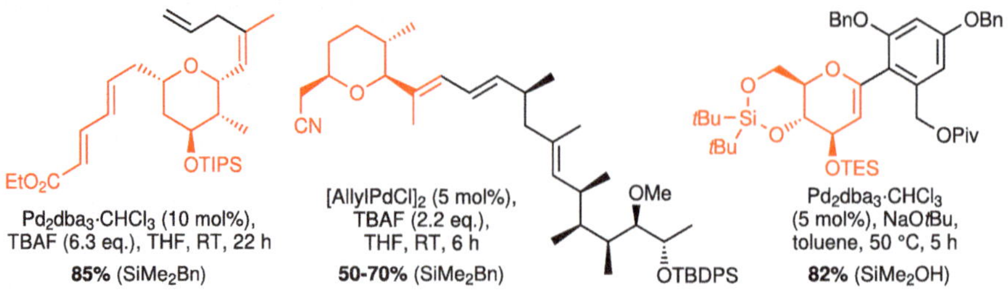

Fig. 6.70 Application of the Hiyama cross-coupling in total syntheses. The nucleophile fragment is highlighted in red [301–303]

The **Sonogashira–Hagihara** coupling, despite being considered part of the cross-coupling canon, somewhat deviates from the repertoire of "standard" methods: The organometallic component is usually generated *in situ* and only terminal alkynes can be converted as nucleophiles. Since metal acetylides are commonly excluded in other cross-couplings, the Sonogashira reaction can be considered complementary to the remaining variants. It is **the** standard method for the construction of alkenyl and aryl alkynes. Its procedure is technically simple, efficient, usually affords very high yields of more than 90% and the tolerance of functional groups is good to excellent despite its slightly basic conditions [144, 304]. In contrast to the related *Stephens–Castro reaction*, which necessitates stoichiometric amounts of copper(I) salts, the Sonogashira reaction can be performed in the presence of catalytic copper(I) salts, usually CuI. If oxidizing agents such as oxygen have not been purged from the reaction, the formation of alkyne homocoupling products from the *in situ* formed Cu acetylides poses one problem of the transformation. Accordingly, the reaction tolerates water but is moderately air-sensitive. When using triflates as electrophiles, the reaction is occasionally referred to as *Cacchi coupling*.

The corresponding homocoupling reaction of two terminal alkynes in the presence of elemental oxygen is referred to as *Glaser coupling*, which exclusively yields symmetric diynes as products. Unsymmetric diynes are accessed via the *Cadiot–Chodkiewicz coupling*, in which an alkynyl bromide serves as an electrophile in the conversion with a terminal alkyne. Both variants are Pd-free and proceed solely Cu-mediated (see Fig. 6.71) [305, 306].

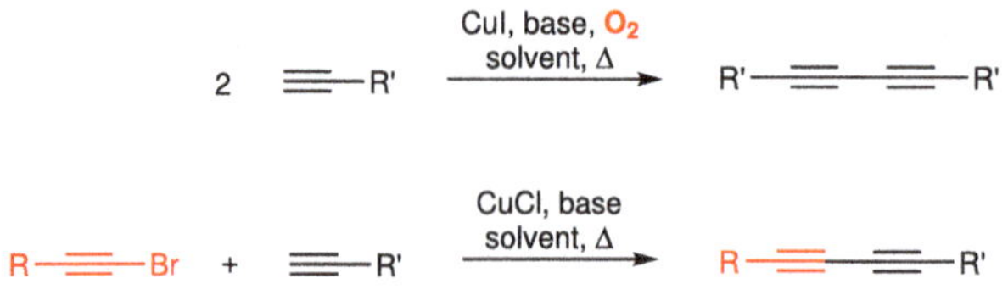

Fig. 6.71 Glaser and Cadiot–Chodkiewicz coupling [305, 306]

Excellent yields of the coupling product can be obtained even with simple phosphines (PPh$_3$, etc.) as ligands, while an amine (NEt$_3$, EtN*i*Pr$_2$, etc.), often used as a solvent or cosolvent, typically serves as base. Aryl bromides possess borderline reactivity and necessitate the use of elevated temperatures, which can be exploited for the selective functionalization of a C-I bond in the presence of a C-Br bond (see Fig. 6.72).

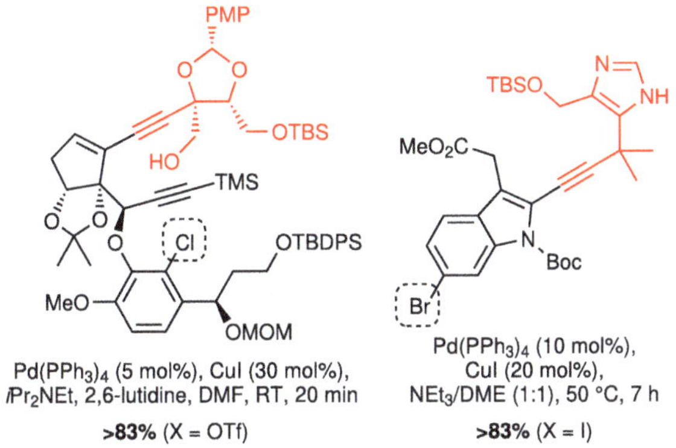

Fig. 6.72 Selective Sonogashira couplings. The alkyne fragment is shown in red, competing coupling sites are highlighted [307, 308]

Electron-poor alkynes (R' = CO_2R, CO_2H, CN, etc.) cannot be coupled or only afford low yields. One way to encourage these substrates to react is the use of preformed metal acetylides [144, 309]. Zinc, tin, and magnesium acetylides have proven to be efficient nucleophiles, with zinc acetylides typically displaying the highest reactivity out of the various metal homologues.[VI] The conversion of other challenging substrates such as acetylene as nucleophile or *ortho*-disubstituted aryl electrophiles often results in high yields under these conditions. If preformed metal acetylides are used, the coupling reaction can be carried out in the absence of copper salts and base (see Fig. 6.73).

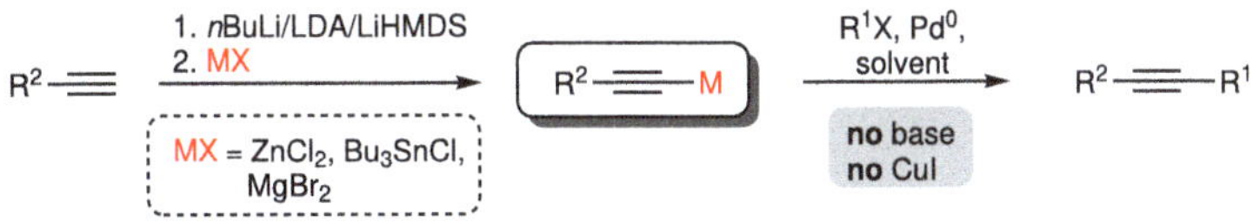

Fig. 6.73 General reaction scheme with metal acetylides [144]

One of the advantages of the standard Sonogashira reaction, its high tolerance towards water, is exchanged for an increased reactivity and yield in the cross-coupling. A direct comparison proves the superiority of the Negishi variant in the conversion of propionic acid derivatives (see Fig. 6.74).

[VI] Although the corresponding tin- and zinc-mediated cross-couplings are also referred to as Negishi and Stille reactions in the literature, the term Sonogashira reaction will be used for simplicity.

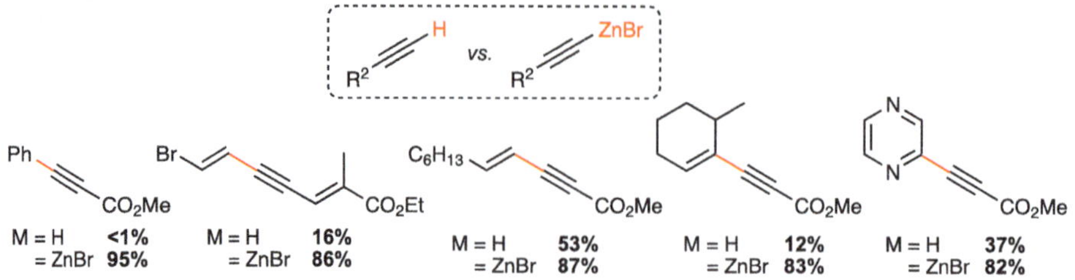

Fig. 6.74 Comparison of the reactivity of standard and Negishi conditions in the Sonogashira cross-coupling [310, 311]

The elimination of the copper co-catalyst is advantageous from several points of view: In addition to lower costs and easier purification (only one metal needs to be recycled), the proportion of homocoupling product decreases, as its formation due to unwanted exposure to oxygen is also linked to the presence of the Cu salt. Depending on the electrophile used and reaction conditions, a Cu-free reaction can often be realized. The Cu-free Sonogashira reaction was discovered in the mid-1990s and allows the coupling of aryl iodides, albeit with high catalyst amounts of at least 5 mol%. Common additives are ammonium halides (Bu$_4$NBr, Bu$_4$NF). The introduction of electron-rich, sterically demanding ligands was also able to significantly expand the substrate scope of the electrophile in the Cu-free Sonogashira cross-coupling (see Fig. 6.75) [312].

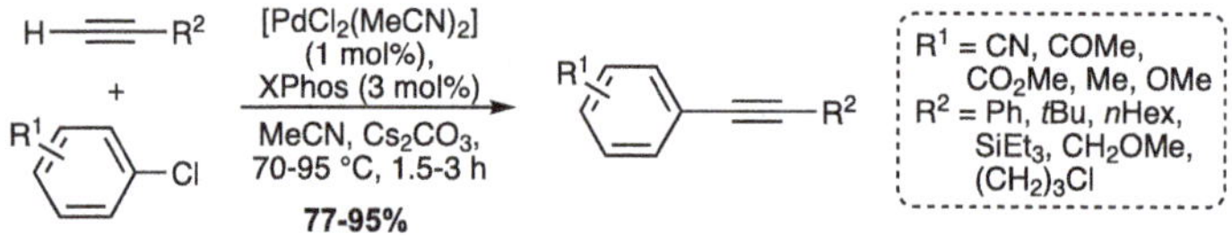

Fig. 6.75 Cu-free Sonogashira reaction developed by Buchwald *et al.* [313]

While the mechanism of the classic Sonogashira cross-coupling is based on coupled Pd/Cu catalytic cycles, the situation becomes more convoluted in the absence of a co-catalyst [147]. The oxidative addition and reductive elimination proceed in the same manner for all cross-couplings. The conceivable variations of the transmetalation and the possibility of further ligand exchange reactions, however, do not necessarily simplify an elucidation of the operative pathway. Recent studies suggest a mechanism involving two interlinked Pd/Pd catalytic cycles, in which the first undergoes oxidative addition and elimination, while the second ensures the activation of the alkyne (see Fig. 6.35) [148]. Previous investigations postulated a solely monometallic transmetalation via a cationic (**103**) or anionic path (**104**), by means of a carbopalladation (**105**) or by amine-halogen exchange prior to the coordination of the alkyne (**106**, see Fig. 6.76) [147].

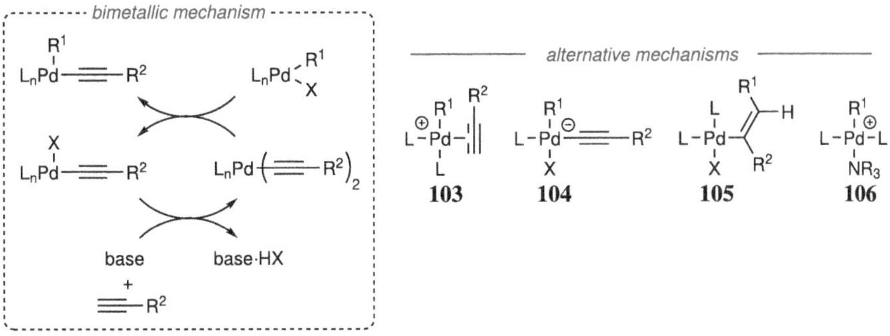

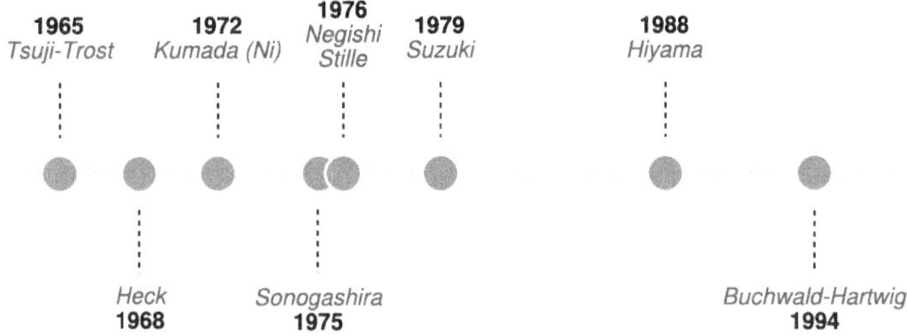

Fig. 6.76 Postulated mechanisms of the Cu-free Sonogashira reaction [147, 148]

6.2.3 Modern Developments and Selected Applications

Fig. 6.77 Development history of the most important coupling reactions

Despite the many decades since their inception, the various cross-coupling methods are still being continuously developed and improved up until today. While the Heck reaction originally required stoichiometric amounts of Pd salts, this problem was already solved in 1971 by Mizoroki and co-workers [314]. The substrate scope of the Kumada cross-coupling could also be significantly enhanced in 1975 by Murahashi through the introduction of a Pd-catalyzed process [315]. The further developments primarily address the most urgent disadvantages of the reactions such as a limited substrate scope (especially the leaving group of the electrophiles), a laborious work-up/purification, the wanting tolerance of functional groups as well as the restricted stability of the organometallic reagents. An increased catalyst activity is directly associated with the possibility of reducing the catalyst load to the ppm range in some cases.

All methods exploited the development of new ligand motifs to enable the conversion of electrophiles beyond aryl and alkenyl iodides, bromides, and triflates, such as the corresponding chlorides and tosylates. In this context, the improved understanding of the available optimization handles offered by the catalytic cycle was capitalized on for rational ligand design. Especially the introduction of biaryl phosphines, N-heterocyclic carbenes and PtBu$_3$ have substantially shaped these developments (see Fig. 6.5). In addition, the problem of reproducibly and rapidly generating a catalytically active species under mild conditions was largely solved by using clearly defined precatalysts, and the requisite amount of catalyst was reduced to below 1% in most cases (see Fig. 6.4). Furthermore, the use of transition metal complexes based on Ni, Cu, Fe, and other less noble homologues was advanced [19].

For the **Stille** cross-coupling, the separation of the stoichiometrically formed tin halide is a particular problem. Apart from its toxicity and thus very low tolerance limits, the difficulty to fully remove R_3SnX by-products even by chromatographic means can be attributed to its nonpolar character. Given the high fluorophilicity of stannanes, fluoride additives are a common approach for their activation via transient, pentacoordinated species. An welcome effect of the fluoride addition is the generation of Bu_3SnF as by-product, which polymerizes and can thus be easily removed in the purification step. To facilitate the Bu_3SnF formation, KF solution can be added in the aqueous work-up [316] or the silica gel for chromatographic purification can be mixed with powdered KF [317]. Alternatively, the addition of $AlMe_3$ (to form Bu_3SnMe) or NaOH (to form Bu_3SnOH) can facilitate the separation, depending on the polarity of the products formed [318]. An accompanying combination with modern catalyst systems allows the isolation of even complex substrates in high yield with a small amount of catalyst and also the more sustainable use of aqueous conditions (see Fig. 6.78) [233, 234, 319–322].

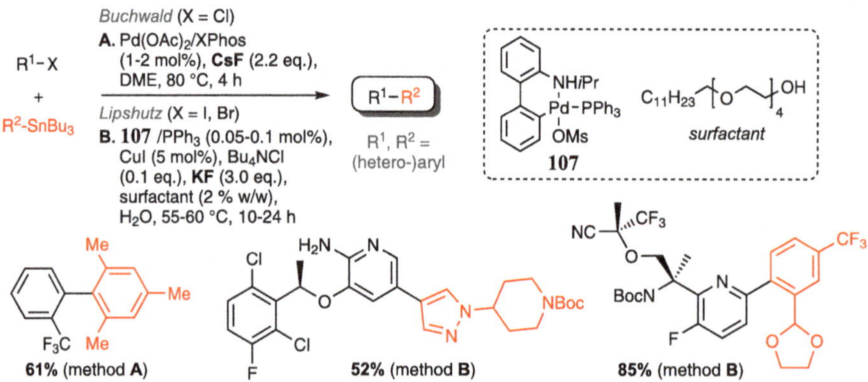

Fig. 6.78 Typical examples of highly efficient Stille–Migita cross-couplings [233, 322]

In addition to the use of fluorides, newer variants also focus on the use of less toxic stannyl derivatives, polymer-bound nucleophiles, and the catalytic use of tin in domino hydrostannylation/Stille cross-coupling reactions to facilitate separation or to reduce the amount of tin from the outset [105]. Despite their promise, these developments have not yet found their way into the canon of commonly encountered Stille methods.

The use of a co-catalyst such as Cu, but also Au, has meanwhile become established. The Stille reaction is often used in academic syntheses on particularly complex substrates in late stages of a sequence, and regularly relies on the addition of Cu salts (see Fig. 6.79).

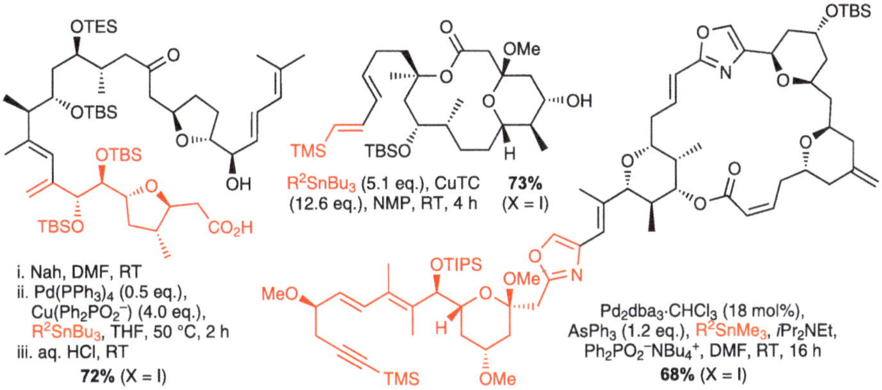

Fig. 6.79 Stille coupling of complex intermediates in total syntheses [211, 323, 324]

The **Suzuki** cross-coupling has arguably received the most attention of all transition metal-mediated coupling methods [325]. The lack of access to organoboron reagents and their limited stability have led to the introduction of the range of boron reagents available today (boronic acid esters, trifluoroborates, MIDA boronates, trialkoxyborates) [176, 326]. In addition, the use of precatalysts has also enabled the use of highly sensitive nucleophiles, which routinely suffer from protodeboronation [327] but are highly desirable for inclusion in active pharmaceutical ingredients due to their structure (see Figs. 6.80 and 6.81).

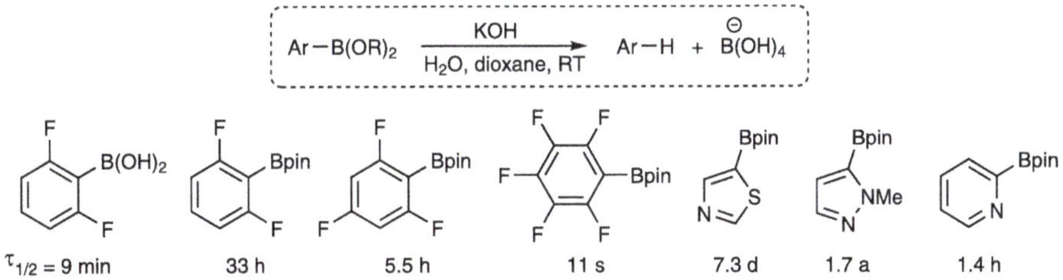

Fig. 6.80 Stability of organoboron nucleophiles against protodeboronation in basic-aqueous medium [247]

Successively disclosed improvements are iterative reactions, usually based on MIDA boronates, as well as the synthesis and use of enantiomerically pure secondary alkyl organoboron reagents [330]. Since the reductive elimination proceeds stereospecifically, the mechanism of transmetalation (open versus cyclic) determines whether the overall reaction occurs with stereoretention or inversion. The synthetic access to chiral alkylboronic acid derivatives remains challenging, and stereoinduction often relies on the presence of directing groups. The introduction of *axial* chirality is also possible with biaryl products bearing

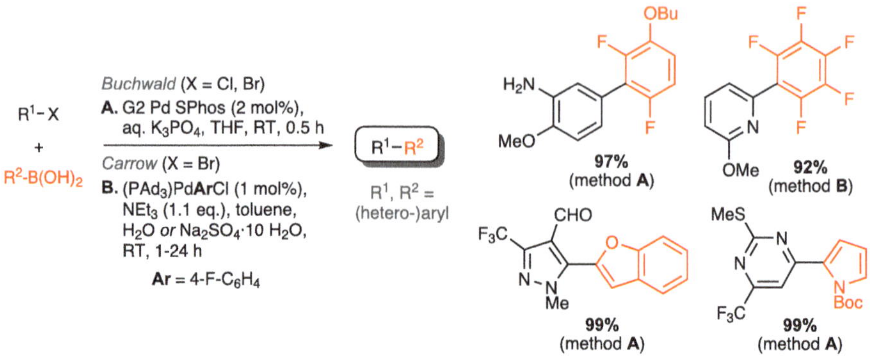

Fig. 6.81 Selected methods for coupling of labile boronic acids [328, 329]

multiple *ortho*-substituents. The pronounced steric interactions during reductive elimination pose a significant hurdle to achieve both an efficient product formation and acceptable enantioinduction by chiral ancillary ligands. Enantiomeric excesses beyond 90% have so far only been attainable in isolated cases [331]. The asymmetric Suzuki–Miyaura coupling does not (yet) play a relevant role on an industrial scale [3]. Boger and co-workers used BINAP as a C_2-symmetric chiral ligand to effect the critical asymmetric biphenyl coupling of **108** and **109** to afford **110**, which was subsequently converted to the antibiotic drug vancomycin. **108** was generated *in situ* and added slowly, which allowed the reaction to be scalable to batches of up to 25 g each (see Fig. 6.82) [332].

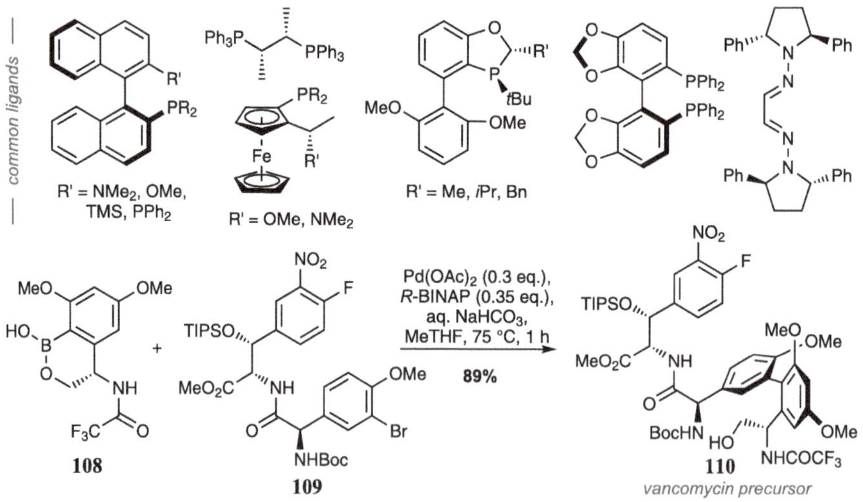

Fig. 6.82 Ligands commonly employed in asymmetric Suzuki cross-couplings and application thereof in Boger's synthesis of vancomycin [331, 332]

The use of alkyl electrophiles, especially in combination with alkyl nucleophiles, remains the focal point of many methodological studies [20, 21, 333]. The suppression of the β-H-elimination was successfully achieved by the use of sterically demanding, electron-rich phosphines and NHCs as ligands. Trifluoroborates constitute the dominating boron derivatives employed as alkyl nucleophiles by virtue of their operatively simple preparation. The alkyl-alkyl Suzuki coupling yet remains underrepresented both in an academic or industrial context, where alkyl organoboron reagents preferentially serve to effect C_{sp^3}-C_{sp^2} bond formation reactions, with Ni-catalyzed Negishi and Fe-catalyzed Kumada-couplings gaining ground [1, 334, 335].

In Evans' synthesis of the cytotoxic metabolite (–)-FR182877, two different Suzuki cross-couplings were used at an earlier stage and very late in the synthetic sequence. In the first Pd-catalyzed reaction, the segments **111** and **112** were stitched together in an intermolecular fashion. The polyene **113** served as precursor for a fully stereocontrolled, intramolecular domino Diels–Alder/Diels–Alder sequence and subsequently comprised the molecular framework of the A-D rings in the natural product. As late-stage C_{sp^3}-C_{sp^2} Suzuki coupling introduced the missing methyl group to afford **115** utilizing trimethyl boroxine as nucleophile. The endgame comprised the concluding ring closure of the lactone to complete the total synthesis of the densely functionalized natural product (see Fig. 6.83) [336].

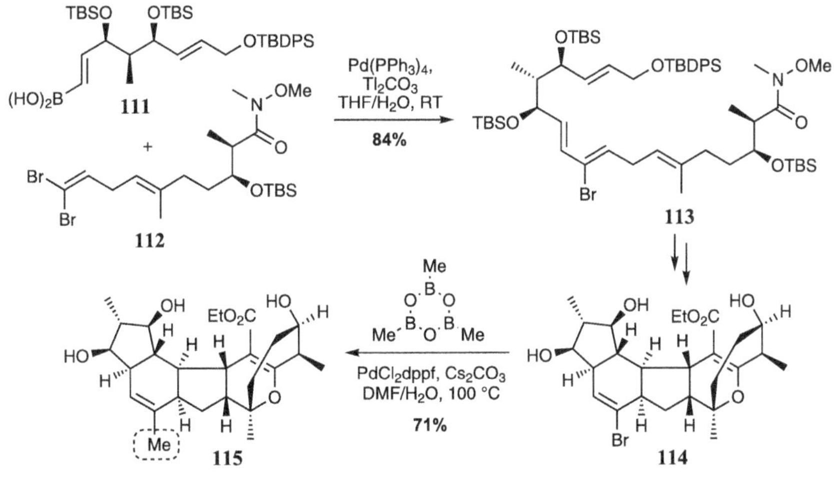

Fig. 6.83 Synthesis of (–)-FR182877 by Evans *et al.* [336]

The **Negishi** cross-coupling has witnessed dramatic leaps on several fronts. The limited stability of organozinc reagents complicates the handling of the water-sensitive nucleophiles, and the use of Ni catalysts is a field with great potential for development [337].

As an extension of their previous work on the preparation of polyfunctionalized organomagnesium and organozinc compounds, the Knochel group disclosed the unprecedented stability of zinc pivalates against moisture and oxygen [338–341]. The reagents can be prepared like standard organozinc compounds via transmetalation, deprotonation/metalation, or insertion into C-halogen bonds. A switch of the counterion of the organometallic compounds (tBuCO$_2^-$ versus Cl/Br/I$^-$) allowed zinc pivalates to be handled outside of a glovebox, open to the air, without significant loss of activity. In addition to their use in Negishi cross-couplings under classic conditions, a Pd-free CC bond formation could also be facilitated by cobalt catalysts (see Fig. 6.84) [342].

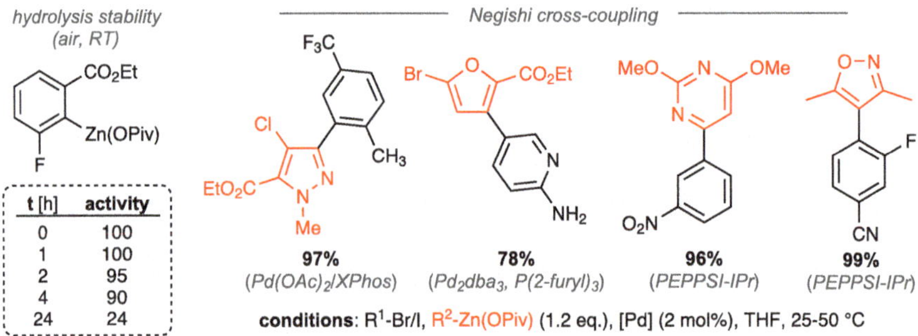

t [h]	activity
0	100
1	100
2	95
4	90
24	24

Fig. 6.84 Stability and reactivity of zinc pivalates [338, 339]

Due to the facile access to alkyl organometallic compounds, the Negishi coupling is optimally suited to utilize these nucleophiles along with the Suzuki–Miyaura reaction [343, 344]. While alkylboranes are foremost prepared by hydroboration, which is suitable for highly complex intermediates, the corresponding alkyl zincates are mainly derived from alkyl halides. Although boronic acids are isolable and stable, they display a comparably lower rate of transmetalation and tend to protodeboronate under the reaction conditions. The respective organozinc reagents are significantly more reactive, but usually have to be freshly prepared due to their limited stability. Secondary alkyl groups additionally suffer from the possibility of isomerization via β-H-elimination and reinsertion to form the n-alkyl coupling product. A catalyst environment that supports reductive elimination over β-H-elimination should accordingly result in high n/r ratios. Unsurprisingly, biaryl phosphines and NHCs have proven to be competent systems, but also bidentate ligands with a certain bonding angle can deliver the desired products with high yields (see Fig. 6.85) [345–347].

The use of first-row transition metal catalysts particularly lends itself for realizing a cost saving potential. Often these catalysts are also less toxic, which makes them ideally suited for the synthesis of pharmaceutical intermediates. While Ni-catalyzed protocols for biaryl synthesis have existed for some time, their extension to alkyl halides has also been achieved. Bidentate ligands (dppe, dppb, DPEPhos, …) are typically used in combination

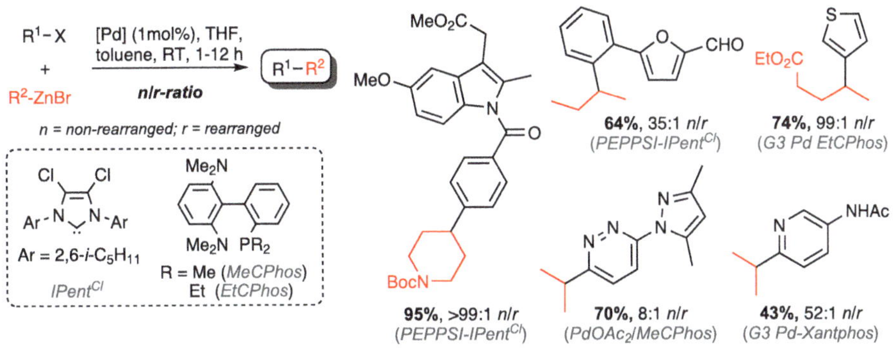

Fig. 6.85 Coupling of secondary alkyl zincates [345–347]

with $Ni(acac)_2$ or an alternative Ni(II) source in etheric solvents. Although catalyst loadings of 2.5–10% are regularly required, the electrophiles are comparable to the scope accessible with Pd catalysts (chloride, tosylate) or even venture beyond (e.g., phosphate) [20]. Under Ni catalysis, the oxidative addition usually proceeds via SET [16]. If chiral catalysts are used, a stereoinduction must be effected by the catalyst system in the transient secondary alkyl radicals. Racemic organozinc compounds can similarly experience an enantioinduction, making the reaction stereoconvergent and ligand-controlled for both components. High enantio- and diastereoselectivities are thus possible (see Fig. 6.86).

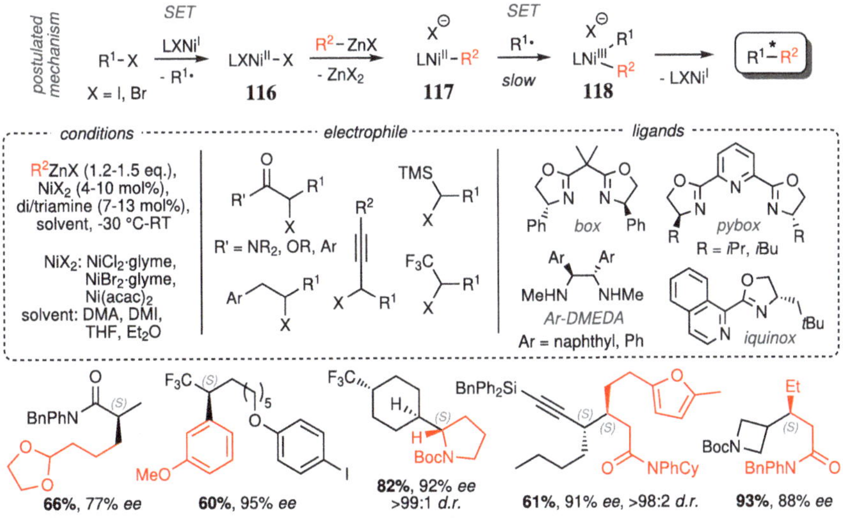

Fig. 6.86 Postulated mechanism, reaction conditions and substrate scope of the Ni-catalyzed, asymmetric Negishi coupling developed by Fu *et al.* [348–352]

The mechanism was investigated for propargyl bromides as electrophiles and appears to proceed via the intermediates **116–118**. Following the halogen abstraction from the electrophile under SET, a racemic radical is obtained. Counterintuitively, the resting state of the catalytic cycle was shown to be the Ni(II) complex **117** resulting after the transmetalation. The rate-determining coordination of the alkyl radical only occurs in the penultimate step prior to reductive elimination from **118**. Since racemic nucleophiles can also be used and enantioinduction occurs through the catalyst, the stereodiscriminating step must be located late in the catalytic cycle [348]. The substrate scope is very broad and includes a range of activated alkyl halides. Unactivated alkyl halides can also be converted under Ni catalysis, but the use of boronic acids as nucleophiles in a Ni-Suzuki cross-coupling is significantly more selective compared to the Negishi system. Corresponding Ni-catalyzed variants of the Kumada and Hiyama cross-coupling have also been realized employing similar conditions [343].

The Negishi reaction is not encountered as frequently in total synthesis as the Stille or Suzuki reaction. Its defining advantage over the other methods is the significantly enhanced reactivity of the organozinc compounds, which do not require additives. The method furthermore allows — barring the intrinsic basicity of the zinc reagent — neutral reaction conditions, so that a large number of functional groups are tolerated (see Fig. 6.87).

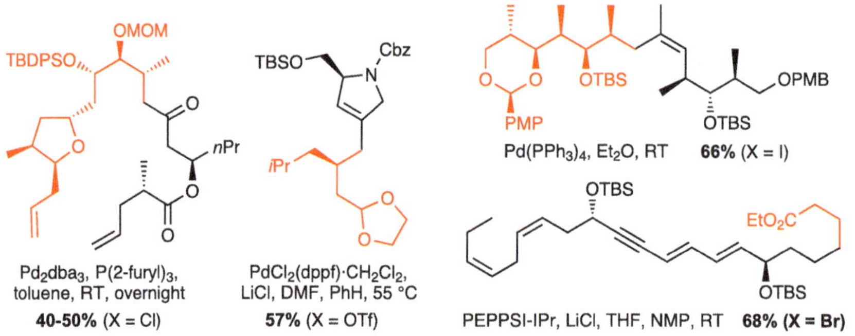

Fig. 6.87 Applications of Negishi cross-couplings in natural product syntheses. The nucleophilic fragment was highlighted [183, 223, 353, 354]

The **Kumada** cross-coupling was encumbered for an extended period by the sparse availability of methods for the preparation of functionalized Mg organyls. This challenge is now largely considered solved by the seminal work in the groups of Knochel and others. The remaining hurdle is the conversion of Mg nucleophiles that possess only a limited temperature stability, as extended reaction times at elevated temperatures result in almost complete decomposition of these reagents (see Fig. 6.88).

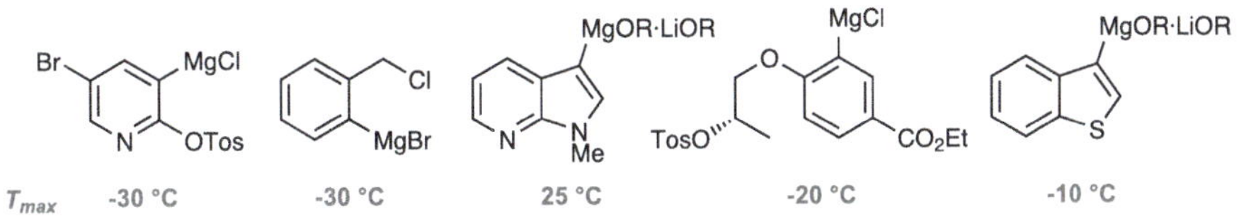

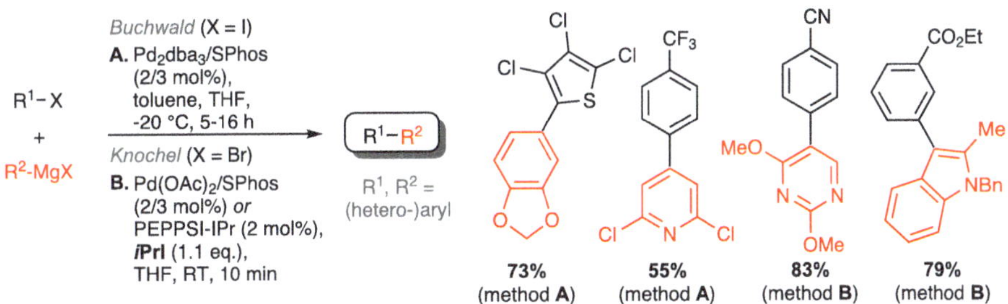

Fig. 6.88 Thermal stability of highly functionalized organomagnesium compounds [184, 190, 355]

Transmetalation to more stable organozinc compounds or boronic acid esters is a viable approach to utilize labile Grignard reagents. However, a direct use would be desirable for the sake of sustainability (atom economy, less handling operations, etc.). Two approaches have proven to be useful in that direction: highly reactive catalysts (biaryl phosphines, NHCs) coupled with either a reaction at low temperatures or a slow addition of the nucleophile at room temperature to slightly elevated temperatures (see Fig. 6.89) [283, 356, 357].

Fig. 6.89 Selected methods for coupling polyfunctional Mg organyls [283, 356]

Despite the considerable progress already achieved, the coupling of very sensitive Mg nucleophiles remains an area where further development is necessary. The situation is yet more challenging when turning to the corresponding organolithium compounds (sometimes also referred to as Murahashi coupling), as they are characterized by an even more pronounced reactivity and thus lower stability compared to the already unstable organomagnesium reagents. Isolated studies for their direct use in the presence of Pd catalysts have been published, but this is still limited to nucleophiles with low functionalization and has so far found little synthetic resonance, despite its appeal [273, 358, 359]. The use of organogelated organolithium nucleophiles may remedy that situation, as they can be handled even in air [360].

The situation is quite different when it comes to research on transition metal catalysts based on Ni, Fe, or other less noble metals. The substrate scope of Ni-catalyzed processes has been significantly expanded beyond classic aryl halides and now includes phosphates, carbamates, sulfamates, esters, and ethers in addition to the traditional substrates [15, 17]. Tricyclohexylphosphine has emerged as a particularly efficient ligand in this context, although the range of methods in Negishi cross-coupling is decidedly more pronounced than the reaction with the corresponding organomagnesium nucleophiles.

Iron catalysts were already identified early on to constitute the optimal match for the conversion of Grignard reagents, which is attributed to a positive, synergistic interaction of both metals. Analogous to Cu-catalyzed CN couplings, the efficiency of iron-catalyzed reactions usually does not depend on a highly specialized precatalyst/ligand combination. Instead, the right "cocktail", i.e., the combination of Fe salt, ligand, additives, solvent, and the starting materials, seems to make the difference. Apart from NMP in the case of vinyl halides, tetramethylethylenediamine (TMEDA) is one of the most common additives; it has proven particularly effective with alkyl halides as electrophiles. Interestingly, the Fe-catalyzed Kumada coupling of alkyl-Mg reagents shows a significantly larger substrate scope than the omnipresent aryl nucleophiles under Pd catalysis. The Fe-catalyzed variant is therefore ideally suited for the conversion of very reactive Grignard reagents, as aryl chlorides can be coupled at very low temperatures of down to $-78\,^\circ$C. Under these conditions, a large number of potentially unstable functionalities remains intact [361]. The application in challenging syntheses emphasizes the relevance of Fe-catalyzed Kumada cross-couplings (see Figs. 6.63 and 6.90).

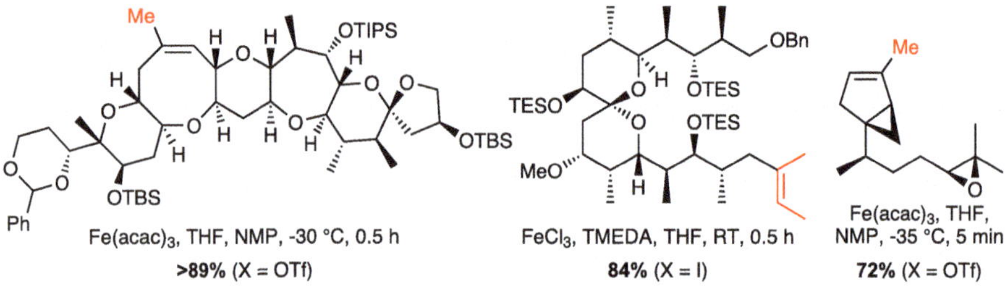

Fig. 6.90 Applications of Fe-catalyzed Kumada couplings in total syntheses [362–364]

The **Hiyama** cross-coupling is, perhaps due to the fate of its late birth, probably the least applied CC coupling in both an academic and industrial context [300]. The greatest development was undoubtedly seen in the expansion of the nucleophile scope, thereby rendering the need for the addition of F^- obsolete [294, 295, 365]. The Denmark group in particular significantly expanded the canon of coupling reagents to include silanes, silanols, and silanolates, which is why the reaction is sometimes also referred to as Hiyama–Denmark coupling. Regarding the electrophilic component, previous methods have mostly been limited to vinyl halides as well as aryl iodides and bromides. Biaryl phosphines and NHCs have so far only been sparingly investigated, which is why aryl chlorides are usually not considered suitable as substrates. A noteworthy exception is the trifluoromethylation developed by Buchwald *et al.*, which relies on TMS-CF$_3$ as nucleophile [366]. In this instance, the reductive elimination to the coupling product rather than the activation of the electrophile proved to be the main challenge, which is attributed to the extremely high stability of the

Pd-CF$_3$ bond. While trifluoromethylations are routinely performed under Cu catalysis, only aryl iodides and to a limited extent bromides afford the desired products (see Fig. 6.91) [367, 368].

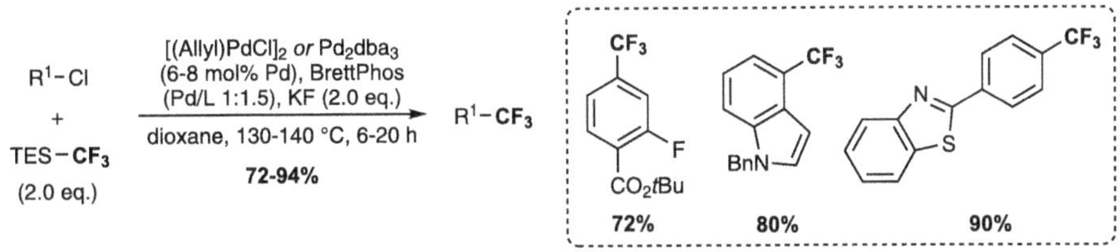

Fig. 6.91 Pd-mediated trifluoromethylation according to Buchwald *et al.* [366]

An synthesis making elegant use of a Hiyama cross-coupling was carried out by the Denmark group. The first total synthesis of the natural product (+)-brasilenyne was achieved in 19 steps starting from malic acid. The oxasilacyclohexene **119**, which was accessed by means of a ring-closing metathesis, afforded the desired tetrahydrooxonine **120** in the intramolecular cross-coupling. The final steps included an extension of the PMB-protected side chain and derivatization to the terminal enyne as well as a concluding Appel reaction for the stereoselective introduction of the missing chlorine substituent to obtain the secondary metabolite brasilenyne (see Fig. 6.92) [369].

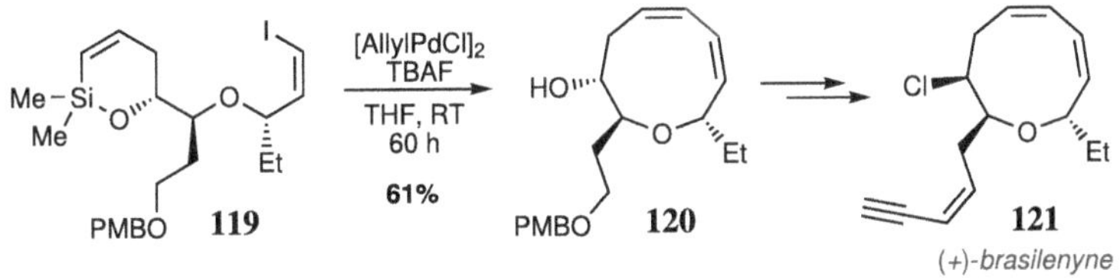

Fig. 6.92 Total synthesis of (+)-brasilenyne (**121**) [369]

The **Sonogashira** reaction has not experienced comparable major advances as the other coupling variants following its expansion of the substrate scope to include aryl chlorides and tosylates through the introduction of modern ligands [312, 370]. In this regard, the reaction may have fallen victim to its own success: The coupling delivers the desired product for most substrates, apart from propionic acid derivatives, in high yields even in the presence of very standard catalyst systems such as Pd(PPh$_3$)$_4$. The previously illustrated Negishi protocol was also able to close the remaining gap for electron-poor alkynes (albeit at the cost of an increased preparative effort) and Cu-free methods have also become successfully established. The maturity of the method has led to a wide range of applications in the synthesis of natural products and functional materials (see Fig. 6.93) [371].

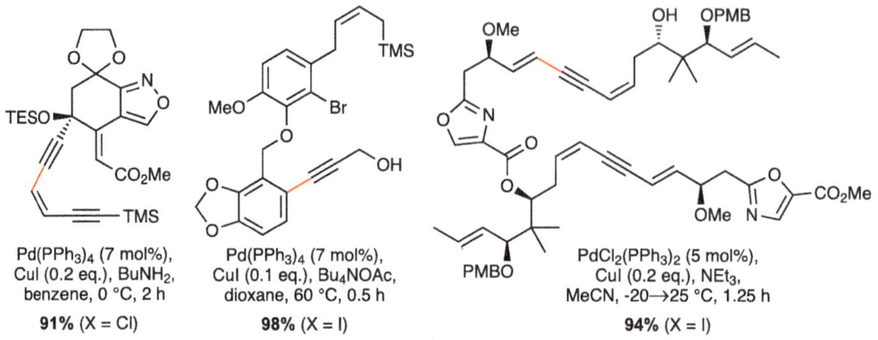

Fig. 6.93 Sonogashira couplings in natural product syntheses [372–374]

To further support its applicability on an industrial scale, one approach is based on the omission or replacement of costly palladium with copper, iron, nickel, and silver salts as a catalyst [375, 376]. Copper-based catalysts in particular have garnered the most attention and shown the greatest promise of the range of alternative metal salts. However, one study observed that contamination of the copper source with palladium, even in the ppb range, had a dramatic impact on the reaction rate and turnover [377]. When using highly pure $[Cu(PPh_3)_2NO_3]$, hardly any turnover was observed in the absence of $[Pd(PPh_3)_2Cl_2]$, while the addition of only 1 ppm of Pd led to full conversion of the starting material. In addition to CuI, Pd traces can also be detected in commercial Cs_2CO_3, phenylacetylene, or even on used magnetic stir bars. The notion of truly Pd-free conversion can therefore be questioned, even though efficient methods have been disclosed that do not require the addition of a Pd salt and which should therefore also find acceptance in an industrial environment [378, 379].

Apart from academic projects, cross-coupling reactions have also proven to be valuable assets in the synthetic toolkit for CC bond formations on an industrial scale. The Suzuki cross-coupling is employed most frequently, followed by Negishi and Sonogashira reactions. The Kumada coupling is rarely encountered for the reasons mentioned above, while the Stille coupling is ruled out based on the toxicity of tin and the Hiyama coupling due to its hitherto scarcely investigated heterocyclic electrophile scope. The core purpose in industrial syntheses is the provision of highly functionalized heteroarenes (see Fig. 6.94) [3, 4, 10, 19, 380, 381].

In addition to the classic cross-coupling of an organometallic nucleophile and an electrophile discussed thus far, reliable methods exploiting alternative reaction modes have also been established in recent years. In addition to CH activation ($\rightarrow$ Sect. 6.5), these comprise the reductive coupling of two electrophiles or of two nucleophiles under oxidative conditions [32–34]. These fields are rapidly developing given the obvious advantages of a reductive coupling: an additional activation of a nucleophile can be omitted, yet at the cost of requiring a stoichiometric redox reagent. The disadvantage, however, is a diminished (chemo-)selectivity, which is why in these reactions both electrophiles or nucleophiles must

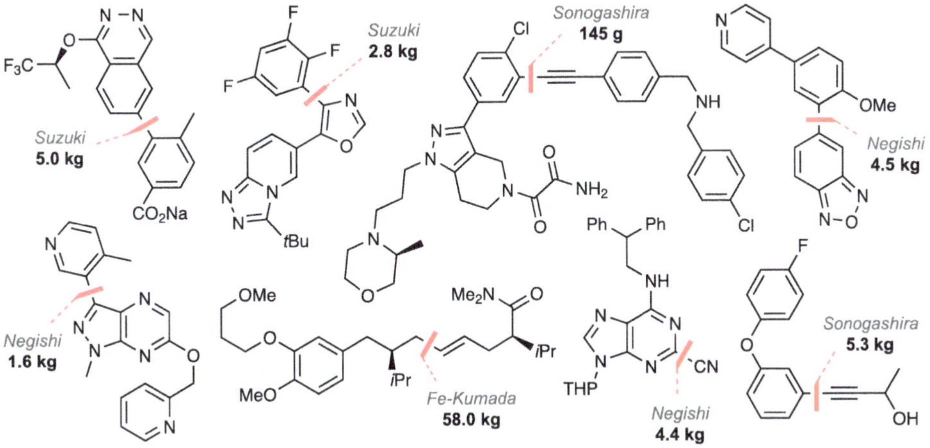

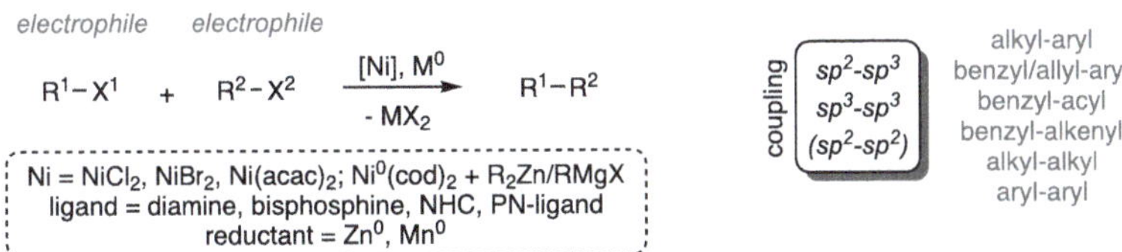

Fig. 6.94 Examples of industrially used CC cross-coupling reactions [382–389]

possess different reactivities to avoid homocouplings and a statistical product composition. This is most easily solved by resorting to two different halides (I/Br) or organometallic reagents. In these cases, a fine-tuning of the reaction conditions accordingly becomes a necessity.

In the more widely studied reductive coupling, metallic zinc or manganese is typically added as reducing agent. Aryl iodides and bromides are most readily reduced. Less noble metals dominate in these methods as the active metal, as most reactions are assumed to proceed via single-electron transfer processes during the substrate activation. Particularly Ni, and to a lesser extent Fe and Co, can perform this chemistry more easily than Pd complexes could (see Fig. 6.95) [34, 390, 391].

Fig. 6.95 Scheme of reductive couplings

The reaction allegedly proceeds via radical steps in most cases, yet so far mechanistic studies have only been carried out on a few selected systems. This is further complicated by the fact that the reactions are inherently heterogeneous by nature due to the employed solid reductants. In addition, some intermediates display a decided sensitivity to air, making them rather elusive to detect or study. Still some problems remain with regard to reproducibility, as the heterogeneous nature of the transformation induces a dependency on the mixing efficiency, the powder size, and its purity. Nevertheless, a wide range of possible coupling types has developed: In addition to the use of functionally complex substrates, the possibility of an aryl-aryl coupling by using two metal catalysts or an enantioconvergent reaction of racemic electrophiles have been reported, as well as select examples of their application in total synthesis (see Fig. 6.96, also cf. Fig. 2.169).

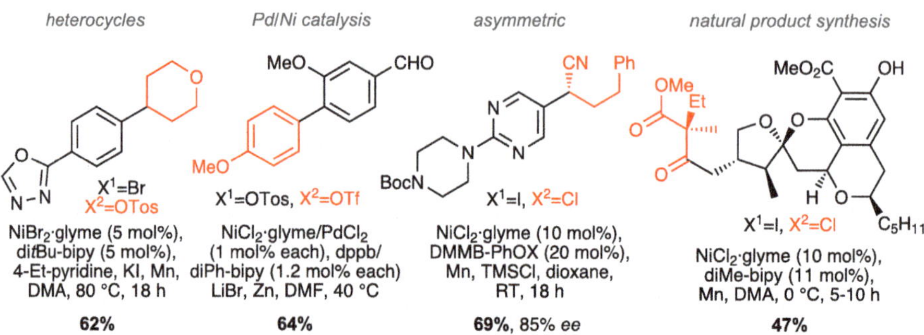

Fig. 6.96 Select examples of Ni-catalyzed reductive CC couplings [392–395]

Furthermore, the activation of catalysts and starting materials by means of photoredox catalysis to achieve a wholly new reactivity has also been successfully integrated [396–398].

Cross-couplings using CO or CO surrogates to access ketones or carboxylic acid derivatives play an important role, especially in the chemical industry. Particularly alkoxy- and aminocarbonylations have been intensely studied as they allow easy access to esters and amides. Mechanistically, the oxidative addition of the electrophile is followed by the coordination of CO, onto which R^1 is subsequently transferred to form an acyl-Pd intermediate. The coordination of a suitable C/N/O nucleophile then allows the reductive elimination of the desired carbonyl derivative (see Fig. 6.97) [399–402].

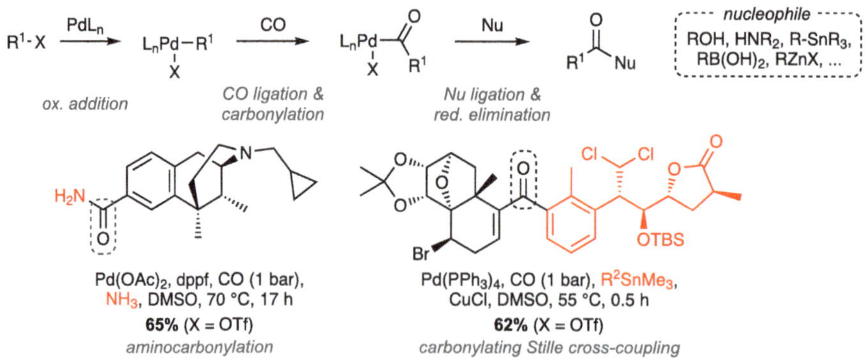

Fig. 6.97 Schematic mechanism of Pd-catalyzed carbonylations and application in the synthesis of bioactive compounds [399, 403, 404]

6.3   The Mizoroki–Heck Reaction and Related Couplings

The Pd-catalyzed arylation and alkenylation of olefins, known as the Heck reaction, was discovered by Heck and co-workers in 1968 and subsequently advanced and expanded in the groups of Mizoroki and Heck [314, 405, 406]. In this coupling, aryl, alkenyl, benzyl, and alkyl halides, triflates, and tosylates react with alkenes (or alkynes) in the presence of a base, usually a sterically hindered amine, and catalytic amounts of a transition metal (see Fig. 6.98) [407, 408].

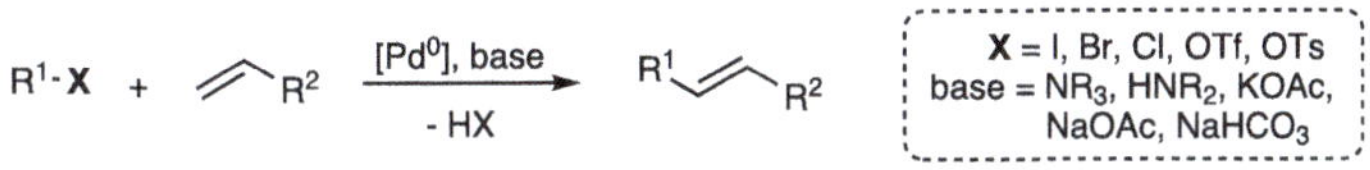

Fig. 6.98 General scheme of the Heck reaction

The Heck reaction proceeds under mild reaction conditions despite the requisite addition of a base and tolerates a broad variety of functional groups, which is why it is performed in both an academic and industrial context for the construction of highly complex compounds (see Fig. 6.99) [3, 407, 408]. Beyond chemo- and regioselective transformations, enantioselective Heck reactions have also been successfully introduced [409, 410]. This can be primarily attributed to the continuous improvement of old and the establishment of new, powerful catalysts.

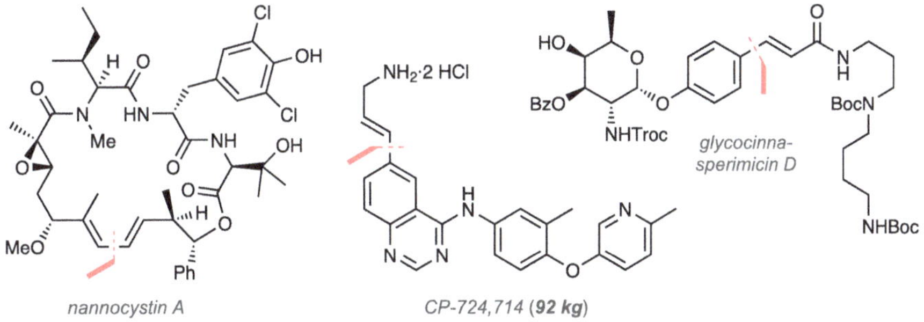

Fig. 6.99 Applications of the Heck coupling in the syntheses of natural products and pharmaceuticals [411–413]

The Heck reaction is best suited (from the standpoint of yield and chemo-, regio-, and stereoselectivity) for the construction of vicinal disubstituted alkenes from terminal olefins. In general, Heck reactions tolerate water and do not need to be carefully degassed, some Heck reactions are even carried out with H_2O as a cosolvent.

6.3.1   Mechanism of the Heck Reaction

Apart from the initial oxidative addition of the electrophile, the mechanism of the Heck reaction deviates from the other CC coupling methods, since no organometallic nucleophile participates as a coupling partner. Its individual details seem to differ depending on the reaction conditions. Especially the precatalyst/ligand combination and the base play a decisive role [414, 415]. Despite these variable elements, a general catalytic cycle can be formulated, which is considered sufficiently proven and generally accepted: Following the oxidative addition, the alkene is coordinated after the prior dissociation of a ligand [416–418]. The carbometalation step ensues, after which the alkyl group reorients itself by a rotation around the CC single bond so that the Pd-hydride species **125** can be received via β-H-elimination, thus affording the product. The base-assisted reductive elimination regenerates the Pd^0 complex and completes the catalytic cycle (see Fig. 6.100) [415].

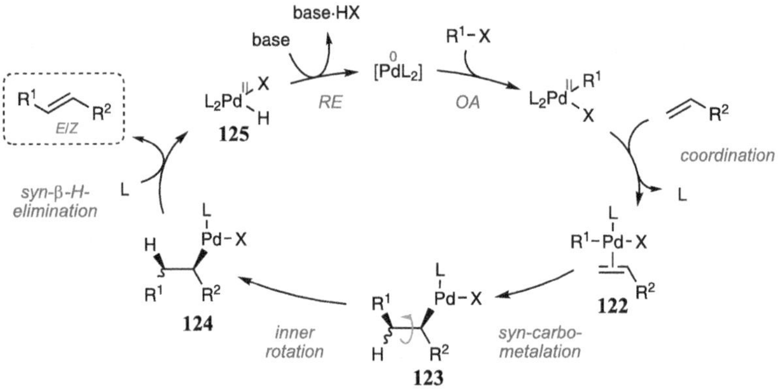

Fig. 6.100 Neutral pathway of the Heck reaction [36, 415]

The formation of the respective *cis*-configured oxidative addition complex or the *cis/trans*-isomerization mechanism corresponds to those of the classic cross-coupling methods (cf. Sect. 6.1).

The catalytic cycle described in Fig. 6.100 depicts the so-called "neutral" pathway, which describes the charge (or rather lack thereof) of the catalyst. The alkene is coordinated by dissociative substitution of a neutral ligand L, possibly assisted by transient coordination of the solvent. Alternatively, the anion X^- can also be initially dissociated and substituted, resulting in a cationic complex in the aptly named "cationic" pathway (cf. Fig. 6.25, **22–24**). The latter path is preferred for electrophiles bearing weakly coordinating leaving groups (e.g., triflate, tosylate, acetate, phosphate, carbonate), whereby facile loss of the anion X^- occurs prior to coordination of the alkene [419, 420]. When using halide electrophiles ($X = I, Br, Cl$), two options are available to force a cationic pathway: either added Ag(I)/Tl(I) salts forcibly abstract the halide from the metal or triflate-based Lewis acids (LiOTf, $Zn(OTf)_2$, etc.) displace the halides with OTf, which subsequently dissociate to the cationic complex. As a third possibility, highly polar solvents can similarly favor the cleavage of the halide without additives through stabilization of the charge by the solvent environment, provided bidentate ligands are utilized. If $Pd(OAc)_2$ constitutes the catalytic species without the addition of phosphines or other ancillary ligands, this is commonly referred to as "ligand free" in the literature. These conditions in turn favor the neutral reaction pathway (see Fig. 6.101) [421, 422].

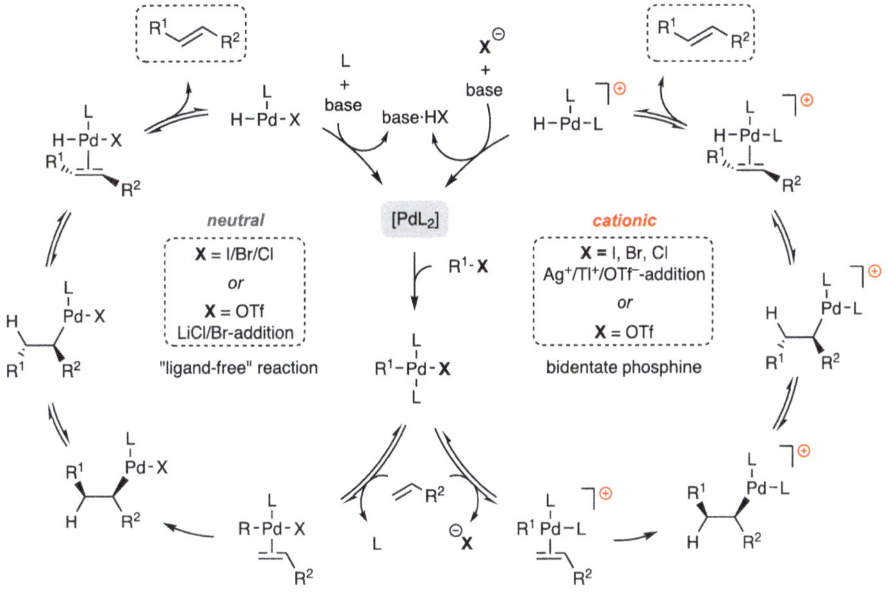

Fig. 6.101 Viable pathways and their dependence on the reaction conditions in the Heck reaction [415, 421]

The carbopalladation can form the Pd-C bond either at the terminal (β-isomer) or internal C-atom of the alkene (α-isomer). The β-regioisomer depicted in Fig. 6.100, 6.101 is preferred for steric reasons, but this preference can reverse on electronic grounds when switching to the cationic path. In the conversion of styrenes under conditions favoring cationic complexes, a build-up of positive charge on the olefin was suggested as rationalization for the observed regioselectivity, which would be stabilized by electron donors in the α-isomer [423]. Indeed, the regioselectivity can reverse when changing the reaction path for monosubstituted olefins. On the neutral pathway, product mixtures favoring the linear alkene usually occur with donor-substituted alkenes. On the cationic path, the branched olefin is preferably obtained. Acceptor-substituted olefins generally favor the linear product [421]. DFT calculations could semi-quantitatively reproduce the experimental trends and also allow a prediction of the selectivity for alkenes bearing multiple substituents (see Fig. 6.102) [424].

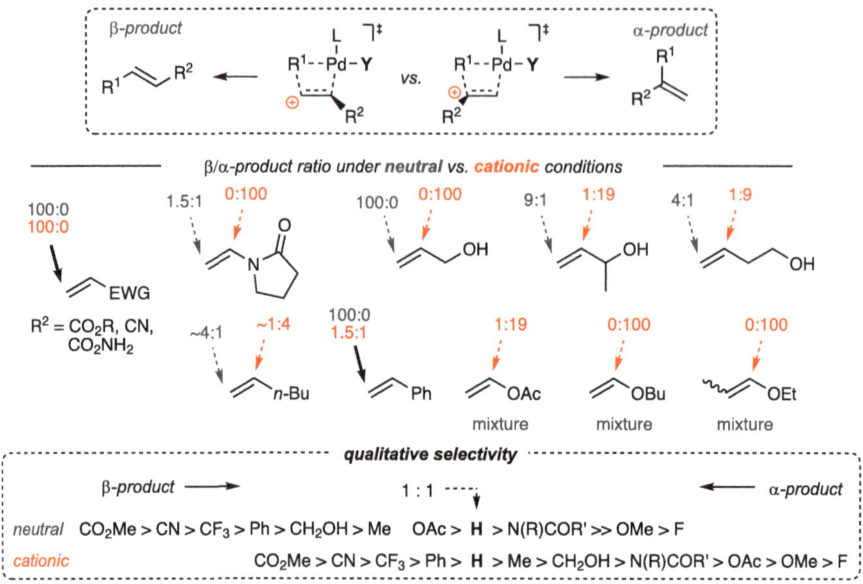

Fig. 6.102 Postulated transition state, experimental regioselectivity, and theoretical prediction depending on the reaction path and the alkene substituents [421, 423, 424]

A pharmaceutical intermediate required the selective access to the α-arylation product in the conversion of an aryl bromide with a vinyl ether. In an extensive solvent, base, ligand and additive screening, only a moderate preference for the internal α-isomer was observed using the bidentate ligand Xantphos in combination with Cs_2CO_3 as base. All other monodentate phosphines and especially alternative bases preferentially yielded the undesired linear alkene. Upon addition of LiOTf, the desired regioisomer could ultimately be isolated with an excellent α/β-selectivity of 24:1, which can be attributed to the cationic path being favored through liberation of Br^- from the metal center. In the subsequent scale-up, the uncontrolled and inferior activation of the precatalyst was identified as the cause for the varying chemo- and regioselectivity, which could be remedied by adding controlled amounts of H_2O in the form of $Na_2SO_4 \cdot 10\,H_2O$ (see Fig. 6.103) [425].

The role of chelate ligands on the different paths is crucial: Since the mechanism of alkene coordination is assumed to proceed in a dissociative fashion, a ligand must be liberated prior to the formation of the alkene complex. With weakly binding or monodentate ancillary ligands, the neutral ligand L would dissociate. With polydentate ligands, no entropic gain for the cleavage of a ligand arm arises ($\Delta S = 0$), making the loss of the anion X^- energetically more favorable than in the respective monodentate system. By resorting to bidentate ligands with a specific bite angle, the cationic path can accordingly be promoted.

The cationic mechanism is of crucial importance in enantioselective transformations, where the use of bidentate ligands is almost obligatory: The cationic path allows the two

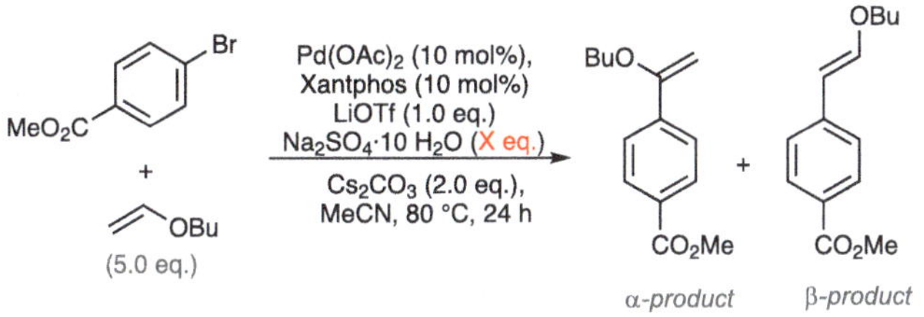

H_2O [eq.]	α:β	yield [%]
–[a]	24 : 1	n.d.
5	3.8 : 1	67
10	9.1 : 1	81
20	1.7 : 1	81

[a] screening scale

Fig. 6.103 Control of regioselectivity by judicious choice of the reaction conditions [425]

arms of the bidentate chiral ligand to remain coordinated to the metal center throughout the entire catalytic cycle, thus achieving an optimal enantioinduction in the substrate. When traversing the neutral pathway, the requisite transient dissociation of a ligand or part of a ligand would lead to a less pronounced induction as a consequence [415]. In enantioselective intramolecular Heck reactions, the facial selectivity upon the coordination of the alkene is assumed to be the enantiodiscriminating step [426].

The binding strength of a ligand to the metal center determines which reactive ligand is transferred. The migration rate of metal-bound ligands decreases in the order [427–429]:

$$\mathrm{H} \;>\; \mathrm{R(C{=}O)} \;\approx\; \mathrm{aryl} \;>\; \mathrm{alkyl} \;\gg\; \mathrm{R_2N}$$

Heteroatom ligands only display a low tendency to migrate, as the metal-ligand bond in these systems can be characterized as a partial multiple bond (interactions of free electron pairs with the metal). In the termination step after the insertion, a β-H-elimination occurs, followed by the dissociation of the alkene from the Pd complex. The catalytic cycle is completed by the base-assisted reductive elimination of HX with concomitant regeneration of the Pd catalyst, which can reenter the reaction. If a β-hydrogen atom is missing, is not accessible, or if the β-hydride elimination proceeds very slowly, the termination can alternatively proceed via nucleophilic substitution at the palladium, transmetalation or reductive elimination of other β-positioned substituents (see Fig. 6.104).

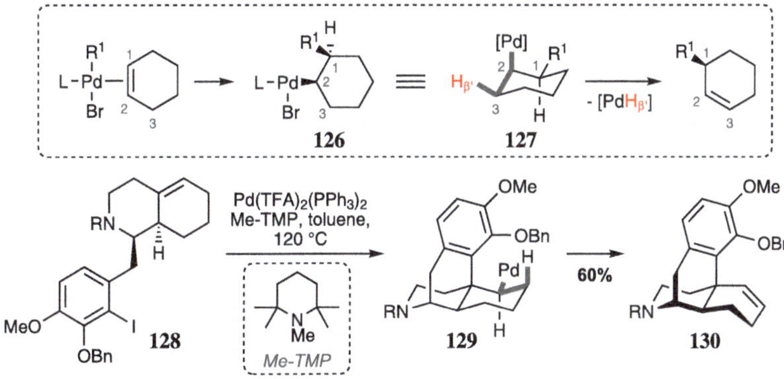

Fig. 6.104 Hindered β-H-elimination in cyclohexene as substrate and application of the β'-H-elimination in Overman's synthesis of morphine [430]

In cyclohexenes, a β-H elimination is not possible due to the anticlinal arrangement in **127**, which is why the synclinal β'-H atom is eliminated instead. This was exploited in Overman's synthesis of morphine to selectively construct the quaternary stereogenic center in the intermediate **130** [430]. Due to their diminished degree of freedom, ring closure reactions usually show high selectivity, which coincides with a good predictability of said chemo-, regio-, and stereoselectivity. The majority of the studied cases follow an *exo-trig* mode, as this is minimizes steric interactions and ring strain. In the formation of small to medium rings, the reaction favors the sterically more stabilized transition structure **131**. An *endo-trig* cyclization necessitates a rotation of the double bond into the newly forming ring for the generation of the requisite π complex **134**, which demands an enhanced freedom of motion of the alkene. The system naturally becomes more flexible with an increase in ring size. Accordingly, an *endo-trig* cyclization via intermediate **135** gains in importance for larger rings (see Fig. 6.105) [407].

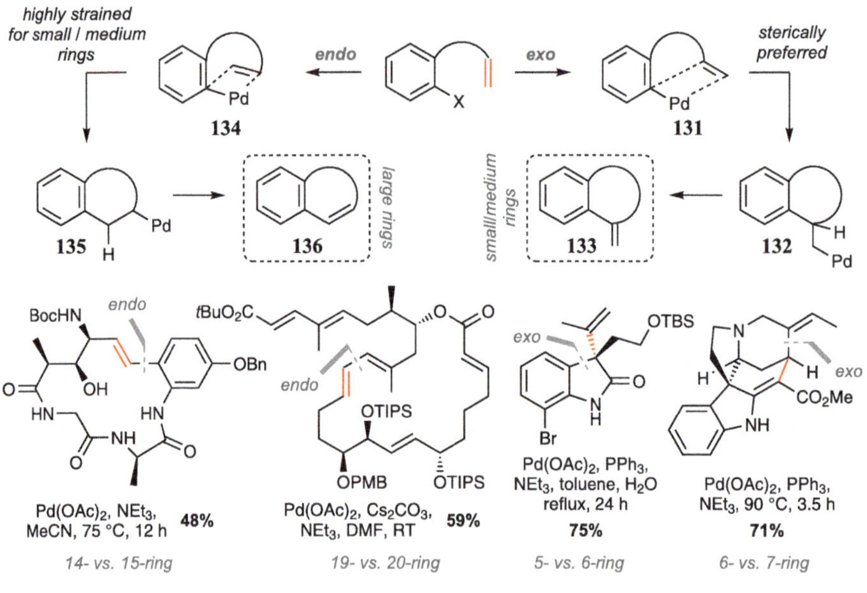

Fig. 6.105 Possible reaction modes of intramolecular Heck reactions [407]. The position of the original double bond is marked in red [431–434]

Apart from its predictable regioselectivity, the Heck coupling is similarly characterized by high diastereoselectivity with a preference for the *E*-configured product. The stereoselectivity is determined by the carbopalladation step, although the subsequent β-hydride elimination has an equally significant influence on the product ratio. The carbopalladation proceeds in a stereospecific manner via **137**. The insertion step is assumed to follow a four-membered, concerted transition state in most cases [407, 435]. In the case of alkenes bearing multiple substituents, the metal and the reactive ligand add from the same face to the double bond with conservation of the stereochemical information. The two possible transition states **138** and **139**, which provide H_a or H_b to the metal center for elimination through clockwise or counterclockwise rotation, determine whether the respective *E*- or *Z*-configured alkene results. Since the rotation has a low energy barrier and constitutes a conformational pre-equilibrium prior to product formation, the product ratio of the alkenes corresponds to the relative energetic position of the transition states **138** versus **139** (*Curtin–Hammett* situation) [435]. The β-H elimination is stereoselective, reversible and proceeds with *syn*-selectivity. Except for sterically very small substituents on the alkene (CN is the most common example), the *E*-isomer is the major product. The transformation is highly stereoselective, which is one of the main advantages over classic olefination methods such as the Wittig reaction. The cationic reaction path proceeds in an analogous fashion (see Fig. 6.106) [407].

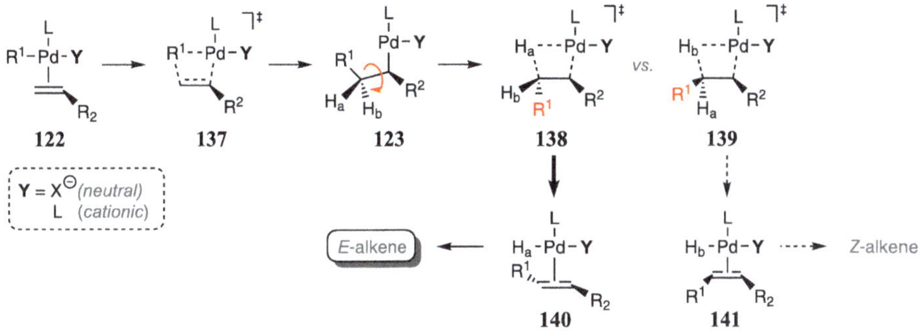

Fig. 6.106 Stereochemical course of the reaction [407]

The efficiency of the termination step is linked to the rate of olefin dissociation from the palladium(II) hydride complex (**140/141**). The β-H-elimination is a reversible process, with a slow dissociation possibly resulting in the formation of product mixtures (β vs. β', *E* vs. *Z*), which are assumed to arise from reinsertion and isomerization of the double bond. The dissociation rate is enhanced in cationic complexes, which can help prevent isomerizations occurring after product formation. Another approach to suppress these undesired side reactions is the utilization of elimination reactions with non-hydrogen atoms. The silyl-terminated Heck reaction, which was first studied by Tietze and co-workers and subsequently expanded by other research groups (Jeffery, etc.), can can accordingly exploit variable pathways depending on the chosen Heck conditions [436–438]. Instead of the customary Pd-H a Pd-SiR$_3$ species is generated via conformer **142**, which can be promoted by suitable reaction conditions (e.g., fluoride sources as an additive) (see Fig. 6.107).

Jeffery also investigated the use of biphasic conditions in the presence of quaternary ammonium salts as phase transfer catalysts for phosphine-free reactions [439–441]. This is now generally referred to as "Jeffery" conditions, and enjoys lasting popularity. The choice of base is crucial in determining whether a biphasic reaction is beneficial: weak bases such as acetate salts show better results under anhydrous conditions (and without

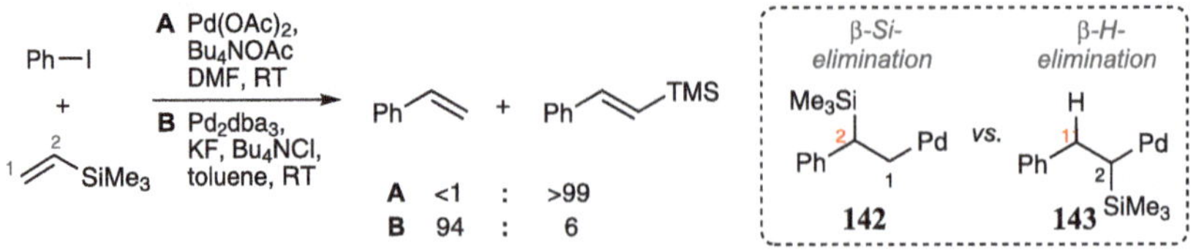

Fig. 6.107 Control of the silyl-terminated Heck reaction according to Jeffery [437]

ammonium additive), while stronger bases like carbonates require the addition (of small amounts) of water. While tertiary amines can afford high yields in the absence of water, the use of aqueous ammonium salts generally increases the reaction rate significantly, so that the reactions can be carried out at low temperatures [407]. In their approach to the synthesis of the molecular skeleton of the cembranoid metabolite family, Hirai and co-workers observed that the reaction provided good yields for the *E*-isomer under standard cationic conditions (Ag_2CO_3, dppe). In contrast, the *Z*-diastereomer unexpectedly did not lead to the desired product. When subjected to Jeffery conditions, the cyclization product was isolated in good yield, whereas the presence of phosphine ligands expectedly countermanded formation of the desired product (see Fig. 6.108) [442].

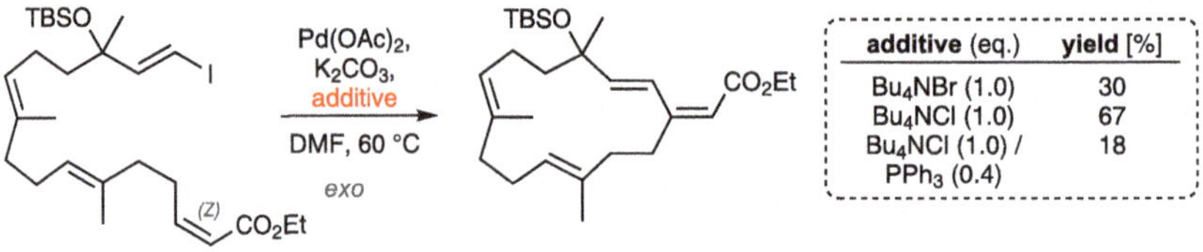

Fig. 6.108 Heck reaction under Jeffery conditions in Hirai's approach to cembranoids [442]

Asymmetric variants of the Heck reaction were introduced in the first half of the 1990s and are usually based on the formation of a stereogenic center within a carbo- or heterocycle with concomitant migration of the double bond [443–447]. Further substrate classes were subsequently realized and also include acyclic alkenes or even alkynes, which afford chiral allenes as products (see Fig. 6.109) [448, 449].

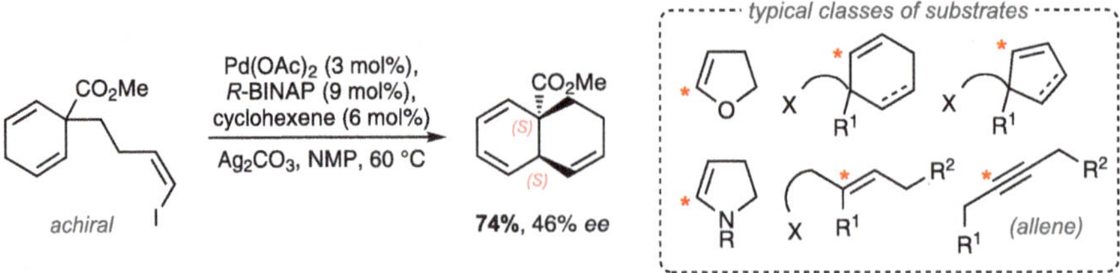

Fig. 6.109 The first asymmetric Heck reaction reported by Shibasaki *et al.* and typical alkene components of enantioselective Heck reactions. The stereocenter is introduced at the C-atom marked with an asterisk [450]

The chiral ligands are usually based on bidentate phosphines. While systems based on BINAP have proven to be particularly effective, planar-chiral or P,N-chelating ligand classes have also been used to great effect [448]. The intramolecular variant was frequently utilized in natural product synthesis (see Fig. 6.110) [2].

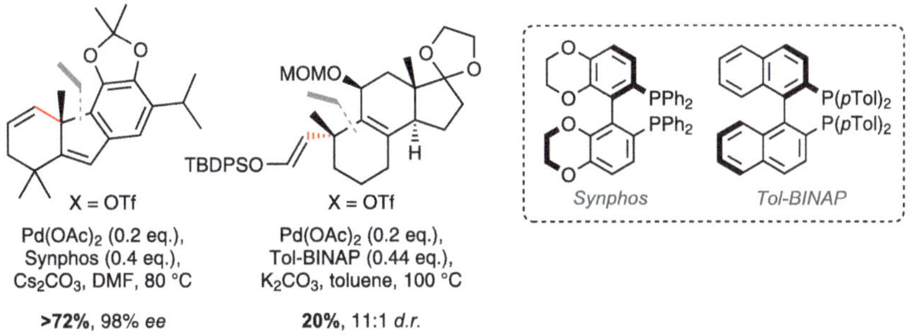

Fig. 6.110 Asymmetric Heck reactions in the syntheses of taiwaniaquinone D and wortmannin. The position of the original double bond is marked in red [451, 452]

Due to the intrinsic mechanistic necessity to reestablish an olefinic double bond in the product, the substrate scope is limited. The asymmetric Heck reaction thus has not become a standard transformation to the same extent as other enantioselective conversions, such as the Sharpless epoxidation or the CBS reduction.

In addition to the mentioned neutral/cationic reaction paths, experimental studies have indicated the existence of anionic intermediates when employing $Pd(OAc)_2$ as precatalyst (e.g., $[PdL_2OAc]^-$ instead of $[PdL_2]$ as the initiating species of the catalytic cycle) [61, 414]. The mechanistic details and the relevance of said species under catalytic conditions are still being controversially discussed [415].

6.3.2   Modern Variants of the Heck Reaction and Application in Synthesis

The methodological development of the Heck reaction has largely focused on a few, select areas: The investigation of modern catalyst systems such as palladacycles as well as the introduction of sterically demanding, electron-rich phosphine and NHC ligands made aryl chlorides and other challenging electrophiles viable as substrates [453, 454]. With the available variants, virtually all standard electrophiles can be converted to the desired product in at least acceptable yield, given the absence of steric complications or very labile functional groups. Despite these advances, the majority of published Mizoroki–Heck reactions in total syntheses still rely on the use of bromides, iodides, and triflates as leaving group (see Fig. 6.111) [455, 456].

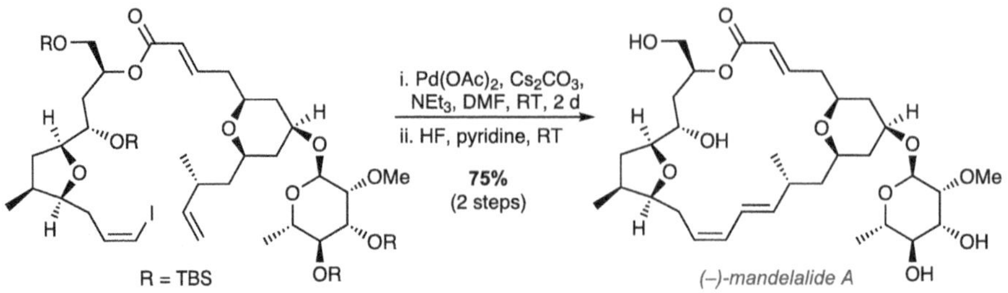

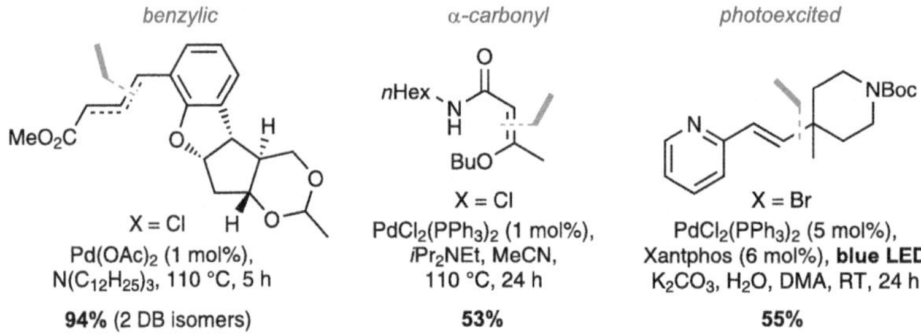

Fig. 6.111 Synthesis of (–)-mandelalide A disclosed by Smith *et al.* [457]

Alkyl electrophiles could also be added to the scope of starting materials by suppressing an unwanted premature β-H elimination. Since the final steps of the catalytic cycle entail a β-H elimination, a complete prevention of this reaction is accordingly undesirable. This precarious balancing act has yielded only a limited number of reports on the conversion of alkyl halides thus far. Most examples cover benzyl halides and α-halogen carbonyl compounds. In the case of unactivated alkyl derivatives, methods relying on a radical pathway have been disclosed in recent years, either employing the support of reducing agents or photoexcitation (see Fig. 6.112) [458].

Fig. 6.112 Conversion of alkyl electrophiles [459–461]

Apart from an expansion of the substrate scope, new types of coupling partners have also been introduced. Especially the use of boronic acids as "electrophiles" under oxidative conditions as well as the activation of CH bonds instead of using an aryl-/alkenyl halide were extensively investigated (see Sect. 6.5 on CH activation) [462, 463]. Mechanistically, the oxidizing agent facilitates the formation of the requisite Pd(II) species, as the boronic acid cannot be oxidatively added (see Fig. 6.113).

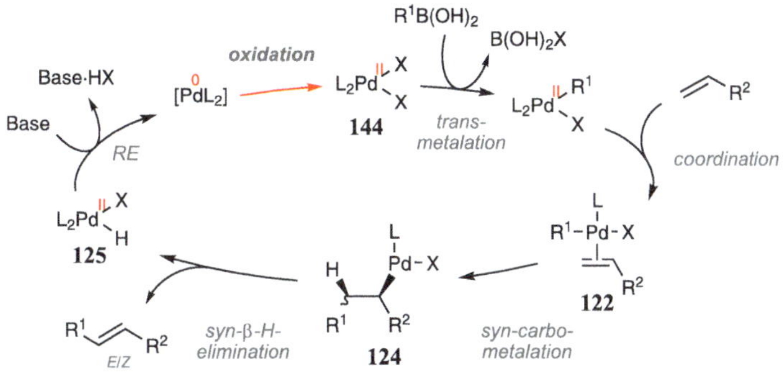

Fig. 6.113 Postulated (neutral) reaction path of the oxidative Heck reaction [462]. Ligands L are omitted as reactant of product for clarity

In addition to the use of stoichiometric oxidants such as Cu or Ag salts, benzoquinone or ketones (affording alcohols as by-product), oxygen can also be utilized as a particularly sustainable, atom-economic, and easy-to-handle oxidant. The reaction can be carried out in air under these conditions, even obviating the use of an inert gas atmosphere (see Fig. 6.114) [462, 463].

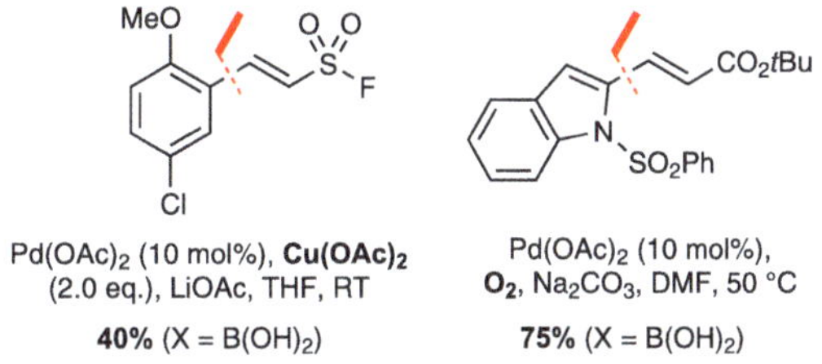

Fig. 6.114 Examples of oxidative boron-Heck reactions [464, 465]

Similar to the various cross-couplings, protocols employing non-noble transition metal complexes as catalysts were also investigated. Nickel complexes lend themselves as catalysts due to their reactivity and possibility to resort to highly optimized ligand classes routinely accessible to palladium [466]. In addition to the formation of carbo- and heterocycles and the arylation of alkenes, even asymmetric Ni-catalyzed reactions have been successfully introduced. The use of alternative catalysts was also reported in some total syntheses (see Fig. 6.115).

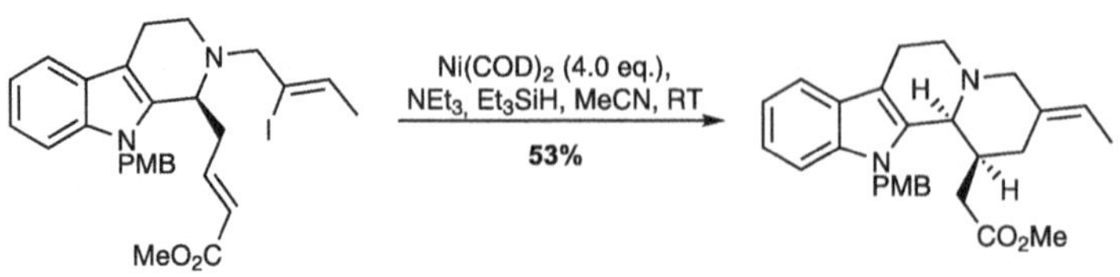

Fig. 6.115 Reductive, stoichiometric Ni-mediated Heck reaction in the synthesis of geissoschizol [467]

The Heck reaction is frequently encountered in the synthesis of complex intermediates, which is partly attributed to its high tolerance towards common functional groups. Intramolecular variants are particularly popular, both in the controlled introduction of quaternary stereogenic centers as well as in macrocyclizations. The requisite double bond may also only be formed *in situ* and have transient character, for example when using an enolate intermediate as alkene component. Overman and co-workers employed a very elegant Heck reaction based on this strategy in their second synthesis of the *Strychnos* alkaloid (+)-minfiensine. The enolate obtained from the α-acidic ketone **145** was converted to **147**, with the intramolecular Heck reaction enabling the final ring closure of the pentacyclic, bridged carbon skeleton of the natural product. **146** could subsequently be converted to (+)-minfiensine in four steps (see Fig. 6.116) [468].

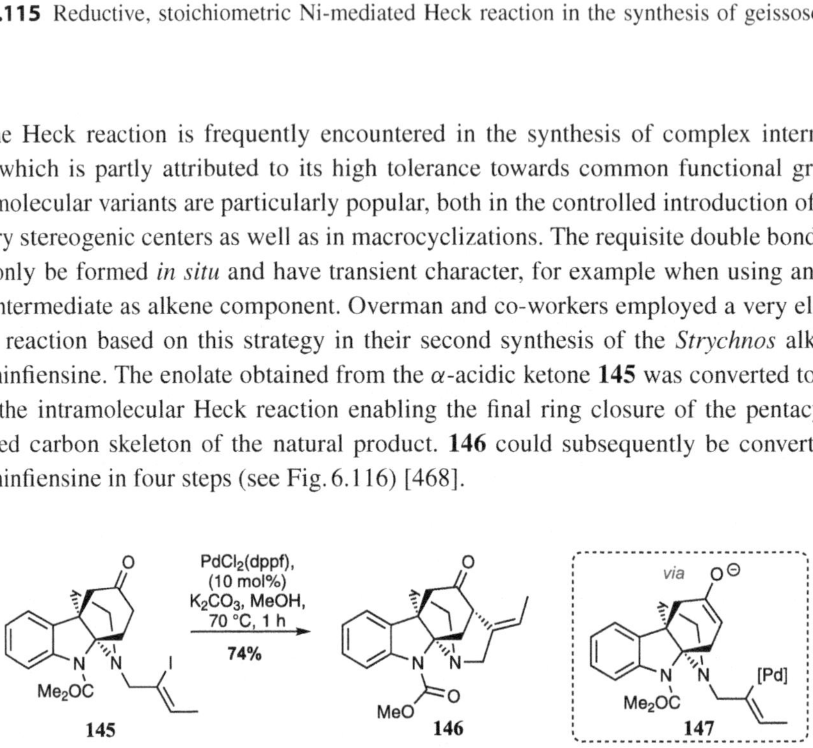

Fig. 6.116 Overman's synthesis of (+)-minfiensine [468]

6.3.3 Carbopalladation of Alkynes

While the majority of reported reactions consists of the C_{sp^2}-C_{sp^2} coupling typical of the Heck reaction, the related C_{sp^2}-C_{sp} couplings have shown to be equally important [469–472]. The primary product of an alkyne carbopalladation, a vinyl palladium species, carries no β-H atoms at the resulting double bond, thus removing a concluding elimination pathway and paving the way for successive reaction steps. They accordingly constitute the most frequently used substrates of domino reactions incorporating the Heck reaction. The carbopalladation of alkynes is a common approach for accessing tetrasubstituted olefins. In general, the introduction of a high degree of complexity is possible with this type of reaction: by virtue of its convergent nature, extensive structural variations can be rapidly achieved in a one-pot approach. However, a good regiocontrol is a key element that must be addressed in order to avoid the isolation of product mixtures (see Fig. 6.117).

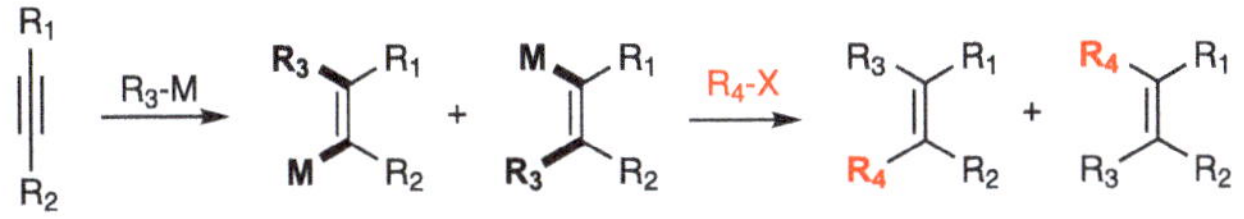

Fig. 6.117 Possible regioisomer formation in the carbopalladations of alkynes

Since carbopalladations are stereospecifically *syn*-selective, only one stereoisomer is obtained. However, in some cases, β-H eliminations can still occur as a side reaction to afford allenes [473].

The problem of regioselectivity can be overcome by the introduction of directing groups or when resorting to symmetric alkynes as substrate. While the use of symmetric alkynes reduces synthetic hurdles, it also significantly limits structural flexibility. In the case of unsymmetric substrates, directing functionalities are routinely used to avoid regioisomer formation. The influence can be both steric and electronic in nature (passive directing groups) as well as encompass chelating functionalities (active directing groups). The Tietze group used a chelate-controlled domino carbopalladation/Stille reaction in their enantioselective synthesis of Feringa-type molecular switches (**151**, see Fig. 6.118) [474].

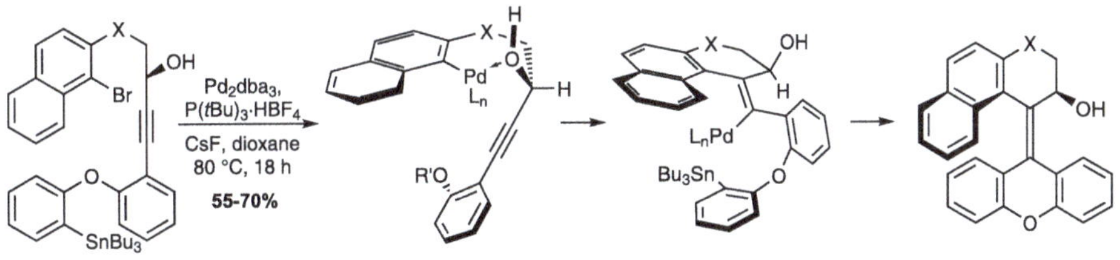

Fig. 6.118 Tietze's synthesis of molecular switches [474]

The tetrasubstituted helical alkenes **151** were obtained in a fully diastereoselective transformation by converting the enantiomerically enriched propargylic alcohols **148**. The observed diastereoselectivity was attributed to a coordination of the free OH group to the metal center following the oxidative addition in intermediate **149**. The subsequent carbopalladation to **150** presumably furnishes a helical structure, which is maintained in the concluding Stille cross-coupling.

For domino reactions of polyenes or polyynes undergoing multiple successive carbopalladations, Negishi defined several reaction modes, which are referred to as *zipper*- and *dumbbell* mode, besides a spirocyclization, to build quaternary stereogenic centers [470].

An extreme example of a zipper process was published by the Negishi group. After the oxidative addition of the vinyl iodide **152**, four carbopalladations ensue, followed by a carbonylation reaction and a terminating esterification with the primary alcohol group. Thus, the pentacyclic lactone **153** could be isolated in 66% yield (see Fig. 6.119) [475].

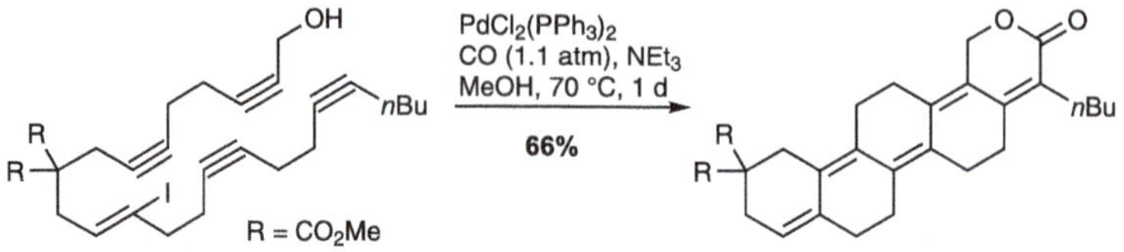

Fig. 6.119 Zipper reaction reported by Negishi [475]

Tietze and co-workers used a domino-carbopalladation-Heck reaction in their synthesis of the lignan linoxepine to assemble the benzonaphthooxepine structure in **156** via **155**. Starting from **154**, the reaction conditions were optimized to allow for the concluding silyl-terminated Heck reaction to become dominant. The final steps to the secondary metabolite included the oxidation of the allylic alcohol to the acid, formation of a methyl ester with TMS-diazomethane, and reductive ozonolysis of the terminal double bond with subsequent lactone formation (see Fig. 6.120) [373].

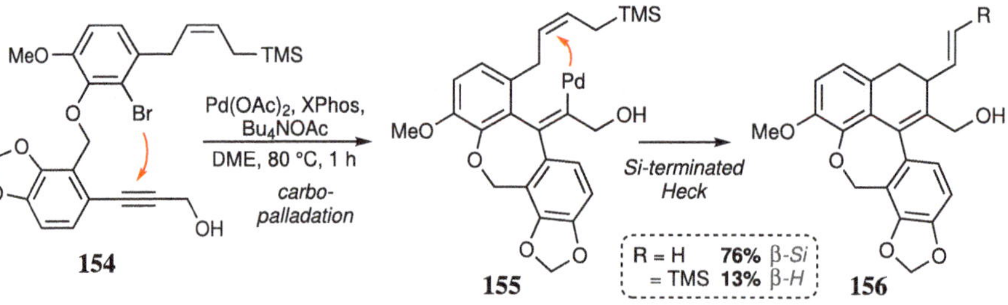

Fig. 6.120 Synthesis of linoxepine by Tietze and co-workers [373]

6.4   C-X Coupling Reactions

In addition to CC coupling reactions, the formation of carbon-heteroatom bonds also constitutes a highly relevant field of transition metal-mediated coupling reactions. The interest in C-X bond formations is rooted in their frequent occurrence in natural products and drugs (see Fig. 6.121).

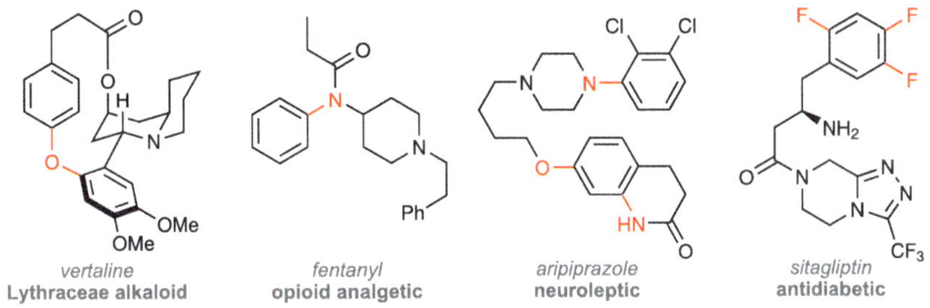

Fig. 6.121 Examples of C-X bonds in natural products and pharmaceuticals

Particularly CN coupling reactions, whose Pd-catalyzed variant is referred to as *Buchwald–Hartwig* coupling, are a key pillar as one of the most frequently employed reactions in the synthesis of pharmaceuticals [242, 476]. While natural products, especially alkaloids, rarely exhibit C_{sp^2}-N bonds, this structural motif can be found in nearly 40% of all drug candidates [10]. Aryl ethers are also contained in a significant proportion of active pharmaceutical ingredients, while thioethers are much less common. The presence of F and CF_3 substituents is likewise strongly pronounced in pharmaceuticals to modulate the lipophilicity, bioavailability, and resistance to enzymatic degradation. Accordingly, the focus of many developments lies on CN bond formations and to a lesser extent on the construction of CO and CF bonds.

In contrast to classic CC coupling reactions that largely rely on Pd catalysts, especially Cu complexes occupy a position of equal importance in CN couplings due to their complementary substrate scope. With exception of the Wacker oxidation or the Tsuji-Trost coupling, CX bond formations are mechanistically derived from the catalytic cycle of the closely related classic CC cross-couplings. Instead of the transmetalation of an organometallic reagent, the coordination of the corresponding heteroatom-containing nucleophile takes place.

6.4.1 Buchwald–Hartwig-Type Couplings

For the longest time, the *Ullmann condensation* exemplified the classic method for the construction of aryl amines and ethers from aryl halides and phenol/aniline derivatives. For the synthesis of amines, the reaction is also occasionally referred to as *Goldberg reaction*. In addition to the use of stoichiometric amounts of Cu salts, temperatures in the range of 100–300 °C are required, which is why the resulting limited functional group tolerance is prohibitive for use in complex substrates under these conditions (see Fig. 6.122) [477].

The subsequently developed variant of the Ullmann condensation relying on aryl halides with nitrogen nucleophiles using catalytic amounts of Pd complexes is referred to as *Buchwald–Hartwig coupling*. In addition to CN couplings, CO and CS couplings of aryl halides are often termed Buchwald–Hartwig-type reactions from the perspective of a com-

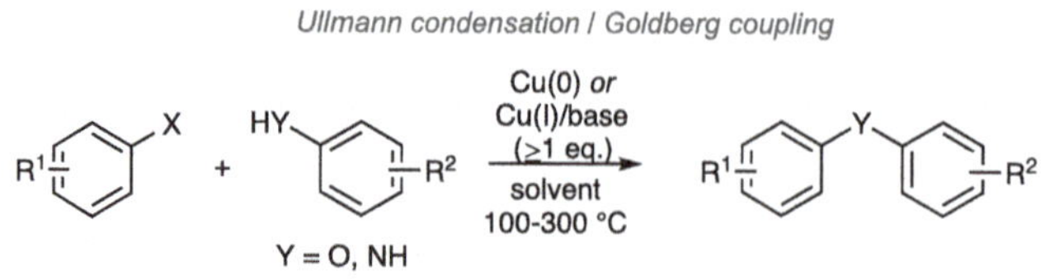

Fig. 6.122 Classic CX couplings relying on stoichiometric Cu amounts [477]

mon mechanistic foundation. The introduction of halogens and other functionalities is usually simply referred to as fluorination, nitration, etc. according to the type of coupled functional group. Mechanistically, these reactions all proceed via a common catalytic cycle, in which the basic steps resemble those of earlier-developed variants of the cross-coupling, in which only the organometallic nucleophile varies (see Fig. 6.123).

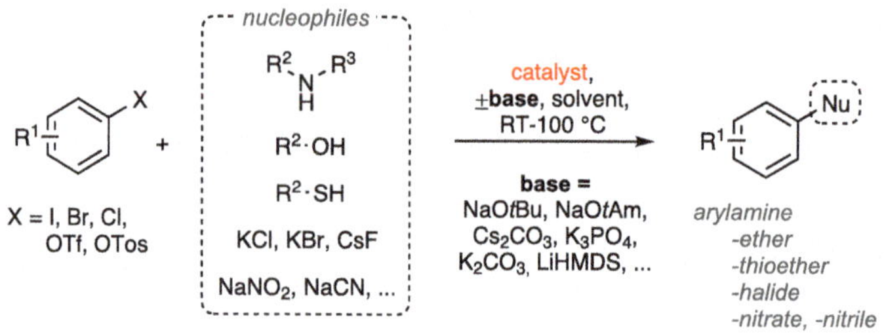

Fig. 6.123 General reaction scheme of Buchwald–Hartwig-type couplings

The bond formation with the nucleophiles is usually facilitated by Pd complexes. In **CN coupling reactions**, highly efficient methods utilizing Cu catalysts are also available, which are derived from classic Ullmann conditions. The substrate scope in Pd- and Cu-catalyzed methods is not only comprehensive with regard to the electrophile — especially the nucleophile can cover a wide scope and range from aliphatic to aromatic amines, (sulfon-) amides and carbamates to hydrazine and even may include ammonia [478]. While Ni-based methods have also seen rapid advances in recent years for use in CN couplings, they typically focus on aryl and alkyl amines as nucleophiles (see Fig. 6.124) [479, 480].

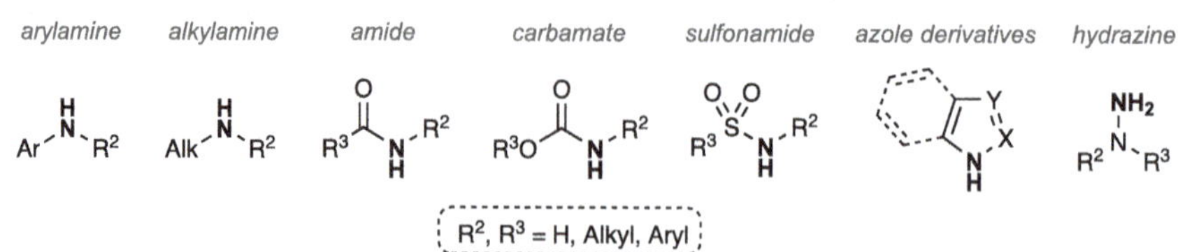

Fig. 6.124 Viable nitrogen nucleophiles in amination reactions [481]

Cu catalysts have been employed early on to promote the desired CN coupling reaction with the right cocktail of base, amine, and solvent for reactive electrophiles. This only became possible for Pd- and subsequently also Ni-based catalysts through the introduction of more efficient ancillary ligands. Pd and Ni complexes differ significantly in their mechanism from Cu catalysts (see Fig. 6.125) [13, 14, 480, 482].

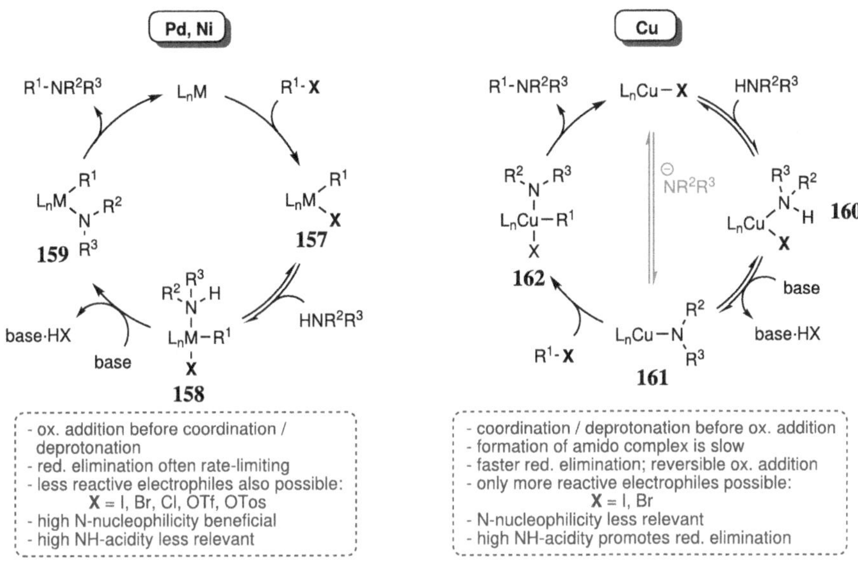

Fig. 6.125 Postulated mechanisms of metal-catalyzed aminations [13, 14, 480, 482]

While the formation of the catalytically active species and oxidative addition to **157** no longer constitutes a challenge for modern Pd precatalysts and ligands, either the formation of the amine complex **158** or the reductive elimination is the rate-determining step, depending on the amine component in question. Ni catalysts, due to their enhanced nucleophilicity and oxophilicity of the metal, also allow the use of phosphates, sulfamates, esters, and even ethers as electrophiles [480]. However, the oxidative addition may proceed very slowly, unlike reactions in the presence of Pd complexes. The active species of the catalytic cycle was proposed to be [PdL] with monodentate phosphines by virtue of their enhanced reactivity, along the lines of what was already discussed for CC coupling reactions (cf. Sect. 6.1) [79]. The experimentally observed enhanced rate of amination with sterically demanding ligands can accordingly be attributed to their stabilizing effect on monoligated Pd complexes [482]. The coordination of the NH-acidic nucleophiles to the metal center results in a significant increase in NH-acidity, thereby greatly facilitating the rapid deprotonation ($\mathbf{158} \rightarrow \mathbf{159}$) of metal-bound nucleophiles for all metals (Pd, Ni, Cu) [13]. Ni catalysts follow the same mechanistic layout as Pd complexes, however the details have not been elucidated as extensively. In addition to a Ni^0/Ni^{II} catalytic cycle, indications of a Ni^I/Ni^{III} pathway have also been discerned. Given the lack of a more detailed picture for these complexes, it is reasonable to assume that the same principles apply as under Pd catalysis [480, 483, 484]. The reductive elimination routinely comprises the rate-determining step with Pd and Ni catalysts, which is why potential side reactions such as β-H eliminations to imines typically can occur at this stage. A high steric demand of the nucleophile and large Lewis basicity accelerate the concluding step of the catalytic cycle with Pd-amido complexes. The more electron-rich and sterically demanding, the faster the reductive elimination [40]:

$$\text{alkyl}_2\text{NH} > \text{alkyl-NH}_2 > \text{aryl-NH}_2 > \text{aryl}_2\text{NH} > \text{amide-NH}_2 \approx \text{azole-NH}$$

Cu complexes possess a smaller coordination sphere and shorter metal-ligand bonds. The effective size of the coordination shell in the vicinity of the metal is smaller, offering less space for large ancillary ligands and thus fewer possibilities for control through sophisticated ligand design. Steric congestion and strain and tuning of the ligand donicity, which are very effective design principles for ligands in Pd catalysis, are not highly useful in copper catalysis [13]. Instead, the success of a given protocol depends on the appropriate interactions in the system as a whole, in which all components (Cu precatalyst, ligands, base, solvent, nucleophile) are important. The reversible coordination of the nucleophile, either as a neutral component or as the corresponding anion, transpires at the beginning of the catalytic cycle. A multitude of possible ligated Cu species exists, the proportion of which is determined by the nucleophilicity of NHR^2R^3 and the ancillary ligand. As in the absence of added ligands, the copper center tends to bind not one but two or more equivalents of the NH nucleophile to form catalytically inactive anionic $Cu(NR_2)_2^-$ species. The ancillary ligand is therefore assumed to compete with the NH nucleophile for coordination sites and thus stabilize the Cu-amido complex for the subsequent productive pathway. The oxidative addition of the Cu(I) species **161** to the electrophile is presumed to occur via the short-lived intermediate **162**. The formation of the highly transient Cu(III) complex can also possibly be omitted if the oxidative addition and reductive elimination proceed in a concerted fashion during the transition state [13]. As opposed to Pd/Ni complexes, the oxidative addition is usually rate-determining in the presence of Cu catalysts. The step has been envisioned to be based on an activating Cu-X interaction with the leaving group of the electrophile, which is why only iodine and bromine as relatively Lewis-basic leaving groups efficiently afford the coupling product. When resorting to very weakly acidic nucleophiles, the formation of the complex **161** in equilibrium can also become the bottleneck for the overall process [13].

When various N-nucleophiles compete for cross-coupling, for example in the successive, twofold or threefold amination of primary amines or in substrates with several acidic NH bonds, a selective CN coupling can usually be achieved (see Fig. 6.126).

The product selectivity in the presence of competing amine components depends on several factors. The equilibrium reaction of amine coordination and subsequent deprotonation (**157→159**) presumably may represent a Curtin–Hammett situation depending on the amine, whereby the rate of reductive elimination or deprotonation determines the selectivity. This applies for the competition of aromatic versus aliphatic amines. The aliphatic amines bind

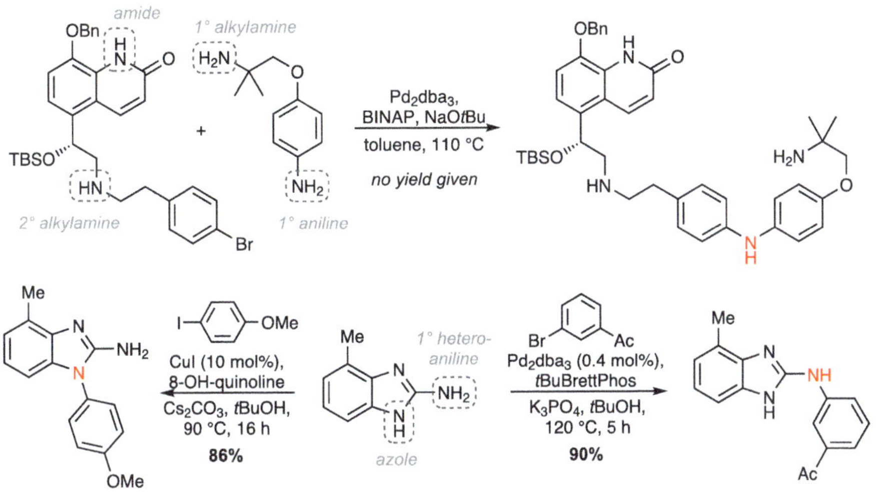

Fig. 6.126 Examples of chemoselective aminations [485, 486]

more strongly, but are less acidic and are deprotonated significantly slower than anilines. The major product results from CN coupling with aniline. In the competition of different anilines, no Curtin–Hammett situation arises: the basicity of the neutral amine rather than the NH acidity determines the selectivity. As the deprotonation becomes more rapid, a pre-equilibrium ceases to occur. The more electron-rich anilines accordingly afford the coupling product faster [487]. The conjugate, anionic form of NH-acidic, electron-rich heterocycles (imidazole, pyrazole, indazole, etc.) can additionally displace phosphine ligands, which is why their concentration should be kept to a minimum. This might be achieved by choosing a weaker base to keep the deprotonation equilibrium on the side of the protonated form and resorting to ligands that are bidentate [482]. Since the oxidative addition in Cu complexes occurs from the anionic amido complex **161** and the subsequent steps are usually fast, the product selectivity for competing amines is determined by the equilibrium position of the corresponding amido complexes. The NH acidity is thus crucial for the rate of product formation when steric influences play only a minor role [488]. Based on these influencing factors, the different metals lend themselves for coupling with a select set of nucleophiles with particularly high efficiency, thus creating "matched" combinations. The substrate scope of Pd and Cu catalysts thus complement each other (see Fig. 6.127) [13].

The optimal reaction conditions, which are inferred from the mechanistic details outlined above, hail from the type of employed nucleophile and whether 5-membered heterocycles are intended to be used as electrophiles. With Pd catalysts, precatalysts in combination with electron-rich, sterically demanding monodentate and bidentate phosphines have become established as a viable reference system that can promote a large proportion of aminations.

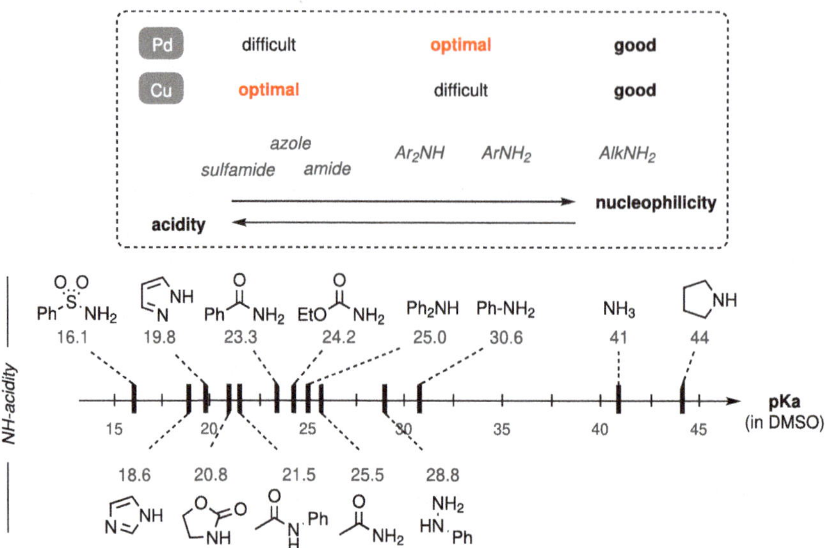

Fig. 6.127 Substrate scope of Pd and Cu catalysts and acidity of selected nucleophiles [13].

The most active Ni catalysts are primarily based on bisphosphines of the dppf or Josiphos type, but the successful use of NHCs has also been reported occasionally (see Fig. 6.128). Cu precatalysts can effect a CN coupling in the absence of complex ligands, provided the solvent solubilizes and stabilizes the intermittent metal complexes. If ligands are added, the catalytic species are typically based on N,N- or N,O-chelates such as bipyridine, phenanthroline, hydroxyquinoline, or oxalic acid amides [12, 13, 37, 483, 484].

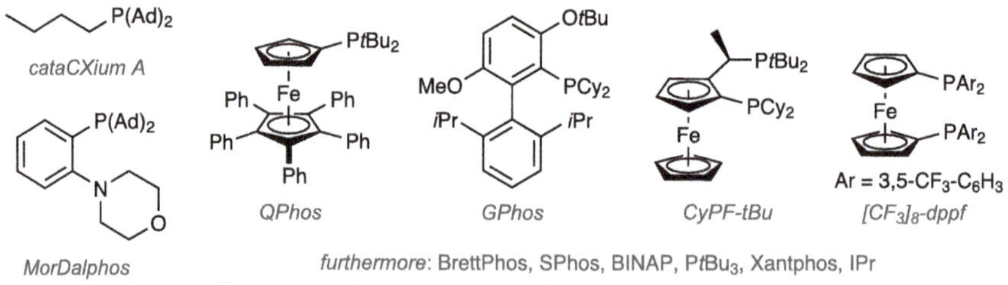

Fig. 6.128 Common ligands of Pd- & Ni-catalyzed aminations [37, 482, 483]

Among the range of possible nucleophiles, particularly weakly binding substrates such as amides, carbamates, azoles, and sulfamides pose a challenge for Pd catalysts. In addition, α-branched primary alkylamines also react slowly due to their high steric demand during coordination. Ammonia proves challenging in the reductive elimination step by virtue of its small size; it can also be arylated multiple times as the coupling product typically becomes successively more reactive than the respective preceding nucleophile. A hallmark of modern ligands such as QPhos, BrettPhos, CyPF-*t*Bu or GPhos lies in their efficient and selective conversion of these demanding nucleophiles [482]. Good overviews with recommendations for certain substrate types have been published for different metal catalysts (Pd, Ni, Cu) [12, 37, 483]. Fitzner and co-workers even evaluated 62,000 Pd-catalyzed reactions from publications and patents to recommend optimal conditions per substrate class using an algorithm-supported data analysis [37, 489].

Due to the large influence of the amination conditions on the reaction yields when converting phenol-based electrophiles (triflate, tosylate), these reactants are rarely utilized under Pd catalysis [482]. Instead, Ni catalysts display a significantly higher reactivity for this substrate class. As precatalysts, very reactive Ni species such as Ni(COD)$_2$ or oxidative addition complexes containing bidentate ligands are routinely reported [483].

Cu catalysts in turn excel with reactive electrophiles (X = I, Br), especially in the coupling of azoles, amides, and carbamates. The use of chelating ligands can also promote the reaction of unbranched primary alkylamines. Free OH groups should be avoided to prevent a competitive CO coupling under Cu catalysis (see below) [13, 490].

For Pd and Ni catalysts, bases such as NaO*t*Bu, LiHMDS, KOH, K$_3$PO$_4$ and occasionally tertiary amines in weakly polar aprotic solvents such as dioxane, DME, or toluene readily effect the desired amination. Cu catalysts are typically reacted in the presence of carbonates as a base and require strongly polar solvents (DMF, NMP, DMSO) to stabilize the Cu intermediates if no solubilizing additives (DMEDA, TMEDA, etc.) are used (see Fig. 6.129) [13, 482, 483].

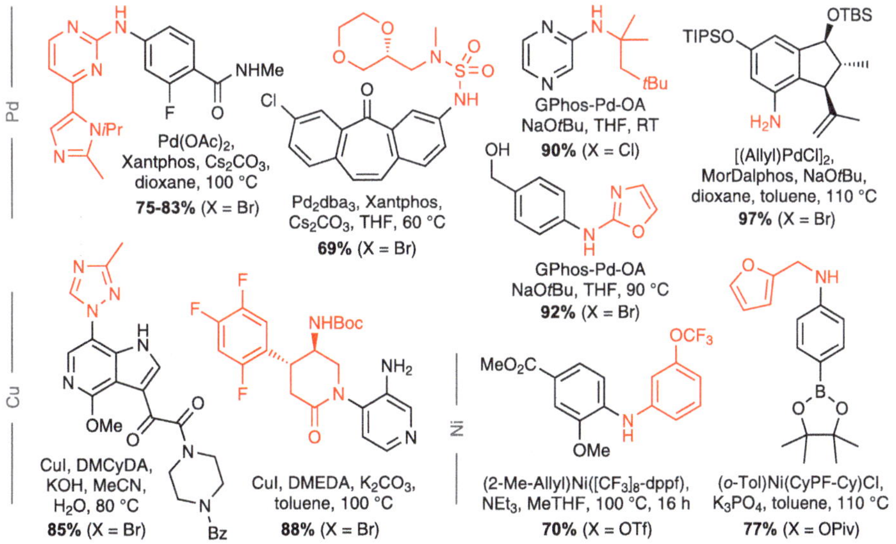

Fig. 6.129 Examples of Pd-, Cu-, and Ni-catalyzed aminations [491–498]

Apart from reductive elimination, in-depth studies for the corresponding **CO** and **CS coupling reactions** remain scarce, although the mechanism of the respective amination has been thoroughly investigated. Given this lack of data, it is assumed that the elementary steps proceed analogously in honor of their mechanistic kinship [499, 500]. These CX coupling modes are dominated by Pd and Cu catalysts, with similar advantages as in the amination of different classes of electrophiles. Ni-based methods have been subsequently introduced as well [501]. Due to the lower nucleophilicity and the formation of stable β-H elimination products (aldehydes or ketones), the C-O bond formation is much more difficult. While early methods only permitted tertiary alcohols as substrates, primary and secondary alcohols and phenols have now also been realized as feasible O-nucleophiles [502, 503]. In addition, further nucleophiles such as (thio)carboxylates can also be coupled (see Fig. 6.130) [12, 477, 504–506].

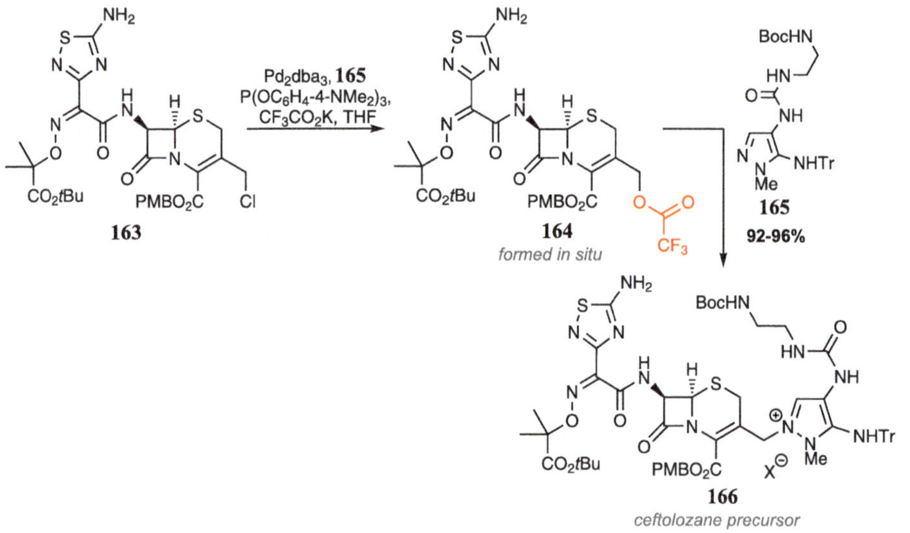

Fig. 6.130 CO coupling in the synthetic route of the cephalosporin antibiotic ceftolozane [507]

Unlike the corresponding amination, the etherification requires significantly elevated temperatures of 80–100 °C to achieve full conversion under Cu catalysis, even if new methods are slowly remedying that drawback [508]. Aryl chlorides have become viable as electrophiles with successively improved protocols, yet even with newer methods the oxidative addition fails to proceed efficiently below 120 °C [12]. Cu catalysts routinely demonstrate a clear preference for CO over CN coupling in case of competing pathways. Aminophenols can thus selectively form either a CO or CN bond by appropriate choice of the active metal and base (see Fig. 6.131) [509].

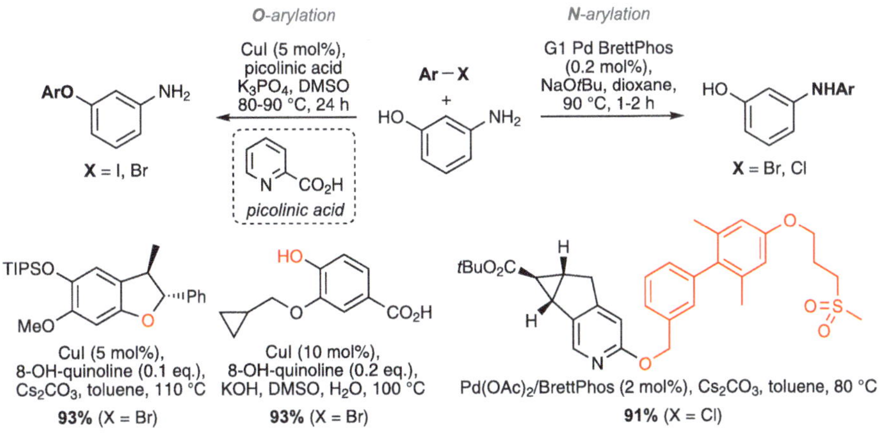

Fig. 6.131 Selective O- or N-arylation of 3-aminophenol using Cu-/Pd catalysis and applications of Cu- and Pd-catalyzed CO couplings [509–512]

The C-S coupling is a very challenging reaction due to the soft character of thiols and the irreversible formation of metal sulfides. In addition, thiols are easily oxidized. While the copper-catalyzed variant was introduced for a relatively wide range of substrates relatively early on, it still requires a high catalyst loading and temperatures well above 100 °C. Pd catalysts have also been reported to promote the CS bond formation for some time. Since the fundamental contributions of the Hartwig group, aryl chlorides and tosylates can now also be efficiently converted (see Fig. 6.132) [505, 513–515].

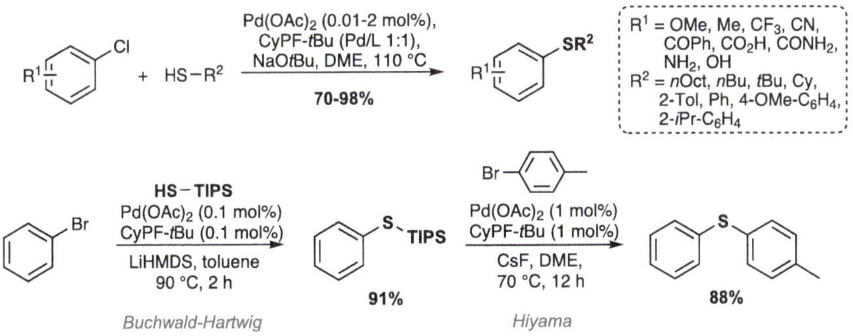

Fig. 6.132 CS coupling of aryl chlorides and sequential Buchwald–Hartwig/Hiyama cross-coupling reported by Hartwig [515]

In addition to the CO and CS coupling reactions, a variety of further functionalities have been installed under Pd catalysis. The conversion of aryl halides to phenols, as well as triflates to aryl bromides or chlorides permits the late-stage introduction of these functional groups and enables the transition from one class of electrophiles to another if so desired [516]. Aryl triflates, which are easily accessible from phenols, can thus be considered as latent aryl halides with these methods and vice versa [517, 518].

Given the frequent occurrence of fluorinated building blocks in pharmaceutically active compounds, a variety of protocols for their introduction has been established [519]. The feasibility of forming a CF bond by reductive elimination from Pd(II) complexes was long controversially debated [520]. The Buchwald group was able to realize this energetically unfavorable elimination by resorting to biarylphosphine ligands and disclosed a practical method for the fluorination of aryl triflates and bromides [166]. The reaction is not yet universally applicable, as regioisomeric mixtures can result from the occurrence of transient Pd-arene intermediates with certain substrates. While the pharmaceutically highly relevant 5-membered heterocycles remain challenging substrates, cyclic vinyl triflates have recently been successfully tackled (see Fig. 6.133) [521, 522].

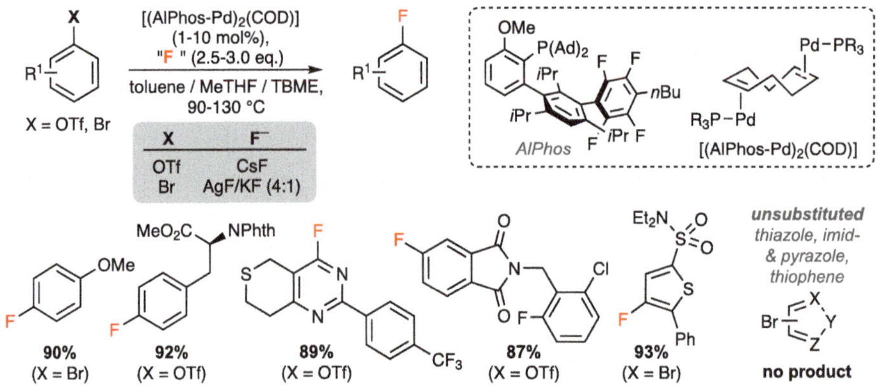

Fig. 6.133 Fluorination of aryl triflates and bromides reported by Buchwald *et al.* [523, 524]

Deviating from the routine electrophile/nucleophile scheme, an organometallic component can also serve as reactant instead of an electrophile, in an analogous manner as the Pd-catalyzed oxidative cross-couplings (cf. Sect. 6.2.3). In the late 1990s, the Cu-catalyzed CN coupling of boronic acids with anilines was introduced by the groups of Chan, Evans, and Lam. In contrast to the Buchwald–Hartwig method, the *Chan–Evans–Lam* coupling can be carried out at room temperature and does not require an inert atmosphere due to the oxidative nature of the transformation. Often, oxygen or air is even employed as the oxidant. Its disadvantage lies in the more laborious synthesis of the boronic acids compared to "simple" aryl halides, and the requisite amounts of catalyst are significantly higher. In addition to amines, alcohols and thiols can also be converted (see Fig. 6.134) [12, 525, 526].

Aryl boronic acids are the standard substrates, other organoboron species possess significantly diminished reactivity. The cause was extensively investigated for pinacol boronates by Watson and co-workers and, based on mechanistic studies, it could be alleviated by

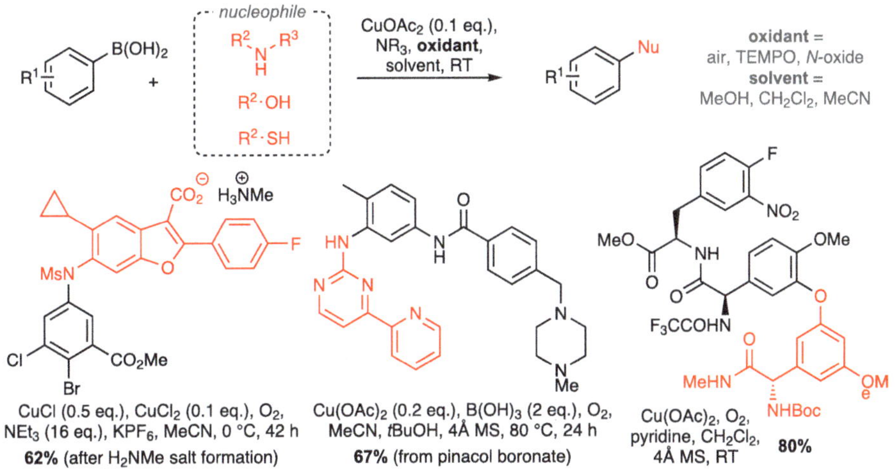

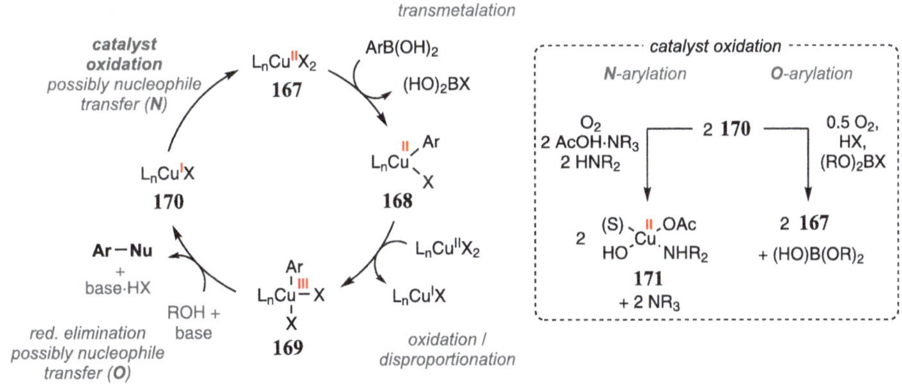

Fig. 6.134 Typical conditions of the Chan–Evans–Lam reaction and its application to CN and CO coupling [527–529]

changing the routine amine additive to $B(OH)_3$ [528]. The mechanism has not yet been fully elucidated, but a catalytic cycle derived from basic investigations for the CN and CO bond formation was envisaged (see Fig. 6.135).

Fig. 6.135 Postulated mechanism of the Chan–Evans–Lam coupling [528, 530, 531]

In contrast to classic couplings, presumably no two-electron transfer takes place. The Cu center is successively oxidized from Cu(I) to Cu(III) via Cu(II), with a Cu-Cu-SET transpiring under disproportionation (**168→169**). The transmetalation and Cu(I)→Cu(II) oxidation proceed in the same way for an amination and alkoxylation. However, the coordination of the nucleophile can either occur before the oxidative addition (CO coupling) or as part of the reoxidation of the catalyst (CN coupling) from **169** to **167** or **171**. The details of the catalytic cycle thus differ slightly depending on the coupling partner and employed reaction conditions [528, 530, 531].

6.4.2 Asymmetric Allylic Alkylation

Although aryl and alkyl halides have thus far primarily been considered as competent electrophiles in CC and CX couplings, allyl derivatives have proven to be equally competent as electrophilic component even prior to the discovery of cross-coupling reactions. Based on initial work by Tsuji, the Trost group intensively studied the mechanism and the feasibility of reacting the intermediate π-allyl metal complexes in an asymmetric fashion. A wide range of starting materials has been reported featuring a variety of leaving groups that ranges from allyl halides to -carbonates, -carbamates, -esters, and -phosphates. The diverse domain of nucleophiles is divided between soft nucleophiles (stabilized carbanions such as enolates or enamines, amines, alkoxides or thiolates) and hard nucleophiles (non-stabilized carbanions of main group metals such as organolithium or Grignard reagents). While Pd complexes have long constituted the focal point of method development, other metal catalysts have now become established, most notably Ir and Mo complexes (see Fig. 6.136) [532–540]. A method already covered is the Krische allylation (cf. Sect. 2.5.1), which represents an Ir-catalyzed allylic alkylation.

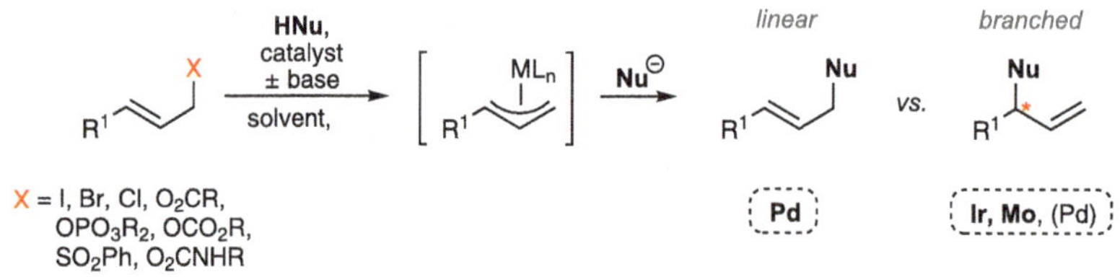

Fig. 6.136 Reaction scheme of the allylic alkylation introduced by Tsuji and Trost

The reaction is routinely referred to as *Tsuji–Trost allylation* when catalyzed by Pd, but the enantioselective variant is often simply termed *asymmetric allylic alkylation*, just as when metal catalysts other than palladium are used. The reaction is usually regio- and stereoselective for chiral substrates. Depending on the nucleophile, either a retention or inversion of configuration is observed (see Fig. 6.137) [541].

Mechanistically, an intermediate allyl-Pd complex **172** is formed by oxidative addition of the allyl system to the metal center, with the metal approaching from the face opposite to the leaving group X. Starting from **172**, soft nucleophiles react directly with the allyl system (via **173**), while hard nucleophiles show a high affinity for the metal center. The transfer of the nucleophile occurs after a transmetalation starting from **174** in an *inner-sphere* mechanism. Soft nucleophiles thus lead stereospecifically to a retention overall and hard nucleophiles to an inversion at the electrophile. For the asymmetric induction at the

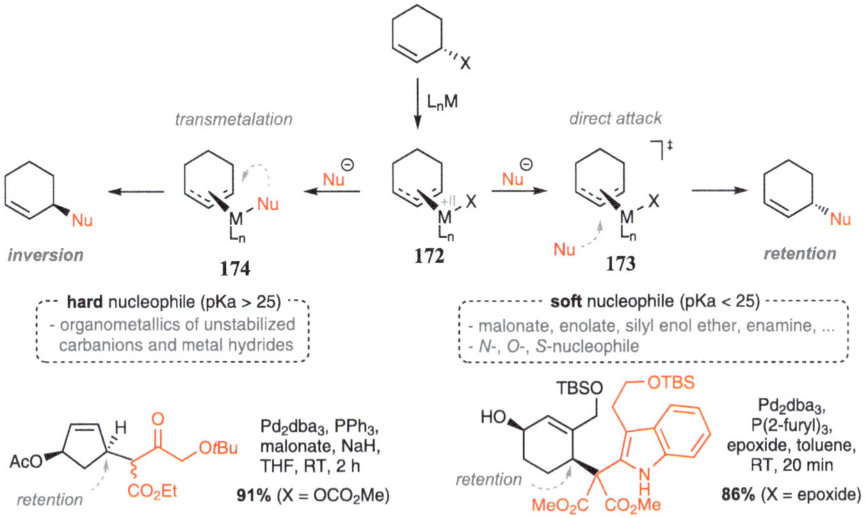

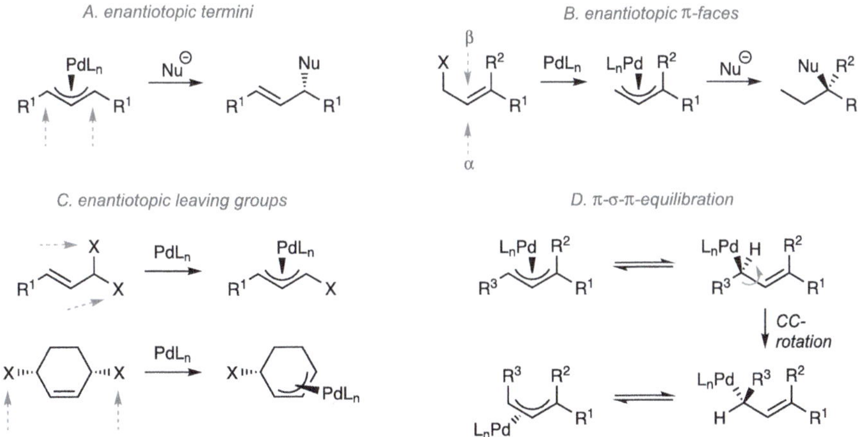

Fig. 6.137 Stereospecificity of the Tsuji-Trost allylation depending on the nucleophile [533, 541–543]

electrophile, four possible modes have been described: The differentiation of enantiotopic termini, enantiotopic π-faces, enantiotopic leaving groups, and a π-σ-π equilibration (see Fig. 6.138) [533].

Fig. 6.138 Avenues for asymmetric induction in electrophiles [533]

If the nucleophile is prochiral, such as an enolate, a discrimination of the two faces of the nucleophile can occur, resulting in the selective introduction of an additional stereogenic center. Achieving good enantioinduction is considerably more challenging, as the outer-sphere mechanism precludes a close interaction with the nucleophile in the vicinity of the inducing metal complex. Especially in recent years, the majority of Pd-catalyzed methods have addressed this shortcoming in order to allow access to tetrasubstituted, acyclic stereocenters (see Fig. 6.139) [533, 538, 544].

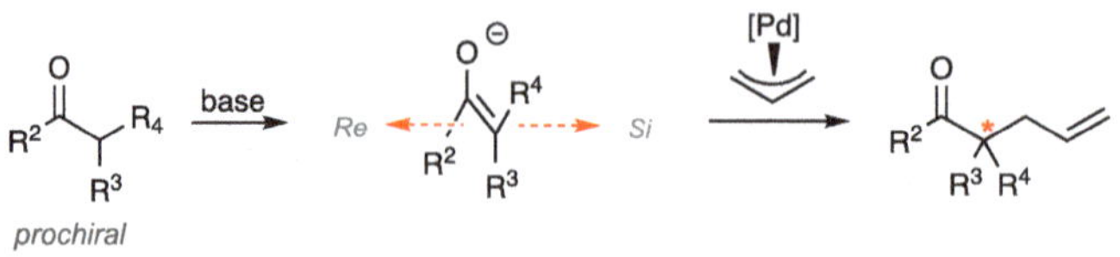

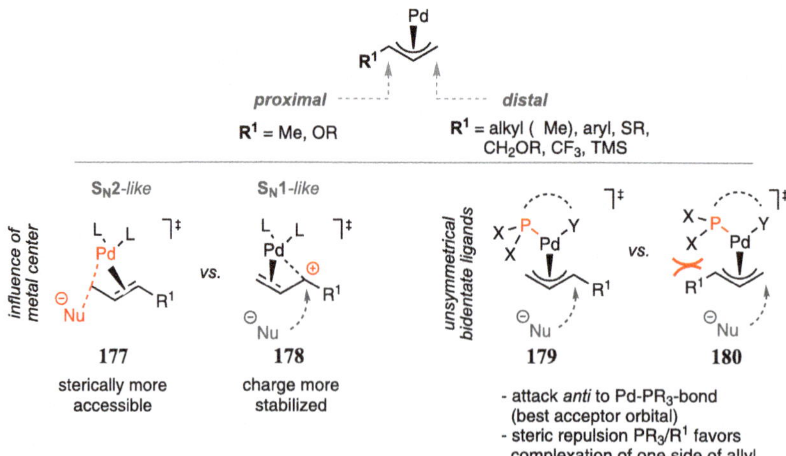

Fig. 6.139 Enantioinduction with prochiral nucleophiles

If the electrophile is asymmetric, either the linear or the branched coupling product can be obtained, as shown in Fig. 6.136. While Pd complexes usually facilitate a proximal reaction, Ir and Mo complexes almost exclusively direct to the distal terminus. The cause of this reversed regioselectivity is not fully understood yet, even though various explanations have been put forth [535, 536]. With Pd complexes, control of regioselectivity is somewhat possible by altering the ligand. Steric and electronic factors are considered as a rationale: The two reaction paths correspond mechanistically to a neutral (S_N2) versus cationic (S_N1) reaction path (**177** versus **178**) and the electronic environment of the metal center can effect a switch of the reaction pathway. Electron-withdrawing ligands, e.g., phosphites or phosphoramidites, favor the S_N1 path (**178**), whereby soft nucleophiles attack the sterically more hindered side [545]. An increased steric demand of the ligand similarly promotes the S_N1 path with soft nucleophiles. This is rationalized by the electronic preference of the nucleophile to attack via a trajectory *trans* to the Pd-P bond [546], whereby the steric repulsion of the phosphane/phosphite favors **179** as transition state (see Fig. 6.140) [547].

Fig. 6.140 Electronic and steric influences on the regioselectivity of the Pd-catalyzed allylation [537, 547]

The scope of possible electrophiles with regard to the leaving group is significantly wider than in classic cross-coupling reactions, which can be attributed to the activation of the allylic position. In contrast to aryl-derived electrophiles, not only halides but also the corresponding esters, carbonates, and phosphates are suitable nucleofuges. While most substrates necessitate basic conditions to deprotonate the nucleophile, the use of allyl carbonates allows the substitution to proceed under neutral conditions. First of all, the carbonate functionality imparts an intrinsically high reactivity. Second of all, the leaving group decomposes upon release of CO_2, thus affording an equivalent of alkoxylate, which serves as an endogenous base. With vinyl epoxides, the respective homoallyl alkoxylate is obtained following cleavage of the allylic CO bond, which can also deprotonate the nucleophile. Whether the insertion of the metal catalysts into the C-X bond proceeds via an oxidative addition or rather by way of a bimolecular nucleophilic substitution with subsequent σ-π isomerization has not yet been fully elucidated (see Fig. 6.141) [548].

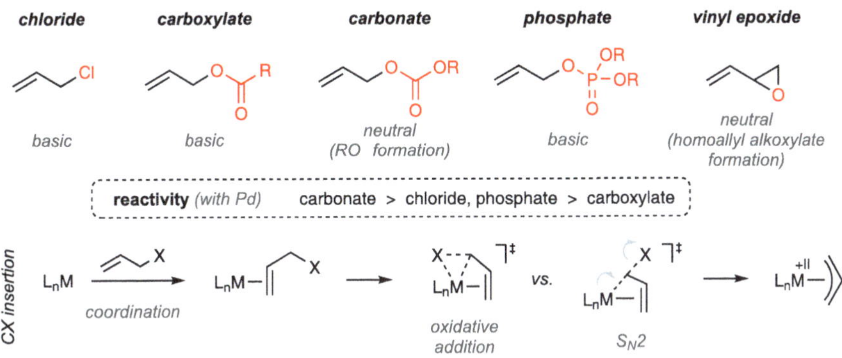

Fig. 6.141 Substrate scope of allylation electrophiles and possible pathways to CX-insertion of the metal [548–550]

The metal-catalyzed allylation is typically used in total synthesis for CC- and CN-bond formation, with the majority of applications relying on an asymmetric protocol in the presence of chiral ligands [1, 532, 551]. By virtue of their high enantioinduction and broad applicability to a variety of substrates, the C_2-symmetric bis(phosphinoamides) referred to as Trost ligand (**181**) have proven to be the ligands of choice in Pd-mediated allylations. In addition, the chiral PHOX ligands (**Pd**), phosphoramidites (**Ir**) and bis(pyridylamides) (**Mo**) have been able to prove their worth, to highlight just a small selection of the investigated structural classes of ligands (see Fig. 6.142) [533–536].

Cuprates can also mediate the conversion of allylic electrophiles with C-nucleophiles [540]. The cuprate-mediated allylation differs from the reactions of other metals in several aspects: First of all, hard nucleophiles such as organozinc, organolithium or organomagnesium species are typically used to generate homo- and heterocuprates (see Sect. 2.5.2).

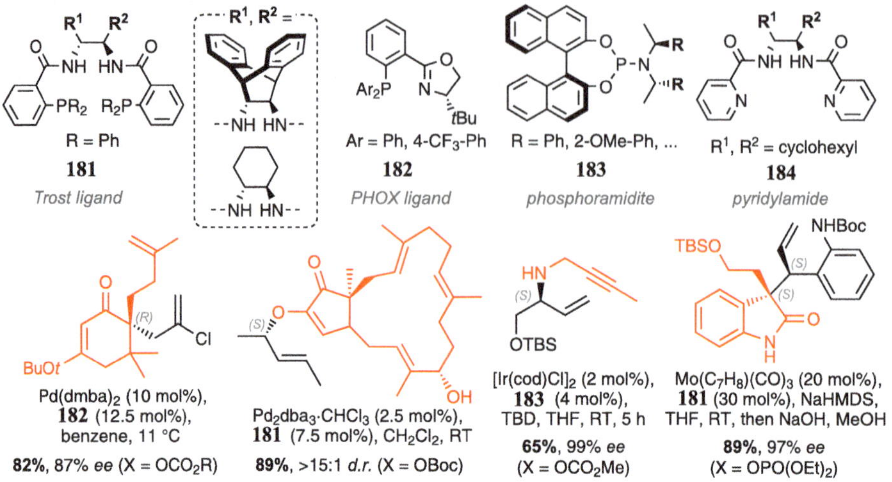

Fig. 6.142 Common chiral ligand classes and application of allylation reactions in total syntheses [552–555]. The nucleophilic fragment was highlighted

Furthermore, stoichiometric amounts of the cuprates are routinely added and the active nucleophile is freshly generated *in situ* prior to dosing of the electrophile to suppress an uncontrolled background reaction. Third of all, the regioselectivity is influenced by the position of the leaving group X: either the bond formation takes place at the original position (α-substitution) or at the opposite end of the allyl system (γ-substitution), respectively. The ratio of α- to γ-attack is a subtle function of substrate structure (steric, electronic situation), the identity of the leaving group, the nature of the organocopper reagent and the presence of directing groups (see Fig. 6.143) [556].

CuMe	α/γ	d.r.
Me$_2$CuLi	~1:1	only *anti*
MeCuCNLi	1:24	only *anti*

OR	α/γ	d.r.
OAc	only α	>99:1 *syn*
OCONHPh	only γ	>99:1 *anti*

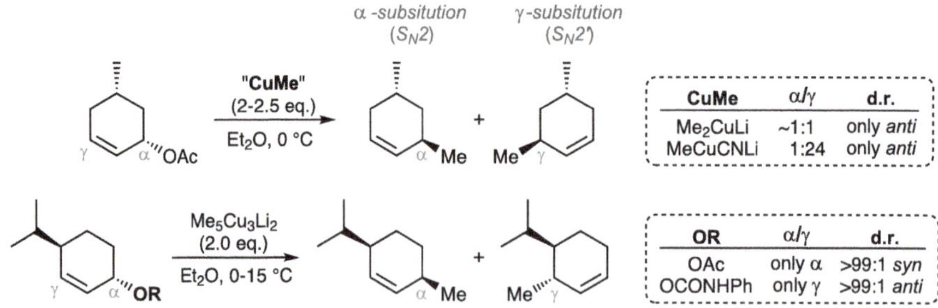

Fig. 6.143 Regio- and stereoselectivity of the Cu-mediated allylic substitution [557–559]

The regio- and stereoselectivity is derived from the mechanism of the transformation. After the initial coordination of the double bond to the bent organocopper species (**185**), the leaving group is irreversibly cleaved, forming a σ-Cu-C bond via transition state **187**. The Cu(III)-allyl species **188** can subsequently undergo reductive elimination and liberate the product from the metal center, starting from complex **190**. The electron transfer from the Cu-3d$_{xz}$ orbital into the mixed π^*_{CC}-σ^*_{CX} orbital is postulated to constitute the dominant orbital interaction of the oxidative addition (**187**). As with the previously discussed metal catalysts for this transformation, the orbital arrangement requires the leaving group to be positioned on the opposite face of the nodal plane and thus *anti* to the nucleophilic metal (**191**). In heterocuprates (Y $\neq$ R^2), two equilibrating isomers **185** and **186** result that can interconvert. Based on theoretical studies, the ensuing oxidative addition for the isomer **185** (Y = CN) is assumed to proceed significantly faster by virtue of a positive *trans*-effect [560]. In the subsequent steps (**188–190**), the CuL$_2$ moiety lacks the freedom to rotate. The reactive ligand R^2 is thus ultimately transferred to the end of the allyl group that is positioned *syn* to it in the transition state **187**. Given the irreversibility of the steps, the regio- and stereoselectivity for heterocuprates is thus determined in the initial oxidative addition. In homocuprates, the selectivity is in contrast induced in the subsequent stages [560]. If coordinating leaving groups such as carbamates are present, the oxidative addition instead proceeds *syn*-selectively via **192/193** (see Fig. 6.144) [556, 561].

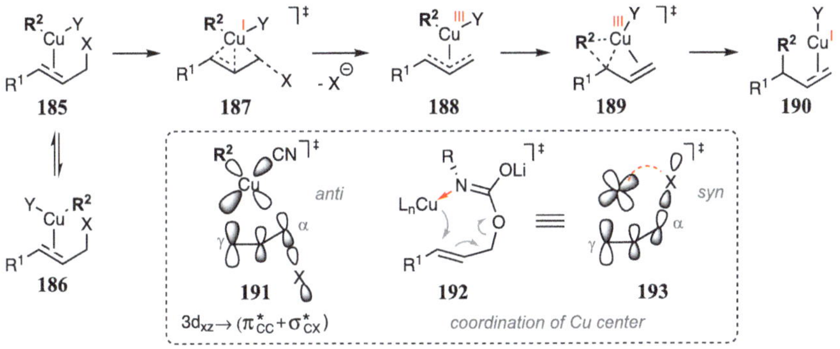

Fig. 6.144 Postulated mechanism of the Cu-mediated allylic substitution [561]

Although π-allyl-Cu complexes of the type **188** have been detected in the reaction of allyl chlorides [562], their intermediacy remains inconclusive when varying the leaving group X. The ratio of α- and γ-substitution products should not be determined by the leaving group in symmetric cuprates, as X was already cleaved in **188**, but only by the characteristics of substituent R^1 in **188**. Nevertheless, complete γ-selectivity can be achieved by using directing leaving groups such as carbamates or phosphates, which argues against the formation of a π-allyl complex and advocates for either a pure $\sigma_{Cu\text{-}C}$ intermediate or a concerted transfer of R^2 and elimination of X [556].

The Cu-mediated allylic alkylation is still underrepresented in natural product synthesis, which may be attributed to the usually requisite stoichiometric amount of cuprates. Nevertheless, there are some selected examples where this reaction has been elegantly incorporated into the synthetic approach (see Fig. 6.145).

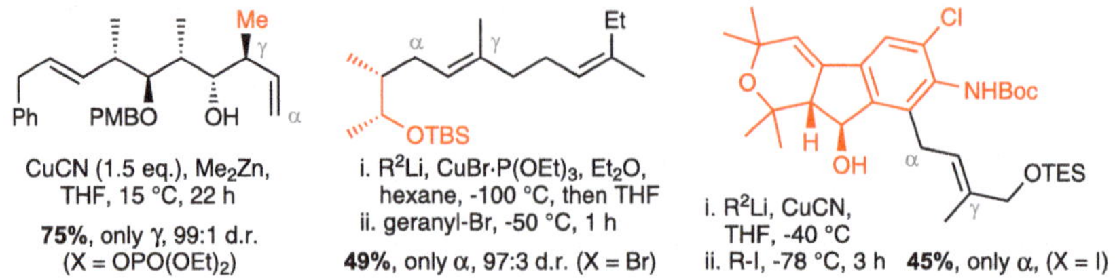

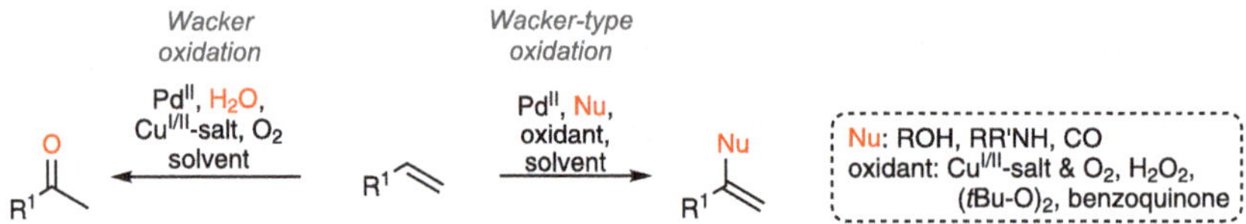

Fig. 6.145 Applications of cuprate-mediated allylic substitution [563–565]

6.4.3 Wacker Oxidation

The Wacker oxidation and Wacker-type oxidation are, in contrast to the methods elaborated so far, Pd(II)-catalyzed processes. In these, terminal alkenes are typically converted with water or other nucleophiles to the corresponding ketones or geminal disubstituted alkenes (see Fig. 6.146) [566–568].

Fig. 6.146 Reaction scheme of the Wacker oxidation

On an industrial scale, O_2 is utilized for the reoxidation of the Pd^0 that is obtained after the reaction, in conjunction with Cu(I) salts for electron transfer [569]. In the Wacker-type process, these conditions are often not compatible with the nucleophiles used, and these transformations resort to alternative oxidants such as peroxides and benzoquinone.

The exact mechanism of the Wacker oxidation is highly complex and is not yet fully understood. The disclosed studies were aimed at elucidating the mechanistic details under the classic conditions of the Wacker oxidation ($PdCl_2$, $Cu(OAc)_2$, O_2, HCl_{aq}, H_2O). While an analogy to the Wacker-type oxidation can be assumed, the transfer of the mechanistic picture to the Wacker-type processes cannot be considered given (see Fig. 6.147). The challenge of an investigation is rooted in the dependence of the reaction path on the reagents and especially their concentration. The most investigated step is the hydroxypalladation of the alkene (**196**→**199**). Either the Pd center and the nucleophile can approach from the same face (*syn*, **197**) or from opposite faces (*anti*, **198**). Prior to the β-H elimination, **197** and **198** can additionally form an equilibrium. Following the elimination (**200**), a hydropalladation to regioisomer **201** ensues if water is the nucleophile, from which the respective ketone is liberated and Pd(0) is generated via a reductive elimination of HX. Other nucleophiles do not necessitate the intermediacy of such a species for tautomerization and directly eliminate the substituted alkene from **200**. The concluding reoxidation to the Pd(II) species completes the catalytic cycle. In the absence of nucleophiles, chlorination to the corresponding vinyl chloride occurs when $CuCl_2$ is used as reoxidation catalyst [570].

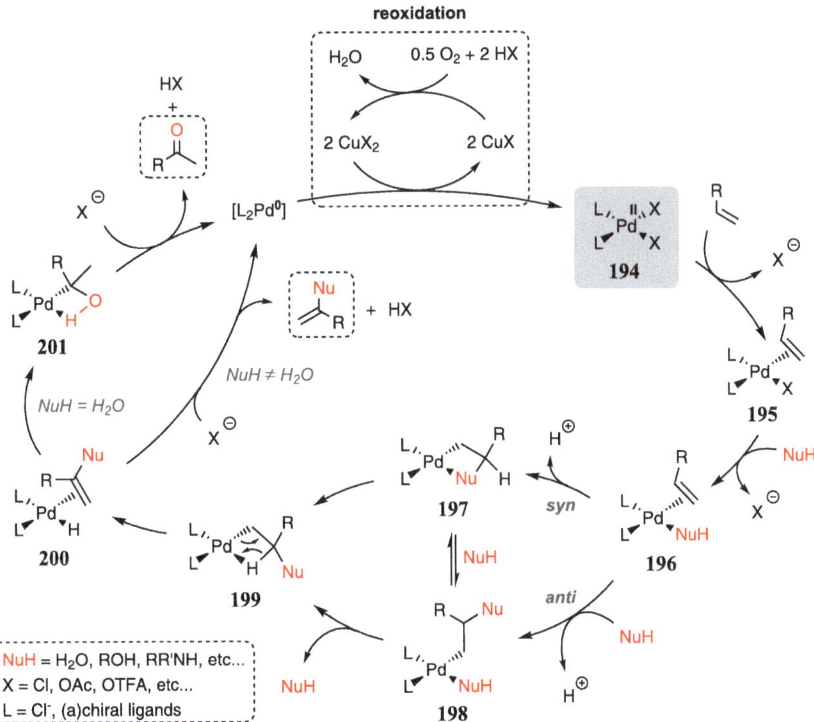

Fig. 6.147 Detailed mechanism of the Wacker and Wacker-type oxidation. The depicted reoxidation step corresponds to the industrial Wacker process [570]

An extension of the Wacker oxidation lies in the possibility of obtaining aldehydes (*anti*-Markovnikov selectivity) as products [571]. In addition, the intermediate **199** can form a Pd-acyl complex in the presence of CO, which delivers β-alkoxy esters with another equivalent of an alcohol (*Semmelhack* reaction) [572].

The Wacker oxidation is occasionally encountered in the synthesis of complex metabolites, usually for the select construction of methyl ketones from terminal alkenes. Mander and co-workers cleverly exploited a Wacker-type reaction to construct the molecular framework of the anticholinergic alkaloid himandrine. The secondary amine in **202** served as an intramolecular nucleophile, whose coupling assembled the six-membered ring of the himandrine structure in the presence of $PdCl_2$. The terminal double bond isomerized to give the internal enamine under the reaction conditions. The advanced intermediate **203** was transformed to the himandrine skeleton via the hydrogenation of the double bond. (see Fig. 6.148) [573].

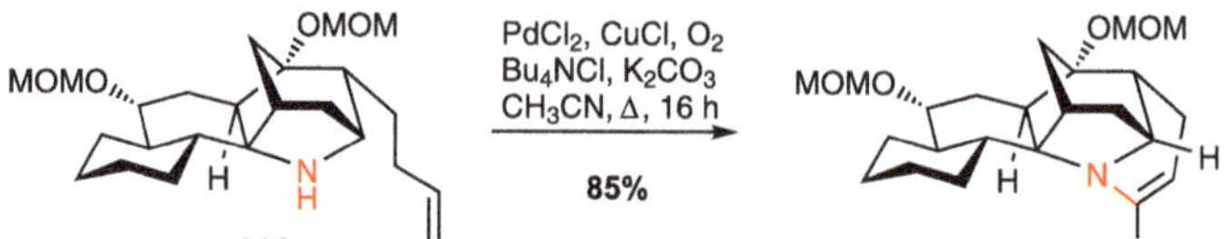

Fig. 6.148 Synthesis of the himandrine skeletal structure by Mander *et al.* [573]

6.5   CH Activation

CH functionalizations entail the derivatization of an otherwise thermodynamically stable CH bond (bond dissociation energies $\geq$90–110 kcal mol^{-1}) and its conversion to a CC or CX bond. The schematic blueprint for these transformations is inspired by nature, where even extremely complex substrates can be selectively functionalized by enzymes, for example mediated by the class of cytochrome P450 oxidases (see Fig. 6.149).

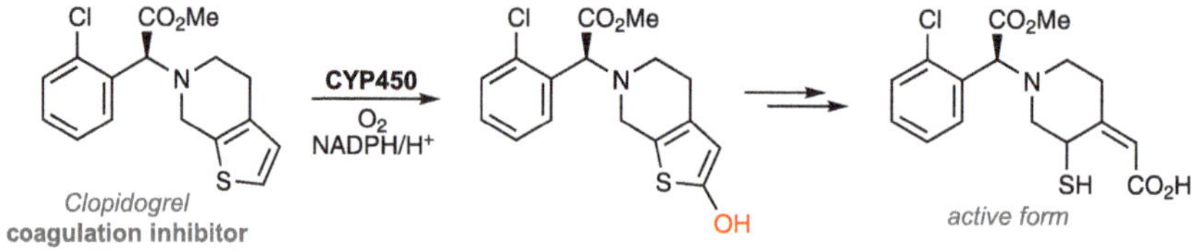

Fig. 6.149 CH-functionalization in the biological activation of the prodrug clopidogrel [574]

Compared to traditional cross-coupling reactions, CH activations are more sophisticated, as the introduction of activating functionalities in the form of carbon-metal or carbon-halogen bonds usually necessitates several steps. A direct coupling thus not only minimizes by-products but also bypasses additional steps, which has a positive effect on the time required for the execution (throughput) and the (atom) economy of the process. Both the α-functionalization of a carbonyl group and the allylic and benzylic oxidation with concomitant formation of a CX or CC bond are often located in the regime of CH functionalizations. Although this is technically correct, these approaches have already been covered in Chaps. 2 and 4 and will therefore not be discussed further here. Radical reactions are also beyond the scope of this chapter. However, it should not go unmentioned that these topics are continuously evolving and are being increasingly incorporated into academic and industrial syntheses (see Fig. 6.150) [575–579].

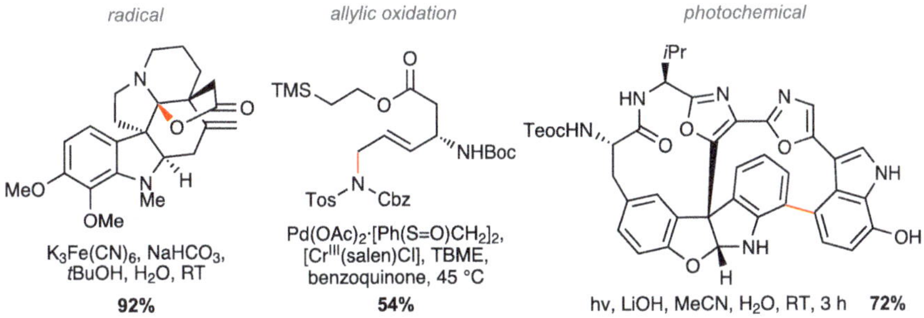

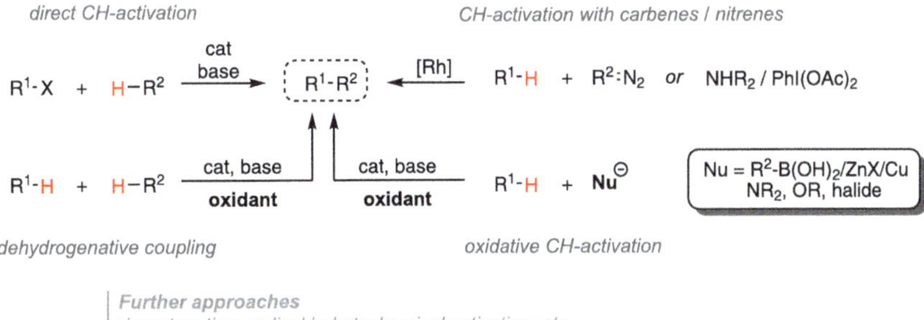

Fig. 6.150 Examples of CH activations outside the scope of this chapter [580–582]

The CH activation or functionalization constitutes an extremely broad field, for which a variety of transition metals have been used in different applications. While the most common incarnation involves a C_{sp^2}-H functionalization, methods to effect increasingly challenging activations have been reported over time, such as C_{sp^3}-H activations or diastereo- and enantioselective conversions. While initial studies heavily relied on iridium and rhodium as active metal, palladium has been dominating both methodological contributions as well as synthetic applications for a long time. However, protocols resorting to rhodium, ruthenium, copper, and even non-noble metals have increasingly gained momentum. Given their oxophilic nature, early transition metals are particularly suited to the conversion of nitrogen-containing building blocks. If the nucleophile is activated in the functionalization, the reaction proceeds redox-neutral and can also be carried out under classic cross-coupling conditions. When activating the electrophile, the stoichiometric use of an oxidizing agent is necessary to regenerate the metal catalyst (see Fig. 6.151) [29, 31, 583, 584].

Fig. 6.151 Possible CH activation modes

6.5.1 Mechanisms of CH Activation

The mechanism of the C-H insertion step in the catalytic cycle has been investigated for individual systems. Its exact nature strongly depends on the substrate, catalyst, and reaction conditions in each case. While early transition metals tend to undergo σ-bond metathesis, the CH bond is activated by oxidative addition in low-valent late transition metals. High-valent complexes in the presence of carboxylate ligands cleave the CH bond with direct involvement of the carboxylate group in the transition state (see Fig. 6.152) [585, 586].

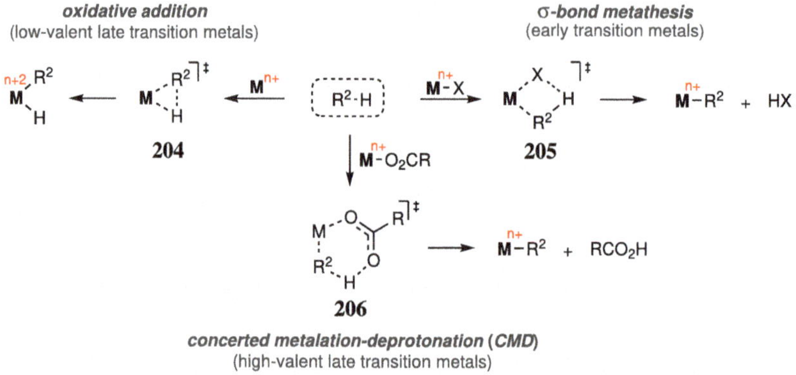

Fig. 6.152 Most common mechanisms of CH insertion [585]

Especially the carboxylate-mediated concerted metalation-deprotonation (CMD) is the most commonly encountered mechanism with Pd catalysts [585]. It was observed that electron-rich and electron-poor substrates show significantly lower activation energy compared to electron-neutral reactants in the CH activation mediated by Pd(II)-phosphine complexes. This was attributed to the fact that electron-rich substrates transfer electron density to the metal center, stabilizing the transition state [587]. A Wheland (or arenium) species can be specifically ruled out since no build-up of charge was observed, indicating that a mechanism fundamentally different from electrophilic aromatic substitution (S_EAr) is operative. A recent theoretical study also came to the conclusion that C_{sp^3}-H bonds are preferentially activated by Pd^0 and Pd^I, whereas Pd^{II}, Pd^{III} enable the functionalization of C_{sp^2}-H bonds, which was largely attributed to the steric environment of the metal center [588].

The extent of metal-C bond formation or CH bond breaking depends primarily on the electronic properties of the metal center and the substrate. A synchronous CMD is rather rare: typically electron-rich d^{10}/d^8 metal centers (Au^I, Ag^I, Cu^I, Pd^{II}-phosphines) abstract the proton prior to M-C bond formation, while electron-poorer d^8/d^6 complexes (Pd^{II}-OTf, Au^{III}, Rh^{III}, Ir^{III}, Ru^{III}) first form the M-C bond (asynchronous transition state). Electron-rich (hetero)arenes show good *polarity matching* and thus low activation energies with electrophilic catalysts (build-up of a partial positive charge in the substrate in the transition state), while electron-poor substrates react faster with nucleophilic catalysts (build-up of a partial negative charge in the substrate in the transition state). If, as in the Heck reaction, reaction conditions are chosen that promote the formation of a cationic intermediate (cf. Sect. 6.3), this similarly enhances the electrophilicity of the metal center. The observed trends suggest a transition state polarization through charge transfer between catalyst and substrate could be an important kinetic contributor to site selectivity. The concept of polarity matching thus can be exploited to enhance the chemo- and regioselectivity of a CH activation, of even overrule the intrinsic reactivity of the substrate (see Fig. 6.153) [589]

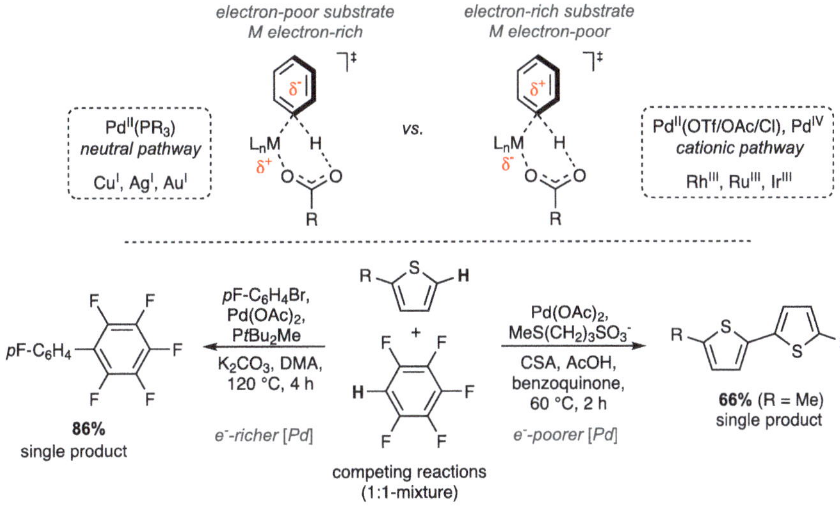

Fig. 6.153 Polarity matching in CH activation via CMD and application thereof in controlling the chemoselectivity [589, 590]

When it comes to the entire catalytic cycle, the modes illustrated in Fig. 6.151 determine which oxidation state the metal center assumes and whether oxidants are needed to regenerate the active catalyst through reoxidation. For example, if an aryl halide is reacted with a CH-reactive arene, the reaction proceeds analogously to a classic cross-coupling — only with a CH activation replacing the transmetalation step. An oxidant is not required, as the metal center can regain its initial oxidation state via reductive elimination. If in turn the electrophile is activated, the use of an oxidizing agent is inevitable (see Fig. 6.154). All Pd-catalyzed CH functionalizations require the metal to assume a Pd(II) oxidation state to effect the insertion into the reacting CH bond.

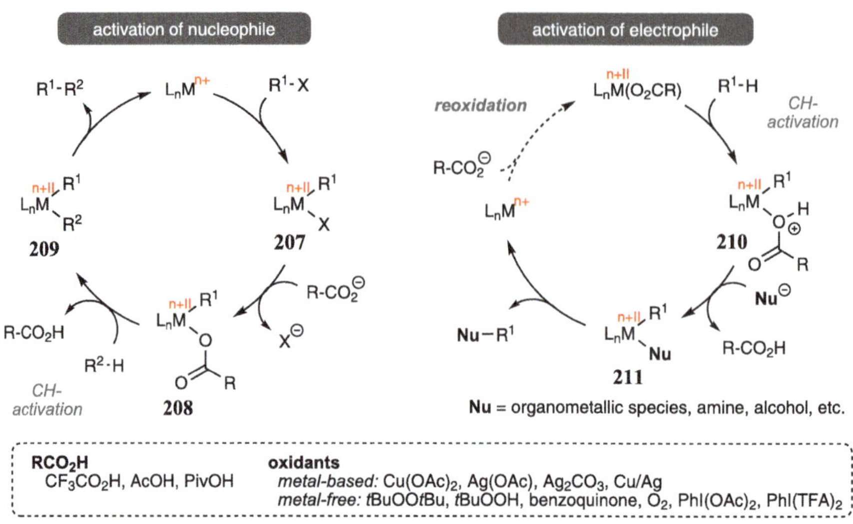

Fig. 6.154 Schematic catalytic cycles of oxidative addition/CH activation and CH activation/transmetalation [585]

In Pd complexes, the catalytic cycles oscillate between oxidation states 0 and +II. However, a series of transformations has also been developed that instead proceed via $Pd^{II/IV}$ if a Pd(II) complex is envisaged to constitute the initial catalytically active species (see Fig. 6.155). On rare occasions, the CH activation can also proceed via the Pd(IV) intermediate if the oxidation step precedes it. In these cases, N-complexed Pd complexes usually serve as catalysts, in which the harder nitrogen can stabilize the high-valent Pd(IV) better than the softer phosphine ligands [591]. Although a Pd(II/IV) mechanism is usually assumed, an alternative Pd(0/II) mechanism has been proposed to be operative during the catalytic cycle for select substrates [592].

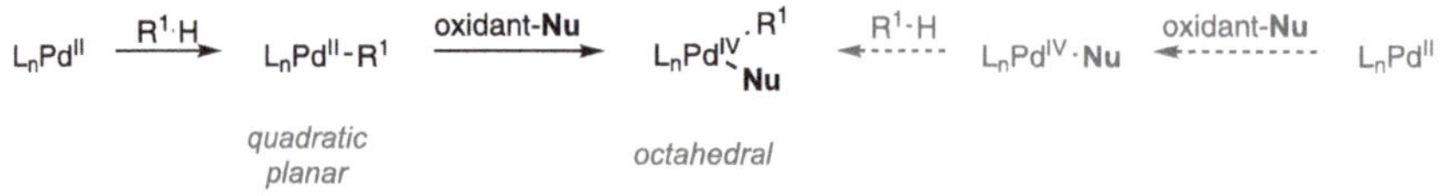

Fig. 6.155 Viable intermediates in $Pd^{II/IV}$ catalysis [591]

When using carbenes or nitrenes as coupling partners, the transformations almost exclusively rely on Rh complexes, with Cu catalysts being used in individual cases. This chemistry differs in that the reactants are metal-bound reagents which are generated *in situ*. Either diazo compounds (CC coupling) or amines in the presence of hypervalent iodine compounds (CN coupling) serve as substrates. Particularly carboxylate-bridged, dimeric Rh catalysts of the type **212** (paddle-wheel structure) are utilized, as the ligands can be chiral and thus offer the option for effecting an asymmetric induction [593]. The mechanism has been extensively studied for the CH functionalization of alkanes and shows some significant differences to classic CH activation: The metal centers of the Rh_2L_4 catalysts do not directly interact with the CH bond in the activation step, but support an asynchronous, concerted transition state, in which the hydride transfer to the carbene carbon atom takes place prior to the CC coupling (**215**). The two metal centers synergistically function as an internal redox system, in which the Rh-Rh bond is initially broken upon binding of the diazo compound and then regenerated by elimination of N_2 (**212**→**213**→**214**). The molecular orbitals of the metal-carbene bond also includes contributions from the second Rh center, so that the Rh→Rh interaction weakens the metal-carbene bond and thus activates it for CH insertion ($4d_{xz} \rightarrow \pi^*_{Rh-C}$). The CH activation step involves the electrophilic attack of the carbene 2p orbital on the CH-σ orbital of the alkane in **215**. The late transition state is characterized by an already pronounced Rh-C bond cleavage and a partial build-up of positive charge on the alkyl C atom. The N_2 expulsion constitutes the rate-determining step of the sequence (see Fig. 6.156) [594].

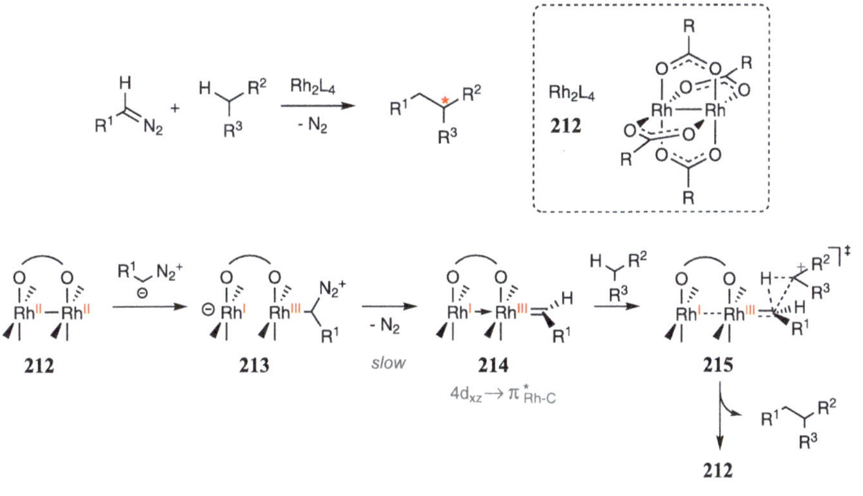

Fig. 6.156 General reaction scheme and postulated mechanism of Rh-mediated CH activations with carbenes [594]

For aminations with nitrenes, an analogous mechanism has been postulated [595]. The activation of the nitrogen is achieved by reaction with hypervalent iodine reagents to form iminoiodinanes (R^2-N $=$ I-Ar). Alternatively, azides or electrophilic nitrogen sources such as chloramines can also be used [596].

6.5.2 Reactivity and Selectivity

A major challenge with C-H activations is rooted in the energetic nature of C-H bonds. They are extremely stable and the formation of a C-C bond by activating two substrates via C-H activation constitutes an endothermic process. For example, the homocoupling of benzene to biphenyl and hydrogen requires 13.8 kJ/mol of energy. While in classic cross-coupling reactions both substrates are activated, e.g., an aryl halide and an arylboronic acid in the Suzuki reaction, the direct coupling of two unactivated substrates can rightfully be considered the ideal process. Since this is only viable for selected substrates and conditions due to thermodynamic reasons in a so-called cross dehydrogenative coupling, one of the reactants is typically prefunctionalized (see Fig. 6.157).

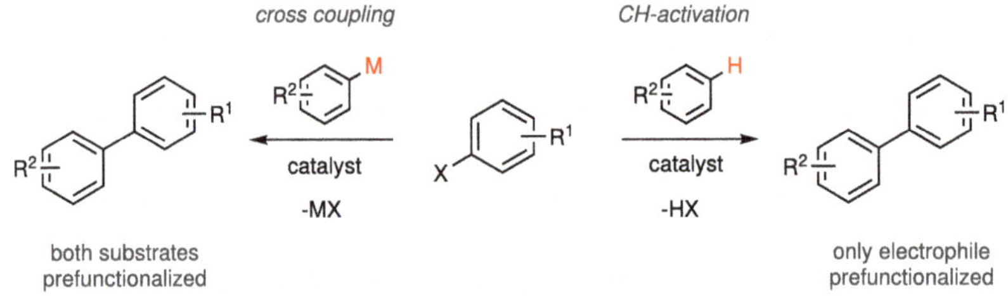

Fig. 6.157 Comparison of cross-coupling and CH activation

Most reactions involve the functionalization of C_{sp^2}-H bonds. Arenes in particular are excellent starting materials, as their bond strength is slightly higher than that of C_{sp^3}-H bonds and therefore more energy is released during the reaction. CH activation therefore particularly lends itself for late-stage functionalization of active pharmaceutical ingredients, which typically contain several (hetero-)aromatic building blocks, for example for diversification in drug discovery. The transition metal-catalyzed CH functionalization has also been successfully applied several times in scale-up and commercial production of pharmaceuticals (see Fig. 6.158) [578].

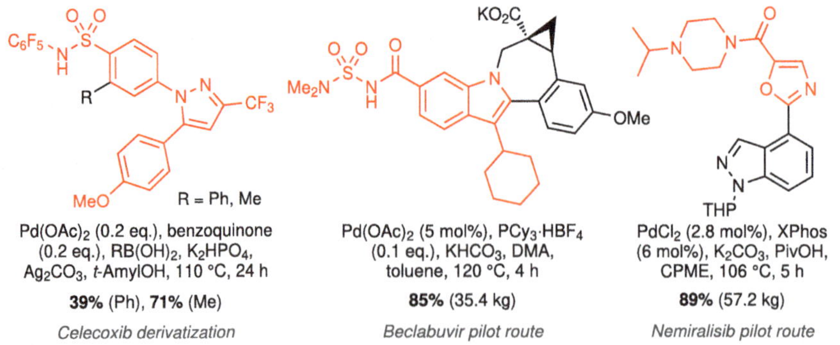

Fig. 6.158 Industrial applications of CH functionalizations [597–599]. The CH-activated component was highlighted

The second and more formidable challenge is the control of chemo- and especially regioselectivity of the transformation [29, 600]. Most substrates contain a significant number of viable reactive bonds, as can be seen from the examples in Fig. 6.158. If a select conversion of the desired position appears questionable or the regioselectivity is unpredictable, many synthetic strategies will likely not opt to rely on such a functionalization as key step. The selective conversion of a CH bond in the presence of others therefore requires careful control. Three possible approaches exist to ensure this: The use of directing groups, which is the dominant approach for aromatic and aliphatic substrates. This also includes the preference for certain ring sizes in cyclizations, either due to steric constraints or relative rate of formation (Baldwin rules, see Chap. 1). Second of all, the nature of the catalyst can impart in some cases a selectivity through coordination (active control), steric interactions (passive control) or modification of the electrophilicity of the metal center (polarity matching, see Fig. 6.153). The third possibility is to exploit the intrinsic reactivity of the substrate governed by electronic and steric factors. Some guidelines for prediction of selectivity have been compiled through comprehensive experimental and theoretical studies, especially for Pd-catalyzed conversions of heteroarenes [601]. Allylic/benzylic and α-heteroatom functionalizations also fall into this category [602]. So-called undirected CH activations are the most difficult because the catalyst does not coordinate to the substrate prior to the CH insertion. As a consequence, no spatial proximity between the CH bond to be functionalized and the metal center exists (see Fig. 6.159). The probably most widely used undirected CH activation is the Ir-catalyzed CH borylation developed by Hartwig, Ishiyama, and Miyaura [603].

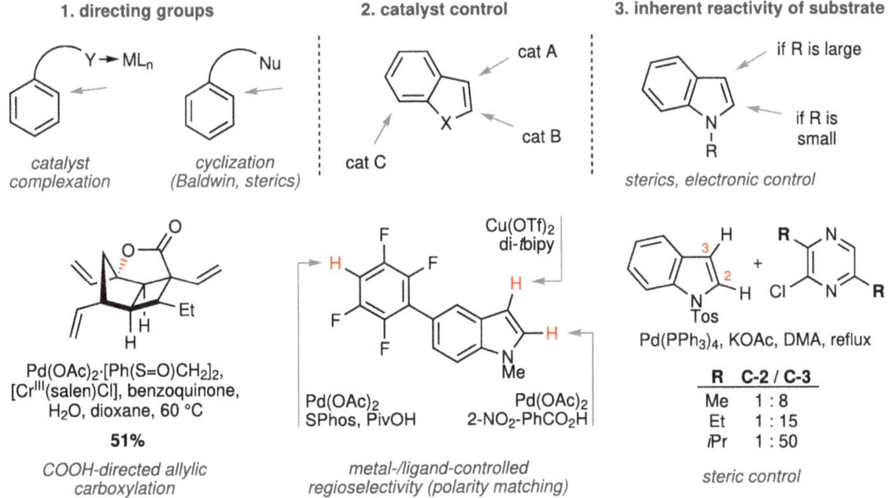

Fig. 6.159 Approaches to control selectivity and representative examples [604–606]

The most common approach in CH functionalizations is the incorporation of directing groups. Their main drawback consists of the need for a subsequent derivatization or cleavage, if this functionality is not present in the target structure. These additional steps can in turn make alternative routes via a traditional cross-coupling significantly more attractive. Coordinating functionalities usually direct to the ortho-position in aromatic systems and to the β-position in aliphatic substrates [607]. In unsymmetric substrates, the sterically less hindered side is functionalized (see Fig. 6.160) [600, 608].

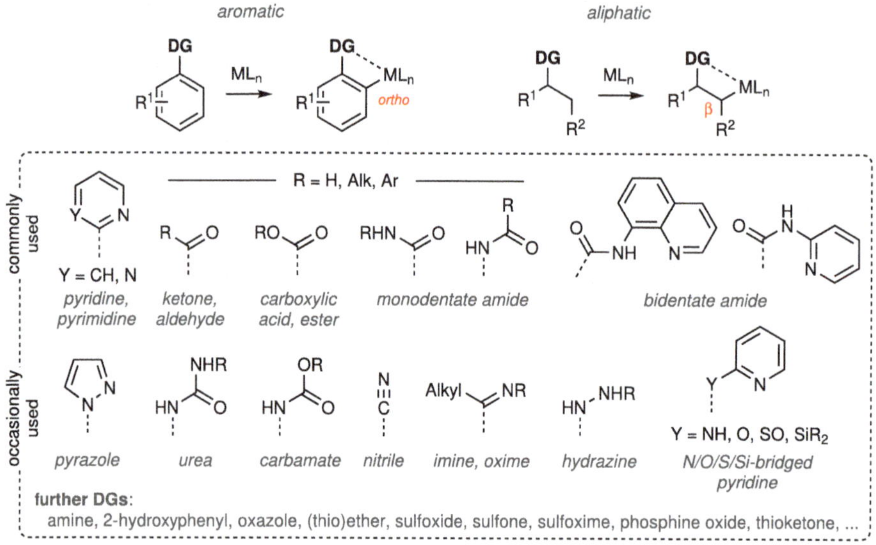

Fig. 6.160 Overview of common directing groups [600, 608]

Early directing groups were based on strongly coordinating functionalities containing nitrogen, sulfur, or phosphorus. However, the metalacycles resulting from a CH bond cleavage are extremely stable, which can be detrimental to a high reaction rate. In contrast, the introduction of weakly binding functionalities such as carboxylic acids, esters, imines, or sulfoxides can promote the desired site-selectivity while pertaining a sufficient rate. In addition to monodentate functionalities, bidentate amides based on 8-aminoquinolines as well as aminopyridine, -oxazoline, and substituted phenyl groups are used for tighter control. Heteroatom-linked pyridine groups can be easily cleaved again [600].

In some cases, the selectivity can be influenced by the electronic and steric characteristics of the catalyst. In addition to the possibility of polarity matching (see Sect. 6.5.1) to preferentially functionalize positions with more pronounced CH-acidity or π-basicity, *meta*-directing functionalizations have been developed by the systematic design of the ligands (see Fig. 6.161) [600].

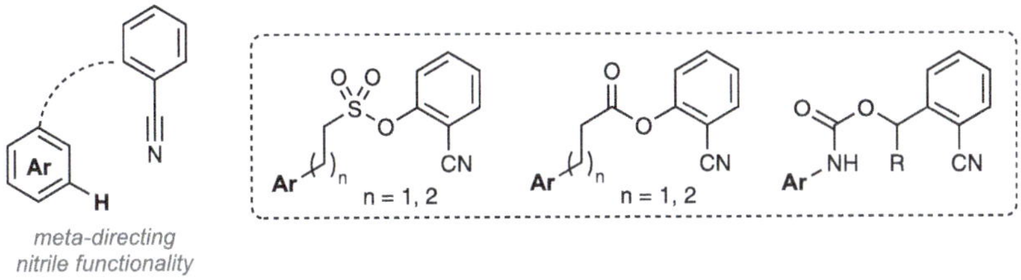

Fig. 6.161 Schematic structure of *meta*-directing nitrile functionalities [600]

A competitive pathway to CH activation occurring with carbenes and nitrenes is the formation of cyclopropanes and aziridines, i.e., a direct reaction with an existing double bond instead of the insertion into CH bonds. The chemoselectivity can be controlled in these cases by modifying the catalyst: Stronger electron-withdrawing carboxylate ligands can shift the equilibrium in favor of the more nucleophilic substrate, while weaker donors such as amides modify the selectivity in favor of less nucleophilic substrates (see Fig. 6.162) [609].

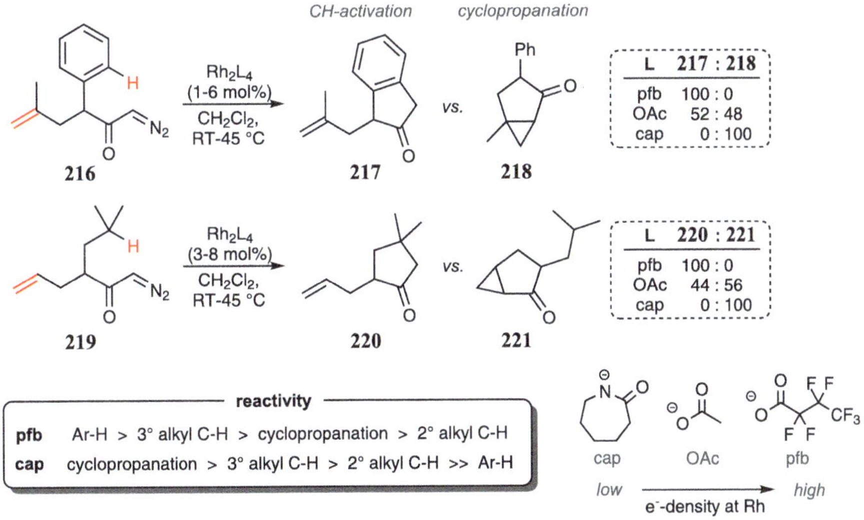

Fig. 6.162 Influence of the catalyst on the chemoselectivity in the reaction of diazoalkenes [610]

The certainly most elegant but also most difficult to predict selectivity is based on the intrinsic reactivity of the substrate. Especially with pharmaceutically relevant heterocycles, extensive studies of various substrate classes have provided a solid foundation that serves as a guide (see Fig. 6.163) [586, 601].

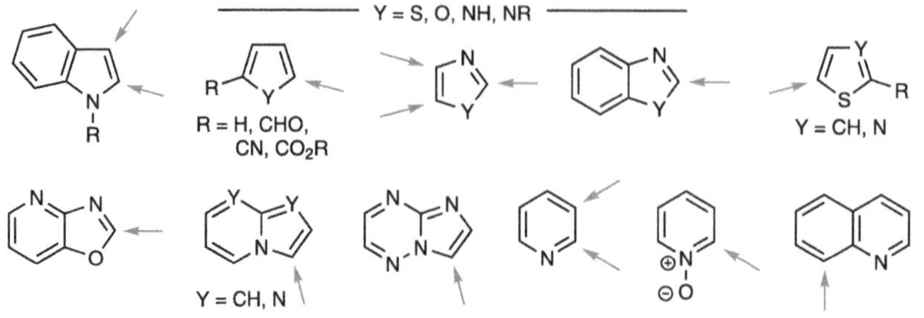

Fig. 6.163 Intrinsic regioselectivity of the Pd-catalyzed CH activation of heteroarenes [586, 601]

In addition to the activation of heteroaromatic CH bonds, allylic and benzylic positions can also be functionalized in a controlled manner by virtue of the lability of these CH bonds and the stability of the resulting carbon-metal bond. The Tsuji-Trost reaction constitutes the corresponding analog in the regime of classic coupling reactions, wherein a proton is substituted via a CH activation instead of the expulsion of a leaving group.

The Rh-catalyzed CH functionalization using carbenes and nitrenes also shows a certain substrate-specific reactivity, which is largely attributed to the stabilization of the positive charge occurring in the transition state (electronic factors) as well as steric repulsion. Unsurprisingly, allylic/benzylic CH bonds and α-heteroatom positions are the most reactive. The competing cyclopropanation often remains the dominant path for terminal, 1,1-disubstituted and *cis*-alkenes. Here, the feasibility of a desired transformation should be checked in advance using model substrates if possible. In the activation of alkanes, tertiary CH bonds are preferred for electronic reasons, whereas secondary and primary ones are favored on the basis of sterics. A detailed analysis of the stereoelectronic properties of the substrate can however enable regiocontrol (see Fig. 6.164) [593, 609].

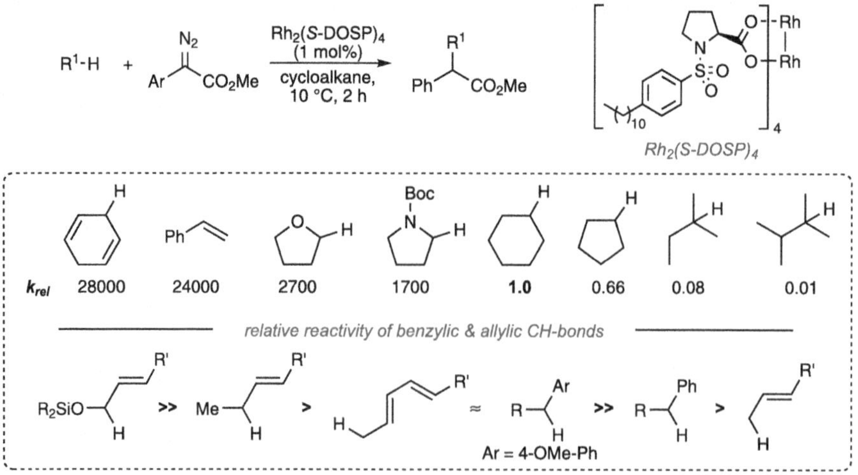

Fig. 6.164 Reactivities of different coupling partners in the reaction with carbenes [593, 611]

CH functionalizations have now become established as important tools in the synthesis of various natural products and pharmaceuticals [576, 577, 612, 613]. The majority of applications are encountered in early stages of lead generation/medicinal chemistry due to the broad range of heteroaromatic building blocks. However, some impressive examples of C_{sp^3}-H functionalizations have also been published in total syntheses (see Fig. 6.165).

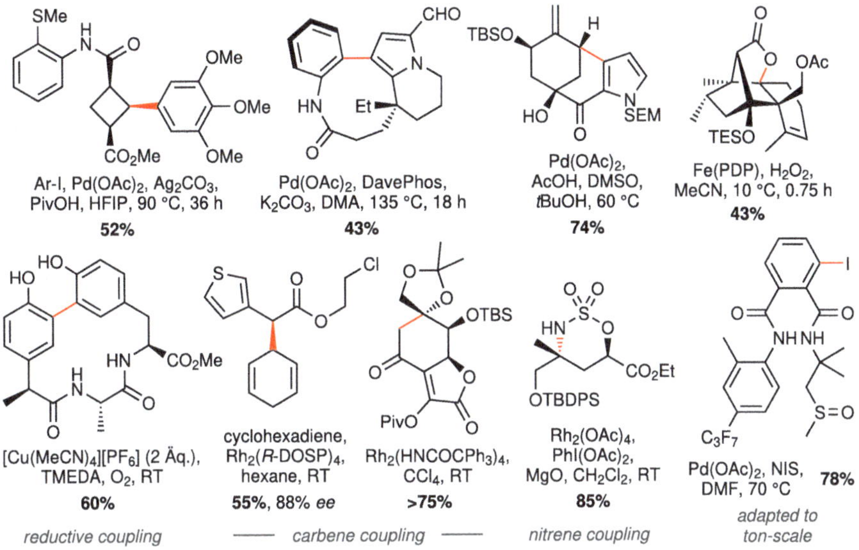

Fig. 6.165 Examples of CH activations in the synthesis of natural substances and pharmaceuticals [614–622]

Due to the breadth of the field, only an introduction to the subject can be provided here. For a more in-depth study, the interested reader is referred to the review articles published in recent years [29, 31, 578, 579, 584, 586, 593, 595, 596, 600, 607, 608, 623–627].

References

1. K. C. Nicolaou, P. G. Bulger, D. Sarlah, *Angew. Chem. Int. Ed.* **2005**, *44*, 4442–4489.
2. A. B. Dounay, L. E. Overman, *Chem. Rev.* **2003**, *103*, 2945–2964.
3. J. Magano, J. R. Dunetz, *Chem. Rev.* **2011**, *111*, 2177–2250.
4. P. Devendar, R.-Y. Qu, W.-M. Kang, B. He, G.-F. Yang, *J. Agric. Food Chem.* **2018**, *66*, 8914–8934.
5. C. Torborg, M. Beller, *Adv. Synth. Catal.* **2009**, *351*, 3027–3043.
6. L. G. Parte, S. Fernández, E. Sandonís, J. Guerra, E. López, *Mar. Drugs* **2024**, *22*, 253.
7. P. Orecchia, D. S. Petkova, R. Goetz, F. Rominger, A. S. K. Hashmi, T. Schaub, *Green Chem.* **2021**, *23*, 8169–8180.
8. K. C. Nicolaou, J. Xu, F. Murphy, S. Barluenga, O. Baudoin, H. Wei, D. L. F. Gray, T. Ohshima, *Angew. Chem. Int. Ed.* **1999**, *38*, 2447–2451.

9. D. Mitchell, K. P. Cole, P. M. Pollock, D. M. Coppert, T. P. Burkholder, J. R. Clayton, *Org. Process Res. Dev.* **2012**, *16*, 70–81.

10. D. G. Brown, J. Boström, *J. Med. Chem.* **2016**, *59*, 4443–4458.

11. X.-F. Wu, P. Anbarasan, H. Neumann, M. Beller, *Angew. Chem. Int. Ed.* **2010**, *49*, 9047–9050.

12. S. Bhunia, G. G. Pawar, S. V. Kumar, Y. Jiang, D. Ma, *Angew. Chem. Int. Ed.* **2017**, *56*, 16136–16179.

13. I. P. Beletskaya, A. V. Cheprakov, *Organometallics* **2012**, *31*, 7753–7808.

14. V. M. Chernyshev, V. P. Ananikov, *ACS Catal.* **2022**, *12*, 1180–1200.

15. M. Tobisu, N. Chatani, *Top. Curr. Chem.* **2016**, *374*, 41.

16. F.-S. Han, *Chem. Soc. Rev.* **2013**, *42*, 5270–5298.

17. B. M. Rosen, K. W. Quasdorf, D. A. Wilson, N. Zhang, A.-M. Resmerita, N. K. Garg, V. Percec, *Chem. Rev.* **2011**, *111*, 1346–1416.

18. T. Zhou, M. Szostak, *Catal. Sci. Technol.* **2020**, *10*, 5702–5739.

19. L.-C. Campeau, N. Hazari, *Organometallics* **2019**, *38*, 3–35.

20. R. Jana, T. P. Pathak, M. S. Sigman, *Chem. Rev.* **2011**, *111*, 1417–1492.

21. N. Kambe, T. Iwasakia, J. Teraob, *Chem. Soc. Rev.* **2011**, *40*, 4937–4947.

22. G. Berionni, B. Maji, P. Knochel, H. Mayr, *Chem. Sci.* **2012**, *3*, 878–882.

23. S. Pratihar, S. Roy, *Organometallics* **2011**, *30*, 3257–3269.

24. *Metal-Catalyzed Cross-Coupling Reactions*, (Eds.: A. de Meijere, F. Diederich), Wiley-VCH, **2004**.

25. J.-P. Corbet, G. Mignani, *Chem. Rev.* **2006**, *106*, 2651–2710.

26. R. B. Bedford, P. B. Brenner, *Top. Organometm Chem.* **2015**, *50*, 19–46.

27. E. Nakamura, T. Hatakeyama, S. Ito, K. Ishizuka, L. Ilies, M. Nakamura, *Org. React.* **2014**, *83*, 1–209.

28. M. S. Jensen, R. S. Hoerrner, W. Li, D. P. Nelson, G. J. Javadi, P. G. Dormer, D. Cai, R. D. Larsen, *J. Org. Chem.* **2005**, *70*, 6034–6039.

29. T. Gensch, M. N. Hopkinson, F. Glorius, J. Wencel-Delord, *Chem. Soc. Rev.* **2016**, *45*, 2900–2936.

30. N. Y. S. Lam, K. Wu, J.-Q. Yu, *Angew. Chem. Int. Ed.* **2021**, *60*, 15767–15790.

31. P. Gandeepan, T. Müller, D. Zell, G. Cera, S. Warratz, L. Ackermann, *Chem. Rev.* **2019**, *119*, 2191–2452.

32. *Transition Metal Catalyzed Oxidative Cross-Coupling Reactions*, (Ed.: A. Lei), Springer, **2019**.

33. A. Lei, W. Shi, C. Liu, W. Liu, H. Zhang, C. He, *Oxidative Cross-Coupling Reactions*, Wiley-VCH, **2016**.

34. C. E. I. Knappke, S. Grupe, D. Gärtner, M. Corpet, C. Gosmini, A. J. von Wangelin, *Chem. Eur. J.* **2014**, *20*, 6828–6842.

35. J. K. Stille, *Angew. Chem. Int. Ed. Engl.* **1986**, *25*, 508–524.

36. L. Xue, Z. Lin, *Chem. Soc. Rev.* **2010**, *39*, 1692–1705.

37. B. T. Ingoglia, C. C. Wagen, S. L. Buchwald, *Tetrahedron* **2019**, *75*, 4199–4211.

38. M. Busch, M. D. Wodrich, C. Corminboeuf, *ACS Catal.* **2017**, *7*, 5643–5653.

39. U. Christmann, R. Vilar, *Angew. Chem. Int. Ed.* **2005**, *44*, 366–374.

40. J. F. Hartwig, *Inorg. Chem.* **2007**, *46*, 1936–1947.

41. C. A. Tolman, *Chem. Rev.* **1977**, *77*, 313–348.

42. H. Clavier, S. P. Nolan, *Chem. Commun.* **2010**, *46*, 841–861.

43. D. J. Durand, N. Fey, *Chem. Rev.* **2019**, *119*, 6561–6594.

44. M.-N. Birkholz, Z. Freixa, P. W. N. M. van Leeuwen, *Chem. Soc. Rev.* **2009**, *38*, 1099–1118.

45. P. Espinet, A. M. Echavarren, *Angew. Chem. Int. Ed.* **2004**, *43*, 4704–4734.

46. A. Echavarren, D. J. Cárdenas in *Metal-Catalyzed Cross-Coupling Reactions* (Eds.: A. de Meijere, F. Diederich), Wiley-VCH, 2nd ed., **2004**, Chapter 1, pp. 1–31.

47. S.-Q. Zhang, X. Hong, *Acc. Chem. Res.* **2021**, *54*, 2158–2171.

48. S. Bajo, G. Laidlaw, A. R. Kennedy, S. Sproules, D. J. Nelson, *Organometallics* **2017**, *36*, 1662–1672.

49. P. M. Pérez-García, A. Darù, A. R. Scheerder, M. Lutz, J. N. Harvey, M.-E. Moret, *Organometallics* **2020**, *39*, 1139–1144.

50. Y. Luo, Y. Li, J. Wu, X.-S. Xue, J. F. Hartwig, Q. Shen, *Science* **2023**, *381*, 1072–1079.

51. A. Fürstner, *ACS Cent. Sci.* **2016**, *2*, 778–789.

52. S. L. Buchwald, C. Bolm, *Angew. Chem. Int. Ed.* **2009**, *48*, 5586–5587.

53. L. Xu, F.-Y. Liu, Q. Zhang, W.-J. Chang, Z.-L. Liu, Y. Lv, H.-Z. Yu, J. Xu, J.-J. Dai, H.-J. Xu, *Nat. Catal.* **2021**, *4*, 71–78.

54. Z. Novák, R. Adamik, J. T. Csenki, F. Béke, R. Gavaldik, B. Varga, B. Nagy, Z. May, J. Daru, Z. Gonda, G. L. Tolnai, *Nat. Catal.* **2021**, *4*, 991–993.

55. M. Avanthay, R. B. Bedford, C. S. Begg, D. Böse, J. Clayden, S. A. Davis, J.-C. Eloi, G. P. Goryunov, I. V. Hartung, J. Heeley, K. A. Khaikin, M. O. Kitching, J. Krieger, P. S. Kulyabin, A. J. J. Lennox, R. Nolla-Saltiel, N. E. Pridmore, B. J. S. Rowsell, H. A. Sparkes, D. V. Uborsky, A. Z. Voskoboynikov, M. P. Walsh, H. J. Wilkinson, *Nat. Catal.* **2021**, *4*, 994–998.

56. J. K. Vinod, A. K. Wanner, E. I. James, K. Koide, *Nat. Catal.* **2021**, *4*, 999–1001.

57. L. Nattmann, R. Saeb, N. Nöthling, J. Cornella, *Nat. Catal.* **2019**, *3*, 6–13.

58. C. Amatore, A. Jutand, M. A. M'Barki, *Organometallics* **1992**, *11*, 3009–3013.

59. L. S. Santos, G. B. Rosso, R. A. Pilli, M. N. Eberlin, *J. Org. Chem.* **2007**, *72*, 5809–5812.

60. J. F. Hartwig, F. Paul, *J. Am. Chem. Soc.* **1995**, *117*, 5373–5374.

61. C. Amatore, A. Jutand, *Acc. Chem. Res.* **2000**, *33*, 314–321.

62. C. S. Wei, G. H. M. Davies, O. Soltani, J. Albrecht, Q. Gao, C. Pathirana, Y. Hsiao, S. Tummala, M. D. Eastgate, *Angew. Chem. Int. Ed.* **2013**, *52*, 5822–5826.

63. C. C. C. Johansson Seechurn, T. Sperger, T. G. Scrase, F. Schoenebeck, T. J. Colacot, *J. Am. Chem. Soc.* **2017**, *139*, 5194–5200.

64. F. Ozawa, A. Kubo, T. Hayashi, *Chem. Lett.* **1992**, 2177–2180.

65. K. H. Shaughnessy, *Isr. J. Chem.* **2020**, *60*, 180–194.

66. B. P. Fors, P. Krattiger, E. Strieter, S. L. Buchwald, *Org. Lett.* **2008**, *10*, 3505–3508.

67. C. Amatore, G. Broeker, A. Jutand, F. Khalil, *J. Am. Chem. Soc.* **1997**, *119*, 5176–5185.

68. S. S. Zalesskiy, V. P. Ananikov, *Organometallics* **2012**, *31*, 2302–2309.

69. S. Ueda, M. Su, S. L. Buchwald, *Angew. Chem. Int. Ed.* **2011**, *50*, 8944–8947.

70. A. G. Sergeev, T. Schulz, C. Torborg, A. Spannenberg, H. Neumann, M. Beller, *Angew. Chem. Int. Ed.* **2009**, *48*, 7595–7599.

71. A. Bruneau, M. Roche, M. Alami, S. Messaoudi, *ACS Catal.* **2015**, *5*, 1386–1396.

72. C. Valente, S. Calimsiz, K. H. Hoi, D. Mallik, M. Sayah, M. G. Organ, *Angew. Chem. Int. Ed.* **2012**, *51*, 3314–3332.

73. P. G. Gildner, T. J. Colacot, *Organometallics* **2015**, *34*, 5497–5508.

74. J. Rio, H. Liang, M.-E. L. Perrin, L. A. Perego, L. Grimaud, P.-A. Payard, *ACS Catal.* **2023**, *13*, 11399–11421.

75. D. J. Jones, M. Lautens, G. P. McGlacken, *Nat. Catal.* **2019**, *2*, 843–851.

76. M. J. Kania, A. Reyes, S. R. Neufeldt, *J. Am. Chem. Soc.* **2024**, *146*, 19249–19260.

77. Z.-B. Dong, G. Manolikakes, L. Shi, P. Knochel, H. Mayr, *Chem. Eur. J.* **2010**, *16*, 248–253.

78. C. Hansch, A. Leo, R. W. Taft, *Chem. Rev.* **1991**, *91*, 165–195.

79. S. J. Firsan, V. Sivakumar, T. J. Colacot, *Chem. Rev.* **2022**, *122*, 16911–17240.

80. F. Barrios-Landeros, J. F. Hartwig, *J. Am. Chem. Soc.* **2005**, *127*, 6944–6945.

81. W.-J. van Zeist, R. Visser, F. M. Bickelhaupt, *Chem. Eur. J.* **2009**, *15*, 6112–6115.

82. K. Vikse, T. Naka, J. S. McIndoe, M. Besora, F. Maseras, *ChemCatChem* **2013**, *5*, 3604–3609.

83. F. Proutiere, F. Schoenebeck, *Angew. Chem. Int. Ed.* **2011**, *50*, 8192–8195.

84. M. Kolter, K. Böck, K. Karaghiosoff, K. Koszinowski, *Angew. Chem. Int. Ed.* **2017**, *56*, 13244–13248.

85. A. Ariafard, Z. Lin, *Organometallics* **2006**, *25*, 4030–4033.

86. H. Kurosawa, S. Ogoshi, Y. Kawasaki, S. Murai, M. Miyoshi, I. Ikeda, *J. Am. Chem. Soc.* **1990**, *112*, 2813–2814.

87. H. Kurosawa, H. Kajimaru, S. Ogoshi, H. Yoneda, K. Miki, N. Kasai, S. Murai, I. Ikeda, *J. Am. Chem. Soc.* **1992**, *114*, 8417–8424.

88. A. Vitagliano, B. Aakermark, S. Hansson, *Organometallics* **1991**, *10*, 2592–2599.

89. M. R. Netherton, G. C. Fu, *Angew. Chem. Int. Ed.* **2002**, *41*, 3910–3912.

90. N. Rodríguez, C. R. de Arellano, G. Asensio, M. Medio-Simón, *Chem. Eur. J.* **2007**, *13*, 4223–4229.

91. I. D. Hills, M. R. Netherton, G. C. Fu, *Angew. Chem. Int. Ed.* **2003**, *42*, 5749–5752.

92. J. Terao, H. Todo, S. A. Begum, H. Kuniyasu, N. Kambe, *Angew. Chem. Int. Ed.* **2007**, *46*, 2086–2089.

93. C.-T. Yang, Z.-Q. Zhang, J. Liang, J.-H. Liu, X.-Y. Lu, H.-H. Chen, L. Liu, *J. Am. Chem. Soc.* **2012**, *134*, 11124–11127.

94. I. Bauer, H.-J. Knölker, *Chem. Rev.* **2015**, *115*, 3170–3387.

95. T. Iwasaki, N. Kambe, *Top. Curr. Chem.* **2016**, *374*, 66.

96. S. L. Zultanski, G. C. Fu, *J. Am. Chem. Soc.* **2013**, *135*, 624–627.

97. A. L. Casado, P. Espinet, *Organometallics* **1998**, *17*, 954–959.

98. E. K. Reeves, E. D. Entz, S. R. Neufeldt, *Chem. Eur. J.* **2021**, *27*, 6161–6177.

99. J. Almond-Thynne, D. C. Blakemore, D. C. Pryde, A. C. Spivey, *Chem. Sci.* **2017**, *8*, 40–62.

100. A. F. Littke, C. Dai, G. C. Fu, *J. Am. Chem. Soc.* **2000**, *122*, 4020–4028.

101. X. Chen, H. Ke, G. Zou, *ACS Catal.* **2014**, *4*, 379–385.

102. E. D. Entz, J. E. A. Russell, L. V. Hooker, S. R. Neufeldt, *J. Am. Chem. Soc.* **2020**, *142*, 15454–15463.

103. V. Palani, M. A. Perea, R. Sarpong, *Chem. Rev.* **2022**, *122*, 10126–10169.

104. J. P. Norman, S. R. Neufeldt, *ACS Catal.* **2022**, *12*, 12014–12026.

105. C. Cordovilla, C. Bartolomé, J. M. Martínez-Ilarduya, P. Espinet, *ACS Catal.* **2015**, *5*, 3040–3053.

106. J. W. Labadie, J. K. Stille, *J. Am. Chem. Soc.* **1983**, *105*, 6129–6137.

107. A. Nova, G. Ujaque, F. Maseras, A. Lledos, P. Espinet, *J. Am. Chem. Soc.* **2006**, *128*, 14571–14578.

108. M. Hervé, G. Lefèvre, E. A. Mitchell, B. U. W. Maes, A. Jutand, *Chem. Eur. J.* **2015**, *21*, 18401–18406.

109. R. Álvarez, O. N. Faza, C. S. López, Ángel R. de Lera, *Org. Lett.* **2006**, *8*, 35–38.

110. A. Ariafard, B. F. Yates, *J. Am. Chem. Soc.* **2009**, *131*, 13981–13991.

111. J. Ye, R. K. Bhatt, J. R. Falck, *J. Am. Chem. Soc.* **1994**, *116*, 1–5.

112. A. A. C. Braga, N. H. Morgon, G. Ujaque, A. Lledós, F. Maseras, *J. Organomet. Chem.* **2006**, *691*, 4459–4466.

113. G. A. Molander, S. R. Wisniewski, *J. Am. Chem. Soc.* **2012**, *134*, 16856–15868.

114. A. A. Thomas, H. Wang, A. F. Zahrt, S. E. Denmark, *J. Am. Chem. Soc.* **2017**, *139*, 3805–3821.

115. A. Olding, C. C. Ho, N. T. Lucas, B. F. Yates, A. J. Canty, A. C. Bissember, *ACS Catal.* **2024**, *14*, 15946–15955.

116. A. J. J. Lennox, G. C. Lloyd-Jones, *Angew. Chem. Int. Ed.* **2013**, *52*, 7362–7370.

117. Y. Shi, J. S. Derasp, T. Maschmeyer, J. E. Hein, *Nat. Commun.* **2024**, *15*, 5436.

118. C. P. Delaney, D. P. Marron, A. S. Shved, R. N. Zare, R. M. Waymouth, S. E. Denmark, *J. Am. Chem. Soc.* **2022**, *144*, 4345–4364.

119. J. J. Fuentes-Rivera, M. E. Zick, M. A. Düfert, P. J. Milner, *Org. Process Res. Dev.* **2019**, *23*, 1631–1637.

120. J. J. Molloy, C. P. Seath, M. J. West, C. McLaughlin, N. J. Fazakerley, A. R. Kennedy, D. J. Nelson, A. J. B. Watson, *J. Am. Chem. Soc.* **2018**, *140*, 126–130.

121. B. P. Carrow, J. F. Hartwig, *J. Am. Chem. Soc.* **2011**, *133*, 2116–2119.

122. C. Amatore, A. Jutand, G. Le Duc, *Chem. Eur. J.* **2011**, *17*, 2492–2503.

123. A. A. C. Braga, N. H. Morgon, G. Ujaque, F. Maseras, *J. Am. Chem. Soc.* **2005**, *127*, 9298–9307.

124. A. Sugiyama, Y. Ohnishi, M. Nakaoka, Y. Nakao, H. Sato, S. Sakaki, Y. Nakao, T. Hiyama, *J. Am. Chem. Soc.* **2008**, *130*, 12975–12985.

125. Y. Hatanaka, T. Hiyama, *J. Am. Chem. Soc.* **1990**, *112*, 7793–7794.

126. C. Amatore, L. Grimaud, G. Le Duc, A. Jutand, *Angew. Chem. Int. Ed.* **2014**, *53*, 6982–6985.

127. S. E. Denmark, R. C. Smith, W.-T. T. Chang, *Tetrahedron* **2011**, *67*, 4391–4396.

128. P. Eckert, S. Sharif, M. G. Organ, *Angew. Chem. Int. Ed.* **2021**, *60*, 12224–12241.

129. P. Eckert, M. G. Organ, *Chem. Eur. J.* **2019**, *25*, 15751–15754.

130. A. B. González-Pérez, R. Alvarez, O. N. Faza, A. R. de Lera, J. M. Aurrecoechea, *Organometallics* **2012**, *31*, 2053–2058.

131. M. V. Polynski, E. A. Pidko, *Catal. Sci. Techn.* **2019**, *9*, 4561–4572.

132. G. T. Achonduh, N. Hadei, C. Valente, S. Avola, C. J. O'Brien, M. G. Organ, *Chem. Commun.* **2010**, *46*, 4109–4111.

133. K. Böck, J. E. Feil, K. Karaghiosoff, K. Koszinowski, *Chem. Eur. J.* **2015**, *21*, 5548–5560.

134. L. C. McCann, M. G. Organ, *Angew. Chem. Int. Ed.* **2014**, *53*, 4386–4389.

135. B. Fuentes, M. G. und Agustí Lledós, F. Maseras, J. A. Casares, G. Ujaque, P. Espinet, *Chem. Eur. J.* **2010**, *16*, 8596–8599.

136. T. Thaler, B. Haag, A. Gavryushin, K. Schober, E. Hartmann, R. M. Gschwind, H. Zipse, P. Mayer, P. Knochel, *Nature Chem.* **2010**, *2*, 125–130.

137. A. Guijarro, R. D. Rieke, *Angew. Chem. Int. Ed.* **2000**, *39*, 1475–1479.

138. J. Skotnitzki, A. Kremsmair, D. Keefer, Y. Gong, R. de Vivie-Riedle, P. Knochel, *Angew. Chem. Int. Ed.* **2020**, *59*, 320–324.

139. E. Gioria, J. M. Martínez-Ilarduya, P. Espinet, *Organometallics* **2014**, *33*, 4394–4400.

140. J. del Pozo, G. Salas, R. Álvarez, J. A. Casares, P. Espinet, *Organometallics* **2016**, *35*, 3604–3611.

141. D. Seyferth, *Organometallics* **2009**, *28*, 1598–1605.

142. B. Hölzer, R. W. Hoffmann, *Chem. Commun.* **2003**, 732–733.

143. K. Matsubara, H. Yamamoto, S. Miyazaki, T. Inatomi, K. Nonaka, Y. Koga, Y. Yamada, L. F. Veiros, K. Kirchner, *Organometallics* **2017**, *36*, 255–265.

144. E. Negishi, L. Anastasia, *Chem. Rev.* **2003**, *103*, 1979–2017.

145. X. Wang, Y. Song, J. Qu, Y. Luo, *Organometallics* **2017**, *36*, 1042–1048.

146. R. J. Oeschger, D. H. Ringger, P. Chen, *Organometallics* **2015**, *34*, 3888–3892.

147. M. Karak, L. C. A. Barbosa, G. C. Hargaden, *RSC Adv.* **2014**, *4*, 53442–53466.

148. M. Gazvoda, M. Virant, B. Pinter, J. Košmrlj, *Nature Commun.* **2018**, *9*, 4814.

149. P. Bertus, F. Fécourt, C. Bauder, P. Pale, *New J. Chem.* **2004**, *28*, 12–14.

150. U. Létinois-Halbes, P. Pale, S. Berger, *J. Org. Chem.* **2005**, *70*, 9185–9190.

151. C. He, J. Ke, H. Xu, A. Lei, *Angew. Chem. Int. Ed.* **2013**, *52*, 1527–1530.

152. T. Ljungdahl, T. Bennur, A. Dallas, H. Emtenäs, J. Mårtensson, *Organometallics* **2008**, *27*, 2490–2498.

153. K. L. Vikse, Z. Ahmadi, C. C. Manning, D. A. Harrington, J. S. McIndoe, *Angew. Chem. Int. Ed.* **2011**, *50*, 8304–8306.

154. R. J. Lundgren, M. Stradiotto, *Chem. Eur. J.* **2012**, *18*, 9758–9769.

155. M. Pérez-Rodríguez, A. A. C. Braga, M. Garcia-Melchor, M. H. Pérez-Temprano, J. A. Casares, G. Ujaque, A. R. de Lera, R. Álvarez, F. Maseras, P. Espinet, *J. Am. Chem. Soc.* **2009**, *131*, 3650–3657.

156. K. J. Bonney, F. Schoenebeck, *Chem. Soc. Rev.* **2014**, *43*, 6609–6638.
157. E. D. Kalkman, Y. Qiu, J. F. Hartwig, *ACS Catal.* **2023**, *13*, 12810–12825.
158. D. A. Culkin, J. F. Hartwig, *Organometallics* **2004**, *23*, 3398–3416.
159. G. Mann, D. Baranano, J. F. Hartwig, A. L. Rheingold, I. A. Guzei, *J. Am. Chem. Soc.* **1998**, *120*, 9205–9219.
160. J. L. Klinkenberg, J. F. Hartwig, *J. Am. Chem. Soc.* **2010**, *132*, 11830–11833.
161. P. Dierkes, P. W. N. M. van Leeuwen, *J. Chem. Soc. Dalton Trans.* **1999**, 1519–1529.
162. K. Tatsumi, R. Hoffmann, A. Yamamoto, J. K. Stille, *Bull. Chem. Soc. Jpn.* **1981**, *54*, 1857–1867.
163. Y. Jean, *Molecular Orbitals of Transition Metal Complexes, S. 78 ff., 178 ff.*, Oxford University Press, **2005**.
164. T. A. Albright, J. K. Burdett, M.-H. Whangbo, *Orbital Interactions in Chemistry, S. 503 ff.*, John Wiley & Sons, **2013**.
165. J. F. Hartwig, *Organotransition Metal Chemistry - From Bonding to Catalysis, S. 323 f.*, University Science Books, **2010**.
166. D. A. Watson, M. Su, G. Teverovskiy, Y. Zhang, J. García-Fortanet, T. Kinzel, S. L. Buchwald, *Science* **2009**, *328*, 1679–1681.
167. P. S. Hanley, S. L. Marquard, T. R. Cundari, J. F. Hartwig, *J. Am. Chem. Soc.* **2012**, *134*, 15281–15284.
168. S. L. Marquard, D. C. Rosenfeld, J. F. Hartwig, *Angew. Chem. Int. Ed.* **2010**, *49*, 793–796.
169. S. L. Marquard, J. F. Hartwig, *Angew. Chem. Int. Ed.* **2011**, *50*, 7119–7123.
170. J. G. de Vries, *Dalton Trans.* **2006**, 421–429.
171. A. S. Kashin, V. P. Ananikov, *J. Org. Chem.* **2013**, *78*, 11117–11125.
172. D. B. Eremin, V. P. Ananikov, *Coord. Chem. Rev.* **2017**, *346*, 2–19.
173. D. Balcells, A. Nova, *ACS Catal.* **2018**, *8*, 3499–3515.
174. M. A. Düfert, K. L. Billingsley, S. L. Buchwald, *J. Am. Chem. Soc.* **2013**, *135*, 12877–12885.
175. M. Wakioka, M. Nagao, F. Ozawa, *Organometallics* **2008**, *27*, 602–608.
176. A. J. J. Lennox, G. C. Lloyd-Jones, *Chem. Soc. Rev.* **2014**, *43*, 412–443.
177. T. Klatt, J. T. Markiewicz, C. Sämann, P. Knochel, *J. Org. Chem.* **2014**, *79*, 4253–4269.
178. G. Dagousset, C. François, T. León, R. Blanc, E. Sansiaume-Dagousset, P. Knochel, *Synthesis* **2014**, *46*, 3133–3171.
179. B. Wei, Y.-H. Chen, P. Knochel, *Acc. Chem. Res.* **2024**, *57*, 1951–1963.
180. R. J. Huntley, R. L. Funk, *Org. Lett.* **2006**, *8*, 4775–4778.
181. D. L. Boger, S. Miyazaki, S. H. Kim, J. H. Wu, S. L. Castle, O. Loiseleur, Q. Jin, *J. Am. Chem. Soc.* **1999**, *121*, 10004–10011.
182. N. K. Garg, D. D. Caspi, B. M. Stoltz, *J. Am. Chem. Soc.* **2004**, *126*, 9552–9553.
183. A. B. Smith III, T. J. Beauchamp, M. J. LaMarche, M. D. Kaufman, Y. Qiu, H. Arimoto, D. R. Jones, K. Kobayashi, *J. Am. Chem. Soc.* **2000**, *122*, 8654–8664.
184. P. Knochel, W. Dohle, N. Gommermann, F. F. Kneisel, F. Kopp, T. Korn, I. Sapountzis, V. A. Vu, *Angew. Chem. Int. Ed.* **2003**, *42*, 4302–4320.
185. D. S. Ziegler, B. Wei, P. Knochel, *Chem. Eur. J.* **2019**, *25*, 2695–2703.
186. V. Dhayalan, V. S. Dodke, M. P. Kumar, H. S. Korkmaz, A. Hoffmann-Röder, P. Amaladass, R. Dandela, R. Dhanusuraman, P. Knochel, *Chem. Soc. Rev.* **2024**, *53*, 11045–11099.
187. A. Krasovskiy, P. Knochel, *Angew. Chem. Int. Ed.* **2004**, *43*, 3333–3336.
188. A. D. H. Pham, J. Bui, K. W. Foreman, *Organometallics* **2023**, *42*, 3266–3274.
189. A. Hermann, R. Seymen, L. Brieger, J. Kleinheider, B. Grabe, W. Hiller, C. Strohmann, *Angew. Chem. Int. Ed.* **2023**, *62*, e202302489.
190. D. S. Ziegler, K. Karaghiosoff, P. Knochel, *Angew. Chem. Int. Ed.* **2018**, *57*, 6701–6704.
191. P. Knochel, M. I. Calaza, E. Hupe in *Metal-Catalyzed Cross-Coupling Reactions* (Eds.: A. de Meijere, F. Diederich), Wiley-VCH, **2004**, Chapter 11, pp. 619–670.

192. A. Krasovskiy, V. Krasovskaya, P. Knochel, *Angew. Chem. Int. Ed.* **2006**, *45*, 2958–2961.

193. S. H. Wunderlich, P. Knochel, *Angew. Chem. Int. Ed.* **2007**, *46*, 7685–7688.

194. C. J. Rohbogner, S. H. Wunderlich, G. C. Clososki, P. Knochel, *Eur. J. Org. Chem.* **2009**, 1781–1795.

195. V. Diemer, F. R. Leroux, F. Colobert, *Eur. J. Org. Chem.* **2011**, 327–340.

196. T. Morita, S. Fuse, H. Nakamura, *Angew. Chem. Int. Ed.* **2016**, *55*, 13580–13584.

197. H. Ren, P. Knochel, *Chem. Commun.* **2006**, 726–728.

198. D. J. Nelson, P. J. Cooper, R. Soundararajan, *J. Am. Chem. Soc.* **1989**, *111*, 1414–1418.

199. D. P. Curran, T. R. McFadden, *J. Am. Chem. Soc.* **2016**, *138*, 7741–7752.

200. H. X. Zhang, F. Guibe, G. Balavoine, *J. Org. Chem.* **1990**, *55*, 1857–1867.

201. L. Ferrié, B. Figadère, *Org. Lett.* **2010**, *12*, 4976–4979.

202. A. B. Smith, III, S. M. Condon, J. A. McCauley, J. L. Leazer, J. W. Leahy, R. E. Maleczka, *J. Am. Chem. Soc.* **1995**, *117*, 5407–5408.

203. P. J. Mohr, R. L. Halcomb, *J. Am. Chem. Soc.* **2003**, *125*, 1712–1713.

204. C. A. Busacca, M. Cerreta, Y. Dong, M. C. Eriksson, V. Farina, X. Feng, J.-Y. Kim, J. C. Lorenz, M. Sarvestani, R. Simpson, R. Varsolona, J. Vitous, S. J. Campbell, M. S. Davis, P.-J. Jones, D. Norwood, F. Qiu, P. L. Beaulieu, J.-S. Duceppe, B. Haché, J. Brong, F.-T. Chiu, T. Curtis, J. Kelley, Y. S. Lo, T. H. Powner, *Org. Process Res. Dev.* **2008**, *12*, 603–613.

205. A. B. Smith, III, K. P. Minbiole, P. R. Verhoest, M. Schelhaas, *J. Am. Chem. Soc.* **2001**, *123*, 10942–10953.

206. K. L. Billingsley, T. E. Barder, S. L. Buchwald, *Angew. Chem. Int. Ed.* **2007**, *46*, 5359–5363.

207. S. Barroso, M. Joksch, P. Puylaert, S. Tin, S. J. Bell, L. Donnellan, S. Duguid, C. Muir, P. Zhao, V. Farina, D. N. Tran, J. G. de Vries, *J. Org. Chem.* **2021**, *86*, 103–109.

208. T. Ishiyama, J. Takagi, K. Ishida, N. Miyaura, N. R. Anastasi, J. F. Hartwig, *J. Am. Chem. Soc.* **2002**, *124*, 390–391.

209. T. Ishiyama, Y. Nobuta, J. F. Hartwig, N. Miyaura, *Chem. Commun.* **2003**, 2924–2925.

210. J. Dufour, L. Neuville, J. Zhu, *Chem. Eur. J.* **2010**, *16*, 10523–10534.

211. A. B. Smith, III, T. M. Razler, J. P. Ciavarri, T. Hirose, T. Ishikawa, *Org. Lett.* **2005**, *7*, 4399–4402.

212. D. F. Fischer, R. Sarpong, *J. Am. Chem. Soc.* **2010**, *132*, 5926–5927.

213. L. F. Tietze, M. A. Düfert, T. Hungerland, K. Oum, T. Lenzer, *Chem. Eur. J.* **2011**, *17*, 8452–8461.

214. G. A. Molander, R. Figueroa, *Org. Lett.* **2006**, *8*, 75–78.

215. G. A. Molander, R. Figueroa, *J. Org. Chem.* **2006**, *71*, 6135–6140.

216. M. H. Becker, P. Chua, R. Downham, C. J. Douglas, N. K. Garg, S. Hiebert, S. Jaroch, R. T. Matsuoka, J. A. Middleton, F. W. Ng, L. E. Overman, *J. Am. Chem. Soc.* **2007**, *129*, 11987–12002.

217. S. E. Denmark, S.-M. Yang, *J. Am. Chem. Soc.* **2002**, *124*, 15196–15197.

218. J. Lee, R. Velarde-Ortiz, A. Guijarro, J. R.Wurst, R. D. Rieke, *J. Org. Chem.* **2000**, *65*, 5428–5430.

219. R. D. Rieke, M. V. Hanson, *Tetrahedron* **1997**, *53*, 1925–1956.

220. R. D. Rieke, *Science* **1989**, *246*, 1260–1264.

221. A. Krasovskiy, V. Malakhov, A. Gavryushin, P. Knochel, *Angew. Chem. Int. Ed.* **2006**, *45*, 6040–6044.

222. S. Huo, *Org. Lett.* **2003**, *5*, 423–425.

223. C. Aissa, R. Riveiros, J. Ragot, A. Fürstner, *J. Am. Chem. Soc.* **2003**, *125*, 15512–15520.

224. C. K. Skepper, T. Quach, T. F. Molinski, *J. Am. Chem. Soc.* **2010**, *132*, 10286–10292.

225. A. Zakarian, A. Batch, R. A. Holton, *J. Am. Chem. Soc.* **2003**, *125*, 7822–7824.

226. M. E. Layton, C. A. Morales, M. D. Shair, *J. Am. Chem. Soc.* **2002**, *124*, 773–775.

227. A. J. Ross, H. L. Lang, R. F. W. Jackson, *J. Org. Chem.* **2010**, *75*, 245–248.

228. V. F. Slagt, A. H. M. de Vries, J. G. de Vries, R. M. Kellogg, *Org. Process Res. Dev.* **2010**, *14*, 30–47.

229. A. L. Casado, P. Espinet, *Organometallics* **2003**, *22*, 1305–1309.

230. A. L. Casado, P. Espinet, *J. Am. Chem. Soc.* **1998**, *120*, 8978–8985.

231. V. Farina, *Pure Appl. Chem.* **1996**, *68*, 73–78.

232. V. Farina, S. Kapadia, B. Krishnan, C. Wang, L. S. Liebeskind, *J. Org. Chem.* **1994**, *59*, 5905–5911.

233. J. R. Naber, S. L. Buchwald, *Adv. Synth. Catal.* **2008**, *350*, 957–961.

234. S. P. H. Mee, V. Lee, J. E. Baldwin, *Angew. Chem. Int. Ed.* **2004**, *43*, 1132–1132.

235. M. Wang, Z. Lin, *Organometallics* **2010**, *29*, 3077–3084.

236. G. D. Allred, L. S. Liebeskind, *J. Am. Chem. Soc.* **1996**, *118*, 2748–2749.

237. A. K. Ghosh, J. R. Born, A. M. Veitschegger, M. S. Jurica, *J. Org. Chem.* **2020**, *85*, 8111–8120.

238. K. C. Nicolaou, G. Bellavance, M. Buchman, K. K. Pulukuri, *J. Am. Chem. Soc.* **2017**, *139*, 15636–15639.

239. M. D. Shair, T. Y. Yoon, K. K. Mosny, T. C. Chou, S. J. Danishefsky, *J. Am. Chem. Soc.* **1996**, *118*, 9509–9525.

240. N. Miyaura, A. Suzuki, *Chem. Rev.* **1995**, *95*, 2457–2483.

241. F. Bellina, A. Carpita, R. Rossi, *Synthesis* **2004**, 2419–2440.

242. B. S. Takale, F.-Y. Kong, R. R. Thakore, *Organics* **2022**, *3*, 1–21.

243. Y. Ashikari, T. Kawaguchi, K. Mandai, Y. Aizawa, A. Nagaki, *J. Am. Chem. Soc.* **2020**, *142*, 17039–17047.

244. D. Qiu, S. Wang, S. Tang, H. Meng, L. Jin, F. Mo, Y. Zhang, J. Wang, *J. Org. Chem.* **2014**, *79*, 1979–1988.

245. N. Miyaura in *Metal-Catalyzed Cross-Coupling Reactions* (Eds.: A. de Meijere, F. Diederich), Wiley-VCH, **2004**, Chapter 2, pp. 41–123.

246. J. C. Anderson, H. Namli, C. A. Roberts, *Tetrahedron* **1997**, *53*, 15123–15134.

247. H. L. D. Hayes, R. Wei, M. Assante, K. J. Geogheghan, N. Jin, S. Tomasi, G. Noonan, A. G. Leach, G. C. Lloyd-Jones, *J. Am. Chem. Soc.* **2021**, *143*, 14814–14826.

248. P. A. Cox, M. Reid, A. G. Leach, A. D. Campbell, E. J. King, G. C. Lloyd-Jones, *J. Am. Chem. Soc.* **2017**, *139*, 13156–13165.

249. Y. Yamamoto, M. Takizawa, X.-Q. Yu, N. Miyaura, *Angew. Chem. Int. Ed.* **2008**, *47*, 928–931.

250. K. L. Billingsley, S. L. Buchwald, *Angew. Chem. Int. Ed.* **2008**, *47*, 4695–4698.

251. W. Shu, L. Pellegatti, M. A. Oberli, S. L. Buchwald, *Angew. Chem. Int. Ed.* **2011**, *50*, 10665–10669.

252. E. P. Gillis, M. D. Burke, *J. Am. Chem. Soc.* **2007**, *129*, 6716–6717.

253. S. J. Lee, K. C. Gray, J. S. Paek, M. D. Burke, *J. Am. Chem. Soc.* **2008**, *130*, 466–468.

254. J. A. Gonzalez, O. M. Ogba, G. F. Morehouse, N. Rosson, K. N. Houk, A. G. Leach, P. H.-Y. Cheong, M. D. Burke, G. C. Lloyd-Jones, *Nat. Chem.* **2016**, *8*, 1067–1075.

255. S. J. Lee, T. M. Anderson, M. D. Burke, *Angew. Chem. Int. Ed.* **2010**, *49*, 8860–8863.

256. G. A. Molander, N. Ellis, *Acc. Chem. Res.* **2007**, *40*, 275–286.

257. G. A. Molander, B. Canturk, *Angew. Chem. Int. Ed.* **2009**, *48*, 9240–9261.

258. R. A. Batey, T. D. Quach, *Tetrahedron Lett.* **2001**, *42*, 9099–9103.

259. M. Butters, J. N. Harvey, J. Jover, A. J. J. Lennox, G. C. Lloyd-Jones, P. M. Murray, *Angew. Chem. Int. Ed.* **2010**, *49*, 5156–5160.

260. A. J. J. Lennox, G. C. Lloyd-Jones, *J. Am. Chem. Soc.* **2012**, *134*, 7431–7441.

261. I. Omari, L. P. E. Yunker, J. Penafiel, D. Gitaari, A. S. Roman, J. S. McIndoe, *Chem. Eur. J.* **2021**, *27*, 3812–3816.

262. G. A. Molander, *J. Org. Chem.* **2015**, *80*, 7837–7848.

263. S. D. Dreher, S.-E. Lim, D. L. Sandrock, G. A. Molander, *J. Org. Chem.* **2009**, *74*, 3626–3631.

264. L. Li, S. Zhao, A. Joshi-Pangu, M. Diane, M. R. Biscoe, *J. Am. Chem. Soc.* **2014**, *136*, 14027–14030.

265. G. A. Molander, I. Shin, *Org. Lett.* **2011**, *13*, 3956–3959.

266. G. A. Molander, N. Fleury-Brégeot, M.-A. Hiebel, *Org. Lett.* **2011**, *13*, 1694–1697.

267. G. A. Molander, I. Shin, *Org. Lett.* **2013**, *15*, 2534–2537.

268. N. Fleury-Brégeot, M. Presset, F. Beaumard, V. Colombel, D. Oehlrich, F. Rombouts, G. A. Molander, *J. Org. Chem.* **2012**, *77*, 10399–10408.

269. F.-L. Haut, K. Speck, R. Wildermuth, K. Möller, P. Mayer, T. Magauer, *Tetrahedron* **2018**, *74*, 3348–3357.

270. V. B. Phapale, D. J. Cárdenas, *Chem. Soc. Rev.* **2009**, *38*, 1598–1607.

271. M. M. Heravi, E. Hashemi, N. Nazari, *Mol. Divers.* **2014**, *18*, 441–472.

272. H. Gong, M. R. Gagné, *J. Am. Chem. Soc.* **2008**, *130*, 12177–12183.

273. S. Hazra, C. C. C. Johansson Seechurn, S. Handa, T. J. Colacot, *ACS Catal.* **2021**, *11*, 13188–13202.

274. C. E. I. Knappke, A. J. von Wangelin, *Chem. Soc. Rev.* **2011**, *40*, 4948–4962.

275. J. D. Firth, P. O'Brien, *ChemCatChem* **2015**, *7*, 395–397.

276. P. Liu, E. N. Jacobsen, *J. Am. Chem. Soc.* **2001**, *123*, 10772–10773.

277. K. Tamao, K. Sumitani, M. Kumada, *J. Am. Chem. Soc.* **1972**, *94*, 4374–4376.

278. R. J. P. Corriu, J. P. Masse, *J. Chem. Soc. Chem. Commun.* **1972**, 144.

279. M. Tamura, J. Kochi, *Synthesis* **1971**, 303–305.

280. J. Terao, N. Kambe, *Acc. Chem. Res.* **2008**, *41*, 1545–1554.

281. G. Bold, A. Fässler, H.-G. Capraro, R. Cozens, T. Klimkait, J. Lazdins, J. Mestan, B. Poncioni, J. Rösel, D. Stover, M. Tintelnot-Blomley, F. Acemoglu, W. Beck, E. Boss, M. Eschbach, T. Hürlimann, E. Masso, S. Roussel, K. Ucci-Stoll, D. Wyss, M. Lang, *J. Med. Chem.* **1998**, *41*, 3387–3401.

282. G. Marzoni, M. D. Varney, *Org. Process Res. Dev.* **1997**, *1*, 81–84.

283. G. Manolikakes, P. Knochel, *Angew. Chem. Int. Ed.* **2009**, *48*, 205–209.

284. V. Wowk, G. Lefèvre, *Dalton Trans.* **2022**, *51*, 10674–10680.

285. G. Cahiez, H. Avedissian, *Synthesis* **1998**, 1199–1205.

286. A. Fürstner, A. Leitner, M. Méndez, H. Krause, *J. Am. Chem. Soc.* **2002**, *124*, 13856–13863.

287. B. Scheiper, M. Bonnekessel, H. Krause, A. Fürstner, *J. Org. Chem.* **2004**, *69*, 3943–3949.

288. O. M. Kuzmina, A. K. Steib, D. Flubacher, P. Knochel, *Org. Lett.* **2012**, *14*, 4818–4821.

289. L. K. Ottesen, F. Ek, R. Olsson, *Org. Lett.* **2006**, *8*, 1771–1773.

290. R. B. Bedford, *Acc. Chem. Res.* **2015**, *48*, 1485–1493.

291. K. Ding, F. Zannat, J. C. Morris, W. W. Brennessel, P. L. Holland, *J. Organomet. Chem.* **2009**, *694*, 4204–4208.

292. A. Fürstner, D. De Souza, L. Parra-Rapado, J. T. Jensen, *Angew. Chem. Int. Ed.* **2003**, *42*, 5358–5360.

293. R. Walsh, *Acc. Chem. Res.* **1981**, *14*, 246–252.

294. Y. Nakao, T. Hiyama, *Chem. Soc. Rev.* **2011**, *40*, 4893–4901.

295. H. F. Sore, W. R. J. D. Galloway, D. R. Spring, *Chem. Soc. Rev.* **2012**, *41*, 1845–1866.

296. S. E. Denmark, R. F. Sweis in *Metal-Catalyzed Cross-Coupling Reactions* (Eds.: A. de Meijere, F. Diederich), Wiley-VCH, **2004**, Chapter 4, pp. 163–216.

297. S. E. Denmark, R. F. Sweis, *Acc. Chem. Res.* **2002**, *35*, 835–846.

298. S. E. Denmark, D. Wehrli, J. Y. Choi, *Org. Lett.* **2000**, *2*, 2491–2494.

299. S. E. Denmark, C. S. Regens, *Acc. Chem. Res.* **2008**, *41*, 1486–1499.

300. S. E. Denmark, J. H.-C. Liu, *Angew. Chem. Int. Ed.* **2010**, *49*, 2978–2986.

301. B. M. Trost, C. E. Stivala, D. R. Fandrick, K. L. Hull, A. Huang, C. Poock, R. Kalkofen, *J. Am. Chem. Soc.* **2016**, *138*, 11690–11701.

302. Y. Zhang, J. S. Panek, *Org. Lett.* **2007**, *9*, 3141–3143.

303. S. E. Denmark, C. S. Regens, T. Kobayashi, *J. Am. Chem. Soc.* **2007**, *129*, 2774–2776.

304. R. Chinchilla, C. Nájera, *Chem. Rev.* **2007**, *107*, 874–922.

305. P. Siemsen, R. C. Livingston, F. Diederich, *Angew. Chem. Int. Ed.* **2000**, *39*, 2632–2657.

306. H. Doucet, J.-C. Hierso, *Angew. Chem. Int. Ed.* **2007**, *46*, 834–871.

307. K. Komano, S. Shimamura, M. Inoue, M. Hirama, *J. Am. Chem. Soc.* **2007**, *129*, 14184–14186.

308. P. S. Baran, R. A. Shenvi, *J. Am. Chem. Soc.* **2006**, *128*, 14028–14029.

309. E. Negishi, *Acc. Chem. Res.* **1982**, *15*, 340–348.

310. E. Negishi, M. Qian, F. Zeng, L. Anastasia, D. Babinski, *Org. Lett.* **2003**, *5*, 1597–1600.

311. L. Anastasia, E. Negishi, *Org. Lett.* **2001**, *3*, 3111–3113.

312. R. Chinchilla, C. Nájera, *Chem. Soc. Rev.* **2011**, *40*, 5084–5121.

313. D. Gelman, S. L. Buchwald, *Angew. Chem. Int. Ed.* **2003**, *42*, 5993–5996.

314. T. Mizoroki, K. Mori, A. Ozaki, *Bull. Chem. Soc. Jpn.* **1971**, *44*, 581.

315. M. Yamamura, I. Moritani, S.-I. Murahashi, *J. Organomet. Chem.* **1975**, *91*, C39–C42.

316. J. E. Leibner, J. Jacobus, *J. Org. Chem.* **1979**, *44*, 449–450.

317. D. C. Harrowven, I. L. Guy, *Chem. Commun.* **2004**, 1968–1969.

318. P. Renaud, E. Lacote, L. Quaranta, *Tetrahedron Lett.* **1998**, *39*, 2123–2126.

319. A. F. Littke, G. C. Fu, *Angew. Chem. Int. Ed.* **1999**, *38*, 2411–2413.

320. A. F. Littke, L. Schwarz, G. C. Fu, *J. Am. Chem. Soc.* **2002**, *124*, 6343–6348.

321. G. A. Grasa, S. P. Nolan, *Org. Lett.* **2001**, *3*, 119–122.

322. B. S. Takale, R. R. Thakore, G. Casotti, X. Li, F. Gallou, B. H. Lipshutz, *Angew. Chem. Int. Ed.* **2021**, *60*, 4158–4163.

323. L. Ferrié, J. Fenneteau, B. Figadère, *Org. Lett.* **2018**, *20*, 3192–3196.

324. H. Fuwa, Y. Okuaki, N. Yamagata, M. Sasaki, *Angew. Chem. Int. Ed.* **2015**, *54*, 868–873.

325. A. Suzuki, Y. Yamamoto, *Chem. Lett.* **2011**, *40*, 894–901.

326. A. B. Pagett, G. C. Lloyd-Jones, *Org. React.* **2020**, *100*, 547–619.

327. A. García-Domínguez, A. G. Leach, G. C. Lloyd-Jones, *Acc. Chem. Res.* **2022**, *55*, 1324–1336.

328. T. Kinzel, Y. Zhang, S. L. Buchwald, *J. Am. Chem. Soc.* **2010**, *132*, 14073–14075.

329. L. Chen, H. Francis, B. P. Carrow, *ACS Catal.* **2018**, *8*, 2989–2994.

330. J. P. G. Rygus, C. M. Crudden, *J. Am. Chem. Soc.* **2017**, *139*, 18124–18137.

331. D. Zhang, Q. Wang, *Coord. Chem. Rev.* **2015**, *286*, 1–16.

332. M. J. Moore, S. Qu, C. Tan, Y. Cai, Y. Mogi, D. J. Keith, D. L. Boger, *J. Am. Chem. Soc.* **2020**, *142*, 16039–16050.

333. R. Kranthikumar, *Organometallics* **2022**, *41*, 667–679.

334. S. R. Chemler, D. Trauner, S. J. Danishefsky, *Angew. Chem. Int. Ed.* **2001**, *40*, 4544–4568.

335. A. T. K. Koshvandi, M. M. Heravi, T. Momeni, *Appl. Organometal. Chem.* **2018**, *32*, e4210.

336. D. A. Evans, J. T. Starr, *Angew. Chem. Int. Ed.* **2002**, *41*, 1787–1790.

337. D. Haas, J. M. Hammann, R. Greiner, P. Knochel, *ACS Catal.* **2016**, *6*, 1540–1552.

338. C. I. Stathakis, S. Bernhardt, V. Quint, P. Knochel, *Angew. Chem. Int. Ed.* **2012**, *51*, 9428–9432.

339. S. Bernhardt, G. Manolikakes, T. Kunz, P. Knochel, *Angew. Chem. Int. Ed.* **2011**, *50*, 9205–9209.

340. Y.-H. Chen, M. Ellwart, V. Malakhov, P. Knochel, *Synthesis* **2017**, *49*, 3215–3223.

341. V. Dhayalan, V. S. Dodke, D. Sharma, R. Dandela, *Eur. J. Org. Chem.* **2024**, *27*, e202301263.

342. J. M. Hammann, F. H. Lutter, D. Haas, P. Knochel, *Angew. Chem. Int. Ed.* **2016**, *56*, 1082–1086.

343. J. Choi, G. C. Fu, *Science* **2017**, *356*, eeaf7230.

344. A. W. Dombrowski, N. J. Gesmundo, A. L. Aguirre, K. A. Sarris, J. M. Young, A. R. Bogdan, M. C. Martin, S. Gedeon, Y. Wang, *ACS Med. Chem. Lett.* **2020**, *11*, 597–604.

345. M. Pompeo, R. D. J. Froese, N. Hadei, M. G. Organ, *Angew. Chem. Int. Ed.* **2012**, *51*, 11354–11357.

346. Y. Yang, K. Niedermann, C. Han, S. L. Buchwald, *Org. Lett.* **2014**, *16*, 4638–4641.

347. A. H. Cherney, S. J. Hedley, S. M. Mennen, J. S. Tedrow, *Organometallics* **2019**, *38*, 97–102.
348. N. D. Schley, G. C. Fu, *J. Am. Chem. Soc.* **2014**, *136*, 16588–16593.
349. C. Fischer, G. C. Fu, *J. Am. Chem. Soc.* **2005**, *127*, 4594–4595.
350. Y. Liang, G. C. Fu, *J. Am. Chem. Soc.* **2015**, *137*, 9523–9526.
351. X. Mu, Y. Shibata, Y. Makida, G. C. Fu, *Angew. Chem. Int. Ed.* **2017**, *56*, 5821–5824.
352. H. Huo, B. J. Gorsline, G. C. Fu, *Science* **2020**, *367*, 559–564.
353. J. D. Mason, D. W. Terwilliger, A. R. Pote, A. G. Myers, *J. Am. Chem. Soc.* **2021**, *143*, 11019–11025.
354. J. E. Tungen, M. Aursnes, J. Dalli, H. Arnardottir, C. N. Serhan, T. V. Hansen, *Chem. Eur. J.* **2014**, *20*, 14575–14578.
355. H. Ila, O. Baron, A. J. Wagner, P. Knochel, *Chem. Commun.* **2006**, 583–593.
356. R. Martin, S. L. Buchwald, *J. Am. Chem. Soc.* **2007**, *129*, 3844–3845.
357. X. Hua, J. Masson-Makdissi, R. J. Sullivan, S. G. Newman, *Org. Lett.* **2016**, *18*, 5312–5315.
358. M. Giannerini, M. Fananás-Mastral, B. L. Feringa, *Nat. Chem.* **2013**, *5*, 667–672.
359. C. Vila, M. Giannerini, V. Hornillos, M. Fananás-Mastral, B. L. Feringa, *Chem. Sci.* **2014**, *5*, 1361–1367.
360. P. Visser, B. L. Feringa, *Chem. Commun.* **2023**, *59*, 5539–5542.
361. A. Fürstner, *Bull. Chem. Soc. Jpn.* **2021**, *94*, 666–677.
362. A. Hamajima, M. Isobe, *Org. Lett.* **2006**, *8*, 1205–1208.
363. C. Gregg, C. Gunawan, A. W. Y. Ng, S. Wimala, S. Wickremasinghe, M. A. Rizzacasa, *Org. Lett.* **2013**, *15*, 516–519.
364. A. Fürstner, A. Schlecker, *Chem. Eur. J.* **2008**, *14*, 9181–9191.
365. F. Foubelo, C. Nájera, M. Yus, *Chem. Rec.* **2016**, *16*, 2521–2533.
366. E. J. Cho, T. D. Senecal, T. Kinzel, Y. Zhang, D. A. Watson, S. L. Buchwald, *Science* **2010**, *328*, 1679–1681.
367. C. Alonso, E. M. de Marigorta, G. Rubiales, F. Palacios, *Chem. Rev.* **2015**, *115*, 1847–1935.
368. O. A. Tomashenko, V. V. Grushin, *Chem. Rev.* **2011**, *111*, 4475–4521.
369. S. E. Denmark, S.-M. Yang, *J. Am. Chem. Soc.* **2004**, *126*, 12432–12440.
370. A. M. Thomas, A. Sujatha, G. Anilkumar, *RSC Adv.* **2014**, *4*, 21688–21698.
371. D. Wang, S. Gao, *Org. Chem. Front.* **2014**, *1*, 556–566.
372. A. L. Smith, C.-K. Hwang, E. Pitsinos, G. R. Scarlato, K. C. Nicolaou, *J. Am. Chem. Soc.* **1992**, *114*, 3134–3136.
373. L. F. Tietze, S.-C. Duefert, J. Clerc, M. Bischoff, C. Maaß, D. Stalke, *Angew. Chem. Int. Ed.* **2013**, *52*, 3191–3194.
374. P. Wipf, T. H. Graham, *J. Am. Chem. Soc.* **2004**, *126*, 15346–15347.
375. H. Zhao, A. K. Ravn, M. C. Haibach, K. M. Engle, C. C. C. Johansson Seechurn, *ACS Catal.* **2024**, *14*, 9708–9733.
376. H. Plenio, *Angew. Chem. Int. Ed.* **2008**, *47*, 6954–6956.
377. Z. Gonda, G. L. Tolnai, Z. Novák, *Chem. Eur. J.* **2010**, *16*, 11822–11826.
378. F. Monnier, F. Turtaut, L. Duroure, M. Taillefer, *Org. Lett.* **2008**, *10*, 3203–3206.
379. L.-H. Zou, A. J. Johansson, E. Zuidema, C. Bolm, *Chem. Eur. J.* **2013**, *19*, 8144–8152.
380. R. Rossi, F. Bellina, M. Lessia, *Adv. Synth. Catal.* **2012**, *354*, 1181–1255.
381. A. Piontek, E. Bisz, M. Szostak, *Angew. Chem. Int. Ed.* **2018**, *57*, 11116–11128.
382. O. R. Thiel, M. Achmatowicz, C. Bernard, P. Wheeler, C. Savarin, T. L. Correll, A. Kasparian, A. Allgeier, M. D. Bartberger, H. Tan, R. D. Larsen, *Org. Process Res. Dev.* **2009**, *13*, 230–241.
383. B. Li, R. A. Buzon, Z. Zhang, *Org. Process Res. Dev.* **2007**, *11*, 951–955.
384. X. Deng, J. T. Liang, M. Peterson, R. Rynberg, E. Cheung, N. S. Mani, *J. Org. Chem.* **2010**, *75*, 1940–1947.
385. P. W. Manley, M. Acemoglu, W. Marterer, W. Pachinger, *Org. Process Res. Dev.* **2003**, *7*, 436–445.

386. A. Gontcharov, J. R. Dunetz, *Org. Process Res. Dev.* **2014**, *18*, 1145–1152.
387. S. Gangula, U. K. Neelam, S. R. Baddam, V. H. Dahanukar, R. Bandichhor, *Org. Process Res. Dev.* **2015**, *19*, 470–475.
388. S. Challenger, Y. Dessi, D. E. Fox, L. C. Hesmondhalgh, P. Pascal, A. J. Pettman, J. D. Smith, *Org. Process Res. Dev.* **2008**, *12*, 575–583.
389. A. V. Thomas, H. H. Patel, L. A. Reif, S. R. Chemburkar, D. P. Sawick, B. Shelat, M. K. Balmer, R. R. Patel, *Org. Process Res. Dev.* **1997**, *1*, 294–299.
390. E. Richmond, J. Moran, *Synthesis* **2018**, *50*, 499–513.
391. K. E. Poremba, S. E. Dibrell, S. E. Reisman, *ACS Catal.* **2020**, *10*, 8237–8246.
392. G. A. Molander, K. M. Traister, B. T. O'Neill, *J. Org. Chem.* **2015**, *80*, 2907–2911.
393. K. Kang, L. Huang, D. J. Weix, *J. Am. Chem. Soc.* **2020**, *142*, 10634–10640.
394. N. T. Kadunce, S. E. Reisman, *J. Am. Chem. Soc.* **2015**, *137*, 10480–10483.
395. Q. Zhou, H.-G. Cheng, Z. Yang, R. Chen, L. Cao, Q. Wei, W.-Y. Tong, Q. Wang, C. Wu, S. Qu, *Angew. Chem. Int. Ed.* **2021**, *60*, 5141–5146.
396. L. Marzo, S. K. Pagire, O. Reiser, B. König, *Angew. Chem. Int. Ed.* **2018**, *57*, 10034–10072.
397. N. A. Romero, D. A. Nicewicz, *Chem. Rev.* **2016**, *116*, 10075–10166.
398. K. L. Skubi, T. R. Blum, T. P. Yoon, *Chem. Rev.* **2016**, *116*, 10035–10074.
399. A. Brennführer, H. Neumann, M. Beller, *Angew. Chem. Int. Ed.* **2009**, *48*, 4114–4133.
400. X.-F. Wu, H. Neumann, M. Beller, *Chem. Soc. Rev.* **2011**, *40*, 4986–5009.
401. C. F. J. Barnard, *Organometallics* **2008**, *27*, 5402–5422.
402. L. Wu, Q. Liu, R. Jackstell, M. Beller, *Angew. Chem. Int. Ed.* **2014**, *53*, 6310–6320.
403. M. P. Wentland, R. Lou, Y. Ye, D. J. Cohen, G. P. Richardson, J. M. Bidlack, *Bioorg. Med. Chem. Lett.* **2001**, *11*, 623–626.
404. S. Gao, Q. Wang, C. Chen, *J. Am. Chem. Soc.* **2009**, *131*, 1410–1412.
405. R. F. Heck, *J. Am. Chem. Soc.* **1968**, *90*, 5518–5526.
406. R. F. Heck, J. P. Nolley, *J. Org. Chem.* **1972**, *37*, 2320–23222.
407. I. P. Beletskaya, A. V. Cheprakov, *Chem. Rev.* **2000**, *100*, 3009–3066.
408. *The Mizoroki-Heck Reaction*, (Ed.: M. Oestreich), Wiley-VCH, **2009**.
409. L. F. Tietze, I. Hiriyakkanavar, H. P. Bell, *Chem. Rev.* **2004**, *104*, 3453–3516.
410. M. Shibasaki, E. M. Vogl, T. Ohshima, *Adv. Synth. Catal.* **2004**, *346*, 1533–1552.
411. Z. Yang, X. Xu, C.-H. Yang, Y. Tian, X. Chen, L. Lian, W. Pan, X. Su, W. Zhang, Y. Chen, *Org. Lett.* **2016**, *18*, 5768–5770.
412. D. H. B. Ripin, D. E. Bourassa, T. Brandt, M. J. Castaldi, H. N. Frost, J. Hawkins, P. J. Johnson, S. S. Massett, K. Neumann, J. Phillips, J. W. Raggon, P. R. Rose, J. L. Rutherford, B. Sitter, A. M. Stewart, III, M. G. Vetelino, L. Wei, *Org. Process Res. Dev.* **2005**, *9*, 440–450.
413. T. Nishiyama, M. Isobe, Y. Ichikawa, *Angew. Chem. Int. Ed.* **2005**, *44*, 4372–4375.
414. A. Jutand in *The Mizoroki-Heck Reaction* (Ed.: M. Oestreich), Wiley-VCH, **2009**, Chapter 1, pp. 1–50.
415. J. P. Knowles, A. Whiting, *Org. Biomol. Chem.* **2007**, *5*, 31–44.
416. M. J. S. Dewar, *Bull. Soc. Chim. Fr.* **1951**, *18*, C79.
417. J. Chatt, L. A. Duncanson, *J. Chem. Soc.* **1953**, 2939–2947.
418. J. Chatt, L. A. Duncanson, L. M. Venanzi, *J. Chem. Soc.* **1955**, 4456–4460.
419. W. Cabri, I. Candiani, S. DeBernardinis, F. Francalanci, S. Penco, R. Santo, *J. Org. Chem.* **1991**, *56*, 5796–5800.
420. G. P. C. M. Dekker, C. J. Elsevier, K. Vrieze, P. W. N. M. Van Leeuwen, *Organometallics* **1992**, *11*, 1598–1603.
421. W. Cabri, I. Candiani, *Acc. Chem. Res.* **1995**, *28*, 2–7.
422. K. S. A. Vallin, M. Larhed, A. Hallberg, *J. Org. Chem.* **2001**, *66*, 4340–4343.
423. P. Fristrup, S. Le Quement, D. Tanner, P.-O. Norrby, *Organometallics* **2004**, *23*, 6160–6165.

424. R. J. Deeth, A. Smith, J. M. Brown, *J. Am. Chem. Soc.* **2004**, *126*, 7144–7151.

425. J. Becica, O. D. Glaze, D. P. Hruszkewycz, G. E. Dobereiner, D. C. Leitch, *React. Chem. Eng.* **2021**, *6*, 1212–1219.

426. M. Shibasaki, E. M. Vogl, *J. Organomet. Chem.* **1999**, *576*, 1–15.

427. C. S. Shultz, J. Ledford, J. M. DeSimone, M. Brookhart, *J. Am. Chem. Soc.* **2000**, *122*, 6351–6356.

428. C. Ehm, P. H. Budzelaar, V. Busico, *J. Organomet. Chem.* **2015**, *775*, 39–49.

429. H. von Schenck, S. Strömberg, K. Zetterberg, M. Ludwig, B. Åkermark, M. Svensson, *Organometallics* **2001**, *20*, 2813–2819.

430. C. Y. Hong, N. Kado, L. E. Overman, *J. Am. Chem. Soc.* **1993**, *115*, 11028–11029.

431. K. Kashinath, G. R. Jachak, P. R. Athawale, U. K. Marelli, R. G. Gonnade, D. S. Reddy, *Org. Lett.* **2016**, *18*, 3178–3181.

432. M. P. Lisboa, D. M. Jones, G. B. Dudley, *Org. Lett.* **2013**, *15*, 886–889.

433. K. Kong, J. A. Enquist, M. E. McCallum, G. M. Smith, T. Matsumaru, E. Menhaji-Klotz, J. L. Wood, *J. Am. Chem. Soc.* **2013**, *135*, 10890–10893.

434. G. Sirasani, T. Paul, W. Dougherty, S. Kassel, R. B. Andrade, *J. Org. Chem.* **2010**, *75*, 3529–3532.

435. C. Bäcktorp, P.-O. Norrby, *Dalton Trans.* **2011**, *40*, 11308–11314.

436. L. F. Tietze, R. Schimpf, *Angew. Chem. Int. Ed. Engl.* **1994**, *33*, 1089–1091.

437. T. Jeffery, *Tetrahedron Lett.* **1999**, *40*, 1673–1676.

438. L. F. Tietze, A. Modi, *Eur. J. Org. Chem.* **2000**, 1959–1964.

439. T. Jeffery, *J. Chem. Soc. Chem. Commun.* **1984**, *19*, 1287–1289.

440. T. Jeffery, *Tetrahedron Lett.* **1985**, *26*, 2667–2670.

441. M. T. Reetz, G. Lohmer, R. Schwickardi, *Angew. Chem. Int. Ed.* **1998**, *37*, 481–483.

442. H. Yokoyama, T. Satoh, T. Furuhata, M. Miyazawa, Y. Hirai, *Synlett* **2006**, 2649–2651.

443. O. Loiseleur, P. Meier, A. Pfaltz, *Angew. Chem. Int. Ed. Engl.* **1996**, *35*, 200–202.

444. L. F. Tietze, K. Thede, F. Sannicolo, *Chem. Commun.* **1999**, 1811–1812.

445. F. Ozawa, A. Kubo, T. Hayashi, *J. Am. Chem. Soc.* **1999**, *113*, 1417–1419.

446. A. B. Machotta, B. F. Straub, M. Oestreich, *J. Am. Chem. Soc.* **2007**, *129*, 13455–13463.

447. W.-Q. Wu, Q. Peng, D.-X. Dong, X.-L. Hou, Y.-D. Wu, *J. Am. Chem. Soc.* **2008**, *130*, 9717–9725.

448. D. M. Cartney, P. J. Guiry, *Chem. Soc. Rev.* **2011**, *40*, 5122–5150.

449. C. Zhu, H. Chu, G. Li, S. Ma, J. Zhang, *J. Am. Chem. Soc.* **2019**, *141*, 19246–19251.

450. Y. Sato, M. Sodeoka, M. Shibasaki, *J. Org. Chem.* **1989**, *54*, 4738–4739.

451. M. Ozeki, M. Satake, T. Toizume, S. Fukutome, K. Arimitsu, S. Hosoi, T. Kajimoto, H. Iwasaki, N. Kojima, M. Node, M. Yamashita, *Tetrahedron* **2013**, *69*, 3841–3846.

452. T. Mizutani, S. Honzawa, S. Tosaki, M. Shibasaki, *Angew. Chem. Int. Ed.* **2002**, *41*, 4680–4682.

453. I. P. Beletskaya, A. V. Cheprakov in *The Mizoroki-Heck Reaction* (Ed.: M. Oestreich), John Wiley & Sons, **2009**, Chapter 2, pp. 51–132.

454. J. Dupont, C. S. Consorti, J. Spencer, *Chem. Rev.* **2005**, *105*, 2527–2572.

455. D. Paul, S. Das, S. Saha, H. Sharma, R. K. Goswami, *Eur. J. Org. Chem.* **2021**, 2057–2076.

456. W. Zhang, *Nat. Prod. Rep.* **2021**, *38*, 1109–1135.

457. M. H. Nguyen, M. Imanishi, T. Kurogi, A. B. Smith, III, *J. Am. Chem. Soc.* **2016**, *138*, 3675–3678.

458. D. Kurandina, P. Chuentragool, V. Gevorgyan, *Synthesis* **2019**, *51*, 985–1005.

459. K. Higuchi, K. Sawada, H. Nambu, T. Shogaki, Y. Kita, *Org. Lett.* **2003**, *5*, 3703–3704.

460. F. Glorius, *Tetrahedron Lett.* **2003**, *44*, 5751–5754.

461. G.-Z. Wang, R. Shang, W.-M. Cheng, Y. Fu, *J. Am. Chem. Soc.* **2017**, *139*, 18307–18312.

462. A.-L. Lee, *Org. Biomol. Chem.* **2016**, *14*, 5357–5366.

463. B. Karimi, H. Behzadnia, D. Elhamifar, P. F. Akhavan, F. K. Esfahani, A. Zamani, *Synthesis* **2010**, 1399–1427.
464. P. K. Chinthakindi, K. B. Govender, A. S. Kumar, H. G. Kruger, T. Govender, T. Naicker, P. I. Arvidsson, *Org. Lett.* **2017**, *19*, 480–483.
465. Y. C. Jung, R. K. Mishra, C. H. Yoon, K. W. Jung, *Org. Lett.* **2003**, *5*, 2231–2234.
466. S. Bhakta, T. Ghosh, *Adv. Synth. Catal.* **2020**, *362*, 5257–5274.
467. Y. Zheng, K. Wei, Y.-R. Yang, *Org. Lett.* **2017**, *19*, 6460–6462.
468. A. B. Dounay, P. G. Humphreys, L. E. Overman, A. D. Wrobleski, *J. Am. Chem. Soc.* **2008**, *130*, 5368–5377.
469. M. Malacria, *Chem. Rev.* **1996**, *96*, 289–306.
470. E. Negishi, C. Copéret, S. Ma, S.-Y. Liou, F. Liu, *Chem. Rev.* **1996**, *96*, 365–393.
471. E. Negishi, *Pure Appl. Chem.* **1992**, *64*, 323–334.
472. I. Marek, N. Chinkov, D. Banon-Tenne in *Metal-Catalyzed Cross-Coupling Reactions* (Eds.: A. de Meijere, F. Diederich), Wiley-VCH, **2004**, Chapter 7, pp. 395–478.
473. A. Arcadi, S. Cacchi, F. Marinelli, *Tetrahedron* **1985**, *41*, 5121–5131.
474. L. F. Tietze, A. Düfert, F. Lotz, L. Sölter, K. Oum, T. Lenzer, T. Beck, R. Herbst-Irmer, *J. Am. Chem. Soc.* **2009**, *131*, 17879–17884.
475. T. Sugihara, C. Coperet, Z. Owczarczyk, L. S. Harring, E. Negishi, *J. Am. Chem. Soc.* **1994**, *116*, 7923–7924.
476. S. D. Roughley, A. M. Jordan, *J. Med. Chem.* **2011**, *54*, 3451–3479.
477. S. V. Ley, A. W. Thomas, *Angew. Chem. Int. Ed.* **2003**, *42*, 5400–5449.
478. P. Ruiz-Castillo, S. L. Buchwald, *Chem. Rev.* **2016**, *116*, 12564–12649.
479. O. R. Taylor, P. J. Saucedo, A. Bahamonde, *J. Org. Chem.* **2024**, *89*, 16093–16105.
480. M. Marín, R. J. Rama, M. C. Nicasio, *Chem. Rec.* **2016**, *16*, 1819–1832.
481. L. Jiang, S. L. Buchwald in *Metal-Catalyzed Cross-Coupling Reactions, Palladium-Catalyzed Aromatic Carbon-Nitrogen Bond Formation* (Eds.: A. de Meijere, F. Diederich), Wiley-VCH, **2004**, Chapter 13, pp. 699–760.
482. J. F. Hartwig, K. H. Shaughnessy, S. Shekhar, R. A. Green, *Org. React.* **2020**, *100*, 853–958.
483. C. M. Lavoie, M. Stradiotto, *ACS Catal.* **2018**, *8*, 7228–7250.
484. V. Ritleng, M. Henrion, M. J. Chetcuti, *ACS Catal.* **2016**, *6*, 890–906.
485. R. M. McKinnell, U. Klein, M. S. Linsell, E. J. Moran, M. B. Nodwell, J. W. Pfeiffer, G. R. Thomas, C. Yu, J. R. Jacobsen, *Bioorg. Med. Chem. Lett.* **2014**, *24*, 2871–2876.
486. S. Ueda, S. L. Buchwald, *Angew. Chem. Int. Ed.* **2012**, *51*, 10364–10367.
487. M. R. Biscoe, T. E. Barder, S. L. Buchwald, *Angew. Chem. Int. Ed.* **2007**, *46*, 7232–7235.
488. L. M. Huffman, S. S. Stahl, *J. Am. Chem. Soc.* **2008**, *130*, 9196–9197.
489. M. Fitzner, G. Wuitschik, R. J. Koller, J.-M. Adam, T. Schindler, J.-L. Reymond, *Chem. Sci.* **2020**, *11*, 13085–13093.
490. F. Monnier, M. Taillefer, *Angew. Chem. Int. Ed.* **2009**, *48*, 6954–6971.
491. C. D. Jones, D. M. Andrews, A. J. Barker, K. Blades, K. F. Byth, M. R. V. Finlay, C. Geh, C. P. Green, M. Johannsen, M. Walker, H. M. Weir, *Bioorg. Med. Chem. Lett.* **2008**, *18*, 6486–6489.
492. G. W. Stewart, K. M. J. Brands, S. E. Brewer, C. J. Cowden, A. J. Davies, J. S. Edwards, A. W. Gibson, S. E. Hamilton, J. D. Katz, S. P. Keen, P. R. Mullens, J. P. Scott, D. J. Wallace, C. S. Wise, *Org. Process Res. Dev.* **2010**, *14*, 849–858.
493. S. D. McCann, E. C. Reichert, P. L. Arrechea, S. L. Buchwald, *J. Am. Chem. Soc.* **2020**, *142*, 15027–15037.
494. P. N. Carlsen, T. J. Mann, A. H. Hoveyda, A. J. Frontier, *Angew. Chem. Int. Ed.* **2014**, *53*, 9334–9338.
495. K. Chen, C. Risatti, M. Bultman, M. Soumeillant, J. Simpson, B. Zheng, D. Fanfair, M. Mahoney, B. Mudryk, R. J. Fox, Y. Hsiao, S. Murugesan, D. A. Conlon, F. G. Buono, M. D. Eastgate, *J. Org. Chem.* **2014**, *79*, 8757–8767.

496. F. Xu, E. Corley, M. Zacuto, D. A. Conlon, B. Pipik, G. Humphrey, J. Murry, D. Tschaen, *J. Org. Chem.* **2010**, *75*, 1343–1353.

497. R. Y. Liu, J. M. Dennis, S. L. Buchwald, *J. Am. Chem. Soc.* **2020**, *142*, 4500–4507.

498. P. M. MacQueen, M. Stradiotto, *Synlett* **2017**, *28*, 1652–1656.

499. M. Palucki, J. P. Wolfe, S. L. Buchwald, *J. Am. Chem. Soc.* **1997**, *119*, 3395–3396.

500. J. P. Stambuli, Z. Weng, C. D. Incarvito, J. F. Hartwig, *Angew. Chem. Int. Ed.* **2007**, *46*, 7674–7677.

501. P. M. MacQueen, J. P. Tassone, C. Diaz, M. Stradiotto, *J. Am. Chem. Soc.* **2018**, *140*, 5023–5027.

502. R. S. Sawatzky, B. K. V. Hargreaves, M. Stradiotto, *Eur. J. Org. Chem.* **2016**, 2444–2449.

503. H. Zhang, P. Ruiz-Castillo, A. W. Schuppe, S. L. Buchwald, *Org. Lett.* **2020**, *14*, 5369–5374.

504. S. Enthaler, A. Company, *Chem. Soc. Rev.* **2011**, *40*, 4912–4924.

505. I. P. Beletskaya, V. P. Ananikov, *Chem. Rev.* **2011**, *111*, 1596–1636.

506. I. P. Beletskaya, V. P. Ananikov, *Chem. Rev.* **2022**, *122*, 16110–16293.

507. D. L. Hughes, *Org. Process Res. Dev.* **2017**, *21*, 430–443.

508. M. J. Strauss, M. E. Greaves, S.-T. Kim, C. N. Teijaro, M. A. Schmidt, P. M. Scola, S. L. Buchwald, *Angew. Chem. Int. Ed.* **2024**, *63*, e202400333.

509. D. Maiti, S. L. Buchwald, *J. Am. Chem. Soc.* **2009**, *131*, 17423–17429.

510. C. Chen, M. Weisel, *Synlett* **2013**, *24*, 189–192.

511. F. Ni, J. Li, *Synthesis* **2012**, *44*, 3598–3602.

512. A. M. Hyde, Z. Liu, B. Kosjek, L. Tan, A. Klapars, E. R. Ashley, Y.-L. Zhong, O. Alvizo, N. J. Agard, G. Liu, X. Gu, N. Yasuda, J. Limanto, M. A. Huffman, D. M. Tschaen, *Org. Lett.* **2016**, *18*, 5888–5891.

513. T. Kondo, T. Mitsudo, *Chem. Rev.* **2000**, *100*, 3205–3220.

514. M. Murata, S. L. Buchwald, *Tetrahedron* **2004**, *60*, 7397–7403.

515. M. A. Fernández-Rodríguez, Q. Shen, J. F. Hartwig, *J. Am. Chem. Soc.* **2006**, *128*, 2180–2181.

516. C. W. Cheung, S. L. Buchwald, *J. Org. Chem.* **2014**, *79*, 5351–5358.

517. J. Pan, X. Wang, Y. Zhang, S. L. Buchwald, *Org. Lett.* **2011**, *13*, 4974–4976.

518. D. A. Petrone, J. Ye, M. Lautens, *Chem. Rev.* **2016**, *116*, 8003–8104.

519. S. Caron, *Org. Process Res. Dev.* **2020**, *24*, 470–480.

520. V. V. Grushin, W. J. Marshall, *Organometallics* **2007**, *26*, 4997–5002.

521. A. C. Sather, S. L. Buchwald, *Acc. Chem. Res.* **2016**, *49*, 2146–2157.

522. Y. Ye, S.-T. Kim, R. P. King, M.-H. Baik, S. L. Buchwald, *Angew. Chem. Int. Ed.* **2023**, *62*, e202300109.

523. A. Sather, H. G. Lee, V. Y. De La Rosa, Y. Yang, P. Müller, S. L. Buchwald, *J. Am. Chem. Soc.* **2015**, *137*, 13433–13438.

524. P. J. Milner, Y. Yang, S. L. Buchwald, *Organometallics* **2015**, *34*, 4775–4780.

525. A. Vijayan, D. N. Rao, K. V. Radhakrishnan, P. Y. S. Lam, P. Das, *Synthesis* **2021**, *53*, 805–847.

526. J. X. Qiao, P. Y. S. Lam, *Synthesis* **2011**, 829–856.

527. K. Arrington, G. A. Barcan, N. A. Calandra, G. A. Erickson, L. Li, L. Liu, M. G. Nilson, I. I. Strambeanu, K. F. VanGelder, J. L. Woodard, S. Xie, C. L. Allen, J. A. Kowalski, D. C. Leitch, *J. Org. Chem.* **2019**, *84*, 4680–4694.

528. J. C. Vantourout, H. N. Miras, A. Isidro-Llobet, S. Sproules, A. J. B. Watson, *J. Am. Chem. Soc.* **2017**, *139*, 4769–4779.

529. D. A. Evans, J. L. Katz, G. S. Peterson, T. Hintermann, *J. Am. Chem. Soc.* **2001**, *123*, 12411–12413.

530. A. E. King, B. L. Ryland, T. C. Brunold, S. S. Stahl, *Organometallics* **2012**, *31*, 7948–7957.

531. S. Bose, S. Dutta, D. Koley, *ACS Catal.* **2022**, *12*, 146.

532. B. M. Trost, M. L. Crawley, *Chem. Rev.* **2003**, *103*, 2921–2043.

533. B. M. Trost, J. E. Schultz, *Synthesis* **2019**, *51*, 1–30.

534. B. M. Trost, *Org. Process Res. Dev.* **2012**, *16*, 185–194.

535. Q. Cheng, H.-F. Tu, C. Zheng, J.-P. Qu, G. Helmchen, S.-L. You, *Chem. Rev.* **2019**, *119*, 1855–1969.

536. O. Belda, C. Moberg, *Acc. Chem. Res.* **2004**, *37*, 159–167.

537. G. Helmchen, U. Kazmaier, S. Förster in *Catalytic Asymmetric Synthesis* (Ed.: I. Ojima), John Wiley & Sons, 3rd ed., **2010**, Chapter 8B, pp. 497–641.

538. J. C. Hethcox, S. E. Shockley, B. M. Stoltz, *ACS Catal.* **2016**, *6*, 6207–6213.

539. F. Panahi, F. Bauer, B. Breit, *Acc. Chem. Res.* **2023**, *56*, 3676–3693.

540. L. Süsse, B. M. Stoltz, *Chem. Rev.* **2021**, *121*, 4084–4099.

541. C. G. Frost, J. Howarth, J. M. J. Williams, *Tetrahedron: Asymmetry* **1992**, *3*, 1089–1122.

542. S. D. Knight, L. E. Overman, G. Pairaudeau, *J. Am. Chem. Soc.* **1995**, *117*, 5776–5788.

543. Y. Kaburagi, H. Tokuyama, T. Fukuyama, *J. Am. Chem. Soc.* **2004**, *126*, 12046–12047.

544. F. Richard, P. Clark, A. Hannam, T. Keenan, A. Jean, S. Arseniyadis, *Chem. Soc. Rev.* **2024**, *53*, 1936–1983.

545. J. Ansell, M. Wills, *Chem. Soc. Rev.* **2002**, *31*, 259–268.

546. B. Goldfuss, U. Kazmaier, *Tetrahedron* **2000**, *56*, 6493–6496.

547. R. Prétôt, A. Pfaltz, *Angew. Chem. Int. Ed.* **1998**, *37*, 323–325.

548. U. Kazmaier, M. Pohlmann in *Metal-Catalyzed Cross-Coupling Reactions* (Eds.: A. de Meijere, F. Diederich), Wiley-VCH, **2004**, Chapter 9, pp. 531–583.

549. J. Tsuji, *Tetrahedron* **1986**, *42*, 4361–4401.

550. Y. Tanigawa, K. Nishimura, A. Kawasaki, S.-I. Murahashi, *Tetrahedron Lett.* **1982**, *23*, 5549–5552.

551. Z. Lu, S. Ma, *Angew. Chem. Int. Ed.* **2008**, *47*, 258–297.

552. D. E. White, I. C. Stewart, R. H. Grubbs, B. M. Stoltz, *J. Am. Chem. Soc.* **2008**, *130*, 810–811.

553. B. M. Trost, G. Dong, J. A. Vance, *Chem. Eur. J.* **2010**, *16*, 6265–6277.

554. A. Farwick, G. Helmchen, *Org. Lett.* **2010**, *12*, 1108–1111.

555. B. M. Trost, M. Osipov, S. Krüger, Y. Zhang, *Chem. Sci.* **2015**, *6*, 349–353.

556. B. Breit, Y. Schmidt, *Chem. Rev.* **2008**, *108*, 2928–2951.

557. H. L. Goering, V. D. Singleton, *J. Org. Chem.* **1983**, *48*, 1531–1533.

558. H. L. Goering, S. S. Kantner, *J. Org. Chem.* **1984**, *49*, 422–426.

559. C. Gallina, *Tetrahedron Lett.* **1982**, *23*, 3093–3096.

560. N. Yoshikai, S.-L. Zhang, E. Nakamura, *J. Am. Chem. Soc.* **2008**, *130*, 12862–12863.

561. N. Yoshikai, E. Nakamura, *Chem. Rev.* **2012**, *112*, 2339–2372.

562. E. R. Bartholomew, S. H. Bertz, S. Cope, M. Murphy, C. A. Ogle, *J. Am. Chem. Soc.* **2008**, *130*, 11244–11245.

563. D. G. Gillingham, A. H. Hoveyda, *Angew. Chem. Int. Ed.* **2007**, *46*, 3860–3864.

564. J. Skotnitzki, L. Spessert, P. Knochel, *Angew. Chem. Int. Ed.* **2019**, *58*, 1509–1514.

565. Y. Zou, X. Li, Y. Yang, S. Berritt, J. Melvin, S. Gonzales, M. Spafford, A. B. Smith, III, *J. Am. Chem. Soc.* **2018**, *140*, 9502–9511.

566. A. Minatti, K. Muniz, *Chem. Soc. Rev.* **2007**, *36*, 1142–1152.

567. R. A. Fernandes, A. K. Jha, P. Kumar, *Catal. Sci. Technol.* **2020**, *10*, 7448–7470.

568. O. N. Temkin, *Kinet. Catal.* **2020**, *61*, 663–720.

569. T. Punniyamurthy, S. Velusamy, J. Iqbal, *Chem. Rev.* **2005**, *105*, 2329–2363.

570. J. A. Keith, P. M. Henry, *Angew. Chem. Int. Ed.* **2009**, *48*, 9038–9049.

571. J. Muzart, *Tetrahedron* **2007**, *63*, 7505–7521.

572. K. R. Holman, A. M. Stanko, S. E. Reisman, *Chem. Soc. Rev.* **2021**, *50*, 7891–7908.

573. P. D. O'Connor, L. N. Mander, M. M. W. McLachlan, *Org. Lett.* **2004**, *6*, 703–706.

574. J.-M. Pereillo, M. Maftouh, A. Andrieu, M.-F. Uzabiaga, O. Fedeli, P. Savi, M. Pascal, J.-M. Herbert, J.-P. Maffrand, C. Picard, *Drug Metab. Dispos.* **2002**, *30*, 1288–1295.

575. T. Cernak, K. D. Dykstra, S. Tyagarajan, P. Vachal, S. W. Krska, *Chem. Soc. Rev.* **2016**, *45*, 546–576.

576. J. Yamaguchi, A. D. Yamaguchi, K. Itami, *Angew. Chem. Int. Ed.* **2012**, *51*, 8960–9009.

577. D. J. Abrams, P. A. Provencher, E. J. Sorensen, *Chem. Soc. Rev.* **2018**, *47*, 8925–8967.

578. R. de Jesus, K. Hiesinger, M. van Gemmeren, *Angew. Chem. Int. Ed.* **2023**, *62*, e202306659.

579. J. H. Docherty, T. M. Lister, G. Mcarthur, M. T. Findlay, P. Domingo-Legarda, J. Kenyon, S. Choudhary, I. Larrosa, *Chem. Rev.* **2023**, *123*, 7692–7760.

580. F. He, Y. Bo, J. D. Altom, E. J. Corey, *J. Am. Chem. Soc.* **1999**, *121*, 6771–6772.

581. S. A. Reed, M. C. White, *J. Am. Chem. Soc.* **2008**, *130*, 3316–3318.

582. A. W. G. Burgett, Q. Li, Q. Wei, P. G. Harran, *Angew. Chem. Int. Ed.* **2003**, *42*, 4961–4966.

583. T. Rogge, N. Kaplaneris, N. Chatani, J. Kim, S. Chang, B. Punji, L. L. Schafer, D. G. Musaev, J. Wencel-Delord, C. A. Roberts, R. Sarpong, Z. E. Wilson, M. A. Brimble, M. J. Johansson, L. Ackermann, *Nat. Rev. Methods Primers* **2021**, *1*, 43.

584. L. Ackermann, *Chem. Rev.* **2011**, *111*, 1315–1345.

585. I. Funes-Ardoiz, F. Maseras, *ACS Catal.* **2018**, *8*, 1161–1172.

586. D. Alberico, M. E. Scott, M. Lautens, *Chem. Rev.* **2007**, *107*, 174–238.

587. S. I. Gorelsky, D. Lapointe, K. Fagnou, *J. Am. Chem. Soc.* **2008**, *130*, 10848–10849.

588. P. Amadeo, B. Bhaskararao, Y.-F. Yang, M. C. Kozlowski, *Organometallics* **2021**, *40*, 2290–2294.

589. B. P. Carrow, J. Sampson, L. Wang, *Isr. J. Chem.* **2020**, *60*, 230–258.

590. L. Wang, B. P. Carrow, *ACS Catal.* **2019**, *9*, 6821–6836.

591. J. J. Topczewski, M. S. Sanford, *Chem. Sci.* **2015**, *6*, 70–76.

592. D. J. Cárdenas, B. Martín-Matute, A. M. Echavarren, *J. Am. Chem. Soc.* **2006**, *128*, 5033–5040.

593. H. M. L. Davies, D. Morton, *Chem. Soc. Rev.* **2011**, *40*, 1857–1869.

594. E. Nakamura, N. Yoshikai, M. Yamanaka, *J. Am. Chem. Soc.* **2002**, *124*, 7181–7192.

595. J. L. Roizen, M. E. Harvey, J. Du Bois, *Acc. Chem. Res.* **2012**, *45*, 911–922.

596. M.-L. Louillat, F. W. Patureau, *Chem. Soc. Rev.* **2014**, *43*, 901–910.

597. H.-X. Dai, A. F. Stepan, M. S. Plummer, Y.-H. Zhang, J.-Q. Yu, *J. Am. Chem. Soc.* **2011**, *133*, 7222–7228.

598. J. Bien, A. Davulcu, A. DelMonte, K. J. Fraunhoffer, Z. Gao, C. Hang, Y. Hsiao, W. Hu, K. Katipally, A. Littke, A. Pedro, Y. Qiu, M. Sandoval, R. Schild, M. Soltani, A. Tedesco, D. Vanyo, P. Vemishetti, R. E. Waltermire, *Org. Process Res. Dev.* **2018**, *22*, 1393–1408.

599. R. N. Bream, H. Clark, D. Edney, A. Harsanyi, J. Hayler, A. Ironmonger, N. Mc Cleary, N. Phillips, C. Priestley, A. Roberts, P. Rushworth, P. Szeto, M. R. Webb, K. Wheelhouse, *Org. Process Res. Dev.* **2021**, *25*, 529–540.

600. C. Sambiagio, D. Schönbauer, R. Blieck, T. Dao-Huy, G. Pototschnig, P. Schaaf, T. Wiesinger, M. F. Zia, J. Wencel-Delord, T. Besset, B. U. W. Maes, M. Schnürch, *Chem. Soc. Rev.* **2018**, *47*, 6603–6743.

601. R. Rossi, F. Bellina, M. Lessi, C. Manzinia, *Adv. Synth. Catal.* **2014**, *356*, 17–117.

602. K. R. Campos, *Chem. Soc. Rev.* **2007**, *36*, 1069–1084.

603. J. F. Hartwig, *Chem. Soc. Rev.* **2011**, *40*, 1992–2002.

604. M. E. McCallum, C. M. Rasik, J. L. Wood, M. K. Brown, *J. Am. Chem. Soc.* **2016**, *138*, 2437–2442.

605. D. Lapointe, T. Markiewicz, C. J. Whipp, A. Toderian, K. Fagnou, *J. Org. Chem.* **2011**, *76*, 749–759.

606. Y. Akita, Y. Itagaki, S. Takizawa, A. Ohta, *Chem. Bull. Pharm.* **1989**, *37*, 1477–1480.

607. S. K. Sinha, S. Guin, S. Maiti, J. P. Biswas, S. Porey, D. Maiti, *Chem. Rev.* **2022**, *122*, 5682–5841.

608. K. M. Engle, T.-S. Mei, M. Wasa, J.-Q. Yu, *Acc. Chem. Res.* **2012**, *45*, 788–802.

609. M. P. Doyle, R. Duffy, M. Ratnikov, L. Zhou, *Chem. Rev.* **2010**, *110*, 704–724.

610. A. Padwa, D. J. Austin, A. T. Price, M. A. Semones, M. P. Doyle, M. N. Protopopova, W. R. Winchester, A. Tran, *J. Am. Chem. Soc.* **1993**, *115*, 8669–8680.
611. H. M. L. Davies, T. Hansen, M. R. Churchill, *J. Am. Chem. Soc.* **2000**, *122*, 3063–3070.
612. D. Y.-K. Chen, S. W. Youn, *Chem. Eur. J.* **2012**, *18*, 9452–9474.
613. W. R. Gutekunst, P. S. Baran, *Chem. Soc. Rev.* **2011**, *40*, 1976–1991.
614. W. R. Gutekunst, P. S. Baran, *J. Am. Chem. Soc.* **2011**, *133*, 19076–19079.
615. A. L. Bowie, D. Trauner, *J. Org. Chem.* **2009**, *74*, 1581–1586.
616. N. K. Garg, D. D. Caspi, B. M. Stoltz, *J. Am. Chem. Soc.* **2004**, *126*, 9552–9553.
617. Q. Ye, P. Qu, S. A. Snyder, *J. Am. Chem. Soc.* **2017**, *139*, 18428–18431.
618. D. S. Peters, F. E. Romesberg, P. S. Baran, *J. Am. Chem. Soc.* **2018**, *140*, 2072–2075.
619. H. M. L. Davies, A. M. Walji, R. J. Townsend, *Tetrahedron Lett.* **2002**, *43*, 4981–4983.
620. A. Hinman, J. Du Bois, *J. Am. Chem. Soc.* **2003**, *125*, 11510–11511.
621. P. M. Wehn, J. Du Bois, *J. Am. Chem. Soc.* **2002**, *124*, 12950–12951.
622. H. Kodama, T. Katsuhira, T. Nishida, T. Hino, K. Tsubata, *EP 1277726 B1* **2007**.
623. L. Ackermann, R. Vicente, A. R. Kapdi, *Angew. Chem. Int. Ed.* **2009**, *48*, 9792–9826.
624. F. Collet, R. H. Dodd, P. Dauban, *Chem. Commun.* **2009**, 5061–5074.
625. X. Chen, K. M. Engle, D.-H. Wang, J.-Q. Yu, *Angew. Chem. Int. Ed.* **2009**, *48*, 5094–5115.
626. I. A. I. Mkhalid, J. H. Barnard, T. B. Marder, J. M. Murphy, J. F. Hartwig, *Chem. Rev.* **2010**, *110*, 890–931.
627. P. Herrmann, T. Bach, *Chem. Soc. Rev.* **2011**, *40*, 2022–2038.

Metathesis

7

The metathesis reaction represents a fundamental tool for the formation of CC-multiple bonds. With the exception of palladium-catalyzed cross-coupling reactions, no other class of transformations has arguably had such a profound impact on how chemists approach carbon–carbon bond formation since the last decade of the twentieth century, with repercussions for the synthesis of polymers as well as the synthesis of natural products (see Fig. 7.1).

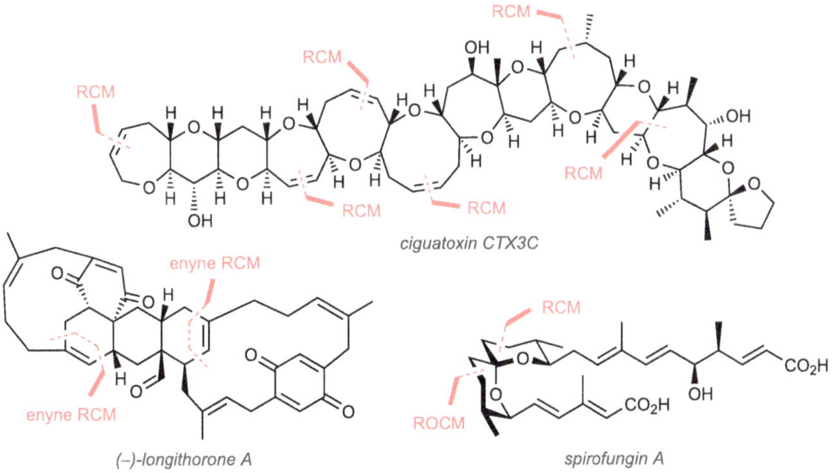

Fig. 7.1 Natural products synthesized via metathesis [1–3]

The name "metathesis" is derived from the Greek *μετάθεσις* (*change of position*), as the reaction involves a metal-catalyzed rearrangement of CC-multiple bonds by cleavage

A. Düfert, *Methods of Organic Synthesis*,
https://doi.org/10.1007/978-3-662-70963-4_7

and subsequent recombination [4]. The process was serendipitously discovered in a study on the Ziegler–Natta polymerization with alternative metal catalysts. It was investigated and expanded in the groups of R. Grubbs and R. Schrock, which subsequently received the Nobel Prize in Chemistry in 2005 in honor of their contributions [5–8]. Apart from the transmetalation of organometallic compounds, which is sometimes also referred to as metathesis, the transformations discussed hereafter comprise three subcategories (see Fig. 7.2).

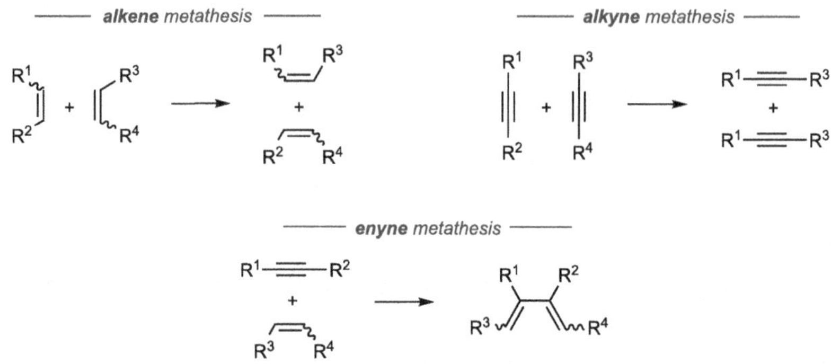

Fig. 7.2 Reaction classes in the metathesis of alkenes and alkynes. See Sect. 7.1.2 for nomenclature of metatheses

The alkene metathesis by far constitutes the most commonly employed variant. The alkyne and especially the enyne metathesis are used to a much lesser extent, which may be due to the fact that the resulting structural elements are not encountered as frequently in intermediates or target molecules: alkenes are simply one of the most commonly utilized platform functionalities (see Sect. 3.3).

7.1 Alkene Metathesis

In alkene metathesis, two olefins react with each other, affording again two, albeit differently substituted, olefins (see Fig. 7.2). The conditions under which metatheses are run require neither bases nor acids. Furthermore, most of these reactions already take place at room temperature. The products are usually obtained as E/Z mixtures. In the case of small rings, however, the reaction yields only the Z-configured product due to the geometric impossibility of E-olefins with small ring sizes. Apart from the extremely mild conditions (neutral medium, room temperature), the tolerance of functional groups is an important issue and the individual catalysts can differ greatly in this respect. (Weakly) polar aprotic solvents (alkanes, benzene/toluene, CH_2Cl_2) are usually used as solvents.

Titanium- and tungsten-based catalysts typically react stoichiometrically with ketones to form olefins. As these reagents are commonly being utilized for the construction of alkenes (e.g., Tebbe reagent), their high oxophilicity disqualifies them for most metathesis processes. In contrast, molybdenum systems are significantly more reactive towards olefins, yet a residual reactivity towards aldehydes, polar and acidic functional groups persists. The ruthenium catalysts introduced by Grubbs can tolerate a broader scope of functional groups, even though modern W and Mo catalysts are slowly closing this gap [9–11]. While carbene complexes based on other transition metals can also be generated, they show a markedly divergent reactivity: Fe, Co, and Rh alkylidenes primarily tend to cyclopropanate olefins. Os- and Ir-based systems do show some activity in olefin metathesis, yet in addition to their lower reactivity, they are also significantly more expensive than the more common metals Ru and Mo [12]. The advantage of molybdenum systems compared to Ru lies in their partly enhanced reactivity as well as their tolerance of some complementary functionalities such as vinyl ethers, S- and P-containing groups, and tertiary amines [13]. Since Ru(2+) (according to HSAB) is a rather soft Lewis acid, ruthenium-based catalysts are often susceptible to sulfur- and phosphorus-containing alkenes (see Fig. 7.3) [13, 14].[1]

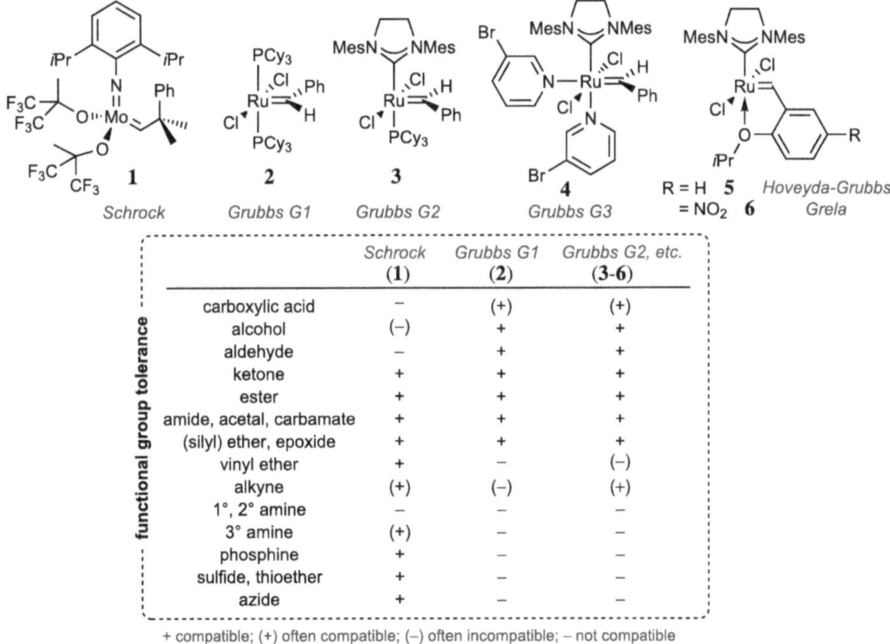

	Schrock (1)	Grubbs G1 (2)	Grubbs G2, etc. (3-6)
carboxylic acid	–	(+)	(+)
alcohol	(–)	+	+
aldehyde	–	+	+
ketone	+	+	+
ester	+	+	+
amide, acetal, carbamate	+	+	+
(silyl) ether, epoxide	+	+	+
vinyl ether	+	–	(–)
alkyne	(+)	(–)	(+)
1°, 2° amine	–	–	–
3° amine	(+)	–	–
phosphine	+	–	–
sulfide, thioether	+	–	–
azide	+	–	–

+ compatible; (+) often compatible; (–) often incompatible; – not compatible

Fig. 7.3 Standard catalysts for alkene metathesis and their substrate scope [13–16]

---

[1] The depiction of the bonding situation in the NHC of the Grubbs catalysts often differs in the literature from that chosen here. Since the NHC is planar (see *Organometallics* **2005**, *24*, 1477–1482), it was adopted hereafter based on said bonding geometry to avoid misunderstandings.

The reactivity of the Schrock catalysts is influenced by all attached ligands, as they do not dissociate from the tetrahedral-coordinated metal center during the catalytic cycle [17]. The Grubbs-type catalysts, on the other hand, mechanistically require the dissociation of a ligand, which is why the Ru catalysts of the second and third generations as well as the Hoveyda–Grubbs and Grela catalyst primarily differ in their activation behavior, but not in their tolerance towards functional groups. The axial carbene ligand is identical in **3-6** [9].

In direct comparison, the second-generation Grubbs catalyst (**3**) is approximately 100 times more reactive than its predecessor (**2**) [18]. The donor properties of the carbene increase the efficiency of the catalyst, either by enhancing the affinity of the metal center towards olefins or by stabilizing reactive carbene ligand conformers versus unproductive ones [19, 20]. The catalyst **5** developed by Hoveyda and Blechert is a phosphine-free variant of **3**. In ring-closing metathesis, it is better suited for the construction of trisubstituted double bonds than the catalyst developed by Grubbs (**3**). Furthermore, highly electron-deficient alkenes can be converted very efficiently. **3** and **5** display a slower initiation rate compared to **2**, which limits their use for polymer synthesis. The third generation of Grubbs' alkylidene catalysts (**4**) relies on the dissociation of hemilabile pyridines, which shortens the initiation phase compared to **2** by up to 10^6 [21]. With the nitro-substituted Grela catalyst **6**, a profoundly accelerated generation of the active species is similarly observed: The acceptor substituent enhances the withdrawal of electron density from the isopropoxy group and thus weakens the O-Ru bond. As a result, bond breaking and accordingly catalyst activation is promoted [22]. The Schrock system **1** is somewhat more reactive than **3** and **5** for some substrates [13]; the use of **1** therefore may afford better yields. A major disadvantage of the Mo catalyst lies in its sensitivity to water and oxygen, which complicates handling in the laboratory. The development of air-stable Mo-alkylidene precatalysts was in turn achieved by coordinatively saturating the metal center with bipyridine, which can be reversibly removed with $ZnCl_2$. Although the active catalyst is still inherently water-sensitive, handling before activation is significantly simplified [10]. However, new investigations have shown that even the Ru alkylidenes considered "water-tolerant" can be impaired by the presence of small amounts of water [23].

From the kinetic studies on different substrates (mono-/di-/tri-/tetrasubstituted, electron-poor or electron-rich) and the partially complementary tolerance towards various functionalities, it becomes clear that there is no single metathesis catalyst that performs exceptionally with every class of substrates. Although the second-generation Grubbs catalyst provides an excellent starting point for a routine transformation, it is worthwhile to test a range of different catalysts in more difficult cases to find the optimal one [22].

7.1.1 Mechanism

The basic features of the generally accepted mechanism of alkene metathesis were postulated as early as 1971 by Chauvin and Hérisson [24]. The reaction proceeds via metallacyclobutane intermediates, which are formed by coordination and subsequent [2+2]-cycloaddition of an olefin to a metal alkylidene. The olefinic substitutents in the substrates and the product are exchanged via a successive bond scission-recombination sequence (see Fig. 7.4) [25].

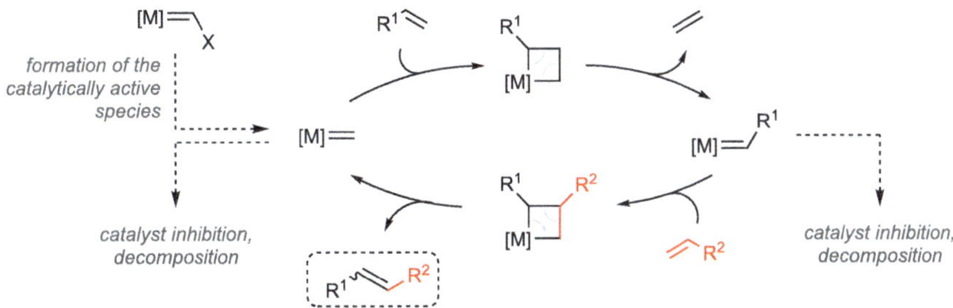

Fig. 7.4 Catalytic cycle of alkene metathesis [25]

Due to the reversibility of all steps in the catalysis cycle, a statistical mixture of all possible products is obtained, the composition of which is dictated by the position of the thermodynamic equilibrium. For a metathesis to preferentially yield only one product, the equilibrium must be shifted in one direction. Apart from the different metathesis products, an *E*/*Z* mixture of each alkene is additionally generated during the reaction, whose ratio can be dramatically influenced by the substrate structure and the properties of the catalyst [9].

The individual steps of the Chauvin mechanism have been thoroughly investigated, particularly for Ru-based catalysts (see Fig. 7.5) [9, 25, 26]. Apart from the initiation, the steps for the corresponding Schrock catalysts proceed analogously [27]. Following the dissociation of a phosphine in **7**, a 14-valence electron species **8** is initially formed [28], whose free coordination site can be occupied by an olefin. After the ligation of the alkene, which presumably coordinates *trans* to the ancillary ligand L, the two square-pyramidal distorted complexes **9** and **10** can result [29]. The formation of the metallacyclobutane (**11, 12**) occurs stereospecifically, with the steric and to a lesser extent also the electronic environment of the metal center (ligands, substrates) defining the dominant reaction path to give **11** or **12** [30]. The Ru-alkylidene finally eliminates the product **13** and can reenter the catalytic cycle. In the elimination step, intermediate **11** is stereospecifically converted to the *E*-alkene and **12** to the *Z*-alkene. The observed diastereoselectivity of the reaction thus roughly reflects the relative contributions of **11** or **12** to the overall reaction. While the dissociative mechanism dominates in about 95% of the reactions, in the presence of an excess of phosphine, an associative pathway can be followed instead [26]. The shown trigonal-bipyramidal geometry of Ru complexes **11, 12** seems to be critical for product formation with the corresponding Mo- and W-based catalysts. A theoretical study indicated that alkene coordination occurs *trans* to the strongly donating imido ligand, whereby the metallacyclobutane can assume two conformations. The trigonal-bipyramidal conformer yields a flat metallacyclobutane, which possesses a carbon-centered, low-lying, vacant p-orbital that rapidly affords the retro-cycloaddition products. The more stable square-pyramidal conformer shows an angled structure lacking a low-lying empty orbital in the ring plane and corresponds to an off-cycle reaction intermediate, susceptible to further catalyst deactivation [31]. With Ru-based catalyst, the coordination of the alkene can also occur equatorially (**14**), resulting in *cis*-metallacyclobutanes (**15**) relative to the ligand L, whose products also occur as *E/Z* mixtures. While theoretical studies disclosed that *trans*-coordination is preferred, both *cis*- (**14**) and *trans*-complexes (**9, 10**) have been isolated and characterized crystallographically [9, 19].

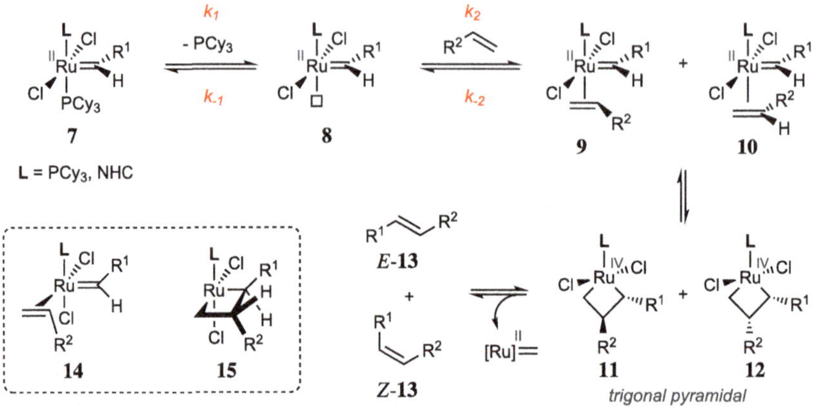

Fig. 7.5 Detailed dissociative mechanism of alkene metathesis for **2, 3** [19, 25]

The initiation in the first- and second-generation catalysts (**2, 3**) proceed dissociatively, i.e., with the loss of the phosphine prior to the coordination of the alkene [25]. There are various explanations for the observed higher reactivity of the second-generation Grubbs catalyst (**3**) compared to the first generation (**2**) [9]. Presumably the effect is not caused by a more efficient phosphine dissociation (k_1, cf. Fig. 7.5), which is actually slower for **3** and **5**. Instead, the 14-valence electron species **8** has a drastically higher binding affinity for alkenes compared to phosphines ($k_2 \gg k_{-1}$) [32]. Alternatively, the strong σ-backbonding of the NHC ligands was used to rationalize a significant reduction in the activation barrier for the formation of metallacyclobutane (**9/10**→**11/12**) based on theoretical studies [19, 20].

With the phosphine-free Ru complexes, the initiation step can vary: The pyridine-coordinated 3rd generation Grubbs catalyst (**4**) first dissociates a pyridine ligand, the second pyridine is subsequently displaced by the alkene in a concerted ligand exchange. The Hoveyda–Grubbs and Grela catalysts (**5, 6**) behave similar to the first- and second-generation Grubbs catalysts and follow a dissociative path with the initial breaking of the isopropoxy-Ru bond. However, a concerted ligand exchange also seems to contribute to the activation of the precatalyst [33, 34]. In contrast to ruthenium, the ligands on molybdenum catalysts do not dissociate during the catalytic cycle, so all ligands influence the properties of the catalytically active species [17]. The presence of two ancillary ligands with different donicity (exempting the imido ligand) in Schrock-type catalysts was actually shown to promote the distortion of the coordination geometry. These complexes can more easily open a coordination site for alkene ligation, thus increasing the reaction rate. One example of such complexes are the monoaryloxide pyrrolide (MAP) complexes (see Sect. 7.1.3) [29].

An incompatibility of the catalysts with certain functional groups is usually caused by the formation of stable intermediates, either for electronic reasons or by affording chelate complexes. This energetic sink reduces the concentration of active catalyst and thus the reaction rate. With Ru alkylidenes, donor substituents (alkoxy, halogen) are particularly problematic as they stabilize the electron-poor metal center in the Fischer carbenes. Alternatively, a decomposition of the catalyst can occur. Especially metal methylidenes, which arise as intermediates in the conversion of terminal alkenes, are susceptible to degradation processes, presumably via a bimolecular pathway (**16**) or β-hydride elimination that is further accelerated by the presence of water (**17**, see Fig. 7.6) [25, 35–37].

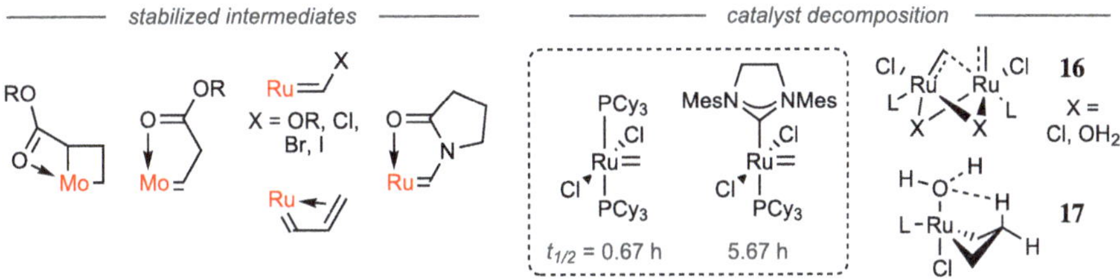

Fig. 7.6 (Meta)stable metal alkylidenes and decomposition of Ru-methylidenes [35, 38–40]

7.1.2 Modes of Alkene Metathesis

Three main variants of the reaction exist, which are derived from the desired products: The ring-closing (RCM), ring-opening (ROM) and cross metathesis (CM). The ring-opening metathesis is primarily used for the formation of polymers from monomeric, highly strained cyclic alkenes, whose conversion is referred to as ring-opening metathesis polymerization (ROMP, see Fig. 7.7) [9].

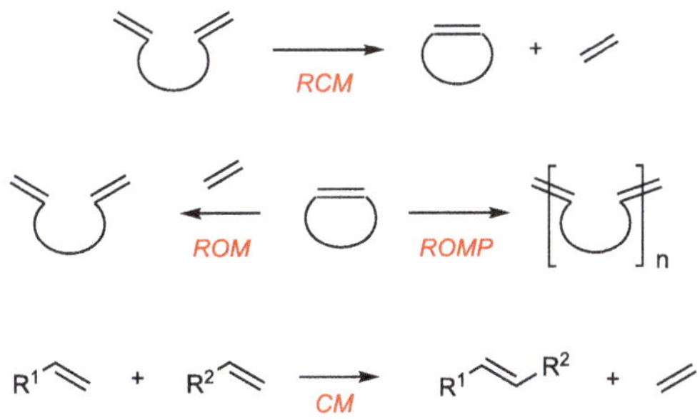

Fig. 7.7 Modes of alkene metathesis [9]

Ring-closing metathesis is entropically favored, as the conversion of the substrate results in the formation of two products. Since one of the target molecules is usually a volatile or even gaseous alkene, the reaction proceeds irreversibly to give the cyclic product if the volatile component can degas from the solution. These supporting effects do not apply to cross metathesis, as there is neither a driving force through the opening of strained rings nor an entropic gain in product formation. For this reason, a selectivity beyond a statistical distribution can only be afforded if the rate of homodimerization of the two reactants differs significantly. The susceptibility of the respective homodimers towards secondary metathesis reactions also plays a role. The driving force in ring-opening metathesis is attributed to the release of ring strain, which also renders this process irreversible [9, 41].

In *ring-closing metathesis* (RCM), cyclizations to normal and large rings compete with oligomerizations. Despite the entropic advantage of forming two products from one reactant, the reaction to medium rings remains challenging.[II] In general, high dilution and the use of a large amount of catalyst favor cyclization over oligomerization [42]. Long-chain α, ω-dienes typically afford the macrocyclic product only in low yield if no preorganization of the linear substrate is ensured. Furthermore, control of the *E/Z* selectivity in macrocyclic products remains problematic with standard catalysts. Especially crown ethers can easily be complexed by choosing a suitable counterion (**20**), whereby the product can be obtained in high yield and with the desired *E/Z* selectivity (see Fig. 7.8) [43].

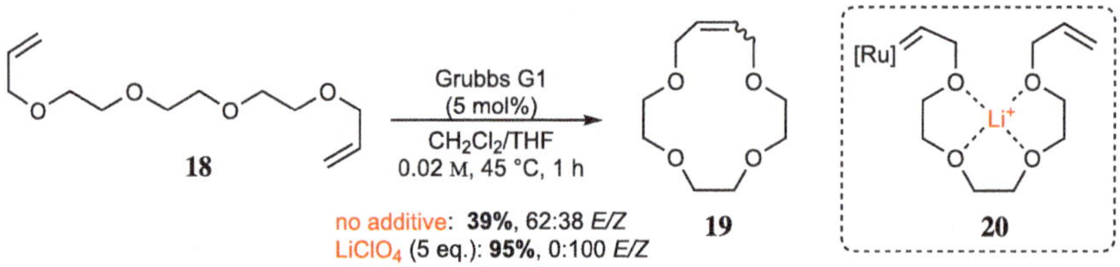

Fig. 7.8 Highly selective synthesis of cyclic crown ethers by template-driven metathesis [43]

The choice of protecting groups on the substrate, if they cannot be avoided, can also influence the *E/Z* ratio of the product to a certain extent, in addition to the yield [42]. For example, by introducing a protecting group at the phenolic OH in Fürstner's synthesis of the cytotoxic natural product (−)-salicylihalamide, the stereoselectivity to the desired *E* isomer could be reversed. It was hypothesized that the unprotected hydroxy functionality restricts the free rotation of the ester group via a hydrogen bond, thus enforcing the formation of the *Z* isomer (see Fig. 7.9) [44].

[II] The cyclization of bifunctional molecules becomes successively unfavorable from an entropic standpoint with increasing chain length, as the number of reactive conformations increases and the probability of the ends meeting diminishes. However, the reaction rates increases for large rings compared to medium rings, as enthalpic effects dominate (loss of strain in the transition state). See *Acc. Chem. Res.* **1981**, *14*, 95–102 and *Angew. Chem. Int. Ed.* **2000**, *39*, 2073–2077.

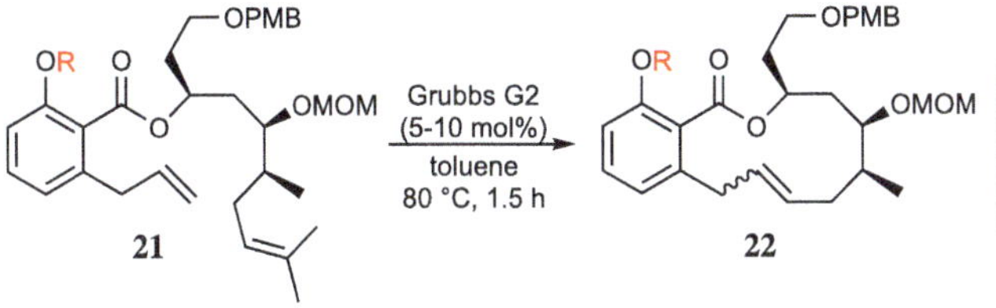

R	yield [%]	E/Z
H	69	0 : 100
TBS	91	40 : 60
Me	93	66 : 34
MOM	91	68 : 32

Fig. 7.9 Synthesis of the key intermediate **22** in Fürstner's approach to (–)-salicylihalamide [44]

If a double bond is unable to form a metal alkylidene with the catalyst due to its steric or electronic properties, the desired product can still be obtained in a domino process by elongating the alkene and attaching a terminal double bond [45]. In this so-called *relay* approach, the metal alkylidene initially formed with the terminal olefin (**24**). The appendage was then cleaved off by eliminating a cyclic alkene, and the catalyst was transferred to the targeted alkene unit (**25**). The concluding metathesis step ultimately yielded the truncated, tetrasubstituted alkene **26**, which is normally not accessible with first-generation Grubbs catalysts (**2**) (see Fig. 7.10).

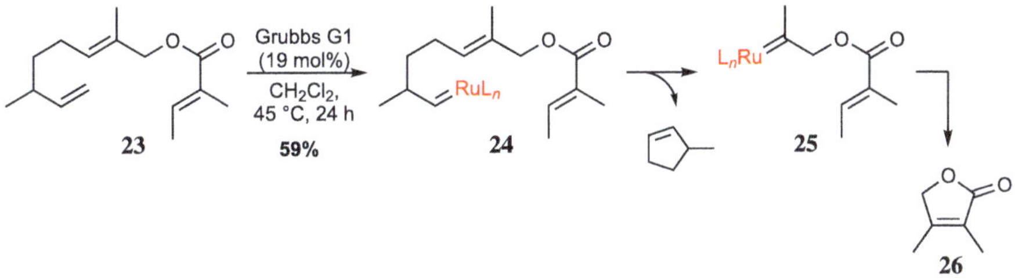

Fig. 7.10 Principle of relay ring-closing metathesis [45]

The challenge of *cross metathesis* (CM) lies in the selective formation of the desired product alongside other statistically possible products. In a non-selective or statistical cross metathesis, the formation of the desired product can be imposed by adding one component in excess (see Fig. 7.11).

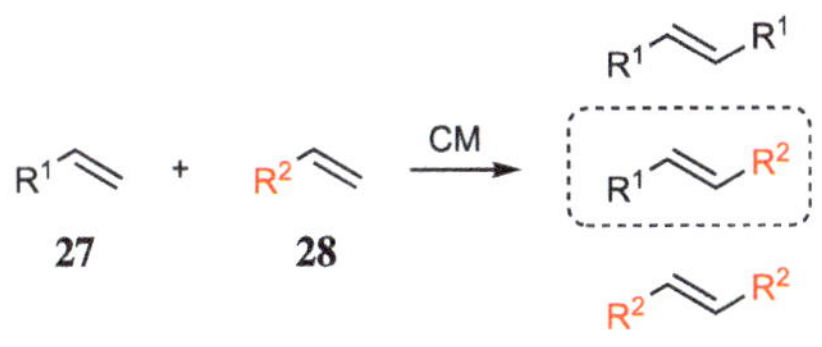

substrate **27** : **28**	CM-selectivity
1 : 1	50%
1 : 2	66%
1 : 4	80%
1 : 10	91%
1 : 20	95%

Fig. 7.11 Statistical product distribution in the cross metathesis of a thermoneutral reaction ($\Delta G_R = 0$)

The homocoupling product of the alkene **28** constitutes the largest amount of obtained product by virtue of the ratio of the reactants. The indicated selectivity thus refers to the cross metathesis, based on the amount of **27**. In a fundamental study of the different categories of olefins and their susceptibility to homodimerization, selection rules for the use of various substrates were established in order to avoid such a large excess of a substrate [41]:

- *Type I* — Rapid homodimerization; homodimers are consumable
- *Type II* — Slow homodimerization; homodimers sparingly consumable
- *Type III* — No homodimerization
- *Type IV* — Alkene inert to cross metathesis but does not lead to catalyst deactivation (spectator)

When two olefins of the same type react with each other, no product is isolated with any preference (statistical or non-selective CM). When alkenes of two different types are combined, the desired CM product is afforded with high selectivity. The classification of which alkenes fall into categories I–IV as substrates depends on the activity of the catalyst, which is why a different catalyst may be appropriate for each substrate combination (Table 7.1).

Table 7.1 Olefin categories for selective cross-metathesis reactions [41]

	Schrock cat.	**Grubbs-I**	**Grubbs-II**
Type I	terminal olefins, allyl silane	terminal olefins, allyl silane, 1° allylic alcohol, ether, ester, allyl boronic ester, allyl halide	terminal olefins, 1° allylic alcohol/ester, allylic boronic ester, allyl halide, allyl phosphonate, allyl silane, allyl amine (protected)
Type II	styrene, allyl stannane	styrene, 2° allylic alcohol, vinyl dioxolane, vinyl boronate	styrene (large ortho-subst.), acrylate, acrylamide, acrylic acid, acrolein, vinyl ketone, vinyl epoxide, 2° allylic alcohol, perfluorinated alkenes
Type III	3° allyl amine, acrylonitrile	vinyl siloxane	1,1-disubstituted olefin, vinyl phosphonate, phenyl vinyl sulfone, 4° allyl carbons
Type IV	1,1-disubstituted olefin	1,1-disubstituted olefin, disubstituted α,β-unsaturated carbonyls, perfluorinated alkenes, 3° allyl amines (protected)	vinyl nitroolefins, trisubstituted allylic alcohols (protected)

With the advent of molybdenum and tungsten monoaryloxide pyrrolide/chloride catalysts (*vide infra*), cross metathesis has also become feasible for trisubstituted alkenes. With these metal alkylidenes the regio- and stereoselectivity is achieved by a combination of steric and electronic effects [46].

One way to take advantage of the benefits of intramolecular reaction is by introducing a temporary connection between both components (*tethering*) [42]. Possible bridging units include silyl ethers or acetals. In Eustache's synthesis of a fragment of okadaic acid (**31**), two alcohols could be connected via a silyl bridge (**29**). **30** could thus be efficiently received in a ring-closing metathesis, and the potentially unselective cross metathesis was foregone (see Fig. 7.12) [47].

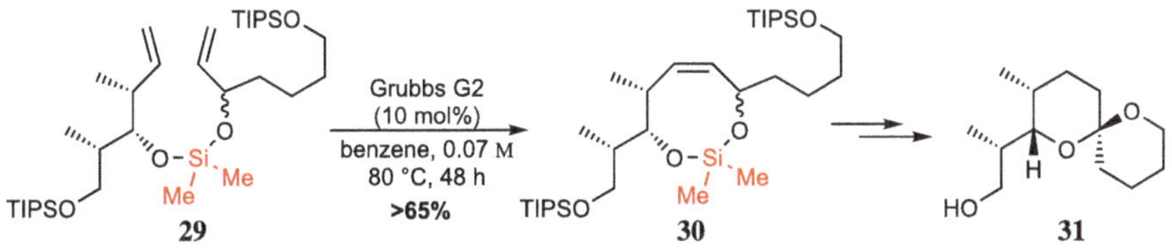

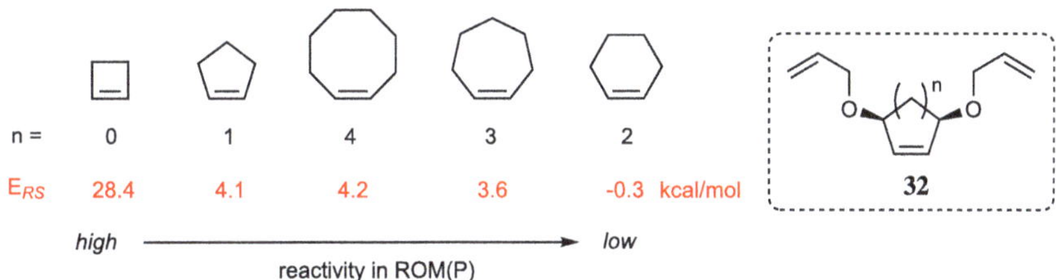

Fig. 7.12 Eustache's synthesis of the C_{28}-C_{38} fragment of okadaic acid utilizing tethered alkenes [47]

The *ring-opening metathesis* (ROM) of highly strained systems such as three-, four-, and eight-membered rings is thermodynamically favored, as the reaction is governed by the release of ring strain. Macrocyclic compounds can also be preferentially opened due to the gain in entropy. If bridged systems are additionally present, the driving force further increases and ΔG becomes more negative (see Fig. 7.13) [48].

Fig. 7.13 Reactivity of cycloalkenes in ROM with substrate **32** [49]. Indication of the ring strain (E_{RS}) of the related cycloalkenes given underneath the ring size [48]

In the case of ROMP, block copolymers can be realized by sequential addition of different monomers, as it is considered a "living" polymerization.

7.1.3 Modern Applications

To further enhance the reactivity and selectivity of alkene metathesis, a variety of newer catalysts, additives, and optimized reaction conditions have been developed. For example, the presence of catalytic amounts of $Ti(OiPr)_4$ can prevent the occurrence of the semi-stable chelates shown in Fig. 7.6, which was developed by the Fürstner group [50]. The addition of the weak Lewis acid prevents the formation of the intermediates **36**, **37** via *reversible* coordination of the titanium species to the impeding ester group, thus preventing the undesired coordination of the ruthenium. Due to the kinetically labile Ti-O bond, the catalytic cycle can still proceed by dissociation of the titanium (see Fig. 7.14).

The incompatibility of Mo catalysts towards acidic functionalities such as hydroxy or carboxy groups has largely been solved through the use of additives. The corresponding

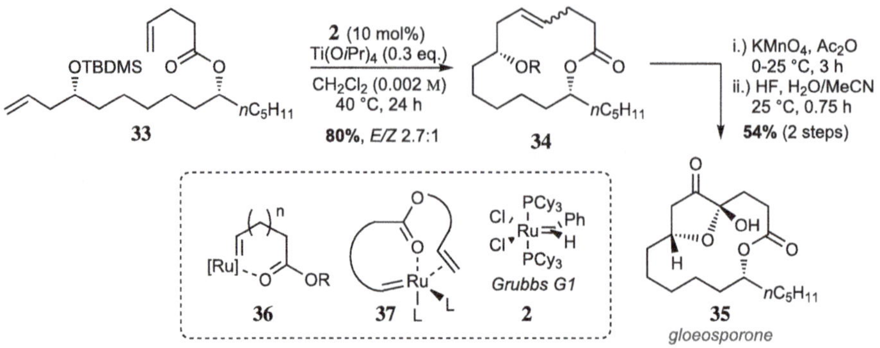

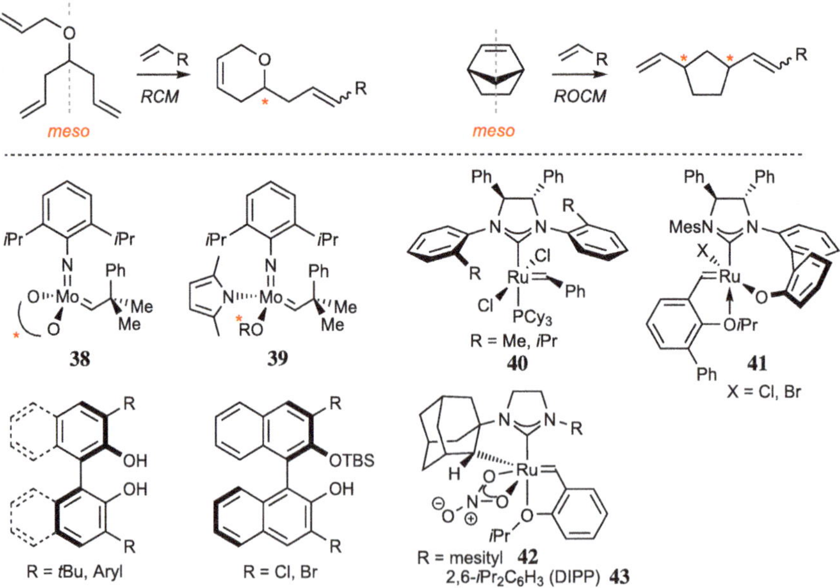

Fig. 7.14 Synthesis of (–)-Gloeosporone (**35**) according to Fürstner *et al.* [50]

alkoxy- or carboxyborate is generated *in situ* by addition of pinacolborane (HBPin), which remains untouched even by Mo complexes with their innate oxygen affinity. A standard work-up and subsequent column chromatographic purification easily and efficiently removes the temporary protective group [51].

The *asymmetric* metathesis entails the construction of chiral alkenes by desymmetrization of prochiral *meso*-compounds [17, 52, 53]. A "direct" enantioselective reaction is not possible, as only sp^2-hybridized C-atoms are connected in the course of alkene metathesis (see Fig. 7.15).

Fig. 7.15 Desymmetrization of prochiral substrates by ring-closing metathesis (RCM) or ring-opening cross metathesis (ROCM). Selected common catalysts for asymmetric metathesis reactions [17, 52, 53]

The development of chiral molybdenum catalysts for asymmetric metathesis reactions started in the mid-1990s and was initially based on the use of biphenolates or binaphtholates as chelate ligands for enantioinduction (**38**). In 2008, Schrock and Hoveyda introduced the class of monoaryloxide pyrrolide complexes (**39**, MAP), which show a significantly increased reaction rate, on par with achiral complexes of type **1**, with equivalent or even superior enantioselectivity [54]. The analogous ruthenium complexes were developed in the early 2000s and either rely on a desymmetrized structural motif on the carbene fragment (usually C_2-symmetric, **40**) or a chelate with its backbone attached to the NHC ligand (**41**-**43**).

Hoveyda and Schrock jointly developed the monoaryloxide pyrrolide complexes for their asymmetric synthesis of the *Aspidosperma* alkaloid (+)-quebrachamine. Both the achiral (**47**) and the chiral complex (**48**) showed outstanding activity and selectivity compared to the previous standard catalysts (**1**, **5**, **38**, **41**). With **48**, the desired product **45** was obtained enantioselectively from **44**, whereafter a subsequent Pt-mediated hydrogenation afforded the natural product **46** in excellent 95% *ee* over a total of 12 synthetic steps (see Fig. 7.16) [55].

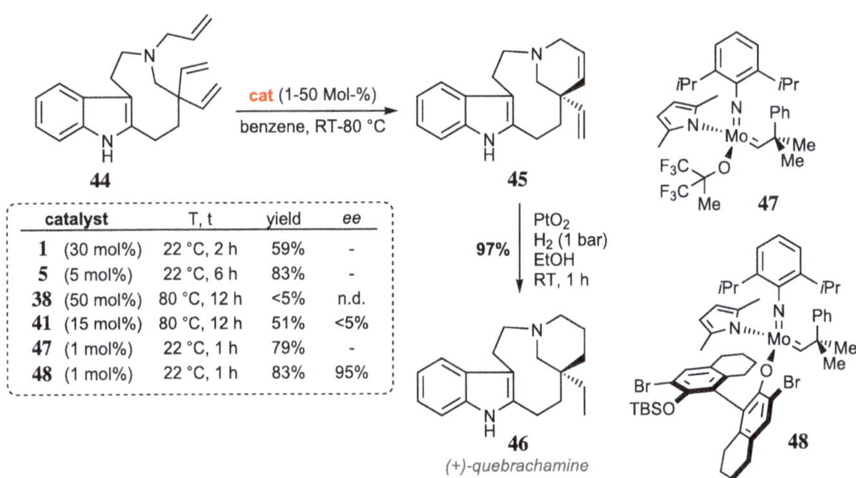

catalyst		T, t	yield	ee
1	(30 mol%)	22 °C, 2 h	59%	-
5	(5 mol%)	22 °C, 6 h	83%	-
38	(50 mol%)	80 °C, 12 h	<5%	n.d.
41	(15 mol%)	80 °C, 12 h	51%	<5%
47	(1 mol%)	22 °C, 1 h	79%	-
48	(1 mol%)	22 °C, 1 h	83%	95%

Fig. 7.16 Synthesis of (+)-quebrachamine (**46**) disclosed by Schrock and Hoveyda [55]

In the context of total synthesis, it was observed through the comparison of the reactivity of several catalysts that complexes bearing only monodentate ligands resulted in a higher rate of reaction. This phenomenon was rationalized by the high-energy, highly rigid metal-lacyclobutane intermediates. Complexes containing bidentate ligands presumably afford a more strained transient metallacyclobutane, leading to an increased reaction barrier and a consequent decrease in reaction rate (see Fig. 7.17) [55].

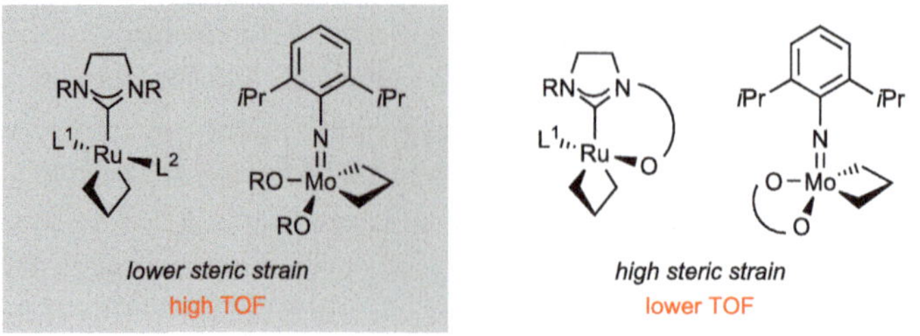

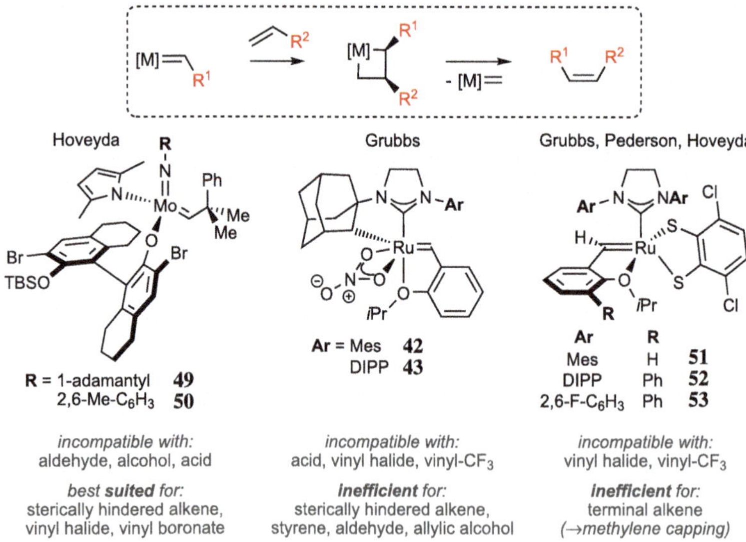

Fig. 7.17 Influence of catalyst structure on reaction rate (TOF) [52]

Although significant improvements in enantioselectivity and reactivity have been achieved in recent years, the realized enantioselectivities obtained are highly substrate dependent, which is why the example shown in Fig. 7.16 is illustrative, but not necessarily representative for other classes of reactants.

While the field of enantioselective metathesis has been extensively explored for some time, the possibility of an *E/Z*-diastereoselective reaction was only realized significantly later. The relevant metallacyclobutane intermediates of a metathesis of two terminal olefins can display a *trans-* or *cis*-configuration of the substituents R^1 and R^2, with only the *cis*-stereoisomer leading to the formation of the *Z*-alkene (see Fig. 7.18) [52, 53, 56–60].

Fig. 7.18 *Z*-Selective olefin metathesis catalysts [58, 59, 61, 62]

The design principle of *Z*-selective catalysts is based on the use of two sterically very different ligands. The steric repulsion of the substituents of the metallacyclobutane and the large ligand energetically favors formation of the *cis*-isomer (**54**). In the case of the MAP catalysts by Hoveyda and Schrock (**49, 50**), the freely rotating monodentate alkoxy group causes the substituents R^1, R^2 to be oriented towards the sterically less demanding imidoadamantyl. In the case of the ruthenium-based catalysts **42, 43, 52, 53**, the *N*-heterocyclic carbene constitutes the largest ligand. Both the bridged adamantyl group and the catechothiolate favor an equatorial coordination of an alkene (cf. **14**, Fig. 7.5 and Fig. 7.17). This arrangement is required for a pronounced energetic differentiation of the *cis*- and *trans*-metallacyclobutane (**56, 57**), otherwise the metallacyclobutane would be *trans*-positioned to the NHC (see Fig. 7.19) [63–65].

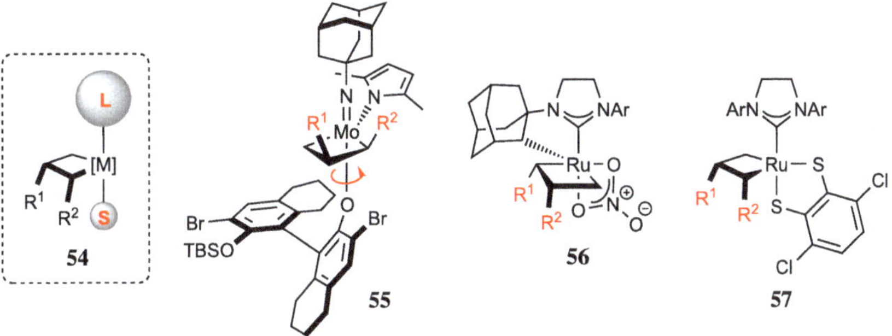

Fig. 7.19 Origin of *Z*-diastereoselectivity and postulated intermediates of **42, 43, 49–53** [63–65]

A metathesis of terminal alkenes can deliver the corresponding *Z*-alkenes highly selectively, partially supported by the removal of ethylene or *in situ* methylene capping (see below). However, the formation of *E*-configured alkenes from the coupling of terminal alkenes is only possible to a limited extent. Interestingly, it was observed that *internal* alkenes afford the cross-metathesis product with stereoretention when employing *Z*-selective catalysts: *Z*-alkenes furnished the *Z*-configured product, while *E*-alkenes yielded the *E*-configured product. This principle of stereoretentive metathesis has been extensively studied and can also be rationalized by the steric repulsion of the intermittent metallacyclobutanes. When using *E*-alkenes, **59** is obtained in which the energetically preferred metallacyclobutane possesses a *trans,trans*-configuration. *Z*-alkenes, on the other hand, form the *cis,cis*-metallacyclobutane **60**, which subsequently liberates *Z*-configured products. This principle has already been illustrated in the synthesis of select complex natural products or their derivatives, whereby *Z*-selective examples dominate thus far (see Fig. 7.20) [62, 66].

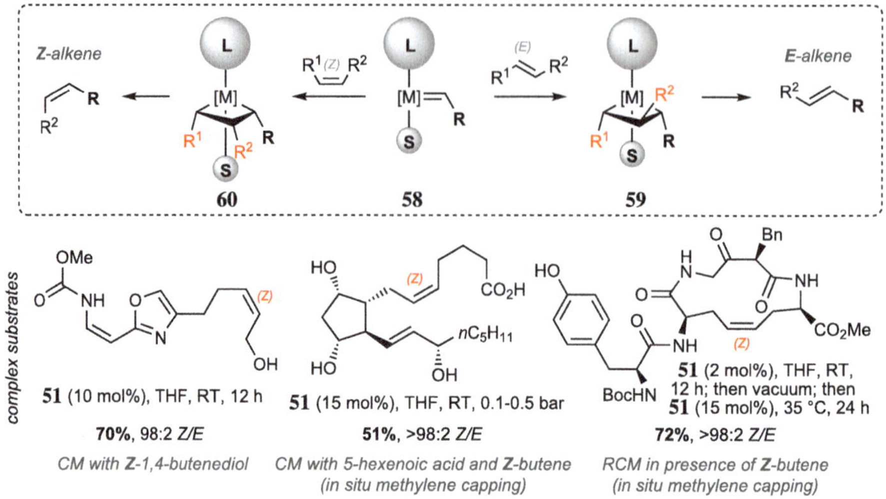

Fig. 7.20 Stereochemical model of stereoretentive reactions and their application [61, 62, 67]

Hoveyda and Schrock examined the application scope of available metathesis catalysts, revisiting ring-closing metatheses of published total syntheses of the natural products nakadomarin A and epothilone C for a selectivity benchmark of modern catalysts. The E/Z-selectivity induced by the catalyst can be impeded during the cyclization and even countermanded, owing to its susceptibility to the steric induction of the substrate (see Fig. 7.21) [68].

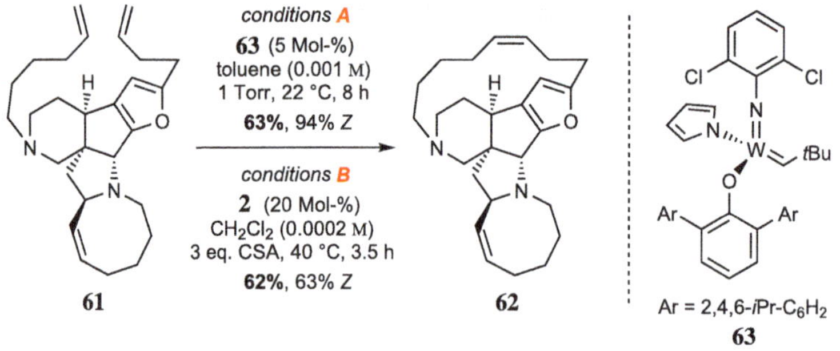

Fig. 7.21 Z-selective ring-closing metathesis in the synthesis of nakadomarin A [69]

Although the authors had previously also reported on MAP catalysts derived from tungsten instead of molybdenum as the active metal, no application in the synthesis of more complex substrates was known to date. Surprisingly, **63** exceeded the selectivity of the commonly used catalyst **49**, which only afforded up to 90% Z-selectivity. This is probably due to the fact that **63** shows a lower activity than **49**. In the case of ring-closing metathesis, the product can potentially isomerize in the presence of the catalyst following the elimination of ethene. With a catalyst that is significantly less active towards disubstituted double bonds, isomerization through a renewed ring opening is efficiently prevented. However, a low pressure was necessary to effect the removal of ethylene from the reaction solution. The applied vacuum prevented a stalling of the cyclization towards the end of the reaction and ensured full conversion. A direct comparison with the Grubbs G1 catalyst (**2**) vividly illustrates the difficulty of the transformation prior to the introduction of modern Z-selective methods [69].

The observed ethylene effect deserves a more differentiated analysis to put it into perspective. Ethylene constitutes an often ignored by-product in the metathesis of terminal olefins. The volatility of the gas is one reason why this reversible transformation benefits from a high driving force towards the desired products. Due to the high reactivity and inherent instability of the metal methylidenes, which are generated as intermediates in the presence of ethylene (see Fig. 7.6), the presence of ethylene can have either a positive or negative effect: It can lead to a rapid catalyst activation and the avoidance of unproductive homocoupling of the reactants or products, which is beneficial for the overall selectivity. On the other hand, a methylidene's low stability and high activity may lead to an erosion of efficiency and/or kinetic selectivity. In this instance, a removal of the generated ethylene is the best course of action. If methylidene decomposition is so fast that there is little or no product formation, its formation should be avoided altogether by resorting to internal olefins as substrates (see Fig. 7.22) [70].

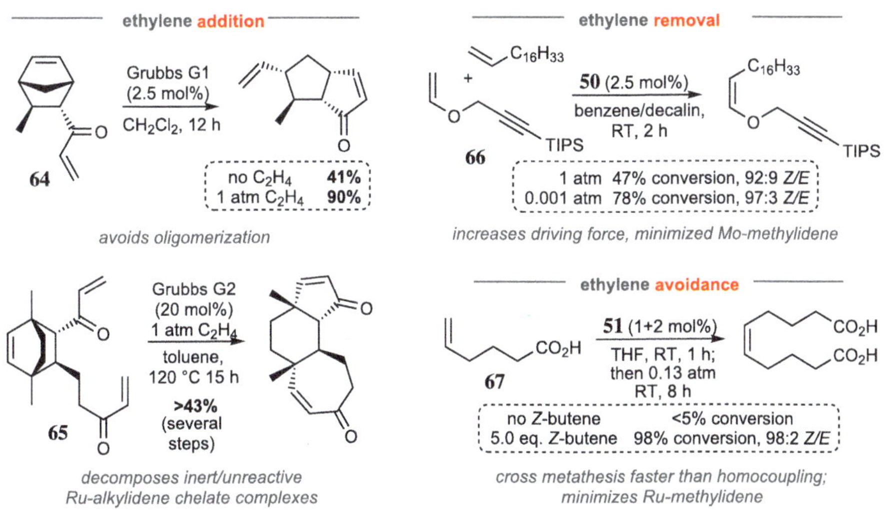

Fig. 7.22 Influence of ethylene on various metathesis reactions [56, 61, 71, 72]

The ring-opening–ring-closing metathesis of the strained alkene **64** displayed low selectivity in the absence of ethylene. This was attributed to the tendency of the primary ring-opened product to react with another equivalent of **64** under oligomerization, rather than undergo the desired ring closure. The addition of ethylene afforded the corresponding unstrained terminal triene by ROCM, which then selectively yielded the desired bicyclic ketone. The use of the less active Grubbs G1 catalyst helped to further minimize oligomerization [71]. Phillips and co-workers also exploited the addition of ethylene in their synthesis of cyanthiwigin U. **65** can undergo an initial reaction at the electron-rich endocyclic double bond or either of the two enone groups. Since the carbonyl group in the Ru-alkylidene can coordinate to the metal center, stabilize the resulting alkylidene intermediate, and thus deactivate the catalyst, the addition of ethylene helps by enabling the reverse reaction to the substrate. The alternative pathway via attack on the strained ring system is the productive route, which favors this irreversible path [72]. Particularly stereoselective variants can benefit from the adjustment of the ethylene concentration if one of the paths is accelerated by it. The cross metathesis of **66** with octadecane was supported by removing the ethylene generated in the process in vacuum. First of all, it can no longer compete with the substrates for the catalyst. Secondly, a subsequent *Z/E* isomerization, which is favored by terminal alkenes, is prevented by removing possible reactants [56]. In the last example, **67** was intended to participate in a homocoupling. The intermediate Ru-methylidene, however, proved to be extremely unstable with high tendency for decomposition. A counterintuitive approach to circumvent formation of ethylene is the use of highly reactive *Z*-2-butene as additive. The presence of butene initially affords 5-heptenoic acid *in situ* with the concomitant release of propene (*in situ methylene capping*). Since the reaction of a Ru-alkylidene with 2-butene outcompetes the conversion of a terminal alkene and butene is used in excess, the formation of methylidene complexes can be largely avoided. In addition, the subsequent *Z/E* isomerization is also suppressed [61].

The general properties of the presented Grubbs and Schrock catalysts with regard to the tolerance of functional groups can routinely be transferred to the more recently developed classes of catalysts. This simplification should help in the preselection of a suitable catalyst system to effect a desired transformation (see Figs. 7.3, 7.15, and 7.18).

7.2 Enyne Metathesis

In the presence of both double and triple bonds in the substrate, a reaction of the alkyne with the olefin can occur in an enyne metathesis instead of a classic alkene metathesis. Given that the intermediate vinyl carbenes are significantly more stable due to the conjugation and additional complexation of the metal center than "normal" metal carbenes of the corresponding olefin metathesis, enyne metathesis proceeds more slowly than the respective alkene metathesis (see Fig. 7.6). This more pronounced minimum of the vinyl carbene in the potential hypersurface heightens the activation barrier of the subsequent product formation and reduces the reaction rate. Terminologically, a distinction is made between the enyne ring-closing metathesis (*enyne RCM*) and the corresponding cross metathesis (*enyne CM*). In addition to the possibility of forming *E/Z* diastereomeric mixtures, constitution isomers can also result in accordance with the envisaged mechanism (see Fig. 7.23) [73–77].

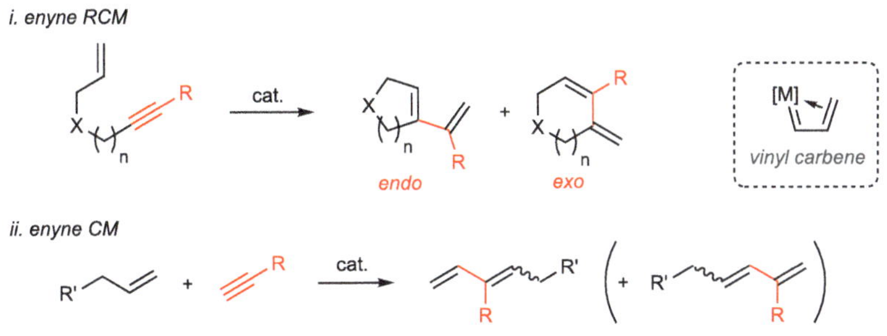

Fig. 7.23 Reaction modes of enyne metathesis [74]

In addition to a high dilution that helps suppress cross metathesis, Mori and co-workers have shown that an ethylene atmosphere could positively influence the yield as well as the conversion rate in the enyne RCM of substrates bearing terminal alkyne groups [78]. Molybdenum alkylidenes tend to undergo enyne metathesis much less frequently and preferentially polymerize the alkyne component instead [79], which is why published methods almost exclusively rely on the use of Ru alkylidenes [74]. So far, there is no model to reliably predict the major product in the enyne CM. In contrast, the preferred product (*endo* vs. *exo*) in the enyne RCM is reliably governed by the resultant ring size: normal and medium rings form the *exo*-product up to a ring size of 10. Eleven-membered rings can follow both the *exo*- and the *endo*-path. Macrocyclic systems (>12) primarily afford the *endo*-product (see Fig. 7.24).

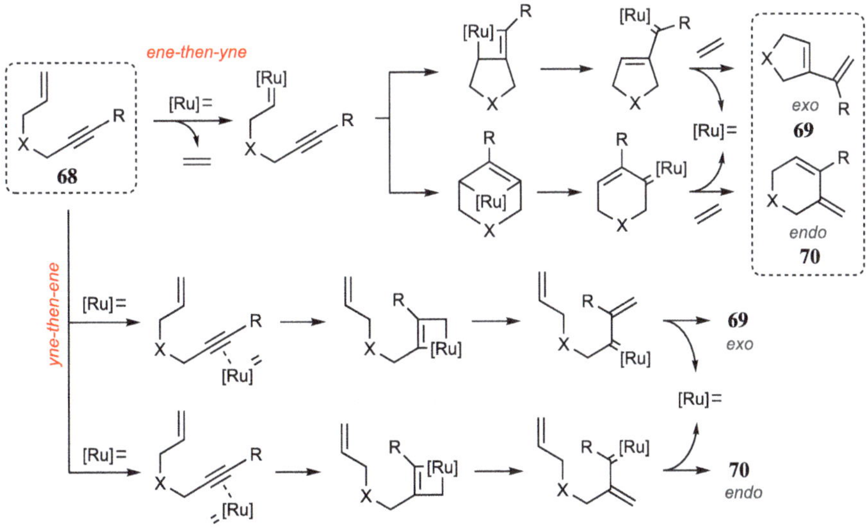

Fig. 7.24 Competing reaction paths of the enyne ring-closing metathesis [73, 74, 77]

In contrast to alkene metathesis, the mechanism of C-C bond formation is much less clearly defined, as several competing reaction paths exist: On the one hand, the initial interaction of the catalyst with the double bond and then the triple bond can occur (*ene-then-yne* mechanism). Alternatively, the triple bond can first generate a metallacyclobutene and a concluding ring-closing metathesis provides the product (*yne-then-ene* mechanism). So far, it has not been elucidated which of the two competing paths is preferentially followed. In general, experimental evidence for both paths has been obtained [78]. The dominant path seems to depend strongly on the chosen reaction conditions as well as the steric environment of the alkyne and the reaction mode (RCM *vs.* CM) [80]. In the intermolecular cross metathesis, evidence was found that the yne-then-ene path seems to be energetically more favorable [81]. Regardless of the initial step of the transformation, the two constitutional isomers **69** and **70** both constitute feasible products. Thus, by different coordination to the alkyne, either the *exo-* (**69**) or the *endo*-product (**70**) can be obtained. The insertion into the alkyne is the only irreversible step of the catalytic cycle and determines the regioselectivity leading to the respective constitutional isomers [82]. In the ene-then-yne mechanism, the *endo/exo*-selectivity is a direct consequence of the ring strain of the intermediate metallacyclobutenes. The ring size accordingly governs the choice of the dominant reaction path.

Based on the current understanding, the use of an ethylene atmosphere has two positive effects on the enyne metathesis: It supports the conversion of the vinyl carbene intermediate to the desired metathesis products and prevents alkyne oligomerization, which can lead to catalyst decomposition as a side reaction (see Fig. 7.25) [83–85].

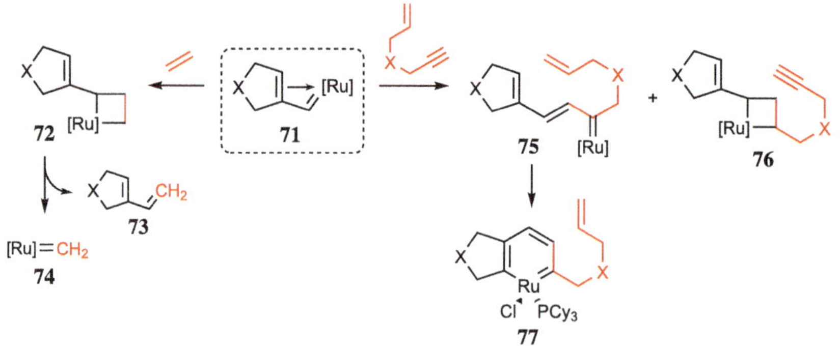

Fig. 7.25 Influence of ethylene on the catalytic cycle of enyne metathesis [83, 85]

The formation of the *exo*-product **73** proceeds via the metal carbene **71**. In the presence of ethylene, the corresponding methylidene catalyst **74** is obtained via **72**, which is significantly more reactive than the vinyl carbene **71** (see Fig. 7.6). **74** subsequently follows the catalytic cycle depicted in Fig. 7.24, which is *catalytic* with respect to ethylene. In the absence of ethylene, either **75** or **76** can result from the reaction of **71** with another equivalent of enyne. **75** rearranges under dissociation of a ligand to give ruthenabenzene **77** or a structural analog thereof. Given the inherent stabilization of the metal center in **77**, this competing pathway can be considered to be irreversible for practical purposes [85]. In contrast, **76** constitutes a competitive intermediate of enyne metathesis and can enter an alternative, ethene-free catalytic cycle [83]. However, the possibility of coordination by alkyne and alkene to the ruthenium in **76** could potentially slow down this catalytic cycle. Substrates possessing steric crowding around the alkyne that prevents the reaction of second equivalent of enyne with **71**, such as internal alkynes, do not lead to formation **75** and **76**. Furthermore, the substrate cannot oligomerize. The addition of ethylene does not increase the rate of reaction in these cases [85]. Previous studies on the ethylene effect employed phosphine-containing metathesis catalysts. Cationic, phosphine-free catalyst system sometimes even exhibit a retardation of the reaction in the presence of ethylene [86].

In the synthesis of the cytotoxic macrolide amphidinolide K (**81**), an enyne metathesis was used to connect the two western fragments and assemble the left hemisphere of the lactone metabolite (see Fig. 7.26) [87].

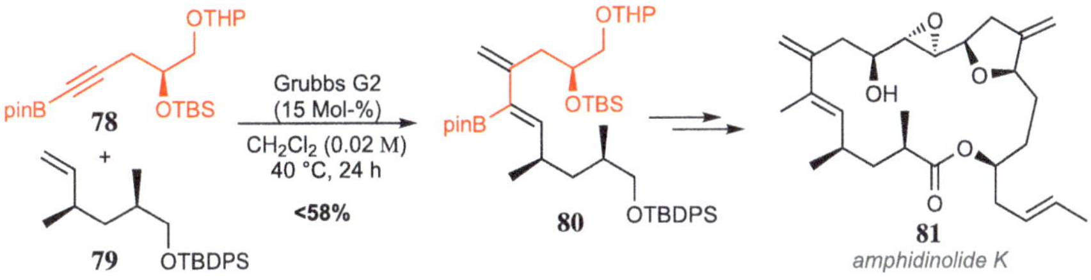

Fig. 7.26 Lee's synthesis of amphidinolide K (**81**) [87]

The cross metathesis of pinacolborane **78** and the olefin **79** yielded the desired product as a 7.5:1 mixture of diastereomers, in which the desired *E*-isomer **80** comprised the major product in the presence of the Grubbs G2 catalyst (**3**). The use of the alkynyl boronate was critical, as the corresponding methyl-substituted alkyne did not furnish any significant amounts of product. The natural product **81** could thus be obtained in 18 linear steps and a total yield of 6.8%.

In addition to the enyne metathesis illustrated here, which is mediated by Ru-/Mo-/W-alkylidene complexes, skeletal rearrangements have been observed to occur in the presence of other transition metal catalysts (e.g., Pd, Pt, Ru, Au, Ir, Rh). While the isolated products can be identical to those received via metathesis, they can also show a different stereoselectivity or be constitutional isomers [88, 89].

7.3   Alkyne Metathesis

The alkyne metathesis can be considered a complementary reaction to olefin metathesis, as olefins are generally inert to alkyne metathesis catalysts and vice versa. Given the challenges in obtaining a high *E/Z*-selectivity with alkene metathesis, the advantage of the alkyne variant lies in the possibility to exclusively generate a *Z*-double bond by subsequent Lindlar reduction or to further derivatize the alkyne, for example, by hydrometallations. The majority of (pre)catalysts are not commercially available and must be synthesized via multi-step routes (see Fig. 7.27) [57, 90–93].

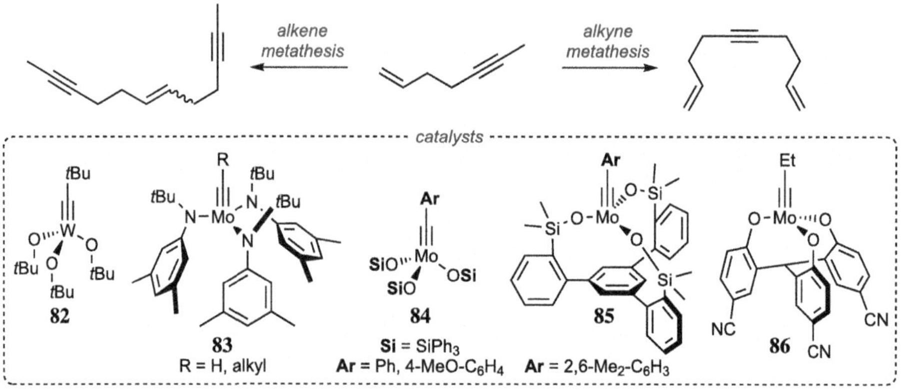

Fig. 7.27 Competition of alkyne and alkene metathesis and frequently used catalysts of alkyne metathesis [90]

The handling of W- and Mo-alkylidines is extremely delicate and requires strictly inert conditions due to their high reactivity towards oxygen and sometimes even dinitrogen, preferably in an Ar-atmosphere. Fürstner succeeded in forming crystalline, air- and moisture-stable precatalysts of **84** by complexation with phenanthroline or $KOSiPh_3$, which can be easily liberated again under reaction conditions [94, 95].

Katz proposed a mechanism for the conversion of alkynes analogous to alkene metathesis, which was subsequently experimentally corroborated and is now considered generally accepted [92, 96]. Following the formal [2+2]-cycloaddition of the alkyne to the metal-carbon triple bond to obtain metallacyclobutadiene **88**, the cycloreversion from **89** to **90** occurs with the elimination of the product. Subsequently, an equivalent of the second alkyne is converted to **91**, which then regenerates the initial catalytic species **87** (see Fig. 7.28).

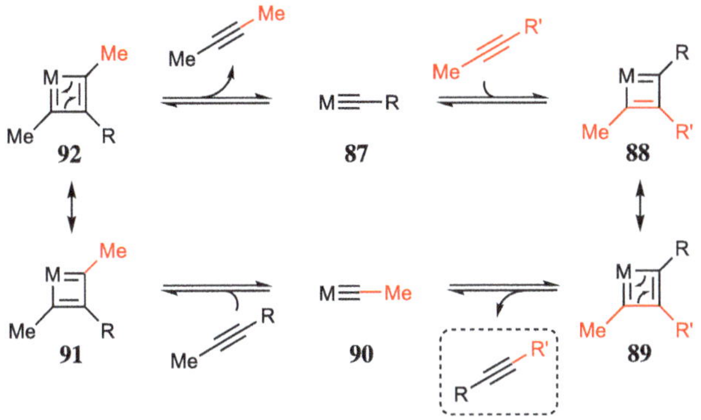

Fig. 7.28 Katz mechanism of alkyne metathesis [92, 96]

In many metathesis reactions, molecular sieves are used as a standard additive [92, 93]. In the reaction of methyl-substituted alkyne, they serve as scavengers of the butyne by-product formed during the regeneration of **87**, thus ensuring a high conversion. Apart from shifting the reaction equilibrium towards the products, their addition is furthermore intended to prevent a deactivation of the catalyst. The polymerization of small alkynes, for example 2-butyne, is one reason why some substrates cannot or can hardly be converted in alkyne metathesis. This deactivation pathway can only be efficiently avoided if the deleterious alkyne is sterically highly hindered, insoluble in the reaction medium or is removed from it (see Fig. 7.29) [97].

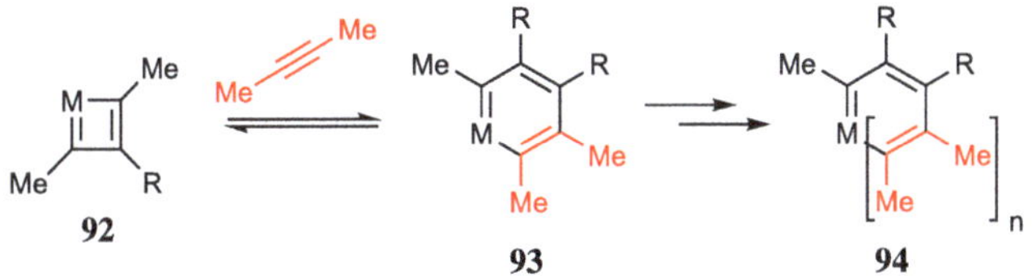

Fig. 7.29 Catalyst deactivation by polymerization of 2-butyne [92, 97]

The efficiency of metathesis is, as already concluded in the case of alkene metathesis, largely determined by the ability to circumvent a reaction outcome comprising a statistical distribution of products. Alkyl-substituted alkynes are more reactive than the respective aryl-substituted substrates, which is why the presence of small alkynes favors the reverse reaction and promotes a statistical product mixture, in addition to undesired catalyst deactivation. An intramolecular reaction constitutes the easiest approach to circumvent these issues, which disfavors the reverse reaction on entropic grounds. Additionally, a product can also be continuously removed from the reaction. The presence of molecular sieves thus can not only help shift the equilibrium towards the products, but also accelerate the reaction by preventing the reverse reaction (see Fig. 7.30) [94].

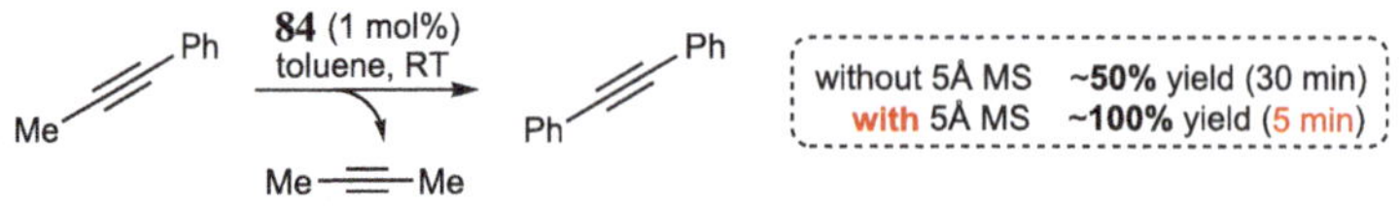

Fig. 7.30 Impact of molecular sieves on alkyne metathesis [94]

The Fürstner group disclosed an example of an intramolecular ring-closing alkyne metathesis. In their synthesis of the actin-binding natural product latrunculin B (**97**), the key macrocyclic ring closure was achieved by connecting the appended alkynes in the presence of the catalyst **83** in 70% yield (see Fig. 7.31) [98].

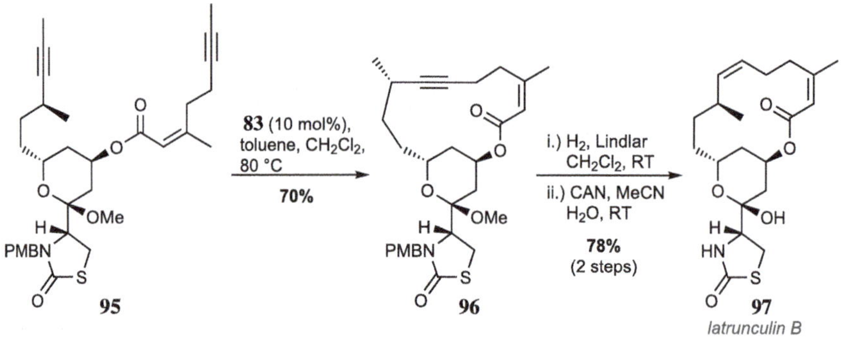

Fig. 7.31 Fürstner's synthesis of latrunculin B (**97**) [98]

The absence of molecular sieves in the reaction of the alkyne **95** to **96** may be attributed to the volatility of 2-butyne under the temperatures used. The first-generation molybdenum catalyst developed by Fürstner (**83**) still required elevated temperatures and an extended reaction time. Nevertheless, even with the now outdated catalyst the tolerance towards functional groups is generally broader than with comparable alkene metathesis catalysts (see Fig. 7.32) [90, 94, 99]. Not only the α, β-unsaturated ester, but also an acetal and a thiocarbamate group remained intact. The subsequent Z-selective reduction of the triple bond with Lindlar catalyst and deprotection of the carbamate and the glycosidic OH group with cerium ammonium nitrate (*CAN*) afforded the desired natural product **97**, which could thus be synthesized in 16 steps in a total yield of 6%.

functional group tolerance	82	83	84	85	86
alkene	+	+	+	+	+
acetal	+	+	+	+	
ester	+	+	+	+	+
(silyl) ether	+	+	+	+	+
sulfone	+	+	+	+	
amide-NH	+	−	+	+	
furan	+		+	+	
aldehyde	−	+	+	+	+
alkyl chloride		+	+	+	
nitrile		+	+	+	+
NO_2		+	+	+	+
pyridine	−	+	+	+	+
epoxide	−	+	+	+	
thioether	−	+	+	+	
aryl halide			+	+	+
CF_3			+	+	
phenol-OH	−	−	−	+	+
alkyl-OH	−	−	−	+	

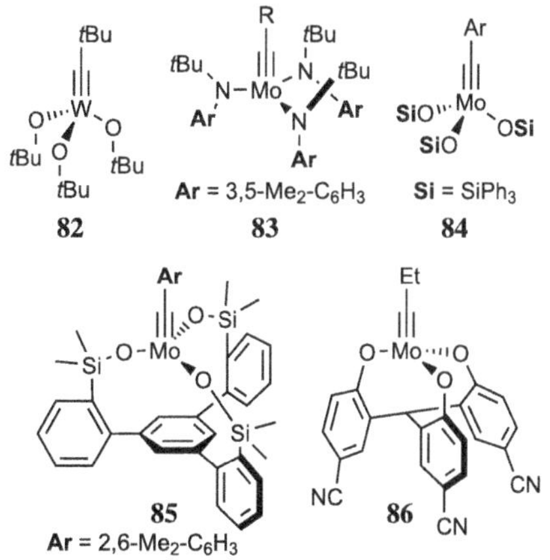

Fig. 7.32 Substrate scope of common alkyne metathesis catalysts [94, 99–101]

Just as in the cross metathesis of olefins, the *intermolecular* alkyne metathesis demonstrates a pronounced selectivity if the two substrates exhibit large differences in their steric and/or electronic properties [92].

The coupling of two terminal alkynes constituted a long sought-after dream reaction. Apart from the possible polymerization of the substrate (cf. Fig. 7.29), the resulting acetylene by-product is a strong catalyst poison [102]. The reaction of **98** via **99–101** represents the "normal" catalytic cycle. The formation of acetylene, which is highly soluble in organic solvents, outcompetes the substrate in its reaction with the catalyst due to its excellent donor properties, thereby depleting the active catalyst species **98**. Both tungsten and molybdenum alkylidines **98** and **100** can generate the dimeric adduct of type **102** under equilibrium conditions. Depending on the substituent R, the equilibrium position lies either on the side of the monomer or the dimer. Sterically demanding substituents effectively prevent the formation of **102** starting from the alkylidene. In the case of R=H (**100**), **102** is thermodynamically strongly favored, which almost completely halts the catalytic cycle due to this energetic catalyst sink [93]. μ-bridged alkylidene complexes **102** can also insert further acetylene and polymerize. Similarly, instead of in a productive metathesis via **99**, the metallacyclobutadiene **103** can be received in a competing pathway. The latter is prone to transannular C-H activation with concomitant loss of one of the ancillary ligands that picks up the proton and hence formally acts as a base. This pathway is particularly favorable in the presence of external donor ligands L. The η^3-bound resultant deprotiometallacycle **104** only requires a hapticity change to reveal alkynyl-alkylidene **105**, which can trigger rapid polymerization of the substrate (see Fig. 7.33) [102].

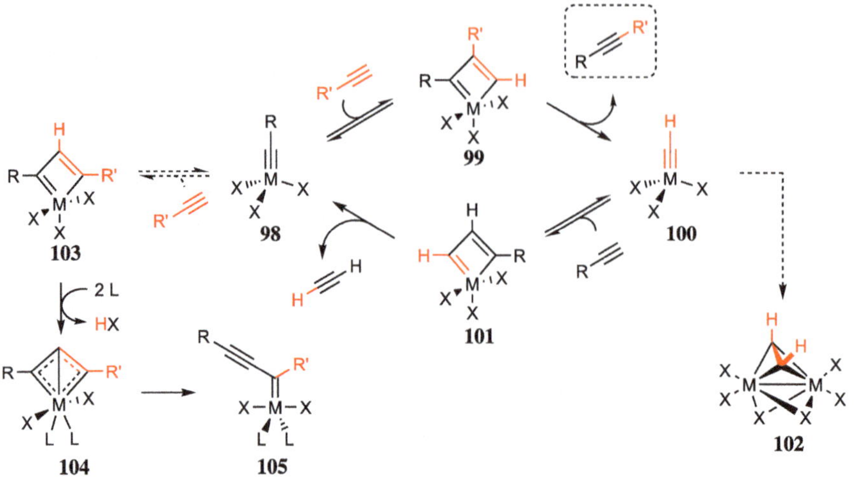

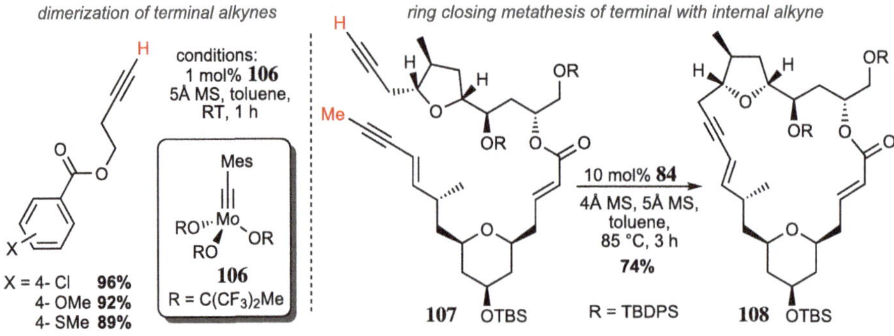

Fig. 7.33 Decomposition pathways when reacting terminal alkynes [102]

As a result, terminal alkynes are generally avoided as substrates in favor of methyl-substituted, internal alkynes. The routinely added molecular sieves can only inadequately sequester the acetylene by-product in the conversion of terminal alkynes, which culminates in a less favorable equilibrium position and deceleration of the reaction — the potential complications described above notwithstanding.

The Tamm group disclosed the first metathesis of terminal alkynes. Their solution to this challenge relied on the use of Schrock-like, Mo-based alkylidene complexes, in which the molybdenum was complexed by the hardly basic hexafluoro-*tert*-butoxy groups (**106**) [103]. Subsequently, it could also be shown that the silanol-ligated catalyst **84** developed by Fürstner could efficiently couple a terminal with an internal alkyne (see Fig. 7.34). However, the respective metathesis of two terminal alkynes failed, presumably due to the formation of acetylene and its reaction with **84** [102, 104].

Fig. 7.34 Metathesis of terminal alkynes [103, 104]

Apart from alternative olefinations covered in Chap. 3, synthesis of Z-alkenes was traditionally only feasible through an alkyne metathesis and subsequent *cis*-hydrogenation due to the lack of a Z-selective alkene metathesis. The development of new catalysts for alkene metathesis that promote selective formation of Z-alkenes has introduced an alternative synthetic access to this class of substances. Nonetheless, alkynes, unless they themselves pose as target of a synthesis, are often exploited as a synthetic platform to obtain mono- to tetrasubstituted alkenes with defined stereochemistry (see Fig. 7.35). Besides the classic Lindlar hydrogenation to access Z-olefins, methods for a *trans*-hydrogenation of the alkyne have also been reported ($\rightarrow$ Sect. 4.2) [105–107]. In addition, a C-C or C-X coupling can be mediated by transition metals, for example a carbometallation, hydroamination, hydrosilylation, or through π-acid catalysis (see Sect. 3.3) (see Fig. 7.35).

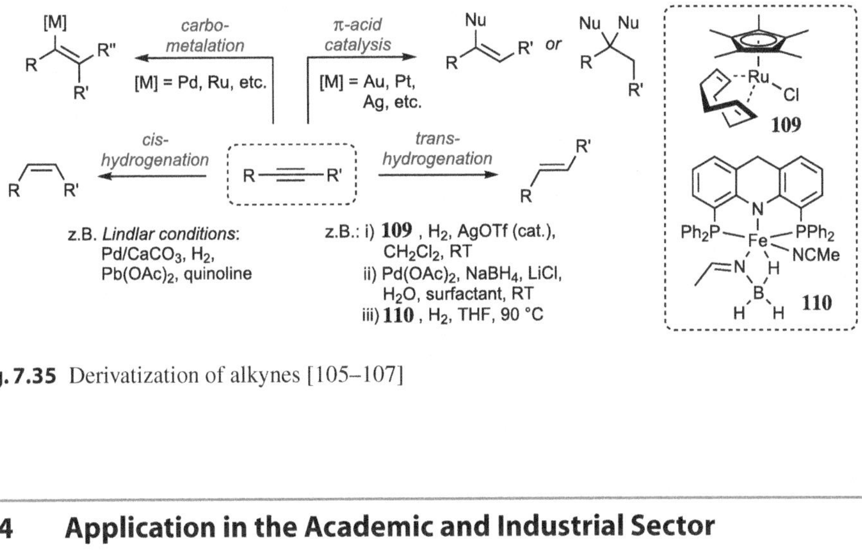

Fig. 7.35 Derivatization of alkynes [105–107]

7.4   Application in the Academic and Industrial Sector

The main academic application of the metathesis reaction, measured by sheer number of publications, is its use in the synthesis of complex molecular structures, for example in total synthesis [15, 16, 56, 68]. However, some applications of particular industrial relevance have also been reported. Barring its use in the field of medicinal chemistry, where metathesis is just on the verge of becoming established [108], alkene metathesis is mainly utilized on a large scale for the production of polymers and unfunctionalized alkenes [109, 110].

In Parker's synthesis of the natural product (–)-englerin A (**116**), a relay enyne/alkene metathesis was used to assemble the bicyclic backbone in a domino reaction (see Fig. 7.36) [111]. The advanced intermediate **111** was synthesized in seven steps starting from geraniol. In the presence of the Grubbs–Hoveyda-derived catalyst **117**, **111** was converted to **114** in 87% yield. The ruthenium alkylidene could successively be directed along the present C-C multiple bonds in the order a–d highlighted in Fig. 7.36. Following the initial conversion of the terminal alkene functionality, elimination of dihydrofuran gave **112**, which subsequently afforded the diene **113** in an enyne metathesis. As concluding step, the trisubstituted double bond of the second ring could be closed and **114** was obtained. The conversion to **115** completed the formal total synthesis, as **115** constituted an advanced intermediate in Echavarren's total synthesis of the natural product [112].

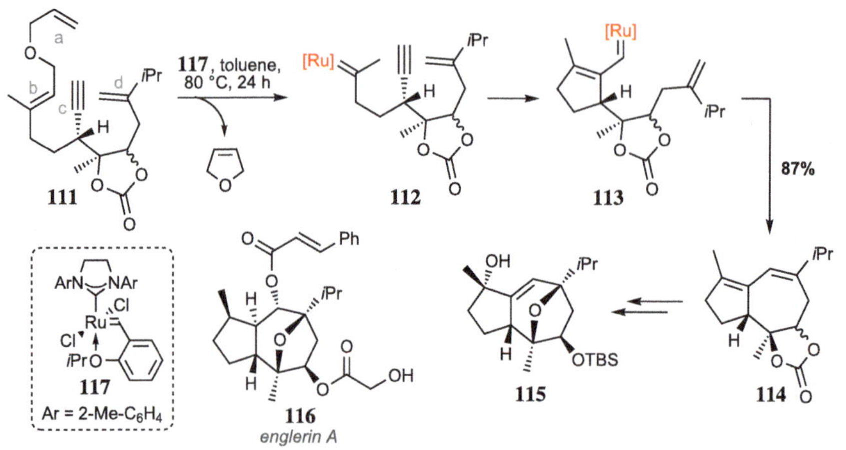

Fig. 7.36 Parker's formal synthesis of (–)-englerin A (**116**)

An impressive testament to the ring-closing metathesis on a larger scale was reported by researchers from GlaxoSmithKline in their approach to SB-462795 (**120**). The active pharmaceutical ingredient (API) inhibits the enzyme cathepsin K, which regulates bone degradation/resorption and allows treatment of diseases such as osteoporosis or other types of bone loss. Due to the start of clinical trials, larger quantities of the API were required, which resulted in the scale-up of the synthetic route to convert kilogram quantities (see Fig. 7.37) [113].

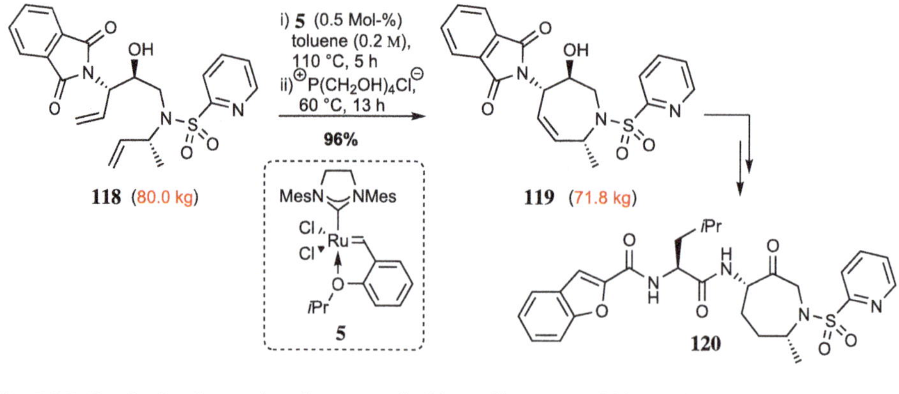

Fig. 7.37 Synthesis of an active pharmaceutical ingredient on an 80 kg scale [113]

Phthalimide **118** was converted to azepine **119** in the presence of the Hoveyda–Grubbs catalyst **5** as a concentrated solution in toluene, from which the product crystallized after formation and thus ensured the complete conversion of the substrate. Crystallization and subsequent addition of the ruthenium scavenger $^{+}P(CH_2OH)_4Cl^{-}$ reduced metal traces in the product to 359 ppm, which were depleted to an acceptable level for clinical studies in the following steps. Basic hydrolysis of the phthalimide to the amine, hydrogenation of the double bond, amide coupling with the left molecular fragment, and final oxidation to the ketone on the azepine enabled access to SB-462795 (**120**) on an 80 kg scale.

An example of an asymmetric domino metathesis reaction was reported by Hoveyda and Schrock. The molecular framework (**124**) of the sesquiterpenoid natural product (+)-africanol (**125**) could be obtained in a ring-opening–ring-closing metathesis in just one step starting from norbornene **121** via **122** and **123** as intermediates. Final modification of the ring periphery then delivered the natural product in another nine steps via an oxidative cleavage of the terminal double bond, hydroformylation of the endocyclic olefin, conversion into a trisubstituted alkene, and a concluding cyclopropanation (see Fig. 7.38) [114].

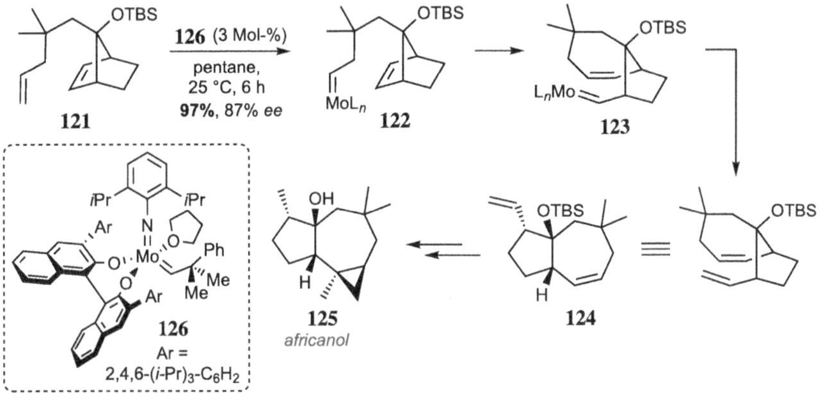

Fig. 7.38 Synthesis of (+)-africanol by Hoveyda and Schrock [114]

Olefin metathesis is used industrially on a million-ton scale for the conversion of petrochemical products. In contrast to academic applications, simple homogeneous precatalysts (mostly WCl_6 and AlR_3) or heterogeneous systems are used, in which rhenium, molybdenum, or tungsten oxides as active component are supported on SiO_2 or Al_2O_3. The advantage compared to well-defined homogeneous catalysts lies in a very cost-effective catalyst preparation, as the molecular structure does not have to be precisely controlled to ensure high activity. The rather harsh reaction conditions constitute their major drawback, which is why the target compounds usually contain no or hardly any functional groups [109, 110].

In the commercially applied SHOP process (*Shell **higher olefin** process*) established in the 1970s, a heterogeneous olefin metathesis is used to produce C_{12}-C_{18} linear alkenes, from which aldehydes and alcohols are subsequently derived (see Fig. 7.39) [115–118].

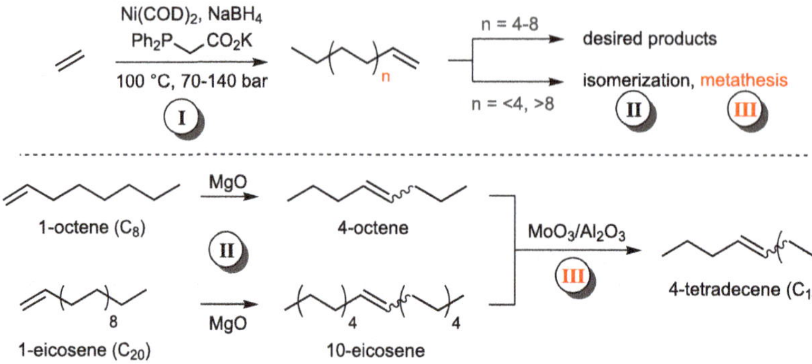

Fig. 7.39 SHOP process [115–118]

The 1-alkenes obtained in the ethylene polymerization are separated by distillation to give a light-boiling (C_4-C_{10}), a heavy-boiling ($\geq C_{20}$), and a product fraction (C_{12}-C_{18}). The light and heavy boilers are reunited and initially subjected to a basic isomerization to afford internal alkenes. Subsequently, a molybdenum catalyst supported on Al_2O_3 produces internal alkenes with a similar chain length as the desired products from both fractions. The higher and lower internal olefins are distillatively separated and reintroduced into the metathesis step. In the subsequent hydroformylation, the cobalt catalyst used can convert the desired terminal alkenes directly to the corresponding aldehyde and also convert the internal alkenes to the respective terminal alkenes, so that the same products result. However, due to a change in demand for C_4-C_{10} olefins, which are used in polyethylene and lubricants, SHOP can now also be operated without isomerization and metathesis, as the product distribution of ethylene oligomerization (*Schulz-Flory* distribution) favors short-chain alkenes [115].

Another classic industrial process employing a cross metathesis is the Phillips triolefin process, which is used for the production of propene from ethylene and butene. The changed availability of the raw material led to a temporary discontinuation of this process. The reverse reaction starting from propene has, however, come back into focus [109].

The ring-opening metathesis polymerization (ROMP) is used on a large scale to produce polynorbornenes or polydicyclopentadienes. The transformation is not limited to opening of the highly strained double bond in the bicyclic substrate to afford **128**. Under certain conditions, the fused cyclopentene can also polymerize (**129**) and introduce cross-linking into the polymer (see Fig. 7.40) [109]. The resulting product is a solid thermosetting polymer with excellent impact strength, from which larger objects can also be produced by reactive injection molding.

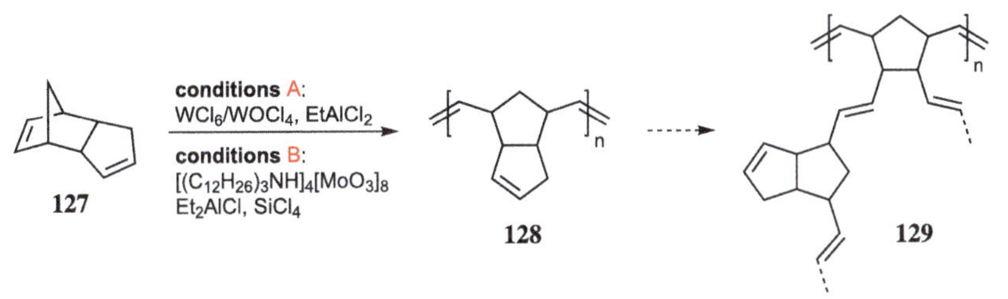

Fig. 7.40 ROMP of dicyclopentadiene [109]

The derivatization of oils and fats as sustainable platform chemicals constitutes another large-scale application. The long-chain, single, or polyunsaturated fatty acids can either be derivatized at the polar carboxylic acid group (reduction, esterification) or the double bonds can be altered by a variety of methods (epoxidation, hydrogenation, hydroformylation, hydroaminomethylation) [119, 120]. Olefin metathesis is another very common method, which can fundamentally change the properties of the fatty acids, as functional termini can be easily incorporated into the backbone through cross metathesis with acrylates, acrylonitrile, or butenediol. In addition, dimerizations or the cleavage of the olefin by reaction with ethylene can also be realized [121, 122].

References

1. M. Inoue, K. Miyazaki, H. Uehara, M. Maruyama, M. Hirama, *Proc. Nat. Acad. Sci.* **2004**, *101*, 12013–12018
2. C. A. Morales, M. E. Layton, M. D. Shair, *Proc. Nat. Acad. Sci.* **2004**, *101*, 12036–12041
3. J. Marjanovic, S. A. Kozmin, *Angew. Chem. Int. Ed.* **2007**, *46*, 8854–8857
4. N. Calderon, H. Y. Chen, K. W. Scott, *Tetrahedron Lett.* **1967**, *8*, 3327–3329
5. R. H. Grubbs, *Angew. Chem. Int. Ed.* **2006**, *45*, 3760–3765
6. R. R. Schrock, *Angew. Chem. Int. Ed.* **2006**, *45*, 3748–3759
7. Y. Chauvin, *Angew. Chem. Int. Ed.* **2006**, *45*, 3740–3747
8. R. H. Grubbs, *Tetrahedron* **2004**, *60*, 7117–7140
9. G. C. Vougioukalakis, R. H. Grubbs, *Chem. Rev.* **2010**, *110*, 1746–1787
10. J. Heppekausen, A. Fürstner, *Angew. Chem. Int. Ed.* **2011**, *50*, 7829–7832
11. M. R. Buchmeiser, S. Sen, J. Unold, W. Frey, *Angew. Chem. Int. Ed.* **2014**, *53*, 9384–9388
12. *Handbook of Metathesis, Vol. 1-3*, (Eds.: R. H. Grubbs, D. J. Leary), Wiley-VCH, 2nd ed., **2015**
13. R. R. Schrock, A. H. Hoveyda, *Angew. Chem. Int. Ed.* **2003**, *42*, 4592–4633
14. T. M. Trnka, R. H. Grubbs, *Acc. Chem. Res.* **2001**, *34*, 18–29
15. K. C. Nicolaou, P. G. Bulger, D. Sarlah, *Angew. Chem. Int. Ed.* **2005**, *44*, 4490–4527
16. A. Fürstner, *Chem. Commun.* **2011**, *47*, 6505–6511
17. S. Kress, S. Blechert, *Chem. Soc. Rev.* **2012**, *41*, 4389–4408
18. J. A. Love, M. S. Sanford, M. W. Day, R. H. Grubbs, *J. Am. Chem. Soc.* **2003**, *125*, 10103–10109
19. B. F. Straub, *Adv. Synth. Catal.* **2007**, *349*, 204–214
20. B. F. Straub, *Angew. Chem. Int. Ed.* **2005**, *44*, 5974–5978
21. J. A. Love, J. P. Morgan, T. M. Trnka, R. H. Grubbs, *Angew. Chem. Int. Ed.* **2002**, *41*, 4035–4037

22. A. Kajetanowicz, K. Grela, *Angew. Chem. Int. Ed.* **2021**, *60*, 13738–13756
23. C. O. Blanco, J. Sims, D. L. Nascimento, A. Y. Goudreault, S. N. Steinmann, C. Michel, D. E. Fogg, *ACS Catal.* **2021**, *11*, 893–899
24. J.-L. Hérisson, Y. Chauvin, *Makromol. Chem.* **1971**, *141*, 161–176
25. D. J. Nelson, S. Manzini, C. A. Urbina-Blanco, S. P. Nolan, *Chem. Commun.* **2014**, *50*, 10355–10375
26. E. L. Dias, S. T. Nguyen, R. H. Grubbs, *J. Am. Chem. Soc.* **1997**, *119*, 3887–3897
27. R. R. Schrock, *Chem. Rev.* **2009**, *109*, 3211–3226
28. C. Adlhart, C. Hinderling, H. Baumann, P. Chen, *J. Am. Chem. Soc.* **2000**, *122*, 8204–8214
29. C. Copéret, Z. J. Berkson, K. W. Chan, J. de Jesus Silva, C. P. Gordon, M. Pucino, P. A. Zhizhko, *Chem. Sci.* **2021**, *12*, 3092–3115
30. P. E. Romero, W. E. Piers, *J. Am. Chem. Soc.* **2005**, *127*, 5032–5033
31. A. Poater, X. Solans-Monfort, E. Clot, C. Copéret, O. Eisenstein, *J. Am. Chem. Soc.* **2007**, *129*, 8207–8216
32. M. S. Sanford, J. A. Love, R. H. Grubbs, *J. Am. Chem. Soc.* **2001**, *123*, 6543–6554
33. V. Forcina, A. García-Domínguez, G. C. Lloyd-Jones, *Faraday Discuss.* **2019**, *220*, 179–195
34. T. Vorfalt, K.-J. Wannowius, H. Plenio, *Angew. Chem. Int. Ed.* **2010**, *49*, 5533–5536
35. C. O. Blanco, D. E. Fogg, *ACS Catal.* **2023**, *13*, 1097–1102
36. K. Młodzikowska-Pieńko, B. Trzaskowski, *Organometallics* **2022**, *41*, 3627–3635
37. W. L. McClennan, S. A. Rufh, J. A. M. Lummiss, D. E. Fogg, *J. Am. Chem. Soc.* **2016**, *138*, 14668–14677
38. J. Feldman, J. S. Murdzek, W. M. Davis, R. R. Schrock, *Organometallics* **1989**, *6*, 2260–2265
39. J. Louie, R. H. Grubbs, *Organometallics* **2002**, *21*, 2153–2164
40. S. H. Hong, M. W. Day, R. H. Grubbs, *J. Am. Chem. Soc.* **2004**, *126*, 7414–7415
41. A. K. Chatterjee, T.-L. Choi, D. P. Sanders, R. H. Grubbs, *J. Am. Chem. Soc.* **2003**, *125*, 11360–11370
42. S. Kotha, M. K. Dipak, *Tetrahedron* **2012**, *68*, 397–421
43. M. J. Marsella, H. D. Maynard, R. H. Grubbs, *Angew. Chem. Int. Ed. Engl.* **1997**, *36*, 1101–1103
44. A. Fürstner, T. Dierkes, O. R. Thiel, G. Blanda, *Chem. Eur. J.* **2001**, *7*, 5286–5298
45. T. R. Hoye, C. S. Jeffrey, M. A. Tennakoon, J. Wang, H. Zhao, *J. Am. Chem. Soc.* **2004**, *126*, 10210–10211
46. A. H. Hoveyda, C. Qin, X. Z. Sui, Q. Liu, X. Li, A. Nikbakht, *Acc. Chem. Res.* **2023**, *56*, 2426–2446
47. J.-G. Boiteau, P. van de Weghe, J. Eustache, *Tetrahedron Lett.* **2001**, *42*, 239–242
48. K. B. Wiberg, *Angew. Chem. Int. Ed. Engl.* **1986**, *25*, 312–322
49. W. J. Zuercher, M. Hashimoto, R. H. Grubbs, *J. Am. Chem. Soc.* **1996**, *118*, 6634–6640
50. A. Fürstner, K. Langemann, *J. Am. Chem. Soc.* **1997**, *119*, 9130–9136
51. Y. Mu, T. T. Nguyen, F. W. van der Mei, R. R. Schrock, A. H. Hoveyda, *Angew. Chem. Int. Ed.* **2019**, *58*, 5365–5370
52. A. H. Hoveyda, *J. Org. Chem.* **2014**, *79*, 4763–4792
53. A. H. Hoveyda, S. J. Malcolmson, S. J. Meek, A. R. Zhugralin, *Angew. Chem. Int. Ed.* **2010**, *49*, 34–44
54. S. J. Malcolmson, S. J. Meek, E. S. Sattely, R. R. Schrock, A. H. Hoveyda, *Nature* **2008**, *456*, 933–937
55. E. S. Sattely, S. J. Meek, S. J. Malcolmson, R. R. Schrock, A. H. Hoveyda, *J. Am. Chem. Soc.* **2009**, *131*, 943–953
56. S. J. Meek, R. V. O'Brien, J. Llaveria, R. R. Schrock, A. H. Hoveyda, *Nature* **2011**, *471*, 461–466
57. A. Fürstner, *Science* **2013**, *341*, 1357 (UNSP 1229713)
58. S. Shahane, C. Bruneau, C. Fischmeister, *ChemCatChem* **2013**, *5*, 3436–3459

59. K. M. Dawood, K. Nomura, *Adv. Synth. Catal.* **2021**, *363*, 1970–1997

60. J. Morvan, A. Del Vecchio, J. Talcik, D. Bouëtard, M. Mauduit, *Eur. J. Org. Chem.* **2023**, *26*, e202300671

61. C. Xu, X. Shen, A. H. Hoveyda, *J. Am. Chem. Soc.* **2017**, *139*, 10919–10928

62. T. P. Montgomery, T. S. Ahmed, R. H. Grubbs, *Angew. Chem. Int. Ed.* **2017**, *56*, 11024–11036

63. I. Ibrahem, M. Yu, R. R. Schrock, A. H. Hoveyda, *J. Am. Chem. Soc.* **2009**, *131*, 3844–3845

64. P. Liu, X. Xu, X. Dong, B. K. Keitz, M. B. Herbert, R. H. Grubbs, K. N. Houk, *J. Am. Chem. Soc.* **2012**, *134*, 1464–1467

65. R. K. M. Khan, S. Torker, A. H. Hoveyda, *J. Am. Chem. Soc.* **2013**, *135*, 10258–10261

66. D. S. Müller, O. Baslé, M. Mauduit, *Beilstein J. Org. Chem.* **2018**, *14*, 2999–3010

67. M. Yu, R. R. Schrock, A. H. Hoveyda, *Angew. Chem. Int. Ed.* **2015**, *54*, 215–220

68. M. Yu, C. Wang, A. F. Kyle, P. Jakubec, D. J. Dixon, R. R. Schrock, A. H. Hoveyda, *Nature* **2011**, *479*, 88–93

69. C. Wang, M. Yu, A. F. Kyle, P. Jakubec, D. J. Dixon, R. R. Schrock, A. H. Hoveyda, *Chem. Eur. J.* **2013**, *19*, 2726–2740

70. A. H. Hoveyda, Z. Liu, C. Qin, T. Koengeter, Y. Mu, *Angew. Chem. Int. Ed.* **2020**, *59*, 22324–22348

71. J. A. Henderson, A. J. Phillips, *Angew. Chem. Int. Ed.* **2008**, *47*, 8499–8501

72. M. W. B. Pfeiffer, A. J. Phillips, *J. Am. Chem. Soc.* **2005**, *127*, 5334–5335

73. H. Villar, M. Fringsa, C. Bolm, *Chem. Soc. Rev.* **2007**, *36*, 55–66

74. S. T. Diver, A. J. Giessert, *Chem. Soc. Rev.* **2004**, *104*, 1317–1382

75. S. P. Nolan, H. Clavier, *Chem. Soc. Rev.* **2010**, *39*, 3305–3316

76. M. Mori, *J. Mol. Cat. A: Chemical* **2004**, *213*, 73–79

77. C.-I. Mitan, P. Filip, L. Delaude, V. Dragutan, *J. Organomet. Chem.* **2024**, *1014*, 123190

78. M. Mori, N. Sakakibara, A. Kinoshita, *J. Org. Chem.* **1998**, *63*, 6082–6083

79. Y. Zhao, A. H. Hoveyda, R. R. Schrock, *Org. Lett.* **2011**, *13*, 784–787

80. F. Nunez-Zarur, X. Solans-Monfort, L. Rodríguez-Santiago, M. Sodupe, *ACS Catal.* **2013**, *3*, 206–218

81. B. R. Galan, A. J. Giessert, J. B. Keister, S. T. Diver, *J. Am. Chem. Soc.* **2005**, *127*, 5762–5763

82. J. J. Lippstreu, B. F. Straub, *J. Am. Chem. Soc.* **2005**, *127*, 7444–7457

83. G. C. Lloyd-Jones, R. G. Margue, J. G. de Vries, *Angew. Chem. Int. Ed.* **2005**, *44*, 7442–7447

84. A. J. Giessert, N. J. Brazis, S. T. Diver, *Org. Lett.* **2003**, *5*, 3819–3822

85. A. G. D. Grotevendt, J. A. M. Lummiss, M. L. Mastronardi, D. E. Fogg, *J. Am. Chem. Soc.* **2011**, *133*, 15918–15921

86. T. M. Gregg, J. B. Keister, S. T. Diver, *J. Am. Chem. Soc.* **2013**, *135*, 16777–16780

87. H. M. Ko, C. W. Lee, H. K. Kwon, H. S. Chung, S. Y. Choi, Y. K. Chung, E. Lee, *Angew. Chem. Int. Ed.* **2009**, *48*, 2364–2366

88. J. Li, D. Lee in *Handbook of Metathesis*, *Vol. 2* (Eds.: R. H. Grubbs, D. J. Leary), Wiley-VCH, 2nd ed., **2015**, Chapter 5, pp. 381–444

89. G. C. Lloyd-Jones, *Org. Biomol. Chem.* **2003**, *1*, 215–236

90. A. Fürstner, *J. Am. Chem. Soc.* **2021**, *143*, 15538–15555

91. H. Ehrhorn, M. Tamm, *Chem. Eur. J.* **2018**, *25*, 3190–2308

92. A. Fürstner, *Angew. Chem. Int. Ed.* **2013**, *52*, 2794–2819

93. W. Zhang, J. S. Moore, *Adv. Synth. Catal.* **2007**, *349*, 93–120

94. J. Heppekausen, R. Stade, R. Goddard, A. Fürstner, *J. Am. Chem. Soc.* **2010**, *132*, 11045–11057

95. J. Heppekausen, R. Stade, A. Kondoh, G. Seidel, R. Goddard, A. Fürstner, *Chem. Eur. J.* **2012**, *18*, 10281–10299

96. T. J. Katz, J. McGinnis, *J. Am. Chem. Soc.* **1975**, *97*, 1592–1594

97. W. Zhang, S. Kraft, J. S. Moore, *J. Am. Chem. Soc.* **2004**, *126*, 329–335

98. A. Fürstner, D. D. Souza, L. Parra-Rapado, J. T. Jensen, *Angew. Chem. Int. Ed.* **2003**, *42*, 5358–5360

99. A. Fürstner, P. W. Davies, *Chem. Commun.* **2005**, 2307–2320

100. J. Hillenbrand, M. Leutzsch, E. Yiannakas, C. P. Gordon, C. Wille, N. Nöthling, C. Copéret, A. Fürstner, *J. Am. Chem. Soc.* **2020**, *142*, 11279–11294

101. Y. Ge, S. Huang, Y. Hu, L. Zhang, L. He, S. Krajewski, M. Ortiz, Y. Jin, W. Zhang, *Nat. Commun.* **2021**, *12*, 1136

102. R. Lhermet, A. Fürstner, *Chem. Eur. J.* **2014**, *20*, 13188–13193

103. B. Haberlag, M. Freytag, C. G. Daniliuc, P. G. Jones, M. Tamm, *Angew. Chem. Int. Ed.* **2012**, *51*, 13019–13022

104. J. Willwacher, B. Heggen, C. Wirtz, W. Thiel, A. Fürstner, *Chem. Eur. J.* **2015**, *21*, 10416–10430

105. K. Radkowski, B. Sundararaju, A. Fürstner, *Angew. Chem. Int. Ed.* **2013**, *52*, 355–360

106. E. D. Slack, C. M. Gabriel, B. H. Lipshutz, *Angew. Chem. Int. Ed.* **2014**, *53*, 14051–14054

107. D. Srimani, Y. Diskin-Posner, Y. Ben-David, D. Milstein, *Angew. Chem. Int. Ed.* **2013**, *52*, 14131–14134

108. S. D. Roughley, A. M. Jordan, *J. Med. Chem.* **2011**, *54*, 3451–3479

109. J. C. Mol, *J. Mol. Cat. A: Chemical* **2004**, *213*, 39–45

110. S. Lwin, I. E. Wachs, *ACS Catal.* **2014**, *4*, 2505–2520

111. J. Lee, K. A. Parker, *Org. Lett.* **2012**, *14*, 2682–2685

112. K. Molawi, N. Delpont, A. M. Echavarren, *Angew. Chem. Int. Ed.* **2010**, *49*, 3517–3519

113. H. Wang, H. Matsuhashi, B. D. Doan, S. N. Goodman, X. Ouyang, W. M. Clark, Jr., *Tetrahedron* **2009**, *65*, 6291–6303

114. G. S. Weatherhead, G. A. Cortez, R. R. Schrock, A. H. Hoveyda, *Proc. Nat. Acad. Sci.* **2004**, *101*, 5805–5809

115. W. Keim, *Angew. Chem. Int. Ed.* **2013**, *52*, 12492–12496

116. B. Reuben, H. Wlttcoff, *J. Chem. Educ.* **1988**, *65*, 605–607

117. E. F. Lutz, *J. Chem. Educ.* **1986**, *63*, 202–203

118. W. Keim, *Chem. Ing. Tech.* **1984**, *56*, 850–853

119. U. Biermann, U. T. Bornscheuer, I. Feussner, M. A. R. Meier, J. O. Metzger, *Angew. Chem. Int. Ed.* **2021**, *60*, 20144–20165

120. A. Behr, A. Westfechtel, J. Pérez Gomes, *Chem. Eng. Technol.* **2008**, *31*, 700–714

121. S. Chikkali, S. Mecking, *Angew. Chem. Int. Ed.* **2012**, *51*, 5802–5808

122. V. Yelchuri, K. Srikanth, R. B. Prasad, M. S. L. Karuna, *J. Chem. Sci.* **2019**, *131*, 39

Organocatalysis

8

Organocatalysis entails the acceleration of chemical reactions through the addition of a substoichiometric quantity of an organic, metal-free compound instead of metal organyls, metal complexes, as well as metal salts or metal oxides. This approach does not constitute a new class of reaction per se, but rather describes an alternative avenue to established catalytic methods. For this reason, a plethora of reactions can be promoted by the presence of an organocatalyst, which enables the isolation of the desired products as efficiently as with standard (metal-based) methods or even with superior yield and selectivity.

In addition to new methods as an alternative to already established metal-catalyzed variants, there are common transformations that have been mediated by organocatalytic systems since their inception (see Fig. 8.1). "Modern" organocatalysis usually refers to asymmetric methods for this reason.

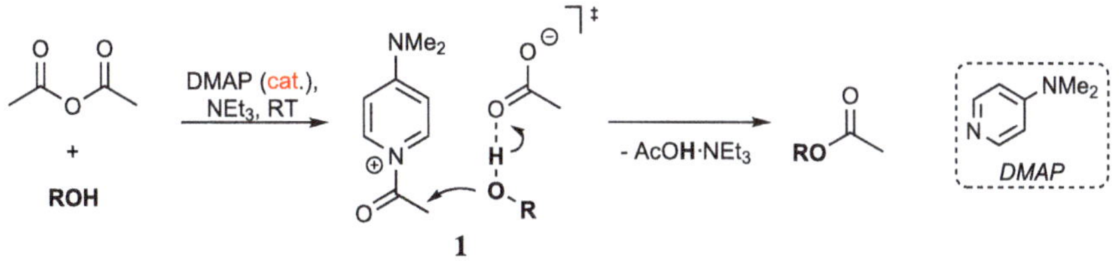

Fig. 8.1 DMAP-catalyzed esterification of alcohols [1]

The advantages and disadvantages of organocatalytic variants compared to classic methods are highly catalyst- and method-specific: While enamine and iminium catalysis tolerate water and oxygen, the use of TMS-triflate in a Lewis acid-catalyzed reaction in contrast requires the rigorous exclusion of traces of water. The frequently mentioned low toxicity

A. Düfert, *Methods of Organic Synthesis*,
https://doi.org/10.1007/978-3-662-70963-4_8

when using organocatalysts strictly applies only to proline. Moreover, the substrate and especially the solvent are often much more critical in the context of toxicity, as they are present in considerably higher quantities than the catalyst. The much-cited high catalyst load of organocatalytic methods applies primarily to covalently bound catalysts, as the formation of the catalytically active species proceeds much slower than a "simple" and very fast activation via hydrogen bonds. Even when using Cinchona alkaloids, higher amounts of the catalyst are required as the alkaloid is prone to decomposition by Hofmann elimination. Many metal-catalyzed methods originally necessitated comparable catalyst loads of >5 mol%, and only years of successive optimization studies led to an increase in efficiency. Organocatalysis can thus be considered complementary to biocatalysis and metal catalysis in many cases [2, 3].

To enable a reaction, at least one of the substrates needs to be activated. For better overview and classification, the reactions will henceforth be classified *based on the activated substrate*. If a ketone attacks a nitroalkene, for example, the reaction could be characterized either as an α-functionalization of a carbonyl (activation of the ketone) or as a Michael addition (activation of the nitroalkene).

α-functionalization
or
Michael addition

Given the nature of the organocatalysts, certain "privileged" transformations exist that particularly lend themselves for an organocatalytic approach. The C=O bond constitutes the most commonly utilized functional group to effect an activation, both for α-functionalizations and 1,2-additions, as well as 1,4-additions in α, β-unsaturated substrates. In addition, CN double bonds can also be activated, and CO and CN single bonds are ideally suited for nucleophilic substitution reactions (see Table 8.1).

The intramolecular conversion of the ketone **2** in the presence of L-proline, independently developed at Hoffmann-La Roche and Schering, is cited as a pivotal milestone of modern organocatalysis [6–8]. Depending on the reaction conditions, either the hydroxyketone **3** or, via concluding elimination of water, the aldol condensation product **4** are afforded with high enantiomeric excess (see Fig. 8.2). The transformation is referred to as the Hajos–Parrish–Eder–Sauer–Wiechert reaction in honor of the original developers, and plays an important role in the synthesis of steroid frameworks.

Table 8.1 Overview of common organocatalytic transformations [4, 5]

reaction category	examples
α-functionalization of carbonyls	aldol reaction
	Mannich reaction
	α-alkylation
	Michael addition
	α-aminoxylation
	α-amination
	α-halogenation
	α-arylation
1,2-addition	imine reduction (e.g. Hantzsch ester)
	Mannich reaction
	Strecker reaction
	nucleophilic substitution (2° alcohols)
	epoxide/aziridine opening
	CBS reduction
	Corey-Chaykovsky reaction
	Knoevenagel condensation
alkene activation	Diels-Alder reaction
	epoxidation
1,4-addition	Michael addition (C/N/S nucleophiles)
	1,4-reduction (e.g. Hantzsch ester)
Umpolung	ester formation / transesterification
	acyl transfer (CC coupling)
H_2 activation	reduction (C=C, C=O, C=N, C≡C)

CO$_2$H

± Säure

2   3   4

Hajos-Parrish	L-proline (3%), DMF, RT, 20 h	**100%**, 93% *ee*
Eder-Sauer-Wiechert	L-proline (48%), 1 N HClO$_4$, MeCN, 80 °C, 22 h	**87%**, 84% *ee*

Fig. 8.2 Intramolecular asymmetric aldol reaction [7, 8]

Building on these preceding contributions, the groups of List and MacMillan disclosed the development of an intermolecular aldol reaction and a Diels–Alder reaction in the presence of secondary amines as organocatalysts in 2000 (see Fig. 8.3), renewing interest in organocatalysis approach and subsequently expanding its scope [9, 10]. Their contributions were ultimately acknowledged with the 2021 Nobel prize in chemistry.

The secondary amines used in these seminal studies opened the field of modern catalyst systems. While many current applications use these types of organocatalysts (via enamines or imines as intermediates), current research has shifted its focus on other classes of catalysts and reaction modes: After the emergence of Brønsted acids and H-bond donors, newer studies

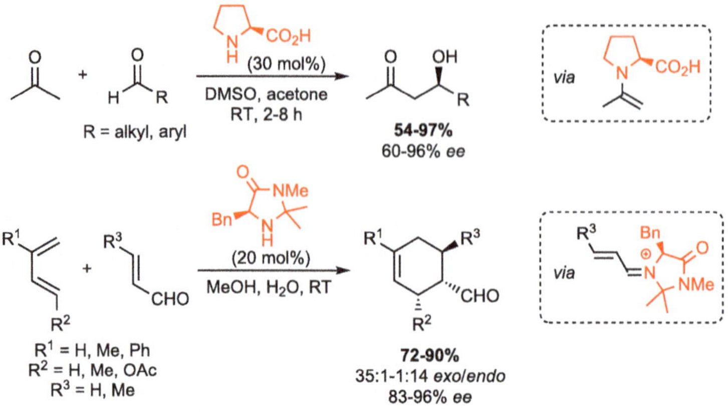

Fig. 8.3 Organocatalytic aldol and Diels–Alder reaction [9, 10]

investigate an asymmetric induction through non-covalent interactions of chiral counterions and the synergistic combination of organocatalysts with metal or photoredox catalysts.

8.1 Reaction Modes, Catalyst Classes, and Applications

Given that many catalysts are particularly suited for a specific set of transformations and the various classes of catalysts operate through very emblematic mechanisms, the scientific literature often uses the class of catalyst and the mechanism synonymously. The most common reaction modes can be roughly divided as follows[1]:

- enamine catalysis
- iminium catalysis
- Umpolung reactions
- activation via H-bonds
- activation through ionic interactions
- other mechanisms

The division into these categories is (unfortunately) rarely as clear-cut as this list might suggest. The enamine intermediates obtained with proline, for example, additionally activate the electrophile via a non-covalent hydrogen bond. Brønsted acids can also cover a wide range between pure hydrogen bond interactions and a protonation of the reactant and thus an ionic interaction. In addition, they often act in a bifunctional manner. The reaction modes are therefore only intended to help with an overall orientation and should be seen more

---

[1] A significantly more holistic but less intuitively understandable classification of organocatalytic reactions is based on the type of initial interaction of the catalyst with the reactant. See *Org. Biomol. Chem.* **2005**, *3*, 719–724.

as an ideal, prototypical guidance. Based on the mechanisms presented in Fig. 8.4 and the illustrated catalysts considered primed to promote them, the mechanistic intertwining of the catalyst classes becomes evident.

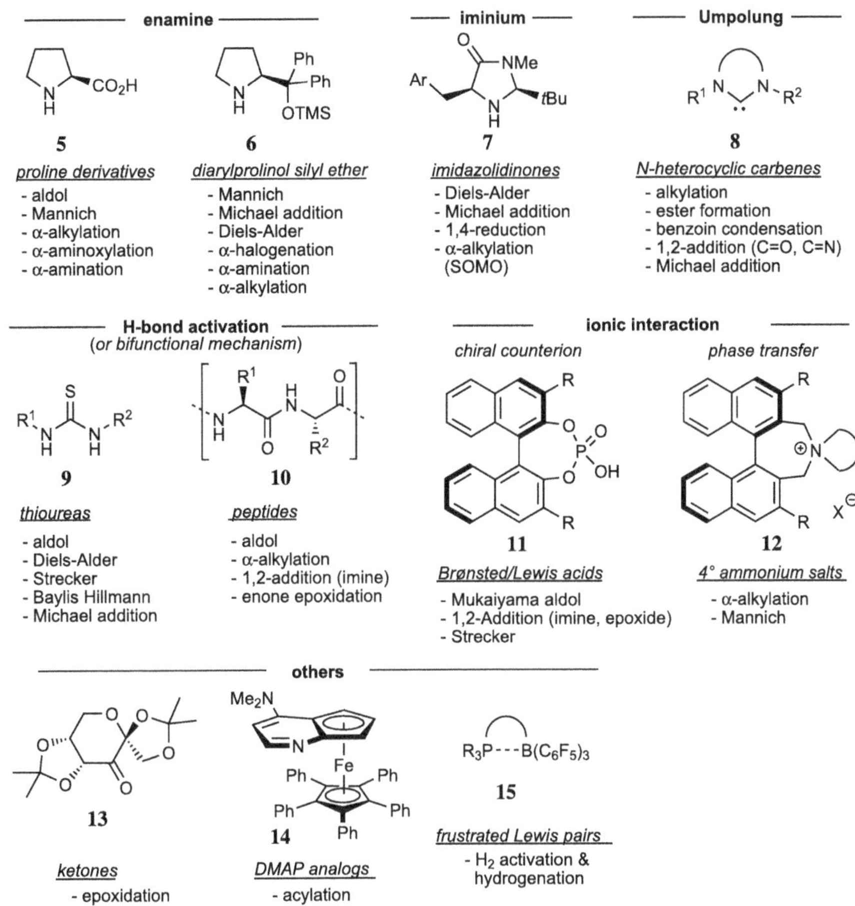

Fig. 8.4 Mechanisms, common catalysts and privileged reactions [4, 5, 11–20]

Since the first seminal contributions on secondary amines as an organocatalyst by Stork *et al.* [21] and proline (**5**) in particular, the proline scaffold has been extensively varied, especially the carboxylate moiety (tetrazole, sulfonamide, ester, etc.). A particularly noteworthy class of derivatives are the prolinol silyl ethers (**6**), which catalyze a very wide range of reactions. To overcome the challenges of sterically demanding substrates, primary amines were subsequently introduced. The MacMillan group developed a series of imidazolidinones (**7**), which are the most common catalysts for activating α, β-unsaturated carbonyl derivatives via iminium intermediates. In addition to these specific methods, enamine and iminium catalysis can generally be achieved with a wide range of amines (see below). *N*-heterocyclic carbenes (**8**) are particularly suited for the Umpolung of aldehydes and the oxidative conversion of these to various ketones, esters, or amides. Thioureas (**9**) and peptide catalysts (**10**) act via non-covalent hydrogen bonds. Many bifunctional catalysts, which can activate both reactants simultaneously, also stem from this category of catalysts. Depending on the exact catalyst structure, these function in a completely non-covalent manner, or one substrate can be activated covalently and the other via hydrogen bonds. The catalytic activation of Brønsted and Lewis acids can rely at one end of the spectrum on a purely ionic interaction with the substrate, yet often they also act via H-bond interactions or bifunctionally. While BINOL-derived phosphoric acids (**11**) are the most extensively studied catalysts of the class of hydrogen bond donors, spiro derivatives and other acidic functionalities such as sulfonamides have also been exploited. Quaternary ammonium catalysts (**12**), which are commonly encountered in asymmetric phase transfer reactions, usually rely on a binaphthyl backbone for chiral induction. Their application is almost exclusively limited to alkylation reactions, where they display an outstanding efficiency. Other noteworthy catalysts include ketones, which effect an asymmetric epoxidation as corresponding dioxiranes (e.g., Shi catalyst **13**), as well as various chiral DMAP derivatives for the desymmetrization of *meso* compounds. Even though the planar-chiral system **14** developed by Fu and co-workers contains ferrocene, the metal center only has a supporting, structural role and does not participate in the catalytic cycle. Other DMAP-derived catalysts efficiently function without containing a metal atom, but are employed less widely for very specific applications. Frustrated Lewis pairs comprise the combination of electron-poor triarylboranes with electron-rich Lewis bases, often derived from phosphines, which are structurally prevented from forming a stable donor-acceptor bond (**15**). These systems serve to activate H_2 for hydrogenations of alkenes, alkynes, imines, and occasionally also ketones.

Organocatalysts can act either **covalently** or **non-covalently** via *ionic interaction* or through *hydrogen bond formation*. Most organocatalytic activation modes in the recently published studies are based on non-covalent interactions, while previous contributions accordingly relied on the then-available covalent methods using enamine/iminium intermediates. Modern, bifunctional catalysts, however, can also act in both ways (see Fig. 8.5).

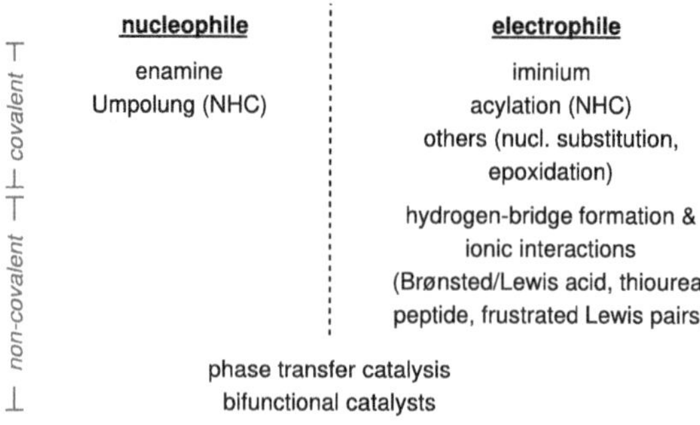

Fig. 8.5 Reaction modes and types of catalyst/reactant interaction

The range of available methods for organocatalytic transformations is exceedingly broad. If one approaches the subject from the frequency of use in total synthesis [22–27] or an industrial [2, 28–31] context, the scope of methods noticeably narrows and some methods particularly lend themselves for application beyond a rather narrowly defined area. Since organocatalysts are usually chiral, the majority of published methods have been developed for asymmetric transformation since the original contributions by List and MacMillan. Although the use on very complex intermediates [32] has room for improvement in absolute terms, the reported examples convincingly demonstrate the advantage of the organocatalytic approach (see Fig. 8.6).

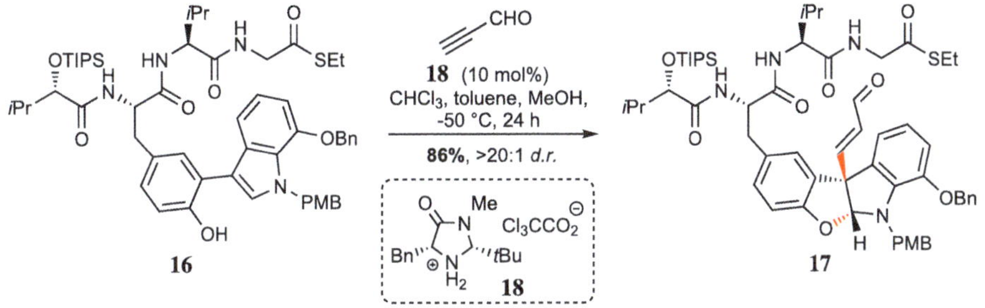

Fig. 8.6 Key step in the total synthesis of diazonamide A reported by MacMillan [33]

The majority of published examples report proportionately high catalyst loads. This may be attributed to the relatively small amount of starting material in total syntheses and the costs of individual steps are generally not subject to commercial criteria. However, there are some methods, especially in the area of cationic phase transfer catalysis, where less than 1% of the catalyst is required [34]. On an industrial scale, individual examples have been carried out in the presence of 10–15 mol% of catalyst, despite the consideration of catalyst costs. *Commercial* production routes of pharmaceuticals so far do not include any organocatalytic steps — the only notable exception being the synthesis of the cholesterol-lowering drug ezetimibe (see Fig. 8.7) [35, 36]. However, details regarding commercial manufacturing routes are only rarely disclosed so it is likely some examples exist that are not known to the scientific community.

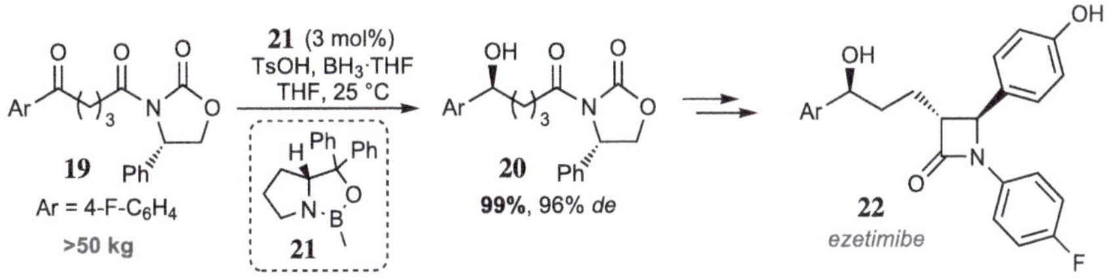

Fig. 8.7 CBS reduction in the manufacturing route of ezetimibe (**22**) [35, 36]

The use of the CBS reduction, which is rarely considered to be an organocatalytic transformation in the literature, enables the selective construction of the benzylic stereocenter in **20** with excellent diastereoselectivity, starting from ketone **19**. The addition of small amounts of acid increases the stereoselectivity by neutralizing trace amounts of nonselective $NaBH_4$, which is contained in technical BH_3 as a stabilizer.

In the synthesis of complex natural products and active pharmaceutical ingredients, the following reaction modes are mainly encountered:

- enamine catalysis (α-alkylation)
- iminium catalysis (Diels–Alder reaction and Michael addition) as well as
- phase transfer catalysis (α-alkylation)

In addition to these prevalent approaches, further selected methods and syntheses are presented below. Nevertheless, the given selection cannot in any way representatively cover this continuously evolving field. The interested reader is hereby referred to the relevant review articles and monographs [11–20, 27, 30, 31, 37–40, 40–62].

8.1.1 Enamine Catalysis

The organocatalytic conversion of carbonyl derivatives in the presence of secondary or primary amines mechanistically relies on the formation of enamine intermediates. These derivatives are characterized by a raised HOMO compared to the respective substrates. Initially, proline derivatives dominated as catalysts, but their limitations subsequently led to the development of methods based on further amines. While secondary amines like proline are suitable for the conversion of α-unsubstituted aldehydes and ketones, α-aryl-substituted aldehydes require sterically less demanding primary amines. α-Substituted ketones can only be efficiently converted with Brønsted acids, which, however, operate via a non-covalent mechanism (see Fig. 8.8).

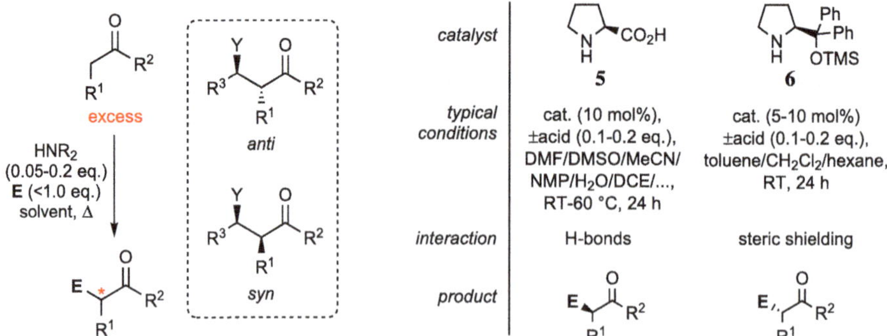

Fig. 8.8 General reaction scheme of enamine catalysis and common reaction conditions

The amines generate enamines with ketones or aldehydes, which are more reactive than the corresponding enol of the original carbonyl substrate by virtue of an increased HOMO and thus allow the desired α-functionalization to occur under very mild conditions. This reaction mode was initially developed using proline as an organocatalyst for an enantioselective reaction. The stoichiometric application of Enders SAMP/RAMP reagent for α-alkylations of ketones and aldehydes is based on a similar principle (see Sect. 2.4). While most theoretical and experimental mechanistic studies focused on proline due to its central role in early method development, many conclusions can be transferred to other amines as catalysts.

The reversible formation of enamines from a carbonyl compound in the presence of catalytic amounts of an amine constitutes the mechanistic foundation of enamine catalysis (see Fig. 8.9) [11].

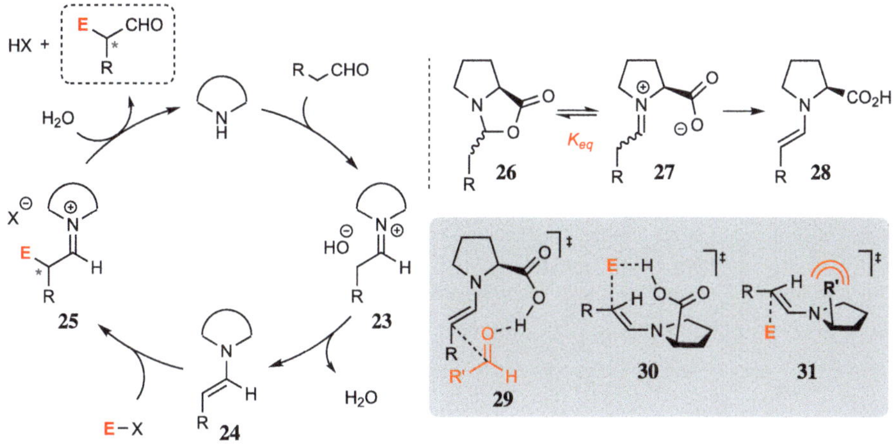

Fig. 8.9 Postulated mechanism of the α-functionalization of carbonyls as well as intermediates and transition states [63–66]

The formation of an iminium ion **23** lowers the LUMO of the C=X bond and increases the α-acidity through a π^*-σ^* interaction, [12] resulting in the formation of enamine **24** upon elimination of H_2O. If stereoinducing elements are present on the amine, an approach from either the *Re*- or *Si*-face of the enamine is favored, depending on the individual catalyst in question. The subsequent reaction with an electrophile by preferential addition from one of the enantio- or diastereotopic faces thus yields the desired product **25** and regenerates the amine. In contrast to the classic aldol reaction, where a defined order of addition may be required to suppress a successive, multiple functionalization of the α-acidic position, organocatalytic aldol reactions stop at the stage of the aldol product.

In the proline-mediated reaction, diverse mechanistic variations have been postulated for the formation of the imine **23** and enamine **24** derived from in-depth experimental and theoretical studies. A direct formation of the enamine **24** from **23** does not seem to occur [67]. The zwitterionic species **26** is assumed to be in equilibrium with oxazolidinone **27**. The position of the equilibrium and the role of **26** as either a resting state or parasitic species in the catalytic cycle have been controversially discussed [64, 65]. The dominant path seems to be able to change with the set of conditions and the presence and strength of acidic/basic additives. **29** is generally assumed to be the operative transition state for the proline-catalyzed aldol reaction, in which the acid functionality directs the electrophile to the *Re*-face of the enamine **28**. In addition to carbonyl compounds, other electrophiles add to the *Re*-face directed by the acid in a similar fashion (via **30**). In contrast to an active stereoinduction, proline analogs without Brønsted-acidic functional groups add *anti* to the substituent R' on the basis of steric repulsion (via transition state **31**). Whether the diarylprolinol silyl ethers exclusively differentiate on the basis of steric repulsion can be questioned, as the silyl ether is prone to cleavage depending on the employed reaction conditions. Nevertheless, the paradigm should apply to the majority of reactions (see Fig. 8.10).

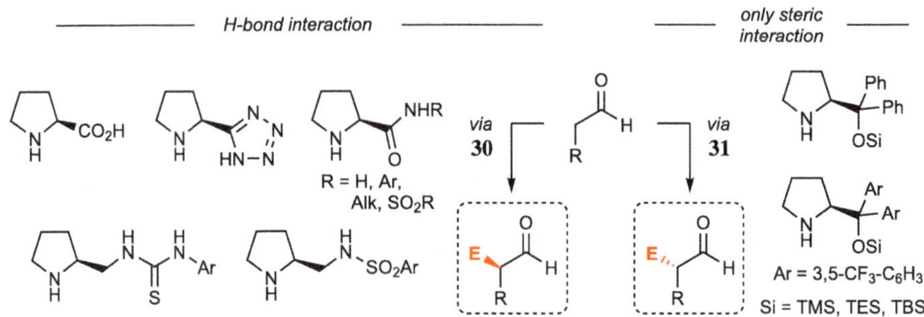

Fig. 8.10 Key structures and observed diastereoselectivity of proline derivatives [11, 40, 68]

In addition to proline derivatives, a variety of other amines has been reported, which also affords the corresponding products via enamines: amino acids, peptides as well as non-amino acid-derived primary or secondary amines [37]. The observed stereoselectivity of these systems typically cannot be rationalized by the transition states **30** and **31**, if they have been elucidated at all. These catalysts are less frequently used with standard substrates, while more densely substituted carbonyl derivatives demand the use of advanced methods [11].

Catalyst loadings of 10–20 mol% are typical for reactions proceeding through enamine intermediates. Although there are isolated examples with lower catalyst loadings, most catalysts are inefficient under these conditions, which, in addition to lower reactivity, might suggest a possible catalyst deactivation [67]. The addition of a Brønsted acid or base can positively influence the reaction rate and diastereoselectivity for some electrophiles or select proline derivatives. This can be attributed to an accelerated iminium formation if acids are added and a faster α-deprotonation and thus more rapid enamine formation when bases are present. The reaction can be significantly accelerated, especially for lethargic substrates, if the acid/base properties of the substrate and catalyst are aligned with additives [67]. Catalysts that utilize hydrogen bonds as directing element necessitate polar aprotic solvents. Proton sources such as H_2O or alcohols are usually added with defined stoichiometry. Amines relying on steric repulsion for facial discrimination, on the other hand, are best matched with apolar to weakly polar aprotic solvents (cf. Fig. 8.8). Excepting intramolecular transformations, the ketone (or the aldehyde) which subsequently furnishes the enamine (so-called donor) is typically used in excess to shift equilibria and selectivities to endorse the formation of the desired product. For this reason, structurally simple and cost-effective substrates are preferably used as donors (see Fig. 8.11).

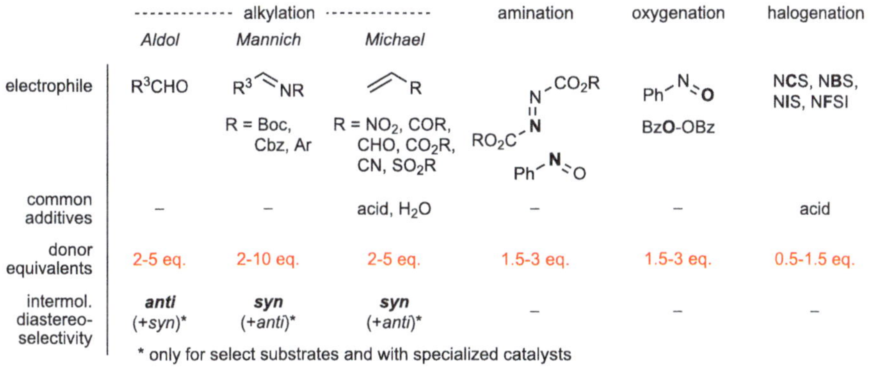

	---------- alkylation ----------			amination	oxygenation	halogenation
	Aldol	*Mannich*	*Michael*			
electrophile	R^3CHO	R^3⁔NR R = Boc, Cbz, Ar	⟋R R = NO_2, COR, CHO, CO_2R, CN, SO_2R	RO_2C–N=N–CO_2R Ph–N=O	Ph–N=O BzO-OBz	NCS, NBS, NIS, NFSI
common additives	–	–	acid, H_2O	–	–	acid
donor equivalents	2-5 eq.	2-10 eq.	2-5 eq.	1.5-3 eq.	1.5-3 eq.	0.5-1.5 eq.
intermol. diastereo-selectivity	**anti** (+*syn*)*	**syn** (+*anti*)*	**syn** (+*anti*)*	–	–	–

* only for select substrates and with specialized catalysts

Fig. 8.11 Substrate scope, equivalents of nucleophile and diastereoselectivity of intermolecular reactions [38, 69]

Catalysts based on an active facial control via H-bonds enable particularly high yield and selectivity in reactions where the electrophile can be additionally activated by H-bonding, such as aldol and Mannich reactions, α-aminations and α-oxygenations. Catalysts in which steric repulsion governs facial selectivity are best suited for transformations where intramolecular H-bonds are either not required or largely inconsequential, such as α-alkylations or α-halogenations [69]. Most reactions are carried out on functionally simple substrates. Due to the requisite excess of donor discussed above for intermolecular reactions, these functionalizations are typically placed very early in a synthetic sequence. Often, these first stereocenters serve as anchors for the subsequent introduction of further stereoinformation, for example, in the syntheses of callipeltoside C (**32**), chloptosin, and *Galbulimima* alkaloid (−)-GB17 (see Fig. 8.12).

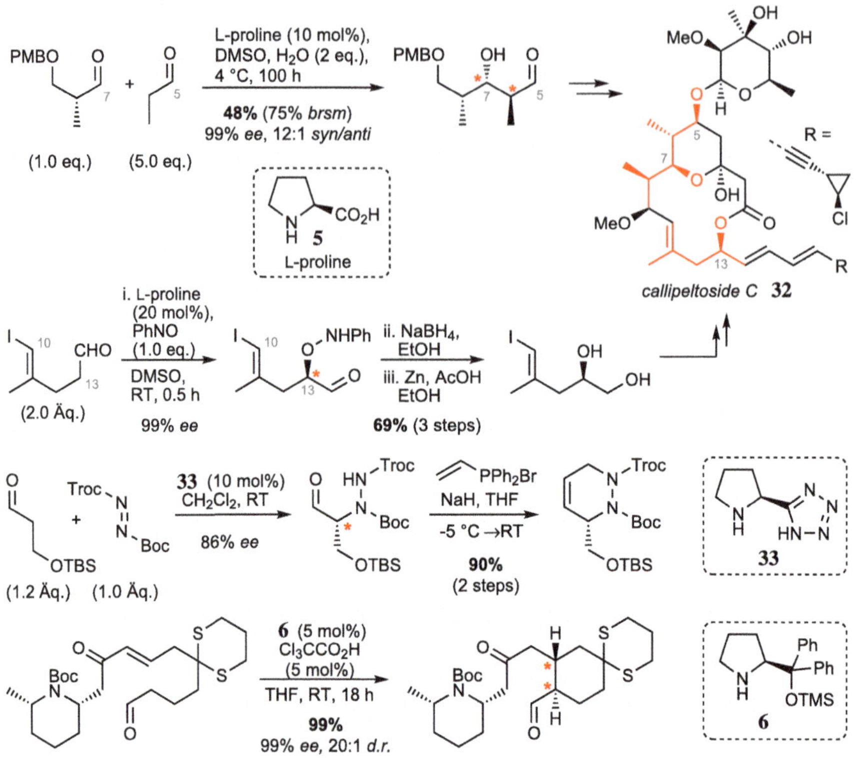

Fig. 8.12 Examples of enamine-catalyzed aldol, oxygenation, amination, and Michael reactions in total syntheses [70–72]

A complementary approach is the formation of radical intermediates by one-electron oxidation of the intermediate enamines, which as a result undergo an α-functionalization. The reaction mode referred to by MacMillan as *SOMO activation* was originally initiated using stoichiometric oxidants such as ceric ammonium nitrate (CAN) or iron(III) salts, but alternative oxidation methods, such as photoredox catalysts, are now preferred. The use of organocatalytic SOMO reactions has not yet been widely established in synthesis [73–75].

8.1.2 Iminium Ion Catalysis

Reactions in which iminium ions constitute intermediate species typically rely on the utilization of imidazolidinones and proline derivatives as catalysts. Both secondary and primary amines are suitable, with secondary amines tending to dominate the field. Primary amines require an acid as a co-catalyst (via **35**) — however, this is also a typical additive when using secondary amines to facilitate the formation of the iminium species. The covalent activation facilitates the conversion of electrophiles in cycloadditions, 1,4-additions and α, β-derivatizations (epoxidation, aziridination, cyclopropanation).

While the majority of catalysts are based on secondary amines, which were introduced in the groups of MacMillan (imidazolidinones) and Jørgensen (prolinol silyl ether), there is a growing number of methods that resort to primary amines derived from Cinchona alkaloids. Particularly sterically more congested substrates such as enones (instead of enals) and α-substituted acroleins can be efficiently converted with good to excellent enantioselectivity. The reported pericyclic reactions usually encompass classic [4+2]-cycloadditions as well as a small contingent of 1,3-dipolar cycloadditions. The variability of the nucleophiles in the 1,4-additions also comprises a wide range: Apart from the reaction of C-nucleophiles, the formation C-X bonds with hydroxylamines, oximes, peracids as well as P- and S-nucleophiles is feasible. Finally, the CC double bond in α, β-position can be reduced to afford the corresponding saturated carbonyl compounds. Most variants employ Hantzsch esters as hydride equivalents (see Fig. 8.13) [76, 77].

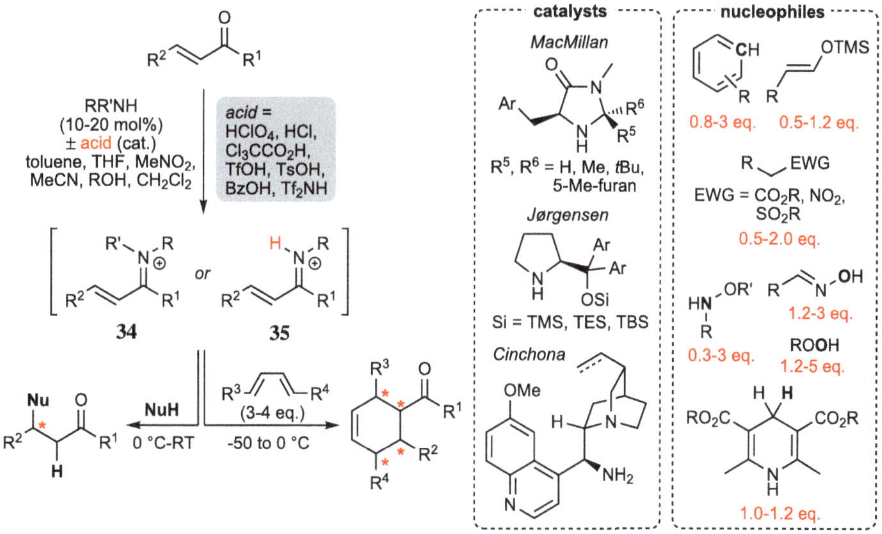

Fig. 8.13 Typical reaction conditions and reaction classes in iminium-catalyzed conversions of α, β-unsaturated carbonyls. The equivalents mark the commonly used amount of nucleophile [76, 77]

The choice of solvent ranges from nonpolar (toluene, infrequently hexane) to polar-protic. Alcohols are very rarely used, as the formation of *N,O*- or *O,O*-acetals could ensue. In cycloadditions, cryogenic conditions and a significant excess of diene or 1,3-dipole are usually required. In the addition of nucleophiles, either the carbonyl or the nucleophile can constitute the stoichiometrically limiting component — this depends on the respective method and the individual nucleophile. A slight to moderate excess of the nucleophile is common for nitroalkanes (C-nucleophile) as well as oximes, peracids, and Hantzsch esters.

1,4-additions are usually carried out at room temperature. Sluggish substrates may require slightly elevated temperatures around 50–80 °C [76, 77].

Mechanistically, the mode of activation is based on a condensation of the secondary amines with α, β-unsaturated carbonyl derivatives to give transient iminium ions, which possess a lower LUMO. The resulting increased reactivity compared to the parent aldehydes and ketones activates the CC double bond for a 1,4-addition or pericyclic reaction (see Fig. 8.14) [67, 78]. To support the generation of the iminium ion, the use of Brønsted-acidic additives is common. If the amine catalysts are used as the respective salt, the counterion can also play an important role in the catalytic cycle.

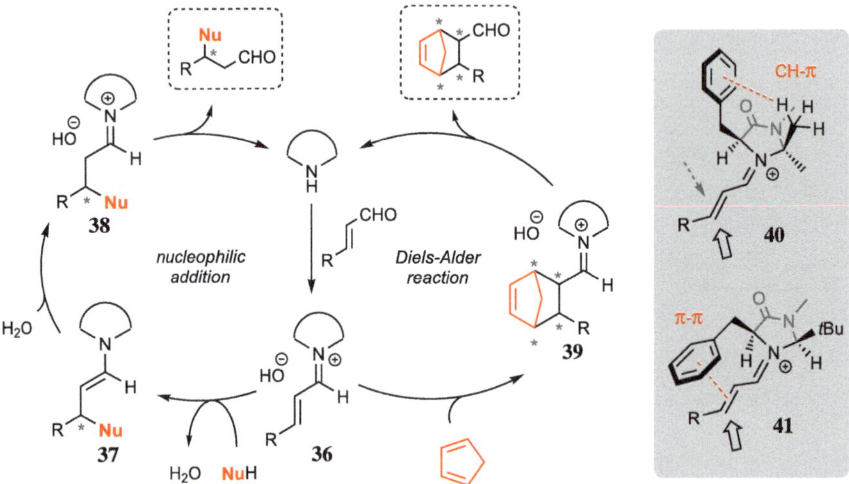

Fig. 8.14 Postulated catalytic cycles in the conversion of α, β-unsaturated carbonyls and purported reactive conformers of **36** [67, 78, 79]

Following the formation of the iminium ion **36** with (usually cyclic) secondary amines, the activated intermediate serves as electrophile. A 1,4-attack of a nucleophile (C/N/O/H) affords enamine **37**, which after protonation of the double bond reforms an iminium species of type **38**. After the elimination of the amine catalyst, the desired addition product is obtained, whereby the configuration of its newly introduced stereocenter is derived from the catalyst-controlled stereoinducing Michael addition. In the case of a pericyclic reaction, **36** acts as a dienophile. The addition product **39** can comprise up to four stereocenters depending on the substitution pattern, which can be selectively synthesized. Following the regeneration of the organocatalyst, the catalytic cycle can commence again.

In the case of the imidazolidinones introduced by MacMillan, two different, reactive conformers were postulated depending on the substitution pattern, to rationalize the observed facial selectivity of the reaction. For the geminal disubstituted dimethyl derivative, conformer **40** stabilized by a CH-π interaction was determined to be energetically favored based on theoretical and experimental studies. The subsequently introduced *t*butyl-derived imidazolidinone **7** was indicated to preferentially assume **41** as reactive conformer in theoretical studies, in which a π-π interaction between the benzyl group and the CC double bond was postulated. In both cases, the stereoselectivity results from the shielding of the (upper) α-*Si*-face by the benzyl group [79, 80].

In cycloadditions, the α, β-unsaturated carbonyl is activated by lowering the LUMO and serves as dienophile or dipolarophile. This type of reaction often benefits from the presence of catalytic amounts of an acid as an additive. In Diels–Alder reactions, a superior reactivity was observed under imidazolidinone catalysis in direct comparison to diarylprolinol silyl ethers [81]. The MacMillan systems preferentially provide the *endo*-product in [4+2]-cycloadditions with the exception of cyclopentadiene, while diarylprolinol silyl ethers afford the *exo*-product (see Fig. 8.15) [76, 82].

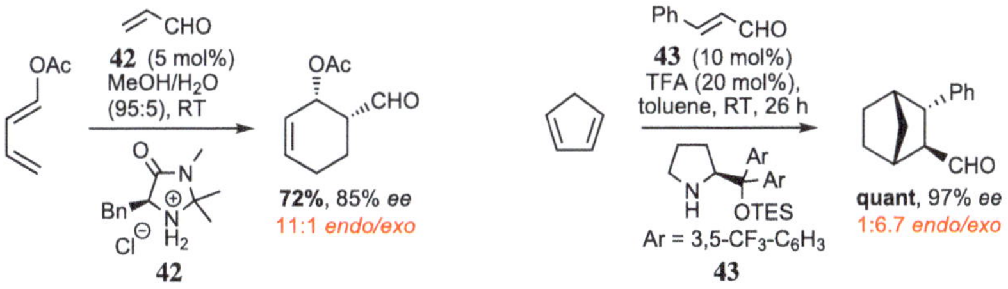

Fig. 8.15 Diastereoselectivity of organocatalytic Diels–Alder reactions [10, 83]

In the majority of published methods, α, β-carbonyls are employed as substrates, which are derived from crotonaldehyde or cinnamaldehyde (by virtue of the higher reactivity of the enals) as well methyl vinyl ketone in some instances. The conversion α-branched aldehydes can now be routinely realized, which further broadens the substrate scope. Cyclopentadiene is often considered as benchmark substrate for dienes. The diversity of functional groups tends to be rather conservative [76, 82].

Some applications are depicted in Fig. 8.16. Holmes' intramolecular [4+2]-cycloaddition, starting from **44** in the presence of the imidazolidinone **42**, provided the framework of the diterpene eunicellin. The synthesis of the fragrance β-santalol by Fehr and co-workers relied on an organocatalytic Diels–Alder reaction. The reaction of cyclopentadiene and crotonaldehyde in the presence of a very low catalyst load of the prolinol silyl ether **45** afforded the cyclohexene with excellent enantiomeric excess, which was subsequently converted to the sandalwood odorant. The synthesis of the *Strychnos* alkaloid (+)-minfiensine by the MacMillan group employed a one-pot reaction to assemble the ABCD ring skeleton of the natural product. Indole **46** first reacted with the propargylaldehyde activated as an iminium ion to **48**. Following protonation of the isolated olefinic double bond, the Boc-protected amine connected the D-ring in an *exo-5-trig*-cyclization and the aldehyde was converted to the enol **50** in a Luche reduction. The natural product could be synthesized via this key sequence in only 9 steps starting from commercial starting materials [84–86].

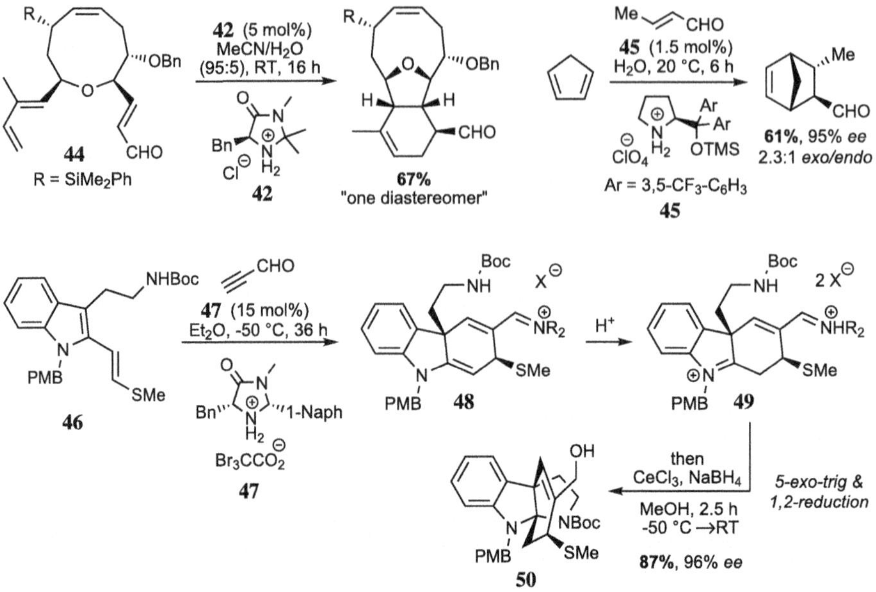

Fig. 8.16 Organocatalytic cycloadditions in total syntheses [84–86]

Judging by the range of viable substrates, the much broader scope of 1,4-additions encompasses the reaction of C-, N-, O-, S-, and P-nucleophiles with the respective iminium ion intermediates of enals and enones. Particularly the Friedel–Crafts alkylation of electron-rich heteroarenes and the addition of 1,3-dicarbonyls (malonic esters, β-oxocarboxylic esters, 1,3-diketones, etc.) are prominently featured in the reported methods. Carbon nucleophiles heavily rely on the use of malonates and occasionally employ nitroalkanes. The stoichiometry and thus the identity of the limiting component can vary from method to method. It is not uncommon to use one of the components as a solvent and thus in great excess. This is common practice with low-cost nitroalkanes such as nitromethane and nitroethane. Silyl enol ethers can also be utilized as a starting material in a Mukaiyama–Michael addition (see Fig. 8.17) [76, 82].

O-protected hydroxylamines usually serve as nitrogen nucleophiles. In addition, NH-acidic azaheterocycles such as tetrazoles, triazoles, and pyrazoles can be used. When resorting to alcohols as nucleophiles in C-O bond formations, the problem of competitive *O,O*- and *N,O*-acetal formation arises. In the case of oximes as nucleophiles, acetal formation is suppressed. Since some iminium-catalyzed reactions can also be carried out in alcoholic solvents, the substrate scope with respect to the O-nucleophile likely is broader than the starting materials in published methodological studies might suggest. When using peroxides or peracids, these can either "only" add to the β-position or also lead to the formation of the corresponding epoxide via nucleophilic attack of the resulting enamine on the oxygen of the newly formed CO bond under extrusion of the original O-substituent (*vide infra*, cf. Fig. 8.21). The reaction conditions (catalyst, addition of acid, temperature, solvent) deter-

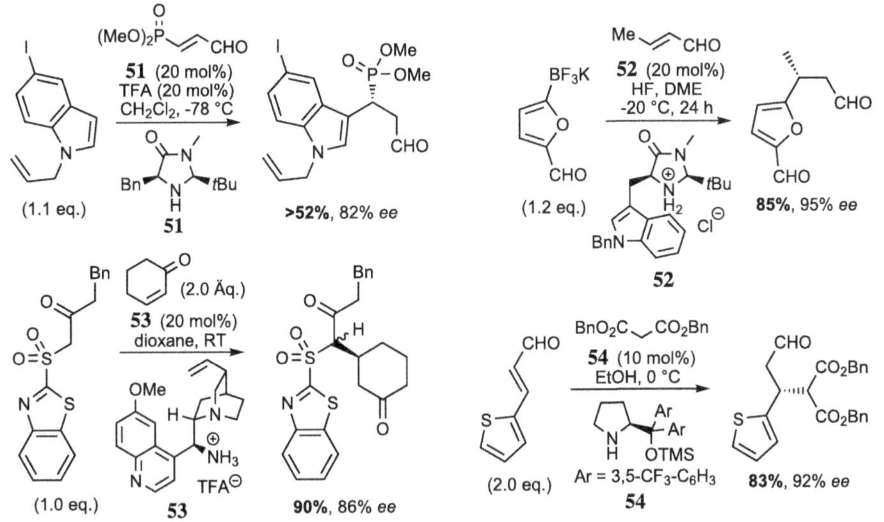

Fig. 8.17 Addition of C-nucleophiles to α, β-unsaturated carbonyls [87–90]

mine the ratio of addition and epoxidation product. An excess of O-nucleophile is required in either case (see Fig. 8.18) [76, 82].

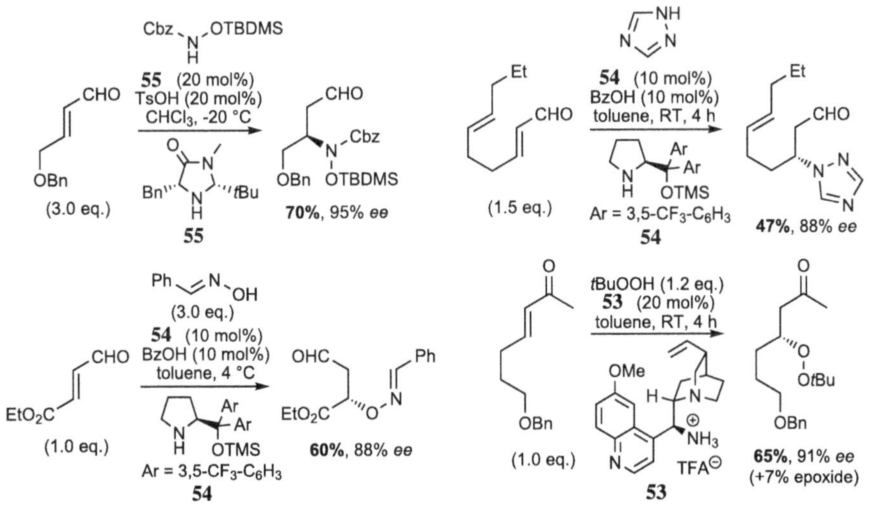

Fig. 8.18 Addition of N- and O-nucleophiles to α, β-unsaturated carbonyls [91–94]

The above methods have also been successfully incorporated into the synthesis of bioactive compounds (see Fig. 8.19). Enders and Tang were able to access several members of the marine plakortin polyketides with the construction of the key intermediate **57** from **56**. Of the examined imidazolidones, proline derivatives, and prolinol silyl ethers, the C_2-symmetric amine **58** provided the requisite high yields and enantioselectivities in the Mukaiyama–Michael addition. Starting from **57**, the natural products hippolachnin A and gracilioether A, E, and F were successfully synthesized. The Michael addition of the substituted pyrazole **59** was used for access to the Janus kinase inhibitor INCB018424. The desired β-aminoaldehyde **60** was subsequently converted into the corresponding nitrile and deprotected in the final step. A particularly impressive example of an organocatalytic 1,4-addition constitutes a key step in the pilot route for the production of the drug candidate telcagepant (**63**), indicated for the acute treatment and prevention of migraine. Enal **61** was converted on a $>100\,\text{kg}$ scale with nitromethane in the presence of the Jørgensen–Hayashi catalyst **54** with good yield and excellent enantioselectivity. The synthetic route to obtain kilogram quantities of **63** for clinical studies comprised 13 steps in the longest sequence, but only three isolations of intermediates and proceeded with a 27% overall yield [95–97].

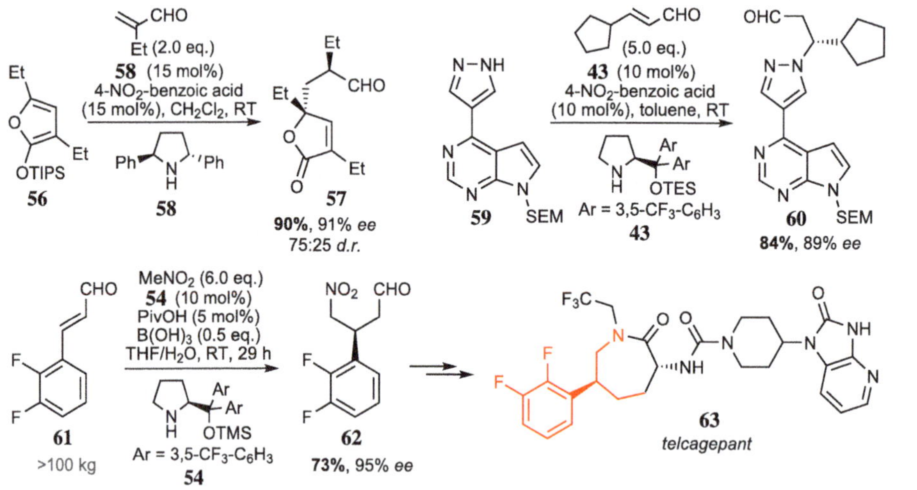

Fig. 8.19 Construction of complex synthesis intermediates under iminium catalysis [95–97]

The use of hydride nucleophiles in Michael addition has largely focused on Hantzsch esters (**64**) as reductants. Protocols have been published that solely rely on amine catalysts (iminium formation) as well as a combination of amines for iminium formation coupled with Brønsted acids (bicatalytic mechanism). In the former case, the enantioinduction originates from the chiral amine in the iminium intermediate **65**, while in the latter approach achiral amines (morpholine, etc.) activate the Michael acceptor and the stereoinformation is induced via coordination of the chiral Brønsted acid in **66**. Common chiral amines are MacMillan systems in combination with a carboxylic acid in polar protic solvents such as ethers or chlorinated alkanes (see Fig. 8.20) [76, 82]. An application of this method can be found in Lear's approach to (−)-platensimycin, in which **67** was diastereoselectively reduced in the presence of the phenylalanine derivative **69**. Alternative reduction methods (Pd–C/H$_2$, Ir(COD)Py(PCy$_3$)/H$_2$, Cu-Hydride, etc.) either resulted in the undesired overreduction of both olefinic bonds or gave the wrong diastereomer in **68** [98].

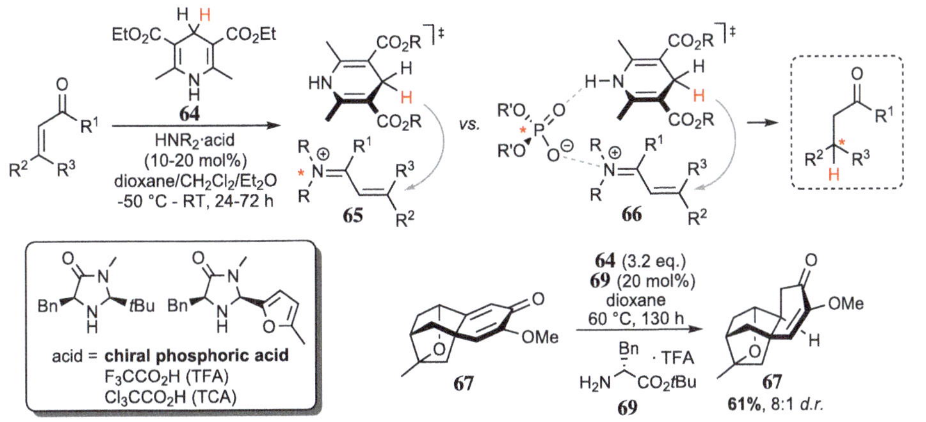

Fig. 8.20 Iminium and Brønsted acid/iminium-catalyzed 1,4-reduction as well as the key step in Lear's synthesis of platensimycin [76, 98]

Although organocatalytic epoxidations are dominated by ketone-based methods (Shi epoxidation, cf. Sect. 3.3.1), particularly α, β-unsaturated carbonyl compounds as starting materials usually necessitate very strong oxidants and thus harsh conditions. This can be rationalized by the typical reactivity of the alkene, which acts as a nucleophile and therefore requires the double bond to be electron-rich, which enals and enones are not. Iminium catalysis, as well as the use of peroxide/base (cf. Fig. 3.107, Sect. 3.3.1), offers a simple approach for *selective* functionalization of electron-poor double bonds in polyenes by participating in a 1,4-addition as an alternative mechanism. The organocatalytic route permits not only an epoxidation but also enables the formation of aziridines and cyclopropanes (see Fig. 8.21) [76].

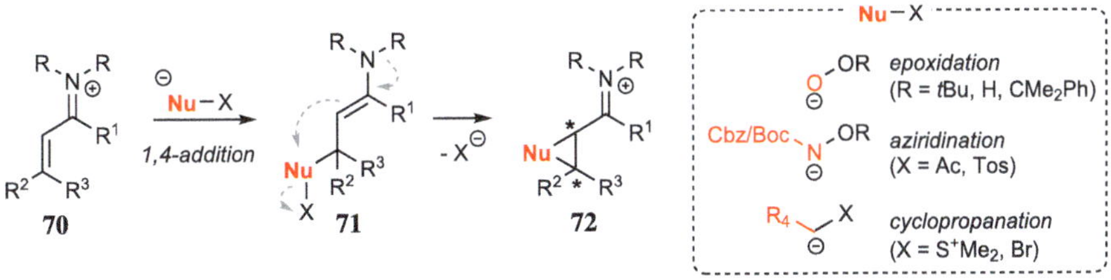

Fig. 8.21 Schematic mechanism of the functionalization of α, β-unsaturated carbonyls [76]

Mechanistically, an initial addition of the nucleophile to **70** affords enamine intermediate **71**. The subsequent nucleophilic attack of the enamine **72** is obtained by the elimination of a leaving group via intramolecular cyclization, which liberates the desired product following hydrolysis of the transient iminium functionality. Suitable catalysts for epoxidation comprise imidazolidines and prolinol silyl ethers as well as Cinchona derivatives. The oxidant is used in slight excess, and the reaction medium is unpolar to polar aprotic. To avoid a stalling at the stage of Michael addition, the reactions are usually carried out at room temperature or at slightly elevated temperatures (see Fig. 8.22) [76].

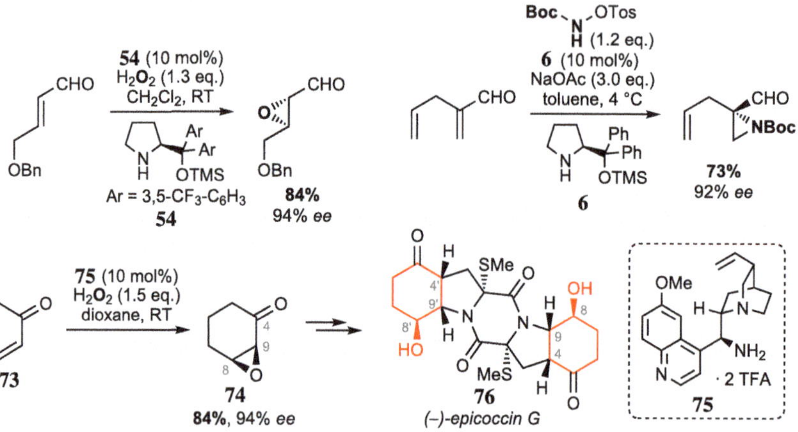

Fig. 8.22 α, β-Functionalization of enals and application in the total synthesis of (–)-epicoccin G (**76**) according to Baudoin *et al.* [99–101]

8.1.3 Umpolung Reactions

The organocatalytic reversal of reactivity in carbonyl derivatives from an electrophilic to a nucleophilic carbon atom (*Umpolung*) [102] takes a cue from nature [13]. The employed organocatalysts are variants of *N*-heterocyclic carbenes, which are a common sight in biosynthetic reactions. In the enzyme-catalyzed benzoin condensation of the pentose phosphate pathway, which is carried out by transketolase enzymes, the reaction is facilitated by the coenzyme catalyst thiamine (vitamin B_1) in the absence of active metal centers (see Fig. 8.23).

Enol **77**, which is referred to as *Breslow* intermediate, constitutes the key intermediate of the biosynthetic transformation. It also plays a central role in organocatalytic Umpolung reactions [104, 105]. The Breslow intermediate can be functionalized by a wide range of electrophiles in a subsequent step. Apart from the benzoin condensation, the reactions widely studied in this context comprise the 1,4-addition of aldehydes to α, β-unsaturated carbonyl compounds (*Stetter* reaction) and transesterifications. In addition to aldehydes, enals can

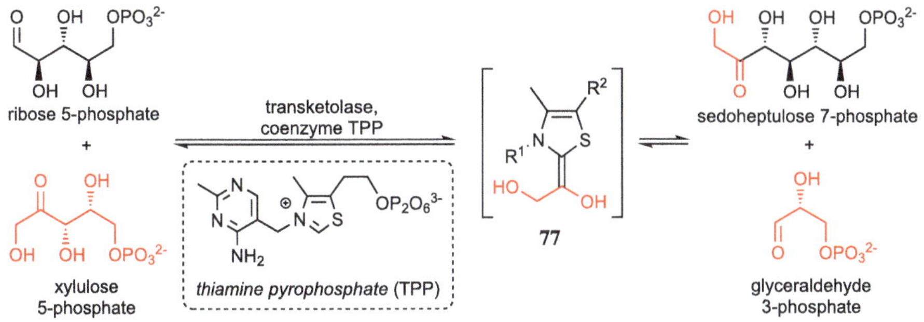

Fig. 8.23 Transketolase-mediated C_2 transfer (benzoin condensation) in the pentose phosphate pathway [103]

also be utilized as nucleophiles, which might also function as homoenolates and thus enable a reaction with electrophiles at either the α- or β-position (see Fig. 8.24).

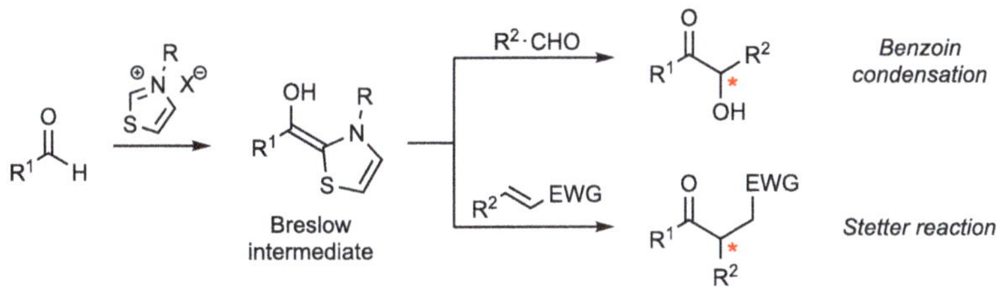

Fig. 8.24 Reactions dominating under NHC catalysis [106–109]

In the benzoin reaction, either an aldehyde can react with itself (homo-benzoin reaction) or two different aldehydes are used (cross-benzoin reaction). Chemo- and enantioselectivity of the latter remain a significant challenge. The two carbonyl substrates must possess a sufficiently different reactivity to not deliver a statistical mixture of the four possible products (two homo- and two cross-benzoin products). Both variants are further compounded by the commonly observed reversibility of the reaction. Apart from intramolecular and intermolecular variants, carbonyl derivatives such as imines or acylsilanes can also be employed as coupling partners. The reaction conditions include the use of a base, usually a tertiary amine, for deprotonation of the precatalyst, and moderate temperatures from $0\,°C$ to room temperature or slightly above [106–109]. Mechanistically, the deprotonation of the precatalyst **78** to the free carbene **79** occurs as the initiating step. Following the attack on the aldehyde, the zwitterionic addition product **80** tautomerizes to give the Breslow intermediate **81**. Depending on the second component, the concluding steps of the catalytic cycle differ. In a cross-benzoin condensation, the acceptor aldehyde is attacked and the CC bond in **82** is formed, from which the desired α-hydroxyketone can be liberated [13, 106]. For certain bases, a concerted deprotonation of **78** under simultaneous CC bond formation with the aldehyde **80** (via transition state **83**) was postulated based on theoretical studies, instead of proceeding via the free carbene **79** [110, 111]. If an electrophile other than a carbonyl compound is employed, such as a Michael system, intermediate **84** would replace **82** as the ultimate species of the catalytic cycle (see Fig. 8.25).

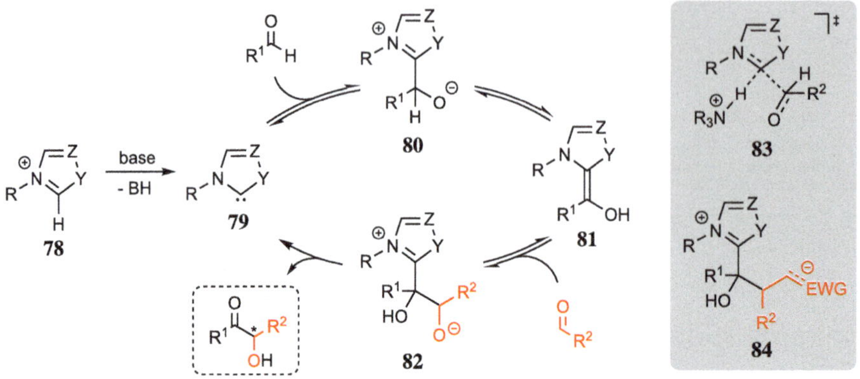

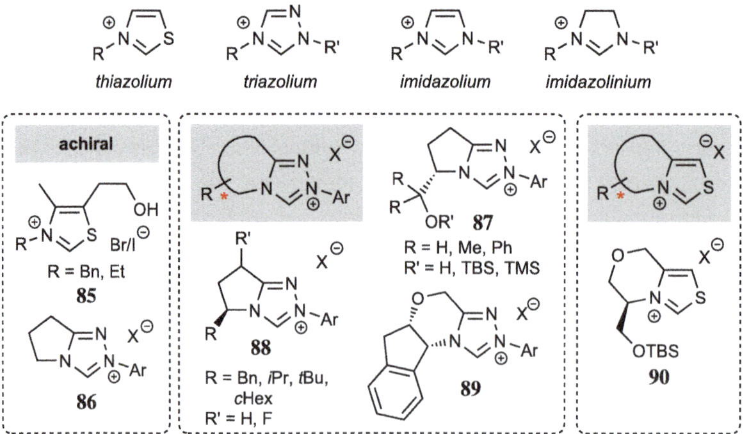

Fig. 8.25 Postulated Breslow mechanism of the benzoin condensation with NHC catalysts [106, 111]

In addition to achiral catalysts, where the nature-derived thiazole **85** still dominates, good stereoinduction is usually achieved by fusing the heterocyclic NHC with other carbocycles and the concomitant introduction of a stereocenter in the appended ring. This chiral backbone is usually derived from the (extended) chiral pool, so that amino acids or readily accessible motifs such as aminoindanol easily lend themselves as building blocks. In the case of triazoles, catalysts of type **87** and **88** dominate, while chiral thiazoles like **90** are only sporadically encountered (see Fig. 8.26) [106–109].

Fig. 8.26 Classes of parent carbene heterocycles and common NHC organocatalysts [106–109]

The enantio- and diastereoselectivity of NHC-catalyzed Umpolung reactions have only been investigated for individual catalyst systems and mainly focused on the benzoin condensation, which is why a more general understanding is only possible to a limited extent (see Fig. 8.27) [5]. The facial differentiation of the electrophile is achieved by the different orientation of the reactant in the transition states *Re*-**90** or *Si*-**91**. The transition state is sensitive to both steric and electronic factors, which is reflected in the slightly differing transition structures from Fig. 8.27. Theoretical studies consistently postulate an orientation of the acceptor double bond synperiplanar to the CO bond of the enol in the Breslow intermediate (cf. **92**, **94**, **96**, **97**). The orientation of the hydroxy group and the substituent R^1 in **81** has not yet been investigated in-depth. In the 1,2,4-triazoles **93** and **98**, the C-O bond is oriented antiperiplanar to the N2-C3 bond (cf. TS **92**, **96**, **97**), while for the thiazole **95**, the hydroxy group is aligned synperiplanar to the endocyclic C2-N3 bond (TS **94**) [5, 105, 112].

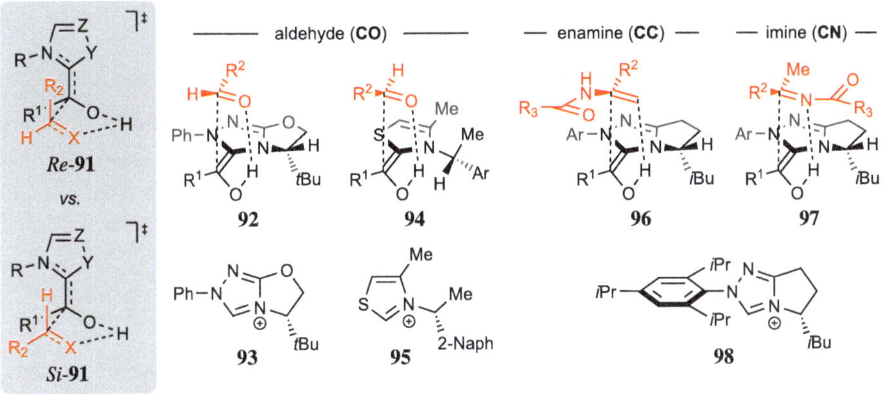

Fig. 8.27 Competing transition states and examples with different acceptors [5, 105, 112]

While the diastereoselectivity is governed by the orientation of the acceptor component, control of the hydroxy group (*E* vs. *Z*) of the Breslow intermediate is also necessary for achieving enantioselectivity. Apart from practical experience with common catalysts, there are currently no universally valid models to reliably predict both orientations.

In the synthesis of complex natural products, the benzoin condensation has so far only been used in the intramolecular reaction of aldehydes with ketones (see Fig. 8.28).

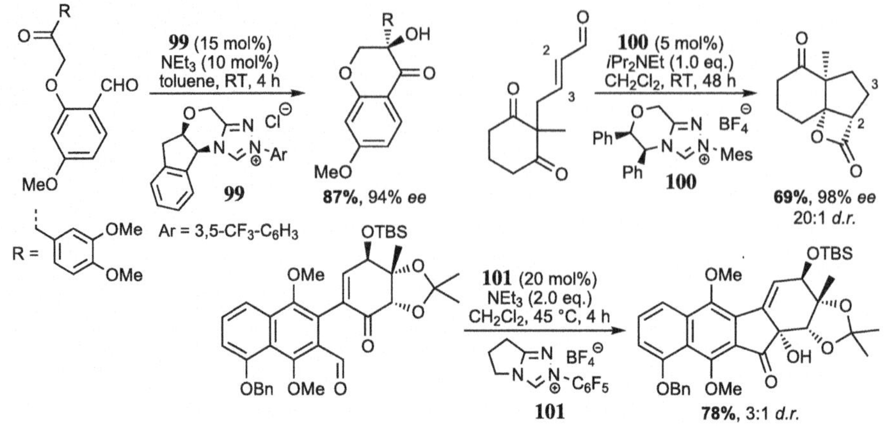

Fig. 8.28 Examples of benzoin condensations in total syntheses [113–115]

The Stetter reaction occupies a significantly broader spectrum, as the electrophiles can bear aldehydes as well as ester, amide, phosphonate, and nitro functionalities as electron-withdrawing group on the alkene. Examples of the NHC-catalyzed Stetter reaction typically are restricted to the use of achiral catalysts (see Fig. 8.29).

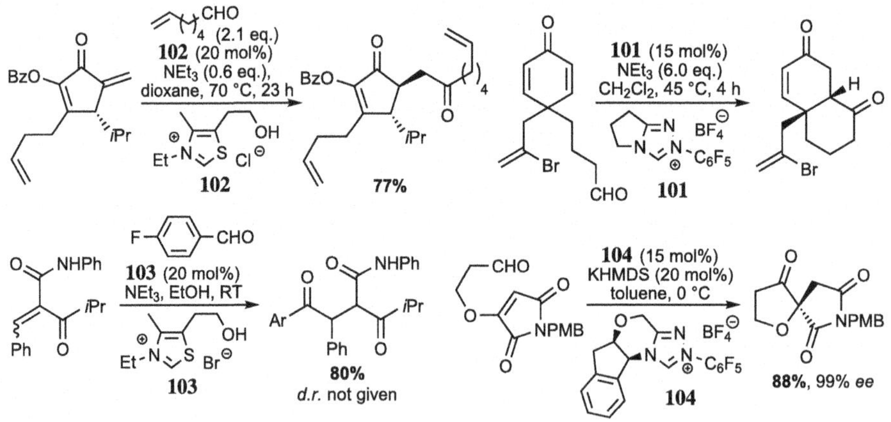

Fig. 8.29 Applications of the NHC-catalyzed Stetter reaction [116–119]

8.1.4   Hydrogen Bond-Based Activation and Bifunctional Catalysis

The activation of substrates by means of hydrogen bonding is borrowed from the modus operandi of enzymes in nature. For the development of stabilizing interactions, substrates require the presence of polar double or single bonds. This accordingly qualifies carbonyl derivatives, imines, and nitro compounds as privileged starting materials [14–16]. The strength of the interaction depends on the Lewis basicity of the substrates, as well as the structure and stereoelectronic (acceptor) properties of the catalyst: The hydrogen bond interaction can transition into a full protonation resulting in a pure ionic interaction between catalyst and substrate,[II] which is why a separation between these mechanisms can only yield an inclination for either ideal for any given reaction. Both extremes should be considered as borderline cases, with most examples lying on the continuum between these poles (see Fig. 8.30) [120–122].

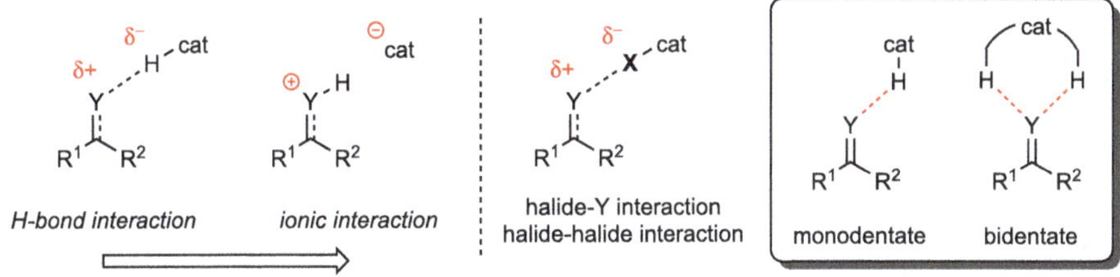

Fig. 8.30 Types of non-covalent interactions in organocatalysis

In addition to the activation of CY double bonds by Y-H interactions, polarization by means of Y-halogen interactions (typically with iodine as the halogen) has also proven to be a viable approach. When using hydrogen or halogen donors in these non-covalent activations, either a monodentate or bidentate interaction with the substrates can occur. The activation of the CY bond lowers the LUMO and increases its reactivity towards nucleophiles.

H-donors can be structurally very diverse, but there are some motifs that are frequently encountered in catalysts: (thio)ureas (**105-107**), diols (**108**) and phosphoric acids (**109**). Particularly the systems developed in the groups of Schreiner (**110**), Takemoto (**111**) and Jacobsen (**112**) are routine catalysts (see Fig. 8.31) [123, 124].

[II] When using Brønsted acids, the ideal case of pure hydrogen bond interaction is referred to as general Brønsted acid catalysis, while a (partial) protonation is categorized as specific Brønsted acid catalysis.

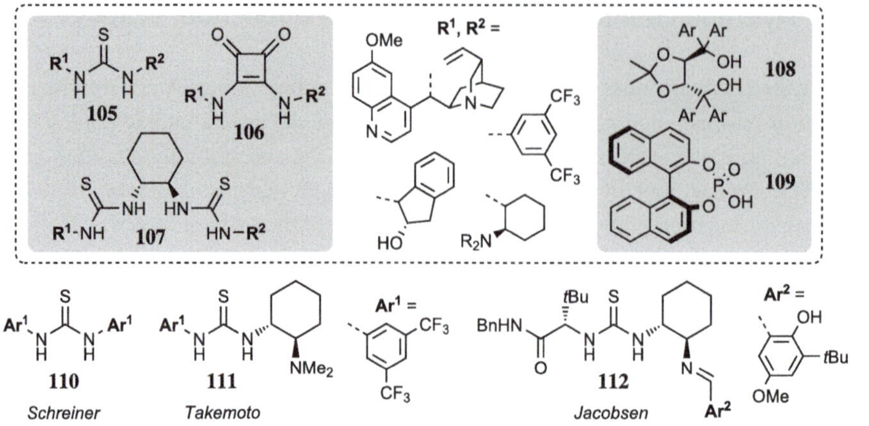

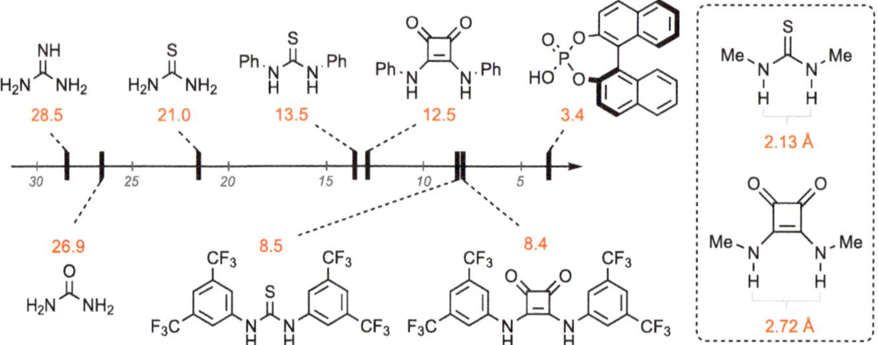

Fig. 8.31 Overview of catalysts based on H-bond interactions

The strength of the donors and thus the activation correlates on one hand with the acidity of the catalysts. However, a consideration of only the pKa values neglects steric effects and bonding geometries, which are equally important. In addition, the stabilization of the transition states can also differ from those of the ground states, which is the principle of action in enzyme catalysis. Both the catalyst structure and its donor strength must therefore be taken into account (see Fig. 8.32) [121, 125].

Fig. 8.32 pKa values (in DMSO) and H-H bond distances in thioureas, squaramides. The pKa value of phosphoric acid **109** is included for comparison [125–128]

The donor properties can be further modified by several endogenous and exogenous factors: If other H-donors or other Lewis acids are present in the side chain of the catalyst, these can interact with the primary H-donor functionality and increase the acidity. This concept is exploited in particular by thiourea and peptide catalysts. Alternatively, the same effect can be achieved by adding an external Lewis acid, either in the presence of a Brønsted acid or via cooperative catalysis through the additional use of transition metal complexes [41, 120].

Thioureas, the most common class of H-bond donors, and squaramides activate carbonyl derivatives, imines, nitro compounds, and epoxides through double hydrogen bond interactions. Diols, phosphoric acids (if they do not follow an ionic mechanism), and peptides only contain one H-donor site for activation of a substrate (see Fig. 8.33) [5].

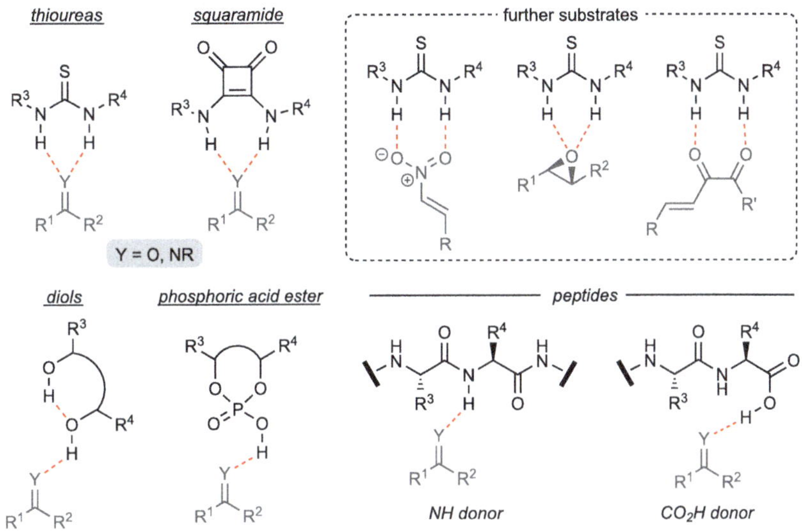

Fig. 8.33 Postulated binding modes of different H-donors [5, 15, 41, 121]

With peptide catalysts, the secondary structure of the amino acid backbone can often successfully enable a facial differentiation of prochiral reactants. The transition state heavily depends on the individual catalyst structure, which makes an intuitive prediction of the selectivity without extensive quantum chemical calculations very difficult for individual cases [15, 16].

Halogen-based Lewis acids (*halogen donors*) usually rely on aryl halides, with aryl iodides occupying a prominent spot by virtue of their high polarizability. Their activation proceeds analogously to an H-bond interaction and serves to polarize C-X single and double bonds to enable an efficient derivatization. The thus activated electrophiles participate in nucleophilic substitutions, additions to carbonyl derivatives and imines as well as pericyclic reactions (see Fig. 8.34) [43, 45, 46, 129].

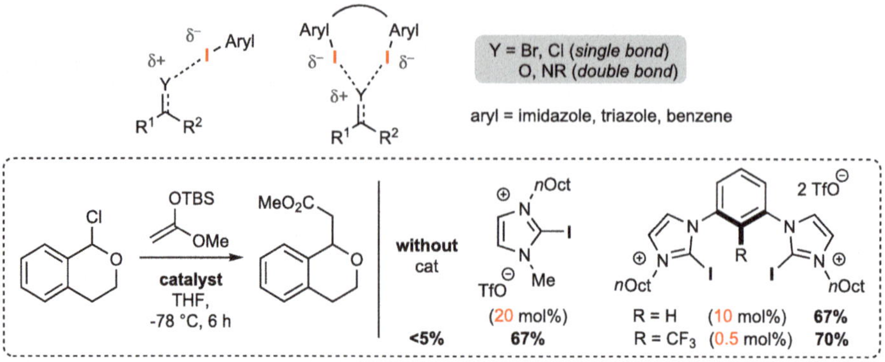

Fig. 8.34 Non-covalent organocatalysis with aryl iodides and comparison of different cationic imidazolium iodides [44]

Many thioureas and the majority of peptide-based catalysts operate via a **bifunctional** mechanism. Here, both reactants are activated, either in a completely non-covalent fashion (transition states **113**-**115**) or via a combination of non-covalent and covalent interactions (transition states **116**, **117**) (see Fig. 8.35) [5, 130–132].

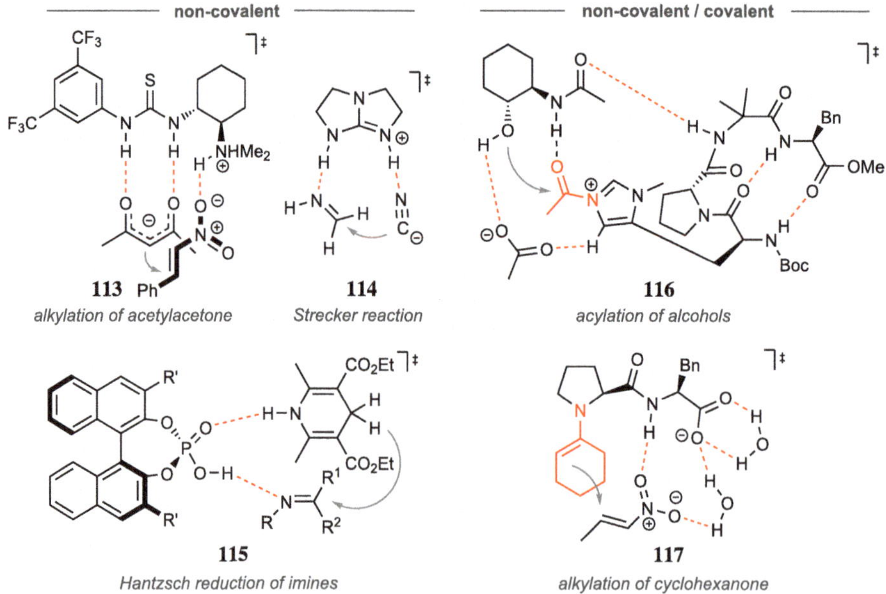

Fig. 8.35 Examples of transition states in bifunctional activation [5, 130–132]

In the case of thioureas, typically one reactant is activated by interaction with the two urea NH protons, while the second substrate interacts with a functionality located in the side chain (**113**). Guanidines behave analogously (**114**). Phosphoric acids combine a Lewis basic and acidic coordination site in their central functionality, so that both electrophile and nucleophile can be electronically activated and brought into close spatial proximity (**115**). Peptides possess a complex secondary structure incorporating intramolecular H-bonds. This structure is normally exploited jointly for stereoinduction as well as in its capacity for substrate activation. If peptides with added functionalities are used, a covalent interaction can also be utilized in addition to the common non-covalent interaction: for example, the use of peptides with proline situated at the C-terminus can induce enamine formation, while the presence of imidazolium moieties allows a covalent acyl transfer via Umpolung (**117** and **116**).

A hallmark of H-donors lies in their aptitude for activating carbonyl and nitro groups. For this reason, the Strecker synthesis and Michael additions of various nucleophiles to nitroalkenes and enones have been particularly intensively studied. Further transformations such as the Pictet–Spengler reaction and other 1,2-additions have also been reported to successfully occur with high selectivities. In hydrogen bond interactions, the solvent constitutes a pivotal element in enabling or disrupting activation, which is why mostly nonpolar (toluene, xylene, hexane) and occasionally polar aprotic solvents (CH_2Cl_2, ethers) are employed (see Fig. 8.36) [123, 124].

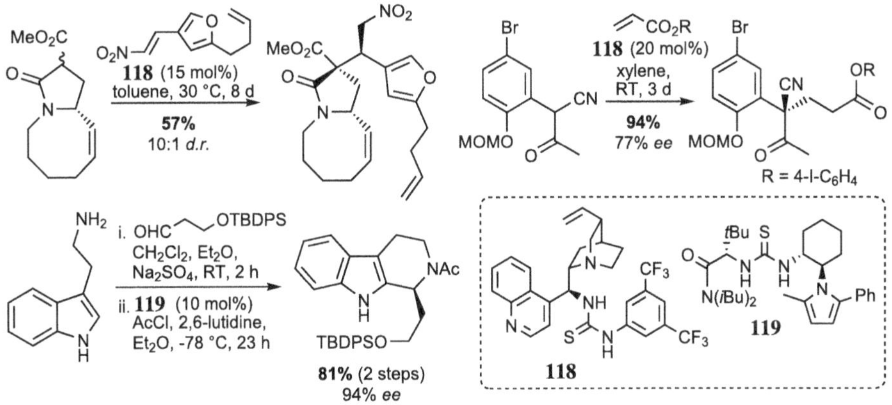

Fig. 8.36 Application of H-donor catalysts in total syntheses [133–135]

Thioureas have also been utilized in reactions carried out on an industrial scale: dihydropyridinone **124** belongs to the group of $P2X_7$ receptor antagonists, which can treat chronic inflammation and pain as well as neurodegenerative diseases. **121** was converted to **122** on a kg scale in the presence of the thiourea **123** with acceptable enantioselectivity of 80%. An alternative enzymatic desymmetrization only led to partial conversion (4% after 5 days) and an asymmetric methanolysis in the presence of stoichiometric amounts of quinine achieved only moderate *ee* values of 48–67%. The recrystallization of the ester **122** in toluene/hexane raised the *ee* value to 97% and enabled isolation of the desired dihydropyridinone **124** in 43–44 and 100% *ee* starting from **121** (see Fig. 8.37) [136].

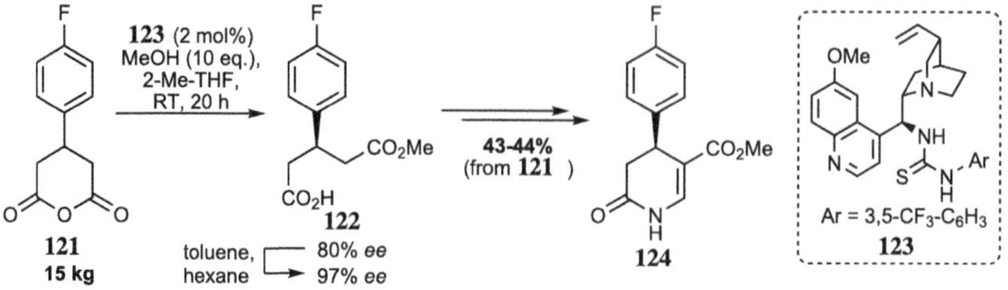

Fig. 8.37 Pilot synthesis of the P2X$_7$ receptor antagonist **124** [136]

While thioureas have been incorporated into syntheses of natural products and drugs, the remaining catalysts rely on non-covalent interactions (hydrogen or halogen donors) such as diols, phosphoric acids, and aryl iodides yet lack notable applications in the synthesis of complex intermediates. Possibly more common than the activation of just one of the two reactants is the introduction of a second active functional group for effecting a *bifunctional* catalysis. Amino functionalities are often the first choice for incorporation into the catalyst structure. They either serve as a base for the deprotonation of acidic nucleophiles as well as for complexation or hydrogen bonding with Lewis acids or enable the covalent formation of enamine or iminium intermediates. A range of other functionalities such as hydroxy, phosphine, or sulfinamide groups have been exploited in bifunctional catalysts apart from amino functionalities (see Fig. 8.38) [137].

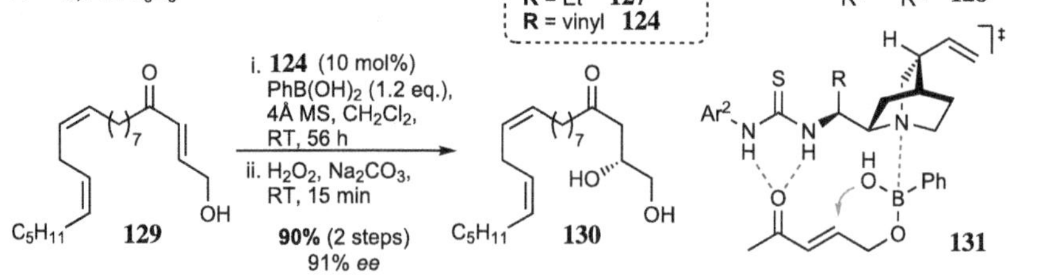

Fig. 8.38 Non-covalent bifunctional catalysis in the synthesis of bioactive substances and possible transition states [138, 139]

The Michael addition of nitromethane to **125** was patented by Syngenta to synthesize various dihydropyrrole derivatives for use as insecticides. While a mechanism for the formation of **126** in the presence of the Cinchona-based catalyst **127** was not provided, **128** might constitute a possible transition state. The intramolecular β-oxidation of the triene **129** presumably proceeds via a dual activation of the α, β-ketone and the borane (transition state **131**) to afford the antifungal and hepatoprotective secondary metabolite of the avocado. In addition, bifunctional catalysts have also been used to access other bioactive substances [140–143].

8.1.5   Ionic Interactions—Phase Transfer Catalysis and Chiral Counterions

Practically all chemical reactions involve some degree of charge polarization and many transformations proceed via the intermediacy of a charged species. The second class of organocatalysts relying on non-covalent binding exploits an ionic interaction between the substrates and the catalyst [47, 48]. The catalytic species, regardless of whether it provides the anionic or cationic counterion, can either be charged itself or it is paired as a neutral species with the counterion of the substrate (see Fig. 8.39). In both cases, it acts as a template for stereochemical induction during the reaction, provided the stabilizing electrostatic interaction ensures sufficient proximity and half-life of the transient species.

	type I	*type II*	*type III*	*type IV*
	chiral cation-directed	chiral anion-directed	cation-binding	anion-binding
	$R^{\ominus}\ ^{\oplus}cat^*$	$R^{\oplus}\ ^{\ominus}cat^*$	$R^{\ominus}\ ^{\oplus}M\text{-}cat^*$	$R^{\oplus}\ ^{\ominus}X\text{-}cat^*$
typical catalysts	4° ammonium salts	Brønsted acids e.g. phosphoric acids	polyether, crown ether	thioureas
common reactions	- α-alkylation	- 1,2-addition - epoxidation - α-halogenation	- α-alkylation	- Strecker reaction - 1,2-addition - nucleophilic substitution

Fig. 8.39 Reactions types based on prevalent ionic interactions [47]

Phase transfer catalysts based on ammonium salts (Type I) are particularly relevant for industrial applications, while in academic syntheses of natural products, BINOL-derived phosphoric acids (Type II) have been used to great effect. Catalysts of type III (crown ethers) and IV (neutral catalyst-complexed anions) play (yet) no significant role as stereoinducing elements.

The spatial proximity between a stereoinducing species and a prochiral substrate is the basis for high stereoselectivity. Depending on the reaction conditions and thus the type of interaction, high diastereo- and enantioinduction can be observed. Solvents with low dielectric constants favor the formation of contact ion pairs with good induction, while more polar solvents lead to solvent-separated ion pairs with significantly weaker induction (see Fig. 8.40).

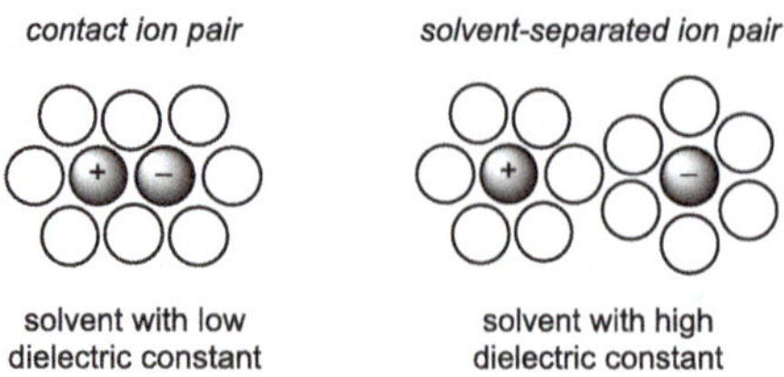

Fig. 8.40 Types of ion pairs [47]

Quaternary ammonium salts constitute a popular motif as cationic catalysts, which can effect a phase transfer catalysis [50–52, 144]. Class III catalysts are frequently conducted under biphasic conditions as well [47, 53, 55, 56]. In addition, chiral bases are used to generate a protonated cat-H species as a counterion (Class I) by combination with acidic substrates [54, 145]. For anionic catalysts, the large group of Brønsted acids based on phosphates, phosphoramides, sulfonic acids, and sulfonimides dominates the field. The complexation of anions by neutral catalysts has been efficiently realized with H-bond donors, foremost thioureas.

Under biphasic conditions (usually organic/aqueous), a **phase transfer catalyst** (PTC) is employed to accelerate the transformation, especially for the coupling of two substrates with different solubility profiles. The "classic" asymmetric reaction is the α-functionalization of carbonyl derivatives, mainly the α-alkylation. Two classes of catalysts are preferably used: *N*-alkylated Cinchona derivatives (**132**) and BINOL-derived C_2-symmetric *spiro*-ammonium compounds (**133, 134**), which are referred to as *Maruoka* catalysts (see Fig. 8.41). Alternative catalysts are based on substituted biphenyls or tartaric acid, among others [18, 29, 50, 51].

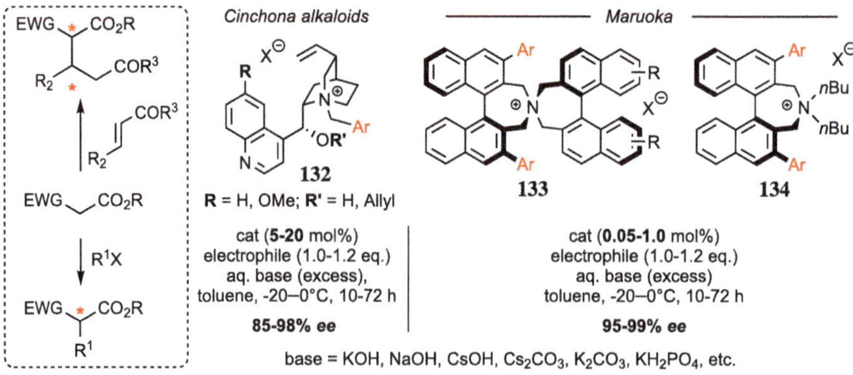

Fig. 8.41 Typical reaction conditions and common catalysts for phase transfer catalysis [18, 29, 50, 51]

Cinchona-type catalysts are readily available and thus relatively inexpensive, while Maruoka catalysts sometimes need to be laboriously synthesized, which makes them significantly more expensive. Compared to Cinchona alkaloid-based catalysts, the Maruoka systems possess superior stability, even under strongly basic reaction conditions, as the Cinchona alkaloids can suffer from a Hofmann elimination under basic-alkylating conditions. The robustness and increased catalyst activity, including the possibility of catalyst recycling, can compensate for the higher costs of **133**, **134** compared to **132** in many cases. Reactions have also been published that require catalyst loads of only 0.01%. The order of addition can play a critical role, especially on an industrial scale [18, 29]. In most cases, aqueous KOH or NaOH serves as the base. The electrophile is only used in slight excess to prevent a possible multiple, successive alkylation. Toluene typically constitutes the organic solvent of choice, but more tailored solvent systems such as mesitylene or cyclopentyl methyl ether have also been used occasionally [146].

While the exact mechanism continues to be controversially discussed, the preferred reaction path naturally includes a deprotonation of the nucleophile in the *boundary layer*. Following cation exchange with the phase transfer catalyst NR_4^+, the subsequent reaction with an electrophile can take place in the organic phase (see Fig. 8.42) [144].

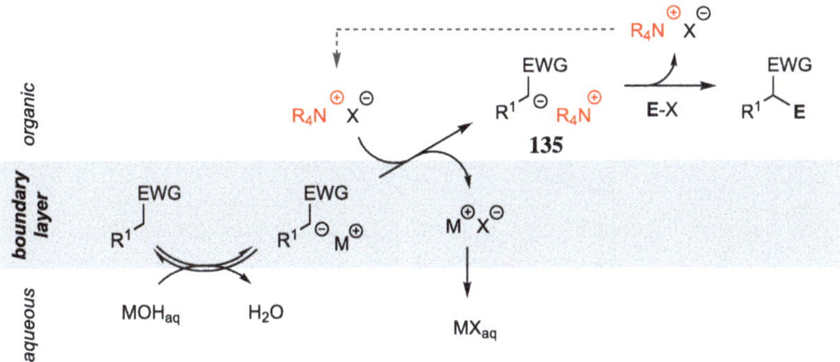

Fig. 8.42 Phase transfer mechanism (*boundary layer mechanism*) according to Makosza [144]

The PTC carbanion species **135** constitutes the key intermediate of the postulated mechanism. In the case of cationic onium salts (type I), the reaction proceeds as shown in Fig. 8.42. However, if neutral catalysts are paired with the cation of an enolate or nitronate (class III), the formation of MX becomes part of the concluding reaction of **135** with the electrophile. The size of the boundary layer, the basicity of the inorganic salt, and the physicochemical properties of the catalyst govern the amount of available phase transfer catalyst [144].

Despite an attractive but non-directional ionic interaction with their counterions, chiral ammonium ions can effect very good facial selectivity. The underlying cause was investigated by theoretical means and several secondary interactions were postulated to be decisive: steric shielding, π-π, and H-donor interactions. The quaternary ammonium ion of the Cinchona and Maruoka catalysts can be envisioned at the center of a tetrahedron, composed of the four substituents attached to the central N-atom. One of the four distinct faces is sufficiently open to allow for close contact and thus generates the requisite asymmetric environment for the anionic substrate (see Fig. 8.43) [147].

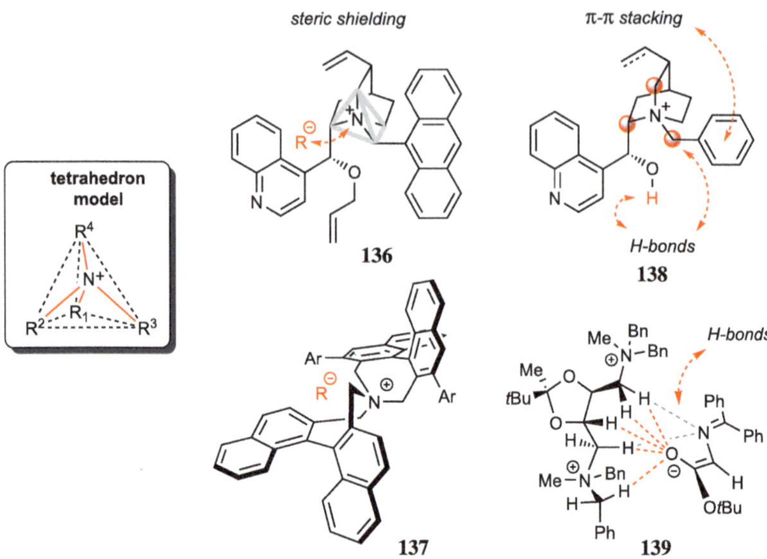

Fig. 8.43 Corey's tetrahedron model and postulated interactions for controlling facial selectivity in PTCs [147–152]

Under PTC conditions, Schiff-base protected α-amino acid esters comprise the nucleophilic component in the majority of published methodologies and synthetic applications. They are commonly α-alkylated. Applications in natural product syntheses are rather rare, [153] but industrially, the access to a multitude of active pharmaceutical ingredients relies on phase transfer catalysis as a key bond-forming step (see Figs. 8.44, 8.45) [28, 29].

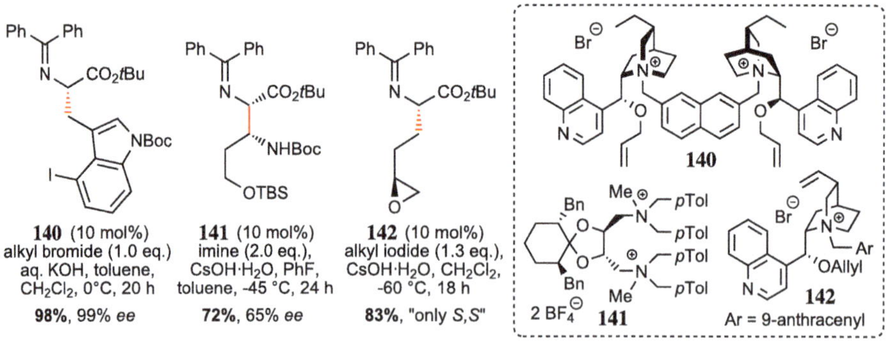

Fig. 8.44 Products of PTC in natural product syntheses [154–156]

Phase transfer conditions are encountered in pharmaceutical chemistry both on a smaller scale (Discovery/Med Chem route) and for the provision of kg quantities for clinical studies (Pilot route). Despite their advantages, Maruoka catalysts have only been used sparingly, which may not only be due to the effort and thus the cost of their preparation but also be related to their patent/IP status: licensing fees may still apply for their use and thus could prove discouraging [28, 29].

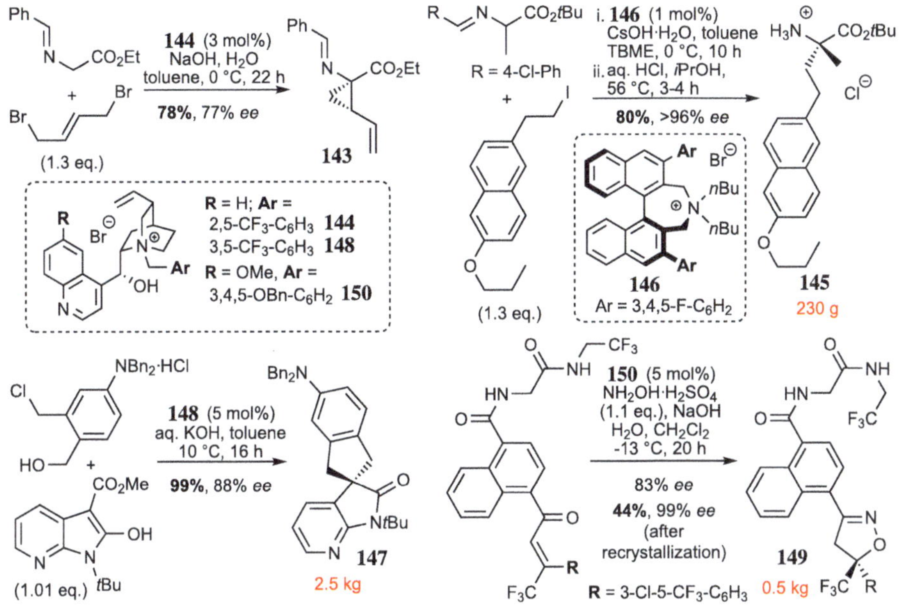

Fig. 8.45 Industrial PTC reactions for synthesis of active pharmaceutical ingredients [157–160]

The examples shown in Fig. 8.45 demonstrate both the versatility of feasible reactions and their scalability. The protected cyclopropanoic acid **143** represents an important building block in the structure of a variety of hepatitis C drugs: grazoprevir, vaniprevir, ciluprevir, danoprevir, and asunaprevir all contain this motif or a variation thereof. The double α-alkylation of the substrate to afford cyclopropane **143** shows only moderate enantioselectivity [157]. Nevertheless, the enantiomeric purities can often be significantly increased by subsequent recrystallization to give useful levels for an industrial application. The class of S1P receptor agonists are important drugs as immunomodulators in organ transplants and for the investigation of autoimmune diseases. **145**, whose amino alcohol is a chiral analog of S1P receptor drug FTY720, was successfully synthesized at Novartis under PTC conditions on a gram to kilogram scale. The simplified Maruoka catalyst **146** provided excellent enantioselectivity, which could even be increased to >98% by subsequent recrystallization [158]. Merck, Sharp & Dohme resorted to a decarboxylating bisalkylation to access **147** under biphasic conditions in their pilot route for the synthesis of the CGRP receptor antagonist MK-3207. The hydroxyazaindole coupled once with the aniline HCl salt in the presence of the Cinchona derivative **148**. The ester group of the respective intermediate was saponified under the reaction conditions and decarboxylated. It afforded spirooxindole

147 almost quantitatively and with an acceptable enantiomeric excess. The product **147** was recrystallized to enrich the desired enantiomer (95% *ee*). Subsequent steps then furnished the enantiomerically pure active ingredient, useful for the treatment of migraine [159]. The isoxazoline insecticide afoxolaner (**149**) could also be obtained under PTC conditions on a kilogram scale: Merial Inc. patented the sequential formation of the desired heterocycle starting from the respective α, β-unsaturated ketone. Mechanistically, the *O*-Michael addition of the hydroxylamine initiates the sequence, followed by the condensation of the ketone with the amino group. The crude product was isolated with an enantiomeric excess of 83%. Subsequent recrystallization afforded the active ingredient practically enantiomerically pure [160].

Xiang and Yasuda at Merck disclosed the development of a new class of PTCs, doubly quaternized Cinchona salts of the type **153**. These novel catalysts achieved excellent enantioselectivity in the alkylation of azaoxindoles with minimum catalyst loads. Under optimized conditions, the conversion of **151** to **152** required only 0.3 mol% of the Cinchona salt to obtain the product with 99% yield and in >94% *ee* (see Fig. 8.46) [161]. A synthesis of the CGRP antagonist ubrogepant (**154**) based on **152** was later also realized on a pilot scale ($\geq$100 kg) [162]. Doubly quaternized Cinchona derivatives were also used in the large-scale synthesis of the commercial antiviral drug letermovir [163].

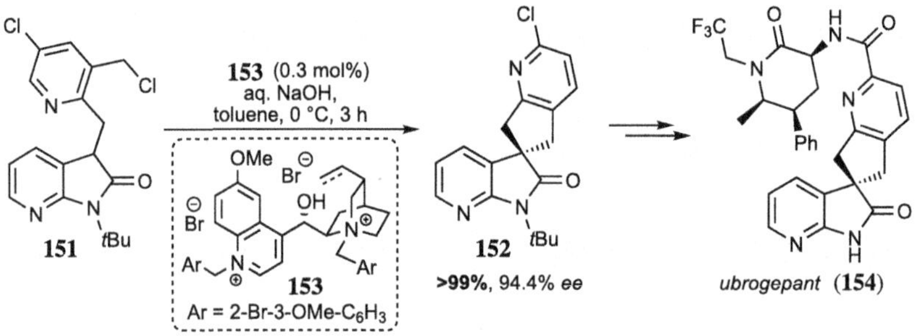

Fig. 8.46 Double quaternary cinchona salts in phase transfer catalysis [161]

The mode of action of **Brønsted acids** can be placed, as initially mentioned in Sect. 8.1.4, in the continuum ranging from H-bonding interactions to a purely electrostatic interactions (see Fig. 8.30). If the ionic character is dominant in the non-covalent binding between substrate and catalyst responsible for asymmetric induction, this type of reaction is referred to as *asymmetric counterion-directed catalysis* (ACDC) according to List [48]. A clear classification as a purely electrostatic interaction with Brønsted acids is only possible for certain substrates that proceed via the intermediacy of a carbocation, iminium, or oxocarbenium species.

The activation of CX double bonds, usually by protonation and thus lowering of the LUMO, allows a nucleophilic 1,2- and 1,4-addition as well as the participation of the activated substrate in pericyclic reactions, especially in cycloadditions. In addition, substitution reactions can also be facilitated — in the majority of cases, the substrate forms a transient cationic intermediate (see Fig. 8.47) [49].

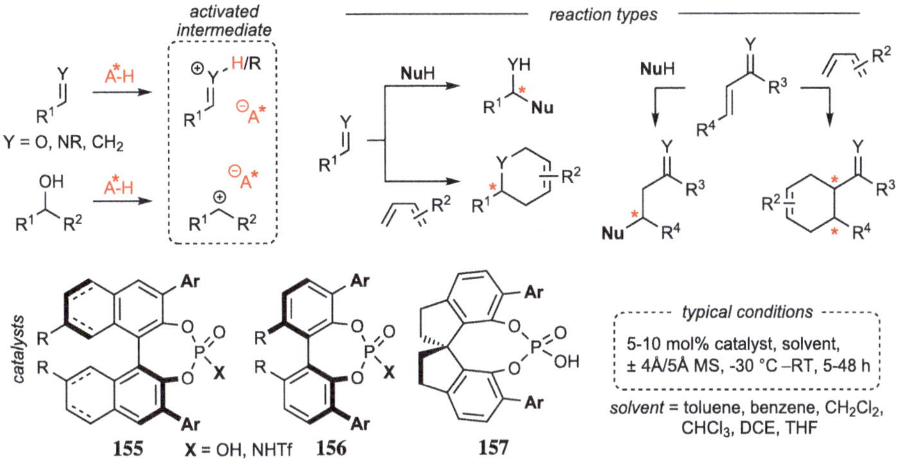

Fig. 8.47 Reactive intermediates, common transformations, and typical reaction conditions under Brønsted acid catalysis

Chiral Brønsted acids, especially the BINOL-based phosphoric acid diesters introduced by Akiyama and Terada, rank among the more frequently encountered organocatalysts. In addition to phosphoric acid esters, phosphortriflimides, as well as bis(sulfuryl)imides, sulfonic acids, and diols can also serve as acidic catalysts (see Fig. 8.48). In reactions that do not involve proton transfer, the corresponding anion of the chiral catalyst can be used as a salt to enable an enantioinduction (type II). In condensation reactions, such as the formation of imines, molecular sieves can be added to the reaction to sequester the by-product and facilitate the conversion. The solvent system ranges between nonpolar and polar aprotic media, with toluene and chlorinated alkanes being the most common.

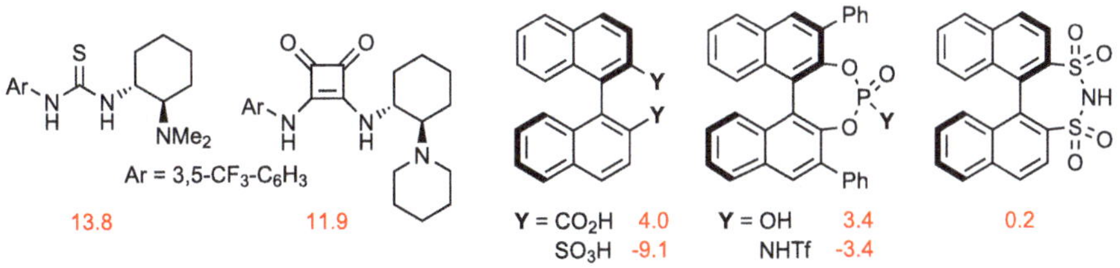

Fig. 8.48 pKa values of relevant Brønsted acids (in DMSO) [125, 126, 128]

When using Brønsted acids, the substrate is protonated by the catalyst or converted to a cation via other paths (cf. Fig. 8.47). The catalyst in the ion pair **158** acts as a chiral template, whereby a nucleophile selectively generates a new stereocenter following the addition to the activated substrate (see Fig. 8.49, left catalytic cycle). In addition to the direct utilization of

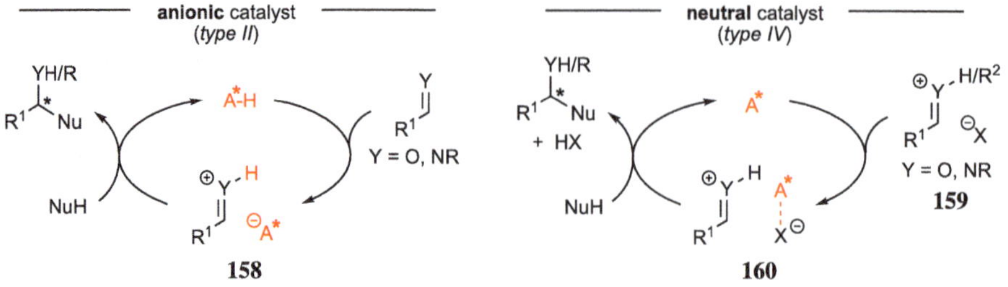

Fig. 8.49 Postulated catalytic cycles for ACDC [47, 48]

an acid (A-H), a common variant is to resort to the corresponding ammonium salt (R_2NH_2A) of the catalyst. Besides Brønsted, Lewis acids can of course also serve as activators. In the case of a neutral catalyst (type IV), the substrate can pursue various ways to arrive at the ion pair **159**. The anion is then bound by the catalyst and the activated complex **160** is obtained, which can subsequently afford the desired product upon reaction with a nucleophile [48, 164, 165].

The electrostatic interaction in the transition states is usually accompanied by secondary interactions such as steric repulsion or hydrogen bonding, which provide the requisite chiral environment for stereoinduction (see Fig. 8.50). In the depicted transition structures, hydrogen bonding (**161**, **162**, **164**) and π-π interactions (**163**) play a role besides steric interactions.

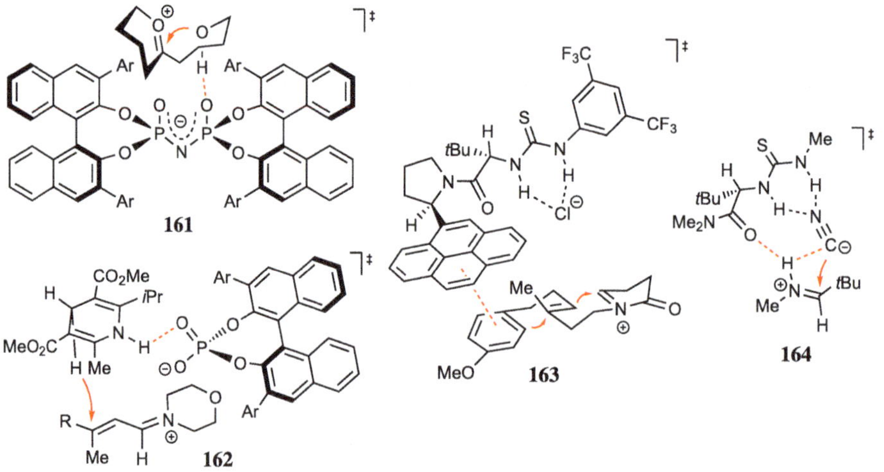

Fig. 8.50 Transition states of anion-directed reactions. Secondary interactions are marked in red [47, 166–169]

In the case of BINOL-derived catalysts, the 3,3'-substitution pattern is particularly crucial for obtaining a high enantioselectivity: Sterically demanding aryl groups such as 2,4,6-

triisopropylphenyl (**165**, TRIP), mesityl or anthracenyl, possibly paired with a modulation of the electronic properties by introducing CF_3 groups, ensure a chiral, pocket-like environment of the reacting substrate by steric shielding (see Fig. 8.51) [170].

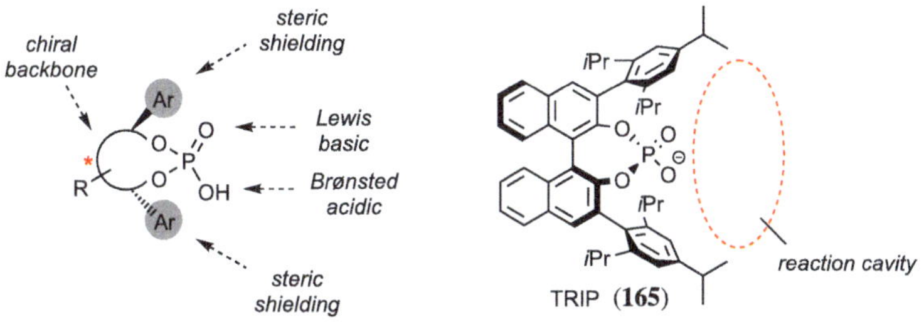

Fig. 8.51 Stereochemical model of catalysis by chiral phosphoric acid esters [170]

Nucleophilic components often rely on electron-rich arenes (Friedel–Crafts alkylation), Hantzsch esters (reduction), but also peracids (epoxidation), alcohols, and amines serve as reactants. Silyl enol ethers (Mukaiyama-Aldol) cannot be converted in the presence of phosphoric acid derivatives, as the catalyst would be silylated and thus deactivated. In these cases, more acidic and thus less Lewis basic systems are required, such as sulfonimides, triflimides, or TMSOTf (see Fig. 8.52) [170–172].

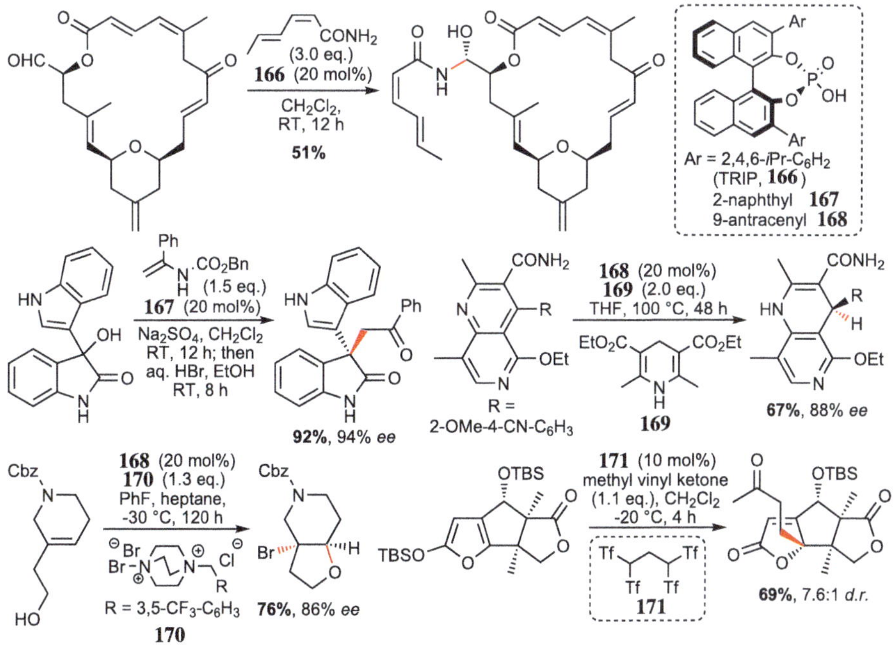

Fig. 8.52 Application of Brønsted acid-catalyzed reactions in natural product syntheses [173–177]

Brønsted bases accept a proton from acidic nucleophiles and also act as a cationic counterion (**172**, see Fig. 8.53), thereby functioning like a chiral proton equivalent [54, 145]. Usually, the corresponding anions are used directly as nucleophiles and the chiral base "only" serves for enantioinduction. Many bases are also capable of acting via a bifunctional mechanism using H-donor functionalities such as hydroxy, amino, or thiourea groups. Typical bases are derived from Cinchona alkaloids (**173**), aminothioureas (**174**), and cyclopropenimines (**175**). The exact origin of a diastereo- or enantioinduction is difficult to generalize due to the breadth of employed catalyst motifs and therefore must be considered on a case-by-case basis.

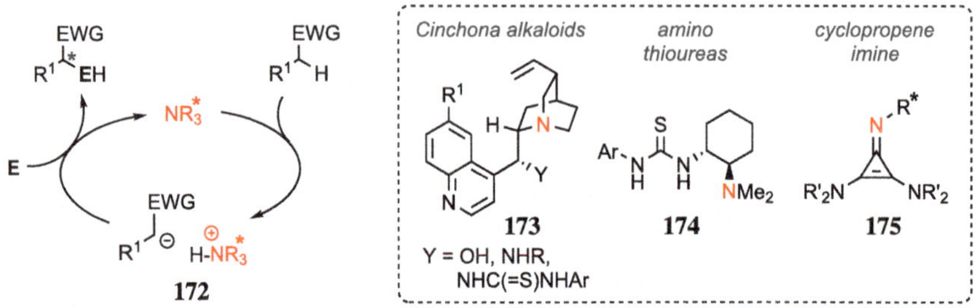

Fig. 8.53 Catalytic cycle of Brønsted bases and structure of common bases

For substrates that are not sufficiently acidic, either stronger bases (usually phosphazenes[III] such as the Schwesinger bases) are needed, or a carbonyl function is activated by means of a co-catalyst, for example, a Lewis acid or alternatively an amine, to enable generation of the more acidic iminium ion [54]. Chiral phosphazenes and guanidines have also been developed as viable organocatalysts for these transformations [178].

So far, there are only select applications in complex natural product syntheses. In Deng's synthesis of manzacidin A, the Cinchona-derived chiral base **177** was used in an asymmetric Michael addition to afford **176**. Dixon and co-workers resorted to catalyst **179** for the deprotonation of the β, γ-ketone and subsequent enantioselective reprotonation at C4 following the key pericyclic reaction. The stereocenter in the phosphazene directs diene and dienophile in the intramolecular cycloaddition to obtain **178** as major product (see Fig. 8.54) [140, 179].

8.1.6 Other Mechanisms and Methods

There remains a multitude of different organocatalysts whose mechanism of action does not fall into the reaction categories discussed thus far. The ketone-catalyzed epoxidation of alkenes, which most commonly employs catalyst **13** developed by Shi and co-workers, is worthy of particular note (see Sect. 3.3.1 for details on the mechanism).

[III] In many publications, these are also referred to as iminophosphoranes.

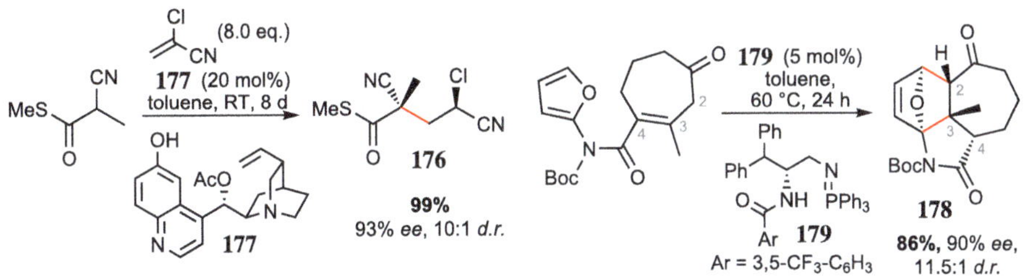

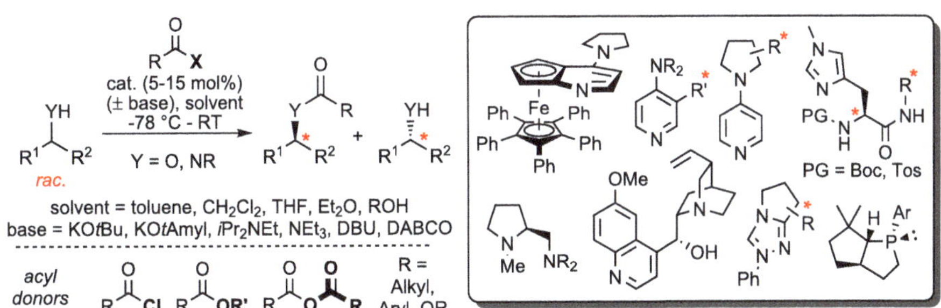

Fig. 8.54 Base-catalyzed key steps in the syntheses of manzacidin A and himalensine A [140, 179]

The **activation of acyl groups** or related functionalities by dimethylaminopyridine derivatives (DMAP) or *n*-alkyl-imidazoles is a frequently encountered method, even though it is rarely considered to be part of the organocatalytic repertoire. The asymmetric variant was first enabled by ferrocene-based catalysts, which were developed independently in the groups of Fu and Vedejs. The mechanism of acylation is shown in Fig. 8.1.

In DMAP derivatives, the chiral induction can be mediated either by the stereoelement of planar chirality originating from η^6-coordination, or a stereogenic center located at the C3 position or placed at the amino substituent. Imidazoles are often incorporated into the side chain of peptide-based catalysts, which enable the chirality transfer of the catalyst to the substrate through either their secondary or tertiary structure. In addition, tertiary amines (e.g., Cinchona alkaloids), phosphines, or *N*-heterocyclic carbenes can facilitate asymmetric acylations. The catalyzed reactions typically involve a kinetic resolution, for example, the esterification of alcohols, the amidation of amines, or various rearrangements but also might encompass the reaction of diverse nucleophiles with acyl acceptors. Apart from the resolution of chiral nucleophiles, [2+2] cycloadditions as well as range of other conversions to chiral products are furthermore catalyzed (see Fig. 8.55) [60, 61, 180].

Fig. 8.55 Reaction conditions, catalysts, and acyl donors in the kinetic resolution of secondary alcohols or amines [180]

The transfer of acyl groups is also applied to the syntheses of secondary metabolites to assemble complex intermediates. In Fürstner's synthesis of formosalides A and B, the hydroxy group at C6 was selectively introduced by TMS-quinidine **182** [181]. The mechanism proceeds via the activation of the acyl chloride with prior HCl-elimination to give ketene. Subsequent formation of the enolate **183**, followed by the aldol reaction to **184** and ring closure to the lactone afforded **181** [182]. The acetylation of erythromycin A (**185**) is particularly challenging due to the number of reactive hydroxy groups. The order of reactivity (2'-OH > 4"-OH > 11-OH) favors a derivatization at the hydroxy groups of the carbohydrates. By using the peptide **187**, the least reactive secondary OH group at C-11 could be selectively converted to **186** [183]. The antiviral drug remdesivir, which was also approved for COVID-19 treatment, was successfully synthesized by Zhang *et al.* The asymmetric formation of the phosphoric acid amido ester **190** starting from **188** and **189** constituted the key step of the sequence. The fused imidazole derivative **191** served as a catalyst to introduce the stereogenic center at phosphorus with excellent diastereoselectivity. The diol functionality in **190** was unmasked as concluding step, yielding the desired active pharmaceutical ingredient (see Fig. 8.56) [184].

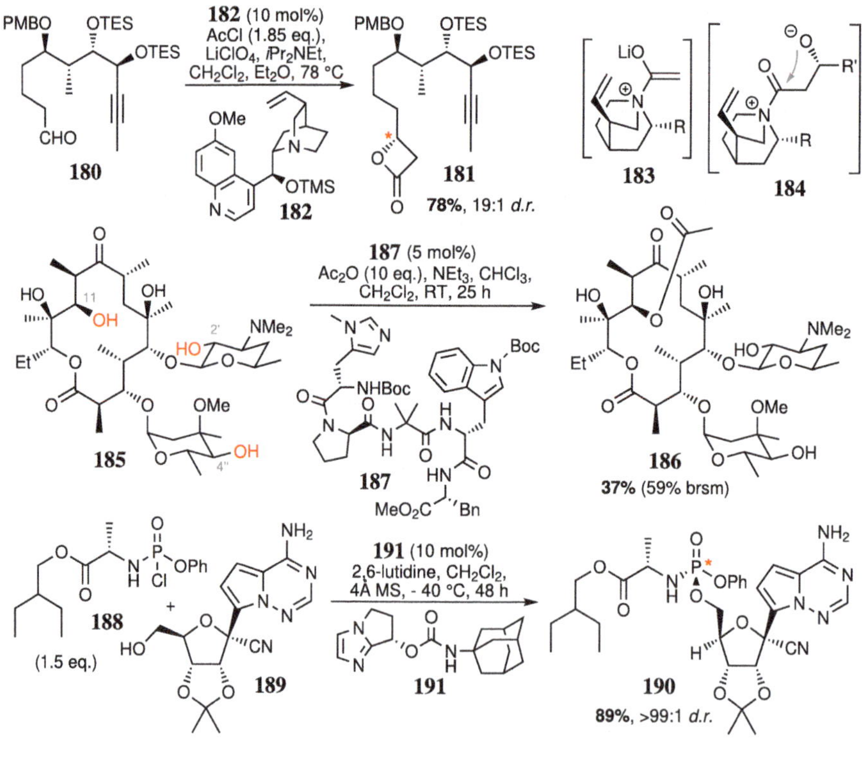

Fig. 8.56 Organocatalytic acylations in syntheses of secondary metabolites and drugs

As previously stated, **frustrated Lewis pairs** (FLPs) are characterized by the presence of a Lewis acid (acceptor A) and Lewis base (donor D), which are prevented from coming into close proximity, whereby they would neutralize and effect a strong acid-base interaction. This state of arrested interaction enables the activation and heterolytic cleavage of H_2

and its subsequent use in hydrogenation reactions. So far, there are only a few examples of asymmetric hydrogenations and no applications in the synthesis of complex natural products by FLPs [185]. The challenge of these reductions is especially rooted in the judicious selection of the catalyst to (i.) avoid an irreversible binding of the reduced product to the catalyst in the case of C-X double bonds and (ii.) enable the initial protonation in the case of unfunctionalized alkenes and alkynes (cf. Fig. 8.59, *vide infra*). This is achieved by the modular structure of the catalyst consisting of the combination of a suitable donor/acceptor couple. While the pairing of electron-poor boranes and electron-rich phosphines has proven to be particularly effective, boranes have also been combined with other Lewis bases such as Et_2O or NHCs in recent years to expand the substrate scope. In addition to hydrogen, other selected molecules such as CO_2 and alkynes can also be activated (see Fig. 8.57) [19, 57–59, 186]. To increase the H_2O tolerance of the classic borane acceptors, various metal-based systems have also been developed in addition to other metal-free FLPs [187].

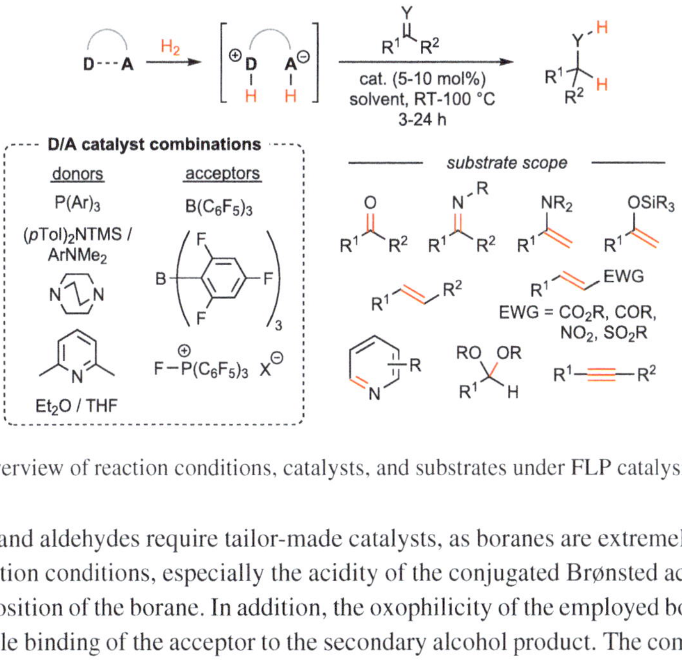

Fig. 8.57 Overview of reaction conditions, catalysts, and substrates under FLP catalysis [59]

Ketones and aldehydes require tailor-made catalysts, as boranes are extremely oxophilic and the reaction conditions, especially the acidity of the conjugated Brønsted acid, result in the decomposition of the borane. In addition, the oxophilicity of the employed boranes leads to a reversible binding of the acceptor to the secondary alcohol product. The combination of $B(C_6F_5)_3$ with an etheric solvent, which can act as a Lewis base, was independently developed in the groups of Ashley and Stephan and successfully prevents catalyst decomposition and coordination to the product (see Fig. 8.58) [188, 189].

To be effective for hydrogen activation, the two functionalities contained within the catalyst must be able to interact with H_2 simultaneously [191]. The dominant orbital interaction in the initial step is a $n \rightarrow \sigma_{HH}^*$ interaction, which allows a heterolytic cleavage of the H-H bond mediated by the active catalyst **192**. The bond scission to give **193** is, depending on the stereoelectronic properties of the chosen catalyst, either reversible or irreversible (e.g.,

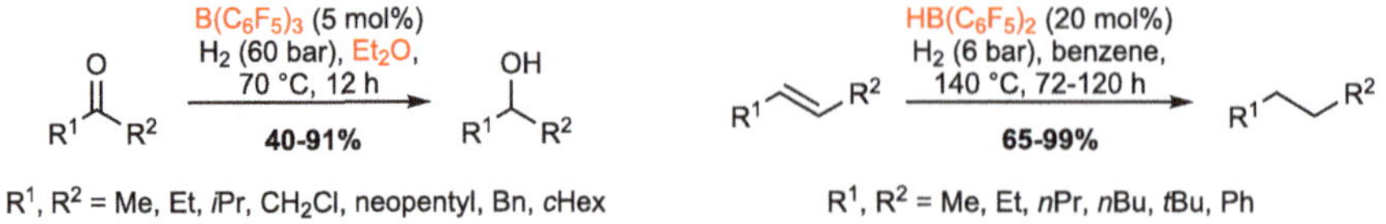

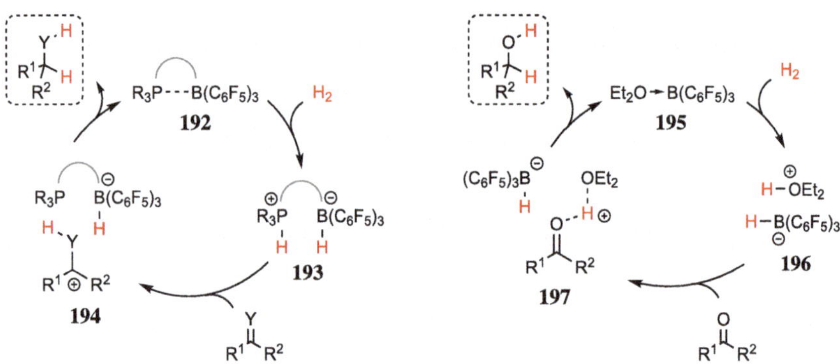

Fig. 8.58 FLP-mediated hydrogenation of ketones and alkenes [189, 190]

in the case of $B(C_6F_4H)_3$-$P(o$-Tol$)_3$ *vs.* $B(C_6F_5)_3$-$P(o$-Tol$)_3$) [192]. In hydrogenations of heteroatomic CX- and activated CC double bonds (imines, oximes, ketones, enamines, and enol ethers), the substrate is protonated following the H_2 cleavage step. The final H^- transfer from borohydride in **194** regenerates the catalyst and allows the catalytic cycle to be initiated once more. In the case of unfunctionalized double bonds, free electron pairs are missing as supporting donor functionality in the protonation. A protolytic mechanism thus displays a high activation barrier in step **193**→**194**. This conundrum requires the use of tailor-made Lewis base/Lewis acid combinations, which are just sufficiently active to cause the H_2 cleavage. Following the H_2 activation, the Lewis base acts as a conjugated Brønsted acid and must possess a sufficiently high acidity to carry out the protonation step of the unfunctionalized substrate (see Fig. 8.59) [192].

Fig. 8.59 Postulated mechanism of FLP-catalyzed hydrogenation—generic scheme and reduction of ketones with $B(C_6F_5)_3$/Et_2O [191–193]

The mechanism of the ketone hydrogenation proceeds in a similar fashion as the general hydrogenation of CX double bonds [193]. The solvent serves as the Lewis base, although the actual number of involved ether molecules in **196** remains unclear. While a protonated species like in **194** was postulated only as a short-lived intermediate or even a true transition state due to the low activation barrier, the related intermediate **197** is assumed for ketone hydrogenations, in which the proton pre-complexes the substrate [193]. In the concluding step, the desired alcohol is obtained, and the FLP **195** reenters the catalytic cycle again.

When using boranes HBR_2, the reduction of alkynes and alkenes follows a different reaction path (see Fig. 8.60). It is assumed that an initial hydroboration of the olefinic or acetylenic CC bond to **198** occurs. The reduction of the respective alkenylborane presumably proceeds either via a σ-bond metathesis (**199**) or through an intermolecular hydride transfer (**200**), depending on the catalyst system and substrate [192, 194].

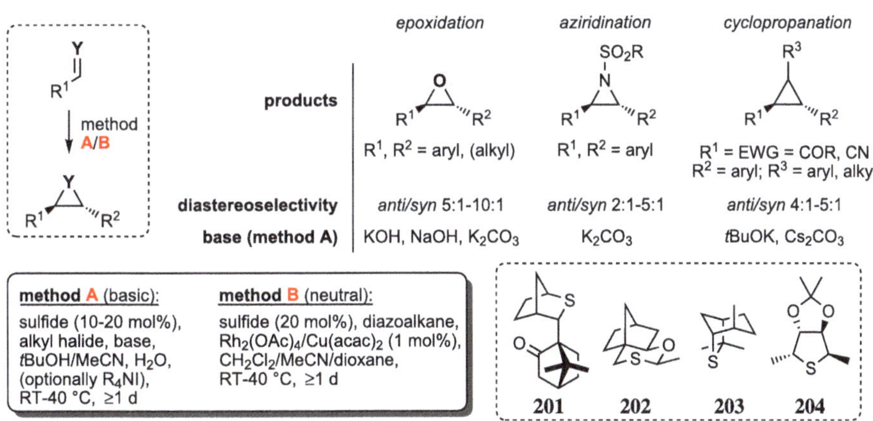

Fig. 8.60 Mechanism of diarylborane-mediated reduction of alkynes and alkenes [192, 194]

Various asymmetric transformations such as epoxidations, aziridinations, and cyclopropanations have been realized using covalent catalysis in the presence of **chalcogen ylides**. Sulfur ylides dominate this field due to their reactivity. Furthermore, chiral sulfides are relatively easily accessible among the chalcogens and thus cheaper compared to corresponding selenides (see Fig. 8.61) [195, 196].

	epoxidation	*aziridination*	*cyclopropanation*
products	R^1, R^2 = aryl, (alkyl)	R^1, R^2 = aryl	R^1 = EWG = COR, CN; R^2 = aryl; R^3 = aryl, alkyl
diastereoselectivity	*anti/syn* 5:1-10:1	*anti/syn* 2:1-5:1	*anti/syn* 4:1-5:1
base (method A)	KOH, NaOH, K_2CO_3	K_2CO_3	tBuOK, Cs_2CO_3

method A (basic): sulfide (10-20 mol%), alkyl halide, base, tBuOH/MeCN, H_2O, (optionally R_4NI), RT-40 °C, ≥1 d

method B (neutral): sulfide (20 mol%), diazoalkane, $Rh_2(OAc)_4$/$Cu(acac)_2$ (1 mol%), CH_2Cl_2/MeCN/dioxane, RT-40 °C, ≥1 d

201   202   203   204

Fig. 8.61 Common reaction conditions and substrate scope in sulfide-catalyzed reactions [195, 196]

For the epoxidation, aziridination, and cyclopropanation, two classes of substrates are generally considered suitable, which generate the respective ylide from the sulfide catalyst: alkyl halides (method A) or diazo compounds (method B). Method A requires the use of a base in a biphasic regime, while diazo compounds can facilitate a reaction under neutral conditions — however, $Cu(acac)_2$ or $Rh_2(OAc)_4$ are needed as a co-catalyst. To achieve high yields, a slow and controlled addition of the diazo compound to the reaction mixture is critical. Instead of their direct addition, N-tosylhydrazone salts can alternatively be used for the *in situ* generation of the diazo compound. This requires biphasic conditions and the use of a phase transfer catalyst. For cyclopropanations, only activated Michael systems such as α, β-unsaturated ketones or nitriles are viable substrates thus far. The employed catalysts can be roughly divided into unsymmetric (**201-203**) and C_2-symmetric sulfides (**204**, and analogs thereof) [195, 196].

The *stoichiometric* sulfur ylide-mediated conversion is also known as the *Corey–Chaykovsky reaction* (see Sect. 3.3.1). Under catalytic conditions, the ylides must be formed as part of the four-step catalytic cycle (see Fig. 8.62). After the initial alkylation of the catalyst to give **205**, the ylide **206** is generated *in situ* by abstraction of the α-acid proton by the added base. The alkylation step is slow and reversible when using alkyl halides [197]. The addition of the ylide to aldehydes/ketones or imines to afford **207** enables the formation of the two vicinal stereocenters and constitutes the enantioinducing step. Depending on the β-substituents present in the substrate, their relative configuration (*syn/anti*), and the employed solvent system, the CC bond formation can be reversible [198]. The stereospecific ring closure of the intermediate betaine **207** furnishes the desired product, eliminates the catalyst, and thus completes the catalytic cycle. Instead of alkyl halides, diazo compounds can also be utilized for generating the transient sulfur ylides. As the transfer necessitates the presence of metal complexes based on Rh or Cu, the reaction can no longer be considered organocatalytic in nature. The advantage of this variant is the avoidance of a base under these conditions, allowing the conversion of base-sensitive substrates to the desired products in high yield. With aliphatic aldehydes or imines, a competing aldol addition or condensation can also be successfully suppressed. Sulfides offer the best combination of nucleophilicity (formation of **205**) and reactivity as a leaving group (ring closure of the betaine **207**) among the chalcogen catalysts [195, 196].

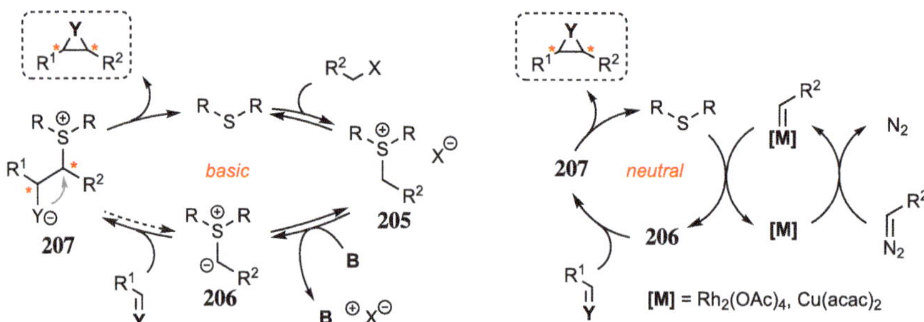

Fig. 8.62 Catalytic cycles of sulfur ylide-mediated epoxidation, aziridination, and cyclopropanation under basic and neutral conditions [195, 199]

The enantioinduction of the addition **206**→**207** is strongly dependent on the identity of the catalyst and a prediction of the stereoselectivity is therefore difficult to realize. Nevertheless, given the obvious analogy to the similar steps of the related and extensively studied Wittig (and Julia) olefination, some conclusions can be drawn, especially in the case of epoxidations

[200]. The ring closure of the betaine **207** to the epoxide with concomitant elimination of the sulfide reliably proceeds in an antiperiplanar manner. Both the diastereo- and the enantioselectivity are accordingly determined during the betaine formation step and the subsequent conformational interconversions. The *diastereoselectivity* is determined by the facial differentiation of the aldehyde. The diastereomeric betaines *sc*-**208** and *ap*-**208** only slowly convert to the respective conformers *anti*-**209** and *syn*-**209**, whose antiperiplanar arrangement of the oxy-anion and the sulfonium ion allows the formation of the oxirane upon sulfide elimination. The rotational barrier **208**→**209** depends on the nature of substituents R^1, R^2 and is different for *sc*-**208** and *ap*-**208**: assuming a rapid equilibration of the two competing diastereomers via **206**, one reacts to the corresponding desired epoxide while the conversion of the other is too slow. The high diastereocontrol therefore results from the *anti*-betaine diastereomer undergoing bond rotation and ring closure more rapidly than the *syn*-isomer, which reverts to the starting materials. The *enantioselectivity* is consequently determined by the equilibrium of the diastereomeric ylides **210** and **211** according to the Curtin–Hammett principle. In addition to the high thermodynamic preference of one ylide, a pronounced enantiomeric excess also requires the addition to be governed by a tight control of facial selectivity (**212** *vs.* **213**). Ideally, the betaine formation should be irreversible for the desired diastereomer and enantiomer, i.e., purely kinetically controlled. The ratio of **212** and **213** and their relative rate of generating the corresponding epoxide reflects the enantioselectivity of the reaction (see Fig. 8.63) [200].

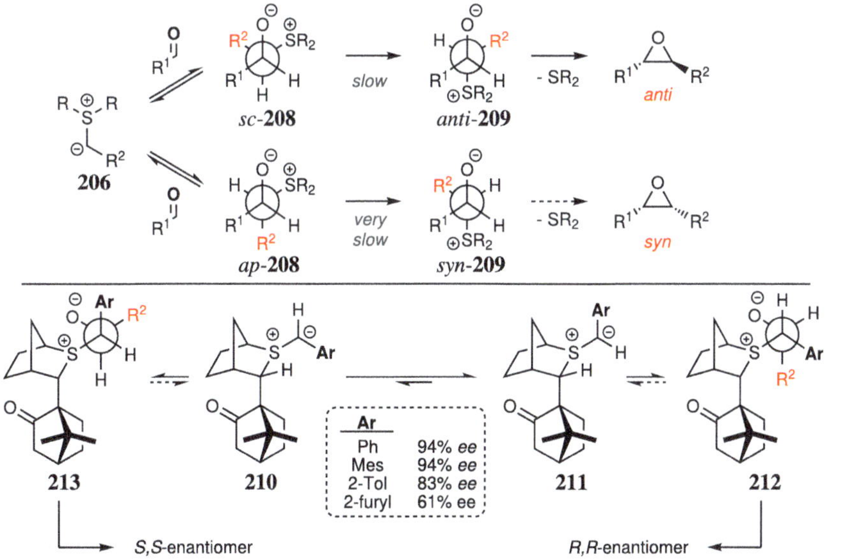

Fig. 8.63 Origin of diastereo- and enantioselectivity in sulfur ylide-catalyzed epoxidations [200]

A high reversibility in the formation of the betaine **207** allows for good diastereoselectivity but only moderate enantioselectivity, while an irreversible reaction to **207** with thermodynamic preference of one addition product offers high enantiomeric excesses but low diastereoselectivity. To enable high diastereo- **and** enantioselectivity, the reversible formation of one betaine (e.g., *sc-***208**) and irreversible reaction to the other betaine (e.g., *ap-***208**) is necessary. This may be controlled to an extent by the chosen reaction conditions (solvent, temperature, metal ions) [200].

The use of sulfur ylides under catalytic conditions, despite the multitude of available methods, has so far found no application in the synthesis of complex natural products [195, 196]. This is attributed to two factors: There exists only a limited range of substrates, as many catalysts only deliver 1,2-diaryl epoxides with high diastereo- and enantioselectivity [201]. Second of all, the chiral sulfides are either not commercially available or only to a limited extent and must be laboriously produced in multi-step syntheses [202]. While newer catalysts can address both issues, the tolerance of functional groups remains a blind spot without extensive studies of the viable substrate scope, which is why proven methods (Sharpless-, Jacobsen-, Shi-epoxidation) are often preferred instead. Additionally, classic (=basic) conditions do not permit the use of base-labile substrates.

Under Corey–Chaykovsky conditions, i.e., using stoichiometric amounts of sulfur ylides, *diastereoselective* epoxidations have already been successfully incorporated into the synthesis of secondary metabolites. An example of a stoichiometric asymmetric aziridination in the presence of a sulfide, which can also be used catalytically in other transformations, was published by Aggarwal *et al.* in the synthesis of the indole alkaloid α-cyclopiazonic acid. In accordance with catalytic conditions, the *anti*-aziridine **216** was preferentially obtained with high diastereoselectivity and excellent enantiomeric excess starting from the *N*-tosyl-substituted indole **214** and sulfonium salt **215** (see Fig. 8.64) [203].

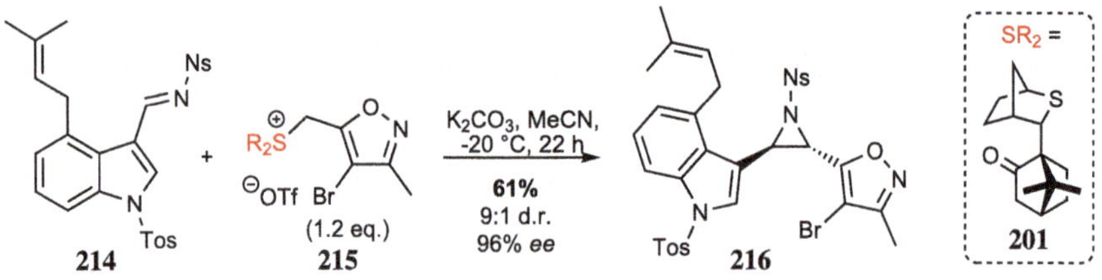

Fig. 8.64 Asymmetric aziridination in the synthesis of α-cyclopiazonic acid [203]

Another noteworthy development is the organocatalytic nucleophilic substitution at C_{sp^3} centers. The *Mitsunobo* reaction has received much attention in this context, as (i.) stoichiometric amounts of an activator are required for the reaction and (ii.) these are difficult to separate from the product [204]. The most common approach is to use a secondary reducing agent, which decreases the requisite amount of phosphine [205]. In addition, a method relying on a redox-neutral catalyst was recently disclosed that does not require additional additives and can also be applied to non-activated, bisalkyl-substituted alcohols (see Fig. 8.65) [206].

The nucleophiles, which must be strongly acidic to effect a conversion under the reaction conditions given the absence of a base, range from O- to N- to S-nucleophiles. The catalyst

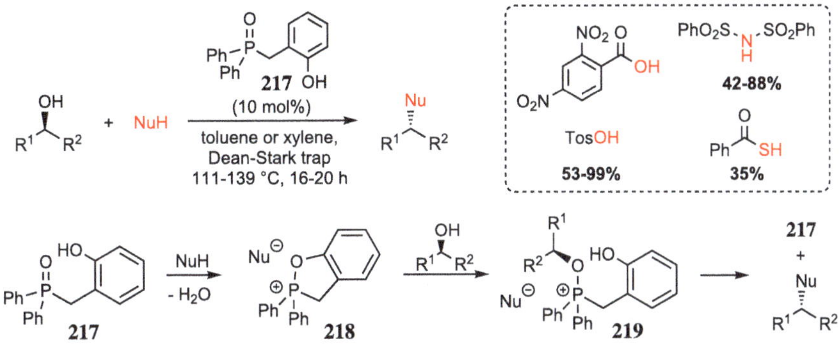

Fig. 8.65 Organocatalytic Mitsunobu reaction by Denton *et al.* [206]

217 can directly convert them to give **218**. The obtained phosphonium salt then reacts with the substrate to **219**, which finally decomposes to afford the desired product and the catalyst **217** via attack of the nucleophile. The method allows a wide range of substrates, especially when employing O-nucleophiles, and shows a stereospecific reaction with little to no enantioerosion in enantiomerically pure substrates.

Corey–Bakshi–Shibata reductions utilize oxazaborolidines, which of course constitute metal-free catalysts, and thus fall under the regime of organocatalytic transformations. The mechanism has been extensively studied and is well understood; the applications of the methodology in the industrial sector are also well documented (see Sect. 4.2.5).

Apart from the approaches discussed so far, organocatalysts can also be combined with metal catalysts. A very prominent example is the use of radical intermediates, which can be obtained through the use of photoredox catalysts (see Sect. 10.2) [62, 74, 207, 208]. This dual catalysis can compensate for the weaknesses of a purely organocatalytic approach in many cases and enable new modes of reactivity. The breadth of the transformations studied shall not be outlined in detail, however, by virtue of their rapid development [209].

References

1. S. Xu, I. Held, B. Kempf, H. Mayr, W. Steglich, H. Zipse, *Chem. Eur. J.* **2005**, *11*, 4751–4757
2. P. G. Bulger in *Comprehensive Chirality*, *Vol. 9* (Ed.: D. L. Hughes), Elsevier, **2012**, Chapter 9.10, pp. 228–252
3. B. M. Sahoo, B. K. Banik, *Curr. Organocat.* **2019**, *6*, 92–105
4. M. C. Holland, R. Gilmour, *Angew. Chem. Int. Ed.* **2015**, *54*, 3862–3871
5. P. H.-Y. Cheong, C. Y. Legault, J. M. Um, N. Celebi-Ölcum, K. N. Houk, *Chem. Rev.* **2011**, *111*, 5042–5137
6. Z. G. Hajos, D. R. Parrish (F. Hoffmann-La Roche), *DE 2102623 A1* **1971**
7. Z. G. Hajos, D. R. Parrish, *J. Org. Chem.* **1974**, *39*, 1615–1621
8. U. Eder, G. Sauer, R. Wiechert, *Angew. Chem. Int. Ed. Engl.* **1971**, *10*, 496–497
9. B. List, R. A. Lerner, C. F. Barbas, *J. Am. Chem. Soc.* **2000**, *122*, 2395–2396
10. K. A. Ahrendt, C. J. Borths, D. W. C. MacMillan, *J. Am. Chem. Soc.* **2000**, *122*, 4243–4244

11. S. Mukherjee, J. W. Yang, S. Hoffmann, B. List, *Chem. Rev.* **2007**, *107*, 5471–5569
12. A. Erkkilä, I. Majander, P. M. Pihko, *Chem. Rev.* **2007**, *107*, 5416–5470
13. D. Enders, O. Niemeier, A. Henseler, *Chem. Rev.* **2007**, *107*, 5606–5655
14. A. G. Doyle, E. N. Jacobsen, *Chem. Rev.* **2007**, *107*, 5713–5743
15. A. J. Metrano, A. J. Chinn, C. R. Shugrue, E. A. Stone, B. Kim, S. J. Miller, *Chem. Rev.* **2020**, *120*, 11479–11615
16. E. A. C. Davie, S. M. Mennen, Y. Xu, S. J. Miller, *Chem. Rev.* **2007**, *107*, 5759–5812
17. T. Akiyama, *Chem. Rev.* **2007**, *107*, 5744–5758
18. K. Maruoka, *Org. Process Res. Dev.* **2008**, *12*, 679–697
19. D. W. Stephan, *J. Am. Chem. Soc.* **2015**, *137*, 10018–10032
20. E. Reyes, L. Prieto, A. Milelli, *Molecules* **2023**, *28*, 271
21. G. Stork, A. Brizzolara, H. Landesman, J. Szmuszkovicz, R. Terrell, *J. Am. Chem. Soc.* **1963**, *85*, 207–222
22. E. Marqués-López, R. P. Herrera, M. Christmann, *Nat. Prod. Rep.* **2010**, *27*, 1138–1167
23. M. E. Abbasov, D. Romo, *Nat. Prod. Rep.* **2014**, *31*, 1318–1327
24. E. Marqués-López, R. P. Herrera in *Comprehensive Enantioselective Organocatalysis: Catalysts, Reactions, and Applications* (Ed.: P. I. Dalko), Wiley-VCH, **2013**, Chapter 44, pp. 1359–1383
25. M. Shoji, Y. Hayashi in *Modern Tools for the Synthesis of Complex Bioactive Molecules* (Eds.: J. Cossy, S. Arseniyadis), John Wiley & Sons, **2012**, Chapter 6, pp. 189–212
26. G. Zhao, Z. Q. Ye, X. Y. Wu in *Efficiency in Natural Product Total Synthesis* (Eds.: Q. Huang, Z.-J. Yao, R. P. Hsung.), John Wiley & Sons, **2018**, Chapter 7, pp. 297–317
27. R. Parella, S. Jakkampudi, J. C.-G. Zhao, *ChemistrySelect* **2021**, *6*, 2252–2280
28. D. L. Hughes, *Org. Process Res. Dev.* **2018**, *22*, 574–584
29. J. Tan, N. Yasuda, *Org. Process Res. Dev.* **2015**, *19*, 1731–1746
30. A. Carlone, L. Bernardi, *Phys. Sci. Rev.* **2019**, *4*, 20180097
31. D. L. Hughes, *Org. Process Res. Dev.* **2022**, *26*, 2224–2239
32. A Scifinder search on December 8th 2020 only gave approx. 100 hits for publications containing the keywords "*organocatalysis*" and "*total synthesis*" or "*natural product synthesis*".
33. R. R. Knowles, J. Carpenter, S. B. Blakey, A. Kayano, I. K. Mangion, C. J. Sinz, D. W. C. MacMillan, *Chem. Sci.* **2011**, *2*, 308–311
34. F. Giacalone, M. Gruttadauria, P. Agrigento, R. Noto, *Chem. Soc. Rev.* **2012**, *41*, 2406–2447
35. X. Fu, T. L. McAllister, T. K. Thiruvengadam, C.-H. Tann, *WO 02/079174 A2* **2001**
36. X. Fu, T. L. McAllister, T. K. Thiruvengadam, C.-H. Tann, D. Su, *Tetrahedron Lett.* **2003**, *44*, 801–804
37. *Comprehensive Enantioselective Organocatalysis: Catalysts, Reactions, and Applications*, (Ed.: P. I. Dalko), Wiley-VCH, **2013**
38. *Science of Synthesis, Asymmetric Organocatalysis*, (Eds.: B. List, K. Maruoka), Thieme, **2012**
39. *Enantioselective Organocatalyzed Reactions I & II*, (Ed.: R. Mahrwald), Springer, **2011**
40. G. J. Reyes-Rodríguez, N. M. Rezayee, A. Vidal-Albalat, K. A. Jørgensen, *Chem. Rev.* **2019**, *119*, 4221–4260
41. T. J. Auvil, A. G. Schafer, A. E. Mattson, *Eur. J. Org. Chem.* **2014**, 2633–2646
42. T. N. Nguyen, P.-A. Chen, K. Setthakarn, J. A. May, *Molecules* **2018**, *23*, 2317
43. P. Nagorny, Z. Sun, *Beilstein J. Org. Chem.* **2016**, *12*, 2834–2848
44. S. H. Jungbauer, S. M. Huber, *J. Am. Chem. Soc.* **2015**, *137*, 12110–12120
45. R. L. Sutar, S. M. Huber, *ACS Catal.* **2019**, *9*, 9622–9639
46. J. Bamberger, F. Ostler, O. G. Mancheno, *ChemCatChem* **2019**, *11*, 5198–5211
47. K. Brak, E. N. Jacobsen, *Angew. Chem. Int. Ed.* **2013**, *52*, 534–561
48. M. Mahlau, B. List, *Angew. Chem. Int. Ed.* **2012**, *51*, 518–533

49. T. Akiyama, K. Mori, *Chem. Rev.* **2015**, *115*, 9277–9306
50. S. Shirakawa, K. Maruoka, *Angew. Chem. Int. Ed.* **2013**, *52*, 4312–4348
51. T. Hashimoto, K. Maruoka, *Chem. Rev.* **2007**, *107*, 5656–5682
52. J. Schörgenhumer, M. Tiffner, M. Waser, *Beilstein J. Org. Chem.* **2017**, *13*, 1753–1769
53. T. Ooi, K. Maruoka, *Angew. Chem. Int. Ed.* **2007**, *46*, 4222–4266
54. B. Teng, W. C. Lim, C.-H. Tan, *Synlett* **2017**, *28*, 1272–1277
55. M. T. Oliveira, J.-W. Lee, *ChemCatChem* **2017**, *9*, 377–384
56. R. Schettini, M. Sicignano, F. D. Riccardis, I. Izzo, G. Della Sala, *Synthesis* **2018**, *50*, 4777–4795
57. D. W. Stephan, G. Erker, *Angew. Chem. Int. Ed.* **2010**, *49*, 46–76
58. D. W. Stephan, *Acc. Chem. Res.* **2015**, *48*, 306–316
59. J. Lam, K. M. Szkop, E. Mosaferi, D. W. Stephan, *Chem. Soc. Rev.* **2019**, *48*, 3592–3612
60. H. Mandai, K. Fujii, S. Suga, *Tetrahedron Lett.* **2018**, *59*, 1787–1803
61. R. P. Wurz, *Chem. Rev.* **2007**, *107*, 5570–5595
62. T. Bortolato, S. Cuadros, G. Simionato, L. Dell'Amico, *Chem. Commun.* **2022**, *58*, 1263–1283
63. M. B. Schmid, K. Zeitler, R. M. Gschwind, *Angew. Chem. Int. Ed.* **2010**, *49*, 4997–5003
64. M. H. Haindl, J. Hioe, R. M. Gschwind, *J. Am. Chem. Soc.* **2015**, *137*, 12835–12842
65. M. A. Ashley, J. S. Hirschi, J. A. Izzo, M. J. Vetticatt, *J. Am. Chem. Soc.* **2016**, *138*, 1756–1759
66. P. Renzi, J. Hioe, R. M. Gschwind, *Acc. Chem. Res.* **2017**, *50*, 2936–2948
67. M. Klussmann in *Science of Synthesis, Asymmetric Organocatalysis 2* (Ed.: K. Maruoka), Thieme, **2012**, Chapter 2.3.4, pp. 633–671
68. H. Kotsuki, N. Sasakura in *Comprehensive Enantioselective Organocatalysis: Catalysts, Reactions, and Applications* (Ed.: P. I. Dalko), Wiley-VCH, **2013**, Chapter 1, pp. 3–31
69. P. M. Pihko, I. Majander, A. Erkkilä, *Top. Curr. Chem.* **2010**, *291*, 29–75
70. J. Carpenter, A. B. Northrup, dM Chung, J. J. M. Wiener, S.-G. Kim, D. W. C. MacMillan, *Angew. Chem. Int. Ed.* **2008**, *47*, 3568–3572
71. A. J. Oelke, F. Antonietti, L. Bertone, P. B. Cranwell, D. J. France, R. J. M. Goss, T. Hofmann, S. Knauer, S. J. Moss, P. C. Skelton, R. M. Turner, G. Wuitschik, S. V. Ley, *Chem. Eur. J.* **2011**, *17*, 4183–4194
72. R. T. Larson, M. D. Clift, R. J. Thomson, *Angew. Chem. Int. Ed.* **2012**, *51*, 2481–2484
73. T. D. Beeson, A. Mastracchio, J.-B. Hong, K. Ashton, D. W. C. MacMillan, *Science* **2007**, *316*, 582–585
74. M. Meciarová, P. Tisovský, R. Sebesta, *New J. Chem.* **2016**, *40*, 4855–4864
75. L. Zhu, D. Wang, Z. Jia, Q. Lin, M. Huang, S. Luo, *ACS Catal.* **2018**, *8*, 5466–5484
76. D. W. C. MacMillan, A. J. B. Watson in *Science of Synthesis, Asymmetric Organocatalysis 2* (Ed.: B. List), Thieme, **2012**, Chapter 1.1.7, pp. 309–401
77. Y. Liu, P. Melchiorre in *Science of Synthesis, Asymmetric Organocatalysis 2* (Ed.: B. List), Thieme, **2012**, Chapter 1.1.8, pp. 403–438
78. I. Pápai in *Science of Synthesis, Asymmetric Organocatalysis 2* (Ed.: K. Maruoka), Thieme, **2012**, Chapter 2.3.3, pp. 601–632
79. R. Gordillo, J. Carter, K. N. Houk, *Adv. Synth. Catal.* **2004**, *346*, 1175–1185
80. M. A. Hellinghuizen, P. Franceschi, J. Roithová, *Chem. Eur. J.* **2024**, *43*, e202400294
81. J. B. Brazier, G. P. Hopkins, M. Jirari, S. Mutter, R. Pommereuil, L. Samulis, J. A. Platts, N. C. O. Tomkinson, *Tetrahedron Lett.* **2011**, *52*, 2783–2785
82. J. B. Brazier, N. C. Tomkinson, *Top. Curr. Chem.* **2010**, *291*, 281–347
83. H. Gotoh, T. Uchimaru, Y. Hayashi, *Chem. Eur. J.* **2015**, *21*, 12337–12346
84. R. Gilmour, T. J. Prior, J. W. Burton, A. B. Holmes, *Chem. Commun.* **2007**, 3954–3956
85. C. Fehr, I. Magpantay, J. Arpagaus, X. Marquet, M. Vuagnoux, *Angew. Chem. Int. Ed.* **2009**, *48*, 7221–7223
86. S. B. Jones, B. Simmons, D. W. C. MacMillan, *J. Am. Chem. Soc.* **2009**, *131*, 13606–13607

87. Y.-C. Guo, D.-P. Li, Y.-L. Li, H.-M. Wang, W.-J. Xiao, *Chirality* **2009**, *21*, 777–785

88. S. Lee, D. W. C. MacMillan, *J. Am. Chem. Soc.* **2007**, *129*, 15438–15439

89. M. W. Paixao, N. Holub, C. Vila, M. Nielsen, K. A. Jørgensen, *Angew. Chem. Int. Ed.* **2009**, *48*, 7338–7342

90. S. Brandau, A. Landa, J. Franzén, M. Marigo, K. A. Jørgensen, *Angew. Chem. Int. Ed.* **2006**, *45*, 4305–4309

91. Y. K. Chen, M. Yoshida, D. W. C. MacMillan, *J. Am. Chem. Soc.* **2006**, *128*, 9328–9329

92. P. Dinér, M. Nielsen, M. Marigo, K. A. Jørgensen, *Angew. Chem. Int. Ed.* **2007**, *46*, 1983–1987

93. S. Bertelsen, P. Dinér, R. L. Johansen, K. A. Jørgensen, *J. Am. Chem. Soc.* **2007**, *129*, 1536–1537

94. X. Lu, Y. Liu, B. Sun, B. Cindric, L. Deng, *J. Am. Chem. Soc.* **2008**, *130*, 8134–8135

95. Q. Li, K. Zhao, A. Peuronen, K. Rissanen, D. Enders, Y. Tang, *J. Am. Chem. Soc.* **2018**, *140*, 1937–1944

96. Q. Lin, D. Meloni, Y. Pan, M. Xia, J. Rodgers, S. Shepard, M. Li, L. Galya, B. Metcalf, T.-Y. Yue, P. Liu, J. Zhou, *Org. Lett.* **2009**, *11*, 1999–2002

97. F. Xu, M. Zacuto, N. Yoshikawa, R. Desmond, S. Hoerrner, T. Itoh, M. Journet, G. R. Humphrey, C. Cowden, N. Strotman, P. Devine, *J. Org. Chem.* **2010**, *75*, 7829–7841

98. S. T.-C. Eey, M. J. Lear, *Chem. Eur. J.* **2014**, *20*, 11556–11573

99. M. Marigo, J. Franzén, T. B. Poulsen, W. Zhuang, K. A. Jørgensen, *J. Am. Chem. Soc.* **2005**, *127*, 6964–6965

100. L. Deiana, G.-L. Zhao, S. Lin, P. Dziedzic, Q. Zhang, H. Leijonmarck, A. Córdova, *Adv. Synth. Catal.* **2010**, *352*, 3201–3207

101. P. Thesmar, O. Baudoin, *J. Am. Chem. Soc.* **2019**, *141*, 15779–15783

102. D. Seebach, *Angew. Chem. Int. Ed. Engl.* **1979**, *18*, 239–258

103. D. Voet, J. G. Voet, C. W. Pratt, *Lehrbuch der Biochemie*, Wiley-VCH, **2019**

104. R. Breslow, *J. Am. Chem. Soc.* **1958**, *80*, 3719–3726

105. M. Pareek, Y. Reddi, R. B. Sunoj, *Chem. Sci.* **2021**, *12*, 7973–7992

106. K. Thai, E. Sánchez-Larios, M. Gravel in *Comprehensive Enantioselective Organocatalysis: Catalysts, Reactions, and Applications* (Ed.: P. I. Dalko), Wiley-VCH, **2013**, Chapter 18, pp. 495–522

107. K. Suzuki, H. Takikawa in *Science of Synthesis, Asymmetric Organocatalysis 2* (Ed.: B. List), Thieme, **2012**, Chapter 1.1.13, pp. 591–618

108. D. A. DiRocco, T. Rovis in *Science of Synthesis, Asymmetric Organocatalysis 2* (Ed.: B. List), Thieme, **2012**, Chapter 1.1.14, pp. 619–637

109. J. L. Moore, T. Rovis, *Top. Curr. Chem.* **2010**, *291*, 77–144

110. S. Gehrke, O. Hollóczki, *Angew. Chem. Int. Ed.* **2017**, *56*, 16395–16398

111. O. Hollóczki, *Chem. Eur. J.* **2020**, *26*, 4885–4894

112. Y. Qiao, X. Chen, D. Wei, J. Chang, *Sci. Rep.* **2016**, *6*, 38200

113. H. Takikawa, K. Suzuki, *Org. Lett.* **2007**, *9*, 2713–2716

114. E. M. Phillips, J. M. Roberts, K. A. Scheidt, *Org. Lett.* **2010**, *12*, 2830–2833

115. K. C. Nicolaou, H. Li, A. L. Nold, D. Pappo, A. Lenzen, *J. Am. Chem. Soc.* **2007**, *129*, 10356–10357

116. P. E. Harrington, M. A. Tius, *J. Am. Chem. Soc.* **2001**, *123*, 8509–8514

117. K. C. Nicolaou, Y. Tang, J. Wang, *Chem. Commun.* **2007**, 1922–1923

118. K. L. Baumann, D. E. Butler, C. F. Deering, K. E. Mennen, A. Millar, T. N. Nanninga, C. W. Palmer, B. D. Roth, *Tetrahedron Lett.* **1992**, *33*, 2283–2284

119. A. Orellana, T. Rovis, *Chem. Commun.* **2008**, 730–732

120. M. Concepción Gimeno, R. P. Herrera, *Eur. J. Org. Chem.* **2020**, 1057–1068

121. M. Žabka, R. Šebesta, *Molecules* **2015**, *20*, 15500–15524

122. N. Sorgenfrei, J. Hioe, J. Greindl, K. Rothermel, F. Morana, N. Lokesh, R. M. Gschwind, *J. Am. Chem. Soc.* **2016**, *138*, 16345–16354

123. K. Hof, K. M. Lippert, P. R. Schreiner in *Science of Synthesis, Asymmetric Organocatalysis 2* (Ed.: K. Maruoka), Thieme, **2012**, Chapter 2.2.4, pp. 297–412

124. D. Uraguchi, T. Ooi in *Science of Synthesis, Asymmetric Organocatalysis 2* (Ed.: K. Maruoka), Thieme, **2012**, Chapter 2.2.5, pp. 413–435

125. G. Jakab, C. Tancon, Z. Zhang, K. M. Lippert, P. R. Schreiner, *Org. Lett.* **2012**, *14*, 1724–1727

126. X. Ni, X. Li, Z. Wang, J.-P. Cheng, *Org. Lett.* **2014**, *16*, 1786–1789

127. S. Ingemann, H. Hiemstra in *Comprehensive Enantioselective Organocatalysis: Catalysts, Reactions, and Applications* (Ed.: P. I. Dalko), Wiley-VCH, **2013**, Chapter 6, pp. 119–160

128. P. Christ, A. G. Lindsay, S. S. Vormittag, J.-M. Neudörfl, A. Berkessel, A. C. O'Donoghue, *Chem. Eur. J.* **2011**, *17*, 8524–8528

129. F. V. Singh, S. E. Shetgaonkar, M. Krishnan, T. Wirth, *Chem. Soc. Rev.* **2022**, *51*, 8102–8139

130. J. Merad, C. Lalli, G. Bernadat, J. Maury, G. Masson, *Chem. Eur. J.* **2018**, *24*, 3925–3943

131. R.-Z. Liao, S. Santoro, M. Gotsev, T. Marcelli, F. Himo, *ACS Catal.* **2016**, *6*, 1165–1171

132. M. Freund, S. Schenker, S. B. Tsogoeva, *Org. Biomol. Chem.* **2009**, *7*, 4279–4284

133. P. Jakubec, D. M. Cockfield, D. J. Dixon, *J. Am. Chem. Soc.* **2009**, *131*, 16632–16633

134. P. Chen, X. Bao, L.-F. Zhang, M. Ding, X.-J. Han, J. Li, G.-B. Zhang, Y.-Q. Tu, C.-A. Fan, *Angew. Chem. Int. Ed.* **2011**, *50*, 8161–8166

135. D. J. Mergott, S. J. Zuend, E. N. Jacobsen, *Org. Lett.* **2008**, *10*, 745–748

136. X. Huang, S. Broadbent, C. Dvorak, S.-H. Zhao, *Org. Process Res. Dev.* **2010**, *14*, 612–616

137. T. Inokuma, Y. Takemoto in *Science of Synthesis, Asymmetric Organocatalysis 2* (Ed.: K. Maruoka), Thieme, **2012**, Chapter 2.2.6, pp. 437–497

138. M. El Qacemi, H. Smits, J. Y. Cassayre, N. P. Mulholland, P. Renold, E. Godineau, T. Pitterna, *US 2016/0073631 A1* **2016**

139. D. R. Li, A. Murugan, J. R. Falck, *J. Am. Chem. Soc.* **2008**, *130*, 46–48

140. Y. Wang, X. Liu, L. Deng, *J. Am. Chem. Soc.* **2006**, *128*, 3928–3930

141. O. Bassas, J. Huuskonen, K. Rissanen, A. M. P. Koskinen, *Eur. J. Org. Chem.* **2009**, 1340–1351

142. F. Xu, E. Corley, M. Zacuto, D. A. Conlon, B. Pipik, G. Humphrey, J. Murry, D. Tschaen, *J. Org. Chem.* **2010**, *75*, 1343–1353

143. Y. Hoashi, T. Yabuta, P. Yuan, H. Miyabe, Y. Takemoto, *Tetrahedron* **2006**, *62*, 365–374

144. S. Shirakawa, K. Maruoka in *Comprehensive Enantioselective Organocatalysis: Catalysts, Reactions, and Applications* (Ed.: P. I. Dalko), Wiley-VCH, **2013**, Chapter 14, pp. 365–379

145. A. Ting, S. E. Schaus in *Comprehensive Enantioselective Organocatalysis: Catalysts, Reactions, and Applications* (Ed.: P. I. Dalko), Wiley-VCH, **2013**, Chapter 13, pp. 343–363

146. S. Shirakawa, K. Maruoka in *Science of Synthesis, Asymmetric Organocatalysis 2* (Ed.: K. Maruoka), Thieme, **2012**, Chapter 2.3.2, pp. 551–599

147. E. J. Corey, F. Xu, M. C. Noe, *J. Am. Chem. Soc.* **1997**, *119*, 12414–12415

148. C. Hofstetter, P. S. Wilkinson, T. C. Pochapsky, *J. Org. Chem.* **1999**, *64*, 8794–8800

149. K. B. Lipkowitz, M. W. Cavanaugh, B. Baker, M. J. O'Donnell, *J. Org. Chem.* **1991**, *56*, 5181–5192

150. C. E. Cannizzaro, K. N. Houk, *J. Am. Chem. Soc.* **2002**, *124*, 7163–7169

151. T. Ooi, M. Kameda, K. Maruoka, *J. Am. Chem. Soc.* **2003**, *125*, 5139–5151

152. T. Ohshima, T. Shibuguchi, Y. Fukuta, M. Shibasaki, *Tetrahedron* **2004**, *60*, 7743–7754

153. G. N. Gururaja, M. Waser, *Studies in Natural Products Chemistry* **2014**, *43*, 409–435

154. J.-M. Ku, B.-S. Jeong, S. Jew, H. Park, *J. Org. Chem.* **2007**, *72*, 8115–8118

155. T. Shibuguchi, H. Mihara, A. Kuramochi, T. Ohshima, M. Shibasaki, *Chem. Asian J.* **2007**, *2*, 794–801

156. R. K. Boeckman, T. J. Clark, B. C. Shook, *Org. Lett.* **2002**, *4*, 2109–2112

157. K. M. Belyk, B. Xiang, P. G. Bulger, W. R. Leonard, Jr., J. Balsells, J. Yin, C. Chen, *Org. Process Res. Dev.* **2010**, *14*, 692–700

158. X. Jiang, B. Gong, K. Prasad, O. Repic, *Org. Process Res. Dev.* **2008**, *12*, 1164–1169

159. K. M. Belyk, P. G. Bulger, X. Linghu, K. M. Maloney, M. Mclaughlin, J. Pan, B. Xiang, Y. Xu, J. Yin, *WO 2011/005731 A2* **2011**

160. C. Yang, L. P. Le Hir de Fallois, C. Q. Meng, A. Long, R. J. G. De Vries, B. Baillon, S. Lafont, M. G. de Saint Michel, S. Kozlovic, *US 2017/0311601 B2* **2017**

161. B. Xiang, K. M. Belyk, R. A. Reamer, N. Yasuda, *Angew. Chem. Int. Ed.* **2014**, *53*, 8375–8378

162. N. Yasuda, E. Cleator, B. Kosjek, J. Yin, B. Xiang, F. Chen, S.-C. Kuo, K. Belyk, P. R. Mullens, A. Goodyear, J. S. Edwards, B. Bishop, S. Ceglia, J. Belardi, L. Tan, Z. J. Song, L. DiMichele, R. Reamer, F. L. Cabirol, W. L. Tang, G. Liu, *Org. Process Res. Dev.* **2017**, *21*, 1851–1858

163. G. R. Humphrey, S. M. Dalby, T. Andreani, B. Xiang, M. R. Luzung, Z. J. Song, M. Shevlin, M. Christensen, K. M. Belyk, D. M. Tschaen, *Org. Process Res. Dev.* **2016**, *20*, 1097–1103

164. Z. Zhang, P. R. Schreiner, *Chem. Soc. Rev.* **2009**, *38*, 1187–1198

165. M. D. Visco, J. Attard, Y. Guan, A. E. Mattson, *Tetrahedron Lett.* **2017**, *58*, 2623–2628

166. I. Coric, B. List, *Nature* **2012**, *483*, 315–318

167. L. Simón, J. M. Goodman, *J. Am. Chem. Soc.* **2008**, *130*, 8741–8747

168. R. R. Knowles, S. Lin, E. N. Jacobsen, *J. Am. Chem. Soc.* **2010**, *132*, 5030–5032

169. S. J. Zuend, E. N. Jacobsen, *J. Am. Chem. Soc.* **2009**, *131*, 15358–15374

170. K. Mori, T. Akiyama in *Comprehensive Enantioselective Organocatalysis: Catalysts, Reactions, and Applications* (Ed.: P. I. Dalko), Wiley-VCH, **2013**, Chapter 11, pp. 289–314

171. T. Akiyama in *Science of Synthesis, Asymmetric Organocatalysis 2* (Ed.: K. Maruoka), Thieme, **2012**, Chapter 2.2.1, pp. 169–217

172. M. Terada, N. Momiyama in *Science of Synthesis, Asymmetric Organocatalysis 2* (Ed.: K. Maruoka), Thieme, **2012**, Chapter 2.2.2, pp. 219–278

173. A. K. Ghosh, X. Cheng, *Org. Lett.* **2011**, *13*, 4108–4111

174. C. Guo, J. Song, J.-Z. Huang, P.-H. Chen, S.-W. Luo, L.-Z. Gong, *Angew. Chem. Int. Ed.* **2012**, *51*, 1046–1050

175. A. Lerchen, N. Gandhamsetty, E. H. E. Farrar, N. Winter, J. Platzek, M. N. Grayson, V. K. Aggarwal, *Angew. Chem. Int. Ed.* **2020**, *59*, 23107–23111

176. K. Yoshida, K. Okada, H. Ueda, H. Tokuyama, *Angew. Chem. Int. Ed.* **2020**, *59*, 23089–23093

177. J. Chen, P. Gao, F. Yu, Y. Yang, S. Zhu, H. Zhai, *Angew. Chem. Int. Ed.* **2012**, *51*, 5897–5899

178. H. Krawczyk, M. Dziegielewski, D. Deredas, A. Albrecht, L. Albrecht, *Chem. Eur. J.* **2015**, *21*, 10268–10277

179. H. Shi, I. N. Michaelides, B. Darses, P. Jakubec, Q. N. N. Nguyen, R. S. Paton, D. J. Dixon, *J. Am. Chem. Soc.* **2017**, *139*, 17755–17758

180. J. I. Murray, Z. Heckenast, A. C. Spivey in *Lewis Base Catalysis in Organic Synthesis* (Eds.: E. Vedejs, S. E. Denmark), Wiley-VCH, **2016**, Chapter 12, pp. 459–526

181. S. Schulthoff, J. Y. Hamilton, M. Heinrich, Y. Kwon, C. Wirtz, A. Fürstner, *Angew. Chem. Int. Ed.* **2021**, *60*, 446–454

182. C. Zhu, X. Shen, S. G. Nelson, *J. Am. Chem. Soc.* **2014**, *126*, 5352–5353

183. C. A. Lewis, S. J. Miller, *Angew. Chem. Int. Ed.* **2006**, *45*, 5616–5619

184. M. Wang, L. Zhang, X. Huo, Z. Zhang, Q. Yuan, P. Li, J. Chen, Y. Zou, Z. Wu, W. Zhang, *Angew. Chem. Int. Ed.* **2020**, *59*, 20814–20819

185. X. Feng, W. Meng, H. Du, *Chem. Soc. Rev.* **2023**, *52*, 8580–8598

186. D. J. Scott, M. J. Fuchter, A. E. Ashley, *Chem. Soc. Rev.* **2017**, *46*, 5689–5700

187. V. Fasano, M. J. Ingleson, *Synthesis* **2018**, *50*, 1783–1795

188. D. J. Scott, M. J. Fuchter, A. E. Ashley, *J. Am. Chem. Soc.* **2014**, *136*, 15813–15816

189. T. Mahdi, D. W. Stephan, *J. Am. Chem. Soc.* **2014**, *136*, 15809–15812

190. Y. Wang, W. Chen, Z. Lu, Z. H. Li, H. Wang, *Angew. Chem. Int. Ed.* **2013**, *52*, 7496–7499

191. A. R. Jupp, *Dalton Trans.* **2022**, *51*, 10681–10689

192. J. Paradies, *Eur. J. Org. Chem.* **2019**, 283–294
193. S. K. Pati, S. Das, *Chem. Eur. J.* **2017**, *23*, 1078–1085
194. K. Chernichenko, Ádám Madarász, I. Pápai, M. Nieger, M. Leskelä, T. Repo, *Nat. Chem.* **2013**, *5*, 718–723
195. S. Liao, P. Wang, Y. Tang in *Comprehensive Enantioselective Organocatalysis: Catalysts, Reactions, and Applications* (Ed.: P. I. Dalko), Wiley-VCH, **2013**, Chapter 20, pp. 547–577
196. E. M. McGarrigle, E. L. Myers, O. Illa, M. A. Shaw, S. L. Riches, V. K. Aggarwal, *Chem. Rev.* **2007**, *107*, 5841–5883
197. K. Julienne, P. Metzner, V. Henryon, *J. Chem. Soc. Perkin Trans. 1* **1999**, 731–735
198. V. K. Aggarwal, S. Calamai, J. G. Ford, *J. Chem. Soc. Perkin Trans. 1* **1997**, 593–599
199. V. K. Aggarwal, H. Abdel-Rahman, L. Fan, R. V. H. Jones, M. C. H. Standen, *Chem. Eur. J.* **1996**, *2*, 1024–1030
200. V. K. Aggarwal, J. Richardson, *Chem. Commun.* **2003**, 2644–2651
201. O. Illa, M. Namutebi, C. Saha, M. Ostovar, C. C. Chen, M. F. Haddow, S. Nocquet-Thibault, M. Lusi, E. M. McGarrigle, V. K. Aggarwal, *J. Am. Chem. Soc.* **2013**, *135*, 11951–11966
202. O. Illa, M. Arshad, A. Ros, E. M. McGarrigle, V. K. Aggarwal, *J. Am. Chem. Soc.* **2010**, *132*, 1828–1830
203. O. Zhurakovskyi, Y. E. Türkmen, L. E. Löffler, V. A. Moorthie, C. C. Chen, M. A. Shaw, M. R. Crimmin, M. Ferrara, M. Ahmad, M. Ostovar, J. V. Matlock, V. K. Aggarwal, *Angew. Chem. Int. Ed.* **2018**, *57*, 1346–1350
204. K. C. K. Swamy, N. N. B. Kumar, E. Balaraman, K. V. P. P. Kumar, *Chem. Rev.* **2009**, *109*, 2551–2651
205. M. Dryzhakov, E. Richmond, J. Moran, *Synthesis* **2016**, *48*, 935–959
206. R. H. Beddoe, K. G. Andrews, V. Magné, J. D. Cuthbertson, J. Saska, A. L. Shannon-Little, S. E. Shanahan, H. F. Sneddon, R. M. Denton, *Science* **2019**, *365*, 910–914
207. W. Yao, E. A. Bazan-Bergamino, M.-Y. Ngai, *ChemCatChem* **2022**, *14*, e202101292
208. C. K. Prier, D. C. MacMillan in *Visible Light Photocatalysis in Organic Chemistry* (Eds.: C. R. J. Stephenson, T. P. Yoon, D. C. MacMillan), Wiley-VCH, **2018**, Chapter 10, pp. 299–333
209. *Science of Synthesis: Dual Catalysis in Organic Synthesis 2*, (Ed.: G. A. Molander), Thieme, **2020**

Click Reactions

9

Click reactions, similar to organocatalysis, do not encompass a specific type of reaction, but rather constitute a fundamental concept and its requirements to attain an ideal reaction. This concept was introduced by Sharpless in 2001 and aimed at quickly obtaining a variety of diverse products stereoselectively with highly efficient reactions [1]. Particular attention was paid to a broad substrate scope and pronounced modularity, as well as the avoidance of toxic solvents and/or offensive by-products, in line with the idea of "Green Chemistry". An additional aim was simple handling and execution of the reaction, ideally a tolerance toward water and oxygen, and the ambition of orthogonal reactivity, so that in the presence of many potential reaction partners, only the desired product would be obtained.

These prerequisites should be understood as an ideal to be aspired, with a few reactions being part of the click concept that come very close to these guidelines (see Fig. 9.1).

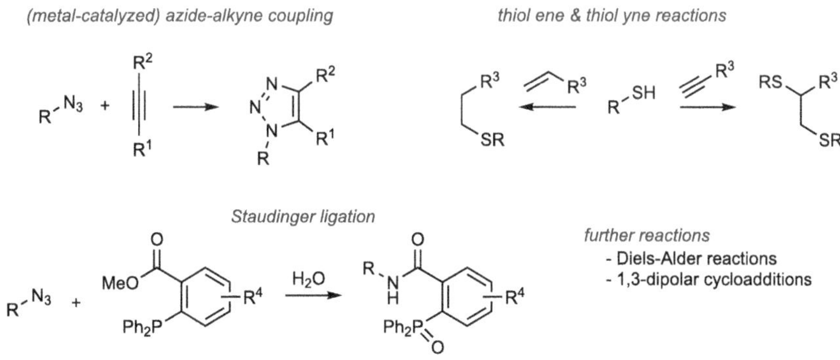

Fig. 9.1 Commonly used click reactions

A. Düfert, *Methods of Organic Synthesis*,
https://doi.org/10.1007/978-3-662-70963-4_9

Especially azides, thiols, and alkynes comprise common reactive components in many click reactions [2, 3]. This may be attributed to an innate reactivity that is only accessible under specific conditions, such as hydrogenations, and thus survives many transformations unchanged. They can therefore often meet the aim for orthogonal reactivity. Since click reactions are routinely used under biochemical conditions (in vitro, in vivo), for example, for attaching biomarkers, their compatibility with biological systems is crucial [4, 5]. This is also referred to as *bioorthogonality*, indicating that these conversions neither interact with a biological system nor disturb it [6].

The sulfur-fluoride exchange reaction constitutes the conversion of sulfonyl fluorides with amines and has only recently been included in the click canon [7]. The Diels–Alder reaction with inverse electron demand also belongs to the repertoire of click reactions, even if it is only mentioned in passing here. The coupling of tetrazines with activated olefins such as *E*-cyclooctene or cyclopropene is also gaining in popularity. Since their mechanistic background has already been extensively illuminated in Chap. 5, only their importance in the context of click chemistry needs to be emphasized at this point [8].

9.1   Metal-Catalyzed Huisgen Cycloaddition

The term metal-catalyzed Huisgen cycloaddition of azides with alkynes is often used synonymously with the term click reaction in the scientific literature. The copper-catalyzed variant was developed independently by the groups of Meldal and Sharpless/Fokin. Fokin subsequently disclosed the ruthenium-catalyzed modification, which offers a complementary regioselectivity to the Cu(I)-mediated method (see Fig. 9.2) [9–11].

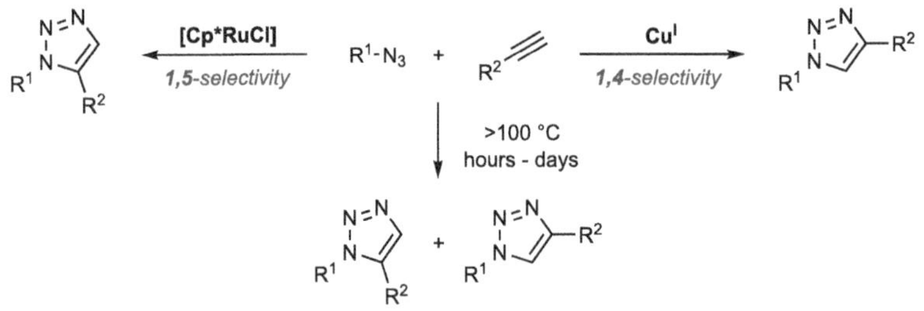

Fig. 9.2 Regioselectivity of the thermal, Cu- and Ru-catalyzed Huisgen cycloaddition [9–11]

Both methods are characterized by a wide substrate scope and tolerate water and oxygen. In particular, copper-catalyzed azide-alkyne cycloadditions have proven to possess the greatest range of applications, as they allow an increase in reaction rate by up to eight orders of magnitude compared to the respective thermal pericylic reaction [11]. The Cu-mediated couplings are used for the synthesis of bioactive substances and in drug discovery, for the derivatization of cellular surfaces (e.g., tagging with fluorophores, etc.), for an *in situ* access to enzyme inhibitors, for the production of dendrimers and functionalized block copolymers, for the construction of structured hydrogels, and many other applications [12–16].

9.1.1   Mechanism of the Catalyzed Cycloaddition

The mechanism of the copper-catalyzed azide-alkyne cycloaddition has been intensively studied, especially with regard to the observed regioselectivity of the transformation [9, 10, 17]. Based on DFT studies and the experimentally determined rate law, a mechanism involving several copper species was discussed early on [18]. Since monomeric Cu acetylides are unreactive towards azides in the absence of external copper sources, the purported mechanism shown in Fig. 9.3 could be derived from studies with isotopically pure organocuprates, which postulated dimeric cuprates as reactive intermediates [19, 20]. The exact structure, including the oxidation state of the heteroleptic Cu dimers, has not yet been conclusively elucidated. Especially the coordination mode of the azide remains elusive. The intermediacy of structures **4** and **5** accordingly is (still) speculative, even though they are cited in many publications in this way or as an analog thereof (see Fig. 9.3) [21–23]. The existence of a competing monometallic catalytic cycle was recently postulated. It was presumed to be operative in addition to the regular, bis-copper catalytic cycle, albeit significantly slower in comparison to the latter [24, 25].

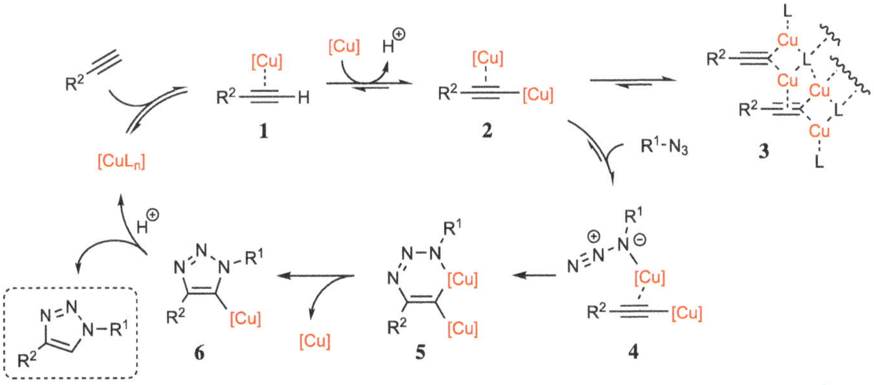

Fig. 9.3 Postulated mechanism of the copper-catalyzed azide-alkyne cycloaddition [20, 25]

The Cu atom of the acetylide species merely acts as a strongly σ-bound Lewis acid, whereas the second copper atom experiences a π-complexation and effects a reciprocal activation of the triple bond. The formation of polymeric copper species (**3**), which can be generated especially at higher Cu concentrations, drastically reduces the reaction rate. As a result, polar solvents (e.g., H_2O, MeOH), which facilitate the dissociation of such complexes, are strongly preferred or act as accelerators compared to apolar solvents, which in contrast promote aggregation.

The ruthenium-catalyzed variant allows the synthesis of the 1,5-triazoles complementary to the Cu route. The observed regioselectivity can be understood in the context of the postulated mechanism shown in Fig. 9.4 [11, 17, 26].

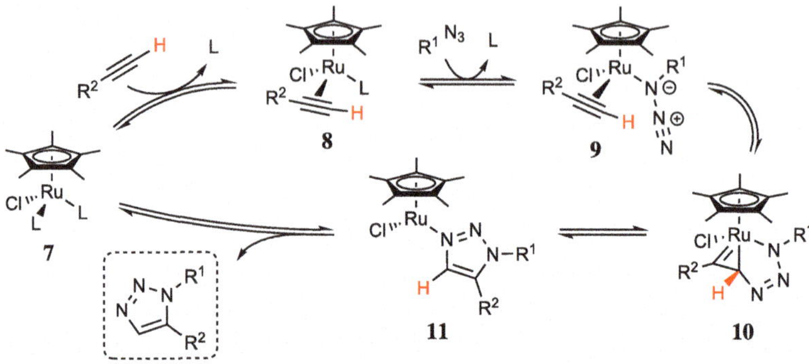

Fig. 9.4 Postulated mechanism of the ruthenium-catalyzed azide-alkyne cycloaddition [11, 17, 26]

Following the coordination of the metal center to the triple bond (**8**) and the azide (**9**), the new C-N bond is formed at the more electronegative and sterically less hindered C-atom of the alkyne with the terminal nitrogen atom of the azide. Azides can coordinate to the metal center with either the terminal nitrogen or the N-atom proximal to the methyl group. Complexes with both modes are known, although the proximal N-atom should display a higher donicity [27]. The intermediate metallacycle **10** can reductively eliminate the observed 1,4-regioisomer via **11** in the concluding step by taking up two ligands L to reenter the catalytic cycle. The rate-determining step is assumed to be the reductive elimination (**10**→**11**) [11].

In general, many Ru salts are catalytically active. However, a hallmark of pentamethylcyclopentadienyl ligands (Cp*) is their pronounced 1,5-regioselectivity and high reaction rate, which renders them ideal for achieving the desired transformation (see Fig. 9.5). The observed increase in catalytic activity is largely attributed to two causes: The increased lability of the nonreactive ligands L during the dissociative coordination of the alkyne and azide and a superior reductive elimination from **10** to **11** facilitated by the steric environment [11].

The advantage of the Ru-mediated variant, apart from the 1,5-regioselectivity, lies in its improved substrate scope: In addition to primary and secondary azides, some tertiary azides are also successfully converted. In addition, internal alkynes are also reactive, which do not afford the addition product in the presence of Cu catalysts [17, 26]. Apart from Cu and Ru complexes, catalysts based on other metal catalysts have also been developed [28].

[Ru]	yield [%] 1,5- / 1,4-product
RuCl$_2$(COD)	– / –
RuCl$_2$(PPh$_3$)$_2$	– / <5
Ru(OAc)$_2$(PPh$_3$)$_2$	– / 46
CpRuCl(PPh$_3$)$_2$	13 / 1
Cp*RuCl(PPh$_3$)$_2$	100 / –

Fig. 9.5 Yield of various Ru catalysts in the reaction of benzyl azide with phenylacetylene [11]

9.1.2 Cu-Free Azide-Alkyne Cycloadditions

Although the Cu-catalyzed Huisgen cycloaddition can be used in cell lysates, it is often not applicable to living systems, as Cu(I) salts are considered cytotoxic. As a consequence, the Bertozzi group developed "Cu-free azide-alkyne cycloadditions" based on initial work by Wittig from the 1960s [29, 30]. The cyclic alkyne component **12** is characterized by a very high ring strain, which delivers sufficient driving force for the thermal Huisgen cycloaddition to **13** to proceed under very mild conditions. The requisite temperatures can also be realized in cell cultures. The strain of the cyclooctyne derivatives (160° vs. 180° in acyclic alkynes) raises the ground state and accordingly narrows the energetic gap to the transition state (*preformation of the transition state*). Apart from the distortion energy, also a smaller HOMO–LUMO gap and an enhanced orbital overlap have been postulated to contribute to the observed high reactivity in the cycloaddition [31]. Optimization of the substitution pattern on the alkyne ultimately led to a further reduction of the activation energy. In particular, a geminal disubstitution with fluorine or a double fusing with benzene showed a pronounced increase in the reaction rate by up to two orders of magnitude (see Fig. 9.6, cf. Chap. 5 for a more detailed mechanistic discussion).

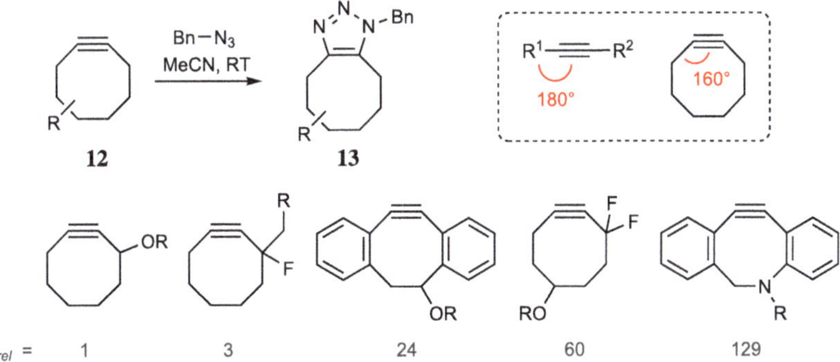

Fig. 9.6 Cu-free azide-alkyne cycloaddition, overview of common cyclic alkynes and their relative reaction rate [29, 30]

9.2   Thiol-Ene and Thiol-Yne Reactions

The thiol-ene and thiol-yne reactions represent the second major group of click reactions. The C-S bond formation can occur either via a radical or ionic pathway, with most variants pursuing a radical reaction [15, 32–34]. Given the possibility of a photoinitiation, this class of reactions is frequently utilized in polymerizations. The resultant highly uniform polymer networks allow a tuning of the material properties, which result from the spatial and temporal control of the click reaction (see Fig. 9.7) [35].

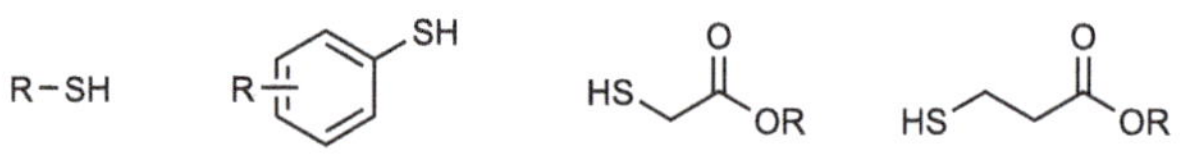

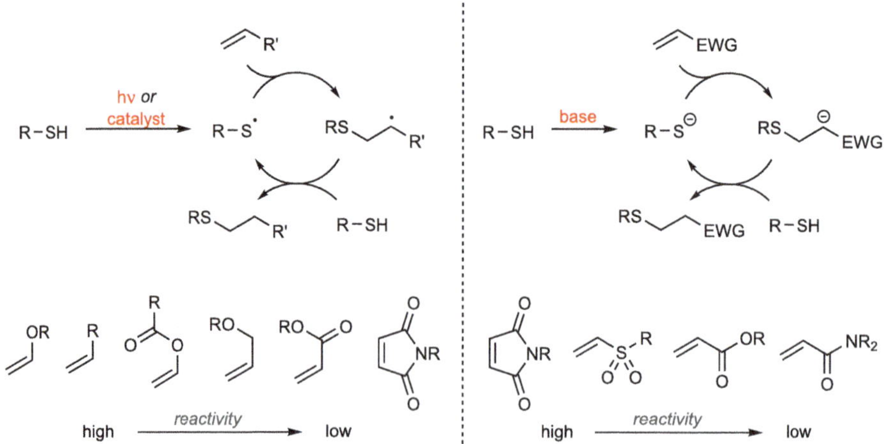

Fig. 9.7 Commonly used thiols in the thiol-ene reaction

Mechanistically, the radical click reaction and the ionic Michael variant show certain similarities. After the formation of the thiol radical or anion, a reaction with the alkene substrate takes place. Following the reaction with a second equivalent of thiol, the original reactive species is regenerated and the addition product is obtained. The afforded thioether shows high *anti*-Markovnikov selectivity under both radical and anionic conditions. Apart from the method used to generate the active species and the temperature, the rate of the reaction mainly depends on the identity of the alkene component. Ideally, only the desired bond formation occurs without any homodimerization of the alkene. In the radical variant, electron-rich or sterically strained alkenes (e.g., norbornene) react most rapidly. In contrast, particularly electron-poor alkenes are preferred substrates in the ionic Michael variant (see Fig. 9.8). Terminal olefins show a higher reaction rate in the radical reaction than substrates with internal double bonds [36].

Fig. 9.8 Mechanism of the radical and Michael-thiol-ene reaction and reactivities of the alkenes under the corresponding conditions [33, 35]

The C-S bond formation is exothermic, the reaction enthalpy varies between -10.5 kcal/mol (vinyl ether) and -22.6 kcal/mol (N-alkyl maleimide) [15]. Under anionic conditions, strong bases, metals, Lewis acids, organometallic compounds as well as amines or phosphines are traditionally employed, which have established themselves as efficient catalysts in recent years.

The thiol-yne reaction was rediscovered in 2009 with previous studies having been disclosed in the middle of the last century. Since then, the transformation has proven itself especially adept at the synthesis of polymers, similar to the corresponding thiol-ene reaction [37, 38]. It is also carried out under radical conditions or as a classic Michael addition. Its advantage over the thiol-ene reaction lies in the possibility to achieve cross-linking even by using only one equivalent of alkyne, as two equivalents of the nucleophile can react with the CC-triple bond. Mechanistically, it is related to the thiol-ene reaction: The radical or anion intermediate obtained after the first addition to the triple bond can subsequently react again with a sulfide, whereby the thiol proton is transferred and a vinyl thioether initially results. This species can react once more with a sulfide radical and thus continues in the chain reaction of the thiol-ene addition. Apart from thiols as nucleophiles, amines and alcohols can also be added to alkynes.

9.3   Staudinger Ligation

The Staudinger ligation is almost exclusively encountered in biochemical applications. Similar to the azide-alkyne cycloaddition, the azide group constitutes an inert functionality under biological conditions, which precludes undesired degradation reactions in the biological medium. The main application lies in the attachment of labels to peptides, proteins, lipids, or carbohydrates in cellulo [39–42]. Taking an inspiration from the Staudinger reaction of azides with phosphines, the Bertozzi group developed an intramolecular variant in which an amide is formed by the elimination of nitrogen (see Fig. 9.9) [30].

In the Staudinger ligation, an azide reacts with phosphine **14** via **15** and **16** with concomitant liberation of dinitrogen. The resultant aza-ylide **17** can subsequently be converted into a variety of different products. When the iminophosphorane **17** reacts with an ester, an amide is obtained. "Traceless" variants have also been developed, in which a native amide bond is formed upon elimination of the phosphine by-product.

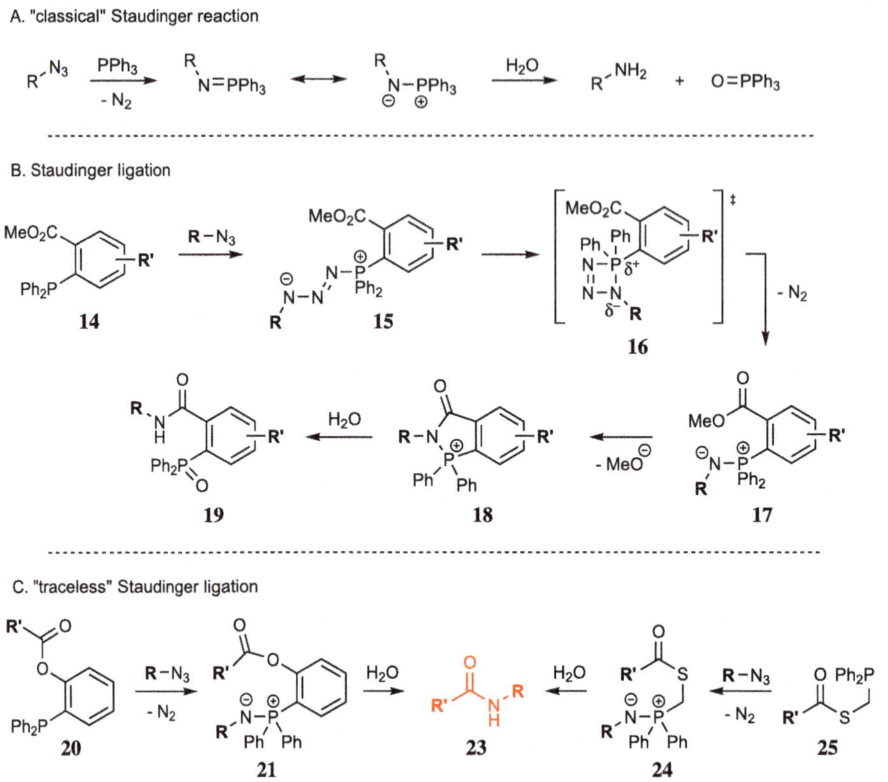

A. "classical" Staudinger reaction

B. Staudinger ligation

C. "traceless" Staudinger ligation

Fig. 9.9 Comparison of bioorthogonal Staudinger variants [40, 42]

9.4   Selected Applications

Click reactions are primarily utilized today in biochemistry and polymer synthesis, far beyond their originally foreseen field of application. Thus, the importance of click reactions was justly recognized in 2022 with the Noble prize in Chemistry for Sharpless, Bertozzi and Meldal. Due to this much wider utilization, the adherence to a strict (bio-)orthogonality of the reactive groups, i.e., the absence of a background reaction of the substrates with other potentially reactive components, such as water, oxygen, salts, proteins, etc., is fundamental.

For select cases of drug research in the discovery phase, Cu-catalyzed azide-alkyne cycloadditions (CuAAC) are still used very successfully. In the search for new small molecule drugs that can selectively modulate proteins, the principle of fragment-based drug discovery was followed. According to this principle, only small fragments are incubated with the target protein, and the strength of the interaction is determined. The interacting fragments are subsequently connected by click chemistry to allow several fragments to react with a protein and thus increase their effectiveness. This quickly yields a series of highly potent drugs that can be assembled rapidly and efficiently (see Fig. 9.10) [13].

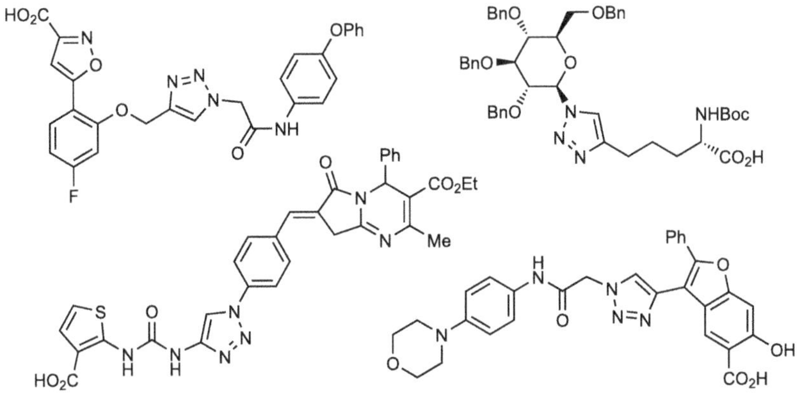

Fig. 9.10 Protein tyrosine phosphatase inhibitors synthesized by CuAAC [13]

A similar concept was followed for the assembly of so-called proteolysis-targeting chimeras (PROTACs). These multifunctional molecules contain a ligand for a target protein tethered to an E3 ubiquitin ligase recruiting group. The bioactivity thus is based on latching on to the desired target protein and facilitating its proteasomal degradation. In order to assemble the binding moieties of these bi- or trifunctional drugs *in situ*, azide-alkyne cycloadditions have been employed to forge the linker unit [43].

One application of the thiol-ene reaction is the construction of hydrogels, which can be used, for example, for the controlled dosage of drugs. Since cysteine contains a thiol group, biofunctionality can easily be incorporated into a polymer backbone (**29**). In a study by Hubbell, a biodegradable hydrogel was linked to albumin (**28**) through a Michael thiol-ene reaction. As albumin constitutes a typical carrier protein for hydrophobic drugs in human blood, this approach can easily achieve loadings with pharmaceutically relevant compounds, similar to the distribution in the human organism (Fig. 9.11) [44].

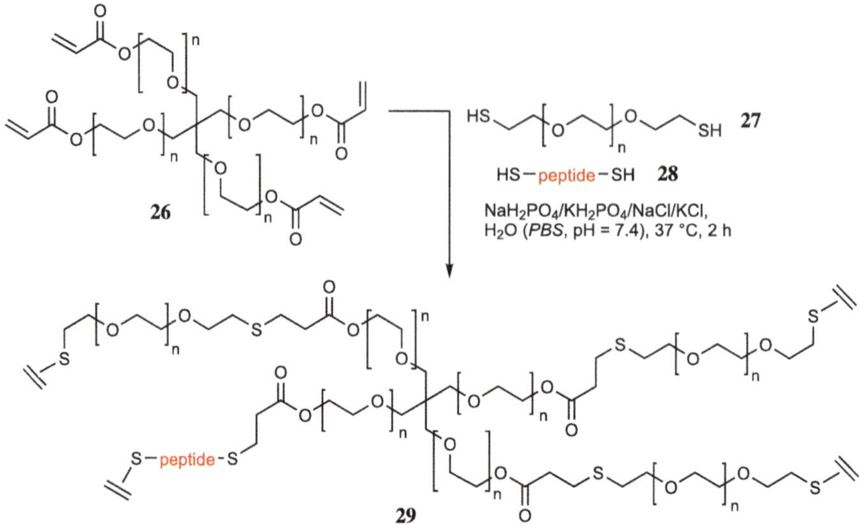

Fig. 9.11 Synthesis of biogels incorporating albumin (**28**) via Michael-thiol-ene reactions (*PBS* = phosphate-buffered aq. NaCl solution) [44]

References

1. H. C. Kolb, M. G. Finn, K. B. Sharpless, *Angew. Chem. Int. Ed.* **2001**, *40*, 2004–2021.
2. J. C. Worch, C. J. Stubbs, M. J. Price, A. P. Dove, *Chem. Rev.* **2021**, *121*, 6744–6776.
3. Z. Geng, J. J. Shin, Y. Xi, C. J. Hawker, *J. Polym. Sci.* **2021**, *59*, 963–1042.
4. A. Pasieka, E. Diamanti, E. Uliassi, M. L. Bolognesi, *ChemMedChem* **2023**, *18*, e202300422.
5. D. Schauenburg, T. Weil, *Adv. Sci.* **2024**, *11*, 2303396.
6. E. M. Sletten, C. R. Bertozzi, *Angew. Chem. Int. Ed.* **2009**, *48*, 6974–6998.
7. A. S. Barrow, C. J. Smedley, Q. Zheng, S. Li, Dong, J. E. Moses, *Chem. Soc. Rev.* **2019**, *48*, 4731–4758.
8. T. Carell, M. Vrabel, *Top. Curr. Chem. (Z)* **2016**, *374*, 9.
9. J. E. Hein, V. V. Fokin, *Chem. Soc. Rev.* **2010**, *39*, 1302–1315.
10. M. Meldal, C. W. Tornøe, *Chem. Rev.* **2008**, *108*, 2952–3015.
11. B. C. Boren, S. Narayan, L. K. Rasmussen, L. Zhang, H. Zhao, Z. Lin, G. Jia, V. V. Fokin, *J. Am. Chem. Soc.* **2008**, *130*, 8923–8930.
12. J. E. Moses, A. D. Moorhouse, *Chem. Soc. Rev.* **2007**, *36*, 1249–1262.
13. P. Thirumurugan, D. Matosiuk, K. Jozwiak, *Chem. Rev.* **2013**, *113*, 4905–4979.
14. C. Barner-Kowollik, F. E. D. Prez, P. Espeel, C. J. Hawker, T. Junkers, H. Schlaad, W. V. Camp, *Angew. Chem. Int. Ed.* **2011**, *50*, 60–62.
15. C. E. Hoyle, C. N. Bowman, *Angew. Chem. Int. Ed.* **2010**, *49*, 1540–1573.
16. S. Neumann, M. Biewend, S. Rana, W. H. Binder, *Macromol. Rapid Commun.* **2020**, *41*, 1900359.
17. C. Wang, D. Ikhlef, S. Kahlal, J.-Y. Saillard, D. Astruc, *Coord. Chem. Rev.* **2016**, *316*, 1–20.
18. V. O. Rodionov, S. I. Presolski, D. D. Díaz, V. V. Fokin, M. G. Finn, *J. Am. Chem. Soc.* **2007**, *129*, 12705–12712.

19. C. Nolte, P. Mayer, B. F. Straub, *Angew. Chem. Int. Ed.* **2007**, *46*, 2101–2103.
20. B. T. Worrell, J. A. Malik, V. V. Fokin, *Science* **2013**, *340*, 457–460.
21. M. S. Ziegler, K. V. Lakshmi, T. Don Tilley, *J. Am. Chem. Soc.* **2017**, *139*, 5378–5386.
22. A. Makarem, R. Berg, F. Rominger, B. F. Straub, *Angew. Chem. Int. Ed.* **2015**, *54*, 7431–7435.
23. C. Iacobucci, S. Reale, J.-F. Gal, F. De Angelis, *Angew. Chem. Int. Ed.* **2015**, *54*, 3065–3068.
24. L. Jin, D. R. Tolentino, M. Melaimi, G. Bertrand, *Sci. Adv.* **2015**, *1*, e1500304.
25. R. Chung, A. Vo, V. V. Fokin, J. E. Hein, *ACS Catal.* **2018**, *8*, 7889–7897.
26. J. R. Johansson, T. Beke-Somfai, A. S. Stålsmeden, N. Kann, *Chem. Rev.* **2016**, *116*, 14726–14768.
27. S. Cenini, E. Gallo, A. Caselli, F. Ragaini, S. Fantauzzi, C. Piangiolino, *Coord. Chem. Rev.* **2006**, *250*, 1234–1253.
28. R. S. Gomes, G. A. M. Jardim, R. L. de Carvalho, M. H. Araujo, E. N. da Silva Júnior, *Tetrahedron* **2019**, *75*, 3697–3712.
29. J. C. Jewett, C. R. Bertozzi, *Chem. Soc. Rev.* **2010**, *39*, 1272–1279.
30. E. M. Sletten, C. R. Bertozzi, *Acc. Chem. Res.* **2011**, *44*, 666–676.
31. T. A. Hamlin, B. J. Levandowski, A. K. Narsaria, K. N. Houk, F. M. Bickelhaupt, *Chem. Eur. J.* **2019**, *25*, 6342–6348.
32. C. E. Hoyle, A. B. Loweb, C. N. Bowman, *Chem. Soc. Rev.* **2010**, *39*, 1355–1387.
33. D. P. Nair, M. Podgorski, S. Chatani, T. Gong, W. Xi, C. R. Fenoli, C. N. Bowman, *Chem. Mater.* **2014**, *26*, 724–744.
34. A. Dondoni, A. Marra, *Chem. Soc. Rev.* **2012**, *41*, 573–586.
35. M. A. Tasdelen, Y. Yagci, *Angew. Chem. Int. Ed.* **2013**, *52*, 5930–5938.
36. T. M. Roper, C. A. Guymon, E. S. Jönsson, C. E. Hoyle, *J. Polym. Sci. A Polym. Chem.* **2004**, *42*, 6283–6298.
37. A. B. Lowe, *Polymer* **2014**, *55*, 5517–5549.
38. A. Massi, D. Nanni, *Org. Biomol. Chem.* **2012**, *10*, 3791–3807.
39. M. Köhn, R. Breinbauer, *Angew. Chem. Int. Ed.* **2004**, *43*, 3106–3116.
40. C. I. Schilling, N. Jung, M. Biskup, U. Schepers, S. Bräse, *Chem. Soc. Rev.* **2011**, *40*, 4840–4871.
41. S. S. van Berkel, M. B. van Eldijk, J. C. M. van Hest, *Angew. Chem. Int. Ed.* **2011**, *50*, 8806–8827.
42. C. Bednarek, I. Wehl, N. Jung, U. Schepers, S. Bräse, *Chem. Rev.* **2020**, *120*, 4301–4354.
43. C. Yang, R. Tripathi, B. Wang, *RSC Chem. Biol.* **2024**, *5*, 189–197.
44. D. L. Elbert, A. B. Pratt, M. P. Lutolf, S. Halstenberg, J. A. Hubbell, *J. Controlled Release* **2001**, *76*, 11–25.

Modern Radical and Redox Chemistry **10**

While the use of redox processes, especially radical reactions, can be counted among the earliest efficient synthetic methods, the generation of radicals was primarily limited to the use of chemical methods. Particularly tin-, samarium-, and hypervalent iodine reagents as well as organic radical initiators like AIBN or peroxides constituted a pivotal component and can accordingly be encountered in many methods (see Fig. 10.1).

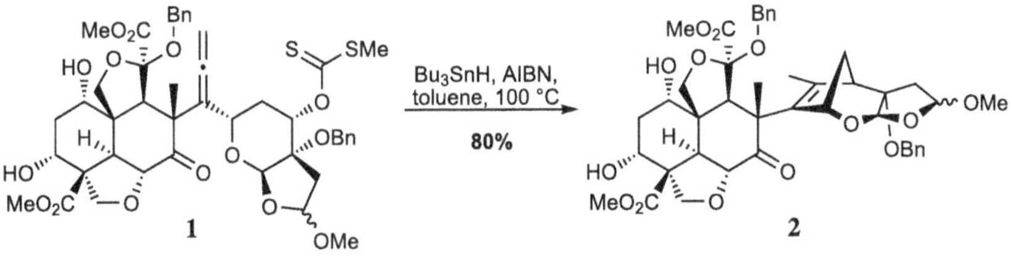

Fig. 10.1 Bu$_3$SnH-mediated radical reaction in Ley's approach to azadirachtin [1]

A similarly mature, but yet underrepresented approach is the use of photochemical or electrochemical energy to convert molecules into reactive intermediates (see Fig. 10.2) [2, 3]. In recent years, there has been an exponential increase in publications in this field. This renewed interest was triggered by a few key publications hailing from the field of method development and natural product synthesis [4–6].

The manifold challenges ailing classic photo- and electrochemistry, such as problems with reproducibility and selectivity, lack of suitable equipment or even of experience on the part of the users, have now largely been solved or addressed. In many aspects, electrosynthetic and

© The Author(s), under exclusive license to Springer-Verlag GmbH, DE, part of Springer Nature 2026
A. Düfert, *Methods of Organic Synthesis*,
https://doi.org/10.1007/978-3-662-70963-4_10

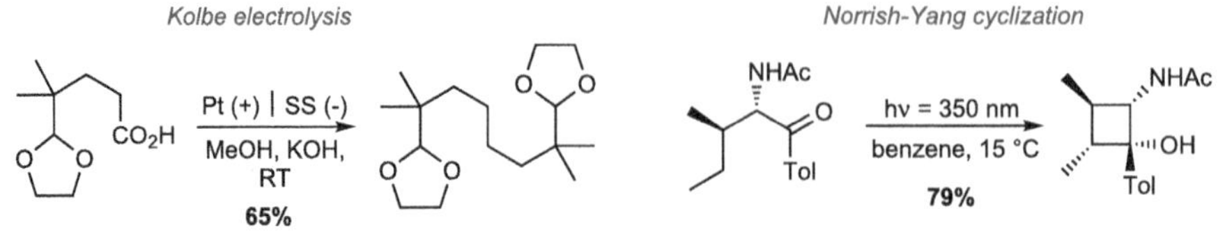

Kolbe electrolysis

Pt (+) | SS (-)
MeOH, KOH,
RT
65%

Norrish-Yang cyclization

NHAc
Tol
hv = 350 nm
benzene, 15 °C
79%
NHAc
OH
Tol

Fig. 10.2 Classic methods using electricity or light for radical generation [7, 8]

photoredox methods share a certain kinship: The reactions require an external stimulus, the energy input of which must be aligned with the redox potential of the desired transformation. The reaction can occur directly or may be facilitated by a catalyst/mediator. Furthermore, barring classic photochemical transformations ([2+2]-cycloadditions, Norrish reactions) and focusing on photoredox chemistry, the reactions proceed through the intermediacy of radicals in both cases.

Given the rapid development in these two fields, only an insight into the fundamental principles and selected examples can be provided to enable an assessment of the potential for the interested user and give an introduction to the relevant literature [2, 3, 9–23].

10.1 Electrosynthesis

The derivatization of organic compounds exploiting single-electron processes is a concept that classic synthetic chemistry routinely relies on, such as in the Birch reduction or related reductive methods, only to name a few. An elemental metal serves as the source of electrons in these cases. The use of an externally applied voltage to generate electrons is governed by similar principles. Nevertheless, it possesses some peculiarities that mandate closer examination and discussion. As with a chemical redox reaction, an (anodic) oxidation and a (cathodic) reduction always occur in parallel. However, normally only one reaction can be exploited for a given substrate as the chemical redox reagent is typically consumed in the transformation. In electrosynthesis, only electrons serve as the reagents and the electrodes are ideally not altered in the process. This makes the reaction both straightforward and more sustainable, since stoichiometric quantities of redox reagents are avoided and no waste is generated. Another difference also lies in the use of both the anode and the cathode reactions to simultaneously carry out two desired transformations. Direct conversion at the electrodes only requires an electroactive functionality to be present in the molecule, a so-called *electrophore*, which is the analog of chromophores necessary for photochemical activation (see Fig. 10.3).[I]

[I] The use of a single cell for electrosynthesis shall be indicated in the reaction conditions of the figures by a single line ("|"), whereas a divided cell is indicated by two parallel lines ("||").

desired reaction	electrode material	
	anode (+)	**cathode (-)**
oxidation	Pt, graphite, GC, RVC	Pt, graphite, GC, RVC, Zn, Mg, Ni, Fe
reduction	*inert*: Pt, graphite, GC, RVC *sacrificial*: Zn, Mg, Ni, Cu	graphite, GC, RVC

GC = glassy carbon; RVC = reticulated vitreous carbon

typical conditions

temperature	-78 °C to 100 °C
solvent	MeCN, MeOH, HFIP, acetone, THF, CH_2Cl_2, DMA, H_2O, ethylene carbonate
electrolyte	NR_4PF_6, NR_4BF_4, $LiPF_6$, $LiBF_4$, $LiClO_4$, etc.

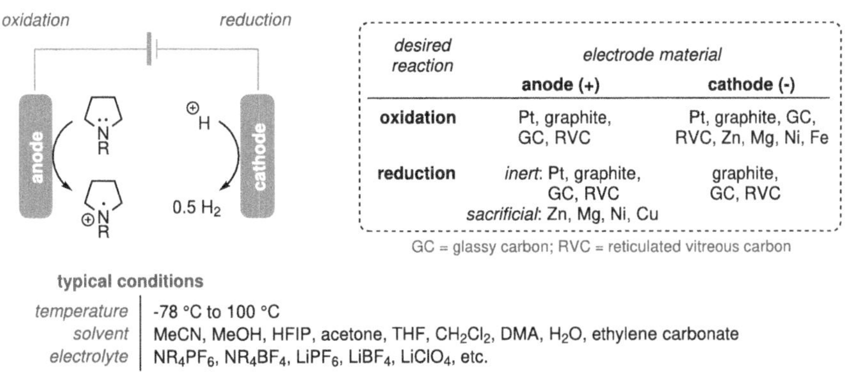

Fig. 10.3 Schematic structure of a direct electrochemical reaction and common electrode materials and reaction conditions [14, 15]

The advantages of electrosynthetic reactions include very mild conditions (usually at room temperature), good scalability, precise control of the potential and, when using redox-active catalysts, the additional possibility of an enantioselective reaction. The disadvantage mainly lies in the sometimes poor predictability of a reaction when several potentially reactive functional groups are present in the substrate. This makes their use in complex natural product syntheses more difficult compared to a purely chemical approach despite the similar challenges, as often many related examples relying on chemical oxidants/reductants have been reported. However, some general principles exist that can be drawn on for a quick assessment. In addition, the number of available methods and use cases is constantly increasing at a rapid pace (see Fig. 10.4) [24].

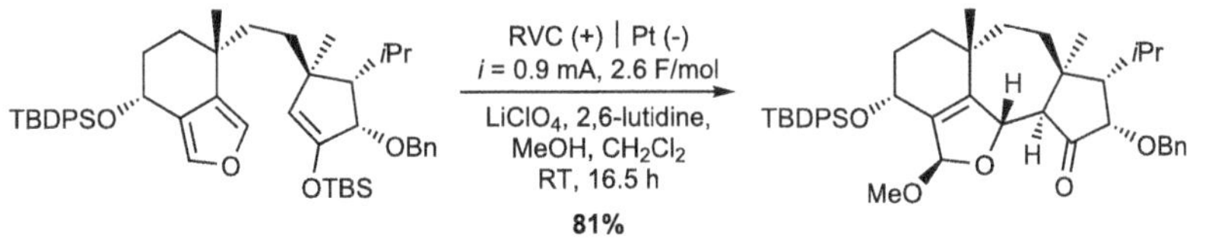

Fig. 10.4 Electrochemical, oxidative Umpolung in Trauner's synthesis of guanacastepene E [25]

Typical electrochemical reactions include reductions and oxidations of CX bonds (alcohol↔aldehyde↔ester; amine↔imine↔nitrile), the α-functionalization of heteroatoms (amine, ether, thioether) and carbonyl groups, the derivatization of olefins as well as a benzylic and allylic oxidation/functionalization [9]. In addition, the oxidation states of transition metal catalysts can be modulated, which can result in an altered reactivity compared to that observed under classic conditions [26].

10.1.1 Basic Principles of Reactivity and Selectivity

For an assessment in a complex synthetic context, the concept of frontier orbitals and their energy levels can be drawn on: The more electron-rich a compound or an electrophore, the higher the HOMO and the easier electrons can be emitted/released. The more electron-poor and the lower the LUMO, the easier a reduction can occur. The ease of oxidation of lone electron pairs on heteroatoms can be attributed to this principle, followed by conjugated CC double/triple bonds or aromatic systems. In contrast, carbonyl compounds and especially quinones can be readily reduced by virtue of their low-lying LUMO [12]. These trends are reflected in Fig. 10.5.

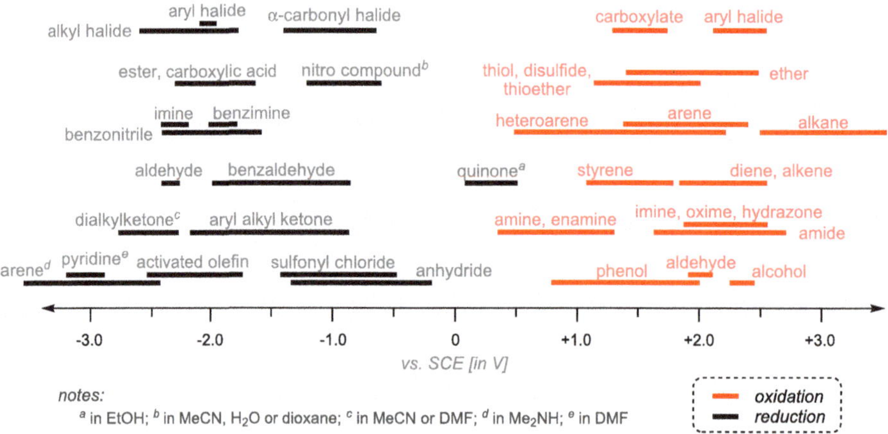

Fig. 10.5 Redox potentials ($E^\circ_{1/2}$, in volts) of common functional groups (in MeCN) [16, 27, 28]. SCE = saturated calomel electrode

In the presence of multiple electrophores with similar redox potential, low selectivity can occur resulting in the formation of undesired products. Generally speaking, functionalities with the lowest redox potential are converted first. However, the selectivity can be modified to a certain extent by the solvent, the identity and total surface of the electrode/current density, and the identity of the electrolytes. Saturated hydrocarbons are rarely redox active, and olefins often show a high barrier to oxidation. The presence of heteroatoms, as in enamines or enol ethers, significantly lowers the redox potential. In most cases, the transfer of a second electron has a higher activation barrier than that of the first and there is no difficulty in preventing the occurrence of the second reaction by an appropriate adjustment of the electrode potential [29]. In addition, control over single-electron versus two-electron processes can be further influenced by the identity of the electrode material, the current density, the electrolyte, or the concentration of the reactant.

When using the tabulated redox potentials, the reference electrode and the solvent always must be taken into account, as the potentials are governed by these aspects. The cited values can be interconverted [30], but a direct comparison is not meaningful without specifying the details of the reference potential.

Since the applied voltage leads to an oxidation at the anode and a reduction at the cathode, the corresponding counterpart must always be considered in addition to the desired reaction as they are intrinsically linked. Both oxidative and reductive processes must occur. The electrode at which the intended reaction takes place is referred to as the *working electrode*, while the second electrode is called the *counter-electrode*. Since electron neutrality must be adhered to, one reaction cannot proceed faster than the other. The reaction at the counter-electrode should therefore proceed as rapidly as possible so as not to become rate-limiting for the overall process. The counter-reaction usually comprises the degradation of the solvent or of a stoichiometric additive. In the case of a counter-reaction at the anode, a sacrificial electrode can also be employed, in which metal ions of the anode are solubilized. Sacrificial anodes are typically based on Zn, Mg, Cu, or Ni [14, 15]. If **both** electrodes are used productively, this is referred to as a *paired* electrosynthesis or a redox-combined electrochemical reaction. In a paired electrosynthesis, both reactions are desired but can be unrelated, whereas a redox-combined synthesis constitutes a coupled process where the product formed on one electrode reacts further on the other electrode. Typical examples of the latter are redox processes in which the oxidation state of a metal catalyst is modulated by the oxidation and reduction at different stages of the catalytic cycle, thus rendering it operative (electrocatalysis). The electron transfer at the electrodes is heterogeneous, as it is a reaction at the solid/liquid or solid/gas phase boundary.

The required energy to obtain the product corresponds to the thermodynamic driving force $\Delta G°$, which can be easily linked to the standard redox potential $\Delta E°$.[II]

$$\Delta G° = -nF\Delta E° \tag{10.1}$$

$$\Delta E = \Delta E° - \frac{RT}{nF} \, ln \, \frac{[\text{product}]}{[\text{reactant}]} \tag{10.2}$$

The redox potential $\Delta E°$ describes the energy to afford a 1:1 equilibrium of product and reactant under standard conditions, if $\Delta G° = 0$ applies (so-called *ergoneutral* reaction). This accordingly signifies only the thermodynamic feasibility of a reaction. The Nernst equation (10.2) provides an explanation as to why the actual necessary voltage can sometimes deviate drastically from the thermodynamic (ideal) potential under synthetic conditions. Assuming full conversion, the product concentration is several orders of magnitude greater than the reactant concentration. At 99% conversion (product:reactant $= 99:1$), this leads to an additional potential of 118 mV for a single-electron process, since a pronounced shift of the

---

[II] Electrochemical potentials measured experimentally by cyclic voltammetry typically correspond to potentials needed to initiate single-electron transfer, which are very different from potentials for net two-electron redox reactions typically of interest to organic chemists [12].

product/reactant equilibrium towards the product is desired. This driving force of the reaction is one of the main reasons for the phenomenon referred to as overpotential. Furthermore, the energy required to excite the substrate or mediator to cross the activation barrier will be greater than the thermodynamic difference between reactant and product ($\Delta G^{\ddagger}$ vs. ΔG°). This additionally imparts an overpotential. Further contributions are often governed by the electrode materials used [12]. The Nernst equation shows the redox potential as a function of the corresponding concentration of oxidized and reduced states of a molecule/atom, etc. If the product continues to react, for example through dimerization, cyclization, etc., its concentration decreases, which consequently also changes the redox potential of the reaction. The faster the subsequent reaction, the lower the concentration of the respective product will be and thus the deviation from the voltage under equilibrium conditions. Therefore, the necessary voltage is usually lower than originally measured, which is referred to as *Nernstian shift*, and runs counter to the overpotential as outlined above [13].

The role of subsequent reactions becomes even more pivotal for substrates bearing multiple reactive functionalities, as they directly affect which of the competitive reaction paths becomes the operative one. Even a reversed selectivity, contrary to the isolated redox potentials, can be observed [31]. In the conversion of **3**, the thioketal moiety should be oxidized more easily than the enol ether by virtue of its lower redox potential. However, the corresponding tetrahydropyran **4** resulting from enol ether excitation was isolated as the major product. This can be rationalized by a Curtin–Hammett situation occurring in the transformation [32]: The two reactive radicals **6** and **7** are in equilibrium, but the reaction of **7** is significantly faster. Since the product formation is largely governed by the greater rate constant k_1 due to the relative position of the activation energies $\Delta\Delta G^{\ddagger}$, the unfavorable equilibrium between **6** and **7** can be overcompensated (see Fig. 10.6) [33].

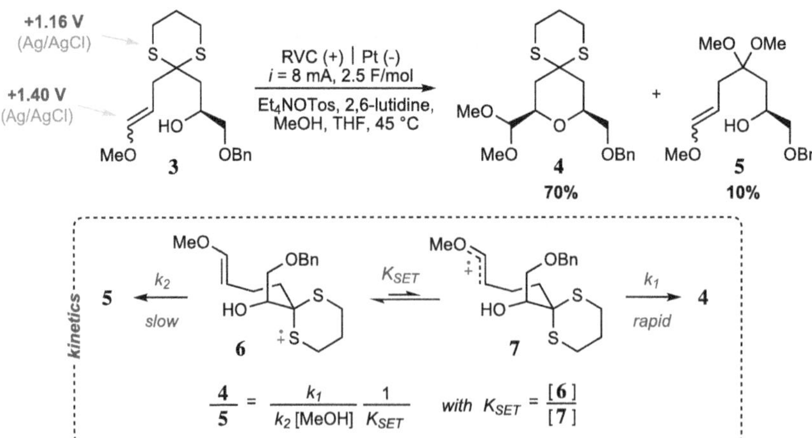

Fig. 10.6 Curtin–Hammett-controlled cyclization in an anodic oxidation [33]

Electrochemical reactions can be carried out either directly or may be mediated. Given the lack of tunable electrophores, direct electrochemical transformations rely predominantly on substrate engineering to regulate reactivity, whereas redox mediators may be fine-tuned independently of the substrate to optimize a reaction. When employing a redox mediator or catalyst, the voltage necessary for the reaction may be significantly decreased. Since the applied voltage determines the driving force of the electron transfer, a high voltage indicates significantly stronger oxidative or reductive conditions — with corresponding implications such as a lower selectivity. Analogous to classical catalysis, a redox mediator potentially allows an alternative reaction path to be taken by forming a catalyst-substrate complex and thus enabling a reactivity independent of the redox potential of the starting material. A typical example is the transfer of a hydrogen, proton, or hydride by the redox catalyst, whereby chemical information is transferred in addition to the electrons (see Fig. 10.7) [10, 11, 34].

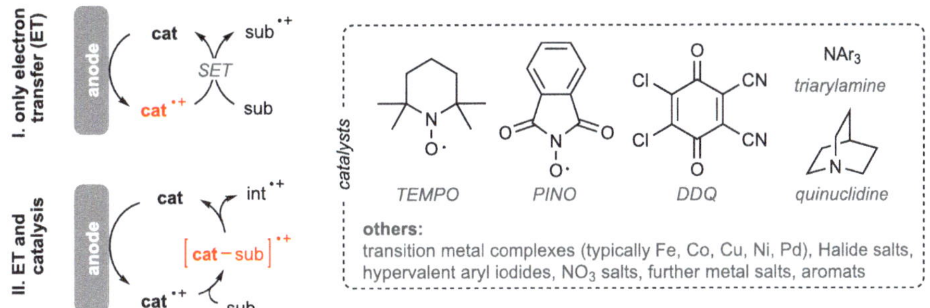

Fig. 10.7 Types of redox-mediated reactions and common catalysts/redox mediators [11]

Due to the lower voltage, catalyzed reactions are often milder and thus more selective. If chiral catalysts are used as mediators, for example transition metal complexes bearing chiral ligands, an enantio- or diastereoinduction and thus an asymmetric reaction may additionally become feasible. Third of all, the electron transfer between redox mediator and the substrate constitutes a homogeneous process. Formation of a passive layer on the electrode by undesired degradation processes (for example a decomposition of the starting material) can thus be avoided.

The type of application is characterized by the innate reactivity of the different catalysts: electron transfers are often mediated by ferrocenes, triarylamines, halide, cerium and cobalt salts, as well as hypervalent aryl iodides. *N*-Oxy radicals, trialkylamines, and metal oxides are capable of transferring hydrogen atoms and are therefore used in anodic oxidations, while *N*-oxides and quinones can serve as hydride sources, for example in CH activations [11, 34, 35]. In addition, the oxidation state of a metal catalyst can be modified (see Fig. 10.8) [34].

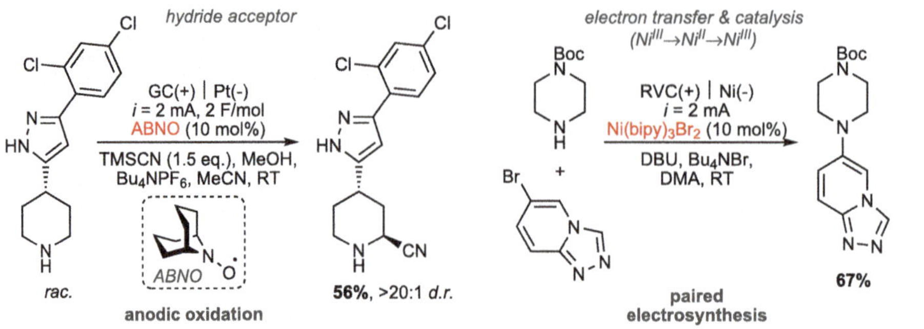

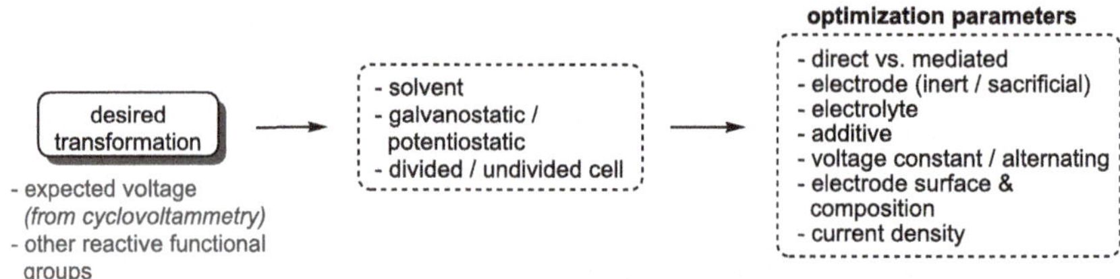

Fig. 10.8 Examples of different types of redox-catalyzed reactions [36, 37]

10.1.2 Practical Aspects and Applications

The reaction conditions of an electrochemical reaction offer many handles besides the voltage to optimize a given transformation. The judicious selection has an influence on the selectivity and reactivity, with several excellent review articles offering practical advice and serving as an easy introduction to the topic (see Fig. 10.9) [14, 15, 38].

Fig. 10.9 Overview of typical optimization parameters [14, 15, 38]

Based on the substrate to be functionalized, the presence of several electrophores in the substrate entails different voltages at which each should react. Two possible approaches need to be considered: Either the reaction is carried out at a certain, constant voltage (*potentiostatic* conditions). The substrate reacts and thus provides for charge compensation of the overall redox process. However, as its concentration decreases steadily over the course of the transformation, the current also decreases with successive progression of the reaction. As the current is proportional to the rate of reaction, a lower current will result in a longer reaction time needed to pass the total charge desired. The reaction is thus very selective under potentiostatic conditions, but becomes almost prohibitively slow at high conversions. In contrast, the current can be kept constant (*galvanostatic* conditions). This results in a constant reaction rate, which allows the transferred charge to be easily calculated accurately, thus rendering the requisite time to fully convert the substrate predictable. However, the voltage continues to rise over the course of the conversion, thus increasing the thermo-dynamic driving force, in order to maintain the desired current at the decreasing substrate concentration. The selectivity may accordingly decrease during the reaction, as the rising voltage successively allows alternative, undesired reaction paths to occur. Both parameters are interrelated and depending on which is kept constant (voltage or current), the other adjusts accordingly. Most reactions are carried out under galvanostatic conditions, which is why the applied current and often the transferred number of electrons (in Faraday per mol) are cited in the reaction conditions. Another, but rather rarely used possibility is the use of alternating polarization of the electrodes: cathode and anode alternate back and forth when their voltage sign is regularly reversed. This can prevent deposition processes on the electrodes and increase the reaction rate in strongly diffusion-limited reactions, possibly resulting in different selectivities due to different reaction rates [39]. A reference electrode is needed for potentiostatic reaction control to be able to set the voltage, which makes the experimental setup slightly more complex.

If the product or the intermediates are unstable towards contact with the counter-electrode, the use of a divided cell is recommended. The two chambers are separated by a semiperme-able membrane/diaphragm or salt bridge. This prevents free diffusion between the chambers, yet guarantees charge transfer. Alternatively, a sacrificial counter-electrode or the addition of sacrificial reagents can be used, which can react at a lower redox potential than the unwanted redox reaction of the substrate, intermediate, or product (often $H^+ \rightarrow H_2$).

The expected voltage, which can be determined by cyclic voltammetry, then dictates the viable solvents.[III] Since solvents decompose at a certain voltage, this provides an operational voltage range for each solvent (*solvent window*). Polar aprotic solvents typically dominate. They ensure sufficient intrinsic conductivity on the one hand, while anions derived from polar protic solvents (conjugate base) can act as nucleophiles in undesired side reactions. If protic solvents are employed, alcohols such as MeOH, EtOH, hexafluoroisopropanol (HFIP), or trifluoroethanol (TFE) are typically used. Fluorinated alcohols furthermore stabilize radical intermediates. To support the current flow, electrolytes are normally added to the solution to decrease the resistance [40]. In practice, quaternary ammonium salts with PF_6^- or BF_4^- as counterions have proven to be particularly effective due to their high solubility in organic solvents. Nevertheless, lithium-derived electrolytes such as $LiClO_4$, $LiPF_6$, or $LiBF_4$ are also routinely encountered. Halides such as Br^- or Cl^- can be used as counterions, yet they are prone to undergo oxidation at elevated voltages. While electrolytes are often needed in stoichiometric amounts, a range of substoichiometric to catalytic additives can also be utilized. For oxidation reactions, for example, a Brønsted acid might be added to facilitate hydrogen formation at the cathode.

Any given selection from the broad range of available methods can only provide an introduction to the topic at this point (see Fig. 10.10). The Lin group primarily studies

---

[III] E^0 is, as outlined above, highly solvent dependent, which renders this an iterative process and several cyclic voltammograms in different solvents might need to be tested in advance.

the functionalization of olefins (hydrocyanation, chlorotrifluoromethylation, oxyamination, dichlorination). One of their contributions focused on double azidation in the presence of *N*-oxy radicals as mediators, thereby tolerating reactive functionalities such as alkyl bromides, epoxides, α,β-ketones, aldehydes, and thioethers [41]. A classic example of an electrochemical reaction is the oxidation of alcohols. Instead of relying on a stoichiometric oxidant, electric current was used under potentiostatic conditions in the TEMPO-mediated reaction (example B) [42]. Yoshida published a series of studies on the amination of electron-rich arenes. The CH activation resorted to pyridine as amine source, which generates a pyridinium ion with the radical cation. The subsequent reaction with piperidine then afforded the primary arylamine (example C) [43]. The concept of generating a reactive cation before adding derivatizing reagents was introduced by Yoshida and is termed the *"cation pool"* method. Unlike normal electrosynthesis, the reactive intermediate does not react immediately; rather it is first generated stoichiometrically before it can subsequently be functionalized upon addition of a nucleophile. However, a divided cell is required to prevent the cations from

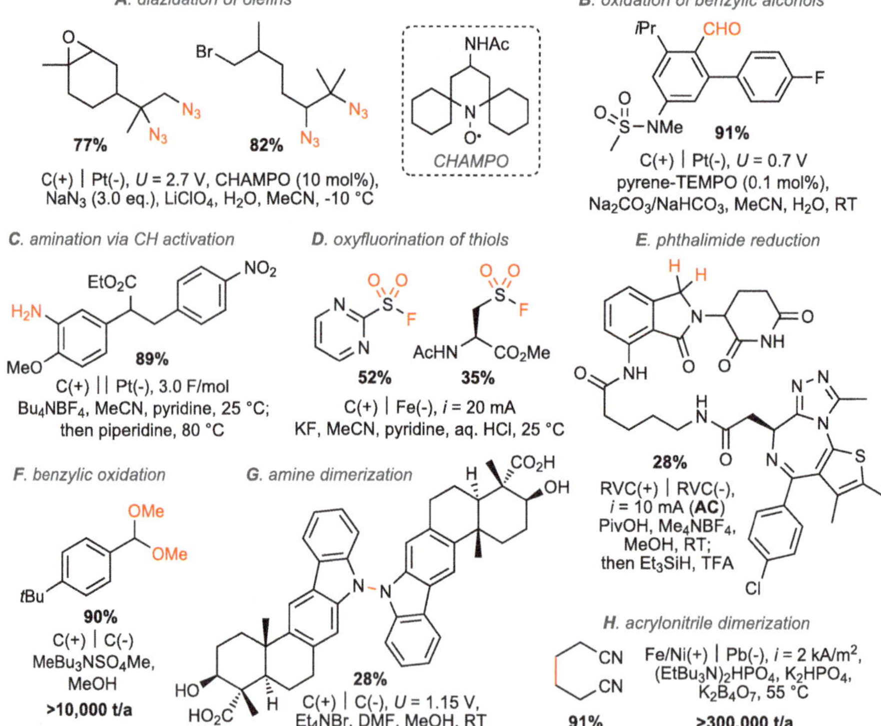

Fig. 10.10 Select examples of electrochemical methods and applications in industrial and academic settings [4, 41–46]

reacting with the cathode [47]. Similar to Lin's work (example A), Noël *et al.* also developed a selective derivatization of the substrate with a spectrum of different reagents. In their anodic oxidation of aromatic and aliphatic thiols, water serves as the oxygen source while KF provides the halide for obtaining the desired sulfonyl fluorides (example D) [44]. As mentioned above, apart from potentiostatic or galvanostatic conditions, the polarity of the electrodes can also be deliberately reversed at a certain frequency. Based on previous work [48, 49], the Baran group investigated alternating polarization in the millisecond range with precise control of the waveform. This enabled a series of substituted phthalimides to be converted to the corresponding lactams. Both oxidation- (alcohol, amine, guanidine) and reduction-sensitive functionalities (azide, ester, amide) remained intact during the transformation. The chosen example E (pomalidomide-JQ1 conjugate) taken from this study, however, had to be reduced in two steps (chemically, then electrochemically) due to the presence of the highly sensitive heterocyclic moiety [45]. The application of electrosynthetic reactions is not limited to academic (total) syntheses but encompasses industrial examples as well [50]. The benzylic oxidation of 4-*t*butyl-toluene to the corresponding acetal is carried out by BASF on a kiloton scale (example F) [46]. The paired electrosynthesis simultaneously produces phthalide at the cathode by reduction of phthalic acid diethyl ester, thus achieving an overall balance in terms of electrons as well as MeOH equivalents. Probably the largest industrial use of electrochemistry lies in the dimerization of acrylonitrile to adiponitrile. The process was developed by Dow and is a classic example of an Umpolung reaction. The electrophilic β-terminus becomes nucleophilic through cathodic reduction and can subsequently react with a second equivalent of the starting material (example H) [46]. In the context of total synthesis, only a few applications in more complex intermediates have been reported. Baran and co-workers employed potentiostatic conditions to dimerize the amine of the monomeric xiamycin A and assemble the secondary metabolite dixiamycin B [4].

10.2 VIS-Photoredox Catalysis

The use of UV-mediated transformations has not been widely established as a standard tool in the synthetic repertoire, exempting some selected applications such as Norrish-type reactions or pericyclic reactions. UV photoreactors are expensive and the technology is complex. For example, the reaction flasks require special UV-transparent glass. In addition, the high energy content beyond near UV (UV-A) can lead to a higher proportion of undesired side reactions. These challenges have barred the widespread adoption of UV-photochemical steps in synthetic endeavors. The use of light with wavelengths located in the visible part of the spectrum (**VIS**) is, on the other hand, technically simple and significantly cheaper to implement [51]. Since a photocatalyst for selective reactions should not absorb in the same region as the substrate, VIS photocatalysts are advantageous because most organic molecules are not receptive towards photons at wavelengths in the range of 400–700 nm. In the part of the electromagnetic spectrum that encompasses increasingly longer wavelengths, the energy content of the photons becomes eventually too low to allow synthetically relevant excitations. This results in a operative window of 300–800 nm, in which photochemical conversions typically occur (see Fig. 10.11) [23].

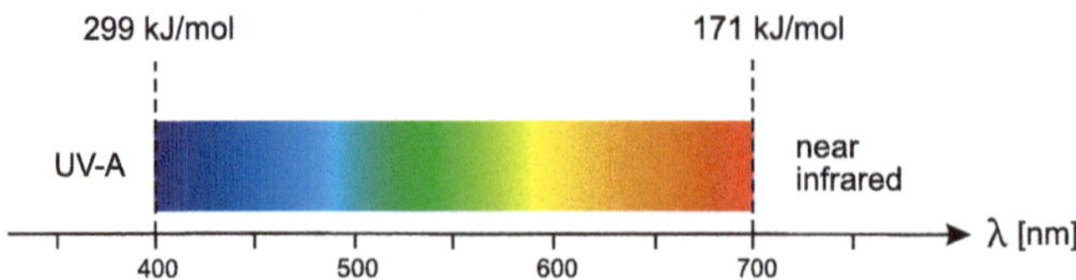

Fig. 10.11 Wavelengths and energy content of visible light

For a long time, the presence of a suitable chromophore in the substrates was necessary to enable a photochemical excitation. This changed with the introduction of photoactive transition metal complexes, which can absorb in the visible range [5, 6]. Two approaches can generally be conceived: Either a molecule is excited by a sensitizer and can subsequently react in the excited state. Typical examples include cycloadditions and *E/Z* isomerizations of alkenes. Alternatively, photosensitizers can serve as redox catalysts. Depending on the substrate, the catalyst can either take up or give off electrons. An additional oxidizing or reducing agent can be employed for the electron transfer. The photocatalyst ensures in the latter case that the oxidant's/reductant's redox potential exactly matches the desired conversion of the substrate (see Fig. 10.12).

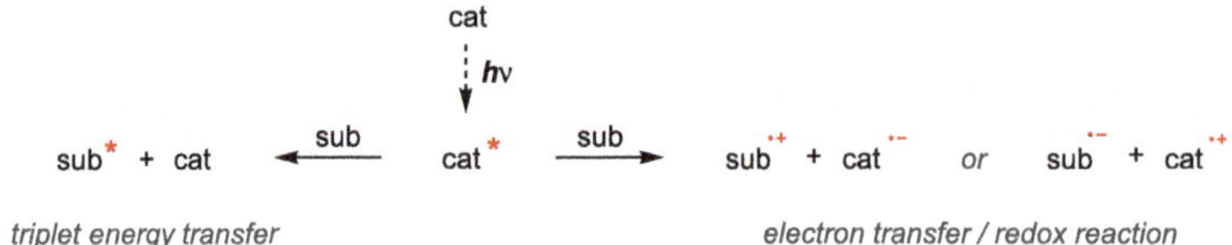

Fig. 10.12 Common approaches to the use of photoactive catalysts

10.2.1 Mechanistic Principles

The direct or mediated photochemical excitation of a substrate can be compared to the mode of action of a catalyst: A prohibitively slow reaction in the ground state becomes potentially possible through the input of energy as the excited state lies on a different potential hypersurface. The excitation can similarly drive a reaction away from their equilibrium position to access the thermodynamically disfavored product. Furthermore, excited states may even result in the omission of an activation barrier when transitioning to the ground state (see Fig. 10.13) [52–54].

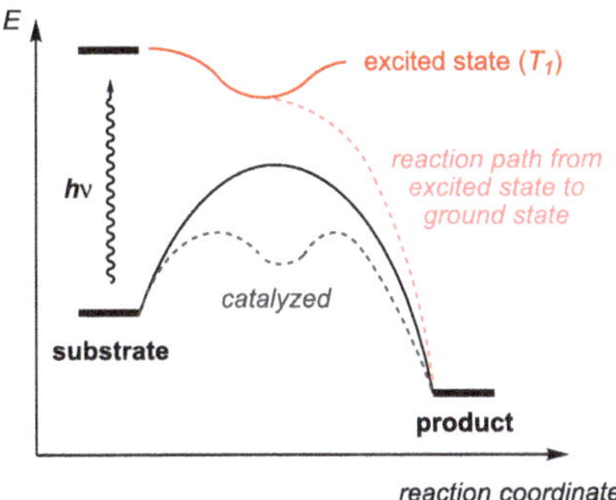

Fig. 10.13 Comparison of the reaction course of (un-)catalyzed and photochemically excited reactions [52, 53]

The mechanism of the photocatalyzed reaction follows a similar pathway to direct excitation. In the uncatalyzed process, an $S_0 \rightarrow S_1$ transition initially occurs, followed by *intersystem crossing* (ISC) to the excited triplet state T_1, which can generate the product.[IV] Given the spin-forbidden relaxation from the T_1 manifold to the ground state, triplet states are relatively long-lived, which renders the interception of a triplet intermediate in an intermolecular process the most likely pathway. When a triplet sensitizer is used, the same processes outlined above occur in the catalyst. The subsequent energy transfer from the triplet state of the catalyst to the substrate allows the reactive state T_1 to be reached in the reactant. This energy transfer (EnT) preferentially follows a *Dexter electron transfer mechanism* involving a non-radiative exchange of two electrons. It requires the substrate and catalyst to come into close proximity to ensure the necessary orbital overlap. The more energetically matched the triplet levels of the sensitizer and substrate are, the stronger the interaction. At the same time, an exergonic reaction is only ensured if the triplet energy level of the sensitizer energetically lies above that of the substrate. A catalyst tailored to a given substrate thus increases the probability of the desired transformation. There is a vast array of metal and organic photoredox catalysts with different excited state redox potentials, typically Ru/Ir complexes or extended aromatic systems. The redox potential of metal catalysts can further be modified with the selection of ligands around the metal center (see Fig. 10.14) [55, 56].

[IV] Further non-radiative relaxation processes (*internal conversion*) and the re-emission of photons as well as the consideration of vibrationally excited states are omitted here for simplicity. These processes still occur and can significantly reduce the quantum yield in individual cases. The effects of different lifetimes of singlet and triplet states are also not considered here. See Nicewicz *et al.*, *Chem. Rev.* **2016**, *116*, 10075–10166 and Bach *et al.*, *Chem. Rev.* **2022**, *122*, 1626–1653.

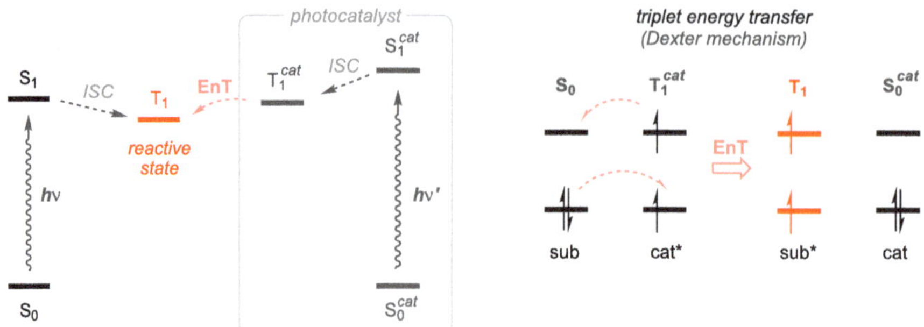

Fig. 10.14 Direct and sensitizer-mediated photoexcitation as well as Dexter mechanism of triplet energy transfer [56]

In a redox process, on the other hand, only one electron is exchanged (electron transfer, ET): Depending on whether it is an oxidation or reduction, the substrate accepts an electron from the catalyst or donates it (see Fig. 10.15).

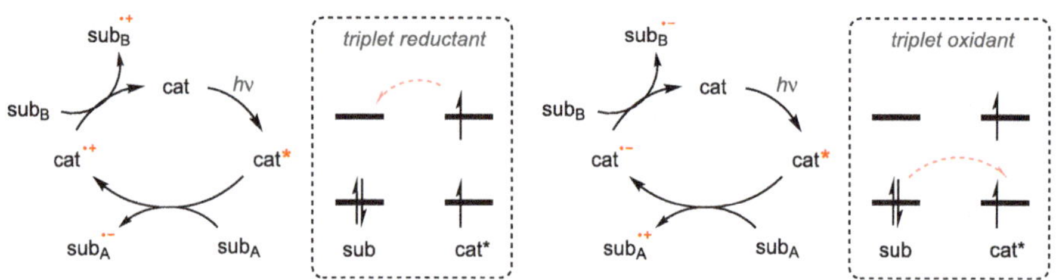

Fig. 10.15 Mechanism of oxidative or reductive photoredox catalysts [18]

As becomes evident in Fig. 10.15, an oxidizing or reducing agent is needed in addition to the substrate to complete the catalytic cycle. In the case of photoinduced redox processes of transition metal catalysts, the "substrate" can be the transition metal catalyst for both steps, as it is first oxidized and then reduced at a later stage. The same principle was exploited in paired electrosynthesis employing a Ni catalyst (see Fig. 10.8). The Ni complex is initially reduced at the anode for the CN coupling, facilitates the transformation, and is reoxidized at the cathode. A chemical oxidant or reducing agent can also be used to effect the reaction. While O_2 constitutes the oxidant in many cases, the substrate or a subsequent intermediate may likewise directly react with the photocatalyst and regenerate it. Finally, combination with an electrochemical cell is also feasible, which is referred to as photoelectrosynthesis or electrophotocatalysis [18, 57]. An illustrative example is provided by Cordero-Vargas' approach to the tetrahydropyran moiety of the marine natural product aspergillide A. Following the initial excitation of the Ru catalyst, it is subsequently reduced by ascorbate. The reduction of iodoacetic acid (**8**) liberates iodide and returns the catalyst to the electronic ground state. The concomitantly obtained acetic acid radical can react with alkene **9** and maintain the radical chain reaction. **10** was cyclized *in situ* upon addition of TFA to **11** and subsequently converted to the natural product (see Fig. 10.16) [58].

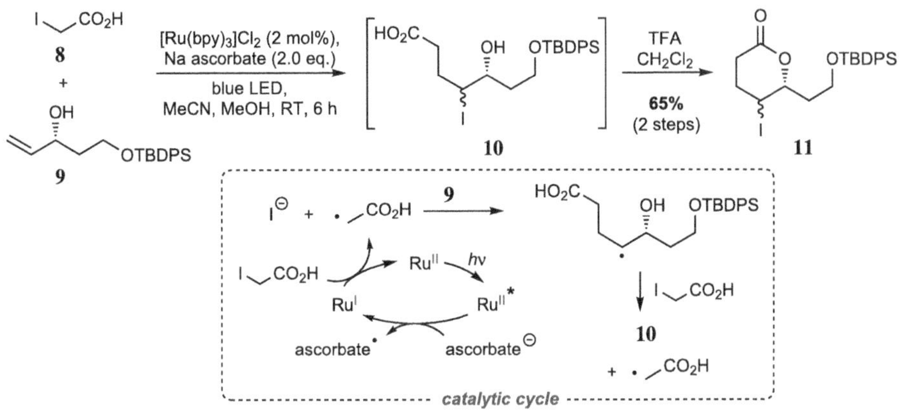

Fig. 10.16 Synthesis of aspergillide A by Cordero-Vargas *et al.* and postulated photoredox catalytic cycle of the key cyclization [3, 58]

From a thermodynamic point of view, the Gibbs energy of the photoinduced electron transfer (*PET*) can be described [18] by

$$\Delta G_{PET} = -F\,[E_{ox}(A) - E_{red}(D)] - w_{Coulomb} - E^* \tag{10.3}$$

A and D correspond to the electron donor and electron acceptor, and F is the Faraday constant. In addition to the redox potential of the donor and acceptor (depending on whether one considers an oxidation or reduction, either the former or the latter term represent the photocatalyst), the equation furthermore contains the excitation energy of the excited donor or acceptor and an electrostatic work term w to account for the Coulombic interaction of charge separation. Since w is usually relatively small, Eq. 10.3 can be simplified by introducing the excitation energy to

$$\Delta G_{PET} = -F\left[E^*_{ox}(\mathrm{cat}) - E_{red}(\mathrm{sub})\right] \tag{10.4}$$

for an oxidative and

$$\Delta G_{PET} = -F\left[E_{ox}(\mathrm{sub}) - E^*_{red}(\mathrm{cat})\right] \tag{10.5}$$

for a reductive PET. For a photoredox catalyst acting as an excited state oxidant, E^*_{red} is positive, while a photoredox catalyst acting as an excited state reductant requires E^*_{ox} to be negative. If a photoinduced oxidation of a substrate is to be feasible (assuming an exergonic reaction with $\Delta G_{PET} < 0$), the oxidation potential $E^*_{ox}(\mathrm{cat})$ must be greater than $E_{red}(\mathrm{sub})$. In a reductive PET, $E^*_{red}(\mathrm{cat})$ must be more negative than the redox potential $E_{ox}(\mathrm{sub})$ of the substrate for the reaction to be thermodynamically favorable [18]. While the energy of the triplet state is decisive for a triplet energy transfer from the catalyst to the substrate, photoredox catalysis is governed by the corresponding redox potentials of the *excited* states (see Fig. 10.17).

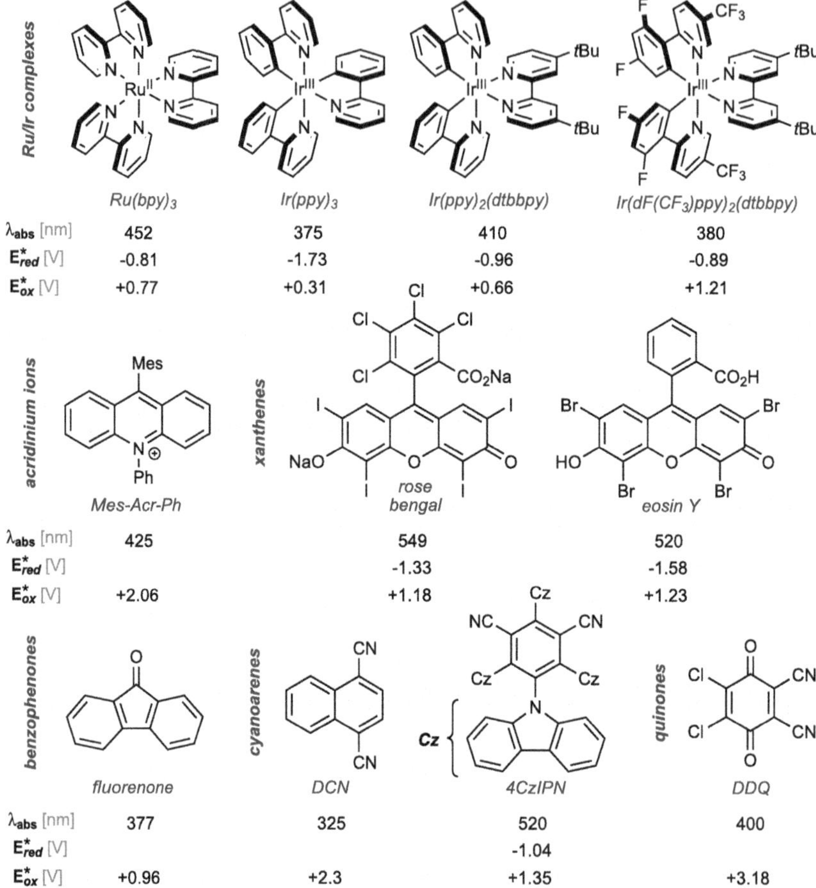

	Ru(bpy)$_3$	Ir(ppy)$_3$	Ir(ppy)$_2$(dtbbpy)	Ir(dF(CF$_3$)ppy)$_2$(dtbbpy)
λ_{abs} [nm]	452	375	410	380
E^*_{red} [V]	-0.81	-1.73	-0.96	-0.89
E^*_{ox} [V]	+0.77	+0.31	+0.66	+1.21

	Mes-Acr-Ph	rose bengal	eosin Y
λ_{abs} [nm]	425	549	520
E^*_{red} [V]		-1.33	-1.58
E^*_{ox} [V]	+2.06	+1.18	+1.23

	fluorenone	DCN	4CzIPN	DDQ
λ_{abs} [nm]	377	325	520	400
E^*_{red} [V]			-1.04	
E^*_{ox} [V]	+0.96	+2.3	+1.35	+3.18

Fig. 10.17 Common photoredox catalysts [18–20, 59]. In the absence of a value for E^*_{red}, the systems are not suitable for a reductive PET

Since various transformations require different excitation energies, a plethora of (metal) organic catalysts has been developed, which absorb in different regions of the VIS spectrum and possess a broad range of redox potentials to match the requisite energy input.

10.2.2 Selected Applications of Photoredox Catalysis

A hallmark of CC or CN single bond formation via radical reactions is its tolerance towards typical polar functionalities such as alcohols, thiols, carboxylic acids, and primary and secondary amines, which are often degraded under basic conditions. The generation of radicals via photoredox catalysis, in contrast to methods using stoichiometric radical initiators, typically does not require special functionalities in the substrate to enable its formation. Thus, additional steps for the introduction of said functional groups can be omitted, which qualifies photoredox catalysis for late-stage functionalizations in the life science sector and in natural product synthesis (see Fig. 10.1) [23, 60, 61]. Another advantage of photoredox catalysis over classic radical methods lies in the access to reactivities that would otherwise not be possible or proceed with low selectivity. Arguably one of the most prominent examples may be redox-neutral reactions, similar to paired electrosynthesis (see Fig. 10.7). Since both oxidizing and reducing agents can be generated *in situ* in the same reaction vessel, electrons can be taken up from substrates and/or intermediates and also be given off by them in photoredox reactions [20]. From the perspective of possible derivatizations, some reactions are encountered more frequently in such a context, as they lend themselves to radical processes by virtue of the straightforward radical formation or a high radical stability. The α-functionalization of amines, derivatizations of benzyl building blocks, functionalizing decarboxylations, the cleavage of diazo groups, or modifications of olefins fall into this category. An illustrative example can be found in Rychnovsky's synthesis of the complex alkaloid himeradine A. Based on a method developed by MacMillan [62], Boc-protected glycine reacted with the advanced key intermediate **12** in the presence of $Ir(dF(CF_3)_2ppy)_2(dttbpy)PF_6$. The reaction mechanism presumably proceeds via initial oxidation of the deprotonated carboxylic acid, cleavage of CO_2, and formation of an α-amino radical species. Following the reaction of the radical with the enone moiety in **12**, the Ir(III) photocatalyst was regenerated upon reduction of the transient radical addition product to the enolate. The concluding protonation yielded **13**. Three further steps subsequently afforded the natural product himeradine A, which could be accessed in overall 17 steps (longest linear sequence) (see Fig. 10.18) [63].

Fig. 10.18 Rychnovsky's synthesis of himeradine A [63]

Initial methodological studies in the field addressed the direct oxidation and reduction of substrates to generate radical intermediates. Subsequently, the possibility of employing two catalysts emerged, in which the photoredox catalyst merely serves to modify the oxidation states of another (transition metal) catalyst [21, 22, 64–66]. Another key area of research is devoted to effecting hydrogen atom transfer (HAT) reactions. Mechanistically, an excited catalyst activates the substrate for functionalization of RH (R = C, Si, S) bonds in the starting material. Either a direct strategy, based on the intrinsic reactivity of a photocatalyst in the excited state, or an indirect one, in which a photocatalytic cycle is used for the generation of a thermal hydrogen abstractor, can be exploited. Typical functionalizations comprise C-C, C-N, C-S of C-F bond formation (see Fig. 10.19) [67–69].

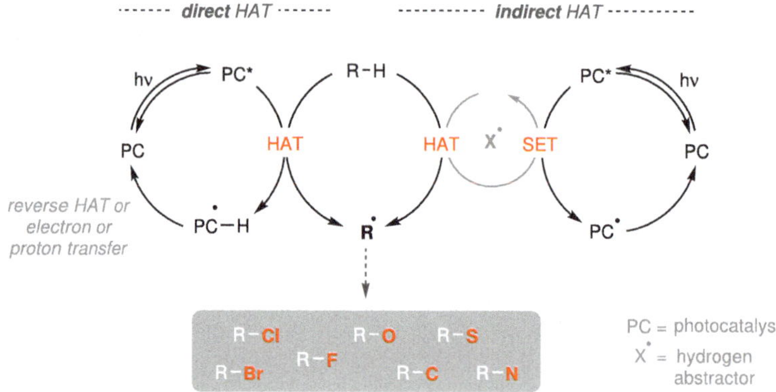

Fig. 10.19 Principle of photocatalyzed direct and indirect hydrogen atom transfer reactions [67–69]

Two commendable examples are encountered in Stephenson's Minisci-like CH activation of electron-poor heterocycles and Buchwald's and MacMillan's Ni-catalyzed CN cross-coupling reaction. The complex substrate **14** was intended for the synthesis of the antineoplastic tyrosine kinase inhibitor gandotinib (LY2784544), whereby the aminomethylene moiety could be varied late-stage. Mechanistically, **16** was postulated as an intermediate in the formation of **15** following the initial attack of the amine radical [70]. The amination of aryl bromides constitutes an established methodology both under Pd and Cu catalysis. The Ni-mediated variant using a photoredox catalyst, which is assumed to modulate the oxidation state of the Ni complex to facilitate its reactivity, proved to be one of the most broadly applicable methods compared to other CN cross-coupling protocols. Several products of a high-throughput screening comprising 18 complex, drug-like substrates showed acceptable yields, which could then serve as a starting point for further optimization studies (see Fig. 10.20) [71].

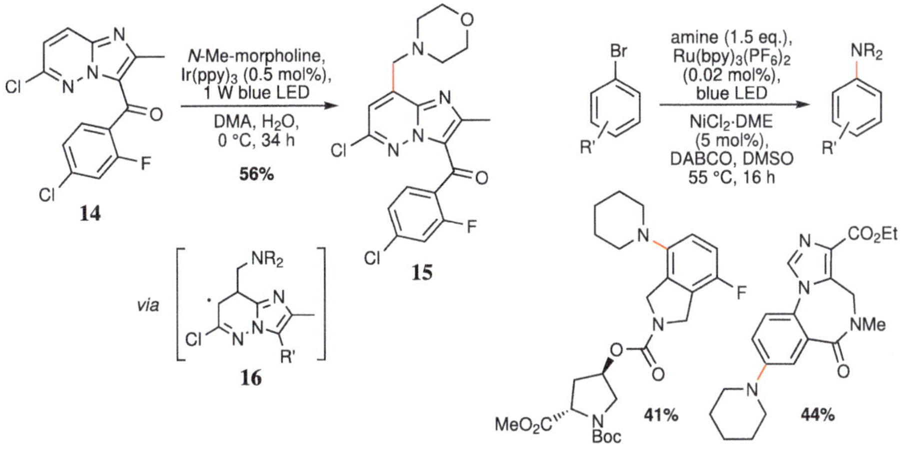

Fig. 10.20 Methods for CC and CN coupling under photoredox catalysis [70, 71]

In general, photoredox-mediated couplings can be more powerful for certain applications than already established and extensively optimized methods. For example, in a broad comparative study of various C_{sp^2}-C_{sp^3} cross-couplings, the Ni/photoredox-alkyl-BF$_3$K coupling emerged as the most powerful protocol for α-oxy and α-amino nucleophiles [72].

In natural product synthesis, some ingenious applications have also been disclosed (see Fig. 10.21) [3, 23, 52]. Overman and co-workers resorted to N-hydroxyphthalimide ester **18** as a radical precursor in their total synthesis of the diterpene aplyviolene. After initial formation of the tertiary radical **20** upon cleavage of CO_2, a Michael addition with the substituted cyclopentenone **17** ensued. A Hantzsch ester served as the reducing agent, which was oxidized in the process to the corresponding pyridinium ion [73]. The Yang group thereafter utilized the same approach in their access to haperforin G via **24**. In this case, an alkyl iodide was converted with a bridged, exocyclic enone in the presence of [Ir(ppy)$_2$(dttbbpy)]PF$_6$. While the transformation could also be carried out under classic radical conditions (AIBN, Bu$_3$SnH), they consistently afforded lower yields [74].

The C3–C3' bond formation of the two heteroaryl units constituted the key step in synthesis of the hexahydropyrroloindoline alkaloid (+)-gliocladin C by Stephenson *et al.*, which was achieved via a photoredox-initiated, radical CC coupling. The aryl bromide **22** generated a tertiary benzylic radical, which subsequently underwent the desired CC connection with indole **21** in the presence of tributylamine as reducing agent. The temporary blockade of the C2 position in **21** by the aldehyde group was necessary to prevent the reaction occurring at this site. The ensuing Rh-catalyzed decarbonylation of the northern fragment, triketopiperazine formation, and final Cbz deprotection provided the natural product [75].

The last example hails from the Yang group, who tackled the synthesis of the marine cembranoid pavidolide B. The formal [3+2]-cycloaddition to assemble the ABC ring skeleton could be achieved with the aid of a thiyl radical mediator. Following its addition to the alkenylcyclopropane moiety in **25**, the strained ring system was opened and **27** was

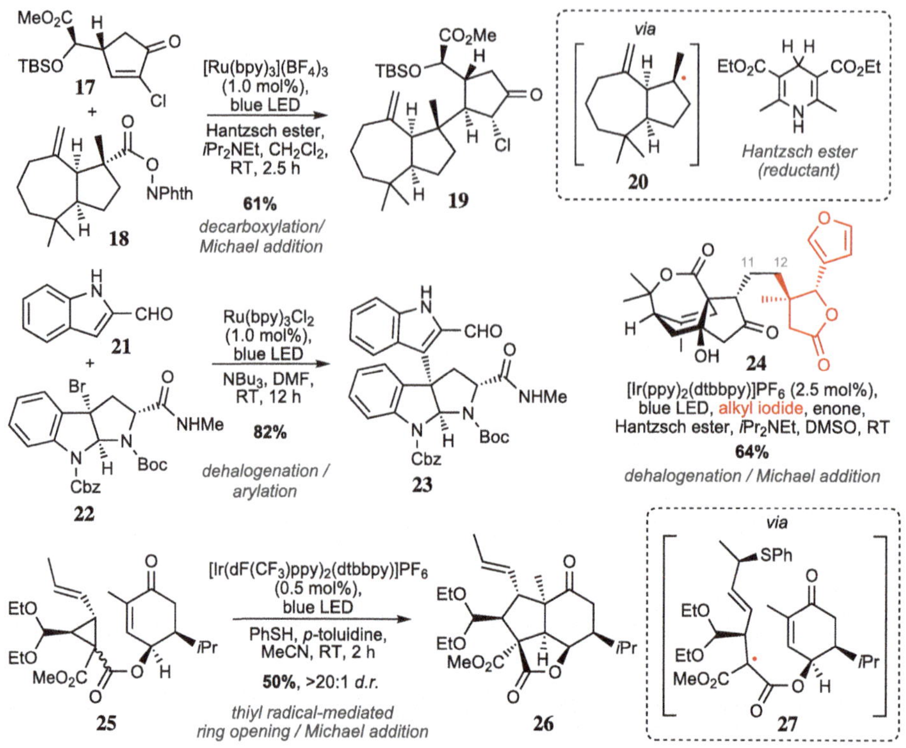

Fig. 10.21 Applications of photoredox-initiated radical reactions in total syntheses [73–76]

presumably generated as intermediate. The following step comprised the intramolecular 1,4-addition to the enone functionality, thereby closing lactone ring. The concluding CC bond formation to **26** then took place with elimination of PhS$^\bullet$ and regeneration of the double bond [76].

References

1. G. E. Veitch, E. Beckmann, B. J. Burke, A. Boyer, S. L. Maslen, S. V. Ley, *Angew. Chem. Int. Ed.* **2007**, *46*, 7629–7632.
2. N. E. S. Tay, D. Lehnherr, T. Rovis, *Chem. Rev.* **2022**, *122*, 2487–2649.
3. M. S. Galliher, B. J. Roldan, C. R. J. Stephenson, *Chem. Soc. Rev.* **2021**, *50*, 10044–10057.
4. B. R. Rosen, E. W. Werner, A. G. O'Brien, P. S. Baran, *J. Am. Chem. Soc.* **2014**, *136*, 5571–5574.
5. D. A. Nicewicz, D. W. C. MacMillan, *Science* **2008**, *322*, 77–80.
6. M. A. Ischay, M. E. Anzovino, J. Du, T. P. Yoon, *J. Am. Chem. Soc.* **2008**, *130*, 12886–12887.
7. H. J. Schäfer, *Top. Curr. Chem.* **1990**, *152*, 91–151.
8. A. G. Griesbeck, H. Heckroth, *J. Am. Chem. Soc.* **2002**, *124*, 396–403.
9. M. Yan, Y. Kawamata, P. S. Baran, *Chem. Rev.* **2017**, *117*, 13230–13319.

10. J. C. Siu, N. Fu, S. Lin, *Acc. Chem. Res.* **2020**, *53*, 547–560.

11. R. Francke, R. D. Little, *Chem. Soc. Rev.* **2014**, *43*, 2492–2521.

12. J. E. Nutting, J. B. Gerken, A. G. Stamoulis, D. L. Bruns, S. S. Stahl, *J. Org. Chem.* **2021**, *86*, 15875–15885.

13. R. D. Little, *J. Org. Chem.* **2020**, *85*, 13375–13390.

14. C. Schotten, T. P. Nicholls, R. A. Bourne, N. Kapur, B. N. Nguyen, C. E. Willans, *Green Chem.* **2020**, *22*, 3358–3375.

15. C. Kingston, M. D. Palkowitz, Y. Takahira, J. C. Vantourout, B. K. Peters, Y. Kawamata, P. S. Baran, *Acc. Chem. Res.* **2020**, *53*, 72–83.

16. L. M. Reid, T. Li, Y. Cao, C. P. Berlinguette, *Sust. En. Fuel* **2018**, *2*, 1905–1927.

17. L. F. T. Novaes, J. Liu, Y. Shen, L. Lu, J. M. Meinhardt, S. Lin, *Chem. Soc. Rev.* **2021**, *50*, 7941–8002.

18. N. A. Romero, D. A. Nicewicz, *Chem. Rev.* **2016**, *116*, 10075–10166.

19. N. Holmberg-Douglas, D. A. Nicewicz, *Chem. Rev.* **2022**, *122*, 1925–2016.

20. C. K. Prier, D. A. Rankic, D. W. C. MacMillan, *Chem. Rev.* **2013**, *113*, 5322–5363.

21. K. L. Skubi, T. R. Blum, T. P. Yoon, *Chem. Rev.* **2016**, *116*, 10035–10074.

22. C.-S. Wang, P. H. Dixneuf, J.-F. Soulé, *Chem. Rev.* **2018**, *118*, 7532–7585.

23. S. P. Pitre, L. E. Overman, *Chem. Rev.* **2022**, *122*, 1717–1751.

24. C. E. Hatch, W. J. Chain, *ChemElectroChem* **2023**, *10*, e202300140.

25. A. K. Miller, C. C. Hughes, J. J. Kennedy-Smith, S. N. Gradl, D. Trauner, *J. Am. Chem. Soc.* **2006**, *128*, 17057–17062.

26. C. Zhu, N. W. J. Ang, T. H. Meyer, Y. Qiu, L. Ackermann, *ACS Cent. Sci.* **2017**, *7*, 415–431.

27. H. G. Roth, N. A. Romero, D. A. Nicewicz, *Synlett* **2015**, *27*, 714–723.

28. T. Fuchigami, M. Atobe, S. Inagi in *Fundamentals and Applications of Organic Electrochemistry: Synthesis, Materials, Devices*, John Wiley & Sons, 1st ed., **2015**, pp. 217–221.

29. C. Costentin, J.-M. Savéant, *Proc. Nat. Acad. Sci.* **2019**, *116*, 11147–11152.

30. V. V. Pavlishchuk, A. W. Addison, *Inorg. Chim. Acta* **2000**, *298*, 97–102.

31. A. Redden, K. D. Moeller, *Org. Lett.* **2011**, *13*, 1678–1681.

32. J. I. Seeman, *Chem. Rev.* **1983**, *83*, 83–134.

33. S. Duan, K. D. Moeller, *J. Am. Chem. Soc.* **2002**, *124*, 9368–9369.

34. C. A. Malapit, M. B. Prater, J. R. Cabrera-Pardo, M. Li, T. D. Pham, T. P. McFadden, S. Blank, S. D. Minteer, *Chem. Rev.* **2022**, *122*, 3180–3218.

35. F. Wang, S. S. Stahl, *Acc. Chem. Res.* **2020**, *53*, 561–574.

36. A. J. J. Lennox, S. L. Goes, M. P. Webster, H. F. Koolman, S. W. Djuric, S. S. Stahl, *J. Am. Chem. Soc.* **2018**, *140*, 11227–11231.

37. Y. Kawamata, J. C. Vantourout, D. P. Hickey, P. Bai, L. Chen, Q. Hou, W. Qiao, K. Barman, M. A. Edwards, A. F. Garrido-Castro, J. N. deGruyter, H. Nakamura, K. Knouse, C. Qin, K. J. Clay, D. Bao, C. Li, J. T. Starr, C. Garcia-Irizarry, N. Sach, H. S. White, M. Neurock, S. D. Minteer, P. S. Baran, *J. Am. Chem. Soc.* **2019**, *141*, 6392–6402.

38. D. Pollok, B. Gleede, A. Stenglein, S. R. Waldvogel, *Aldrichim. Acta* **2021**, *54*, 3–15.

39. L. Zeng, J. Wang, D. Wang, H. Yi, A. Lei, *Angew. Chem. Int. Ed.* **2023**, *62*, e202309620.

40. L. G. Gombos, J. Nikl, S. R. Waldvogel, *ChemElectroChem* **2024**, *11*, e202300730.

41. J. C. Siu, J. B. Parry, S. Lin, *J. Am. Chem. Soc.* **2019**, *141*, 2825–2831.

42. A. Das, S. S. Stahl, *Angew. Chem. Int. Ed.* **2017**, *56*, 8892–8897.

43. T. Morofuji, A. Shimizu, J. Yoshida, *J. Am. Chem. Soc.* **2013**, *135*, 5000–5003.

44. G. Laudadio, A. de A. Bartolomeu, L. M. H. M. Verwijlen, Y. Cao, K. T. de Oliveira, T. Noel, *J. Am. Chem. Soc.* **2019**, *141*, 11832–11836.

45. Y. Kawamata, K. Hayashi, E. Carlson, S. Shaji, D. Waldmann, B. J. Simmons, J. T. Edwards, C. W. Zapf, M. Saito, P. S. Baran, *J. Am. Chem. Soc.* **2021**, *143*, 16580–16588.

46. T. Fuchigami, M. Atobe, S. Inagi in *Fundamentals and Applications of Organic Electrochemistry: Synthesis, Materials, Devices*, John Wiley & Sons, 1st ed., **2015**, Chapter 8.

47. J. Yoshida, S. Suga, *Chem. Eur. J.* **2002**, *8*, 2650–2659.

48. D. E. Blanco, B. Lee, M. A. Modestino, *Proc. Nat. Acad. Sci.* **2019**, *116*, 17683–17689.

49. S. Rodrigo, C. Um, J. C. Mixdorf, D. Gunasekera, H. M. Nguyen, L. Luo, *Org. Lett.* **2020**, *22*, 6719–6723.

50. D. Lehnherr, L. Chen, *Org. Process Res. Dev.* **2024**, *28*, 338–366.

51. L. Buglioni, F. Raymenants, A. Slattery, S. D. A. Zondag, T. Noël, *Chem. Rev.* **2022**, *122*, 2752–2906.

52. M. D. Kärkäs, J. A. Porco, C. R. J. Stephenson, *Chem. Rev.* **2016**, *116*, 9683–9747.

53. J. C. Scaiano, *Chem. Soc. Rev.* **2023**, *52*, 6330–6343.

54. A. Lin, S. Lee, R. R. Knowles, *Acc. Chem. Res.* **2024**, *57*, 1827–1838.

55. J. D. Bell, J. A. Murphy, *Chem. Soc. Rev.* **2021**, *50*, 9540–9685.

56. J. Großkopf, T. Kratz, T. Rigotti, T. Bach, *Chem. Rev.* **2022**, *122*, 1626–1653.

57. M. C. Lamb, K. A. Steiniger, L. K. Trigoura, J. Wu, G. Kundu, H. Huang, T. H. Lambert, *Chem. Rev.* **2024**, *124*, 12264–12304.

58. J. B. Mateus-Ruiz, A. Cordero-Vargas, *J. Org. Chem.* **2019**, *84*, 11848–11855.

59. T. Neveselý, M. Wienhold, J. J. Molloy, R. Gilmour, *Chem. Rev.* **2022**, *122*, 2650–2694.

60. P. Bellotti, H.-M. Huang, T. Faber, F. Glorius, *Chem. Rev.* **2023**, *123*, 4237–4352.

61. L. Candish, K. D. Collins, G. C. Cook, J. J. Douglas, A. Gómez-Suárez, A. Jolit, S. Keess, *Chem. Rev.* **2022**, *122*, 2907–2980.

62. L. Chu, C. Ohta, Z. Zuo, D. W. C. MacMillan, *J. Am. Chem. Soc.* **2014**, *136*, 10886–10889.

63. A. Burtea, J. DeForest, X. Li, S. D. Rychnovsky, *Angew. Chem. Int. Ed.* **2019**, *58*, 16193–16197.

64. A. Y. Chan, I. B. Perry, N. B. Bissonnette, B. F. Buksh, G. A. Edwards, L. I. Frye, O. L. Garry, M. N. Lavagnino, B. X. Li, Y. Liang, E. Mao, A. Millet, J. V. Oakley, N. L. Reed, H. A. Sakai, C. P. Seath, D. W. C. MacMillan, *Chem. Rev.* **2022**, *122*, 1485–1542.

65. L.-C. Campeau, N. Hazari, *Organometallics* **2019**, *38*, 3–35.

66. T. Bortolato, S. Cuadros, G. Simionato, L. Dell'Amico, *Chem. Commun.* **2022**, *58*, 1263–1283.

67. L. Capaldo, D. Ravelli, M. Fagnoni, *Chem. Rev.* **2022**, *122*, 1875–1924.

68. H. Cao, X. Tang, H. Tang, Y. Yuan, J. Wu, *Chem. Catal.* **2021**, *1*, 523–598.

69. L. Capaldo, D. Ravelli, *Eur. J. Org. Chem.* **2017**, 2056–2071.

70. J. J. Douglas, K. P. Cole, C. R. J. Stephenson, *J. Org. Chem.* **2014**, *79*, 11631–11643.

71. E. B. Corcoran, M. T. Pirnot, S. Lin, S. D. Dreher, D. A. DiRocco, I. W. Davies, S. L. Buchwald, D. W. C. MacMillan, *Science* **2016**, *353*, 279–283.

72. A. W. Dombrowski, N. J. Gesmundo, A. L. Aguirre, K. A. Sarris, J. M. Young, A. R. Bogdan, M. C. Martin, S. Gedeon, Y. Wang, *ACS Med. Chem. Lett.* **2020**, *11*, 597–604.

73. M. J. Schnermann, L. E. Overman, *Angew. Chem. Int. Ed.* **2012**, *51*, 9576–9580.

74. W. Zhang, Z. Zhang, J.-C. Tang, J.-T. Che, H.-Y. Zhang, J.-H. Chen, Z. Yang, *J. Am. Chem. Soc.* **2020**, *142*, 19487–19492.

75. L. Furst, J. M. R. Narayanam, C. R. J. Stephenson, *Angew. Chem. Int. Ed.* **2011**, *50*, 9655–9659.

76. P. Zhang, Y. Li, Z. Yan, J. Gong, Z. Yang, *J. Org. Chem.* **2019**, *84*, 15958–15971.

Principles of Synthetic Planning

11

Total synthesis is widely considered the pinnacle and prime art of organic chemistry [1–4]. The analysis of the target molecule and its breakdown to simple, commercially available precursors is a formidable challenge that necessitates a high degree of skill, experience, and chemical intuition to realize an efficient and selective synthesis. The *retrosynthetic* or *antithetic* analysis constitutes a methodical approach to the deconstruction molecules of both low and high complexities, such as oseltamivir (**1**) and palytoxin (**2**) (see Fig. 11.1) [5–7].

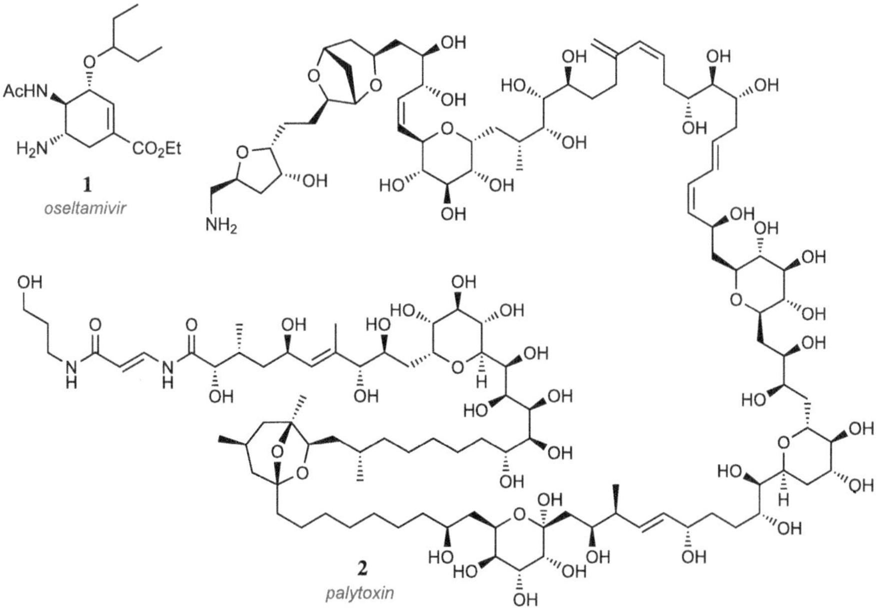

1
oseltamivir

2
palytoxin

Fig. 11.1 Structures of oseltamivir (**1**) and palytoxin (**2**)

A. Düfert, *Methods of Organic Synthesis*,
https://doi.org/10.1007/978-3-662-70963-4_11

Depending on the motivation for a synthetic endeavor, a synthesis plan needs to take these requirements into account: In natural product syntheses, the target structure, while often intricate and densely functionalized, remains clearly defined. Their high complexity requires a retrosynthetic disconnection that allows easily accessible alternative routes if a planned step fails to deliver the desired intermediate at any given point. In contrast, when exploring structure–activity relationship studies of bioactive compounds, a large variation in the corresponding analogs is required. It is thus imperative to incorporate a divergence point at an advanced intermediate that allows the desired late derivatizations without having to progress through too many stages to the diverse set of products. If access to a class of several natural products is the target, the introduction of a common key intermediate can be immensely advantageous to simplify the synthetic efforts and improve the efficiency of these divergent routes.

While the initial steps of a successful synthesis commence with the retrosynthesis of a target molecule or even a single functionality, the synthetic maneuvering does not end with that retrosynthesis. Routinely, corrections to the initially proposed synthetic route need to be incorporated [8]. The more menial aspects of a synthesis (reaction conditions, work-up, etc.) are typically not mentioned when discussing the general approach, yet any practitioner of the art should consider them in order to avoid additional corrections to the route along the way (see Fig. 11.2).

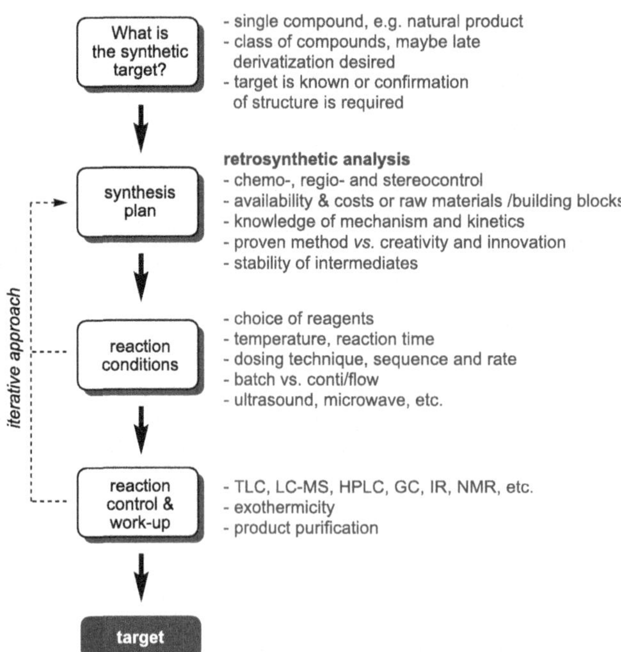

Fig. 11.2 Decision points in synthesis planning

11.1 **Retrosynthesis**

In synthetic chemistry, there is not **one** correct way to access a given target molecule. During the course of retrosynthetic analysis the various "fragments" are evaluated differently by each chemist. An approach heavily depends on a practitioner's familiarity with certain disconnections, be they routine or unique in nature, as well as the knowledge of a spectrum of synthetic methods and what repertoire of transformations has actually been previously used in the laboratory by them. The most elegant syntheses usually rely either on a methodology frequently utilized or even developed in the respective research group, on a biosynthesis postulated for this class of substances, or on known syntheses of structurally related molecules. Especially the first and last points necessitate a good grasp on the reactivity of all common functionalities and how they influence each other. Despite the ingenuity of introducing new ways to assemble a molecule, some structural elements lend themselves for retrosynthetic disconnection at very distinct positions of a target structure (see Fig. 11.3).

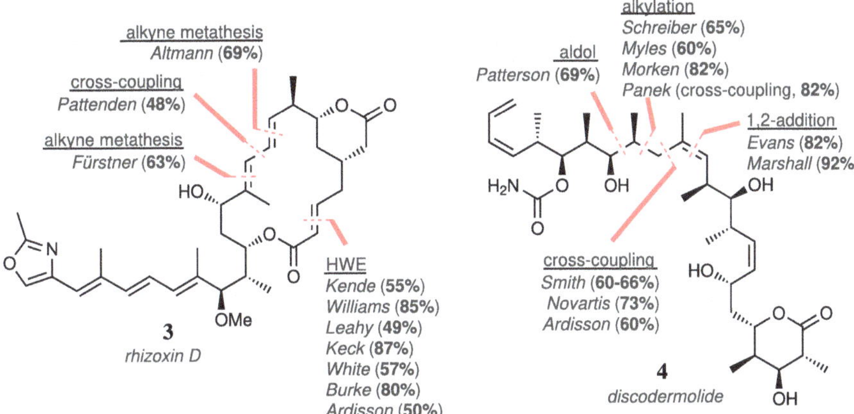

Fig. 11.3 Comparison of the central disconnections in the total syntheses of rhizoxin D (**3**) and discodermolide (**4**) [9–11]

In the various total syntheses of rhizoxin D (**3**), the macrocyclization was realized in the majority of cases by means of Horner–Wadsworth–Emmons olefination at the α,β-unsaturated ester site. In the synthetic approaches to discodermolide (**4**), the coupling of the central and the left-hand fragment took place at different bonds. However, the range of utilized methods did not vary widely once a position was selected for bond formation, as the structural patterns possess a sort of "natural" reactivity that is mirrored by the selected transformations [9–11].

Stereogenic elements, molecular size, sterically congested molecules, strained ring systems, fused structures, chemical reactivity, and unstable functionalities can pose considerable hurdles in a synthesis. The better one "understands" the corresponding parts of a molecule — this includes insights into possible biosynthetic pathways and the behavior of the molecule under various chemical conditions — the easier it becomes to circumvent undesired side reactions and predict critical points of the synthetic strategy. An expedient retrosynthetic analysis should therefore start with a detailed examination of all structural subtleties and functional groups.

An example of the particular lability of certain structural features can be encountered in the cytostatic drug dynemycin A (**5**, see Fig. 6.55, Chap. 6). The active pharmaceutical ingredient contains a reactive enediyne motif, which forms a DNA-alkylating benzene-1,4-diradical in a Bergman cyclization. Preliminary studies had shown that the enediyne is stabilized by the presence of the structurally rigid epoxide, thus preventing a premature cycloaromatization. However, an acidic environment leads to the opening of the epoxide in the presence of nucleophiles, requiring the strict omission of acidic conditions. Second of all, anthraquinone is very prone towards reduction, which can also trigger the cleavage of the epoxide in a domino sequence [12]. The introduction of the anthraquinone moiety should therefore occur as late as possible if the enediyne-epoxide functionality is already installed.

In the case of the insecticidal terpenoid azadirachtin (**6**, see Fig. 5.138, Sect. 5.3.1.2), the coupling of the central CC bond between the eastern and western fragments constituted the most formidable challenge of the sequence, the general instability of the molecule towards basic and acidic conditions or UV light notwithstanding. Despite positive indications on structurally simplified model substrates, all coupling attempts with the fully functionalized intermediates failed due to the sterically crowded environment of both substrates. Following an extensive screening of available coupling methods, it was found that an intramolecular, acetylenic Claisen rearrangement could ultimately provide the critical C8-C14 connection in a relay approach (see Fig. 11.4) [13].

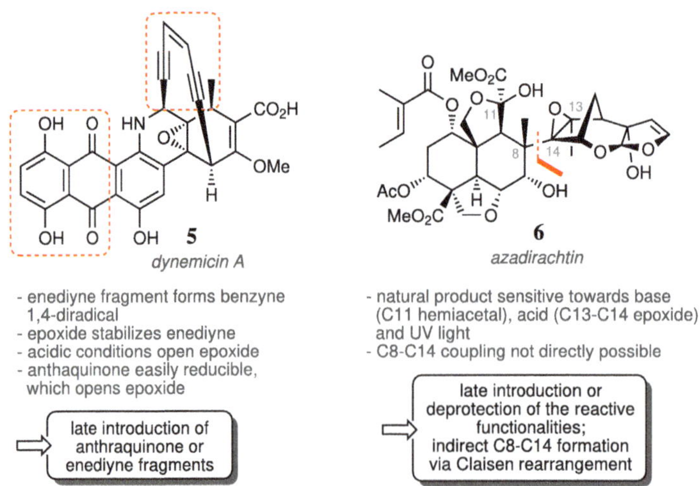

Fig. 11.4 Challenges in accessing dynemycin A and azadirachtin and derived synthesis strategies [12, 13]

For a disconnection, several options are usually available. The example shown in Fig. 11.5 is of course trivial, as 1,3-cyclohexadiene is commercially available. Nonetheless, this essentially illustrates that one can easily deviate from the obvious E_1 strategy through a different repertoire of transformations. If elements of further complexity are added to the target molecule, e.g., additional substituents or functional groups that are base-labile, the left path can no longer be pursued as easily. The most difficult tasks frequently do not lie in the assembly of large, complex molecular frameworks, but surprisingly often rather in the construction of basic building blocks. Thus, a simple target such as cyclohexadiene can quickly lose its triviality.

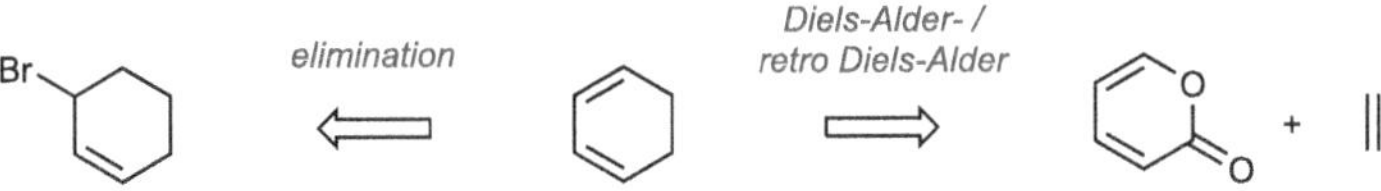

Fig. 11.5 Retrosynthetic analysis of 1,3-cyclohexadiene

A systematic approach of antithetic synthesis was in large part initiated and formalized by *E. J. Corey*, which was honored with the Nobel Prize in Chemistry in 1990 [14, 15]. The ideas introduced by Corey were supplemented and further developed in the subsequent years by the contributions of many researchers [16–22]. In general, there is no silver bullet for systematically reducing a target molecule to commercially available starting materials, as the number of feasible methods for such disconnections continuously increases over time. The formation of cycloalkenes or conjugated dienes, for example, has been fundamentally simplified by the introduction of alkene/alkyne metathesis and CC cross-coupling reactions (see Fig. 11.3). The "rules" of retrosynthesis should therefore only be viewed as indicative guidelines, yet often prove highly helpful in this formal approach [14].

While continuously improving antithetic analysis, several algorithms have been developed that allow a complete reduction of a molecule to possible precursors using computer programs (e.g., LHASA, Synthia/Chematica). There now also exist *machine learning* approaches that can be trained on the available datasets [23–28]. However, sometimes machine learning may simply capture literature trends as the data can be biased [29]. Computer-aided approaches have now become so advanced that a distinction between AI-devised and "human" synthetic routes is no longer obvious or sometimes possible, even for very complex natural products [30].

Regardless of the target structure to be synthesized, the following guidelines can followed, some of which will be discussed in more detail over the course of this chapter:

- Disconnect bonds between two functional groups
- Disconnect the molecule at central points
- Disconnect at branch points
- Disconnect side chains

- Use symmetry elements if possible
- Use a convergent rather than a linear strategy if possible
- Analyze oxidation states and possible functional group interconversions
- Exploit the natural reactivity of functional groups if possible
- Install most reactive functional groups last
- Consider compatibility of functional groups with regard to chemo-, regio-, and stereoselectivity

11.1.1 Nomenclature

The transformation of a molecule into a synthetic precursor is referred to as *transform* and accordingly represents the reverse of a synthetic reaction. The hypothetical fragments or *synthons* are thought of as idealized molecules or reagents and do not necessarily have to correspond to stable compounds. Synthons typically comprise anions or cations, which may also constitute intermediates occurring during the course of a synthetic reaction. A minimal keying element or *retron* is the substructural unit of functionalities required for enabling a transform. The retron of a Diels–Alder transformation, for example, is a six-membered ring containing a π bond. A reagent is the compound or intermediate actually used in place of the hypothetical synthon to carry out the synthetic operation and must exhibit the same reactivity as the associated retron. Synthetically equivalent reagents perform identical transformations (see Fig. 11.6) [15].

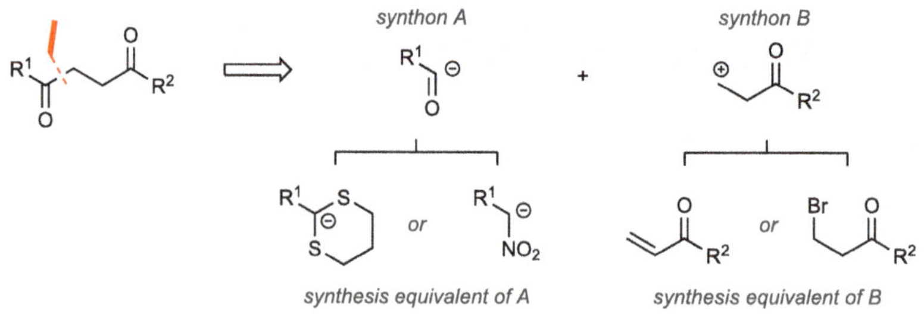

Fig. 11.6 Retrosynthetic disconnection of a 1,4-diketone

The bonds chosen for disconnection in a retrosynthetic step are designated as the bond-set for the intended synthesis. A nomenclature introduced by *Seebach* [31] considers the positions and polarity of the functional groups involved in a disconnection, and is most commonly employed with carbonyl-bearing substrates. The root functional group is assigned the position 1, the remaining atoms are assigned numbers along the carbon chain.

Upon disconnection, two fragments are obtained. The nucleophilic synthon is referred to as a **d**-fragment (*donor*), the electrophilic as an **a**-fragment (*acceptor*). For the substrate shown in Fig. 11.7, a scission between atoms 3 and 4 can generate either an a^3-fragment or a d^3-fragment (*i*). For the indicated disconnection *ii*, an a^2- or a d^2-synthon results.

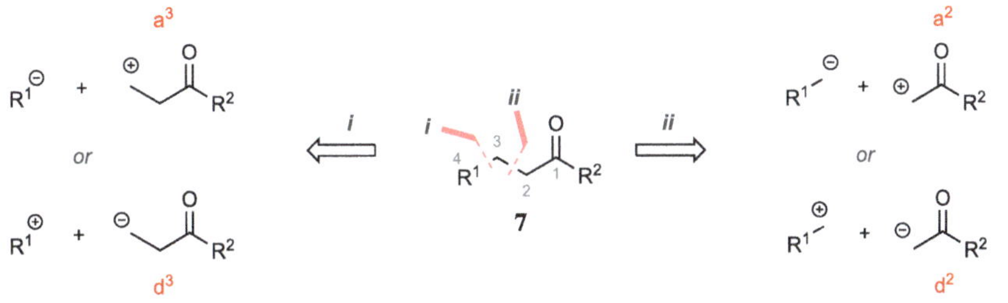

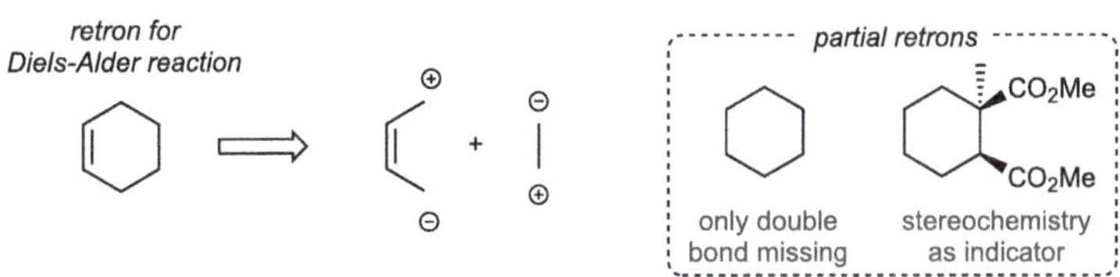

Fig. 11.7 Possible disconnections of **7** and application of the Seebach nomenclature

This nomenclature is extremely helpful in determining the feasible reagents or their synthetic equivalents, as the designation can be utilized to infer which synthons especially lend themselves for use in a given transform. *Even-numbered* donor synthons ($\mathbf{d^2}$, $\mathbf{d^4}$, etc.) and *odd-numbered* acceptor synthons ($\mathbf{a^1}$, $\mathbf{a^3}$, etc.) usually correspond to the natural polarity of a compound. For *odd-numbered* donor and *even-numbered* acceptor synthons, an *Umpolung* usually has to take place due to their unnatural polarity, as can be seen from the disconnection of a 1,4-diketone into a d^1- and an a^3-synthon shown in Fig. 11.6: While the electrophilic synthon B suggests a Michael system as substrate, one of the corresponding precursors designated by synthon A is a reagent with reversed polarity (dithiane anion).

In a *partial* retron, not all prerequisites of a certain transform are fulfilled. The retron of a Diels–Alder reaction, for example, is a cyclohexene; a cyclohexane would be a partial retron due to its lack of a π-bond. Additional key elements can also help to identify partial retrons. These include functional groups, stereocenters or ring structures, which are typically associated with a certain reaction (see Fig. 11.8).

Fig. 11.8 Diels–Alder retron and partial retrons

The major goal of any transform is to simplify the complex molecular structure of a target. However, non-simplifying steps can also be applied if they help to convert a partial retron into a complete retron, thus enabling a strongly simplifying transformation. These include C-C coupling reactions or a rearrangement of the carbon skeleton, a conversion or introduction of functional groups, and a configurational change of stereocenters, either by inversion or stereotransfer. In Fig. 11.9, both the introduction and conversion of functional groups as well as the inversion of a stereocenter can be considered as non-simplifying steps. However, they transform **8** into the retron **10** and enable the strongly simplifying cyclization to **13**.

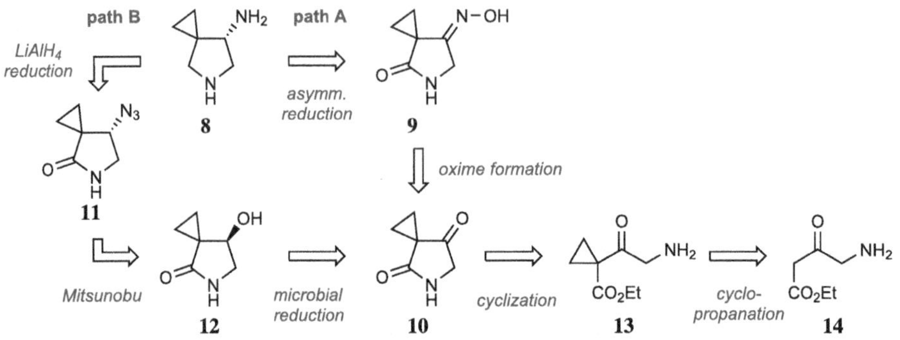

Fig. 11.9 Retrosynthetic analysis of **8** [28, 32]

11.1.2 Classic Approaches

Since many possible paths can be generated by disconnections at different sites, the antithetic analysis typically produces a retrosynthesis tree, with each intermediate potentially being the target molecule of further disconnection. The skill of an experienced synthetic chemist is then expressed in excluding branches of the tree from the outset, thus reducing the retrosynthesis to a few strategies that should be as flexible as possible for subsequent adaptation. If one considers, for example, a relatively simple aniline derivative, the classic (and probably also in most cases the most efficient and cost-effective) route is a straightforward reduction of a nitroarene precursor, which can be obtained by nitration. Alternatively, a Buchwald–Hartwig coupling of the respective aryl bromide can be pursued. Since a direct CN coupling with ammonia can still pose a challenge depending on the substrate, a potential ammonia equivalent is required, which must subsequently be deprotected or unmasked. In an academic context, the second variant may have its charm. For an industrial large-scale synthesis, however, the aryl halide building block will rarely be used as a starting material. With further application of the antithetic analysis, the aryl halide can be accessed by an electrophilic aromatic bromination, direct deprotonation, and subsequent reaction with Br_2 as well as a Sandmeyer reaction of the respective diazo compound. However, the corresponding aniline is needed for the Sandmeyer reaction to obtain the diazo compound, so the last option requires the final target molecule as starting material, which can be thus excluded from the retrosynthesis tree. The problem is thus reduced to questions of regioselectivity and functional group tolerance during nitration, bromination, or deprotonation (see Fig. 11.10). Especially with heterocycles, there are now good and very simple approaches to predict the regioselectivity for these transformations, even with unknown structures [33].

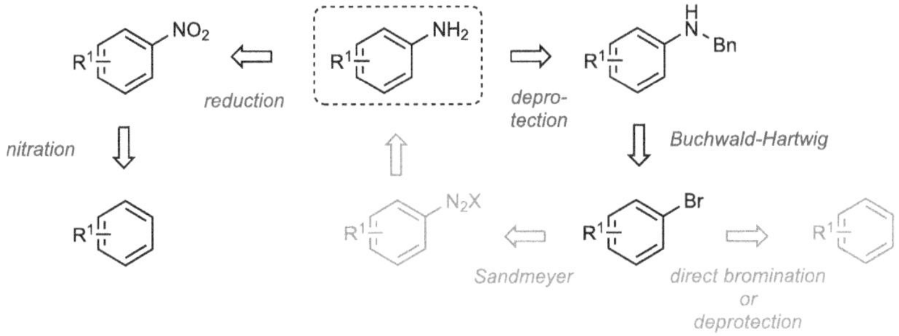

Fig. 11.10 Retrosynthetic analysis of a substituted aniline

The above example illustrates the various, classic approaches to accessing certain target structures. Either the molecular skeleton can dictate where plausible disconnections are to be placed—such as the introduction of the aryl ring as a building block without building it up *de novo*. In addition, the type and arrangement of functional groups, certain key reactions, or the stereochemical elements of the target structure can determine a retrosynthetic strategy. They are accordingly referred to as

- structure-goal (*S-goal*),
- functional group-based (*FG*),
- transform-goal (*T-goal*), and
- stereotopological strategies.

A complete retrosynthesis typically combines several strategies at different stages, whereby the fragments are assembled and subsequently combined. Both the identification of potential starting materials and the identification of subunits containing partial or complete retrons allow for a significant simplification of the synthetic problem. In this way, a combination of several strategies can be achieved or a bidirectional analysis can be carried out, in which both the retrosynthesis from the target molecule and a hypothetical forward synthesis from the starting materials can be performed.

Regardless of the selected strategy, a retrosynthetic disconnection always proceeds in the direction of energetically increasing synthons/reagents. The generation of more stable intermediates runs counter to the endergonic retrosynthesis direction and must be avoided at all costs. Thus, it is not uncommon to proceed via the intermediacy of strained ring systems if the subsequent forward steps exploit this energetic build-up to enable otherwise unfeasible transformations. The use of medium ring systems, however, usually does not afford the desired product in a forward sense, barring only few exceptions.

An example of a combined S-goal/T-goal strategy can be found in Shair's synthesis of longithorone A (**15**). Based on the cyclohexene retrons, a successive Diels–Alder strategy (transannular and intermolecular Diels–Alder reaction of **16** and **17**) was pursued, whose precursors **18** and **19** can be assembled via an enyne metathesis. The disconnection of the natural product to **16** and **17** was derived from a hypothetical biosynthesis, which was postulated already several years prior [34, 35]. The convergent–divergent approach allowed the diene **17** and the dienophile **16** to be disconnected to a common starting material **20**. The synthetic approach therefore also ingeniously exploited a hidden symmetry element (see Fig. 11.11) [36].

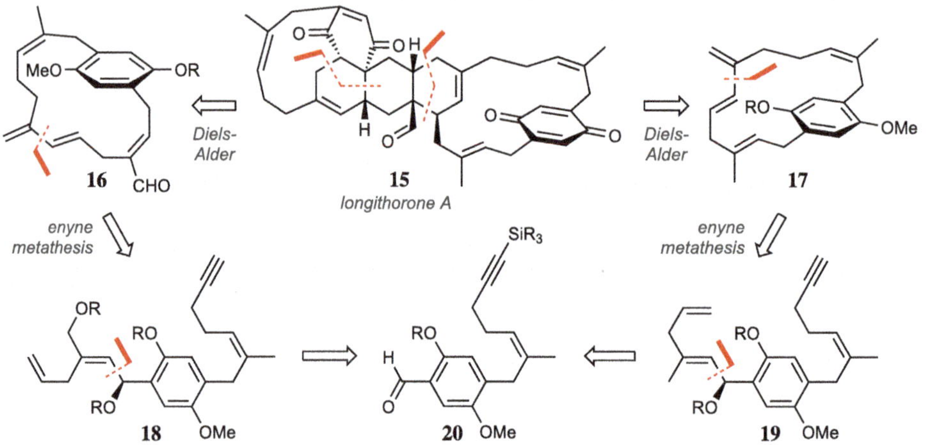

Fig. 11.11 Retrosynthesis of longithorone A (**15**) according to Shair *et al.* [36]

In a **structure-goal** strategy, the bonds of the molecular skeleton are analyzed and classified according to two categories if possible: *strategic* bonds, whose disconnection brings a high degree of simplification, and *preserved* bonds, whose preservation is desirable, e.g., because they are located in a part of the molecule that can be derived from a building block as a whole. The majority of the CC bonds of a molecule will remain retrosynthetically ambiguous and fall into neither of these categories unfortunately. A case-by-case evaluation is therefore required to assess whether the molecule should be disconnected at a certain bond. C-heteroatom bonds are, with few exceptions, always considered strategic. Aromatic rings were categorized as static by the original guidelines proposed by Corey, suggesting their introduction from a precursor as a whole. However, some very efficient methods have been developed for constructing even densely substituted arenes *de novo* by transition metal catalysis, especially fused systems and heteroarenes. An approach that also falls into the S-goal category is the use of naturally occurring and cost-effective chiral building blocks from the chiral pool (*vide infra*). These precursors are also referred to as *chirons* (chiral synthons), a term first introduced by Hanessian [37, 38].

Acyclic structures are commonly accessible via a FG- or transform-based strategy. In the case of a (hetero-)cyclic core structure, acyclic side chains are retrosynthetically disconnected near the branching point and usually attached to the cyclic moiety over the course of the synthesis as one fragment. Regarding fused and bridged carbo- and heterocyclic ring systems on the other hand, guidelines have been disclosed on how they can be most efficiently build up.

For simple carbo- and heterocycles, Baldwin's rules can be consulted, relying on both thermodynamic and kinetic aspects of ring closure (see Sect. 1.1). Kinetic factors are primarily understood as the spatial proximity of the reacting termini, while the thermodynamic driving force is usually based on the extent of transaxial and angular strain present in the resulting carbo- or heterocycles (see Table 11.1) [39].

Table 11.1 Overview of kinetic and thermodynamic factors favoring (+) or disfavoring (−) cyclization reactions [39, 40]

ring size	kinetic	thermodynamic
3	+++	-
4	-	-
5	++	+
6	+	++
7	+	+
8+	-	+

In medium and large rings, the two ends are located so far apart that a folding of the molecule is required before the ring closure can take place in the conformation leading to the desired product. Cyclization in these systems is therefore significantly retarded despite the thermodynamic stability of the corresponding products. For this reason, the intermediacy of medium rings during a synthesis is strongly discouraged unless they are part of the target structure.

Especially bridged, polycyclic systems yield a quick simplification due to their high topological complexity if the right strategic bonds are chosen for retrosynthetic disconnection. An analysis of the ring system allows the identification of the ring with *maximum bridging*. The maximum bridging rings of a molecule are bridged at the greatest number of different sites/atoms. Bridging sites are defined as the atoms from which a linked ring with more than one shared bond departs, both ends of which also lie within a considered carbo- or heterocycle (see Fig. 11.12) [41].

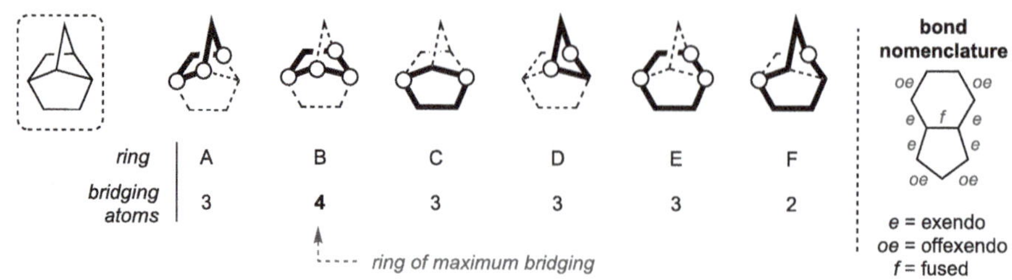

Fig. 11.12 Primary rings of a bridged tricyclononane. Bridging atoms are indicated with ○ [41]

The positions indicated with *e* are *exendo* bonds, which are located *endo* to/within one ring but *exo* to another. In polycyclic systems, in addition to the core bond where the molecules are fused (*f*), *offexendo* bonds (*oe*) may be encountered, which run off or from an exendo bond. Strategic bonds are located in the ring of maximum bridging and are exendo. An f-bond is considered a *core* bond and not strategic (even if it meets the two previous criteria) if its disconnection results in a ring containing more than seven members [41].

Looking at some synthetic targets such as the class of CP molecules, the conscious or unconscious implementation of these criteria can also be observed. Of the four total syntheses disclosed by the groups of Nicolaou, Shair, Fukuyama, and Danishefsky, three incorporated a disconnection in the ring of maximum bridging (see Fig. 11.13) [42–47].

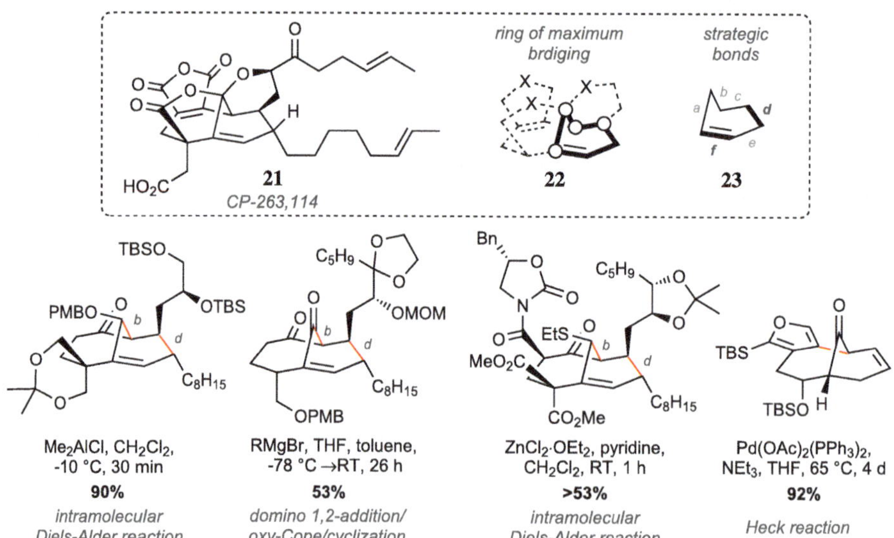

Fig. 11.13 (+)-CP-263,114 (**21**), ring of maximum bridging (**22**) and the strategic bonds in the ring (**23**). Approaches to assembling the core structure in the total syntheses by Nicolaou, Shair, Fukuyama, and Danishefsky [42–47]

Following the determination of the ring of maximum bridging, the bonds within that ring are classified according to the above guidelines. Bonds a, b, c, d, and f are all located *exendo*. Of these, bonds a, b, and c are excluded for one-bond disconnection as they constitute core bonds which would give cyclic precursors with more than seven-membered rings.[I] In the depicted total syntheses, Nicolaou, Shair, and Fukuyama used bonds b and d to construct the molecular skeleton [42–45]. Only the approach of the Danishefsky group deviated in this regard and relied on a different strategy to synthesize the carbocyclic framework [46, 47]. These guidelines originally introduced by Corey can indicate which antithetic disconnection leads to the greatest simplification for a given cyclic structure. The intermediates of the synthesis of CP-263,114 shown in Fig. 11.13 have a simpler structure compared to that of the parent natural product, which is why their synthesis does not necessarily have to be in line with the analysis proceeding via **22** and **23**. It is nevertheless intriguing that a good match between the syntheses and the original analysis of the natural product exists, further corroborating the viability of the antithetic logic.

Comparable guidelines also exist for fused structures on how to most efficiently disconnect them. For a more in-depth study of the topic, the reader is referred to the original publications [14, 15].

If a transformation is not feasible for thermodynamic reasons, the desired conversion can often still be achieved by resorting to high-energy intermediates. In nature, a phosphorylation or the formation of thioesters is a proven method for realizing such intermediates, thereby enabling a process that normally is considered to be endergonic. In a synthetic context, an identical effect is accomplished through the formation of highly strained intermediates [18]. Typical examples involve the cleavage of cyclopropane or cyclobutane moieties to enable energetically unfavorable reaction paths by exploiting the loss of ring strain as a driving force [48]. Seven- and eight-membered rings, which represent challenging motifs from both thermodynamic and kinetic perspectives, are routinely assembled via the intermediacy of small rings (see Fig. 11.14; cf. Table 11.1).

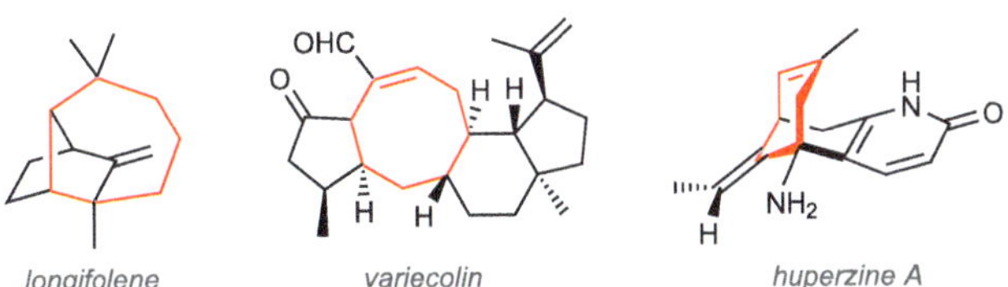

Fig. 11.14 Examples of natural products that were synthesized using high-energy cyclobutane intermediates [49–51]

[I] Bond-pair disconnections involving core bonds are strategic if both rings are assembled in one step, e.g., via a cycloaddition [15].

An impressive example of inverse complexity planning was reported in Snapper's synthesis of pleocarpene (**24**). The construction of the 3/4/4/5-fused tetracyclic skeleton in **26** was accomplished by an intramolecular [2+2]-cycloaddition of the stabilized butadiene **27**, followed by cyclopropanation under Cu(acac)$_2$ catalysis. The advanced key intermediate **26** subsequently rearranged to unveil the 5/7-fused carbon framework at 200 °C in the presence of catalytic amounts of DBU. Modification of the ring periphery and saturation of the cycloheptadiene afforded the desired secondary metabolite **24** (see Fig. 11.15) [52].

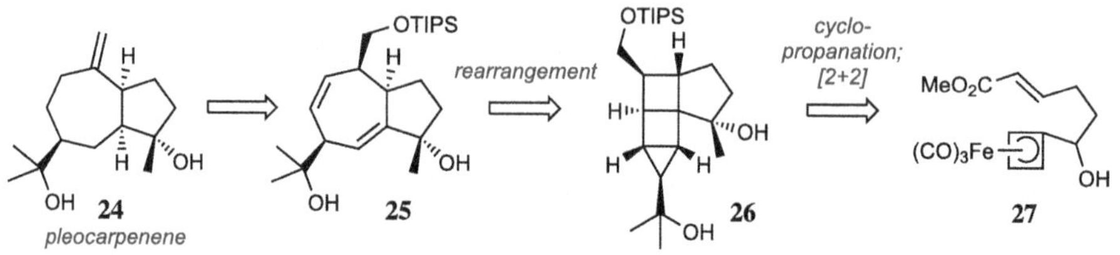

Fig. 11.15 Retrosynthesis of pleocarpene (**24**) according to Snapper *et al.* [52]

While the size and complexity of the carbon skeleton have a profound impact on whether and how long a molecule can resist all attempts at total synthesis, the density of functionalities present plays an equally important and often superior role. The second approach, in which the type and arrangement of the functionalities in the molecule govern a corresponding disconnection, is referred to as **functional group-based** retrosynthesis (*FG* based). A functionality usually serves either as the target of a fragment itself (if it is contained in the product) or as a key element of a retron (e.g., the construction of a cyclohexene for a Diels–Alder reaction). Especially in highly functionalized molecules, the density and type of contained functional groups can dominate a synthesis route (see Fig. 11.16).

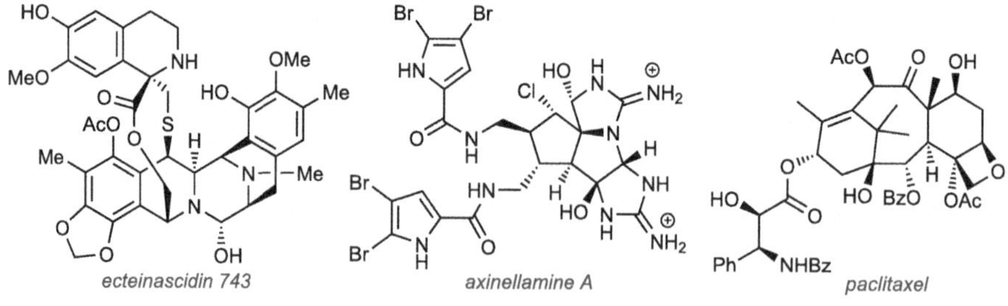

Fig. 11.16 Examples of highly functionalized natural products

An analysis of a molecule's functionalities and their oxidation state may directly provide an approach for their synthesis. Thus, (hemi-)ketals have the same oxidation state as ketones, while orthoesters and carboxylic acid derivatives can also be considered equivalent.

A functional group is referred to as *latent* or *masked* if its reactivity differs significantly from the target functionality, but it can be easily converted to the latter. In addition to the masking of carbonyl derivatives, alkynes and aldehydes/ketones can be considered masked alkenes due to a multitude of methods available for their interconversion (see Chap. 3 and 4) [53].

In Woodward's synthesis of vitamin B_{12}, **28** was one of the target fragments of the corrin ring. An analysis of the oxidation states in **28** traced the ketal group back to the acyclic ketone **29**. By (retrosynthetically) reconnecting the longest, most highly oxidized chain, the retron of a Diels–Alder reaction was obtained (see Fig. 11.17) [54].

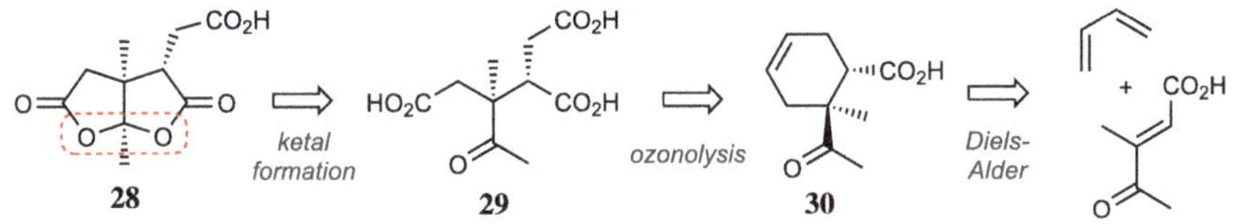

Fig. 11.17 Retrosynthesis of **28** by Woodward [54]

Since reliable methods for the diastereo- and enantioselective construction of acyclic molecules have only become widely available in the last decades of the twentieth century, previous syntheses relied on structurally rigid, cyclic compounds to introduce the desired stereocenters. The use of cyclic intermediates, which can be subsequently cleaved by ozonolysis, retroaldol, or pericyclic reactions, accordingly represents a common motif in total syntheses of the 1960s and 1970s.

In Storck's antithetic analysis of lycopodine (**31**), the retron for the intramolecular addition of an enolate to an imine was identified. The use of a direct enolate was not deigned feasible: the required precursor **32** contains an eight-membered ring, for the construction of which only a few reliable methods existed at the time of synthesis. Instead, Storck opted for the use of a methoxyphenyl group as a masked enolate, which could be cleaved following the cyclization by a Birch reduction and ensuing ozonolysis (see Fig. 11.18) [55].

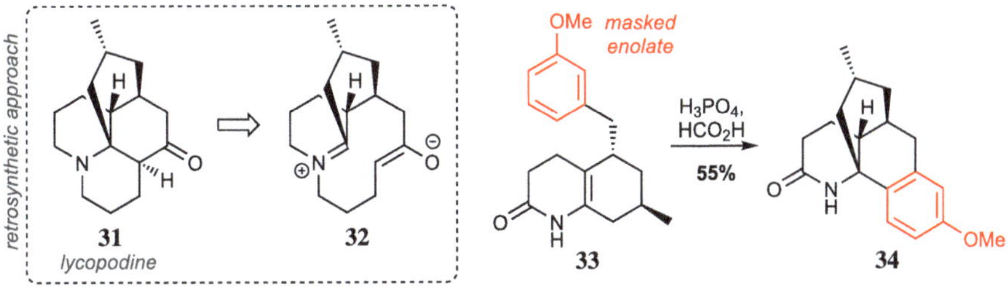

Fig. 11.18 Synthesis of lycopodine by Storck [55]

Just as the presence of stereocenters in the target compound demands a high degree of stereoselectivity, the interconversion and introduction of functional groups requires excellent chemoselectivity. If an incompatibility of certain functional groups with the desired reaction conditions is expected, suitable *protecting groups* are used. Nevertheless, the chemoselectivity gained through the use of protecting groups is purchased at the expense of a series of non-productive reactions (protection, deprotection). When planning a synthesis, the functionalities designated for protection should be identified up front, which reaction conditions these protecting groups must withstand, and when which protecting group should be removed. Especially an *orthogonal* protection, i.e., the independent introduction and removal of different protecting groups, is often crucial [56].

The ideal of a protecting group-free synthesis of complex target compounds is not new, but rather has been a long-existing goal that could thus far only be realized in selected synthetic endeavors [57–61]. The key to success is the introduction of sensitive functionalities after all reactions, against which the functional group might be labile, have taken place.

Protecting groups are roughly divided into three categories, whose reactivities should be orthogonal to each other.

- *short-term protection*, which blocks a functionality over one or a maximum of two steps;
- *mid-term protection*, which is removed during the course of the synthesis; and
- *long-term protection*, which is typically cleaved at the end of a synthesis.

The majority of functionalities to be protected are oxygen based, with hydroxy functions possibly representing the most common class of protected functionalities in natural product syntheses due to their ubiquity. Since the introduction of silyl-based protecting groups, most long-term protecting groups are derived from this class of functionalities. Silyl ethers can even be selectively deprotected in the presence of other silyl ethers by judicious choice of the reaction conditions [62]. TMS ethers are very sensitive to protonolysis, while TIPS ethers often even require forcing conditions for deprotection. Ketals are mainly employed for masking carbonyl groups, while amines can be protected as the corresponding carbamates. A selection of the most common protecting groups is shown in Fig. 11.19 [63, 64].

If protecting groups cannot be avoided, the requisite number of unproductive steps for introduction and removal should at least be minimized. Especially towards the end of a synthesis, it is most efficient to choose a convergent protecting group strategy, so that all blocked functionalities can ideally be unmasked in one step. Two representative examples can be found in the syntheses of streptide and chivosazole F. The fully protected peptide-based macrolactam streptide was deprotected by Boger and co-workers in the final step. In addition to the protonolysis of the Boc and *t*butyl ester groups, an amide and a C-Si bond were concomitantly cleaved. Paterson's approach to the complex polyene macrolide chivosazole F mainly relied on silyl-based protecting groups for the secondary hydroxy groups. In the concluding sequence, a variant of the Wittig olefination was first used, followed by a Stille cross-coupling to effect the macrocyclization and the global deprotection of the

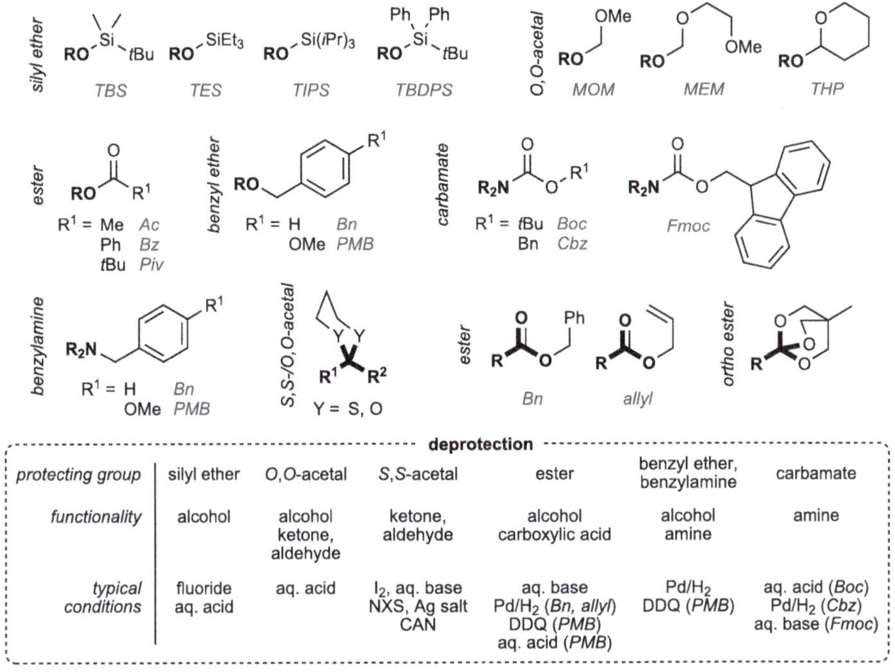

Fig. 11.19 Standard protecting groups and typical deprotection conditions [63, 64]

three different silyl ethers (see Fig. 11.20) [65, 66]. An outstanding example can be found in Kishi's synthesis of palytoxin (**2**), in which a total of 42 (!) protecting groups of eight different types was removed at the end of the synthesis [5, 6].

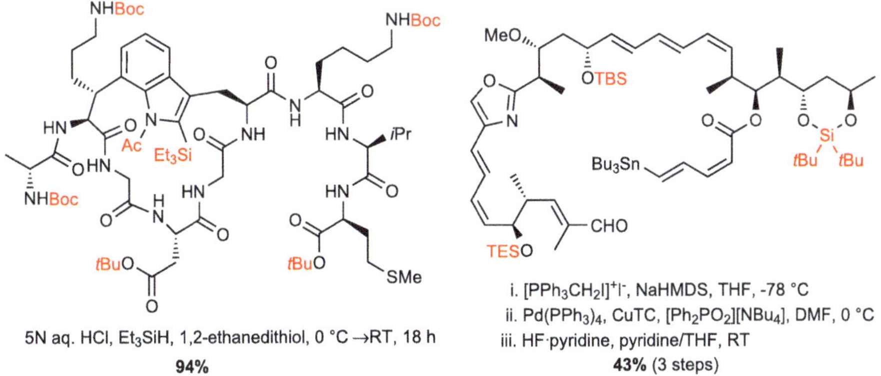

Fig. 11.20 Global deprotection in the syntheses of streptide and chivosazole F [65, 66]. The affected protecting groups in the substrates are highlighted

The use of certain, highly efficient reactions can enable an easy access to specific structural motifs. A large part of modern total syntheses follows a **transform-based** approach in addition to a structure-goal strategy. A key step usually encompasses the introduction of a considerable part of the functionalities or the assembly of the molecular skeleton in a highly efficient manner. Such syntheses often rely on methodologies that have been developed or frequently used in the respective research groups.

The synthesis of the cytotoxic secondary metabolide (–)-pseudolaric acid B, which also exhibits antifungal and fertility-inhibiting activity, falls into this category. Trost and co-workers accomplished the total synthesis via a key ruthenium-catalyzed [5+2]-cycloaddition [67], which had been developed and tested on several structures of variable complexity by the Trost group prior to the natural product synthesis [68–70]. After the retrosynthetic disconnection of the side chain and a radical cyclization, a 5/7-fused bicyclic precursor is obtained. The 1,4-diene **37** is received after adjustment of the oxidation state and isomerization of the double bonds, which can be accessed by a [5+2]-cycloaddition (see Fig. 11.21).

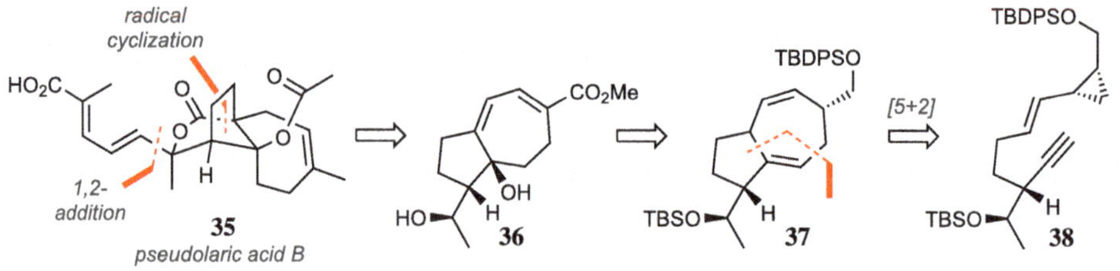

Fig. 11.21 Retrosynthesis of (–)-pseudolaric acid B (**35**) by Trost and co-workers [67]

There are numerous examples of such transform-based approaches in the scientific literature, which can easily be attributed to the fact that a research group frequently relies on a methodology developed in-house. The major challenge of such retrosyntheses is thus reduced to the recognition of structural features that can be assembled by the transformation in question. Often only partial retrons are present, leading to the introduction of the corresponding full retrons as a sub-strategy.

The Wood group exploited the inherent symmetry of the product in addition to a transform-based strategy in their synthesis of the bacchopetiolone core (**42**). The construction of the dimeric structure was achieved via the intermediacy of the monomer units **40**. The dimerization proceeded in an oxidation/Diels–Alder sequence starting from **39**. Unfortunately, all efforts at the concluding decarbonylation to afford the natural product proved fruitless, which nonetheless does not detract from the high convergence and elegance of the overall synthetic sequence (see Fig. 11.22) [71].

These rather complex examples illustrate that without a good understanding of the corresponding reaction mechanisms as well as thermodynamic and kinetic principles, the prospect of successfully completing a transform-based strategy is greatly diminished. This

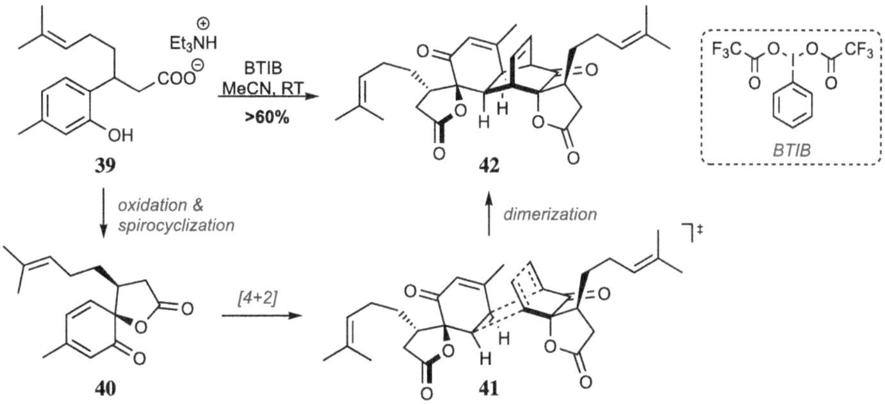

Fig. 11.22 Access to the bacchopetiolone core by Wood *et al.* [71]

fact explains why key steps almost exclusively comprise reactions whose functional group scope, robustness, and limitations are sufficiently known. These may either be reactions that have been developed in the respective research groups or which have already been applied to a large number of substrates. Prominent examples include transformations such as the Diels–Alder reaction, cross-coupling, or macrolactonization reactions among many others.

11.1.3 Control of Stereochemistry

Stereogenic elements continue to occupy a pivotal role in a retrosynthetic analysis, despite the constantly increasing number of available asymmetric methods for their selective introduction. While their presence is more the rule than the exception in natural products, a considerable number of small molecule active pharmaceutical ingredients are being introduced that do not contain stereogenic elements. Even though data on commercial synthetic routes is only sporadically available, the approaches for the introduction of said stereoinformation into the active drug have been evaluated for the period up to 2003 (see Fig. 11.23) [72].

Asymmetric methods are constantly gaining traction, yet for a large part of the active ingredients, the stereoinformation still originates from naturally abundant chiral building blocks (*chiral pool*) or from a racemic transformation followed by separation of the enantiomers (*resolution*) [73].

The surprisingly low proportion of asymmetric methods in Fig. 11.23 can be explained by two circumstances. First of all, the development of commercial manufacturing processes requires up to a decade, thus exhibiting a certain delay compared to the current state of the art. The period illustrated above thus only covers methods that were available until the end of the 1990s. Second of all, only a certain proportion of the developed methods

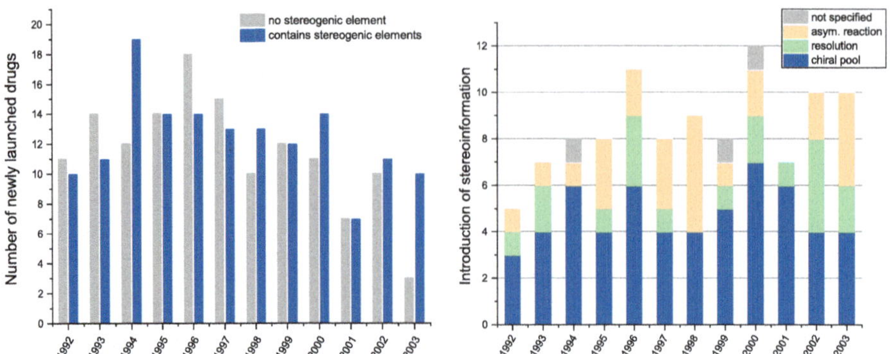

Fig. 11.23 Proportion of newly approved active pharmaceutical ingredients without and with stereocenters as well as methods for their introduction [72]

are amenable for an industrial application for various reasons (scalability of the method, availability of catalysts/reagents on a large scale, costs, safety aspects, etc.). Nevertheless, some methods have established themselves as belonging to the standard repertoire routinely used for the production of active ingredients, such as asymmetric hydrogenations, Sharpless epoxidations and dihydroxylations, asymmetric phase transfer alkylations, and enzyme-catalyzed reactions, just to name a few (see Sect. 11.2) [72]. In an analysis of the source of chirality of FDA-approved drugs in 2016–2020, approximately half relied on the chiral pool, with a chiral resolution constituting the second most popular approach (30%). Chiral catalysts, reagents, or auxiliaries made up the remaining 20% [74].

Upon closer inspection, the planning of a stereotopological strategy is largely governed by the recognition of retrons for a desired stereoinduction and the search for methods or precursors capable of enabling it. The constant development of **enantioselective reactions** and their ever-increasing abundance in an academic setting has been sufficiently demonstrated within the scope of this book. They probably represent the most efficient access from a pure feasibility perspective, as long as the enantioinduction is good and the yield is significantly above 50%. The use of chiral auxiliaries or reagents has been declining sharply, as they are needed in stoichiometric amounts and, in the case of auxiliaries, require unproductive steps for their introduction and ensuing removal following the desired diastereoselective transformation.

In enantio- or diastereoselective reactions, a differentiation of the possible products can only occur through *diastereomorphic* transition states. Depending on the substrates and reaction conditions, transformations can either be *substrate-controlled* or *reagent-controlled*. Given the presence of stereogenic elements in the substrate and the use of chiral reagents/catalysts, they can either synergistically favor the same product or antagonistically steer towards different enantio- or diastereomers (*matched* or *mismatched* situation) [75, 76]. Despite the occurrence of a mismatched situation, a catalyst-induced selectivity can override the inherent substrate selectivity, thus resulting in an improved diastereoinduction than with the sole use of the chirality of the starting material (see Fig. 11.24).

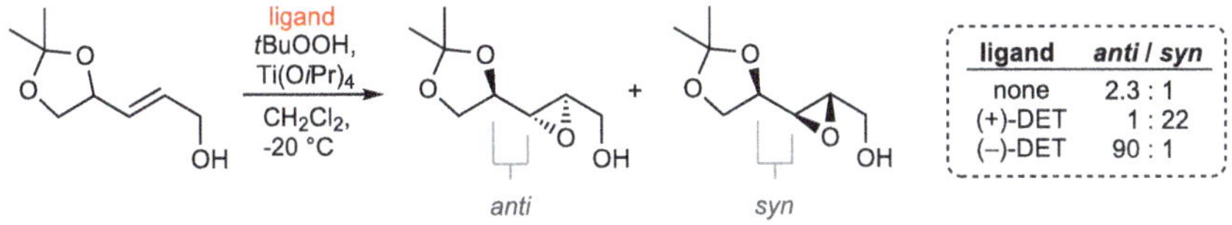

Fig. 11.24 Double stereoinduction in the asymmetric epoxidation of an allylic alcohol [75]

In the absence of a ligand, the *anti*-configured product is favored. The (+)-diethyl tartrate completely overrides the substrate's innate selectivity in the mismatched case, leading to a significantly enhanced product ratio than without a ligand, although the opposite diastereomer is the major product. In the matched situation, an excellent selectivity of 90 : 1 is afforded [75].

A substrate-dependent induction can either proceed *actively* through coordination of the reagent/catalyst or *passively* through steric repulsion. The guiding principles during facial discrimination for active or passive induction by a chiral starting material are usually based on stereoelectronic effects such as Felkin–Anh control, 1,2-/1,3-allylic strain, the anomeric effect, or a conformational control in the case of cyclic substrates (see Sect. 1.1).

In Evans' synthesis of rutamycin B (**45**), the double anomerically-stabilized spiroketal **44** constituted a key advanced intermediate. The acyclic precursor **43** afforded the cyclization product in good yield in a hydrolysis/ketalization sequence in the presence of HF in acetonitrile/water. Formation of the two singly anomerically stabilized side products was not observed under the employed conditions (see Fig. 11.25) [77, 78].

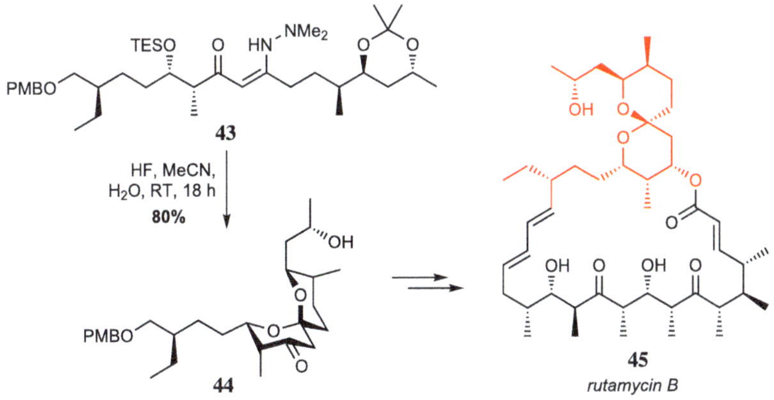

Fig. 11.25 Synthesis of the spiroketal key fragment of rutamycin B (**45**) by Evans *et al.* [77]

The use of the **chiral pool** as source of stereoinformation combined with a subsequent diastereomeric resolution remains an important tactic that is somewhat underrepresented in academic endeavors. Especially the pharmaceutical industry resorts to amino acids, amino alcohols, or sugars as starting materials for the synthesis of bioactive ingredients (chiron approach, *vide supra*). The advantage of this approach is evident: The molecules are available as sources of stereogenic centers in enantiomerically pure form, they are usually very inexpensive and accessible in large quantities. The requisite increased effort for their functionalization has reduced their occurrence in academic syntheses. If they are encountered, however, their use is typically highly efficient. A selection of the most common representatives of the chiral pool is depicted in Fig. 11.26: L-amino acids, α-hydroxy acids, amino alcohols as well as various carbohydrates and fatty acids. Glycidol, epichlorohydrin, cyclohexanediamine, and aminoindanol are not derived from natural sources. However, asymmetric methods allow easy access on a ton scale which is why they are also routinely employed as a source of stereogenic centers and considered members of the "expanded" chiral pool (see Fig. 11.26) [79–83].

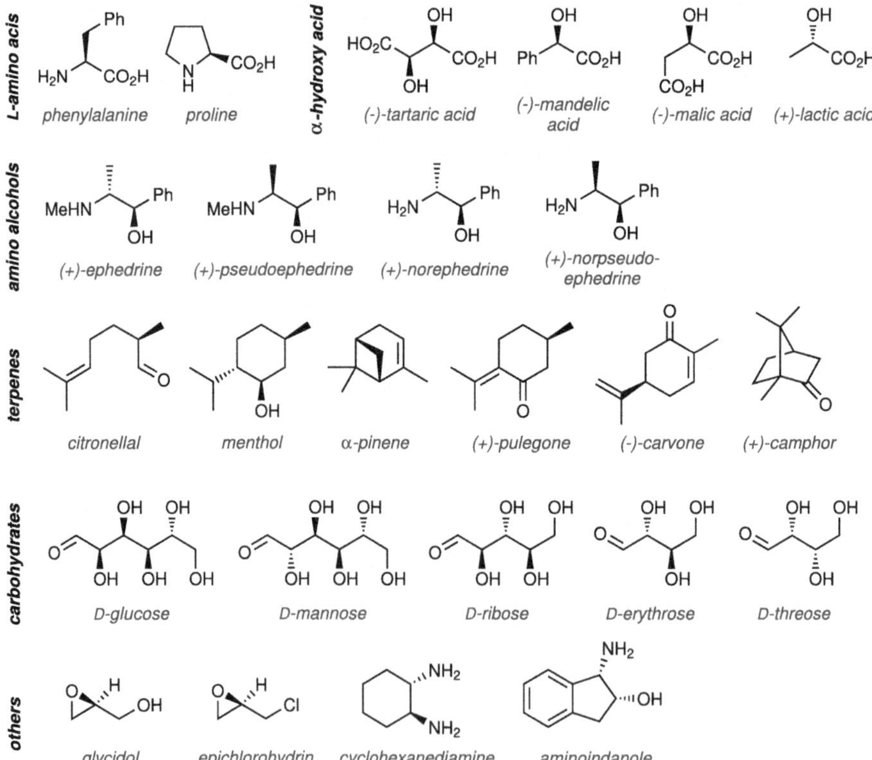

Fig. 11.26 Selection of molecules from the expanded chiral pool [79–83]

In the synthesis of platensimycin (**48**) by the Nicolaou group, the stereocenter of (–)-carvone (**51**) was exploited as a handle to introduce the further stereogenic elements of the bridged tetracylic moiety. Only a few functionalization steps of the terpene were required and it was seamlessly integrated into the molecule framework with high efficiency (see Fig. 11.27). Compound **49** serves as a key intermediate for the access to **48** also in other syntheses of the natural product (see Fig. 11.27) [84].

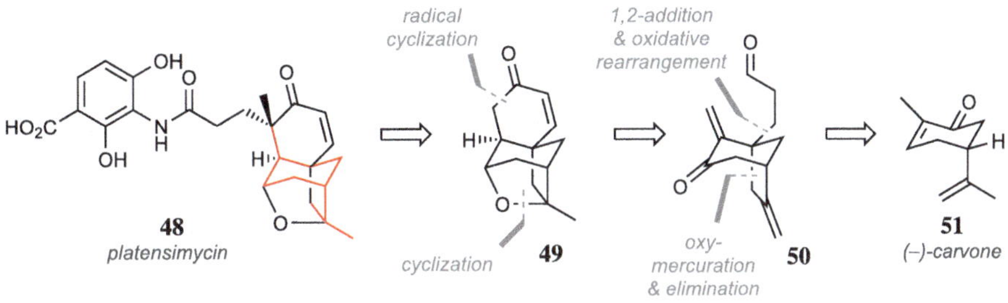

Fig. 11.27 Retrosynthesis of platensimycin (**48**) according to Nicolaou *et al.* [84]

Another illustrative example is provided by the synthesis of the unnatural enantiomer of 6-*epi*-ophioboline N (**52**). Maimone and co-workers traced the 5/8/5-fused framework of the sesterpene back to linalool (**53**) and farnesol (**54**). Following the coupling of both building blocks, a reductive, radical cyclization constituted the key reaction, which was based on the postulated biosynthesis of the secondary metabolite. Subsequent modification of functionalities finally yielded the natural substance in only nine steps in the longest linear sequence. The chiral pool starting materials thus not only served as source of the target molecule's stereoinformation but concomitantly supplied a major proportion of the carbon skeleton (see Fig. 11.28) [85].

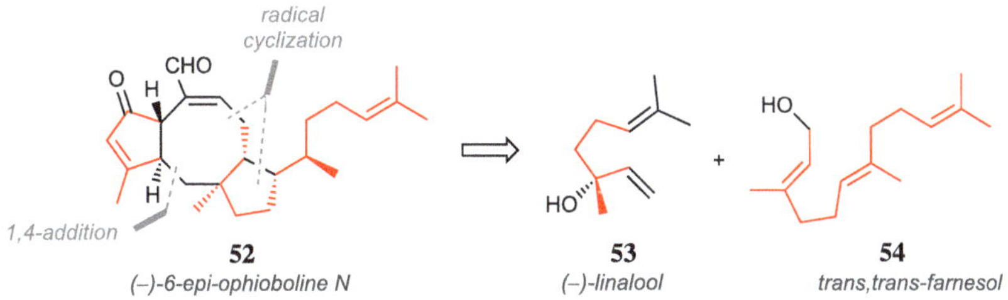

Fig. 11.28 Maimone's retrosynthesis of (–)-6-*epi*-ophioboline N (**52**) [85]

The **resolution** of an enantiomeric or diastereomeric mixture is rarely encountered in academic syntheses, but is a common practice in an industrial synthetic environment [86, 87]. In "unreceptive" substrates, the subsequent separation of the enantiomers or diastereomers might even prove advantageous over an enantioselective reaction: As soon as the asymmetric variant proceeds with a yield of 50% or lower while the racemic route retains high yields, a resolution of the mixture can afford the desired stereoisomere in a comparable or even higher yield. Furthermore, additional steps required to build up the substrates for the asymmetric reaction are avoided. Often the reagents for an enantioselective variant are also more expensive as a consequence of their more laborious preparation.

The principle of resolution can be traced back to Louis Pasteur's manual racemate resolution of tartaric acid [88]. More than a hundred years later, the field of stereoisomer resolution has developed into a powerful method for the separation of both enantiomers or diastereomers [89]. It is based either on the formation of diastereomeric salts/complexes with subsequent crystallization of one isomer or a selective reaction of only one stereoisomer (*kinetic* resolution).

The formation of diastereomeric salts constitutes the simplest approach. The separation is based on the different solubility of the diastereomers (see Fig. 11.29). If conditions for a selective crystallization are known, this method can easily be scaled up to range from multi-kilogram to multi-ton batches and is very efficient [90, 91].

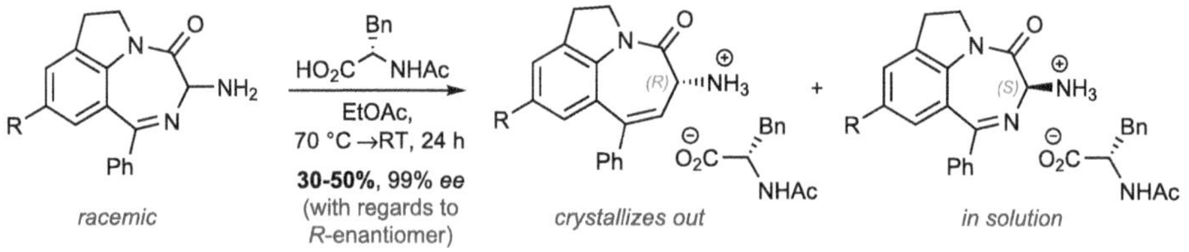

Fig. 11.29 Resolution of racemic benzodiazepine derivatives by formation of diastereomeric salts [92]

The counterions usually originate from the chiral pool: As anions, derivatives of tartaric acid are often employed [93–95]. Other acids, mostly α-hydroxy acids, are also routinely used in co-crystal formation [96, 97]. For chiral cations, amines such as α-phenylethylamine, naturally occurring amino alcohols of the ephedrine family, amino alcohols derived from amino acids, or the pseudoenantiomers quinine and quinidine are utilized, whose derivatives also serve as ligands in asymmetric catalysis (see Fig. 11.30) [98].

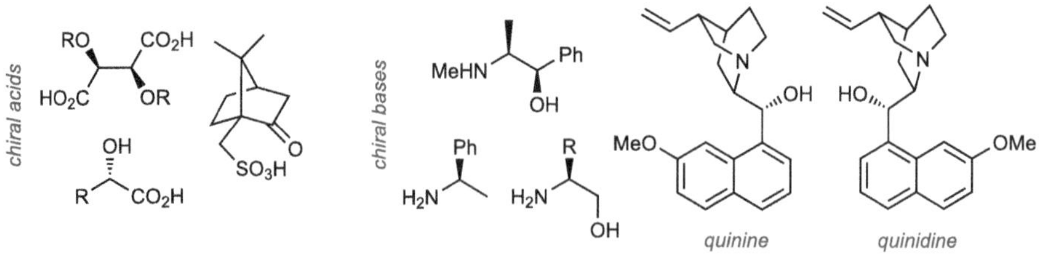

Fig. 11.30 Common acids and bases for diastereomeric salt resolutions [98]

The *kinetic* resolution is based on the discrimination of both enantiomers due to their different reaction rates in a chiral environment. The quality of a separation is determined by the influence that the stereogenic center can exert on the rate of the reaction. Ideally only one enantiomer should react, while the other remains unaffected (see Fig. 11.31) [99].

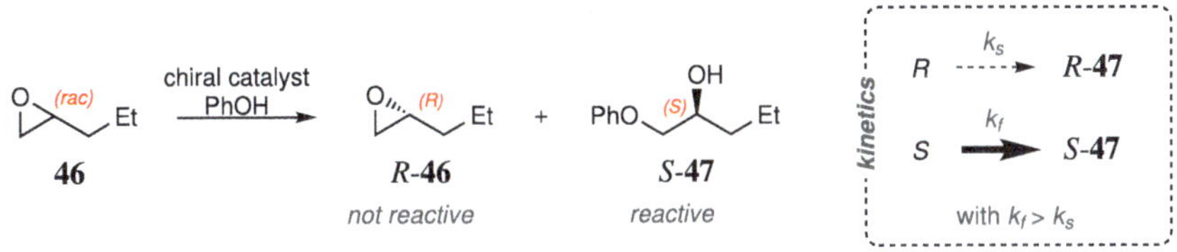

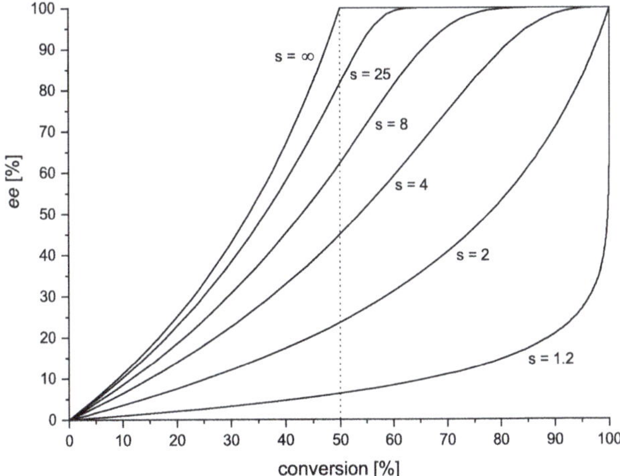

Fig. 11.31 Principle of kinetic resolution [99]

The selectivity s, defined as the ratio of the two reaction rates k_f (faster reaction) and k_s (slower reaction), should therefore be as large as possible. In the limit case $k_f/k_s \to \infty$, 100% ee is achieved at a conversion of 50%.[II] As can be easily seen, however, ratios of $k_f/k_s = 25 : 1$ are already sufficient to closely approach the ideal scenario. The lower the selectivity s, the higher the required conversion to obtain the starting material in high enantiomeric excesses (see Fig. 11.32).

Fig. 11.32 Dependence of the ee values of the remaining substrate on selectivity s and the conversion [100, 101]

[II] The relationships of the ee values of the starting material (ee_{SM}) and product (ee_P), the conversion C, and the ratio of the rate constants can be expressed by the following equations [100, 101]:

$$s = \frac{k_f}{k_s} = \frac{ln[(1 - C)(1 - ee_{SM})]}{ln[(1 - C)(1 + ee_{SM})]} \tag{11.1}$$

$$s = \frac{k_f}{k_s} = \frac{ln[1 - C(1 + ee_P)]}{ln[1 - C(1 - ee_P)]} \tag{11.2}$$

The acylation of secondary alcohols and amines can be considered a prototypical kinetic resolution, which is inspired by enzyme catalysis. Vinyl acetate typically comprises the acyl donor, as the enol received after transesterification tautomerizes to acetaldehyde and is thus removed from a potential equilibrium (backward) reaction as reactant. Acid anhydrides can also be used. Fu developed a kinetic resolution of secondary alcohols based on the activation of the acyl donor by a chiral DMAP derivative (see Fig. 11.33) [102, 103].

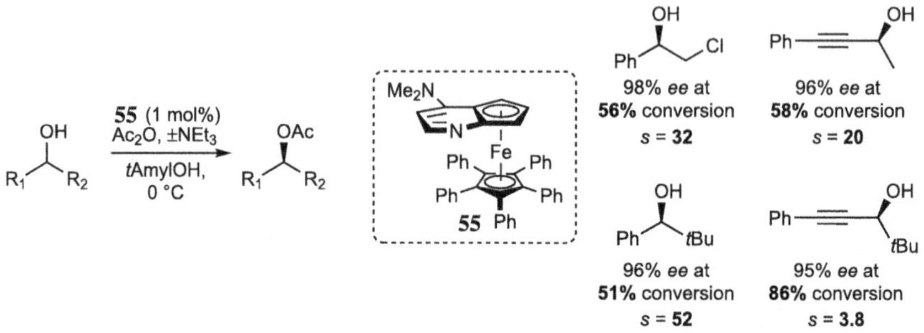

Fig. 11.33 Kinetic resolution of secondary alcohols according to Fu *et al.* [102, 103]

In addition to acylations, the most commonly encountered metal-catalyzed resolutions are achieved by epoxidations (Sharpless and Jacobsen epoxidations), Sharpless dihydroxylations, and nucleophilic epoxide openings (using the Co/Cr-Salen complexes developed by Jacobsen), as these reaction types allow a very broad (and often complementary) substrate scope and the observed *ee* values are usually beyond 90% (see Sect. 3.3.1) [99].

When *meso*-compounds are employed as substrates, yields of up to 100% can even be accomplished [104]. An enantiotopic distinction of functional groups in *meso*-compounds is a convenient method to selectively synthesize one enantiomer, while avoiding the separation and disposal of the undesired opposite stereoisomer. This is also referred to as the "*meso-trick*", which was first reported in the 1970s for the asymmetric synthesis of prostaglandins [105, 106].

In addition to the successes in metal-catalyzed resolutions, biobased catalysts in particular excel at this type of transformation. It can be carried out by a wide range of different *enzymes* or whole organisms, including lipases/esterases (ester hydrolysis or formation), amidases (amide hydrolysis/formation), and dehydrogenases (oxidoreduction of alcohols/ketones) [107–109]. The rate laws or the relationship between conversion, *ee* values and selectivities follow the same principles and equations as their metal-catalyzed counterparts. An enzyme can only selectively convert a few, structurally closely related substrates. Finding a suitable enzyme for a specific reaction is often associated with considerable effort. The handling and the sometimes exotic reaction and work-up conditions (for organic chemists) also have somewhat limited their general establishment as part of the synthetic toolbox. However, enzymes can catalyze reactions that are often beyond the capabilities of chemical reagents. In addition, the achieved enantioselectivities are typically very high due to the precisely defined structural environment around the active center.

Acylation or deacylation reactions of hydroxyl groups or amines dominate the scientific literature [110–113]. The most important and best-studied enzymes are lipases, which catalyze the formation and cleavage of ester bonds. Predicting an enzyme's selectivity towards a substrate is challenging and can vary greatly even with minor changes in the structure of the starting material or the used solvent system. This was ambly demonstrated in the hydrolysis of the dihydropyridine moiety **56** of amlodipine, a nifedipine-type calcium channel antagonist (see Fig. 11.34) [114]. Despite these challenges, empirical models have been developed by Kazlauskas, Prelog, and others that provide a good match between prediction and the actual selectivity observed with lipases (**57** and **58**) [115–118].

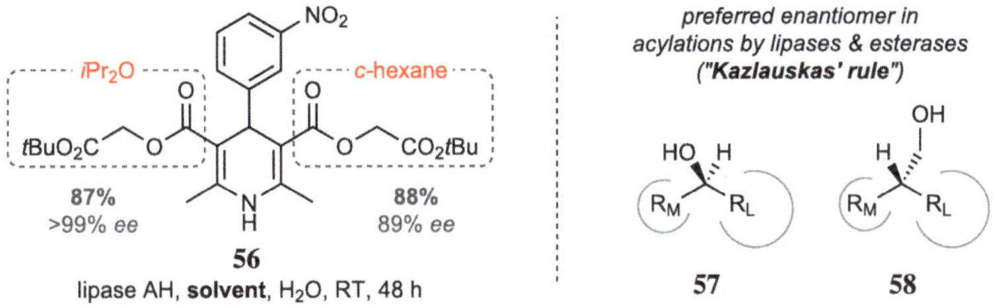

Fig. 11.34 Reversal of enantiopreference upon change of the solvent and empirical model according to Kazlauskas [114, 117]

Especially the pharmaceutical industry resorts to enzymatic reactions to synthesize key intermediates in enantiomerically pure form, as these are usually very cost-effective processes [119, 120]. This leads to a particular emphasis of their use on an industrial scale. Schering's synthesis of posaconazole (**62**, formerly known as SCH 56592), an antifungal triazole derivative, includes an enzyme-catalyzed resolution by a lipase (Novozym 435) as key step. Following addition of iodine the newly-generated stereocenter directs the free hydroxyl group to the β-face of the transient iodonium intermediate derived from the double bond to stereospecifically afford **61** in enantiomerically pure form. The batch size was successfully adapted to a multi-kilogram scale (see Fig. 11.35) [121]. Other commercial syntheses that rely on enzymes for kinetic resolutions include highly complex active ingredients such as paclitaxel [122], epothilones [123, 124], or β-lactam antibiotics, among others [125].

The *dynamic kinetic* resolution (DKR) combines a reagent/catalyst for discriminating the antipodal stereoisomers while simultaneously racemizing the starting material. The remaining unreactive enantiomer is thereby continuously transformed to the reactive one, which can subsequently be converted. Thus, as with the kinetic resolution of *meso*-compounds, yields of up to 100% can be achieved. A prerequisite for this type of resolution is the presence of acidic protons at the stereogenic center, and the racemization must occur more rapidly than the reaction of the preferred enantiomer. Therefore, benzylic and allylic alcohols or amines as well as α-functionalized carbonyls constitute commonly used substrates.

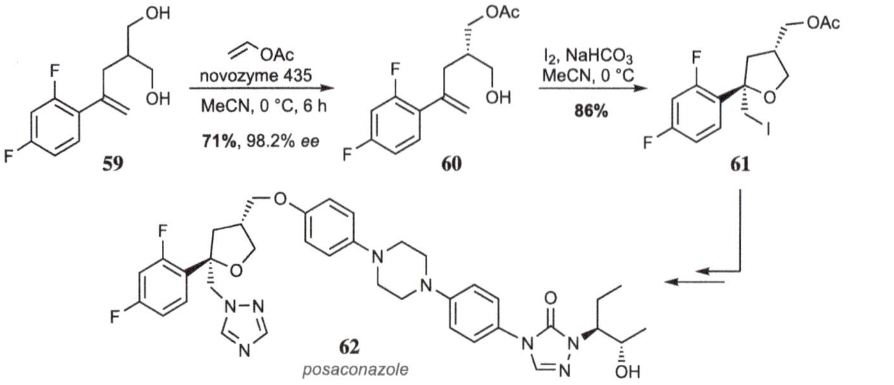

Fig. 11.35 Schering's approach to posaconazole (**62**) [121]

For a dynamic kinetic resolution, non-enzymatic catalysts can be utilized for racemization and derivatization. However, there is also a wide range of chemoenzymatic methods that combine both enzymatic catalysts for the enantioselective conversion and classic chemical catalysts for the racemization [126–128]. A (pre)catalyst frequently used for the latter is the dimeric ruthenium complex developed by Shvo **63** that operates via transfer hydrogenation (see Fig. 11.36) [129].

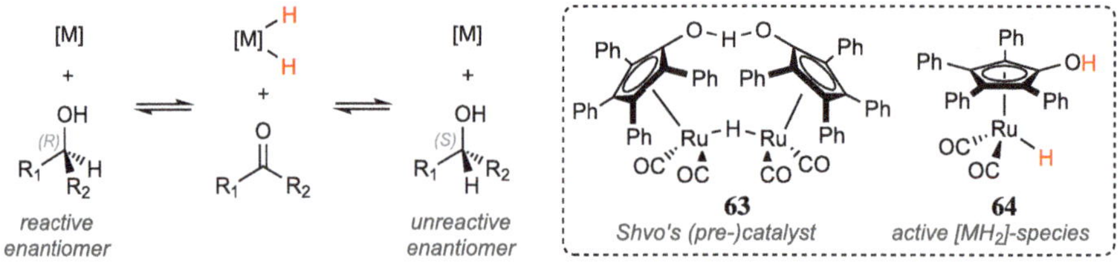

Fig. 11.36 General mechanism of DKR via transfer hydrogenation and structures of the Shvo catalyst **63** and its active form [126, 129]

11.1.4 Advanced Concepts

In addition to the approaches discussed so far for assembling complex target structures, some further concepts can be exploited that result in a significant increase in the efficiency of a synthesis. These include the use of

- domino and multicomponent reactions,
- iterative reactions,
- bridging or intramolecular reactions,

- (hidden) symmetry,
- biomimetic reactions as well as
- divergent syntheses or the use of common key intermediates in product families.

The use of **domino-** [130, 131] or **multicomponent** processes (e.g., the *Ugi* or *Passerini* reaction) [132–134] leads to high convergence, flexibility, and efficiency of a synthesis. However, identifying the corresponding retrons or partial retrons requires a high degree of experience and keen insight.

Iterative reactions lend themselves for use when repeating structural features are present in the target molecule. To build these motifs, a series of reactions can be used at which end the creation of a functionality results that allows the sequence to be repeated [135]:

$$ A \quad \xrightarrow{\ ''B''\ } \quad A\text{–}B \quad \xrightarrow[\text{repeat}]{\ ''B''\ } \quad A\text{–}B\text{–}B \quad \xrightarrow[\text{repeat}]{\ ''B''\ } \quad A\text{–}B\text{–}B\text{–}B \quad \xrightarrow[\text{repeat}]{\ ''B''\ } \quad \ldots $$

Classic examples of such repeating reactions comprise the construction of peptides or oligosaccharides, the synthesis of polyketides via aldol reactions, or of polyenes by Suzuki cross-coupling reactions (see Fig. 11.37).

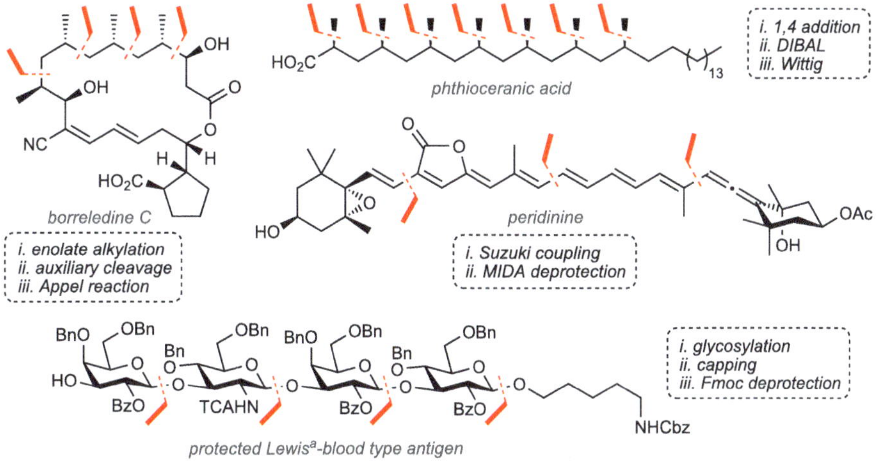

Fig. 11.37 Examples of natural products in whose syntheses iterative reactions were used [136–139]

Tethering describes a technique in which two molecules are connected to enable an intramolecular reaction. Such an approach can have multiple positive effects on yield and selectivity: First of all, an intramolecular reaction typically increases the rate of conversion. From an entropic point of view, the spatial proximity of the reactants is beneficial for enhancing the rate of product formation. Second of all, the rigidification of the substrates can additionally improve the regio- and stereoselectivity significantly by avoiding undesired reaction paths and reducing the number of possible conformations (see Fig. 11.38).

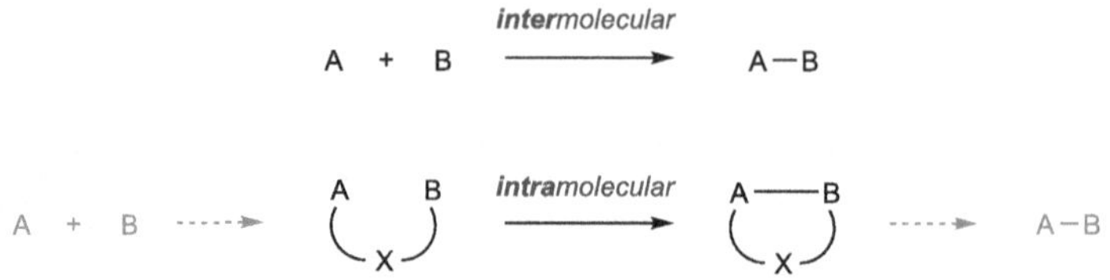

Fig. 11.38 Principle of tethering

Most commonly, two hydroxy groups are chosen as attachment points, from which the corresponding bridging functionality can be subsequently cleaved again. Typical reagents for introducing the tether are silicon-derived (resulting in silyl ethers), yet occasionally boron- and phosphate-based linkers can be encountered as well [140, 141]. Kozmin and Marjanovic doubly capitalized on the silyl tether in their synthesis of the antiproliferative natural product spirofungin A. In the metathesis of the enone moiety with the terminal alkene in **65**, an intramolecular reaction was used instead of a cross metathesis with potentially purely statistical product distribution. As the cyclization is first order with respect to the substrate concentration and an intermolecular side reaction is of second order, the selectivity of the desired product **66** could be increased by high dilution conditions. In the ensuing hydrogenolytic debenzylation and enone hydrogenation with subsequent spirocyclization, only the required spiroketal **67** was obtained. Although **67** is doubly anomerically-stabilized and should be favored for electronic reasons, it is sterically highly congested, which is why a 1:1 mixture of the double and the single anomerically-stabilized ketal is afforded without tethering (see Fig. 11.39) [142]. The same approach was also exploited in further natural product syntheses [143–147].

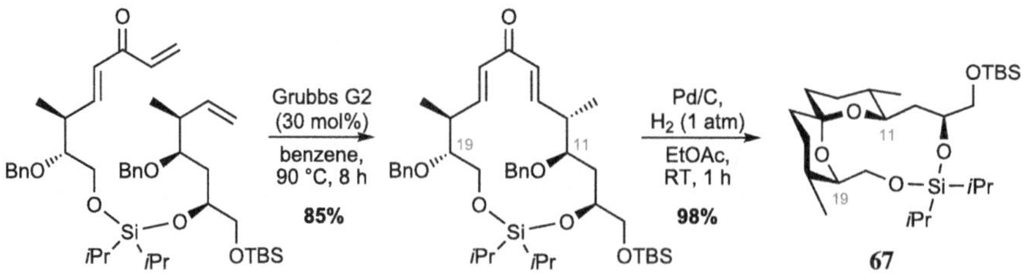

Fig. 11.39 Si-tethering in Kozmin's synthesis of spirofungin A [142]

The regioselectivity of Diels–Alder reactions may in select cases even be completely reversed, as Fortin and co-workers impressively demonstrated by comparing intra- and intermolecular reactions. The required *meta*-regioisomer could not be isolated in the intermolecular cycloaddition. In contrast, the intramolecular variant with **69** as substrate not only yielded the *meta*-regioisomer **70** but afforded the desired *endo*-stereoisomer as the major product as well (see Fig. 11.40) [148].

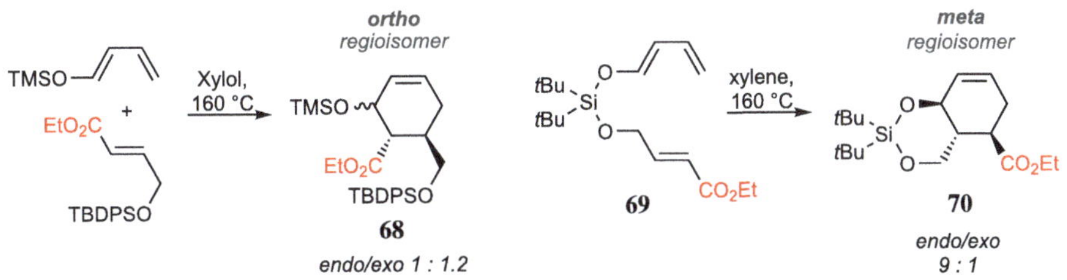

Fig. 11.40 Product composition of inter- *vs.* intramolecular Diels–Alder reactions [148]

The beginning of a **symmetry-based** retrosynthesis unavoidably starts with the recognition of the underlying symmetry. This may be obvious or could be hidden axes of rotation, a dimeric structure, or the possible occurrence of achiral transition states/intermediates in chiral substrates (see Fig. 11.41) [149–151].

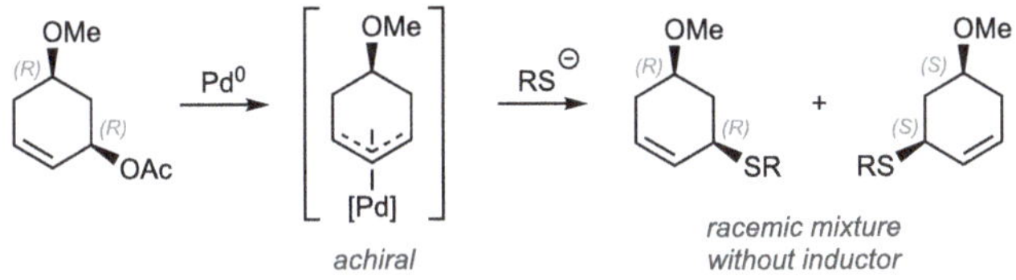

Fig. 11.41 Impact of achiral intermediates in a Tsuji–Trost reaction of chiral substrates

While this triviality may be easy to perform in simple substrates, complex structures do not easily yield their inherent symmetry to the chance observer. If their discovery is successful, such an approach usually leads to a significant simplification of the synthetic problem. Three approaches of a symmetry-based synthesis design dominate:

- antithetic disconnection to a symmetric precursor;
- use of a *meso*-precursor;
- utilization of a di-/oligomeric structure or of a C_n-axis of the target molecule.

The use of symmetric precursors is comparatively simple. Often these building blocks are C_n-symmetric arenes or aliphatic (carbo/hetero)cycles. In their synthesis of hemibrevetoxin B (**71**), the Nelson group resorted to a *meso*-strategy [152]. By tracing the key fragment **72** back to the epoxide **73**, which contains an inversion center, a previous synthesis [153] which had not utilized the latent symmetry, could be significantly shortened. The late-stage, symmetry-breaking step was achieved by a kinetic resolution in the presence of a Co(II)-Jacobsen catalyst (see Fig. 11.42, cf. Sect. 3.3) [154].

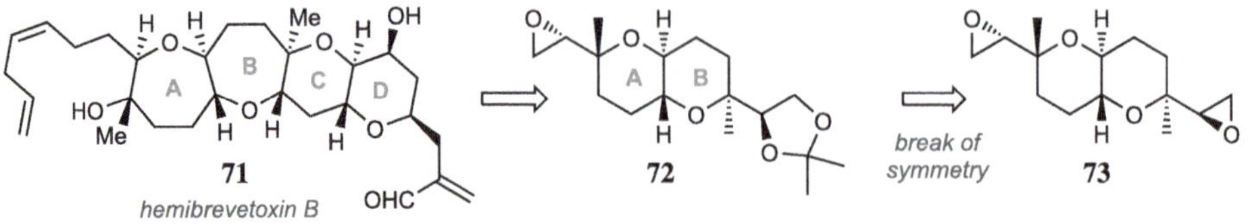

Fig. 11.42 Retrosynthesis of hemibrevetoxin B by Nelson *et al.* [154]

While the opportunity to maintain the symmetry of a *meso*-compound for an extended period in a synthesis is rare, C_n-symmetric or dimeric key intermediates have firmly established themselves in total synthesis [149, 150]. The advantage of a C_n-symmetry lies in the homotopic nature of the functional groups. Therefore, it does not matter which of the reacting termini is converted. A desymmetrization can thus be carried out with *achiral* reagents, as long as it is ensured that only one functional group is transformed.

While recognizing the symmetry elements in carpanone [155] and yuehchukene [156] is still relatively simple (see Fig. 11.43), there are also natural products where only a thoroughly executed retrosynthetic analysis brought the symmetry to light [157].

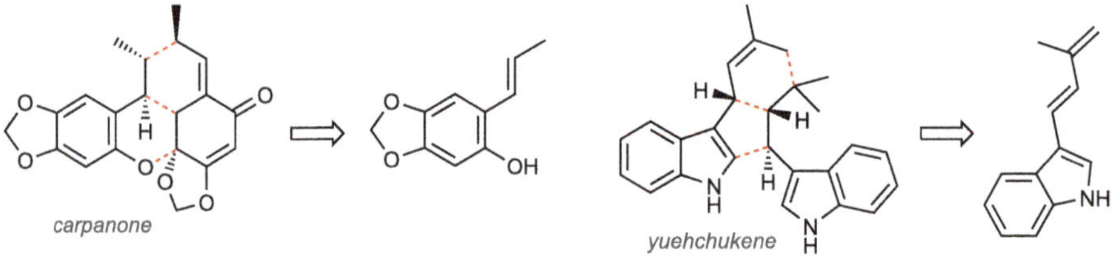

Fig. 11.43 Retrosynthesis of the dimeric natural products carpanone and yuehchukene [155, 156]. The dashed lines indicate the disconnections

One of the most complex examples can be encountered in the secondary metabolite spongistatin 1, a marine macrolide containing 24 stereogenic centers with exceptional antiproliferative activity. None of the previous syntheses [158–164] exploited the latent C_2-symmetry, which was insightfully achieved by the Ley group and led to a dramatic reduction of the requisite steps compared to their original synthetic plan (see Fig. 11.44) [165].

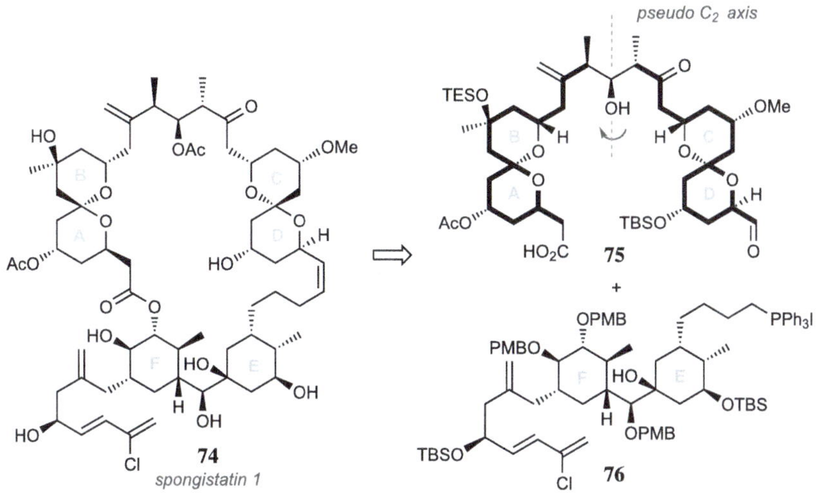

Fig. 11.44 Retrosynthesis and pseudo-C_2-symmetry of the ABCD fragment of spongistatin 1 (**74**) [165]

In this approach, a divergent–convergent method was pursued, in which the stereocenter at C-5 in **79** was purposely introduced as a 1:1 mixture of the *R/S*-diastereomers. The AB (**77**) and CD fragments (**78**) were then obtained from each of the two synthesized stereoisomers. The derived ketals could subsequently be recombined at a later stage of the synthesis. The synthesis impressively underlines the advantage of a well-designed retrosynthesis, through which the construction of the ABCD fragment **75** could be shortened from 65 to 46 steps (see Fig. 11.45) [165].

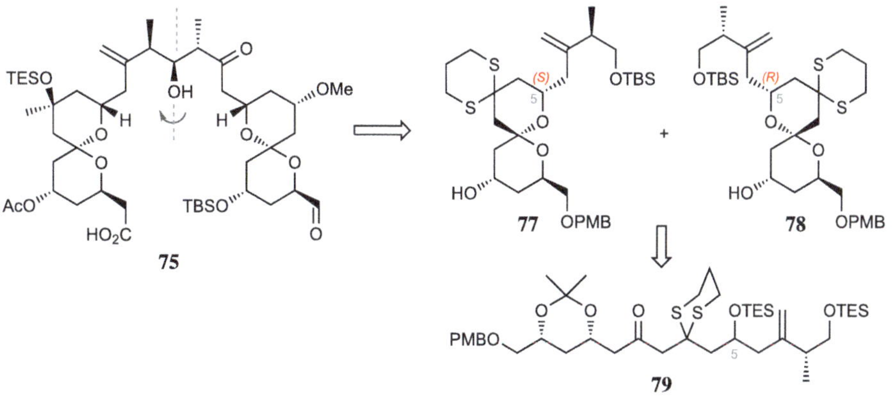

Fig. 11.45 Retrosynthetic disconnection of the ABCD fragment to **79** [165]

If a total synthesis uses the same or a similar bond formation pattern that is carried out by enzymes in the biosynthesis of a natural substance, this is referred to as a **biomimetic** approach. Knowledge of a biosynthetic pathway can be of considerable help in a retrosynthetic analysis: nature thus already indicates where the preferred disconnections are to be placed. In addition, considerable insight into the properties of the intermediates and target molecules can be gained (see Fig. 11.4), and the natural reactivity of the functional groups can be utilized.

A vivid example of the high efficiency of this approach is provided by the access to the daphniphyllum alkaloids [166]. Initial model studies of the daphniphylline skeleton were conducted using a regular synthetic probe [167] followed by investigations with partly pre-formed ring structures [168]. In the concluding total synthesis of ($\pm$)-proto-daphniphylline, the biosynthetic precursor of the daphniphyllum alkaloids, an impressive biomimetic domino reaction from an acyclic precursor afforded the desired target (see Fig. 11.46) [169]. As the biosynthetic precursor to proto-daphniphylline (**81**) was elucidated to be squalene (**80**) during these synthetic efforts [170], the retrosynthesis yielded the squalene derivative **82** (also a C_{30} body) as starting material. Upon reaction with ammonia and acetic acid, the domino reaction of **82** generated five rings with eight stereogenic centers, including three quaternary carbon atoms, and afforded the desired alkaloid in one step.

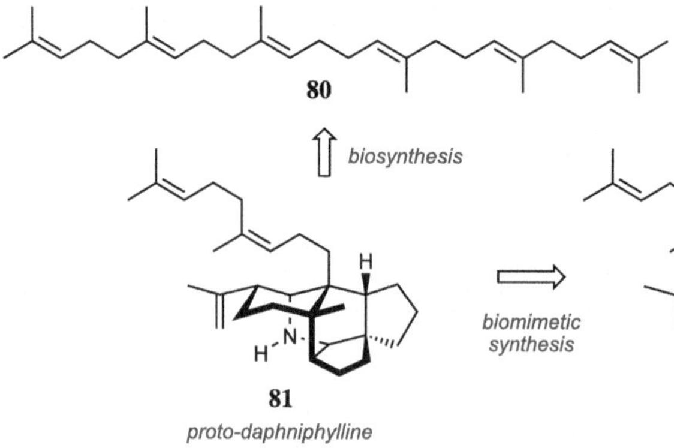

Fig. 11.46 Incorporation of biosynthesis into the retrosynthesis of proto-daphniphylline [169]

Based on a postulated biosynthesis, the antimalarial monoterpene cardamom peroxide (**83**) was synthesized by Maimone and Hu in only four steps. The hypothetical conversion, consisting of two molecules of pinene and multiple units of molecular oxygen as the sole building blocks, could be successfully translated into a biomimetic sequence. It additionally exploited the pseudo-dimeric structure of the natural product. Starting from the structurally related myrtenal, the key enedione intermediate **84** was assembled by a McMurry coupling, a reaction with singlet oxygen, and a subsequent Dess–Martin oxidation. The manganese hydride-mediated oxidative functionalization of the two enone double bonds led, as in the purported biosynthesis, to the desired endoperoxide **83** via diperoxide **86** (see Fig. 11.47) [171].

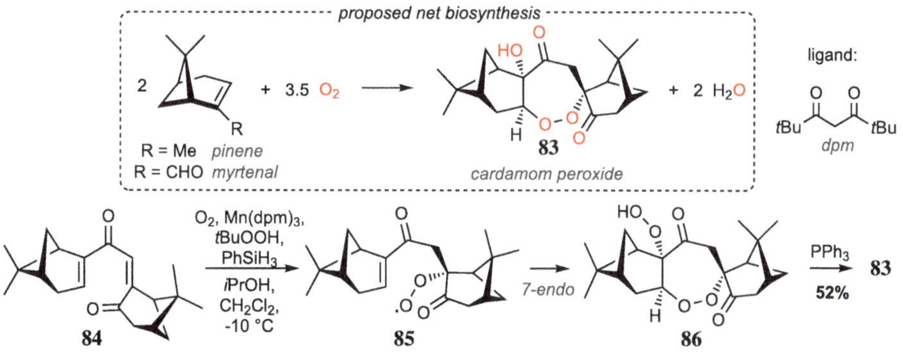

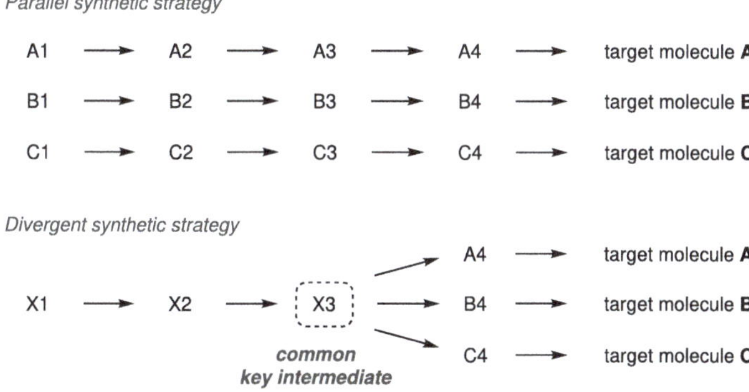

Fig. 11.47 Biomimetic synthesis of cardamom peroxide (**83**) by Maimone and Hu [171]

If access to a structurally related family of natural products or derivatives of a given lead structure is desired, one can potentially resort to a strategy relying on a **common key intermediate** [172–175]. This category also formally includes dimeric or trimeric natural products, which have already been mentioned in the context of capitalizing on inherent molecular symmetry [149–151]. In contrast to a parallel synthesis, where all target molecules are constructed via individual routes, a *divergent* synthetic strategy yields all natural product analogs or derivatives starting from one intermediate towards the end of the sequence (see Fig. 11.48).

Fig. 11.48 Principle of divergent syntheses [172]

The approach is significantly more challenging than "merely" conceiving a route to a single natural product. The structural patterns of the substance family to be synthesized serve as a blueprint in the search for a common starting point. In addition to the basic molecular skeleton, such an advanced intermediate must take into account the location of all stereocenters as well as the functionality distribution and substitution pattern. The biosynthesis of the natural product family may also provide further clues. However, the biosynthetic disconnections are usually understood more as an inspiration and can rarely be used directly due to challenges related to stability, chemoselectivity, or polarity, which are much simpler to overcome with highly specialized enzymes (see Fig. 11.49).

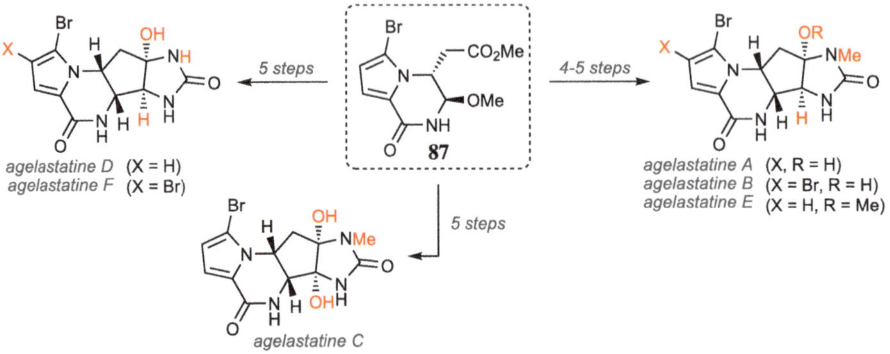

Fig. 11.49 Synthesis of *agelastatin* alkaloids according to Movassaghi *et al.* [176]

Starting from these privileged intermediates, the introduction of various functionalities (see Fig. 11.49), the attachment of different side chains or a structural reorganization usually ensues to gain access to various target molecules (see Fig. 11.50).

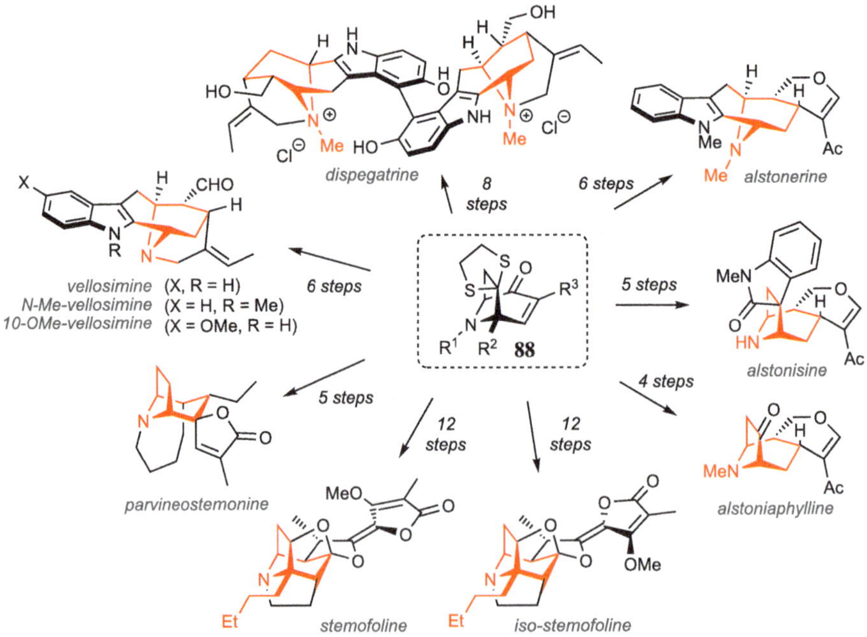

Fig. 11.50 Gaich's approach to *sarpagine*, *macroline*, and *stemona* alkaloids [173]

These examples can only provide a very limited insight into this rather complex topic. A multitude of further, inspiring approaches for the syntheses of alkaloids, terpenes, polyketides, and other classes of natural products can be found in the corresponding literature [172–174]. Since the common intermediate (in the illustrated examples **87** and **88**) is specifically tailored to allow an entry to the desired substance class, a specific precursor cannot be used for access to a different family of natural products. However, the approach can be conceptually well transferred. Thus, many natural product families become accessible with less synthetic effort, compared to target-oriented syntheses of each individual natural product.

11.2   Requirements of Academic and Industrial Syntheses

> *It does [...] no good to offer an elegant, difficult and expensive process to an industrial manufacturing chemist, whose ideal is something to be carried out in a disused bathtub by a one-armed man who cannot read, the product being collected continuously through the drain hole in 100% purity and yield* [177].
>
> Sir John Cornforth

Although syntheses in both academic and industrial settings ideally afford the desired product in high yield, there are some significant differences in each field's requirements, which are also related to the corresponding objectives. Basic academic research in the synthetic field often deals with access to rare and structurally very complex natural products. This can involve the synthesis of a challenging target molecule, the confirmation of a postulated structure [178], or the application of a specially developed method in the key step of the sequence. Studies on the potential application of a bioactive metabolite, such as the structure–activity relationship, are much less common but their frequency is steadily increasing. In industrial syntheses, it is necessary to differentiate between active pharmaceutical ingredients and crop protection agents, fine and bulk chemicals. Even in a pharmaceutical synthesis, there exists a stark difference in the requirements, depending on whether the objective is lead discovery or lead optimization, the provision of a few kilograms of an active ingredient candidate for clinical studies, or the commercial production of an approved active ingredient (see Fig. 11.51).[III]

The commercialization of an active ingredient often takes 10 years or more and incurs costs of 1–2 billion dollars for pharmaceutical drugs [182]. The quantities required for studies range from a few milligrams to hundreds of kilograms in the third clinical phase.

[III] The time periods shown can vary greatly depending on the individual case: While the search for active ingredients can take up to 6 years, the clinical phases are relatively standardized at 6–7 years. The registration phase until approval can take another 2–5 years. For agrochemicals, the search for active ingredients and lead optimization, as well as registration, can also take significantly longer, and the entire process can extend to over more than 10 years.

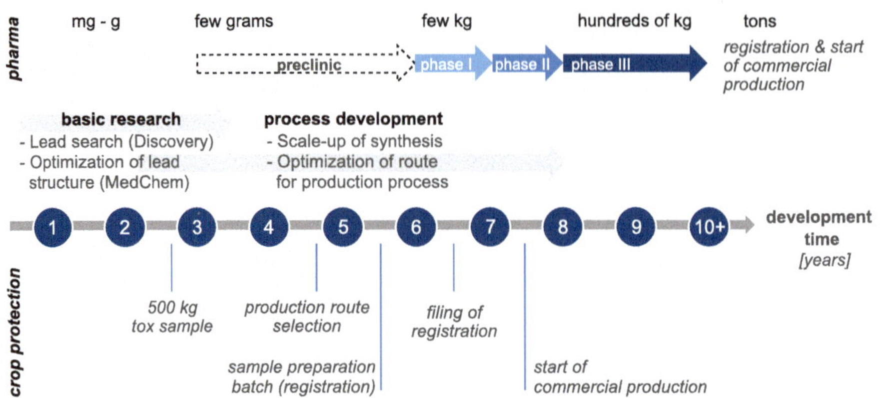

Fig. 11.51 Rough comparison of the development periods and phases of pharmaceutical drugs and crop protection agents [179–181]

Accordingly, the pilot route for the first kilogram quantities is important, yet route selection occurs relatively late in the development process, approximately after 5–7 years. The situation is rather different for crop protection agents. Especially toxicity studies require a large amount of substance early on, and the final commercial production route needs to be selected much sooner [179, 180].

The typical length of commercial syntheses of active pharmaceutical ingredients comprises on average 4–10 steps [183]. This has successively increased to an average of 12 steps in the early 2010s [184]. The complexity of the active ingredients (molecular weight, density of functional groups, number of stereocenters, number of carbo-/heterocycles) is continuously expanding with the improved understanding of the mode of action and the resulting optimized molecular structure (see Fig. 11.52) [185].

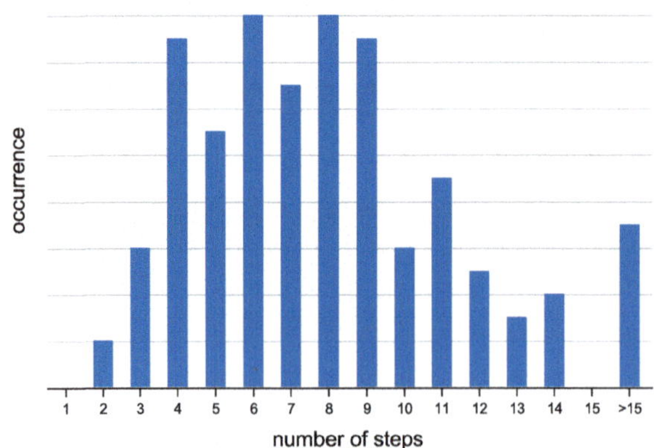

Fig. 11.52 Analysis of the syntheses of active ingredient candidates [183]

Heterocycles occupy a central role, particularly in a pharmaceutical context. A statistical evaluation of the most common heterocyclic building blocks revealed that 84% of all active substances approved in the USA up to 2012 contain a nitrogen atom, the majority of which are attributable to nitrogen-containing heterocycles (59% of all active substances) [186]. Sulfur and fluorine, also typical elements in pharmaceutical ingredients, are in comparison much less common in approved drugs (26 and 13%, respectively) [187]. Given the significant differences in target structures of industrial and academic routes and the quantities required at different clinical phases, it is not surprising that industrial syntheses rely on a select repertoire of key reactions [184, 188]. Depending on the stage of process development (Med. Chem. routes *vs.* pilot routes on a multi-kg scale), an analysis of the reactions shows that protection and deprotection steps as well as acylations (especially amidations) occur much less frequently on a larger scale, while resolutions via formation of diastereomeric salts play no significant role in drug discovery (see Fig. 11.53) [86, 183, 189].

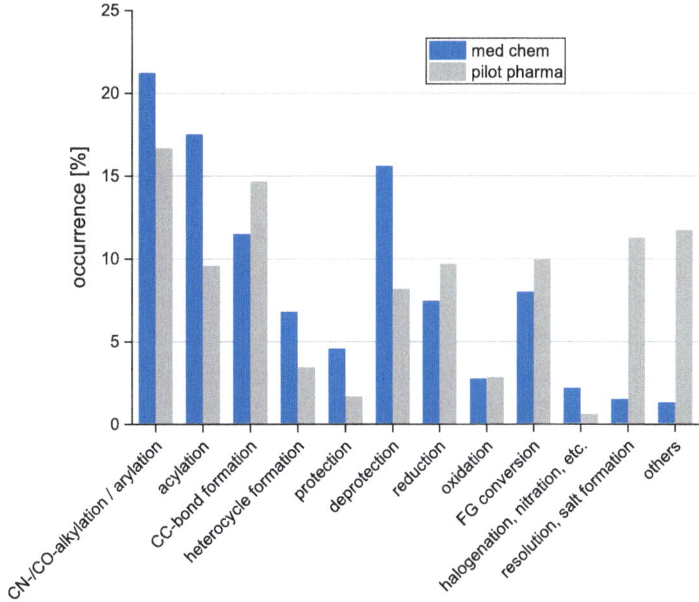

Fig. 11.53 Analysis of reaction types in Med. Chem. and pilot routes of active pharmaceutical ingredients [86, 183, 189]

The basic, trivial-sounding premise of an industrial process is its economic viability — if the process were unprofitable, it simply would not be carried out. All synthetic approaches or methods in which the starting materials are more expensive than the products therefore can be considered nonsensical for an economically sustainable production. For bulk chemicals such as solvents, simple building blocks like piperazine or acrylic acid, the production costs typically comprise less than 10 $/kg. For fine chemicals (catalysts, fragrances, vitamins, etc.), the produced quantities are smaller and the production costs correspondingly higher (between 10 and 100 \$/kg). Agrochemicals and pharmaceuticals sometimes cost significantly over 100 \$/kg, many pharma routes even result in production costs of 1000 \$/kg and beyond [180, 190]. In large-scale processes whose scope extends to multi-ton batches, additional factors also play a role, which need less consideration on a laboratory scale (Table 11.2) [179–181, 191, 192].

Table 11.2 Requirements/targets of academic and industrial syntheses

academic synthesis	industrial process
yield	(space/time) yield
elegance / number of steps	scalability
complexity of the target molecule	heat removal / supply
(± application aspects)	by-product control
	kinetics / reaction network
	purification concept
	production costs
	availability of starting materials
	materials of construction / corrosion
	waste disposal / recycling
	safety concept
	patent situation / *freedom-to-operate*
	customer demand
	logistics of starting materials and products

The direct comparison elucidates why an industrial synthesis often is not commercialized, as many aspects have to be considered. The targeted amount of the product may not match the availability of raw materials on this scale. In the case of particularly exothermic reactions or when using highly reactive or toxic reagents, the necessary safety concept can potentially make the process prohibitively expensive. The patent situation must also be considered, as a competing patent can prohibit its production.[IV]

In addition to these factors, which are largely irrelevant for the technical feasibility of the synthesis, the kinetics and the resulting spectrum of by-products can play a significant role from a chemical point of view. The entire reaction profile, i.e., the kinetics of the step in question and the development of by-products (especially gases or solids), and temperature

---

[IV] A patent is a form on intellectual property that does **not** give the patent owner the right to exploit the invention subject to the patent. Patent law is an **exclusionary** right, which only grants its owner the legal right to prohibit others from commercially executing a process. If there are no obstructive patents to execute a process, this is referred to as *"freedom-to-operate"* (FTO). FTO can only ever be a momentary snapshot but some countries recognize customary rights to keep a process operating, even if obstructive patents have been granted subsequently.

profile, should ideally be fully elucidated. In kinetics of higher than zeroth order, the reaction rate depends on the concentration of the reactants. The rate thus inevitably decreases over time and with conversion (see Fig. 11.54).[V]

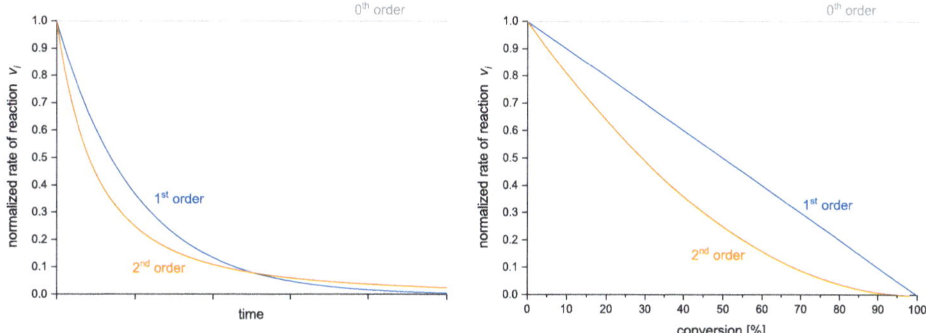

Fig. 11.54 Dependence of reaction rates on time and conversion

The efficiency of utilizing the available equipment is referred to as *space-time yield* (STY). The rate of product formation relative to the reactor volume (expressed in $kg_{product}$ L^{-1} h^{-1}) can be extremely low in highly diluted reactions despite a good yield, as the necessary reactor volume is very large due to the dilution factor. An internal study by AstraZeneca estimated that the throughput of an active ingredient decreases exponentially with the number of (linear) synthetic steps (see Fig. 11.55) [191, 193].

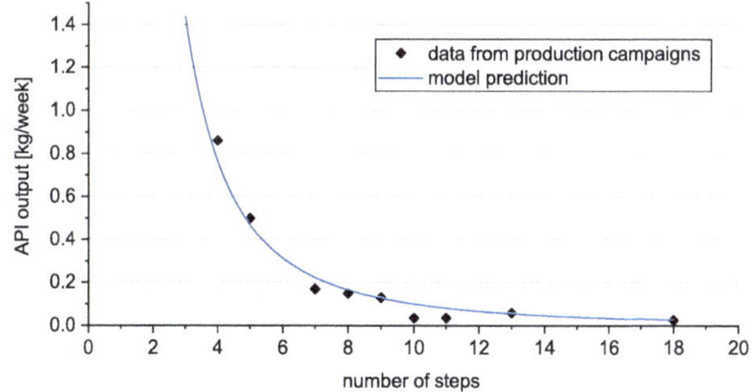

Fig. 11.55 Impact of the number of steps on production output according to AstraZeneca [191]

 At conversions above >98%, the reaction rate also decreases exponentially in first-order reactions. A "full conversion" is therefore not feasible on an industrial scale as the reaction time would become prohibitively long.

A low reaction throughput can certainly be a reason to change an existing route. Omeprazole (**89**) is a proton pump inhibitor used in reflux esophagitis to regulate the pH of gastric acid. Given the enhanced bioavailability of esomeprazole (**91**), the corresponding *S*-enantiomer of racemic omeprazole, it was developed as an improved drug by AstraZeneca. Material required for initial clinical studies relied on the mandelic acid derivative **90**. It was synthesized from omeprazole and laboriously separated into its diastereomers using preparative column chromatography. Subsequent studies required ten times the amount of active ingredient, which would have led to an estimated consumption of 60,000 L of eluent and several months of time for the chromatographic separation alone. To circumvent the bottleneck enantiomer separation, a direct asymmetric oxidation of the thioether precursor **92** to the sulfoxide was developed. The desired enantiomer was thereby obtained directly with 94% *ee* instead of requiring separate six steps, thus significantly shortening the production time from 14 weeks to 14 days (Fig. 11.56) [191].

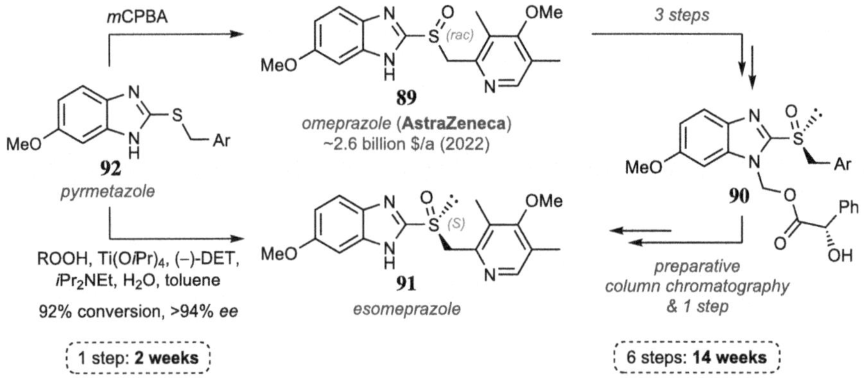

Fig. 11.56 Optimized production route to esomeprazole [191]

Continuous production is generally much more efficient than batch operation, as a batch approach involves additional work-up, filling, emptying, and possibly cleaning steps in addition to the actual reaction time. Additionally, dosage and especially temperature control are more challenging and sluggish, which can lead to a reduced selectivity. The disadvantage is the often necessary customized equipment for continuous processes, which brings high specific investment costs and usually prevents its use for other types of reactions. The larger the production throughput, the more worthwhile a continuous operation becomes, which is why many fine chemicals and practically all bulk chemicals are synthesized continuously. Particularly with efficient heating/cooling/mixing, the occurrence of labile intermediates, the use of highly reactive and/or dangerous reagents, or in heterogeneous reactions, continuous operation can fully play to its strengths from the perspective of process control and safety (see Fig. 11.57) [194–198].

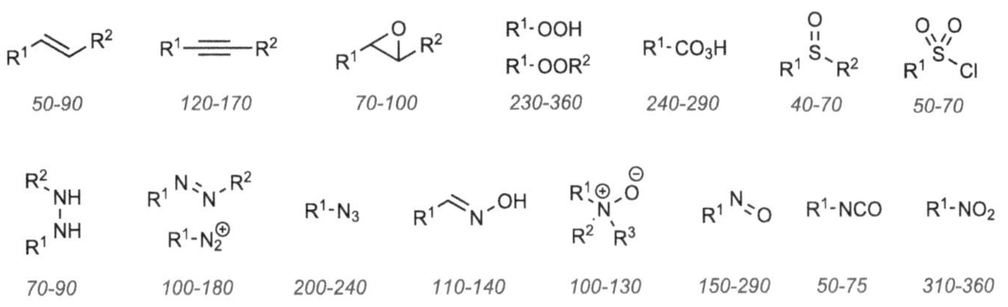

Fig. 11.57 Decomposition energies of selected functional groups (in kJ/mol) [199]

In the academic field, this approach is usually referred to as *continuous flow* synthesis. A homogeneous continuous reaction is prevalent in academic syntheses or methods, where the substrates and catalyst are in solution and are quenched after having passed through the reactor. In contrast, fixed bed reactors dominate in the industrial production of bulk chemicals. The reactants pass in liquid form or as a gas over a heterogeneous catalyst and are then either purified by distillation or by continuous crystallization. The high local concentrations of the starting materials on the catalyst surface can enable rapid reactions and continuous operation may be the most efficient way to produce large quantities with a small physical space requirement/footprint, i.e., high space-time yield [200, 201].

Continuous operation can also be used in highly complex syntheses to simplify the reaction conditions and thus reduce variable costs. Eribulin (**97**) is a structurally simplified cytostatic derived from halichondrin B, which is synthesized on industrial scale in 64 (!) steps [202, 203]. The mechanism of action via tubulin aggregation differs from the widely employed drug classes of taxanes, *Vinca* alkaloids, and epothilones. Given its high efficacy, only quantities in the kg range are needed annually. However, the complexity of the target structure and the length of the synthetic route place a pronounced emphasis on efficiency in terms of selectivity, space-time yield, and simple technical execution of each step. For this purpose, a continuous operation was investigated for the DIBAL reduction of the ester **93** and the coupling of the two hemispheres **94** and **95** instead of a classic batch reaction at -70 °C. Under flow conditions, the lithiation of **95** and subsequent 1,2-addition to **94** even showed the highest yield at 10 °C, reducing the reaction time of the crucial CC coupling to a few seconds. The preceding DIBAL reduction of **93** could also be optimized to yield a significantly increased throughput at a lower excess of reductant without affecting the yield (see Fig. 11.58) [204].

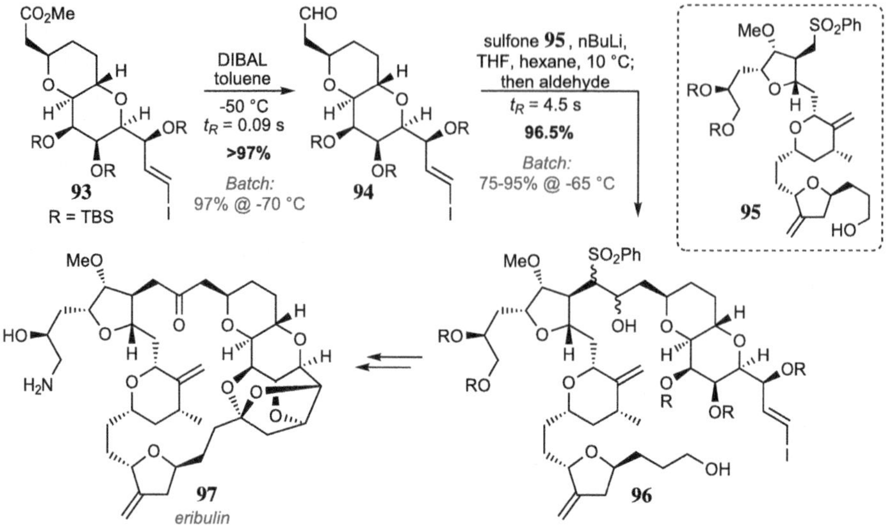

Fig. 11.58 Optimization of the commercial route to eribulin through a continuous process [204]

Regardless of whether a reaction is operated continuously or as a batch process — if it is possible to continue the synthetic sequence with the crude product, this not only increases throughput but also potentially reduces waste. In addition, non-crystalline, oily intermediates that are toxic or malodorous can be handled relatively safely. The crude product is thereby typically utilized as a solution in the following steps. This principle is referred to as *telescoping*. The disadvantage of this approach is the potential carryover of undesired components into subsequent steps, which can lead to a reduction in yield or a change in the spectrum of by-products. An investigation of these effects is therefore essential.

An illustrative example was reported by Bristol-Myers Squibb in their pilot synthesis of the hepatitis C drug beclabuvir (**104**). The chiral cyclopropane fragment **103** was accessed from benzaldehyde **98** over four telescoped steps (**99**→**102**). The Ph_3PO generated in the formation of styrene **99** had to be strictly avoided in subsequent steps, necessitating a filtration of the reaction mixture over basic Al_2O_3 following precipitation of the majority of Ph_3PO. The asymmetric cyclopropanation with the diazo compound (also formed *in situ*) proceeded smoothly. A solvent switch for the subsequent ozonolysis with reductive workup afforded the alcohol **101**, whose ester group was saponified. The final crystallization in *i*PrOH/H_2O with the chiral amine *R*-AMBA allowed the isolation of the nearly enantiomerically pure building block **103**. The active drug **104** was thus synthesized in a total of 12 linear steps with only 5 purifications (see Fig. 11.59) [205].

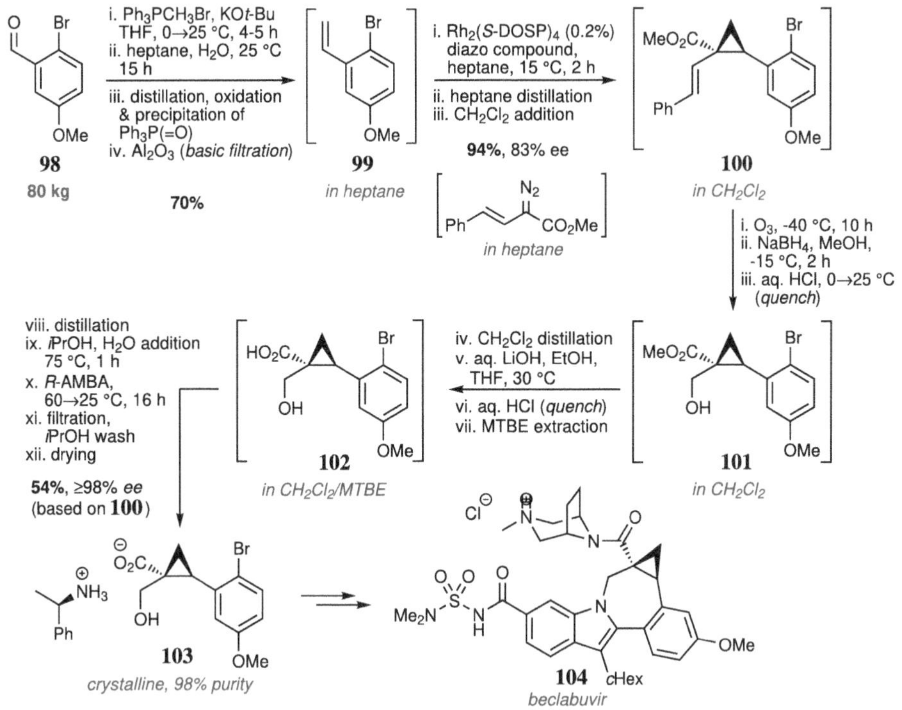

Fig. 11.59 Telescoping in the synthesis of beclabuvir (**104**) [205]

In addition to environmental aspects [206] including waste minimization, the chosen reaction conditions heavily impact the control of the selectivity and thus the spectrum of generated by-products. Not only the purity of the product, but also the identity and ratio of trace impurities is one of the central aspects of the manufacturing process for pharmaceuticals and crop protection ingredients: The toxicity studies for agrochemicals and the clinical studies for pharmaceutical drugs are conducted based on the given composition of organic or inorganic substances present in the active ingredient due to the chosen manufacturing route (see Table 11.3).[VI]

When the route is changed, these studies must be repeated if a significant deviation of the by-products in the product results. This aspect illustrates the rationale of the high hurdles for changing certain reaction parameters in the production of active ingredients. Contrary to popular belief, commercially marketed active ingredients do not have to possess 100% purity — however, their composition may only vary within narrowly defined limits. The

---

[VI] The *International Council for Harmonisation of Technical Requirements for Pharmaceuticals for Human Use* (ICH) has published a series of guidelines as an industry standard for the production, development, quality control, risk management, etc. of pharmaceuticals. See www.ich.org.

Table 11.3 Limit exposures of selected metals in daily dosed pharmaceuticals (ICH Q3D) [207]

metal	oral intake [ppm]	parenteral dosage [ppm]
As, Cd, Hg, Pb	0.5-3	0.2-1.5
Co	5	0.5
Ni	20	2
Pd, Pt, Rh, Ru	10	1
Cu	300	30
Sn	600	60
Cr	1100	110

more robust a process is to variations in the quality of the starting materials and deviations in manufacturing, the more likely it is to be favored over alternative routes.

To effectively control product quality, several approaches are pursued: Particularly the purification steps of a synthetic sequence allow the prevention of unwanted changes in the composition. Many active ingredients are solids or salts, which is why a crystallization is the most common purification method of pharmaceutical intermediates of active ingredients on an industrial scale. The control of product purity by a final crystallization in the last step is a central element of almost all manufacturing routes. The second approach is the close monitoring of online analytics for process control. The monitoring of a conversion by means of gas chromatography, HPLC, or thin layer chromatography, techniques very common in the academic field, is often too slow to ensure an effective control.[VII] For process analytics, therefore, quasi-instantaneous methods such as temperature, pH, pressure and conductivity measurements as well as (automated) IR/UV/VIS, NMR or Raman spectroscopy, which allow real-time observation, are preferred on large scale [208]. Detectability is the fundamental aspect: If not all components can be detected, complementary analytical methods must be used to obtain the complete mass balance.[VIII] The third aspect is the systematic investigation of a reaction step for its susceptibility to variations in dosing rate, temperature, mixing efficiency, etc. This way, an operating window can be devised in which the product quality can be ensured, which is referred to as *"quality-by-design"* [209].

In contrast to an unwanted variation in the product, the role of trace impurities in the starting materials may be both positive or negative in nature: The Nozaki–Hiyama–Kishi coupling, for example, was only able to become established as a reliable standard method once the role of the occasional Ni(II) "impurities" had been elucidated. Prior to that under-

---

[VII] As long as the state of the reaction does not change during sample preparation, measurement, and analysis, almost any analytical method can serve for reaction control. With a typical GC or HPLC analysis duration of 30 min or more, this requirement is not always met.

[VIII] Decomposition reactions can lead to gas formation, for example. Many polar functionalities can also prevent their detection by GC or HPLC when resorting to standard columns. Oligo- and polymers also cannot be detected and quantified by GC or HPLC. In any case, a calibrated analytical method should be used. The measured proportion in area percent can be dramatically influenced by the present functionalities.

standing, the yield varied depending on the traces of other metals in the employed chromium salt [210]. The "right" impurity thus finally enabled a reproducible transformation. Just as in the case of the previously mentioned Cu-catalyzed CN coupling by Bolm and co-workers, originally an Fe-catalyzed process was assumed as operative mechanism (cf. Fig. 6.12, Sect. 6.1). Even reused stir bars can lead to a false-positive result, as metal contaminants can be present on their surface [211]. Although the awareness for transition metal impurities has been raised in the academic environment, little thought is usually given to the purity of the organic reactants. Kadyrov and co-workers demonstrated the dramatic effect of the identity of the impurities in a metathesis reaction. The (percentage) purity of the reactant had only a marginal effect. With specially synthesized material and by the selective addition of the identified trace compounds, the conversion and selectivity[IX] for the desired metathesis product could be modulated. In the process of the investigation, dozens of trace components were elucidated and their influence as an inhibitor or promoter identified (see Fig. 11.60) [212].

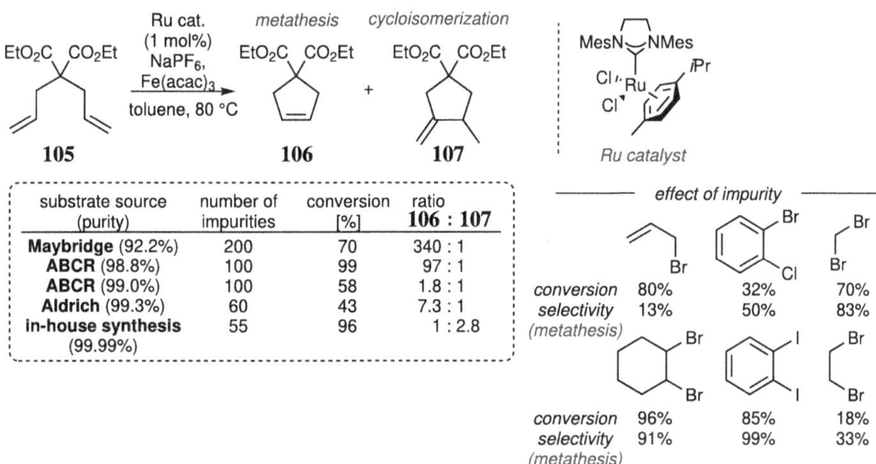

substrate source (purity)	number of impurities	conversion [%]	ratio **106 : 107**
Maybridge (92.2%)	200	70	340 : 1
ABCR (98.8%)	100	99	97 : 1
ABCR (99.0%)	100	58	1.8 : 1
Aldrich (99.3%)	60	43	7.3 : 1
in-house synthesis (99.99%)	55	96	1 : 2.8

Fig. 11.60 Influence of reactant composition on a ring-closing metathesis [212]

In the following section, the evaluation of different synthesis routes bears a closer look. Since multiple routes are routinely developed and evaluated, especially during industrial process development, examples from the pharmaceutical and agro sectors are particularly suitable for weighing the pros and cons of different (industrial) criteria.

[IX] Selectivity was defined as the ratio of the yield of the metathesis product to the conversion.

11.3 Evaluation and Optimization of Synthesis Routes

An "ideal" synthesis is inevitably dependent on a subjective assessment, as the targets or requirements differ from case to case (see Table 11.2). When considering the efficiency and environmental sustainability of a sequence, various definitions of ideality have been published by Wender, Baran, and others [213–218]. For example, the number of strategic bond formations and redox reactions in relation to all steps may constitute the key metric. Further desirable traits comprise the lowest possible cost of the starting materials, a high throughput, as well as environmentally friendly and safe reaction conditions, among many others.

Depending on the perspective of an academic, medicinal, or process chemist, analyzing different syntheses would likely come to contrasting conclusions. For example, an academic chemist pursuing a first total synthesis of a topographically complex target has vastly different goals than a process chemist aiming at a simple and cheap access to an active pharmaceutical ingredient on a multi-ton scale with defined economic and/or time constraints. Even when comparing some of the more than 20 published academic syntheses to strychnine [219], a different route may be considered optimal from various points of view: The number of steps of the longest linear sequence (*LLS*),[X] the average yield per step, the total yield or the ideality [214] of a route, all could potentially point to a different synthesis as preferred. For example, Overman's approach is similar in total yield to Vanderwal's route despite the higher number of steps. Yet most chemists probably would favor the six-step sequence in terms of its "efficiency", even though it only provides the racemic natural product in contrast to Overmann's asymmetric route (see Fig. 11.61).

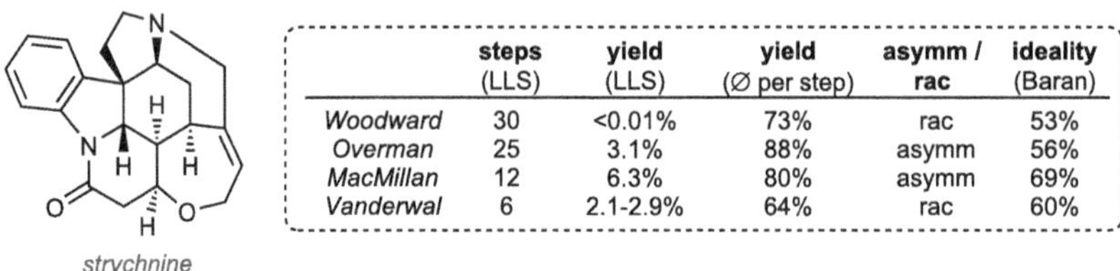

	steps (LLS)	yield (LLS)	yield (Ø per step)	asymm / rac	ideality (Baran)
Woodward	30	<0.01%	73%	rac	53%
Overman	25	3.1%	88%	asymm	56%
MacMillan	12	6.3%	80%	asymm	69%
Vanderwal	6	2.1-2.9%	64%	rac	60%

Fig. 11.61 Comparison of selected syntheses of strychnine [220–224]

In particular, the use of catalysts has had a fundamental influence on the development of industrial and academic reactions — an estimated 90% of chemicals can be traced back directly or indirectly to catalyzed reactions [184]. The use of catalytic methods is therefore

---

[X] Despite its routine use in the description of total syntheses, the use of the longest linear sequence as a metric only allows a comparison from a limited perspective. By referencing the number of steps starting from known intermediates instead of commercially available starting materials, the sequence can be artificially shortened. Nevertheless, these steps need to be considered for a more accurate comparison.

preferable in most cases to an uncatalyzed variant for various reasons (environmental aspects, economic efficiency, process safety, etc.) [206, 225, 226].

The calculation of the non-recyclable waste generated in a route provides a possible quantitative criterion by which synthesis routes can be assessed. The so-called *E-factor*, the ratio of waste to product formed, has become widely established as one standard even outside an industrial setting. Depending on the industry segment and tonnage and thus the complexity of a synthesis (number of stages, atom economy, type of reagents, proportion of reused starting materials, etc.), the mass of waste can exceed the mass of the desired product by several orders of magnitude (see Table 11.4) [227].

Table 11.4 E-factors in the chemical industry [227]

industry segment	product Quantity [t/a]	E-factor [kg_{waste} / $kg_{product}$]
oil refining	10^6–10^8	<0.1
bulk chemicals	10^4–10^6	<1.5
fine chemicals	100–10^4	5–50
pharmaceuticals	10–1000	>25

Furthermore, other concepts such as the combined mass of the reactants used per kg of product are also common (*process mass intensity*, PMI) [228]. The evaluation of the E-factor and PMI is extremely complex in detail, as all reagents and catalysts along with the respective stoichiometry and their production route, by-products, and solvents must be taken into account. A comparison of academic and industrial routes to oseltamivir was able to show significant differences between the E-factors of the individual synthetic approaches (see Fig. 11.62) [229, 230].

AcHN
H_2N CO_2Et
oseltamivir

	industrial / academic	steps	# of substrates	yield [%]	E-factor [10^3]
Roche (shikimic acid)	indust.	13	19	39	0.23
Roche (quinic acid)	indust.	14	20	22	0.31
Gilead	indust.	12	21	6.3	0.94
Fang	acad.	18	35	13	2.6
Trost	acad.	9	17	30	2.7
Corey	acad.	11	17	22	3.3
Fukuyama	acad.	13	22	5.5	4.0
Roche (Diels-Alder)	indust.	9	17	1.1	5.1
Kann	acad.	15	25	3.4	13.6
Shibasaki G1	acad.	15	34	1.4	16.2
Shibasaki G2	acad.	16	32	4.5	20.2
Okamura-Corey	acad.	13	25	2.6	22.4
Shibasaki G3	acad.	11	23	1.4	26.5

Fig. 11.62 Comparison of various syntheses of oseltamivir [229]

The proper disposal of a waste stream is estimated to approximately cost 1 $ per kg [231]. Assuming an E-factor of only 50, this means that with an annual production of 10 tons of an active ingredient, the disposal costs amount to 0.5 million $ per year. Even small process improvements can thus result in a high impact.

Pfizer's commercial manufacturing route of the antidepressant sertraline (**111**) was only slightly varied, which, however, led to an annual saving of more than 100,000 $, a reduction of the waste stream, and increased process safety. The original route required stoichiometric amounts of highly reactive TiCl₄ to sequester (as TiO₂) the water released during imine formation from **108** to **109**. The stereoselective hydrogenation to the amine **110** and subsequent crystallization furnished the mandelic acid derivative of the active ingredient. In addition to a relatively high solvent use, the mixture of the resulting solvents in the mother liquor was rather complex (EtOH, EtOAc, THF, toluene, hexane), which complicated its distillative work-up for recycling. The optimized route, on the other hand, only relied on EtOH as a solvent. The amount of solvent could thus be reduced overall and the composition of the mother liquor for work-up could be simplified (EtOH, EtOAc). The low solubility of **109** in alcoholic solvents allowed for a complete conversion of the starting material **108**, and a very time-consuming filtration of the TiO₂/MeNH₃Cl salt mixture could be omitted. The crude product was therefore not isolated in the optimized route, but instead directly hydrogenated to give the amine **112** in the presence of a modified catalyst, which resulted in an increased stereoselectivity and a lower proportion of dehalogenation. The ensuing second telescoping step allowed the direct crystallization of the mandelate **111** (see Fig. 11.63) [232].

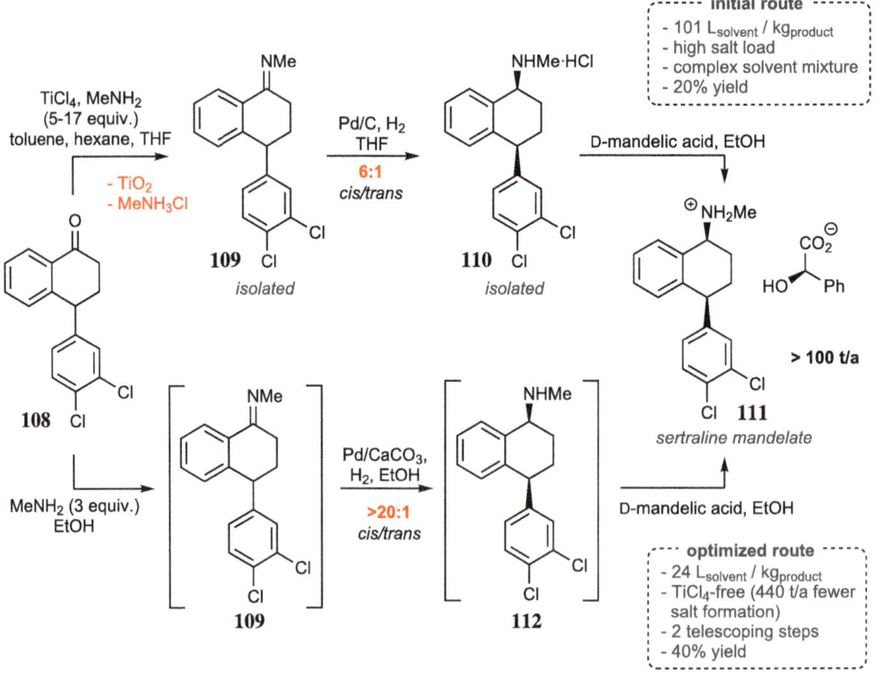

Fig. 11.63 Optimization of the commercial route of sertraline [232]

A frequently cited ideal in challenging syntheses is the concept of **convergence**, i.e., the construction of a target molecule from several building blocks of similar complexity. An example of a highly convergent synthesis is provided in Fürstner's approach to amphidinolide H (**113**). The retrosynthesis envisages a coupling of four nearly equal-sized fragments, with a different reaction being used for each connection: a metathesis, a Stille coupling, a macrolactonization, and an aldol reaction (see Fig. 11.64) [233].

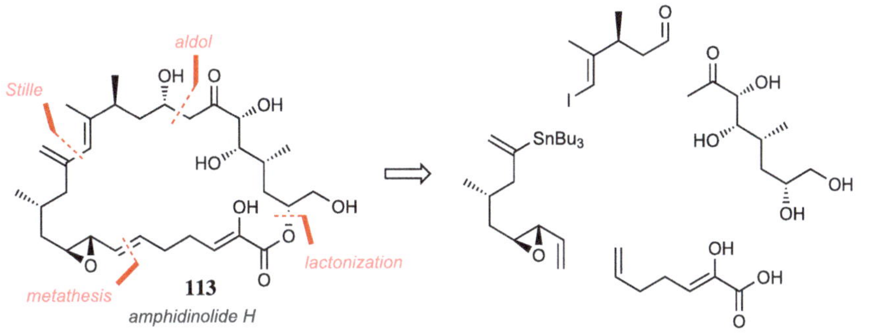

Fig. 11.64 Fürstner's retrosynthesis of amphidinolide H [233]

Herzon and his team attempted to approach a quantification of convergence by examining their previous total syntheses for the number of steps after fragment coupling and for the construction of each fragment, as well as analyzing the number of bonds, stereocenters, and rings that were formed in the coupling step [234]. A convergent route does not necessarily mean that the number of steps is reduced compared to a linear approach, as can easily be illustrated by the hypothetical sequence shown in Fig. 11.65. Both the linear and the convergent routes encompass a total of seven steps.

The above model reactions exemplify that the definition of the longest linear sequence, often cited in academic projects, can obscure the true effort of a synthesis. The starting point in a convergent synthesis typically cannot be clearly pinpointed: beginning of the longest sequence, structurally largest building block, most expensive starting material, etc. If the costs of a certain reactant, for example of building block B in Fig. 11.65, are decisive, the yield based on this can be significantly increased by switching to a convergent strategy: assuming an average yield of 80% per step an improvement of 21 to 51% ensues![XI]

In general, a highly convergent synthesis results in a reduction of the mass of employed reactants and intermediates, which has a positive effect on costs and throughput. For example, the synthesis of the octapeptide **114** necessitates seven coupling steps in a fully linear approach. If instead the last step constitutes the attachment of a longer building block, thus increasing the convergence of the synthesis, an additional subsequent step is required

---

[XI] According to Fig. 11.65, building block B proceeds through seven steps in the linear route, but only requires three in the convergent case, hence 0.8^3 *vs.* 0.8^7.

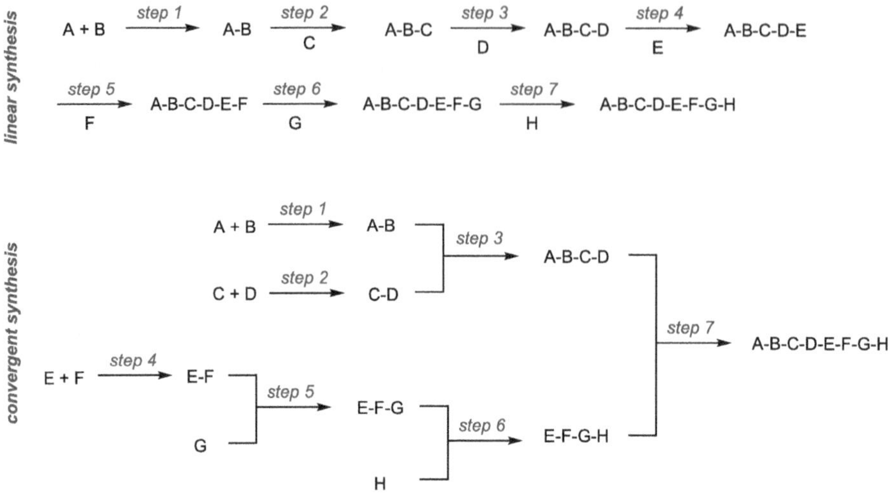

Fig. 11.65 diagram:

linear synthesis

A + B —step 1→ A-B —step 2 (C)→ A-B-C —step 3 (D)→ A-B-C-D —step 4 (E)→ A-B-C-D-E

—step 5 (F)→ A-B-C-D-E-F —step 6 (G)→ A-B-C-D-E-F-G —step 7 (H)→ A-B-C-D-E-F-G-H

convergent synthesis

A + B —step 1→ A-B

C + D —step 2→ C-D

A-B + C-D —step 3→ A-B-C-D

E + F —step 4→ E-F

E-F + G —step 5→ E-F-G

E-F-G + H —step 6→ E-F-G-H

A-B-C-D + E-F-G-H —step 7→ A-B-C-D-E-F-G-H

Fig. 11.65 Comparison of the number of steps in a hypothetical linear and convergent synthesis

(Boc-deprotection). However, this additional step can quickly be overcompensated and a convergent approach has a positive effect overall if the average yield per step lies above 75–80%. When using a tri- or tetrapeptide as the last building block (route 3 or 4), a high average yield accomplishes a decrease in the amount of reactants by more than half (see Fig. 11.66).

Despite an increased yield with regard to a single building block, a switch to a convergent strategy does not infer that the total yield is higher compared to a linear approach. Apart from a lower reactant input, the further advantages comprise an increased flexibility, an enhanced compatibility of a reaction sequence with the present functional groups, as well as an improved robustness against challenges of chemo-, regio-, and diastereoselectivity [181].

First of all, a convergent synthesis is significantly more robust against perturbations in the synthetic process. The possibility of one or more steps failing, especially when resorting to novel chemistry whose broad applicability has not yet been fully elucidated, is significant and should ideally be considered in the synthetic approach. Thus, the incorporation of alternative methods or branching points for each questionable and/or critical step that can be resorted to only seems prudent. If a key step in a linear synthesis proves unsuccessful, the entire synthesis must be started from scratch. If a key step in one of the fragments fails, only the assembly of the problematic fragment needs to be revised. Secondly, the fragments are usually bearing fewer functional groups than in a linear sequence prior to coupling, resulting in the omission of some side reactions from the outset. Also, in most cases, each fragment is subjected to fewer transformations, which also often improves chemoselectivity of the entire sequence. In the best case, such an approach can even be used to completely avoid the use of protecting groups (see Sect. 11.1). Third of all, due to the more flexible approach, a variation of the molecular structure often proves easier, which is particularly advantageous in divergent syntheses or drug discovery/lead optimization. Last of all, as the total weight of

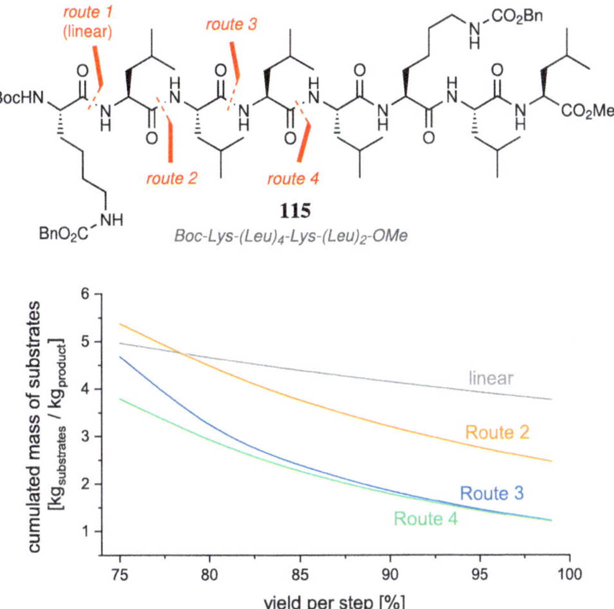

Fig. 11.66 Effect of increasing convergence on reactant use

material processed in a convergent strategy is reduced, this increases to overall throughput as the space-time yield improves. Also, parallel processing becomes viable, especially in an industrial setting.

Just as a convergent approach can lead to a reduction in the amount of starting material and thus the production costs, synthesis steps with i. low yield, ii. high catalyst/starting material costs, and iii. high dilution can also be strategically placed [181]. For example, if a resolution step via diastereomeric salt formation is required, said separation of the stereoisomers could theoretically be incorporated at any point in the synthetic sequence. The overall yield will not change. However, if a step with *low yield* is placed early in the route, all subsequent steps would necessitate less reagents, which has a positive effect on the cost of the overall process (see Fig. 11.67).

Given the illustrated impact of starting material quantities shown in Fig. 11.67, a 58% increase in total weight of substrates would be required if the low-yielding step were placed at the end of a sequence instead of at its beginning.

Reactions that utilize *cost-intensive* reagents, e.g., expensive starting materials or catalysts, instead require the opposite placement. The later a step relying on cost-intensive reagents is carried out, the less of them is needed for the transformation, as the amount of intermediates inevitably decreases during the synthesis through less than perfect yield. The same principle applies to *highly diluted* reactions: The smaller the amount of a substrate, the less solvent will be needed, which is why these reactions should ideally be placed late in a synthetic route if they cannot be avoided altogether.

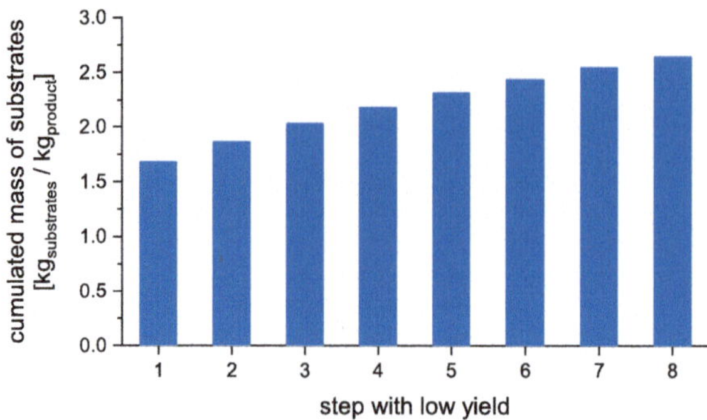

Fig. 11.67 Impact of the placement of a step with 50% yield in an eight-step linear sequence on the cumulative reagent weight. *Assumptions*: The remaining steps proceed with 90% yield. The molar mass of each building block is 50 g/mol, the product thus amounts to 450 g/mol

In addition to economic (raw material costs, space-time yield) and sustainability factors (waste quantity and identity), a number of other criteria play a decisive role in industrial syntheses: The reproducibility of the reaction and the product quality, but also, as touched upon in the previous example of sertraline, process safety as well as a reliable and long-term access to raw materials are aspects that can tip the balance for or against a route. An implementation in existing facilities is also desirable, as an invest into new assets can thus be avoided [181, 235].

Bristol-Myers Squibb required larger quantities of the IRAK4 kinase inhibitor BMS-986236 for its development as a potential drug candidate (**115**, see Fig. 11.68) [236].

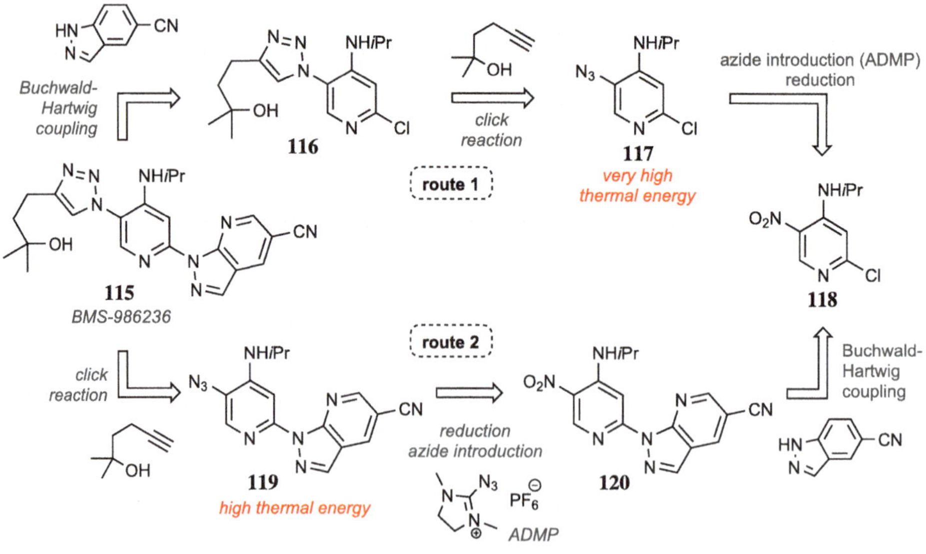

Fig. 11.68 Optimization of the synthesis of BMS-986236 by reorganization of the key steps [236]

The original route 1 showed a low overall yield (18%) and a high trace Pd contamination in the final API of BMS-986236 due to the low yield of the Buchwald–Hartwig coupling (**116**→**115**) in the final step. Apart from the need for a proprietary ligand for the CN coupling (Me$_4t$Bu-XPhos), the energy content of **117** (1111 J/g) and its sensitivity to impact carried a considerable safety risk. In the optimized route 2, the thermal stability of the key azide was increased and the safety hazard mitigated: the energy content of the azide **119** (650 J/g) was reduced compared to **117** due to the higher C/N ratio in the molecule, and it no longer displayed a shock sensitivity. Additionally, the CN coupling could be carried out with a generic ligand (XantPhos), reducing the cost of the reaction by 90% and concomitantly increasing the yield to 84%. The overall yield was thus improved to an acceptable 41%. Despite extensive screening, ADMP remained the best reagent for azide introduction. The transfer reagent is hampered by a low shelf life and therefore difficult to procure on large scale. As a consequence, an in-house preparation of ADMP was pursued, enabling the isolation of almost 500 g of **115** in total [236].

The development of the virostatic agent fostemsavir (**128**) by Bristol-Myers Squibb spanned almost two decades from identification of the first lead structures to its approval as an oral anti-HIV drug in 2020. Since the resorption of the free azaindole (active form: temsavir, **136**) is largely governed by the solubility and dissolution rate of the drug, a phosphonoxymethylation was employed to furnish the prodrug fostemsavir possessing a significantly improved bioavailability [237]. The pilot route was largely based on the initial,

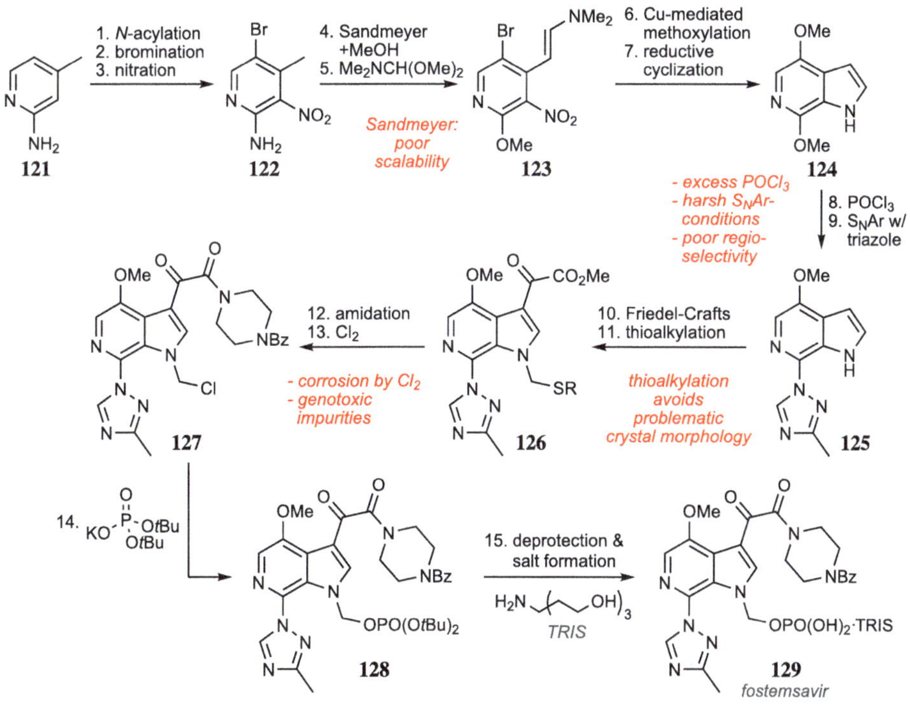

Fig. 11.69 Pilot route to fostemsavir (**129**) [238]

medicinal chemistry synthesis of the active ingredient (see Fig. 11.69). However, to ensure scalability for the preparation of over a metric ton of API needed for clinical development, some modifications and additional steps were required in the pilot route. Furthermore, a possible contamination of the active ingredient by genotoxic intermediates or reagents was a realistic concern. Highly corrosive reagents such as chlorine gas were also needed [238].

The above challenges necessitated a complete redesign of the route already established for >100 kg quantities in the transition to commercialization. The key approach relied on assembling the azaindole starting from a pyrrole (**130**) instead of a pyridine (**121**), as this strategy significantly eased the build-up of the desired substituent pattern and positively impacted the reactivity of the intermediates. The *N*-alkylation of the azaindole in the commercial route mirrored the initial Med Chem approach. However, the use of the preformed Li salt **136** instead of the free amine reduced the filtration resistance, allowing a dramatic improvement in filtration rates. The final control of impurity contamination was achieved by crystallizing the protected prodrug **128**. In-depth analysis and optimization of the processing parameters in the subsequent TRIS salt formation resulted in the generation of the target compound with consistent quality, solids properties and yield. In addition to solving the technical challenges during scale-up, resorting to starting materials available on large scale, as well as increasing (space-time) yield, the requisite quantities of reagents used (PMI, see above) were also minimized (see Fig. 11.70) [239].

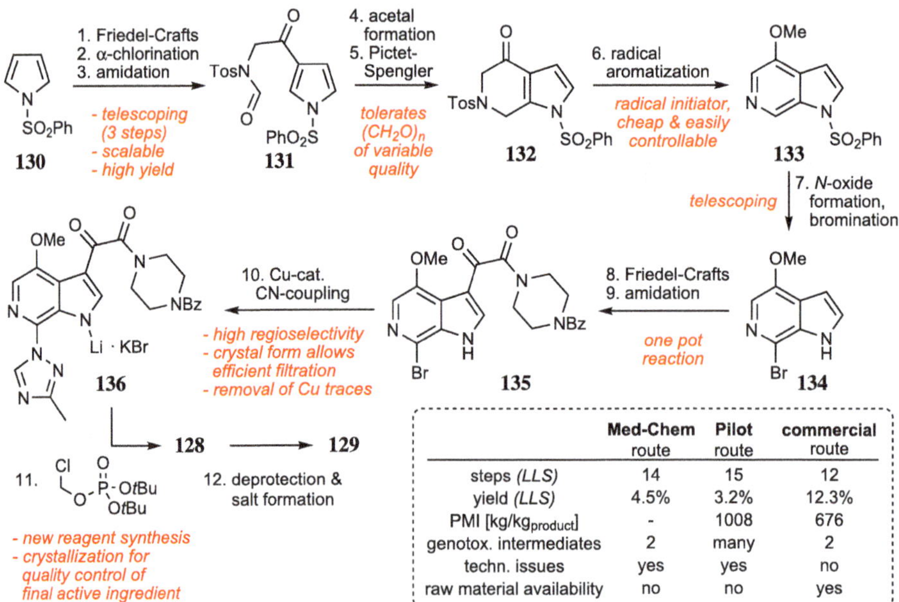

	Med-Chem route	Pilot route	commercial route
steps *(LLS)*	14	15	12
yield *(LLS)*	4.5%	3.2%	12.3%
PMI [kg/kg$_{product}$]	-	1008	676
genotox. intermediates	2	many	2
techn. issues	yes	yes	no
raw material availability	no	no	yes

Fig. 11.70 Commercial route to fostemsavir (**129**) [239]

In an academic setting, optimizations of already published syntheses by a research group in the shape of a second- or third-generation approach are encountered much less frequently. However, some noteworthy examples have been published, in which the optimization was sometimes necessary for the provision of larger quantities of a target molecule, for example for (pre-)clinical testing. Alternatively, either the impact of incorporating a newly developed method could be demonstrated to great effect or the previous route shortened and simplified in the subsequent syntheses [236, 240–246]. The groups of Hayakawa and Kigoshi revisited their synthesis of the potent antitumor marine macrolide aplyronine A (**137**) to optimize steps with low yield and only moderate stereoselectivity. While their original synthesis relied on a passably stereoselective Julia olefination and a Yamaguchi macrolactonization as key steps, the optimized route resorted to an intermolecular esterification to first link the two fragments and a Nozaki–Hiyama–Kishi coupling to afford the macrocyclic framework. This approach reduced the number of steps (absolute and LLS) and more than tripled the overall yield (see Fig. 11.71) [247, 248].

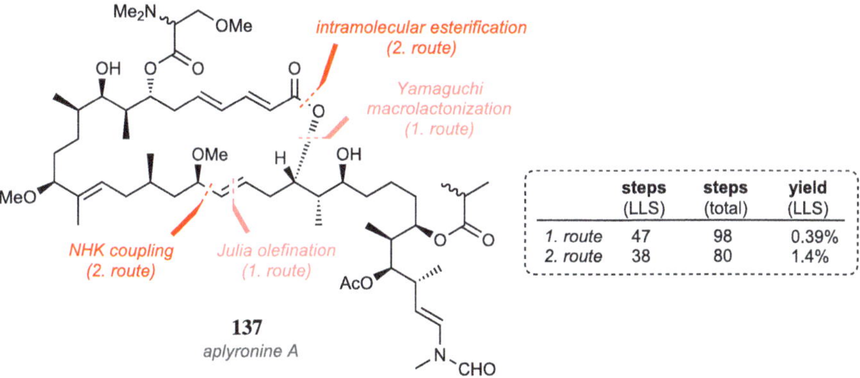

	steps (LLS)	steps (total)	yield (LLS)
1. route	47	98	0.39%
2. route	38	80	1.4%

Fig. 11.71 Optimization of the total synthesis of aplyronine A (**137**) by Hayakawa and Kigoshi [247, 248]

Dow published an overview of the various process routes to several sulfonamide herbicides and disclosed the evaluation criteria which were considered during the route selection phase for the commercial syntheses (see Fig. 11.72) [179].

The sulfonamide bond was consistently formed as final step in the highlighted three active ingredients from the corresponding sulfonyl chlorides and amines. Access to the sulfonyl chlorides (**141**, **148**, **154**) thus proved paramount to the commercial production of the herbicides **138-140**.

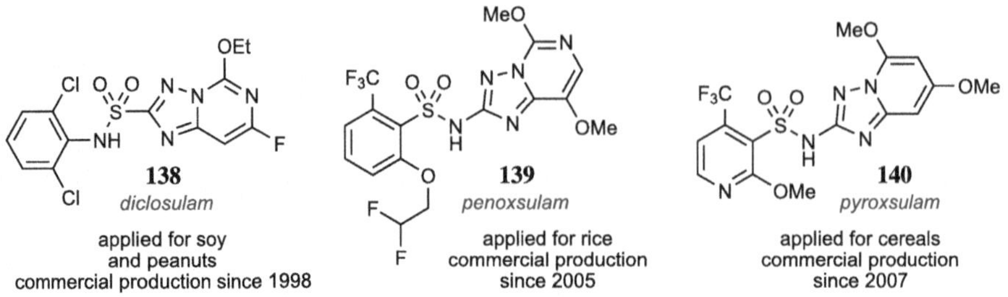

Fig. 11.72 Sulfonamide herbicides developed and marketed by Dow/Corteva

In the synthesis of diclosulam (**138**), four competing routes to **138** were developed, one of which was preferred on the basis of process safety, atom economy, and the availability of the starting materials (see Fig. 11.73). The fused triazolopyrimidine heterocycle in **141** could either be assembled at a late stage in the synthesis via rearrangement (**142**, **144**, **145**) or directly from the 1,2,4-triazole (**143**). The product sample required for toxicity studies was prepared via route 1, which is why a significant deviation in by-products, particularly occurring in route 2, posed a considerable additional hurdle. The use of CS_2 to build the triazolopyrimidine framework was identified as a safety concern, as it has an autoignition temperature of only 90 °C and a very broad flammability envelope (1.3–50 vol% in air). Additional safety precautions and mitigation measures imply additional expenditures in terms of capital and operating costs. Route 2 was CS_2-free, but displayed a lower overall yield, afforded a higher waste stream, and thus required an added effort in the purification of the crude product. Since intermediate **146** from route 1 exhibited significant dermal toxicity, its formation was avoided in the further routes. The third route similarly constructed the heterocyclic moiety relatively late, providing a comparable spectrum of by-products as in the production of the tox lot. The atom economy of the process was not optimal due to the necessary protection of the labile sulfur in the sulfide intermediates **142-144**. The superfluous steps could be circumvented by directly transforming the disulfide **145** as the corresponding protected thiol to sulfonyl chloride **141**. This additionally enabled the use of crude **141** as part of a telescoping sequence, which further improved the efficiency of the route. The lower toxicity of the intermediates, the similarity of the by-products to the tox sample, and the higher yield and efficiency as well as the avoidance of protection/deprotection steps ultimately tipped the balance in favor of route 4. Despite the safety concerns, the decision was taken to use CS_2 as source of sulfur. However, the risks associated with its storage and transport could be circumvented by generating it *in situ*. A challenge identified following the selection of route 4 revolved around the use of anhydrous cyanamide (**147**). The reagent was only available in limited quantities at considerable cost from a single supplier, which was assessed to be an exclusion criterion given the possible supply bottleneck and raw material expenditures. The substitution with aqueous cyanamide, which was much more readily and cheaply available, was subsequently incorporated into the commercial synthesis [179].

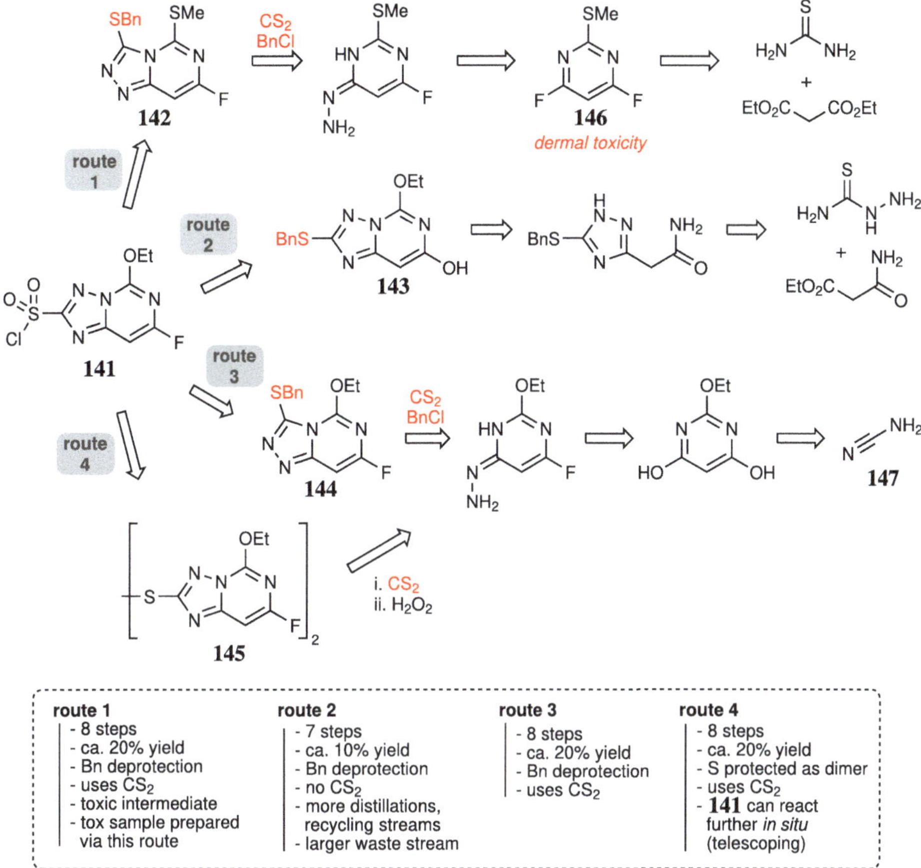

route 1	route 2	route 3	route 4
- 8 steps	- 7 steps	- 8 steps	- 8 steps
- ca. 20% yield	- ca. 10% yield	- ca. 20% yield	- ca. 20% yield
- Bn deprotection	- Bn deprotection	- Bn deprotection	- S protected as dimer
- uses CS₂	- no CS₂	- uses CS₂	- uses CS₂
- toxic intermediate	- more distillations,		- **141** can react
- tox sample prepared	recycling streams		further *in situ*
via this route	- larger waste stream		(telescoping)

Fig. 11.73 Evaluated routes to diclosulam (**138**) [179]

In the process development of penoxsulam (**139**), three synthetic routes were evaluated to be viable (see Fig. 11.74). Early stage cost estimates showed that all routes considered had roughly the same cost of manufacturing despite the different number of stages, which is why other criteria became instrumental for route selection. The sulfonyl chloride **148** could be disconnected to three differently substituted benzotrifluorides. The sulfur in **148** was incorporated by reacting the regioselectively lithiated arenes with elemental sulfur or dipropyl disulfide. Route 3 could quickly be ruled out due to the thermal lability of **152**. The elimination of LiF generates a highly reactive benzyne, which could lead to an uncontrolled thermal runaway. In addition, as the tox sample was produced via route 1, a commercialization via route 3 would have induced an additional expenditure and delayed production. Furthermore, an additional crystallization and the use of potentially toxic solvents were necessary on this path to ensure the purity and yield of **148**. The advantage of route 2 over the first route lay in the avoidance of a protection of the phenolic hydroxy group as acetal, which reduced the total number of steps. However, it relied on the odorous reagent dipropyl disulfide as sulfur source, which also possesses high water toxicity. The methyl propyl thioether generated via demethylation of the anisole constitutes a similarly malodorous by-product and is volatile. Since control of fugitive odors can be a challenge on the production scale and experience with the chemistry inherent to route 1 had already been gathered on a pilot scale, the longest route was chosen for commercial production counter to intuition [179].

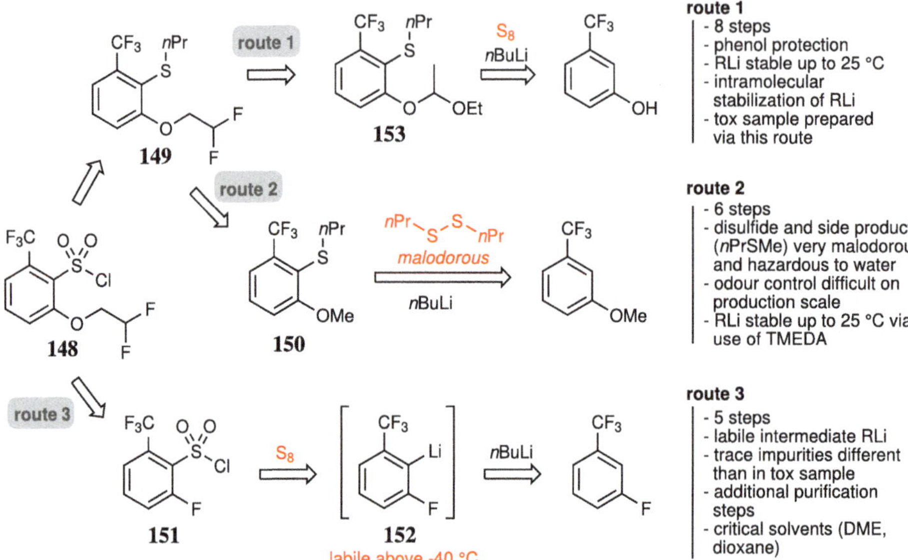

Fig. 11.74 Evaluated routes to penoxsulam (**139**) [179]

In the approach to pyroxsulam (**140**), the key building block **154** was disconnected to fluorinated anisole **155** using known chemistry (lithiation and reaction with S_8, followed by oxidative chlorination). The primary route selection decision revolved around two very different synthetic strategies: route 1 was shorter and had superior yield. In addition, γ-picoline **156** was readily available on a large scale and correspondingly cheap. However, the handling of elemental chlorine and especially of HF required specialized equipment/technology for the transformation of **156** to **155**. In the alternative cyclization route 2, an additional step was necessary, the overall yield was lower, and the material costs of the building block **157** were significantly higher than those of **156**. Nevertheless, the decision was ultimately taken to pursue the longer and more cost-intensive route 2: since the execution of the chemistry required for route 1 could only take place in-house in the existing facility, a potential added cost related to the projected loss of production time of other high-demand, exceedingly profitable products needed to be considered. The more conventional chemistry of route 2 allowed it to be run at standard contract facilities without significant capital investment.

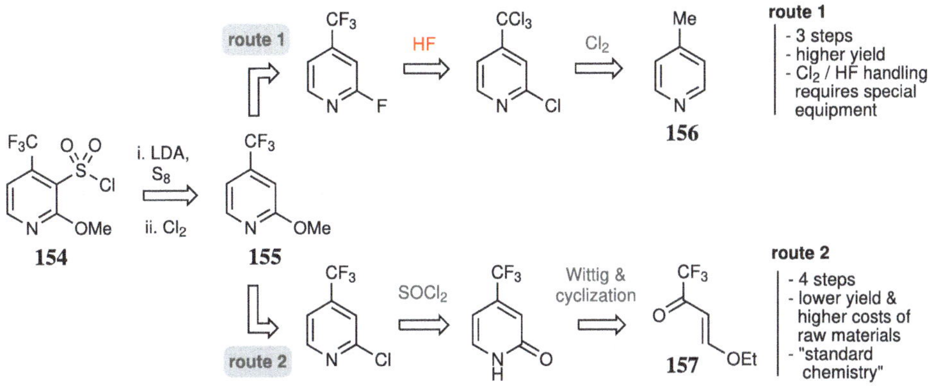

Fig. 11.75 Viable routes to pyroxsulam (**140**) [179]

The two routes were compared in a 10-year cost/revenue analysis, revealing that even with modest yield assumptions and expensive starting materials, route 2 had a significant cost advantage if carried out at a contract manufacturer. This strategy minimized the upfront capital risk when the commercial success of a new product yet is unproven, as investments can thus be avoided or postponed to a later date (see Fig. 11.75). Route 2 could be subsequently optimized following its selection, further reducing the manufacturing costs [179].

These three industrial examples certainly do not cover the entire range of decision criteria that may be considered for selecting a commercial route. However, they do provide a rare and very instructive insight into why the choice for a longer and significantly more expensive production route can be justified.

References

1. K. C. Nicolaou, D. Vourloumis, N. Winssinger, P. S. Baran, *Angew. Chem. Int. Ed.* **2000**, *39*, 44–122
2. L. W. Hernandez, D. Sarlah, *Chem. Eur. J.* **2019**, *25*, 13248–13270
3. K. C. Nicolaou, C. R. H. Hale, *Natl. Sci. Rev.* **2014**, *1*, 233–252
4. A. M. Armaly, Y. C, DePorre, E. J. Groso, P. S. Riehl, C. S. Schindler, *Chem. Rev.* **2015**, *115*, 9232–9276
5. R. W. Armstrong, J.-M. Beau, S. H. Cheon, W. J. Christ, H. Fujioka, W.-H. Ham, L. D. Hawkins, H. Jin, S. H. Kang, Y. Kishi, M. J. Martinelli, W. W. McWhorter, M. Mizuno, M. Nakata, A. E. Stutz, F. X. Talamas, M. Taniguchi, J. A. Tino, K. Ueda, J. Uenishi, J. B. White, M. Yonaga, *J. Am. Chem. Soc.* **1989**, *111*, 2525–2530
6. R. W. Armstrong, J.-M. Beau, S. H. Cheon, W. J. Christ, H. Fujioka, W.-H. Ham, L. D. Hawkins, H. Jin, S. H. Kang, Y. Kishi, M. J. Martinelli, W. W. McWhorter, M. Mizuno, M. Nakata, A. E. Stutz, F. X. Talamas, M. Taniguchi, J. A. Tino, K. Ueda, J. Uenishi, J. B. White, M. Yonaga, *J. Am. Chem. Soc.* **1989**, *111*, 7530–7533
7. V. Farina, J. D. Brown, *Angew. Chem. Int. Ed.* **2006**, *45*, 7330–7334

8. M. A. Sierra, M. C. de la Torre, *Dead Ends and Detours: Direct Ways to Successful Total Synthesis*, Wiley VCH, **2004**.

9. A. Fürstner, P. Karier, F. Ungeheuer, A. Ahlers, F. Anderl, C. Wille, *Angew. Chem. Int. Ed.* **2019**, *58*, 248–253

10. A. B. Smith, III, B. S. Freeze, *Tetrahedron* **2008**, *64*, 261–298

11. Z. Yu, R. J. Ely, J. P. Morken, *Angew. Chem. Int. Ed.* **2014**, *53*, 9632–9636

12. K. C. Nicolaou, W.-M. Dai, *Angew. Chem. Int. Ed. Engl.* **1991**, *30*, 1387–1416

13. S. V. Ley, A. Abad-Somovilla, J. C. Anderson, C. Ayats, R. Bänteli, E. Beckmann, A. Boyer, M. G. Brasca, A. Brice, H. B. Broughton, B. J. Burke, E. Cleator, D. Craig, A. A. Denholm, R. M. Denton, T. Durand-Reville, L. B. Gobbi, M. Göbel, B. L. Gray, R. B. Grossmann, C. E. Gutteridge, N. Hahn, S. L. Harding, D. C. Jennens, L. Jennens, P. J. Lovell, H. J. Lovell, M. L. de la Puente, H. C. Kolb, W.-J. Koot, S. L. Maslen, C. F. McCusker, A. Mattes, A. R. Pape, A. Pinto, D. Santafianos, J. S. Scott, S. C. Smith, A. Q. Somers, C. D. Spilling, F. Stelzer, P. L. Toogood, R. M. Turner, G. E. Veitch, A. Wood, C. Zumbrunn, *Chem. Eur. J.* **2008**, *14*, 10683–10704

14. E. J. Corey, *Chem. Soc. Rev.* **1988**, *17*, 111–133

15. E. J. Corey, X.-M. Cheng, *The Logic of Chemical Synthesis*, Wiley, New York, **1995**.

16. S. Warren, *Designing Organic Syntheses: A programmed Introduction to the Synthon Approach*, Wiley Blackwell, **1978**.

17. S. Warren, *Organic synthesis: The Disconnection Approach*, John Wiley & Sons, **1983**

18. R. W. Hoffmann, *Elemente der Syntheseplanung*, Elsevier, München, **2006**.

19. P. Wyatt, S. Warren, *Organic Synthesis: Strategy and Control*, Wiley, **2007**.

20. T. Hudlicky, J. Reed, *The Way of Synthesis: Evolution of Design and Methods for Natural Products*, Wiley VCH, **2007**.

21. S. Hanessian, S. Giroux, B. L. Merner, *Design and Strategy in Organic Synthesis: From the Chiron Approach to Catalysis*, Wiley-VCH, **2013**.

22. N. A. Doering, R. Sarpong, R. W. Hoffmann, *Angew. Chem. Int. Ed.* **2020**, *59*, 10722–10731

23. Z. Tu, T. Stuyver, C. W. Coley, *Chem. Sci.* **2023**, *14*, 226–244.

24. S. Szymkuc, E. P. Gajewska, T. Klucznik, K. Molga, P. Dittwald, M. Startek, M. Bajczyk, B. A. Grzybowski, *Angew. Chem. Int. Ed.* **2016**, *55*, 5904–5937

25. C. W. Coley, W. H. Green, K. F. Jensen, *Acc. Chem. Res.* **2018**, *51*, 1281–1289

26. K. Molga, S. Szymkuc, B. A. Grzybowski, *Acc. Chem. Res.* **2021**, *54*, 1094–1106

27. A. Cook, A. P. Johnson, J. Law, M. Mirzazadeh, O. Ravitz, A. Simon, *WIREs Comput. Mol. Sci.* **2012**, *2*, 79–107

28. M. H. Todd, *Chem. Soc. Rev.* **2005**, *34*, 247–266

29. W. Beker, R. Roszak, A. Wołos, N. H. Angello, V. Rathore, M. D. Burke, B. A. Grzybowski, *J. Am. Chem. Soc.* **2022**, *144*, 4819–4827

30. B. Mikulak-Klucznik, P. Golebiowska, A. A. Bayly, O. Popik, T. Klucznik, S. Szymkuc, E. P. Gajewska, P. Dittwald, O. Staszewska-Krajewska, W. Beker, T. Badowski, K. A. Scheidt, K. Molga, J. Mlynarski, M. Mrksich, B. A. Grzybowski, *Nature* **2020**, *588*, 83–88

31. D. Seebach, *Angew. Chem. Int. Ed. Engl.* **1979**, *18*, 283–336

32. I. Hayakawa, S. Atarashi, M. Imamura, Y. Kimura, *EP357047A1* **1990**.

33. M. Kruszyk, M. Jessing, J. L. Kristensen, M. Jørgensen, *J. Org. Chem.* **2016**, *81*, 5128–5134

34. X. Fu, M. B. Hossain, D. van der Helm, F. J. Schmitz, *J. Am. Chem. Soc.* **1994**, *116*, 12125–12126

35. X. Fu, M. B. Hossain, D. van der Helm, F. J. Schmitz, *J. Org. Chem.* **1997**, *62*, 3810–3819

36. M. E. Layton, C. A. Morales, M. D. Shair, *J. Am. Chem. Soc.* **2002**, *124*, 773–775

37. S. Hanessian, J. Franco, B. Larouche, *Pure Appl. Chem.* **1990**, *62*, 1887–1990

38. S. Hanessian, *J. Org. Chem.* **2012**, *77*, 6657–6688

39. G. Illuminati, L. Mandolini, *Acc. Chem. Res.* **1981**, *14*, 95–102
40. K. B. Wiberg, *Angew. Chem. Int. Ed. Engl.* **1986**, *25*, 312–322
41. E. J. Corey, W. J. Howe, H. W. Orf, D. A. Pensak, G. Petersson, *J. Am. Chem. Soc.* **1975**, *97*, 6116–6124
42. K. C. Nicolaou, P. S. Baran, Y.-L. Zhong, H.-S. Choi, W. H. Yoon, Y. He, K. C. Fong, *Angew. Chem. Int. Ed.* **1999**, *38*, 1669–1675
43. K. C. Nicolaou, P. S. Baran, Y.-L. Zhong, K. C. Fong, Y. He, H.-S. Choi, *Angew. Chem. Int. Ed.* **1999**, *38*, 1676–1678
44. C. Chem, M. E. Layton, S. M. Sheehan, M. D. Shair, *J. Am. Chem. Soc.* **2000**, *122*, 7424–7425
45. N. Waizumi, T. Itoh, T. Fukayama, *J. Am. Chem. Soc.* **2000**, *122*, 7825–7826
46. Q. Tan, S. J. Danishefsky, *Angew. Chem. Int. Ed.* **2000**, *39*, 4509–4511
47. D. Meng, Q. Tan, S. J. Danishefsky, *Angew. Chem. Int. Ed.* **1999**, *38*, 3197–3201
48. W. Oppolzer, *Acc. Chem. Res.* **1982**, *15*, 135–141
49. W. Oppolzer, T. Godel, *J. Am. Chem. Soc.* **1978**, *100*, 2583–2584
50. M. R. Krout, C. E. Henry, T. Jensen, K.-L. Wu, S. C. Virgil, B. M. Stoltz, *J. Org. Chem.* **2018**, *83*, 6995–7009
51. J. D. White, Y. Li, J. Kim, M. Terinek, *J. Org. Chem.* **2015**, *80*, 11806–11817
52. M. J. Williams, H. L. Deak, M. L. Snapper, *J. Am. Chem. Soc.* **2007**, *129*, 486–487
53. L. Call, *Chemie in unserer Zeit* **1978**, *12*, 122–133
54. R. B. Woodward, *Pure Appl. Chem.* **1968**, *17*, 519–547
55. G. Stork, R. A. Kretchmer, R. H. Schlessinger, *J. Am. Chem. Soc.* **1968**, *90*, 1647–1648
56. M. Schellhaas, H. Waldmann, *Angew. Chem. Int. Ed. Engl.* **1996**, *35*, 2057–2083
57. R. W. Hoffmann, *Synthesis* **2006**, 3531–3541
58. I. S. Young, P. S. Baran, *Nat. Chem.* **2009**, *1*, 193–205
59. R. N. Saicic, *Tetrahedron* **2014**, *70*, 8183–8218
60. C. Hui, F. Chen, F. Pu, J. Xu, *Nat. Rev. Chem.* **2019**, *3*, 85–107
61. R. A. Fernandes, P. Kumar, P. Choudhary, *Chem. Commun.* **2020**, *56*, 8569–8590
62. T. D. Nelson, R. D. Crouch, *Synthesis* **1996**, 1031–1069.
63. P. J. Kocienski, *Protecting Groups*, Thieme, 3rd ed., **2005**.
64. P. G. M. Wuts, *Greene's Protective Groups in Organic Synthesis*, John Wiley & Sons, 5th ed., **2014**.
65. N. A. Isley, Y. Endo, Z.-C. Wu, B. C. Covington, L. B. Bushin, M. R. Seyedsayamdost, D. L. Boger, *J. Am. Chem. Soc.* **2019**, *141*, 17361–17369
66. S. Williams, J. Jin, S. B. J. Kan, M. Li, L. J. Gibson, I. Paterson, *Angew. Chem. Int. Ed.* **2017**, *56*, 645–649
67. B. M. Trost, J. Waser, A. Meyer, *J. Am. Chem. Soc.* **2007**, *129*, 12556–14557
68. P. A. Wender, H. Takahashi, B. Witulski, *J. Am. Chem. Soc.* **1995**, *117*, 4720–4721
69. B. M. Trost, F. D. Toste, H. Shen, *J. Am. Chem. Soc.* **2000**, *122*, 2379–2380
70. B. M. Trost, Y. Hu, D. B. Horne, *J. Am. Chem. Soc.* **2007**, *129*, 11781–11790
71. A. Bérubé, I. Drutu, J. L. Wood, *Org. Lett.* **2006**, *8*, 5421–5424
72. V. Farina, J. T. Reeves, C. H. Senanayake, J. J. Song, *Chem. Rev.* **2006**, *106*, 2734–2793
73. R. Siedlecka, *Tetrahedron* **2013**, *69*, 6331–6363
74. R. Tamatam, D. Shin, *Pharmaceuticals* **2023**, *16*, 339
75. S. Masamune, W. Choy, J. S. Petersen, L. R. Sita, *Angew. Chem. Int. Ed. Engl.* **1985**, *24*, 1–3
76. O. I. Kolodiazhnyi, *Tetrahedron* **2003**, *59*, 5953–6018
77. D. A. Evans, D. L. Rieger, T. K. Jones, S. W. Kaldor, *J. Org. Chem.* **1990**, *55*, 6260–6268
78. D. A. Evans, H. P. Ng, D. L. Rieger, *J. Am. Chem. Soc.* **1993**, *115*, 11446–11459
79. J. Kühlborn, J. Groß, T. Opatz, *Nat. Prod. Rep.* **2020**, *37*, 380–424
80. Z. G. Brill, M. L. Condakes, C. P. Ting, T. J. Maimone, *Chem. Rev.* **2017**, *117*, 11753–11795

81. S.-M. Paek, M. Jeong, J. Jo, Y. M. Heo, Y. T. Han, H. Yun, *Molecules* **2016**, *21*, 951
82. G. Casiraghi, F. Zanardi, G. Raw, P. Spanu, *Chem. Rev.* **1995**, *95*, 1677–1716
83. H.-U. Blaser, *Chem. Rev.* **1992**, *92*, 935–952
84. K. C. Nicolaou, D. Pappo, K. Y. Tsang, R. Gibe, D. Y.-K. Chen, *Angew. Chem. Int. Ed.* **2008**, *47*, 944–946
85. Z. G. Brill, H. K. Grover, T. J. Maimone, *Science* **2016**, *352*, 1078–1082
86. R. W. Dugger, J. A. Ragan, D. H. B. Ripin, *Org. Process Res. Dev.* **2005**, *9*, 253–258
87. L.-C. Yang, H. Deng, H. Renata, *Org. Process Res. Dev.* **2022**, *26*, 1925–1943
88. L. Pasteur, *Ann. Chim. Phys.* **1850**, *28*, 56–99
89. E. Fogassy, M. Nógrádi, D. Kozma, G. Egri, E. Pálovics, V. Kiss, *Org. Biomol. Chem.* **2006**, *4*, 3011–3030
90. K. G. Gadamasetti, *Process Chemistry in the Pharmaceutical Industry*, CRC Press, **1999**
91. K. G. Gadamasetti, T. Braish, *Process Chemistry in the Pharmaceutical Industry. Volume 2. Challenges in an Ever-changing Climate*, CRC Press, **2007**
92. C. Bournouf, E. Auclair, N. Avenel, B. Bertin, C. Bigot, A. Calvet, K. Chan, C. Durand, V. Fasquelle, F. Féru, R. Gilbertsen, H. Jacobelli, A. Kebsi, E. Lallier, J. Maignel, B. Martin, S. Milano, M. Ouagued, Y. Pascal, M.-P. Pruniaux, J. Puaud, M.-N. Rocher, C. Terrasse, R. Wrigglesworth, A. M. Doherty, *J. Med. Chem.* **2000**, *43*, 4850–4867
93. J. F. Larrow, E. N. Jacobsen, Y. Gao, Y. Hong, X. Nie, C. M. Zepp, *J. Org. Chem.* **1994**, *59*, 1939–1942
94. D. Kozma, Z. Madarász, C. Kassai, E. Fogassy, *Chirality* **1999**, *11*, 373–375
95. K. S. Kim, J. S. Sack, J. S. Tokarski, L. Qian, S. T. Chao, L. Leith, Y. F. Kelly, R. N. Misra, J. T. Hunt, S. D. Kimball, W. G. Humphreys, B. S. Wautlet, J. G. Mulheron, K. R.Webster, *J. Med. Chem.* **2000**, *43*, 4126–4134
96. K. Sakai, R. Sakurai, A. Yuzawa, N. Hirayama, *Tetrahedron: Asymmetry* **2003**, *14*, 3713–3718
97. J. Bálint, Z. Hell, I. Markovits, L. Párkányi, E. Fogassy, *Tetrahedron: Asymmetry* **2001**, *11*, 1323–1329
98. J. Qiu, J. M. Stevens, *Org. Process Res. Dev.* **2020**, *24*, 1725–1734
99. H. Pellissier, *Adv. Synth. Catal.* **2011**, *353*, 1613–1666
100. V. S. Martin, S. S. Woodard, T. Katsuki, Y. Yamada, M. Ikeda, K. B. Sharpless, *J. Am. Chem. Soc.* **1981**, *103*, 6237–6240
101. C.-S. Chen, Y. Fujimoto, G. Girdaukas, C. J. Sih, *J. Am. Chem. Soc.* **1982**, *104*, 7294–7299
102. J. C. Ruble, J. Tweddell, G. C. Fu, *J. Org. Chem.* **1998**, *63*, 2794–2795
103. B. Tao, J. C. Ruble, D. A. Hoic, G. C. Fu, *J. Am. Chem. Soc.* **1999**, *121*, 5091–5092
104. E. N. Jacobsen, F. Kakiuchi, R. G. Konsler, J. F. Larrow, M. Tokunaga, *Tetrahedron Lett.* **1997**, *38*, 773–776
105. A. Fischli, M. Klaus, H. Mayer, P. Schönholzer, R. Rüegg, *Helv. Chim. Acta* **1975**, *58*, 564–584
106. M. Nada, S. Terashima, S. Yamada, *Tetrahedron* **1980**, *36*, 3161–3170
107. E. Schoffers, A. Golebiowski, C. R. Johnson, *Tetrahedron* **1995**, *52*, 3769–3826
108. G. Carrea, S. Riva, *Angew. Chem. Int. Ed.* **2000**, *39*, 2226–2254
109. K. M. Koeller, C.-H. Wong, *Nature* **2001**, *409*, 232–240
110. M. T. Reetz, *Curr. Opin. Chem. Biol.* **2002**, *6*, 145–150
111. V. Gotor-Fernández, R. Brieva, V. Gotor, *J. Mol. Cat. B* **2006**, *40*, 111–120
112. A. Ghanem, *Tetrahedron* **2007**, *63*, 1721–1754
113. I. Alfonso, V. Gotor, *Chem. Soc. Rev.* **2004**, *33*, 201–209
114. Y. Hirose, K. Kariya, I. Sasaki, Y. Kurono, H. Ebiike, K. Achiwa, *Tetrahedron Lett.* **1992**, *33*, 7157–7160
115. V. Prelog, *Pure Appl. Chem.* **1964**, *9*, 119–130
116. R. J. Kazlauskas, A. N. E. Weissfloch, A. T. Rappaport, L. A. Cuccia, *J. Org. Chem.* **1991**, *56*, 2656–2665

117. A. N. E. Weissfloch, R. J. Kazlauskas, *J. Org. Chem.* **1995**, *60*, 6959–6969

118. R. J. Kazlauskas, A. N. E. Weissfloch, *J. Mol. Catal. B: Enzym.* **1997**, *3*, 65–72

119. V. Gotor, *Org. Process Res. Dev.* **2002**, *6*, 420–426

120. A. Zaks, D. R. Dodds, *Drug Discovery Today* **1997**, *2*, 513–531

121. A. K. Saksena, V. M. Girijavallabhan, R. G. Lovey, R. E. Pike, H. Wang, A. K. Ganguly, B. Morgan, A. Zaks, M. S. Puar, *Tetrahedron Lett.* **1995**, *36*, 1787–1790

122. R. N. Patel, A. Banerjee, R. Y. Ko, J. M. Howell, W. S. Li, F. T. Comezoglu, R. A. Partyka, F. T. Szarka, *Biotechnol. Appl. Biochem.* **1994**, *20*, 23–33

123. T. D. Machajewski, C.-H. Wong, *Synthesis* **1999**, 1469–1472

124. S. C. Sinha, J. Sun, G. Miller, C. F. Barbas III, R. A. Lerner, *Org. Lett.* **1999**, *1*, 1623–1626

125. M. J. Zmijewski, B. S. Briggs, A. R. Thompson, I. G. Wright, *Tetrahedron Lett.* **1991**, *32*, 1621–1622

126. H. Pellissier, *Tetrahedron* **2011**, *67*, 3769–3802.

127. O. Pàmies, J.-E. Bäckvall, *Chem. Rev.* **2003**, *103*, 3247–3261

128. O. Verho, J.-E. Bäckvall, *J. Am. Chem. Soc.* **2015**, *137*, 3996–4009

129. D. G. Gusev, D. M. Spasyuk, *ACS Catal.* **2018**, *8*, 6851–6861

130. L. F. Tietze, U. Beifuss, *Angew. Chem. Int. Ed. Engl.* **1993**, *32*, 131–163

131. L. F. Tietze, G. Brasche, K. M. Gerricke, *Domino-Reactions in Organic Synthesis*, Wiley-VCH, Weinheim, **2006**.

132. A. Dömling, I. Ugi, *Angew. Chem. Int. Ed.* **2000**, *39*, 3168–3210

133. J. D. Sunderhaus, S. F. Martin, *Chem. Eur. J.* **2009**, *15*, 1300–1308

134. H. Eckert, *Molecules* **2017**, *22*, 349

135. J. W. Lehmann, D. J. Blair, M. D. Burke, *Nat. Rev. Chem.* **2018**, *2*, 0115

136. B. G. Vong, S. H. Kim, S. Abraham, E. A. Theodorakis, *Angew. Chem. Int. Ed.* **2004**, *43*, 3947–3951

137. B. ter Horst, B. L. Feringa, A. J. Minnaard, *Org. Lett.* **2007**, *9*, 3013–3015

138. E. M. Woerly, A. H. Cherney, E. K. Davis, M. D. Burke, *J. Am. Chem. Soc.* **2010**, *132*, 6941–6943

139. Y. Yu, A. Kononov, M. Delbianco, P. H. Seeberger, *Chem. Eur. J.* **2018**, *24*, 6075–6078

140. S. Bracegirdle, E. A. Anderson, *Chem. Soc. Rev.* **2010**, *39*, 4114–4129

141. P. R. Hanson, S. Jayasinghe, S. Maitra, J. L. Markley, *Top. Curr. Chem.* **2014**, *361*, 253–271

142. J. Marjanovic, S. A. Kozmin, *Angew. Chem. Int. Ed.* **2007**, *46*, 8854–8857

143. G. L. Nattrass, E. Díez, M. M. McLachlan, D. J. Dixon, S. V. Ley, *Angew. Chem. Int. Ed.* **2005**, *44*, 580–584

144. P. K. Park, S. J. O'Malley, D. R. Schmidt, J. L. Leighton, *J. Am. Chem. Soc.* **2006**, *128*, 2796–2797

145. S. E. Denmark, S.-M. Yang, *J. Am. Chem. Soc.* **2002**, *124*, 15196–15197

146. A. D. Brosius, L. E. Overman, L. Schwink, *J. Am. Chem. Soc.* **1999**, *21*, 700–709

147. K. C. Nicolaou, J. J. Liu, C.-K. Hwang, W.-M. Daia, R. K. Guy, *J. Chem Soc. Chem. Commun.* **1992**, 1118–1120

148. J. Gillard, R. Fortin, E. Grimm, M. Maillard, M. Tjepkema, M. Bernstein, R. Glaser, *Tetrahedron Lett.* **1991**, *31*, 1145–1148

149. W.-J. Bai, X. Wang, *Nat. Prod. Rep.* **2017**, *34*, 1345–1358

150. J. Sun, H. Yang, W. Tang, *Chem. Soc. Rev.* **2021**, *50*, 2320–2336

151. S. A. Snyder, A. M. ElSohly, F. Kontes, *Nat. Prod. Rep.* **2011**, *28*, 897–924

152. R. W. Hoffmann, *Angew. Chem. Int. Ed.* **2003**, *42*, 1096–1109

153. T. Nakata, *J. Synth. Org. Chem. Jpn.* **1998**, *56*, 940–951

154. J. M. Holland, M. Lewis, A. Nelson, *Angew. Chem. Int. Ed.* **2001**, *40*, 4082–4084

155. O. L. Chapman, M. R. Engel, J. P. Springer, J. C. Clardy, *J. Am. Chem. Soc.* **1971**, *93*, 6696–6698

156. K.-F. Cheng, Y.-C. Kong, T.-Y. Chan, *J. Chem. Soc. Chem. Commun.* **1985**, 48–49.

157. D. J. Critcher, S. Connolly, M. Willis, *J. Org. Chem.* **1997**, *62*, 6638–6657
158. D. A. Evans, B. W. Trotter, B. Côté, P. J. Comelan, L. C. Dias, A. N. Tyler, *Angew. Chem. Int. Ed. Engl.* **1997**, *36*, 2744–2747
159. J. Guo, K. J. Duffy, K. L. Stevens, P. I. Dalko, R. M. Roth, M. M. Hayward, Y. Kishi, *Angew. Chem. Int. Ed.* **1998**, *37*, 187–190
160. A. B. Smith, Q. Lin, V. A. Doughty, L. Zhuang, M. D. McBriar, J. K. Kerns, C. S. Brook, N. Murase, K. Nakayama, *Angew. Chem. Int. Ed.* **2001**, *40*, 196–199
161. I. Paterson, D. Y.-K. Chen, M. J. Coster, J. L. Acena, J. Bach, K. R. Gibson, L. E. Keown, R. M. Oballa, T. Trieselmann, D. J. Wallace, A. P. Hodgson, R. D. Norcross, *Angew. Chem. Int. Ed.* **2001**, *40*, 4055–4060
162. M. T. Crimmins, J. D. Katz, D. G. Washburn, S. P. Allwein, L. F. McAtee, *J. Am. Chem. Soc.* **2002**, *124*, 5661–5663
163. C. H. Heathcock, M. McLaughlin, J. Medina, J. L. Hubbs, G. A. Wallace, R. Scott, M. M. Claffey, C. J. Hayes, G. R. Ott, *J. Am. Chem. Soc.* **2003**, *125*, 12844–12849
164. I. Paterson, D. Y.-K. Chen, M. J. Coster, J. L. Acena, J. Bach, D. J. Wallace, *Org. Biomol. Chem.* **2005**, *3*, 2431–2440
165. M. Ball, M. J. Gaunt, D. F. Hook, A. S. Jessiman, S. Kawahara, P. Orsini, A. Scolaro, A. C. Talbot, H. R. Tanner, S. Yamanoi, S. V. Ley, *Angew. Chem. Int. Ed.* **2005**, *44*, 5433–5438
166. C. H. Heathcock, *Angew. Chem. Int. Ed. Engl.* **1992**, *31*, 665–681
167. C. H. Heathcock, S. K. Davidsen, S. G. Mills, M. A. Sanner, *J. Org. Chem.* **1992**, *57*, 2531–2544
168. C. H. Heathcock, M. M. Hansen, R. B. Ruggeri, J. C. Kath, *J. Org. Chem.* **1992**, *57*, 2544–2553
169. C. H. Heathcock, S. Piettre, *Science* **1990**, *248*, 1532–1534
170. C. H. Heathcock, S. Piettre, R. B. Ruggeri, J. A. Ragan, J. C. Kath, *J. Org. Chem.* **1992**, *57*, 2554–2566
171. X. Hu, T. J. Maimone, *J. Am. Chem. Soc.* **2014**, *136*, 5287–5290
172. L. Li, Z. Chen, X. Zhang, Y. Jia, *Chem. Rev.* **2018**, *118*, 3752–3852
173. C. K. G. Gerlinger, T. Gaich, *Chem. Eur. J.* **2019**, *25*, 10782–10791
174. J. Shimokawa, *Tetrahedron Lett.* **2014**, *55*, 6156–6162
175. R. A. Fernandes, *Chem. Commun.* **2023**, *59*, 12205–12230
176. M. Movassaghi, D. S. Siegel, S. Han, *Chem. Sci.* **2010**, *1*, 561–566
177. J. W. Cornforth, *Chem. Br.* **1975**, *11*, 432
178. K. C. Nicolaou, S. A. Snyder, *Angew. Chem. Int. Ed.* **2005**, *44*, 1012–1044
179. R. B. Leng, M. V. M. Emonds, C. T. Hamilton, J. W. Ringer, *Org. Process Res. Dev.* **2012**, *16*, 415–424
180. N. G. Anderson, *Practical Process Research & Development*, Academic Press, 2nd ed., **2012**
181. T. Y. Zhang, *Chem. Rev.* **2006**, *106*, 2583–2595
182. L. M. Jarvis, *Chem. Eng. News* **2011**, *88 (23)*, 13
183. J. S. Carey, D. Laffan, C. Thomson, M. T. Williams, *Org. Biomol. Chem.* **2006**, *4*, 2337–2347
184. C. A. Busacca, D. R. Fandrick, J. J. Song, C. H. Senanayake, *Adv. Synth. Catal.* **2011**, *353*, 1825–1864
185. S. Caille, S. Cui, M. M. Faul, S. M. Mennen, J. S. Tedrow, S. D. Walker, *J. Org. Chem.* **2019**, *84*, 4583–4603
186. E. Vitaku, D. T. Smith, J. T. Njardarson, *J. Med. Chem.* **2014**, *57*, 10257–10274
187. E. A. Ilardi, E. Vitaku, J. T. Njardarson, *J. Med. Chem.* **2014**, *57*, 2832–2842
188. D. J. Ager, *Synthesis* **2015**, *47*, 760–768
189. S. D. Roughley, A. M. Jordan, *J. Med. Chem.* **2011**, *54*, 3451–3479
190. P. Métivier, *Stud. Surf. Sci. Catal.* **2000**, *130*, 167–176
191. M. Butters, D. Catterick, A. Craig, A. Curzons, D. Dale, A. Gillmore, S. P. Green, I. Marziano, J.-P. Sherlock, W. White, *Chem. Rev.* **2006**, *106*, 3002–3027

192. T. Schaub, *Chem. Eur. J.* **2021**, *27*, 1865–1869

193. Y. Hayashi, *J. Org. Chem.* **2021**, *86*, 1–23

194. L. Capaldo, Z. Wen, T. Nöel, *Chem. Sci.* **2023**, *14*, 4230–4247

195. P. Poechlauer, J. Manley, R. Broxterman, B. Gregertsen, M. Ridemark, *Org. Process Res. Dev.* **2012**, *16*, 1586–1590

196. M. B. Plutschack, B. Pieber, K. Gilmore, P. H. Seeberger, *Chem. Rev.* **2017**, *117*, 11796–11893

197. M. Movsisyan, E. I. P. Delbeke, J. K. E. T. Berton, C. Battilocchio, S. V. Ley, C. V. Stevens, *Chem. Soc. Rev.* **2016**, *45*, 4892–4928

198. B. Gutmann, D. Cantillo, C. O. Kappe, *Angew. Chem. Int. Ed.* **2015**, *54*, 6688–6728

199. S. M. Rowe, *Org. Process Res. Dev.* **2002**, *6*, 877–883

200. A. Domokos, B. Nagy, B. Szilágyi, G. Marosi, Z. K. Nagy, *Org. Process Res. Dev.* **2021**, *25*, 721–739

201. S. K. Teoh, C. Rathi, P. Sharratt, *Org. Process Res. Dev.* **2015**, *20*, 414–431

202. B. C. Austad, T. L. Calkins, C. E. Chase, F. G. Fang, T. E. Horstmann, Y. Hu, B. M. Lewis, X. Niu, T. A. Noland, J. D. Orr, M. J. Schnaderbeck, H. Zhang, N. Asakawa, N. Asai, H. Chiba, T. Hasebe, Y. Hoshino, H. Ishizuka, T. Kajima, A. Kayano, Y. Komatsu, M. Kubota, H. Kuroda, M. Miyazawa, K. Tagami, T. Watanabe, *Synlett* **2013**, *24*, 333–337

203. A. Bauer, *Top. Heterocycl. Chem.* **2016**, *44*, 209–270

204. T. Fukuyama, H. Chiba, H. Kuroda, T. Takigawa, A. Kayano, K. Tagami, *Org. Process Res. Dev.* **2016**, *20*, 503–509

205. J. Bien, A. Davulcu, A. J. DelMonte, K. J. Fraunhoffer, Z. Gao, C. Hang, Y. Hsiao, W. Hu, K. Katipally, A. Littke, A. Pedro, Y. Qiu, M. Sandoval, R. Schild, M. Soltani, A. Tedesco, D. Vanyo, P. Vemishetti, R. E. Waltermire, *Org. Process Res. Dev.* **2018**, *22*, 1393–1408

206. P. T. Anastas, J. C. Warner, *Green Chemistry: Theory and Practice*, Oxford University Press, **1998**

207. ICH Harmonised Guideline for Elemental Impurities (Q3D R1), **2022**

208. A. Chanda, A. M. Daly, D. A. Foley, M. A. LaPack, S. Mukherjee, J. D. Orr, G. L. Reid, III, D. R. Thompson, H. W. Ward, II, *Org. Process Res. Dev.* **2015**, *19*, 63–83

209. N. M. Thomson, K. D. Seibert, S. Tummala, S. Bordawekar, W. F. Kiesman, E. A. Irdam, B. Phenix, D. Kumke, *Org. Process Res. Dev.* **2015**, *19*, 925–934

210. K. Takai, *Bull. Chem. Soc. Jpn.* **2015**, *88*, 1511–1529

211. E. O. Pentsak, D. B. Eremin, E. G. Gordeev, V. P. Ananikov, *ACS Catal.* **2019**, *9*, 3070–3081

212. C. Lübbe, A. Dumrath, H. Neumann, M. Schäffer, R. Zimmermann, M. Beller, R. Kadyrov, *ChemCatChem* **2014**, *6*, 684–688

213. P. A. Wender, *Chem. Rev.* **1996**, *96*, 1–2

214. T. Gaich, P. S. Baran, *J. Org. Chem.* **2010**, *75*, 4657–4673

215. D. S. Peters, C. R. Pitts, K. S. McClymont, T. P. Stratton, C. Bi, P. S. Baran, *Acc. Chem. Res.* **2021**, *54*, 605–617

216. J. B. Hendrickson, *J. Am. Chem. Soc.* **1975**, *97*, 5784–5800

217. B. M. Trost, *Science* **1991**, *254*, 1471–1477

218. P. L. Fuchs, *Tetrahedron* **2001**, *57*, 6855–6875

219. J. S. Cannon, L. E. Overman, *Angew. Chem. Int. Ed.* **2012**, *51*, 4288–4311

220. J. Li, M. D. Eastgate, *Org. Biomol. Chem.* **2015**, *13*, 7164–7176

221. R. B. Woodward, M. P. Cava, W. D. Ollis, A. Hunger, H. U. Daeniker, K. Schenker, *J. Am. Chem. Soc.* **1954**, *76*, 4749–4751

222. S. D. Knight, L. E. Overman, G. Pairaudeau, *J. Am. Chem. Soc.* **1993**, *115*, 9293–9294

223. S. B. Jones, B. Simmons, A. Mastracchio, D. W. C. MacMillan, *Nature* **2011**, *475*, 183–188

224. D. B. C. Martin, C. D. Vanderwal, *Chem. Sci.* **2011**, *2*, 649–651

225. J. D. Hayler, D. K. Leahy, E. M. Simmons, *Organometallics* **2019**, *38*, 36–46

226. D. J. Ager, A. H. M. de Vries, J. G. de Vries, *Chem. Soc. Rev.* **2012**, *41*, 3340–3380
227. R. A. Sheldon, *Green Chem.* **2007**, *9*, 1273–1283
228. C. Jimenez-Gonzalez, C. S. Ponder, Q. B. Broxterman, J. B. Manley, *Org. Process Res. Dev.* **2011**, *15*, 912–917
229. J. Andraos, *Org. Process Res. Dev.* **2009**, *13*, 161–185
230. J. Magano, *Tetrahedron* **2011**, *67*, 7875–7899
231. B. W. Cue, Jr., *Chem. Eng. News* **2005**, *83 (39)*, 46
232. G. P. Taber, D. M. Pfisterer, J. C. Colberg, *Org. Process Res. Dev.* **2004**, *8*, 385–388
233. A. Fürstner, L. C. Bouchez, J.-A. Funel, V. Liepins, F.-H. Porée, R. Gilmour, F. Beaufils, D. Laurich, M. Tamiya, *Angew. Chem. Int. Ed.* **2007**, *46*, 9265–9270
234. I. T. Hsu, M. Tomanik, S. B. Herzon, *Acc. Chem. Res.* **2021**, *54*, 903–916
235. R. Dach, J. J. Song, F. Roschangar, W. Samstag, C. H. Senanayake, *Org. Process Res. Dev.* **2012**, *16*, 1697–1706
236. P. N. Arunachalam, P. Kuppusamy, S. Ganesan, S. Krishnamoorthy, R. Y. Nimje, L. B. Jarugu, N. K. Chikkananjaiah, C. A. Reddy, P. Anjanappa, M. Botlagunta, S. Vanteru, N. Maddala, M. Shankar, S. Nair, J. Hynes, Jr., J. B. Santella, III, P. H. Carter, R. Rampulla, M. Vetrichelvan, A. Gupta, A. K. Gupta, A. Mathur, *Org. Process Res. Dev.* **2019**, *23*, 912–918
237. N. A. Meanwell, M. R. Krystal, B. Nowicka-Sans, D. R. Langley, D. A. Conlon, M. D. Eastgate, D. M. Grasela, P. Timmins, T. Wang, J. F. Kadow, *J. Med. Chem.* **2017**, *61*, 62–80
238. R. J. Fox, J. C. Tripp, M. J. Schultz, J. F. Payack, D. D. Fanfair, B. M. Mudryk, S. Murugesan, C.-P. H. Chen, T. E. La Cruz, S. E. Ivy, S. Broxer, R. Cullen, D. Erdemir, P. Geng, Z. Xu, A. Fritz, W. W. Doubleday, D. A. Conlon, *Org. Process Res. Dev.* **2017**, *21*, 1095–1109
239. K. Chen, C. Risatti, J. Simpson, M. Soumeillant, M. Soltani, M. Bultman, B. Zheng, B. Mudryk, J. C. Tripp, T. E. La Cruz, Y. Hsiao, D. A. Conlon, M. D. Eastgate, *Org. Process Res. Dev.* **2017**, *21*, 1110–1121
240. A. B. Smith, III, T. J. Beauchamp, M. J. LaMarche, M. D. Kaufman, Y. Qiu, H. Arimoto, D. R. Jones, K. Kobayashi, *J. Am. Chem. Soc.* **2000**, *122*, 8654–8664
241. A. B. Smith, III, B. S. Freeze, I. Brouard, T. Hirose, *Org. Lett.* **2003**, *5*, 4405–4408
242. I. Paterson, O. Delgado, G. J. Florence, I. Lyothier, M. O'Brien, J. P. Scott, N. Sereinig, *J. Org. Chem.* **2005**, *70*, 150–160
243. D. L. Boger, S. H. Kim, Y. Mori, J.-H. Weng, O. Rogel, S. L. Castle, J. J. McAtee, *J. Am. Chem. Soc.* **2001**, *123*, 1862–1871
244. S. Benson, M.-P. Collin, A. Arlt, B. Gabor, R. Goddard, A. Fürstner, *Angew. Chem. Int. Ed.* **2011**, *50*, 8739–8744
245. M. Inoue, K. Miyazaki, H. Uehara, M. Maruyama, M. Hirama, *Proc. Nat. Acad. Sci.* **2004**, *101*, 12013–12018
246. K. C. Nicolaou, M. O. Frederick, E. Z. Loizidou, G. Petrovic, K. P. Cole, T. V. Koftis, Y. M. A. Yamada, *Chem. Asian J.* **2006**, *1*, 245–263
247. I. Hayakawa, K. Saito, S. Matsumoto, S. Kobayashi, A. Taniguchi, K. Kobayashi, Y. Fujii, T. Kaneko, H. Kigoshi, *Org. Biomol. Chem.* **2017**, *15*, 124–131
248. H. Kigoshi, M. Ojika, T. Ishigaki, K. Suenaga, T. Mutou, A. Sakakura, T. Ogawa, K. Yamada, *J. Am. Chem. Soc.* **1994**, *116*, 7443–7444

Index

Symbols

[2+2]-cycloaddition, **518**, *see also* cycloaddition

1,2,4-trioxolane, 279

1,2-Addition, *see* Carbonyl-Addition

1,3-dipolar cycloaddition, **509**, *see also* cycloaddition

1,4-addition, **145**
  asymmetric, 148
  cause of selectivity, 127
  cuprate, 145
  enolate formation, 71
  reduction, 150

9-BBN, 65, **272**, 630

A

α-alkylation, *see* enolate

α-amination, *see* enolate

α-halogenation, *see* enolate

α-hydroxylation, *see* enolate

A-values, 3

Acceptorless alcohol dehydrogenation, 157

Acetal
  O,O-acetal, 11, 876
  S,S-acetal, 26

Acidity, **27**, 679
  effect of hybridization, 30
  kinetic acidity, 28
  thermodynamic acidity, 28

Activated DMSO, **326**

Activation volume, 497

Acylation, 797

Adam's catalyst, 416

AD Mix, 261

Aggregation
  of organolithium compounds, 60, 189

AIBN, 390

Albright–Goldman oxidation, 329

Aldehyde, *see* carbonyl group

Alder *endo*-rule, 490

Aldol reaction, **74**
  1,4-selectivity, 87
  1,5-selectivity, 89
  acetate/propionate aldol reaction, 75, 99
  catalytic, 97
  chiral boron reagents, 96
  direct, 102
  diastereoselectivity, 76
  Evans product, 92
  homoaldol reaction, 56
  Mukaiyama aldol reaction, 81, 98
  open *vs.* cyclic transition state, 76
  organocatalytic, 102
  reductive, 104
  vinylogous aldol reaction, 84
  Zimmerman-Traxler model, 76

Alkene, **237**
  aminohydroxylation, 263
  aziridination, 257
  dihydroxylation, 258
  epoxidation, 238

© The Editor(s) (if applicable) and The Author(s), under exclusive license to Springer Verlag GmbH, DE, part of Springer Nature 2026
A. Düfert, *Methods of Organic Synthesis*,
https://doi.org/10.1007/978-3-662-70963-4

formation from carbonyls, 179
halogenation, 266
hydroboration, 272
hydrogenation, 403
metathesis, 724
Alkene metathesis, **724**
asymmetric, 734
cross metathesis, 731
ethylene effect, 739
mechanism, 726
ring-closing metathesis, 730
substrate scope, 725
Z-selective, 736
Alkylidenation, 181
Alkyne, **286**
acidity, 30
carbometalation, 292
click reaction, 814
formation, 226, 644, 744
hydroboration, 289
hydrometalation, 630
hydrozirconation, 287
metathesis, 744
π-acid activation, 294
partial reduction, 299, 388, 411, 423
Alkyne metathesis, **744**
substrate scope, 746
Allylation of carbonyls, **136**
Allyl borane, 136
Allylic alcohol, 238, *see also* Sharpless epox-
idation
Allylic oxidation, 348, 356
Allylic strain, **13**
Allyl-Pd complex, 686
Allyl silane, 136
Allyl stannane, 136
Alpine-Borane, 428
Aluminum acetal, 369
Amide formation, 156
Amide, *see* carboxylic acid derivatives
Amino acid, 760, 868
Amino alcohol, 263, 285
Ancillary ligand, 595
Anelli oxidation, *see* TEMPO oxidation
Anode, 826
Anomeric effect, **7**
generalized anomeric effect, 8
kinetic anomeric effect, 10
of spiroacetals, 11

reverse anomeric effect, 13
solvent dependence, 9
Antarafacial, 460
Arbuzov reaction, 196
Asymmetric allylic alkylation, **686**
Cu-mediated substitution, 689
regioselectivity, 688
Asymmetric counterion-directed catalysis,
792
Au-catalysis, **294**
Auxiliary
Abiko-Masamune auxiliary, 90
cleavage, 94
Coltart auxiliary, 117
Crimmins auxiliary, 90
Evans auxiliary, 90, 117, 502
Myers auxiliary, 117
Oppolzer auxiliary, 90, 117, 502
SAMP/RAMP, 117
Yan auxiliary, 90
Azide, 510, 814
Aziridine, 257
Azodicarboxylate, 154
Azomethine ylide, 511

B
B-enolate, **64**, 79
diastereoselectivity, 64
Baeyer–Villiger oxidation, **350**
BAIB, 330
Baldwin's rules, **18**, 857
exceptions, 19
kinetics, 21
Barton ester, 393
Barton–Kellogg olefination, 225
Barton–McCombie deoxygenation, 392
Beckmann rearrangement, 354
Benkeser reaction, 302, 386
Benzoin condensation, 776
Benzylic oxidation, 348
Betaine, 182, 803
Bifunctional catalysis, 784
Bimetallic catalysis, 53
BINAL-H, 428
BINAP, 431, 593
BINOL, 139
Biomimetic synthesis, 569, 880
Bioorthogonality, 814

Bipyridine (bipy), 593
Birch reduction, 301, **384**
  alkylating Birch reduction, 387
  of alkynes, 388
Bisoxazoline (box), 100, 506
BOP, 153
Borane
  as reducing agent, 371, 429
  derivatization, 272
  reactivity, 273
Boronate ester, 637
Boronic acid, 637
Borylation, Ir-catalyzed, 631
Bouveault–Blanc reaction, 389
Breslow intermediate, 776
BrettPhos, 593
Bromination, 125, 269
Brønsted acid, **792**
Brønsted base, 796
Brown allylation, **139**
Buchwald–Hartwig coupling, 591, **675**
  CO/CS coupling, 682
  Pd versus Cu/Ni catalysis, 679
  substrate scope, 676
Buchwald precatalyst, 592
Burgess reagent, 224
Bürgi-Dunitz trajectory, 40

C
Cacchi coupling, 644
Cadiot–Chodkiewicz coupling, 644
Caprolactam synthesis, 354
Carbene, 295, 530, 698
Carbometalation, 292
Carbonyl addition, **40**, 129
  allylation, 136
  asymmetric, 132
  Cieplak model, 47
  competition of 1,2- and 1,3-control, 50
  Cram chelate model, 42, 376
  Evans model of 1,3-addition, 49
  Felkin–Anh model, 42, 375
  Midland model, 376
  organomagnesium reagents, 53
  organozinc reagents, 54
  Reetz model of 1,3-addition, 49
Carbonylation, 281, 660

Carbonyl-ene reaction, 571
Carbonyl group, **37**, 152
  α-functionalization, 108, 764
  aldol reaction, 74
  deprotonation, 56
  reactivity, 38
  redox potential, 348, 828
Carbonyl reduction, **362**, 396, 409, 426
  asymmetric, 428
  in cyclohexanones, 378
  with silanes, 381
Carbonyl ylide, 512
Carbopalladation, 293, 663, 672
Carboxylic acid
  activation, 153
  formation, 324, 335, 339, 342
Carboxylic acid derivatives, **152**
  conversion, 153, 757, 797
  formation, 350
CataCXium A, 680
Catecholborane, 273
Cathode, 826
CBS reduction, **429**, 764
CH activation, **694**
  control of selectivity, 701
  mechanisms, 696
Chan–Evans–Lam coupling, 684
Cheletropic reaction, 459, **528**
Chiral pool, 868
Chiron, 857
Chlorination, 125, 269
Cieplak model, 47
Cinchona alkaloid, 122, 261, 758, 769, 788, 796
Claisen rearrangement, 545
Clemmensen reduction, 396
Click reaction, **813**
  azide-alkyne cycloaddition, 814
  Staudinger ligation, 819
  thiol-ene/thiol-yne reaction, 818
Co-catalysis, 252, 281
CO coupling, *see* Buchwald–Hartwig coupling
Collins reagent, 322
Colvin rearrangement, 232
Comins' reagent, 401
Comproportionation, 332

Configuration interaction, 467
Conformers, **2**, 31, 115, 502, 512, 544, 770
  energies, 7
  gauche effect, 7
Conia-ene reaction, 296
Conjugation, 4
Conrotatory, 461
Continuous flow, 889
Convergent synthesis, 897
Cope rearrangement, 543
Copper chromite, 416
Copper hydride, 150
Corey–Bakshi–Shibata reduction, *see* CBS
    reduction
Corey borane, 96
Corey–Chaykovsky reaction, 258, 802
Corey–Fuchs reaction, **228**
Corey–Kim oxidation, 326
Corey-Nicolaou macrolactonization, 158
Corey–Winter olefination, 221
Correlation diagram, 462, 623
Co salen complex, 252
Cr-catalysis, 252
Cr-mediated oxidation, **322**
Crabtree catalyst, 403
Cram chelate model, 42
Criegee intermediate, 351
Criegee mechanism, 279
Cross-coupling, **626**
    mechanism, 595
    method comparison, 599, 626
    substrate scope, 590
Crotylation, 136
Cr salen complex, 252
CS coupling, *see* Buchwald-Hartwig coupling
Cu-catalysis, 291, 590, 600, 676, 692, 814
    1,4-addition, 145
Cuprate, 145, 689
Curtin–Hammett principle, 436, 667, 803,
    830
Cycloaddition, 459, **476**, 771, 814
  asymmetric, 500
  cheletropic reaction, 528
  diastereoselectivity, 477
  electron demand, 481, 510
  endo-/exo-selectivity, 490
  high pressure, 497
  Lewis acid additives, 499
  periselectivity, 487

  photochemical, 518
  regioselectivity, 482, 512
  retro-cycloaddition, 494, 517
Cyclopropanation, **530**, 704

D

Danishefsky's diene, 489
Davis reagent, 123
DCC, 153, 327
DDQ, 348, 831
DEAD, 154
Debenzylation, *see* hydrogenolysis
Defunctionalization, 373, **390**, 425
De-Mayo reaction, 524
Dess–Martin oxidation, 330
Dess–Martin periodinane, *see* Dess–Martin
    oxidation
DET, 244
Dexter electron transfer mechanism, 837
(DHQ)$_2$PHAL, 261
(DHQD)$_2$PHAL, 261
DIAD, 154
Diazoalkane, 512
Diazomethane, 155
DIBAL, 365
DIC, 153
Diels-Alder reaction, **489**, *see also* cycloaddition
Dihalogenation of alkenes, **267**
Dihydroxylation of alkenes, **258**
Diimide, 300, 576
DIOP, 437
Dioxirane, *see* epoxidation of alkenes
DIPAMP, 435
DIP-Cl, 428
DIPT, 244
Directing group, 702
Disproportionation, 366
Disrotatory, 461
Divergent synthesis, 881
DMAP, 757, 797, 872
DMDO, 238
DMPU
  influence on reactivity, 61
DPEPhos, 282, 598
dppe, 593
dppf, 593
dppp, 593

Drug development, 883
DuPhos, 435

E
E-factor, 895
EDC, 153
Electrochemical reaction, 826
Electrocyclic reaction, 460, **556**
Electrode, 826
Electronically controlled reaction, 2
Electrophore, 826
Elimination, **218**
  alkyne formation, 234
  olefin formation, 218
Enamine, 55, 117, **764**
Ene reaction, **571**
Enolate, **55**
  α-alkylation, 109
  α-amination, 125
  α-halogenation, 125
  α-hydroxylation, 123
  aggregation of Li enolates, 60
  C-/O-alkylation, 109
  kinetic enolate, 57
  reactivity towards electrophiles, 55
  thermodynamic enolate, 57
Enolate formation
  Birch reduction, 73
  diastereoselectivity, 59
  influence of HMPA/DMPU, 61
  Ireland model, 59
  of amides, 59
  of esters, 59
  of ketones, 59
  Reformatsky reaction, 73
  soft enolization, 63
Enyne cycloisomerization, 296
Enyne metathesis, **740**
Enzyme, 872
Epoxidation of alkenes, **238**
  directing groups, 242
  of allylic alcohols, 242
  of enones, 243
Epoxide opening, 19, 284, 373
Ester, *see* carboxylic acid derivatives
Ethylene effect, 739
Evans model of 1,3-addition, 49

Evans–Saksena reduction, 378

F
Favorskii rearrangement, 562
Fe-catalysis, 590, 601, 656
Felkin-Anh model, 42
Fétizon reagent, 346
Fischer esterification, 153
Fluorination, 125, 269
Frontier orbital interaction, 470
Frustrated Lewis pair, 396, 416, 798
Fürst–Plattner rule, **16**, 113, 546

G
Galvanostatic conditions, 832
gauche effect, 7
Geminal dialkyl effect, *see* Thorpe-Ingold
    effect
Gilman cuprate, 145
Glaser coupling, 644
Goldberg coupling, 675
Grieco elimination, 222
Grignard reagent, 53, 129, 148, 618, 628
Group transfer reaction, 460, **571**
Grubbs catalyst, 725
Gusev catalyst, 410

H
Hajos–Parrish–Eder–Sauer–Wiechert reac-
    tion, 758
Hantzsch ester, 769, 774
HATU, 153
HBTU, 153
Heck reaction, 591, **661**
  Jeffery conditions, 667
  mechanism, 662
  oxidative coupling, 670
  regioselectivity, 663
Henbest effect, 242
Henry reaction, 107
Hetero-Diels-Alder reaction, 493, *see also*
    cycloaddition
Heyns oxidation, 342
Hiyama cross-coupling, 591, 615, 642, 656
HMPA

influence on reactivity, 61, 109, 129
HOAt, 153
HOBt, 153
Hofmann product, *see* elimination
Homocoupling, 56, 618
Horiuti–Polanyi mechanism, 420
Horner–Wadsworth–Emmons reaction, **191**
    Ando modification, 193
    diastereoselectivity, 192
    Masamune–Roush modification, 194
    Still–Gennari modification, 193
Hosomi-Sakurai allylation, 139
Hoveyda–Grubbs catalyst, 726
Huisgen cycloaddition, 814
Hydrazone, 117, 397
Hydroamination, 282
Hydroboration, 15, **272**
    asymmetric, 277
    catalytic, 275
    of alkynes, 289
    reactivity, 273, 630
    regioselectivity, 273
Hydroformylation, 281
Hydrogenation, **401**
    heterogeneous, 416
    homogeneous, 402
    kinetics, 421
Hydrogen atom transfer reaction, 284, 842
Hydrogen bond interaction, 762
Hydrogenolysis, 425
Hydrometalation, 301, 405, 630
Hydrozirconation, **287**
Hyperconjugation, 4, 47
Hypervalent iodine, **330**

I
IBX, 330
Ideality of syntheses, 894
Imine, 38, 105, 566, 790, 799
    redox potential, 828
    reduction, 366, 409, 433
Iminium ion, 38, 105, **768**
Infinite dilution, *see* pseudo-dilution
Intersystem crossing, 520, 837
Iodolactonization, 270
Ion pair, 110, 129, 787
Ir-catalysis, 142, 438, 631, 689, 840

Ireland–Claisen rearrangement, 549
Ireland model, 59
Isopinocampheylborane, 65, 96, 139, 276
Isoxazole, 514
Iterative reactions, 638, 875

J
Jacobsen epoxidation, **248**
Jacobsen kinetic resolution, **252**
Jones oxidation, 323
Jørgensen catalyst, *see* prolinol silyl ether
JosiPhos, 402, 435
Julia–Kocienski olefination, *see* Julia–
    Lythgoe olefination
Julia–Lythgoe olefination, **197**
    Barbier conditions, 202
    classic conditions, 197
    modified reaction, 200
    synthesis of the sulfones, 205

K
Keck allylation, 140
Keck macrolactonization, 158
Ketene, 523, 525, 566
Ketone, *see* carbonyl group
Kinetic resolution, 17, 870
Klopman–Salem equation, 469
Krische allylation, 142
Kumada cross-coupling, 591, 618, 639, 654
    Ni-/Fe-catalyzed, 640

L
Lactam, *see* carboxylic acid derivatives
Lactone, *see* carboxylic acid derivatives
Latent functionality, 861
Lawesson's reagent, 426
LDA, **57**
Leighton allylation, 139
Ley oxidation, 340
Li-enolate, 56, 79
LiHMDS, 57
Lindlar catalyst, 416
Lindlar hydrogenation, 300, 423
LiTMP, 57
Lombardo olefination, 210
Luche reduction, 144, 368

M
MacMillan catalyst, 769
Macrocyclization, 23, 158
Macrolactamization, 23, 159
Macrolactonization, 23, **158**
Mannich reaction, 105
Martin's sulfurane, 224
Maruoka catalyst, 122, 788
Masamune borane, 96, 276
McMurry coupling, **214**
    mechanism, 216
*m*CPBA, 239, 351
Meerwein–Ponndorf–Verley reduction, 412
Menthol synthesis, 575
meso-compound, 877
Metal hydride, **363**
    hydride equivalents, 367
    method comparison, 364
    nucleophilicity and electrophilicity, 363
    reactivity, 365
Metallacyclobutane, 726
Metathesis, **723**
    alkene metathesis, 724
    alkyne metathesis, 744
    enyne metathesis, 740
Methenylation, 181
Mg-enolate, 70, 79
Midland model, 376
Milstein catalyst, 158, 410
Mislow–Evans rearrangement, 550
Mitsunobu reaction, 154, 205, 804
Miyaura borylation, 631
Mn-catalysis, 248
Mn salen complex, 248
Mo-catalysis, 689, 725, 735, 744
Molozonide, 279
MoOPH, 123
MorDalphos, 680
Mukaiyama macrolactonization, 158
Mukaiyama reagent, 153
Murahashi cross-coupling, 639, 655

N
N-heterocyclic carbene
    Catalyst, 777
    ligand, 151, 291, 295, 592, 726
Nazarov cyclization, 563
Negishi cross-coupling, 591, 617, 638, 651

Nernst equation, 829
Nernstian shift, 830
NFSI, 123
Ni-catalysis, 133, 590, 599
Nitrene, 698
Nitrile oxide, 512
Nitrone, 511
Nitroxyl radicals, **335**, 831
Non-nucleophilic base, 56
Normant cuprate, 145
Noyori catalyst, **431**
Noyori–Ikariya catalyst, 433
Nozaki-Hiyama-Kishi reaction, 133
Nysted olefination, 211

O
Ohira–Bestmann modification, *see* Seyferth–
    Gilbert homologation
Olefination of carbonyls, **179**
    comparison of methods, 181
    elimination, 218
    phosphorus ylide, 181
    sulfur ylides, 197
    transition metal-based, 205
Oppenauer oxidation, 348
Orbital
    acceptor capacity, 4
    donor capacity, 5
Orbital interactions, **4**, 31, 40, 43, 47, 64, 462,
    481, 510, 530, 559
Organititanium compounds
    olefination, 214
Organoboron reagent, *see* Suzuki cross-
    coupling
Organocatalysis, **757**
    acyl transfer, 797
    catalyst classes, 761
    enamine catalysis, 764
    hydrogen bond interaction, 781
    iminium ion catalysis, 768
    ionic interaction, 788
Organochromium reagents, 133, 211
Organolithium base
    basicity, 56
Organometallic reagent
    reactivity, 591
    synthesis, 628, 654
Organosilicon reagent, 642

Organotitanium compound
  olefination, 206
Organozinc reagent, 130, 211, 652
Orthogonal protection, *see* protecting group
Oxaphosphetane, 182
Oxasiletane, 219
Oxazaborolidine, 429, 504
Oxaziridine, 123
Oxazolidinone, 67, 90, 118, 502
Oxidation of alcohols, **320**
  activated DMSO, 326
  chromium-based, 322
  hypervalent iodine, 330
  method comparison, 322
  nitroxyl radicals, 335
  ruthenium-based, 339
  selective oxidation of primary alcohols, 346
  selective oxidation of secondary alcohols, 347
Oxidation state, 319
OximaPure, 153
Oxime, 353
Oxone, 238
Oxo synthesis, *see* hydroformylation
Oxy-Cope rearrangement, 547
Ozonolysis, **278**, 510
  indicator dye, 279

P
π-acid activation of alkynes, 294
Parikh–Doering oxidation, 326
Partial reduction, *see* alkyne
Passive volume, 87, 115
Paternò–Büchi reaction, 520
PCC, 322
Pd-catalysis, **590**
  CH activation, 694
  CN coupling, 676
  cross-couplings, 626
  elementary reactions, 594
  Heck reaction, 661
Pd black, 624
PDC, 322
Pearlman catalyst, 416
PEPPSI, 592
Peracid, *see* epoxidation of alkenes

Pericyclic reaction, **457**
  conservation of symmetry, 462
  correlation diagram, 463
  cycloaddition, 476
  electrocyclic reaction, 556
  group transfer reaction, 571
  orbital interaction, 470
  sigmatropic rearrangement, 537
  Woodward–Hoffmann rules, 474
Persistent stereogenic center, 86, 501
Petasis olefination, 207
Peterson olefination, 218
Pfaltz catalyst, 438
Pfitzner–Mofatt oxidation, 326
Phase transfer reaction, 108, 122, **788**
Phosphine ligand, 148, 151, 291, 295, 435, **592**, 679
  steric and electronic properties, 597
Phosphinooxazoline (PHOX), 406, 438, 689
Phosphoramidite, 148, 689
Phosphorus ylide, 182
  synthesis, 196
Photochemical reaction, 519, 835
Photoredox catalyst, 840
PIDA, *see* BAIB
Pinnick-Oxidation, 344
pK_a value, *see* acidity
PMHS, 151, 283, 394
Polarity matching, 696
Polyketide, 74
Potentiostatic conditions, 832
Precatalyst, 592, 602
Prilezhaev reaction, *see* epoxidation of alkenes
Principle of least motion, 33
Principle of non-perfect synchronization, 33
Process mass intensity, 895
Proline, 761
Prolinol silyl ether, 764, 766, 769
Protecting group, 862
Protecting group-free synthesis, 862
Protodeboronation, 637
Pseudo-dilution, 23
Pummerer rearrangement, 328
PyBOP, 153
PyOxim, 153
Pyridine bisoxazoline (pybox), 100

Q

QPhos, 680
Quasi-enantiomers, 261
Quinidine, 261
Quinine, 261

R

Radical reaction, 825
Ramberg–Bäcklund rearrangement, 225, 532
Raney nickel, 416, 426
Reaction throughput, *see* space-time yield
Rearrangement, 351, 354, 357, 526, 723, 819
Red-Al, 365
Redox potential, 384, 828
Reetz model of 1,3-addition, 49
Reformatsky reaction, 73
Resolution, 248, 252, **869**
Retron, 852
Retrosynthesis, **849**
  functional group based, 860
  structure-goal strategy, 856
  transform based, 864
Rh-catalysis, 275, 281, 435, 698
Rieke metal, 632
Riley oxidation, *see* selenium dioxide oxidation
Ring closure reaction, *see* Baldwin's rules; macrocyclization
Ring strain, 733
Rosenmund catalyst, 416
Rosenmund reduction, 426
Roush allylation, **139**
Ru-catalysis, 302, 431, 435, 437, 725, 816
Ru-MACHO, 410
Rubottom-Oxidation, 123

S

Saegusa–Ito oxidation, 360
Sarett reagent, *see* Collins reagent
Saudan catalyst, 410
Saytzeff product, *see* elimination
Schlenk equilibrium, 53, 617
Schrock catalyst, 725
Schrock/Osborn catalyst, 403
Schwartz reagent, 287, *see also* hydrozirconation

Secondary orbital interaction, 490
Secondary ozonide, *see* 1,2,4-trioxolane
Selectfluor, 123
Selectride, 365
Selenium dioxide oxidation, 356
Semireduction, *see* partial reduction
SET, *see* single electron transfer
Seyferth–Gilbert homologation, 230
  Colvin rearrangement, 232
  Ohira–Bestmann modification, 230
Sharpless aminohydroxylation, **263**
Sharpless dihydroxylation, **261**
Sharpless epoxidation, **244**
Shi epoxidation, 255
Shiina-Macrolactonization, 158
SHOP process, 752
Shvo catalyst, 415, 874
Sigmatropic rearrangement, 460, **537**
  [1,n]-rearrangement, 540
  [2,3]-rearrangement, 550
  [3,3]-rearrangement, 543
  selection rules, 539
Silanol(ate), 643
Siletane, 643
Simmons–Smith cyclopropanation, 530
Single electron transfer, 199, 375, 383, 390, 397, 601, 653, 826, 838
Smiles rearrangement, 200
Sn-enolate, 69, 79
Solvent window, 833
Sonogashira coupling, 591, 644, 657
  metal acetylides, 645
Space-time yield, 887
SPhos, 593
Spirocyclization, 11, 876
Squaramide, 782
Stahl oxidation, 339, 361
Staudinger ligation, 819
Staudinger reaction, 566
Stephens–Castro reaction, 644
Stereochemical drift, 186
Stereoisomers, 1
  properties, 1
Sterically controlled reaction, 2
Steric demand, *see* A-values
Stetter reaction, 776
Stevens oxidation, 347

Stille cross-coupling, 591, 612, 634, 648
  copper effect, 634
Stille–Kelly coupling, 631
Strain release Lewis acidity, 643
Stryker's reagent, 71, 150
Substitution, 205, 373, 686
Sulfonamide, 160
Sulfur ylide, 197, 801
Superhydride, 365
Supersilane, 395
Suprafacial, 460
Suzuki cross-coupling, 591, 614, 635, 649
  comparison of boron reagents, 637
Swern oxidation, 326
Symmetry-based synthesis, 877
Symmetry breaking, 877
Synthon, 852

T
Takai olefination, 210
Takai–Utimoto olefination, 211
Takeda olefination, 207
TBHP, 238, 244
Tebbe olefination, **207**
Telescoping, 890
TEMPO, *see* nitroxyl radicals
TEMPO oxidation, **335**
Tethering, 732, 875
Thiol-ene/thiol-yne reaction, 818
Thiourea, 781
Thorpe-Ingold effect, **24**
Ti-catalysis, 139, 244, 271
Ti-enolate, 67, 79
Tolman electronic parameter, 597
Torquoselectivity, **559**
TPAP, 339
trans-diaxial effect, *see* Fürst-Plattner rule
Transfer hydrogenation, **411**, 433
trans-hydrogenation, 302, 411
Trialkylaluminum, 294, 377
Tributylstannane, 390
Trifluoroborate, 637
Trimethylsilyl diazomethane, 156
TRIP, 795
Triplet state, 520, 837
Trost ligand, 689
Tsuji–Trost allylation, 120, **686**
Turbo-Grignard reagent, 629

U
Ullmann condensation, 675
Umpolung, **776**, 853

V
Vilsmeier reagent, 355
Vinyl halide
    via hydrozirconation, 288
    via Takai–Utimoto olefination, 213
Vitamin A synthesis, 187

W
Wacker oxidation, 692
Wagner–Meerwein rearrangement, 542
W-catalysis, 738, 744
Weinreb amide, 160, 371
Westheimer rule, 183, 191
White oxidation, 357
Wilkinson catalyst, 403
Wittig reaction, **181**
    betaine, 182, 188
    diastereoselectivity, 183, 190
    Schlosser modification, 187
    stereochemical drift, 186
Wittig rearrangement, 550
Wolff–Kishner reduction, **397**
    from tosylhydrazones, 399
    Huang–Minlon modification, 398
    Myers modification, 399
Woodward–Hoffmann rules, *see* pericyclic
    reaction

X
Xanthate, 392
XantPhos, 282, 598
XPhos, 593

Y
Yamaguchi macrolactonization, 158
Ylide, 182

Z
Ziegler-Natta polymerization, 54
Zimmerman-Traxler model, 76
Zipper reaction, 673

The manufacturer's authorised representative in the EU is Springer
Nature Customer Service Centre GmbH, Europaplatz 3, 69115 Heidelberg,
Germany. If you have any concerns regarding our products, please
contact ProductSafety@springernature.com

Printed and bound by CPI Group (UK) Ltd, Croydon, CR0 4YY

26/05/2026

02118120-0002